Your Learning Style

To discover what kind of learner you are, complete the Learning Styles Inventory on page xiv (or in MyMathLab). Then in the textbook, watch for the Learning Strategy boxes and the accompanying icons that provide ideas for maximizing your own learning style.

_____ 1. I remember information better if I write it down or draw a picture of it.

_____ 2. I remember things better when I hear them instead of just reading or seeing them.

_____ 3. When I receive something that has to be assembled, I just start doing it. I don't read the directions.

_____ 4. If I am taking a test, I can visualize the page of text or lecture notes where the answer is located.

_____ 5. I would rather have the professor explain a graph, chart, or diagram to me instead of just showing it to me.

_____ 6. When learning new things, I want to do it rather than hear about it.

_____ 7. I would rather have the instructor write the information on the board or overhead instead of just lecturing.

_____ 8. I would rather listen to an audiobook than read the book.

_____ 9. I enjoy making things, putting things together, and working with my hands.

_____ 10. I am able to conceptualize quickly and visualize information.

_____ 11. I learn best by hearing words.

_____ 12. I have been called hyperactive by my parents, spouse, partner, or professor.

_____ 13. I have no trouble reading maps, charts, or diagrams.

_____ 14. I can usually pick up on small such as like bells, crickets, or frogs or on distant sounds such as train whistles.

_____ 15. I use my hands and gesture a lot when I speak to others.

Your Learning Strategies

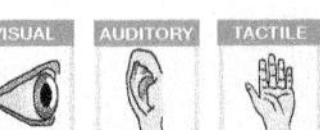
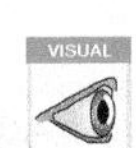
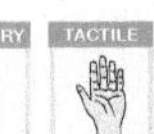

VISUAL AUDITORY TACTILE

Learning Strategy

Developing a good study system and understanding how you best learn is essential to academic success. Make sure you familiarize your-self with the study system outlined in the To the Student section at the beginning of the text. Also take a moment to complete the Learning Styles Inventory found at the end of that section to discover your personal learning style. In these Learning Strategy boxes, we offer tips and suggestions on how to connect the study system and your learning style to help you be successful in the course.

VISUAL AUDITORY TACTILE

Learning Strategy

In the To the Student section, we suggest that when taking notes, you use a red pen for definitions and a blue pen for rules and procedures. Notice that we have used those colors in the design of the text to connect with your notes.

Definition **Additive inverses:** Two numbers whose sum in zero.

Examples of additive inverses: 15 and -15 because $15 + (-15) = 0$

-9 and 9 because $-9 + 9 = 0$

0 and 0 because $0 + 0 = 0$

Notice that 0 is its own additive inverse and that for numbers other than 0, the additive inverses have the same absolute value but *opposite* signs.

Rules

The sum of two additive inverse is zero.

The additive inverse of 0 is 0.

The additive inverse of a nonzero number has the same absolute value but opposite sign.

The Carson Math Study System

The Carson Math Study System is designed to help you succeed in your math course. You will discover your own learning style, and you will use study strategies that match the way you learn best. You will also learn how to organize your course materials, manage your time efficiently, and study and review effectively.

Your Math Notebook

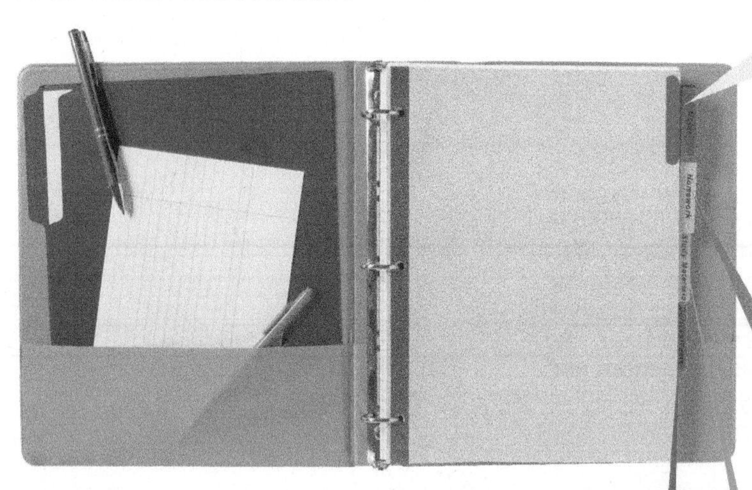

Notes

(see pages xvii–xviii)

Section # 9/20

We can simplify an expression by combining like terms.

p. 37 def. Like terms: constant terms or variable terms that have the same variable(s) raised to the same powers.

Ex 1) $2x$ and $3x$ are like terms.

Ex 2) $5x$ and $7y$ are not like terms.

Homework

(see pages xviii–xxi)

Section # Homework 9/21

#1 – 15 odd

1. $5^2 + 3 \cdot 4 - 7$
$= 25 + 3 \cdot 4 - 7$
$= 25 + 12 - 7$
$= 37 - 7$
$= 30$

Quizzes/Tests

(see page xxv)

Chapter # Quiz 10/10

For 1-4, simplify.

1. $|-8|$ $= 8$ ✓
2. $|9|$ $= 9$ ✓
3. $-15 + 5$ $= -10$ ✓
4. $-8 + -6$ $= -14$ ✓

4/4 = 100% Nice work!

Study Materials

(see pages xxii–xxiv)

Chapter # Study Sheets 9/10

To graph a number on a number line, draw a dot on the mark for the number. ex) Graph -2.

−4 −3 −2 −1 0 1

The absolute value of a positive number is positive. ex) $|7| = 7$

The absolute value of a negative number is positive. ex) $|-12| = 12$

Fourth Edition

Intermediate Algebra

Tom Carson

Bill Jordan

Seminole State College of Florida

Boston Columbus Indianapolis New York San Francisco Upper Saddle River
Amsterdam Cape Town Dubai London Madrid Milan Munich Paris Montréal Toronto
Delhi Mexico City São Paulo Sydney Hong Kong Seoul Singapore Taipei Tokyo

Editorial Director: Chris Hoag
Editor in Chief: Maureen O'Connor
Executive Editor: Cathy Cantin
Content Editor: Christine Whitlock
Editorial Assistant: Chase Hammond
Senior Managing Editor: Karen Wernholm
Associate Managing Editor: Tamela Ambush
Digital Assets Manager: Marianne Groth
Media Producer: Aimee Thorne
QA Manager, Assessment Content: Marty Wright
Executive Content Manager: Rebecca Williams
Senior Content Developer: John Flanagan
Senior Marketing Manager: Rachel Ross
Marketing Assistant: Kelly Cross
Liaison Manager, Text Permissions Group: Joseph Croscup
Image Manager: Rachel Youdelman
Procurement Specialist: Debbie Rossi
Associate Director of Design, USHE North and West: Andrea Nix
Program Design Lead: Heather Scott
Text Design: Tamara Newnam
Production Coordination: PreMediaGlobal
Composition: PreMediaGlobal
Illustrations: PreMediaGlobal
Cover Design: Tamara Newnam
Cover Image: Papajka/Shutterstock

For permission to use copyrighted material, grateful acknowledgment is made to the copyright holders on page P-1, which is hereby made part of this copyright page.

Library of Congress Cataloging-in-Publication Data

Carson, Tom, 1967–
Intermediate algebra / Tom Carson, Franklin Classical School, Bill Jordan, Seminole State College of Florida—Fourth edition.
pages cm.
title: Intermediate algebra
Includes index.
ISBN-13: 978-0-321-91587-0
ISBN-10: 0-321-91587-9
1. Algebra—Textbooks. I. Jordan, Bill. II. Title.
QA154.3.C37 2015
512.9—dc23 2013009378

6 17

www.pearsonhighered.com

ISBN 13: 978-0-321-91587-0
ISBN 10: 0-321-91587-9

Contents

Appendixes

Preface

The *Why* behind Understanding Algebra
+
The *Carson Math Study System* with *Learning Styles Inventory*

From the Authors

Welcome to the fourth edition of *Intermediate Algebra*! Revising our program has been both exciting and rewarding, and it has given us the opportunity to respond to valuable instructor and student feedback. With great pride, we share with you the improvements to this edition, as well as the key features and proven style of our approach.

Intermediate Algebra, Fourth Edition, is one title in a series of four that also includes *Prealgebra*, Fourth Edition; *Elementary Algebra*, Fourth Edition; and *Elementary and Intermediate Algebra*, Fourth Edition. We have designed our program to be versatile enough for use in a variety of teaching and learning formats, including standard lecture, self-paced lab, hybrid, online, and even independent study.

We write in a relaxed, nonthreatening style, taking great care to ensure that students who have struggled with math in the past will be comfortable with our subject matter. Throughout the text, we explain *why* an algebraic process works the way it does, instead of just showing how to follow steps to solve problems. In addition, through problems from science, engineering, accounting, health, the arts, and everyday life, we link algebra to the real world.

Finally, to help students succeed in these courses, we offer the complete Carson Study System that includes a note To the Student, a Learning Styles Inventory, a Math Study System plan for developing a notebook as a personalized organizational and study tool, and integrated Learning Strategies from both the authors and students.

Tom Carson
Bill Jordan

The Carson Math Study System

The Carson Math Study System is designed to help students develop the skills (for example, time management, test prep, and note-taking) that students need to succeed in math, their college careers, and life.

1. **To the Student** (page xiii) focuses on why math is important and what students need to do to succeed in their math courses.
2. Taking the **Learning Styles Inventory** (page xv) will help students assess their particular style of learning and use that knowledge to identify helpful study skills.
3. **The Math Study System** (page xvii) offers students abundant suggestions for developing a class notebook to reflect their personal learning styles, taking effective notes, doing homework, and reviewing for quizzes and tests.
4. Throughout the text, **Learning Strategy** boxes offer advice on implementing the study system effectively based on a student's learning style (pages xv–xvi).

New to this Edition

In response to feedback from instructors and students, the presentation was refined; examples, exercises, Your Turns, and Instructor Notes were added; and real-data applications were updated or replaced with current data and topics. In addition, the order and content of some topics was improved for student comprehension and retention.

Content Changes

Chapter 1 The real numbers diagram in Section 1.1 has been revised to match the treatment in the revised diagram for complex numbers, which appears in Section 8.7.

Chapter 2 Application problems involving two unknowns have been eliminated from this chapter. In the authors' experience, students find it easier and more intuitive to solve those problems using systems of equations; so the topic is now covered with systems.

Chapter 3 Absolute value functions have been added to Section 3.5.

Chapter 4 Problems involving two unknowns are now introduced in this chapter rather than in an earlier chapter using single-variable equations as in previous editions. Solving these problems using systems of equations is more intuitive to students. Cramer's Rule has been moved to Appendix D.

Chapter 5 The exposition in Section 5.2 has been streamlined.

Chapter 6 In Section 6.4, the order of Objectives 2 and 3 have been switched from the previous edition so that finding intercepts of quadratic and cubic functions precedes solving problems involving quadratic equations.

Chapter 7 The discussion of LCD and the examples of variation were streamlined and are now covered in a more visual presentation.

Chapter 8 The development of complex numbers is now more precise.

New and Revised Features

Student Learning Strategies and helpful tips written by successful students and recent college graduates were added throughout the text. These can be identified by the student's name at the end of the strategy.

The **Chapter Openers** now provide a brief topical overview of the chapter. In the instructor's edition, each opener now includes teaching suggestions for the chapter.

Each section now begins with **Warm-up Exercises** that review material presented previously and are helpful to understanding the concepts in the section.

Section Exercises are now grouped by objective to make it easier for students to connect examples with related exercises. In addition, new **Prep Exercises** help students focus on the terminology, rules, and processes corresponding to the exercises that immediately follow.

The **Chapter Summary and Review** is now interactive, with Review Exercises integrated within the summary to encourage active learning and review. For each key topic, students are prompted to complete the corresponding definitions, rules, and procedures.

Calculator tips have been removed to help reduce text length. Also, we find that students now tend to be fairly calculator savvy and with the variety of calculators on the market it is difficult to present accurate keystrokes without appearing to endorse a particular brand.

All **collaborative exercises** have been relocated to an appendix. They retain the section references so that students know where to find the related material.

Finally, we now offer two versions of MyMathLab: **Standard MyMathLab courses** allow instructors to build the course their way, offering maximum flexibility and control over all aspects of assignment creation. **New Ready-to-Go courses** provide students with all the same great MyMathLab features, but make it easier for instructors to get started.

Key Features

In addition to the new and revised features just described, the following key components round out the comprehensive guided learning approach.

An Algebra Pyramid is used throughout the text to help students see how the topic they are learning relates to the big picture of algebra—focusing particularly on the relationship between constants, variables, expressions, and equations and inequalities (page 3). In the end-of-section Review Exercises, Chapter Review Exercises, and Cumulative Review Exercises, an Algebra Pyramid icon indicates the level of the pyramid that correlates to a particular group of exercises. This helps students determine what actions are appropriate with these exercises (for example, whether to "simplify" or "solve" (pages 74, 75–76, 187–188).

Connection Boxes help students understand how math concepts are related and build on each other (pages 190, 291).

Your Turn Practice Exercises, found after most examples, give students an opportunity to work problems similar to the examples they just saw. This practice step engages students and provides immediate feedback to help them develop confidence in their problem-solving skills (pages 5, 337).

Real, Relevant, and Interesting Applications in the examples and exercises reflect real-world situations in science, engineering, health, finance, the arts, and other areas. These applications illustrate the everyday use of basic algebraic concepts and encourage students to apply mathematical concepts to solve problems (pages 94, 247).

Thorough Explanations are key to student understanding. The authors take great care to explain not only *how* to do the math but also *why* the math works the way it does, how concepts are related, and how the math is relevant to students' everyday lives. Knowing all of this gives students a context in which to learn and remember math concepts.

BREAKTHROUGH
To improving results

MyMathLab

Ties the Complete Learning Program Together

MyMathLab® Online Course (access code required)

MyMathLab from Pearson is the world's leading online resource in mathematics, integrating interactive homework, assessment, and media in a flexible, easy-to-use format. MyMathLab delivers **proven results** in helping individual students succeed. It provides **engaging experiences** that personalize, stimulate, and measure learning for each student. And it comes from an **experienced partner** with educational expertise and an eye on the future.

MyMathLab® for Developmental Mathematics

Prepared to go wherever you want to take your students.

Personalized Support for Students

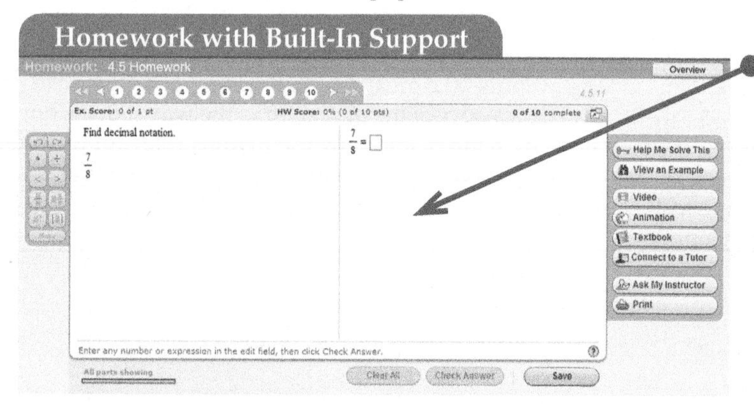

Exercises: The homework and practice exercises in MyMathLab are correlated to the exercises in the textbook, and they regenerate algorithmically to give students unlimited opportunity for practice and mastery. The software offers immediate, helpful feedback when students enter incorrect answers.

Multimedia Learning Aids: Exercises include guided solutions, sample problems, animations, videos, and eText access for extra help at point of use.

Expert Tutoring: Although many students describe the whole of MyMathLab as "like having your own personal tutor," students using MyMathLab do have access to live tutoring from Pearson, from qualified math instructors.

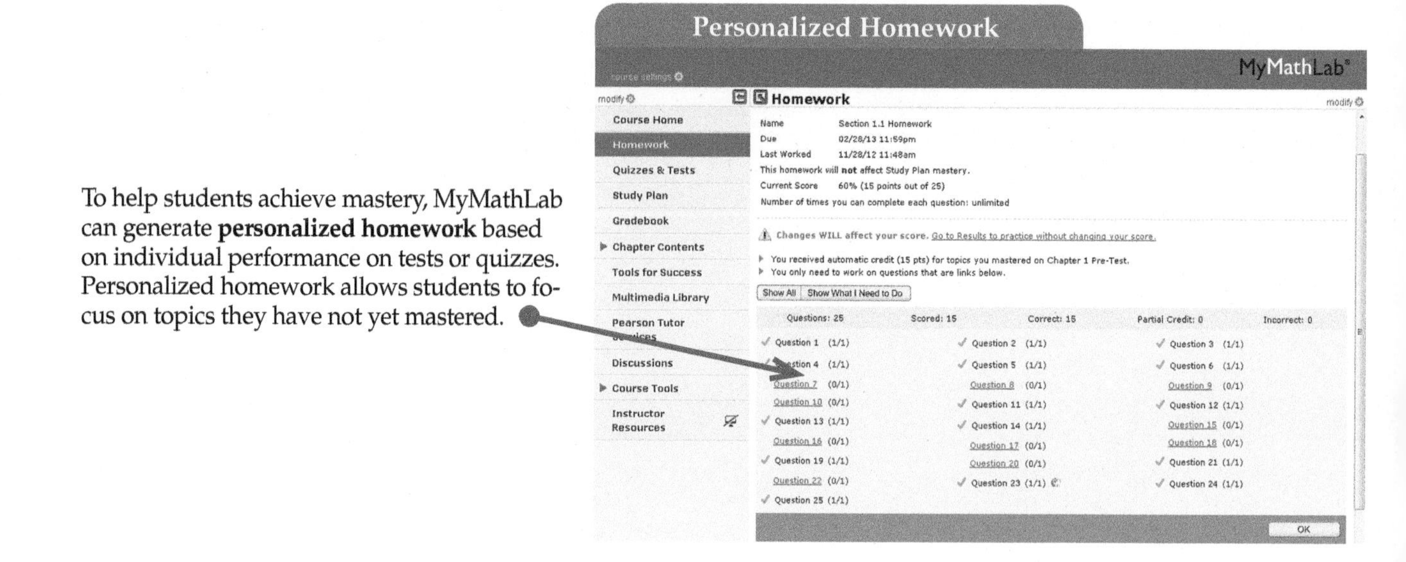

To help students achieve mastery, MyMathLab can generate **personalized homework** based on individual performance on tests or quizzes. Personalized homework allows students to focus on topics they have not yet mastered.

The **Adaptive Study Plan** makes studying more efficient and effective for every student. Performance and activity are assessed continually in real time. The data and analytics are used to provide personalized content—reinforcing concepts that target each student's strengths and weaknesses.

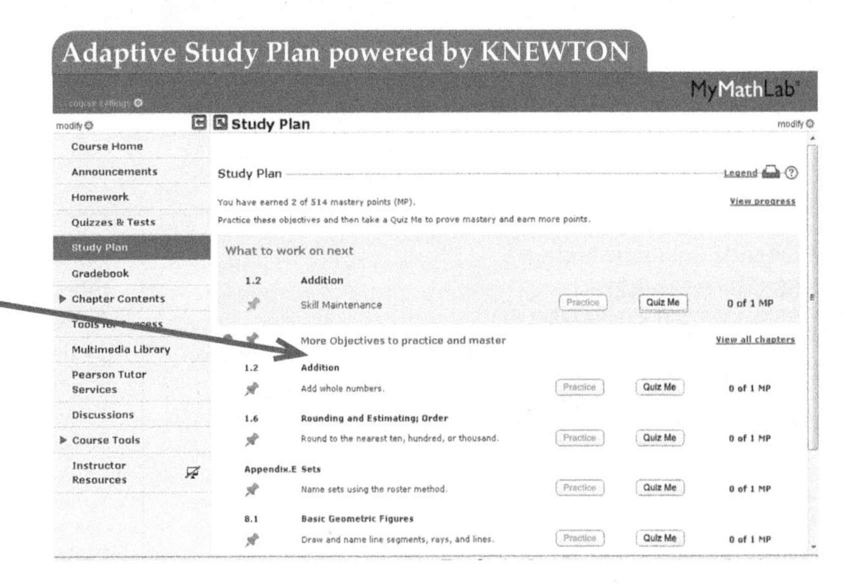

Flexible Design, Easy Start-up, and Results for Instructors

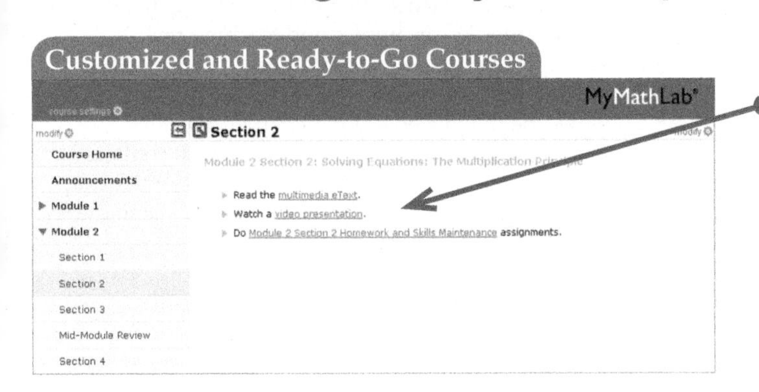

Instructors can modify the left-hand nav and insert their own directions onto course-level landing pages, and a **custom MyMathLab** course can be built to reorganize material and structure the course by chapters, modules, units—whatever the needs may be.

Ready-to-Go courses include pre-assigned homework, quizzes, and tests to make it even easier for instructors to get started.

The **comprehensive online gradebook** automatically tracks students' results on tests, quizzes, homework, and in the study plan. You can use the gradebook to intervene quickly if students have trouble or to provide positive feedback on a job well done. The data within MyMathLab is easily exported to a variety of spreadsheet programs such as Microsoft Excel. Instructors can determine which points of data to export and then analyze the results to determine success.

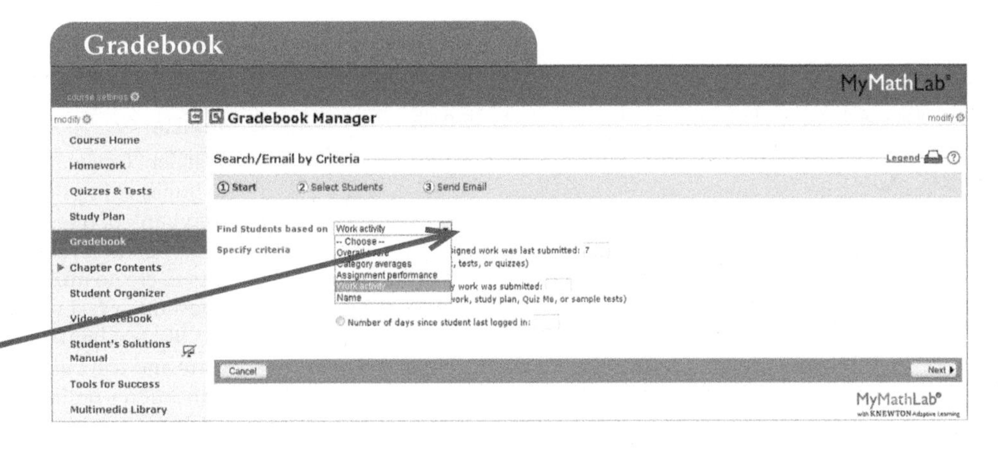

New features such as **Search/Email by criteria** make the gradebook a powerful tool for instructors. With this feature, instructors can easily communicate with both at-risk and successful students. Instructors can search by score on specific assignments, noncompletion of assignments within a given time frame, last log-in date, or overall score.

Additional Resources in MyMathLab

*In addition to the robust course delivery, the course also includes the **full Carson eText, additional Carson Program Features,** and the **entire set of instructor and student resources** in one easy place to access online.*

For Students

Student's Solutions Manual*

ISBN-10: 0-321-91228-4
ISBN-13: 978-0-321-91228-2

- Contains complete solutions to the odd-numbered section exercises and solutions to all of the section-level Review Exercises, Chapter Review Exercises, Practice Tests, and Cumulative Review Exercises.

MyWorkBook for Intermediate Algebra*

ISBN-10: 0-321-92231-X
ISBN-13: 978-0-321-92231-1

MyWorkbook can be packaged with the textbook or with the MyMathLab access kit and includes the following resources for each section of the text:

- Key vocabulary terms and vocabulary practice problems.
- Guided Examples with stepped-out solutions and similar Practice Exercises, keyed to the text by Learning Objective.
- References to textbook Examples and Section Lecture Videos for additional help.
- Additional Exercises with ample space for students to show their work, keyed to the text by Learning Objective.

Video Resources

Within MyMathLab, students can access short lectures for each section of the text, plus Chapter Test Prep Videos, which allow students to watch an instructor work through step-by-step solutions for all of the Chapter Test exercises from the textbook. All videos include optional English and Spanish captions.

*Printed supplements are also available for separate purchase through MyMathLab, MyPearsonStore.com, or other retail outlets. They can also be value-packed with a textbook or MyMathLab code at a discount.

For Instructors

Annotated Instructor's Edition**

ISBN-10: 0-321-91599-2
ISBN-13: 978-0-321-91599-3

- Includes answers to all exercises, including Puzzle Problems and Collaborative Exercises, printed in bright blue near the corresponding problems.
- Useful teaching tips are printed in the margin.
- A ★ icon, found in the AIE only, indicates especially challenging exercises in the exercise sets.

Instructor's Resource Manual with Tests and Mini Lectures** (download only)

ISBN-10: 0-321-91594-1
ISBN-13: 978-0-321-91594-8

- A mini-lecture for each section of the text, organized by objective, includes key examples and teaching tips.
- Designed to help both new and adjunct faculty with course preparation and classroom management.
- Contains one diagnostic test per chapter; four free-response test forms per chapter, one of which contains higher-level questions; one multiple-choice test per chapter; a midchapter check-up for each chapter; one midterm exam; and two final exams.

Instructor's Solutions Manual** (Download only)

ISBN-10: 0-321-91592-5
ISBN-13: 978-0-321-91592-4

- Contains complete solutions to all even-numbered section exercises, Puzzle Problems, and Collaborative Exercises.

PowerPoint® Lecture Slides** (download only)

Present key concepts and definitions from the text.

TestGen®

TestGen®(www.pearsoned.com/testgen) enables instructors to build, edit, print, and administer tests using a computerized bank of questions developed to cover all the objectives of the text. TestGen is algorithmically based, allowing instructors to create multiple but equivalent versions of the same question or test with the click of a button. Instructors can also modify test bank questions or add new questions. The software and test bank are available for download from Pearson Education's online catalog.

**Also available in print or for download from the Instructor Resource Center (IRC) on www.pearsonhighered.com.

To learn more about how MyMathLab combines proven learning applications with powerful assessment, visit **www.mymathlab.com** or contact your Pearson representative.

Acknowledgments

Many people gave of themselves in so many ways during the development of this text. Mere words cannot contain the fullness of our gratitude. Although the words of thanks that follow may be few, please know that our gratitude is great.

We would like to thank the following people who have given of their time in reviewing this and previous editions of this textbook. Their thoughtful input was vital to the development of the text.

Marwan Abu-Sawwa, *Florida Community College at Jacksonville*
Ahmed Adala, *Metropolitan Community College*
Janet Archibald, *Ventura College*
Daniel Bacon, *Massasoit Community College*
Sandra Belcher, *Midwestern State University*
Connie Buller, *Metropolitan Community College*
Nancy Carpenter, *Johnson County Community College*
Mary Deas, *Johnson County Community College*
Sharon Edgmon, *Bakersfield College*
Dolen Freeouf, *Southeast Community College*
Elise Fischer, *Johnson County Community College*
Kathy Garrison, *Clayton College and State University*
Haile K. Haile, *Minneapolis Community and Technical College*
Pauline Hall, *Iowa State University*
James W. Harris, *John A. Logan College*
Darlene Hatcher, *Metropolitan Community College*
Kristy Hill, *Hinds Community College–Raymond*
Charles Glenn Hudson, *Columbia State Community College*
Nancy R. Johnson, *Manatee Community College*
Tracey L. Johnson, *University of Georgia*
Jeff Kroll, *Brazosport College*
Peter Lampe, *University of Wisconsin–Whitewater*
Sandra Lofstock, *California Lutheran University*
Stephanie Logan, *Lower Columbia College*
John Long, *Jefferson Community College*
Brenda Moore, *Hutchinson Community College*
Janis Orinson, *Central Piedmont Community College*
Merrel Pepper, *Southeast Technical Institute*
William Plemmons, *University of Central Florida*
Kim Ramsey-Chin, *University of Akron*
Sherri Rankin, *Hutchinson Community College*
Vivian Reinhard, *Johnson County Community College*
Patrick Riley, *Hopkinsville Community College*
Reynaldo Rivera, *Estrella Mountain Community College*
Mary Robinson, *University of New Mexico, Valencia Campus*
Rebecca Schantz, *Prairie State College*
James Smith, *Columbia State Community College*
Kay Stroope, *Phillips County Community College*
John Thoo, *Yuba College*
Bettie Truitt, *Black Hawk College*
Natalie Weaver, *Daytona Beach Community College*
Judith A. Wells, *University of Southern Indiana*
Joe Westfall, *Carl Albert State College*
Peter Willett, *Diablo Valley College*
Tom Williams, *Rowan-Cabarrus Community College*
Tom Worthing, *Hutchinson Community College*

In addition, we would like to thank the students who provided their successful learning strategies for inclusion in the book. We hope their tips will help other students succeed as well.

Ellyn G., *Stanly Community College*
Jason J., *University of Alabama*
Judah G., *Columbus State Community College*
Lauren H., *University of Central Florida*
Biridiana N.
Zheng Alick Z., *University of Southern California*

We would like to extend a heartfelt thank you to Cathy Cantin, Kerianne Okie, Christine Whitlock, Kari Heen, Tamela Ambush, Michelle Renda, Rachel Ross, Alicia Frankel, Kelly Cross, Aimee Throne, Rebecca Williams, Greg Tobin, and all of the wonderful people at Pearson Arts & Sciences who helped so much in the revision of the text. Special thanks to Janis Cimperman and Paul Lorczak for checking the manuscript and pages for accuracy and to Becky Troutman for researching application problems and preparing the Glossary.

To Erin Donahue and all of the fabulous people at PreMediaGlobal, thank you for your attention to detail and for working so hard to put together the finished pages.

To Melinda McLaughlin and her team at GEX for their work on the solutions manuals, the Instructor's Resource Manual, and the MyWorkBook.

Finally, we'd like to thank our families for their support and encouragement during the process of developing and revising this text.

Tom Carson
Bill Jordan

To the Student

Why Do I Have to Take This Course?

Often, this is one of the first questions students ask when they find out they must take an algebra course. What a great question! But why focus on math alone? What about English, history, psychology, or science? Does anyone really use *every* topic of *every* course in the curriculum? Most jobs do not require that we write essays on Shakespeare, discuss the difference between various psychological theories, or analyze the cell structure of a frog's liver. So what's the point? The issue comes down to recognizing that general education courses are not job training. The purpose of those courses is to stretch and exercise the mind so that the educated person can better communicate, analyze situations, and solve problems, which are all valuable skills in life and any job.

Professional athletes offer a good analogy. They usually have an exercise routine apart from the sport designed to build and improve their body. They may seek a trainer to design exercises intended to improve strength, stamina, or balance and then push them in ways they would not normally push themselves. That trainer may have absolutely no experience with his or her client's sport, but can still be quite effective in designing an exercise program because the trainer is focused on building basic skills useful for any athlete in any sport. Education is similar: it is exercise for the mind. A teacher's job is like that of a physical trainer. A teacher develops exercises intended to improve communication skills, critical-thinking skills, and problem-solving skills. Different courses are like different types of fitness equipment. Some courses may focus more on communication through writing papers and through discussion and debate. Other courses, such as mathematics, focus more on critical thinking and problem solving.

Another similarity is that physical exercise must be challenging for your body to improve. Similarly, mental exercise must be challenging for the mind to improve. Expect course assignments to challenge you and push you mentally in ways you wouldn't push yourself. That's the best way to grow. So as you think about the courses you are taking and the assignments in those courses, remember the bigger picture of what you are developing: your mind. When you are writing papers, responding to questions, analyzing data, and solving problems, you are developing skills important to life and any career out there.

What Do I Need to Do to Succeed?

☑ **Adequate Time** To succeed, you must have adequate time and be willing to use that time to perform whatever is necessary. To determine if you have adequate time, use the following guide.

Step 1. Calculate your work hours per week.
Step 2. Calculate the number of hours in class each week.
Step 3. Calculate the number of hours required for study by doubling the number of hours you spend in class.
Step 4. Add the number of hours from steps 1–3.

Adequate time: If the total number of hours is below 60, then you have adequate time.

Inadequate time: If the total number of hours is 60 or more, then you do not have adequate time. You may be able to hang in there for a while, but eventually, you will find yourself overwhelmed and unable to fulfill all of your obligations. Remember, the above calculations do not consider other likely elements of life such as a commute, family, and recreation. The wise thing to do is cut back on work hours or drop some courses.

Assuming you have adequate time available, choosing to use that time to perform whatever is necessary depends on your attitude, commitment, and self-discipline.

☑ **Positive Attitude** We do not always get to choose our circumstances, but we do get to choose our reaction and behavior. A positive attitude is choosing to be cheerful, hopeful, and encouraging no matter the situation. A benefit of a positive attitude is that it tends to encourage people around you. As a result, they are more likely to want to help you achieve your goal. A negative attitude, on the other hand, tends to discourage people around you. As a result, they are less likely to want to help you. The best way to maintain a positive attitude is to keep life in perspective, recognizing that difficulties and setbacks are temporary.

☑ **Commitment** Commitment means binding yourself to a course of action. Remember, expect difficulties and setbacks, but don't give up. That's why a positive attitude is important; it helps you stay committed in the face of difficulty.

☑ **Self-Discipline** Self-discipline is choosing to do what needs to be done—even when you don't feel like it. In pursuing a goal, it is normal to get distracted or tired. It is at those times that your positive attitude and commitment to the goal help you discipline yourself to stay on task.

Thomas Edison, inventor of the lightbulb, provides an excellent example of all of these principles. Edison tried over 2000 different combinations of materials for the filament before he found a successful combination. When asked about all his failed attempts, Edison replied, "I didn't fail once, I invented the lightbulb. It was just a 2000-step process." He also said, "Our greatest weakness lies in giving up. The most certain way to succeed is always to try just one more time." In those two quotes, we see a person who obviously had time to try 2000 experiments, had a positive attitude about the setbacks, never gave up, and had the self-discipline to keep working.

Behaviors of Strong Students and Weak Students

The four requirements for success can be translated into behaviors. The following table compares the typical behaviors of strong students with the typical behaviors of weak students.

Strong Students ...	Weak Students ...
• are relaxed, patient, and work carefully.	• are rushed, impatient, and hurry through work.
• almost always arrive on time and leave the classroom only in an emergency.	• often arrive late and often leave class to "take a break."
• sit as close to the front as possible.	• sit as far away from the front as possible.
• pay attention to instruction.	• ignore instruction, chit-chat, draw, fidget, etc.
• use courteous and respectful language, encourage others, make positive comments, are cheerful and friendly.	• use disrespectful language, discourage others, make negative comments, are grumpy and unfriendly. Examples of unacceptable language include: "I hate this stuff!" (or even worse!) "Are we doing anything important today?" "Can we leave early?"
• ask appropriate questions and answer instructor's questions during class.	• avoid asking questions and rarely answer instructor's questions in class.
• take a lot of notes, have organized notebooks, seek out and use study strategies	• take few notes, have disorganized notebooks, do not use study strategies.
• begin assignments promptly and manage time wisely and almost always complete assignments on time.	• procrastinate, manage time poorly, and often complete assignments late.
• label assignments properly and show all work neatly.	• show little or no work and write sloppily.
• read and work ahead.	• rarely read or work ahead.
• contact instructors outside of class for help, and use additional resources such as study guides, solutions manuals, computer aids, videos, and tutorial services.	• avoid contacting instructors outside of class and rarely use additional resources available.

Assuming you have the prerequisites for success and understand the behaviors of a good student, our next step is to develop two major tools for success:

1. **Learning Style:** Complete the Learning Styles Inventory to determine how you tend to learn.
2. **The Study System:** This system describes a way to organize your notebook, take notes, and create study tools to complement your learning style. We've seen students transform their mathematics grades from D's and F's to A's and B's by using the Study System that follows.

Learning Styles Inventory

What Is Your Personal Learning Style?

A learning style is the way in which a person processes new information. Knowing your learning style can help you make choices in the way you focus on and study new material. Below are 15 statements that will help you assess your learning style. After reading each statement, rate your response to the statement using the scale below. There are no right or wrong answers.

3 = Often applies **2** = Sometimes applies **1** = Never or almost never applies

_____ 1. I remember information better if I write it down or draw a picture of it.

_____ 2. I remember things better when I hear them instead of just reading or seeing them.

_____ 3. When I receive something that has to be assembled, I just start doing it. I don't read the directions.

_____ 4. If I am taking a test, I can visualize the page of text or lecture notes where the answer is located.

_____ 5. I would rather have the professor explain a graph, chart, or diagram to me instead of just showing it to me.

_____ 6. When learning new things, I want to do it rather than hear about it.

_____ 7. I would rather have the instructor write the information on the board or overhead instead of just lecturing.

_____ 8. I would rather listen to an audiobook than read the book.

_____ 9. I enjoy making things, putting things together, and working with my hands.

_____ 10. I am able to conceptualize quickly and visualize information.

_____ 11. I learn best by hearing words.

_____ 12. I have been called hyperactive by my parents, spouse, partner, or professor.

_____ 13. I have no trouble reading maps, charts, or diagrams.

_____ 14. I can usually pick up on small sounds such as bells, crickets, frogs or on distant sounds such as train whistles.

_____ 15. I use my hands and gesture a lot when I speak to others.

Write your score for each statement beside the appropriate statement number below. Then add the scores in each column to get a total score for that column.

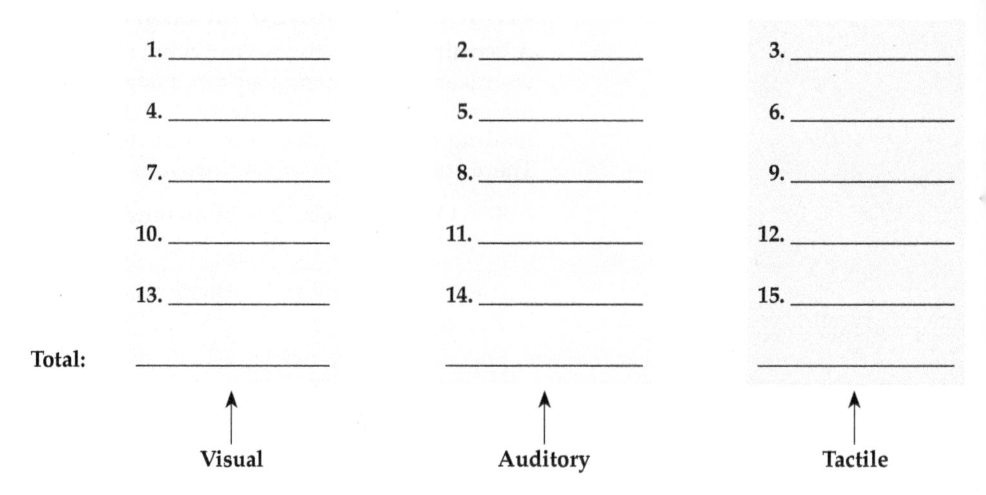

	1. ______	2. ______	3. ______
	4. ______	5. ______	6. ______
	7. ______	8. ______	9. ______
	10. ______	11. ______	12. ______
	13. ______	14. ______	15. ______
Total:	______	______	______
	↑ **Visual**	↑ **Auditory**	↑ **Tactile**

The largest total of the three columns indicates your dominant learning style.

VISUAL

Visual learners learn best by seeing. If this is your dominant learning style, then you should focus on learning strategies that involve seeing. The color coding in the study system (see page xvii–xviii) will be especially important. The same color coding is used in the text. Draw diagrams, arrows, and pictures in your notes to help you see what is happening. Reading your notes, study sheets, and text repeatedly will be an important strategy.

AUDITORY

Auditory learners learn best by hearing. If this is your dominant learning style, then you should use learning strategies that involve hearing. After getting permission from your instructor, bring a recorder to class to record the discussion. When you study your notes, play back the recording. Also, when you learn rules, say the rule over and over. As you work problems, say the rule before you do the problem. You may also find the videotapes to be beneficial because you can hear explanations of problems taken from the text.

TACTILE

Tactile (also known as kinesthetic) learners learn best by touching or doing. If this is your dominant learning style, you should use learning strategies that involve doing. Doing a lot of practice problems will be important. Make use of the Your Turn exercises in the text. These are designed to give you an opportunity to do problems that are similar to the examples as soon as a topic is discussed. Writing out your study sheets and doing your practice tests repeatedly will be important strategies for you.

Note that the study system developed in this text is for all learners. Your learning style will help you decide what aspects and strategies in the study system to focus on, but being predominantly an auditory learner does not mean that you shouldn't read the textbook, do a lot of practice problems, or use the color-coding system in your notes. Auditory learners can benefit from seeing and doing, and tactile learners can benefit from seeing and hearing. In other words, do not use your dominant learning style as a reason for not doing things that are beneficial to the learning process. Also remember that the Learning Strategy boxes presented throughout the text provide tips to help you use your personal learning style to your advantage.

The Carson Math Study System

Organize the notebook into four parts using dividers shown:

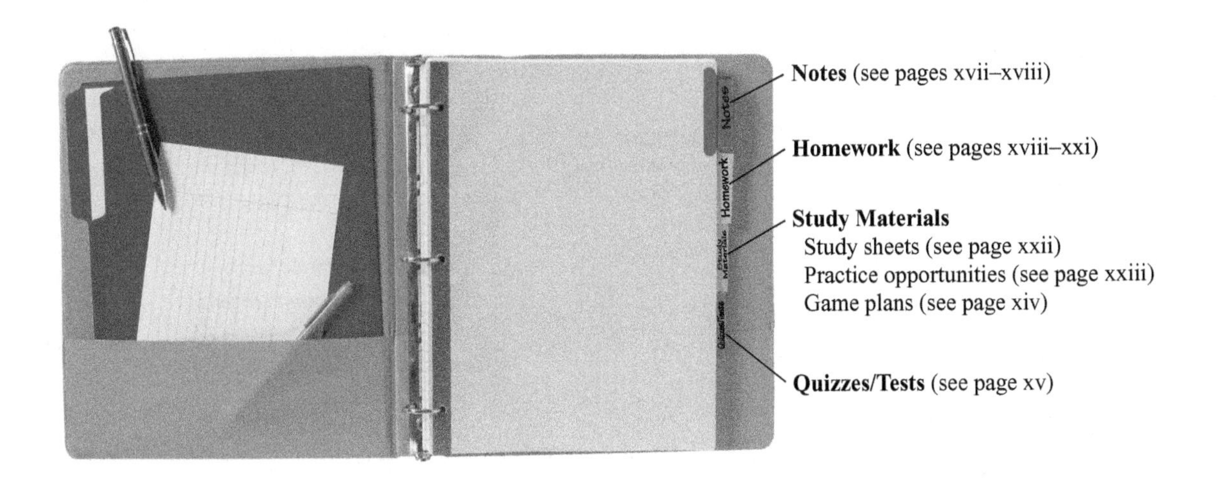

Notes

- Use a color code: red for definitions, blue for rules or procedures, and pencil for all examples and other notes.
- Begin notes for each class on a new page (front and back for that day is okay). Include a topic title or section number and the date on each page.
- Try to write your instructor's spoken explanations along with the things he or she writes on the board.
- Mark examples your instructor emphasizes in some way to give them a higher priority. These problems often appear on quizzes and tests.
- Write warnings your instructor discusses about a particular situation.
- Include common errors that your instructor illustrates, but mark them clearly as errors so that you do not mistake them for correct.
- To speed note taking, eliminate unnecessary words such as *the* and use codes for common words such as + for *and* and $\therefore$ for *therefore*. Also, instead of writing complete definitions, rules, or procedures, write the first few words and place the page reference from the text so that you can copy from the text later.

Sample Notes with Color Code

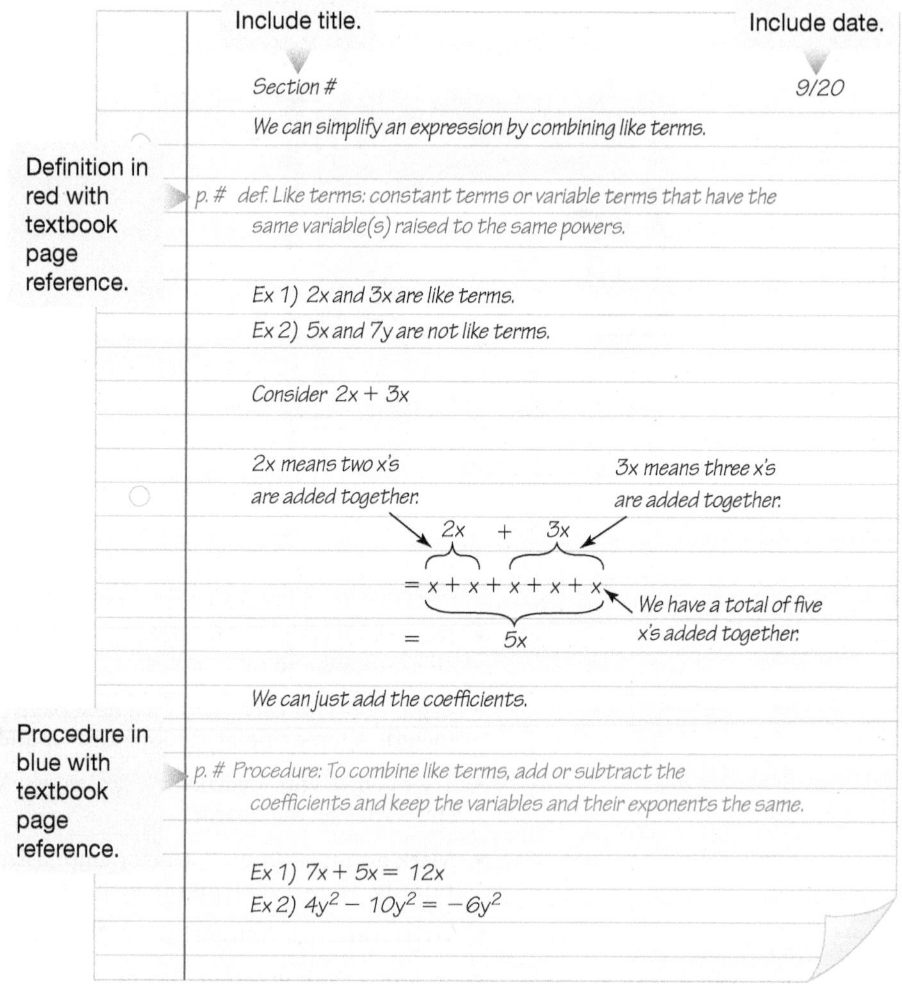

Include title. Include date.

Section # *9/20*

We can simplify an expression by combining like terms.

Definition in red with textbook page reference.

p. # def. Like terms: constant terms or variable terms that have the same variable(s) raised to the same powers.

Ex 1) 2x and 3x are like terms.

Ex 2) 5x and 7y are not like terms.

Consider $2x + 3x$

2x means two x's are added together. *3x means three x's are added together.*

$$2x + 3x$$

$$= x + x + x + x + x$$

We have a total of five x's added together.

$$= 5x$$

We can just add the coefficients.

Procedure in blue with textbook page reference.

p. # Procedure: To combine like terms, add or subtract the coefficients and keep the variables and their exponents the same.

Ex 1) $7x + 5x = 12x$

Ex 2) $4y^2 - 10y^2 = -6y^2$

Homework

This section of the notebook contains all homework. Use the following guidelines whether your assignments are from the textbook, a handout, or a computer program such as MyMathLab or MathXL.

- Use pencil so that mistakes can be erased (scratching through mistakes is messy and should be avoided).
- Label according to your instructor's requirements. Usually, at least include your name, the date, and the assignment title. It is also wise to write the assigned problems at the top as they were given. For example, if your instructor writes "Section 1.5 #1–15 odd," write it that way at the top. Labeling each assignment with this much detail shows that you take the assignment seriously and leaves no doubt about what you interpreted the assignment to be.
- For each problem you solve, write the problem number and show all solution steps neatly.

Why do I need to show work and write all the steps? Isn't the right answer all that's needed?

- Mathematics is not just about getting correct answers. You really learn mathematics when you organize your thoughts and present those thoughts clearly using mathematical language.
- You can arrive at correct answers with incorrect thinking. Showing your work allows your instructor to verify that you are using correct procedures to arrive at your answers.
- Having a labeled, well-organized, and neat hard copy is a good study tool for exams.

What if I submit my answers in MyMathLab or MathXL? Do I still need to show work?

Think of MyMathlab or MathXL as a personal tutor who provides the exercises, offers guided assistance, and checks your answers before you submit the assignment. For the same reasons as those listed above, you should still create a neatly written hard copy of your solutions, even if your instructor does not check the work. Following are some additional reasons to show your work when submitting answers in MyMathlab or MathXL.

- If you have difficulties that are unresolved by the program, you can show your instructor. Without the written work, your instructor cannot see your thinking.
- If you have a correct answer but have difficulty entering that correct answer, you have a record of it and can show your instructor. If correct, your instructor can override the score.

Sample Homework: Simplifying Expressions or Solving Equations

Suppose you are given the following exercise:

For Exercises 1–30 simplify.

1. $5^2 + 3 \cdot 4 - 7$

Your homework should look something like the following:

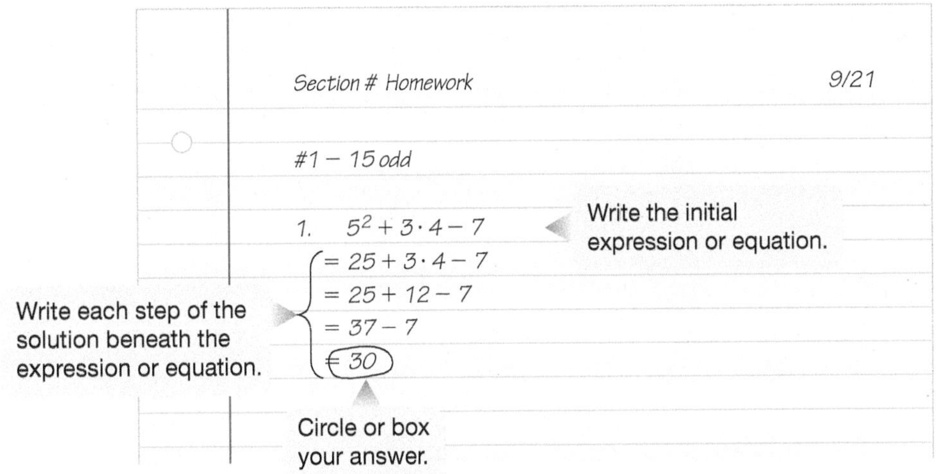

Section # Homework 9/21

#1 – 15 odd

1. $5^2 + 3 \cdot 4 - 7$

$= 25 + 3 \cdot 4 - 7$

$= 25 + 12 - 7$

$= 37 - 7$

$= 30$

Write the initial expression or equation.

Write each step of the solution beneath the expression or equation.

Circle or box your answer.

Sample Homework: Solving Application Problems

Example: Suppose you are given the following two problems.

For Exercises 1 and 2, solve.

1. Find the area of a circle with a diameter of 10 feet.
2. Two cars are traveling toward each other on the same highway. One car is traveling 65 miles per hour, and the other is traveling at 60 miles per hour. If the two cars are 20 miles apart, how long will it be until they meet?

Your homework should look something like the following:

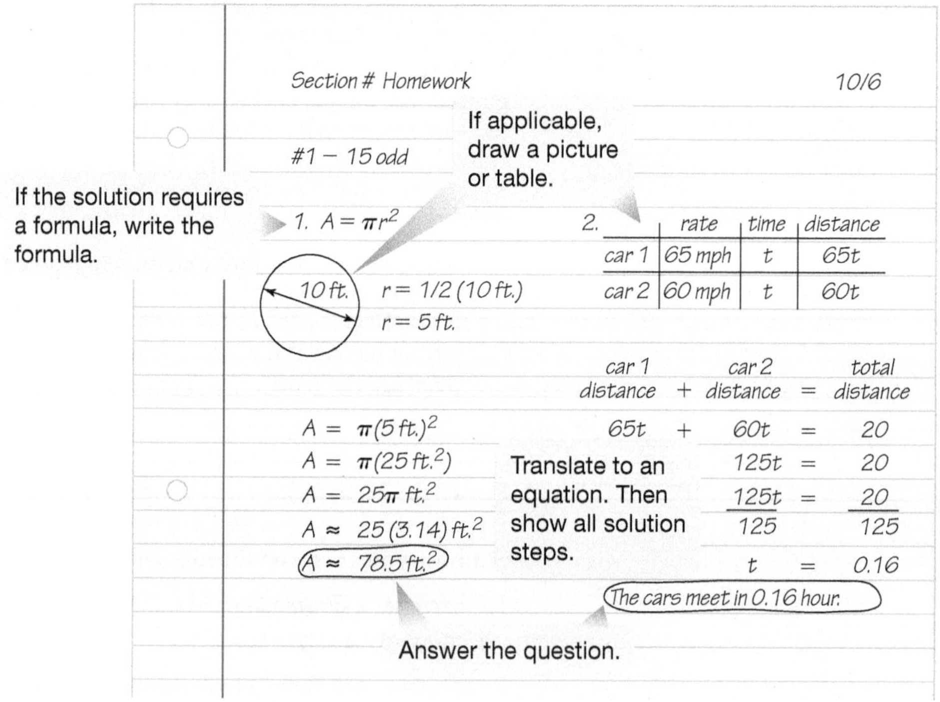

Section # Homework 10/6

#1 – 15 odd

If applicable, draw a picture or table.

If the solution requires a formula, write the formula.

1. $A = \pi r^2$

10 ft.

$r = 1/2\,(10\text{ ft.})$

$r = 5\text{ ft.}$

$A = \pi(5\text{ ft.})^2$

$A = \pi(25\text{ ft.}^2)$

$A = 25\pi\text{ ft.}^2$

$A \approx 25(3.14)\text{ ft.}^2$

$A \approx 78.5\text{ ft.}^2$ (circled)

2.

	rate	time	distance
car 1	65 mph	t	$65t$
car 2	60 mph	t	$60t$

car 1 distance + car 2 distance = total distance

$65t + 60t = 20$

$125t = 20$

$\frac{125t}{125} = \frac{20}{125}$

$t = 0.16$

Translate to an equation. Then show all solution steps.

The cars meet in 0.16 hour. (circled)

Answer the question.

Sample Homework: Graphing

Example: Suppose you are given the following two problems.

For Exercises 1 and 2, graph the equation.

1. $y = 2x - 3$ **2.** $y = -2x + 1$

Your homework should look something like the following:

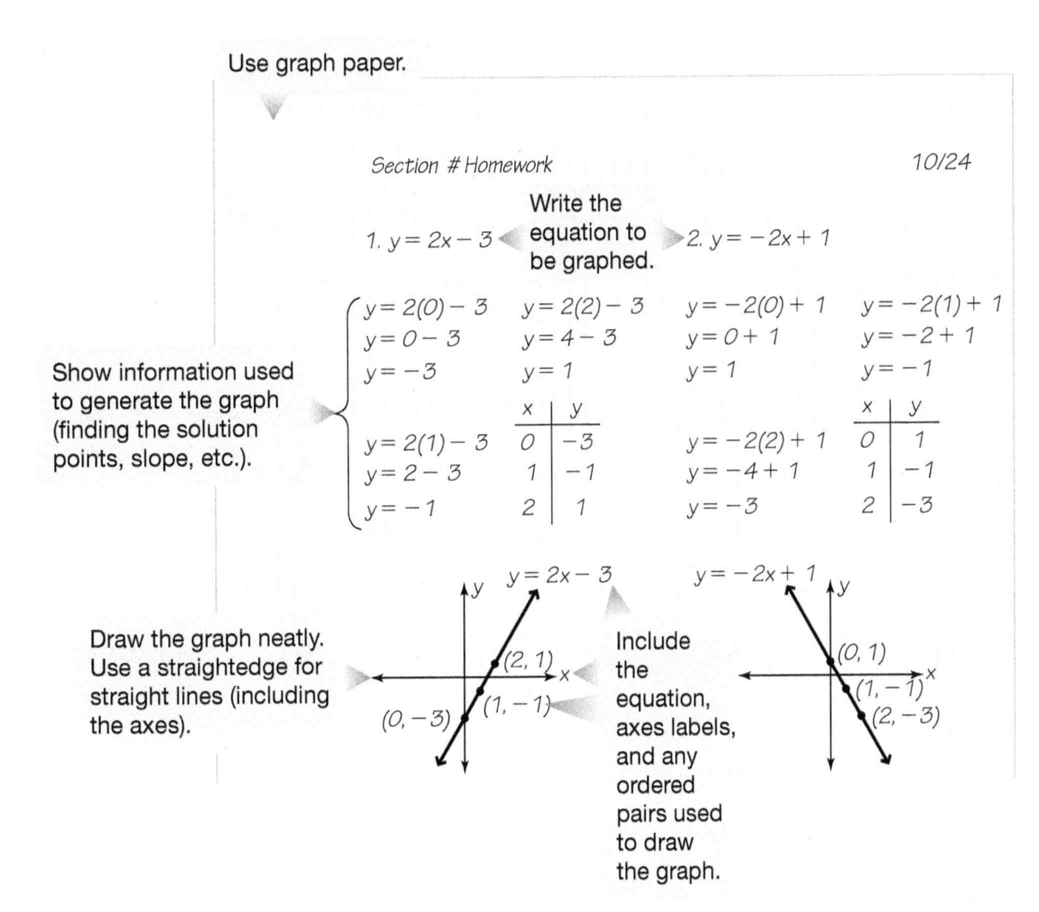

Study Materials

This section of the notebook contains three types of study materials for each chapter.

Study Material 1: The Study Sheet A study sheet contains *every* rule or procedure in the current chapter.

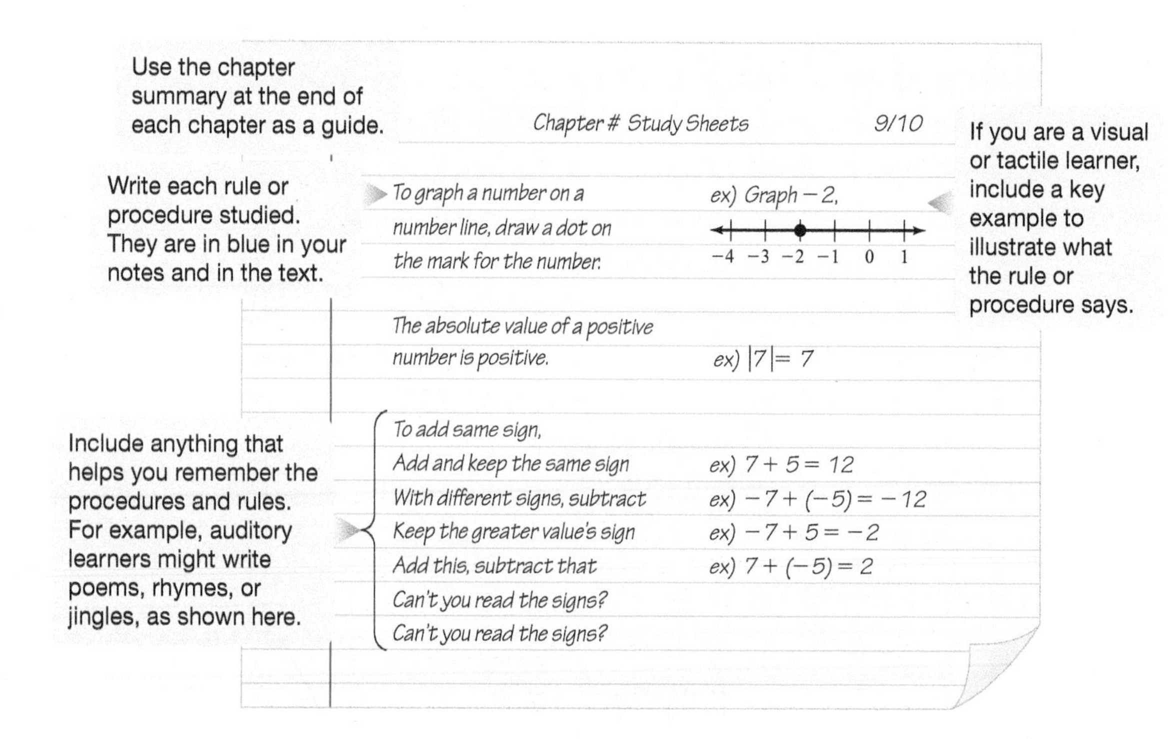

Study Material 2: The Practice Test If your instructor gives you a practice test, proceed to the discussion of creating a game plan.

If your instructor does not give you a practice test, use your notes to create your own practice test from the examples given in class. Include only the instructions and the problem, not the solutions. The following sample practice test was created from examples in the notes for Chapter 2 in a prealgebra course.

Chapter # Practice Test 10/10

For # 1 and 2, graph on a number line.

1. 4
2. -3

For each example in your notes, write the directions and the problem but not the solution.

For #3 and 4, simplify.

3. $|-8|$
4. $|9|$

After working through the practice test, use your notes to check your solutions.

For #5 and 6, find the additive inverse of the number.

5. 13
6. -15

For #7 – 10, add.

7. $13 + (-9)$
8. $-20 + (-6)$
9. $-15 + 8$
10. $3 + (-24)$

Study Material 3: The Game Plan The game plan refines the study process further. It is your plan for the test based on the practice test. For each problem on your practice test, write the definition, rule, or procedure used to solve the problem.

The sample shown gives the rule or procedure used to solve each problem on the preceding sample practice test. The rules and procedures came from the sample study sheet.

Chapter # Game Plan *9/10*

Multiple problems that use the same rule or procedure can be grouped together.

#1 and 2: Draw a dot on the mark for the number.

#3 and 4: The absolute value of a positive number is a positive number.
The absolute value of a negative number is a positive number.
The absolute value of 0 is 0.

Write the rule or procedure used to solve the problems on the practice test.

#5 and 6: Additive inverses are numbers whose sum is 0.

#7 – 10: To add same sign,
Add and keep the same sign
With different signs
Subtract and keep the greater value's sign
Add this, subtract that
Can't you read the signs?
Can't you read the signs?

Quizzes/Tests

Archive all returned quizzes and tests in this section of the notebook.

- Midterm and final exam questions are often taken from the quizzes and tests, so they make excellent study tools for those cumulative exams.
- Keeping all graded quizzes and tests offers a backup system in the unlikely event your instructor should lose any of your scores.
- If a dispute arises about a particular score, you have the graded test to show your instructor.

Chapter # Quiz *10/10*

For 1-4, simplify.

1. $|-8|$ $= 8$ ✓
2. $|9|$ $= 9$ ✓
3. $-15 + 5$ $= -10$ ✓
4. $-8 + -6$ $= -14$ ✓

4/4 = 100% Nice work!

CHAPTER

1

Real Numbers and Expressions

Chapter Overview

Algebra is like a building. It has a foundation, and various topics build upon that foundation and connect to one another to form a beautiful structure. In this chapter, we briefly review the foundation, which is arithmetic. More specifically, we review the following topics:

- Number sets
- Properties of real numbers
- Evaluating numeric and variable expressions
- Simplifying expressions

1.1 Sets and the Structure of Algebra

Objectives

1. Understand the structure of algebra.
2. Classify number sets.
3. Determine the absolute value of a number.
4. Compare numbers.

Learning Strategy

It is important to sit up front in class since it helps you avoid distractions and makes it easier to hear any hints the professor might be giving.

—Jason J.

Connection Think of expressions as phrases and equations as complete sentences. The expression $2 + 6$ is read "two plus six," which is not a complete sentence. The equation $2 + 6 = 8$ can be read as "Two plus six equals eight." Notice that the equal sign translates to the verb "equals"; thus, we have a complete sentence.

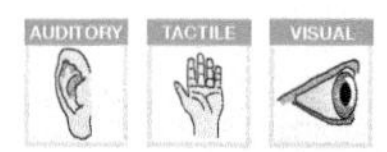

Learning Strategy

Developing a good study system and understanding how you best learn is essential to academic success. If you haven't already done so, read the study system in the To the Student section at the beginning of the text. Also take a moment to complete the Learning Styles Inventory found on page xv to discover your learning style. In these learning strategy boxes, we offer tips on how to connect the study system and your learning style to help you be successful in this course.

Warm-up *Refer to the To the Student Section*

1. What are the four sections of the notebook?
2. What is the color code for notes?
3. Complete the Learning Styles Inventory. What is your learning style?

Objective 1 Understand the structure of algebra.

In algebra, our basic components are **variables** and **constants,** which we use to build **expressions, equations,** and **inequalities.**

Definitions **Variable:** A symbol varying in value.
Constant: A symbol that does not vary in value.
Expression: A constant; a variable; or any combination of constants, variables, and arithmetic operations.
Equation: Two expressions set equal to each other.
Inequality: Two expressions separated by $\neq$, $<$, $>$, $\leq$, **or** $\geq$.

Variables are usually letters of the alphabet such as x or y. Constants are symbols for specific values such as $1, 2, \frac{3}{4}, 6.74$, or π. Following are some examples of expressions.

$$2 + 6 \qquad 4x - 5 \qquad \frac{1}{3}\pi r^2 h$$

Note Expressions do *not* contain an equal sign or inequality symbol.

Examples of equations are as follows:

$$2 + 6 = 8 \qquad 4x - 5 = 12 \qquad V = \frac{1}{3}\pi r^2 h.$$

The following table contains examples of inequalities and their verbal translations.

Inequality Symbols and Their Translations

Symbolic Form	Translation
$8 \neq 3$	Eight is not equal to three.
$5 < 7$	Five is less than seven.
$7 > 5$	Seven is greater than five.
$x \leq 3$	x is less than or equal to three.
$y \geq 2$	y is greater than or equal to two.

The following Algebra Pyramid illustrates how variables, constants, expressions, equations, and inequalities relate to one another. Constants and variables form

Answers to Warm-up

1. notes, homework, study materials, graded work
2. red = definitions; blue = rules/procedures; pencil = all other notes
3. Answers will vary.

the foundation on which we build expressions, which in turn form equations and inequalities.

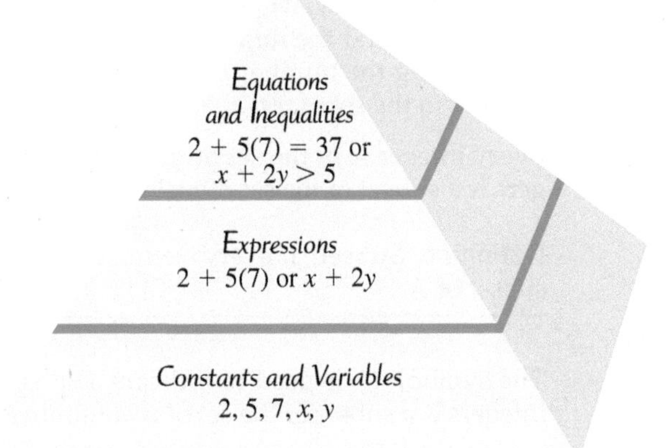

Note During this course, we move back and forth between expressions, equations, and inequalities. When we change topics, we use this Algebra Pyramid as a visual aid so that it's clear what we are working on.

Objective 2 Classify number sets.

In mathematics, we often group numbers, variables, or other objects in **sets**.

Definition **Set:** A collection of objects.

Braces are used to indicate a set. For example, the set containing the numbers 1, 2, 3, and 4 can be written $\{1, 2, 3, 4\}$. The numbers 1, 2, 3, and 4 are called the *members* or *elements* of this set.

Example 1 Write the set containing the last five letters of the English alphabet.

Answer: $\{V, W, X, Y, Z\}$

Your Turn 1 Write the set containing the first four months of the year.

Sets can contain a finite (countable) number of elements, an infinite number of elements, or no elements at all. A set with no elements is an *empty set,* which we write using empty braces { } or the symbol $\varnothing$. The set in Example 1 is a finite set with five elements. Some examples of infinite sets are the **set of natural numbers**, the **set of whole numbers**, and the **set of integers**.

Definitions **The set of natural numbers:** $\{1, 2, 3, \ldots\}$
The set of whole numbers: $\{0, 1, 2, 3, \ldots\}$
The set of integers: $\{\ldots, -3, -2, -1, 0, 1, 2, 3, \ldots\}$

Note The three dots are known as an *ellipsis* and indicate that the numbers continue forever in the same pattern.

So far, we have written sets in *roster form,* which means that we have listed each element or, in the case of infinite sets, the pattern of the set. We can also write sets in *set-builder notation,* where we describe what the set contains. Here we write the whole numbers from 0 to 4 using both notations.

Roster form: $\{0, 1, 2, 3, 4\}$

Set builder: $\{x \mid x \text{ is a whole number and } x \leq 4\}$

Note The set-builder notation is read "the set of all x such that x is a whole number and x is less than or equal to 4."

The symbol $\in$ is read "is an element of" and indicates that an object is a member of a set. For example, $-3 \in \{x \mid x \text{ is an integer}\}$ is a true statement indicating that -3 "is an element of" the set of integers. The symbol $\notin$ is translated "is not an element of." The statement $-3 \notin \{x \mid x \text{ is a whole number}\}$ is true because -3 is not an element of the set of whole numbers.

Answer to Your Turn 1
{January, February, March, April}

Number lines are often used in mathematics. The following number line has the integers from −6 to 6 marked and the integer 2 plotted.

Although we can view only a portion of a number line, the arrows at the ends indicate that the line and the numbers on it continue forever in both directions. If we were to travel along the number line forever in both directions, we would encounter every number in the set of *real numbers*.

Every integer is in the set of real numbers. Consequently, we say that the set of integers is a **subset** of the set of real numbers.

Definition **Subset:** If every element of a set B is an element of a set A, then B is a subset of A.

Note The following letter symbols are often used to indicate the given number sets.

Real numbers	$\boldsymbol{R}$
Irrational numbers	$\boldsymbol{I}$
Rational numbers	$\boldsymbol{Q}$
Integers	$\boldsymbol{Z}$
Whole numbers	$\boldsymbol{W}$
Natural numbers	$\boldsymbol{N}$

The symbol $\subseteq$ is used to indicate a subset. For example, to indicate that the set of integers is a subset of the set of real numbers, we write $\boldsymbol{Z} \subseteq \boldsymbol{R}$.

The set of real numbers also contains numbers that are not integers, such as fractions. For example, the following number line has the numbers $\frac{5}{4}$ and −0.3 plotted, which are not integers.

Numbers such as $\frac{5}{4}$ and −0.3, along with the integers, are in a subset of the real numbers called the set of **rational numbers**.

Note In the definition for a rational number, the notation $b \neq 0$ is important because if the denominator b were to equal 0, the fraction would be undefined. The reason it is undefined will be explained later.

Definition **Rational number:** Any real number that can be expressed in the form $\frac{a}{b}$, where a and b are integers and $b \neq 0$.

The number $\frac{5}{4}$ is a rational number because 5 and 4 are integers. Similarly, −0.3 is rational because it can be written as $-\frac{3}{10}$. All of the integers are rational numbers because they can be written in the form $\frac{a}{b}$, where b is 1. For example, $7 = \frac{7}{1}$. Some less obvious rational numbers are decimal numbers such as $0.\overline{6}$. The bar over the 6 in $0.\overline{6}$ is called a repeat bar and means that the digit 6 repeats forever, as in 0.6666 . . . . It is a rational number because $0.\overline{6}$ can be written as $\frac{2}{3}$.

Note We will learn how to write numbers with repeating decimal digits as fractions in Appendix B. For now, you just need to be able to recognize that these types of decimal numbers are rational numbers.

Some real numbers are not rational numbers. For example, the exact value of the real number π cannot be expressed as a ratio of integers or as a repeating decimal. Real numbers such as π are called **irrational numbers**.

Definition **Irrational number:** Any real number that is not rational.

Some other irrational numbers are $\sqrt{2}$ and $\sqrt{3}$. Square roots are explained in more detail in Section 1.3. Because an irrational number cannot be written as a ratio of integers, if a calculation involves an irrational number, we must leave it in symbolic form or use a rational number approximation. We can approximate π with rational numbers such as 3.14 or $\frac{22}{7}$.

The following diagram illustrates how the set of real numbers and all of its subsets are organized. Note the letters used to indicate each set.

Real Numbers

Rational Numbers: Real numbers that can be expressed in the form $\frac{a}{b}$, where a and b are integers and $b \neq 0$, such as $-4\frac{3}{4}$, $-\frac{2}{3}$, 0.018, $0.\overline{3}$, and $\frac{5}{8}$.

Integers: $\ldots, -3, -2, -1, 0, 1, 2, 3, \ldots$

Whole Numbers: $0, 1, 2, 3, \ldots$

Natural Numbers: $1, 2, 3, \ldots$

Irrational Numbers: Any real number that is not rational, such as $-\sqrt{2}$, $-\sqrt{3}$, $\sqrt{0.8}$, and π.

Another set of numbers that we often use in mathematics is the set of **prime numbers**, which is a subset of the natural numbers.

Definition **Prime number:** A natural number with exactly two different factors, 1 and the number itself.

There are an infinite number of prime numbers. The first ten prime numbers are 2, 3, 5, 7, 11, 13, 17, 19, 23, and 29.

Example 2 Determine whether the statement is true or false.

a. $-5 \in \{x \mid x \text{ is an integer}\}$

Answer: True because -5 is a member of the set of integers.

b. $\frac{2}{3} \in W$

Answer: False because $\frac{2}{3}$ is not a whole number.

c. November $\notin \{n \mid n \text{ is a month}\}$

Answer: False because November is a member of the set containing the months.

d. Given $A = \{1, 2, 3, 4, 5\}$ and $B = \{2, 3, 5\}$, then $B \subseteq A$.

Answer: True; B is a subset of A because every element of B is in A.

e. $0.4 \in Q$

Answer: True because 0.4 is the decimal equivalent of $\frac{2}{5}$; so 0.4 is a rational number.

Your Turn 2 Determine whether the statement is true or false.

a. Friday $\in \{n \mid n \text{ is a day of the week}\}$
b. $0.2 \notin I$
c. Given $A = \{1, 2, 3, 4, 5\}$ and $B = \{2, 3, 5\}$, then $A \subseteq B$.

Answers to Your Turn 2
a. true **b.** true **c.** false

Objective 3 Determine the absolute value of a number.

In mathematics, we often discuss the magnitude, or **absolute value**, of a number.

Definition **Absolute value:** A number's distance from zero on a number line.

For example, the absolute value of 5, written $|5|$, is 5 because it is 5 units from 0 on a number line. Similarly, $|-5| = 5$ because -5 is also 5 units from zero.

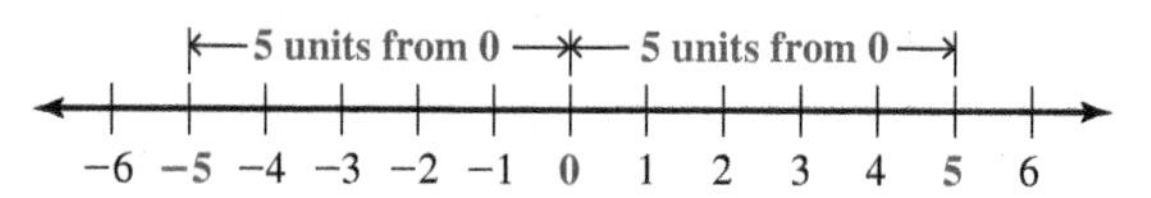

Consider $|0|$. Because 0 is 0 units from itself, $|0| = 0$.

Rule Absolute Value

The absolute value of every real number is either positive or 0 according to the following rule:

$$|n| = \begin{cases} n \text{ when } n \geq 0 \\ -n \text{ when } n < 0 \end{cases}$$

Example 3 Determine the absolute value.

a. $|-2.3|$

Answer: $|-2.3| = 2.3$

b. $\left|\frac{1}{4}\right|$

Answer: $\left|\frac{1}{4}\right| = \frac{1}{4}$

Your Turn 3 Determine the absolute value.

a. $\left|-4\frac{1}{2}\right|$

b. $|9.8|$

Objective 4 Compare numbers.

We can also use a number line to determine which of two numbers is greater. Because numbers increase from left to right on a number line, the number farther to the right is the greater of two numbers.

Rule Comparing Numbers

For any two real numbers a and b, a is greater than b if a is to the right of b on a number line.

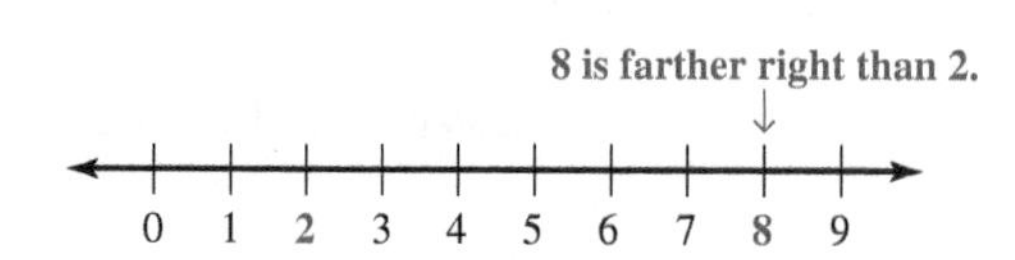

Because the number 8 is farther to the right on a number line than the number 2, we say that 8 is greater than 2, or in symbols, $8 > 2$.

Answers to Your Turn 3

a. $4\frac{1}{2}$ b. 9.8

Example 4 Use =, <, or > to write a true statement.

a. 6 $\quad$ −8

Answer: $6 > -8$ because 6 is farther right on a number line than −8.

b. −3.2 $\quad$ −3.1

Answer: As we see on the following number line, $-3.2 < -3.1$ because −3.1 is farther right than −3.2.

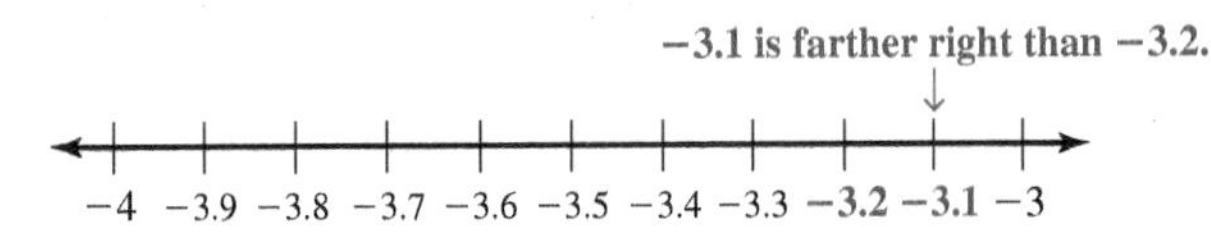

c. $\left|-2\frac{3}{4}\right| \quad 2.75$

Answer: $\left|-2\frac{3}{4}\right| = 2.75$ because the absolute value of $-2\frac{3}{4}$ is $2\frac{3}{4}$, which is 2.75 when written as a decimal number.

Answers to Your Turn 4
a. $-14 > -25$ b. $-2.4 < 0$
c. $3\frac{5}{6} > 3\frac{1}{4}$ d. $\left|-9\frac{1}{2}\right| = 9.5$

Your Turn 4 Use =, <, or > to write a true statement.

a. −14 $\quad$ −25 **b.** −2.4 $\quad$ 0 **c.** $3\frac{5}{6} \quad 3\frac{1}{4}$ **d.** $\left|-9\frac{1}{2}\right| \quad 9.5$

1.1 Exercises

For Extra Help MyMathLab®

Note: Exercises marked with a ★ represent challenging exercises.

Objective 1

Prep Exercise 1 Explain the difference between a constant and a variable.

Prep Exercise 2 An expression is a constant, a variable, or any combination of constants, variables, and arithmetic operations that describes a ______________.

Prep Exercise 3 An equation is a mathematical relationship that contains an ______________.

Prep Exercise 4 List the inequality symbols.

Prep Exercise 5 Complete the name of each level of the Algebra Pyramid.

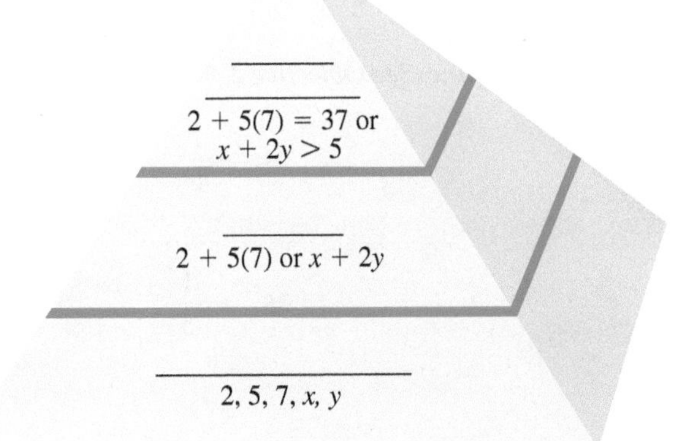

Prep Exercise 6 What is a *set*?

Prep Exercise 7 To write a set, write the members or elements of the set separated by commas within ______________, ______________.

Prep Exercise 8 What is the difference between a finite set and an infinite set?

Prep Exercise 9 We can indicate an empty set using { } or $\varnothing$, but not $\{\varnothing\}$. Why is $\{\varnothing\}$ incorrect?

For Exercises 1–14, write a set representing each statement. See Example 1.

1. The name of the weekend days

2. The vowels of the English language

3. The months whose name begin with J

4. The months containing only 30 days

5. The states in the United States whose names begin with the word North

6. The states in the United States whose names begin with *W*

7. The whole numbers less than 5

8. The prime numbers less than 10

9. The natural-number multiples of 3

10. The natural-number multiples of 5

11. The integers greater than $-\frac{5}{4}$ and less than $\frac{4}{3}$

12. The integers greater than $-\frac{13}{6}$ and less than $\frac{7}{3}$

13. The integers greater than $-\frac{3}{4}$ and less than $-\frac{1}{4}$

14. The integers greater than 2 and less than -3

For Exercises 15–22, write the set in set-builder notation. See Objective 2.

15. $\{\ldots, -2, -1, 0, 1, 2, \ldots\}$

16. $\{1, 2, 3, \ldots\}$

17. $\{a, b, c, \ldots, x, y, z\}$

18. $\{a, e, i, o, u\}$

19. {Monday, Tuesday, Wednesday, Thursday, Friday, Saturday, Sunday}

20. {January, February, March, April, May, June, July, August, September, October, November, December}

21. $\{5, 10, 15, \ldots\}$

22. $\{4, 8, 12, \ldots\}$

For Exercises 23–30, graph each number on a number line. See Objective 2.

23. $\frac{2}{3}$

24. $-\frac{3}{4}$

25. $1\frac{3}{8}$

26. $-8\frac{1}{5}$

27. -2.1

28. 5.8

29. 3.62

30. -1.22

Objective 2

Prep Exercise 10 What is a subset of a set?

Prep Exercise 11 Explain the difference between a rational number and an irrational number.

Prep Exercise 12 Every real number is either ____________ or ____________.

For Exercises 31–56, answer true or false. See Example 2.

31. $n \in \{a, e, i, o, u\}$

32. $m \in \{l, m, n, o, p\}$

33. "go" $\in \{n \mid n \text{ is a verb}\}$

34. "eat" $\in \{n \mid n \text{ is a noun}\}$

35. $4 \in \{x \mid x \text{ is a rational number}\}$

36. $\pi \in \{x \mid x \text{ is an irrational number}\}$

37. "James" $\notin \{n \mid n \text{ is the last name of a U.S. president}\}$

38. "Florida" $\notin \{n \mid n \text{ is the last name of a state in the United States}\}$

39. $-0.6 \notin R$

40. $\sqrt{3} \notin Q$

41. $\{a, e, i, o, u\} \subseteq \{n \mid n \text{ is a letter of the English alphabet}\}$

42. $\{\text{red, blue, yellow}\} \subseteq \{n \mid n \text{ is a color}\}$

43. $\{1, 3, 5, 7, \ldots\} \subseteq Z$

44. $\{2, 4, 6, 8, \ldots\} \subseteq Q$

45. $Z \subseteq Q$

46. $Q \subseteq Z$

47. $I \subseteq R$

48. $N \subseteq Q$

49. A number exists that is both rational and irrational.

50. All integers are irrational numbers.

51. The only difference between the set of whole numbers and the set of natural numbers is that the set of whole numbers contains 0.

52. The set of integers contains every negative number.

53. Every rational number can be written as a fraction.

54. A number expressed as a repeating decimal can be written in the form $\frac{a}{b}$, where a and b are integers and b is not equal to zero.

★ **55.** The set of rational numbers contains the set of natural numbers.

★ **56.** All real numbers are either positive or negative.

Objective 3

Prep Exercise 13 The absolute value of every real number is either ________ or ________.

For Exercises 57–64, determine the absolute value. See Example 3.

57. $|2.6|$ **58.** $|-2.8|$ **59.** $\left|-1\frac{2}{5}\right|$ **60.** $\left|1\frac{1}{4}\right|$ **61.** $|-1|$ **62.** $|0|$ **63.** $|-8.75|$ **64.** $|18|$

Objective 4

Prep Exercise 14 When comparing two numbers, the number farthest to the ________ on a number line is the greater number.

For Exercises 65–76, use =, <, or > to write a true statement. See Example 4.

65. $-6 \quad -8$

66. $-19 \quad -7$

67. $0 \quad -1.8$

68. $-3.7 \quad -1.6$

69. $3\frac{4}{5} \quad 3\frac{3}{4}$

70. $2\frac{3}{5} \quad 2\frac{1}{4}$

71. $|-3| \quad |3|$

72. $6.2 \quad |-6.2|$

73. $6.7 \quad |6.7|$

74. $|10.4| \quad 10.4$

75. $\left|-\frac{2}{3}\right| \quad \left|-\frac{4}{3}\right|$

76. $\left|-\frac{7}{4}\right| \quad \left|-\frac{5}{3}\right|$

For Exercises 77–80, list the given numbers in order from least to greatest. See Example 4.

77. $0.4, -0.6, 0, 3\frac{1}{4}, \left|-0.02\right|, -0.44, \left|1\frac{2}{3}\right|$

78. $-2.56, 5.4, |8.3|, \left|-7\frac{1}{2}\right|, -4.7$

79. $2.9, 1, \left|-2\frac{3}{4}\right|, -12.6, |-1.3|, -9.6$

80. $-1, \pi, |-0.05|, -1.3, \left|4\frac{2}{3}\right|, 0.\overline{4}$

For Exercises 81–84, use the graph provided.

81. Write a set containing the three years from 2002 to 2012 that had the greatest rate of inflation.

82. Write a set containing the three years from 2002 to 2012 that had the least rate of inflation.

83. Write a set containing the years from 2002 to 2012 that had an inflation rate of at least 3%.

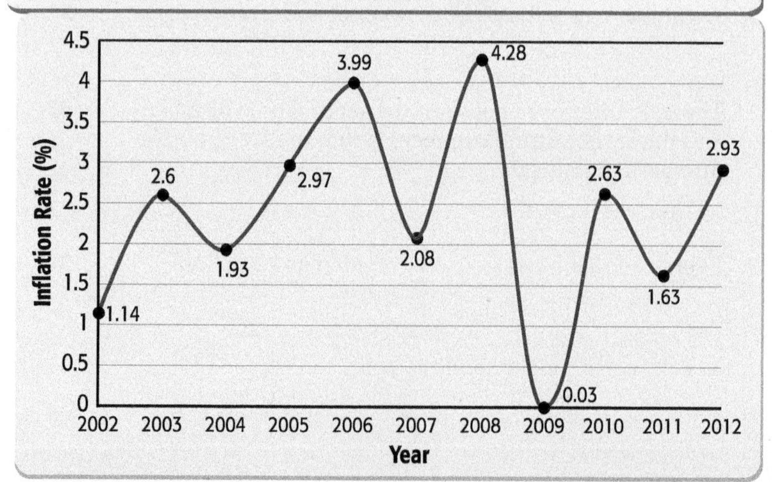

84. Write a set containing the years from 2002 to 2012 that had an inflation rate less than 2%.

For Exercises 85–88, use the following graph, which shows the percent of the market using each type of browser on their desktop computer.

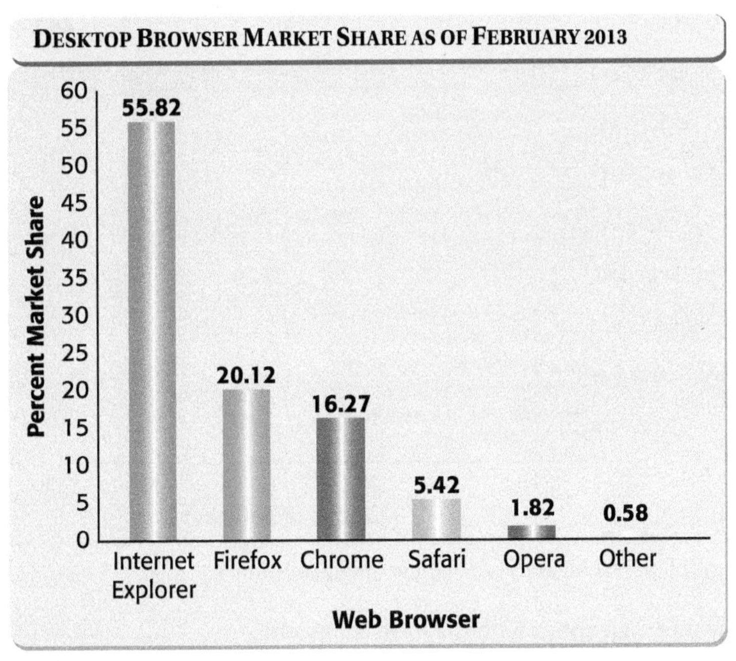

Source: Market Share Reports

85. Write a set containing the three most popular desktop browsers.

86. Write a set containing the two least popular desktop browsers named.

87. Write a set containing the browsers that have more than 20% of the market share.

88. Write a set containing the browser with exactly 10% of the market share.

Of Interest

On mobile devices, Safari is the most popular browser with 55.41% of the market.

For Exercises 89–92, use the following graph which shows the ten celebrities with the greatest Twitter followings.

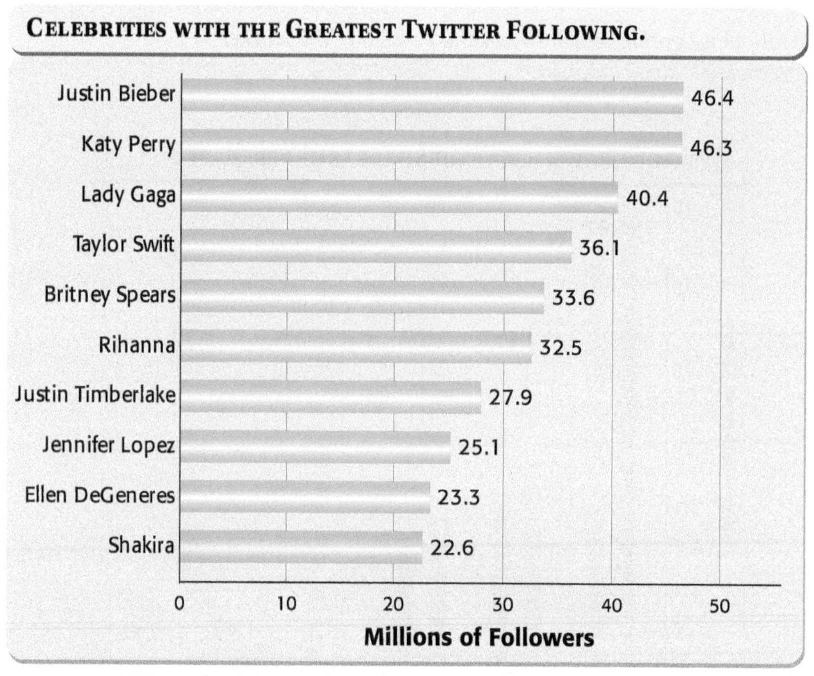

Source: http://socialbakers.com/twitter/group/celebrities

89. Write a set containing the three celebrities with the most followers.

90. Write a set containing the three of the top ten celebrities with the least followers.

91. Write a set containing the celebrities that have more than 35 million Twitter followers.

92. Write a set containing the celebrity with exactly 50 million Twitter followers

1.2 Operations with Real Numbers; Properties of Real Numbers

Objectives

1 Add real numbers.

2 Subtract real numbers.

3 Multiply real numbers.

4 Divide real numbers.

Warm-up

[1.1] **1.** Use $=$, $<$, or $>$ to write a true statement.

a. $-12 \quad 12$ **b.** $|-12| \quad 12$ **c.** $-12 \quad -16$

[1.1] For Exercises 2 and 3, answer true or false.

2. $\sqrt{2}$ is a rational number.

3. The natural numbers are a subset of the integers.

Answers to Warm-up

1. a. $<$ **b.** $=$ **c.** $>$
2. false **3.** true

Objective 1 Add real numbers.

Addition Properties of Real Numbers

First, let's consider some properties of addition that are true for all real numbers. The following table summarizes these properties.

Property of Addition	Symbolic Form	Word Form
Identity property of addition	$a + 0 = a$	The sum of a number and 0, the additive identity, is that number.
Commutative property of addition	$a + b = b + a$	Changing the order of addends does not affect the sum.
Associative property of addition	$a + (b + c) = (a + b) + c$	Changing the grouping of three or more addends does not affect the sum.

Example 1 Identify the property of addition that is illustrated in the given equation.

a. $3(a + 2) = 3(2 + a)$

Solution: Commutative property. The order of the addends a and 2 changed.

b. $8 + 0 = 8$

Solution: Identity property. 0 is the additive identity.

c. $(2 + x) + 3 = 2 + (x + 3)$

Solution: Associative property. The grouping of the addends changed.

Your Turn 1 Identify the property of addition that is illustrated in the given equation.

a. $x + 2 = 2 + x$ **b.** $(3 + m) + n = 3 + (m + n)$ **c.** $t + 0 = t$

Adding Signed Numbers

Now we consider how to add signed numbers.

Note We can relate the addition of signed numbers to money. Example 2(a) illustrates a sum of two debts, which increases the amount of debt. Example 2(b) illustrates a credit with a debt where the value of the credit is more than the debt; so the result is a credit. In Example 2(c), we see that the value of the debt is more than the credit; so the result is a debt.

Procedure Adding Real Numbers

To add two numbers that have the *same sign*, add their absolute values and keep the same sign.

To add two numbers that have *different signs*, subtract the smaller absolute value from the greater absolute value and keep the sign of the number with the greater absolute value.

Example 2 Add.

a. $-15 + (-21)$

Solution: $-15 + (-21) = -36$

b. $26 + (-17)$

Solution: $26 + (-17) = 9$

c. $-28 + 5$

Solution: $-28 + 5 = -23$

Answers to Your Turn 1

a. commutative property
b. associative property
c. identity property

d. $-\frac{4}{5} + \frac{2}{3}$

Solution: $-\frac{4}{5} + \frac{2}{3} = -\frac{4(3)}{5(3)} + \frac{2(5)}{3(5)}$ Write equivalent fractions with their least common denominator (LCD), 15.

$= -\frac{12}{15} + \frac{10}{15}$

$= \frac{-12 + 10}{15}$ Add numerators and keep the common denominator. Because the addends have different signs, we subtract and keep the sign of the number with the greater absolute value.

$= -\frac{2}{15}$

e. $-12.5 + (-16.4)$

Solution: $-12.5 + (-16.4) = -28.9$

Your Turn 2 Add.

a. $-52 + (-16)$ b. $37 + (-42)$ c. $-29 + 12$

d. $-\frac{1}{6} + \left(-\frac{5}{8}\right)$ e. $-21.6 + 15.2$

Note Additive inverses add to equal the additive identity.

Note that adding two numbers that have the same absolute value but different signs gives us 0. For example, $6 + (-6) = 0$. We say that 6 and -6 are **additive inverses** or opposites.

Definition **Additive inverses:** Two numbers whose sum is 0.

Example 3 Find the additive inverse of the given number.

a. -1.6

Answer: 1.6 because $-1.6 + 1.6 = 0$

b. $\frac{7}{12}$

Answer: $-\frac{7}{12}$ because $-\frac{7}{12} + \frac{7}{12} = 0$

Your Turn 3 Find the additive inverse of the given number.

a. -7 b. 0.8 c. $-\frac{3}{4}$

We sometimes indicate the additive inverse, or opposite, of a number using a negative sign. For example, the additive inverse of -7 can be written symbolically as $-(-7)$, which when simplified is 7.

Rule **Double Negative Property**

For any real number n, $-(-n) = n$.

Example 4 Simplify.

a. $-(-(-6))$

Solution: $-(-(-6)) = -(6)$ Simplify $-(-6)$, which is 6.

$= -6$ Find the additive inverse of 6.

Answers to Your Turn 2

a. -68 b. -5 c. -17 d. $-\frac{19}{24}$ e. -6.4

Answers to Your Turn 3

a. 7 b. -0.8 c. $\frac{3}{4}$

Warning Don't confuse $-|-15|$ with $-(-15)$. The expression $-|-15|$ indicates the additive inverse of the absolute value of -15, which is -15. The expression $-(-15)$ indicates the additive inverse of -15, which is 15.

b. $-|-15|$

Solution: $-|-15| = -(15)$ Simplify $|-15|$, which is 15.

$= -15$ Find the additive inverse of 15.

Your Turn 4 Simplify.

a. $-(-(5))$ **b.** $-|13|$

Objective 2 Subtract real numbers.

Recall that when we add two numbers that have different signs, we actually subtract, which suggests that subtraction and addition are related. In fact, every subtraction statement can be written as an equivalent addition statement. For example, the subtraction statement $8 - 5 = 3$ and the addition statement $8 + (-5) = 3$ are equivalent. Note that in writing $8 - 5$ as $8 + (-5)$, we change the operation sign and change the subtrahend (number being subtracted) 5 to its additive inverse.

$$8 - 5$$

Change the operation from minus to plus. Change the subtrahend to its additive inverse.

$$= 8 + (-5)$$

$$= 3$$

Procedure Rewriting Subtraction

To write a subtraction statement as an equivalent addition statement, change the operation symbol from a minus sign to a plus sign and change the subtrahend to its additive inverse.

Example 5 Subtract.

Connection Notice the use of the double negative property when changing $-(-6)$ to $+6$.

a. $-15 - (-6)$

Solution: $-15 - (-6) = -15 + 6$ Write as an equivalent addition.

$= -9$

b. $-\frac{2}{3} - \frac{3}{4}$

Note Just because we *can* rewrite subtraction does not mean we *have* to rewrite a subtraction as an equivalent addition. For example, it is easy enough to determine the result of $9 - 2$ without rewriting it as $9 + (-2)$.

Solution: $-\frac{2}{3} - \frac{3}{4} = -\frac{2}{3} + \left(-\frac{3}{4}\right)$ Write as an equivalent addition.

$$= -\frac{2(4)}{3(4)} + \left(-\frac{3(3)}{4(3)}\right)$$

$$= -\frac{8}{12} + \left(-\frac{9}{12}\right)$$

$$= -\frac{17}{12}$$

c. $5.04 - 8.01$

Solution: $5.04 - 8.01 = 5.04 + (-8.01)$

$= -2.97$

Warning Subtraction is neither commutative nor associative. For example, $5.04 - 8.01 = -2.97$ whereas $8.01 - 5.04 = 2.97$.

Your Turn 5 Subtract.

a. $25 - (-17)$ **b.** $-\frac{3}{8} - \left(-\frac{1}{3}\right)$ **c.** $0.06 - 4.02$

Answers to Your Turn 4

a. 5 b. -13

Answers to Your Turn 5

a. 42 b. $-\frac{1}{24}$ c. -3.96

Objective 3 Multiply real numbers.

Properties of Multiplication

Like addition, multiplication has properties, which we list in the following table.

Property of Multiplication	Symbolic Form	Word Form
Multiplicative property of 0	$0 \cdot a = 0$	The product of a number and 0 is 0.
Identity property	$1 \cdot a = a$	The product of a number and 1, the multiplicative identity, is the number.
Commutative property of multiplication	$ab = ba$	Changing the order of factors does not affect the product.
Associative property of multiplication	$a(bc) = (ab)c$	Changing the grouping of three or more factors does not affect the product.
Distributive property of multiplication over addition	$a(b + c) = ab + ac$	A sum multiplied by a factor is equal to the sum of that factor multiplied by each addend.

Example 6 Identify the property of multiplication that is illustrated in the given equation.

a. $1 \cdot 7 = 7$

Solution: Identity property. 1 is the multiplicative identity.

b. $xy = yx$

Solution: Commutative property. The order of the factors changed.

c. $(2 \cdot 3) \cdot 5 = 2 \cdot (3 \cdot 5)$

Solution: Associative property. The grouping of the factors changed.

Your Turn 6 Identify the property of multiplication that is illustrated in the given equation.

a. $xyz = yxz$ **b.** $5 \cdot 0 \cdot n = 0$ **c.** $2(x + 3) = 2x + 6$

Multiplying Signed Numbers

Now let's consider multiplying signed numbers.

Rules **Multiplying Two Signed Numbers**

The product of two numbers that have the *same sign* is positive.
The product of two numbers that have *different signs* is negative.

Example 7 Multiply.

a. $(-8)(-4)$

Solution: $(-8)(-4) = 32$

Answers to Your Turn 6
a. commutative property
b. multiplicative property of 0
c. distributive property

Warning Don't confuse $6(-7)$ with $6 - 7$. The expression $6(-7)$ indicates multiplication, whereas $6 - 7$ indicates subtraction.

b. $6(-7)$

Solution: $6(-7) = -42$

c. $-\frac{3}{4} \cdot \left(\frac{5}{6}\right)$

Solution: $-\frac{3}{4} \cdot \left(\frac{5}{6}\right) = -\frac{3 \cdot 5}{4 \cdot 6} = -\frac{15}{24} = -\frac{5}{8}$

d. $(-2.3)(-0.07)$

Solution: $(-2.3)(-0.07) = 0.161$

Your Turn 7 Multiply.

a. $-12(-6)$ **b.** $-4(9)$ **c.** $\left(-\frac{5}{8}\right)\left(-\frac{7}{10}\right)$ **d.** $(-0.6)(7.82)$

When multiplying more than two factors, we multiply from left to right. However, the following examples suggest that to determine the sign of the answer, we can simply consider the number of negative factors.

$$(-2)(-3)(4) = 6(4) = 24 \qquad (-2)(-3)(-4) = 6(-4) = -24$$

Even number of negative factors → Positive result

Odd number of negative factors → Negative result

Rule **Multiplying with Negative Factors**

The product of an even number of negative factors is positive, whereas the product of an odd number of negative factors is negative.

Warning Remember that we multiply in order from left to right. A common error is to "distribute" the first factor over the remaining factors. For example,

$4(-3)(5) \neq (-12)(20)$.

$4(-3)(5) = (-12)(5) = -60$.

Example 8 Multiply.

a. $(-1)(-1)(-3)(8)$

Solution: $(-1)(-1)(-3)(8) = -24$ Because there is an odd number of negative factors, the result is negative.

b. $(-1)(-2)(-5)(3)(-4)$

Solution: $(-1)(-2)(-5)(3)(-4) = 120$ Because there is an even number of negative factors, the result is positive.

Your Turn 8 Multiply.

a. $(-1)(6)(-4)(2)$ **b.** $(7)(-2)(-1)(3)(-1)$

Every nonzero real number has a **multiplicative inverse**.

Definition **Multiplicative inverses:** Two numbers whose product is 1.

For example, $\frac{2}{3}$ and $\frac{3}{2}$ are multiplicative inverses because their product is 1.

$$\frac{2}{3} \cdot \frac{3}{2} = \frac{6}{6} = 1$$

Connection Additive inverses *add* to equal the additive identity, 0, and multiplicative inverses *multiply* to equal the multiplicative identity, 1.

Notice that to write a number's multiplicative inverse, we simply invert the numerator and denominator. Multiplicative inverses are also known as *reciprocals*.

Answers to Your Turn 7

a. 72 **b.** -36 **c.** $\frac{7}{16}$ **d.** -4.692

Answers to Your Turn 8

a. 48 **b.** -42

Example 9 Find the multiplicative inverse of the given number.

a. $-\frac{3}{8}$

Answer: $-\frac{8}{3}$ because $-\frac{3}{8}\cdot\left(-\frac{8}{3}\right) = 1$

b. 6

Answer: $\frac{1}{6}$ because $6\cdot\frac{1}{6} = 1$

Your Turn 9 Find the multiplicative inverse of the given number.

a. $\frac{4}{5}$ b. $-\frac{1}{7}$ c. 9

Objective 4 Divide real numbers.

Recall that we can write a subtraction statement as an equivalent addition statement using the additive inverse. Similarly, we can write a division statement as an equivalent multiplication statement using the multiplicative inverse. We use this rule when dividing fractions.

Change the operation from division to multiplication.

$$\frac{1}{3} \div \frac{2}{5} = \frac{1}{3}\cdot\frac{5}{2} = \frac{5}{6}$$

Change the divisor to its multiplicative inverse.

Procedure Rewriting Division

To write a division statement as an equivalent multiplication statement, change the operation symbol from division to multiplication and change the divisor to its multiplicative inverse.

The rules for determining the sign of a quotient, which is the result of a division problem, are the same as for determining the sign of a product.

Rules Dividing Signed Numbers

The quotient of two numbers that have the *same sign* is positive.
The quotient of two numbers that have *different signs* is negative.

Warning Division is neither commutative nor associative. For example, $-45 \div (-5) = 9$ whereas $-5 \div (-45) = 1/9$.

Example 10 Divide.

a. $-45 \div (-5)$

Solution: $-45 \div (-5) = 9$

b. $56 \div (-8)$

Solution: $56 \div (-8) = -7$

c. $-\frac{2}{3} \div \frac{5}{6}$

Solution:

$-\frac{2}{3} \div \frac{5}{6} = -\frac{2}{3}\cdot\frac{6}{5}$ Write an equivalent multiplication.

$= -\frac{12}{15}$ Multiply.

$= -\frac{4}{5}$ Simplify.

d. $-12.6 \div (-0.5)$

Solution: $-12.6 \div (-0.5) = 25.2$

Answers to Your Turn 9

a. $\frac{5}{4}$ b. -7 c. $\frac{1}{9}$

Your Turn 10 Divide.

a. $(-60) \div 12$ b. $-42 \div (-6)$ c. $\frac{5}{6} \div \left(-\frac{3}{4}\right)$ d. $9.03 \div (-4.3)$

Division Involving 0

If 0 is involved in division, we must be careful how we evaluate the result. If we divide 0 by a nonzero number, the quotient is 0. For example $0 \div 9 = 0$, which we can check using multiplication: $0 \cdot 9 = 0$.

Now consider division by 0, as in $8 \div 0$. To check this division, the quotient must multiply by 0 to equal 8, which is impossible because any number multiplied by 0 equals 0, not 8. Consequently, we say that dividing a nonzero number by 0 is *undefined*.

Now consider $0 \div 0$. To check this division, the quotient must multiply by 0 to equal 0. Notice the quotient could be any number because any number times 0 equals 0. Because we cannot determine a unique quotient, we say that $0 \div 0$ is *indeterminate*.

Following is a summary of these rules:

Rules **Division Involving 0**

$0 \div n = 0$ when $n \neq 0$.
$n \div 0$ is undefined when $n \neq 0$.
$0 \div 0$ is indeterminate.

Answers to Your Turn 10

a. -5 b. 7 c. $-\frac{10}{9}$ d. -2.1

1.2 Exercises

For Extra Help MyMathLab®

Objective 1

Prep Exercise 1 Why is 0 called the additive identity and 1 the multiplicative identity?

For Exercises 1–12, indicate whether the given equation illustrates the additive identity property, multiplicative identity property, additive inverses, or multiplicative inverses. See Examples 1, 3, 6, and 9.

1. $3 + (-3) = 0$
2. $4 \cdot \frac{1}{4} = 1$
3. $-6.1 + 0 = -6.1$
4. $-8\frac{1}{2} + 8\frac{1}{2} = 0$
5. $(-8.1)1 = -8.1$
6. $-\frac{2}{3}\left(-\frac{3}{2}\right) = 1$
7. $-\frac{1}{5}(-5) = 1$
8. $2\frac{3}{5} + 0 = 2\frac{3}{5}$
9. $-5.2(1) = -5.2$
10. $9\frac{1}{7} + 0 = 9\frac{1}{7}$
11. $-6 + 6 = 0$
12. $-\frac{2}{3} \cdot 1 = -\frac{2}{3}$

Prep Exercise 2 Explain why 8 and -8 are additive inverses.

Prep Exercise 3 What are multiplicative inverses?

For Exercises 13–20, find the additive inverse and multiplicative inverse. See Examples 3 and 9.

13. 8

14. 12

15. -7

16. -9

17. $-\frac{5}{8}$

18. $\frac{7}{2}$

19. 0.3

20. -2.5

Prep Exercise 4 Explain the difference between the commutative property of addition and the associative property of addition.

For Exercises 21–32, indicate whether the equation illustrates the commutative property of addition, commutative property of multiplication, associative property of addition, associative property of multiplication, or distributive property. See Examples 1 and 6.

21. $3 + 2 = 2 + 3$

22. $2 \cdot n \cdot 5 = 2 \cdot 5 \cdot n$

23. $5(x + y) = 5x + 5y$

24. $-\frac{1}{3} + \frac{2}{5} = \frac{2}{5} + \left(-\frac{1}{3}\right)$

25. $\frac{3}{4}\left(\frac{2}{9} \cdot \frac{5}{7}\right) = \left(\frac{3}{4} \cdot \frac{2}{9}\right)\frac{5}{7}$

26. $0.5 + [-8.1 + (-9)] = [0.5 + (-8.1)] + (-9)$

27. $m + (2m + 5) = (m + 2m) + 5$

28. $6t + 6u = 6(t + u)$

29. $2x + (3y + 5x) = 2x + (5x + 3y)$

30. $x(5 + y) = (5 + y)x$

31. $(x - 3)(x + 4) = (x + 4)(x - 3)$

32. $-3(mn) = (-3m)\,n$

Prep Exercise 5 Explain how to add two numbers that have the same sign.

Prep Exercise 6 Explain how to add two numbers that have different signs.

For Exercises 33–48, add. See Example 2.

33. $9 + (-16)$

34. $23 + (-29)$

35. $-27 + (-13)$

36. $-8 + (-12)$

37. $-15 + 9$

38. $-21 + 14$

39. $14 + (-19)$

40. $-32 + 16$

41. $-\frac{3}{4} + \frac{1}{6}$

42. $\frac{1}{4} + \left(-\frac{5}{6}\right)$

43. $-\frac{1}{8} + \left(-\frac{2}{3}\right)$

44. $-\frac{2}{5} + \left(-\frac{3}{4}\right)$

45. $-0.18 + 6.7$

46. $-0.81 + 4.28$

47. $-3.28 + (-4.1)$

48. $-7.8 + (-9.16)$

Prep Exercise 7 For any real number n, $-(-n) =$ ________.

For Exercises 49–56, simplify. See Example 4.

49. $-(-7)$

50. $-(-3)$

51. $-(-(-2.7))$

52. $-(-(-4.2))$

53. $-|12|$

54. $-|23|$

55. $-\left|-\frac{3}{4}\right|$

56. $-\left|-\frac{7}{8}\right|$

Objective 2

Prep Exercise 8 Explain how to write a subtraction statement as an equivalent addition statement.

For Exercises 57–72, subtract. See Example 5.

57. $2 - (-3)$

58. $4 - (-9)$

59. $10 - (-2)$

60. $8 - (-1)$

61. $7 - 11$

62. $6 - 21$

63. $8 - 3$

64. $17 - 13$

65. $\frac{7}{10} - \left(-\frac{3}{5}\right)$

66. $-\frac{1}{2} - \left(-\frac{1}{3}\right)$

67. $-\frac{1}{5} - \left(-\frac{1}{5}\right)$

68. $-\frac{2}{5} - \left(-\frac{4}{10}\right)$

69. $6.2 - 3.65$

70. $8.1 - 4.76$

71. $-6.1 - (-4.5)$

72. $-7.1 - (-2.3)$

Objectives 3 and 4

Prep Exercise 9 What is the sign of the product or quotient of two numbers with different signs?

Prep Exercise 10 The product of an even number of negative factors is ________.

The product of an odd number of negative factors is ________.

For Exercises 73–92, multiply or divide. See Examples 7, 8, and 10.

73. $4(-3)$

74. $5(-2)$

75. $(-2)(-1)$

76. $(-5)(-1)$

77. $-2 \cdot \frac{1}{4}$

78. $3 \cdot \left(-\frac{2}{9}\right)$

79. $-25 \div (-5)$

80. $-28 \div (-4)$

81. $-\frac{1}{4} \div \frac{2}{3}$

82. $\frac{3}{5} \div \left(-\frac{1}{3}\right)$

83. $-12 \div 0.3$

84. $-14 \div (-0.7)$

85. $-1(-2)(-3)$

86. $(4)(-2)(-3)$

87. $2.7(-0.1)(-2)$

88. $(-0.6)(-2.5)(-3)$

89. $-2(5)(-2)(-3)$

90. $(4)(-2)(-5)(-3)$

91. $(-1)(-3)(-5)(2)(3)$

92. $(4)(-2)(3)(-1)(2)$

For Exercises 93–96, solve.

93. On March 29, 2008, the NASDAQ index closed at 2261.18, which was a change of −19.65 from the previous day's closing value. What was the closing value on March 28, 2008? (*Source:* Motley Fool)

94. The temperature at which molecular motion is at a minimum is −273.15°C. A piece of metal is cooled to −256.5°C. Write an equation that describes the difference in the temperatures; then calculate the difference.

95. A district manager has a balance of −$1475.84 on her business credit card. In one week, the transactions shown occurred. Find the new balance.

Description	Amount
Payment	$1200.00
Italian Café	−$124.75
Starbucks	−$12.50
Office Maxx	−$225.65
Kinko's	−$175.92

96. A small company has a balance of $10,450.75 in the bank. The table lists income and expenses for one week. Find the new balance.

Description	Income	Description	Expense
Sales	$3400	Electricity	−$234.45
		Sewer	−$150.00
		Payroll	−$8500.00
		Materials	−$2400.00
		Office Supplies	−$142.75
		Phone Service	−$225.80

Of Interest

The temperature −273.15°C is also known as *absolute zero* on the Kelvin temperature scale. The Kelvin scale is the scale most scientists use to measure temperatures. Increments on the Kelvin scale are equal to increments on the Celsius scale, so a change of 1 K corresponds to a change of 1°C.

For Exercises 97 and 98, find the resultant force on each object. The resultant force on an object is the sum of all of the forces acting on the object.

97.

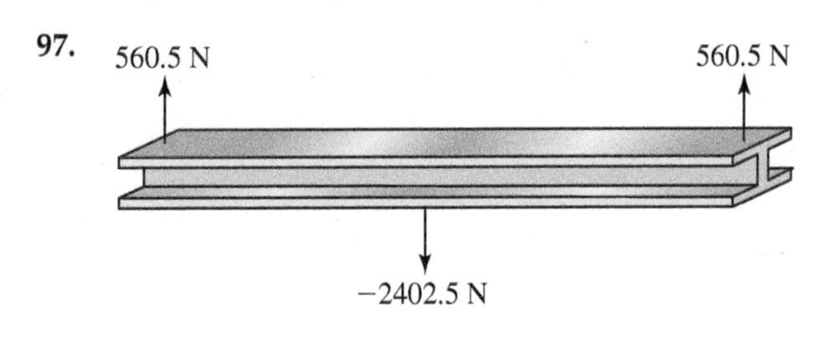

98.

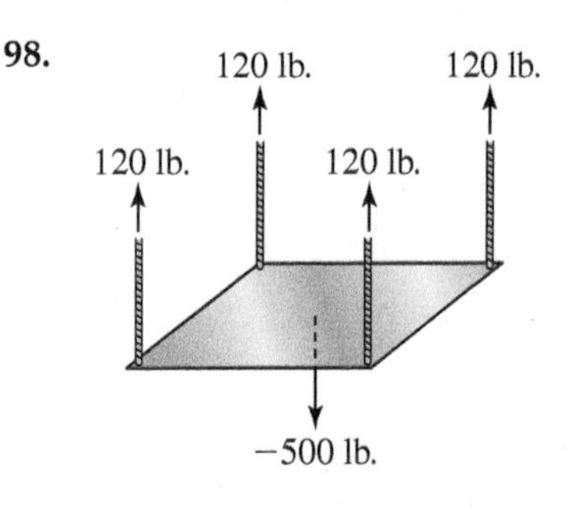

Of Interest

N is an abbreviation for the metric unit of force called the newton. The unit is named after Sir Isaac Newton (1642–1727) in honor of his development of a theory of forces.

99. The budget for a company has a −3.5% change from last year's budget. If last year's budget was $4.8 million, what is the new budget?

100. A company experiences a −2.5% change in profit from the previous year. If the profit for the previous year was $1.4 million, what is this year's profit?

For Exercises 101 and 102, use the fact that an object's weight is a downward force due to gravity and is calculated by multiplying the object's mass by the acceleration due to gravity, which is a constant (force = mass · acceleration). The following table lists the gravitational acceleration constants in American units for various bodies in our solar system. American mass units are slugs (s), *and force units are pounds* (lb.).

Earth	-32.2 ft./sec.^{2}
Moon	-5.5 ft./sec.^{2}
Mars	-12.3 ft./sec.^{2}

101. a. Find the force due to gravity on a person with a mass of 5.5 slugs on Earth, the Moon, and Mars.

b. On which of the Earth, the Moon, or Mars does a person experience the greatest force due to gravity?

Of Interest

On July 20, 1969, during the *Apollo 11* mission, Neil Armstrong and Buzz Aldrin became the first people to walk on the surface of the Moon. Armstrong was first out of the lunar lander and stated the now famous words, "That's one small step for man, one giant leap for mankind."

102. a. During the NASA Apollo missions, the lunar lander had a mass of about 271 slugs (without fuel and equipment). Find the force due to gravity on the lander on Earth and on the Moon.

b. What gravitational force would that same lander have experienced on Mars?

For Exercises 103 and 104, use the fact that in an electrical circuit, voltage is equal to the product of the current, measured in amperes (A), and the resistance of the circuit, measured in ohms (Ω) (voltage=current $\times$ resistance).

103. Suppose the current in a circuit is -6.4 A and the resistance is 8 Ω. Find the voltage.

104. An electrical technician measures the voltage in a circuit to be -15 V and the current to be -8 A. What is the resistance of the circuit?

Puzzle Problem Two books in an algebra series, *Elementary Algebra* and *Intermediate Algebra*, are placed on a bookshelf as shown. Suppose each hardback cover is $\frac{1}{8}$ of an inch thick and each book without its covers is 2 in. thick. If a bookworm begins at the first page of *Elementary Algebra* and bores straight through to the last page of *Intermediate Algebra*, how far has it traveled?

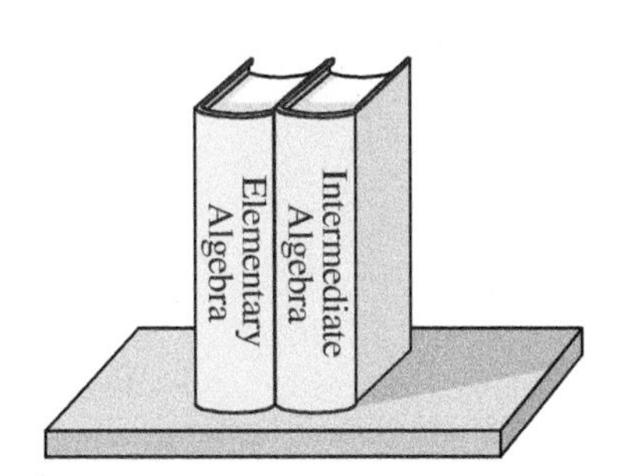

Review Exercises

Exercises 1–5 Constants and Variables

[1.1] **1.** Write a set containing the last names of the first four presidents of the United States.

[1.1] **2.** If $A = \{1, 3, 5, 7, 9\}$ and $B = \{3, 5, 6\}$, is B a subset of A? Why or why not?

[1.1] **3.** Is the set of whole numbers a finite set or an infinite set?

[1.1] **4.** Explain why -6 is a rational number.

[1.1] **5.** Evaluate $|-25|$.

Exercise 6 Equations and Inequalities

[1.1] **6.** Use $<, >$, or $=$ to make a true statement. $-\frac{5}{6}\quad -\frac{15}{18}$

1.3 Exponents, Roots, and Order of Operations

Objectives

1. Evaluate numbers in exponential form.
2. Evaluate roots.
3. Use the order of operations agreement to simplify numerical expressions.

Warm-up

[1.2] For Exercises 1–3, evaluate.

1. $\left(-\frac{3}{4}\right)\left(-\frac{3}{4}\right)\left(-\frac{3}{4}\right)$
2. $(0.6)(0.6)$
3. $-\frac{1}{8}(-8) = -8\left(-\frac{1}{8}\right)$ illustrates which property?

Objective 1 Evaluate numbers in exponential form.

An expression such as 2^4 is in *exponential form* where the number 2 is called the *base* and 4 is an *exponent*. The expression 2^4 is read "two to the fourth power," or simply "two to the fourth." To evaluate 2^4, we write 2 as a factor four times, then multiply.

$$2^4 = \underbrace{2 \cdot 2 \cdot 2 \cdot 2}_{\text{Four 2's}} = 16$$

Base: 2; Exponent: 4

Rule Evaluating an Exponential Form

For any real number b and any natural number n, the nth power of b, or b^n, is found by multiplying b as a factor n times.

$$b^n = \underbrace{b \cdot b \cdot b \cdot \cdots \cdot b}_{b \text{ used as a factor } n \text{ times}}$$

Example 1 Evaluate.

a. $(-9)^2$

Solution: $(-9)^2 = (-9)(-9) = 81$

b. $\left(-\frac{3}{4}\right)^3$

Solution: $\left(-\frac{3}{4}\right)^3 = \left(-\frac{3}{4}\right)\left(-\frac{3}{4}\right)\left(-\frac{3}{4}\right) = -\frac{27}{64}$

c. -2^4

Solution: $-2^4 = -(2 \cdot 2 \cdot 2 \cdot 2) = -16$ The base is 2, not -2.

Note A base raised to the second power, as in $(-9)^2$, is said to be *squared*. A base raised to the third power, as in $\left(-\frac{3}{4}\right)^3$, is said to be *cubed*.

Warning The expressions $(-2)^4$ and -2^4 are different. The expression $(-2)^4$ means that the number -2 is being raised to the fourth power, and -2^4 means the additive inverse of 2 to the fourth power.

Examples 1(a) and 1(b) suggest the following sign rules.

Rules Evaluating Exponential Forms with Negative Bases

If the base of an exponential form is a negative number and the exponent is even, the product is positive.
If the base is a negative number and the exponent is odd, the product is negative.

Answers to Warm-up

1. $-\frac{27}{64}$
2. 0.36
3. commutative property of multiplication

Your Turn 1 Evaluate.

a. $\left(-\frac{1}{2}\right)^5$ b. $(-0.3)^4$ c. -3^2

Objective 2 Evaluate roots.

Square Roots

The square root of a number is a number whose square is the given number. For example, one square root of 25 is 5 because $5^2 = 25$. Another square root of 25 is -5 because $(-5)^2 = 25$. In fact, every positive real number has *two* real-number square roots, a *positive* root and a *negative* root. For convenience, we can write both the positive and negative roots in a compact expression, ± 5. The number 0 has only one square root, 0.

Example 2 Find all real-number square roots of the given number.

a. 64

Answer: ± 8

b. -36

Answer: No real-number square roots exist.

Explanation: The square of a real number is always positive or zero.

Note We do not say that this situation is undefined or indeterminate because we will eventually define the square root of a negative number. Such roots will be defined using imaginary numbers.

Your Turn 2 Find all real-number square roots of the given number.

a. 121 b. -49

The Principal Square Root

The symbol $\sqrt{}$ is called a *radical* and indicates the *principal* (nonnegative) square root of a number.

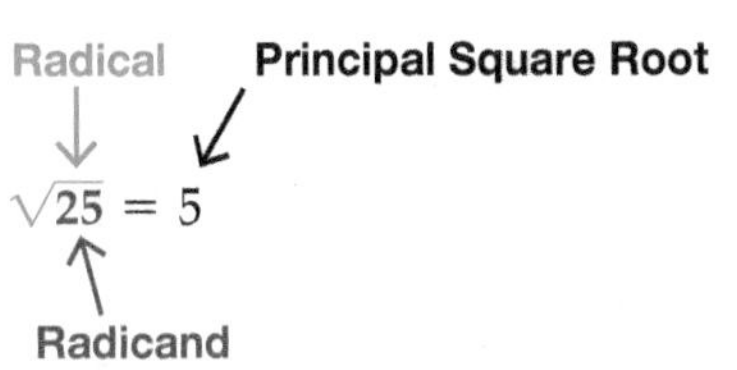

Note For now, we consider only simple rational roots. In Chapter 8, we consider irrational roots.

Example 3 Evaluate.

a. $\sqrt{49}$

Answer: $\sqrt{49} = 7$

b. $\sqrt{\frac{25}{81}}$

Answer: Because $\left(\frac{5}{9}\right)^2 = \frac{25}{81}$, we can say that $\sqrt{\frac{25}{81}} = \frac{5}{9}$.

Note This suggests that $\sqrt{\frac{25}{81}} = \frac{\sqrt{25}}{\sqrt{81}} = \frac{5}{9}$.

c. $\sqrt{0.36}$

Answer: $\sqrt{0.36} = 0.6$

d. $\sqrt{-100}$

Answer: Not a real number.

We can summarize what we have learned so far with the following rules for square roots.

Answers to Your Turn 1

a. $-\frac{1}{32}$ b. 0.0081 c. -9

Answers to Your Turn 2

a. ± 11
b. No real-number square roots exist.

Rules Square Roots

Every positive number has two square roots, a positive root and a negative root.

The square root of 0 is 0.

Negative numbers have no real-number square roots.

The radical symbol $\sqrt{\ }$ denotes the principal (nonnegative) square root.

$\sqrt{\frac{a}{b}} = \frac{\sqrt{a}}{\sqrt{b}}$, where $a \geq 0$ and $b > 0$.

Your Turn 3 Evaluate.

a. $\sqrt{121}$ b. $\sqrt{\frac{81}{169}}$ c. $\sqrt{-0.09}$ d. $\sqrt{0.81}$

Other Roots

Each natural-number exponent has a corresponding root, which we indicate as an *index* in a radical. For example, to indicate the cube root of a number a, we write $\sqrt[3]{a}$. To find a cube root, we find a number that when cubed, equals the radicand.

The index is 3. ⟶ $\sqrt[3]{64} = 4$ because $4^3 = 64$.

Higher roots follow a similar pattern.

Note If no index is shown, as in $\sqrt{16}$, the radical is understood to mean a square root.

Example 4 Evaluate.

a. $\sqrt[4]{81}$

Answer: $\sqrt[4]{81} = 3$ because $3^4 = 81$.

b. $\sqrt[5]{32}$

Answer: $\sqrt[5]{32} = 2$ because $2^5 = 32$.

c. $\sqrt[3]{-125}$

Answer: $\sqrt[3]{-125} = -5$ because $(-5)^3 = -125$.

Note If the index is odd and the radicand is negative, the root will be negative because a negative number raised to an odd power is negative.

Your Turn 4 Evaluate.

a. $\sqrt[3]{125}$ b. $\sqrt[4]{16}$ c. $\sqrt[5]{-243}$

Objective 3 Use the order of operations agreement to simplify numerical expressions.

If an expression contains a mixture of operations, we agree to perform the operations in a specific order because using other orders can result in different answers.

Procedure Order of Operations Agreement

Perform operations in the following order:

1. Within grouping symbols: parentheses ( ), brackets [ ], braces { }, absolute value | |, above and/or below fraction bars, and radicals $\sqrt{\ }$.
2. Exponents/Roots from left to right, in order as they occur.
3. Multiplication/Division from left to right, in order as they occur.
4. Addition/Subtraction from left to right, in order as they occur.

Answers to Your Turn 3

a. 11 b. $\frac{9}{13}$

c. not a real number d. 0.9

Answers to Your Turn 4

a. 5 b. 2 c. −3

Note Recall that the expressions $(-5)^2$ and -5^2 are different. The expression $(-5)^2$ means the square of the number -5, whereas -5^2 means the additive inverse of the square of 5.

$$(-5)^2 = (-5)(-5) = 25$$
$$-5^2 = -(5 \cdot 5) = -25$$

Example 5 Evaluate.

a. $18 - 12 \div 3(-2) + (-5)^2$

Solution:
$= 18 - 12 \div 3(-2) + 25$ Simplify the exponential form.
$= 18 - 4(-2) + 25$ Multiply or divide from left to right: $-12 \div 3$ is first.
$= 18 + 8 + 25$ Multiply: $-4(-2) = 8$.
$= 51$ Add, in order, left to right.

b. $(-2)^3 - 3|5 - (4 + 3)| - \sqrt{16 + 9}$

Solution: When grouping symbols are embedded one within another, as in $|5 - (4 + 3)|$, we work from the innermost set of grouping symbols outward.

$= (-2)^3 - 3|5 - 7| - \sqrt{25}$ Simplify the innermost parentheses: $4 + 3 = 7$; add inside the radical: $16 + 9 = 25$.
$= -8 - 3|-2| - 5$ Evaluate the exponential form, subtract within the absolute value, and find the square root.
$= -8 - 3(2) - 5$ Simplify the absolute value: $|-2| = 2$.
$= -8 - 6 - 5$ Multiply: $-3(2) = -6$.
$= -19$ Subtract.

c. $\dfrac{(-3)^4 - 3(-5)}{-\sqrt{4 \cdot 9}}$

Connection The expression in Example 5(c) is equivalent to the expression

$$[(-3)^4 - 3(-5)] \div [-\sqrt{4 \cdot 9}].$$

Solution: Simplify the numerator and denominator separately; then divide the results.

$= \dfrac{81 - 3(-5)}{-\sqrt{36}}$ Evaluate the exponential form in the numerator and simplify within the radical in the denominator.
$= \dfrac{81 + 15}{-6}$ Multiply in the numerator and evaluate the square root in the denominator.
$= \dfrac{96}{-6}$ Add in the numerator.
$= -16$ Divide.

Your Turn 5 Evaluate.

a. $15 - [3 - 5]^2 - \sqrt{25 - 9}$ **b.** $24 \div (-3)(-4) + |9 - 16| + (-4)^3$

c. $\dfrac{8(4 - 5)^5 - 1}{15 - 3(12)}$

Now let's consider some applications that require the use of the order of operations agreement.

Example 6 A cumulative grade point average (GPA) is found by dividing the total number of grade points by the total number of course credits. Grade points are calculated by multiplying the numerical value of the letter grade ($A = 4, B = 3, C = 2, D = 1, F = 0$) by the number of credits for the course. For the student's grade report in the margin, write a numerical expression that describes the GPA; then calculate the GPA.

Course	Course Credits	Grade
ENG 102	3	A
MAT 102	3	B
BIO 110	4	C
SOC 101	3	F

Solution: $\text{GPA} = \dfrac{3 \cdot 4 + 3 \cdot 3 + 4 \cdot 2 + 3 \cdot 0}{3 + 3 + 4 + 3}$

$= \dfrac{12 + 9 + 8 + 0}{13}$ Multiply in the numerator; add in the denominator.
$= \dfrac{29}{13}$ Add in the numerator.
≈ 2.23 Divide and round the quotient to the nearest hundredth.

Answers to Your Turn 5

a. 7 **b.** -25 **c.** $\dfrac{3}{7}$

Your Turn 6 John leases a car with the following agreement: He pays \$895 down, monthly payments of \$279 for 36 months, and \$0.18 per mile in excess of 36,000 miles. When he returned the car, it had 42,500 miles on it. What was his total cost for leasing the car?

Counting problems often generate expressions with exponents that require using order of operations. Consider counting the number of different numbers that can be expressed using two LED displays if a single digit (0–9) is shown in each display.

Note Each display can contain 0–9, so ten different digits are possible in each display. ▶

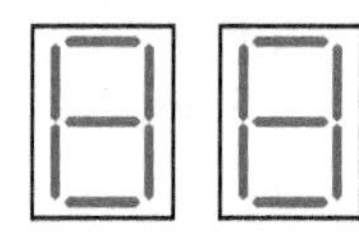

We could list every possible number using two digits—00, 01, 02, 03, 04, . . . , 99—to see that there are 100 possible numbers. Or because each of the ten digits in the first display is paired with ten digits in the second display, we could multiply $10 \cdot 10$ (or 10^2) to find out that there are 100 possible numbers. In general, if a problem involves counting the total number of possibilities with items in a set of positions, the result will be the product of the number of items possible in each position.

Example 7 How many license plates are possible if each license plate has three capital letters (A–Z) followed by three digits (0–9)?

Solution: Multiply the number of letters or digits that are possible in each position.

Note Each of the three letter positions has 26 different letter possibilities. Each of the three digit positions has 10 different digit possibilities. ▶

$$26^3 \cdot 10^3 = 17{,}576 \cdot 1000 = 17{,}576{,}000$$

Answer to Your Turn 6
\$12,109

Answer to Your Turn 7
12,000

Your Turn 7 How many different combinations are possible if a combination lock has four dials, one dial with the letters A–L and each of the other three dials with the digits 0–9 on them?

1.3 Exercises

For Extra Help MyMathLab®

Objective 1

For Exercises 1–4, identify the base and the exponent; then translate the expression into words. See Objective 1.

1. $(-4)^3$

2. 5^2

3. -1^7

4. -2^8

Prep Exercise 1 Explain how to evaluate b^n, where b is any real number and n is a natural number.

Prep Exercise 2 Is $(-8)^6$ positive or negative? Why?

Prep Exercise 3 Is -8^6 positive or negative? Why?

For Exercises 5–22, evaluate. See Example 1.

5. 5^4 **6.** 2^7 **7.** $(-3)^4$ **8.** $(-7)^2$ **9.** $(-4)^3$ **10.** $(-2)^5$

11. -6^2 **12.** -3^4 **13.** $(-1)^{10}$ **14.** $(-1)^{12}$ **15.** $\left(-\frac{3}{8}\right)^2$ **16.** $\left(-\frac{1}{9}\right)^2$

17. $\left(-\frac{5}{6}\right)^3$ **18.** $\left(-\frac{2}{3}\right)^3$ **19.** $(0.4)^3$ **20.** $(0.1)^4$ **21.** $(-2.1)^3$ **22.** $(-1.4)^3$

Objective 2

Prep Exercise 4 What is the difference between a number's square and its square root?

Prep Exercise 5 How many square roots does each positive real number have?

Prep Exercise 6 How many real number square roots does a negative real number have?

For Exercises 23–30, find all real-number square roots of the given number. See Example 2.

23. 225 **24.** 81 **25.** 256 **26.** 121 **27.** $\frac{4}{9}$ **28.** $\frac{25}{36}$ **29.** -25 **30.** -400

Prep Exercise 7 What does the radical symbol, $\sqrt{\ }$, denote?

Prep Exercise 8 Why is an odd root of a negative number a negative number?

For Exercises 31–44, evaluate the root. See Examples 3 and 4.

31. $\sqrt{25}$ **32.** $\sqrt{100}$ **33.** $\sqrt[5]{32}$ **34.** $\sqrt[4]{81}$ **35.** $\sqrt{0.36}$ **36.** $\sqrt{0.01}$ **37.** $\sqrt[3]{-27}$

38. $-\sqrt[3]{-64}$ **39.** $\sqrt[3]{\frac{8}{27}}$ **40.** $\sqrt[4]{\frac{1}{16}}$ **41.** $\sqrt{-36}$ **42.** $\sqrt{-16}$ **43.** $\sqrt[3]{\frac{24}{3}}$ **44.** $\sqrt{\frac{50}{2}}$

Objective 3

Prep Exercise 9 If a numerical expression contains embedded grouping symbols, such as $\{15 - 4[3 - (5 + 2)]\}$, explain the order in which you simplify the grouping symbols.

Prep Exercise 10 What is the general plan for simplifying a numerical expression containing a fraction line, such as $\frac{12 + 8(16 - 4)}{3^2 - 12}$?

For Exercises 45–68, evaluate using the order of operations. See Example 5.

45. $-4 + 3(-1)^4 + 18 \div 3$

46. $2(-2)^3 - 8 + 24 \div 4$

47. $-8^2 + 36 \div (9 - 5)$

48. $-7^2 - 25 \div (8 - 3)$

49. $12^2 \div \sqrt{45 - 9} - 8$

50. $6 + 4^2 \div \sqrt{13 + 3}$

51. $-2|-8 - 2| \div (-5)(2)$

52. $-3|-6 - 8| \div 2(-3)$

53. $-24 \div (-6)(2) + \sqrt{169 - 25}$

54. $\sqrt{100 - 64} + 18 \div (-3)(-2)$

55. $-18 \cdot \frac{2}{9} \div (-2) + |9 - 5(-2)|$

56. $24\left(-\frac{3}{8}\right) \div (-3) + |-6 + 2(-3)|$

57. $13.02 \div (-3.1) + 6^2 - \sqrt{25}$

58. $4^2 + \sqrt{49} - 9.03 \div (-4.3)$

59. $(1 - 0.8)^2 + 2.4 \div (0.3)(-0.5)$

60. $(0.1)^3 - (-6)(3) + 5(2 - 8)$

61. $\frac{4}{5} \div \left(-\frac{1}{10}\right) \cdot (-2) + \sqrt[5]{16 + 16}$

62. $-\frac{3}{4} \div \frac{1}{8}(-3) + \sqrt[4]{96 - 15}$

63. $\frac{9}{8} \cdot \left(-\frac{2}{3}\right) + \left(\frac{1}{5} - \frac{2}{3}\right) \div \sqrt{\frac{125}{5}}$

64. $\frac{6}{5} \cdot \left(-\frac{2}{3}\right) + \left(\frac{2}{5} - \frac{2}{5}\right) \div \sqrt{\frac{36}{4}}$

65. $\frac{12 - 2^3}{5 - 3 \cdot 2}$

66. $\frac{6^2 + 4}{8 \div (-2) \cdot 5}$

67. $\frac{6^2 - 3(4 + 2^5)}{5 + 20 - (2 + 3)^2}$

68. $\frac{4[5 - 8(2 + 1)]}{10 + 6 - (-4)^2}$

In Exercises 69–72, a property of arithmetic was correctly used as an alternative to the order of operations. Determine what property of arithmetic was applied and explain how it is different from the order of operations agreement.

69. $13 - 1 \cdot 3 \cdot 3 + 8^2$
$= 13 - 1 \cdot 9 + 64$
$= 13 - 9 + 64$
$= 68$

70. $-3(2 + 5) - \sqrt{36}$
$= -6 - 15 - 6$
$= -27$

71. $-6(-1 + 6^2) - \sqrt{14 + 11}$
$= -6(-1 + 36) - \sqrt{25}$
$= 6 + (-216) - 5$
$= -215$

72. $(-3)^3 + 2[-11 + 8 + (-2)]$
$= -27 + 2(-13 + 8)$
$= -27 + 2(-5)$
$= -27 + (-10)$
$= -37$

Find the Mistake *For Exercises 73–76, explain the mistake. Then simplify correctly.*
See Example 5.

73. $12 \div 4 \cdot 3 - 11$
$= 12 \div 12 - 11$
$= 1 - 11$
$= -10$

74. $25 - 3(1 - 8)$
$= 25 - 3(-7)$
$= 22(-7)$
$= -154$

75. $30 \div 2 + \sqrt{16 + 9}$
$= 15 + \sqrt{16 + 9}$
$= 15 + 4 + 3$
$= 22$

76. $-2^4 + 16 \div 4 - (5 - 7)$
$= 16 + 4 - (-2)$
$= 20 + 2$
$= 22$

77. Jennifer is preparing for a 10-mile minimarathon. To qualify, she must have an average time of 72 minutes or less on four 10-mile runs. She records her time on each of four practice runs, which are listed here.

April 2, 2010	76.5 minutes	April 16, 2010	71.4 minutes
April 9, 2010	74.5 minutes	April 23, 2010	69.2 minutes

a. Find her average time for the four 10-mile practice runs.

b. Does she qualify?

78. Harold receives an offer from his electric and gas provider where he can pay a fixed rate of \$200 each month to avoid drastic variations in his bill from month to month. To determine whether the offer is reasonable, he decides to compare the \$200 fixed monthly rate with his average monthly cost from the previous year. The following list contains his monthly charges from 2010.

January	\$310	July	\$120
February	\$324	August	\$122
March	\$210	September	\$125
April	\$185	October	\$188
May	\$160	November	\$248
June	\$128	December	\$298

a. Find the average of the monthly bills from 2010.

b. Is the \$200 fixed monthly rate a reasonable deal?

For Exercises 79 and 80, use the following bar graph that shows the projected number of graduates from District of Columbia public high schools for each school year 2013–2014 to 2020–2021.

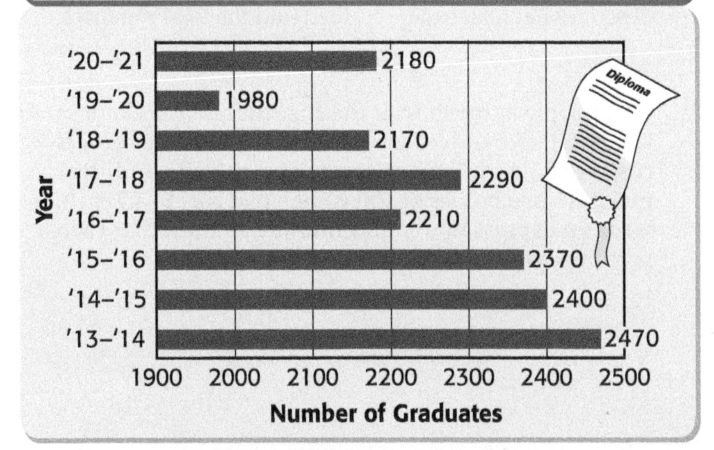

Source: Office of Evaluating and Reporting, Florida Department of Education.

79. Find the average projected number of graduates from 2013–2014 through 2016–2017.

80. Find the average projected number of graduates from 2018–2019 through 2020–2021.

For Exercises 81 and 82, use the following line graph, which shows the closing value of the S&P 500 index each day from July 30 through August 10.

81. Find the average closing value for the S&P 500 for the week of July 30–August 3.

82. Find the average closing value for the S&P 500 for the week of August 6–August 10.

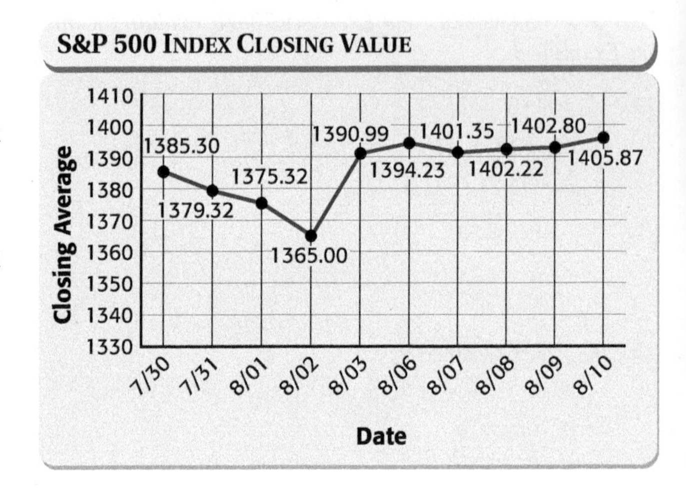

For Exercises 83 and 84, find the GPA. See Example 6.

83.

Course	Course Credits	Grade
CHM 110	4	B
MAT 140	4	A
PSY 201	3	C
ECO 101	3	D

84.

Course	Course Credits	Grade
MAT 110	3	B
ENG 201	3	A
PHI 101	3	C
ART 101	3	A
PHY 101	4	D

For Exercises 85–92, solve. See Example 6.

85. A wireless service charges $35.50 for 500 minutes in the primary calling zone, then $0.10 for each additional minute in the calling zone. Calls outside the calling zone cost $0.12 per minute. Suppose a subscriber uses 658 minutes in the calling zone and 45 minutes outside the calling zone. Write a numerical expression that describes the total cost for the month; then calculate the total.

86. A wireless service charges $49.00 for 1000 minutes in the primary calling zone, then $0.05 for each additional minute in the calling zone. Calls outside the calling zone cost $0.15 per minute. Suppose a subscriber uses 1258 minutes in the calling zone and 72 minutes outside the calling zone. Write a numerical expression that describes the total cost for the month; then calculate the total.

87. Dona gets reimbursed by her company for travel expenses. Her company pays for fuel and $0.35 per mile when employees use their own vehicle, which she did. She drove 814 miles and spent $54.50 on gas. She spent three nights in a hotel at $89.90 per night and spent $112.45 on food. Write a numerical expression that describes her total expenses; then find the total expenses.

88. Jackson gets reimbursed by his company for travel expenses. His company pays for fuel and $0.30 per mile when employees use their own vehicle, which he did. He drove 127 miles and spent $21.26 on gas. He spent two nights in a hotel at $112.80 per night and spent $131.08 on food. Write a numerical expression that describes his total expenses; then find the total expenses.

89. Five people agree to split the cost equally for Chinese takeout. If two orders cost $8.95 each, two other orders cost $10.95 each, and one order costs $6.95 and they purchase two bottles of soft drinks that cost $1.45 each, write an expression for and find the amount that each person pays.

90. If eight sorority sisters agree to split the cost equally for two pizzas that cost $10.98 each, two pizzas that cost $12.98 each, and four drinks that cost $1.68 each, write an expression for and find the amount that each person pays.

91. Caleb leased a car with the following terms: He agreed to put $1200 down. He also agreed to pay $349.00 per month for three years and $0.15 per mile for all miles in excess of 45,000. When he returned the car at the end of the lease period, it had 52,000 miles on the odometer. Write an expression for and find the total amount that Caleb paid for the lease.

92. Jo Wie leased a car with the following terms: She agreed to put $1500 down. She also agreed to pay $499.00 per month for four years and $0.20 per mile for all miles in excess of 60,000. When she returned the car at the end of the lease period, it had 68,000 on the odometer. Write an expression for and find the amount that Jo paid for the lease.

For Exercises 93–98, solve. See Example 7.

93. How many combinations are possible on a combination lock that has four dials, each with the letters A–H?

94. A slot machine has three wheels that spin. Each wheel has an orange, an apple, a cherry, a strawberry, a banana, and a joker. How many different ways can the three wheels stop?

95. The memory locations on a MIDI device are coded using a 7-bit memory chip. If each of the 7 bits can have a value of 0 or 1, how many memory locations are possible?

96. How many five-letter "words" can be formed using the five lowercase vowels (a, e, i, o, u)?

97. How many seven-digit phone numbers are possible if the first three digits can be 7, 8, or 9 and the other four digits can be any numeral 0–9?

98. How many ten-digit phone numbers are possible with the area codes 801, 802, and 803?

Review Exercises

Exercise 1 Constants and Variables

[1.1] **1.** Write a set containing the whole numbers less than 9.

Exercises 2–5 Expressions

[1.2] *For Exercises 2–4, simplify.*

2. $19 - (-54)$

3. $-2(-5)(-6)$

4. $\frac{-38}{2}$

[1.1] **5.** Is $5x^2 - 4y$ an expression or an equation? Explain.

Exercise 6 Equations and Inequalities

[1.1] **6.** What property is being applied in $4(9 + 8) = (9 + 8)4$?

1.4 Evaluating and Rewriting Expressions

Objectives

1. Translate word phrases to variable expressions.
2. Evaluate variable expressions.
3. Rewrite expressions using the distributive property.
4. Rewrite expressions by combining like terms.

Warm-up

[1.3] For Exercises 1–3, evaluate.

1. $\dfrac{4(-3)^2}{-3-6}$ **2.** $-(-10)$ **3.** $\sqrt{-16}$

We have reviewed how to simplify numeric expressions. Now we will review basic concepts of variable expressions.

Objective 1 Translate word phrases to variable expressions.

To translate a word phrase, select a variable to represent the unknown amount; then use the key words to translate to the appropriate operation. The following table contains some basic phrases and their translations.

Addition	Translation
the sum of x and three	$x + 3$
h plus k	$h + k$
t added to seven	$7 + t$
three more than a number	$n + 3$
y increased by two	$y + 2$

Note Because addition is a commutative operation, it doesn't matter in what order we write the translation.

For "the sum of x and three," we can write

$$x + 3 \text{ or } 3 + x.$$

Subtraction	Translation
the difference of three and x	$3 - x$
h minus k	$h - k$
seven subtracted from t	$t - 7$
three less than a number	$n - 3$
two decreased by y	$2 - y$

Note Subtraction is not a commutative operation; therefore, the order in which we write the translation matters. We must translate each key phrase exactly as shown. Notice that when we translate "less than" or "subtracted from," the translation is in reverse order from what we read.

Multiplication	Translation
the product of x and three	$3x$
h times k	hk
twice a number	$2n$
triple the number	$3n$
two-thirds of a number	$\frac{2}{3}n$

Note Multiplication is a commutative operation, so it doesn't matter in what order we write the translation.

h times k can be hk or kh.

Division	Translation
the quotient of x and three	$x \div 3$ or $\frac{x}{3}$
h divided by k	$h \div k$ or $\frac{h}{k}$
h divided into k	$k \div h$ or $\frac{k}{h}$
the ratio of a to b	$a \div b$ or $\frac{a}{b}$

Note Division is not a commutative operation; therefore, we must translate division phrases exactly as shown. Notice how "divided into" is translated in reverse order of what we read.

Exponents	Translation
c squared or the square of c	c^2
k cubed or the cube of k	k^3
n to the fourth power	n^4
y raised to the fifth power	y^5

Roots	Translation
the square root of x	$\sqrt{x}$
the cube root of y	$\sqrt[3]{y}$
the fifth root of n	$\sqrt[5]{n}$

Answers to Warm-up
1. -4
2. 10
3. does not exist as a real number

Example 1 Translate the phrase to an algebraic expression.

a. four more than three times a number

Translation: $4 + 3n$ or $3n + 4$

Note The commutative property of addition allows us to write the expression in either order.

b. six less than the square of a number

Translation: $n^2 - 6$

Note Because subtraction is not commutative, we must translate the key words carefully. The phrase "six *less than*" indicates that the 6 comes *after* the subtraction sign.

c. the sum of h raised to the fifth power and fifteen

Translation: $h^5 + 15$

d. the ratio of m times n to r cubed

Translation: $\frac{mn}{r^3}$

Note When coupled with the word *ratio*, the word *to* translates into the fraction line. The amount to the left of the word *to* goes in the numerator, and the amount to the right of the word *to* goes in the denominator.

e. one-half of v divided by the square root of t

Translation: $\frac{1}{2}v \div \sqrt{t}$

Note When the word *of* is preceded by a fraction, it means multiply.

f. five times the sum of x and y

Translation: $5(x + y)$

Note Without the parentheses, the expression is $5x + y$, which is "the sum of five times x and y."

g. the square root of the difference of the square of x and the square of y

Translation: $\sqrt{x^2 - y^2}$

Note "The square root of the difference" indicates that the entire subtraction is under the radical sign.

h. the product of x and y divided by the sum of x^2 and five

Translation: $xy \div (x^2 + 5)$ or $\frac{xy}{x^2 + 5}$

Your Turn 1 Translate the phrase to an algebraic expression.

a. two-thirds subtracted from the product of nine and a number
b. −6.2 increased by 9.8 times a number
c. a number minus twice the sum of the number and seven
d. the difference of m and n, all raised to the fourth power
e. the sum of x and five divided by the square root of the difference of a and three

Sometimes it is necessary to translate expressions in terms of a given variable or variable(s).

Example 2 Translate the indicated expression.

a. The base of a triangle is three more than the height. If the height is represented by h, write an expression for the base.

Translation: $h + 3$ or $3 + h$

b. If a ribbon is 12 inches long and x inches are cut off, write an expression for the length of the remaining piece.

Translation: $(12 - x)$ inches

c. The area of a rectangle is the product of its length, l, and width, w. Write an expression for finding the area of a rectangle.

Translation: lw

d. The volume of a pyramid is one-third the product of the area of the base, A, and the height h. Write an expression for finding the volume of a pyramid.

Translation: $\frac{1}{3}Ah$

Answers to Your Turn 1

a. $9n - \frac{2}{3}$ b. $-6.2 + 9.8y$
c. $x - 2(x + 7)$ d. $(m - n)^4$
e. $\frac{x + 5}{\sqrt{a - 3}}$

Your Turn 2 Translate the indicated expression.

a. The longer base of a trapezoid is three times as long as the shorter base. If the shorter base is represented by x, write an expression for the longer base.
b. Heinz received an end-of-year bonus of \$25,000, which he invested in mutual funds and municipal bonds. If he invested n dollars in mutual funds, write an expression for the amount invested in municipal bonds.
c. The area of a triangle is one-half the product of the base, b, and the height, h. Write an expression for finding the area of a triangle.
d. The area of a circle is found by multiplying the square of the radius, r, by π. Write an expression for finding the area of a circle.

Objective 2 Evaluate variable expressions.

Recall that variables represent numbers. When we replace the variables in an expression with numbers, we are *evaluating* the expression.

Procedure Evaluating a Variable Expression

1. Replace each variable with its corresponding given value.
2. Simplify the resulting numerical expression.

Example 3 Evaluate $\frac{4x^2}{x - 6}$ when

Connection We evaluate expressions when we check solutions to equations, as we will see in Chapter 2.

a. $x = -3$

Solution: $\frac{4(-3)^2}{-3 - 6}$ Replace x with -3.

$= \frac{36}{-9} = -4$ Simplify.

b. $x = 6$

Solution: $\frac{4(6)^2}{6 - 6}$ Replace x with 6; then simplify.

$= \frac{144}{0}$, which is undefined.

Your Turn 3

a. Evaluate $2x^3 - 6y$ when $x = -3$ and $y = 1$.

b. Evaluate $\frac{2n}{n + 5}$ when $n = -5$.

In Example 3(b), we found that $\frac{4x^2}{x - 6}$ was undefined when $x = 6$. With any variable expression, it is important to identify values for the variable that cause the expression to be undefined. Remember that if a denominator is 0, the expression is undefined.

Answers to Your Turn 2
a. $3x$ **b.** $\$25{,}000 - n$
c. $\frac{1}{2}bh$ **d.** πr^2

Answers to Your Turn 3
a. -60 **b.** undefined

Example 4 Determine all values for the variable that causes the expression to be undefined.

a. $\dfrac{x - 3}{x}$

Answer: If $x = 0$, this expression is undefined because the denominator is 0.

b. $\dfrac{y}{y - 7}$

Answer: If $y = 7$, this expression is undefined because the denominator is 0.

Your Turn 4 Determine all values for the variable that causes the expression to be undefined.

a. $\dfrac{2a + 3}{a}$ **b.** $\dfrac{k + 2}{k - 8}$

Objective 3 Rewrite expressions using the distributive property.

Recall from Section 1.2 the distributive property of multiplication over addition.

Rule The Distributive Property of Multiplication over Addition

$$a(b + c) = ab + ac$$

We can use the distributive property to rewrite an expression in another form that is equivalent to the original form.

Example 5 Use the distributive property to write an equivalent expression.

a. $3(x + 4)$

Solution: $3(x + 4) = 3 \cdot x + 3 \cdot 4$ Distribute 3.

$= 3x + 12$ Multiply.

Note You might find it helpful to think of $-5(t - 2)$ as $-5[t + (-2)]$ so that multiplying -5 and -2 gives positive 10 directly without your having to use the double negative property.

b. $-5(t - 2)$

Solution: $-5(t - 2) = -5 \cdot t - (-5) \cdot 2$ Distribute -5.

$= -5t - (-10)$ Multiply.

$= -5t + 10$ Rewrite $-(-10)$ as $+10$.

c. $\dfrac{3}{8}\left(2m + \dfrac{4}{5}\right)$

Solution: $\dfrac{3}{8}\left(2m + \dfrac{4}{5}\right) = \dfrac{3}{8} \cdot 2m + \dfrac{3}{8} \cdot \dfrac{4}{5}$ Distribute $\dfrac{3}{8}$.

$= \dfrac{3}{4}m + \dfrac{3}{10}$ Multiply and simplify.

Your Turn 5 Use the distributive property to write an equivalent expression.

a. $-4(6 - 5y)$ **b.** $\dfrac{4}{5}\left(\dfrac{1}{2}y - 10\right)$

Objective 4 Rewrite expressions by combining like terms.

Some expressions such as $5x^2 + 9x + 7$ contain addition of expressions called **terms**.

Answers to Your Turn 4
a. 0 **b.** 8

Answers to Your Turn 5
a. $-24 + 20y$ **b.** $\dfrac{2}{5}y - 8$

Definition **Term:** A number or the product of a number and one or more variables raised to powers.

For example, the terms in $5x^2 + 9x + 7$ are $5x^2, 9x$, and 7. An expression containing subtraction, such as $3x^2 - 9x + 2$, can be rewritten as a sum to identify its terms. So $3x^2 - 9x + 2$ becomes $3x^2 + (-9x) + 2$ and we see that its terms are $3x^2$, $-9x$, and 2. A term that has no variable, such as 2, is called a *constant term*.

The numerical factor in a term is called the **numerical coefficient**, or simply the coefficient of the term.

Definition **Numerical coefficient:** The numerical factor in a term.

For example, the coefficients of the terms in $\frac{2}{3}x^2 - 7x + 2$ are $\frac{2}{3}$, -7, and 2.

Sometimes expressions contain **like terms**.

Definition **Like terms:** Variable terms that have the same variable(s) raised to the same exponents, or constant terms.

Examples of like terms:	Examples of unlike terms:	
$3x$ and $5x$	$0.2x$ and $8y$	(different variables)
$\frac{3}{4}y^2$ and $9y^2$	$4t^2$ and $4t^3$	(different exponents)
$7xy$ and $3xy$	x^2y and xy^2	(different exponents)
6 and 15	12 and $12x$	(different variables)

Combining Like Terms

By combining the like terms, we can rewrite expressions containing like terms so they have fewer terms. The process of rewriting an algebraic expression with fewer terms is called *simplifying*. An algebraic expression is simplified when grouping symbols have been removed and like terms have been combined.

To simplify an expression by combining like terms, we use the distributive property. Consider the expression $3x + 5x$. Notice the common factor of x in both terms. Using the distributive property, we write the common factor x outside the parentheses and write the remaining factors, which are the coefficients 3 and 5, as a sum inside the parentheses.

$3x + 5x = (3 + 5)x$ Because the parentheses contain a sum of two numbers, we can simplify by adding them.

$= 8x$

Notice that when combining like terms, we simply add the coefficients and keep the variable the same.

Procedure **Combining Like Terms**

To combine like terms, add or subtract the coefficients and keep the variables and their exponents the same.

Example 6 Combine like terms.

a. $8x + 7x$

Solution: $8x + 7x = 15x$

Note Think "eight x's plus seven x's is fifteen x's."

b. $15k^2 - k^2$

Solution: $15k^2 - k^2 = 14k^2$

Note The coefficient of $-k^2$ is -1.

Your Turn 6 Combine like terms.

a. $2.4n + 6.9n$

b. $\frac{1}{4}y^3 - \frac{2}{3}y^3$

Sometimes expressions are more complex and contain different sets of like terms or may require that the distributive property be used to simplify parentheses before like terms can be combined.

Example 7 Simplify.

Note When we use the commutative property to rearrange the expressions so that like terms are together, we say that we are *collecting* the like terms.

a. $10x + 7y - x + 6y$

Solution:

$10x + 7y - x + 6y$ Use the commutative property to rearrange the terms.

$= 10x - x + 7y + 6y$

$= 9x + 13y$ Combine like terms: $10x - x = 9x$ and $7y + 6y = 13y$.

b. $0.2m - 1.5mn - 2.5m + 9 - 0.3mn$

Solution:

$0.2m - 1.5mn - 2.5m + 9 - 0.3mn$

$= 0.2m - 2.5m - 1.5mn - 0.3mn + 9$ Use the commutative property to collect like terms.

$= -2.3m - 1.8mn + 9$ Combine like terms.

c. $3h + k - 2(5h + 3k) - 9 + 5k$

Solution:

$3h + k - 2(5h + 3k) - 9 + 5k$

$= 3h + k - 10h - 6k - 9 + 5k$ Distribute -2.

$= 3h - 10h + k - 6k + 5k - 9$ Collect like terms.

$= -7h - 9$ Combine like terms.

Learning Strategy

Repetition is Key!! Sometimes homework is done quickly for the grade or because of time constraints, but it should be taken very seriously because it's the closest thing to the test. Do the Practice Test and use all materials that Pearson has to offer.

—Lauren H.

Your Turn 7 Combine like terms.

a. $3.2x^2 - 9x + 12 - 0.4x - 12$

b. $\frac{5}{6}h - k + 4 - \frac{3}{4}h + 5k$

c. $3(t + 5) - (5t - 3u) - 1$

Answers to Your Turn 6

a. $9.3n$ b. $-\frac{5}{12}y^3$

Answers to Your Turn 7

a. $3.2x^2 - 9.4x$

b. $\frac{1}{12}h + 4k + 4$

c. $-2t + 3u + 14$

1.4 Exercises

For Extra Help MyMathLab®

Note: Exercises marked with a ★ represent challenging exercises.

Objective 1

Prep Exercise 1 Why can we translate the phrase "eight more than x," which indicates addition, as $8 + x$ or $x + 8$?

Prep Exercise 2 The phrase "nine subtracted from n" translates to $n - 9$, which is in reverse order of what we read. What other key words for subtraction translate in reverse order?

For Exercises 1–22, translate the phrase to an algebraic expression. See Example 1.

1. five times a number
2. the product of a number and eight
3. two less than three times a number
4. the difference of five times a number and four
5. five less p
6. seven less than m
7. the quotient of a number to the fourth power and 8
8. the quotient of the square of y and 9
9. twenty less than the product of a number and two
10. a number subtracted from twice another number
11. x to the fourth power times y squared
12. the product of x cubed and y
13. one-half subtracted from the quotient of p and q
14. two-thirds less than the quotient of m and n
15. m minus three times the sum of a number and 5
16. four times the difference of seven and a number, decreased by nine
17. the difference of four and t, all raised to the fifth power
18. six less than y, all raised to the fourth power
19. six-sevenths of a number increased by seven
20. seven-tenths of a number added to 12
21. the sum of x and y less than the difference of m and n
22. the difference of a and b subtracted from the sum of c and d

Find the Mistake *For Exercises 23–26, explain the mistake. Then translate the phrase correctly.*

23. six less than y; translation $6 - y$
24. four subtracted from the square root of m; translation: $4 - \sqrt{m}$

25. four times the sum of r and seven; translation: $4r + 7$

26. seven divided by the product of two and r; translation: $2r \div 7$

For Exercises 27–36, translate the indicated expression. See Examples 2(a) and 2(b).

27. The length of a rectangle is five times the width. If the width is represented by w, write an expression that describes the length.

28. The width of a rectangle is twice the length. If the length is represented by l, write an expression that describes the width.

29. The length of a rectangle is three times the width subtracted from two. If the width is represented by the variable w, write an expression that describes the length.

30. The width of a rectangle is five less than four times the length. If the length is represented by l, write an expression that describes the width.

31. The diameter of a circle is twice the radius. If r represents the radius, write an expression for the diameter.

32. The radius of a circle is half the diameter. If d represents the diameter, write an expression for the radius.

33. Mickie has 17 coins in her change purse, all of which are either nickels or quarters. If n represents the number of nickels she has, write an expression in terms of n that describes the number of quarters.

34. Millie split \$2500 between two savings accounts. If n represents the amount in dollars in one account, write an expression in terms of n for the amount in the other account.

35. Zelda passes a rest stop on the highway. One-half of an hour later, Scott, traveling in the same direction, passes the same rest stop. If t represents the amount of time in hours it takes Scott to catch up to Zelda, write an expression in terms of t that describes the amount of time Zelda has traveled since passing the rest stop.

36. Renee is on her way home and passes a sign. One-fourth of an hour later, Gary, who is traveling in the same direction, passes the same sign. If t represents the amount of time in hours it takes Gary to catch up to Renee, write an expression in terms of t that describes the amount of time Renee has traveled since passing the sign.

Exercises 37–48 contain word descriptions of expressions from mathematics and physics. Translate each description to a variable expression. See Examples 2(c) and 2(d).

37. The circumference of a circle can be found by multiplying the diameter, d, by π. Write an expression for finding the circumference.

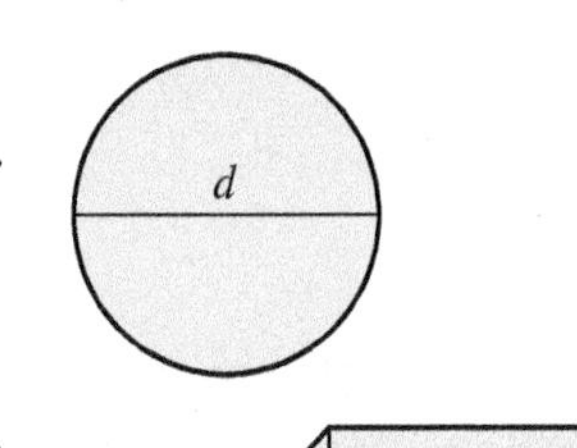

38. The area of a parallelogram can be found by finding the product of the base, b, and the height, h. Write an expression for finding the area of a parallelogram.

39. The area of a trapezoid is one-half of the product of the height, h, and the sum of the lengths of the sides a and b. Write an expression for finding the area of a trapezoid.

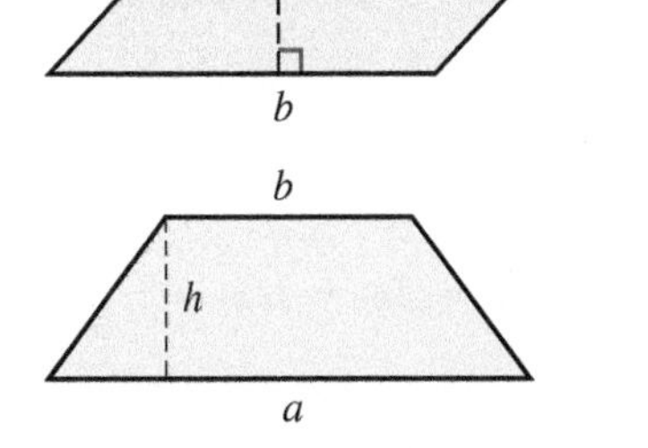

40. The surface area of a right circular cylinder is found by multiplying the sum of the radius, r, and the height, h, by the product of 2π and the radius. Write an expression for the surface area of a right circular cylinder.

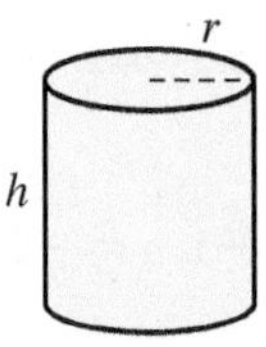

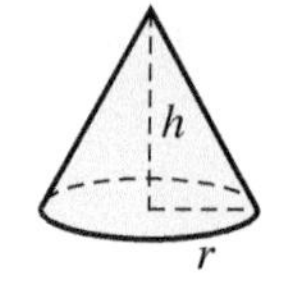

41. The volume of a cone is one-third of the product of π; the square of the radius, r; and the height, h, of the cone. Write an expression for finding the volume of a cone.

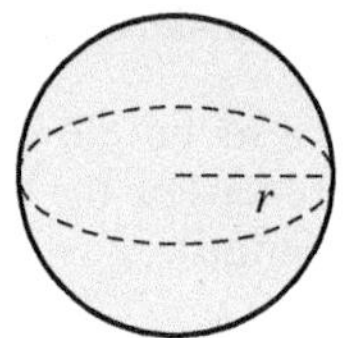

42. The volume of a sphere is four-thirds of the product of π and the cube of the radius, r. Write an expression for finding the volume of a sphere.

43. In physics, energy is calculated by multiplying the product of the mass, m, and the square of the velocity, v, by one-half. Write an expression for finding energy.

44. When an object accelerates from rest for an amount of time, the distance the object travels is the product of one-half of the acceleration, a, and the square of the time, t. Write an expression for finding the distance.

45. Isaac Newton discovered that the relationship for gravitational attraction between two bodies is the product of their masses M and m divided by the square of the distance, d, between them. Write an expression for finding the gravitational attraction between two bodies.

46. Albert Einstein developed an expression that describes the energy of a particle at rest, which is the product of the mass, m, of the particle and the square of the speed of light, c. Write an expression for finding the rest energy.

★ 47. Albert Einstein's theory of relativity includes a mathematical expression that is the square root of the difference of 1 and the ratio of the square of the velocity, v, of an object in motion to the square of the speed of light, c. Translate to an algebraic expression.

★ 48. René Descartes developed an expression for the distance between two points in the coordinate plane. The distance between two points is the square root of the sum of the square of the difference of x_2 and x_1 and the square of the difference of y_2 and y_1. Write an expression for finding the distance between two points in the coordinate plane.

Isaac Newton is considered to be one of the greatest mathematicians ever. He received a BA from Trinity College, Cambridge, in 1664. Over the next two years, he developed calculus, discovered the universal law of gravitation, and proved that white light is made up of all colors in the spectrum (all before he was 25). Newton's mathematical model for gravity held until the early 1900s, when Albert Einstein revolutionized physics with his theory of relativity.

Objective 2

Prep Exercise 3 Explain how to evaluate a variable expression.

For Exercises 49–56, evaluate the expression using the given values. See Example 3.

49. $-0.4(x + 2) - 5; x = 3$

50. $1.6(y - 3) + 4; y = -2$

51. $-3m^2 + 5m + 1; m = -\frac{2}{3}$

52. $y^2 - 2y + 3; y = -\frac{1}{2}$

53. $|2x^2 - 3y| + x; x = 5, y = -1$

54. $|8m^2 - 3n| + n; m = 4, n = -2$

55. $\sqrt[3]{c} - 2ab^2; a = -1, b = -2, c = 8$

56. $\sqrt{m - 9} + 3n^2; m = 13, n = -2$

57. The expression $ad - bc$ is used to calculate the determinant of a matrix, which we will study in detail in Section 4.5. Find the determinant given each set of values.

a. $a = 5, b = 0.2, c = -3, d = 7$ **b.** $a = -8, b = \frac{2}{3}, c = 2, d = -\frac{5}{6}$

58. The expression $b^2 - 4ac$ is called the *discriminant* and is used to determine the types of solutions for quadratic equations. We will study the discriminant in detail in Section 9.2. Find the value of the discriminant of each set of values.

a. $a = -2, b = 4, c = 3$ **b.** $a = -1, b = 2, c = -4.$

59. The expression $\frac{y_2 - y_1}{x_2 - x_1}$ is used to calculate the slope of a line, which we will discuss fully in Section 3.2. Find the slope given each set of values.

a. $x_1 = 3, y_1 = -1, x_2 = 5, y_2 = -7$ **b.** $x_1 = 3, y_1 = -1, x_2 = -1, y_2 = -2$

60. The expression $\sqrt{(x_2 - x_1)^2 + (y_2 - y_1)^2}$ is used to calculate the distance between two points in the coordinate plane, which we will discuss in Section 11.1. Evaluate the expression using the given values.

a. $x_1 = 5, y_1 = -2, x_2 = -7, y_2 = 3$ **b.** $x_1 = 1, y_1 = -2, x_2 = 9, y_2 = 4$

For Exercises 61–68, determine all values for the variable that cause the expression to be undefined. See Example 4.

61. $\frac{5 + x^2}{x}$ **62.** $\frac{6 - u^2}{u}$ **63.** $\frac{-7}{y - 6}$ **64.** $\frac{8}{m + 2}$

65. $\frac{-5y}{(y + 5)(y - 1)}$ **66.** $\frac{2m}{(m + 2)(m - 1)}$ ★ **67.** $\frac{x + 1}{4x + 1}$ ★ **68.** $\frac{3y}{5y - 3}$

Objective 3

Prep Exercise 4 Complete the distributive property of multiplication over addition.
$a(b + c) =$ ____________

For Exercises 69–76, use the distributive property to write an equivalent expression. See Example 5.

69. $9(3x - 5)$ **70.** $4(6x + 2)$ **71.** $-5(m + 2)$ **72.** $-8(2x - 7)$

73. $\frac{3}{8}\left(\frac{2}{9}x - 24\right)$ **74.** $-\frac{4}{7}\left(-14k - \frac{1}{8}\right)$ **75.** $-2.1(3x + 2.4)$ **76.** $4.2(2.1x - 5)$

Objective 4

Prep Exercise 5 What are like terms?

Prep Exercise 6 Explain how to combine like terms.

For Exercises 77–86, simplify by combining like terms. See Examples 6 and 7.

77. $2x - 13x$

78. $5p - 8p$

79. $\frac{6}{7}b^2 - \frac{8}{7}b^2$

80. $\frac{1}{2}y - \frac{5}{6}y$

81. $4x - 9y - 12 + y + 3x$

82. $n - 12m + 7 + 2m + 8n$

83. $1.5x + y - 2.8x + 0.3 - y - 0.7$

84. $0.4t^2 - 0.9t - 2.8 - t^2 + 0.9t - 4$

85. $2.6h^2 + \frac{5}{3}h - \frac{2}{5}h^2 + h + 7$

86. $-0.3t + \frac{3}{4}t^2 - 8 + t + \frac{1}{2}t - \frac{1}{3}t^2$

For Example 87–90, simplify. See Example 7(c).

87. $2(5n - 6) + 4(n + 1) - 8$

88. $3(x - 7) + 6(2x - 3) + 9$

89. $7a + 5b - 3(4a + 2b) - 12 + 8b$

90. $t - 2(t + 9u) + 7t - 9 + 18u$

91. **a.** Translate to an algebraic expression: fourteen plus the difference of six times a number and eight times the same number.

b. Simplify the expression.

c. Evaluate the expression when the number is -3.

92. **a.** Translate to an algebraic expression: the sum of negative five times a number and the difference of eight and two times the same number.

b. Simplify the expression.

c. Evaluate the expression when the number is 0.2.

Review Exercises

Exercises 1–4 Expressions

[1.1] **1.** Write the set containing all integers greater than or equal to -2 in set-builder notation.

[1.3] *For Exercises 2–4, evaluate.*

2. $-5^2 + 3(6 - 8) - \sqrt{16}$

3. $(20 - 24)^3 - 4|3 - 8|$

4. -7^2

Exercises 5 and 6 Equations and Inequalities

[1.2] **5.** What property of addition is represented by $-9 + (6 + 2) = -9 + (2 + 6)$?

[1.2] **6.** What property is represented by $-1(5 - 2) = -5 + 2$?

Chapter 1 Summary and Review Exercises

Complete each incomplete definition, rule, or procedure; study the key examples; and then work the related exercises.

The Real-Number System

Real Numbers

Rational Numbers: Real numbers that can be expressed in the form $\frac{a}{b}$, where a and b are integers and $b \neq 0$, such as $-4\frac{3}{4}$, $-\frac{2}{3}$, 0.018, $0.\overline{3}$, and $\frac{5}{8}$.

Integers: ..., −3, −2, −1, 0, 1, 2, 3, ...

Whole Numbers: 0, 1, 2, 3, ...

Natural Numbers: 1, 2, 3, ...

Irrational Numbers: Any real number that is not rational, such as $-\sqrt{2}$, $-\sqrt{3}$, $\sqrt{0.8}$, and π.

Definitions

[1.1]

A ____________ is a symbol varying in value.

A ____________ is a symbol that does not vary in value.

An **expression** is a ____________; a ____________ or any combination of ____________, ____________, and arithmetic operations.

An **equation** is two expressions ____________.

An **inequality** is two expressions separated by ________, ________, ________, ________, or ________.

A ____________ is a collection of objects.

The set of **natural numbers** is ____________.

The set of **whole numbers** is ____________.

The set of **integers** is ____________.

If every element of set B is an element of set A, then B is a ____________ of A.

[1.2]

A **rational number** is any real number that can be expressed in the form ____________, where a and b are ____________ and $b \neq 0$.

An ____________ is any real number that is not rational.

A **prime number** is a natural number with exactly two factors, ____________ and ____________.

The **absolute value** of a number is its distance from ____________ on the number line.

Two numbers are **additive inverses** if their ____________ is ____________.

Two numbers are **multiplicative inverses** if their ____________ is ____________.

[1.4]

A ____________ is number or the product of a number and one or more variables raised to powers.

The **numerical coefficient** is the ____________ of a term.

Variable terms that have the same variable(s) raised to the same exponents, or constant terms are ____________.

1.1 Sets and the Structure of Algebra

Definitions/Rules/Procedures	Key Example(s)
To write a set, write the ________ or ________ of the set within braces.	Write a set containing the odd natural numbers. **Answer:** $\{1, 3, 5, 7, \ldots\}$

Exercises 1–12 Expressions

[1.1] *For Exercises 1–4, write a set representing each description.*

1. The states outside the 48 contiguous states of the United States

2. The odd integers

3. The natural-number multiples of 5

4. The letters in the word *simplify*

[1.1] *For Exercises 5–8, write the set in set-builder notation.*

5. $\{3, 6, 9, \ldots\}$

6. $\{0, 1, 2, 3, \ldots\}$

7. $\{2, 3, 5, 7, 11, \ldots\}$

8. {Monday, Tuesday, Wednesday, Thursday, Friday, Saturday, Sunday}

[1.1] *For Exercises 9–12, answer true or false.*

9. March $\notin \{n \mid n \text{ is a day of the week}\}$

10. If $A = \{3, 4, 5\}$ and $B = \{2, 3, 5\}$, then $A \subseteq B$.

11. $\frac{2}{3} \in I$

12. $\sqrt{2} \notin Z$

Definitions/Rules/Procedures	Key Example(s)
The **absolute value** of every real number is either positive or zero according to the following rule: $\lvert n \rvert = \begin{cases} \_\_\_\_ & \text{for } n \geq 0 \\ \_\_\_\_ & \text{for } n < 0 \end{cases}$	Find the absolute value. **a.** $\lvert 12 \rvert = 12$ **b.** $\lvert -9 \rvert = 9$ **c.** $\lvert 0 \rvert = 0$
For any two real numbers a and b, a is greater than b if a is to the ________ of b on a number line.	Use $<$ or $>$ to write a true statement. **a.** $-15 \quad -10$ **b.** $8 \quad -5$ **Answers:** **a.** $-15 < -10$ **b.** $8 > -5$

Exercises 13–16 Equations and Inequalities

[1.1] *For Exercises 13–16, use =, <, or > to write a true statement.*

13. $-(-5) \quad \lvert -5 \rvert$

14. $\left\lvert -3\frac{1}{4} \right\rvert \quad -3\frac{1}{4}$

15. $0.5 \quad \frac{1}{2}$

16. $-\lvert -3 \rvert \quad -4$

1.2 Operations with Real Numbers; Properties of Real Numbers

Definitions/Rules/Procedures	Key Example(s)
Properties of Arithmetic In each of the following, a, b, and c represent real numbers. [1.2] **Additive Identity** $a + 0 =$ ______ **Commutative Property of Addition** $a + b =$ ______ **Associative Property of Addition** $(a + b) + c =$ ______ **Multiplication Property of 0** $a \cdot 0 =$ ______ **Multiplicative Identity** $a \cdot 1 =$ ______ **Commutative Property of Multiplication** $ab =$ ______ **Associative Property of Multiplication** $(ab)c =$ ______ [1.4] **Distributive Property** $a(b + c) =$ ______	Identify the property illustrated by each of the following. **a.** $-2 + 8 = 8 + (-2)$ **Answer:** Commutative Property of Addition **b.** $-2(8 \cdot 4) = (-2 \cdot 8)4$ **Answer:** Associative Property of Multiplication **c.** $-9 + 0 = -9$ **Answer:** Additive identity **d.** $-6(-4 + 2) = -6(-4) + -6(2)$ **Answer:** Distributive property **e.** $4(-6) = -6 \cdot 4$ **Answer:** Commutative Property of Multiplication **f.** $8 \cdot 1 = 8$ **Answer:** Multiplicative Identity **g.** $(-9 + 3) + 7 = -9 + (3 + 7)$ **Answer:** Associative Property of Addition **h.** $-13 \cdot 0 = 0$ **Answer:** Multiplication Property of 0

Exercises 17–26 Expressions

[1.2] *For Exercises 17–20, indicate whether the given equation illustrates the additive identity, multiplicative identity, additive inverse, or multiplicative inverse.*

17. $5.8 + (-5.8) = 0$

18. $-\frac{2}{5} \cdot -\frac{5}{2} = 1$

19. $0 + (-9) = -9$

20. $8.7(1) = 8.7$

[1.2] *For Exercises 21–26, indicate whether the given equation illustrates the commutative property of addition, commutative property of multiplication, associative property of addition, associative property of multiplication, or distributive property.*

21. $6(7 + 5) = 6 \cdot 7 + 6 \cdot 5$

22. $-3(4 \cdot 5) = (-3 \cdot 4)5$

23. $-6 + 5 = 5 + (-6)$

24. $3 \cdot (2 \cdot 4) = 3 \cdot (4 \cdot 2)$

25. $-7 + (1 + 8) = (-7 + 1) + 8$

26. $-2(3 + 5) = -2 \cdot 3 - 2 \cdot 5$

Definitions/Rules/Procedures	Key Example(s)
To **add** two numbers that have the **same sign**, add their __________ and keep the same sign. To **add** two numbers that have **different signs**, __________ the smaller absolute value from the greater absolute value and keep the sign of the number with the __________.	Add. **a.** $3 + 9 = 12$ **b.** $-3 + (-9) = -12$ **c.** $-3 + 9 = 6$ **d.** $3 + (-9) = -6$
For any real number n, $-(-n) =$ __________.	Simplify $-(-6)$. **Answer:** $-(-6) = 6$
To write a **subtraction statement** as an equivalent **addition statement**, change the operation symbol from a __________ to a __________ and change the subtrahend to its __________.	Subtract. **a.** $3 - 9 = 3 + (-9) = -6$ **b.** $-3 - 9 = -3 + (-9) = -12$ **c.** $3 - (-9) = 3 + 9 = 12$ **d.** $-3 - (-9) = -3 + 9 = 6$

Exercises 27–34 Expressions

[1.2] *For Exercises 27–30, add.*

27. $6 + (-7)$

28. $-4 + 9$

29. $-15 + (-2)$

30. $-2 + (-5)$

[1.2] *For Exercises 31–34, subtract.*

31. $7 - 9$

32. $-2 - 8$

33. $15 - (-2)$

34. $-8 - (-1)$

Definitions/Rules/Procedures	Key Example(s)
The **product** of two numbers that have the **same sign** is __________. The **product** of two numbers that have **different signs** is __________. The **product** of an **even number** of negative factors is __________, whereas the product of an **odd number** of negative factors is __________.	Multiply. **a.** $(3)(4) = 12$ **b.** $(-3)(-4) = 12$ **c.** $(3)(-4) = -12$ **d.** $(-3)(4) = -12$ **e.** $(-1)(-2)(-3)(-4) = 24$ **f.** $(-2)(-3)(-4) = -24$
To write a **division statement** as an equivalent **multiplication statement**, change the operation symbol from __________ to __________ and change the divisor to it's __________.	Divide. $\frac{2}{5} \div \frac{3}{5} = \frac{2}{5} \cdot \frac{5}{3} = \frac{2}{3}$
The **quotient** of two numbers that have the **same sign** is __________. The **quotient** of two numbers that have **different signs** is __________. $0 \div n =$ __________ when $n \neq 0$. $n \div 0$ is __________ when $n \neq 0$. $0 \div 0$ is __________.	Divide. **a.** $24 \div 8 = 3$ **b.** $-24 \div (-8) = 3$ **c.** $-24 \div 8 = -3$ **d.** $24 \div (-8) = -3$ **e.** $0 \div 7 = 0$ **f.** $35 \div 0$ is undefined.

Exercises 35–43 Expressions

[1.2] *For Exercises 35–42, multiply or divide.*

35. $-2(4)$

36. $-3(-5)$

37. $7(-8)$

38. $25 \div (-5)$

39. $-10 \div (-5)$

40. $-8 \div 4$

41. $-1(-2)(-3)$

42. $-50 \div [4 \div (-2)]$

For Exercise 43, solve.

[1.2] **43.** Leanne has a balance of $-\$245.85$ on her credit card. Find her new balance if she makes the following transactions.

Transaction	Amount
Payment	$125.00
Walmart	−$72.34
Starbucks	−$12.50
Krispy Kreme	−$14.75

1.3 Exponents, Roots, and Order of Operations

Definitions/Rules/Procedures	Key Example(s)
For any real number b and any natural number n, the nth power of b, or b^n, is found by multiplying ________ as a factor ________ times. $b^n = \underbrace{b \cdot b \cdot b \cdot \cdots \cdot b}_{b \text{ used as a factor } n \text{ times}}$	Evaluate. **a.** $3^4 = 3 \cdot 3 \cdot 3 \cdot 3 = 81$ **b.** $(-5)^4 = (-5)(-5)(-5)(-5) = 625$ **c.** $(-2)^5 = (-2)(-2)(-2)(-2)(-2) = -32$ **d.** $-8^2 = -(8 \cdot 8) = -64$
If the **base** of an exponential form is a **negative** number and the exponent is **even**, the product is ________.	
If the **base** of an exponential form is a **negative** number and the exponent is **odd**, the product is ________.	

Exercises 44–49 Expressions

[1.3] *For Exercises 44 and 45, identify the base and the exponent, translate the expression to words, and then simplify.*

44. -2^7

45. $(-1)^4$

[1.3] *For Exercises 46–49, evaluate.*

46. -3^2

47. -2^3

48. $(-4)^2$

49. $\left(-\frac{2}{5}\right)^3$

Definitions/Rules/Procedures	Key Example(s)
Every **positive number** has two square roots, a ________ root and a ________ root.	Find all square roots of 36. **Answer:** ± 6
The square root of 0 is ________. **Negative numbers** have ______ real-number square roots. The **radical symbol** $\sqrt{\ }$ denotes the ____________ square root. $\sqrt{\frac{a}{b}} =$ ________, where $a \geq 0$ and $b > 0$. **Roots** other than **square roots** are indicated by using a root ________.	Simplify. **a.** $\sqrt{100} = 10$ **b.** $\sqrt{-64}$ is not a real number. **c.** $\sqrt{\frac{25}{81}} = \frac{\sqrt{25}}{\sqrt{81}} = \frac{5}{9}$ **d.** $\sqrt[3]{64} = 4$

Exercises 50–53 Expressions

[1.3] *For Exercises 50–53, evaluate the root.*

50. $\sqrt{121}$ **51.** $\sqrt[4]{81}$ **52.** $\sqrt[3]{-8}$ **53.** $\sqrt[8]{1}$

Definitions/Rules/Procedures	Key Example(s)
Perform operations in the following order: **1.** Within grouping symbols: parentheses ( ), brackets [ ], braces { }, above and/or below fraction bars, and radicals $\sqrt{\ }$. **2.** ________ from left to right, in order as they occur. **3.** ________ from left to right, in order as they occur. **4.** ________ from left to right, in order as they occur.	Simplify. $-4^3 - 3\|16 - (4 + 2\cdot 9)\| + \sqrt{49}$ $= -4^3 - 3\|16 - (4 + 18)\| + 7$ $= -4^3 - 3\|16 - 22\| + 7$ $= -64 - 3\|-6\| + 7$ $= -64 - 3(6) + 7$ $= -64 - 18 + 7$ $= -75$

Exercises 54–62 Expressions

[1.3] *For Exercises 54–59, evaluate using the order of operations.*

54. $-7|8 - 2^4| + 7 - 3^2$ **55.** $8(1 - 3^2) + \sqrt{16 - 7}$ **56.** $\sqrt[3]{8} - 7\cdot 3^2$

57. $\sqrt{16} + \sqrt{9} - 3(2 - 7)$ **58.** $5^2(3 - 8)^2$ **59.** $\dfrac{3^4 - 5[3 - 4(-2)]}{16 \div 8(-5)}$

For Exercises 60–62, solve.

[1.3] **60.** Find the GPA.

Course	Course Credits	Grade
NUR 110	2	B
MAT 140	4	A
PSY 201	3	B
ECO 101	3	D

[1.3] 61. A wireless service charges $59 for 2000 minutes in the primary calling zone, then $0.10 for each additional minute in the calling zone. Calls outside the calling zone cost $0.15 per minute. Suppose a subscriber uses 1568 minutes in the calling zone and 37 minutes outside the calling zone. Write a numerical expression that describes the total cost for the month; then calculate the total.

[1.3] 62. A store owner wants to use a code system to label his inventory. How many different codes are possible if each code label has three letters (A–E) followed by five digits (0–9)?

1.4 Evaluating and Rewriting Expressions

Definitions/Rules/Procedures	Key Example(s)
To **translate** a word phrase to an expression, identify the ________, ________, and ________; then write the corresponding symbolic form.	Translate the phrase to an expression. **a.** Three less than four times a number **Answer:** $4n - 3$ **b.** Eight times the difference of a number and five **Answer:** $8(n - 5)$

Exercises 63–70 Expressions

[1.4] *For Exercises 63–68, translate the phrase to an algebraic expression.*

63. fourteen minus a number times eight

64. twice the difference of a number and two

65. the sum of a number and one-third of the difference of the number and four

66. the ratio of m and the square root of n

67. sixteen subtracted from half of the difference of a number and eight

68. twenty more than the sum of a number and five

[1.4] *For Exercises 69 and 70, translate the indicated phrase.*

69. The length of a rectangle is twice the width. If the width is represented by w, write an expression that describes the length.

70. Laura is traveling south on a highway and passes a grocery store. One-third of an hour later, Tom, who is traveling in the same direction on the highway, passes the same store. If t represents the amount of time in hours it takes Tom to catch up to Laura, write an expression in terms of t that describes the amount of time Laura has traveled since passing the store.

Definitions/Rules/Procedures	Key Example(s)
To **evaluate** an **algebraic** expression, **1.** Replace each ________ with its corresponding given value. **2.** ________ the resulting numerical expression. A **expression** is **undefined** for all value(s) of the variable(s) that make the ________________.	Evaluate $5x^3 - 4x$ when $x = -2$. $5(-2)^3 - 4(-2) = 5(-8) - 4(-2)$ $= -40 + 8$ $= -32$ Determine all values for the variable that makes $\frac{x-2}{x+5}$ undefined. **Solution:** If $x = -5$ the expression is undefined because the denominator is equal to 0.

Exercises 71–76 Expressions

[1.4] *For Exercises 71–74, evaluate each expression using the given values.*

71. $3a^2 - 4a + 2; a = -1$

72. $-4|-b + 2ac|; a = 1, b = -2, c = -1$

73. $15(x + y)^2 + x^2; x = 1, y = -2$

74. $\sqrt{a - b} + 3^0 - a; a = 25, b = 16$

[1.4] *For Exercises 75 and 76, determine all values for the variable that cause the expression to be undefined.*

75. $\frac{2}{x - 3}$

76. $\frac{5x - 3}{2x - 1}$

Definitions/Rules/Procedures	Key Example(s)
Rewrite expressions using the distributive property. $a(b + c) =$ ________ To **combine** like **terms**, add or subtract the ________ and keep the ________ and their ________ the same.	Distribute. $-3(x - 5) = -3 \cdot x - (-3) \cdot 5$ $= -3x + 15$ Combine like terms. $6x^2 - 9x + x^2 - 12 + 9x + 7$ $= 6x^2 + x^2 - 9x + 9x - 12 + 7$ $= 7x^2 - 5$

Exercises 77–82 Expressions

[1.4] *For Exercises 77 and 78, use the distributive property to write an equivalent expression.*

77. $-2(5x + 1)$

78. $4(2a + 3b - 4)$

[1.4] *For Exercises 79–82, simplify by combining like terms.*

79. $3x + 2x^2 - 4x - x - 3x^2$

80. $5m^5 + 3mn - 2mn^2 - mn - 4m^5$

81. $-7ab + 3ab^2 + 2ab + 3a - 7a^2 - 8$

82. $6r - 3 - r - 2r - 7$

Chapter 1 Practice Test

For Extra Help

Step-by-step test solutions are found on the Chapter Test Prep Videos available in MyMathLab® *or on* YouTube.

For Exercises 1–5, simplify.

1. $|8.1|$

2. $-\left|-\frac{11}{4}\right|$

3. $\sqrt{169}$

4. $\sqrt[3]{125}$

5. $\sqrt[5]{\frac{1}{32}}$

For Exercises 6 and 7, indicate whether the expression illustrates the commutative property of addition, the associative property of addition, the commutative property of multiplication, the associative property of multiplication, or the distributive property.

6. $8(1 + 3) = 8(3 + 1)$

7. $2(5 \cdot 3) = (2 \cdot 5)3$

For Exercises 8–19, simplify.

8. $9 + (-1)$

9. $\frac{2}{3} - \left(-\frac{1}{4}\right)$

10. $(-3)(2.5)$

11. $(-5)^2$

12. $-\frac{2}{5} \div \frac{5}{2}$

13. $\sqrt[4]{16}$

14. $8 \div 4 \cdot 2$

15. $-3^2 + 7 - 2(5 - 1)$

16. $\sqrt[4]{16} + 8 - (3 + 1)^2$

17. $4 \div |8 - 6| + 2^4$

18. $(5 - 4)^5 + (2 - 3)^3$

19. $\sqrt{9 + 16} + [-2^2 + 3(2 - 5)]$

20. Collin has a balance of $-\$423.75$ on a credit card. Find his balance after the transactions listed here.

Description	Amount
Outback Steakhouse	−$84.50
Texaco	−$24.80
Payment	$500.00
Best Buy	−$356.45

21. Listed are Internet reference websites. Find the average number of visitors per site. Each site is listed with the number of unique visitors in millions.

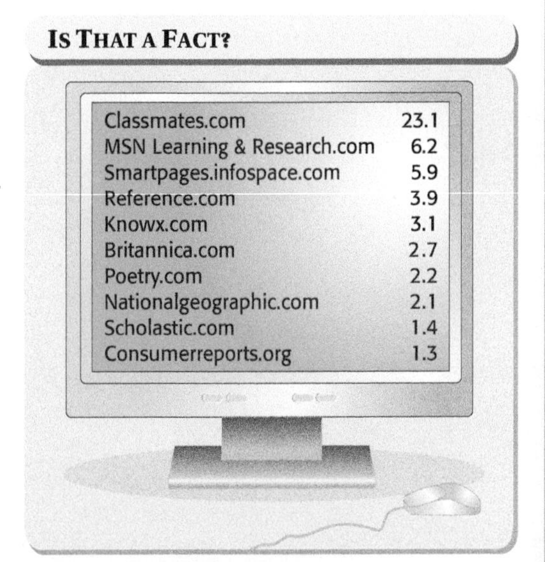

Source: Comscore Media Metrix

22. Evaluate $-2|3 - 4xy^2|$ when $x = 2$ and $y = -3$.

23. Evaluate $\frac{a}{b} - \sqrt{a} + \sqrt{b}$ when $a = 16$ and $b = 4$.

24. Use the distributive property to write an equivalent expression: $-7(3x + 5)$

25. Simplify $\frac{2}{5}x + \frac{3}{4}y - 5x + 6y + 2.7$.

CHAPTER

2

Linear Equations and Inequalities in One Variable

Chapter Overview

In Chapter 1, we reviewed number sets, arithmetic operations and properties, and how to evaluate and rewrite expressions. In this chapter, we build on that foundation and learn to solve equations and inequalities that have one variable. Specifically, we will explore the following topics:

- Solve linear equations and inequalities.
- Solve absolute value equations and inequalities.
- Solve application problems.

2.1 Linear Equations and Formulas

Objectives

1. Solve linear equations in one variable.
2. Identify identities and contradictions.
3. Solve formulas for specified variables.

Warm-Up

[1.4] **1.** Translate to an expression. nine less than five times a number

[1.4] **2.** Evaluate. $|x^3 - 5x|; x = -3$

[1.4] **3.** Simplify. $2(3m - 4) + 5(m + 1)$

In Chapter 1, we reviewed real numbers and their properties and defined constants, variables, expressions, equations, and inequalities. We now move up the Algebra Pyramid and consider equations. Recall that an equation has an equal sign.

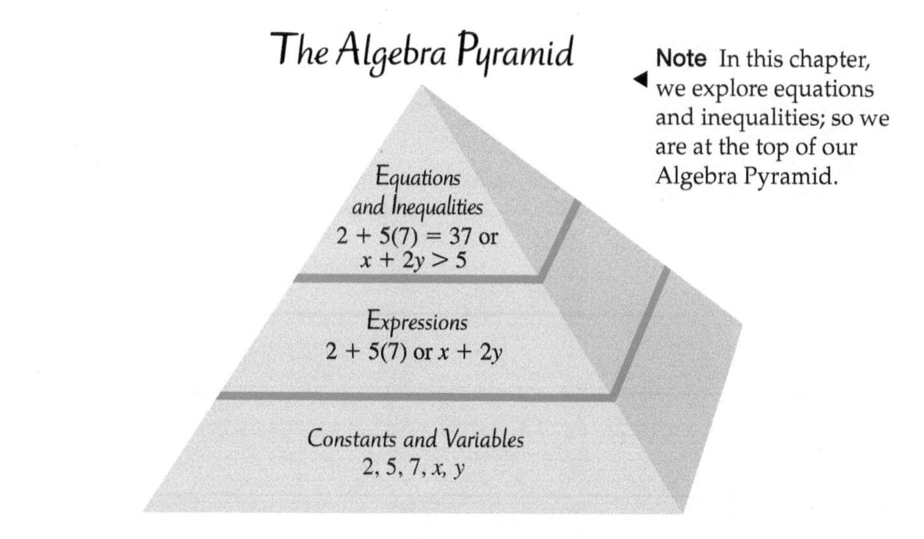

Note In this chapter, we explore equations and inequalities; so we are at the top of our Algebra Pyramid.

Learning Strategy (Auditory, Tactile, Visual)

As suggested in the To the Student section on page xviii, arrange your notebook with four sections: Notes, Homework, Study Sheets, and Practice Tests. Remember to use color in your notes: red for definitions and blue for rules and procedures. Organization and color coding will help you locate important items faster.

An equation can be true or false. For example, the equation $4 + 1 = 5$ is true and the equation $8 - 2 = 7$ is false. If an equation contains a variable for a missing number, such as $x + 3 = 9$, our goal is to find its **solution(s)**. We can write those solutions in a **solution set**.

Definitions **Solution:** A number that makes an equation true when it replaces the variable in the equation.
Solution set: A set containing all of the solutions for a given equation.

For example, 6 is the solution to the equation $x + 3 = 9$ because replacing x with 6 makes the equation true.

$$x + 3 = 9$$
$$\downarrow$$
$$6 + 3 = 9 \quad \text{True}$$

Note By showing that the equation $x + 3 = 9$ is a true statement when x is replaced with 6, we have **checked** that 6 is the solution for the equation.

The solution set for this equation is $\{6\}$, or in set-builder notation, $\{x | x = 6\}$. We focus on writing solution sets when we learn about inequalities in Section 2.3.

Objective 1 Solve linear equations in one variable.

In this chapter, we solve **linear equations in one variable**.

Definition **Linear equation in one variable:** An equation that can be written in the form $ax + b = c$, where a, b, and c are real numbers and $a \neq 0$.

Answers to Warm-up
1. $5n - 9$ **2.** 12 **3.** $11m - 3$

Note Except for two special cases, linear equations have exactly one solution.

The equation $2x + 3 = 7$ is a linear equation in one variable. Notice that the exponent of x is 1, which is an important feature of linear equations. If an equation in one variable has a variable raised to an exponent other than 1, the equation is not linear. For example, $x^2 - 2x + 5 = 8$ is not linear because it contains x^2.

Two principles of equality are used to solve linear equations in one variable: the *addition principle of equality* and the *multiplication principle of equality*.

Note Because subtraction is defined in terms of addition $(x - y = x + (-y))$ and division is defined in terms of multiplication $\left(\frac{x}{y} = x \cdot \frac{1}{y}\right)$, these properties allow us to either subtract the same number from both sides of an equation or to divide both sides of an equation by the same nonzero number.

Rules **The Addition Principle of Equality**
If $a = b$, then $a + c = b + c$ is true for all real numbers a, b, and c.

The Multiplication Principle of Equality
If $a = b$, then $ac = bc$ is true for all real numbers a, b, and c, where $c \neq 0$.

The addition principle of equality says that adding the same term to both sides of an equation gives an equivalent equation, meaning its solution set is the same. So we can add the additive inverse of the term that we want to move to the other side of an equation to both sides of the equation. Likewise, the multiplication principle says that multiplying both sides of an equation by the same nonzero expression gives an equivalent equation. So when we have isolated a term on one side of an equation, we can multiply both sides of the equation by the multiplicative inverse of the coefficient of that term in order to isolate the variable.

When we solve an equation using these principles, our goal is to write an equation equivalent to the original that has the variable isolated. In other words, we want the variable alone on one side of the equation.

Example 1 Solve and check: $2x - 9 = 5$

Solution: Use the addition principle to isolate the variable term; then use the multiplication principle to isolate the variable.

$$2x - 9 + 9 = 5 + 9$$ Add 9 to both sides to isolate $2x$.

$$2x = 14$$ Simplify.

$$\frac{1}{2} \cdot \frac{2x}{1} = \frac{14}{2} \cdot \frac{1}{1}$$ Multiply both sides by $\frac{1}{2}$ to isolate x.

$$x = 7$$

Check: $2(7) - 9 \stackrel{?}{=} 5$ Replace x in the original equation with 7 and verify that the equation is true.

$$14 - 9 \stackrel{?}{=} 5$$

$$5 = 5$$ True; therefore, 7 is correct.

Note Dividing by 2 is equivalent to multiplying by $\frac{1}{2}$. Also, dividing both sides by 2 looks cleaner.

$$\frac{2x}{2} = \frac{14}{2}$$

$$x = 7$$

From now on, if the coefficient is an integer, we will divide both sides by that integer instead of multiplying by its reciprocal.

Your Turn 1 Solve and check.

a. $4y - 19 = 1$ **b.** $9 = -12t + 3$

Solving Equations with Variable Terms on Both Sides

If variable terms appear on *both* sides of the equal sign, we use the addition principle to get the variable terms on one side of the equal sign and the constant terms on the other side. To avoid a negative coefficient on the remaining variable term, eliminate the variable term with the lesser coefficient.

Answers to Your Turn 1

a. 5 **b.** $-\frac{1}{2}$

Example 2 Solve and check: $5y - 6 = 2y - 30$

Note Had we added $-5y$ to both sides, after simplifying, the right side of the equation would contain $-3y$, which has a negative coefficient. By adding $-2y$, we avoided a negative coefficient after simplifying.

Solution: $5y - 2y - 6 = 2y - 2y - 30$ Add $-2y$ to both sides to get the variable terms together.

$3y - 6 + 6 = -30 + 6$ Add 6 to both sides to isolate $3y$.

$\frac{3y}{3} = \frac{-24}{3}$ Divide both sides by 3 to isolate y.

$y = -8$

Check: $5(-8) - 6 \stackrel{?}{=} 2(-8) - 30$ Replace y in the original equation with -8 and verify that the equation is true.

$-40 - 6 \stackrel{?}{=} -16 - 30$

$-46 = -46$ True; therefore, -8 is correct.

Your Turn 2 Solve and check: $6x + 17 = 2x - 7$

Simplifying First

If an equation contains parentheses or like terms that appear on the same side of the equal sign, we first eliminate the parentheses and combine like terms.

Example 3 Solve and check: $18 - (3n - 5) = 11n - 4(n + 3)$

Solution: $18 - 3n + 5 = 11n - 4n - 12$ Distribute to eliminate parentheses.

$23 - 3n = 7n - 12$ Combine like terms.

$23 - 3n + 3n = 7n + 3n - 12$ Because $-3n$ has the lesser coefficient, we add $3n$ to both sides.

$23 + 12 = 10n - 12 + 12$ Add 12 to both sides to isolate $10n$.

$\frac{35}{10} = \frac{10n}{10}$ Divide both sides by 10 to isolate n.

$3.5 = n$

Check: $18 - (3(3.5) - 5) \stackrel{?}{=} 11(3.5) - 4(3.5 + 3)$ Replace n in the original equation with 3.5 and verify that the equation is true.

$18 - (10.5 - 5) \stackrel{?}{=} 38.5 - 4(6.5)$

$18 - 5.5 \stackrel{?}{=} 38.5 - 26$

$12.5 = 12.5$ True; therefore, 3.5 is correct.

Your Turn 3 Solve and check.

a. $6y - 12 + 3y = 15 + 4y - 7$

b. $11 + 5(k - 1) = 3k - (6k - 6)$

Eliminating Fractions or Decimals from an Equation

We can use the multiplication principle of equality to eliminate fractions or decimals from an equation so that it contains only integers. Although equations can be solved without eliminating fractions or decimals, most people find equations that contain only integers easier to solve.

If the equation contains fractions, we multiply both sides by a number that will eliminate all of the denominators. Although any multiple common to all denominators will work, using the LCD (least common denominator) results in the simplest equations.

Answer to Your Turn 2
-6

Answers to Your Turn 3
a. 4 **b.** 0

Example 4 Solve and check: $\frac{1}{5}x - \frac{3}{4} = \frac{1}{2}x + 1$

Solution: $20\left(\frac{1}{5}x - \frac{3}{4}\right) = 20\left(\frac{1}{2}x + 1\right)$ Eliminate the fractions by multiplying both sides by the LCD, 20.

$20 \cdot \frac{1}{5}x - 20 \cdot \frac{3}{4} = 20 \cdot \frac{1}{2}x + 20 \cdot 1$ Distribute.

$4x - 15 = 10x + 20$ Simplify.

$4x - 4x - 15 = 10x - 4x + 20$ Subtract $4x$ from both sides.

$-15 - 20 = 6x + 20 - 20$ Subtract 20 from both sides.

$\frac{-35}{6} = \frac{6x}{6}$ Divide both sides by 6 to isolate x.

$-\frac{35}{6} = x$

Check: $\frac{1}{5}\left(-\frac{35}{6}\right) - \frac{3}{4} \stackrel{?}{=} \frac{1}{2}\left(-\frac{35}{6}\right) + 1$ Replace x in the original equation with $-\frac{35}{6}$ and verify that the equation is true.

$-\frac{7}{6} - \frac{3}{4} \stackrel{?}{=} -\frac{35}{12} + 1$ Multiply.

$-\frac{14}{12} - \frac{9}{12} \stackrel{?}{=} -\frac{35}{12} + \frac{12}{12}$ Write equivalent fractions with their LCD.

$-\frac{23}{12} = -\frac{23}{12}$ True; therefore, $-\frac{35}{6}$ is correct.

Eliminating Decimals

To eliminate decimal numbers in an equation, we multiply by an appropriate power of 10 as determined by the decimal number with the most decimal places.

Note 0.25 has the most decimal places of the numbers in the equation. Because it has two decimal places, we multiply by 10^2, which is 100.

Connection Because decimal numbers represent fractions with denominators that are powers of 10, we are still multiplying by the LCD to eliminate them. For example,

$0.25 = \frac{25}{100}$ and $0.4 = \frac{4}{10}$

The LCD for these fractions is 100.

Example 5 Solve and check: $0.4(n - 3) = 0.25n + 6$

Solution: $100 \cdot 0.4(n - 3) = 100(0.25n + 6)$ Multiply both sides by 100.

$40(n - 3) = 100 \cdot 0.25n + 100 \cdot 6$ Multiply $100 \cdot 0.4$ and distribute 100.

$40n - 120 = 25n + 600$ Distribute 40 and multiply.

$40n - 25n - 120 = 25n - 25n + 600$ Subtract $25n$ from both sides.

$15n - 120 + 120 = 600 + 120$ Add 120 to both sides.

$\frac{15n}{15} = \frac{720}{15}$ Divide both sides by 15.

$n = 48$

Check: $0.4(48 - 3) \stackrel{?}{=} 0.25(48) + 6$ Replace n in the original equation with 48 and verify that the equation is true.

$0.4(45) \stackrel{?}{=} 12 + 6$ Simplify.

$18 = 18$ True; therefore, 48 is correct.

The following outline summarizes the process of solving linear equations in one variable.

Procedure Solving Linear Equations

To solve linear equations:

1. Simplify both sides of the equation as needed.
   a. Distribute to eliminate parentheses.
   b. Eliminate fractions or decimals by multiplying through by the LCD. In the case of decimals, the LCD is the power of 10 with the same number of 0 digits as decimal places in the number with the most decimal places. (Eliminating fractions and decimals is optional.)
   c. Combine like terms.
2. Use the addition principle so that all variable terms are on one side of the equation and all constants are on the other side. (Eliminate the variable term with the lesser coefficient to avoid negative coefficients.) Then combine like terms.
3. Use the multiplication principle to eliminate any remaining coefficient.

Your Turn 5 Solve and check.

a. $\frac{1}{3}(x - 1) = \frac{3}{4}x + \frac{1}{6}$ **b.** $4.8t - 2.46 = 0.3t - 14.16$

Objective 2 Identify identities and contradictions.

In general, a linear equation in one variable has only one real-number solution. We say that such an equation is a **conditional equation**.

Definition Conditional linear equation in one variable: A linear equation with exactly one solution.

However, there are two special cases that we need to consider, identities and contradictions:

Identities

If the left and right sides of an equation are *identical*, as in $2x + 1 = 2x + 1$, it is an **identity**. Because the expressions are identical, we can replace x with any real number and the equation will check; so every real number is a solution for an identity.

Definition Identity: An equation in which every real number (for which the equation is defined) is a solution.

Sometimes, we have to simplify the expressions in an equation to see that it is an identity.

Example 6 Solve and check: $4(3x - 2) + 5 = 9x - 3(1 - x)$

Solution: $12x - 8 + 5 = 9x - 3 + 3x$ Distribute.

$12x - 3 = 12x - 3$ Combine like terms.

Because the left and right sides are *identical*, the equation is an *identity* and every real number is a solution. The solution set is $\{x \mid x \text{ is a real number}\}$, or $\mathbb{R}$.

Although we stopped solving the equation at $12x - 3 = 12x - 3$, we could have continued the process. Each step would affirm that the equation is an identity.

$12x - 12x - 3 = 12x - 12x - 3$ Subtract $12x$ from both sides.

$-3 = -3$ It is still clear that the equation is an identity.

Answers to Your Turn 5

a. $-\frac{6}{5}$ **b.** -2.6

Contradictions

After simplifying both sides of an equation, if the variable terms are identical but the constant terms are not identical, as in $2x + 1 = 2x + 5$, the equation is a **contradiction**. Replacing x in $2x + 1$ and $2x + 5$ with the same real number always yields different results, suggesting that contradictions have no solution.

Definition **Contradiction:** An equation that has no solution.

Example 7 Solve and check: $4x + 2x - 7 = 5 + 6x - 8$

Solution: $6x - 7 = 6x - 3$ Combine like terms.

Because the variable terms are identical and the constant terms are not, the equation is a contradiction and has no solution. The solution set is empty, which we indicate by { } or $\varnothing$.

As we saw with the identity in Example 6, we can continue the process of solving a contradiction and each step will affirm that it is a contradiction.

$$6x - 6x - 7 = 6x - 6x - 3$$ Subtract $6x$ from both sides.

$$-7 = -3$$ This false equation with no variable terms affirms that the equation is a contradiction.

Warning Do not write the symbol for empty set within braces, as in $\{\varnothing\}$. This no longer indicates an empty set because the braces now contain something; so they are not truly empty.

Your Turn 7 Solve and check.

a. $3(2x - 5) - 4x = 2(x - 3) - 9$ **b.** $3(x + 1) - 2 = 3x - 5$

Objective 3 Solve formulas for specified variables.

If an equation contains more than one variable, such as in a formula, we often can use the methods we have learned to solve for one of the variables. For example, suppose we are asked to solve for w in the formula for the area of a rectangle, $A = lw$. Because w is multiplied by l, we view l as a coefficient of w; so to solve for w, we divide both sides by l.

$$\frac{A}{l} = \frac{\cancel{l}w}{\cancel{l}}$$

$$\frac{A}{l} = w$$

Our example suggests the following procedure:

Procedure **Isolating a Variable in a Formula**

To isolate a particular variable in a formula, treat all other variables like constants and isolate the desired variable using the outline for solving equations.

Example 8 Solve the formula for the indicated variable.

a. $P = R - C$; solve for R.

Solution: $P + C = R - C + C$ Add C to both sides to isolate R.

$P + C = R$ Simplify.

b. $I = Prt$; solve for r.

Solution: $\frac{I}{Pt} = \frac{\cancel{P}r\cancel{t}}{\cancel{Pt}}$ The product Pt can be treated like a coefficient of r; so we divide both sides by Pt to isolate r.

$\frac{I}{Pt} = r$ Simplify.

Of Interest

The formula $P = R - C$ is used to calculate profit, given revenue R and cost C. The formula $I = Prt$ is used to calculate interest I, given principal P at a rate r over a time t.

Answers to Your Turn 7

a. $\{x \mid x \text{ is a real number}\}$ (identity)

b. $\varnothing$ (contradiction)

c. $P = 2l + 2w$; solve for w.

Solution: $P - 2l = 2l - 2l + 2w$ Subtract $2l$ from both sides to isolate $2w$.

$\frac{P - 2l}{2} = \frac{2w}{2}$ Divide both sides by 2 to isolate w.

$\frac{P - 2l}{2} = w$ Simplify.

Note We could rewrite $\frac{P - 2l}{2}$.

$$w = \frac{P - 2l}{2} = \frac{P}{2} - \frac{2l}{2} = \frac{P}{2} - l$$

d. $C = \frac{5}{9}(F - 32)$; solve for F

Solution: $\frac{9}{5} \cdot C = \frac{9}{5} \cdot \frac{5}{9}(F - 32)$ Multiply both sides by $\frac{9}{5}$ to isolate $F - 32$.

$\frac{9}{5}C + 32 = F - 32 + 32$ Add 32 to both sides.

$\frac{9}{5}C + 32 = F$ Simplify.

Of Interest

The formula

$$C = \frac{5}{9}(F - 32)$$

is used to convert degrees Fahrenheit to degrees Celsius.

Your Turn 8 Solve the formula for the indicated variable.

a. $V = ir$; solve for i.

b. $A = \frac{1}{2}h(a + b)$; solve for h.

c. $B = P + Prt$; solve for r.

Answers to Your Turn 8

a. $i = \frac{V}{r}$ **b.** $h = \frac{2A}{a + b}$

c. $r = \frac{B - P}{Pt}$ or $r = \frac{B}{Pt} - \frac{1}{t}$

2.1 Exercises

For Extra Help MyMathLab®

Note: Exercises marked with a ★ represent challenging exercises.

Objective 1

Prep Exercise 1 What is a solution for an equation?

Prep Exercise 2 What is a solution set?

Prep Exercise 3 Explain how to eliminate fractions from an equation.

Prep Exercise 4 What is the least power of 10 that could be used to eliminate all of the decimal numbers in the equation $0.4x + 1.58 = 0.2x - 0.436$?

For Exercises 1–22, solve and check. See Examples 1–5.

1. $2x - 2 = 10$

2. $3x - 3 = 15$

3. $9 - 3m = 12$

4. $7 - 5u = 17$

5. $2x + 2 = x + 3$

6. $3p + 4 = 4p - 4$

7. $4t + 3 = 2t + 9$

8. $6q - 5 = 3q + 4$

9. $2(3z + 1) = 8$

10. $3(h + 2) = 12$

11. $n + 2(3n + 1) = 9$

12. $2b + 3(b - 2) = 4$

13. $6u - 17 = 4(u + 3) - 3$

14. $2a + 9 = 11 + 7(a - 1)$

15. $\frac{1}{2}z - 1 = 4z - 3 - 3z$

16. $\frac{1}{3}y - 5 = 7y + 10 - 5y$

17. $\frac{1}{4}x + \frac{1}{3}x + \frac{5}{6} = \frac{15}{4}$

18. $\frac{2}{3}v + \frac{1}{6}v - \frac{1}{4} = \frac{7}{12}$

19. $0.5x + 0.95 = 0.2x - 1$

20. $0.2p - 0.7 = 0.7p - 3.6$

21. $1.6x - 18 + 0.9x = 0.1x + 6$

22. $0.3z + 1.4z + 0.7 = 2.9 - 0.5z$

Objectives 1 and 2

Prep Exercise 5 Why is $5x + 1 = 9x - 6 - 4x + 7$ an identity?

Prep Exercise 6 What is the solution set for an identity?

Prep Exercise 7 Why is $x - 3 + 2x = 3x + 4$ a contradiction?

Prep Exercise 8 What is the solution set for a contradiction?

For Exercises 23–42, solve and check. Note that some equations may be identities or contradictions. See Examples 3–7.

23. $17a - 5 - 7a = 2a + 19$

24. $7d - 42 - 27d = -64 - 9d$

25. $8r - 2(r + 5) = 5(2r - 6) - 8$

26. $12c - 2(c - 3) = 11 - 5(2c - 7)$

27. $5 - 5(3x - 2) = 38 - 2(x - 8)$

28. $18 - 8(3 - s) = -5(s + 4) + 27$

29. $8h - 27 + 5(2h - 3) = 62 - (3h + 8)$

30. $5m + 3 - (4m - 5) = 5(m + 5) - 3$

31. $\frac{2}{3}q - 4 = \frac{1}{5}q + 10$

32. $\frac{4}{5}k - 6 = 7 - \frac{1}{2}k$

33. $\frac{2}{7}(x - 5) = -2 + \frac{2}{5}x$

34. $\frac{2}{3}(d - 6) = 3 + \frac{3}{4}d$

35. $0.5(x - 2) + 1.76 = 0.3x + 0.8$

36. $3.3 - 0.6a = 1.1(4 - a) - 2$

37. $2x - 3.24 + 2.4x = 6.2 + 0.08x + 3.52$

38. $x - 2.28 + 1.6x = 4.6 + 0.05x - 0.25$

39. $6(m + 3) - 5 + 2m = 3(3m + 1) - m$

40. $16(q - 3) + 2q - 1 = 5(q + 1) + 13q + 2$

41. $7(n + 2) - 3n = 4 + 4n + 10$

42. $-5(p + 2) - 3p + 5 = -5 - 8p$

Find the Mistake *For Exercises 43–48, explain the mistake; then solve correctly.*

43. $2(x + 3) = 7x - 1$
$2x + 3 = 7x - 1$
$3 = 5x - 1$
$4 = 5x$
$\frac{4}{5} = x$

44. $4 + 3(p - 1) = 14$
$4 + 3p - 1 = 14$
$3 + 3p = 14$
$3p = 11$
$p = \frac{11}{3}$

45. $4 - 5(x + 1) + 4x = 7 - 11$
$-1(x + 1) + 4x = 7 - 11$
$-x - 1 + 4x = -4$
$3x - 1 = -4$
$3x = -3$
$x = -1$

46. $7 - 2(d + 2) - 3 = 5$
$5(d + 2) - 3 = 5$
$5d + 10 - 3 = 5$
$5d + 7 = 5$
$5d = -2$
$d = -\frac{2}{5}$

47. $-2(x + 4) + 6x = 2x + 6$
$-2x + 8 + 6x = 2x + 6$
$4x + 8 = 2x + 6$
$2x + 8 = 6$
$2x = -2$
$x = -1$

48. $3x - 4(2x - 3) = -2x + 6$
$3x - 8x - 12 = -2x + 6$
$-5x - 12 = -2x + 6$
$-3x - 12 = 6$
$-3x = 18$
$x = -6$

Objective 3

Prep Exercise 9 Explain how you would isolate t in the distance formula, $d = rt$.

Prep Exercise 10 What is the first step for isolating x in the slope formula, $y = mx + b$?

For Exercises 49–68, solve the formula for the indicated variable. See Example 8.

49. $P = R - C$; solve for C

50. $c^2 = a^2 + b^2$; solve for b^2

51. $A = bh$; solve for b

52. $I = Prt$; solve for t

53. $A = 2\pi pw$; solve for p

54. $A = 2\pi rh$; solve for h

55. $A = \frac{1}{2}\theta r^2$; solve for θ

56. $V = \frac{1}{3}\pi r^2 h$; solve for h

57. $F = \frac{kMm}{d^2}$; solve for M

58. $t = -\frac{2\omega}{\alpha}$; solve for ω

59. $A = \pi s(R + r)$; solve for s

60. $B = P(1 + rt)$; solve for P

61. $P = 2l + 2w$; solve for l

62. $P = 2\pi r + 2d$; solve for d

63. $3x + 2y = 6$; solve for y

64. $4x + 3y = 9$; solve for y

65. $F = \frac{9}{5}C + 32$; solve for C

66. $h = -16t^2 + h_0$; solve for t^2

★ **67.** $x = vt + \frac{1}{2}at^2$; solve for a

★ **68.** $E = \frac{1}{2}mv^2 + mgy$; solve for v^2

Find the Mistake *For Exercises 69–72, explain the mistake; then solve correctly.*

69.
$$V = lwh;\text{ solve for } h$$
$$V - lw = lwh - lw$$
$$V - lw = h$$

70.
$$A = bh;\text{ solve for } h$$
$$A - b = bh - b$$
$$A - b = h$$

71.
$$P = 2l + 2w;\text{ solve for } l$$
$$P - 2w = 2l + 2w - 2w$$
$$P - 2w = 2l$$
$$P - 2w - 2 = 2l - 2$$
$$P - 2w - 2 = l$$

72.
$$A = \frac{1}{2}bh;\text{ solve for } b$$
$$A = 2 \cdot \frac{1}{2}bh$$
$$A = bh$$
$$\frac{A}{h} = b$$

Review Exercises

Exercises 1–6 Expressions

[1.1] 1. Write a set containing the odd natural numbers less than 14.

[1.2] 2. Simplify: $14 - 12[6 - 8(3 + 2^5)] + \sqrt{9 \cdot 16}$

[1.4] *For Exercises 3 and 4, translate to an expression.*

3. nine less than the product of seven and n

4. negative three times the sum of a number and eight

5. Combine like terms: $8x - y - 10x + 9 - 5y - 4$

6. Distributive: $-6(9m - 4)$

2.2 Solving Problems

Objectives

1. Use formulas to solve problems.
2. Translate words to equations.

Warm-up

[2.1] ***For Exercises 1–3, solve.***

1. $-5 = \frac{9}{5}x + 22$ **2.** $\frac{3}{4}(x + 5) = 1 - \frac{1}{3}x$ **3.** $29.95n + 19.95(2700 - n) = 61{,}365$

A primary purpose of studying mathematics is to develop and improve problem-solving skills. George Polya formulated a problem-solving process that follows a four-step outline: (1) understand the problem, (2) devise a plan for solving the problem, (3) execute the plan, and (4) check the results. The following outline for problem solving is based on Polya's four-step process. We'll see Polya's process illustrated throughout the rest of the text in application problems.

Procedure Problem-Solving Outline

1. **Understand** the problem.
   a. Read the question(s) (not the whole problem, just the question at the end) and make a note of what you are to find.
   b. Read the whole problem, underlining the key words.
   c. If possible or useful, draw a picture, make a list or table to organize what is known and unknown, simulate the situation, or search for a related example problem.
2. **Plan** your solution by searching for a formula or using the key words to translate to an equation.
3. **Execute** the plan by solving the equation/formula.
4. **Answer** the question. Look at your note about what you were to find and make sure you answered that question. Include appropriate units.
5. **Check** results.
   a. Try finding the solution in a different way, reversing the process, or estimating the answer and making sure the estimate and actual answer are reasonably close.
   b. Make sure the answer is reasonable.

Objective 1 Use formulas to solve problems.

We will explore the various strategies listed in the outline. First, we focus on using formulas to solve problems.

Example 1 The average temperature on Mars is −81°F. Use $F = \frac{9}{5}C + 32$ to convert this temperature to degrees Celsius.

Understand We are given a temperature in degrees Fahrenheit and the formula $F = \frac{9}{5}C + 32$.

Plan In the formula, replace F with −81 and solve for C.

Execute

$-81 = \frac{9}{5}C + 32$ Replace F with −81.

$-81 - 32 = \frac{9}{5}C + 32 - 32$ Subtract 32 from both sides.

$\frac{5}{9} \cdot (-113) = \frac{5}{9} \cdot \frac{9}{5}C$ Simplify and then multiply both sides by $\frac{5}{9}$.

$-62.\overline{7} = C$

Answer The average temperature on Mars in degrees Celsius is $-62.\overline{7}$°C.

Check To check, we could convert the answer back to degrees Fahrenheit, which we will leave to the reader.

Answers to Warm-up

1. $x = -15$

2. $x = -\frac{33}{13}$ or $-2\frac{7}{13}$

3. $n = 750$

Your Turn 1 The formula $B = P + Prt$ can be used to calculate the balance in an account if simple interest is added to a principal P at an interest rate of r after t years. Suppose an account earns 4% interest and after one year the balance is \$4368. Find the principal that was invested.

Example 2 The face of one wing of a new house is to be covered with brick. To order the correct number of bricks, the builder needs to know the area to be covered. Calculate the area that will be covered.

Understand We are to calculate the area that will be covered with brick.

Plan Since the window will not be covered, the area in question is a combination of a large rectangle and a triangle minus the area of the window.

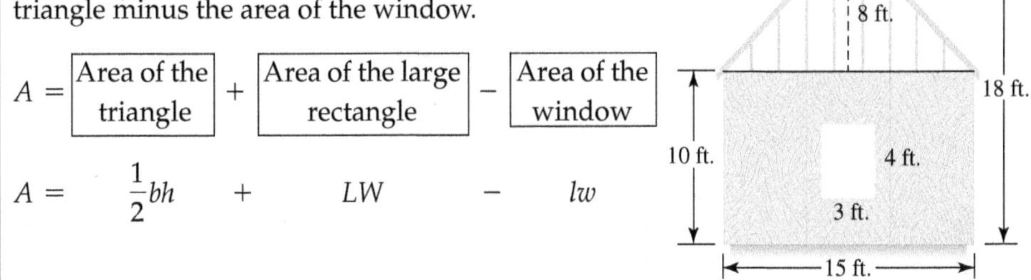

$$A = \boxed{\text{Area of the triangle}} + \boxed{\text{Area of the large rectangle}} - \boxed{\text{Area of the window}}$$

$$A = \frac{1}{2}bh + LW - lw$$

Note Uppercase L and lowercase l indicate different values. Similarly, uppercase W and lowercase w indicate different values.

Execute Replace the variables with the corresponding values and calculate.

$$A = \frac{1}{2}bh + LW - lw$$

$$A = \frac{1}{2}(15)(8) + (15)(10) - (3)(4)$$

$$A = 60 + 150 - 12$$

$$A = 198 \text{ ft.}^2$$

Answer The area to be covered by brick is 198 ft.^{2}

Check Verify the calculations. We will leave the check to the reader.

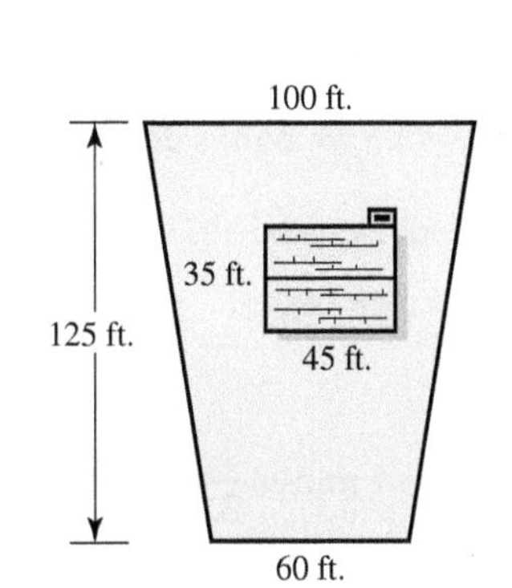

Your Turn 2 A house is to be built on a lot as shown in the figure in the margin. Once the house is built, the remaining area will be landscaped. Calculate the area to be landscaped. Recall the formula for the area of a trapezoid is $A = \frac{1}{2}h(a + b)$.

Objective 2 Translate words to equations.

In Section 1.4, we translated word phrases to expressions. Now we will translate sentences to equations. The key words to look for that indicate an equal sign are listed here.

Note You may want to review the key words for various operations in Section 1.4 before proceeding with this material.

Key words for an equal sign: is equal to, is the same as, is, produces, yields, results in

Procedure Translating Word Sentences

To translate a word sentence to an equation, identify the variable(s), constants, and key words; then write the corresponding symbolic form.

Answer to Your Turn 1
\$4200

Answer to Your Turn 2
8425 ft.^{2}

Example 3 Three-fourths of the sum of a number and five is equal to one minus one-third of the number. Translate to an equation and solve.

Understand The phrase *three-fourths of the sum* indicates to multiply *the sum of the number and five* by $\frac{3}{4}$. Because the sum is being multiplied, we write it in parentheses. *Is equal to* means an equal sign, *minus* means subtract, and *one-third of the number* indicates multiplication.

Plan Let n represent the unknown number, use the key words to translate to an equation, and solve.

Execute Translation:

Three-fourths of the sum of a number and five is equal to one minus one-third of the number.

$$\frac{3}{4} \cdot \quad (n + 5) \quad = \quad 1 \quad - \quad \frac{1}{3} \cdot n$$

$$\frac{3}{4}(n + 5) = 1 - \frac{1}{3}n$$

Solve: $12 \cdot \frac{3}{4}(n + 5) = 12 \cdot \left(1 - \frac{1}{3}n\right)$ Multiply both sides by the LCD, 12, to eliminate the fractions.

$9(n + 5) = 12 \cdot 1 - 12 \cdot \frac{1}{3}n$ Multiply $12 \cdot \frac{3}{4}$ on the left; distribute 12 on the right.

$9n + 45 = 12 - 4n$ Distribute 9 and multiply.

$9n + 4n + 45 = 12 - 4n + 4n$ Add $4n$ to both sides.

$13n + 45 - 45 = 12 - 45$ Subtract 45 from both sides.

$\frac{13n}{13} = \frac{-33}{13}$ Divide both sides by 13.

Answer $n = -\frac{33}{13}$ or $-2\frac{7}{13}$

Check Verify that $\frac{3}{4}$ of the sum of $-\frac{33}{13}$ and 5 is equal to 1 minus $\frac{1}{3}$ of $-\frac{33}{13}$. We will leave this check to the reader.

Your Turn 3 Translate to an equation and solve. One-half of the difference of a number and six is equal to two-fifths plus three-fourths of the number.

Example 4 A clothing store determines retail prices by adding 20% of the wholesale price to the wholesale price. If the retail price of a suit is $84.90, find the wholesale price the store paid.

Understand Because $84.90 is a retail price, it is the result of adding 20% of the wholesale price to the wholesale price.

Plan Let p represent the wholesale price, use the key words to translate to an equation, and solve.

Answer to Your Turn 3

$\frac{1}{2}(n - 6) = \frac{2}{5} + \frac{3}{4}n$;
$n = -\frac{68}{5}$ or $-13\frac{3}{5}$

Note Because addition is commutative, we can translate "Adding 20% of the wholesale price to the wholesale price" as $p + 0.2p$ or $0.2p + p$.

Execute Translation:

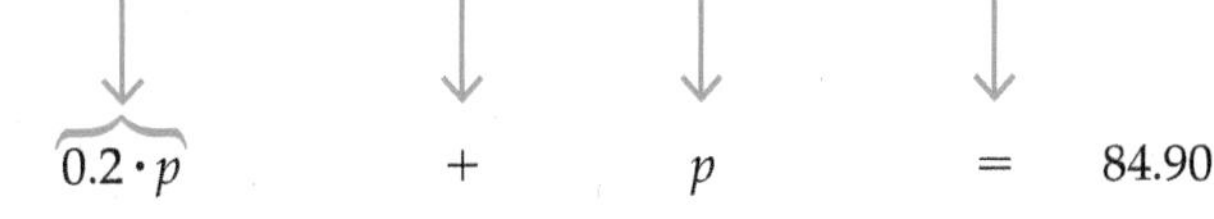

$$0.2 \cdot p \quad + \quad p \quad = \quad 84.90$$

$$0.2p + p = 84.90$$

Solve: $1.2p = 84.90$ Combine like terms.

$$\frac{1.2p}{1.2} = \frac{84.90}{1.2}$$ Divide both sides by 1.2.

$$p = 70.75$$

Answer The wholesale price is $70.75.

Check Verify that adding 20% of $70.75 to $70.75 results in a price of $84.90.

$$0.2(70.75) + 70.75 = 14.15 + 70.75 = 84.90$$

Your Turn 4 A telemarketer selling children's books tells a customer that after she receives a 15% discount for ordering over the phone, the price will be $5.27 per book. What was the original price for each book?

Example 5 Two cars are traveling toward each other on the same highway. One car is traveling 65 miles per hour; the other, 60 miles per hour. If the two cars are 20 miles apart, how long until they meet?

Understand Draw a picture of the situation.

Note Car 1 will travel the greater distance because it is going faster.

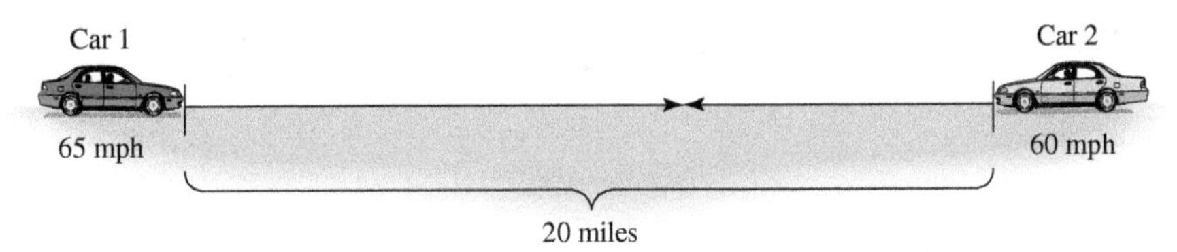

From the diagram, we see that the sum of the individual distances traveled will be the total distance separating the two cars, which is 20 miles. Because we are given their rates, we can use $d = rt$ to write expressions for their individual distances.

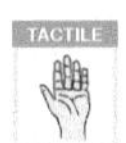
TACTILE

Learning Strategy

If you are a tactile learner, you may find it helpful to simulate the situations in rate problems with model cars, pencils, or small pieces of paper.

Car	Rate	Time	Distance
Car 1	65	t	$65t$
Car 2	60	t	$60t$

We were given these rates.

Both cars start at the same time and meet at the same moment in time, so they travel the same amount of time, t.

To describe each distance, we use the fact that $d = rt$ and multiply the rate and time.

Plan Use the sum of the individual distances traveled (20 miles) to write an equation.

Execute Car 1's distance + Car 2's distance = 20

$$65t + 60t = 20$$

$$125t = 20$$

$$\frac{125t}{125} = \frac{20}{125}$$

$$t = 0.16$$

Note If solving for time, the time unit will match the unit of time in the rate. In this problem, the rate is in miles per hour, so the time unit is hours.

Answer to Your Turn 4
$6.20

Answer The cars will meet in 0.16 hours, which is 9.6 minutes or 9 minutes and 36 seconds.

Check Use $d = rt$ to verify that in 0.16 hours, the cars will travel a combined distance of 20 miles.

$$\text{Car 1: } d = 65(0.16) = 10.4 \text{ miles}$$
$$\text{Car 2: } d = 60(0.16) = 9.6 \text{ miles}$$

The combined distance is $10.4 + 9.6 = 20$ miles.

Your Turn 5 Deon and Kayla are jogging along the same trail. Kayla crosses a bridge at 5:00 P.M. Deon crosses the same bridge at 5:10 P.M. Kayla is traveling at 4 miles per hour, while Deon is traveling at 6 miles per hour. What time will Deon catch up to Kayla?

Answer to Your Turn 5
5:30 P.M.

2.2 Exercises

For Extra Help MyMathLab®

Objective 1

Prep Exercise 1 What are the four steps to problem solving according to George Polya?

Prep Exercise 2 List four strategies suggested as aids to understanding a problem.

For Exercises 1–14, solve. See Examples 1 and 2.

1. If you cool carbon dioxide down to −109.3°F at sea level, you get dry ice. Use $F = \frac{9}{5}C + 32$ to convert this temperature to degrees Celsius.

2. The coldest spot on Earth was found in Antarctica at −89°C. Use $C = \frac{5}{9}(F - 32)$ to convert this temperature to degrees Fahrenheit.

3. A sheet of cardboard with an area of 2280 square inches will be formed into a box 24 inches long by 15 inches wide. Use the formula for surface area $A = 2lw + 2lh + 2wh$ to find the height of the box.

4. A sheet of metal with an area of 30 square feet will be formed into a box 6 feet long by 1.5 feet wide. Use the formula for surface area $A = 2lw + 2lh + 2wh$ to find the height of the box.

5. The area of the trapezoidal piece of plastic shown is 370 square centimeters. Find the length of the side labeled a. $\left(\text{Use } A = \frac{1}{2}h(a + b).\right)$

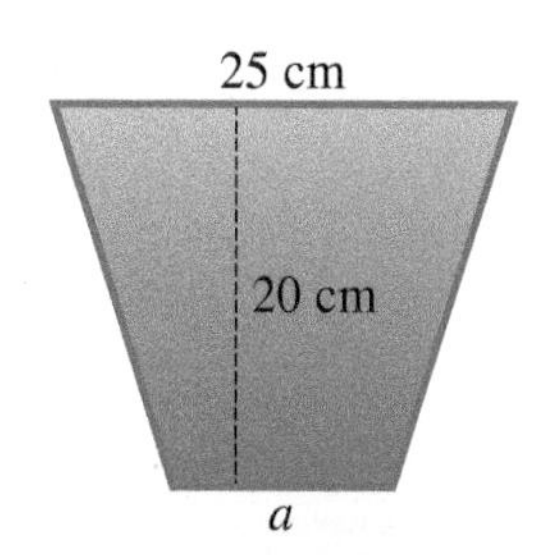

6. A county is in the shape of a trapezoid. If the area is 1296 square miles, find the length of the border labeled b. $\left(\text{Use } A = \frac{1}{2}h(a + b).\right)$

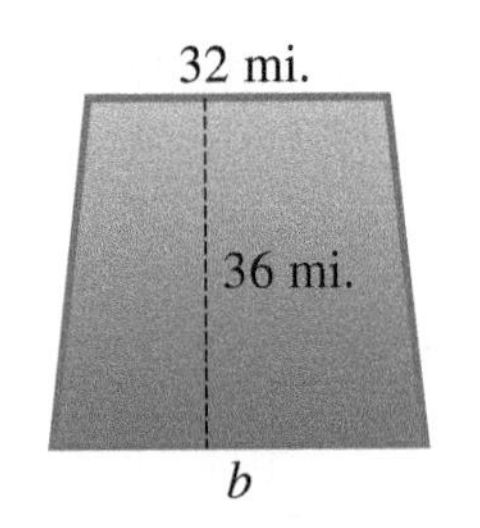

7. A carpenter installs exactly 84 feet of crown molding on the walls of a rectangular room. If the length is 6 feet more than the width, find the dimensions of the room. (Use $P = 2l + 2w$.)

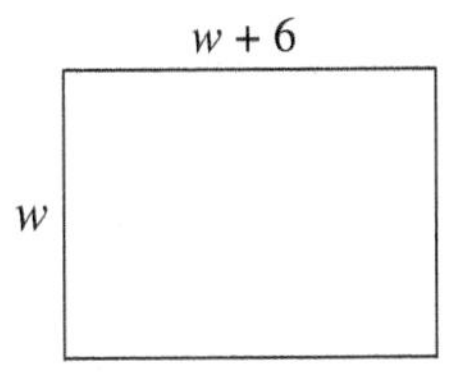

8. Seven wood planks each 8 feet long are to be used to build the walls of a box garden. If the length of the garden is to be 4 feet more than the width, find the dimensions of the garden. (Use $P = 2l + 2w$.)

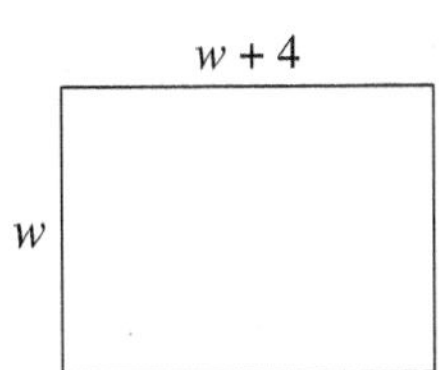

9. The figure to the right shows the wall of a room that is to be painted. The two windows are 4 feet by 5 feet. To purchase the correct amount of paint, the homeowner needs to calculate the area to be painted. Calculate the area to be covered.

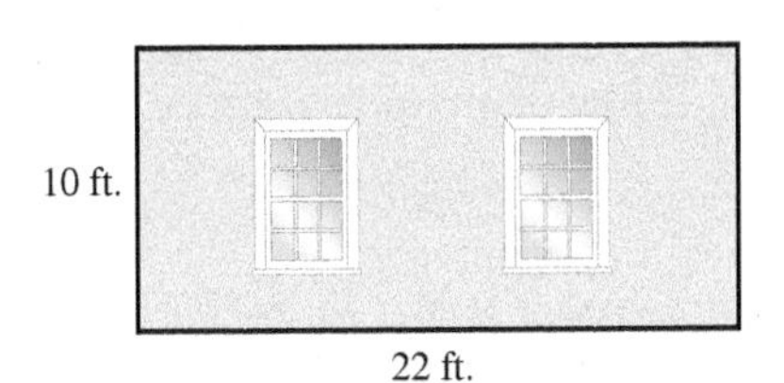

10. The figure to the right shows a lot plan for a new house. After the house is built, the remainder of the lot will be landscaped. Find the area to be landscaped.

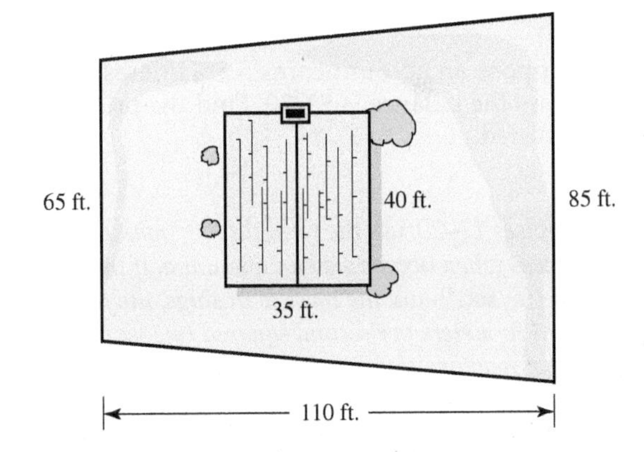

11. The following parking lot has an area of 2020 square feet. Find the length of the two sides labeled x and $x + 10$.

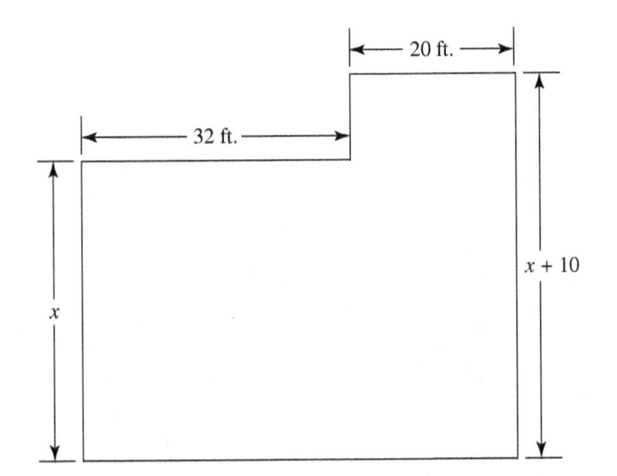

12. The following building covers an area of 3250 square feet. Find the length of the two sides labeled x and $x + 30$.

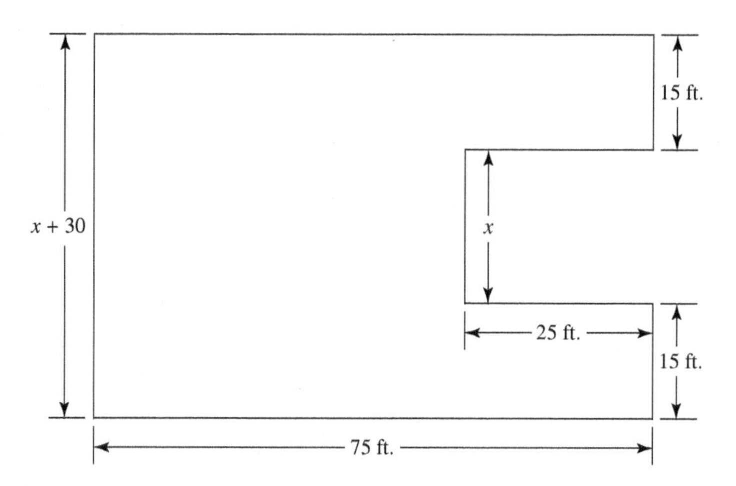

13. A business breaks even if the revenue and cost are equal. Suppose the expression $50n - 1200$ describes the monthly revenue where n represents the number of units of product sold. The expression $24n + 750$ describes the monthly cost of producing those n units. How many units must be sold to break even?

14. A business breaks even if the revenue and cost are equal. Suppose the expression $45n - 600$ describes the monthly revenue where n represents the number of units of product sold. The expression $15n + 300$ describes the monthly cost of producing those n units. How many units must be sold to break even?

For Exercises 15 and 16, use the formula $B = P + Prt$, which is used to calculate the balance in an account if simple interest is added to a principal P at an interest rate of r after t years. See Example 1.

15. Suppose an account earns 5.5% interest and after two years the balance is \$3330. Find the principal that was invested.

16. Suppose an account earns 3% interest and after five years the balance is \$1725. Find the principal that was invested.

For Exercises 17–20, use the formula $F = ma$, which describes the force an object with mass m experiences when accelerated an amount a. If the acceleration is measured in feet per second squared (ft./sec.^{2}) and the mass is in slugs, the force is in pounds (lb.). If the acceleration is measured in meters per second squared (m/sec.^{2}) and the mass is in kilograms (kg), the force is in units of newtons (N). See Example 1.

17. A woman weighs 135 pounds so that the downward force due to gravity is -135 pounds. If the acceleration due to gravity is -32.2 feet per second squared, find the woman's mass rounded to the nearest tenth of a slug.

18. A man weighs 180 pounds so that the downward force due to gravity is -180 pounds. If the acceleration due to gravity is -32.2 feet per second squared, find the man's mass rounded to the nearest tenth of a slug.

19. The world record for the heaviest deadlift is 4512.9 newtons set by Benedikt Magnusson on April 4, 2011. If the downward force was -4512.9 newtons and the acceleration due to gravity was -9.8 meters per second squared, find the mass he lifted rounded to the nearest tenth of kilogram. (Source: Powerliftingwatch.com)

20. The world record for the heaviest unequipped bench press is 3209.5 newtons set by Eric Spoto on May 19, 2013. If the downward force was -3209.5 newtons and the acceleration due to gravity was -9.8 meters per second squared, find the mass he pressed rounded to the nearest tenth of kilogram. (Source: Powerliftingwatch.com)

Of Interest

Benedikt Magnusson's world record deadlift was 1015 pounds. Eric Spoto's world record bench press was 722 pounds.

For Exercises 21–24, use the formula $d = v_i t + \frac{1}{2}at^2$*, which describes the distance an object travels if the object has an initial velocity of* v_i *and an acceleration of* a *for a time* t*. See Example 1.*

21. An object travels 260 meters after being accelerated at 25 meters per second squared for 4 seconds. Find the object's initial velocity.

22. An object travels 318 feet after being accelerated at 15 feet per second squared for 6 seconds. Find the object's initial velocity.

23. A car travels 525 feet after accelerating for 6 seconds. If the initial velocity was 50 feet per second, what was the acceleration?

24. A boat travels 53 meters after accelerating for 2 seconds. If the initial velocity was 18 meters per second, what was the acceleration?

Objective 2

Prep Exercise 3 What key words translate to an equal sign?

Prep Exercise 4 When solving problems involving two objects traveling in opposite directions, what can you conclude about the amount of time until they meet?

For Exercises 25–42, solve. See Examples 3–7.

25. Three subtracted from five times a number is thirty-seven. What is the number?

26. Eleven is equal to seven less than six times a number. What is the number?

27. Eight more than five times a number is equal to ten less than triple the number. What is the number?

28. Eleven minus four times a number is equal to twenty-three plus twice the number. What is the number?

29. Three times the sum of a number and four is negative six. What is the number?

30. Twice the difference of a number and three is negative eight. What is the number?

31. Eight less than the product of five and the difference of a number and two is twice the number. What is the number?

32. Six minus three times the difference of a number and five is equal to four times the number. What is the number?

33. A bookstore determines the retail price by marking up the wholesale price 25%. The retail price of a chemistry book is $135.68. What was the wholesale price?

34. The manufacturer's suggested retail price (MSRP) of a particular car is $22,678. If the MSRP represents a markup of 15% of the wholesale price, what is the wholesale price?

35. A computer store is selling older stock computers at a discount of 30% off the original price. If the price after the discount is $699.95, what was the original price?

36. A clothing store has a special sale with every item discounted 40%. If the price of a dress after the discount is $24.50, what was the original price?

37. An annual report for a company shows that 45,000 units of a particular product were sold, which is an increase of 20% over the previous year. How many units of that product were sold the previous year?

38. A report shows that revenue for one month was $24,360, which is a decrease of 12.5% from the previous month. What was the previous month's revenue?

39. Two cars are traveling toward each other on the same road. One car is traveling 65 miles per hour; the other, at 50 miles per hour. If the two cars are 230 miles apart, how long until they meet?

40. Two cyclists ride along the same path towards each other. One cyclist is traveling 14 miles per hour; the other 18 miles per hour. If they are 12 miles apart, how long until they meet?

41. Two boats traveling in opposite directions on a river meet and pass each other. The northbound boat is traveling 9 miles per hour; the southbound boat at 15 miles per hour. If they maintain those rates, how long will it take them to be 42 miles apart?

42. Two cars traveling in opposite directions meet and then pass each other. The eastbound car is traveling 60 miles per hour; the westbound car 48 miles per hour. If they maintain those rates, how long will it take them to be 270 miles apart?

Review Exercises

Exercises 1 and 2 Constants and Variables

[1.1] **1.** Place the following values in order from least to greatest.

$$0.02, -\frac{1}{6}, 4.5\%, |-15.8|, \sqrt{48}, -5\frac{3}{8}$$

[1.1] **2.** What is the smallest prime number?

Exercises 3–6 Equations and Inequalities

[1.1] *For Exercises 3 and 4, use $<$, $>$, or $=$ to write a true statement.*

3. $-|5.8| \quad -|-5.8|$

4. $-(-6) \quad -(-(-8))$

[2.1] *For Exercises 5 and 6, solve.*

5. $6x - 19 = 4x - 31$

6. $-\frac{4}{9}n = \frac{5}{6}$

2.3 Solving Linear Inequalities

Objectives

1 Represent solutions to linear inequalities in one variable using set-builder notation, interval notation, and graphs.
2 Solve linear inequalities in one variable.
3 Solve problems involving linear inequalities.

Warm-ups

[2.1] *For Exercises 1 and 2, solve.*

1. $5(t - 3) = 7 - (t + 4) + 6t$

2. $\frac{1}{3}(9h + 5) - 2h = \frac{1}{4}h + 1$

[2.2] **3.** A business breaks even if its revenue and costs are equal. Suppose the expression $30n + 200$ describes the monthly revenue where n is the number of units of product sold. The expression $10n + 1800$ describes the monthly cost of producing those n units of the product. How many units must be sold to break even?

Answers to Warm-up

1. No solution **2.** $h = -\frac{8}{9}$ **3.** 80

Objective 1 Represent solutions to linear inequalities in one variable using set-builder notation, interval notation, and graphs.

Not all problems translate to equations. Sometimes a problem can have a range of values as solutions. We often use inequalities to describe situations where a range of solutions is possible. Inequality symbols and their meanings are as follows:

$<$ is less than	$>$ is greater than
$\leq$ is less than or equal to	$\geq$ is greater than or equal to

In this section, we focus on how to solve **linear inequalities.**

Definition Linear inequality in one variable: An inequality that can be written in the form $ax + b < c$, $ax + b > c$, $ax + b \leq c$, or $ax + b \geq c$, where a, b, and c are real numbers and $a \neq 0$.

Here are some examples of linear inequalities in one variable:

$$x > 5 \qquad n + 2 < 6 \qquad 2(y - 3) \leq 5y - 9$$

A solution for an inequality is any number that can replace the variable(s) in the inequality and make the inequality true. Because inequalities can have a range of solutions, we often write those solutions in set-builder notation, which we introduced in Section 1.1. For example, the solution set for the inequality $x > 5$ is $\{x | x > 5\}$.

Another popular notation used to indicate ranges of values is *interval notation.* With interval notation, parentheses are used for end values that are not included in the interval and brackets are used for end values that are included. For example, because the solution set for $x > 5$ does not include 5, we write $(5, \infty)$, whereas the solution set for $x \geq 5$ is written as $[5, \infty)$. The symbol ∞ means infinity and is never included as an end value.

Solution sets can also be represented graphically using a number line. Two styles are common. One style uses open or solid circles and the other style uses the parentheses or brackets from interval notation.

$x > 5$ is represented by

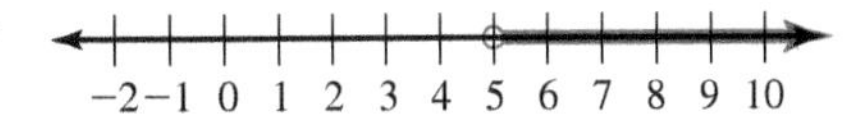

or

Note This form reinforces the interval notation $(5, \infty)$. ▶

−2 −1 0 1 2 3 4 5 6 7 8 9 10

$x \geq 5$ is represented by

−2 −1 0 1 2 3 4 5 6 7 8 9 10

or

Note This form reinforces the interval notation $[5, \infty)$. ▶

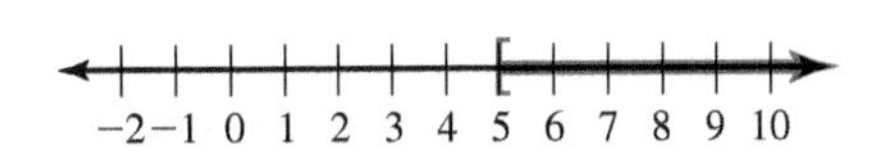

Your instructor may have a preference of style. We use the parentheses and brackets because they reinforce interval notation.

Procedure Graphing Inequalities

To graph an inequality in one variable of the form $x \leq a$, $x \geq a$, $x < a$ or $x > a$ on a number line:

1. If the symbol is $\leq$ or $\geq$, draw a bracket (or solid circle) on the number line at the indicated number open to the left for $\leq$ and to the right for $\geq$. If the symbol is $<$ or $>$, draw a parenthesis (or open circle) on the number line at the indicated number open to the left for $<$ and to the right for $>$.
2. If the variable is greater than the indicated number, shade to the right of the indicated number. If the variable is less than the indicated number, shade to the left of the indicated number.

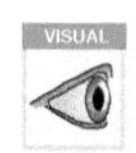

Learning Strategy

If you are a visual learner, you may find it helpful to graph the solution set first and then use your graph to write the interval and set-builder notations.

Example 1 Write the solution set in set-builder notation and interval notation; then graph the solution set.

a. $x \leq -4$
Set-builder notation: $\{x | x \leq -4\}$
Interval notation: $(-\infty, -4]$

Graph: 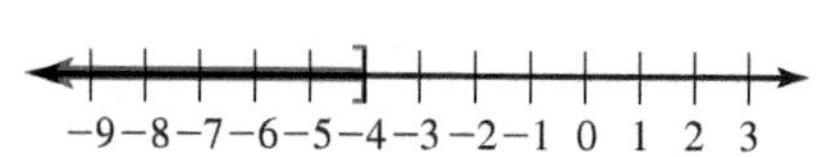

Note When writing interval notation, we write from the leftmost end value to the rightmost end value.

b. $n < 0$
Set-builder notation: $\{n | n < 0\}$
Interval notation: $(-\infty, 0)$

Graph: 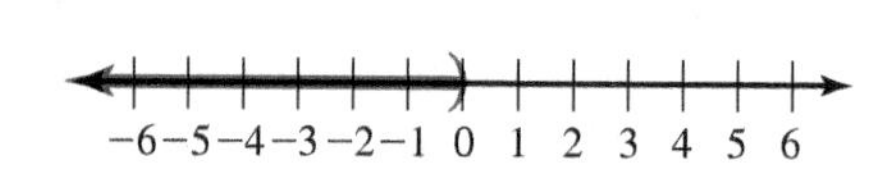

Connection The set $\{n | n < 0\}$ is a way of expressing the set of all negative real numbers.

Your Turn 1 Write the solution set in set-builder notation and interval notation; then graph the solution set.

a. $x < 3$ **b.** $h \geq -2$

Objective 2 Solve linear inequalities in one variable.

To solve linear inequalities such as $n + 2 < 6$ and $2(x - 3) \leq 5x - 9$, we follow essentially the same process as for solving linear equations. The principles for inequalities, summarized next, are very similar to the principles for equations. Be careful to note the special condition in the multiplication principle of inequality.

Rules The Addition Principle of Inequality

If $a < b$, then $a + c < b + c$ is true for all real numbers a, b, and c.

The Multiplication Principle of Inequality

If a and b are real numbers where $a < b$, then $ac < bc$ is true if c is a *positive* real number and $ac > bc$ is true if c is a *negative* real number. Note that $c \neq 0$.

Note Although we have written the principles in terms of the $<$ symbol, the principles are true for any inequality symbol.

Notice that the addition principle of inequality indicates that adding the same amount to (or subtracting the same amount from) both sides of an inequality does not affect the inequality. Similarly, multiplying (or dividing) both sides of an inequality by the same *positive* real number does not affect an inequality. But if we multiply (or divide) both sides by the same *negative* real number, we must *reverse the direction of the inequality symbol* to keep the inequality true. The following procedure summarizes the process of solving a linear inequality.

Procedure Solving Linear Inequalities in One Variable

To solve a linear inequality in one variable:

1. Simplify both sides of the inequality as needed.
   a. Distribute to eliminate parentheses.
   b. Eliminate fractions or decimals by multiplying through by the LCD. (This step is optional.)
   c. Combine like terms.
2. Use the addition principle so that all variable terms are on one side of the inequality and all constants are on the other side. Then combine like terms.
3. Use the multiplication principle to eliminate any remaining coefficient. If you multiply (or divide) both sides by a negative number, reverse the direction of the inequality symbol.

Note By eliminating the term with the lesser coefficient, you can avoid a negative coefficient on the remaining variable term.

Answers to Your Turn 1

a. Set-builder notation: $\{x | x < 3\}$
Interval notation: $(-\infty, 3)$
Graph:

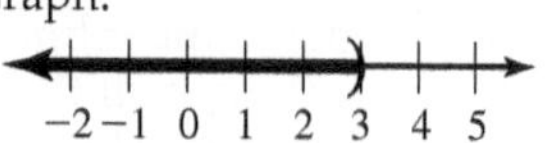

b. Set-builder notation: $\{h | h \geq -2\}$
Interval notation: $[-2, \infty)$
Graph:

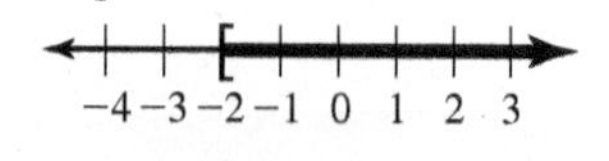

Example 2 For each inequality, solve and write the solution set in set-builder notation and in interval notation. Then graph the solution set.

a. $-6x > 18$

Solution: $\frac{-6x}{-6} < \frac{18}{-6}$ Divide both sides by -6 and change the inequality.

$x < -3$

Set-builder notation: $\{x \mid x < -3\}$
Interval notation: $(-\infty, -3)$
Graph:

-6 -5 -4 -3 -2 -1 0 1 2 3 4 5 6

Note Because we divided both sides by a negative number, we reversed the direction of the inequality symbol.

b. $-10 \le 3m + 2$

Solution: $-10 - 2 \le 3m + 2 - 2$ Subtract 2 from both sides.

$\frac{-12}{3} \le \frac{3m}{3}$ Divide both sides by 3.

$-4 \le m$

Set-builder notation: $\{m \mid m \ge -4\}$
Interval notation: $[-4, \infty)$
Graph:

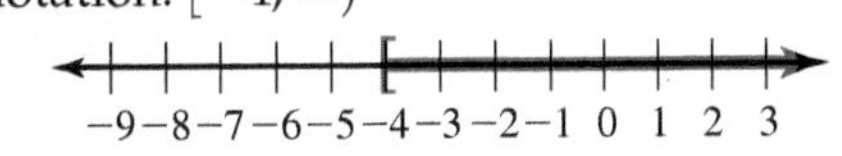

Note Dividing both sides by a positive number does not affect the inequality. Also, $-4 \le m$ and $m \ge -4$ are the same.

c. $\frac{1}{3}(9h + 5) - 2h > \frac{1}{4}h + 1$

Solution: $3h + \frac{5}{3} - 2h > \frac{1}{4}h + 1$ Distribute $\frac{1}{3}$.

$h + \frac{5}{3} > \frac{1}{4}h + 1$ Combine like terms.

$12\left(h + \frac{5}{3}\right) > 12\left(\frac{1}{4}h + 1\right)$ Multiply both sides by the LCD, 12, to eliminate the fractions.

$12h + 20 > 3h + 12$ Distribute 12.

$12h - 3h + 20 > 3h - 3h + 12$ Subtract $3h$ from both sides.

$9h + 20 - 20 > 12 - 20$ Subtract 20 from both sides.

$\frac{9h}{9} > \frac{-8}{9}$ Divide both sides by 9.

$h > -\frac{8}{9}$

Note By subtracting $3h$ from both sides, we get a positive 9 coefficient after combining like terms; therefore, we won't have to reverse the inequality when we divide both sides by 9.

Set-builder notation: $\left\{h \mid h > -\frac{8}{9}\right\}$

Interval notation: $\left(-\frac{8}{9}, \infty\right)$

Graph:

$-\frac{8}{9}$

-1 0 1

Note We could have multiplied by the LCD, 12, first.

$12\left[\frac{1}{3}(9h + 5) - 2h\right] > 12\left(\frac{1}{4}h + 1\right)$

$4(9h + 5) - 24h > 3h + 12$

Continuing the solution process yields the same solution, $h > -\frac{8}{9}$.

Answers to Your Turn 2

a. $x < -2.5$
Set-builder: $\{x \mid x < -2.5\}$
Interval: $(-\infty, -2.5)$
Graph:

-4 -3 -2.5 -2 -1

b. $n \ge \frac{24}{5}$ or $4\frac{4}{5}$

Set-builder: $\left\{n \mid n \ge \frac{24}{5}\right\}$

Interval: $\left[\frac{24}{5}, \infty\right)$

Graph:

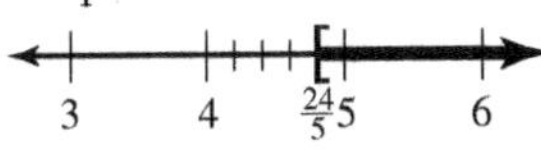

c. $k \le 4$
Set-builder: $\{k \mid k \le 4\}$
Interval: $(-\infty, 4]$
Graph:

-1 0 1 2 3 4 5

Your Turn 2 For each inequality, solve and write the solution set in set-builder notation and in interval notation. Then graph the solution set.

a. $-0.2x > 0.5$ **b.** $\frac{3}{4}n - 1 \ge \frac{1}{2}n + \frac{1}{5}$ **c.** $2(k - 5) + 4k \le 7k + 6 - 5k$

Linear inequalities also may have no solution or every real number as a solution.

Example 3 For each inequality, solve and write the solution set in set-builder notation and in interval notation. Then graph the solution set.

a. $9x - 8 < 4(x + 3) + 5x$

Solution:

$9x - 8 < 4x + 12 + 5x$ Distribute 4.

$9x - 8 < 9x + 12$ Combine like terms.

$9x - 9x - 8 < 9x - 9x + 12$ Subtract 9x from both sides.

$-8 < 12$

Because there are no variable terms left in the inequality and $-8 < 12$ is true, every real number is a solution for the original inequality.

Set-builder notation: $\{x \mid x \text{ is a real number}\}$, or $\mathbb{R}$
Interval notation: $(-\infty, \infty)$

Graph:

−6 −5 −4 −3 −2 −1 0 1 2 3 4 5 6

b. $5(t - 3) \geq 7 - (t + 4) + 6t$

Solution:

$5t - 15 \geq 7 - t - 4 + 6t$ Distribute.

$5t - 15 \geq 5t + 3$ Combine like terms.

$5t - 5t - 15 \geq 5t - 5t + 3$ Subtract 5t from both sides.

$-15 \geq 3$

Because there are no variable terms left in the inequality and $-15 \geq 3$ is false, there are no solutions for the original inequality.

Set-builder notation: { } or $\varnothing$
Interval notation: We do not write interval notation when there are no solutions.

Graph:

−6 −5 −4 −3 −2 −1 0 1 2 3 4 5 6

Objective 3 Solve problems involving linear inequalities.

Problems requiring inequalities can be translated using key words much like we used key words to translate sentences to equations. The following table lists some common key words that indicate inequalities.

Words That Mean ≥	Words That Mean >	Words That Mean ≤	Words That Mean <
at least	is greater than	at most	is less than
minimum of	more than	maximum of	smaller than
not less than		not more than	

Example 4 Solve.

a. Sasha's grade in her math course is calculated by the average of four tests. To receive an A for the course, she needs an average of at least 89.5. If her current test scores are 84, 92, and 94, what range of scores can she make on the last test to receive an A for the course?

Understand We are given three out of four test scores and must find the range of values for the fourth score that will give her an average for the four scores of 89.5 or higher.

Plan Because we are to find a range of scores, we will write an inequality with n representing the score on the last test, then solve.

Execute Sasha's average must be at least 89.5.

$$\frac{84 + 92 + 94 + n}{4} \geq 89.5$$

Note n represents Sasha's score on the fourth test.

$$\frac{270 + n}{4} \geq 89.5 \quad \text{Simplify in the numerator.}$$

$$270 + n \geq 358 \quad \text{Multiply both sides by 4.}$$

$$n \geq 88 \quad \text{Subtract 270 from both sides.}$$

Note If the highest possible score on the last test is 100, any score in the interval [88, 100] will result in an A.

Answer A score of 88 or greater on the last test will give her an average of at least 89.5.

Check Verify that a score of 88 will give her an average of 89.5; then choose a value greater than 88 and verify that the average will be greater than 89.5. We will leave this check to the reader.

b. A painter charges $80 plus $1.50 per square foot. If a family is willing to spend no more than $500, what range of square footage can they afford?

Understand We are given the painter's prices and the maximum amount of money.

Plan Because we are looking for a range of square footage, we will write an inequality with n representing the number of square feet and then solve.

Execute Translate: The total cost has to be less than or equal to $500.

$$80 + 1.5n \leq 500$$

Solve: $$80 - 80 + 1.5n \leq 500 - 80 \quad \text{Subtract 80 from both sides.}$$

$$\frac{1.5n}{1.5} \leq \frac{420}{1.5} \quad \text{Divide both sides by 1.5.}$$

$$n \leq 280$$

Answer They can afford up to 280 square feet.

Check Verify that if they paid for exactly 280 square feet, the painter would charge exactly $500.

c. A business breaks even if its revenue and costs are equal and makes a profit if the revenue exceeds costs. Suppose the expression $30n + 2000$ describes the monthly revenue for a company, where n is the number of units of product sold. The expression $10n + 18,000$ describes the monthly cost of producing those n units of the product. How many units must be sold to break even or make a profit?

Understand We are given expressions for revenue and cost and are to find the number of units that the company must sell to break even or make a profit.

Plan Write an inequality and then solve.

Execute Translate: To break even or make a profit,
the revenue has to be greater than or equal to the cost.

$$30n + 2000 \geq 10n + 18{,}000$$

Solve: $$30n - 10n + 2000 \geq 10n - 10n + 18{,}000 \quad \text{Subtract } 10n \text{ from both sides.}$$

$$20n + 2000 \geq 18{,}000$$

$$20n + 2000 - 2000 \geq 18{,}000 - 2000 \quad \text{Subtract 2000 from both sides.}$$

$$\frac{20n}{20} \geq \frac{16{,}000}{20} \quad \text{Divide both sides by 20.}$$

$$n \geq 800$$

Answer The company must sell 800 units of its product to break even and more than 800 units to make a profit.

Check Verify that selling exactly 800 units causes the company to break even and that selling more than 800 units causes the company to make a profit. We will leave this to the reader.

Your Turn 4 Solve.

a. A wireless company offers a calling plan that charges \$35.00 for the first 500 minutes and \$0.06 for each additional minute. Find the range of total minutes Polina could use with this plan if she budgets a maximum of \$50 per month for her wireless charges.

b. Pecan flooring costs \$11.50 per square foot, \$3.50 per square foot for installation, and \$900 for supplies. If the Smiths have budgeted at most \$18,900 for the flooring, what range of square footage can they afford?

Answers to Your Turn 4

a. $t \leq 750$ minutes (500 minutes plus 250 additional minutes)

b. 1200 ft.^{2} or less

2.3 Exercises

For Extra Help MyMathLab®

Note: Exercises marked with a ★ represent challenging exercises.

Objective 1

Prep Exercise 1 What is a solution for an inequality?

Prep Exercise 2 Explain the difference between $x < a$ and $x \leq a$, where a is a real number.

Prep Exercise 3 How do you read $\{x|x \geq 2\}$? What does it mean?

Prep Exercise 4 Explain what $(-\infty, 4]$ means.

Prep Exercise 5 Explain what $(-2, \infty)$ means.

Prep Exercise 6 The following graph is a solution set for an inequality. What does it show?

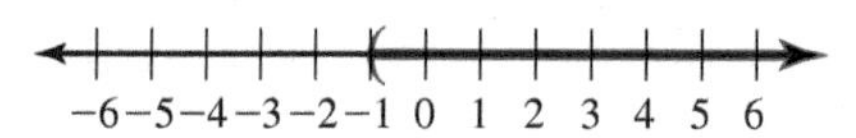

Prep Exercise 7 The following graph shows a solution set for an inequality. What does it show?

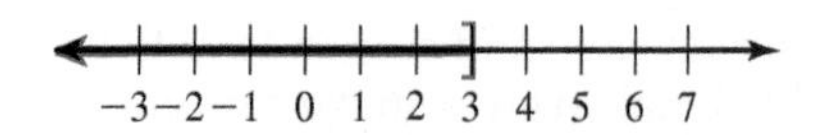

For Exercises 1–8: *a. Write in set-builder notation.* *b. Write in interval notation.* *c. Graph the solution set.* *See Example 1.*

1. $x \geq 5$

2. $n \leq 7$

3. $q < -1$

4. $x > -3$

5. $p < \frac{1}{5}$

6. $a \geq -\frac{2}{3}$

7. $r \leq 1.9$

8. $s \leq -3.2$

Objective 2

Prep Exercise 8. When solving an inequality, what action causes the direction of an inequality symbol to change?

For Exercises 9–38, solve:
a. Write the solution set using set-builder notation.
b. Write the solution set using interval notation.
c. Graph the solution set.
See Examples 2 and 3.

9. $r - 6 < -12$

10. $x + 6 \geq -1$

11. $-4y \leq -16$

12. $-4x < -24$

13. $3p + 9 < 21$

14. $9y - 7 > 11$

15. $6 - 5x > 29$

16. $-3c + 4 < 17$

17. $\frac{a}{8} + 1 < \frac{3}{8}$

18. $4 > \frac{1}{5} + \frac{q}{5}$

19. $7 + 2x < -2 + x$

20. $5b - 3 \geq 13 + 4b$

21. $-11k - 8 > -16 - 9k$

22. $4 - 7k < -2k + 19$

23. $2(3w - 4) - 5 \leq 17$

24. $6(3a + 2) - 10 \geq 2$

25. $4(y + 2) + 5 > 3(2y - 1)$

26. $4(v - 10) \leq 17(v + 3) + 13$

27. $\frac{1}{6}(5x + 1) < -\frac{1}{6}x - \frac{5}{3}$

28. $\frac{1}{7}(3x - 28) > -\frac{4}{7}x + \frac{1}{4}$

29. $\frac{1}{6}(7m + 2) - \frac{1}{6}(11m - 7) \geq 0$

30. $\frac{1}{5}(p + 10) - \frac{1}{2}(p + 5) > 0$

31. $0.7x - 0.3 \leq 0.8x + 0.7$

32. $1.2b - 1.4 \geq 1.5b - 0.5$

33. $0.09z + 20.34 < 3(1.4z - 1.5) + 2.1z$

34. $3.2f + 3.6 - 1.8f \leq 0.3(f + 6) - 0.4$

35. $3(x + 2) - 4 < x + 5 + 2x$

36. $2(2a + 4) - 11 \leq 4(a - 2) + 6$

37. $4b - 3(b + 5) \geq 2(2b + 3) - 3(b + 5)$

38. $3(2x - 6) - 2(x - 4) > 5(x + 3) - x$

Objective 3

Prep Exercise 9 Which inequality does "at most" indicate?

Prep Exercise 10 Which inequality does "at least" indicate?

For Exercises 39–46, translate to an inequality and then solve. See Objective 3.

39. Three-fourths of a number is less than negative six.

40. Two-thirds of a number is greater than negative eight.

41. Five times a number decreased by one is greater than fourteen.

42. Two less than three times a number is less than ten.

43. Four times a number less than one is at most twenty-five.

44. Three minus two times a number is at least seventeen.

45. Six more than twice the difference of a number and five will not exceed twelve.

★ **46.** Eight more than three times the difference of a number and four is at least negative sixteen.

For Exercises 47–62, solve. See Example 4.

47. To earn an A in her math course, Yvonne needs the average of five tests to be at least 90. Her current test scores are 92, 100, 83, and 81. What range of scores on the fifth test would earn her at least an A in the course?

48. In an introduction to theater course, the final grade is determined by the average of four papers. The department requires any student whose average falls below 70 to repeat the course. Eric's scores on the first three papers are 70, 62, and 75. What range of scores on the fourth paper would result in him having to repeat the course?

49. For an upcoming golf tournament, Blake must show his scores for his most recent four rounds of golf. He wants the average of those scores to be 84 or less. If he has played three rounds and scored 86, 82, and 88, what range of scores must he have on his fourth round to meet his goal?

50. Paulette wants to average at least 175 during a bowling tournament in which bowlers roll three games. So far, she has scored 186 and 178. What range of values must she score on the third game to meet her goal?

51. An architect is designing a house. The master bedroom is to be rectangular, and the area of the room cannot exceed 234 square feet. If the length is 18 feet, what range of values can the width have?

52. Dante needs to make a rectangular spa cover that is at least 14 square feet. If the spa length is 4 feet, what range of values can the width have?

53. Susan makes ceramic flower pots. One particular flower pot requires her to glue a strip of mosaic around the circumference of the top. She is designing a new flower pot and has enough mosaic to cover 150 inches. What range of values can the radius have?

54. Frank is designing a drainpipe for a storage tank. He has a gasket that can vary in size up to a maximum circumference of 12 centimeters. Find the range of values for the diameter of the pipe.

55. Juan wants to meet his girlfriend at the airport before she gets on a plane. If the airport is 180 miles away and she is leaving in 3 hours, what is the minimum average rate that he must drive to see her?

56. Therese does not drive over 75 miles per hour on the interstate. If she plans a trip of 480 miles, find the range of values that describes her time to complete the trip.

57. A company produces lamps. The expression $12.5n + 3000$ describes the monthly revenue if n lamps are sold. The expression $8n + 21{,}000$ describes the monthly cost of producing those n lamps. How many lamps must be sold per month to break even or make a profit?

58. The expression $65n + 10{,}000$ describes the monthly revenue for a cleaning service, where n is the number of homes cleaned. The expression $25n + 15{,}000$ describes the monthly cost of operating the service. How many homes must be cleaned per month to break even or make a profit?

59. Gold is a solid up to a temperature of 1064.58°C. At 1064.58°C or hotter, gold is liquid.

a. Find the range of temperatures in degrees Fahrenheit that gold remains solid.

b. Find the range of temperatures in degrees Fahrenheit that gold is liquid.

60. Iron is a solid up to a temperature of 2795°F. At 2795°F or hotter, iron is liquid.

a. Find the range of temperatures in degrees Celsius that iron remains solid.

b. Find the range of temperatures in degrees Celsius that iron is liquid.

★ **61.** The design of a circuit specifies that the voltage cannot exceed 15 V. If the resistance of the circuit is 6 Ω, find the range of values that the current, in amps, can have. (Use the formula voltage = current × resistance.)

★ **62.** A label on a forklift indicates that the maximum safe load is 24,500 N; that is, the maximum downward force is −24,500 N. If the acceleration due to gravity is -9.8 m/sec.^2, find the range of mass, in kilograms, that can be safely lifted. (Use the formula force = mass × acceleration.)

Puzzle Problem A lady and a gentleman are sister and brother. We do not know who is older.

Someone asked them, "Who is older?"

The sister said, "I am older."

The brother said, "I am younger."

At least one of them was lying. Who is older?

Review Exercises

Exercises 1–4 Expressions

[1.1] **1.** If $A = \{1, 2, 3, 4\}$ and $B = \{0, 1, 2, 3, 4, 5, 6\}$, is $A \subseteq B$?

[1.2] **2.** What property is illustrated by $-5(2 + 7) = -5(7 + 2)$?

[1.4] **3.** Evaluate $0.5r^2 - t$ when $r = -6$ and $t = 8$.

[1.4] **4.** Evaluate $|x^3 - y|$ when $x = -4$ and $y = -19$.

Exercises 5 and 6 Equations and Inequalities

[2.1] *For Exercises 5 and 6, solve and check.*

5. $6x + 9 = 13 + 4(3x + 2)$

6. $\frac{3}{4}n - 1 = \frac{1}{5}(2n + 3)$

2.4 Compound Inequalities

Objectives

1 Solve compound inequalities involving *and*.

2 Solve compound inequalities involving *or*.

Learning Strategy

When taking notes, always write down steps and tips the instructor tells you. Write *all* example problems. Don't worry about writing everything down. Instead, concentrate on the instructor; the basics are in the book.

—Judah G.

Warm-up

[2.3] *For Exercises 1 and 2, solve the inequality.*

1. $2x + 4 \leq 14$ **2.** $-3x + 7 > -2$

In the previous section we learned to solve linear inequalities and represent their solution sets with set-builder notation, interval notation and graphically on number lines. For example, the solution set for $x > 3$ is represented in those three ways here.

Set-builder notation: $\{x \mid x > 3\}$

Interval notation: $(3, \infty)$

Graph:

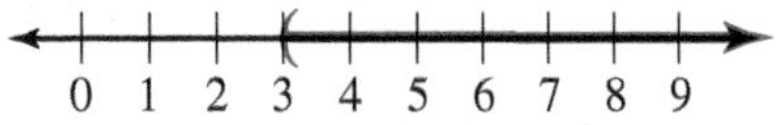

Note Remember that the parenthesis indicates that 3 is **not** included in the solution set.

In this section, we build on this foundation and explore **compound inequalities.**

Definition **Compound inequality:** Two inequalities joined by either *and* or *or.*

Examples of compound inequalities are as follows:

$$x > 3 \text{ and } x \leq 8 \qquad -2 \geq x \text{ or } x > 4$$

Objective 1 Solve compound inequalities involving *and.*

Interpreting Compound Inequalities with *and*

First, let's consider inequalities involving *and*, such as $x > 3$ *and* $x \leq 8$. The word *and* indicates that the solution set contains only values that satisfy *both* inequalities. Therefore, the solution set is the **intersection**, or overlap, of the two inequalities' solution sets.

Definition **Intersection:** For two sets A and B, the intersection of A and B, symbolized by $A \cap B$, is a set containing only elements that are in both A and B.

For example, if $A = \{1, 2, 3, 4, 5\}$ and $B = \{3, 4, 5, 6, 7\}$, then $A \cap B = \{3, 4, 5\}$ because 3, 4, and 5 are the only elements in both A and B. Let's look at our compound inequality $x > 3$ and $x \leq 8$. First, we will graph the two inequalities separately, then consider their intersection.

$x > 3$

$x \leq 8$

Intersection of $x > 3$ and $x \leq 8$:

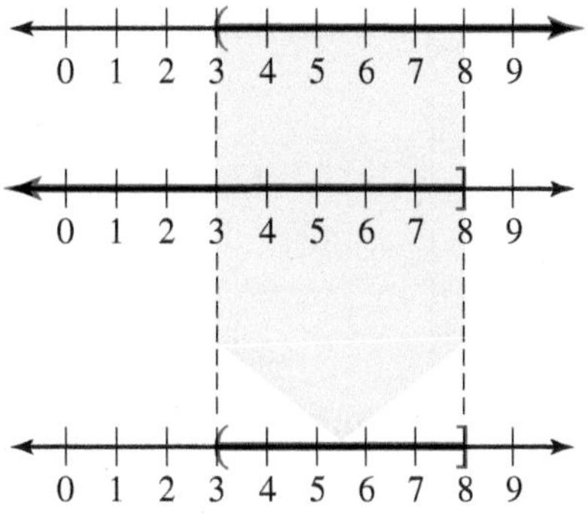

Note We use the blue field to indicate the intersection and the dashed lines to indicate the intersection boundaries.

Note The intersection of the two inequalities contains elements that are only in **both** of their solution sets. Notice that 3 is excluded in the intersection because it was excluded in one of the individual graphs.

So $x > 3$ and $x \leq 8$ means that x is any number greater than 3 *and* less than or equal to 8. Because this inequality indicates that x is between two values, we can write it without the word *and*: $3 < x \leq 8$. In set-builder notation, we write $\{x \mid 3 < x \leq 8\}$. Using interval notation, we write $(3, 8]$.

Answers to Warm-up

1. $x \leq 5$ **2.** $x < 3$

Example 1 For the compound inequality $x > -2$ and $x < 3$, graph the solution set and write the compound inequality without *and* if possible. Then write in set-builder notation and in interval notation.

Solution: The solution set is the region of intersection.

$x > -2$

$x < 3$

Solution set graph:

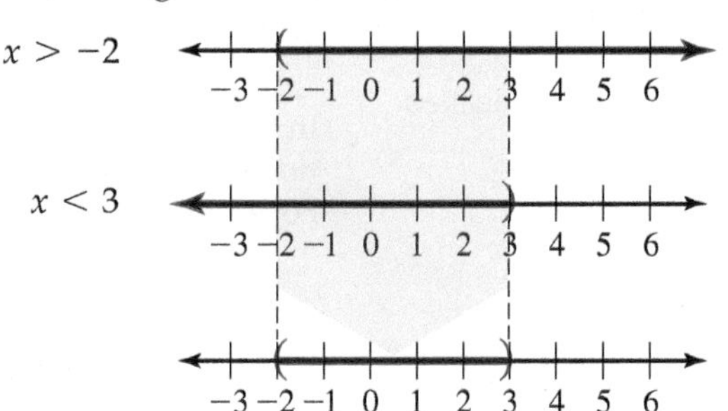

Without *and*: $-2 < x < 3$

Set-builder notation: $\{x \mid -2 < x < 3\}$

Interval notation: $(-2, 3)$

Warning Be careful not to confuse the interval notation $(-2, 3)$ with an ordered pair.

Your Turn 1 For each compound inequality, graph the solution set and write the compound inequality without *and* if possible. Then write in set-builder notation and in interval notation.

a. $x > 4$ and $x < 7$

b. $x \geq -8$ and $x \leq -2$

Solving Compound Inequalities with *and*

Now let's consider solution sets for more complex inequalities involving *and*.

Example 2 For the inequality $3x - 1 > 2$ and $2x + 4 \leq 14$, graph the solution set. Then write the solution set in set-builder notation and in interval notation.

Solution:

$3x - 1 > 2$ and $2x + 4 \leq 14$

$3x > 3$ and $2x \leq 10$ Use the addition principle to isolate each x term.

$x > 1$ and $x \leq 5$ Divide out each coefficient.

The solution set is the intersection of the two individual solution sets.

$x > 1$

$x \leq 5$

Solution set graph:

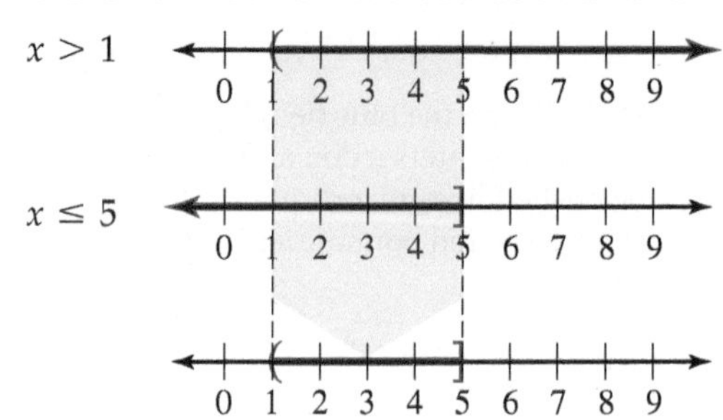

Set-builder notation: $\{x \mid 1 < x \leq 5\}$

Interval notation: $(1, 5]$

Example 2 suggests the following procedure:

Procedure Solving Compound Inequalities Involving *and*

To solve a compound inequality involving *and*:

1. Solve each inequality in the compound inequality.
2. Find the intersection of the individual solution sets.

Answers to Your Turn 1

a. Without *and*: $4 < x < 7$
Set-builder: $\{x \mid 4 < x < 7\}$
Interval: $(4, 7)$

b. Without *and*: $-8 \leq x \leq -2$
Set-builder: $\{x \mid -8 \leq x \leq -2\}$
Interval: $[-8, -2]$

Finally, we consider some special situations with compound inequalities involving *and*.

Example 3 For the compound inequality, graph the solution set. Then write the solution set in set-builder notation and in interval notation.

a. $-2 \le -3x + 7 \le 1$

Solution:

$-3x + 7 \ge -2$ and $-3x + 7 \le 1$ Write $-2 \le -3x + 7 \le 1$ using *and*. Remember, $-2 \le -3x + 7$ is the same as $-3x + 7 \ge -2$.

$-3x \ge -9$ and $-3x \le -6$ Subtract 7 from both sides of each inequality.

$x \le 3$ and $x \ge 2$ Divide both sides of each inequality by -3, which changes the direction of each inequality.

Note Remember, when we multiply or divide both sides of an inequality by a negative number, we must change the direction of the inequality symbol.

Remember, $x \le 3$ and $x \ge 2$ is the same as $2 \le x \le 3$.

Solution set graph: 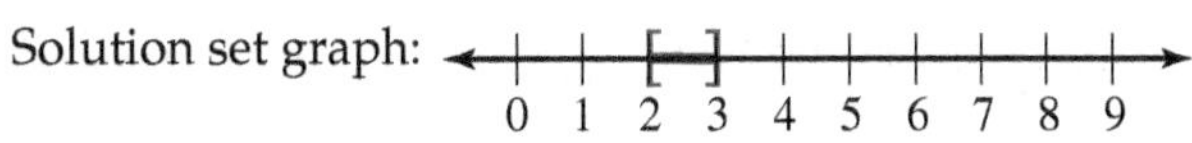

Set-builder notation: $\{x \mid 2 \le x \le 3\}$

Interval notation: $[2, 3]$

The compound inequality can also be solved in its original form.

$-2 \le -3x + 7 \le 1$

$-9 \le -3x \le -6$ Subtract 7 from all three parts of the compound inequality.

$3 \ge x \ge 2$ Divide all three parts of the compound inequality by -3.

Note that $3 \ge x \ge 2$ is the same as $2 \le x \le 3$.

b. $2x + 11 \ge 5$ and $2x + 11 > 3$

Solution:

$2x + 11 \ge 5$ and $2x + 11 > 3$

$2x \ge -6$ and $2x > -8$ Subtract 11 from both sides of each inequality.

$x \ge -3$ and $x > -4$ Divide both sides of each inequality by 2.

$x \ge -3$

−5 −4 −3 −2 −1 0 1 2 3 4

$x > -4$

−5 −4 −3 −2 −1 0 1 2 3 4

Solution set graph: 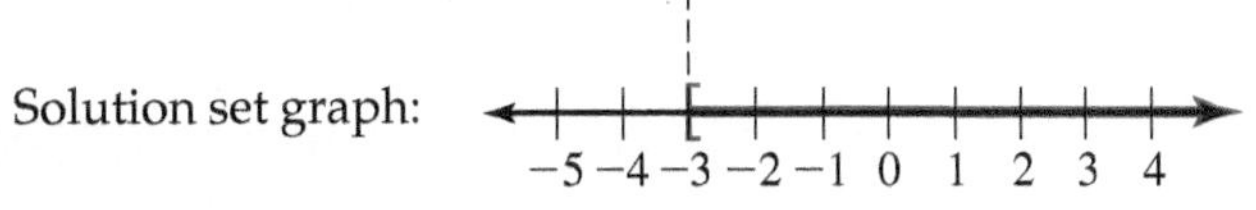

Note The region of intersection for these graphs is from -3 to ∞.

Set-builder notation: $\{x \mid x \ge -3\}$

Interval notation: $[-3, \infty)$

c. $-4x - 3 > 1$ and $-4x - 3 < -11$

Solution:

$-4x - 3 > 1$ and $-4x - 3 < -11$

$-4x > 4$ and $-4x < -8$ Add 3 to both sides of each inequality.

$x < -1$ and $x > 2$ Divide both sides of each inequality by -4, which changes the direction of each inequality.

$x < -1$ 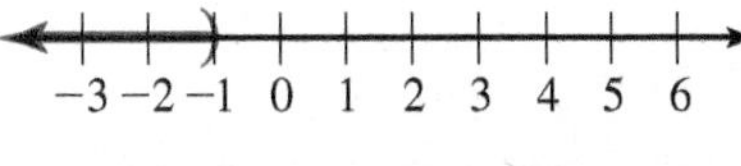

$x > 2$

−3 −2 −1 0 1 2 3 4 5 6

Note There is no region of intersection for these graphs, so the solution set contains no values and is said to be empty.

To graph an empty set, we draw a number line with no shading.

Solution set graph:

−3 −2 −1 0 1 2 3 4 5 6

Set-builder notation: { } or ∅

Interval notation: We do not write interval notation because there are no values in the solution set.

Your Turn 3 For each compound inequality, graph the solution set. Then write the solution set in set-builder notation and in interval notation.

a. $-9 \le -5x + 11 < 21$ **b.** $3x - 2 \le 4$ and $5x + 7 \le 32$

c. $-4x - 3 > 1$ and $-4x - 3 < -5$

Objective 2 Solve compound inequalities involving *or*.

Interpreting Compound Inequalities with *or*

In a compound inequality such as $x > 2$ or $x \le -1$, the word *or* indicates that the solution set contains values that satisfy *either* inequality. Therefore, the solution set is the **union**, or joining, of the two inequalities' solution sets.

Definition **Union:** For two sets A and B, the union of A and B, symbolized by $A \cup B$, is a set containing every element in A or in B.

For example, if $A = \{1, 2, 3, 4, 5\}$ and $B = \{3, 4, 5, 6, 7\}$, then $A \cup B = \{1, 2, 3, 4, 5, 6, 7\}$. Consider our compound inequality $x > 2$ or $x \le -1$. Let's graph the two inequalities separately, then consider their union.

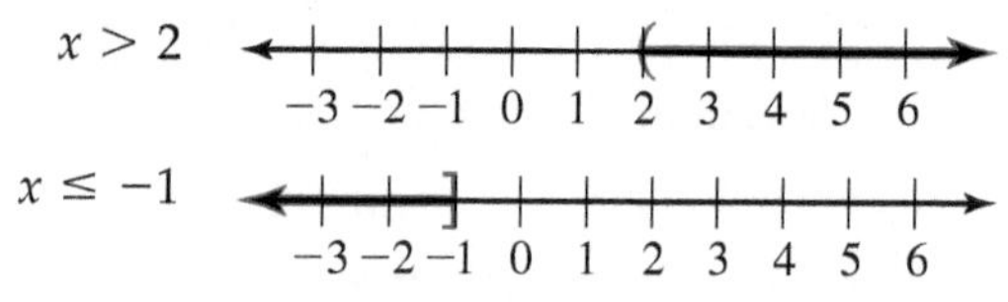

Solution set:
$x > 2$ or $x \le -1$

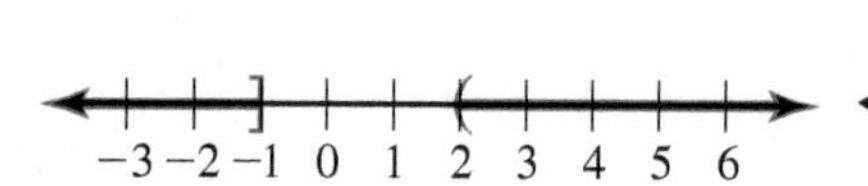

Note This graph is the union of the two individual solution sets because it contains every element in either of the individual solution sets.

Using set-builder notation, we write the solution set as $\{x \mid x > 2 \text{ or } x \le -1\}$. Interval notation is a little more challenging for compound inequalities such as $x > 2$ or $x \le -1$ because the solution set is a combination of two intervals. As we scan from left to right, the first interval is $(-\infty, -1]$. The second interval is $(2, \infty)$. To indicate the union of these two intervals, we write $(-\infty, -1] \cup (2, \infty)$.

Solving Compound Inequalities with *or*

Now let's solve more complex inequalities involving *or*. The process is similar to the process we used for solving compound inequalities involving *and*.

Procedure **Solving Compound Inequalities Involving *or***

To solve a compound inequality involving *or*:

1. Solve each inequality in the compound inequality.
2. Find the union of the individual solution sets.

Answers to Your Turn 3

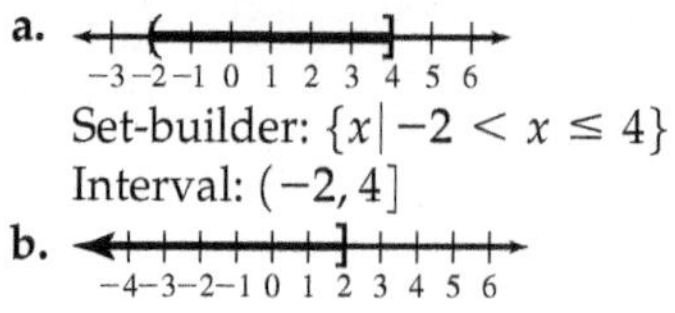

Set-builder: $\{x \mid x \le 2\}$
Interval: $(-\infty, 2]$

c. −3 −2 −1 0 1 2 3 4 5 6

Set-builder: { } or ∅
No interval notation

Example 4 For each compound inequality, graph the solution set. Then write the solution set in set-builder notation and in interval notation.

a. $\frac{2}{3}x - 5 \le -7$ or $\frac{2}{3}x - 5 \ge -1$

Solution:

$\frac{2}{3}x - 5 \le -7$ or $\frac{2}{3}x - 5 \ge -1$

$\frac{2}{3}x \le -2$ or $\frac{2}{3}x \ge 4$ Add 5 to both sides of each inequality.

$x \le -3$ or $x \ge 6$ Multiply both sides by $\frac{3}{2}$ in each inequality.

The solution set is the union of the two individual solution sets.

$x \le -3$

$x \ge 6$

Solution set graph:

Set-builder notation: $\{x | x \le -3 \text{ or } x \ge 6\}$

Interval notation: $(-\infty, -3] \cup [6, \infty)$

Learning Strategy

If you are a visual learner, imagine placing one graph on top of the other to form their union.

b. $-3x + 8 < -1$ or $-3x + 8 \le 20$

Solution: $-3x + 8 < -1$ or $-3x + 8 \le 20$

$-3x < -9$ or $-3x \le 12$ Subtract 8 from both sides of each inequality.

$x > 3$ or $x \ge -4$ Divide both sides of each inequality by -3, which changes the direction of each inequality.

Now we graph the individual solution sets and consider their union.

$x > 3$

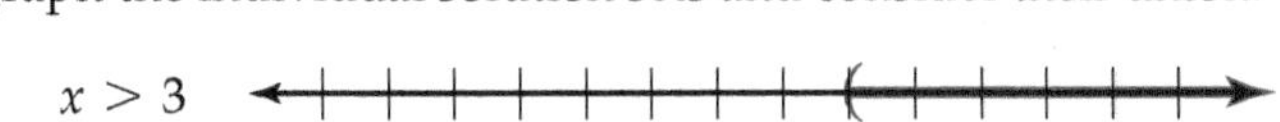

$x \ge -4$

Solution set graph:

Note When we join the two graphs, the graph of $x \ge -4$ covers all of the graph of $x > 3$; so the union of the two graphs is actually $x \ge -4$.

Set-builder notation: $\{x | x \ge -4\}$

Interval notation: $[-4, \infty)$

c. $9x - 16 > -34$ or $9x - 16 \le 29$

Solution: $9x - 16 > -34$ or $9x - 16 \le 29$

$9x > -18$ or $9x \le 45$ Add 16 to both sides of each inequality.

$x > -2$ or $x \le 5$ Divide both sides of each inequality by 9.

Now we graph the individual solution sets and consider their union.

$x > -2$

$x \le 5$

Solution set graph:

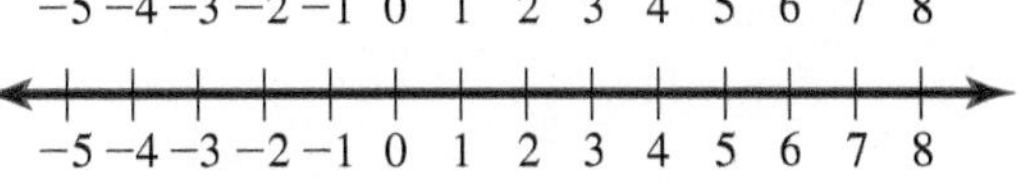

Note When we join the two graphs, the entire number line is covered; so the union of the two graphs is the set of real numbers.

Answers to Your Turn 4

a. Number line: −6 −5 −4 −3 −2 −1 0 1 2 3
Set-builder: $\{x \mid x < -4 \text{ or } x \geq -1\}$
Interval: $(-\infty, -4) \cup [-1, \infty)$

b. Number line: −3 −2 −1 0 1 2 3 4 5 6
Set-builder: $\{x \mid x < 4\}$
Interval: $(-\infty, 4)$

c. Number line: −4 −3 −2 −1 0 1 2 3 4
Set-builder: $\{x \mid x \text{ is a real number}\}$, or $\mathbb{R}$
Interval: $(-\infty, \infty)$

Set-builder notation: $\{x \mid x \text{ is a real number}\}$, or $\mathbb{R}$.
Interval notation: $(-\infty, \infty)$

Your Turn 4 For each compound inequality, graph the solution set. Then write the solution set in set-builder notation and in interval notation.

a. $-5x - 8 > 12$ or $-5x - 8 \leq -3$

b. $\frac{3}{4}x + 5 < -1$ or $\frac{3}{4}x + 5 < 8$

c. $5 - 3x \geq 11$ or $5 - 3x < 17$

2.4 Exercises

For Extra Help MyMathLab®

Note: Exercises marked with a ★ represent challenging exercises.

Objective 1

Prep Exercise 1 What is a compound inequality?

Prep Exercise 2 What two key words are used to denote compound inequalities?

Prep Exercise 3 Describe the intersection of two sets.

Prep Exercise 4 Describe how to graph a compound inequality involving *and*.

For Exercises 1–4, find the intersection of the given sets. See Objective 1.

1. $A = \{1, 3, 5\}$
$B = \{1, 3, 5, 7, 9\}$

2. $A = \{2, 4, 6\}$
$B = \{2, 4, 6, 8, 10\}$

3. $A = \{5, 6, 7, 8\}$
$B = \{7, 8, 9\}$

4. $A = \{8, 10, 12, 14, 16\}$
$B = \{14, 16, 18\}$

For Exercises 5–12, write each inequality without and. *See Example 1.*

5. $x > -4$ and $x < 5$

6. $n \geq -3$ and $n < 7$

7. $y > 0$ and $y \leq 2$

8. $m \geq 0$ and $m < 3$

9. $w > -7$ and $w < 3$

10. $r > -1$ and $r \leq 1$

11. $u \geq 0$ and $u \leq 2$

12. $t > 7$ and $t < 15$

For Exercises 13–20, graph the compound inequality. See Examples 1–3.

13. $x > 2$ and $x < 7$

14. $x > 3$ and $x < 9$

15. $x > -1$ and $x \leq 5$

16. $x \geq -4$ and $x < 0$

17. $1 \le x \le 10$

18. $-2 < x < -1$

19. $-3 \le x < 4$

20. $0 < x \le 8$

For Exercises 21–40: *a. Graph the solution set.*
b. Write the solution set using set-builder notation.
c. Write the solution set using interval notation.
} *See Examples 1–3.*

21. $x > -3$ and $x < -1$

22. $x > 4$ and $x < 7$

23. $x + 2 > 5$ and $x - 4 \le 2$

24. $x - 3 \ge 1$ and $x + 2 < 10$

25. $-x > 1$ and $-2x \le -10$

26. $-4x \ge 8$ and $-3x \le -15$

27. $3x + 8 \ge -7$ and $4x - 7 < 5$

28. $3x + 5 \ge -1$ and $5x - 2 < 13$

29. $-3x - 8 > 1$ and $-4x + 5 \le -3$

30. $-6x + 4 < -14$ and $-x - 3 \ge -2$

31. $-3 < x + 4 < 1$

32. $-2 \le x - 2 \le 2$

33. $-7 \le 4x - 3 \le 5$

34. $8 < 3x - 4 < 14$

35. $0 \le 2 + 3x < 8$

36. $0 \le -2 + 5x \le 13$

37. $-1 < -2x + 5 \le 5$

38. $-3 < 5 - 2x < 1$

39. $3 \le 6 - x \le 6$

40. $-4 \le -4 - x \le 2$

Objective 2

Prep Exercise 5 Describe the union of two sets.

Prep Exercise 6 Describe how to graph a compound inequality involving *or*.

For Exercises 41–44, find the union of the given sets. See Objective 2.

41. $A = \{c, a, t\}$
$B = \{d, o, g\}$

42. $A = \{l, o, v, e\}$
$B = \{m, a, t, h\}$

43. $A = \{x, y, z\}$
$B = \{w, x, y, z\}$

44. $A = \{a, b, c, d, e\}$
$B = \{c, d, e, f\}$

For Exercises 45–52, graph each inequality. See Objective 2.

45. $x < -2$ or $x > 6$

46. $n \le 3$ or $n > 4$

47. $y < -3$ or $y \ge 0$

48. $m \le 2$ or $m > 8$

49. $w > -3$ or $w > 2$

50. $r > -3$ or $r \ge -1$

51. $u \ge 0$ or $u \le 2$

52. $t > 6$ or $t < 13$

For Exercises 53–68: **a.** *Graph the solution set.*
b. *Write the solution set using set-builder notation.*
c. *Write the solution set using interval notation.*
See Example 4.

53. $y + 2 < -7$ or $y + 2 > 7$

54. $a - 4 \le -2$ or $a - 4 \ge 2$

55. $4r - 3 < -11$ or $2r - 3 > -1$

56. $3t - 5 \le -14$ or $5t + 5 \ge -5$

57. $-w + 2 \le -5$ or $-w + 2 \ge 3$

58. $-2x - 3 \le -1$ or $-2x - 3 \ge 5$

59. $7 - 4k \le -5$ or $6 - 2k \ge 2$

60. $8 - 2q \le 2$ or $8 - 2q \ge 4$

61. $\frac{2}{3}x - 4 \geq -2$ or $\frac{2}{5}x - 2 \leq -4$

62. $\frac{3}{4}x + 3 \geq 6$ or $\frac{3}{5}x - 3 \leq -6$

63. $-2(c - 1) < -4$ or $-2(c - 2) < -6$

64. $-3(y - 1) \leq 6$ or $-4(y - 2) \leq 4$

65. $2(x + 3) + 3 \leq 1$ or $2(x + 1) + 7 \leq -3$

66. $5(d - 1) + 8 < 8$ or $5(d + 1) - 2 < 18$

67. $-3(x - 1) - 1 \leq -1$ or $-3(x + 1) + 5 \geq -2$

68. $-4(k - 2) - 1 \geq -1$ or $-4(k - 2) - 1 \leq 5$

Objectives 1 and 2

For Exercises 69–80: **a.** *Graph the solution set.* **b.** *Write the solution set using set-builder notation.* **c.** *Write the solution set using interval notation.* } *See Examples 2, 3, and 4.*

69. $-3 < x + 4$ and $x + 4 < 7$

70. $0 < x - 1$ and $x - 1 \leq 3$

71. $4x + 3 < -5$ or $3x + 2 > 8$

72. $6x - 2 < -8$ or $4x - 3 > 9$

73. $4 < -2x < 6$

74. $9 \leq -3x \leq 15$

75. $7 \leq 4x - 5 \leq 19$

76. $-9 < 2x - 7 < -3$

77. $-5 < 1 - 2x < -1$

78. $4 \leq -3x + 1 < 7$

★ 79. $x + 3 < 2x + 1 < 3x$

★ 80. $2x - 2 \leq x + 1 \leq 2x + 5$

For Exercises 81–90, solve. Then:
a. Graph the solution set.
b. Write the solution set using set-builder notation.
c. Write the solution set using interval notation.

81. If Andrea's current long-distance bill is from \$36 to \$45 in a month, she can save money by switching to another company. Her current rate is \$0.09 per minute. What is the range of minutes that Andrea must use long distance to justify switching to the other company?

82. A mail-order music club offers one bonus CD if the total of an order is from \$75 to \$100. Dayle decides to buy the lowest-priced CDs, which are \$12.50 each. In what range would the number of CDs he orders have to be for him to receive a bonus CD?

83. Juan has taken four of the five tests in his history course. To get a B in the course, his average needs to be at least 80 and less than 90. If his scores on the first four tests are 95, 80, 82, and 88, in what range of values can his score on the fifth test be so that he has a B average?

84. Students in a chemistry course receive a C if the average of four tests is at least 70 and less than 80. Suppose a student has the following scores: 76, 72, and 84. What range of scores on the fourth test would result in the student receiving a C?

85. To conserve energy, it is recommended that a home's thermostat be set at 5° above 73°F during summer or 5° below 73° during winter. If the heat pump does not run when the temperature is at or between those values, in what range of temperatures is the heat pump off?

86. A house thermostat is set so that the heat pump/air conditioner comes on if the temperature is 2° or more above or below 70°. In what range of temperatures is the heat pump off?

87. In the maintenance of a saltwater aquarium, the temperature should be within 4° of 76°. What range of temperatures would be acceptable for saltwater fish? (*Source: The Conscientious Marine Aquarist*, Robert M. Fenner.)

88. When driving in Mississippi, motorists risk a ticket if they drive more than 15 miles per hour over or under the 55 miles per hour speed limit on a state highway. In what range of speeds would a motorist not risk receiving a ticket? (*Source:* Mississippi State Code 63-3-509.)

89. A building is to be designed in the shape of a trapezoid as shown. Find the range of values for the length of the back side of the building so that the square footage is from 6000 to 8000 square feet.

[Use $A = \frac{1}{2}h(a + b)$.]

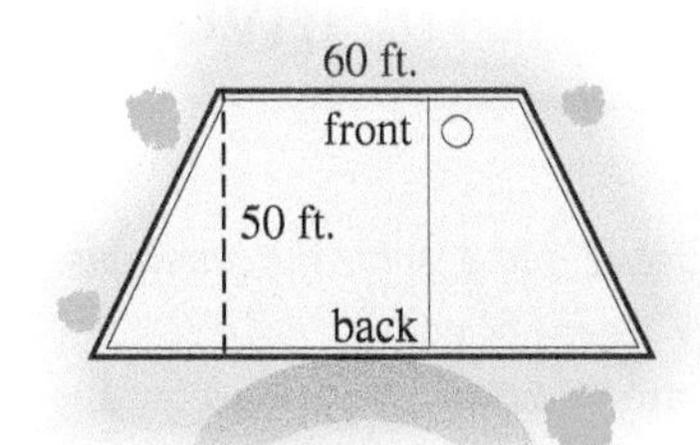

90. Water is a liquid between 32°F and 212°F. Use a compound inequality to describe the range of values in degrees Celsius for water in its liquid state. [Use $C = \frac{5}{9}(F - 32)$.]

Puzzle Problem A cyclist is involved in a multiple-day race. She believes that she needs to complete today's 40 kilometers between 1 hour and 45 minutes and 2 hours. Find the range of values her rate can be to complete this leg of the race in her desired time frame.

Review Exercises

Exercises 1–3 Expressions

[1.1] **1.** Is the absolute value of a number always positive? Explain.

[1.3] *For Exercises 2 and 3, simplify.*

2. $-|2 - 3^2| - |-4|$

3. $-| - |-16||$

Exercises 4–6 Equations and Inequalities

[2.1] *For Exercises 4 and 5, solve.*

4. $3x - 14x = 17 - 12x - 11$

5. $\frac{3}{5}(25 - 5x) = 15 - \frac{3}{5}$

[2.2] **6.** One-half the sum of a number and 2 is zero. Find the number.

2.5 Equations Involving Absolute Value

Objective

1 Solve equations involving absolute value.

Warm-up

[2.1] *For Exercises 1 and 2, solve the equations.*

1. $3x - 7 = -(5 - 3x)$ **2.** $8 - 2(4x - 8) = -16$

[1.4] **3.** Evaluate $-|x - 2| + 1$ for $x = 4$.

Objective 1 Solve equations involving absolute value.

We learned in Section 1.1 that the absolute value of a number is its distance from zero on a number line. For example, $|4| = 4$ and $|-4| = 4$ because both 4 and -4 are 4 units from 0 on a number line.

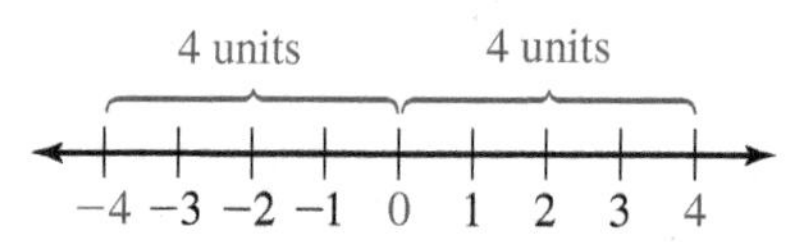

Now we will solve equations in which a variable appears within the absolute value symbols, as in $|x| = 4$. Notice that the solutions are 4 and -4, which suggests the following rule:

Rule Absolute Value Property

If $|x| = a$, where x is a variable or an expression and $a \geq 0$, then $x = a$ or $x = -a$.

Example 1 Solve.

a. $|2x - 5| = 9$

Solution: $2x - 5 = 9$ or $2x - 5 = -9$ Use the absolute value property to separate into a positive case and a negative case.

$2x = 14$ or $2x = -4$ In each case, add 5 to both sides.

$x = 7$ or $x = -2$ In each case, divide both sides by 2.

The solutions are 7 and -2.

Note The absolute value property does not apply if $a < 0$. For example, $|x| = -5$ has no solution because there is no real number whose absolute value is -5.

b. $|4x + 1| = -11$

Solution: This equation has the absolute value equal to a negative number, -11. Because the absolute value of every real number is a positive number or zero, this equation has no solution.

From Example 1(b), we can say that an absolute value equation in the form $|x| = a$, where $a < 0$, has no solution.

Your Turn 1 Solve.

a. $|3x + 4| = 8$ **b.** $|2x - 1| = 0$ **c.** $|5x - 2| = -3$

To use the absolute value property, we need the **absolute value isolated**.

Example 2 Solve $8 - 2|4x - 8| = -16$.

Solution: $-2|4x - 8| = -24$ Subtract 8 from both sides of the equation.

$|4x - 8| = 12$ Divide both sides by -2 to isolate the absolute value.

Answers to Your Turn 1

a. $\frac{4}{3}$ and -4 **b.** $\frac{1}{2}$

c. no solution

Answers to Warm-up

1. No solution or $\varnothing$

2. $x = 5$

3. -1

$4x - 8 = 12$	or	$4x - 8 = -12$	Use the absolute value property.
$4x = 20$	or	$4x = -4$	
$x = 5$	or	$x = -1$	

The solutions are 5 and -1.

Examples 1 and 2 suggest the following procedure:

Procedure Solving Equations Containing a Single Absolute Value

To solve an equation containing a single absolute value:

1. **Isolate the absolute value** so that the equation is in the form $|ax + b| = c$. If $c > 0$, proceed to steps 2 and 3. If $c < 0$, the equation has no solution.
2. Separate the absolute value into two equations: $ax + b = c$ and $ax + b = -c$.
3. Solve both equations.

Your Turn 2 Solve.

a. $12 - 3|3x - 6| = -6$ **b.** $|2x + 1| + 6 = 4$

More Than One Absolute Value

Some equations contain more than one absolute value expression, such as $|x + 3| = |2x - 8|$. If two absolute values are equal, they must contain expressions that are *equal* or *opposites*. Some simple examples follow.

Equal		Opposites
$\lvert 7\rvert = \lvert 7\rvert$	or	$\lvert 7\rvert = \lvert -7\rvert$
$\lvert -5\rvert = \lvert -5\rvert$	or	$\lvert -5\rvert = \lvert 5\rvert$

For the equation $|x + 3| = |2x - 8|$, therefore, the expressions $x + 3$ and $2x - 8$ must be equal or opposites.

Equal		Opposites
$x + 3 = 2x - 8$	or	$x + 3 = -(2x - 8)$

This suggests the following procedure:

Procedure Solving Equations in the Form $|ax + b| = |cx + d|$

To solve an equation in the form $|ax + b| = |cx + d|$:

1. Separate the absolute value equation into two equations: $ax + b = cx + d$ and $ax + b = -(cx + d)$.
2. Solve both equations.

Example 3 Solve.

a. $|3x + 5| = |x - 9|$

Solution: Separate the absolute value equation into two equations.

Equal		Opposites
$3x + 5 = x - 9$	or	$3x + 5 = -(x - 9)$
$2x + 5 = -9$	or	$3x + 5 = -x + 9$
$2x = -14$	or	$4x + 5 = 9$
$x = -7$	or	$4x = 4$
		$x = 1$

Note Think of $-(x - 9)$ as $-1(x - 9)$.

The solutions are -7 and 1.

Answers to Your Turn 2
a. 0 and 4 **b.** no solution

b. $|3x - 7| = |5 - 3x|$

Solution:

Equal		Opposites	
$3x - 7 = 5 - 3x$	or	$3x - 7 = -(5 - 3x)$	Separate into two equations.
$6x - 7 = 5$	or	$3x - 7 = -5 + 3x$	
$6x = 12$	or	$-7 = -5$	
$x = 2$			

Note Subtracting $3x$ from both sides gives us a false equation; so the "opposites" equation is a contradiction and thus has no solution.

This absolute value equation has only one solution, 2.

Your Turn 3 Solve.

a. $|5x - 3| = |3x - 13|$ **b.** $|9 + 4x| = |1 - 4x|$

Answers to Your Turn 3
a. $-5, 2$ **b.** -1

2.5 Exercises

For Extra Help MyMathLab®

Objective 1

Prep Exercise 1 What does absolute value mean in terms of a number line?

Prep Exercise 2 How do we interpret $|x| = 5$?

Prep Exercise 3 State the absolute value property in your own words.

Prep Exercise 4 If given an equation in the form $|ax + b| = c$ and $c < 0$, what can you conclude about its solution(s)?

For Exercises 1–16, solve using the absolute value property. See Example 1.

1. $|x| = 2$ **2.** $|y| = 5$ **3.** $|a| = -4$ **4.** $|r| = -1$

5. $|x + 3| = 8$ **6.** $|w - 1| = 4$ **7.** $|2m - 5| = 1$ **8.** $|3s + 6| = 12$

9. $|6 - 5x| = 1$ **10.** $|2 - 3x| = 5$ **11.** $|4 - 3w| = 6$ **12.** $|1 - 2x| = 2$

13. $|4m - 2| = -5$ **14.** $|6p + 5| = -1$ **15.** $|4w - 3| = 0$ **16.** $|3m - 5| = 0$

For Exercises 17–32, isolate the absolute value; then use the absolute value property. See Example 2.

17. $|2y| - 3 = 5$ **18.** $|3r| + 7 = 10$ **19.** $|y + 1| + 2 = 4$ **20.** $|x + 3| - 1 = 5$

21. $|b - 4| - 6 = 2$ **22.** $|v - 2| + 2 = 4$ **23.** $3 + |5x - 1| = 7$ **24.** $2 + |5t - 2| = 5$

25. $1 - |2k + 3| = -4$ **26.** $-3 = 1 - |4u - 8|$ **27.** $4 - 3|z - 2| = -8$ **28.** $3 - 2|x - 5| = -7$

29. $6 - 2|3 - 2w| = -18$ **30.** $15 - 2|5 - 2x| = -5$ **31.** $|3x - 2(x + 5)| = 10$ **32.** $|4y - 2(6 - y)| = 18$

Prep Exercise 5 Explain the first step in solving an equation containing two absolute values, such as $|2x + 1| = |3x - 5|$.

Prep Exercise 6 How can you tell that an absolute value equation containing two absolute values has only one solution?

For Exercises 33–42, solve by separating the two absolute values into equal and opposite cases. See Example 3.

33. $|2x + 1| = |x + 5|$ **34.** $|2p + 5| = |3p + 10|$ **35.** $|x + 3| = |2x - 4|$ **36.** $|p - 1| = |2p + 8|$

37. $|3v + 4| = |1 - 2v|$ **38.** $|3 - 6c| = |2c + 3|$ **39.** $|2n + 3| = |3 + 2n|$ **40.** $|2r - 1| = |1 - 2r|$

41. $|2k + 1| = |2k - 5|$ **42.** $|4 + 2q| = |2q + 8|$

For Exercises 43–50, solve. See Examples 1 and 2.

43. $|10 - (5 - h)| = 8$ **44.** $|7 - (2 - n)| = 17$ **45.** $\left|\frac{b}{2} - 1\right| = 4$ **46.** $\left|\frac{u}{3} - 2\right| = 3$

47. $\left|\frac{4 - 3x}{2}\right| = \frac{3}{4}$ **48.** $\left|\frac{6 - 5w}{6}\right| = \frac{2}{3}$ **49.** $\left|2y + \frac{3}{2}\right| - 2 = 5$ **50.** $\left|p + \frac{2}{3}\right| - 1 = 7$

Review Exercises

Exercises 1–6 Equations and Inequalities

[2.3] *For Exercises 1 and 2, solve.*

1. $6 - 4x > -5x - 1$

2. $-4x \leq 12$

[2.3] **3.** Translate to a linear inequality and solve: Two less than a number is less than five.

[2.4] **4.** Use interval notation to represent the solution set shown on the graph to the right.

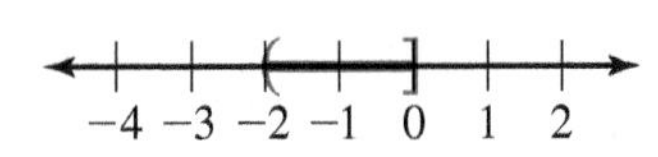

[2.4] **5.** Graph the compound inequality: $-1 < x \leq 0$

[2.4] **6.** Solve the compound inequality: $-1 < \frac{3x + 2}{4} < 4$

2.6 Inequalities Involving Absolute Value

Objectives

1 Solve absolute value inequalities involving less than.

2 Solve absolute value inequalities involving greater than.

Warm-up

For Exercises 1 and 2, solve.

[2.4] 1. $-3 < -0.5x + 1 < 3$

[2.5] 2. $|2x - 1| + 3 = 16$

Objective 1 Solve absolute value inequalities involving less than.

Now let's solve inequalities that contain absolute value. Think about $|x| \le 3$. A solution for this inequality is any number whose absolute value is less than or equal to 3. This means that the solutions are a distance of 3 units or less from 0 on a number line.

Note Solutions for the **equal to** part of $|x| \le 3$ are 3 and -3. The **less than** part of $|x| \le 3$ means all numbers between 3 and -3 because their absolute values are less than 3.

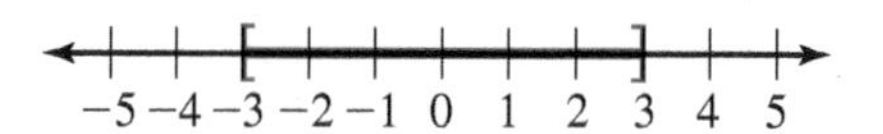

Notice that the solution region corresponds to the compound inequality $x \ge -3$ and $x \le 3$, which we can write as $-3 \le x \le 3$. In set-builder notation, the solution is $\{x | -3 \le x \le 3\}$, and in interval notation, we write $[-3, 3]$.

Our examples suggest the following procedure:

Procedure Solving Inequalities in the Form $|x| < a$, where $a > 0$

To solve an inequality in the form $|x| < a$, where $a > 0$:

1. Rewrite the inequality as a compound inequality involving *and*: $x > -a$ and $x < a$ (or use $-a < x < a$).
2. Solve the compound inequality.

Similarly, to solve $|x| \le a$, we write $x \ge -a$ and $x \le a$ (or $-a \le x \le a$).

Example 1 For each inequality, solve, graph the solution set, and write the solution set in both set-builder and interval notation.

a. $|x| < 4$

Solution: $x > -4$ **and** $x < 4$ Rewrite as a compound inequality.

Solution set graph: −5 −4 −3 −2 −1 0 1 2 3 4 5

Note $x > -4$ and $x < 4$ means $-4 < x < 4$.

Set-builder notation: $\{x | -4 < x < 4\}$

Interval notation: $(-4, 4)$

b. $|x - 4| \le 3$

Solution: $x - 4 \ge -3$ **and** $x - 4 \le 3$ Rewrite as a compound inequality.

$x \ge 1$ **and** $x \le 7$ Add 4 to both sides of each inequality.

Solution set graph: −1 0 1 2 3 4 5 6 7 8 9

Note The solution set for $|x - 4| \le 3$ contains every number whose distance from 4 is 3 units or less.

Set-builder notation: $\{x | 1 \le x \le 7\}$

Interval notation: $[1, 7]$

c. $|3x + 1| < 5$

Solution: Instead of $3x + 1 > -5$ and $3x + 1 < 5$, we use the more compact form.

$-5 < 3x + 1 < 5$ Rewrite as a compound inequality.

$-6 < 3x < 4$ Subtract 1 from all three parts of the inequality.

$-2 < x < \frac{4}{3}$ Divide all three parts of the inequality by 3 to isolate x.

Answers to Warm-up

1. $-4 < x < 8$

2. $7, -6$

Solution set graph: 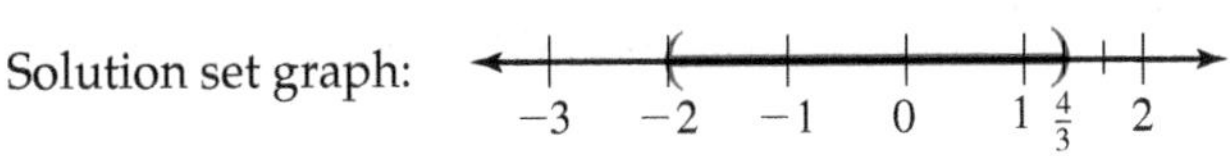

Set-builder notation: $\left\{x \mid -2 < x < \frac{4}{3}\right\}$

Interval notation: $\left(-2, \frac{4}{3}\right)$

d. $|-0.5x + 1| - 2 < 1$

Solution: Notice that the equation is not in the form $|x| < a$; so our first step is to isolate the absolute value.

$|-0.5x + 1| < 3$ Add 2 to both sides to isolate the absolute value.

$-3 < -0.5x + 1 < 3$ Rewrite as a compound inequality.

$-4 < -0.5x < 2$ Subtract 1 from all three parts of the inequality.

$8 > x > -4$ Divide all three parts of the inequality by -0.5, which changes the direction of each inequality.

Solution set graph: 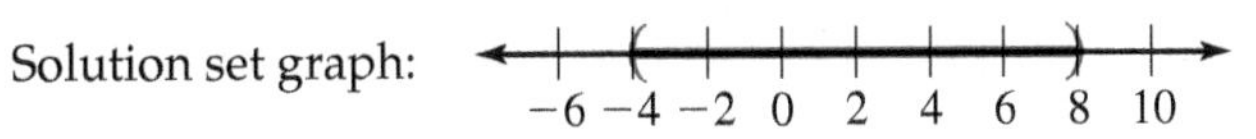

Set-builder notation: $\{x \mid -4 < x < 8\}$

Interval notation: $(-4, 8)$

e. $\left|\frac{1}{3}x - 4\right| + 6 < -1$

Note Any inequality of the form $|x| < a, a \le 0$, or $|x| \le a, a < 0$, has no solution because the absolute value of any real number is nonnegative and, therefore, can only be zero or positive. ▶

Solution: $\left|\frac{1}{3}x - 4\right| < -7$ Subtract 6 from both sides to isolate the absolute value.

Because absolute values cannot be negative, this inequality has no solution; so the solution set is empty.

Solution set graph: 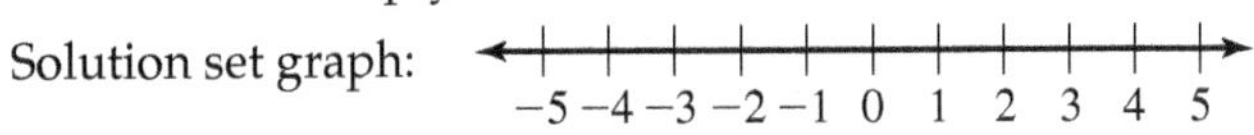

Set-builder notation: { } or $\varnothing$

Interval notation: We do not write interval notation because there are no values in the solution set.

Your Turn 1 For each inequality, solve, graph the solution set, and write the solution set in both set-builder notation and interval notation.

a. $|x + 5| < 2$

b. $|2x - 3| \le 7$

c. $\left|-\frac{1}{3}x + 1\right| - 2 \le 1$

d. $\left|\frac{1}{2}x - 1\right| + 3 < -2$

Answers to Your Turn 1

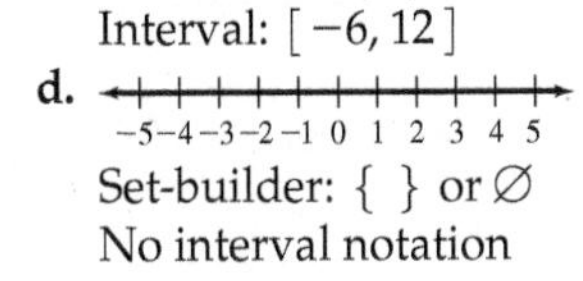

a. Set-builder: $\{x \mid -7 < x < -3\}$
Interval: $(-7, -3)$

b. Set-builder: $\{x \mid -2 \le x \le 5\}$
Interval: $[-2, 5]$

c. Set-builder: $\{x \mid -6 \le x \le 12\}$
Interval: $[-6, 12]$

d. Set-builder: { } or $\varnothing$
No interval notation

Objective 2 Solve absolute value inequalities involving greater than.

Now we consider inequalities with greater than, such as $|x| \ge 5$. A solution for $|x| \ge 5$ is any number whose absolute value is greater than or equal to 5. As the graph shows, the solution set contains all values that are a distance of 5 units or more from 0.

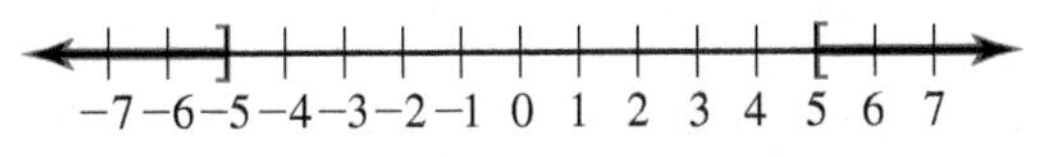

Notice that the solutions are equivalent to $x \le -5$ or $x \ge 5$.

◀ **Note** Solutions for the **equal to** part of $|x| \ge 5$ are 5 and -5. The **greater than** part means all values that are farther from 0 than 5 and -5 because their absolute values are greater than 5.

An inequality such as $|x| > 2$ uses parentheses.

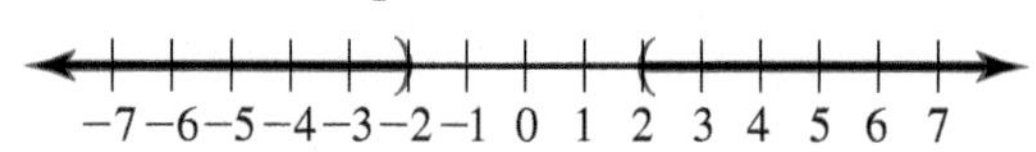

The solutions are equivalent to $x < -2$ or $x > 2$. Our examples suggest that we can split these inequalities into a compound inequality involving *or*.

Procedure Solving Inequalities in the Form $|x| > a$, where $a > 0$

To solve an inequality in the form $|x| > a$, where $a > 0$:

1. Rewrite the inequality as a compound inequality involving *or*: $x < -a$ or $x > a$.
2. Solve the compound inequality.

Similarly, to solve $|x| \geq a$, we write $x \leq -a$ or $x \geq a$.

Warning When solving absolute value inequalities, look carefully at the order symbol. The rules for $<$ and $>$ are very different.

Example 2 For each inequality, solve, graph the solution set, and write the solution set in both set-builder notation and interval notation.

a. $|x - 2| \geq 5$

Solution: $x - 2 \leq -5$ **or** $x - 2 \geq 5$ — Rewrite as a compound inequality.

$x \leq -3$ **or** $x \geq 7$ — Add 2 to both sides of each inequality.

Solution set graph:

−5 −4 −3 −2 −1 0 1 2 3 4 5 6 7 8

Set-builder notation: $\{x|x \leq -3 \text{ or } x \geq 7\}$

Interval notation: $(-\infty, -3] \cup [7, \infty)$

b. $|-3x - 4| > 5$

Solution: $-3x - 4 < -5$ or $-3x - 4 > 5$ — Rewrite as a compound inequality.

$-3x < -1$ or $-3x > 9$ — Add 4 to both sides of each inequality.

$x > \frac{1}{3}$ or $x < -3$ — Divide both sides of each inequality by -3.

Solution set graph:

−4 −3 −2 −1 0 $\frac{1}{3}$ 1 2

Set-builder notation: $\left\{x|x < -3 \text{ or } x > \frac{1}{3}\right\}$

Interval notation: $(-\infty, -3) \cup \left(\frac{1}{3}, \infty\right)$

c. $\left|\frac{3}{4}x - 2\right| + 1 \geq 6$

Solution: $\left|\frac{3}{4}x - 2\right| \geq 5$ — Subtract 1 from both sides to isolate the absolute value.

$\frac{3}{4}x - 2 \leq -5$ or $\frac{3}{4}x - 2 \geq 5$ — Rewrite as a compound inequality.

$\frac{3}{4}x \leq -3$ or $\frac{3}{4}x \geq 7$ — Add 2 to both sides of each inequality.

$x \leq -4$ or $x \geq \frac{28}{3}$ — Multiply both sides of each inequality by $\frac{4}{3}$.

Solution set graph:

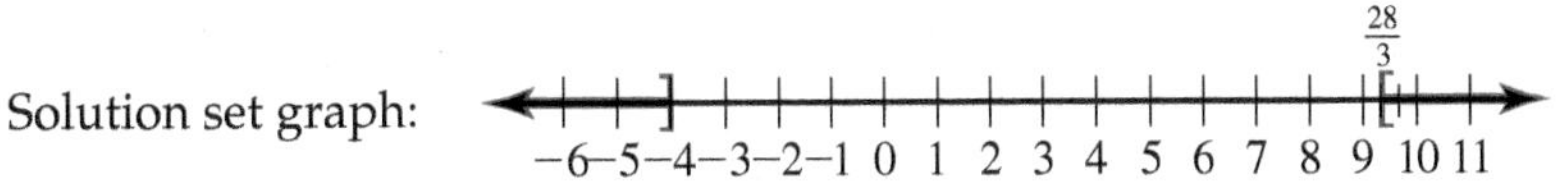

Set-builder notation: $\left\{x \mid x \le -4 \text{ or } x \ge \frac{28}{3}\right\}$

Interval notation: $(-\infty, -4] \cup \left[\frac{28}{3}, \infty\right)$

d. $|0.2x + 1| - 3 > -7$

Solution: $|0.2x + 1| > -4$ Add 3 to both sides to isolate the absolute value.

This inequality indicates that the absolute value is greater than a negative number. Because the absolute value of every real number is either positive or 0, the solution set is $\mathbb{R}$. The graph is the entire number line.

Note The solutions of any inequality of the form $|x| > a$, $a < 0$, is all real numbers because the absolute value of any real number is nonnegative and, therefore, greater than any negative number.

Solution set graph:

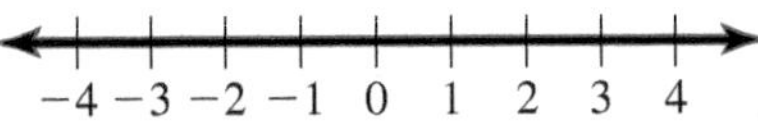

Set-builder notation: $\{x \mid x \text{ is a real number}\}$, or $\mathbb{R}$

Interval notation: $(-\infty, \infty)$

Your Turn 2 For each inequality, solve, graph the solution set, and write the solution set in both set-builder notation and interval notation.

a. $|x - 3| > 2$

b. $|4x + 1| \ge 9$

c. $|-0.4x + 1| + 2 \ge 3$

d. $\left|\frac{1}{3}x - 4\right| - 2 > -9$

Answers to Your Turn 2

a. Set-builder: $\{x \mid x < 1 \text{ or } x > 5\}$
Interval: $(-\infty, 1) \cup (5, \infty)$

b. Set-builder: $\left\{x \mid x \le -\frac{5}{2} \text{ or } x \ge 2\right\}$
Interval: $\left(-\infty, -\frac{5}{2}\right] \cup [2, \infty)$

c.

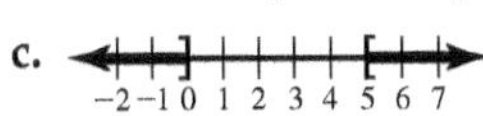

Set-builder: $\{x \mid x \le 0 \text{ or } x \ge 5\}$
Interval: $(-\infty, 0] \cup [5, \infty)$

d.

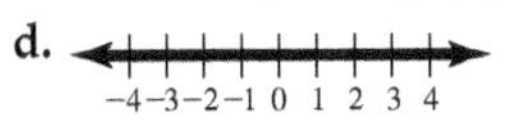

Set-builder: $\{x \mid x \text{ is a real number}\}$
Interval: $(-\infty, \infty)$

2.6 Exercises

For Extra Help MyMathLab®

Note: Exercises marked with a ★ represent challenging exercises.

Objective 1

For Exercises 1–4, assume that a is a positive number.

Prep Exercise 1 What compound inequality is related to $|x| \le a$, where $a > 0$?

Prep Exercise 2 How would you characterize the graph of $|x| < a$, where $a > 0$?

Prep Exercise 3 Under what conditions does $|x| < a$ have no solution?

For Exercises 1–14, solve the inequality. Then:
a. Graph the solution set.
b. Write the solution set using set-builder notation.
c. Write the solution set using interval notation.
See Example 2.

1. $|x| < 5$

2. $|y| \le 3$

3. $|x + 3| \leq 7$

4. $|m - 5| < 2$

5. $|s + 3| + 6 < 9$

6. $|p - 3| + 4 \leq 8$

7. $|2m - 5| - 3 < 6$

8. $|4x + 12| + 2 < 10$

9. $|-3k + 5| + 7 \leq 8$

10. $|-5h - 1| - 6 < 8$

11. $2|x| + 7 \leq 3$

12. $3|u| + 2 < -7$

13. $2|w - 3| + 4 < 10$

14. $4|n + 2| - 3 \leq 9$

Objective 2

Prep Exercise 4 What compound inequality is related to $|x| \geq a$, where $a > 0$?

Prep Exercise 5 How would you characterize the graph of $|x| > a$, where $a > 0$?

Prep Exercise 6 Under what conditions does the solution set for $|x| > a$ contain all real numbers?

For Exercises 15–28, solve the inequality. Then: **a.** *Graph the solution set.* **b.** *Write the solution set using set-builder notation.* **c.** *Write the solution set using interval notation.* *See Example 2.*

15. $|c| > 12$

16. $|h| \geq 6$

17. $|y + 2| \geq 7$

18. $|a - 4| > 3$

19. $|p - 6| - 3 > 5$

20. $|x + 5| - 3 \geq 6$

21. $|3x + 6| - 3 \geq 9$

22. $|2x - 5| - 1 \geq 10$

23. $|-4n - 5| + 3 > 8$

24. $|-6h + 1| - 7 > 4$

25. $4|v| + 3 \geq 7$

26. $5|m| + 2 > 12$

27. $4|y + 2| - 1 > 3$

28. $3|x - 2| - 5 \geq 1$

Objectives 1 and 2

For Exercises 29–48, solve the inequality. Then:
a. Graph the solution set.
b. Write the solution set using set-builder notation.
c. Write the solution set using interval notation.
See Examples 1 and 2.

29. $|4m + 8| - 2 > 10$

30. $|2x + 4| - 2 \geq 10$

31. $|-3x + 6| < 6$

32. $|-2y + 3| \leq 3$

33. $|2r - 3| > -3$

34. $|3b + 7| > -2$

35. $4 - 2|x + 3| > 2$

36. $5 - 2|u + 4| \geq 1$

37. $6|2x - 1| - 3 < 3$

38. $4|3p + 6| - 2 < 22$

39. $5 - |w + 4| > 10$

40. $4 - |5 + k| > 7$

41. $\left|2 - \frac{3}{2}k\right| \le 5$

42. $\left|3 - \frac{1}{2}x\right| \le 7$

43. $|0.25x - 3| + 2 > 4$

44. $|0.5y - 3| + 4 > 5$

45. $\left|2.4 - \frac{3}{4}y\right| \le 7.2$

46. $\left|5.3 - \frac{2}{3}w\right| \ge 5.3$

47. $|2p - 8| + 5 > 1$

48. $|8p + 7| + 4 > 3$

★ *For Exercises 49–58, write an inequality involving absolute value that describes the graph shown.*

49. 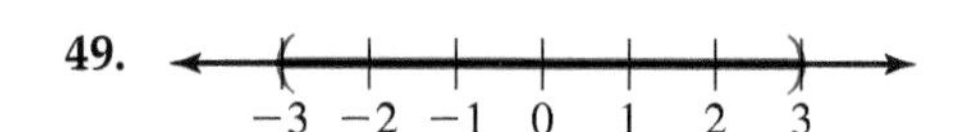

50. 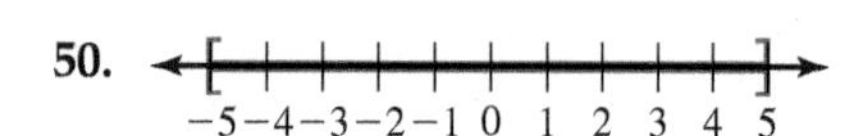

51. 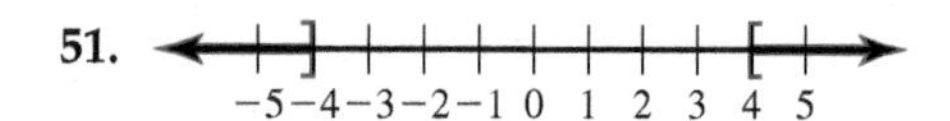

52. 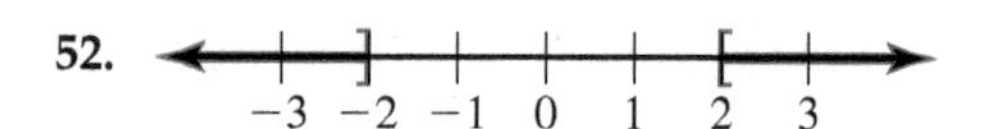

53. 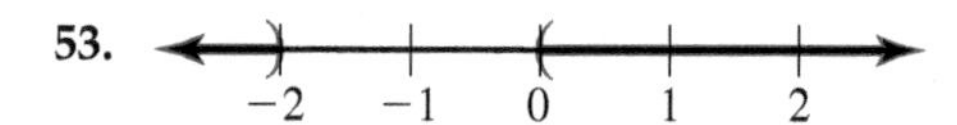

54. 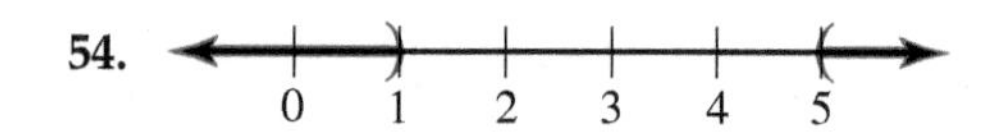

55. 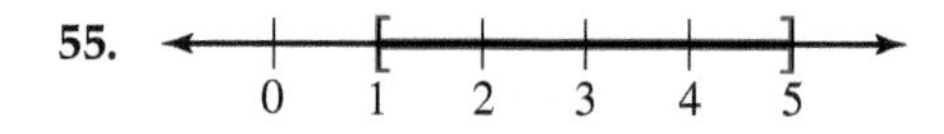

56. −6 −5 −4 −3 −2 −1 0 1 2 3 4

57. −3 −2 −1 0 1

58. 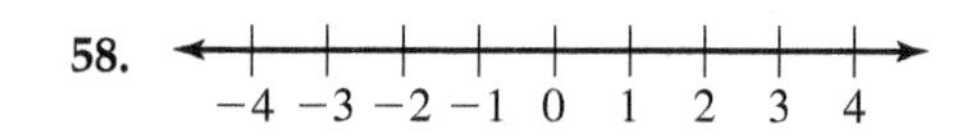

Review Exercises

Exercises 1 and 2 Constants and Variables

[1.1] ***For Exercises 1 and 2, indicate whether the statement is true or false.***

1. $-\frac{2}{3} \in \{x \mid x \text{ is a rational number}\}$

2. $\{1, 3, 5, 7\} \subseteq N$

Exercises 3–6 Equations and Inequalities

[2.1] ***For Exercises 3 and 4, solve.***

3. $23 - 5(2 - 3x) = 12x - (4x + 1)$

4. $\frac{5}{6}x - \frac{3}{4} = \frac{2}{3}x + \frac{1}{2}$

[2.1] **5.** Solve $Ax + By = C$ for y.

[2.2] **6.** Two-fifths of the sum of a number and five is five less than three-fourths of the number. Find the number.

Chapter 2 Summary and Review Exercises

Complete each incomplete definition, rule, or procedure; study the key examples; and then work the related exercises.

2.1 Linear Equations and Formulas

Definitions/Rules/Procedures	Key Example(s)
The Addition Principle of Equality If $a = b$, then ______ = ______ is true for all real numbers a, b, and c. **The Multiplication Principle of Equality** If $a = b$, then ______ = ______ is true for all real numbers a, b and c, where $c \neq 0$. **To solve linear equations:** 1. Simplify both sides of the equation as needed. a. Distribute to eliminate parentheses. b. Eliminate fractions or decimals by multiplying through by the ______. In the case of decimals, the ______ is the power of ______ with the same number of 0 digits as decimal places in the number with the most decimal places. (Eliminating fractions and decimals is optional.) c. Combine like terms. 2. Use the ______ principle so that all variable terms are on one side of the equation and all constants are on the other side. (Eliminate the variable term with the lesser coefficient.) Then combine like terms. 3. Use the ______ principle to eliminate any remaining coefficient.	Solve and check: $\frac{2}{3}x - \frac{1}{2}(x + 1) = \frac{5}{6}x - 2$ $6 \cdot \frac{2}{3}x - 6 \cdot \frac{1}{2}(x + 1) = 6 \cdot \frac{5}{6}x - 6 \cdot 2$ Multiply both sides by the LCD 6. $4x - 3(x + 1) = 5x - 12$ Simplify. $4x - 3x - 3 = 5x - 12$ Distribute -3. $x - 3 = 5x - 12$ Combine like terms. $x - x - 3 = 5x - x - 12$ Subtract x from both sides. $-3 + 12 = 4x - 12 + 12$ Add 12 to both sides. $\frac{9}{4} = \frac{4x}{4}$ Divide both sides by 4. $\frac{9}{4} = x$ **Check:** Substitute $\frac{9}{4}$ for x in the original equation; then verify that the equation is true. We will leave this to the reader.

Exercises 1–12 Equations and Inequalities

[2.1] *For Exercises 1–12, solve and check. Note that some equations may be identities or contradictions.*

1. $3x - 9 = 12$

2. $7m - 3 = 4m + 9$

3. $6(3a - 4) + (3a + 4) = 5(a - 2) - 10$

4. $2 - 2y - 6(y + 3) = 5 - 7y$

5. $\frac{1}{5}d - 2 = -3 + d$

6. $\frac{1}{2}k + \frac{5}{8} = \frac{1}{4}$

7. $2w + 2(7 - w) = 2(5w + 10) + 4$

8. $5r - 2(5r - 3) = 6 - 5r$

9. $0.53 - 0.2z = 0.2(2z - 13) - 0.47$

10. $18(v + 2) - 6v = 30 + 12v - 2$

11. $\frac{2}{5}p - \frac{3}{20} = \frac{13}{20} - \frac{3}{10}(6 - 2p)$

12. $0.5(l + 4) = 0.3(3l - 1) - 0.9$

Definitions/Rules/Procedures	Key Example(s)
To solve for a particular variable in a formula, treat all other variables like ________ and solve for the desired variable using the outline for solving equations.	Solve the formula $d = vt + x$ for t. $d - x = vt + x - x$ Subtract x from both sides. $\frac{d - x}{v} = \frac{vt}{v}$ Divide both sides by v. $\frac{d - x}{v} = t$

Exercises 13–16 Equations and Inequalities

[2.1] *For Exercises 13–16, solve the formula for the indicated variable.*

13. $I = Prt$; solve for t

14. $P = 2l + 2w$; solve for w

15. $A = \frac{1}{2}bh$; solve for h

16. $P = a + b + c$; solve for a

2.2 Solving Problems

Definitions/Rules/Procedures	Key Example(s)
Problem-Solving Outline 1. ________ the problem. a. Read the question(s) (not the whole problem, just the question at the end) and write a note to yourself about what you are to find. b. Read the whole problem, underlying the key words. c. If possible or useful, draw a picture, make a list or table to organize what is known and unknown, simulate the situation, or search for a related example problem. 2. ________ your solution by searching for a formula or using the key words to translate to an equation. 3. ________ the plan by solving the equation/formula. 4. ________ the question. Look at the note about what you were to find and make sure you answer the question. Include appropriate units. 5. ________ results. a. Try finding the solution in a different way, reversing the process, or estimating the answer making sure the estimate and actual answer are reasonably close. b. Make sure the answer is reasonable.	The surface area of an enclosed box is 148 square feet. Use the formula SA $= 2lw + 2wh + 2lh$ to find the height if the length is 5 feet and the width is 4 feet. Understand We are given the surface area, the formula SA $= 2lw + 2wh + 2lh$, the length is 5 feet and the width is 4 feet. We are asked to find the height. Plan Replace SA with 148, l with 5, w with 4, and solve for h. Execute $148 = 2(5)(4) + 2(4)h + 2(5)h$ Replace SA with 148, l with 5 and w with 4. $148 = 40 + 8h + 10h$ Simplify. $148 = 40 + 18h$ Combine like terms. $148 - 40 = 40 - 40 + 18h$ Subtract 40 from both sides. $108 = 18h$ Simplify. $6 = h$ Divide both sides by 18. Answer The height of the box is 6 feet. Check Find the surface area of a box 5 feet long, 4 feet wide, and 6 feet high and verify that it is 148 ft.^{2}.
To **translate** a word sentence to an equation, identify the ________, ________, and ________; then write the corresponding symbolic form.	Translate to an equation. a. Three less than eight times a number is thirteen. **Answer:** $8n - 3 = 13$ b. Four times the sum of a number and five is equal to negative three times the difference of the number and two. **Answer:** $4(n + 5) = -3(n - 2)$

Exercises 17–23 Equations and Inequalities

[2.2] *For Exercises 17–23, solve.*

17. Liquid nitrogen is often used to freeze objects quickly. It is so cold that when a room temperature object is immersed, the liquid nitrogen immediately boils at a temperature of about −196°C. Using $C = \frac{5}{9}(F - 32)$, convert this temperature to degrees Fahrenheit.

18. The formula $B = P + Prt$ can be used to determine the final balance, B, in an account if a principal P earns interest at a rate r for t years. If the balance in an account is $4368 after earning interest at a rate of 4% for one year, find the principal that was invested.

19. Two-thirds of a number is equal to two plus one-half of the number. Find the number.

20. Twice the difference of a number and seven is negative six. What is the number?

21. A clothing store has a special sale with every item discounted 30%. If the price of a dress after the discount is $44.94, what was the original price?

22. An annual report for a company shows that 37,791 units of their top-selling product were sold during the year, which is an increase of 10.5% over the previous year's sales. How many units of that product were sold the previous year?

23. Two cars are traveling toward each other on the same road. One car is traveling 50 miles per hour; the other, 55 miles per hour. If the two cars are 315 miles apart, how long until they meet?

2.3 Solving Linear Inequalities

Definitions/Rules/Procedures	Key Example(s)
To **graph an inequality** in one variable of the form $x \le a, x \ge a, x < a$, or $x > a$ on a number line: **1.** If the symbol is $\le$ or $\ge$, draw a __________ (or __________) on the number line at the indicated number open to the __________ for $\le$ and to the __________ for $\ge$. If the symbol is $<$ or $>$, draw a __________ (or __________) on the number line at the indicated number open to the __________ for $<$ and to the __________ for $>$.	Graph $x \le 3$ on a number line. −3 −2 −1 0 1 2 3 4
2. If the variable is greater than the indicated number, shade to the __________ of the indicated number. If the variable is less than the indicated number, shade to the __________ of the indicated number.	Graph $x > -4$ on a number line. −5 −4 −3 −2 1 0 1 2

Definitions/Rules/Procedures	Key Example(s)
The Addition Principle of Inequality If $a < b$, then $a + c$ ________ $b + c$ is true for all real numbers a, b, and c. **The Multiplication Principle of Inequality** If a and b are real numbers where $a < b$, then $ac < bc$ is true if c is a ________ real number and $ac > bc$ is true if c is a ________ real number. **Note:** Although we have written the principles in terms of the $<$ symbol, the principles are true for any inequality symbol.	Solve. Then write the solution set in set-builder notation and in interval notation. **a.** $-3x \geq 12$ $\frac{-3x}{-3} \leq \frac{12}{-3}$ Divide both sides by -3, which changes the direction of the inequality symbol. $x \leq -4$ Set-builder notation: $\{x \mid x \leq -4\}$ Interval notation: $(-\infty, -4]$
To **solve a linear inequality** in one variable: **1.** Simplify both sides of the inequality as needed. **a.** Distribute to eliminate ________. **b.** Eliminate fractions or decimals by multiplying through by the ________ just as we did for equations. (This step is optional.) **c.** Combine like terms. **2.** Use the ________ principle so that all the variable terms are on one side of the inequality and all constants are on the other side. Then combine like terms. **3.** Use the ________ principle to eliminate any remaining coefficient. If you multiply (or divide) both sides by a negative number, ________ the direction of the inequality symbol.	**b.** $12m - (2m - 9) > 5m - 7 + 3m$ $12m - 2m + 9 > 8m - 7$ Distribute -1. $10m - 8m + 9 > 8m - 8m - 7$ Subtract $8m$ from both sides. $2m + 9 - 9 > -7 - 9$ Subtract 9 from both sides. $\frac{2m}{2} > \frac{-16}{2}$ Divide both sides by 2. $m > -8$ Set-builder notation: $\{m \mid m > -8\}$ Interval notation: $(-8, \infty)$

Exercises 24–30 Equations and Inequalities

[2.3] *For Exercises 24–29, solve and then:*
a. Write the solution set using set-builder notation.
b. Write the solution set using interval notation.
c. Graph the solution set.

24. $-8n \geq -32$

25. $3x + 5 > -1$

26. $9 - 2m > 15$

27. $5h - 11 \geq 9h + 9$

28. $\frac{2}{3}t - 1 \leq \frac{1}{4}t + \frac{1}{2}$

29. $12 - (u + 7) < 3(u - 1)$

For Exercise 30, solve using inequalities.

[2.3] **30.** A homeowner is planning to put a pool in the backyard. The area of a rectangular pool may not exceed 300 square feet. If the width is 12 feet, what range of values can the length have?

2.4 Compound Inequalities

Definitions/Rules/Procedures	Key Example(s)
A **compound inequality** is two inequalities joined by either ________ or ________.	Examples of compound inequalities are $x > 5$ or $x < 7$, $x \geq -4$ and $x < 8$.
The intersection of two sets A **and** B, symbolized by $A \cap B$, is a set containing only elements that are in ________.	If $A = \{-2, -1, 0, 3, 7\}$ and $B = \{-3, -1, 1, 3\}$, then $A \cap B = \{-1, 3\}$ and $A \cup B = \{-3, -2, -1, 0, 1, 3, 7\}$.
The union of two sets A **and** B, symbolized by $A \cup B$, is a set containing every element in ________.	

Exercises 31 and 32 Expressions

[2.4] *For Exercises 31 and 32, find the intersection and union of the given sets.*

31. $A = \{1, 5, 9\}$
$B = \{1, 5, 7, 9\}$

32. $A = \{1, 2, 3\}$
$B = \{4, 5, 6, 7\}$

Definitions/Rules/Procedures	Key Example(s)
To solve a compound inequality involving ***and*****:** **1.** Solve each ________ in the compound inequality. **2.** Find the ________ of the individual solution sets.	Solve the compound inequality. Then: **a.** Graph the solution set. **b.** Write the solution set using set-builder notation. **c.** Write the solution set using interval notation. $-3 < -2x - 5 \leq 1$ Solution: **a.** $-2x - 5 > -3$ and $-2x - 5 \leq 1$ $-2x > 2$ and $-2x \leq 6$ $x < -1$ and $x \geq -3$ Graph: (number line from −6 to 1 with [at −3 and) at −1) **b.** Set-builder notation: $\{x \mid -3 \leq x < -1\}$ **c.** Interval notation: $[-3, -1)$

Exercises 33–40 Equations and Inequalities

[2.4] *For Exercises 33–38, solve the compound inequality. Then:*
a. Graph the solution set.
b. Write the solution set using set-builder notation.
c. Write the solution set using interval notation.

33. $4 < -2x < 6$

34. $9 \leq -3x \leq 15$

35. $-3 < x + 4 < 7$

36. $0 < x - 1 \leq 3$

37. $2x + 3 \geq -1$ and $2x + 3 < 3$

38. $3x - 2 > 1$ and $3x - 2 < -8$

[2.4] *For Exercises 39 and 40, solve. Then:* *a. Graph the solution set.*
b. Write the solution set using set-builder notation.
c. Write the solution set using interval notation.

39. Students in a biology course receive a B if the average of four tests is 80 or higher and less than 90. Suppose a student has the following scores: 80, 89, and 83. What range of scores on the fourth test would cause the student to receive a B?

40. The width of a rectangular building is to be 80 feet. What range of values can the length have so that the area of the base of the building is from 12,000 to 16,000 square feet?

Definitions/Rules/Procedures	Key Example(s)
To solve a compound inequality involving *or*: 1. Solve each ________ in the compound inequality. 2. Find the ________ of the individual solution sets.	Solve the compound inequality. Then: **a.** Graph the solution set. **b.** Write the solution set using set-builder notation. **c.** Write the solution set using interval notation. $5x - 1 \leq -16$ or $5x - 1 > 9$ Solution: **a.** $5x - 1 \leq -16$ or $5x - 1 > 9$ $5x \leq -15$ or $5x > 10$ $x \leq -3$ or $x > 2$ Graph: (number line from −5 to 4) **b.** Set-builder notation: $\{x \mid x \leq -3 \text{ or } x > 2\}$ **c.** Interval notation: $(-\infty, -3] \cup (2, \infty)$

Exercises 41–46 Equations and Inequalities

[2.4] *For Exercises 41–46, solve the compound inequality. Then:* *a. Graph the solution set.*
b. Write the solution set using set-builder notation.
c. Write the solution set using interval notation.

41. $w + 4 \leq -2$ or $w + 4 \geq 2$

42. $4w - 3 < 1$ or $4w - 3 > 0$

43. $2m - 5 < 0$ or $2m - 5 > 5$

44. $3x + 2 \leq -2$ or $3x + 2 \geq 8$

45. $-x - 6 \le -2$ or $-x - 6 \ge 3$

46. $-4w + 1 \le -3$ or $-4w + 1 \ge 5$

2.5 Equations Involving Absolute Value

Definitions/Rules/Procedures	Key Example(s)
Absolute Value Property If $\|x\| = a$, where x is a variable or an expression and $a \ge 0$, then $x =$ ________ or $x =$ ________. **To solve an equation containing a single absolute value:** **1.** Isolate the absolute value so that the equation is in the form ________. If $c > 0$, proceed to steps 2 and 3. If $c < 0$, the system has no solution. **2.** Separate the absolute value into two equations: ________ or ________. **3.** Solve both equations.	Solve $\|4x + 1\| - 3 = 2$. **Solution:** $\|4x + 1\| = 5$ Isolate the absolute value. $4x + 1 = 5$ or $4x + 1 = -5$ Use the absolute value property. $4x = 4$ or $4x = -6$ $x = 1$ or $x = -\frac{3}{2}$

Exercises 47–54 Equations and Inequalities

[2.5] *For Exercises 47–54, solve.*

47. $|x| = 4$

48. $|x - 4| = 7$

49. $|2w - 1| = 3$

50. $|5r + 8| = -3$

51. $|q - 4| - 3 = 8$

52. $|5w| - 2 = 13$

53. $2|3x - 4| = 8$

54. $4 - 2|r - 5| = -8$

Definitions/Rules/Procedures	Key Example(s)
To solve an equation in the form $\|ax + b\| = \|cx + d\|$: **1.** Separate the absolute value equation into two equations: $ax + b =$ ________ or $ax + b =$ ________. **2.** Solve both equations.	Solve $\|5x - 7\| = \|3x + 1\|$. **Solution:** Separate into two equations. **Equal** **Opposites** $5x - 7 = 3x + 1$ or $5x - 7 = -(3x + 1)$ $2x - 7 = 1$ or $5x - 7 = -3x - 1$ $2x = 8$ or $8x = 6$ $x = 4$ or $x = \frac{3}{4}$

Exercises 55–58 Equations and Inequalities

[2.5] *For Exercises 55–58, solve.*

55. $|3x - 2| = |x + 2|$

56. $|2x - 1| = |3x + 2|$

57. $|-3x + 1| = |3 - 2x|$

58. $|9 - 4x| = |7 - 2x|$

2.6 Inequalities Involving Absolute Value

Definitions/Rules/Procedures	Key Example(s)
To solve an inequality in the form $\|x\| < a$, where $a > 0$: 1. Rewrite the inequality as a compound inequality ________ and ________ (or use $-a < x < a$). 2. Solve the compound inequality. Similarly, to solve $\|x\| \le a$, we write ________ and ________ (or use $-a \le x \le a$).	For each example, solve the inequality. Then: **a.** Graph the solution set. **b.** Write the solution set using set-builder notation. **c.** Write the solution set using interval notation. $\|2x - 3\| - 1 < 4$ **Solution:** **a.** $\|2x - 3\| < 5$ Isolate the absolute value. $2x - 3 > -5$ and $2x - 3 < 5$ Separate $2x > -2$ and $2x < 8$ using *and*. $x > -1$ and $x < 4$ Solution graph: (number line from −2 to 5, open at −1 and 4) **b.** Set-builder notation. $\{x\|-1 < x < 4\}$ **c.** Interval notation: $(-1, 4)$

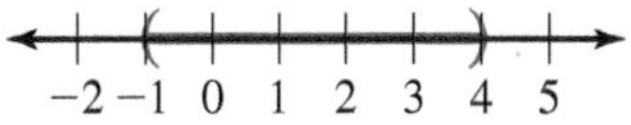

Exercises 59–62 Equations and Inequalities

[2.6] *For Exercises 59–62, solve the inequality. Then:* *a. Graph the solution set.* *b. Write the solution set using set-builder notation.* *c. Write the solution set using interval notation.*

59. $|x| < 5$

60. $|2m + 6| < 4$

61. $|3s - 1| \le -2$

62. $7|m + 3| \le 21$

Definitions/Rules/Procedures	Key Example(s)
To solve an inequality in the form $\|x\| > a$, where $a > 0$: 1. Rewrite the inequality as a compound inequality ________ or ________. 2. Solve the compound inequality.	For each example, solve the inequality. Then: **a.** Graph the solution set. **b.** Write the solution set using set-builder notation. **c.** Write the solution set using interval notation. $\|3x - 5\| - 7 \ge 4$

Definitions/Rules/Procedures	Key Example(s)
Similarly, to solve $\|x\| \geq a$, we write ______ or ______.	**Solution:** **a.** $\|3x - 5\| \geq 11$ Isolate the absolute value. $3x - 5 \leq -11$ or $3x - 5 \geq 11$ Separate using *or*. $3x \leq -6$ or $3x \geq 16$ $x \leq -2$ or $x \geq \frac{16}{3}$ Solution graph: (number line from −3 to 7, closed at −2 shading left, closed at $\frac{16}{3}$ shading right) **b.** Set-builder notation: $\left\{x \mid x \leq -2 \text{ or } x \geq \frac{16}{3}\right\}$ **c.** Interval notation: $(-\infty, -2] \cup \left[\frac{16}{3}, \infty\right)$

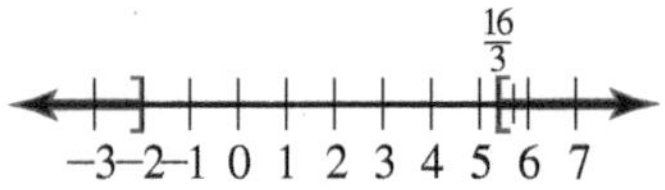

Exercises 63–68 Equations and Inequalities

[2.6] *For Exercises 63–68, solve the inequality. Then:*
a. *Graph the solution set.*
b. *Write the solution set using set-builder notation.*
c. *Write the solution set using interval notation.*

63. $|p| \geq 4$

64. $|x - 3| > 7$

65. $5|b| - 2 > 3$

66. $-2|t - 5| < -10$

67. $5 - 2|2k - 3| \leq -15$

68. $3 - 7|2p + 4| \leq 24$

Chapter 2 Practice Test

For Extra Help

Step-by-step test solutions are found on the Chapter Test Prep Videos available in MyMathLab® *or on* YouTube.

1. Find the intersection and union of the given sets.

$A = \{h, o, m, e\}$

$B = \{h, o, u, s, e\}$

For Exercises 2–7, solve.

2. $2(w + 2) - 3 = 3(w - 1)$

3. $\frac{2}{3}q - 4 = \frac{1}{5}q + 10$

4. $|x + 3| = 5$

5. $3 - |2x - 3| = -6$

6. $|2x + 3| = |x - 5|$

7. $|5x - 4| = -3$

8. Solve for b: $A = \frac{1}{2}h(b + B)$.

For Exercises 9–16, solve. Then: *a. Graph the solution set.*
b. Write the solution set using set-builder notation.
c. Write the solution set using interval notation.

9. $-3 < x + 4 \le 7$

10. $4 < -2x \le 6$

11. $|x + 4| < 9$

12. $2|x - 1| > 4$

13. $3 - 2|x + 4| > -3$

14. $|3y - 2| < -2$

15. $|8t + 4| \ge -12$

16. $2|3x - 4| \le 10$

17. A sheet of metal with an area of 1500 square inches will be formed into a box 15 inches long by 8 inches high. Find the width of the box.

18. A clothing store has a special sale in which items are discounted 30%. If the price after the discount for a dress is \$31.36, what was the original price?

19. Two cars are traveling in opposite directions along the same highway. One car is traveling 65 miles per hour and the other is traveling 70 miles per hour. If they are 54 miles apart, how long until they meet?

20. A mail-order book club offers one bonus book if the total of an order is between \$40 and \$50. Johnson decides to buy the least expensive books, which are \$8 each. What range would the number of books he orders have to be for him to receive a bonus book?

Chapters 1–2 Cumulative Review Exercises

For Exercises 1–3, answer true or false.

[1.1] **1.** Every rational number is an integer.

[1.2] **2.** Subtraction is commutative.

[2.4] **3.** The expressions $x > a$ and $x < -a$ can be written more compactly as $-a < x < a$.

For Exercises 4–6, fill in the blank.

[2.1] **4.** An equation that is true for all replacements of the variable(s) for which it is defined is called a(n) ________.

[1.3] **5.** $\sqrt{-16}$ is not a(n) ________.

[2.3] **6.** If you multiply or divide both sides of an inequality by a negative number, you must ________________.

Exercises 7–20 Expressions

For Exercises 7 and 8, evaluate.

[1.1] **7.** $-\left|-\frac{5}{9}\right|$

[1.3] **8.** $\sqrt[3]{-125}$

For Exercises 9–14, evaluate.

[1.2] **9.** $-16 + 9$

[1.2] **10.** $(-4)(-5)(4)$

[1.3] **11.** -9^2

[1.3] **12.** $14 - 4(-2)^2$

[1.3] **13.** $|7 - 3(8 - 2) \div 6| + [4(2 - 4^2) \div 2]$

[1.3] **14.** $[-3^2 - 2(-6 + 2)] + \sqrt{25 + 144}$

[1.4] **15.** Evaluate $\dfrac{4y - 5x^2}{3y^3 - z^2}$ for $x = -3, y = 2$, and $z = -4$.

[1.4] **16.** Use the distributive property to write $-3(2x^2 - 4x + 3)$ as an equivalent expression.

[1.4] **17.** For what value(s) of x is $\dfrac{x - 5}{x + 4}$ undefined?

[1.4] *For Exercises 18 and 19, simplify by combining like terms.*

18. $4d^2 - 4 + 4d + 4d^2 - 9 - 3d$

19. $4(2y - 5) - 2(3y - 6) + 6y$

[1.4] **20.** Translate to an algebraic expression and simplify: 12 minus the product of 4 and the difference of 5 and a number.

Exercises 21–30 Equations and Inequalities

[2.1] *For Exercises 21–23, solve and check.*

21. $8(2z - 3) - z = 5z - 34$

22. $1.4u - 0.5(44 - 6u) = 66 + 2.4u$

23. $\frac{3x}{4} - \frac{2}{5} = \frac{3x}{10} + \frac{1}{2}$

For Exercises 24–26, solve and then:
a. Write the solution set using set-builder notation.
b. Write the solution set using interval notation.
c. Graph the solution set.

[2.3] **24.** $7u - 10(2u + 3) \geq 22$

[2.6] **25.** $|2x - 5| - 1 \leq 4$

[2.4] **26.** $3 < -2y + 5 \leq 7$

[2.5] **27.** Solve: $|3x - 9| + 6 = 12$

[2.2] *For Exercises 28–30, solve.*

28. Three minus twice the difference of a number and six is five less than three times the number.

29. A company breaks even if the revenue and cost are equal. Suppose the expression $7n - 250$ describes the monthly revenue where n represents the number of units sold. The expression $2n + 1200$ describes the cost of producing those n units. How many units must be sold for the company to break even?

30. A commercial passenger jet leaves Miami, Florida, traveling toward Reno, Nevada, at an average speed of 560 miles per hour. At the same time, a private jet leaves Reno, traveling toward Miami at an average rate of 440 miles per hour. How long will it take until the planes meet if the distance from Miami to Reno is 3000 miles?

CHAPTER 3

Equations and Inequalities in Two Variables; Functions

Chapter Overview

In this chapter, we find solution sets for equations and inequalities that have two variables and graph those solution sets in the *coordinate plane*. More specifically, we consider the following topics:

- Graphing linear equations and inequalities
- Writing equations of lines
- Recognizing functions
- Graphing linear, absolute value and quadratic functions

3.1 Graphing Linear Equations

Objectives

1. Plot points in the coordinate plane.
2. Graph linear equations.

Of Interest

René Descartes was born in 1596 at La Haye, near Tours, France. While serving in the military, Descartes had a series of dreams in which he became aware of a new way of viewing geometry using algebra. In 1637, after some urging by his friends, he reluctantly allowed one work known as the *Method* to be printed. It was in this book that the rectangular coordinate system and analytical geometry was given to the world.

Learning Strategy

If you are a tactile learner, when you plot a point, move your pencil from the origin along the x-axis first and then move up or down to the point location, just as we've done with the arrows.

Answers to Warm-up

1. $x = -6$
2. $y = 4$
3. $x > 2$ or $x < -\dfrac{2}{3}$

Warm-up

For Exercises 1–3, solve.

[2.1] **1.** $0 = \frac{1}{3}x + 2$ [2.1] **2.** $6 - 3y = -6$ [2.6] **3.** $|3x - 2| > 4$

Objective 1 Plot points in the coordinate plane.

In 1619, René Descartes, the French philosopher and mathematician, recognized that positions of points in a plane could be described using two perpendicular number lines, called *axes*. The axes form what is called a *rectangular coordinate system*, or the *Cartesian coordinate system*, named in honor of René Descartes. Usually, we call the horizontal axis the *x-axis* and the vertical axis the *y-axis*.

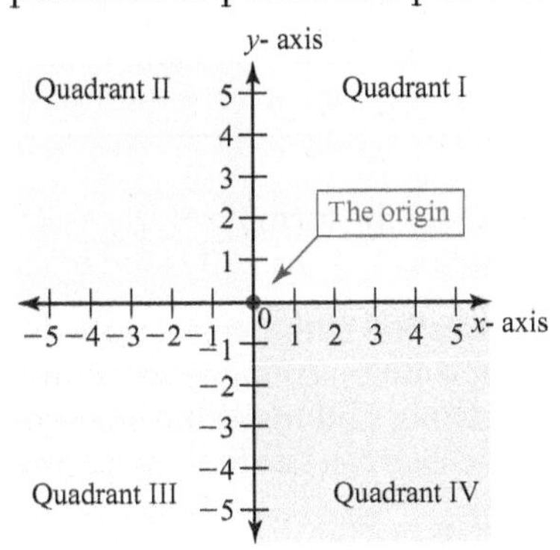

The point where the axes intersect is called the *origin* and has the value 0 for both the x-axis and y-axis. Also, the axes divide the plane into four regions, which we call *quadrants*. The quadrants are numbered using Roman numerals, as shown in the figure to the left. Points on the axes are not in any quadrant.

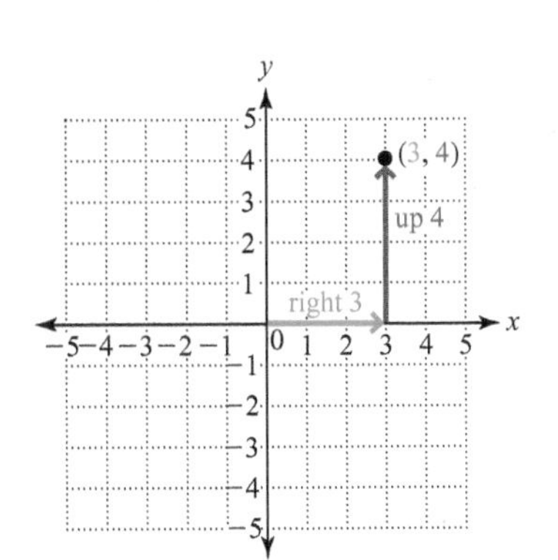

Any point in the plane can be described using one number from each axis written in a specific order. The horizontal axis value is given first and then the vertical axis value. Because the order is specific, we say that the two numbers form an *ordered pair*. Each number in an ordered pair is called a *coordinate* of the ordered pair.

The notation for writing ordered pairs is (horizontal coordinate, vertical coordinate). For example, the ordered pair for the point shown in the coordinate plane to the left is $(3, 4)$ because it is 3 units to the right and 4 units up from the origin. The origin has coordinates $(0, 0)$.

Example 1 Plot the point described by the coordinates and identify the point's quadrant. If the point is on an axis, state which axis.

a. $(-3, -2)$ **b.** $(0, 4)$ **c.** $(2, -4)$

Solution:

Note The ordered pair $(0, 4)$ means the point is 0 units from the origin along the x-axis and 4 units up the y-axis. It is not in a quadrant; it is on the y-axis.

Note The ordered pair $(-3, -2)$ means 3 units to the left of the origin and 2 units down, placing the point in quadrant III.

Note The ordered pair $(2, -4)$ means 2 units to the right of the origin and 4 units down, placing the point in quadrant IV.

Your Turn 1 Plot the point described by the coordinates and identify the point's quadrant. If the point is on an axis, state which axis.

a. $A: (4, 3)$ **b.** $B: (-3, 2)$ **c.** $C: (-5, -2)$ **d.** $D: (2, 0)$

Objective 2 Graph linear equations.

We can graph the solution set of an equation or inequality in the coordinate plane. In this section, we focus on linear equations. Because the coordinate plane has an x- and y-axis, we need to discuss solutions to **linear equations in two variables** such as $x + y = 3$ and $y = 2x - 3$.

Definition Linear equation in two variables: An equation that can be written in the form $Ax + By = C$, where A, B, and C are real numbers and A and B are not both 0.

An equation written in the form $Ax + By = C$, such as $x + y = 3$, is said to be in *standard form*. The equation $y = 2x - 3$ is not in standard form.

A solution for an equation with two variables is an ordered pair of numbers like we plotted in Example 1. A solution contains one number for each variable that can replace the corresponding variables and make the equation true. For example, $(1, 2)$ is a solution for $x + y = 3$, which we can verify by replacing x with 1 and y with 2.

$$x + y = 3$$

$$1 + 2 = 3$$ This is true, so the ordered pair is a solution.

Note that $(1, 2)$ is just one solution out of an infinite number of possible solutions for $x + y = 3$. An infinite number of solutions are possible because every x-value has a corresponding y-value that will add to the x-value to equal 3 and vice versa. We list some other solutions in the following table.

x	y	Ordered Pair
−2	5	$(-2, 5)$
−1	4	$(-1, 4)$
0	3	$(0, 3)$
1	2	$(1, 2)$
2	1	$(2, 1)$
3	0	$(3, 0)$
4	−1	$(4, -1)$
5	−2	$(5, -2)$

Note To find these solutions, we choose a value for x or y, then use the equation to find the corresponding value of the other variable. For example, if we choose x to be 2, then y has to be 1 so that the sum is 3.

Our exploration suggests the following procedure for finding solutions to equations with two variables.

Procedure Finding Solutions to Equations with Two Variables

To find a solution to an equation in two variables:

1. Replace one of the variables with a chosen number (any number).
2. Solve the equation for the other variable.

Because these equations have an infinite number of ordered pairs as solutions, a graph offers a geometric way to represent all of the ordered pairs in the solution set. For example, on the accompanying grid, we plot each of the solutions that we listed for $x + y = 3$.

Answers to Your Turn 1

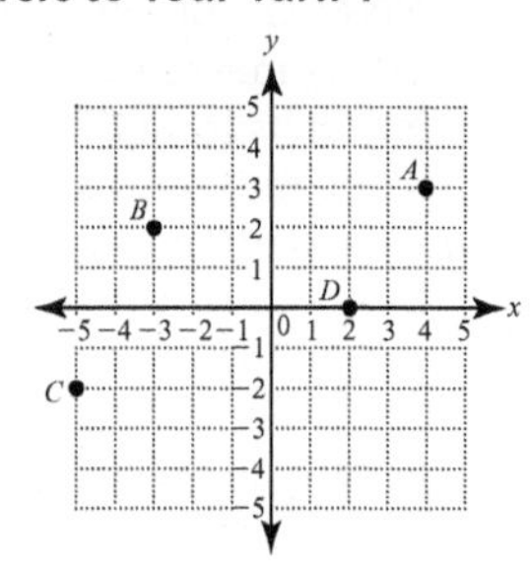

a. A: quadrant I
b. B: quadrant II
c. C: quadrant III
d. D: not in a quadrant; it is on the x-axis.

Note The arrows on either end of the line indicate that the solutions continue indefinitely beyond our grid.

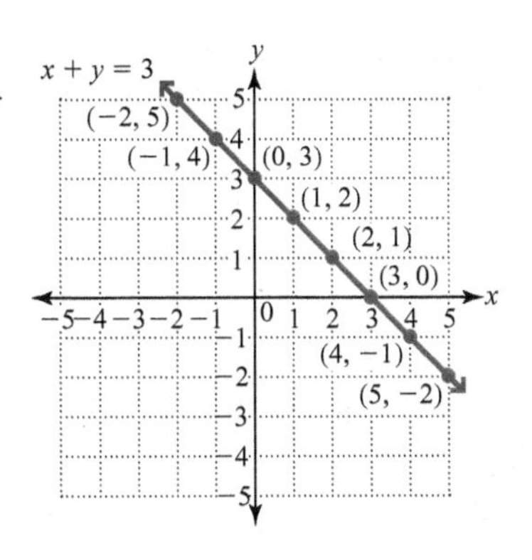

Notice that the ordered pair solutions lie on a straight line. In fact, every ordered pair along that line is a solution for $x + y = 3$; so the line represents the solution set for the equation.

The graph of the solutions of any linear equation is a straight line. Because two points determine a line in a plane, we need at least two ordered pair solutions to draw the graph of a linear equation. It is wise, however, to find three solutions, using the third solution as a check. If we plot all three points and they cannot be connected with a straight line, we know something is wrong.

Procedure Graphing Linear Equations

To graph a linear equation:

1. Find at least two solutions to the equation.
2. Plot the solutions as points in the rectangular coordinate system.
3. Draw a line through the points to form a straight line.

Note The line is drawn with arrowheads on both ends to indicate that it continues indefinitely in both directions, unless otherwise specified.

Example 2 Graph $2x + y = 4$.

Solution: We will find three solutions.

If $x = 0$,

$2(0) + y = 4$

$y = 4$

First solution
$(0, 4)$

If $x = 1$,

$2(1) + y = 4$

$2 + y = 4$

$y = 2$

Second solution
$(1, 2)$

If $x = 2$,

$2(2) + y = 4$

$4 + y = 4$

$y = 0$

Third solution
$(2, 0)$

Choose a value for *x*.

In the equation, replace *x* with the chosen value. Then solve for *y*.

Note We also can choose values for y and solve for x. For example, if $y = -4$,

$$2x + (-4) = 4$$
$$2x = 8$$
$$x = 4$$

Note that the point $(4, -4)$ is also on the graph of $2x + y = 4$.

A table helps organize the solutions. To graph, we plot our solutions and connect the points to form a straight line.

x	y	Ordered Pair
0	4	$(0, 4)$
1	2	$(1, 2)$
2	0	$(2, 0)$

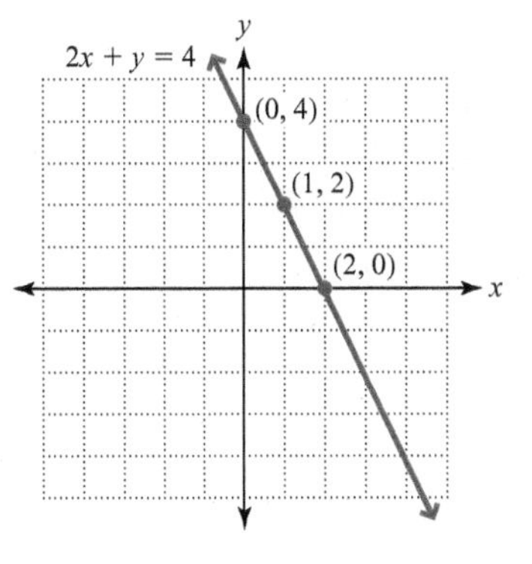

Your Turn 2 Graph $x + 3y = 3$.

Answer to Your Turn 2

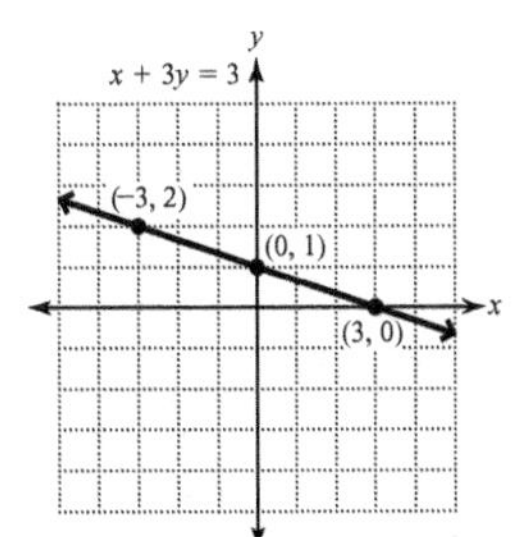

Notice that the graph of $2x + y = 4$ intersects the x-axis at $(2, 0)$ and the y-axis at $(0, 4)$. Points where a graph intersects an axis are called **intercepts.**

Definitions ***x*-intercept:** A point where a graph intersects the x-axis.
***y*-intercept:** A point where a graph intersects the y-axis.

Note that x-intercepts always have the form $(a, 0)$ and y-intercepts always have the form $(0, b)$, where a and b are real numbers.

Procedure Finding the *x*- and *y*-intercepts

To find an x-intercept:
1. Replace y with 0 in the given equation.
2. Solve for x.

To find a y-intercept:
1. Replace x with 0 in the given equation.
2. Solve for y.

Because the resulting equations are usually easier to solve when x or y is replaced with 0, many people find it easy to graph equations by finding the intercepts. This method is best used when the equation is in standard form $(Ax + By = C)$.

VISUAL

Learning Strategy

If you are a visual learner, to find an intercept, cover the term containing the other variable and then solve the remaining portion of the equation mentally. For example, to find the x-intercept of $2x - 3y = -6$, cover $3y$ and then solve for x.

$$2x - 3y = -6$$

Covering $3y$ gives

$$2x = -6$$
$$x = -3.$$

Example 3 Find the x- and y-intercepts and then graph.

a. $2x - 3y = -6$

Solution:

$2x - 3(0) = -6$ For the x-intercept, replace y with 0.
$2x - 0 = -6$ Simplify.
$2x = -6$
$x = -3$ Divide both sides by 2.

x-intercept: $(-3, 0)$

$2(0) - 3y = -6$ For the y-intercept, replace x with 0.
$0 - 3y = -6$ Simplify.
$-3y = -6$
$y = 2$ Divide both sides by -3.

y-intercept: $(0, 2)$

Note Although two points determine a line, it is wise to find a third point as a check. To find a third point, we will let $x = 3$, ▶

$2(3) - 3y = -6$ Replace x with 3.
$6 - 3y = -6$ Simplify.
$-3y = -12$ Subtract 6.
$y = 4$ Divide by -3.

Notice that $(3, 4)$ does indeed lie on the line.

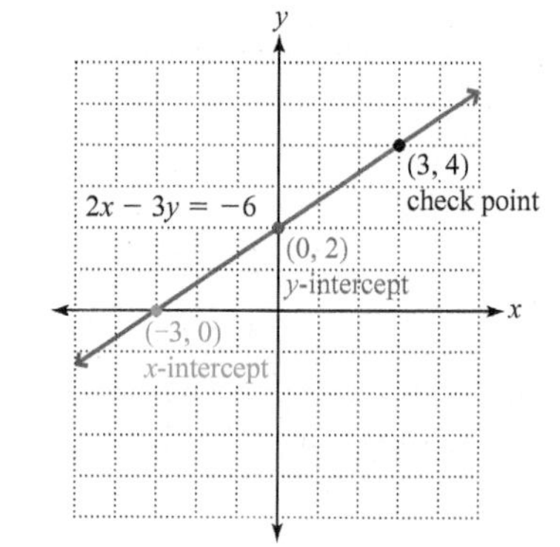

b. $y = -\frac{1}{3}x + 2$

Solution:

$0 = -\frac{1}{3}x + 2$ For the x-intercept, replace y with 0.
$-2 = -\frac{1}{3}x$ Subtract 2 from both sides.
$6 = x$ Multiply both sides by -3.

x-intercept: $(6, 0)$

$y = -\frac{1}{3}(0) + 2$ For the y-intercept, replace x with 0.
$y = 2$ Simplify.

y-intercept: $(0, 2)$

▲

Note In the equation $y = -\frac{1}{3}x + 2$, we can see that 2 is the y-coordinate of the y-intercept because replacing x with zero eliminates the x term, leaving $y = 2$.

Note We will find a third point as a check. To find a third point, let $x = 3$.

$y = -\frac{1}{3}(3) + 2$ Replace x with 3.

$y = 1$ Simplify.

Notice that $(3, 1)$ is on the line.

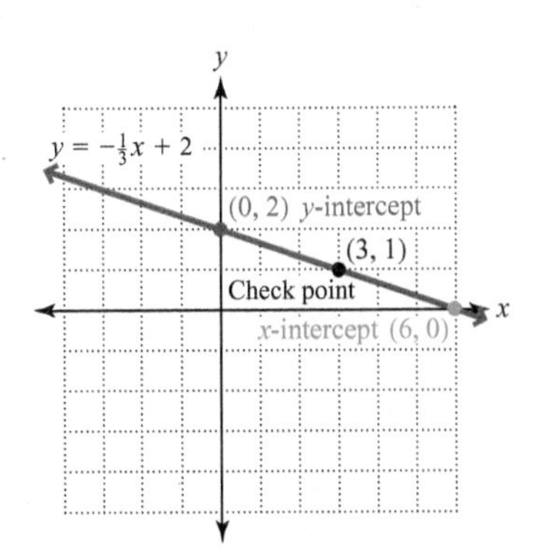

c. $y = 2x$

Solution: $0 = 2x$ For the x-intercept, replace y with 0. $\qquad y = 2(0)$ For the y-intercept, replace x with 0.

$0 = x$ Divide both sides by 2. $\qquad y = 0$ Simplify.

Note The origin is the only point in the coordinate plane that can be both the x- and y-intercept.

Because the x- and y-intercepts are the same point, $(0, 0)$, we need at least one more point to be able to draw the graph. We will find another point by choosing a value for x and solving for y.

Let $x = 1$.

$y = 2(1)$ Replace x with 1.

$y = 2$ Simplify.

Second solution: $(1, 2)$

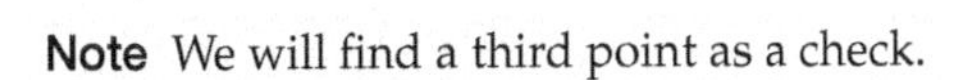

Note We will find a third point as a check.

Let $x = 2$.

$y = 2(2)$ Replace x with 2.

$y = 4$ Simplify.

Notice that $(2, 4)$ is on the line.

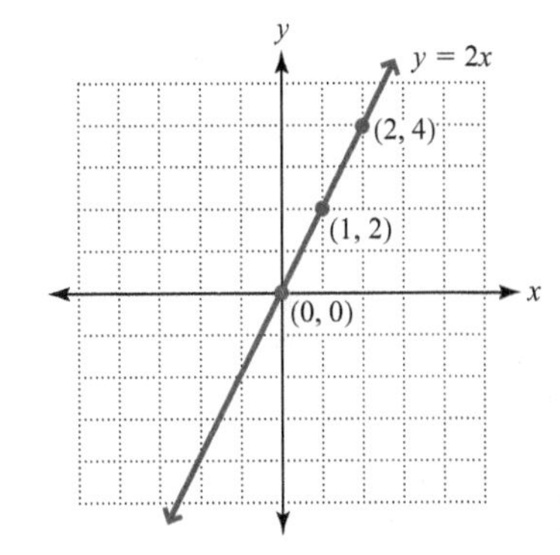

The equations $y = -\frac{1}{3}x + 2$ and $y = 2x$ from Example 3 are in the form $y = mx + b$. A nice feature of equations in the form $y = mx + b$ is that b is the y-coordinate of the y-intercept. If we replace x with 0 in $y = mx + b$, we have $y = m(0) + b = b$; so the y-intercept is $(0, b)$. We will learn more about equations of the form $y = mx + b$ in Section 3.2.

Rule The y-intercept of $y = mx + b$

Given an equation in the form $y = mx + b$, the coordinates of the y-intercept are $(0, b)$.

Your Turn 3 Find the x- and y-intercepts and then graph.

a. $x + 3y = 3$ **b.** $y = 2x - 1$ **c.** $y = -3x$

We have graphed linear equations in two variables. Now let's consider equations with one variable, such as $y = 3$ and $x = 4$, in which the variable is equal to a constant.

Answers to Your Turn 3

a. x-intercept: $(3, 0)$; y-intercept: $(0, 1)$

b. x-intercept: $\left(\frac{1}{2}, 0\right)$; y-intercept: $(0, -1)$

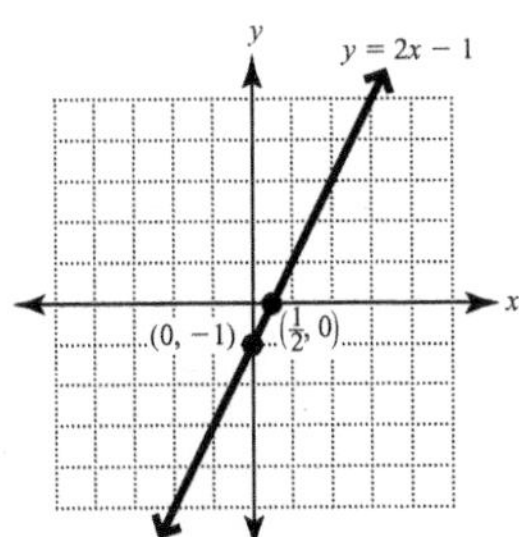

c. x- and y-intercepts: $(0, 0)$

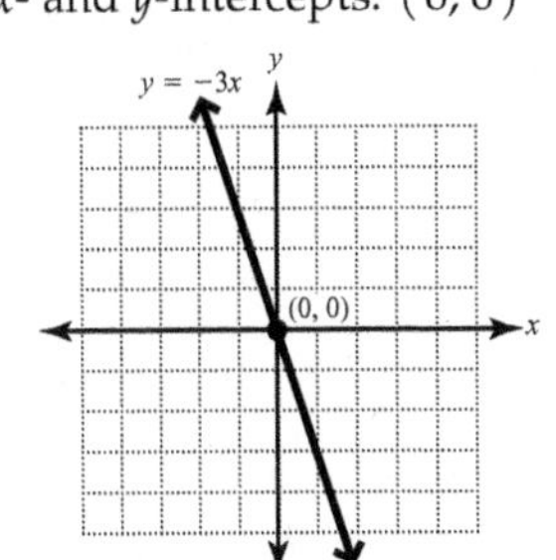

Example 4 Graph.

a. $y = 3$

Solution: To find solutions, we can think of $y = 3$ as $0x + y = 3$. Because the coefficient of x is 0, y is always 3 no matter what we choose for x.

	x	y	Ordered Pair
If we let x equal 0, then y equals 3.	0	3	$(0, 3)$
If we let x equal 2, then y is 3.	2	3	$(2, 3)$
If we let x equal 4, then y is still 3.	4	3	$(4, 3)$

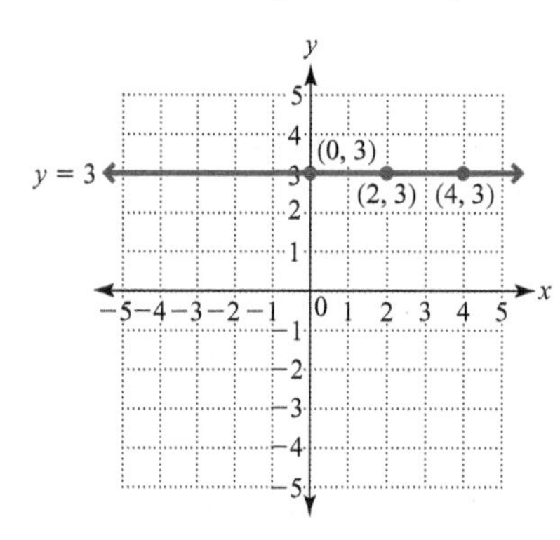

Note The graph of $y = 3$ is a horizontal line parallel to the x-axis. It has no x-intercept, and its y-intercept is $(0, 3)$.

b. $x = -4$

Solution: The equation $x = -4$ indicates that x is equal to a constant, -4. We can write $x = -4$ as $x + 0y = -4$ and complete a table of solutions.

	x	y	Ordered Pair
If we let y equal 0, then x equals -4.	-4	0	$(-4, 0)$
If we let y equal 1, then x is -4.	-4	1	$(-4, 1)$
If we let y equal 2, then x is still -4.	-4	2	$(-4, 2)$

Note The graph of $x = -4$ is a vertical line parallel to the y-axis. It has no y-intercept, and its x-intercept is $(-4, 0)$.

Rules Horizontal Lines and Vertical Lines

The graph of $y = c$, where c is a real-number constant, is a *horizontal line* parallel to the x-axis with a y-intercept at $(0, c)$.

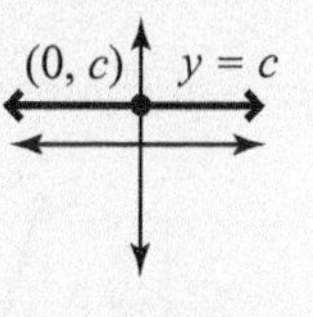

The graph of $x = c$, where c is a real-number constant, is a *vertical line* parallel to the y-axis with an x-intercept at $(c, 0)$.

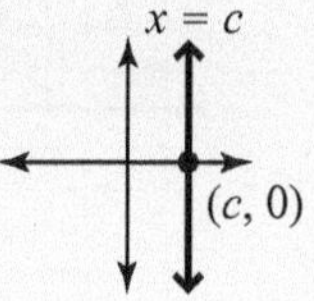

Your Turn 4 Graph.

a. $y = -2$ **b.** $x = 5$

Answers to Your Turn 4

a.

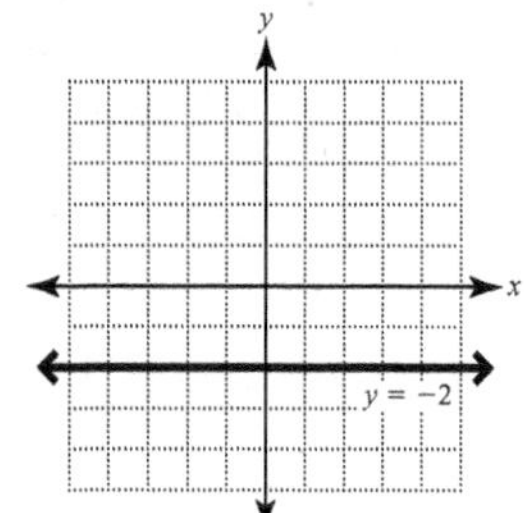

b.

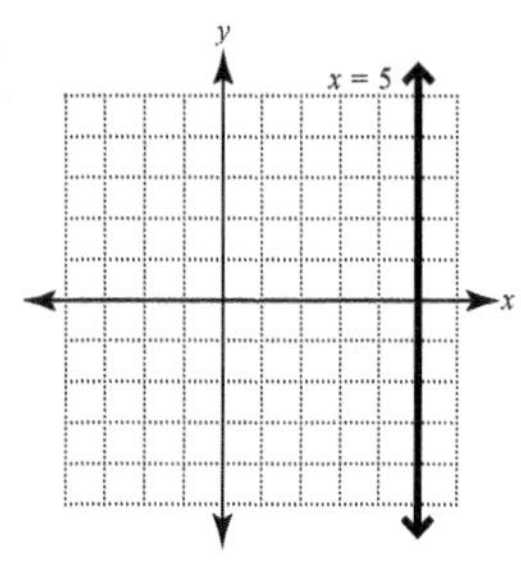

3.1 Exercises

For Extra Help MyMathLab®

Note: Exercises marked with a ★ represent challenging exercises.

Objective 1

Prep Exercise 1 Describe how to locate the point described by $(-4, 3)$ in the rectangular coordinate system.

Prep Exercise 2 What signs would the coordinates of a point in quadrant I have? quadrant II? quadrant III? quadrant IV?

For Exercises 1 and 2, write the coordinates for each point. See Objective 1.

1.

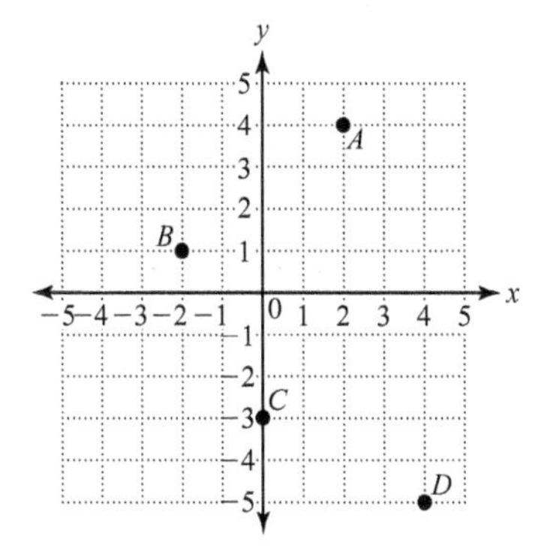

2.

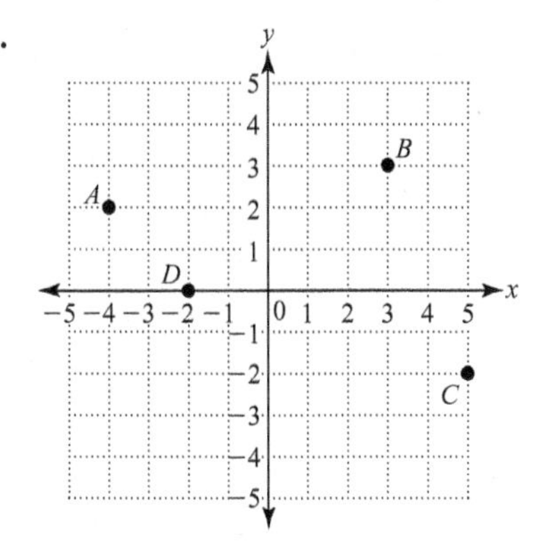

For Exercises 3–6, plot and label the points indicated by the coordinate pairs and give the quadrant in which the point is located. If the point is on an axis, state which axis. See Example 1.

3. a. $(4, 1)$ **b.** $(3, -1)$ **c.** $(-5, -2)$ **d.** $(0, 2)$

4. a. $(-1, 4)$ **b.** $(2, 5)$ **c.** $(-2, 0)$ **d.** $(4, -1)$

5. a. $(-3, 0)$ **b.** $(5, -3)$ **c.** $(-2, -4)$ **d.** $(1, 1)$

6. a. $(2, -3)$ **b.** $(0, -4)$ **c.** $(5, -1)$ **d.** $(-3, 1)$

7. The ordered pairs $(-4, 2)$, $(-1, 2)$, $(-1, 4)$, $(2, 4)$, $(2, -3)$, and $(-4, -3)$ form the vertices of a figure.
a. Plot the points and connect them to form the figure.
b. Find the perimeter of the figure.

c. Find the area of the figure.

8. The ordered pairs $(-5, 2), (-2, 2), (-2, 5), (1, 5), (1, 3), (4, 3), (4, -3)$, and $(-5, -3)$ form the vertices of a figure.
a. Plot the points and connect them to form the figure.
b. Find the perimeter of the figure.

c. Find the area of the figure.

Objective 2

Prep Exercise 3 Explain how to determine whether an ordered pair is a solution for a given equation.

For Exercises 9–16, determine whether the ordered pair is a solution for the equation.

9. $(4, 2); 2x + 3y = 14$

10. $(-1, -2); 3x - 2y = 1$

11. $\left(-\frac{1}{3}, \frac{2}{3}\right); x - y = -1$

12. $\left(\frac{3}{5}, -\frac{1}{4}\right); 2x - y = 7$

13. $(5, -1); \frac{2}{3}x - y = 8$

14. $(-2, -4); -y = \frac{1}{2}x + 3$

15. $(2.5, -2.1); 5.2x = 6.7 - 3y$

16. $(-2.1, -0.2); -3.4 - 2x = 4y$

Prep Exercise 4 Explain how to find a solution for a given equation with two unknowns.

Prep Exercise 5 What does the graph of an equation represent?

Prep Exercise 6 What is the minimum number of points needed to draw a straight line? Explain.

Prep Exercise 7 After finding at least two solutions, explain the process of graphing a linear equation.

For Exercises 17–24, graph by finding three ordered pairs that solve the equation. Note that these equations are in standard form. See Examples 2 and 3.

17. $x - y = 4$

18. $x - y = 5$

19. $2x + y = 6$

20. $3x + y = 9$

21. $x - 2y = 10$

22. $x + 4y = 12$

23. $2x - 5y = 10$

24. $3x + 4y = 12$

For Exercises 25–48, graph by finding three ordered pairs that solve the equation. Note that these equations are not in standard form. See Examples 2–4.

25. $y = x$

26. $y = 2x$

27. $y = 3x$

28. $y = 5x$

29. $y = -x$

30. $y = -2x$

31. $y = -3x$

32. $y = -5x$

33. $y = \frac{1}{3}x$

34. $y = \frac{1}{2}x$

35. $y = -\frac{2}{3}x$

36. $y = -\frac{2}{5}x$

37. $y = 3x + 4$

38. $y = 2x - 1$

39. $y = -2x + 1$

40. $y = -5x - 3$

41. $y = \frac{3}{4}x + 2$

42. $y = \frac{2}{3}x + 1$

43. $y = -\frac{1}{3}x - 1$

44. $y = -\frac{3}{5}x - 4$

45. $y = 4$

46. $y = -5$

47. $x = 2$

48. $x = -3$

Prep Exercise 8 Explain how to find x- and y-intercepts.

For Exercises 49–60, find the x- and y-intercepts and then graph. See Example 3.

49. $2x + y = 4$

50. $x + 3y = 6$

51. $3x + 2y = 6$

52. $3x + 4y = 12$

53. $2x - 3y = -6$

54. $x - 2y = -4$

55. $y = x - 3$

56. $y = 2x + 1$

57. $y = -3x$

58. $y = 2x$

59. $y = -\frac{3}{4}x + 1$

60. $y = -\frac{2}{3}x - 2$

61. Compare the graphs of $y = x, y = 2x, y = 3x$, and $y = 5x$ from Exercises 25–28. For an equation in the form $y = mx$, what effect does increasing m seem to have on the graph?

62. Compare the graphs of and $y = x$ and $y = -x$ from Exercises 25 and 29. Then compare the graphs of $y = 2x$ and $y = -2x$ from Exercises 26 and 30. For an equation in the form $y = mx$, what can you conclude about the graph when m is positive versus when m is negative?

63. Compare the graphs of $y = 3x$ and $y = 3x + 4$ from Exercises 27 and 37. Then compare the graphs of $y = -2x$ and $y = -2x + 1$ from Exercises 30 and 39. For an equation in the form $y = mx + b$, what effect does adding b (where $b > 0$) to mx seem to have on the graph of $y = mx + b$?

64. Compare the graphs of $y = 2x$ and $y = 2x - 1$ from Exercises 26 and 38. Then compare the graphs of $y = -5x$ and $y = -5x - 3$ from Exercises 32 and 40. For an equation in the form $y = mx - b$, what effect does subtracting b (where $b > 0$) from mx seem to have on the graph of $y = mx - b$?

For Exercises 65 and 66, a polygon has been moved from its original position, shown with the dashed lines, to a new position, shown with solid lines.
a. List the coordinates of each vertex of the original position.
b. List the coordinates of each vertex of the new position.
c. Write a rule that describes in mathematical terms how to move any point on the polygon from its original position to any point on the polygon in its new position.

65.

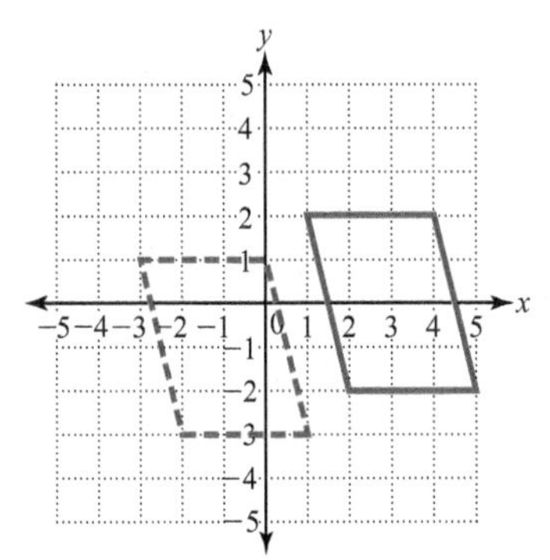

★66.

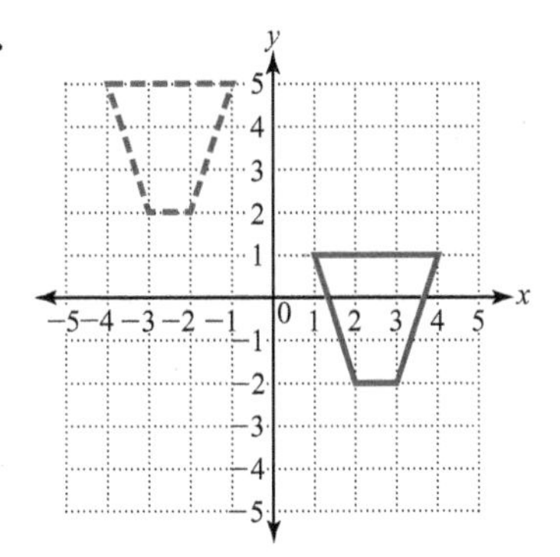

For Exercises 67–72, solve. In each situation, $n \geq 0$ and $c \geq 0$.

67. An academic tutor charges \$60 per month plus \$5 per hour of tutoring. The equation $c = 5n + 60$ describes the total he would charge for tutoring a month, where n represents the number of hours of tutoring and c is the total cost.
a. Find the total cost if the tutor works 7 hours.
b. If the tutor's total charges are \$90, for how many hours of labor is the tutor charging?
c. Graph the equation with n along the horizontal axis and c along the vertical axis.

68. A plumber charges \$120 plus \$60 per hour of labor. The equation $c = 60n + 120$ describes the total he would charge for a service visit, where n represents the number of hours of labor and c is the total cost.
a. Find the total cost if labor is 2.5 hours.

b. If the plumber's total charges are \$240, for how many hours of labor is the plumber charging?

c. Graph the equation with n along the horizontal axis and c along the vertical axis.

69. A wireless company charges \$40.00 per month for 450 anytime minutes plus \$0.05 per minute above 450. The equation $c = 0.05n + 40$ describes the total, c, the company would charge for a monthly bill, where n represents the number of minutes used above 450.
a. Find the total cost if a customer uses 600 anytime minutes.

b. If the customer's total charges are \$41.75, for how many additional minutes was the customer charged?

c. Graph the equation with n along the horizontal axis and c along the vertical axis.

70. A stock begins a ten-day decline in price at \$47 per share. The equation $c = -n + 47$ describes the total for which the stockbroker could sell the stock, where n represents the number of days into the decline and c is the total cost.

a. Find the total cost if the stock is at day 4.

b. If the stock is worth \$39 per share, how many days into the decline is the price?

c. Graph the equation with n along the horizontal axis and c along the vertical axis.

71. A copy center charges \$5.00 for the first 200 copies plus \$0.03 per copy after that. The equation $c = 0.03n + 5$ describes the total charges, c, for copying, where n represents the number of copies after the first 200 copies.

a. Find the total cost for 300 copies.

b. If a customer is charged a total of \$11, how many copies did the customer request?

c. Graph the equation with n along the horizontal axis and c along the vertical axis.

72. A mechanic charges \$100 plus \$50 per hour of labor. The equation $c = 50n + 100$ describes the total she would charge for a service visit, where n represents the number of hours of labor and c is the total cost.

a. Find the total cost if labor is 1.5 hours.

b. If the mechanic's total charges are \$275, for how many hours of labor is the mechanic charging?

c. Graph the equation with n along the horizontal axis and c along the vertical axis.

Review Exercises

Exercises 1–4 Expressions

[1.2] **1.** What property is illustrated by $3x + 5 - 6x + 9 = 5 + 3x - 6x + 9$?

[1.3] **2.** Simplify: $\dfrac{3(5 - 8) + 21}{8 - 3(6)}$

[1.4] *For Exercises 3 and 4, evaluate the expression* $\dfrac{y_2 - y_1}{x_2 - x_1}$ *using the given values.*

3. $x_1 = 4, x_2 = 1, y_1 = -8$, and $y_2 = -2$

4. $x_1 = -10, x_2 = 2, y_1 = 13$, and $y_2 = 5$

Exercises 5 and 6 Equations and Inequalities

[2.1] **5.** Solve: $12 - 3(n - 2) = 6n - (2n + 1)$

[2.3] **6.** Solve $7x - 5 > 4x + 13$ and write the solution set in set-builder notation and in interval notation. Then graph the solution set.

3.2 The Slope of a Line

Objectives

1 Compare lines with different slopes.
2 Graph equations in slope–intercept form.
3 Find the slope of a line given two points on the line.

Warm-up

[2.1] 1. Solve $5x - 2y = 8$ for y

[1.4] 2. Evaluate $\frac{a-b}{c-d}$ for $a = 3, b = 7, c = 5, d = -1$

[3.1] 3. Find the x- and y-intercepts of the graph of $4x - 5y = 20$.

In Section 3.1, we stated that the standard form for a linear equation in two variables is $Ax + By = C$. We also graphed equations in the form $y = mx + b$, which is not in standard form, and discovered that their y-intercept is $(0, b)$. In this section, we further explore equations in the form $y = mx + b$ and consider how the coefficient m affects the graph.

Objective 1 Compare lines with different slopes.

First, consider graphs of equations in which b is 0. If $b = 0$, the equation $y = mx + b$ becomes $y = mx$ and the graph has the origin $(0, 0)$ as its x- and y-intercepts.

Example 1 Graph on the same grid.

a. $y = x, y = 3x$, and $y = \frac{1}{3}x$

Solution: In the following table, the same choice for x has been substituted into each equation and the corresponding y-coordinate has been found.

x	y = x	y = 3x	$y = \frac{1}{3}x$
0	0	0	0
1	1	3	$\frac{1}{3}$
2	2	6	$\frac{2}{3}$

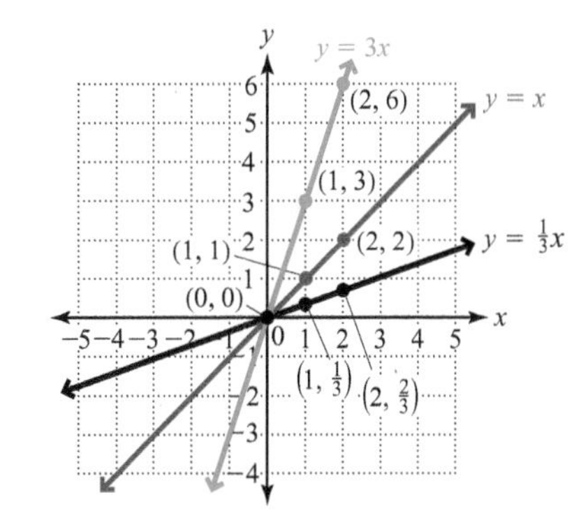

Connection When an equation is written with y isolated, like those in Example 1, we can think of x-values as inputs and the corresponding y-values as outputs. In Example 1, we chose x-input values of 0, 1, and 2 and used the equations to find the corresponding y-output values. This is the fundamental way of thinking about a mathematical concept called a *function*, which we will discuss in Section 3.5.

b. $y = -x, y = -2x$, and $y = -\frac{1}{2}x$

Solution:

x	y = −x	y = −2x	$y = -\frac{1}{2}x$
0	0	0	0
1	−1	−2	$-\frac{1}{2}$
2	−2	−4	−1

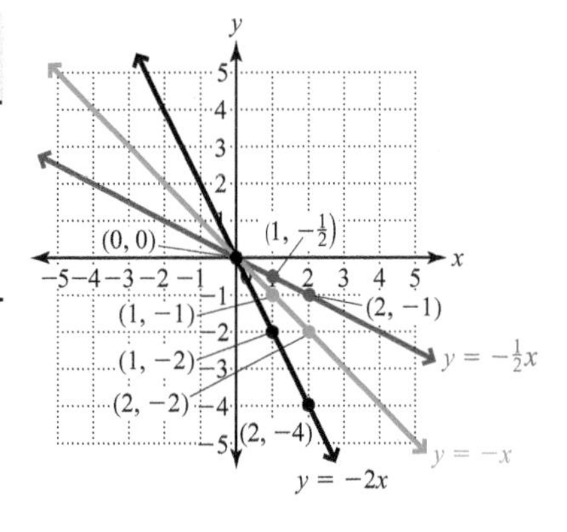

Note In math, the convention is to visually scan a graph from left to right just as we would read a sentence. ▶

Answers to Warm-up

1. $y = \frac{5}{2}x - 4$ **2.** $-\frac{2}{3}$ **3.** $(5, 0), (0, -4)$

Of Interest

No one knows for sure why the letter m is used to represent slope. One belief is that it comes from the French word *monter* which means "to climb."

Example 1 suggests that for equations of the form $y = mx$, the coefficient m affects the incline of the line. Consequently, m is called the *slope* of the line.

For $y = x$, the slope is 1 $(m = 1)$. For $y = -x$, the slope is -1 $(m = -1)$.

For $y = 3x$, the slope is 3 $(m = 3)$. For $y = -2x$, the slope is -2 $(m = -2)$.

For $y = \frac{1}{3}x$, the slope is $\frac{1}{3}$ $\left(m = \frac{1}{3}\right)$. For $y = -\frac{1}{2}x$, the slope is $-\frac{1}{2}$ $\left(m = -\frac{1}{2}\right)$.

Notice that lines with positive slope incline upward from left to right and lines with negative slope incline downward from left to right. Also notice that regardless of whether the incline is upward or downward, the greater the absolute value of the slope, the steeper the incline. We summarize our discoveries about slope in Objective 2.

Your Turn 1

a. Which has a steeper graph, $y = 4x$, or $y = 2x$? Why?

b. Does the graph of $y = -\frac{1}{3}x$ incline upward or downward from left to right? Why?

Objective 2 Graph equations in slope–intercept form.

Note A precise formula for the slope of a line is given in Objective 3.

Let's explore more specifically how slope determines the incline of a graph. Notice that if we isolate m in $y = mx$, we get $m = \frac{y}{x}$, which suggests that **slope** is a ratio of the amount of vertical change (y) to the amount of horizontal change (x).

Definition Slope: The ratio of the vertical change (change in y) to the horizontal change (change in x) between any two points on a line.

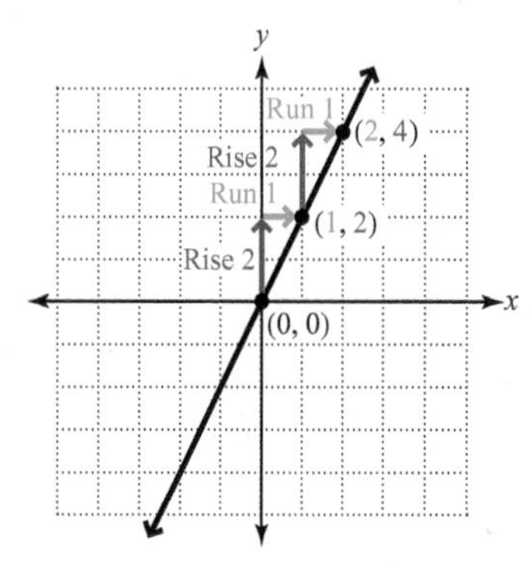

For example, the slope of $y = 2x$ is 2, which can be written as $\frac{2}{1}$. This slope value means that rising vertically 2 units and then running horizontally 1 unit to the right from any point on the line ends up at another point on the same line. For example, rising 2 units and running 1 unit from $(0, 0)$ gives us $(1, 2)$. Similarly, rising 2 units and running 1 unit from $(1, 2)$ gives us $(2, 4)$.

This discovery about slope suggests another approach to graphing. In an equation of the form $y = mx + b$, the slope is m and the y-intercept is $(0, b)$. We can use the y-intercept as a starting point and then find other points on the line using the slope.

Example 2 For the equation $y = -\frac{2}{3}x + 1$, determine the slope and the y-intercept. Then graph the equation.

Solution: The slope is $-\frac{2}{3}$, and the y-intercept is $(0, 1)$. To graph the line using the slope, we plot the y-intercept and then rise -2 (move down two) and run 3 (move right three), arriving at $(3, -1)$, or rise 2 and run -3, arriving at $(-3, 3)$.

Answers to Your Turn 1

a. The graph of $y = 4x$ is steeper than $y = 2x$ because a slope of 4 is greater than a slope of 2.

b. The graph's incline is downward from left to right because the slope is negative.

Note A positive rise means go up, and a negative rise means go down. A positive run means go right, and a negative run means go left.

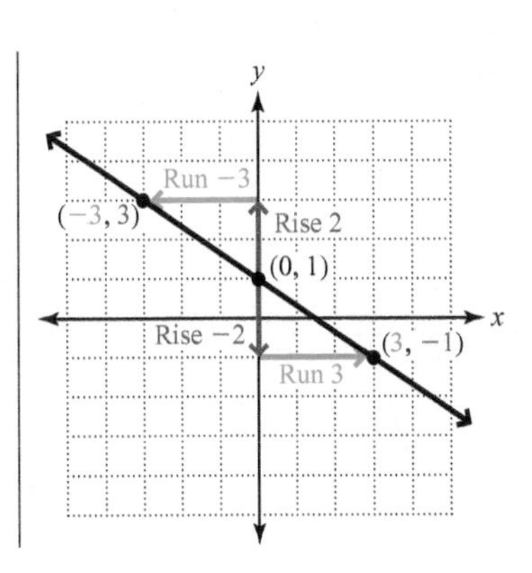

The points we found using the slope can be verified in the equation. We will show the check for $(-3, 3)$ and leave the check for $(3, -1)$ to the reader.

$$y = -\frac{2}{3}x + 1$$

$$3 \stackrel{?}{=} -\frac{2}{3}(-3) + 1$$

$$3 \stackrel{?}{=} 2 + 1$$

$$3 = 3 \qquad \text{This checks.}$$

Your Turn 2 For the equation $y = \frac{2}{5}x - 2$, determine the slope and the y-intercept. Then graph the equation.

The equations $y = -\frac{2}{3}x + 1$ and $y = \frac{2}{5}x - 2$ have the form $y = mx + b$, where m is the slope and b is the y-coordinate of the y-intercept. Because we can easily determine the slope and y-intercept of an equation in the form $y = mx + b$, we call it *slope–intercept form.*

Rules Graphs of Equations in Slope–Intercept Form

The graph of an equation in the form $y = mx + b$ (slope–intercept form) is a line with slope m and y-intercept $(0, b)$. The following rules indicate how m affects the graph:

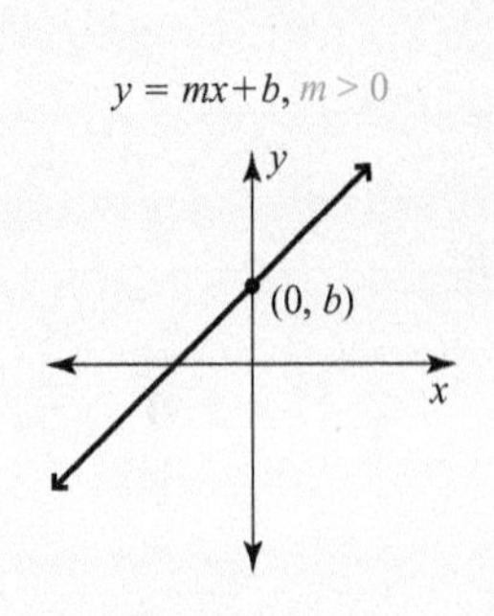

If $m > 0$, the line slants *upward* from left to right.

If $m < 0$, the line slants *downward* from left to right.

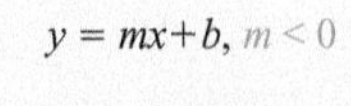

The greater the absolute value of m, the steeper the line.

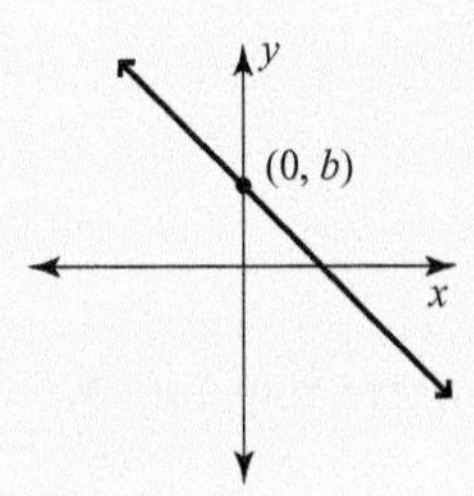

If an equation is not in slope–intercept form and we need to determine the slope and y-intercept, we can write the equation in slope–intercept form by solving for y.

Answer to Your Turn 2

$m = \frac{2}{5}$; y-intercept: $(0, -2)$

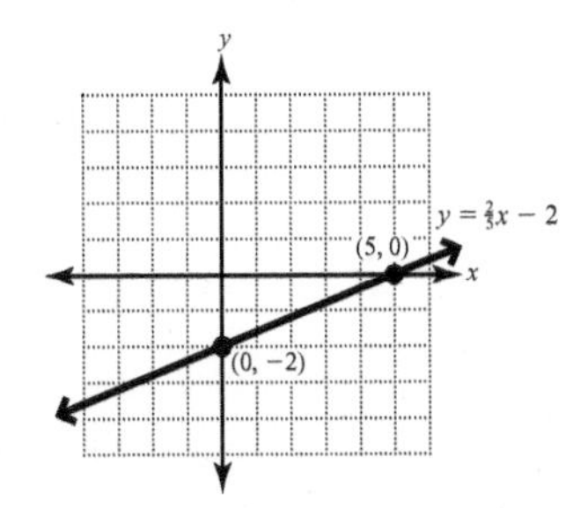

Example 3 For the equation $5x - 2y = 8$, determine the slope and the y-intercept. Then graph the equation.

Solution: Write the equation in slope–intercept form by isolating y.

$$-2y = -5x + 8 \qquad \text{Subtract } 5x \text{ from both sides.}$$

$$y = \frac{-5}{-2}x + \frac{8}{-2} \qquad \text{Divide both sides by } -2 \text{ to isolate } y.$$

$$y = \frac{5}{2}x - 4 \qquad \text{Simplify.}$$

The slope is $\frac{5}{2}$, and the y-intercept is $(0, -4)$. To graph the line, we begin at $(0, -4)$ and then rise 5 and run 2, which gives the point $(2, 1)$. Rising 5 and running 2 from $(2, 1)$ gives the point $(4, 6)$.

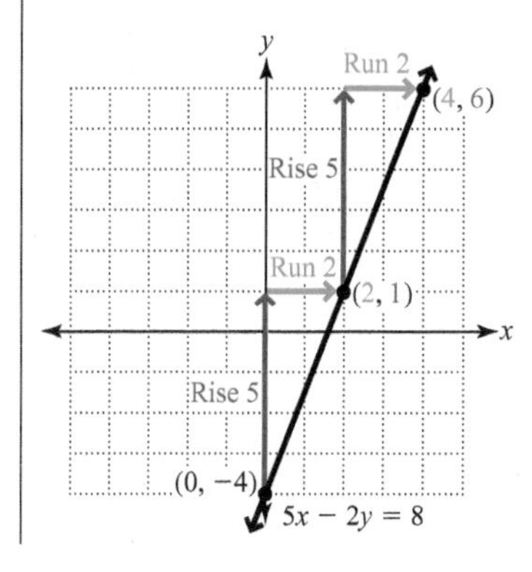

We can check the ordered pairs in the equation.

For (2, 1):	For (4, 6):
$5(2) - 2(1) \stackrel{?}{=} 8$	$5(4) - 2(6) \stackrel{?}{=} 8$
$10 - 2 \stackrel{?}{=} 8$	$20 - 12 \stackrel{?}{=} 8$
$8 = 8$	$8 = 8$

Both ordered pairs check.

Your Turn 3 For the equation $3x + 4y = 12$, determine the slope and the y-intercept. Then graph the equation.

Objective 3 Find the slope of a line given two points on the line.

Given two points on a line, we can determine the slope of the line. Consider the points $(4, 6)$ and $(2, 1)$ on the graph of $5x - 2y = 8$ from Example 3. Remember that we found the slope of this line to be $\frac{5}{2}$.

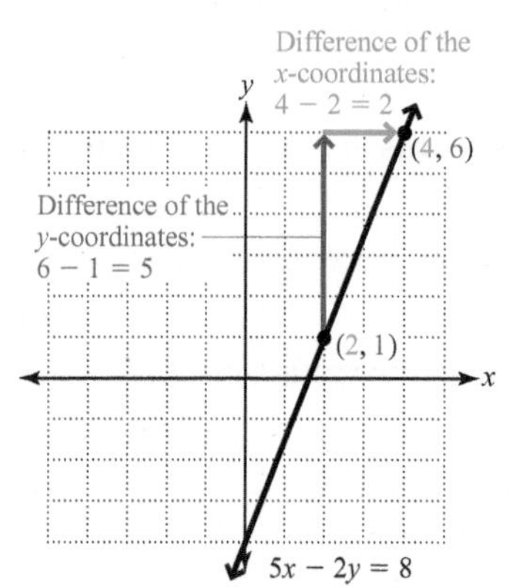

Notice that the amount of rise is equal to the difference in the y-coordinates of the two points and the amount of run is the difference of the x-coordinates.

$$m = \frac{6 - 1}{4 - 2} = \frac{5}{2}$$

In general, we can write a formula for slope.

Answer to Your Turn 3

$m = -\frac{3}{4}$; y-intercept: $(0, 3)$

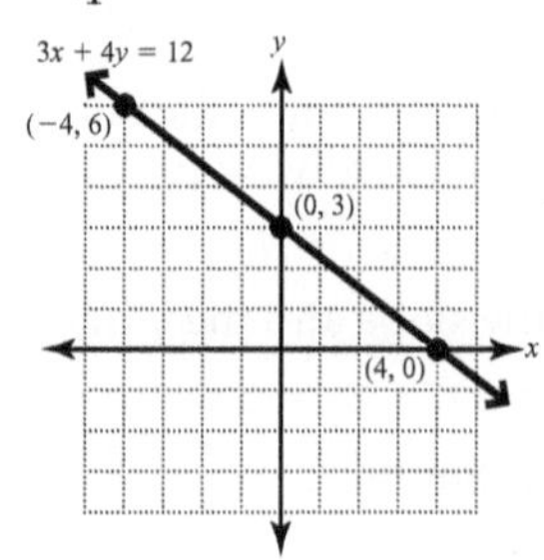

Rule The Slope Formula

Given two points (x_1, y_1) and (x_2, y_2), where $x_2 \neq x_1$, the slope of the line connecting the two points is given by the formula $m = \frac{y_2 - y_1}{x_2 - x_1}$.

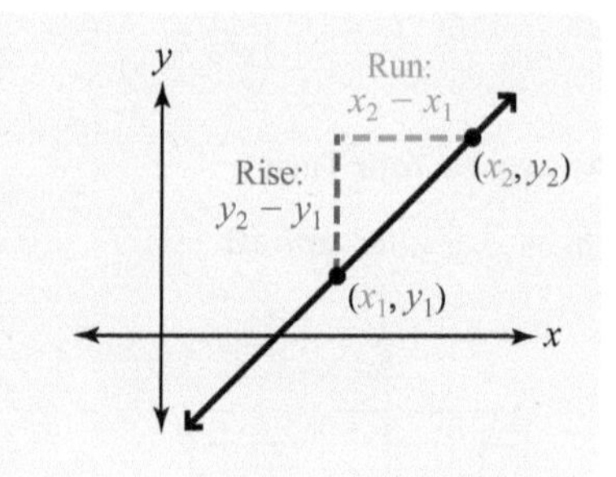

Example 4 Find the slope of the line connecting $(-1, 7)$ and $(5, 3)$.

Solution: Using $m = \dfrac{y_2 - y_1}{x_2 - x_1}$, replace the variables with their corresponding values and then simplify.

Note Some people find it helpful to label the coordinates first.

$(-1, 7)$ $(5, 3)$
(x_1, y_1) (x_2, y_2)

$$m = \frac{3 - 7}{5 - (-1)} = \frac{-4}{6} = -\frac{2}{3}$$

It doesn't matter which ordered pair is (x_1, y_1) and which is (x_2, y_2). If we let $(5, 3)$ be (x_1, y_1) and $(-1, 7)$ be (x_2, y_2), we get the same slope.

$$m = \frac{7 - 3}{-1 - 5} = \frac{4}{-6} = -\frac{2}{3}$$

Your Turn 4 Find the slope of the line connecting the pair of points.

a. $(2, 5)$ and $(6, 1)$ **b.** $(3, -2)$ and $(-4, 3)$

Example 5 Graph the line connecting the given points and find its slope.

a. $(-2, -3)$ and $(4, -3)$

Solution Because the y-coordinates are the same, the graph is a horizontal line.

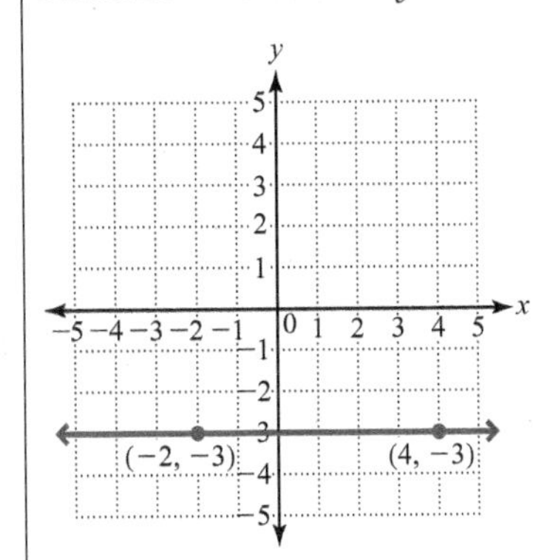

Calculating the slope:

$$m = \frac{-3 - (-3)}{4 - (-2)} = \frac{0}{6} = 0$$

Note The equation of this line is $y = -3$.

b. $(3, 4)$ and $(3, -2)$

Solution Because the x-coordinates are the same, the graph is a vertical line.

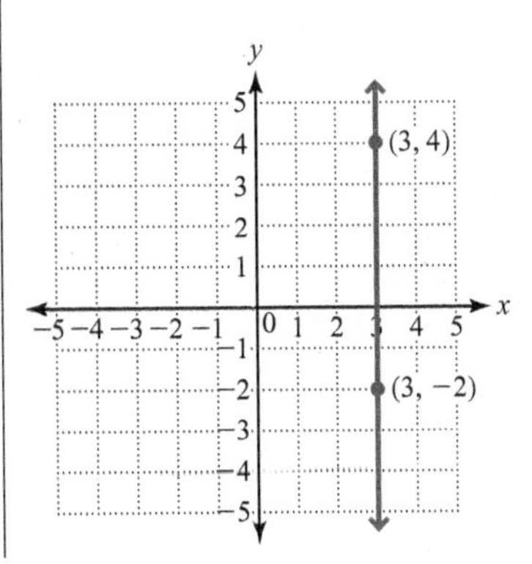

Calculating the slope:

$$m = \frac{4 - (-2)}{3 - 3} = \frac{6}{0}, \text{ which is undefined}$$

Note The equation of this line is $x = 3$.

We can summarize what we learned in Example 5 with the following rules.

Rules Slopes of Horizontal and Vertical Lines

Two points with different x-coordinates and the same y-coordinates, (x_1, c) and (x_2, c), form a horizontal line with slope 0 and equation $y = c$.

Two points with the same x-coordinates and different y-coordinates, (c, y_1) and (c, y_2) form a vertical line with undefined slope and equation $x = c$.

Answers to Your Turn 4

a. -1 **b.** $-\dfrac{5}{7}$

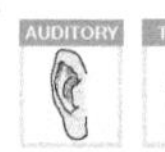 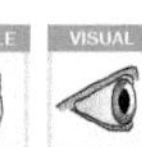

Learning Strategy

To best do your homework, work practice problems similar to the homework before attempting to do it and keep working on a problem even if it is difficult. Make yourself think about what you know and how to apply it to the problem.

—Stella P.

Your Turn 5 Graph the line connecting the given points and find its slope.

a. $(4, -1)$ and $(-3, -1)$ **b.** $(-2, 2)$ and $(-2, -3)$

Example 6 The following graph shows the number of product units a company produced and sold each year from 2000 to 2010. An analyst determines that the red line reasonably describes the trend shown in the data.

Note The graph contains values from the first quadrant only since the year and the number of units must both be nonnegative.

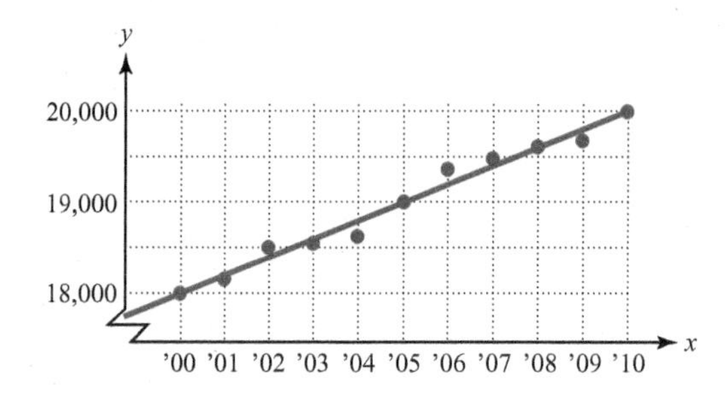

a. Find the slope of the line.

Solution: $m = \dfrac{19{,}000 - 18{,}000}{2005 - 2000}$ Use $m = \dfrac{y_2 - y_1}{x_2 - x_1}$ with two data points on the line such as (2000, 18,000) and (2005, 19,000).

$= \dfrac{1000}{5} = 200$

b. What does the slope indicate?

Answer: The slope indicates that demand for the product increased by about 200 units each year.

c. If the trend continued, what was the demand in 2011?

Solution: $20{,}000 + 200 = 20{,}200$ Add the slope, 200, to the demand in 2010, which was 20,000.

Your Turn 6 The following graph shows the closing price for one share of a particular stock each day for one week. Note that x represents the number of days after May 10. The prices seem to show a linear pattern.

a. Find the slope of the line from the point on the graph at May 10 ($x = 0$) to the point on the graph at May 14 ($x = 4$).
b. What does the slope indicate?
c. If the trend continues, predict the price of a share of the stock on the next day of trading.

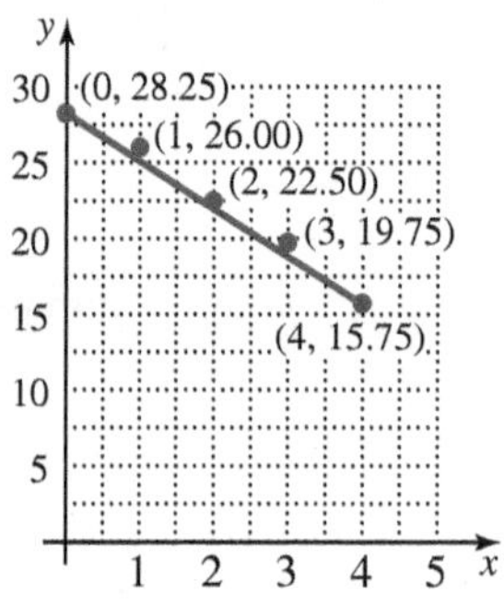

Answers to Your Turn 5

a. $m = 0$

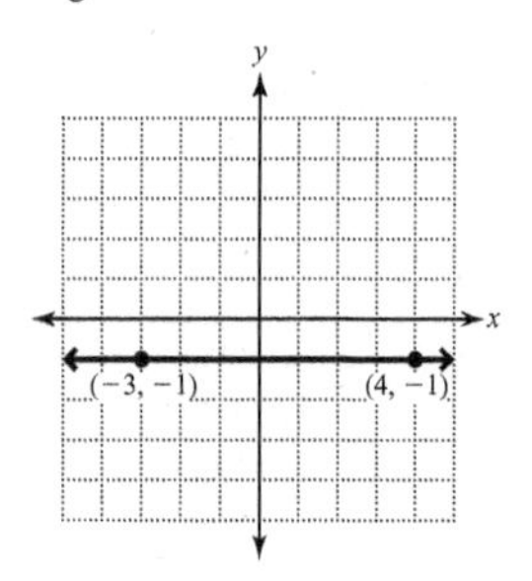

b. m is undefined.

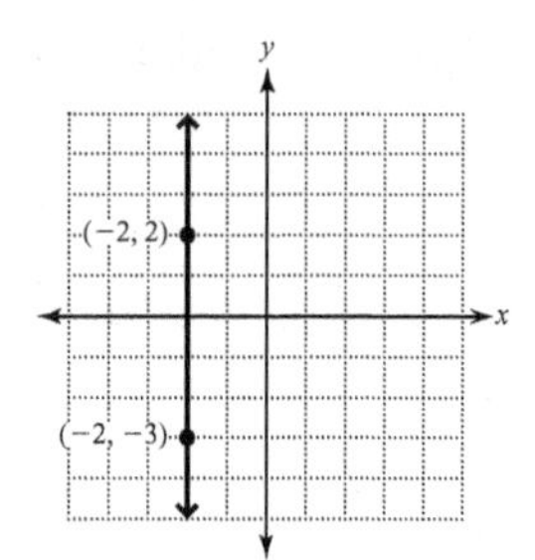

Answers to Your Turn 6

a. -3.125
b. The price decreased by about \$3.125 each day.
c. $\approx \$12.63$

3.2 Exercises

For Extra Help MyMathLab®

Note: Exercises marked with a ★ represent challenging exercises.

Objective 1

Prep Exercise 1 In your own words, what is the slope of a line?

For Exercises 1–4, which equation's graph has the steeper slope? See Example 1.

1. $y = 3x + 1$ or $y = 2x - 7$

2. $y = 3x - 2$ or $y = 4x + 3$

3. $y = 0.2x + 3$ or $y = x - 4$

4. $y = \frac{3}{4}x - 1$ or $y = \frac{2}{3}x + 5$

Prep Exercise 2 Given an equation in the slope–intercept form $y = mx + b$, if $m > 0$, what does that indicate about the incline of the graph?

Prep Exercise 3 Given an equation in the slope–intercept form $y = mx + b$, if $m < 0$, what does that indicate about the incline of the graph?

For Exercises 5–8, indicate whether the graph of the equation inclines upward or downward from left to right. See Example 1.

5. $y = 3x + 2$

6. $y = -4x + 1$

7. $y = -\frac{4}{5}x - 3$

8. $y = \frac{2}{3}x - 5$

Objective 2

Prep Exercise 4 Given a graph of a line, explain how to determine the line's slope.

Prep Exercise 5 Given a graph of a line, explain how to identify the y-intercept.

Prep Exercise 6. Describe a line with slope 0. Explain why a slope of 0 produces this type of line.

Prep Exercise 7 Describe a line with undefined slope. Explain why undefined slope produces this type of line.

For Exercises 9–12, determine the slope and the y-intercept of the line.

9.

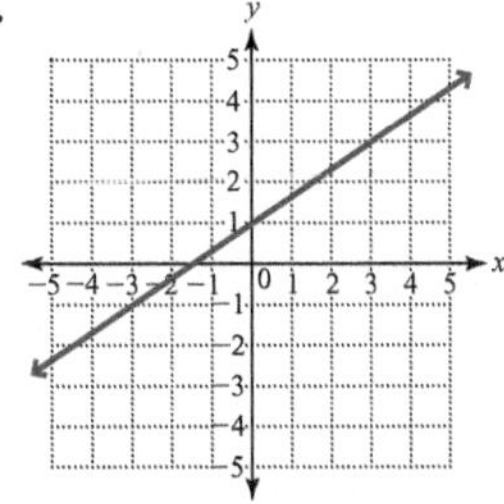

10.

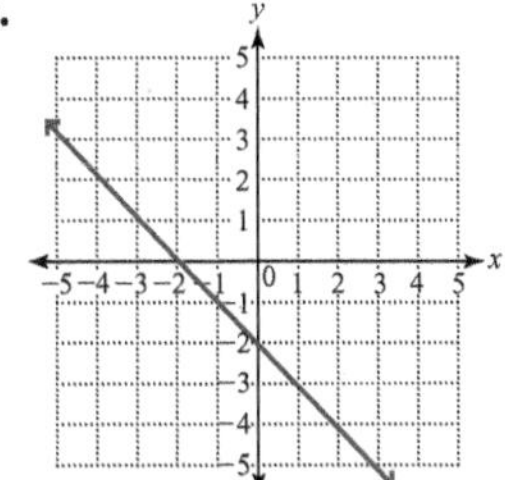

11.

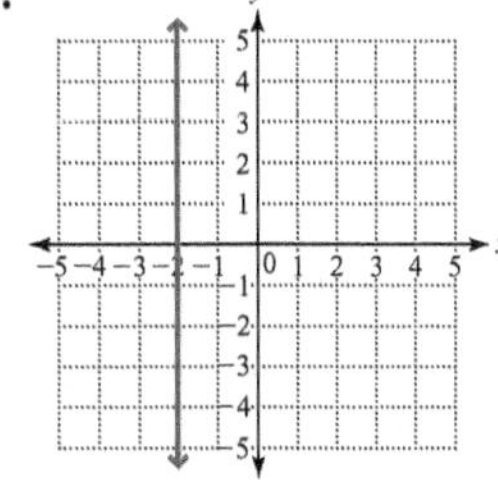

12.

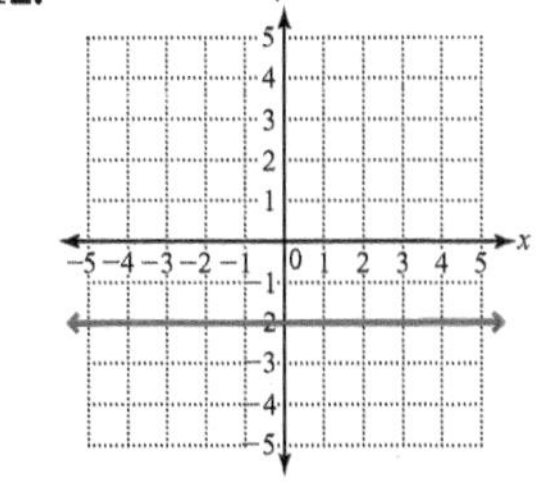

For Exercises 13–20, match the equation with its graph.

13. $y = 4x - 1$

14. $y = 2x + 3$

15. $y = -2x + 3$

16. $y = -4x + 2$

17. $y = \frac{2}{3}x - 1$

18. $y = \frac{3}{4}x + 2$

19. $y = -\frac{1}{2}x + 3$

20. $y = -\frac{3}{4}x - 1$

a.

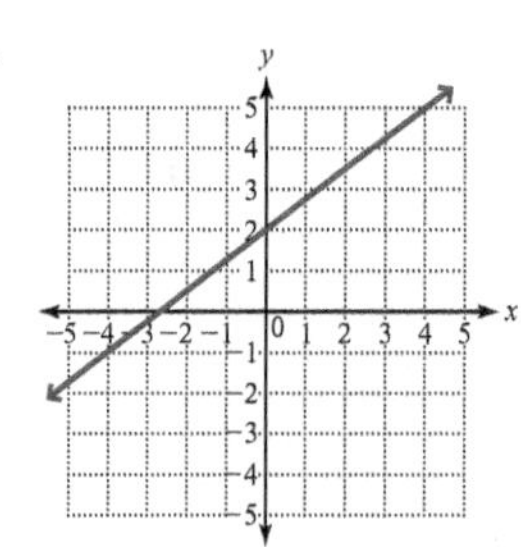

b.

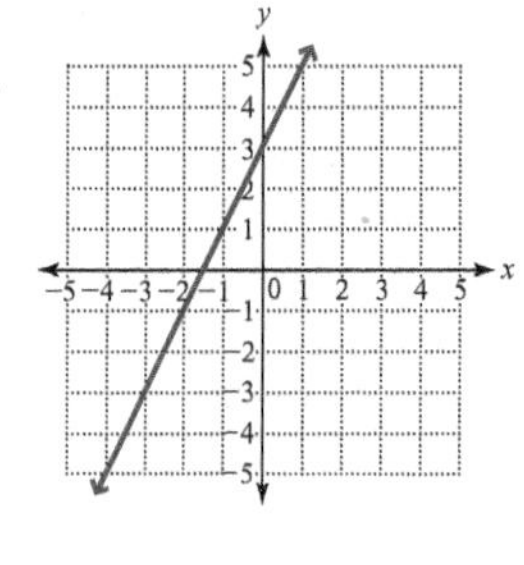

c.

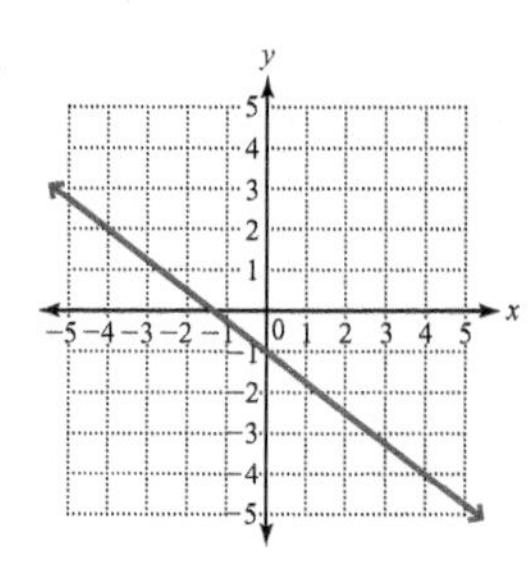

d.

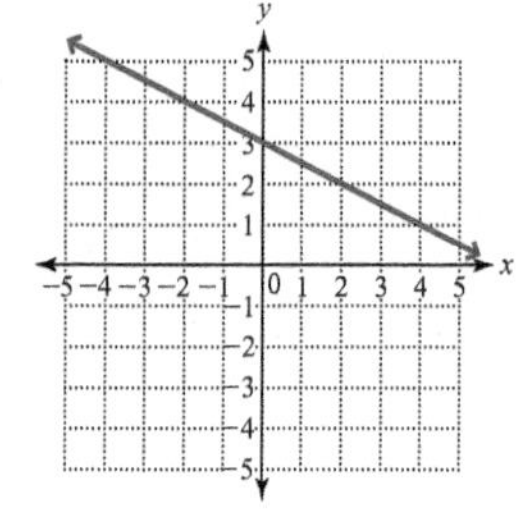

e.

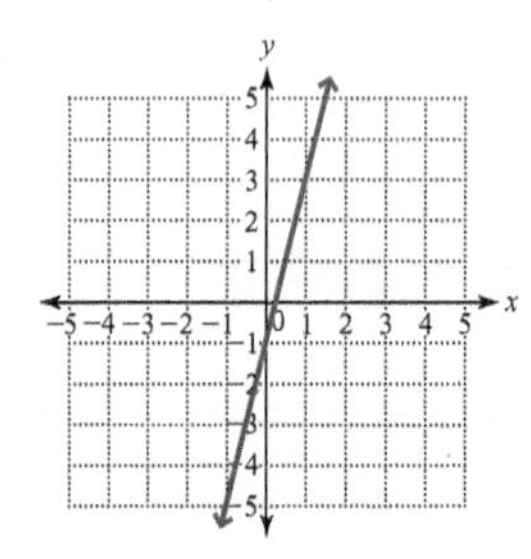

f.

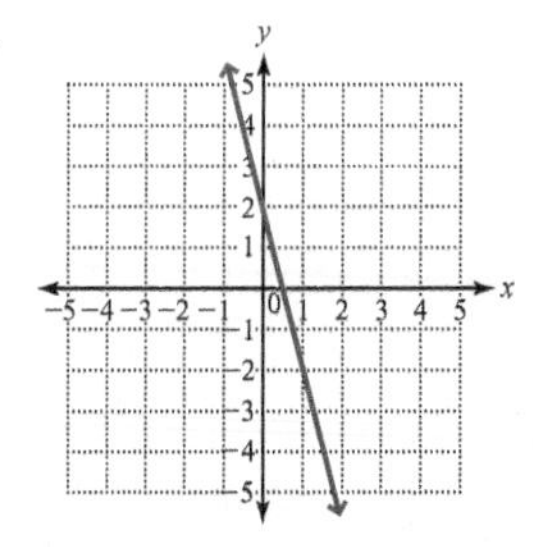

g.

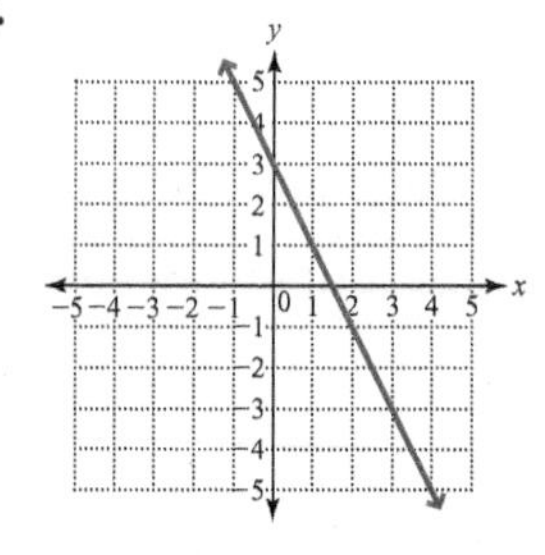

h.

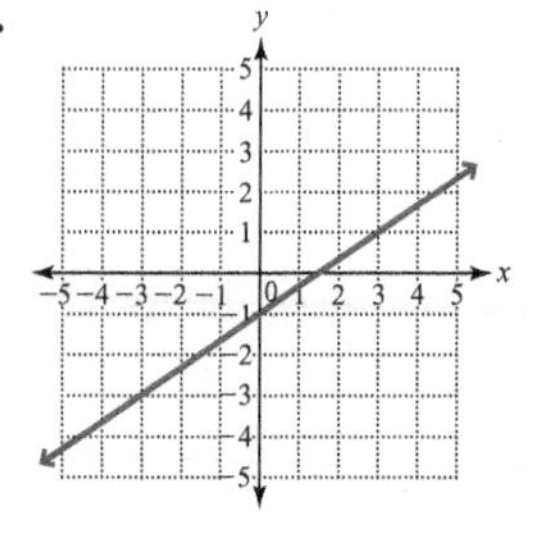

Prep Exercise 8 Explain how to use the slope and y-intercept to graph $y = \frac{3}{4}x - 2$.

For Exercises 21–36, determine the slope and the y-intercept. Then graph the equation. See Examples 2 and 3.

21. $y = \frac{2}{3}x + 5$

22. $y = \frac{3}{5}x - 8$

23. $y = -\frac{3}{4}x - 2$

24. $y = -\frac{4}{5}x - 1$

25. $y = x + 4$

26. $y = x - 3$

27. $y = -5x + \frac{2}{3}$

28. $y = -3x - \frac{4}{5}$

29. $2x + 3y = 6$

30. $5x + 3y = 6$

31. $x - 2y = -7$

32. $3x - 2y = 9$

33. $2x - 7y = 8$

34. $5x - 4y = -9$

35. $-x + y = 0$

36. $2x + y = 0$

Objective 3

Prep Exercise 9 Given two points (x_1, y_1) and (x_2, y_2), where $x_2 \neq x_1$, what is the formula for finding the slope of the line connecting the two points?

Prep Exercise 10 Does it matter which of the two points is (x_1, y_1) and which is (x_2, y_2)? Why or why not?

For Exercises 37–48, find the slope of the line through the given points. See Examples 4 and 5.

37. $(4, 2), (2, 6)$

38. $(1, 7), (3, 1)$

39. $(-1, 3), (4, -6)$

40. $(6, -7), (-2, 4)$

41. $(1, 4), (-3, 10)$

42. $(-7, 9), (2, 3)$

43. $(8, 2), (8, -5)$

44. $(-3, 1), (-3, -7)$

45. $(5, 12), (-1, 12)$

46. $(-5, -1), (-3, -1)$

47. $(0, 0), (10, -8)$

48. $(10, 2), (0, 0)$

For Exercises 49 and 50, solve.

49. A parallelogram has vertices at $(-1, 1)$, $(1, 6)$, $(5, 1)$, and $(7, 6)$.

a. Plot the vertices in a coordinate plane; then connect them to form the parallelogram.

b. Find the slope of each side of the parallelogram.

c. What do you notice about the slopes of the parallel sides?

50. A right triangle has vertices at $(-2, 1)$, $(-4, 4)$, and $(-1, 6)$.

a. Plot the vertices in a coordinate plane; then connect them to form the triangle.

b. Find the slope of each side of the triangle.

c. Two lines that intersect at right angles are perpendicular. What do you notice about the slopes of the perpendicular sides?

For Exercises 51–54, find the indicated slope. Disregard the object's orientation so that each slope is given as a positive number.

51. A wheelchair ramp is to be built with a height of 29 inches and a length of 348 inches. Find the slope of the ramp.

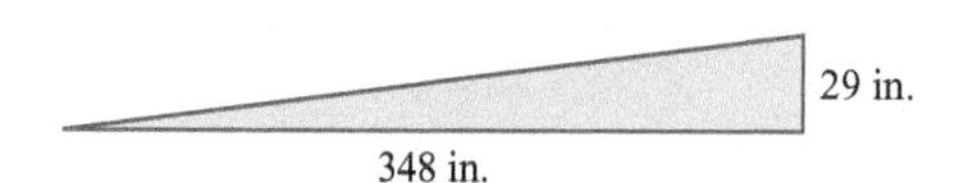

52. A wood frame for a roof is shown. Find the slope of the roof.

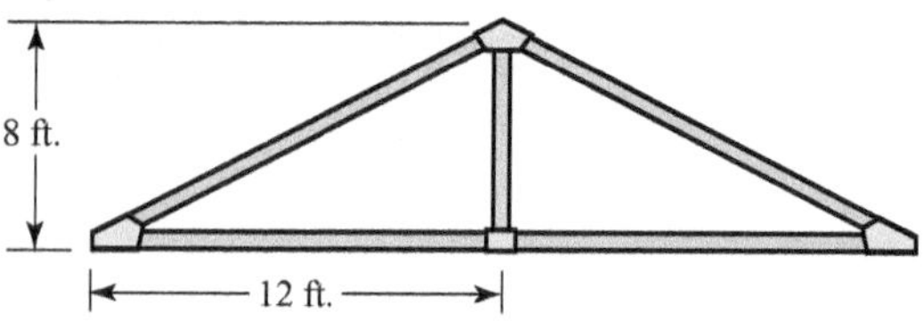

Of Interest

In construction, the slope of a roof is called its *pitch*.

53. An architect consults an elevation plan to determine the slope of a driveway for a new home. The bottom of the driveway will be at an elevation of 210 feet, and the top will be at an elevation of 215 feet. The drive will cover a horizontal distance of 20 feet. Find the slope of the driveway. Write your answer as a decimal number.

54. The four faces of the Great Pyramid in Egypt originally had a smooth stone surface. It was originally 481 feet tall, and each side along the square base was 754 feet long. If the top of the pyramid was centered over the base, find the slope of its original faces. Write your answer as a decimal number rounded to the nearest hundredth.

Of Interest

After losing the stones that formed its smooth outer surface, the Great Pyramid is now 449 feet tall and each side is 745 feet long.

For Examples 55–64, solve. See Example 6.

55. A roller coaster begins on flat elevated track that is 15 feet above the ground, as shown (point B). After traveling 50 feet on the flat stretch of track to point C, it begins climbing a hill. The top of the hill (point D) is 80 feet above the ground and 100 feet horizontally from the beginning position of the ride.

a. If we were to graph the line representing the hill in the coordinate plane with the origin at point A, what would be the coordinates of points C and D?

b. Find the slope of the hill that the coaster climbs.

c. What does the slope indicate?

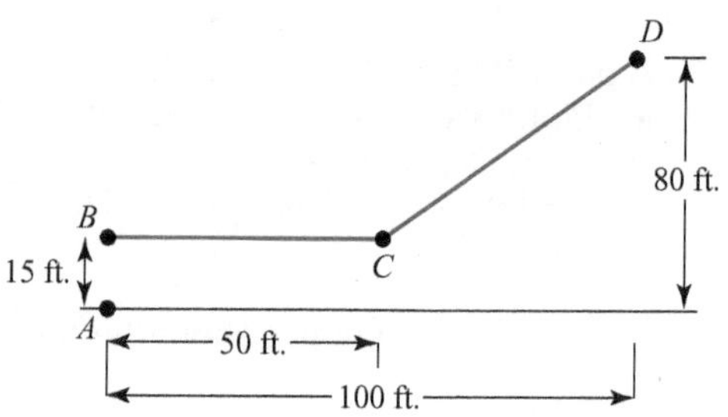

56. Body mass index, BMI, assesses the amount of fat in a person's body. At age 20, June weighed 125 and her BMI was 16.9. Now at age 38, she weighs 150 and has a BMI of 27.4. Suppose the increase in weight and BMI was roughly linear from age 20 to age 38. Plotting her weight along the x-axis and BMI along the y-axis gives the graph to the right.

a. Find the slope of the line shown. Write your answer as a decimal number rounded to the nearest hundredth.

b. What does the slope indicate?

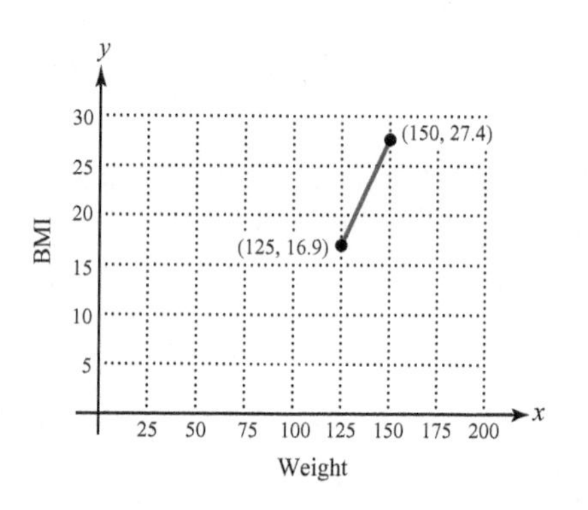

57. The graph shows the closing price for a share of Krispy Kreme doughnut stock on four consecutive days beginning with Tuesday, June 21. Note that x represents the number of days after June 21. The prices seem to show a linear pattern.
   a. Draw a straight line from the point on the graph at June 21 ($x = 0$) to the point on the graph at June 24 ($x = 3$).
   b. Find the slope of this line.
   c. If the same trend continued, predict the price of a share of Krispy Kreme stock on the next day of trading (Monday, June 27).

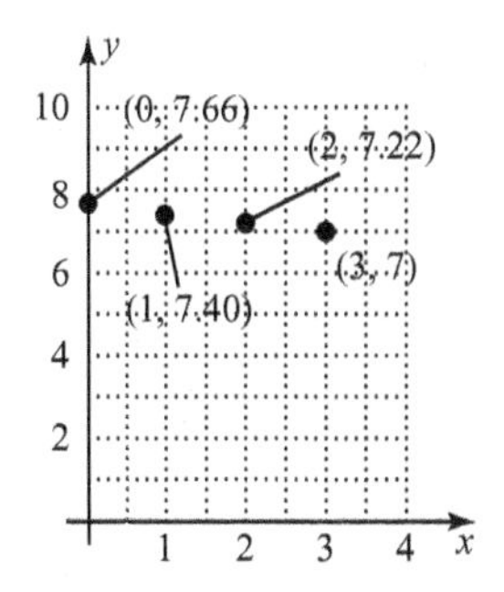

58. The graph shows the S&P 500 closing values each day of the week beginning March 7. Note that x represents the number of trading days after March 7. The closing values show a linear pattern.
   a. Draw a line from the point on the graph at March 7 to the point on the graph at March 11.
   b. Find the slope of this line.
   c. If the same trend continued, find the closing value on March 14, which was the next day of trading.

★ 59. The graph shows the change in NASDAQ closing values from December 15 to January 16, where x represents the number of trading days after December 15. The change can be roughly modeled by a line.
   a. Draw a straight line connecting the NASDAQ closing value on December 15 to the value on January 16.
   b. Find the slope of this line.
   c. If the same trend continued, what was the NASDAQ closing value on January 19, which was the next day the market opened?

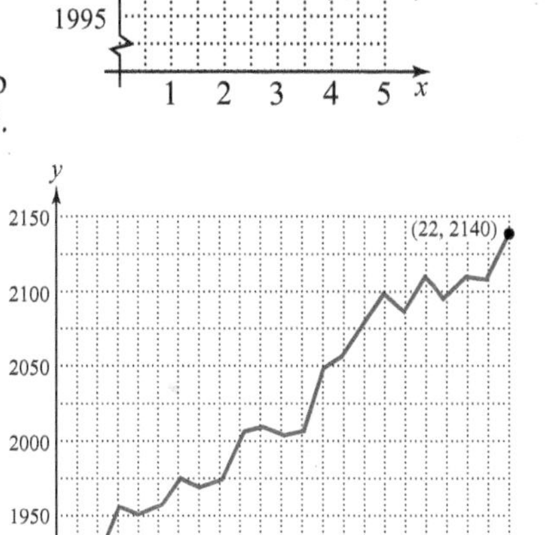

★ 60. The graph shows the change in NASDAQ from May 16 through June 17, where x represents the number of trading days after May 16. The change can be roughly modeled by a line.
   a. Draw a straight line connecting the NASDAQ value on May 16 to the value on June 17.

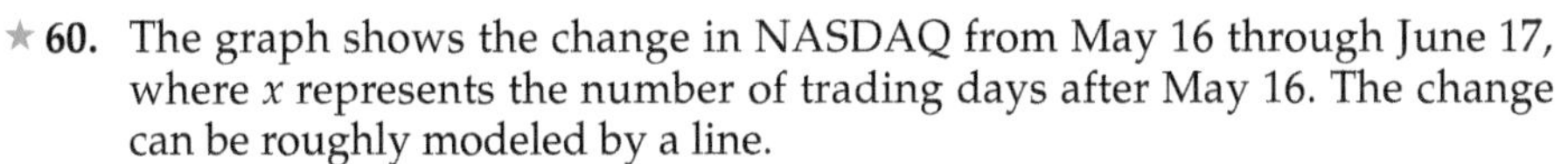

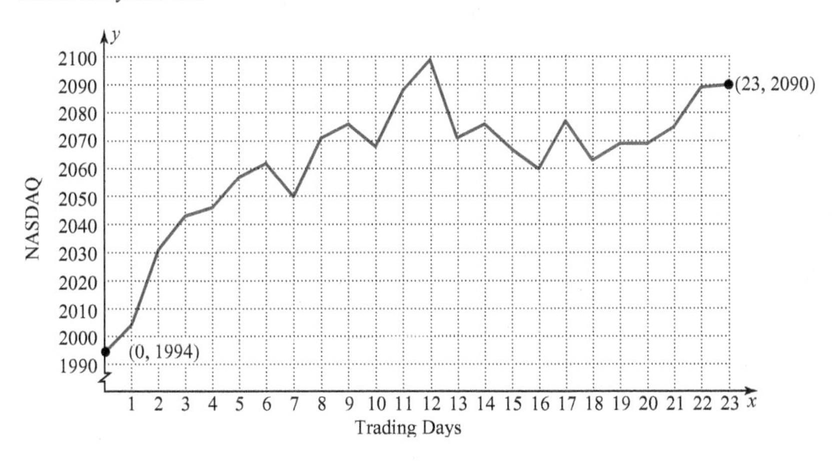

   b. Find the slope of this line rounded to the nearest whole number.
   c. If the same trend continued, what was the NASDAQ closing value on June 20, which was the next day the market opened?

61. In 2000, 47.0 million U.S. residents spoke a native language other than English. In 2010, that number had increased to 58.1 million. (*Source:* U.S. Bureau of the Census.)

a. If x represents the number of years after 2000 (with 2000 being $x = 0$) and y represents the number of residents who spoke a native language other than English, plot the two data points given; then connect them with a straight line.

b. Find the slope of the line connecting the two data points given.

62. In 2000, the average earnings per hour for U.S. construction workers was \$17.48. The average earnings increased steadily each year so that in 2010 the average hourly earnings was \$23.22. (*Source:* Bureau of Labor Statistics, U.S. Department of Labor.)

a. If x represents the number of years after 2000 (with 2000 being $x = 0$) and y represents the average hourly earnings, plot the two data points given; then connect them with a straight line.

b. Find the slope of the line connecting the two data points given.

63. In 1995, the average number of overtime hours that workers producing computer and electronic products in the United States worked per week was 4.9. The number of hours declined each year so that in 2010, the average number of hours was 2.9. (*Source:* Bureau of Labor Statistics, U.S. Department of Labor.)

a. If x represents the number of years after 1995 (with 1995 being $x = 0$) and y represents the average number of hours per week, plot the two data points given; then connect them with a straight line.

b. Find the slope of the line connecting the two data points given.

64. In 2006, the employment rate in the U.S. was 63.1%. The employment rate decreased each year so that in 2010 it was 58.5%.

a. If x represents the number of years after 2006 (with 2006 being $x = 0$) and y represents the employment rate plot the two data points given; then connect them with a straight line.

b. Find the slope of the line segment joining the two data points.

Review Exercises

Exercises 1–4 Expressions

[1.4] *For Exercises 1 and 2, evaluate* $4n^2 - m(n + 2)$ *using the given values.*

1. $m = 7, n = -2$

2. $m = -5, n = 3$

3. Simplify: $16x - 3y - 15 - x + 3y - 12$

4. Distribute: $-5(3x - 2)$

Exercises 5 and 6 Equations and Inequalities

[2.1] *For Exercises 5 and 6, solve the equation for the indicated variable.*

5. $V = lwh$; solve for h

6. $Ax + By = C$; solve for y

3.3 The Equation of a Line

Objectives

1. Use slope–intercept form to write the equation of a line.
2. Use point–slope form to write the equation of a line.
3. Write the equation of a line parallel to a given line.
4. Write the equation of a line perpendicular to a given line.

Warm-up

[3.2] **1.** Find the slope of the line passing through $(0, 5)$ and $(100, 7.5)$.

[2.1, [3.2] **2.** Simplify $y - 6 = -\frac{2}{3}(x - (-3))$ and leave the answer in slope–intercept form.

[3.2] **3.** Find the slope of the line whose equation is $3x + 5y = 10$.

In this section, we explore how to write an equation of a line given information about the line, such as its slope and the coordinates of a point on the line.

Objective 1 Use slope–intercept form to write the equation of a line.

If we are given the y-intercept $(0, b)$ of a line, to write the equation of the line, we need the slope or another point so that we can calculate the slope. First, consider a situation in which we are given the y-intercept and the slope of a line.

Example 1 A line has a slope of $-\frac{3}{5}$. If the y-intercept is $(0, 1)$, write the equation of the line in slope–intercept form.

Solution: We use $y = mx + b$, the slope–intercept form of the equation, replacing m with the slope, $-\frac{3}{5}$, and b with the y-coordinate of the y-intercept, 1.

$$y = -\frac{3}{5}x + 1$$

Your Turn 1 A line has a slope of -5. If the y-intercept is $(0, -2)$, write an equation of the line in slope–intercept form.

Now suppose we are given the y-intercept and a second point on the line. To write the equation in slope–intercept form, we need the slope. So our first step is to calculate the slope using the formula from Section 3.2, $m = \frac{y_2 - y_1}{x_2 - x_1}$, where the two points, (x_1, y_1) and (x_2, y_2), are the y-intercept and the other given point.

Example 2 The following data points show the amount a city charges for the number of cubic feet of water consumed. When plotted, all of the points lie on a straight line.

a. Write the equation of the line in slope–intercept form.

b. Find the price when the consumption is 500 cubic feet.

Note Since the consumption (x) and the price (y) are both nonnegative ($x \geq 0$ and $y \geq 0$), the graph is in the first quadrant only.

x (consumption in cubic feet)	y (price in \$)
0	5.00
100	7.50
200	10.00
300	12.50

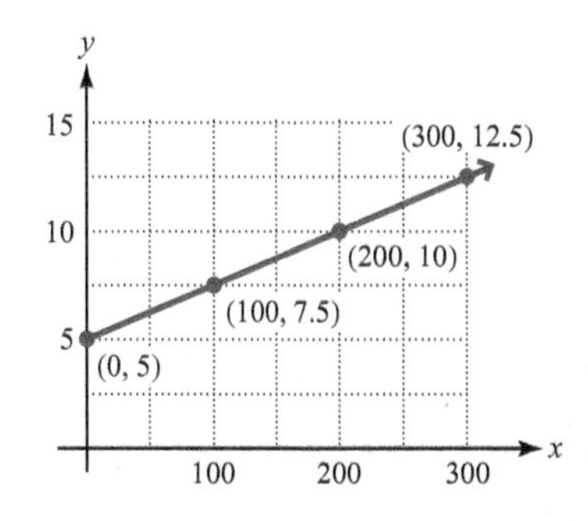

Answer to Your Turn 1
$y = -5x - 2$

Answers to Warm-up
1. 0.025
2. $y = -\frac{2}{3}x + 4$
3. $-\frac{3}{5}$

Solution: a. We can actually find the slope of the line using any two ordered pairs in the slope formula. We will use $(0, 5)$, the y-intercept, and $(100, 7.5)$.

$$m = \frac{7.5 - 5}{100 - 0} = \frac{2.5}{100} = 0.025$$

Using the y-intercept, $(0, 5)$, we can write the equation in slope–intercept form.

$$y = 0.025x + 5$$

b. To find the price when the consumption is 500 cubic feet, replace x with 500.

$$y = 0.025(500) + 5 = 12.5 + 5 = 17.5$$

The cost of 500 cubic feet of water is $17.50.

Examples 1 and 2 suggest the following procedure.

Procedure Equation of a Line Given Its y-intercept

To write the equation of a line given its y-intercept, $(0, b)$, and its slope, m, use the slope–intercept form of the equation, $y = mx + b$. If given a second point and not the slope, calculate the slope using $m = \frac{y_2 - y_1}{x_2 - x_1}$; then use $y = mx + b$.

Your Turn 2 Write the equation of the line passing through $(0, 3)$ and $(4, -2)$ in slope–intercept form.

Objective 2 Use point–slope form to write the equation of a line.

Now let's see how to write the equation of a line given *any* two points on the line. We can use the slope formula to derive the *point–slope* form of the equation of a line. Recall the slope formula, as follows:

$$m = \frac{y_2 - y_1}{x_2 - x_1}$$

$$(x_2 - x_1) \cdot m = \left(\frac{y_2 - y_1}{x_2 - x_1}\right) \cdot (x_2 - x_1)$$ Multiply both sides by $x_2 - x_1$ to isolate the y's.

$$(x_2 - x_1)m = (y_2 - y_1)$$ Rewrite with y's on the left side to resemble slope–intercept form.

$$y_2 - y_1 = m(x_2 - x_1)$$

With this point–slope form, we can write the equation of a line by substituting the slope of the line for m and the coordinates of any point on the line for x_1 and y_1, leaving x_2 and y_2 as variables. To indicate that x_2 and y_2 remain variables, we remove their subscripts so that we have $y - y_1 = m(x - x_1)$, which is called the point–slope form of the equation of a line.

Procedure Using the Point–Slope Form of the Equation of a Line

To write the equation of a line given its slope and any point, (x_1, y_1), on the line, use the point–slope form of the equation of a line, $y - y_1 = m(x - x_1)$. If given a second point, (x_2, y_2), and not the slope, calculate the slope using $m = \frac{y_2 - y_1}{x_2 - x_1}$; then use $y - y_1 = m(x - x_1)$.

Answer to Your Turn 2

$y = -\frac{5}{4}x + 3$

Example 3

a. Write the equation of a line with a slope of -3 that passes through the point $(2, -4)$. Write the equation in slope–intercept form.

Solution: Because we are given the coordinates of a point, $(2, -4)$, and the slope of a line, -3, passing through the point, we use the point–slope formula, $y - y_1 = m(x - x_1)$.

Note We leave x and y as variables and substitute only for x_1 and y_1.

$$y - (-4) = -3(x - 2)$$ Replace m with -3, x_1 with 2, and y_1 with -4.

$$y + 4 = -3x + 6$$ Simplify.

$$y = -3x + 2$$ Subtract 4 from both sides so that you have slope–intercept form.

Check: Verify that the point $(2, -4)$ is a solution for the equation.

$$-4 \stackrel{?}{=} -3(2) + 2$$

$$-4 = -4$$ True

b. Write the equation of a line passing through the points $(-3, 6)$ and $(9, -2)$. Write the equation in slope–intercept form.

Solution: Because we do not have the slope, we calculate it using $m = \frac{y_2 - y_1}{x_2 - x_1}$.

$$m = \frac{-2 - 6}{9 - (-3)} = \frac{-8}{12} = -\frac{2}{3}$$ Replace x_1 with -3, y_1 with 6, x_2 with 9, and y_2 with -2.

Now we can use the point–slope form. We can use either of the two given points for (x_1, y_1) in the point–slope equation. We will use $(-3, 6)$ for (x_1, y_1).

$$y - 6 = -\frac{2}{3}(x - (-3))$$ Using $y - y_1 = m(x - x_1)$, replace m with $-\frac{2}{3}$, x_1 with -3, and y_1 with 6.

$$y - 6 = -\frac{2}{3}x - 2$$ Distribute and simplify.

$$y = -\frac{2}{3}x + 4$$ Add 6 to both sides to get slope–intercept form.

Your Turn 3

a. Write the equation of a line with a slope of -0.4 that passes through the point $(-2, 3)$. Write the equation in slope–intercept form.

b. Write the equation of a line passing through $(-3, -5)$ and $(3, -1)$. Write the equation in slope–intercept form.

Writing Linear Equations in Standard Form

We may also write equations in standard form, which is $Ax + By = C$, where A, B, and C are real numbers. In standard form, it is customary to write the equation so that the x term is first with a positive coefficient and, if possible, A, B, and C are all integers.

Example 4 A line connects the points $(8, 4)$ and $(2, -1)$. Write the equation of the line in standard form $Ax + By = C$, where A, B, and C are integers and $A > 0$.

Answers to Your Turn 3

a. $y = -0.4x + 2.2$

b. $y = \frac{2}{3}x - 3$

Solution: Because we are given two points, we find the slope using $m = \frac{y_2 - y_1}{x_2 - x_1}$.

$$m = \frac{-1 - 4}{2 - 8} = \frac{-5}{-6} = \frac{5}{6}$$ Replace x_1 with 8, y_1 with 4, x_2 with 2, and y_2 with -1.

Because we were not given the y-intercept, we use the point–slope form of the linear equation, $y - y_1 = m(x - x_1)$.

$$y - (-1) = \frac{5}{6}(x - 2)$$ Use the slope and the point $(2, -1)$.

$$y + 1 = \frac{5}{6}(x - 2)$$ Simplify the signs.

Now manipulate the equation to write in the form $Ax + By = C$, where A, B, and C are integers and $A > 0$.

$$6(y + 1) = 6 \cdot \frac{5}{6}(x - 2)$$ Multiply both sides by the LCD, 6.

$$6y + 6 = 5x - 10$$ Simplify and distribute.

$$6y - 5x = -16$$ Subtract $5x$ and 6 from both sides to separate the variable terms and constant terms.

$$-5x + 6y = -16$$ Use the commutative property of addition so that the x term is the first term.

$$5x - 6y = 16$$ Multiply both sides by -1 so that the coefficient of the x term is positive. Now A, B, and C are integers and $A > 0$.

Your Turn 4 A line connects the points $(3, 2)$ and $(-1, 5)$. Write the equation of the line in the form $Ax + By = C$, where A, B, and C are integers and $A > 0$.

Objective 3 Write the equation of a line parallel to a given line.

Consider the graphs of $y = 2x + 1$ and $y = 2x - 3$.

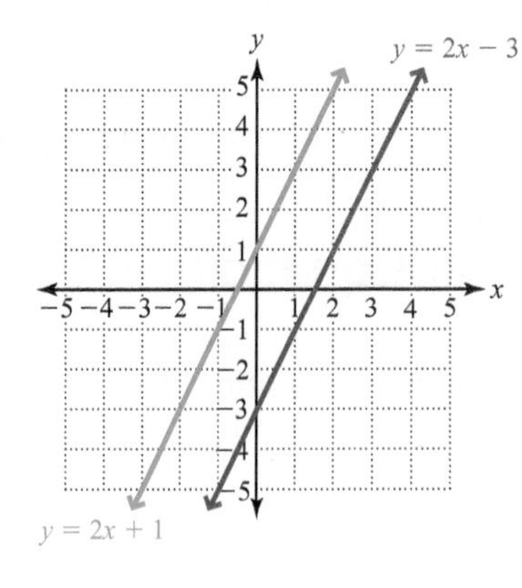

These two lines are parallel, which means that they do not intersect at any point. The graph of $y = 2x + 1$ has a slope of 2 and a y-intercept of $(0, 1)$. The graph of $y = 2x - 3$ has a slope of 2 and a y-intercept of $(0, -3)$. Both lines have the same slope, 2, but different y-intercepts, which suggests the following rule.

Rule Parallel Lines

Nonvertical parallel lines have equal slopes and different y-intercepts. Vertical lines are parallel.

Example 5 Write the equation of the line in slope–intercept form that passes through $(4, -1)$ and is parallel to the graph of $y = -\frac{1}{2}x + 3$.

Answer to Your Turn 4
$3x + 4y = 17$

Solution: The slope of the given equation is $-\frac{1}{2}$, so the slope of the parallel line must also be $-\frac{1}{2}$. We can now write the equation of the parallel line.

$$y - (-1) = -\frac{1}{2}(x - 4)$$ Using $y - y_1 = m(x - x_1)$, replace m with $-\frac{1}{2}$, x_1 with 4, and y_1 with -1.

$$y + 1 = -\frac{1}{2}x + 2$$ Distribute and simplify.

$$y = -\frac{1}{2}x + 1$$ Subtract 1 from both sides to isolate y.

Your Turn 5 Write the equation of a line in slope–intercept form that passes through $(-3, 2)$ and is parallel to the graph of $y = \frac{2}{3}x - 5$.

Objective 4 Write the equation of a line perpendicular to a given line.

Consider the graphs of $y = \frac{2}{3}x - 4$ and $y = -\frac{3}{2}x + 1$.

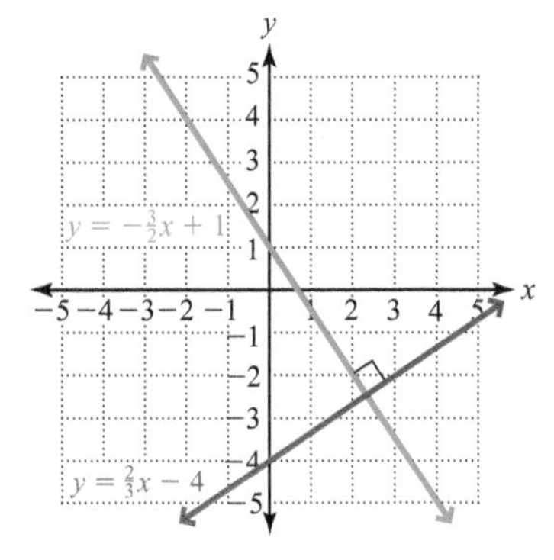

These two lines are perpendicular, which means that they intersect at a 90° angle. In the equations, we see that their slopes, $\frac{2}{3}$ and $-\frac{3}{2}$, are reciprocals with opposite signs, which suggests the following rule.

Rule Perpendicular Lines
The slope of a line perpendicular to a line with a slope of $\frac{a}{b}$ is $-\frac{b}{a}$.
Horizontal and vertical lines are perpendicular.

This means that if a line has a slope of $-\frac{3}{4}$, then the slope of any line perpendicular to it is $\frac{4}{3}$; if the slope is 4, then the slope of any perpendicular line is $-\frac{1}{4}$; and so on.

Example 6 Write the equation of a line in standard form that passes through $(3, 1)$ and is perpendicular to the graph of $3x + 5y = 10$.

Solution: First, we need to determine the slope of the given line; so we will write the given equation in slope–intercept form.

$$5y = -3x + 10$$ Subtract $3x$ from both sides.

$$y = -\frac{3}{5}x + 2$$ Divide both sides by 5 to get slope–intercept form.

Answer to Your Turn 5

$y = \frac{2}{3}x + 4$

We see that the slope of the given line is $-\frac{3}{5}$; so the slope of the perpendicular line must be $\frac{5}{3}$. We can now write the equation of the perpendicular line.

$y - 1 = \frac{5}{3}(x - 3)$ Using $y - y_1 = m(x - x_1)$, replace m with $\frac{5}{3}$, x_1 with 3, and y_1 with 1.

$3y - 3 = 5x - 15$ Multiply both sides by 3, simplify, and distribute.

$3y - 5x = -12$ Subtract $5x$ from and add 3 to both sides.

$-5x + 3y = -12$ Rearrange the x and y terms.

$5x - 3y = 12$ Multiply both sides by -1.

Your Turn 6 Write the equation of a line in standard form that passes through $(-6, 2)$ and is perpendicular to the graph of $3x + y = -1$.

Answer to Your Turn 6
$x - 3y = -12$

3.3 Exercises

For Extra Help MyMathLab®

Objective 1

Prep Exercise 1 In the slope–intercept form of the equation, $y = mx + b$, what do m and b represent?

For Exercises 1–6, use the given slope and coordinate of the y-intercept to write the equation of the line in slope–intercept form. See Example 1.

1. $m = -4; (0, 3)$

2. $m = 3; (0, -1)$

3. $m = \frac{3}{5}; (0, -2)$

4. $m = -\frac{1}{4}; \left(0, \frac{1}{2}\right)$

5. $m = -0.2; (0, -1.5)$

6. $m = 1.6; (0, 4.5)$

For Exercises 7–10, determine the slope and the y-intercept and write the equation of the line in slope–intercept form.

7.

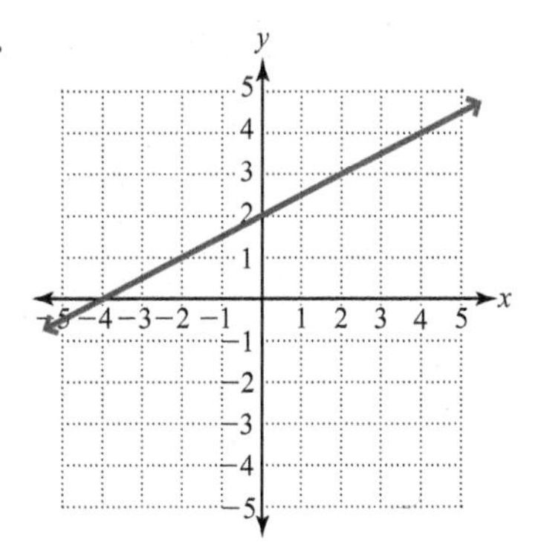

8.

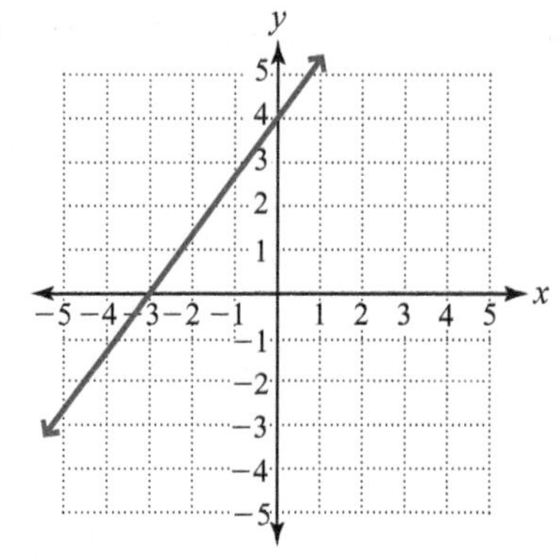

9.

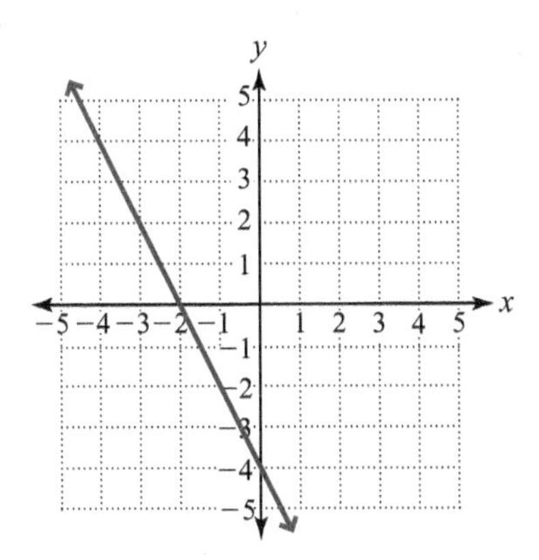

10.

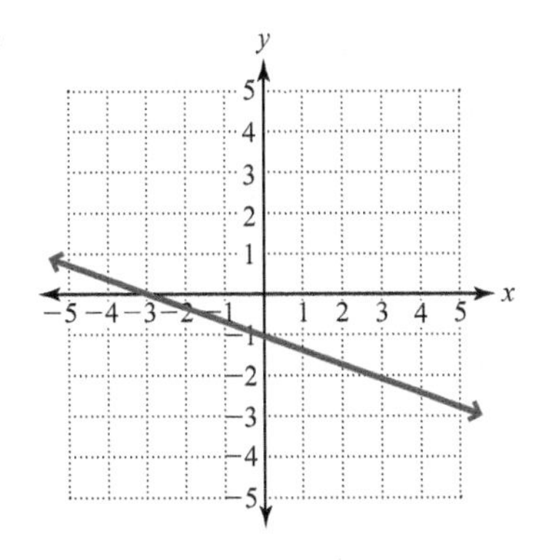

Prep Exercise 2 What is the first step in writing the equation of a line through two points, (x_1, y_1) and (x_2, y_2)?

For Exercises 11–14, write the equation of the line with the given y-intercept that passes through the given point. Leave the answer in slope–intercept form. See Example 2.

11. $(0, 2), (3, 4)$

12. $(0, 4), (5, 7)$

13. $(0, -6), (-2, -3)$

14. $(0, -3), (4, -6)$

Objective 2

Prep Exercise 3 What is the point–slope equation of a line?

For Exercises 15–26, write the equation of a line in slope–intercept form with the given slope that passes through the given point. See Example 3a.

15. $m = 2; (3, 1)$

16. $m = 3; (3, -5)$

17. $m = -2; (-5, 0)$

18. $m = -4; (-3, 0)$

19. $m = -3; (0, 0)$

20. $m = -1; (0, 0)$

21. $m = \frac{2}{5}; (0, -2)$

22. $m = \frac{2}{3}; (0, 3)$

23. $m = \frac{4}{5}; (-2, -2)$

24. $m = \frac{4}{3}; (-5, -9)$

25. $m = -\frac{3}{2}; (-1, 3)$

26. $m = -\frac{5}{2}; (3, -2)$

Prep Exercise 4 Explain how to rewrite an equation that is in standard form in slope–intercept form.

Prep Exercise 5 In standard form, $Ax + By = C$, it is customary to write the equation so that A, B, and C are all integers and A is positive. If A, B, or C is a fraction, explain how to eliminate the fraction.

Prep Exercise 6 In standard form, $Ax + By = C$, it is customary to write the equation so that A, B, and C are all integers and A is positive. If A is a negative integer, explain how to eliminate the negative sign.

For Exercises 27–38: *a. Write the equation of a line through the given points in slope–intercept form. See Example 3b.*
b. Write the equation in standard form, $Ax + By = C$, where A, B, and C are integers and $A > 0$. See Example 4.

27. $(4, -1), (2, 3)$ **28.** $(1, 3), (-2, -9)$ **29.** $(1, -4), (-2, -7)$ **30.** $(3, -4), (-1, 8)$

31. $(-6, -6), (3, 0)$ **32.** $(-4, -7), (2, 8)$ **33.** $(-5, 0), (0, 6)$ **34.** $(-2, 0), (0, 3)$

35. $(4, 2), (-9, 3)$ **36.** $(-9, 0), (4, -1)$ **37.** $(-2, 2), (-5, 9)$ **38.** $(-5, -2), (6, -4)$

Objectives 3 and 4

Prep Exercise 7 If two non-vertical lines are parallel, what is true about their slopes?

Prep Exercise 8 What is the slope of a line perpendicular to a line with a slope of $\frac{a}{b}$?

For Exercises 39–50, determine whether the given lines are parallel, perpendicular, or neither.

39. $y = \frac{1}{3}x - 5$, $y = \frac{1}{3}x + 2$

40. $y = \frac{2}{5}x - 2$, $y = \frac{2}{5}x + 2$

41. $y = \frac{3}{4}x - 2$, $y = -\frac{4}{3}x + 4$

42. $y = -\frac{2}{5}x + 4$, $y = \frac{5}{2}x - 6$

43. $y = 5x + 1$, $y = -5x - 6$

44. $y = -\frac{1}{4}x + 5$, $y = -4x$

45. $2x + 7y = 8$, $4x + 14y = -9$

46. $3x - 4y = 7$, $-9x + 12y = -2$

47. $3x + 5y = 4$, $5x - 3y = 2$

48. $6x - 2y = 7$, $x + 3y = -5$

49. $x = 4$, $y = -2$

50. $y = 3$, $x = 3$

Objective 3

For Exercises 51–60: *a. Write the equation of a line in slope–intercept form that passes through the given point and is parallel to the graph of the given equation. See Example 5.*
b. Write the equation in the form $Ax + By = C$, where A, B, and C are integers and $A > 0$.

51. $(2, -6); y = -5x + 3$ **52.** $(-1, 1); y = -3x + 1$ **53.** $(-5, -2); y = 4x - 2$ **54.** $(-5, -7); y = 2x - 5$

55. $(3, -4); y = \frac{2}{3}x - 5$ **56.** $(-4, 2); y = \frac{5}{2}x - 4$ **57.** $(4, -3); 4x + 6y = 3$ **58.** $(1, 1); 3x - 4y = 8$

59. $(-3, -7); 2x + 5y - 30 = 0$

60. $(-2, -2); x - 3y + 9 = 0$

Objective 4

For Exercises 61–70: a. *Write the equation of a line in slope–intercept form that passes through the given point and is perpendicular to the graph of the given equation.*
b. *Write the equation in the form $Ax + By = C$, where A, B, and C are integers and $A > 0$. See Example 6.*

61. $(2, -1); y = \frac{1}{3}x + 6$

62. $(-1, 5); y = \frac{1}{2}x - 7$

63. $(0, -3); y = -\frac{2}{5}x - 8$

64. $(3, 2); y = -\frac{3}{4}x + 1$

65. $(-2, -9); y = -3x + 4$

66. $(-3, -3); y = 2x - 5$

67. $(-2, -3); x + 4y = -10$

68. $(2, 8); x + 2y = 9$

69. $(1, 4); 2x - 3y = 15$

70. $(-1, -5); 3x + 2y = 1$

71. What is the equation of a horizontal line through $(0, -4)$?

72. What is the equation of a vertical line through $\left(\frac{3}{4}, 0\right)$?

73. What equation describes the x-axis?

74. What equation describes the y-axis?

For Exercises 75–82, solve. See Example 2.

75. On the graph, p represents the U.S. resident population, in millions of people, and t represents the number of years after 2000. The graph shows that the U.S. resident population has increased from 282.4 million people in 2000 (year 0) to 313.9 million people in 2012 (year 12). (*Source:* U.S. Bureau of the Census.)
a. Find the slope of the line.

b. Write the equation of the line in slope–intercept form.

c. If this trend continues, predict the U.S. resident population in 2020.

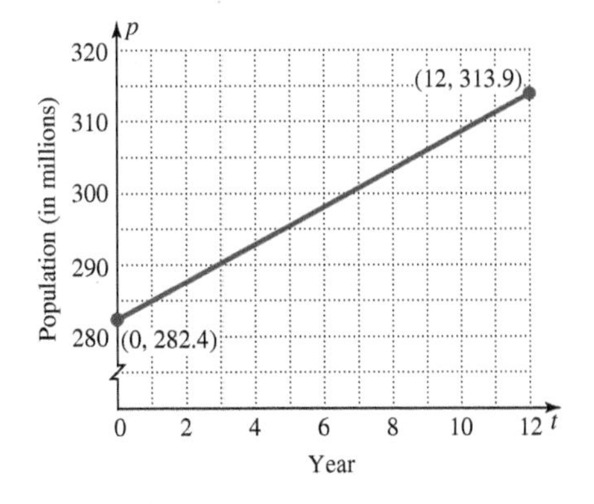

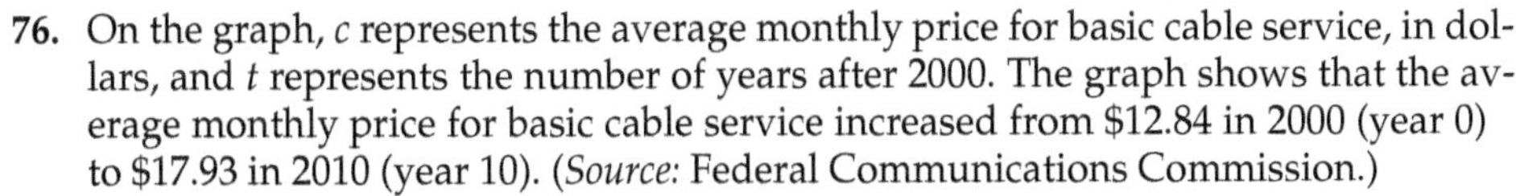

76. On the graph, c represents the average monthly price for basic cable service, in dollars, and t represents the number of years after 2000. The graph shows that the average monthly price for basic cable service increased from \$12.84 in 2000 (year 0) to \$17.93 in 2010 (year 10). (*Source:* Federal Communications Commission.)
a. Find the slope of the line.

b. Write the equation of the line in slope–intercept form.

c. If this trend continues, predict the average monthly bill for basic cable subscribers in 2020.

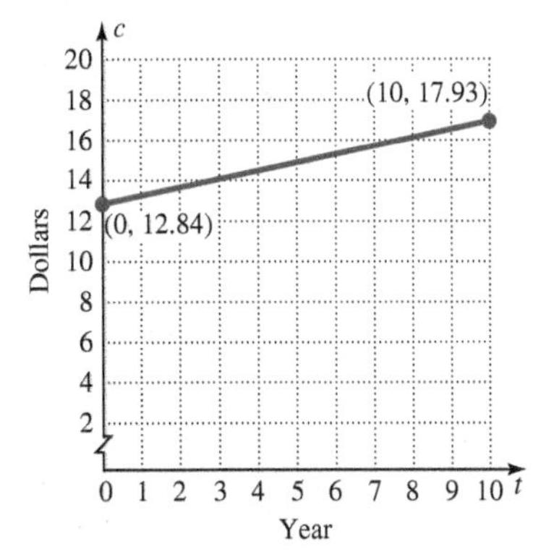

77. On the graph, b represents the average number of barrels of oil produced by Alaska per day, in thousands, and t represents the number of years after 2004. The graph shows that Alaska's production of crude oil decreased from an average of 908 thousand barrels per day in 2004 to 526 thousand barrels per day in 2012. (*Source:* Energy Information Administration, *Monthly Energy Review,* May 2013.)

a. Find the slope of the line.

b. Write the equation of the line in slope–intercept form.

c. If this trend continues, predict the average number of barrels of crude oil expected to be produced per day in Alaska in 2020.

78. On the graph, n represents the number of music CDs manufactured, in millions, and t represents the number of years after 1998. The graph shows that the number of music CDs manufactured decreased from 847 million in 1998 to 253 million in 2010. (*Source:* Recording Industry Association of America.)

a. Find the slope of the line.

b. Write the equation of the line in slope–intercept form.

c. If this trend continues, predict the year in which CD's will no longer be produced. (Hint: $n = 0$)

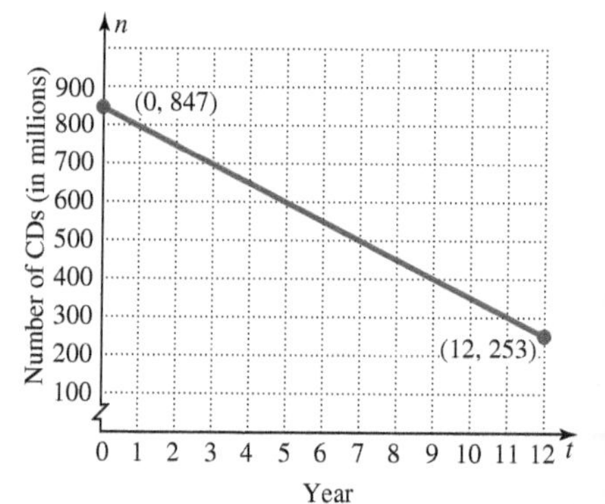

79. Paying discount points can lower a mortgage loan interest rate and monthly payment. According to www.interest.com, paying 1 point will save a home owner \$300 after the first year of payments and \$1500 after 5 years of payments.

a. Make a grid with the horizontal axis labeled t and the vertical axis labeled s so that t represents the number of years payments have been completed with 1 point and s represents the amount of money saved. Plot the two given data points and connect them with a straight line segment. (Hint: $t = 1$ means payments have been made for one year.)

b. Find the slope of the line.

c. Write an equation of the line in slope–intercept form.

d. If the discount savings follow the same trend, predict the savings after 10 years.

80. The consumer price index (CPI) compares the current value of a dollar against what the dollar would buy in a past year. Since 1967, the CPI has steadily risen in a linear pattern. Items that cost \$1.00 in 1967 cost \$6.95 in 2013. (*Source:* Bureau of Labor Statistics, Department of Labor.)

a. Make a grid with the horizontal axis labeled t and the vertical axis labeled p so that t represents the number of years after 1967 ($t = 0$) and p represents the CPI. Plot the two given data points and connect them with a straight line segment.

b. Find the slope of the line connecting the two data points given.

c. Write the equation of the line in slope–intercept form that connects the two data points given.

d. Using the equation from part c, predict the CPI in 2025.

81. The percentage of men 65 years of age and older who are in the labor force has increased from 17.6% in 1990 to 20.8% in 2010. (*Source:* Bureau of Labor Statistics, Department of Labor)

a. Make a grid with the horizontal axis labeled t and the vertical axis labeled p so that t represents the number of years after 1990 ($t = 0$) and p represents the percentage of men 65 years of age and older who are in the labor force. Plot the two given data points and connect them with a straight line segment.

b. Find the slope of the line connecting the two data points given.

c. Write the equation of the line in slope–intercept form that connects the two data points given.

d. Using the equation from part c, predict the percentage of men 65 years of age and older in the labor force in 2025.

82. In 2000, 23.3% of adults aged 18 and over in the United States were cigarette smokers. In 2010, the percent decreased to 19.3% of U.S. adults. (*Source:* Centers for Disease Control, National Center for Health Statistics)

a. Make a grid with the horizontal axis labeled t and the vertical axis labeled p so that t represents the number of years after 2000 ($t = 0$) and p represents the percentage of adults ages 18 and over in the United States who smoke. Plot the two given data points and connect them with a straight line segment.

b. Find the slope of the line.

c. Write an equation of the line in slope–intercept form.

d. If this trend continues, predict the percent of the U.S. adults ages 18 and over expected to be cigarette smokers in 2025.

Review Exercises

Exercises 1–6 Equations and Inequalities

[1.1] **1.** Use $<$, $>$, or $=$ to write a true statement: $-0.8 \quad -\frac{1}{8}$

[2.1] *For Exercises 2 and 3, solve and check.*

2. $4(2x + 3) - 2(x - 6) = 7 - (3x + 1)$

3. $\frac{1}{4}x - \frac{5}{6} = \frac{3}{2}x - 1$

[2.3] *For Exercises 4 and 5, solve and graph the solution set. Then write the solution set in set-builder notation and in interval notation.*

4. $-4x - 7 \geq 13$

5. $6x - 2(x - 1) \geq 9x - 8$

6. A certain breed of dog should have at least 2000 square feet of space to get adequate exercise. If the length of a fence must be fixed at 50 feet, what range of values can the width have so that the area is at least 2000 square feet?

3.4 Graphing Linear Inequalities

Warm-up

[3.1] *For Exercises 1–3, graph.*

1. $3x + 2y = 6$ **2.** $y = 2x$

[3.3] 3. Write the equation of a line in slope–intercept form that passes through $(5, -2)$ with slope of $\frac{3}{5}$.

Objective 1 Graph linear inequalities in the coordinate plane.

Now let's graph **linear inequalities in two variables**, such as $3x + 2y > 6$ and $y \leq x - 2$, in the coordinate plane.

Note Linear inequalities have the same form as linear equations except that they contain an inequality symbol instead of an equal sign.

Definition Linear inequality in two variables: An inequality that can be written in the form $Ax + By > C$, where A, B, and C are real numbers and A and B are not both 0. Note that the inequality could also be $<$, $\leq$, or $\geq$.

A solution to a linear inequality is an ordered pair that makes the inequality true. For example, the ordered pair $(2, 4)$ is a solution for $3x + 2y > 6$.

$$3x + 2y > 6$$
$$3(2) + 2(4) \overset{?}{>} 6 \quad \text{Replace } x \text{ with 2 and } y \text{ with 4.}$$
$$14 > 6 \quad \text{This is true, so } (2, 4) \text{ is a solution.}$$

Recall that the graph of an equation in the coordinate plane is its solution set. This is also true for inequalities. The solution set for a linear inequality, such as $3x + 2y \geq 6$, contains ordered pairs that satisfy $3x + 2y > 6$ *or* $3x + 2y = 6$. The graph of the equation $3x + 2y = 6$ is a line, which we can graph using intercepts $(2, 0)$ and $(0, 3)$. That line is called a *boundary* for the graph of the inequality, and it separates the coordinate plane into two half planes. One of those half planes is the solution region for $3x + 2y > 6$.

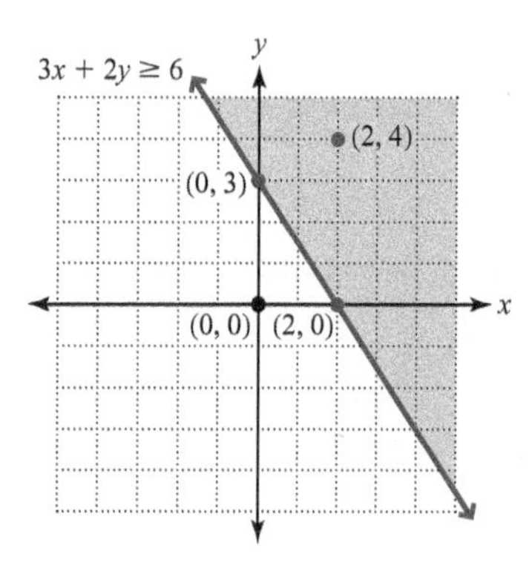

Earlier we showed that $(2, 4)$ is a solution for $3x + 2y > 6$. In fact, every ordered pair in the half plane with $(2, 4)$ is a solution for $3x + 2y > 6$; so we shade that region. Therefore, the solution set for $3x + 2y \geq 6$ contains every ordered pair on the line $3x + 2y = 6$ along with every ordered pair in the half plane containing $(2, 4)$.

Ordered pairs in the other half plane do not satisfy the inequality, which is why it is not shaded. For example, consider $(0, 0)$.

$$3(0) + 2(0) \overset{?}{\geq} 6$$
$$0 \geq 6 \quad \text{This inequality is false, so } (0, 0) \text{ is not a solution.}$$

Answers to warm-up

1.

2.

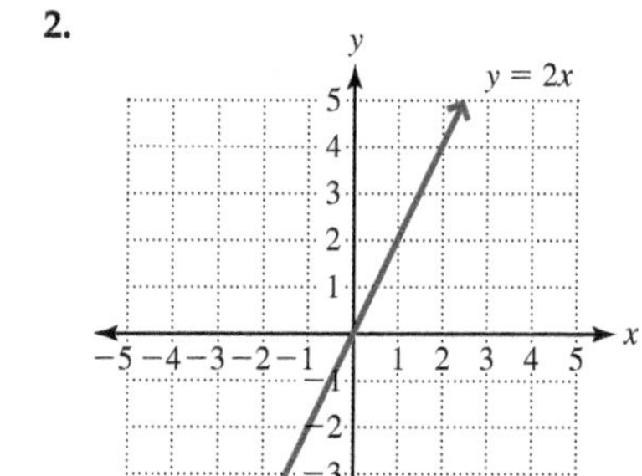

3. $y = \frac{3}{5}x - 5$

If the inequality were $3x + 2y > 6$, we would shade the same region, but because the ordered pairs on the line $3x + 2y = 6$ are *not* in the solution set, we draw a dashed line instead of a solid line as shown to the right.

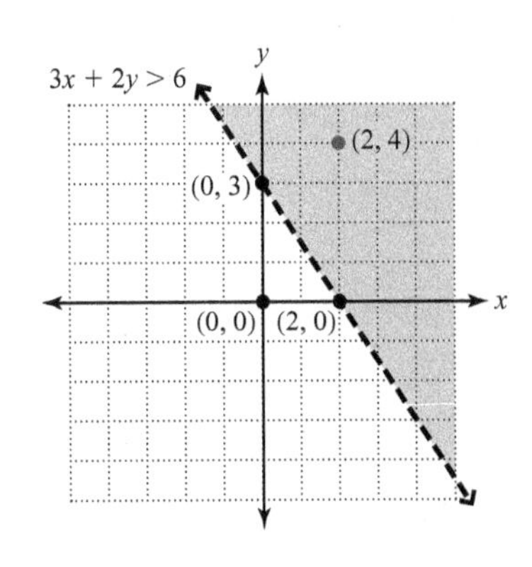

For the inequalities $3x + 2y \leq 6$ and $3x + 2y < 6$, we shade the half plane on the other side of the boundary line, which contains $(0, 0)$.

Connection When graphing a solution set on a number line for inequalities involving $\leq$ or $\geq$, we used brackets for end values. For inequalities involving $<$ or $>$, we used parentheses to show that the end values are not part of the solution set. When solution sets for inequalities are graphed in the coordinate plane, a solid line is like a bracket, whereas a dashed line is like a parenthesis.

The graph of $3x + 2y \leq 6$:

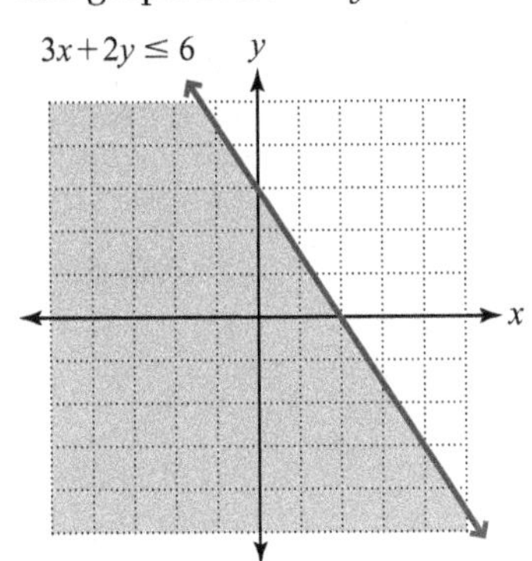

The graph of $3x + 2y < 6$:

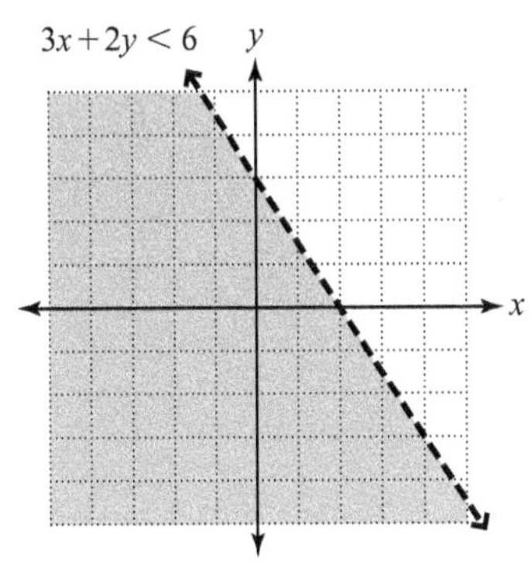

Procedure Graphing Linear Inequalities

To graph a linear inequality in two variables:

1. Graph the related equation (the boundary line). The related equation has an equal sign in place of the inequality symbol. If the inequality symbol is $\leq$ or $\geq$, draw a solid boundary line. If the inequality symbol is $<$ or $>$, draw a dashed boundary line.
2. Choose an ordered pair on one side of the boundary line and test this ordered pair in the inequality. If the ordered pair satisfies the inequality, shade the region that contains it. If the ordered pair does not satisfy the inequality, shade the region on the other side of the boundary line.

Example 1 Graph.

a. $y < -2x$

Solution: First, graph the related equation $y = -2x$.

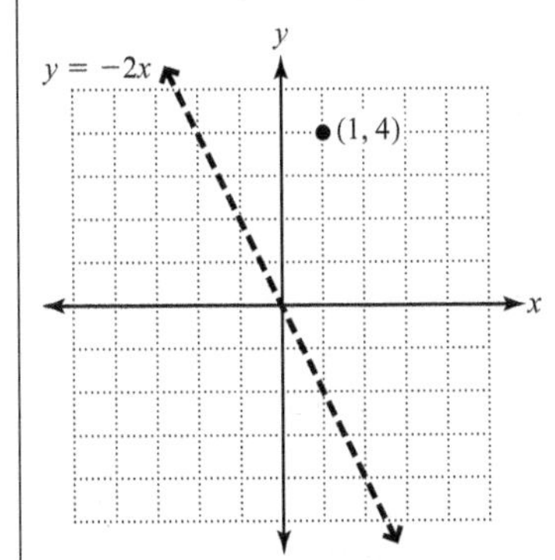

Because the inequality is $<$, we draw a dashed line to indicate that ordered pairs along the boundary line are not in the solution set.

Next, we choose a point, $(1, 4)$, to see if it satisfies the inequality.

$4 \overset{?}{<} -2(1)$ Replace x with 1 and y with 4.

$4 < -2$ This inequality is false, so $(1, 4)$ is not a solution.

Because our chosen point $(1, 4)$ did not satisfy the inequality, we shade the region on the other side of the boundary line.

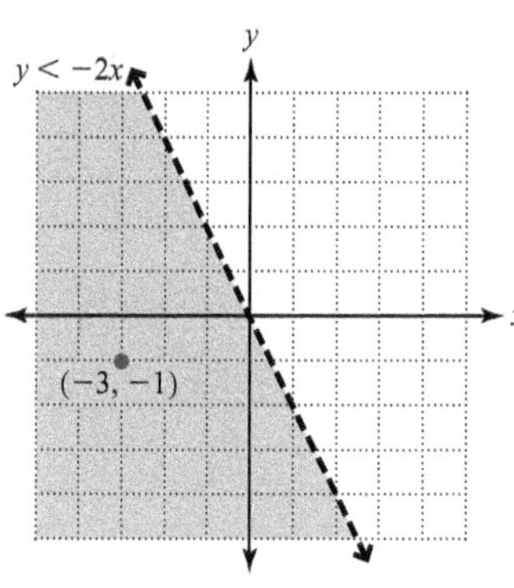

We can confirm our shading by checking an ordered pair in that region. We will choose $(-3, -1)$.

$-1 \overset{?}{<} -2(-3)$ Replace x with -3 and y with -1.

$-1 < 6$ This inequality is true, confirming that we have shaded the correct half plane.

b. $x - 3y \le 6$

Solution:

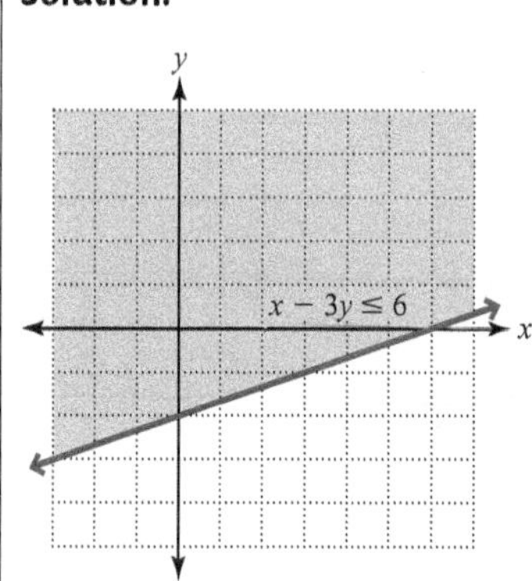

Because the inequality is $\le$, when we graph the related equation $x - 3y = 6$, we draw a solid line to indicate that ordered pairs along the boundary line are in the solution set.

We will choose the origin $(0, 0)$ as our test point.

$$0 - 3(0) \overset{?}{\le} 6$$
$$0 \le 6$$

Because this is true, $(0, 0)$ is a solution and we shade the half plane containing it.

Note The origin is often a good choice for the test point because it's easy to work with. However, the origin cannot be used if it lies on the boundary line.

c. $x > 2$

Solution:

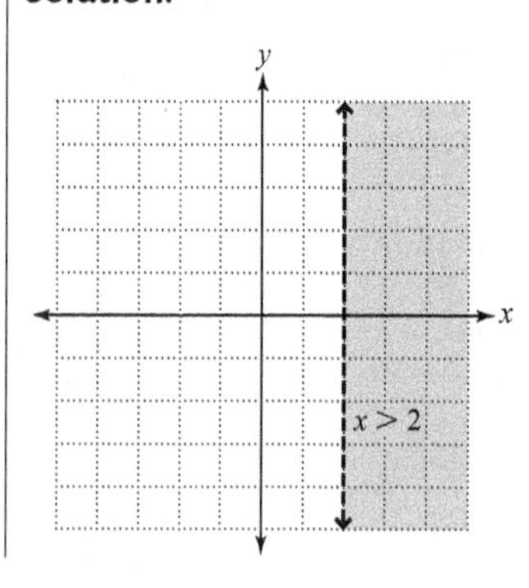

Because the inequality is $>$, when we draw the related equation $x = 2$, we draw a dashed line to indicate that ordered pairs along the boundary line are not in the solution set.

We could use a test point, but it isn't necessary. As on the number line, the ordered pairs whose x-values are greater than 2 are to the right of those whose x-values are equal to 2. So shade the region to the right of the boundary line.

Your Turn 1 Graph.

a. $y < -\dfrac{2}{5}x + 3$ **b.** $y \le -3$

Answers to Your Turn 1

a.

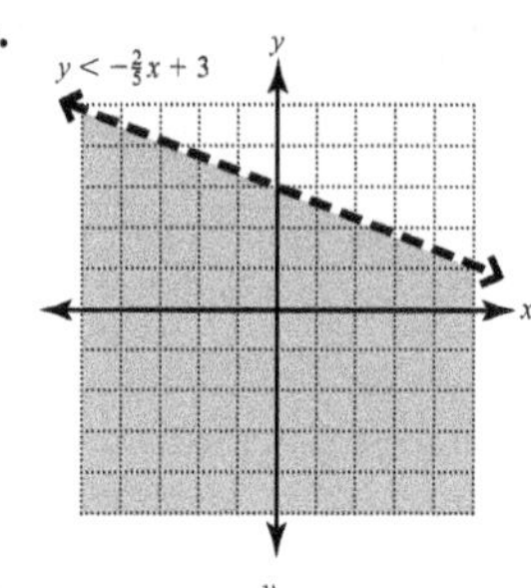

b.

Example 2 A drink stand in an amusement park sells two sizes of drinks. The large size sells for \$3; the smaller, for \$2. Park management believes that the stand needs to have a total revenue from drink sales of at least \$600 each day to be profitable.

a. Write an inequality that describes the amount of revenue the stand must make to be profitable.
b. Graph the inequality.
c. Find two combinations of the number of large and small drinks that must be sold to be profitable.

Solution: **a.** We can use a table like those from Chapter 2.

Category	Price	Number Sold	Revenue
Large	3.00	x	$3x$
Small	2.00	y	$2y$

The total revenue is found by the expression $3x + 2y$. If that total revenue must be at least 600, we can write the following inequality.

$$3x + 2y \geq 600$$

b.

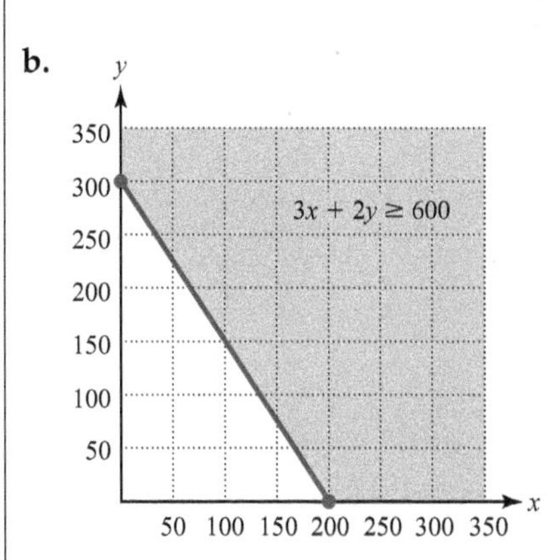

Note We show only the first quadrant because it is impossible to sell a negative number of drinks. The line represents every combination of numbers of large and small drinks sold that will produce exactly $600 in revenue. The shaded region represents all combinations that produce revenue exceeding $600. The shaded region also includes non-integer values that solve the inequality but do not solve the problem because we cannot sell fractions of a drink.

c. Assuming that fractions of a particular size are not sold, we give only whole number combinations that are profitable. One combination is selling 200 large drinks and 0 small drinks, which gives exactly $600. A second combination is selling 100 large drinks and 200 small drinks, which gives a total revenue of $700.

Answers to Your Turn 2

a. $2x + 3y \leq 84$

b.

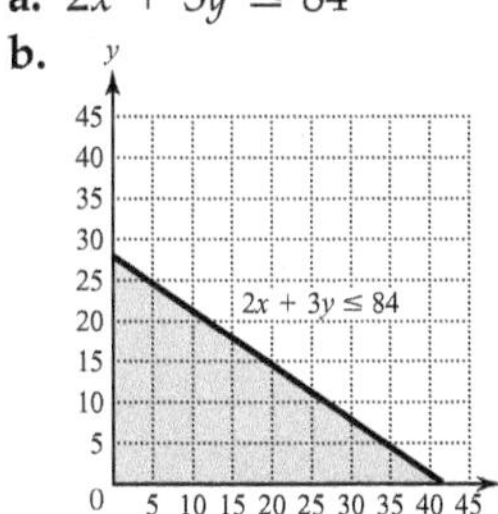

c. no

Your Turn 2 A small company manufactures lawn chairs. It takes 2 hours of labor to make a regular chair and 3 hours of labor to make a lounge chair. A maximum of 84 hours of labor are available per day.

a. Let x represent the number of regular chairs produced and y the number of lounge chairs produced in a day. Write an inequality describing the relationship between the number of each type of chair produced and the labor available.

b. Graph the inequality.

c. Is it possible to produce 20 regular chairs and 20 lounge chairs in a day?

3.4 Exercises

For Extra Help MyMathLab®

Objective 1

Prep Exercise 1 Why is $(3, -1)$ a solution to $x + y < 4$?

For Exercises 1–8, determine whether the ordered pair is a solution for the linear inequality.

1. $(-1, 3); y \geq -x + 7$

2. $(-1, -1); y > x + 15$

3. $(0, 0); 2x + 3y \geq 0$

4. $(7, 2); 3x - 2y > 4$

5. $(3, 4); 3y - x < 8$

6. $(0, -3); y - x \leq -5$

7. $(5, 3); y > \frac{3}{4}x - 5$

8. $(0, 1); y > \frac{2}{3}x + 4$

Prep Exercise 2 When graphing a linear inequality in two variables, what is the first step?

Prep Exercise 3 Explain how to determine the related equation for a given inequality.

Prep Exercise 4 When drawing the boundary line for a linear inequality, use a dashed line if the inequality symbol is ____________ or ____________.

Prep Exercise 5 When drawing the boundary line for a linear inequality, use a solid line if the inequality symbol is ____________ or ____________.

Prep Exercise 6 When graphing a linear inequality, how do you determine which side of the boundary line to shade?

For Exercises 9–34, graph the linear inequality. See Example 1.

9. $y \leq x + 3$

10. $y \leq x - 1$

11. $y \geq -4x + 6$

12. $y \geq -3x + 8$

13. $y > x$

14. $y < 2x$

15. $y > -3x$

16. $y > -x$

17. $y > \frac{2}{5}x$

18. $y < \frac{1}{4}x$

19. $x - y < 2$

20. $x - y < 5$

21. $x + 3y > -9$

22. $x + 4y < -8$

23. $x - 2y \geq -6$

24. $x - 3y \geq -3$

25. $3x - 2y > 6$ **26.** $2x - 5y < 10$ **27.** $5x - y \leq 0$ **28.** $x + 7y \geq -14$

29. $4x + 2y \leq 3$ **30.** $5x + 2y < 9$ **31.** $x > 6$ **32.** $x \leq -1$

33. $y \leq 7$ **34.** $y > 5$

For Exercises 35–40, solve. See Example 2.

35. A company produces two versions of a game. The board game costs \$10 per unit to produce, and the video game costs \$20 per unit to produce. Management plans for the cost of production to be a maximum of \$250,000 for the first quarter. The inequality $10x + 20y \leq 250{,}000$ describes the total cost as prescribed by management.

a. What do x and y represent?

b. Graph the inequality. Because x and y represent nonnegative numbers, the graph should be in the first quadrant only ($x \geq 0$ and $y \geq 0$).

c. What does the line represent?

d. What does the shaded region represent?

e. List two combinations of numbers of units produced of each version of the game that yield a total cost of \$250,000.

f. List two combinations of numbers of units produced of each version of the game that yield a total cost less than \$250,000.

g. In reality, is every combination of units produced that is represented by the line and shaded region a possibility? Explain.

36. A cosmetics company sells two different sizes of foundation. The revenue for each unit of the large bottle is $5. The revenue for each unit of the small bottle is $3.50. For the company to break even during the first quarter, the company must generate $400,000 in revenue. The inequality $5x + 3.5y \geq 400{,}000$ describes the amount of revenue that must be generated from each size bottle for the company to break even or turn a profit.

a. What do x and y represent?

b. Graph the inequality. Because x and y represent nonnegative numbers, the graph should be in the first quadrant only ($x \geq 0$ and $y \geq 0$).

c. What does the boundary line represent?

d. What does the shaded region represent?

e. List two combinations of sales numbers of each size that allow the company to break even.

f. List two combinations of sales numbers of each size that allow the company to turn a profit.

g. In reality, is every combination of units sold that is represented by the line and shaded region a possibility? Explain.

37. A rectangular fence is to be designed so that the maximum perimeter is 200 feet.

a. Letting l represent the length and w the width, write an inequality in which the maximum perimeter is 200 feet.

b. Graph the inequality. Note that $l > 0$ and $w > 0$.

c. What does the line represent?

d. What does the shaded region represent?

e. Find two combinations of length and width that yield a perimeter of exactly 200 feet.

f. Find two combinations of length and width that yield a perimeter of less than 200 feet.

38. A landscaper plans a rectangular garden so that the maximum perimeter is 250 feet.

a. Letting l represent the length and w the width, write an inequality in which the maximum perimeter is 250 feet.

b. Graph the inequality. Note that $l > 0$ and $w > 0$.

c. What does the line represent?

d. What does the shaded region represent?

e. Find two combinations of length and width that yield a perimeter of exactly 250 feet.

f. Find two combinations of length and width that yield a perimeter of less than 250 feet.

39. Roberto is coordinating a fund-raiser for a high school marching band in which students sell two different boxes of fruit. A box of oranges sells for $12; a box of grapefruits, for $15. The goal is to raise at least $18,000.

a. Letting x represent the number of boxes of oranges sold and y the number of boxes of grapefruits sold, write an inequality in which the total sales is at least $18,000.

b. Graph the inequality. Note that $x \geq 0$ and $y \geq 0$.

c. What does the line represent?

d. What does the shaded region represent?

e. Find two combinations of fruit boxes sold that yield exactly $18,000 in total sales.

f. Find two combinations of fruit boxes sold that yield more than $18,000.

40. Veronica visits a home center to purchase new border edging for her garden. She has a gift certificate for $75. Edging comes in two styles: straight edge pieces, which cost $3.50 each, and scalloped edge pieces, which cost $5.50 each.

a. Letting x represent the number of straight edge pieces and y the number of scalloped edge pieces, write an inequality that has her total cost within the amount of the gift certificate.

b. Graph the inequality. Note that $x \geq 0$ and $y \geq 0$.

c. What does the line represent?

d. What does the shaded region represent?

e. Find two combinations of containers she could purchase.

f. In reality, are there any combinations she could purchase that would yield a total in the exact amount of the gift certificate?

Review Exercises

Exercises 1–4 Expressions

[1.1] *For Exercises 1 and 2, use the ordered pairs* $(-2, 7)$, $(0, 1)$, $(3, 5)$, *and* $(4, 2)$.

1. Write a set containing all of the x-coordinates.

2. Write a set containing all of the y-coordinates.

[1.4] **3.** Evaluate $x^2 - 4x + 1$ when $x = -2$.

[1.4] **4.** Find every value for x that makes $\frac{x}{x-3}$ undefined.

Exercises 5 and 6 Equations and Inequalities

[2.1] **5.** Solve and check: $3x - 4(x + 8) = 5(x - 1) + 21$

[3.1] **6.** Find the x- and y-intercepts for $4x - 5y = 8$.

3.5 Introduction to Functions and Function Notation

Objectives

1 Identify the domain and range of a relation and determine whether a relation is a function.
2 Find the value of a function.
3 Graph functions.

Warm-up

[1.4] **1.** Evaluate $-5x + 3$ for $x = 4$.

[1.4] **2.** Evaluate $\frac{x - 3}{2x + 4}$ for $x = -2$.

[3.4] **3.** Graph $y < 3x + 4$.

Objective 1 Identify the domain and range of a relation and determine whether a relation is a function.

We have used ordered pairs to generate graphs of equations and inequalities in two variables. Ordered pairs are the foundation of two important concepts in mathematics, *relations* and *functions*. First, we consider **relations**.

Definition Relation: A set of ordered pairs.

For example, the accompanying relation contains ordered pairs that track the price of a share of a particular stock at the close of the stock market each day for a week.

Day	Closing Price
1	\$6.50
2	\$8.00
3	\$10.00
4	\$9.25
5	\$8.00

Because a relation involves ordered pairs such as $(4, 9.25)$, it has two sets of values called the **domain** and **range**.

Definitions Domain: A set containing initial values of a relation; its input values; the first coordinates in ordered pairs.

Range: A set containing all values that are paired to domain values in a relation; its output values; the second coordinates in ordered pairs.

Note When giving the domain and range, it isn't necessary to relist repeated values.

For example, the domain for our stock price relation contains the number of each day; so the domain is {1, 2, 3, 4, 5}. The range contains all of the prices; so the range is {6.50, 8.00, 10.00, 9.25}.

Note that each day has exactly one closing price, which makes this relation a **function**.

Note Every input in a function has exactly one output.

Definition Function: A relation in which each value in the domain is assigned to exactly one value in the range.

From our stock price example, we see that a function can have different values in the domain assigned to the same value in the range. For example, domain values 2 and 5 correspond to the same range value, 8.00.

Using arrows to map the domain values to the range values can be a helpful technique in determining whether a relation is a function. Such a map might look like this:

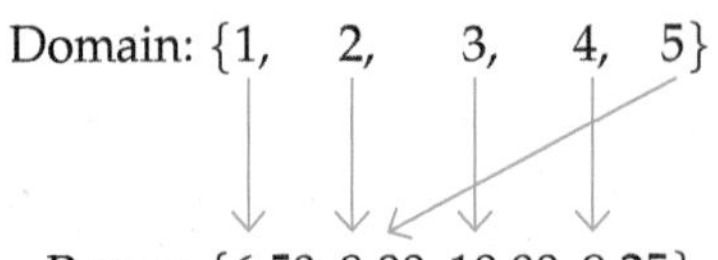

Note Each element in the domain has a single arrow pointing to an element in the range.

A relation is not a function if any value in the domain is assigned to more than one value in the range. For example, suppose we have a domain that contains some players on the national champion 2008 University of Florida football team and a range containing the positions they play. The following relation shows their position assignments.

Answers to Warm-up

1. -17

2. undefined

3.

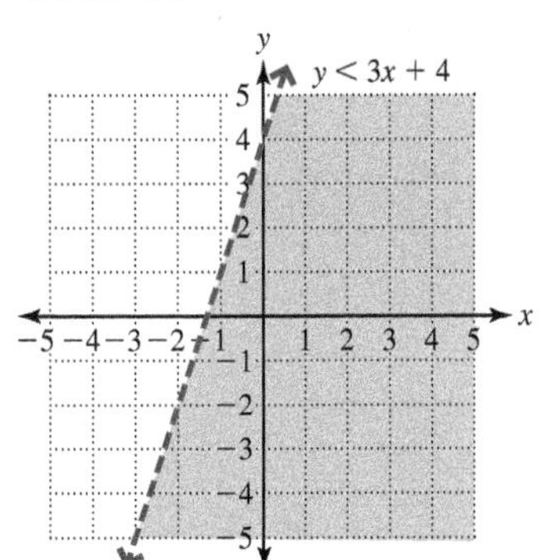

Note We now have an element in the domain, Percy Harvin, assigned to three elements in the range; so the relation is not a function.

Domain: {Tim Tebow, Cornelius Ingram, Kestahn Moore, Percy Harvin}

Range: {quarterback, receiver, tailback, punt returner}

Conclusion: If any value in the domain is assigned to more than one value in the range, the relation is not a function.

Example 1 Identify the domain and range of the relation; then determine whether it is a function.

a. Birthrate in the United States for each year:

Year	Rate per 1000 People
2000	14.7
2001	14.1
2002	13.9
2003	14.1
2004	14.0
2005	14.0

Source: Centers for Disease Control and Prevention, National Center for Health Statistics

Note In this text, relations in table form always have the domain in the left column and the range in the right column.

Solution: Domain: $\{2000, 2001, 2002, 2003, 2004, 2005\}$
Range: $\{14.7, 14.1, 13.9, 14.0\}$

The relation is a function because every element in the domain is paired with exactly one element in the range.

b. Results of the 2013 Masters golf tournament

Place	Player
1	Adam Scott
2	Angel Cabrera
3	Jason Day
4	Marc Leishman, Tiger Woods

Solution: Domain: $\{1, 2, 3, 4\}$
Range: {Adam Scott, Angel Cabrera, Jason Day, Tiger Woods, Marc Leishman}

The relation is not a function because an element in the domain, 4, is paired with more than one element in the range: Marc Leishman *and* Tiger Woods.

c. $\{(1, 6), (-1, 4), (3, 6), (-4, -2)\}$

Solution: Domain: $\{1, -1, 3, -4\}$ Range: $\{6, 4, -2\}$

The relation is a function because every element in the domain is paired with exactly one element in the range.

Your Turn 1 Identify the domain and range of the relation and determine whether it is a function.

a. Unemployment rate as a percent of the population from January to June of 2012:

Month	Unemployment Rate
1	8.3
2	8.3
3	8.2
4	8.1
5	8.2
6	8.2

Source: U.S. Department of Labor, Bureau of Labor Statistics

b. Depth chart for each position on the 2013 Boston Red Sox roster:

Position	Player
Left field	J. Gomes, D. Nava, M. Carp
Center field	J. Ellsbury, S. Victorino, R. Sweeney, R. Kalish
Right field	S. Victorino, R. Sweeney

c. $\{(-1, 4), (2, -7), (3, -5), (-1, 3)\}$

Answers to Your Turn 1

a. Domain: $\{1, 2, 3, 4, 5, 6\}$
Range: $\{8.1, 8.2, 8.3\}$
It is a function

b. Domain: {left field, center field, right field}
Range: {J. Gomes, D. Nava, M. Carp, J. Ellsbbury, S. Victorino, R. Sweeney, R. Kalish}
It is not a function.

c. Domain: $\{-1, 2, 3\}$
Range: $\{4, -7, -5, 3\}$
It is not a function.

Because relations involve ordered pairs, we sometimes see them as graphs in the coordinate plane. Consider a graph of the stock price relation that we explored earlier.

Notice that the domain is plotted along the horizontal axis and the range is plotted along the vertical axis. On a graph in the standard x- and y-coordinate plane, the domain will be the x-coordinates and the range will be the y-coordinates.

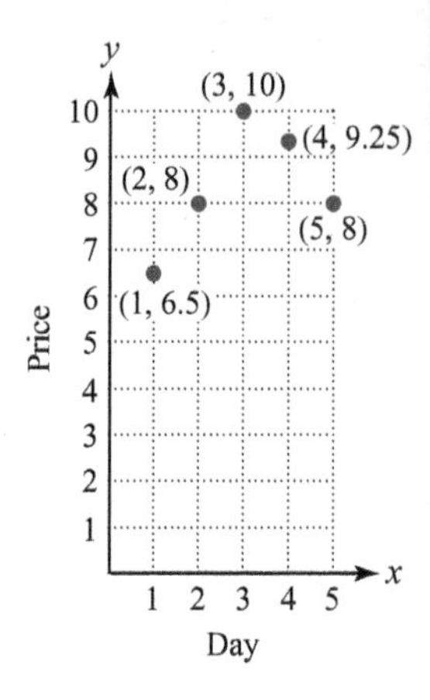

Procedure Determining the Domain and Range of a Graph

The domain is a set containing the first coordinate (x-coordinate) of every point on the graph.
The range is a set containing the second coordinate (y-coordinate) of every point on the graph.

To determine whether a graphed relation is a function, we can perform the *vertical line test*. Remember, on a function's graph, every x-coordinate will correspond to exactly one y-coordinate. Consequently, a vertical line drawn through each x-coordinate in the domain of a function will intersect the graph at only one point.

For example, in our graph of the stock price relation, we have drawn vertical lines, shown in blue, through each x-coordinate in the domain. Notice that every one of those vertical lines intersects only one point; so the relation is a function.

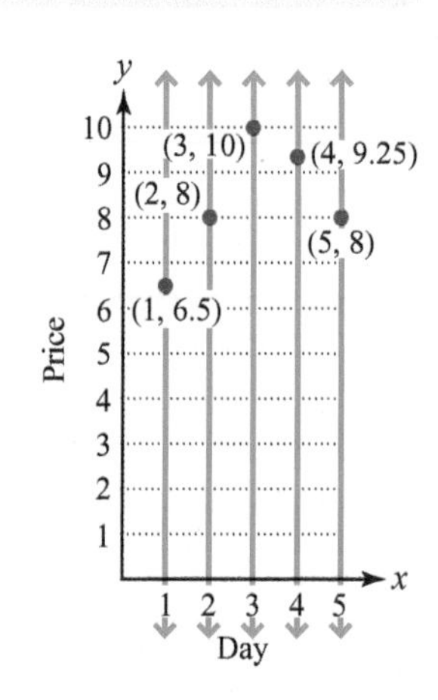

Procedure Vertical Line Test

To determine whether a graphical relation is a function, draw or imagine vertical lines through each value in the domain. If each vertical line intersects the graph at only one point, the relation is a function. If any vertical line intersects the graph more than once, the relation is not a function.

Example 2 Identify the domain and range of the relation; then determine whether it is a function.

Note The arrowheads at the ends of the graph indicate that the graph continues indefinitely. We say that the graph continues to infinity and use the symbol ∞ to represent infinity and $-\infty$ to represent negative infinity.

a.

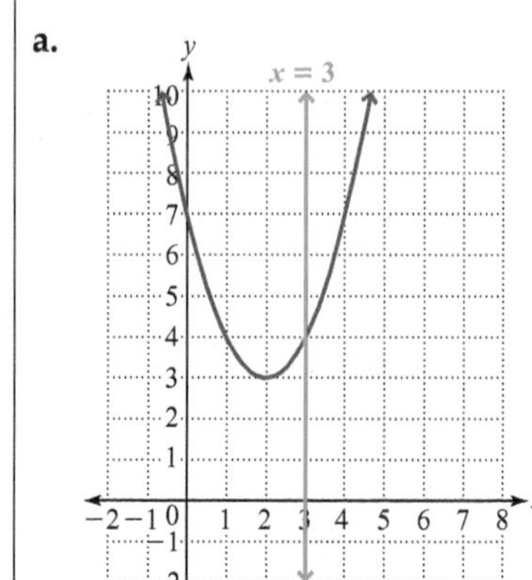

Solution: Domain: $\{x \mid -\infty < x < \infty\}$ or $(-\infty, \infty)$
Range: $\{y \mid 3 \leq y < \infty\}$ or $[3, \infty)$

This relation is a function because a vertical line at each x-value would intersect the graph at only one point. We have shown one such vertical line in blue passing through $x = 3$.

b.

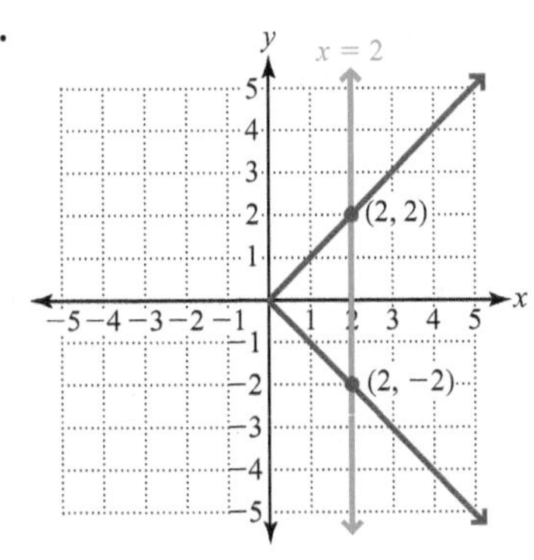

Solution: Domain: $\{x \mid 0 \le x < \infty\}$ or $[0, \infty)$
Range: $\mathbb{R}$ or $(-\infty, \infty)$

This relation is not a function because there are values in the domain that correspond to two values in the range. We can see this with the vertical line test. For example, $x = 2$ (shown in blue) passes through two different points on the graph, $(2, 2)$ and $(2, -2)$.

c. The following relation is from a survey that shows the percent of travelers that choose to use their personal vehicle according to trip distance.

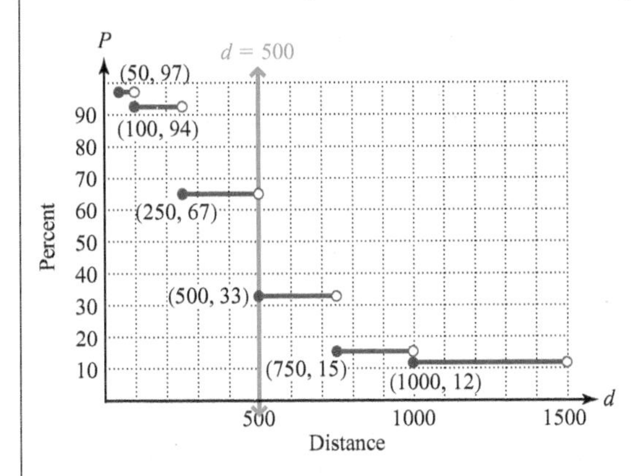

Solution:

Domain: $\{d \mid 50 \le d < 1500\}$ or $[50, 1500)$
Range: $\{12\%, 15\%, 33\%, 67\%, 94\%, 97\%\}$

This relation is a function. Remember that solid circles indicate values that are in the set, whereas open circles are not. Notice that each open circle and solid circle align vertically so that a vertical line through them intersects the graph at only one point, the solid circle. For example, $d = 500$ (shown in blue) intersects the graph only at the solid circle at 33%.

Your Turn 2 Identify the domain and range of the relation; then determine whether it is a function.

a.

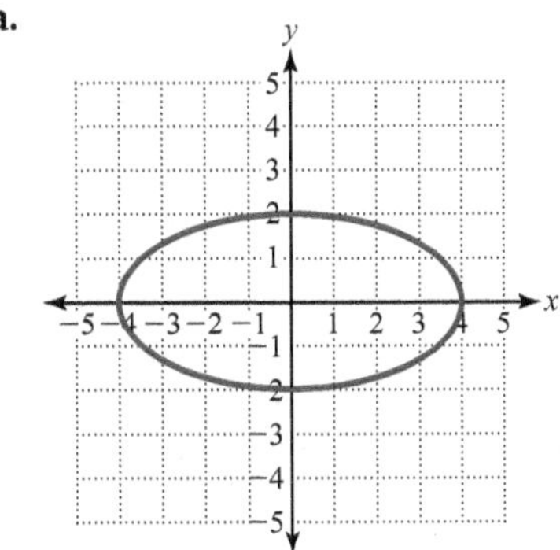

b.

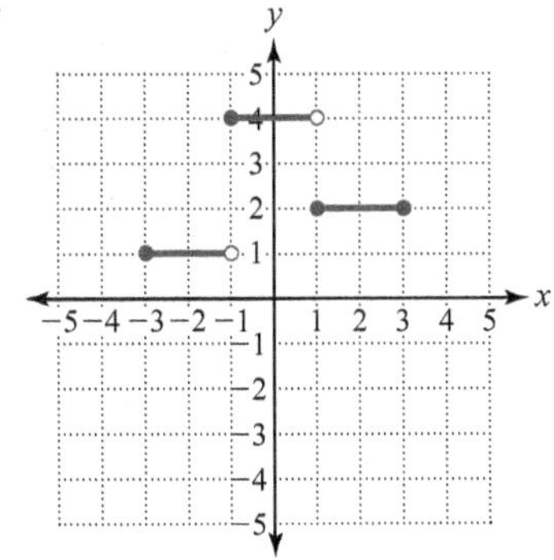

Objective 2 Find the value of a function.

We have seen functions as paired sets and graphs, but they also can take the form of equations. In equation form, functions are essentially the same as equations in two variables. For example, we can write the equation $y = 3x + 1$ as $f(x) = 3x + 1$ in function notation. Notice that the notation $f(x)$, which is read "a function in terms of x," or "f of x," takes the place of the variable y. The following table shows some equations in two variables and their corresponding function form:

Answers to Your Turn 2

a. Domain: $\{x \mid -4 \le x \le 4\}$ or $[-4, 4]$
Range: $\{y \mid -2 \le y \le 2\}$ or $[-2, 2]$
It is not a function.

b. Domain: $\{x \mid -3 \le x \le 3\}$ or $[-3, 3]$
Range: $\{1, 2, 4\}$
It is a function.

Note $f(x)$ does not mean f times x.

Equation in two variables:	Function notation:
$y = -2x + 3$	$f(x) = -2x + 3$
$y = \frac{3}{4}x - 1$	$f(x) = \frac{3}{4}x - 1$

We can think of the x-values in equations such as $y = 3x + 1$ as inputs and the y-values as outputs. Similarly, the domain's x-values in $f(x) = 3x + 1$ are the inputs and the resulting function values, $f(x)$, are its outputs (elements of the range). Compare the following tables.

x	y
0	$3(0) + 1 = 1$
1	$3(1) + 1 = 4$
2	$3(2) + 1 = 7$

x	$f(x)$
0	$3(0) + 1 = 1$
1	$3(1) + 1 = 4$
2	$3(2) + 1 = 7$

Both tables list the same information. Finding the value of a function is the same as finding the y-coordinate of an equation in two variables that is solved for y. In both cases, we input a value for x and calculate the output value, which is y or $f(x)$ depending on the form.

Function notation offers a clever way to indicate that a specific x-value is to be used. For the function $f(x) = 3x + 1$, we indicate that we want the value of the function when $x = 2$ by writing "find $f(2)$."

Procedure Finding the Value of a Function

Given a function $f(x)$, to find $f(a)$, where a is a real number in the domain of f, replace x in the function with a and then evaluate or simplify.

Connection Finding the value of a function is similar to evaluating an expression. Following are function and expression forms that indicate essentially the same procedure.

Function language:

If $f(x) = 3x$, find $f(2)$.

Expression language:

Evaluate the expression $3x$ when $x = 2$.

Example 3 a. For the function $f(x) = -5x + 3$, find $f(4)$.

Solution: $f(4) = -5(4) + 3$ Replace x with 4; then evaluate.
$= -20 + 3$ Multiply.
$= -17$ Add.

b. For the function $f(x) = 4x^2 - 3$, find $f(n)$.

Solution: $f(n) = 4(n)^2 - 3$ Replace x with n; then simplify.
$= 4n^2 - 3$ Square n.

Connection For the function in part a, we found the ordered pair $(4, -17)$. For part b, the ordered pair is $(n, 4n^2 - 3)$.

Your Turn 3

a. For the function $f(x) = 0.5x - 2$, find $f(-6)$.
b. For the function $f(x) = x^2 - 4x - 3$, find $f(a)$.

Answers to Your Turn 3
a. -5 b. $a^2 - 4a - 3$

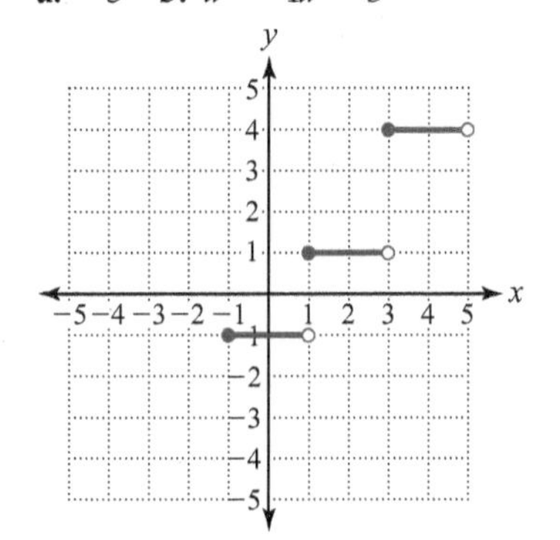

We also can find the value of a function given its graph. For a given value in the domain (x-value), we look on the graph for the corresponding value in the range (y-value).

Example 4 Use the graph in the margin to find the indicated value of the function.

a. $f(0)$

Solution: The notation $f(0)$ means to find the value of the function (y-value) when $x = 0$. On the graph, we see that when $x = 0$, the corresponding y-value is -1; so we say that $f(0) = -1$.

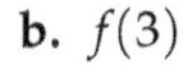

b. $f(3)$

Solution: When $x = 3$, we see that $y = 4$; so $f(3) = 4$.

c. $f(-4)$

Solution: The domain for this function is $\{x \mid -1 \le x < 5\}$. Because -4 is not in the domain, we say that $f(-4)$ is undefined.

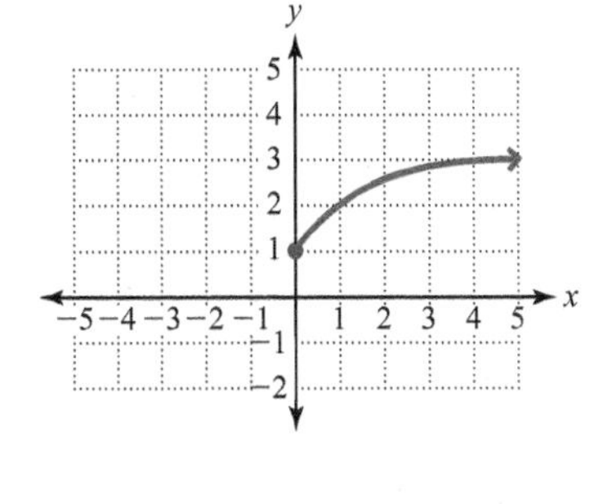

Your Turn 4 Use the graph in the margin to find the indicated value of the function.

a. $f(-2)$ **b.** $f(0)$ **c.** $f(5)$

Objective 3 Graph functions.

We create the graph of a function the same way we create the graph of an equation in two variables. First, we consider linear functions.

Linear Functions

Linear functions are similar to linear equations in slope–intercept form.

Slope–intercept form: $y = mx + b$ Linear function: $f(x) = mx + b$

The graph of a linear function is a straight line with slope m and y-intercept $(0, b)$.

Example 5 Graph $f(x) = -2x + 3$. Give the domain and range.

Solution: Think of the function as the equation $y = -2x + 3$. We could make a table of ordered pairs or use the fact that the slope is -2 and the y-intercept is 3.

Table of ordered pairs:

x	$f(x)$
0	3
1	1
2	−1

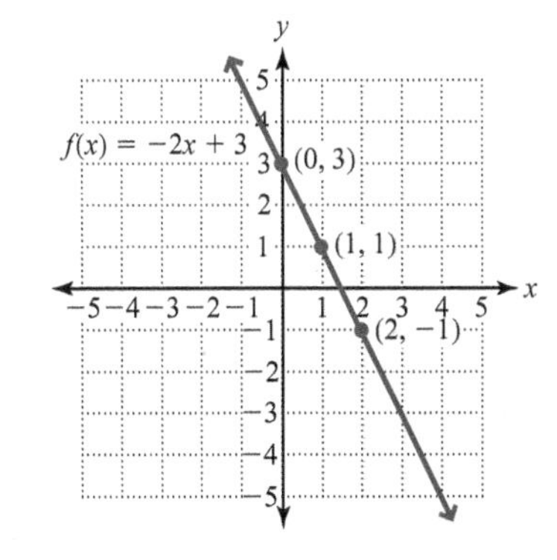

Domain: $\mathbb{R}$ or $(-\infty, \infty)$
Range: $\mathbb{R}$ or $(-\infty, \infty)$

Your Turn 5 Graph $f(x) = 3x + 1$. Give the domain and range.

Example 6 An electric and gas company charges \$10 per month plus \$3.1245 per cubic foot of natural gas used. The function $c(v) = 3.1245v + 10$ describes a customer's monthly cost for natural gas, where v represents the volume in cubic feet of natural gas used during the month.

a. Graph the function.

b. Find the cost if a customer uses 24 cubic feet of natural gas.

Answers to Your Turn 4
a. undefined **b.** 1 **c.** 3

Answer to Your Turn 5

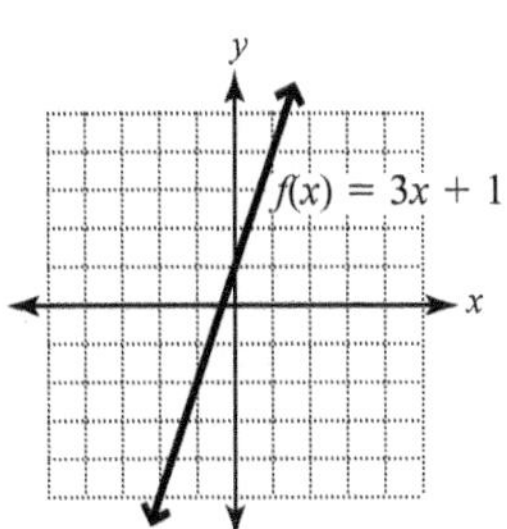

Domain: $\mathbb{R}$ or $(-\infty, \infty)$
Range: $\mathbb{R}$ or $(-\infty, \infty)$

Solution:

a.

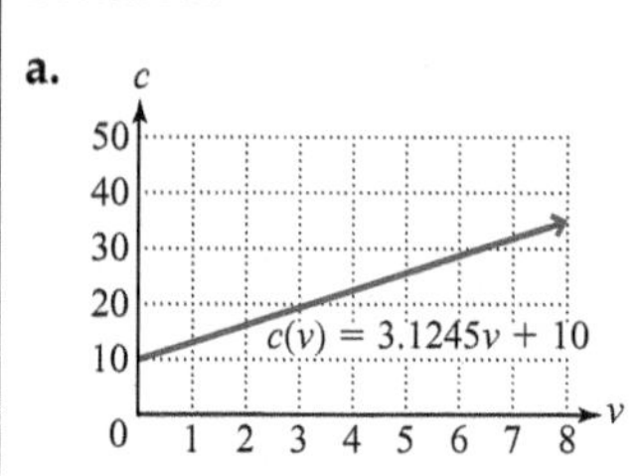

Note The slope of the line, 3.1245, is the cost per cubic foot of natural gas used each month. The y-intercept (or c-intercept in this case) is the $10 flat fee.

b. Finding the cost for using 24 cubic feet means that we are evaluating $c(24)$.

$$c(24) = 3.1245(24) + 10 = 84.988$$

Rounding to the nearest cent, the total cost is $84.99.

Absolute Value Functions

The basic absolute value function is $f(x) = |x|$. It is the same as $y = |x|$. As we learned with linear functions, to graph a function, we plot ordered pairs found by evaluating the function using various values of x. For example, if we let $x = -4$, we have $f(-4) = |-4| = 4$; so the ordered pair is $(-4, 4)$. Below, we complete a table of ordered pairs, plot those ordered pairs, and then connect the points to see that the graph is a V shape.

x	$f(x)$
−4	4
−2	2
0	0
2	2
4	4

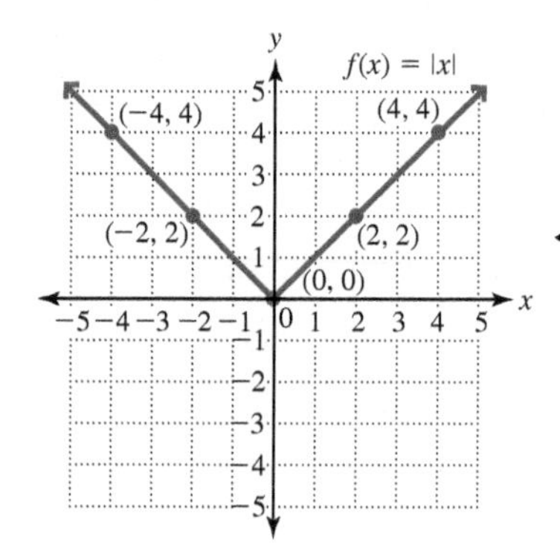

Note $f(x) = |x|$ is equivalent to $f(x) = -x$ when $x < 0$ and to $f(x) = x$ when $x \geq 0$. Notice that $f(x) = -x$ and $f(x) = x$ are linear functions with opposite slopes, which explains the V shape of the graph.

We can see that $f(x) = |x|$ is a function because it passes the vertical line test. Notice that every real number along the x-axis can be used as an input in the function, so the domain is $(-\infty, \infty)$. The range is $[0, \infty)$ because the minimum y-coordinate on the graph is 0 and the graph continues upward without end.

Example 7 Graph each function. Give the domain and range.

a. $f(x) = |x| + 1$

Solution:

x	$f(x)$
−4	5
−2	3
0	1
2	3
4	5

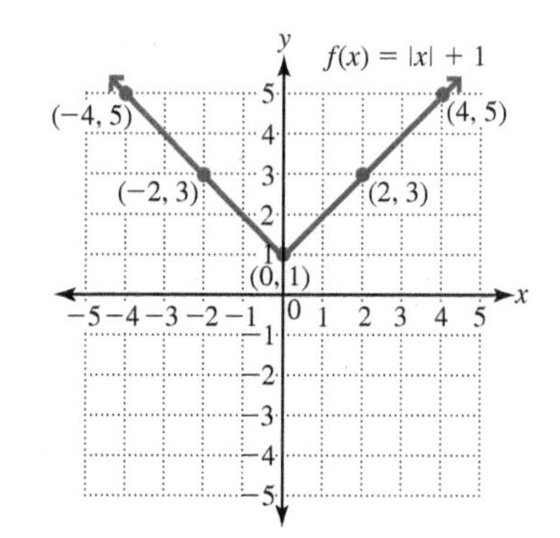

Note Every point on $f(x) = |x| + 1$ is one higher than every point on $f(x) = |x|$. Adding 1 to $|x|$ shifts the graph of $f(x) = |x|$ up 1 unit.

Domain: $(-\infty, \infty)$
Range: $[1, \infty)$

b. $f(x) = |x - 2| + 1$

Solution:

x	$f(x)$
−4	7
−2	5
0	3
2	1
4	3

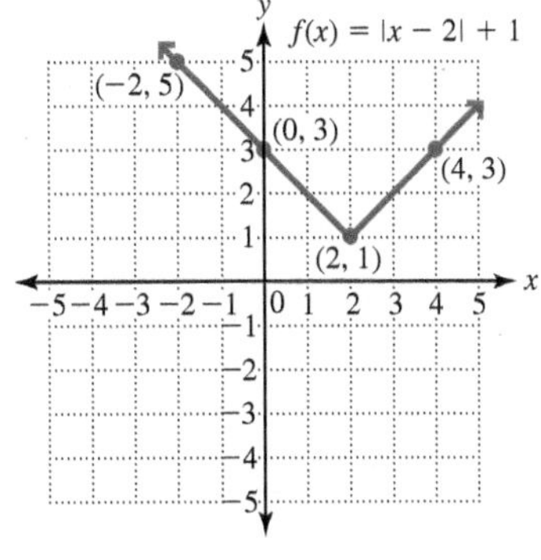

Note This graph is the graph of $f(x) = |x|$ shifted to the right 2 units and up 1 unit. The $x - 2$ inside the absolute value causes the shift to the right 2 units, and the +1 outside the absolute value causes the upward shift 1 unit.

Domain: $(-\infty, \infty)$
Range: $[1, \infty)$

c. $f(x) = -|x - 2| + 1$

Solution:

x	$f(x)$
−4	−5
−2	−3
0	−1
2	1
4	−1

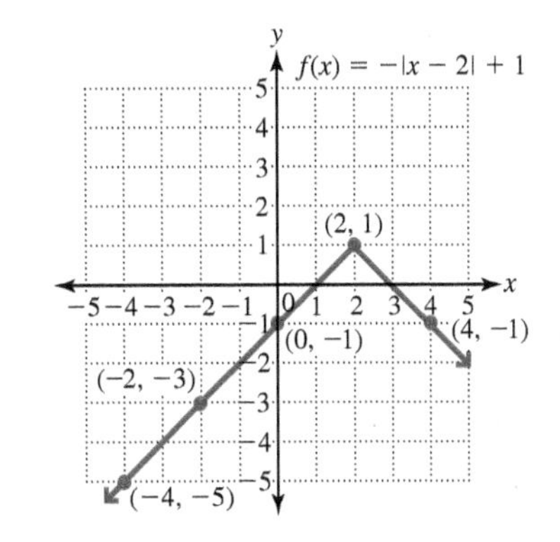

Note The minus sign in front of the absolute value causes the graph to invert so that the **V** is "upside down." This means that the range goes from a low of $-\infty$ up to 1.

Domain: $(-\infty, \infty)$
Range: $(-\infty, 1]$

Your Turn 7 Graph each function. Give the domain and range.

a. $f(x) = |x| - 2$

b. $f(x) = -2|x - 1| + 3$

Quadratic Functions

A quadratic function has the form $f(x) = ax^2 + bx + c$ where $a \neq 0$. The simplest quadratic function is $f(x) = x^2$. We will see that the graph of a quadratic function is a curve called a parabola. Because the graphs of quadratic functions are not straight lines, they are *nonlinear* functions. Let's make a table of ordered pairs for $f(x) = x^2$ to see the graph. We will explore nonlinear functions in greater detail later.

x	$f(x)$
−2	4
−1	1
0	0
1	1
2	4

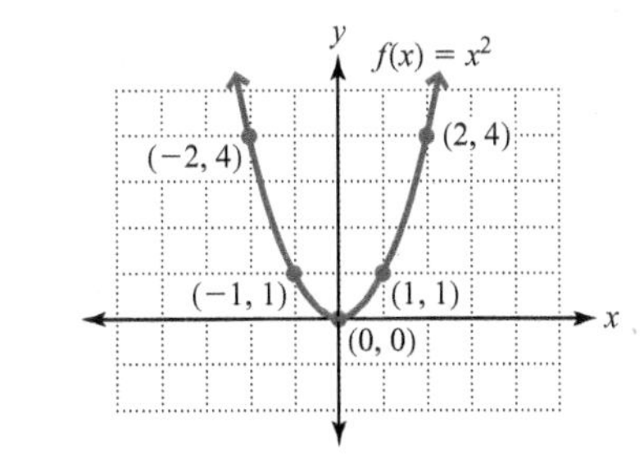

Note The domain is R or $(-\infty, \infty)$ and the range is $\{y | 0 \leq y < \infty\}$ or $[0, \infty)$.

Answers to Your Turn 7

a.

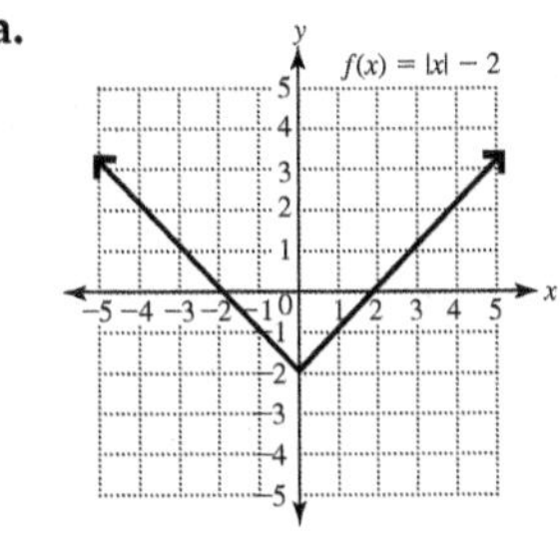

Domain: $\mathbb{R}$ or $(-\infty, \infty)$
Range: $[-2, \infty)$

b.

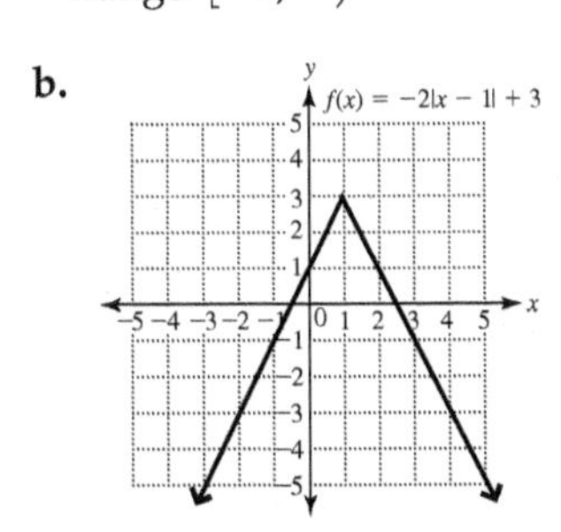

Domain: $\mathbb{R}$ or $(-\infty, \infty)$
Range: $(-\infty, 3]$

Learning Strategy

The best way to prepare for class is to complete all the assigned homework problems and read the material in the text that will be discussed in the next class.

—Paul S.

Example 8 Graph $f(x) = x^2 - 4$. Give the domain and range.

Solution:

x	$f(x)$
−3	5
−2	0
−1	−3
0	−4
1	−3
2	0
3	5

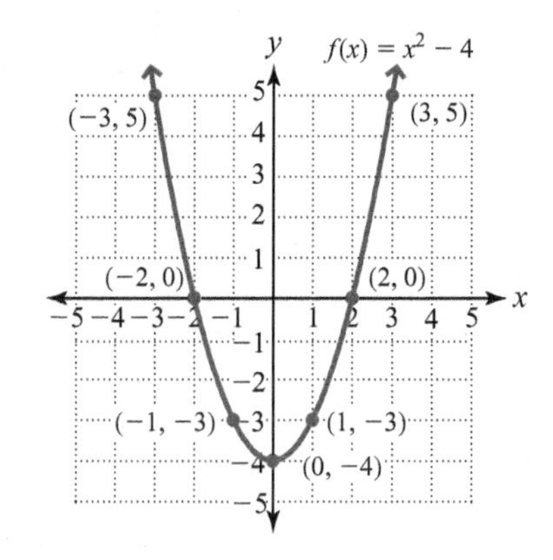

Note Subtracting 4 from x^2 shifts the graph of $f(x) = x^2$ down four units.

Domain: $\mathbb{R}$ or $(-\infty, \infty)$
Range: $\{y \mid -4 \le y < \infty\}$ or $[-4, \infty)$

Your Turn 8 Graph $f(x) = -x^2 + 1$. Give the domain and range.

Answer to Your Turn 8

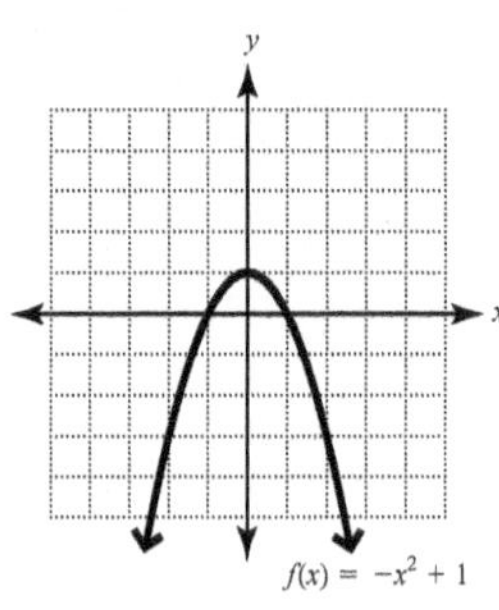

Domain: $\mathbb{R}$ or $(-\infty, \infty)$
Range:
$\{y \mid -\infty < y \le 1\}$ or $(-\infty, 1]$

Note: Exercises marked with a ★ represent challenging exercises.

Objective 1

Prep Exercise 1 What is the domain of a relation?

Prep Exercise 2 What is the range of a relation?

Prep Exercise 3 What makes a relation a function?

For Exercises 1–16, identify the domain and range of the relation; then determine whether it is a function. See Examples 1 and 2.

1. The five most expensive college tuitions according to the *Chronicle of Higher Education*, 2012–13:

Institution	Tuition
Sarah Lawrence College	\$61,236
New York University	\$59,337
Harvey Mudd College	\$58,913
Columbia College at Columbia University	\$58,742
Wesleyan University	\$58,502

2. The five most populated cities in 2013:

City	Population
New York	8,175,133
Los Angeles	3,792,621
Chicago	2,695,598
Houston	2,099,451
Philadelphia	1,526,006

Source: http://en.wikipedia.org

3. Superbowl wins as of December 2013. (*Source:* www.nfl.com)

Number of Wins	NFL Team
6	Pittsburgh Steelers
5	San Francisco 49ers, Dallas Cowboys
4	New York Giants, Green Bay Packers
3	New England Patriots, Oakland/LA Raiders, Washington Redskins
2	Miami Dolphins, Denver Broncos, Indianapolis/Baltimore Colts, Baltimore Ravens
1	Chicago Bears, New York Jets, Tampa Bay Buccaneers, Kansas City Chiefs, St. Louis/LA Rams, New Orleans Saints

4. Olympic gold medals won by an individual:

Number	Individual
14	Michael Phelps
10	Ray Ewry
9	Carl Lewis, Paavo Nurmi, Mark Spitz
8	Matthew Biondi, Bjorn Daehlie, Sawao Kato
7	Nikolay Andrianov, Viktor Churkarin, Aladar Gerevich, Boris Shakhlin

5. Market share of various operating systems:

Percentage	Operating System
42%	Windows XP
35%	Windows 7
8%	Windows Vista
6%	Mac OS X
5%	iOS
1%	Android
1%	Linux
2%	Other

Source: National Association of Realtors

6. The top ten cable television shows for the week of March 11, 2013: (*Source:* Nielson Media Research.)

Rank	Show
1	The Big Bang Theory
2	Person of Interest
3	NCIS
4	American Idol: Wednesday
5	Two-and-a-Half Men
6	NCIS: Los Angeles
7	American Idol-Thursday
8	Elementary
9	Bachelor: After Final Rose(s)
10	The Bachelor, Blue Bloods

7. The top five best selling albums of all time (Source: toptens.com) (*Source: The New York Times*):
1. Thriller
2. Back in Black
3. Dark Side of the Moon
4. Bat out of Hell
5. Their [The Eagles] Greatest Hits 1971–75

8. The top grossing movies of all time (*Source:* Box Office Mojo):
1. Avatar
2. Titanic
3. The Avengers
4. Harry Potter and the Deathly Hallows Part 2
5. Transfomers: Dark of the Moon

9. $\{(-2, 4), (6, -2), (5, 1), (-1, -6)\}$

10. $\{(8, -3), (1, 5), (-2, -6), (5, 4)\}$

11. $\{(-6, 2), (2, -3), (-5, 7), (2, 4)\}$

12. $\{(3, 1)(4, -5), (-4, 2), (3, -3)\}$

13. $\{(2, -4), (-4, 3), (5, 3), (3, -1)\}$

14. $\{(1, 8), (-3, -3), (-2, 8), (4, -3)\}$

15. The graph shows the number of occupational fatalities in the United States for each year.

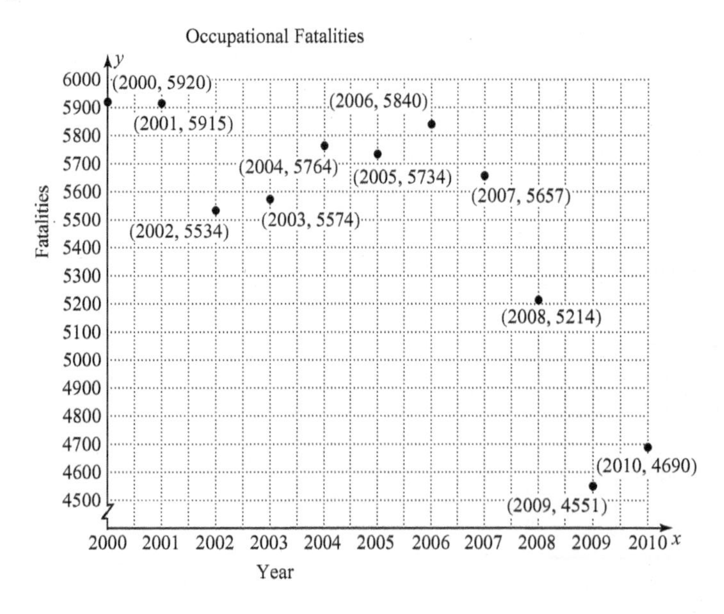

16. The graph shows the number of motorcyclist fatalities in the United States for each year.

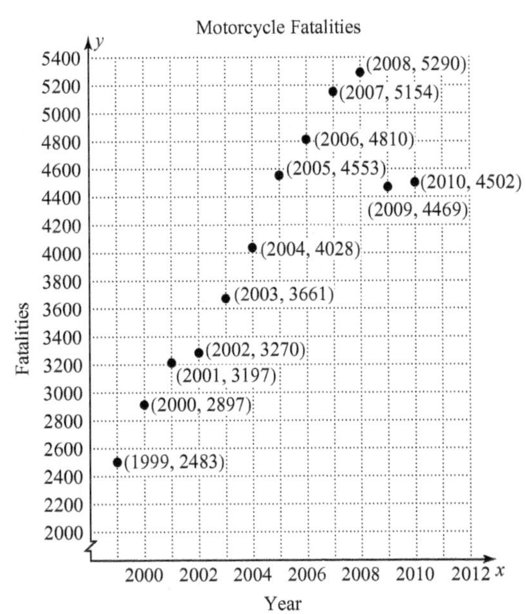

Prep Exercise 4 Explain how the vertical line test shows whether a graph represents a function or not.

For Exercises 17–24, identify the domain and range of the relation, then determine if it is a function. See Example 2.

17.

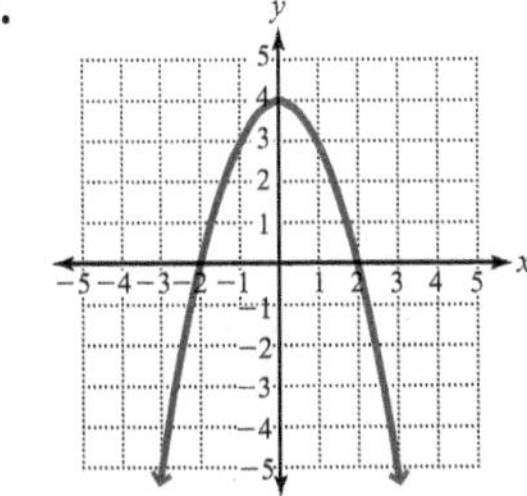

18.

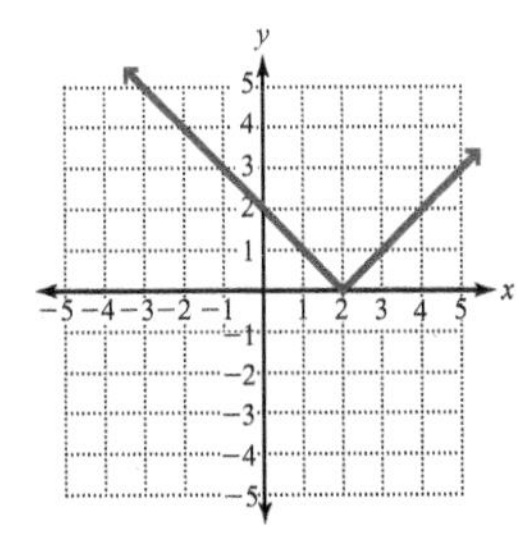

19.

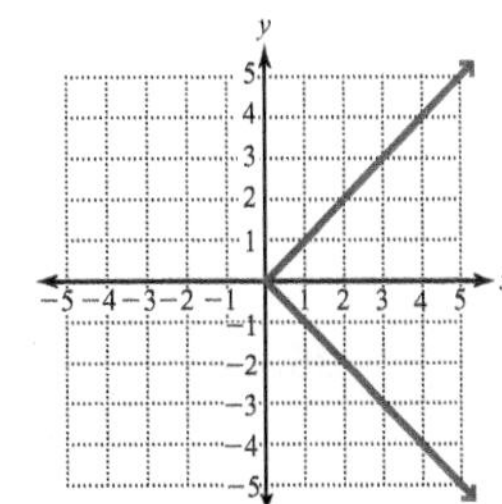

20.

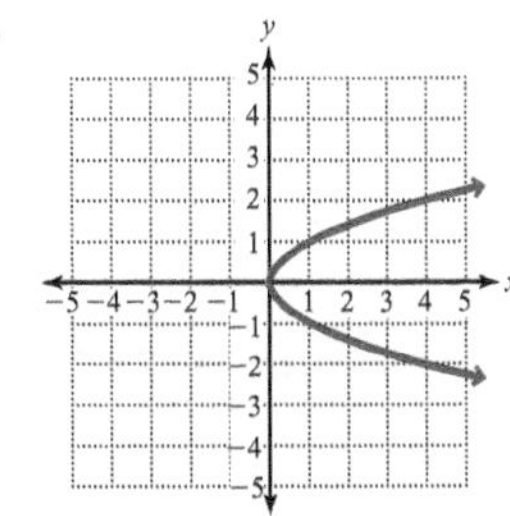

21.

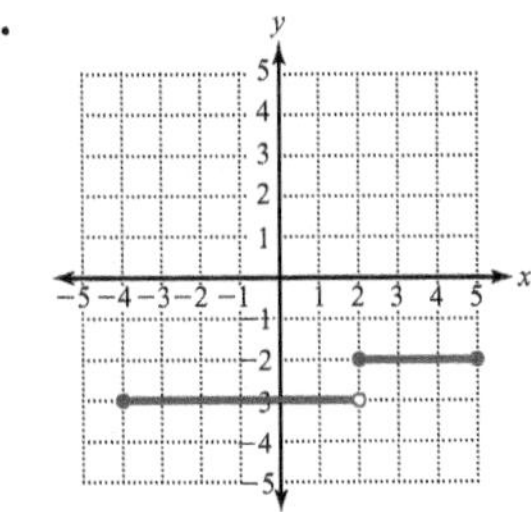

22.

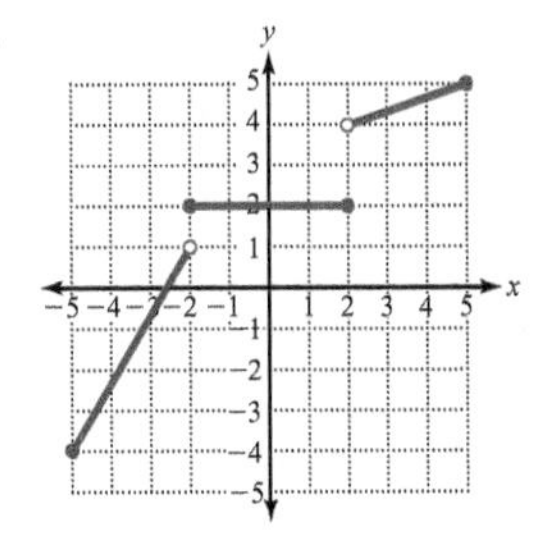

23.

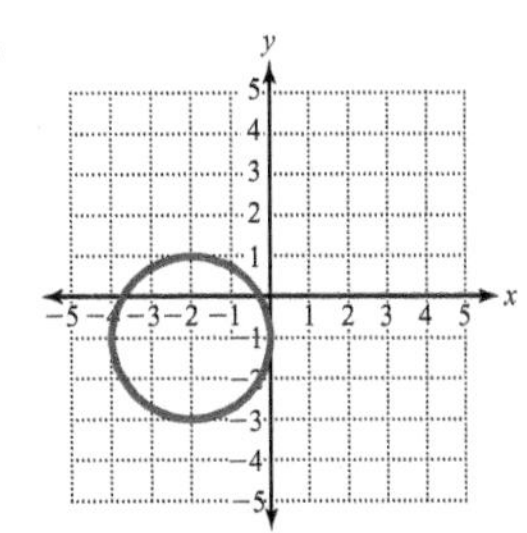

24.

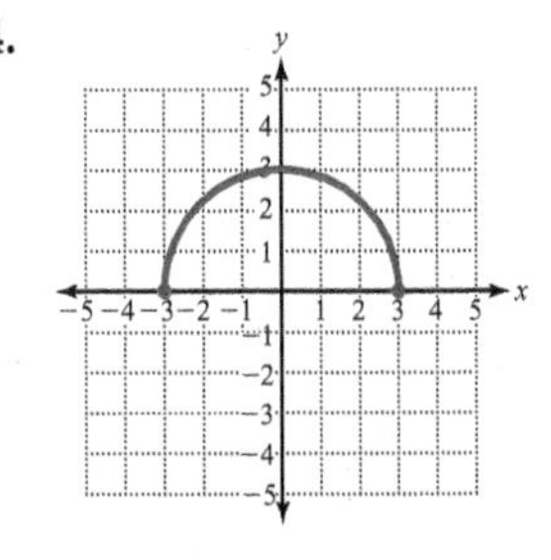

Objective 2

Prep Exercise 5 Given a function, $f(x)$, explain how to find the value of a function $f(a)$, where a is a real number in the domain of f.

For Exercises 25–44, find the indicated value of the function. See Example 3.

25. $f(x) = -2x - 9$
- **a.** $f(0)$
- **b.** $f(1)$
- **c.** $f(-1)$
- **d.** $f(a + 1)$

26. $f(x) = -6x - 2$
- **a.** $f(1)$
- **b.** $f(0)$
- **c.** $f(-1)$
- **d.** $f(t + 3)$

27. $f(x) = 2x^2 - x + 7$
- **a.** $f(0)$
- **b.** $f(1)$
- **c.** $f(-1)$
- **d.** $f(a)$

28. $f(x) = 3x^2 + 2x - 4$
- **a.** $f(-2)$
- **b.** $f(-1)$
- **c.** $f(3)$
- **d.** $f(a)$

29. $f(x) = \sqrt{3 - x}$
- **a.** $f(-1)$
- **b.** $f(12)$
- **c.** $f(-2)$
- **d.** $f(t)$

30. $f(x) = \sqrt{4x - 2}$
- **a.** $f(-1)$
- **b.** $f(1.5)$
- **c.** $f(3)$
- **d.** $f(a)$

31. $f(x) = \frac{2}{5}x + 1$
- **a.** $f(0)$
- **b.** $f(-5)$
- **c.** $f(-1)$
- **d.** $f(r)$

32. $f(x) = \frac{3}{4}x - 3$
- **a.** $f(4)$
- **b.** $f(0)$
- **c.** $f(3)$
- **d.** $f(t)$

33. $f(x) = x^2 - 2.1x - 3$
- **a.** $f(0)$
- **b.** $f(-2.2)$
- **c.** $f\left(\frac{2}{3}\right)$
- **d.** $f(a)$

34. $f(x) = 3.1x^2 + 2x$
- **a.** $f(-1)$
- **b.** $f(0.1)$
- **c.** $f\left(-\frac{1}{2}\right)$
- **d.** $f(a)$

35. $f(x) = \sqrt{x^2 - 4x}$
- **a.** $f(4)$
- **b.** $f(7)$
- **c.** $f(2)$
- **d.** $f(n)$

36. $f(x) = \sqrt{-x^2 + 5x^3}$
- **a.** $f(-2)$
- **b.** $f(1)$
- **c.** $f(0)$
- **d.** $f(w)$

37. $f(x) = |3x^2 + 1|$
- **a.** $f(0)$
- **b.** $f\left(\frac{2}{3}\right)$
- **c.** $f(-1)$
- **d.** $f(-2)$

38. $f(x) = |x - 3x^2|$
- **a.** $f(-2)$
- **b.** $f(-1)$
- **c.** $f\left(\frac{1}{3}\right)$
- **d.** $f(3)$

39. $f(x) = \frac{3 - x}{x - 4}$
- **a.** $f(5)$
- **b.** $f(4)$
- **c.** $f(-2)$
- **d.** $f(3)$

40. $f(x) = \dfrac{1 - 2x}{x + 2}$

a. $f(-3)$

b. $f(-2)$

c. $f(3)$

d. $f(-1)$

41. $f(x) = \dfrac{x}{x^2 - 1}$

a. $f(0)$

b. $f(1)$

c. $f(2)$

d. $f(m)$

42. $f(x) = \dfrac{x - 2}{x^2 - 9}$

a. $f(4)$

b. $f(-1)$

c. $f(3)$

d. $f(a)$

43. $f(x) = \dfrac{2x}{\sqrt{2 - x}}$

a. $f(-2)$

b. $f(1)$

c. $f(2)$

d. $f(6)$

44. $f(x) = \dfrac{4x}{\sqrt{2x - 5}}$

a. $f(3)$

b. $f\left(\dfrac{5}{2}\right)$

c. $f(-2)$

d. $f(a)$

For Exercises 45–50, use the graph to find the indicated value of the function. See Example 4.

45. a. $f(-4)$ **b.** $f(0)$ **c.** $f(2)$

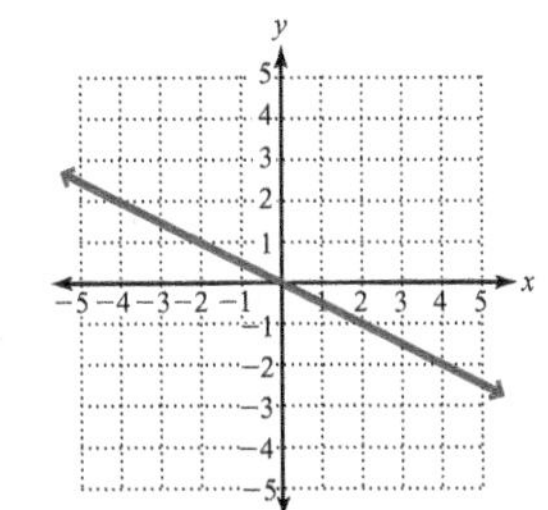

46. a. $f(-4)$ **b.** $f(-2)$ **c.** $f(1)$

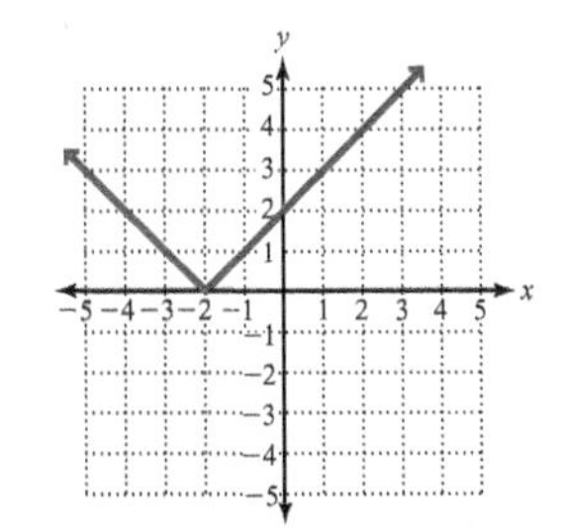

47. a. $f(-2)$ **b.** $f(0)$ **c.** $f(2)$

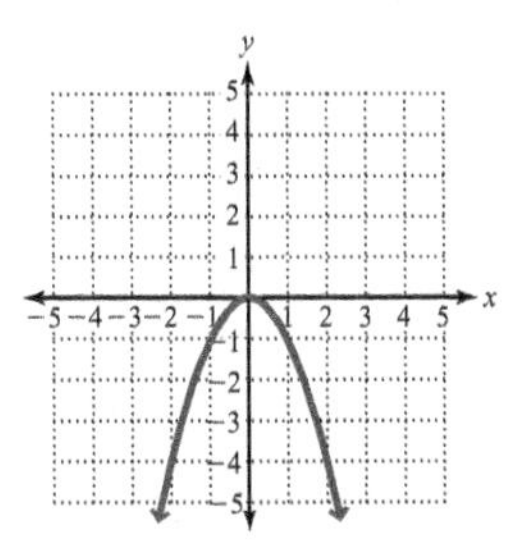

48. a. $f(-2)$ **b.** $f(-1)$ **c.** $f(1)$

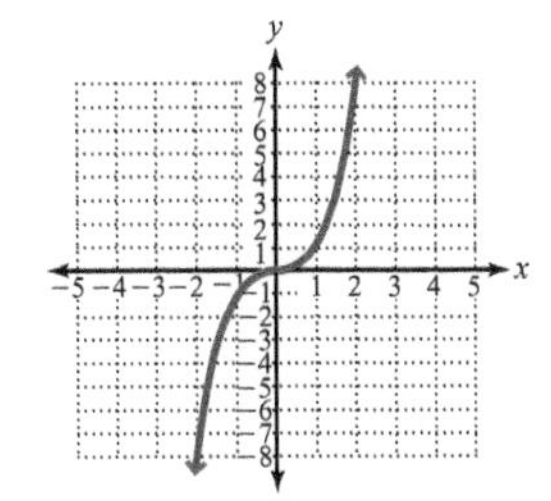

49. a. $f(-4)$ **b.** $f(0)$ **c.** $f(3)$

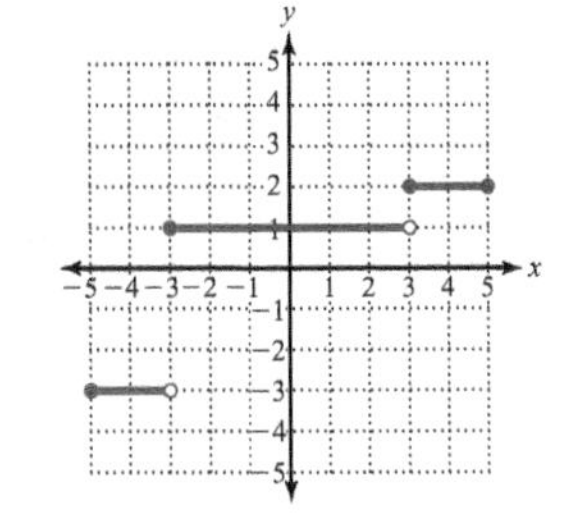

50. a. $f(-5)$ **b.** $f(-1)$ **c.** $f(2)$

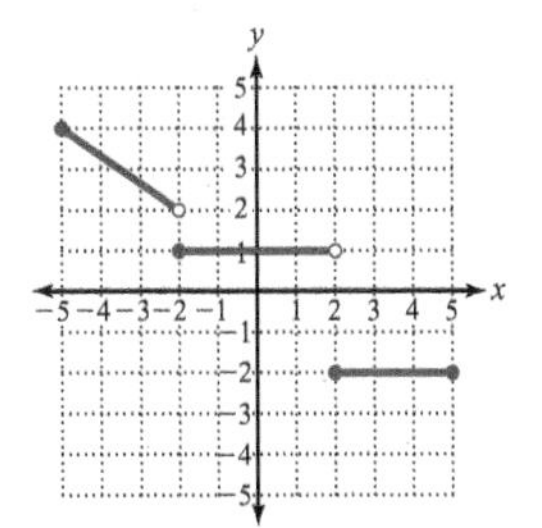

Objective 3

Prep Exercise 6 Is every linear equation of the form $y = mx + b$ a function?

Prep Exercise 7 What is the shape of the graph of every absolute value function? How do you graph an absolute value function?

Prep Exercise 8 What is the shape of every quadratic function?

For Exercises 51–72, graph. Give the domain and range. See Examples 5 and 7.

51. $f(x) = -3x + 2$

52. $f(x) = -4x - 1$

53. $f(x) = \frac{2}{3}x - 1$

54. $f(x) = -\frac{3}{2}x - 5$

55. $f(x) = -5x$

56. $f(x) = \frac{4}{3}x$

57. $f(x) = |x| - 4$

58. $f(x) = |x| + 1$

59. $f(x) = |x + 3| - 2$

60. $f(x) = |x - 1| + 3$

61. $f(x) = 2|x - 1| + 3$

62. $f(x) = 3|x + 1| - 2$

63. $f(x) = -|x| + 3$

64. $f(x) = -|x| - 1$

65. $f(x) = -3|x + 2| - 1$

66. $f(x) = -2|x - 1| + 4$

67. $f(x) = -\frac{1}{2}|x + 1| + 1$

68. $f(x) = -\frac{1}{4}|x - 2| + 2$

69. $f(x) = x^2 - 1$

70. $f(x) = x^2 + 2$

71. $f(x) = -x^2 + 3$

72. $f(x) = -x^2 - 2$

For Exercises 73–80, solve. See Example 6.

73. A factory employee gets paid \$50 per day plus \$10 per hour. The employee punches in and out on a special clock that records the employee's hours. The function $w(t) = 10t + 50$ describes an employee's daily earnings for working t hours.

a. Graph the function. Note that $t \geq 0$ and $w \geq 0$.

b. Find the daily earnings for an employee who works 7.5 hours.

74. An appliance repair service charges a \$75 service charge and \$40 per hour. The function $c(t) = 40t + 75$ describes the charges (excluding parts) for a repair that takes t hours.

a. Graph the function. Note that $t \geq 0$ and $c \geq 0$.

b. Find the charges for a repair that takes 3.2 hours.

75. A city charges \$6 per month plus \$1.225 per cubic foot of water used.

a. If C represents cost in dollars and V represents the volume of water used in cubic feet, write a function for C in terms of V that describes a customer's monthly cost.

b. Graph the function. Note that $V \geq 0$ and $C \geq 0$.

c. Find the monthly cost if 40 cubic feet of water are used.

76. A trucker is paid $200 plus $0.85 per mile traveled.

a. If p represents the total amount the trucker is paid in dollars and d represents the distance traveled in miles, write a function for p in terms of d that describes the trucker's total pay for a trip.

b. Graph the function. Note that $d \geq 0$ and $p \geq 0$.

c. Find the total pay if the trucker travels 250 miles.

★ **77.** The formula for the area of a circle, A, can be expressed in function form as $A(r) = \pi r^2$, where r represents the radius of the circle.

a. Use a graphing calculator to graph the function. Set your window so that the range on the x-axis is from 0 to 3 with a scale of 1 and the range on the y-axis is from 0 to 20 with a scale of 5.

b. Is it a linear function?

c. In the notation $A(1.5)$, to what does the 1.5 refer?

d. Calculate $A(1.5)$. Round to the nearest hundredth.

★ **78.** The formula for the volume of a cylinder, V, with a height of 8 centimeters can be expressed in function form as $V(r) = 8\pi r^2$, where r represents the radius of the cylinder in centimeters.

a. Use a graphing calculator to graph the function. Set your window so that the range on the x-axis is from 0 to 3 with a scale of 1 and the range on the y-axis is from 0 to 50 with a scale of 5.

b. Is it a linear function?

c. In the notation $V(1.2)$, to what does the 1.2 refer?

d. Calculate $V(1.2)$. Round to the nearest tenth.

79. Temperature and wind combine to cause heat loss from the body surface. The wind makes the temperature feel colder than it actually is and this effect is called the windchill temperature. The table at the right gives the actual temperature and the windchill temperature if the wind is blowing 10 miles per hour.

Actual	Windchill
0	−16
5	−10
10	−4
15	3
20	9
25	15

a. Make a grid with the horizontal axis labeled a and the vertical axis labeled w so that a represents the actual temperature and w represents the windchill temperature. Plot the ordered pairs in the coordinate plane.

b. Do the points follow a linear pattern? If so, find the slope of the line that passes through them using the first and last ordered pairs.

c. Write the equation of this line in functional notation, $w(a)$.

d. Find $w(40)$, which is the windchill temperature when the actual temperature is 40°F.

80. The table to the right shows the average price of a movie ticket in the United States.

Year	Price
2000	\$5.39
2002	\$5.80
2004	\$6.21
2006	\$6.55
2008	\$7.18
2010	\$7.89

Source: National Association of Theater Owners

a. Make a grid with the horizontal axis labeled t and the vertical axis labeled p so that t represents the number of years after 2000 and p represents the average price of a movie ticket in the United States. Plot the ordered pairs in the coordinate plane. (Hint: $t = 0$ means 2000.)

b. Do the points follow an approximately linear pattern? If so, using the first and last ordered pairs, find the slope of the line that passes through them.

c. Write the equation of this line in function notation, $p(t)$.

d. Find $p(25)$, which is the predicted average movie ticket price in 2025.

Review Exercises

Exercise 1 Expressions

[1.4] **1.** Simplify: $x + 9y - 3x + 7 - 3y - 9$

Exercises 2–5 Equations and Inequalities

[2.1] *For Exercises 2 and 3, solve.*

2. Solve: $5(x + 1) - x = 3x - (x + 1)$

3. $2\left(2 - \frac{4}{3}y\right) + 6y = -1$

[2.3] **4.** Solve $4x - (6x + 3) \geq 8x - 23$. Write the solution set in set-builder notation and in interval notation. Then graph the solution set.

[3.1] **5.** Is $(3, -2)$ a solution of $x + 4y = -5$? Is it also a solution of $2x - 3y = 12$?

Chapter 3 Summary and Review Exercises

Complete each incomplete definition, rule, or procedure; study the key examples; and then work the related exercises.

3.1 Graphing Linear Equations

Definitions/Rules/Procedures	Key Example(s)
A **linear equation in two variables** is an equation that can be written in the form __________ where A, B, and C are real numbers and A and B are not both 0. To find a **solution** to an equation in two variables, **1.** Replace one of the variables with a chosen __________. **2.** Solve the equation for the other __________.	Find two solutions for the equation $y = 3x - 1$. First solution: Let $x = 0$ $y = 3(0) - 1$ $y = -1$ **Solution:** $(0, -1)$ Second solution: Let $x = 1$ $y = 3(1) - 1$ $y = 3 - 1$ $y = 2$ **Solution:** $(1, 2)$

Exercises 1–4 Equations and Inequalities

For Exercises 1–4, determine whether the given pair of coordinates is a solution for the given equation.

1. $(2, 1);\ 2x - y = -3$

2. $\left(\frac{4}{5}, 2\right);\ 5x + y = 6$

3. $(0.4, 1.2);\ y + 2x = 4.1$

4. $(5, 2);\ y = \frac{2}{5}x$

Definitions/Rules/Procedures	Key Example(s)
To **graph** a linear equation: **1.** Find at least __________ solutions to the equation. **2.** __________ the solutions as points in the rectangular coordinate system. **3.** Draw a line through the __________ to form a straight line. Put __________ on both ends of the line to indicate that the line extends indefinitely in both directions. An ***x*-intercept** is the point(s) where a graph crosses the __________. A ***y*-intercept** is the point(s) where a graph crosses the __________.	Graph $y = 3x - 1$. We found two solutions above: $(0, -1)$ and $(1, 2)$. 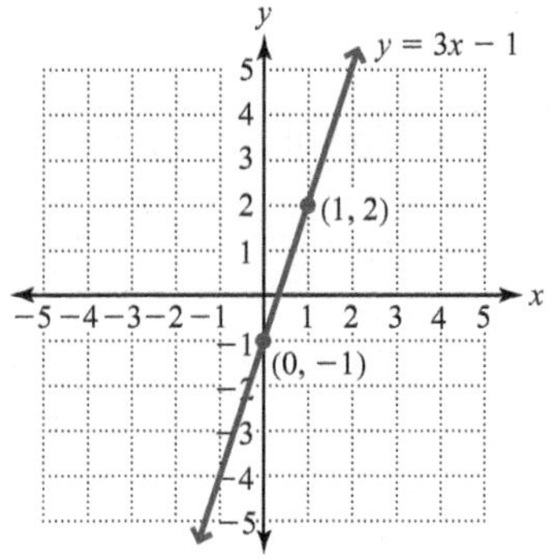

Definitions/Rules/Procedures	Key Example(s)
To find an ***x*-intercept**: 1. Replace ________ with 0 in the given equation. 2. Solve for ________. To find a ***y*-intercept**: 1. Replace ________ with 0 in the given equation. 2. Solve for ________. Given an equation of the form $y = mx + b$, the coordinates of the y-intercept are ________.	Find the x- and y-intercepts for $4x - 3y = 24$. **Solution:** x-intercept: $4x - 3(0) = 24$; $4x = 24$; $x = 6$; x-intercept: $(6, 0)$ y-intercept: $4(0) - 3y = 24$; $-3y = 24$; $y = -8$; y-intercept: $(0, -8)$

Exercises 5–10 Equations and Inequalities

For Exercises 5–10, determine the x- and y-intercepts; then graph.

5. $y = x - 4$

6. $y = 5x$

7. $y = \frac{2}{3}x - 3$

8. $y = -\frac{2}{7}x + 1$

9. $2x - 3y = 6$

10. $3x + 4y = 28$

Definitions/Rules/Procedures	Key Example(s)
The **graph of $y = c$** where c is a real number is a ________ line parallel to the ________-axis and with a y-intercept at ________. The **graph of $x = c$** where c is a real number is a ________ line parallel to the ________-axis and with an x-intercept at ________.	Graph $y = -2$ and $x = 3$. **Solution:** 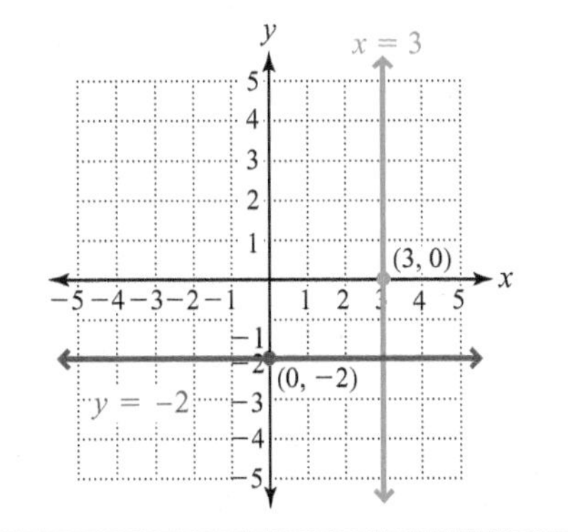

Exercises 11 and 12 Equations and Inequalities

For Exercises 11 and 12, find the x- and y-intercepts; then graph.

11. $x = 3$

12. $y = -4$

3.2 The Slope of a Line

Definitions/Rules/Procedures	Key Example(s)
The **slope** of a line is the ratio of the ____________ ____________ to the ____________ between any two points on a line.	Graph $y = \frac{1}{2}x + 1$ and $y = -\frac{1}{2}x + 1$.
The **graph of an equation** in the form $y = mx + b$ (slope–intercept form) is a line with slope of ____________ and y-intercept of ____________. The following rules indicate how m affects the graph.	
If $m > 0$, the line slants ____________ from left to right.	
If $m < 0$, the line slants ____________ from left to right.	

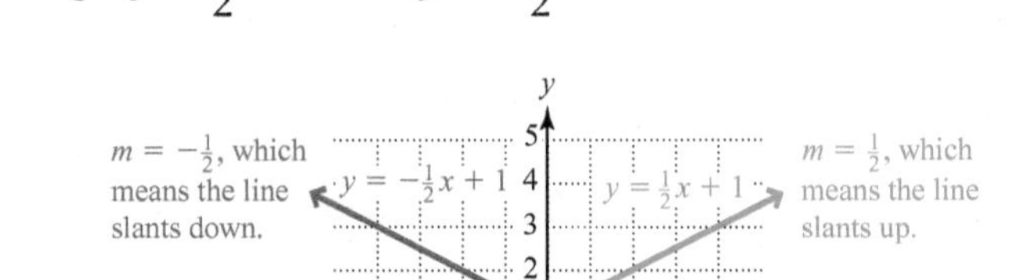

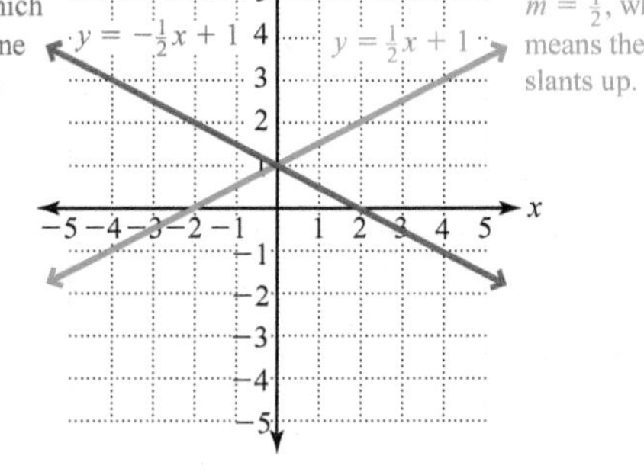

Exercises 13–16 Equations and Inequalities

For Exercises 13–16, determine the slope and the y-intercept; then graph.

13. $y = -3x + 2$

14. $y = \frac{5}{2}x + 3$

15. $2x + 5y = 15$

16. $y - 4x + 3 = 0$

Definitions/Rules/Procedures	Key Example(s)
Given two points (x_1, y_1) and (x_2, y_2), where $x_2 \neq x_1$, the **slope** of the line connecting the two points is given by the formula $m =$ ______________.	Find the slope of a line passing through the points $(4, -2)$ and $(-1, -3)$. **Solution:** $m = \dfrac{-3 - (-2)}{-1 - 4} = \dfrac{-1}{-5} = \dfrac{1}{5}$
Two points with different x-coordinates and the same y-coordinates, (x_1, c) and (x_2, c), form a ______________ line with slope of ______________ and equation ______________.	The slope of a line connecting the points $(1, 3)$ and $(-2, 3)$ is 0, and the equation of the line is $y = 3$.
Two points with the same x-coordinates and different y-coordinates, (c, y_1) and (c, y_2), form a ______________ line with ______________ slope and equation ______________.	The slope of a line connecting the points $(-1, 6)$ and $(-1, -2)$ is undefined, and the equation of the line is $x = -1$.

Exercises 17–22 Equations and Inequalities

For Exercises 17–20, find the slope through the given points.

17. $(1, 8), (4, -1)$

18. $(3, -1), (3, 2)$

19. $(7, -4), (2, -9)$

20. $(-1, -1), (3, -1)$

21. A builder is putting together a staircase where the base of each step is 12 inches and the rise is 10 inches. What is the slope of the staircase?

22. In 2006, 12.2% of 12th-grade students smoked daily. In 2010, the rate had declined to 10.7%. (*Source: Monitoring the Future,* University of Michigan Institute for Social Research and National Institute on Drug Abuse.)

a. If the years are plotted along the x-axis so that 2006 is year 0 and 2010 is year 5, and the percent is plotted along the y-axis, find the slope of the line passing through the two points given.

b. Write the equation of the line in slope–intercept form that connects the two points.

c. Assuming that the trend continues, use the equation from part b to determine the percent of 12th-grade students that will smoke daily in 2020.

3.3 The Equation of a Line

Definitions/Rules/Procedures	Key Example(s)
To write **the equation of a line** given its ***y*-intercept**, $(0, b)$, and its **slope**, m, use the slope–intercept form of the equation, ______________. If given a second point and not the slope, calculate the slope using $m =$ ______________ then use ______________.	Write the slope–intercept form of the equation of a line passing through $(0, 3)$ with slope $-\frac{5}{6}$. **Solution:** $y = -\frac{5}{6}x + 3$

Exercises 23–28 Equations and Inequalities

For Exercises 23–26, write the equation of the line in slope–intercept form given the slope and the coordinates of the y-intercept.

23. $m = 2; (0, -4)$

24. $m = -\frac{2}{5}; (0, 4)$

25. $m = -0.3; (0, -1)$

26. $m = -3; (0, 0)$

For Exercises 27 and 28, determine the slope and the y-intercept. Then write the equation of the line in slope–intercept form.

27.

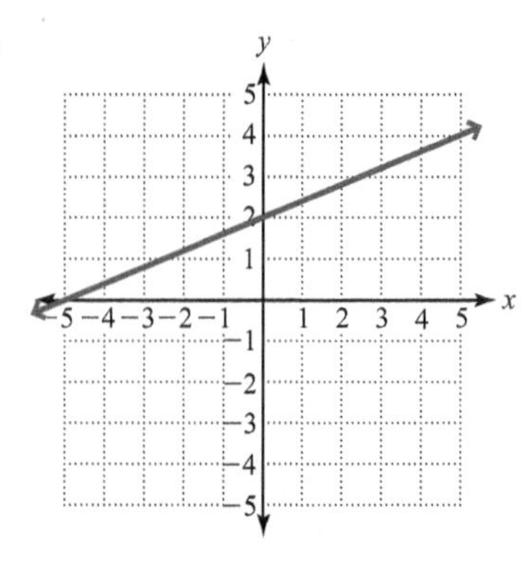

28.

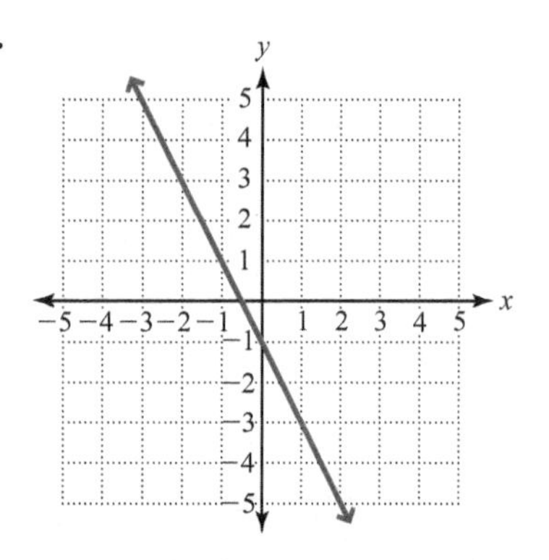

Definitions/Rules/Procedures	Key Example(s)
To write **the equation of a line** given its **slope**, m, and any **point** (x_1, y_1) on the line, use the point–slope form of the equation of a line ____________. If given a second point, (x_2, y_2), and not the slope, calculate the slope using $m =$ __________, then use ____________.	Write the slope–intercept form of the equation of a line passing through $(-4, 1)$ with slope -2. **Solution:** $y - y_1 = m(x - x_1)$ Replace m with -2, x_1 with -4, and y_1 with 1. $y - 1 = -2(x - (-4))$ $y - 1 = -2x - 8$ Distribute -2. $y = -2x - 7$ Add 1 to both sides.
To write the equation of a line in **standard form**, use the preceding rules and manipulate the resulting equation so it is in the form __________ where A, B, and C are integers and $A > 0$.	Write an equation of a line connecting $(1, 6)$ and $(-5, 2)$ in the form $Ax + By = C$, where A, B, and C are integers and $A > 0$. **Solution:** Find the slope: $m = \frac{2-6}{-5-1} = \frac{-4}{-6} = \frac{2}{3}$ Write the equation: $y - 6 = \frac{2}{3}(x - 1)$ $3(y - 6) = 3 \cdot \frac{2}{3}(x - 1)$ Multiply by 3 to clear the fraction. $3y - 18 = 2x - 2$ Simplify and distribute. $-2x + 3y = 16$ Subtract $2x$ from and add 18 to both sides. $2x - 3y = -16$ Multiply by -1 so that the x-term is positive.

Exercises 29–36 Equations and Inequalities

For Exercises 29–32, write the equation of a line in slope–intercept form with the given slope that passes through the given point.

29. $m = -1; (2, 8)$

30. $m = 6.2; (3, -5)$

31. $m = \frac{2}{3}; (4, 0)$

32. $m = -3; (-2, -2)$

For Exercises 33–36: *a. Write the equation of a line through the given points in slope–intercept form.*
b. Write the equation in standard form, $Ax + By = C$, where A, B, and C are integers and $A > 0$.

33. $(3, 3), (5, 9)$

34. $(-2, 2), (-4, -5)$

35. $(-2, -3), (4, -2)$

36. $(6, 0), (0, 6)$

Definitions/Rules/Procedures	Key Example(s)
Nonvertical parallel lines have __________ slopes and __________ y-intercepts. Vertical lines are parallel. The slope of a line perpendicular to a line with a slope of $\frac{a}{b}$ is __________. Horizontal and vertical lines are perpendicular.	Find the slopes of the lines parallel and perpendicular to the graph of $4x - 5y = -10$. **Solution:** Write $4x - 5y = -10$ in slope–intercept form. $-5y = -4x - 10$ Subtract 4x from both sides. $y = \frac{4}{5}x + 2$ Divide both sides by −5. The slope of $4x - 5y = -10$ is $\frac{4}{5}$. The slope of a line parallel to $4x - 5y = -10$ is $\frac{4}{5}$. The slope of a line perpendicular to $4x - 5y = -10$ is $-\frac{5}{4}$.

Exercises 37–44 Equations and Inequalities

For Exercises 37–40, determine whether the graphs of the equations are parallel, perpendicular, or neither.

37. $y = \frac{2}{3}x$
$y = \frac{2}{3}x - 4$

38. $7x + y = -9$
$7x - y = -6$

39. $y = \frac{4}{3}x - 1$
$y = -\frac{3}{4}x + 4$

40. $x + y = -1$
$x - y = 5$

For Exercises 41–44, write the equation of the line in slope–intercept form.

41. Find the equation of a line passing through $(0, -2)$ that is parallel to the line $y = -\frac{3}{5}x + 2$.

42. Find the equation of a line passing through $(2, 5)$ that is parallel to the line $x - 5y = 10$.

43. Find the equation of a line passing through $(-1, -2)$ that is perpendicular to the line $y = -\frac{2}{3}x + 5$.

44. Find the equation of a line passing through $(-3, -5)$ that is perpendicular to the line $y = 3x + 4$.

3.4 Graphing Linear Inequalities

Definitions/Rules/Procedures	Key Example(s)
A **linear inequality in two variables** is an inequality that can be written in the form ________ where A, B, and C are real numbers and A and B are not both 0. Note that the inequality could also be ________, ________, or ________.	
To **graph a linear inequality in two variables:** 1. Graph the related ________ (the boundary line). The related ________ has an equal sign in place of the inequality symbol. If the inequality symbol is $\leq$ or $\geq$ draw a ________ boundary line. If the inequality symbol is $<$ or $>$, draw a ________ boundary line.	Graph $3x + 5y > 15$. **Solution:** Because the inequality is $>$, we graph the related equation $3x + 5y = 15$ using a dashed line. Choose a point on one side of the line to see if it solves the inequality. We choose $(0, 0)$.
2. Choose an ordered pair on one side of the boundary line and test this ordered pair in the inequality. If the ordered pair satisfies the inequality, shade the region ________. If the ordered pair does not satisfy the inequality, shade the region ________.	$3(0) + 5(0) > 15$ $0 > 5$ This inequality is false, so $(0, 0)$ is not a solution. Therefore, we shade the region on the other side of the line. 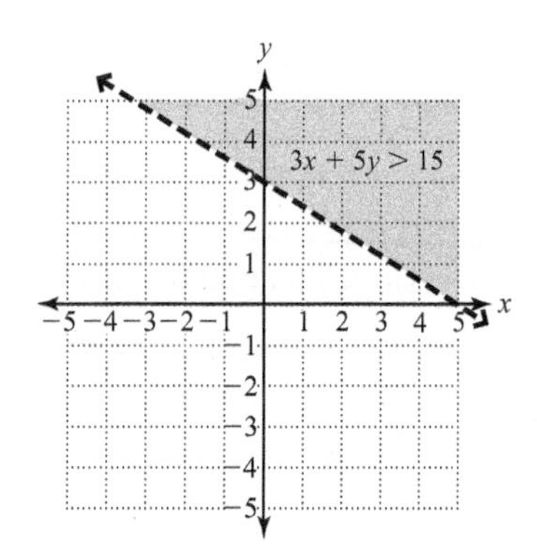

Exercises 45–53 Equations and Inequalities

For Exercises 45 and 46, determine whether the ordered pair is a solution for the linear inequality.

45. $(-3, -1); 3x + 5y > 1$

46. $(-4, 2); y \geq 3x + 1$

For Exercises 47–52, graph the linear inequality.

47. $y < -2x + 5$

48. $3x - 4y > 12$

49. $-2x - 5y \leq -10$

50. $y > \frac{4}{5}x$

51. $x + y \geq 3$

52. $y \geq -3$

53. A furniture company sells two different types of dining room chairs. A basic chair sells for \$35, and an armchair sells for \$50. For the company to break even or make a profit, the combined revenue from those two chairs must be at least \$70,000.

a. Letting x represent the number of basic chairs sold and y the number of armchairs sold, write an inequality that describes the revenue requirement for the company to break even or make a profit.

b. Graph the inequality. Note that $x \geq 0$ and $y \geq 0$.

c. What does the boundary line represent?

d. What does the shaded region represent?

e. List two combinations of sales numbers of each chair that allow the company to break even.

f. List two combinations of sales numbers that give the company a profit.

3.5 Introduction to Functions and Function Notation

Definitions/Rules/Procedures	Key Example(s)
A **relation** is a ________________.	Identify the domain and range of the relation; then determine whether it is a function.
A **function** is a relation in which each value in the domain is assigned to ________________.	
The **domain** of a relation is a set that contains all its ________________ values. When the graph of a relation is given, the domain contains the ________________ of every point on the graph.	
The **range** of a relation is a set that contains all its ________________ values. When the graph of a relation is given, the range contains the ________________ of every point on the graph.	
To determine whether a **graphical relation** is a **function**, draw or imagine ________________ lines through each value in the domain. If each ________________ line intersects the graph at ________________, the relation is a function. If any vertical line intersects the graph ________________, the relation is not a function.	

a.

Date of Birth	Person
March 1	Candice, Ricky
April 7	Sheri
May 19	Berry

Domain: {March 1, April 7, May 19}
Range: {Candice, Ricky, Sheri, Berry}
It is not a function because one element in the domain corresponds to two elements in the range.

b.

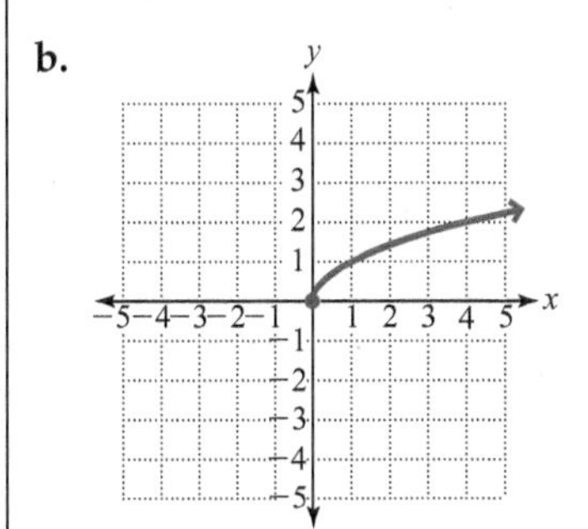

Domain: $\{x \mid x \geq 0\}$ or $[0, \infty)$
Range: $\{y \mid y \geq 0\}$ or $[0, \infty)$
It is a function. (It passes the vertical line test.)

c.

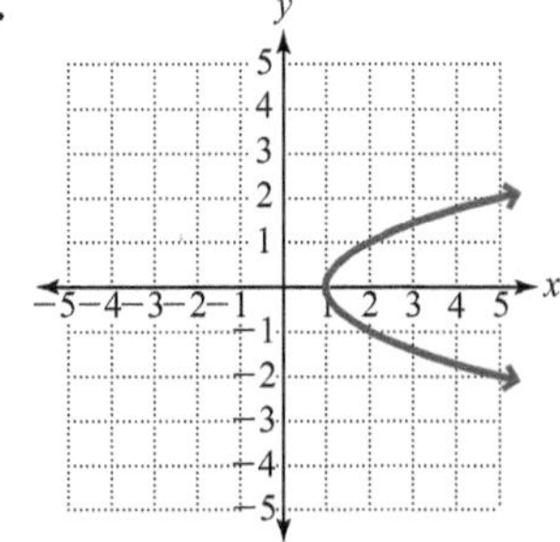

Domain: $\{x \mid x \geq 1\}$ or $[1, \infty)$
Range: $\mathbb{R}$ or $(-\infty, \infty)$
It is not a function. (It fails the vertical line test.)

Exercises 54–59 Equations and Inequalities

For Exercises 54–59, find the domain and range of the relation; then determine whether it is a function.

54. Tallest peaks in North America:

Peak	Height (ft.)
McKinley	20,320
Logan	19,551
Pico de Orizaba	18,555
St. Elias	18,008
Popocatépetl	17,930

55. The number of drive-in theaters in each state:

Number of Drive-ins	State
21	California
23	Indiana
32	New York
35	Ohio, Pennsylvania

Source: United Drive-in Theater Owners Association

56.

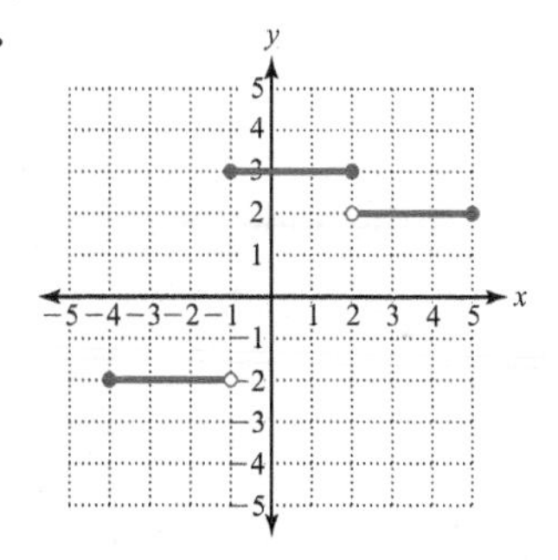

57.

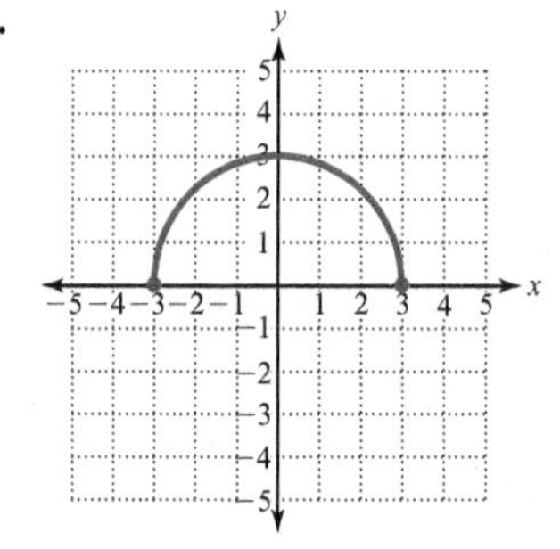

58.

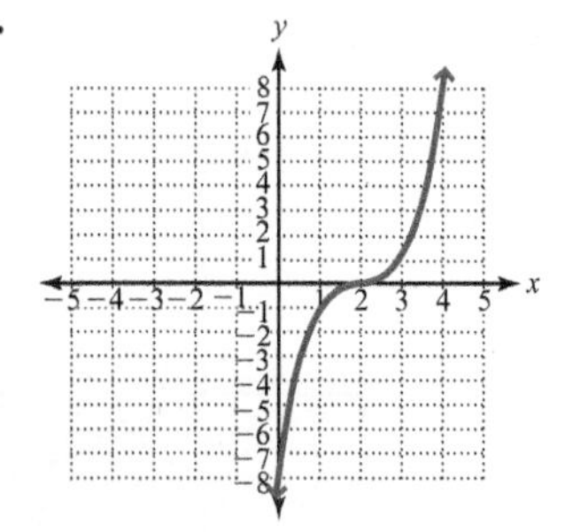

59.

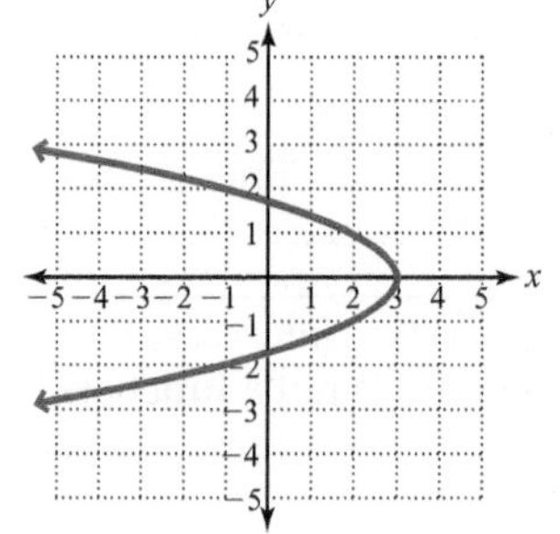

Definitions/Rules/Procedures	Key Example(s)
Given a function $f(x)$, to find $f(a)$, where a is a real number in the domain of f, replace __________ in the function with __________ and then evaluate or simplify.	For the function $f(x) = x^3 - 2x$, find $f(-4)$. $f(-4) = (-4)^3 - 2(-4)$ $= -64 + 8$ $= -56$

Exercises 60–64 Equations and Inequalities

60. Find the indicated value of the function $f(x) = x^2 - 4$.
a. $f(2)$ **b.** $f(0)$ **c.** $f(-3)$ **d.** $f(n)$

61. Find the indicated value of the function $g(x) = \dfrac{x - 3}{x + 5}$.
a. $g(2)$ **b.** $g(3)$ **c.** $g(-5)$

62. Use the graph in Exercise 56 to find the indicated value of the function.
a. $f(-3)$ **b.** $f(2)$ **c.** $f(0)$ **d.** $f(5)$

63. Use the graph in Exercise 57 to find the indicated value of the function.
a. $f(-3)$ **b.** $f(0)$ **c.** $f(3)$ **d.** $f(4)$

64. A landscaper charges $75 per visit plus $25 per hour of labor.
a. If c represents the total cost of a visit and t represents the time in hours she works, write an equation using function notation that describes the total cost.

b. Find the total cost if labor is 1.5 hours.

c. If a client's bill is $150, how many hours of labor was the client charged?

d. Graph the function. Note that $t \geq 0$ and $c \geq 0$.

Definitions/Rules/Procedures	Key Example(s)
A **linear function** has the form $f(x) =$ __________. The graph of a linear function is a __________.	Graph. Give the domain and range. **a.** $f(x) = 2x - 4$ 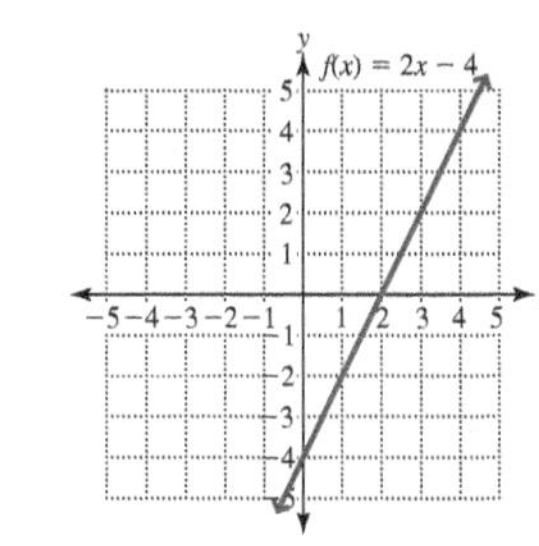 Domain: R Range: R
An **absolute value function** is of the form $f(x) =$ __________. The graph of an absolute value function is a __________ shape. To graph, find enough ordered pairs to generate the __________ shape.	**b.** $f(x) = \lvert x + 2 \rvert$ 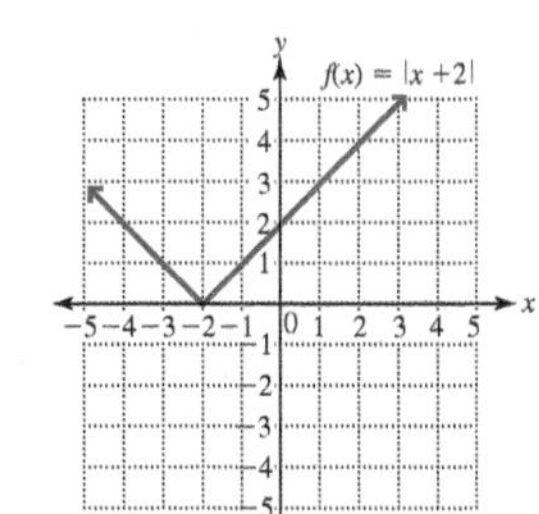 Domain: R Range: $[0, \infty)$
A **quadratic function** has the form $f(x) =$ __________ where $a \neq 0$. The graph of a quadratic function is a __________.	**c.** $f(x) = -x^2 + 4$ 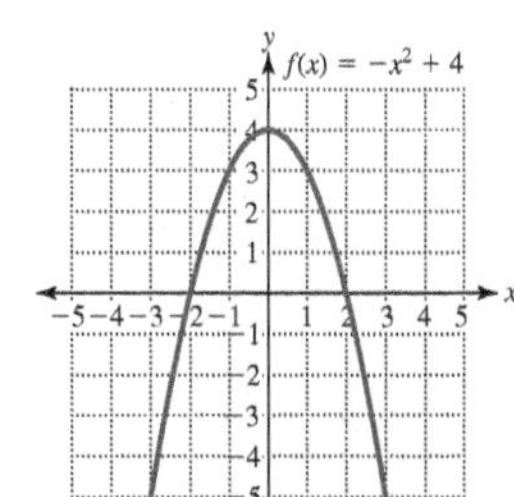 Domain: R Range: $(-\infty, 4]$

Exercises 65–68 Equations and Inequalities

[3.5] *For Exercises 65–68, graph. Give the domain and range.*

65. $f(x) = -\frac{2}{3}x + 4$

66. $f(x) = |x - 2|$

67. $f(x) = -2|x + 1| + 3$

68. $f(x) = x^2 - 3$

Chapter 3 Practice Test

For Extra Help

Step-by-step test solutions are found on the Chapter Test Prep Videos available in MyMathLab® *or on* YouTube.

1. Determine the coordinates for each point in the graph to the right.

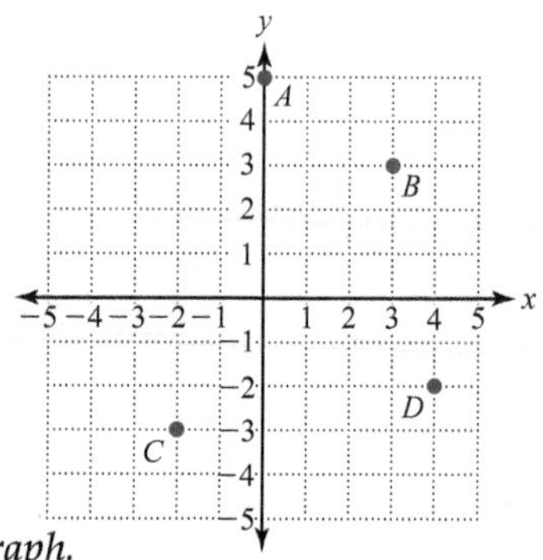

2. State the quadrant in which $\left(3.6, -501\frac{2}{3}\right)$ is located.

3. Determine whether $(-3, 5)$ is a solution for $y = -\frac{1}{4}x + 3$.

For Exercises 4 and 5, determine the slope and the coordinates of the y-intercept; then graph.

4. $y = -\frac{4}{3}x + 5$

5. $x - 2y = -8$

For Exercises 6 and 7, determine the slope of the line through the given points.

6. $(-1, -4), (-2, -4)$

7. $(-3, -9), (4, -1)$

8. Write the equation of a line in slope–intercept form with y-intercept $(0, 5)$ and slope $\frac{2}{7}$.

9. Write the equation of a line in slope–intercept form that passes through the points $(4, 2), (-5, -1)$.

10. Write the equation of a line through the points $(1, 4)$ and $(-3, -1)$ in the form $Ax + By = C$, where A, B, and C are integers and $A > 0$.

11. Are the graphs of $y = \frac{3}{4}x - 2$ and $y = \frac{4}{3}x + 2$ paralle perpendicular, or neither?

For Exercises 12 and 13, graph the linear inequality.

12. $y \geq 3x - 1$

13. $2x - 3y < 6$

For Exercises 14 and 15, identify the domain and range of the relation; then determine whether it is a function.

14. The number of U.S. residents who are not native English speakers:

Language	U.S. Residents (millions)
Spanish	37.6
Chinese	2.9
Tagalog	1.6
Vietnamese	1.4
French	1.3

Source: U.S. Bureau of the Census

15. 

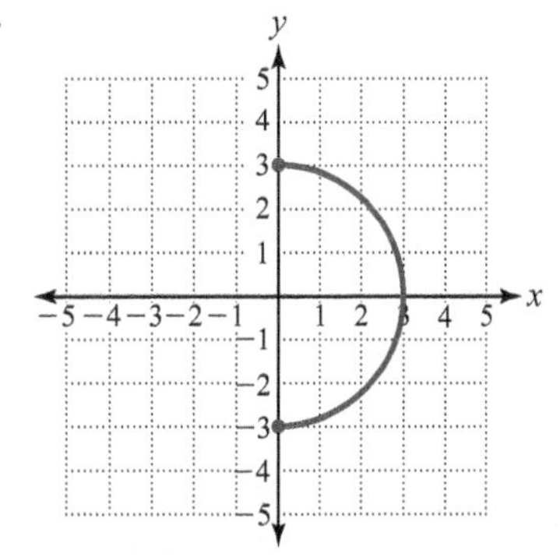

For Exercises 16 and 17, find the indicated value of the function.

16. $f(x) = 2x^2 - 7$

a. $f(-2)$ **b.** $f(3)$ **c.** $f(t)$

17. $f(x) = \dfrac{4x}{x + 5}$

a. $f(-1)$ **b.** $f(5)$ **c.** $f(-5)$

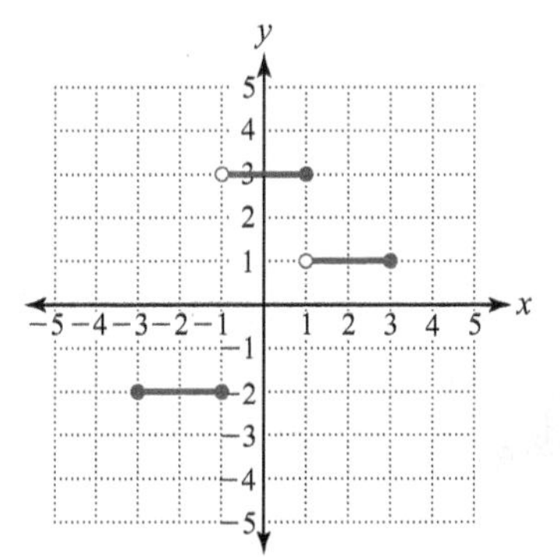

18. **a.** Give the domain and range of the function shown in the graph.

b. Use the graph to find $f(1)$.

For Exercises 19 and 20, graph. Give the domain and range.

19. $f(x) = |x + 1| - 2$

20. $f(x) = -x^2 + 3$

21. Part of a farmer's land is to be enclosed for pasture. Materials limit the perimeter to 1000 feet.

a. Letting l represent the length in feet and w the width in feet, write an inequality in which the maximum perimeter is 1000 feet.

b. Graph the inequality. Note that $l > 0$ and $w > 0$.

c. Find two combinations of length and width that satisfy the inequality.

22. A shipping company charges \$3.00 to use its overnight shipping service plus \$0.45 per ounce of weight of the package being shipped.

a. If c represents the total cost in dollars and w represents the package weight in ounces, write an equation using function notation that describes the cost of shipping the package overnight.

b. Graph the function. Note that $w \geq 0$ and $c \geq 0$.

c. Find the cost of shipping a 32-ounce package.

Chapters 1–3 Cumulative Review Exercises

For Exercises 1–3, answer true or false.

[2.3] **1.** The inequality $x > 0$ can also be expressed as $[0, \infty)$ in interval notation.

[1.3] **2.** If the base is a negative number and the exponent is odd, the product is negative.

[1.3] **3.** $a(b + c) = ab + ac$

For Exercises 4–6, fill in the blank.

[2.1] **4.** To clear decimal numbers in an equation, we multiply by an appropriate power of ______ as determined by the decimal number with the most decimal places.

[1.3] **5.** The radical symbol $\sqrt{\ }$ denotes only the ______________ square root.

[1.3] **6.** List the order in which we perform operations of arithmetic.

Exercises 7–12 Expressions

[2.4] **7.** Find the intersection and union of the given sets.
$A = \{w, e, l, o, v\}$ $B = \{m, a, t, h\}$

[1.3] *For Exercises 8–10, evaluate.*

8. $(-3)^2$

9. $\sqrt[3]{27}$

10. $-|4 + 3| - 8(1 - 5)^2$

[1.4] **11.** Evaluate $\dfrac{3x}{x - 5}$ when $x = 5$.

[1.4] **12. a.** Translate to an algebraic expression: the product of five and n is subtracted from the sum of eight and negative four times n.

b. Simplify the expression.

c. Evaluate the expression when n is 4.

[1.2] *For Exercises 13 and 14, indicate whether the given equation illustrates the additive identity, multiplicative identity, additive inverse, commutative property of addition, commutative property of multiplication, associative property of addition, associative property of multiplication, distributive property, or multiplicative inverse.*

Exercises 13–30 Equations and Inequalities

13. $-\frac{1}{4}(-4) = 1$

14. $6(5 + 2) = 6 \cdot 5 + 6 \cdot 2$

For Exercises 15 and 16, solve.

[2.1] 15. $6(5 + y) - 3y = 2y + 1$

[2.5] 16. $|2x - 1| = |x + 8|$

For Exercises 17 and 18, solve and then:
a. Write the solution set using set-builder notation.
b. Write the solution set using interval notation.
c. Graph the solution set.

[2.3] 17. $3 + 2x \leq 8 - 4x$

[2.6] 18. $3|x + 1| - 7 > 2$

[2.1] 19. Solve $d = rt$ for t.

[3.2] 20. Find the slope of a line passing through $(3, -1)$ and $(-2, -1)$.

[3.3] 21. Determine whether the graphs of $2x + y = 6$ and $y = -\frac{1}{2}x + 1$ are parallel, perpendicular, or neither.

[3.3] 22. Find the equation of a line passing through $(1, 8)$ that is parallel to the line $2x - y = 3$.

[3.4] 23. Graph $2x - y \geq 6$.

[3.5] 24. Given $f(x) = 3x^2 + 2$, find $f(-1)$.

[3.5] 25. Graph $f(x) = \frac{2}{5}x - 1$.

For Exercises 26–30, solve.

[1.3] 26. Find the GPA.

Course	Course Credits	Grade
Math 111	3	A
Eng 212	3	A
Phys 201	4	B

[2.2] 27. The formula $B = P + Prt$ can be used to calculate the balance in an account if simple interest is added to a principal P at an interest rate of r after t years. Suppose an account earns 3.5% interest and after 4 years the balance is \$6840. Find the principal that was invested.

[2.2] **28.** Three-fourths the difference of a number and four is fourteen more than negative two-thirds of the number. Translate to an equation and solve.

[2.2] **29.** Two cars begin at the same location and travel in opposite directions on the same highway. One car is traveling at 45 miles per hour; the other at 55 miles per hour. How long until they are 60 miles apart?

[2.3] **30.** A wireless company has a plan that charges $40.00 for 500 minutes, then $0.04 for each additional minute. Find the range of total minutes a person can use with this plan if the budget is a maximum of $60 per month for wireless charges.

CHAPTER

4

Systems of Linear Equations and Inequalities

Chapter Overview

In this chapter, we solve problems involving two or more unknowns. We assign a different variable to each unknown and translate the problem to several equations, called a system of equations. We learn several techniques for solving a system of equations:

- ▸ Graphing
- ▸ Substitution
- ▸ Elimination
- ▸ Matrices

Finally, in Section 4.5, we solve systems of linear inequalities.

4.1 Solving Systems of Linear Equations in Two Variables

Objectives

1. Determine whether an ordered pair is a solution for a system of equations.
2. Solve systems of linear equations graphically and classify systems.
3. Solve systems of linear equations using substitution.
4. Solve systems of linear equations using elimination.

Warm-up

[3.1] 1. Graph. $y = 3x + 1$

[2.1] 2. Solve for x. $3x - 2y = 12$

[2.1] 3. Solve. $6\left(4 + \frac{2}{3}y\right) - 8y = 16$

[1.4] 4. Distribute. **a.** $-2(x - 5y)$ **b.** $10(0.3x + 0.6y)$

[1.4] 5. Combine like terms. $-2x + 10y + 3x + 6y$

Objective 1 Determine whether an ordered pair is a solution for a system of equations.

A problem with more than one unknown can be represented by a group of equations called a **system of equations**.

Definition **System of equations:** A group of two or more equations.

Consider the following problem: The sum of two numbers is 3. Twice the first number plus three times the second number is 8. What are the two numbers? If x represents the first number and y represents the second number, we can translate each sentence to an equation. Together the two equations form a system of equations that describe the problem.

$$\text{System of equations} \quad \begin{cases} x + y = 3 & \text{(Equation 1)} \\ 2x + 3y = 8 & \text{(Equation 2)} \end{cases}$$

Our goal will be to find the **solution for the system of equations.**

Definition **Solution for a system of equations:** An ordered set of numbers that makes all equations in the system true.

For example, in the preceding system of equations, if $x = 1$ and $y = 2$, both equations are true; so they form a solution to the system of equations. We can write the solution as an ordered pair: $(1, 2)$. We can check by substituting those values in place of the corresponding variables.

Procedure **Checking a Solution to a System of Equations**

To verify or check a solution to a system of equations:

1. Replace each variable in each equation with its corresponding value.
2. Verify that each equation is true.

Example 1 Determine whether each ordered pair is a solution to the system of equations.

$$\begin{cases} x - 2y = 3 & \text{(Equation 1)} \\ y = 3x + 1 & \text{(Equation 2)} \end{cases}$$

a. $(5, 1)$

Solution:

$x - 2y = 3$ (Equation 1)
$5 - 2(1) \stackrel{?}{=} 3$
$5 - 2 \stackrel{?}{=} 3$
$3 = 3$ True

$y = 3x + 1$ (Equation 2)
$1 \stackrel{?}{=} 3(5) + 1$
$1 \stackrel{?}{=} 15 + 1$
$1 = 16$ False

In both equations, replace x with 5 and y with 1.

Because $(5, 1)$ does not satisfy both equations, it is not a solution for the system.

Answers to Warm-up

1.

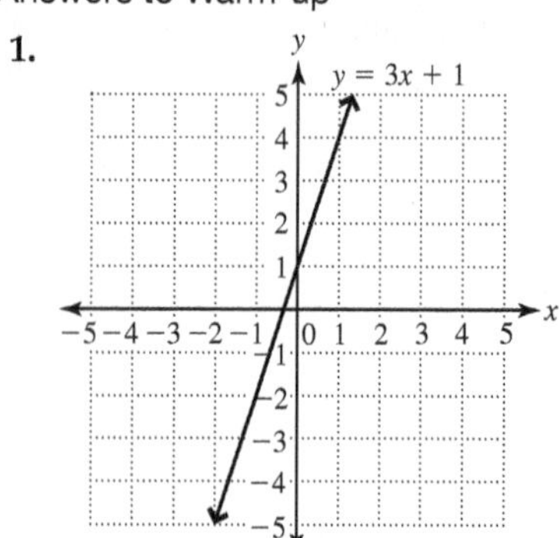

2. $x = 4 + \frac{2}{3}y$
3. $y = 2$
4. **a.** $-2x + 10y$
   b. $3x + 6y$
5. $x + 16y$

b. $(-1, -2)$

Solution: $x - 2y = 3$ (Equation 1) $\qquad y = 3x + 1$ (Equation 2)

$$-1 - 2(-2) \stackrel{?}{=} 3 \qquad -2 \stackrel{?}{=} 3(-1) + 1$$

$$-1 + 4 \stackrel{?}{=} 3 \qquad -2 \stackrel{?}{=} -3 + 1$$

$$3 = 3 \quad \text{True} \qquad -2 = -2 \quad \text{True}$$

In both equations, replace x with -1 and y with -2.

Because $(-1, -2)$ satisfies both equations, it is a solution for the system.

Your Turn 1 Determine whether each ordered pair is a solution to the system of equations.

$$\begin{cases} 3x + y = 1 & \text{(Equation 1)} \\ y = 1 - 2x & \text{(Equation 2)} \end{cases}$$

a. $(2, -5)$ **b.** $(0, 1)$

Objective 2 Solve systems of linear equations graphically and classify systems.

Let's graph the equations in the system from Example 1: $\begin{cases} x - 2y = 3 & \text{(Equation 1)} \\ y = 3x + 1 & \text{(Equation 2)} \end{cases}$.

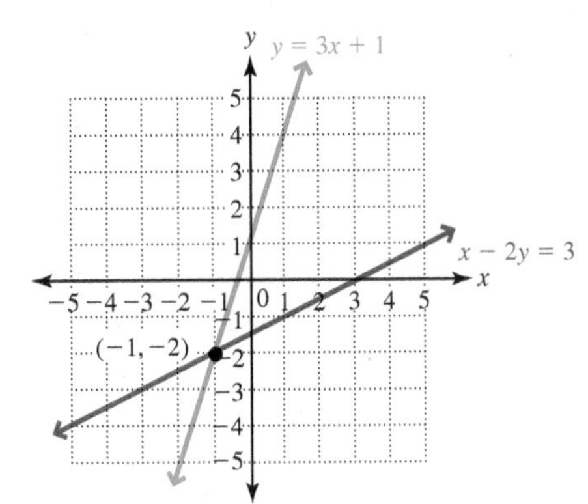

Notice that the graphs of $x - 2y = 3$ and $y = 3x + 1$ intersect at the point $(-1, -2)$. Because $(-1, -2)$ lies on both lines, its coordinates satisfy both equations; so it is the solution for the system. If two linear graphs intersect at a single point, the system has a *single solution* at the point of intersection. Also notice that these lines have different slopes, which is always the case when a system of two linear equations has a single solution.

A system can also have no solution. Look at the graphs of the equations in the system $\begin{cases} y = 2x + 1 & \text{(Equation 1)} \\ y = 2x - 3 & \text{(Equation 2)} \end{cases}$.

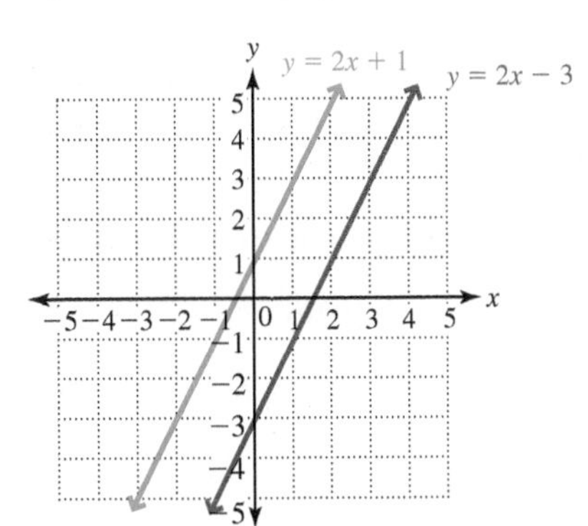

Notice that the equations $y = 2x + 1$ and $y = 2x - 3$ have the same slope, 2, and different y-intercepts; so their graphs are parallel lines. Because the graphs of the equations in this system have no point of intersection, the system has no solution.

Some systems have an infinite number of solutions. Consider the system

$$\begin{cases} 4x + 8y = 16 & \text{(Equation 1)} \\ 2x + 4y = 8 & \text{(Equation 2)} \end{cases}.$$

Answers to Your Turn 1
a. not a solution **b.** solution

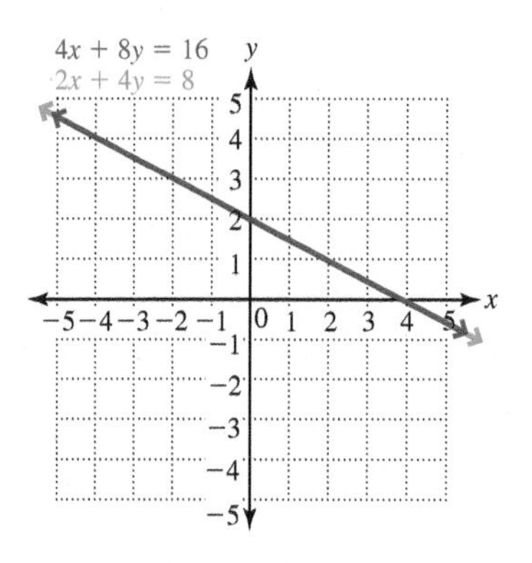

Notice that the graphs of the two equations are identical. Consequently, an infinite number of solutions to this system of equations lie on that line.

Note Having an infinite number of solutions is not the same as having all real numbers as solutions as in the solutions of identities.

Procedure Solving Systems of Linear Equations Graphically

To solve a system of linear equations graphically:

1. Graph each equation.
   a. If the lines intersect at a single point, the coordinates of that point form the solution.
   b. If the lines are parallel, there is no solution.
   c. If the lines are identical, there are an infinite number of solutions, which are the coordinates of all of the points on the line.
2. Check your solution.

Example 2 Solve the system of equations graphically.

a. $\begin{cases} y = -2x + 3 & \text{(Equation 1)} \\ x - 2y = 4 & \text{(Equation 2)} \end{cases}$

Solution: Graph each equation.

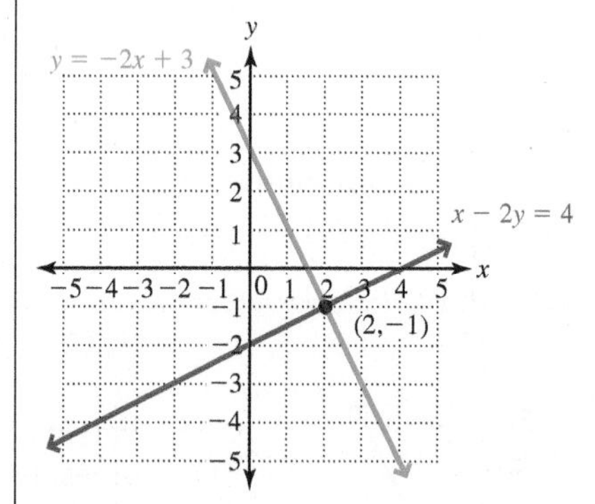

The lines intersect at a single point, which appears to be $(2, -1)$. We can verify that it is the solution by substituting into the equations.

$$\begin{aligned} y &= -2x + 3 \\ -1 &\stackrel{?}{=} -2(2) + 3 \\ -1 &\stackrel{?}{=} -4 + 3 \\ -1 &= -1 \quad \text{True} \end{aligned} \qquad \begin{aligned} x - 2y &= 4 \\ 2 - 2(-1) &\stackrel{?}{=} 4 \\ 2 + 2 &\stackrel{?}{=} 4 \\ 4 &= 4 \quad \text{True} \end{aligned}$$

Warning Graphing by hand can be imprecise, and not all solutions have integer coordinates. So you should check your solutions by substituting them into the original equations.

Answer: Because $(2, -1)$ makes both equations true, it is the solution.

b. $\begin{cases} 3x + 4y = 8 & \text{(Equation 1)} \\ y = -\dfrac{3}{4}x - 3 & \text{(Equation 2)} \end{cases}$

Solution: Graph each equation.

The lines appear to be parallel, which we can verify by comparing the slopes. The slope of Equation 2 is $-\dfrac{3}{4}$. To determine the slope of Equation 1, we rewrite it in slope–intercept form.

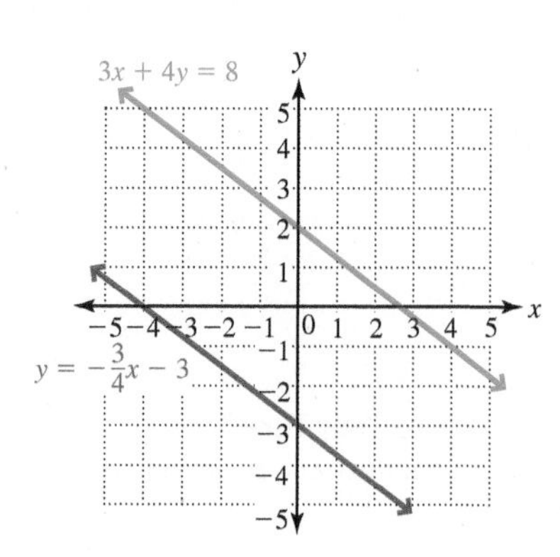

$$3x + 4y = 8$$
$$4y = -3x + 8 \quad \text{Subtract } 3x \text{ from both sides.}$$
$$y = -\frac{3}{4}x + 2 \quad \text{Divide both sides by 4 to isolate } y.$$

Answer: Because the slopes are equal and the y-intercepts are different, the lines are parallel; so the system has no solution.

c. $\begin{cases} 6x + 2y = 4 & \text{(Equation 1)} \\ y = -3x + 2 & \text{(Equation 2)} \end{cases}$

Solution: Graph each equation.

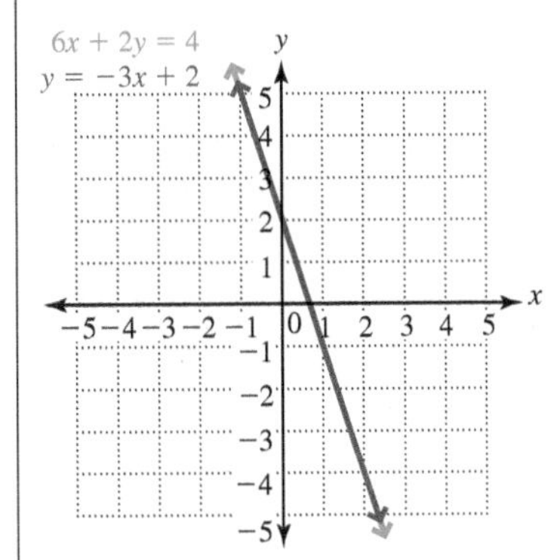

The lines appear to be identical, which we can verify by rewriting $6x + 2y = 4$ in slope–intercept form.

$$6x + 2y = 4$$
$$2y = -6x + 4 \quad \text{Substract } 6x \text{ from both sides.}$$
$$y = -3x + 2 \quad \text{Divide both sides by 2 to isolate } y.$$

Answer: Because the equations are the same, the lines coincide and there are an infinite number of solutions to the system, which are the coordinates of all of the points on the line. We can say that the solution set contains every ordered pair that solves $6x + 2y = 4$ (or $y = -3x + 2$) and can be written as $\{(x, y) \mid 6x + 2y = 4\}$ (or $\{(x, y) \mid y = -3x + 2\}$).

Your Turn 2 Solve the system of equations graphically.

$$\begin{cases} 3x + y = -2 \\ x + 2y = 6 \end{cases}$$

Classifying Systems of Equations

As we have seen, a system of two equations with two variables can have one solution, an infinite number of solutions along a common line, or no solution. Special terms are used to indicate each of these three classifications. The first terms describe whether a system has a solution.

Definitions **Consistent system of equations:** A system of equations that has at least one solution.
Inconsistent system of equations: A system of equations that has no solution.

We have seen two types of consistent systems of equations. One type has a single solution because the two equations produce different lines that intersect at a single point. The other type has an infinite number of solutions because the two equations are identical and therefore produce the same line. To distinguish the two consistent cases, we use additional terms. If the equations in a consistent system produce different lines so that the system has one solution, we say that the equations are *independent*. If the equations in a consistent system produce the same line so that the system has an infinite number of solutions along that line, we say that the equations are *dependent*.

We have learned that linear equations whose graphs are different have different slopes or, in the case of parallel lines, have the same slopes and different y-intercepts. Linear equations whose graphs are identical have the same slope and same y-intercept. Therefore, by observing the slope and y-intercept of the equations in a system of equations, we can determine with which of the three cases we are dealing, which suggests the following procedure.

Answer to Your Turn 2

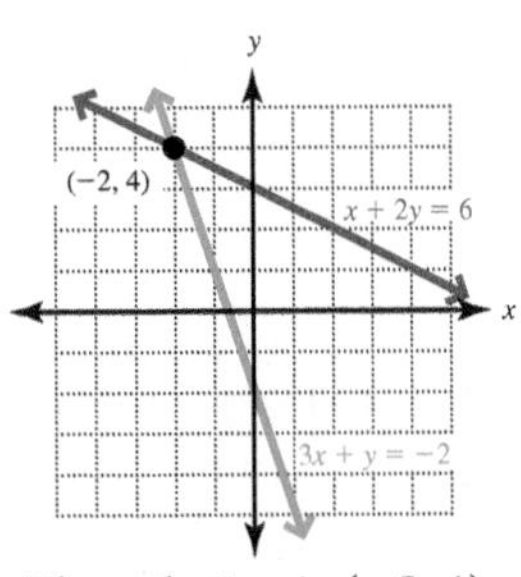

The solution is $(-2, 4)$.

Procedure Classifying Systems of Equations

To classify a system of two linear equations in two unknowns, write the equations in slope–intercept form and compare the slopes and y-intercepts.

Consistent system with independent equations: The system has a single solution at the point of intersection. The graphs are different. They have different slopes.

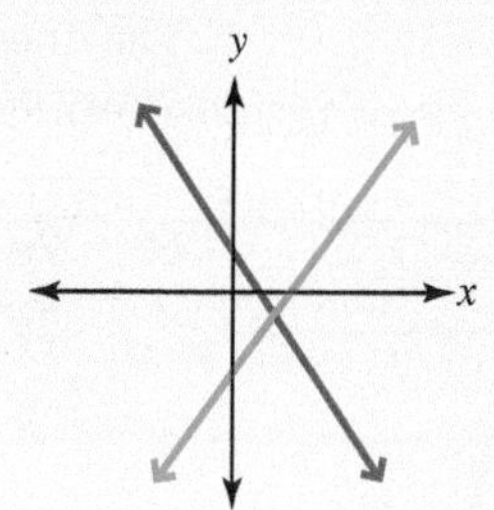

Consistent system with dependent equations: The system has an infinite number of solutions. The graphs are identical. They have the same slope and same y-intercept.

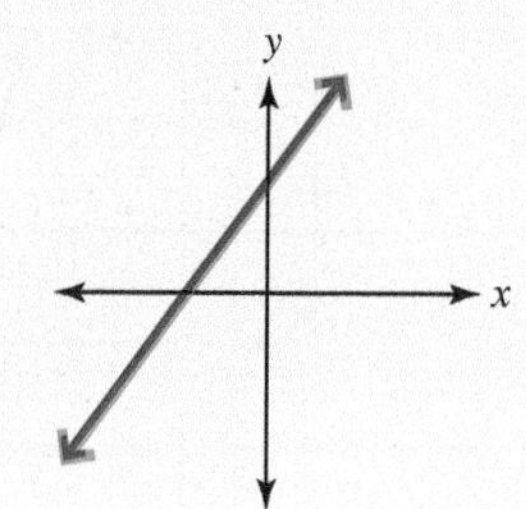

Inconsistent system: The system has no solution. The graphs are parallel lines. They have the same slope, but different y-intercepts.

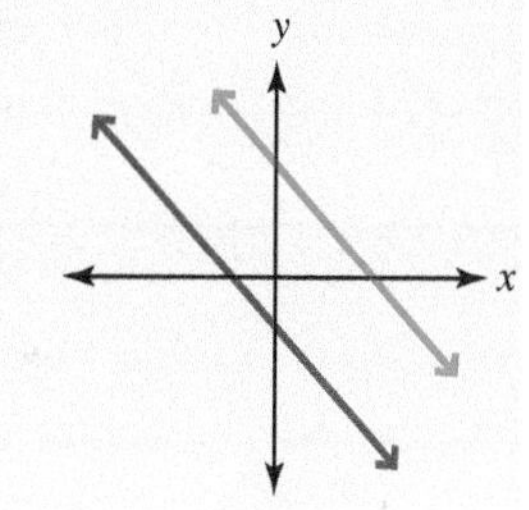

Example 3 For each of the systems of equations in Example 2, is the system consistent or inconsistent, are the equations dependent or independent, and how many solutions does the system have?

Solution: The system $\begin{cases} y = -2x + 3 \\ x - 2y = 4 \end{cases}$ from Example 2(a) is consistent with independent equations and has one solution, $(2, -1)$.

The system $\begin{cases} 3x + 4y = 8 \\ y = -\dfrac{3}{4}x - 3 \end{cases}$ from Example 2(b) is inconsistent and has no solution.

The system $\begin{cases} 6x + 2y = 4 \\ y = -3x + 2 \end{cases}$ from Example 2(c) is consistent with dependent equations and has an infinite number of solutions.

Your Turn 3

a. Determine whether the following system is consistent with independent equations, consistent with dependent equations, or inconsistent.

b. How many solutions does the following system have?

$$\begin{cases} x - 2y = 1 \\ x - 4y = 2 \end{cases}$$

Objective 3 Solve systems of linear equations using substitution.

If the solution to a system of equations contains fractions or decimal numbers, it could be difficult to determine those values using the graphing method. A more practical method is the *substitution* method. Consider the following system.

$$\begin{cases} x + 3y = 10 & \text{(Equation 1)} \\ y = x + 4 & \text{(Equation 2)} \end{cases}$$

Remember that in a system's solution, the same x and y values satisfy both equations. In the preceding system, notice that Equation 2 indicates that y is equal to $x + 4$; so y

Answer to Your Turn 3

a. consistent with independent equations

b. one solution

must be equal to $x + 4$ in Equation 1 as well. Therefore, we can substitute $x + 4$ for y in Equation 1.

Note By replacing y with $x + 4$, the variable y has been eliminated from $x + 3y = 10$.

$$x + 3y = 10$$
$$x + 3(x + 4) = 10 \quad \text{Substitute } x + 4 \text{ for } y.$$

Now we have an equation in terms of a single variable, x, which allows us to solve for x.

$$x + 3x + 12 = 10 \quad \text{Distribute 3.}$$
$$4x + 12 = 10 \quad \text{Combine like terms.}$$
$$4x = -2 \quad \text{Subtract 12 from both sides.}$$
$$x = -\frac{1}{2} \quad \text{Divide both sides by 4 and simplify.}$$

We can find the y-value by substituting $-\frac{1}{2}$ for x in either of the original equations. Equation 2 is easier because y is isolated.

$$y = x + 4$$
$$y = -\frac{1}{2} + 4 \quad \text{Substitute } -\frac{1}{2} \text{ for } x.$$
$$y = 3\frac{1}{2}$$

The solution to the system of equations is $\left(-\frac{1}{2}, 3\frac{1}{2}\right)$.

Procedure Solving Systems of Two Linear Equations Using Substitution

To find the solution of a system of two linear equations using the substitution method:

1. Isolate one of the variables in one of the equations.
2. In the other equation, substitute the expression you found in step 1 for that variable.
3. Solve this new equation. (It will now have only one variable.)
4. Using one of the equations containing both variables, substitute the value you found in step 3 for that variable and solve for the value of the other variable.
5. Check the solution in the original equations.

In step 1, if no variable is already isolated in any of the given equations, we select an equation and isolate one of its variables. Isolating a variable that has a coefficient of 1 or -1 will avoid fractions.

Example 4 Solve the system using substitution.

$$\begin{cases} x + y = 6 & \text{(Equation 1)} \\ 5x + 2y = 8 & \text{(Equation 2)} \end{cases}$$

Solution:

Step 1: In Equation 1, both x and y have a coefficient of 1; so isolating either variable is easy. We will isolate y.

$$y = 6 - x \quad \text{Subtract } x \text{ from both sides to isolate } y.$$

Step 2: Substitute $6 - x$ for y in Equation 2.

$$5x + 2y = 8$$
$$5x + 2(6 - x) = 8$$

Step 3: Solve for x using the equation we found in step 2.

$$5x + 12 - 2x = 8 \quad \text{Distribute 2.}$$
$$3x + 12 = 8 \quad \text{Combine like terms.}$$
$$3x = -4 \quad \text{Subtract 12 from both sides.}$$
$$x = -\frac{4}{3} \quad \text{Divide both sides by 3.}$$

Step 4: Find the value of y by substituting $-\frac{4}{3}$ for x in one of the equations containing both variables. Because we isolated y in $y = 6 - x$, we will use this equation.

$$y = 6 - \left(-\frac{4}{3}\right) \quad \text{Substitute } -\frac{4}{3} \text{ for } x \text{ in } y = 6 - x.$$
$$y = \frac{18}{3} + \frac{4}{3} \quad \text{Write as addition and find a common denominator.}$$
$$y = \frac{22}{3}$$

The solution is $\left(-\frac{4}{3}, \frac{22}{3}\right)$.

Step 5: We will leave the check to the reader.

Your Turn 4 Solve the system using substitution.

$$\begin{cases} x - y = 1 \\ 3x + y = 11 \end{cases}$$

If none of the variables in the system of equations has a coefficient of 1, then for step 1, select the equation that seems easiest to work with and the variable in that equation that seems easiest to isolate. When selecting that variable, recognize that you'll eventually divide by its coefficient; so choose the variable whose coefficient will divide evenly into most, if not all, of the other numbers in its equation.

Example 5 Solve the system of equations using substitution: $\begin{cases} 3x + 4y = 6 \\ 2x + 6y = -1 \end{cases}$

Solution:

Step 1: We will isolate x in $3x + 4y = 6$.

Note We chose to isolate x because its coefficient, 3, goes evenly into one of the other numbers in the equation, 6, whereas y's coefficient, 4, does not go evenly into any of the other numbers.

$$3x = 6 - 4y \quad \text{Subtract } 4y \text{ from both sides.}$$
$$x = 2 - \frac{4}{3}y \quad \text{Divide both sides by 3 to isolate } x.$$

Step 2: Substitute $2 - \frac{4}{3}y$ for x in $2x + 6y = -1$.

$$2x + 6y = -1$$
$$2\left(2 - \frac{4}{3}y\right) + 6y = -1$$

Step 3: Solve for y using the equation we found in step 2.

$$4 - \frac{8}{3}y + 6y = -1 \quad \text{Distribute to clear the parentheses.}$$
$$3 \cdot 4 - 3 \cdot \frac{8}{3}y + 3 \cdot 6y = 3 \cdot (-1) \quad \text{Multiply both sides by 3 to clear the fraction.}$$
$$12 - 8y + 18y = -3 \quad \text{Simplify.}$$
$$12 + 10y = -3 \quad \text{Combine like terms.}$$

Answer to Your Turn 4
(3, 2)

$$10y = -15 \quad \text{Subtract 12 from both sides.}$$
$$y = -\frac{3}{2} \quad \text{Divide both sides by 10 and simplify.}$$

Step 4: Find the value of x by substituting $-\frac{3}{2}$ for y in one of the equations containing both variables. We will use $x = 2 - \frac{4}{3}y$.

$$x = 2 - \frac{4}{3}\left(-\frac{3}{2}\right) \quad \text{Substitute } -\frac{3}{2} \text{ for } y \text{ in } x = 2 - \frac{4}{3}y.$$
$$x = 2 + 2 \quad \text{Simplify.}$$
$$x = 4$$

The solution is $\left(4, -\frac{3}{2}\right)$.

Step 5: We will leave the check to the reader.

Your Turn 5 Solve the system of equations using substitution.

$$\begin{cases} 5x - 2y = 10 \\ 3x - 6y = 2 \end{cases}$$

Inconsistent Systems and Dependent Equations

Let's see what happens when we solve an inconsistent system or a system with dependent equations using the substitution method.

Example 6 Solve the system of equations using substitution.

a. $\begin{cases} x + 2y = 4 \\ y = -\frac{1}{2}x + 1 \end{cases}$

Solution: Because y is isolated in the second equation, we substitute $-\frac{1}{2}x + 1$ in place of y in the first equation.

$$x + 2y = 4$$
$$x + 2\left(-\frac{1}{2}x + 1\right) = 4 \quad \text{Substitute } -\frac{1}{2}x + 1 \text{ for } y.$$

Now solve for x.

$$x - x + 2 = 4 \quad \text{Distribute to eliminate the parentheses.}$$
$$2 = 4 \quad \text{Combine like terms.}$$

Notice that $2 = 4$ no longer has a variable and is false. This false equation with no variable indicates that the system is inconsistent and has no solution.

Connection If we were to solve the system in Example 6a using graphing, we would see that the lines are parallel.

b. $\begin{cases} x = 3y + 4 \\ 2x - 6y = 8 \end{cases}$

Solution: Because x is isolated in the first equation, we substitute $3y + 4$ in place of x in the second equation.

$$2x - 6y = 8$$
$$2(3y + 4) - 6y = 8 \quad \text{Substitute } 3y + 4 \text{ for } x.$$

Answer to Your Turn 5
$\left(\frac{7}{3}, \frac{5}{6}\right)$

Now solve for y.

$$6y + 8 - 6y = 8 \quad \text{Distribute to clear the parentheses.}$$
$$8 = 8 \quad \text{Combine like terms.}$$

Connection If we were to solve the system in Example 6b using graphing, we would see that the lines are identical.

Notice that $8 = 8$ no longer has a variable and is true. This true equation with no variable indicates that the equations in the system are dependent; so there are an infinite number of solutions that are all of the ordered pairs along $x = 3y + 4$ $(\text{or } 2x - 6y = 8)$.

Your Turn 6 Solve the systems of equations using substitution.

a. $\begin{cases} y = 2x - 7 \\ 4x - 2y = -3 \end{cases}$ **b.** $\begin{cases} 3x - y = 4 \\ 2y - 6x = -8 \end{cases}$

Objective 4 Solve systems of linear equations using elimination.

Note This method is also called the elimination by addition method or simply the addition method.

Because substitution can be tedious when no coefficients are 1, we turn to a third method, the *elimination* method. In this method, we use the addition principle of equality to add equations so that a new equation emerges with one of the variables eliminated. We'll work through some examples to get a sense of the method before we write a formal procedure.

Example 7 Solve the system using elimination: $\begin{cases} x + y = 5 & \text{(Equation 1)} \\ 2x - y = 7 & \text{(Equation 2)} \end{cases}$

Solution: If we add the two equations, the resulting equation has one variable.

Note Because y and $-y$ are additive inverses, their sum is 0. Because y is eliminated in the new equation, we can easily solve for the value of x.

$$x + y = 5 \quad \text{(Equation 1)}$$
$$\underline{2x - y = 7} \quad \text{(Equation 2)}$$
$$3x + 0 = 12 \quad \text{Add Equation 1 to Equation 2.}$$
$$3x = 12 \quad \text{Simplify.}$$
$$x = 4 \quad \text{Divide both sides by 3 to isolate } x.$$

Connection The addition principle of equality (Chapter 2) says that adding the same amount to both sides of an equation will not affect its solution(s). Because Equation 1 indicates that $x + y$ and 5 are the same amount, if we add $x + y$ to the left side of Equation 2 and 5 to the right side of Equation 2, we are applying the addition principle of equality.

Now we can find the value of y by substituting 4 for x in one of the original equations. We will use $x + y = 5$.

$$4 + y = 5 \quad \text{Substitute 4 for } x \text{ in } x + y = 5.$$
$$y = 1$$

The solution is $(4, 1)$. We can check by verifying that $(4, 1)$ makes both of the original equations true. We will leave this to the reader.

Your Turn 7 Solve the system using elimination.

$$\begin{cases} 4x + 3y = 8 \\ x - 3y = 7 \end{cases}$$

Answers to Your Turn 6
a. no solution
b. all ordered pairs along $3x - y = 4$ $(\text{or } 2y - 6x = -8)$

Answer to Your Turn 7
$\left(3, -\frac{4}{3}\right)$

Connection The multiplication principle of equality (Chapter 2) says that we can multiply both sides of an equation by the same nonzero amount without affecting its solution(s).

Multiplying One Equation by a Number to Create Additive Inverses

In Example 7, you may have noticed that the expressions $x + y$ and $2x - y$ conveniently had additive inverses y and $-y$. If no such pairs of additive inverses appear in a system of equations, we can use the multiplication principle of equality to multiply both sides of one or both equations by a number to create additive inverse pairs.

Example 8 Solve the system using elimination.

$$\begin{cases} x + y = 6 & \text{(Equation 1)} \\ 2x - 5y = -16 & \text{(Equation 2)} \end{cases}$$

Solution: We will rewrite one of the equations so that it has a term that is the additive inverse of one of the terms in the other equation. We will multiply both sides of Equation 1 by 5 so that the y terms will be additive inverses.

$$5 \cdot x + 5 \cdot y = 5 \cdot 6 \quad \text{Multiply both sides of Equation 1 by 5.}$$
$$5x + 5y = 30 \quad \text{(Equation 1 rewritten)}$$

Note We could have multiplied both sides of Equation 1 by -2. We would get a $-2x$ term in Equation 1 that is the opposite of the $2x$ term in Equation 2.

Now we add the rewritten Equation 1 to Equation 2 to eliminate the y term.

$$5x + 5y = 30 \quad \text{(Equation 1 rewritten)}$$
$$\underline{2x - 5y = -16} \quad \text{(Equation 2)}$$
$$7x + 0 = 14 \quad \text{Add rewritten Equation 1 to Equation 2 to eliminate } y.$$
$$7x = 14 \quad \text{Simplify.}$$
$$x = 2 \quad \text{Divide both sides by 7 to isolate } x.$$

Note Multiplying Equation 1 by 5 made the elimination of the y terms possible.

To finish, we substitute 2 for x in an equation containing both variables. We will use $x + y = 6$.

$$2 + y = 6 \quad \text{Substitute 2 for } x \text{ in } x + y = 6.$$
$$y = 4$$

The solution is $(2, 4)$. We will leave the check to the reader.

Your Turn 8 Solve the system using elimination.

$$\begin{cases} -3x - 5y = 6 \\ x + 2y = -1 \end{cases}$$

Multiplying Each Equation by a Number to Create Additive Inverses

If no coefficient in a system of equations is 1, we may have to multiply each equation by a number to generate a pair of additive inverses.

Example 9 Solve the system using elimination.

$$\begin{cases} 4x - 3y = -2 & \text{(Equation 1)} \\ 6x - 7y = 7 & \text{(Equation 2)} \end{cases}$$

Solution: We will choose to eliminate x, so we will multiply Equation 1 by 3 and Equation 2 by -2, which makes the x terms additive inverses.

$$4x - 3y = -2 \xrightarrow{\text{Multiply by 3}} 12x - 9y = -6$$
$$6x - 7y = 7 \xrightarrow[\text{Multiply by } -2]{} -12x + 14y = -14$$

Note The x terms are now additive inverses, $12x$ and $-12x$.

Now we can add the rewritten equations to eliminate the x term.

Answer to Your Turn 8
$(-7, 3)$

$$\begin{aligned} 12x - 9y &= -6 \\ -12x + 14y &= -14 \\ \hline 0 + 5y &= -20 \quad \text{Add the equations.} \\ y &= -4 \quad \text{Solve for } y \text{ by dividing both sides by 5.} \end{aligned}$$

To finish, we substitute -4 for y in one of the equations and solve for x. We will use $4x - 3y = -2$.

$$\begin{aligned} 4x - 3(-4) &= -2 \quad \text{Substitute } -4 \text{ for } y \text{ in } 4x - 3y = -2. \\ 4x + 12 &= -2 \quad \text{Simplify.} \\ 4x &= -14 \quad \text{Subtract 12 from both sides.} \\ x &= -\frac{7}{2} \quad \text{Divide both sides by 4 and simplify.} \end{aligned}$$

Note It doesn't matter which equation we use because x has the same value in any equation in the system.

The solution is $\left(-\frac{7}{2}, -4\right)$. We will leave the check to the reader.

Fractions or Decimals in a System

We can use the multiplication principle to clear any fractions or decimals from the equations in a system.

Example 10 Solve the system using elimination.

$$\begin{cases} \frac{1}{2}x - y = \frac{3}{4} & \text{(Equation 1)} \\ 0.4x - 0.3y = 1 & \text{(Equation 2)} \end{cases}$$

Note Because this system contains both fractions and decimal numbers, the answer can be written either in fraction or decimal form.

Solution: To eliminate the fractions in Equation 1, we multiply both sides by the LCD, 4. To eliminate the decimals in Equation 2, we multiply both sides by 10.

$$\frac{1}{2}x - y = \frac{3}{4} \xrightarrow{\text{Multiply by 3}} 2x - 4y = 3$$

$$0.4x - 0.3y = 1 \xrightarrow[\text{Multiply by 10}]{} 4x - 3y = 10$$

Now that both equations contain only integers, the system is easier to solve. We eliminate x by multiplying the first equation by -2, which makes the x terms additive inverses; then we add the equations.

$$\begin{aligned} -4x + 8y &= -6 \quad -2(2x - 4y = 3) \text{ gives } -4x + 8y = -6. \\ 4x - 3y &= 10 \\ \hline 0 + 5y &= 4 \quad \text{Add the equations.} \\ y &= \frac{4}{5} \text{ or } 0.8 \quad \text{Solve for } y \text{ by dividing both sides by 5.} \end{aligned}$$

To finish, substitute 0.8 $\left(\text{or } \frac{4}{5}\right)$ for y in one of the equations and solve for x.

$$\begin{aligned} 0.4x - 0.3(0.8) &= 1 \quad \text{Substitute 0.8 for } y \text{ in Equation 2.} \\ 0.4x - 0.24 &= 1 \quad \text{Simplify.} \\ 0.4x &= 1.24 \quad \text{Add 0.24 to both sides.} \\ x &= 3.1 \text{ or } \frac{31}{10} \quad \text{Divide both sides by 0.4.} \end{aligned}$$

The solution is $(3.1, 0.8)$ or $\left(\frac{31}{10}, \frac{4}{5}\right)$. We will leave the check to the reader.

Your Turn 10 Solve the system using elimination.

$$\begin{cases} 0.6x + y = 3 \\ \frac{2}{5}x + \frac{1}{4}y = 1 \end{cases}$$

Rewriting the Equations in the form $Ax + By = C$

In Examples 7–10, all of the equations were in standard form, which we learned in Chapter 3 to be $Ax + By = C$. To use the elimination method, the equations are usually written in standard form. Consider the following system.

$$\begin{cases} y = 2 - x & \text{(Equation 1)} \\ 3x - y = 1 & \text{(Equation 2)} \end{cases}$$

We need to rewrite Equation 1 in standard form before using elimination.

$$\begin{cases} x + y = 2 & \text{(Equation 1)} \\ 3x - y = 1 & \text{(Equation 2)} \end{cases}$$

We can now summarize the elimination method with the following procedure.

Procedure Solving Systems of Two Linear Equations Using Elimination

To solve a system of two linear equations using the elimination method:

1. Write the equations in standard form ($Ax + By = C$).
2. Use the multiplication principle of equality to eliminate fractions or decimals (optional).
3. If necessary, multiply one or both equations by a number (or numbers) so that they have a pair of terms that are additive inverses.
4. Add the equations. The result is an equation in terms of one variable.
5. Solve the equation from step 4 for the value of that variable.
6. Using an equation containing both variables, substitute the value you found in step 5 for the corresponding variable and solve for the value of the other variable.
7. Check your solution in the original equations.

Inconsistent Systems and Dependent Equations

How would we recognize an inconsistent system or a system with dependent equations using the elimination method?

Connection If we were to graph the equations in Example 11(a), we would see that the lines are parallel, confirming that there is no solution for the system.

Example 11 Solve the system using elimination.

a. $\begin{cases} 2x - y = 1 \\ 2x - y = -3 \end{cases}$

Solution: Notice that the left sides of the equations match. Multiplying one of the equations by -1 and then adding the equations will eliminate both variables.

$$\begin{array}{rcl} 2x - y = 1 & \xrightarrow{\text{Multiply by } -1} & -2x + y = -1 \\ 2x - y = -3 & & \underline{2x - y = -3} \\ & & 0 = -4 \end{array}$$

Both variables have been eliminated, and the resulting equation, $0 = -4$, is false; therefore, there is no solution. This system of equations is inconsistent.

Answer to Your Turn 10

$\left(1, \frac{12}{5}\right)$ or $(1, 2.4)$

b. $\begin{cases} 3x + 4y = 5 \\ 9x + 12y = 15 \end{cases}$

Solution: To eliminate x, we could multiply the first equation by -3, then combine the equations.

$$\begin{aligned} 3x + 4y &= 5 \xrightarrow{\text{Multiply by } -3} & -9x - 12y &= -15 \\ 9x + 12y &= 15 & 9x + 12y &= 15 \\ & & 0 &= 0 \end{aligned}$$

Both variables have been eliminated, and the resulting equation, $0 = 0$, is true. This means the equations are dependent, so there are an infinite number of solutions. We can say that the solution set contains every ordered pair that solves $3x + 4y = 5$ (or $9x + 12y = 15$).

Connection If graphed, both equations in Example 11(b) would generate the same line, indicating that the equations are dependent.

Your Turn 11 Solve the system using elimination.

a. $\begin{cases} x - 4y = 2 \\ 5x - 20y = 10 \end{cases}$

b. $\begin{cases} x + 2y = 3 \\ x + 2y = 1 \end{cases}$

Answers to Your Turn 11

a. all ordered pairs that solve $x - 4y = 2$ (dependent equations)

b. no solution (inconsistent)

4.1 Exercises

For Extra Help MyMathLab®

Objective 1

Prep Exercise 1 How do you check a solution for a system of equations?

For Exercises 1–8, determine whether the given ordered pair is a solution to the system of equations. See Example 1.

1. $(-1, 2); \begin{cases} x - y = -3 \\ x + y = 1 \end{cases}$

2. $(1, -2); \begin{cases} 4x - 3y = 10 \\ 3x - 2y = 7 \end{cases}$

3. $(-2, 3); \begin{cases} 3x + 4y = 6 \\ x - 4y = 8 \end{cases}$

4. $(-3, 2); \begin{cases} 2x + 3y = -12 \\ 4x - 5y = 2 \end{cases}$

5. $\left(-\frac{3}{4}, -\frac{2}{3}\right); \begin{cases} 4x + 3y = -5 \\ 12x - 6y = -5 \end{cases}$

6. $\left(\frac{3}{4}, \frac{2}{3}\right); \begin{cases} 4x + 6y = 7 \\ 8x - 3y = 4 \end{cases}$

7. $(3, -4); \begin{cases} 3x + 2y = 1 \\ \frac{1}{3}x + \frac{1}{2}y = 3 \end{cases}$

8. $(4, -5); \begin{cases} \frac{1}{4}x + \frac{3}{5}y = 4 \\ x + 2y = -6 \end{cases}$

For Exercises 9–12, translate into a system of equations. Do not solve. See Objective 1.

9. The sum of two numbers is five. The difference of the same numbers is three. What are the numbers?

10. The difference of two numbers is eight. The sum of the same numbers is twelve. What are the numbers?

11. The perimeter of a rectangle is 50 feet. If the width is 2 feet less than the length, find the dimensions of the rectangle.

12. The perimeter of a rectangle is 65 centimeters. If the length is 5 centimeters more than the width, find the dimensions of the rectangle.

Objective 2

Prep Exercise 2 If a system of linear equations in two variables has one solution, what does this indicate about the graphs of the equations?

Prep Exercise 3 If a system of linear equations in two variables has no solution (inconsistent), what does this indicate about the graphs of the equations?

Prep Exercise 4 If two linear equations in two variables has an infinite number of solutions (dependent), what does this mean about their graphs?

For Exercises 13–22, solve the system graphically. See Example 2.

13. $\begin{cases} x + y = 5 \\ x - y = 3 \end{cases}$

14. $\begin{cases} x - y = 1 \\ 2x + y = 8 \end{cases}$

15. $\begin{cases} y = 2x + 5 \\ y = -x - 4 \end{cases}$

16. $\begin{cases} y = -2x - 4 \\ y = x - 1 \end{cases}$

17. $\begin{cases} 2x - y = 3 \\ 2x - y = 8 \end{cases}$

18. $\begin{cases} 3x + 2y = 6 \\ 6x + 4y = 18 \end{cases}$

19. $\begin{cases} 3x + y = 4 \\ 6x + 2y = 8 \end{cases}$

20. $\begin{cases} 2x + y = -2 \\ 8x + 4y = -8 \end{cases}$

21. $\begin{cases} x = 4 \\ y = -2 \end{cases}$

22. $\begin{cases} x = -3 \\ y = 4 \end{cases}$

For Exercises 23–28: a. *Determine whether the system of equations is consistent with independent equations, consistent with dependent equations, or inconsistent.*
b. *How many solutions does the system have? See Example 3.*

23.

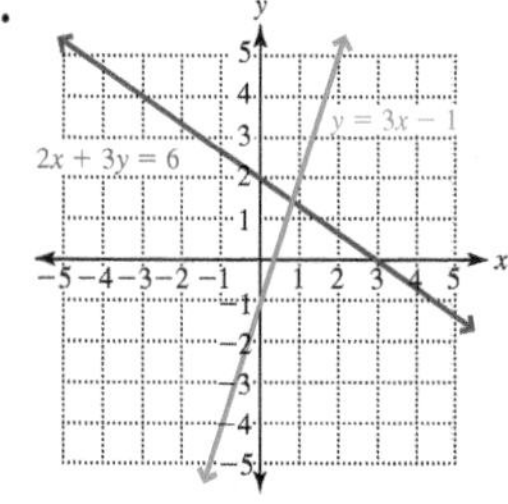

24.

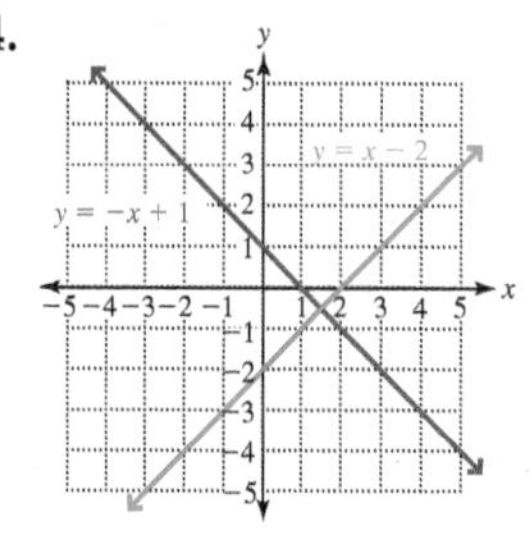

25.

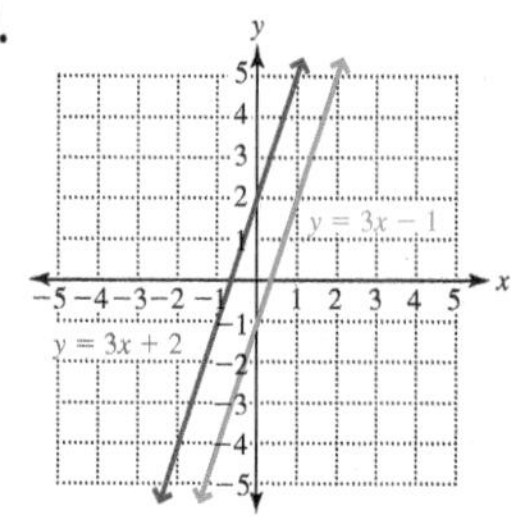

26.

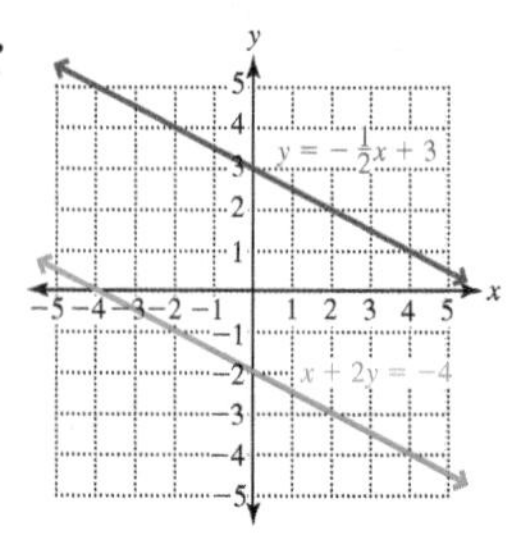

27.

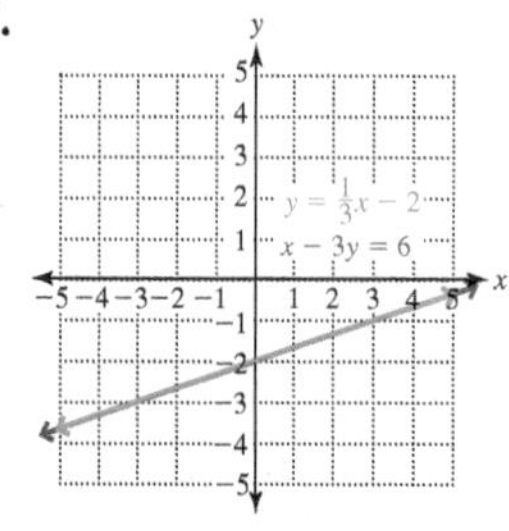

28.

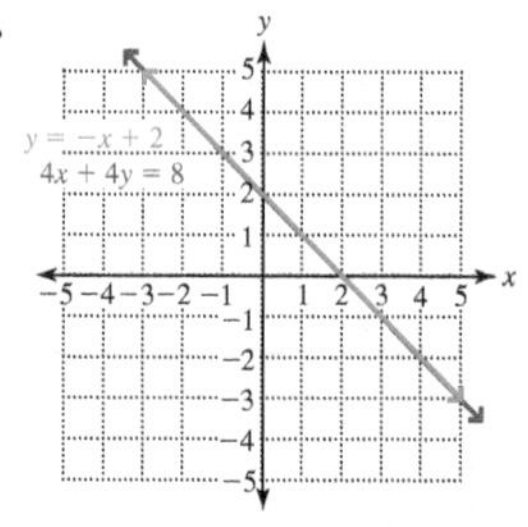

For Exercises 29–34, determine whether the system of equations is consistent with independent equations, consistent with dependent equations or inconsistent. See Example 3.

29. $\begin{cases} y = -x \\ y - x = 6 \end{cases}$

30. $\begin{cases} y = x + 4 \\ 4x + y = 10 \end{cases}$

31. $\begin{cases} x + 3y = 1 \\ 2x + 6y = 2 \end{cases}$

32. $\begin{cases} 6x + 9y = 6 \\ 2x + 3y = 2 \end{cases}$

33. $\begin{cases} 3x + 2y = 12 \\ 6x + 4y = -12 \end{cases}$

34. $\begin{cases} 2x + y = 4 \\ 4x + 2y = 12 \end{cases}$

Objective 3

Prep Exercise 5 What advantage does the method of substitution have over graphing for solving systems of equations?

Prep Exercise 6 Suppose you plan to use substitution to solve the system

$$\begin{cases} y = x + 2 \\ 3x - 4y = -9 \end{cases}$$

a. Which variable would you replace in the substitution process?

b. What expression would replace that variable?

Prep Exercise 7 Explain how to recognize a system with no solution (inconsistent system) when using substitution.

Prep Exercise 8 Explain how to recognize a system with an infinite number of solutions (consistent with dependent equations) when using substitution.

For Exercises 35–50, solve the system using substitution. See Examples 4–6.

35. $\begin{cases} 3x - y = 2 \\ y = 2x \end{cases}$

36. $\begin{cases} 2x - y = 5 \\ y = -3x \end{cases}$

37. $\begin{cases} x + y = 1 \\ y = -2x - 1 \end{cases}$

38. $\begin{cases} x = y + 6 \\ x + y = -2 \end{cases}$

39. $\begin{cases} y = -\frac{3}{4}x \\ x - 8y = -7 \end{cases}$

40. $\begin{cases} y = \frac{2}{3}x \\ 6x + 3y = 4 \end{cases}$

41. $\begin{cases} 3x + 4y = 11 \\ x + 2y = 5 \end{cases}$

42. $\begin{cases} 2x + y = 7 \\ 3x + 2y = 12 \end{cases}$

43. $\begin{cases} 4x - y = -13 \\ 3x - 4y = -13 \end{cases}$

44. $\begin{cases} 5x - 3y = 2 \\ 3x - y = -2 \end{cases}$

45. $\begin{cases} 4x + 3y = 2 \\ 3x + 2y = 2 \end{cases}$

46. $\begin{cases} 2x - 5y = 18 \\ 3x + 2y = 8 \end{cases}$

47. $\begin{cases} x - 3y = -4 \\ 5x - 15y = -6 \end{cases}$

48. $\begin{cases} 2x - y = 1 \\ 4x - 2y = 5 \end{cases}$

49. $\begin{cases} x - 2y = 6 \\ -2x + 4y = -12 \end{cases}$

50. $\begin{cases} x + y = -5 \\ -2x - 2y = 10 \end{cases}$

Find the Mistake *For Exercises 51 and 52, explain the mistake; then find the correct solution.*

51. $\begin{cases} 2x + y = 8 \\ x = y + 1 \end{cases}$

Substitute:

$$2(y + 1) + y = 8$$
$$2y + 1 + y = 8$$
$$3y + 1 = 8$$
$$3y = 7$$
$$y = \frac{7}{3}$$

Solve for x:

$$x = \frac{7}{3} + 1$$
$$x = \frac{10}{3}$$

Solution: $\left(\frac{10}{3}, \frac{7}{3}\right)$

52. $\begin{cases} 3x + 7y = 8 \\ x - y = 6 \end{cases}$

Rewrite: $x = 6 + y$

Substitute: $6 + y - y = 6$
$$6 = 6$$

Solution: Infinite solutions along $x - y = 6$

Objective 4

Prep Exercise 9 What advantage does the method of elimination have over graphing and substitution?

Prep Exercise 10 Suppose you plan to use elimination to solve the system

$$\begin{cases} 4x + 2y = 10 \\ 3x - 2y = 4 \end{cases}.$$

Which variable would you eliminate? Why?

Prep Exercise 11 Explain how to recognize a system with no solution (inconsistent system) when using elimination.

Prep Exercise 12 Explain how to recognize a system with an infinite number of solutions (consistent with dependent equations) when using elimination.

For Exercises 53–72, solve the system using elimination. See Examples 7–11.

53. $\begin{cases} x + y = 1 \\ 2x - y = 2 \end{cases}$

54. $\begin{cases} x - y = 7 \\ x + y = 5 \end{cases}$

55. $\begin{cases} 2x + 3y = 9 \\ 4x + 3y = 15 \end{cases}$

56. $\begin{cases} 3x + 2y = 6 \\ 5x + 2y = 14 \end{cases}$

57. $\begin{cases} 4x + y = 8 \\ 5x + 3y = 3 \end{cases}$

58. $\begin{cases} 5x + y = 12 \\ 5x - 6y = -2 \end{cases}$

59. $\begin{cases} x + 3y = -1 \\ 3x + 6y = -1 \end{cases}$

60. $\begin{cases} x + 4y = 1 \\ 3x + 8y = 4 \end{cases}$

61. $\begin{cases} 3x = 2y + 7 \\ 4x - 3y = 10 \end{cases}$

62. $\begin{cases} 4x = 3y - 23 \\ 5x - 4y = -30 \end{cases}$

63. $\begin{cases} 3x - 4y = 2 \\ 4x + 5y = 6 \end{cases}$

64. $\begin{cases} 2x + 3y = 6 \\ 5x - 2y = 4 \end{cases}$

65. $\begin{cases} \frac{1}{5}x + \frac{1}{2}y = \frac{1}{5} \\ \frac{1}{2}x + \frac{1}{3}y = -\frac{4}{3} \end{cases}$

66. $\begin{cases} \frac{1}{4}x - \frac{1}{2}y = -\frac{1}{4} \\ \frac{1}{3}x + \frac{1}{6}y = \frac{4}{3} \end{cases}$

67. $\begin{cases} 0.4x - 0.3y = 1.3 \\ 0.3x + 0.5y = 1.7 \end{cases}$

68. $\begin{cases} 0.5x - 0.2y = -2.1 \\ 0.2x + 0.5y = 0.9 \end{cases}$

69. $\begin{cases} 2x + 3y = 6 \\ 4x - 18 = -6y \end{cases}$

70. $\begin{cases} 4x = 2y + 7 \\ 2x - y = 4 \end{cases}$

71. $\begin{cases} x + 2y = 4 \\ x - 4 = -2y \end{cases}$

72. $\begin{cases} 4x + 2y = 6 \\ 6x = 9 - 3y \end{cases}$

Find the Mistake *For Exercises 73 and 74, explain the mistake; then find the correct solution.*

73. $\begin{cases} x + y = 8 \\ \underline{x - y = 7} \end{cases}$

$2x = 1$

$x = \frac{1}{2}$

$\frac{1}{2} + y = 8$

$y = \frac{15}{2}$

Solution: $\left(\frac{1}{2}, \frac{15}{2}\right)$

74. $\begin{cases} x + y = 3 \\ x + 2y = 4 \end{cases}$

$x + y = 3$

$-1(x + 2y = 4)$

$x + y = 3$

$\underline{-x - 2y = 4}$

$-y = 7$

$y = -7$

$x - 7 = 3$

$x = 10$

Solution: $(10, -7)$

For Exercises 75–78, solve.

75. A business breaks even when its costs and revenues are equal. The graph shows the cost of a product based on the number of units produced and the revenue based on the number of units sold.

a. How many units must be sold for the business to break even?

b. What amount of revenue and cost is needed for the business to break even?

c. Write an inequality that describes the number of units that must be sold for the business to make a profit.

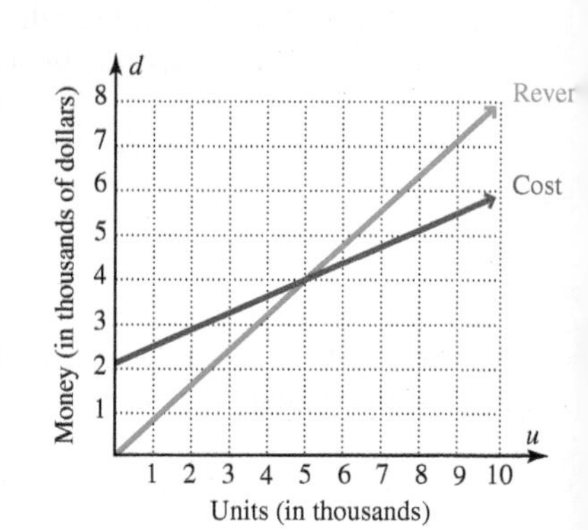

76. The demand for a product is related to the supply available. The graph shows the supply and demand curves for a particular product.

a. What is the number of units for which the supply equals the demand?

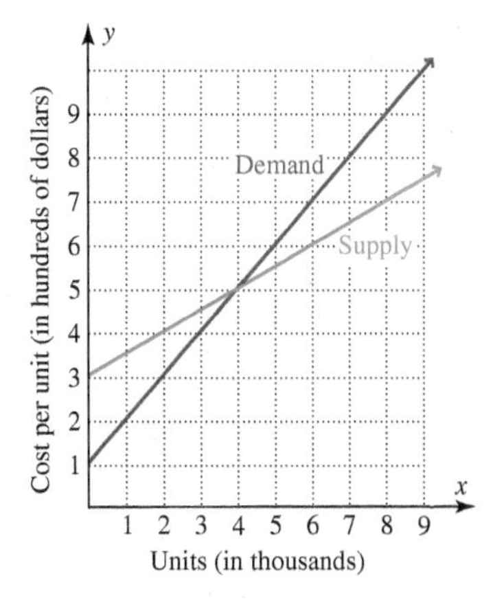

b. What is the cost per unit when the supply and demand are equal?

c. Write an inequality that describes the number of units for which the demand is greater than the supply.

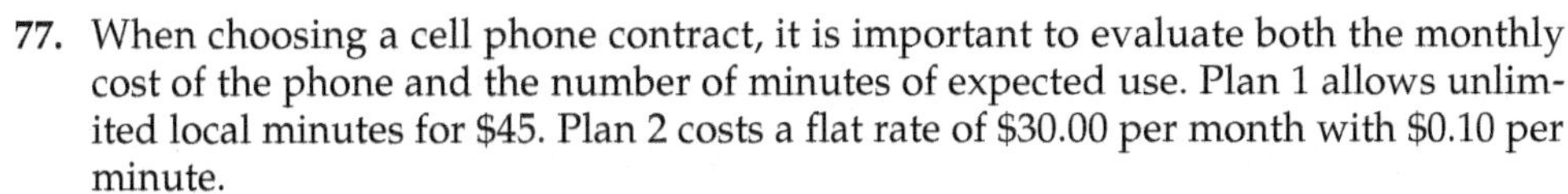

77. When choosing a cell phone contract, it is important to evaluate both the monthly cost of the phone and the number of minutes of expected use. Plan 1 allows unlimited local minutes for \$45. Plan 2 costs a flat rate of \$30.00 per month with \$0.10 per minute.

a. Write a system of equations so that one equation describes the total cost of plan 1 and the other equation describes the total cost of plan 2.

b. On one grid, graph both equations in the system.

c. How many minutes must be used for the two plans to cost the same per month?

d. Which plan is cheaper if the customer plans to use more than 200 minutes per month?

e. If the phone is used only for emergencies, which plan is less expensive?

78. Deciding whether to purchase or lease a car can be a challenging decision. Suppose a dealership offers two deals for a particular car. The first deal is to purchase the car with \$2700 down plus a monthly payment of \$550 for 36 months. The second deal is to lease the car with \$1500 down plus a lease payment of \$350 per month for 36 months. If the vehicle is retured in good condition, the lease agreement can be transferred to a new vehicle at the same payment (\$350) and no additional down payment.

a. Write a system of equations so that one equation describes the total cost of buying a vehicle and the other equation describes the total cost of leasing a car indefinitely.

b. On one grid, graph both equations in the system.

c. After how many months is the total cost of leasing the same as that of purchasing?

d. If the customer wants to drive a new vehicle every three years, is it better to go with the purchase or the lease deal?

e. What other factors should a person consider when deciding whether to purchase or lease a car?

Review Exercises

Exercises 1 and 2 Expressions

[1.4] 1. Use the distributive property: $4(5x + 7y - z)$

[1.4] 2. Simplify: $3x - 4y + z - 2x + y - z$

Exercises 3–6 Equations and Inequalities

[2.1] 3. Solve for y: $x + 3y = 6$

[2.1] 4. Solve for x: $\frac{1}{3}x - 2y = 10$

[1.4] 5. Given $x + y + 2z = 7$, find x if $y = -1$ and $z = 3$.

[1.4] 6. Given $x - 3y + 2z = 6$, find y if $x = -2$ and $z = 4$.

4.2 Solving Systems of Linear Equations in Three Variables

Objectives

1 Determine whether an ordered triple is a solution for a system of equations.
2 Understand the types of solution sets for systems of three equations.
3 Solve a system of three linear equations using the elimination method.

Warm-up

[1.4] **1.** Find the value of $2x + 3y + 2z$ if $x = 3, y = 1$, and $z = -1$.
[1.4] **2.** Find the product: $-2(2x + y + 2z)$
[1.4] **3.** Combine like terms: $-4x - 2y - 4z + 4x + 8y + 5z$

Objective 1 Determine whether an ordered triple is a solution for a system of equations.

In this section, we solve systems of three linear equations with three unknowns. Solutions of these systems are *ordered triples* with the form (x, y, z). We check a solution for a system of three equations the same way we check a system of two equations, by replacing each variable with its corresponding value and verifying that each equation is true.

Example 1 Determine whether each ordered triple is a solution to the system of equations.

$$\begin{cases} x + y + z = 3 & \text{(Equation 1)} \\ 2x + 3y + 2z = 7 & \text{(Equation 2)} \\ 3x - 4y + z = 4 & \text{(Equation 3)} \end{cases}$$

a. $(2, 1, 0)$

Solution: In all three equations, replace x with 2, y with 1, and z with 0.

Equation 1:
$$x + y + z = 3$$
$$2 + 1 + 0 \stackrel{?}{=} 3$$
$$3 = 3 \quad \text{True}$$

Equation 2:
$$2x + 3y + 2z = 7$$
$$2(2) + 3(1) + 2(0) \stackrel{?}{=} 7$$
$$4 + 3 + 0 \stackrel{?}{=} 7$$
$$7 = 7 \quad \text{True}$$

Equation 3:
$$3x - 4y + z = 4$$
$$3(2) - 4(1) + 0 \stackrel{?}{=} 4$$
$$6 - 4 + 0 \stackrel{?}{=} 4$$
$$2 = 4 \quad \text{False}$$

Because $(2, 1, 0)$ does not satisfy all three equations in the system, it is not a solution for the system.

b. $(3, 1, -1)$

Solution:

Equation 1:
$$x + y + z = 3$$
$$3 + 1 + (-1) \stackrel{?}{=} 3$$
$$3 = 3 \quad \text{True}$$

Equation 2:
$$2x + 3y + 2z = 7$$
$$2(3) + 3(1) + 2(-1) \stackrel{?}{=} 7$$
$$6 + 3 - 2 \stackrel{?}{=} 7$$
$$7 = 7 \quad \text{True}$$

Equation 3:
$$3x - 4y + z = 4$$
$$3(3) - 4(1) + (-1) \stackrel{?}{=} 4$$
$$9 - 4 - 1 \stackrel{?}{=} 4$$
$$4 = 4 \quad \text{True}$$

Because $(3, 1, -1)$ satisfies all three equations in the system, it is a solution for the system.

Learning Strategy

The best way to prepare for class is to stay up to date on homework. Seek help immediately if you have a problem that you cannot answer. In math classes, teachers explain problems that will be used again later in more complicated problems.

—Matt D.

Your Turn 1 Determine whether the following ordered triples are solutions to the system of equations.

a. $(4, 1, 1)$ **b.** $(-1, 5, 2)$

$$\begin{cases} x + y + z = 6 & \text{(Equation 1)} \\ 3x - 2y + 3z = -7 & \text{(Equation 2)} \\ 4x - 2y + z = -12 & \text{(Equation 3)} \end{cases}$$

Answers to Your Turn 1
a. not a solution
b. solution

Answers to Warm-up
1. 7
2. $-4x - 2y - 4z$
3. $6y + z$

Objective 2 Understand the types of solution sets for systems of three equations.

Recall that we plot an ordered pair using two axes: x and y. Similarly, we plot an ordered triple such as $(3, 4, 5)$ using three axes—x, y, and z—each of which is perpendicular to the other two, as shown.

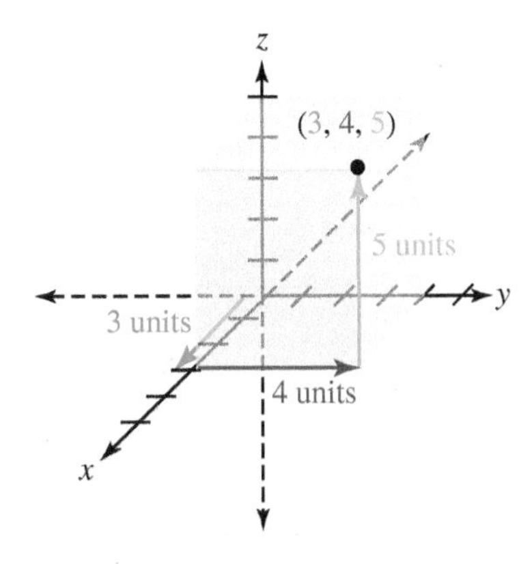

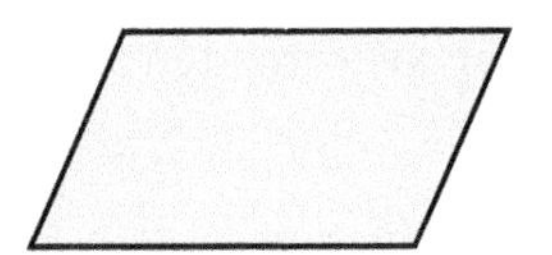

The graph of a linear equation in three variables is a *plane*, which is a flat surface much like a sheet of paper with infinite length and width. We draw a plane as a parallelogram (shown in the margin), but remember that the length and width are infinite.

In Section 4.1, we found that two lines could intersect in one point (consistent system), no points (inconsistent system), or an infinite number of points (consistent system with dependent equations). Similarly, the intersection of the planes of a three-variable system can tell us about the number of solutions for the system.

Connection Recall from Section 4.1 that a system of equations that has a solution (single or infinite) is consistent and a system that has no solution is inconsistent. This is true for systems of three equations as well. We also discussed dependent and independent equations. However, for systems of three equations, defining dependent and independent equations and discussing their graphical representations is too complicated for this course. Here, we simply say that systems with an infinite number of solutions have dependent equations. As we will see, we can identify these systems when solving them using the elimination method.

A Single Solution

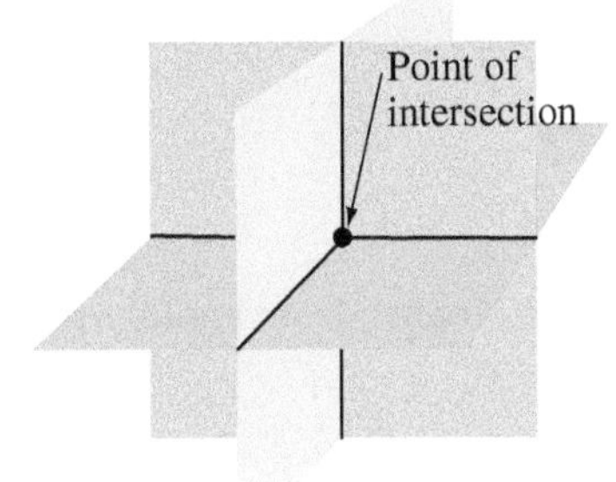

If the planes intersect at a single point, that ordered triple is the solution to the system.

Infinite Number of Solutions

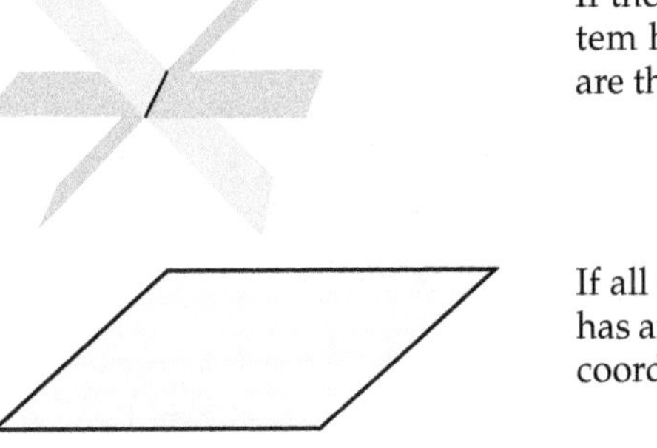

If the three planes intersect along a line, the system has an infinite number of solutions, which are the coordinates of any point along that line.

If all three graphs are the same plane, the system has an infinite number of solutions, which are the coordinates of any point in the plane.

No Solution

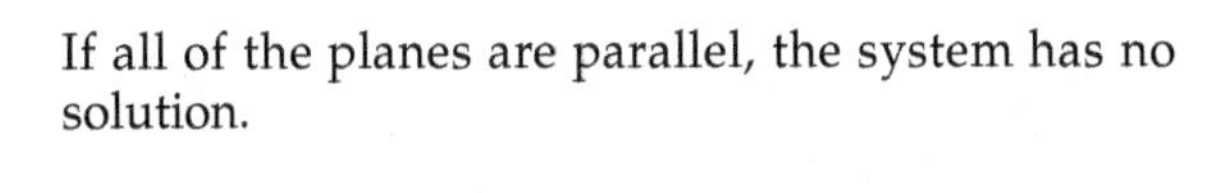

If all of the planes are parallel, the system has no solution.

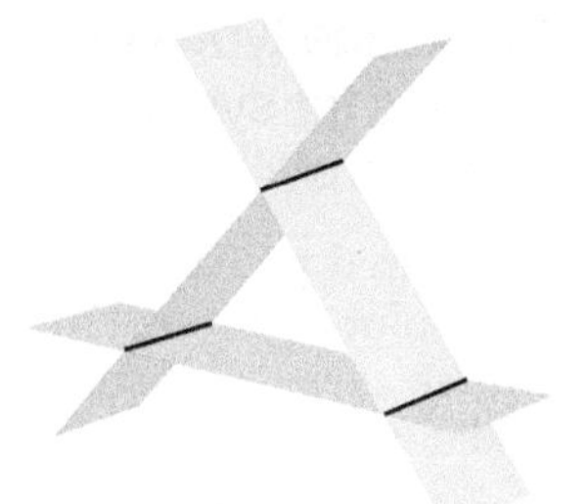

Pairs of planes also can intersect, as shown. However, because all three planes do not have a common intersection, the system has no solution.

Note Planes can be configured in still more ways so that they do not have a common intersection.

Objective 3 Solve a system of three linear equations using the elimination method.

Because graphing systems of three linear equations is usually impractical, we solve them using elimination. The process is much like what we learned in Section 4.1.

Example 2 Solve the system of equations using elimination.

a. $\begin{cases} x + y + z = 2 & \text{(Equation 1)} \\ 2x + y + 2z = 1 & \text{(Equation 2)} \\ 3x + 2y + z = 1 & \text{(Equation 3)} \end{cases}$

Solution: First, we choose a variable and eliminate that variable from two pairs of equations. Let's eliminate y using Equations 1 and 2. We multiply Equation 1 by -1, then add the equations.

(Eq. 1) $x + y + z = 2$ $\xrightarrow{\text{Multiply by } -1.}$ $-x - y - z = -2$

(Eq. 2) $2x + y + 2z = 1$ $\qquad 2x + y + 2z = 1$ Add the equations.

$x + z = -1$ (Equation 4)

Note Our initial goal is to generate two equations with the same two variables.

Now we need to eliminate y from another pair of equations. We will choose Equations 1 and 3.

(Eq. 1) $x + y + z = 2$ $\xrightarrow{\text{Multiply by } -2.}$ $-2x - 2y - 2z = -4$

(Eq. 3) $3x + 2y + z = 1$ $\qquad 3x + 2y + z = 1$ Add the equations.

$-x - z = -3$ (Equation 5)

Equations 4 and 5 form a system of two equations with two variables x and z. We can eliminate z by adding the equations.

(Eq. 4) $x + z = -1$

(Eq. 5) $x - z = -3$ Add the equations.

$2x = -4$

$x = -2$ Divide both sides by 2 to isolate x.

To find z, we substitute -2 for x in Equation 4 or 5. We will use Equation 4.

$x + z = -1$

$-2 + z = -1$ Substitute -2 for x.

$z = 1$ Add 2 to both sides.

To find y, substitute -2 for x and 1 for z in any of the original equations. We will use Equation 1.

$x + y + z = 2$

$-2 + y + 1 = 2$ Substitute -2 for x and 1 for z.

$y - 1 = 2$ Simplify the left side.

$y = 3$ Add 1 to both sides to isolate y.

Note Recall that ordered triples are written in the form (x, y, z).

The solution is $(-2, 3, 1)$. We can check the solution by verifying that the ordered triple satisfies each of the three original equations. We will leave the check to the reader.

b.
$$\begin{cases} 2x + 3y = 3 & \text{(Equation 1)} \\ 2y - 3z = -8 & \text{(Equation 2)} \\ 4x - z = 10 & \text{(Equation 3)} \end{cases}$$

Solution: Notice that a variable is missing in each of the equations, which can simplify solving the system. We will eliminate z from Equations 2 and 3.

Note We chose Equation 3 because z has a coefficient of -1.

$$\begin{array}{lll} \text{(Eq. 2)}\ 2y - 3z = -8 & & 2y - 3z = -8 \\ \text{(Eq. 3)}\ 4x - z = 10 & \xrightarrow{\text{Multiply by } -3.} & -12x + 3z = -30 \quad \text{Add the equations.} \\ & & \overline{-12x + 2y = -38} \quad \text{(Equation 4)} \end{array}$$

Because Equations 1 and 4 do not have z terms, we do not need to eliminate z from another pair of equations. We will use Equations 1 and 4 to eliminate x.

$$\begin{array}{lll} \text{(Eq. 1)}\quad 2x + 3y = 3 & \xrightarrow{\text{Multiply by 6.}} & 12x + 18y = 18 \\ \text{(Eq. 4)}\ -12x + 2y = -38 & & -12x + 2y = -38 \quad \text{Add the equations.} \\ & & \overline{ 20y = -20} \\ & & y = -1 \quad \text{Divide both sides by 20 to isolate } y. \end{array}$$

To find x, substitute -1 for y in Equation 1 or 4. We will use Equation 1.

$$\begin{aligned} 2x + 3(-1) &= 3 && \text{Substitute } -1 \text{ for } y \text{ in Equation 1.} \\ 2x - 3 &= 3 && \text{Simplify.} \\ 2x &= 6 && \text{Add 3 to both sides.} \\ x &= 3 && \text{Divide both sides by 2 to isolate } x. \end{aligned}$$

To find z, substitute in Equation 2 or 3. (Equation 1 has no z term.) We will use Equation 2.

$$\begin{aligned} 2(-1) - 3z &= -8 && \text{Substitute } -1 \text{ for } y \text{ in Equation 2.} \\ -2 - 3z &= -8 && \text{Simplify.} \\ -3z &= -6 && \text{Add 2 to both sides.} \\ z &= 2 && \text{Divide both sides by } -3 \text{ to isolate } z. \end{aligned}$$

The solution is $(3, -1, 2)$. We will leave the check to the reader.

Following is a summary of the process of solving a system of three linear equations.

Procedure Solving Systems of Three Linear Equations Using Elimination

To solve a system of three linear equations with three unknowns using elimination:

1. Write each equation in the form $Ax + By + Cz = D$.
2. Eliminate one variable from one pair of equations using the elimination method.
3. If necessary, eliminate the same variable from another pair of equations.
4. Steps 2 and 3 result in two equations with the same two variables. Solve these equations using the elimination method.
5. To find the third variable, substitute the values of the variables found in step 4 into any of the three original equations that contain the third variable.
6. Check the ordered triple in all three original equations.

Your Turn 2 Solve the system using elimination.

a. $\begin{cases} x + 3y + 2z = 6 \\ 2x - 3y + z = -18 \\ -3x + 2y + z = 12 \end{cases}$

b. $\begin{cases} 2x - 3z = -6 \\ x + 3y = -3 \\ 2y - 3z = -16 \end{cases}$

Note In Section 4.1, we learned that if we get $0 = 0$ when using elimination to solve a system of two equations, then the two equations are dependent (see Example 11b in Section 4.1). Similarly, the equations in a system of three equations are dependent if we get $0 = 0$ when adding any two equations. For example, if Equation 4 of Example 3 were $-7x - z = -18$, the sum of Equations 4 and 5 would be $0 = 0$.

$$\begin{array}{lr} \text{(Eq. 4)} & -7x - z = -18 \\ \text{(Eq. 5)} & \underline{7x + z = 18} \\ & 0 = 0 \end{array}$$

In such a case, we say that the equations in the system are dependent and that the system has an infinite number of solutions.

Let's see what happens when a system of three linear equations has no solution (inconsistent) or has an infinite number of solutions (dependent).

Example 3 Solve the system using elimination.

$$\begin{cases} 2x + y + 2z = -1 & \text{(Equation 1)} \\ -3x + 2y + 3z = -13 & \text{(Equation 2)} \\ 4x + 2y + 4z = 5 & \text{(Equation 3)} \end{cases}$$

Solution: We need to eliminate a variable from two pairs of equations. Let's eliminate y using Equations 1 and 2.

$$\begin{array}{ll} \text{(Eq. 1)}\ 2x + y + 2z = -1 \xrightarrow{\text{Multiply by } -2.} & -4x - 2y - 4z = 2 \\ \text{(Eq. 2)}\ -3x + 2y + 3z = -13 & \underline{-3x + 2y + 3z = -13} \quad \text{Add the equations.} \\ & -7x \quad - z = -11 \quad \text{(Equation 4)} \end{array}$$

Now we eliminate y from Equations 2 and 3.

$$\begin{array}{ll} \text{(Eq. 2)}\ -3x + 2y + 3z = -13 \xrightarrow{\text{Multiply by } -1.} & 3x - 2y - 3z = 13 \\ \text{(Eq. 3)}\ 4x + 2y + 4z = 5 & \underline{4x + 2y + 4z = 5} \quad \text{Add the equations.} \\ & 7x \quad + z = 18 \quad \text{(Equation 5)} \end{array}$$

Equations 4 and 5 form a system in x and z.

$$\begin{array}{lr} \text{(Eq. 4)} & -7x - z = -11 \\ \text{(Eq. 5)} & \underline{7x + z = 18} \quad \text{Add the equations.} \\ & 0 = 7 \end{array}$$

All variables are eliminated and the resulting equation is false, which means that this system has no solution; it is inconsistent.

Your Turn 3 Solve the following systems of equations using elimination.

a. $\begin{cases} 2x + y + 2z = 1 \\ x + y + z = 2 \\ 4x + 2y + 4z = 6 \end{cases}$

b. $\begin{cases} x + 3y - 6z = 9 \\ -7y + 6z = -3 \\ -2x + y + 6z = -15 \end{cases}$

Answers to Your Turn 2
a. $(-2, 4, -2)$ b. $(3, -2, 4)$

Answers to Your Turn 3
a. no solution (inconsistent)
b. infinite number of solutions (dependent equations)

4.2 Exercises

For Extra Help MyMathLab®

Objectives 1 and 2

Prep Exercise 1 When solving a system of three equations with three variables, you eliminate a variable from one pair of equations, then eliminate the same variable from a second pair of equations. Does it matter which equations are chosen?

Prep Exercise 2 When solving a system of three equations with three variables, we eliminate a variable from one pair of equations, then eliminate the same variable from a second pair of equations. Why do we eliminate the same variable from two pairs of equations?

Prep Exercise 3 Suppose you are given this system of equations:

$$\begin{cases} x + y + z = 0 & \text{(Equation 1)} \\ 2x + 4y + 3z = 5 & \text{(Equation 2)} \\ 4x - 2y + 3z = -13 & \text{(Equation 3)} \end{cases}$$

Which variable would you choose to eliminate? Which pairs of equations would you use? Why?

Prep Exercise 4 When using elimination to solve a system of three equations with three variables, how do you know whether it has no solution?

Prep Exercise 5 Two drawings were given in the text of inconsistent systems of equations with three unknowns. Make another drawing of an inconsistent system other than the ones given.

Prep Exercise 6 Where will a point be located if it solves two equations of a system of three equations in three variables but does not solve the third equation?

For Exercises 1–6, determine whether the ordered triple is a solution of the system. See Example 1.

1. $(3, -1, 1)$
$$\begin{cases} x + y + z = 3 \\ 2x - 2y - z = 7 \\ 2x + y - 2z = 3 \end{cases}$$

2. $(2, -2, 1)$
$$\begin{cases} 3x + 2y + z = 3 \\ 2x - 3y - 2z = 8 \\ -2x + 4y + 3z = -9 \end{cases}$$

3. $(1, 0, 2)$
$$\begin{cases} 2x + 3y - 3z = -4 \\ -2x + 4y - z = -4 \\ 3x - 4y + 2z = 5 \end{cases}$$

4. $(3, 4, 0)$
$$\begin{cases} x + 2y + 5z = 11 \\ 3x - 2y - 4z = 1 \\ 2x + 2y + 3z = 12 \end{cases}$$

5. $(2, -2, 4)$
$$\begin{cases} x + 2y - z = -6 \\ 2x - 3y + 4z = 26 \\ -x + 2y - 3z = -18 \end{cases}$$

6. $(0, -2, 4)$
$$\begin{cases} 3x + 2y - 3z = -16 \\ 2x - 4y - z = 4 \\ -3x + 4y - 2z = -16 \end{cases}$$

Objective 3

For Exercises 7–28, solve the systems of equations. See Examples 2 and 3.

7. $\begin{cases} x + y + z = 5 \\ 2x + y - 2z = -5 \\ x - 2y + z = 8 \end{cases}$

8. $\begin{cases} x + y + z = 2 \\ 3x + y - z = -2 \\ 2x - 2y + 3z = 15 \end{cases}$

9. $\begin{cases} x + y + z = 2 \\ 4x - 3y + 2z = 2 \\ 2x + 3y - 2z = -8 \end{cases}$

10. $\begin{cases} x + y - z = 7 \\ -2x + 2y - z = -5 \\ 3x - 3y + 2z = 6 \end{cases}$

11. $\begin{cases} 2x + y + 2z = 5 \\ 3x - 2y + 3z = 4 \\ -2x + 3y + z = 8 \end{cases}$

12. $\begin{cases} 2x + 3y - 2z = -4 \\ 4x - 3y + z = 25 \\ x + 2y - 4z = -12 \end{cases}$

13. $\begin{cases} x + 2y - z = 1 \\ 2x + 4y - 2z = -8 \\ 3x + y - 4z = 6 \end{cases}$

14. $\begin{cases} 3x - 2y + z = 5 \\ 4x - 5y + 2z = 7 \\ 9x - 6y + 3z = 7 \end{cases}$

15. $\begin{cases} 4x - 2y + 3z = 6 \\ 6x - 3y + 4.5z = 9 \\ 12x - 6y + 9z = 18 \end{cases}$

16. $\begin{cases} -8x + 4y + 6z = -18 \\ 2x - y - 1.5z = 4.5 \\ 4x - 2y - 3z = 9 \end{cases}$

17. $\begin{cases} x = 2y + z + 7 \\ y = -3x + 2z + 1 \\ 2x + y - z = 0 \end{cases}$

18. $\begin{cases} z = -3x + y - 10 \\ 2x + 3y - 2z = 5 \\ x = 3y - 3z - 14 \end{cases}$

19. $\begin{cases} x = 4y - z + 1 \\ 3x + 2y - z = -8 \\ x + 6y + 2z = -3 \end{cases}$

20. $\begin{cases} 2x - 3y - 4z = 3 \\ y = -4x + 8z - 1 \\ x = -5y - 2z - 4 \end{cases}$

21. $\begin{cases} 4x + 2y + 3z = 9 \\ 2x - 4y - z = 7 \\ 3x - 2z = 4 \end{cases}$

22. $\begin{cases} 4x + 3y - 2z = 19 \\ 2x + 5z = -4 \\ 3x + 2y + 3z = 5 \end{cases}$

23. $\begin{cases} 3x - 2z = -1 \\ 4x + 5y = 23 \\ y + 2z = -1 \end{cases}$

24. $\begin{cases} 4y + 3z = 2 \\ 3x + 4y = 2 \\ 2x - 5z = -6 \end{cases}$

25. $\begin{cases} 3x + 2y = -2 \\ 2x - 3z = 1 \\ 0.4y - 0.5z = -2.1 \end{cases}$

26. $\begin{cases} 0.2x - 0.3z = -1.8 \\ 3x + 2y = 5 \\ 3x + 2z = -1 \end{cases}$

27. $\begin{cases} \frac{3}{2}x + y - z = 0 \\ 4y - 3z = -22 \\ -0.2x + 0.3y = -2 \end{cases}$

28. $\begin{cases} -0.2x - 0.3y + 0.1z = -0.3 \\ -3x + 2y = 13 \\ \frac{1}{4}y - \frac{1}{2}z = 2 \end{cases}$

Review Exercises

Exercise 1 Expressions

[1.4] **1.** Use the distributive property to rewrite $-0.10(x - y + 200)$

Exercises 2–4 Equations and Inequalities

For Exercises 2–4, solve.

[2.1] **2.** $12x = 16\left(x - \frac{1}{6}\right)$

[2.1] **3.** $12.25x + 15.75(40 - x) = 539$

[4.1] **4.** $\begin{cases} x = y + \frac{1}{30} \\ 6x = 10y \end{cases}$

4.3 Solving Applications Using Systems of Equations

Objectives

1. Solve application problems that translate to a system of two linear equations.
2. Solve application problems that translate to a system of three linear equations.

Warm-up

Solve.

[4.1] 1. $\begin{cases} y = 2x + 6 \\ 3x + y = -19 \end{cases}$

[4.2] 2. $\begin{cases} x + 2y + z = 1 \\ x - y + z = 7 \\ x + y - 3z = -5 \end{cases}$

In this section, we use systems of equations to solve a variety of application problems. The following procedure describes our general approach to solving problems using systems of equations.

Procedure Solving Problems Using Systems of Linear Equations

To solve a problem using a system of linear equations:

1. Select a variable to represent each unknown.
2. Write a system of equations.
3. Solve the system.

Objective 1 Solve application problems that translate to a system of two linear equations.

Geometry Problems

In geometry problems, the definition of a geometry term often provides the needed relationship. For example, a problem might involve angles that are **complementary** or **supplementary**. We indicate a right angle by putting a square at the vertex.

Definitions **Complementary angles:** Two angles are complementary if the sum of their measures is 90°.
Supplementary angles: Two angles are supplementary if the sum of their measures is 180°.

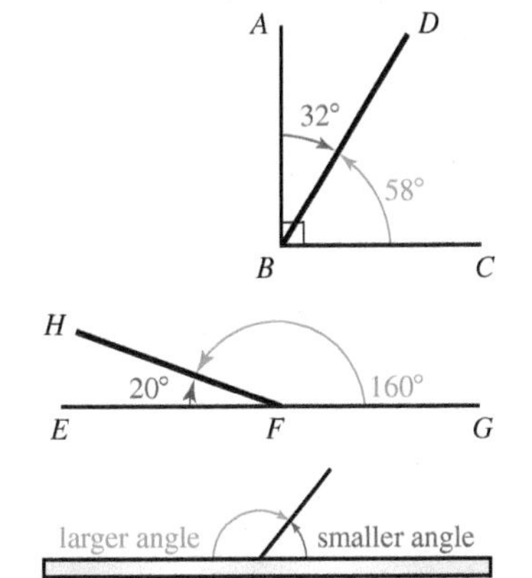

In the figures in the margin, $\angle ABD$ and $\angle DBC$ are complementary because $32° + 58° = 90°$. Also, $\angle EFH$ and $\angle HFG$ are supplementary because $20° + 160° = 180°$.

Now let's consider a problem containing one of these terms.

Example 1 A steel cable is connected to a suspension bridge, creating two angles. If the larger angle is 30° less than twice the smaller angle, what are the angle measurements?

Understand We must find the two angle measurements. We need two relationships. First, the larger angle is 30° less than twice the smaller angle. A sketch shows our second relationship, that the two angles are supplementary.

Plan and Execute Let l represent the larger angle and a the smaller angle.

Relationship 1: The larger angle is 30° less than twice the smaller angle.

Translation: $l = 2a - 30$

Relationship 2: The two angles are supplementary.

Translation: $l + a = 180$

Our system: $\begin{cases} l = 2a - 30 & \text{(Equation 1)} \\ l + a = 180 & \text{(Equation 2)} \end{cases}$

Answers to Warm-up
1. $(-5, -4)$ **2.** $(3, -2, 2)$

Because l is isolated in Equation 1, we will substitute $2a - 30$ for l in Equation 2.

$$(2a - 30) + a = 180 \quad \text{Substitute } 2a - 30 \text{ for } l.$$
$$3a - 30 = 180 \quad \text{Combine like terms.}$$
$$3a - 30 + 30 = 180 + 30 \quad \text{Add 30 to both sides.}$$
$$\frac{3a}{3} = \frac{210}{3} \quad \text{Divide both sides by 3.}$$
$$a = 70$$

Now we can find the value of l by substituting 70 for a in Equation 1.

Note We also could have found the larger angle by subtracting 70° from 180° to get 110°.

$$l = 2(70) - 30 = 110$$

Answer The angle measurements are 70° and 110°.

Check The two angles are supplementary because 110° + 70° = 180° and 110° is 30° less than twice 70°.

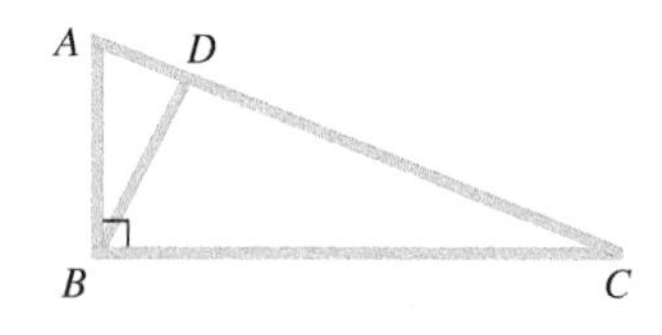

Your Turn 1 Wood joists in a roof are connected as shown in the figure in the margin. The measure of $\angle DBC$ is 9° more than twice the measure of $\angle ABD$. Find the measure of $\angle DBC$ and $\angle ABD$.

Business and Currency Problems

Note The graphing method is the least desirable method because it can be time-consuming and inaccurate. The substitution method is simpler than elimination if one of the equations has an isolated variable or a variable with a coefficient of 1 or −1. If all equations in the system are in standard form, elimination is preferred.

Example 2 An artist produces two types of ornaments: small and large. In reviewing her records for October, she notes that the total revenue was \$1582, but she did not record how many of each size she sold. She remembers selling 75 more large ornaments than small. If the small ornament sells for \$3.50 and the large ornament sells for \$5.00, how many of each size did she sell in October?

Understand We are given the total revenue and the price of each ornament. We also know that she sold 75 more large ornaments than small. We are to find the number of each size sold.

Plan and Execute Let x represent the number of small ornaments sold and y represent the number of large ornaments sold.

Categories	Selling Price	Number Sold	Revenue
Small	3.50	x	$3.50x$
Large	5.00	y	$5.00y$

Relationship 1: She sold 75 more large than small.

Translation: $y = x + 75$

Relationship 2: The total revenue is \$1582.

Translation: $3.50x + 5.00y = 1582$

$$\text{Our system: } \begin{cases} y = x + 75 & \text{(Equation 1)} \\ 3.50x + 5.00y = 1582 & \text{(Equation 2)} \end{cases}$$

Because y is isolated in Equation 1, the substitution method seems easier.

$$3.50x + 5.00y = 1582$$
$$3.50x + 5.00(x + 75) = 1582 \quad \text{Substitute } x + 75 \text{ for } y.$$
$$3.50x + 5.00x + 375 = 1582 \quad \text{Distribute.}$$
$$8.50x = 1207 \quad \text{Combine like terms and subtract 375 from both sides.}$$
$$x = 142 \quad \text{Divide both sides by 8.50.}$$

Now we can find the value of y by substituting 142 for x in Equation 1.

$$y = 142 + 75 = 217$$

Answer to Your Turn 1
27° and 63°

Answer She sold 142 small ornaments and 217 large ornaments.

Check Verify the solution in both given relationships.

Your Turn 2 A broker sells a combined total of 40 shares of two different stocks. The first stock sold for \$12.25 per share, and the second stock sold for \$15.75 per share. If the total sale was \$539, how many shares of each stock were sold?

Rate Problems

Example 3 To gain strength, a rowing crew practices in a stream with a fairly quick current. When rowing against the stream, the team takes 15 minutes to row 1 mile, whereas with the stream, the team rows the same mile in 6 minutes. Find the team's speed in still water and how much the current changes the team's speed.

Understand We assume that the amount the current adds to the speed is the same as the amount subtracted depending on which way the team is traveling.

Plan and Execute Let x represent the team's speed in still water. Let y represent the amount the current changes the team's speed. We can use a table to organize the information.

Direction	Rate	Time (in hours)	Distance
Upstream	$x - y$	0.25	$0.25(x - y)$
Downstream	$x + y$	0.1	$0.1(x + y)$

Relationship 1: The distance upstream is 1 mile.

Translation: $0.25(x - y) = 1$

$x - y = 4$ Divide both sides by 0.25 to clear the decimal (or multiply both sides by 4).

Relationship 2: The distance downstream is 1 mile.

Translation: $0.1(x + y) = 1$

$x + y = 10$ Divide both sides by 0.1 to clear the decimal (or multiply both sides by 10).

Our system: $\begin{cases} x - y = 4 & \text{(Equation 1)} \\ x + y = 10 & \text{(Equation 2)} \end{cases}$

Because the equations are in standard form, we will use the elimination method.

$$x - y = 4$$
$$x + y = 10$$
$$2x + 0 = 14$$ Add the equations to eliminate y.
$$2x = 14$$ Simplify.
$$x = 7$$ Divide both sides by 2.

Now we can find the value of y by substituting 7 for x in one of the equations. We will use $x + y = 10$.

$$7 + y = 10$$ Substitute 7 for x.
$$y = 3$$ Subtract 7 from both sides.

Answer The team can row 7 miles per hour in still water. The stream's current changes the speed by 3 miles per hour depending on the direction of travel.

Answer to Your Turn 2
26 shares at \$12.25,
14 shares at \$15.75

Check Verify the time of travel for each part of the round trip.

Going upstream: $7 - 3 = 4$ miles per hour

$$\text{Time: } t = \frac{d}{r} = \frac{1}{4} = 0.25 \text{ hours, which is 15 minutes}$$

Going downstream: $7 + 3 = 10$ miles per hour

$$\text{Time: } t = \frac{d}{r} = \frac{1}{10} = 0.1 \text{ hours, which is 6 minutes}$$

Your Turn 3 A plane traveling east with the jet stream travels from Salt Lake City to Atlanta in 3 hours. From Atlanta to Salt Lake City, flying against the jet stream, the plane takes 3.75 hours. If the distance from Salt Lake City to Atlanta is 1590 miles, find the plane's speed in still air.

Example 4 Suppose Angel is running in a marathon. He passes the first checkpoint at an average speed of 6 miles per hour. Two minutes later, Jay passes the first checkpoint at an average speed of 10 miles per hour. If both runners maintain their average speed, how long will it take Jay to catch up with Angel?

Understand We are to find the amount of time it takes Jay to catch Angel given their individual rates and the amount of time separating them.

Plan and Execute Let x represent Angel's travel time and y represent Jay's travel time. We can use a table to organize the information.

Runner	Rate	Time (in hours)	Distance
Angel	6	x	$6x$
Jay	10	y	$10y$

Relationship 1: When Jay catches up with Angel, Angel will have run for 2 minutes $\left(\frac{1}{30} \text{ of an hour}\right)$ longer than Jay since passing the checkpoint.

$$\text{Translation: } x = y + \frac{1}{30}$$

Relationship 2: When Jay catches up, they will have traveled the same distance from the start of the race.

$$\text{Translation: } 6x = 10y$$

$$\text{Our system: } \begin{cases} x = y + \frac{1}{30} & \text{(Equation 1)} \\ 6x = 10y & \text{(Equation 2)} \end{cases}$$

Equation 1 is solved for x, so we will use substitution.

$$6\left(y + \frac{1}{30}\right) = 10y \quad \text{Substitute } y + \frac{1}{30} \text{ for } x \text{ in Equation 2.}$$

$$6y + \frac{1}{5} = 10y \quad \text{Multiply.}$$

$$\frac{1}{5} = 4y \quad \text{Subtract } 6y \text{ from both sides.}$$

$$\frac{1}{20} = y \quad \text{Divide both sides by 4.}$$

Answer Jay will catch up in $\frac{1}{20}$, or 0.05, hours, which is 3 minutes.

Answer to Your Turn 3
477 mph

Check Notice that when Jay catches Angel, Angel will have run for 5 minutes $(3 + 2)$ after passing the checkpoint. In that 5 minutes $\left(\frac{1}{12}\text{ of an hour}\right)$ after passing the checkpoint, Angel travels $6\left(\frac{1}{12}\right) = \frac{1}{2}$ mile. In the 3 minutes after the checkpoint, Jay travels $10\left(\frac{1}{20}\right) = \frac{1}{2}$ mile, indicating that Jay is side by side with Angel.

Your Turn 4 Ian and Lindsey are riding bikes in the same direction. Ian is traveling at an average speed of 12 miles per hour and passes a sign at 3:30 P.M. Lindsey, who is traveling at an average speed of 16 miles per hour, passes the same sign at 3:40 P.M. At what time will Lindsey catch up with Ian?

Mixture Problems

Example 5 How many milliliters of a 10% HCl solution and a 30% HCl solution must be mixed together to make 200 milliliters of a 15% HCl solution?

Understand The two unknowns are the volumes of a 10% solution and a 30% solution that are mixed. One relationship involves concentrations of each solution in the mixture, and the other relationship involves the total volume of the final mixture (200 ml).

Plan and Execute Let x and y represent the two amounts to be mixed. We can use a table to organize the information.

Note To find the volume of HCl in each solution (fourth column), multiply the concentration (as a decimal) and the volume.

Solution	Concentration	Volume of Solution	Volume of HCl
10% HCL	0.10	x	$0.10x$
30% HCL	0.30	y	$0.30y$
15% HCL	0.15	200	$0.15(200)$

Relationship 1: The total volume is 200 ml.

Translation: $x + y = 200$

Relationship 2: The combined volumes of HCl in the two mixed solutions are equal to the total volume of HCl in the mixture.

Translation: $0.10x + 0.30y = 0.15(200)$

$$\text{Our system: } \begin{cases} x + y = 200 \\ 0.10x + 0.30y = 30 \end{cases}$$

Because the equations are in standard form, elimination is the better method. We will eliminate x.

$$x + y = 200 \xrightarrow{\text{Multiply by } -0.10.} -0.10x - 0.10y = -20$$

$$0.10x + 0.30y = 30 \qquad\qquad 0.10x + 0.30y = 30$$

$$0.20y = 10 \quad \text{Add the equations.}$$

$$y = 50 \quad \text{Divide both sides by 0.20.}$$

Now we can find x by substituting 50 for y in one of the equations. We will use $x + y = 200$.

$$x + 50 = 200 \quad \text{Substitute 50 for } y.$$

$$x = 150 \quad \text{Subtract 50 from both sides.}$$

Answer Mixing 150 milliliters of the 10% solution with 50 milliliters of the 30% solution gives 200 milliliters of the 15% solution.

Answer to Your Turn 4
4:10 P.M.

Check The mixture volume is $150 + 50 = 200$ milliliters. The volume of HCl is $0.10(150) + 0.30(50) = 15 + 15 = 30$, which means that 30 milliliters out of the 200 milliliters is pure HCl and $\frac{30}{200}$ is 0.15 or 15%.

Your Turn 5 How much 20% HCl solution and 40% HCl solution must be mixed together to make 400 milliliters of 35% HCl solution?

Objective 2 Solve application problems that translate to a system of three linear equations.

We follow the same procedure to solve applications involving three unknowns as we did when solving applications with two unknowns.

Example 6 At a movie theater, John buys one popcorn, one soft drink, and one candy bar, all for \$7. Fred buys two popcorns, three soft drinks, and two candy bars for \$16. Carla buys one popcorn, two soft drinks, and three candy bars for \$12. Find the price of one popcorn, one soft drink, and one candy bar.

Understand We have three unknowns and three relationships, and we are to find the cost of each.

Plan and Execute Let x represent the cost of one popcorn, y the cost of one soft drink, and z the cost of one candy bar.

Relationship 1: One popcorn, one soft drink, and one candy bar cost \$7.

Translation: $x + y + z = 7$

Relationship 2: Two popcorns, three soft drinks, and two candy bars cost \$16.

Translation: $2x + 3y + 2z = 16$

Relationship 3: One popcorn, two soft drinks, and three candy bars cost \$12.

Translation: $x + 2y + 3z = 12$

$$\text{Our system: } \begin{cases} x + y + z = 7 & \text{(Equation 1)} \\ 2x + 3y + 2z = 16 & \text{(Equation 2)} \\ x + 2y + 3z = 12 & \text{(Equation 3)} \end{cases}$$

We will choose to eliminate z from two pairs of equations. We will start with Equations 1 and 2.

$$\begin{array}{llcrl} \text{(Eq. 1)} & x + y + z = 7 & \xrightarrow{\text{Multiply by } -2} & -2x - 2y - 2z = -14 & \\ \text{(Eq. 2)} & 2x + 3y + 2z = 16 & & \underline{2x + 3y + 2z = 16} & \text{Add the equations.} \\ & & & y \qquad = 2 & \text{(Equation 4)} \end{array}$$

Note Although we were trying to eliminate z, we ended up eliminating both x and z.

Equation 4 gives us the value of y, indicating that a soft drink costs \$2. Now we choose another pair of equations and eliminate z again. We will use Equations 1 and 3.

$$\begin{array}{llcrl} \text{(Eq. 1)} & x + y + z = 7 & \xrightarrow{\text{Multiply by } -3} & -3x - 3y - 3z = -21 & \\ \text{(Eq. 3)} & x + 2y + 3z = 12 & & \underline{x + 2y + 3z = 12} & \text{Add the equations.} \\ & & & -2x - y \qquad = -9 & \text{(Equation 5)} \end{array}$$

Answer to Your Turn 5
100 ml of 20% solution with 300 ml of 40% solution

Because we already know that $y = 2$, we can substitute for y in $-2x - y = -9$.

$$-2x - 2 = -9 \quad \text{Substitute 2 for } y.$$
$$-2x = -7 \quad \text{Add 2 to both sides.}$$
$$x = 3.5 \quad \text{Divide both sides by } -2 \text{ to isolate } x.$$

Because x represents the cost of one popcorn, a popcorn costs \$3.50. To find z, substitute for x and y into one of the original equations. We will use Equation 1.

$$3.5 + 2 + z = 7 \quad \text{Substitute 3.5 for } x \text{ and 2 for } y.$$
$$5.5 + z = 7 \quad \text{Simplify.}$$
$$z = 1.5 \quad \text{Subtract 5.5 from both sides to isolate } z.$$

Because z represents the cost of one candy bar, a candy bar costs \$1.50.

Answer Popcorn costs \$3.50, a soft drink costs \$2.00, and a candy bar costs \$1.50.

Check Verify that at these prices—the amounts of money spent by John, Fred, and Carla—are correct. We will leave the check to the reader.

Answer to Your Turn 6
length: 20 in.
width: 12 in.
height: 14 in.

Your Turn 6 A small aquarium is in the shape of a rectangular solid. The sum of the length, width, and height is 46 inches. The sum of twice the length, three times the width, and the height is 90 inches. The sum of the length, twice the width, and three times the height is 86 inches. Find the dimensions of the aquarium.

4.3 Exercises

For Extra Help MyMathLab®

Note: Exercises marked with a ★ represent challenging exercises.

Objective 1

Prep Exercise 1 Suppose x and y represent the measure of two angles.

a. If the angles are complementary, what equation relates their measures?

b. If the angles are supplementary, what equation relates their measures?

Prep Exercise 2 Translate to an equation: The length of a rectangle is ten more than twice the width. Use l to represent the length and w to represent the width.

For Exercises 1–10, translate to a system of two equations; then solve. See Example 1.

1. Two angles are complementary. If the measure of one angle is 40° more than the other angle, find the measure of both angles.

2. Two angles are complementary. If the measure of one angle is 42° less than the other angle, find the measure of both angles.

3. Two angles are supplementary. If the measure of one angle is four times the other angle, find the measure of both angles.

4. Two angles are supplementary. If the measure of one angle is five times the other angle, find the measure of both angles.

5. A laser beam is shined on glass so that it gets bent, or refracted, as it passes through as shown in the figure. Suppose the angle made by the incident ray and the glass boundary measures 24° less than twice the measure of angle of incidence, I. Find the measure of the angle of incidence.

The angle of refraction depends on the density of the material. Materials with greater density refract light more than less dense materials.

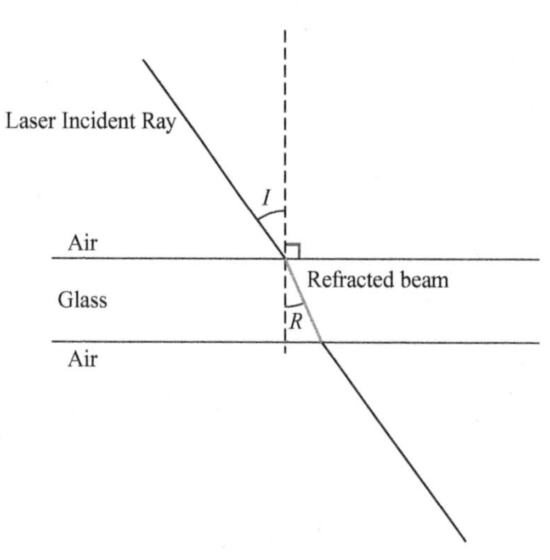

6. Using the figure from Exercise 5, suppose the angle made by the refracted beam and the boundary of the glass measures 6° less than triple the measure of the angle of refraction, R. Find the measure of the angle of refraction.

7. A support beam is attached to a wall and to the bottom of a ceiling truss as shown in the figure to the right. If the measure of the angle made on one side of the support beam is 10° more than four times the angle made on the other side, find the measure of both angles.

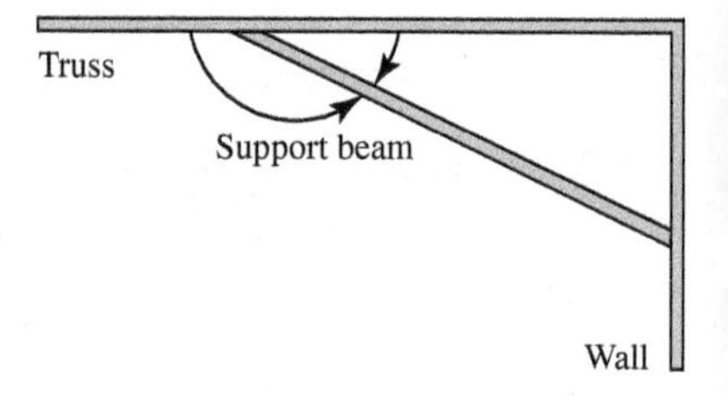

8. Two pieces of metal are welded together as shown. If the measure of the angle formed on one side of the support beam is 18° less than triple the angle on the other side, find the measure of the two angles.

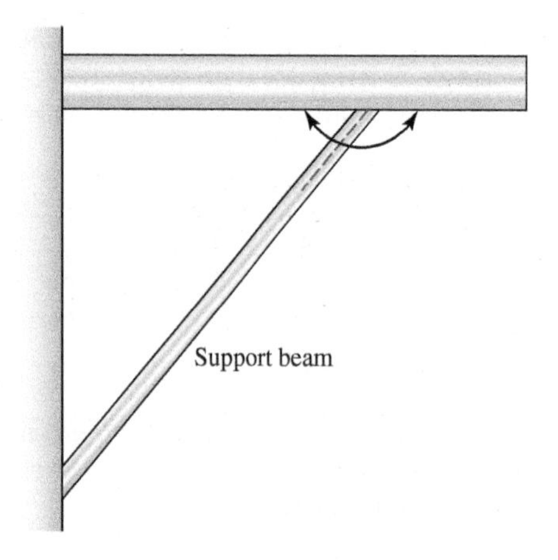

9. The perimeter of the base of the Washington Monument is 220.5 feet. If the width is equal to its length, what are the dimensions of the base?

10. The perimeter of the playing lines for a doubles tennis match is 228 feet. If the width is 42 feet less than the length, what are the dimensions?

Prep Exercise 3 A movie theater sells two different sizes of candy. Small boxes cost $5.50 and large boxes cost $8.00.

a. Complete the following table.

Categories	Value	Number	Amount
Small box	5.50	x	
Large box	8.00	y	

b. If the theater sold 45 boxes of candy in one hour, write an equation relating the number of boxes sold.

c. If the theater earned a total of $285 in that one hour, write an equation relating the total amount earned.

Prep Exercise 4 Suppose there are four more $1 bills than $5 bills in a wallet. If x represents the number of $5 bills and y represents the number of $1 bills, write an equation relating the number of bills.

For Exercises 11–24, translate to a system of two equations; then solve. See Example 2.

11. A vendor at a farmer's market sells a small container of tomatoes for $6 and the large for $10. In one day, she sold 42 containers earning a total of $308. How many of each size did she sell?

12. A store has a clearance sale so that all t-shirts sell for $12 and all shorts sell for $15. The day the sale began, 57 clearance items sold (t-shirts and shorts) for at total of $747. How many t-shirts and how many shorts were sold?

13. As of April 2013, *The Twilight Saga: Breaking Dawn Part 1* and *The Twilight Saga: Breaking Dawn: Part 2* grossed a combined total of $1542 million. If *Part 2* grossed $118 million more than *Part 1*, and how much did each movie gross? (*Source:* The Numbers.)

14. In 2010, the combined budget for the U.S. Senate and House of Representatives was $2326 million. If the budget for the House of Representatives was $474 million more than that of the Senate, how much was allocated to each branch of Congress?

15. Of the 11.6 million students enrolled in community colleges, there are 1.8 million more women than men. How many women are enrolled in a community college? (*Source:* American Association of Community Colleges.)

16. In 2010, there were 3.5 million nurses employed in the United States. If there were 1.9 million more registered nurses than licensed practical nurses, how many registered nurses were employed? (*Source:* U.S. Bureau of Labor Statistics.)

17. The median salary for a postsecondary math teacher is $37,330 less than that of a mathematician in industry. Together the average salaries total $161,430. Find each average annual salary. (*Source:* U.S. Bureau of Labor Statistics.)

18. The civilian labor force in March 2012 was 321 thousand less than in March 2013. The total for the two years was 309,735 thousand. How many civilians were in the labor force in March of each year? (*Source:* U.S. Department of Labor Statistics.)

19. Shark attacks are almost six times as likely to occur in deep water (greater than 30 feet) as in other depths. Find the percentage of attacks occurring in deep water. (*Hint:* Because all attacks occur in the ocean, the combined percent is 100%.) (*Source:* International Shark Attack File; Florida Museum of Natural History, Knight Ridder Tribune, Associated Press.)

20. The number of people infected with West Nile virus in the United States has been decreasing. In 2008, approximately 1.9 times as many people were infected as in 2011. If a total of 2068 people were infected during those two years, find the number infected each year. (*Source:* Centers for Disease Control and Prevention.)

21. The Old-Fashioned Christmas Store sells two sizes of fresh wreaths. An 18-inch wreath costs $20, and a 22-inch wreath costs $25. In one day, the number of 22-inch wreaths sold was six more than triple the number of 18-inch wreaths, for a total of $1100. How many of each were sold?

22. Pier 1 sells two sizes of pillar candles. The larger one sells for $15; the smaller, for $10. One day the number of small candles sold was four more than twice the number of larger candles, for a combined total of $845. How many of each size were sold?

Of Interest

Of the 498,000 soldiers who died in the Civil War, about 215,000 died in battle (140,000 Union and 75,000 Confederate). The rest died from disease and malnutrition.

23. In the Civil War, the Union and Confederate forces lost a combined total of about 498,000 soldiers. If the Union lost 38,000 less than three times the number of soldiers lost by the Confederate forces, how many soldiers died on each side? (*Source:* U.S. Department of Defense.)

24. The unemployment rate in January 2000 for men 20 years and older was 6.4% less than the unemployment rate in January 2010. The rate in 2010 was 2% less than four times the rate in 2000. Find the unemployment rate for men 20 years and older in 2010.

Prep Exercise 5 Josh and Rhonda are traveling north on the same highway. Rhonda, traveling 60 miles per hour, passes Exit 40 at 2:00 P.M. Josh, traveling 75 miles per hour, passes Exit 40 at 2:30 P.M.

a. Complete the following table.

Categories	Rate	Time	Distance
Josh	75	x	
Rhonda	60	y	

b. How much time, in hours, separates Josh and Rhonda?

c. If x represents the amount of time it takes Josh to catch up to Rhonda after passing Exit 40, and y represents the amount of time that Rhonda travels after passing Exit 40, write an equation relating x and y.

d. At the time Josh catches up to Rhonda, what can you conclude about the distances they each have traveled from Exit 40? Write an equation relating their distances.

Prep Exercise 6 If a plane's speed in still air is x miles per hour and the amount the jet stream changes the plane's speed is y miles per hour, write an expression that describes the plane's speed with the jet stream and another expression that describes the plane's speed against the jet stream.

For Exercises 25–32, translate to a system of two equations; then solve. See Example 3 and 4.

25. In a cross-country skiing event, Josh, averaging 12 miles per hour, passes a checkpoint. Three minutes later, Dolph, averaging 15 miles per hour, passes the same checkpoint. If they maintain their average speeds, how long will it take Dolph to catch up?

26. In a NASCAR event, Dale Earnhardt Jr. finds himself 4 seconds behind the leader, who is averaging 180 miles per hour (264 feet per second). If Dale can increase his average speed to 183 miles per hour (268.4 feet per second), how long will it take him to catch up with the leader?

27. Francis and Poloma are traveling south in separate cars on the same interstate. Poloma is traveling at 65 miles per hour; Francis, at 70 miles per hour. Poloma passes a rest stop at 3:00 P.M. Francis passes the same rest stop at 3:30 P.M. At what time will Francis catch up with Poloma?

28. Hale and Austin are jogging in the same direction, Hale passes a waterfall at 9.00 A.M. Austin passes the same waterfall at 9:30 A.M. Austin is jogging at 5 miles per hour, and Hale is jogging at 4 miles per hour. When will Austin catch up with Hale?

29. A boat travels 36 miles upstream in 3 hours. Going downstream, it can travel 48 miles in the same amount of time. Find the speed of the current and the speed of the boat in still water.

30. Going upstream, it takes 2 hours to travel 36 miles in a boat. Downstream, the same distance takes only 1.5 hours. Find the speed of the current and the speed of the boat in still water.

31. If a plane can travel 650 miles per hour with the wind and only 550 miles per hour against the wind, find the speed of the wind and the speed of the plane in still air.

32. If a plane can travel 320 miles per hour with the wind and only 280 miles per hour against the wind, find the speed of the wind and the speed of the plane in still air.

Prep Exercise 7 A 30% sulfuric acid solution is to be added to 500 milliliters of 10% sulfuric acid solution to make a 15% sulfuric acid solution.

a. Complete the following table.

Solutions	Concentration	Volume of Solution	Volume of Sulfuric Acid
30%	0.30	x	
10%	0.10	500	
15%	0.15	y	

b. Write an equation that describes the volume of the resulting 15% solution, y, in terms of x.

c. Write an equation that describes the volume of sulfuric acid in the resulting solution.

For Exercises 33–40, translate to a system of two equations; then solve. See Example 5

33. How many milliliters of a 5% HCl mixture and a 20% HCl mixture must be combined to get 10 milliliters of 12.5% HCl mixture?

34. How much of a 45% saline solution and a 30% solution must be mixed to produce 20 liters of a 39% saline solution?

35. A mechanic has 24 ounces of a mixture that is 20% antifreeze and another mixture that is 10% antifreeze. He wants a 16% antifreeze mixture. How much of the 10% solution should he add to the 20% solution to form a 16% antifreeze solution?

36. A gardener has 10 ounces of a 40% plant fertilizer solution and some 50% plant fertilizer solution. He wants a 46% plant fertilizer solution. How much of the 50% solution should he add to the 40% solution to form a 46% solution?

37. A farmer has 20 gallons of 10% pesticide mixture and some 40% pesticide mixture. He wants to make a 20% pesticide mixture. How much of the 40% mixture should he combine with the 10% mixture to form a 20% mixture?

38. Janice has 50 milliliters of 10% HCl solution and some 25% HCl solution. How much of the 25% solution must be added to the 10% solution to create a 20% concentration?

39. Julio invested some money in an account that pays 9% and the rest in an account that pays 5%. He invested four times as much in the 9% account as in the 5% account. If the combined interest from both investments was $1435, how much did he invest in each?

40. Marta invested twice as much money in an account paying 7% interest than she did in an account paying 4% interest. If the total interest paid was $720, how much did she invest in each?

Objective 2

Prep Exercise 8 If a problem has three unknowns, how many equations would you expect to be in the corresponding system of equations?

Prep Exercise 9 Suppose x, y, and z represent three consecutive integers.

a. Write an equation that describes y in terms of x.

b. Write an equation that describes z in terms of y.

c. Write an equation that describes z in terms of x.

Prep Exercise 10 A problem involves the price of three different sizes of drink: small, medium, and large. If the large drink sells for \$0.20 less than the small and medium drinks combined, write an equation describing the pricing. Use l, m, and s to represent the prices.

For Exercises 41–60, translate to a system of three equations; then solve. See Example 6.

41. The sum of three numbers is 16. One of the numbers is 2 more than twice a second number and 2 less than the third number. Find the numbers.

42. The sum of three numbers is 18. Twice the first number is 1 less than the third number, and twice the second number is 1 more than the third number. Find the numbers.

43. The sum of three consecutive integers is 39. Find the integers.

44. The sum of three consecutive even integers is 54. Find the integers.

45. The sum of the measures of the angles in every triangle is 180°. The measure of one angle of a triangle is three times that of a second angle, and the measure of the second angle is 5° less than the measure of the third. Find the measure of each angle of the triangle.

46. The perimeter of a triangle is 33 inches. The sum of the length of the longest side and twice the length of the shortest side is 31 inches. Twice the length of the longest side minus both of the other side lengths is 12 inches. Find the side lengths.

47. At a fast-food restaurant, one burger, one order of fries, and one drink cost \$5.00; three burgers, two orders of fries, and two drinks cost \$12.50; and two burgers, four orders of fries, and three drinks cost \$14.00. Find the individual cost of one burger, one order of fries, and one drink.

48. James went to the college bookstore and purchased two pens, two erasers, and one pack of paper for \$6. Tamika purchased four pens, three erasers, and two packs of paper for \$11.50. Jermaine purchased three pens, one eraser, and three packs of paper for \$9.50. Find the individual costs of one pen, one eraser, and one pack of paper.

49. Tickets for a high school band concert were \$3 for children, \$5 for students, and \$8 for adults. A total of 500 tickets were sold, and the total money received from the sale of the tickets was \$2500. There were 150 fewer adult tickets sold than student tickets. Find the number of each type of ticket sold.

50. Tickets to a play were \$2 for children, \$4 for students, and \$6 for adults. A total of 300 tickets were sold, and the total money received from the sale of the tickets was \$1250. There were 35 more student tickets sold than children tickets. Find the number of each type of ticket sold.

51. In a basketball game, John scored 14 points with a combination of 3-point field goals, 2-point field goals, and free throws (1 point each). If he scored a total of 7 times and made two more free throws than 2-point field goals, find the number of each that he made.

52. At a track meet, 10 points are awarded for each first-place finish, 5 points for each second place, and 1 point for each third place. Suppose a track team scored a total of 71 points and had two more first-place finishes than seconds and one less first-place finish than third. Find the number of first-, second-, and third-place finishes for the team.

53. Bronze is an alloy that is made of zinc, tin, and copper in a specified proportion. Suppose an order is placed for 1000 pounds of bronze. The amount of tin in this order of bronze is three times the amount of zinc, and the amount of copper is 20 pounds more than 15 times the amount of tin. Find the number of pounds of zinc, tin, and copper. (*Source: World Almanac and Book of Facts.*)

54. When bought in bulk, peanuts sell for \$1.34 per pound, almonds for \$4.36 per pound, and pecans for \$5.88 per pound. Suppose a local specialty shop wants 200 pounds of a mixture of all three kinds of nuts that will cost \$2.70 per pound. Find the number of pounds of each type of nut if the sum of the number of pounds of peanuts and almonds is nine times the number of pounds of pecans.

55. A delicatessen sells Black Forest ham for \$11.96 per pound, turkey breast for \$8.76 per pound, and roast beef for \$9.16 per pound. Suppose the deli makes a 10-pound party tray of these three meats such that the average cost is \$9.80 per pound. Find the number of pounds of each meat if the number of pounds of turkey breast is equal to the sum of the number of pounds of Black Forest ham and roast beef.

56. A coffee shop sells Jamaican Blue Mountain coffee for \$45.99 per pound, Hawaiian Kona for \$36.99 per pound, and Sulawesi Kalossi for \$12.99 per pound. A 25-pound mixture of these three coffees sells for \$30.27 per pound. Find the number of pounds of each coffee if the sum of the number of pounds of Jamaica Blue Mountain and Hawaiian Kona is 5 pounds more than the number of pounds of Sulawesi Kalossi.

57. A total of $8000 is invested in three stocks, which paid 4%, 6%, and 7% dividends, respectively, in one year. The amount invested in the stock that returned 7% dividends is $1500 more than the amount that returned 4% dividends. If the total dividends in one year were $475, find the amount invested in each stock.

58. A total of $5000 is invested in three funds. The money market fund pays 5% annually, the income fund pays 6% annually, and the growth fund pays 8% annually. The total earnings for one year from the three funds are $340. If the amount invested in the growth fund is $500 less than twice the amount invested in the money market fund, find the amount in each fund.

59. The number of calories burned per hour bicycling is 120 more than the number burned from brisk walking. The number of calories burned per hour in climbing stairs is 180 more than from bicycling and is twice that from brisk walking. Find the number of calories burned per hour for each of the three activities. (*Source: Numbers: How Many, How Long, How Far, How Much.*)

60. The intensity of sound is measured in decibels. The decibel reading of a rock concert is 60 less than three times the reading of normal conversation and 10 less than that of a jet at takeoff. The decibel reading of a jet takeoff is 10 more than twice that of normal conversation. Find the decibel reading of each. (*Source: Numbers: How Many, How Long, How Far, How Much.*)

★ **61.** An object is thrown upward with an initial velocity of v_0 from an initial height of h_0. The height of the object, h, is described by an equation of the form $h = at^2 + v_0t + h_0$, where h and h_0 are in feet and t is in seconds. The table shows the height for different values of t. Find the values of a, v_0, and h_0 and write the equation for h.

t	h
1	234
3	306
6	174

★ **62.** Use the same general equation from Exercise 51, $h = at^2 + v_0t + h_0$, where h and h_0 are in feet and t is in seconds. The table shows the distance above the ground for different values of t. Find the values of a, v_0, and h_0 and write the equation for h.

t	h
2	136
3	106
4	44

Puzzle Problem The sum of four positive integers a, b, c, and d is 228, and $a < b < c < d$. The greatest number subtracted from twice the smallest number gives 18. If twice the sum of the third number and fourth number is subtracted from the product of 6 and the second number, the result is 10. If the smallest two numbers are consecutive odd integers, find all of the numbers.

Review Exercises

Exercise 1 Expressions

[1.4] 1. Use the distributive property to rewrite $-3(2x - 4y - z + 9)$.

[2.1] 2. Given the equation $x - 0.5y = 8$, let $x = 6$; then solve for y.

Exercises 2–6 Equations and Inequalities

[3.5] *For Exercises 3–6, use the function* $f(x) = -2x + 1$.

3. Find $f(1)$.

4. Find the y-intercept.

5. Find the slope.

6. Graph the function.

4.4 Solving Systems of Linear Equations Using Matrices

Objectives

1. Write a system of equations as an augmented matrix.
2. Solve a system of linear equations by transforming its augmented matrix to row echelon form.
3. Solve applications problems using matrices.

Warm-up

[2.1] **1.** Given $x + 3y = \frac{1}{2}$ and $y = \frac{1}{2}$, find x.

[2.1] **2.** Given $x + y + 2z = 7$ and $y = -1$ and $z = 3$, find x.

Objective 1 Write a system of equations as an augmented matrix.

Although the elimination method that we learned in Section 4.1 is effective for solving systems of linear equations, we can streamline the method by manipulating just the coefficients and constants in a **matrix**.

Definition **Matrix:** A rectangular array of numbers.

Following are some examples of matrices (plural of *matrix*).

$$\begin{bmatrix} 1 & -2 \\ -3 & 4 \end{bmatrix} \quad \begin{bmatrix} 1 & -5 & 6 \\ -2 & 4 & 0 \end{bmatrix} \quad \begin{bmatrix} -3 & 0 & 9 \\ -2 & 4 & 7 \\ 9 & -2 & 0 \end{bmatrix}$$

Matrices are made up of horizontal rows and vertical columns.

Note One way to remember that columns are vertical is to imagine the columns on a building, which are vertical.

Column 1 Column 2 Column 3

$$\begin{matrix} \text{Row 1} \longrightarrow \\ \text{Row 2} \longrightarrow \end{matrix} \begin{bmatrix} 1 & 4 & -2 \\ 5 & -1 & 3 \end{bmatrix}$$

Column 1 Column 2

$$\begin{matrix} \text{Row 1} \longrightarrow \\ \text{Row 2} \longrightarrow \\ \text{Row 3} \longrightarrow \end{matrix} \begin{bmatrix} 3 & -4 \\ -1 & 2 \\ 5 & -3 \end{bmatrix}$$

The number of rows followed by the number of columns gives the *dimensions* of the matrix. So $\begin{bmatrix} 2 & 5 & -1 \\ 5 & 4 & 3 \end{bmatrix}$ is a 2×3 (read "2 by 3") matrix because it has two rows and three columns; $\begin{bmatrix} 2 & 5 \\ -4 & 6 \\ 3 & -2 \end{bmatrix}$ is a 3×2 matrix because it has three rows and two columns.

Each number in a matrix is called an *element*. To solve a system of equations using matrices, we first rewrite the system as an **augmented matrix**.

Definition **Augmented matrix:** A matrix made up of the coefficients and the constant terms of a system. The constant terms are separated from the coefficients by a dashed vertical line.

For example, to write the system $\begin{cases} 3x - 2y = 7 \\ 4x - 3y = 10 \end{cases}$ as an augmented matrix, we omit the variables and write $\left[\begin{array}{cc|c} 3 & -2 & 7 \\ 4 & -3 & 10 \end{array}\right]$.

Example 1 Write $\begin{cases} 2x + 3y - 4z = 1 \\ 3x - 5y + z = -4 \\ 2x - 6y + 3z = 2 \end{cases}$ as an augmented matrix.

Solution: $\left[\begin{array}{ccc|c} 2 & 3 & -4 & 1 \\ 3 & -5 & 1 & -4 \\ 2 & -6 & 3 & 2 \end{array}\right]$

Note It is helpful to think of the dashed line as the equal sign of each equation.

Answers to Warm-up
1. $x = -1$ **2.** 2

Your Turn 1 Write the augmented matrix for $\begin{cases} x + 3y = -2 \\ -4x - 5y = 7 \end{cases}$.

Objective 2 Solve a system of linear equations by transforming its augmented matrix to row echelon form.

In using the elimination method, we can interchange equations, add equations, or multiply equations by a number and then add the rewritten equations. Because each row of an augmented matrix contains the constants and coefficients of each equation in the system, we can interchange rows, add rows, or multiply rows by a number and then add the rewritten rows.

Rule Row Operations

The solution of a system is not affected by the following row operations in its augmented matrix.

1. Any two rows may be interchanged.
2. The elements of any row may be multiplied (or divided) by any nonzero real number.
3. Any row may be replaced by a row resulting from adding the elements of that row (or multiples of that row) to a multiple of the elements of any other row.

We use these row operations to solve a system of linear equations by transforming the augmented matrix of the system into an equivalent matrix that is in **row echelon form**.

Definition Row echelon form: An augmented matrix whose coefficient portion has 1's on the diagonal from upper left to lower right and 0s below the 1's.

For example, $\left[\begin{array}{cc|c} 1 & 3 & 3 \\ 0 & 1 & 4 \end{array}\right]$ and $\left[\begin{array}{ccc|c} 1 & 2 & -4 & 5 \\ 0 & 1 & -5 & 7 \\ 0 & 0 & 1 & 5 \end{array}\right]$ are in row echelon form.

In the following examples, we let R_1 mean row 1, R_2 mean row 2, and so on.

Example 2 Solve the following linear systems by transforming their augmented matrices into row echelon form.

a. $\begin{cases} x - 2y = -4 & \text{(Equation 1)} \\ -2x + 5y = 9 & \text{(Equation 2)} \end{cases}$

Solution: First, we write the augmented matrix: $\left[\begin{array}{cc|c} 1 & -2 & -4 \\ -2 & 5 & 9 \end{array}\right]$.

Now we perform row operations to transform the matrix into row echelon form. The element in the first row, first column is already 1, which is what we want. Therefore, we need to rewrite the matrix so that −2 in the second row, first column becomes 0. To do this, we multiply the first row by 2 and add it to the second row.

Note The result of the row operations replaces the affected row in the matrix. Other rows are rewritten unchanged.

$$2R_1 + R_2 \longrightarrow \left[\begin{array}{cc|c} 1 & -2 & -4 \\ 0 & 1 & 1 \end{array}\right]$$

$2(1\ -2 \mid -4) = (2\ -4 \mid -8)$ and $(2\ -4 \mid -8) + (-2\ 5 \mid 9) = (0\ 1 \mid 1)$ New $R_2 = 2R_1 + R_2$

The resulting matrix represents the system $\begin{cases} x - 2y = -4 \\ y = 1 \end{cases}$.

Answers to Your Turn 1

$\left[\begin{array}{cc|c} 1 & 3 & -2 \\ -4 & -5 & 7 \end{array}\right]$

Because $y = 1$, we can solve for x using substitution.

$x - 2(1) = -4$ Substitute 1 for y in $x - 2y = -4$.
$x - 2 = -4$ Simplify.
$x = -2$ Add 2 to both sides.

The solution is $(-2, 1)$. The check is left to the reader.

b. $\begin{cases} 2x + 6y = 1 & \text{(Equation 1)} \\ x + 8y = 3 & \text{(Equation 2)} \end{cases}$

Solution: First, we write the augmented matrix: $\left[\begin{array}{cc|c} 2 & 6 & 1 \\ 1 & 8 & 3 \end{array}\right]$.

Now we perform row operations to get row echelon form.

$\frac{1}{2}R_1 \longrightarrow \left[\begin{array}{cc|c} 1 & 3 & \frac{1}{2} \\ 1 & 8 & 3 \end{array}\right]$ We need the 2 in row 1, column 1 to be a 1; so multiply each number in row 1 by $\frac{1}{2}$.

$-R_1 + R_2 \longrightarrow \left[\begin{array}{cc|c} 1 & 3 & \frac{1}{2} \\ 0 & 5 & \frac{5}{2} \end{array}\right]$ We need the 1 in row 2, column 1 to be 0; so multiply row 1 by -1 and add it to row 2.

$\frac{1}{5}R_2 \longrightarrow \left[\begin{array}{cc|c} 1 & 3 & \frac{1}{2} \\ 0 & 1 & \frac{1}{2} \end{array}\right]$ We need the 5 in row 2, column 2 to be 1; so multiply row 2 by $\frac{1}{5}$.

Connection Row operations correspond to equation manipulations in the elimination method. For example, $-R_1 + R_2$ corresponds to multiplying $x + 3y = \frac{1}{2}$ by -1 and then adding the resulting equation to $x + 8y = 3$ so that we get $5y = \frac{5}{2}$.

This matrix represents the system $\begin{cases} x + 3y = \frac{1}{2} \\ y = \frac{1}{2} \end{cases}$.

Because $y = \frac{1}{2}$, we can solve for x using substitution.

$x + 3\left(\frac{1}{2}\right) = \frac{1}{2}$ Substitute $\frac{1}{2}$ for y in $x + 3y = \frac{1}{2}$.
$x + \frac{3}{2} = \frac{1}{2}$ Simplify.
$x = -1$ Subtract $\frac{3}{2}$ from both sides.

The solution is $\left(-1, \frac{1}{2}\right)$. The check is left to the reader.

Your Turn 2 Solve the following system by transforming its augmented matrix into row echelon form.

$$\begin{cases} 3x + 4y = 6 \\ x - 3y = -11 \end{cases}$$

Now let's use row operations to solve systems of three equations.

Example 3 Use the row echelon method to solve the following system.

$$\begin{cases} 2x + y - 2z = -3 & \text{(Equation 1)} \\ x + y + 2z = 7 & \text{(Equation 2)} \\ 4y + 3z = 5 & \text{(Equation 3)} \end{cases}$$

Answer to Your Turn 2
$(-2, 3)$

Note When writing a matrix in row echelon form, if you get a row that is all zeros to the left of the dashed line and nonzero to the right, the system is inconsistent (no solution). If you get a row that is all zeros, the system is dependent (infinite solutions).

Solution: Write the augmented matrix: $\left[\begin{array}{ccc|c} 2 & 1 & -2 & -3 \\ 1 & 1 & 2 & 7 \\ 0 & 4 & 3 & 5 \end{array}\right]$.

Now we perform row operations to get row echelon form.

$$\begin{array}{l} R_2 \longrightarrow \\ R_1 \longrightarrow \end{array} \left[\begin{array}{ccc|c} 1 & 1 & 2 & 7 \\ 2 & 1 & -2 & -3 \\ 0 & 4 & 3 & 5 \end{array}\right]$$

Because we need a 1 in row 1, column 1, we interchange rows 1 and 2.

$$-2R_1 + R_2 \longrightarrow \left[\begin{array}{ccc|c} 1 & 1 & 2 & 7 \\ 0 & -1 & -6 & -17 \\ 0 & 4 & 3 & 5 \end{array}\right]$$

To get a 0 in row 2, column 1, we multiply the new row 1 by -2 and add it to row 2.

$$4R_2 + R_3 \longrightarrow \left[\begin{array}{ccc|c} 1 & 1 & 2 & 7 \\ 0 & -1 & -6 & -17 \\ 0 & 0 & -21 & -63 \end{array}\right]$$

Row 3 needs only a 0 in the second column, so we multiply row 2 by 4 and add it to row 3.

$$\begin{array}{l} -R_2 \longrightarrow \\ -\frac{1}{21}R_3 \longrightarrow \end{array} \left[\begin{array}{ccc|c} 1 & 1 & 2 & 7 \\ 0 & 1 & 6 & 17 \\ 0 & 0 & 1 & 3 \end{array}\right]$$

To get a 1 in row 2, column 2, we multiply row 2 by -1.
To get a 1 in row 3, column 3, we multiply row 3 by $-\frac{1}{21}$.

The resulting matrix represents the system $\begin{cases} x + y + 2z = 7 \\ y + 6z = 17. \\ z = 3 \end{cases}$

To find y, substitute 3 for z in $y + 6z = 17$.

$y + 6(3) = 17$ Substitute 3 for z.
$y + 18 = 17$ Multiply 6 and 3.
$y = -1$ Subtract 18 from both sides.

To find x, substitute -1 for y and 3 for z in $x + y + 2z = 7$.

$x + (-1) + 2(3) = 7$ Substitute -1 for y and **3** for z.
$x + 5 = 7$ Simplify the left side of the equation.
$x = 2$ Subtract 5 from both sides of the equation.

The solution is $(2, -1, 3)$. We can check by verifying that the ordered triple satisfies all three of the original equations. This is left to the reader.

Your Turn 3 Use the row echelon method to solve the following system.

$$\begin{cases} 3x + y + z = 2 \\ x + 2y - z = -4 \\ 2x - 2y + 3z = 9 \end{cases}$$

Objective 3 Solve applications problems using matrices.

We now solve an applications problem using matrices that we previously solved using the elimination method.

Answer to Your Turn 3
$(1, -2, 1)$

Example 4 Jeff invests \$5000 in two accounts. The first account has an annual percentage rate (APR) of 4%. The second account has an APR of 6%. If the total interest earned after one year is \$252, what principal was invested in each account?

Solution: This resulted in the system of equations $\begin{cases} x + y = 5000 \\ 0.04x + 0.06y = 252, \end{cases}$ where x represents the amount invested at 4% and y represents the amount invested at 6%. The augmented matrix for this system is as follows:

$$\left[\begin{array}{cc|c} 1 & 1 & 5000 \\ 0.04 & 0.06 & 252 \end{array}\right]$$

To put the matrix in row-echelon form, we need a 0 in the second row, first column where 0.04 is presently located. So multiply the first row by -0.04, add it to the second row, and replace the second row with the result.

$$-0.04R_1 + R_2 \longrightarrow \begin{bmatrix} 1 & 1 & 5000 \\ 0 & 0.02 & 52 \end{bmatrix}$$

$-0.04(1 \;\; 1 \mid 5000) = (-0.04 \;\; -0.04 \mid -200)$ and $(-0.04 \;\; -0.04 \mid 200) + (0.04 \;\; 0.06 \mid 252) = (0 \;\; 0.02 \mid 52)$

Dividing R_2 by 0.02 gives

$$R_2 \div 0.02 \longrightarrow \begin{bmatrix} 1 & 1 & 5000 \\ 0 & 1 & 2600 \end{bmatrix}, \text{ which gives us the system } \begin{cases} x + y = 5000 \\ y = 2600 \end{cases}.$$

Because $y = 2600$, we can solve for x using substitution.

$$x + y = 5000$$

$$x + 2600 = 5000 \quad \text{Substitute 2600 for } y.$$

$$x = 2400 \quad \text{Subtract 2600 from both sides.}$$

So Jeff invested \$2400 at 4% and \$2600 at 6%.

Your Turn 4 In winning the 2013 NCAA basketball championship, Louisville scored a total of 46 times in its 82 to 76 victory over Michigan. The sum of the number of 3-point field goals and 2-point field goals was ten more than the number of free throws (1 point each). How many of each did Louisville score?

Answer to Your Turn 4
3-point = 8, 2-point = 20, free throws = 18

4.4 Exercises

For Extra Help MyMathLab®

Note: Exercises marked with a ★ represent challenging exercises.

Objective 1

Prep Exercise 1 How many rows and columns are in a 4×2 matrix?

Prep Exercise 2 How do you write a system of equations as an augmented matrix?

Prep Exercise 3 To what does the dashed line in an augmented matrix correspond in a system of equations?

Prep Exercise 4 Explain how solving a system of equations using row operations is similar to solving the system using elimination.

Prep Exercise 5 What is row echelon form?

Prep Exercise 6 Once an augmented matrix is in row echelon form, how do you find the values of the variables?

For Exercises 1–8, write the augmented matrix for the system of equations. See Example 1.

1. $\begin{cases} 14x + 7y = 6 \\ 7x + 6y = 8 \end{cases}$

2. $\begin{cases} 5x + 6y = 2 \\ 10x + 3y = -2 \end{cases}$

3. $\begin{cases} 7x - 6y = 1 \\ \quad -2y = 5 \end{cases}$

4. $\begin{cases} 3x = 6 \\ 9x + 2y = -2 \end{cases}$

5. $\begin{cases} x - 3y + z = 4 \\ 2x - 4y + 2z = -4 \\ 6x - 2y + 5z = -4 \end{cases}$

6. $\begin{cases} 3x + 2y - 3z = 2 \\ 2x - 4y + 5z = -10 \\ 5x - 4y + z = 0 \end{cases}$

7. $\begin{cases} 4x + 6y - 2z = -1 \\ 8x + 3y = -12 \\ -y + 2z = 4 \end{cases}$

8. $\begin{cases} x - 3y = 3 \\ -3x + 4y + 9z = 3 \\ 2x + 7z = -9 \end{cases}$

Objective 2

For Exercises 9–12, given the matrices in row echelon form, find the solution for the system. See Objective 2 and Example 2.

9. $\left[\begin{array}{cc|c} 1 & -3 & -2 \\ 0 & 1 & 2 \end{array}\right]$

10. $\left[\begin{array}{cc|c} 1 & -3 & 7 \\ 0 & 1 & -5 \end{array}\right]$

11. $\left[\begin{array}{ccc|c} 1 & -4 & -8 & 6 \\ 0 & 1 & -2 & -7 \\ 0 & 0 & 1 & 1 \end{array}\right]$

12. $\left[\begin{array}{ccc|c} 1 & -2 & 6 & 16 \\ 0 & 1 & -4 & -13 \\ 0 & 0 & 1 & 3 \end{array}\right]$

For Exercises 13–18, complete the indicated row operation. See Objective 2 and Example 2.

13. Replace R_2 in $\left[\begin{array}{cc|c} 1 & 3 & -1 \\ -2 & 5 & 6 \end{array}\right]$ with $2R_1 + R_2$.

14. Replace R_2 in $\left[\begin{array}{cc|c} 1 & 2 & -2 \\ 3 & 8 & -4 \end{array}\right]$ with $-3R_1 + R_2$.

15. Replace R_3 in $\left[\begin{array}{ccc|c} 1 & -2 & 4 & 6 \\ 0 & 2 & -1 & -5 \\ 0 & 8 & -6 & -3 \end{array}\right]$ with $-4R_2 + R_3$.

16. Replace R_2 in $\left[\begin{array}{ccc|c} 1 & 5 & -3 & 8 \\ -3 & 2 & -4 & -6 \\ 0 & 1 & -2 & 9 \end{array}\right]$ with $3R_1 + R_2$.

17. Replace R_1 in $\left[\begin{array}{cc|c} 4 & 8 & -10 \\ -1 & 3 & 2 \end{array}\right]$ with $\frac{1}{4}R_1$.

18. Replace R_3 in $\left[\begin{array}{ccc|c} 1 & 3 & 4 & 8 \\ 0 & 1 & -2 & -6 \\ 0 & 0 & -6 & 24 \end{array}\right]$ with $-\frac{1}{6}R_3$.

For Exercises 19–22, describe the next row operation that should be performed to make the matrix closer to row echelon form.

19. $\left[\begin{array}{rr|r} 1 & 2 & -1 \\ -3 & 4 & 5 \end{array}\right]$

20. $\left[\begin{array}{rr|r} 1 & -3 & 2 \\ 5 & 6 & -4 \end{array}\right]$

21. $\left[\begin{array}{rrr|r} 1 & -4 & 3 & 6 \\ 0 & 1 & -2 & 4 \\ 0 & 2 & -5 & 1 \end{array}\right]$

22. $\left[\begin{array}{rrr|r} 1 & -2 & 4 & 7 \\ 0 & 1 & -2 & 3 \\ 0 & -5 & 4 & -9 \end{array}\right]$

For Exercises 23–46, solve by transforming the augmented matrix into row echelon form. See Examples 2 and 3.

23. $\begin{cases} x - y = 5 \\ x + y = -1 \end{cases}$

24. $\begin{cases} x + y = 2 \\ x - y = -8 \end{cases}$

25. $\begin{cases} x + y = 3 \\ 3x - y = 1 \end{cases}$

26. $\begin{cases} x + 3y = -9 \\ -x + 2y = -11 \end{cases}$

27. $\begin{cases} x + 2y = -7 \\ 2x - 4y = 2 \end{cases}$

28. $\begin{cases} x - 3y = -13 \\ 3x + 2y = 5 \end{cases}$

29. $\begin{cases} -2x + 5y = 4 \\ x - 2y = -2 \end{cases}$

30. $\begin{cases} 2x + y = 12 \\ x - 3y = 6 \end{cases}$

31. $\begin{cases} 4x - 3y = -2 \\ 2x - 3y = -10 \end{cases}$

32. $\begin{cases} 6x - 3y = 0 \\ 2x + 4y = 0 \end{cases}$

33. $\begin{cases} 5x + 2y = 12 \\ 2x + 3y = -4 \end{cases}$

34. $\begin{cases} 4x + 3y = 14 \\ 3x - 2y = 2 \end{cases}$

35. $\begin{cases} x + y + z = 3 \\ 2x + y - 3z = -10 \\ 2x + 2y + z = 3 \end{cases}$

36. $\begin{cases} x + 2y + z = 2 \\ 3x + y - z = 3 \\ 2x + y + 2z = 7 \end{cases}$

37. $\begin{cases} x - y + z = -1 \\ 2x - 2y + z = 0 \\ x + 3y + 2z = 1 \end{cases}$

38. $\begin{cases} x + 3y - 2z = 4 \\ 2x - 3y + 2z = -7 \\ 3x + 2y - 2z = -1 \end{cases}$

39. $\begin{cases} 2x - y + z = 8 \\ x - 2y + 3z = 11 \\ 2x + 3y - z = -6 \end{cases}$

40. $\begin{cases} 2x + 3y - 2z = -21 \\ 2x - 4y + 3z = 15 \\ 3x + 2y - 3z = -24 \end{cases}$

41. $\begin{cases} 3x + 2y - 3z = 1 \\ -2x + 3y - 4z = 7 \\ 5x - 2y + z = -5 \end{cases}$

42. $\begin{cases} 2x - 2y + z = 6 \\ -2x + 4y + 3z = 4 \\ 4x - 3y - 2z = -7 \end{cases}$

43. $\begin{cases} 3x - 6y + z = -10 \\ 7y - z = 2 \\ 2x + 4z = 14 \end{cases}$

44. $\begin{cases} 6x - y + 5z = -28 \\ 4x + 2y = 10 \\ 5x + 6z = -23 \end{cases}$

45. $\begin{cases} 2x + 5y - z = 10 \\ x - y - z = 14 \\ x - 6y = 20 \end{cases}$

46. $\begin{cases} 2x + 4y - 5z = 16 \\ x - 2y - 11z = 1 \\ 4x + 5y = -6 \end{cases}$

★ *For Exercises 47–54, solve using a matrix on a graphing calculator.*

47. $\begin{cases} 2x + 5y = -14 \\ 6x + 7y = -10 \end{cases}$

48. $\begin{cases} 4x - 3y = 11 \\ 5x + 4y = 68 \end{cases}$

49. $\begin{cases} 4x - 7y = -80 \\ 9x + 5y = -14 \end{cases}$

50. $\begin{cases} 7x - 3y = 23 \\ 4x - 9y = -67 \end{cases}$

51. $\begin{cases} 3x + 2y - 5z = -19 \\ 4x - 7y + 6z = 67 \\ 5x - 6y - 4z = 18 \end{cases}$

52. $\begin{cases} 6x - 7y + 2z = -52 \\ 8x + 5y - 6z = 132 \\ 12x + 4y - 7z = 150 \end{cases}$

53. $\begin{cases} 8x - 7y + 2z = 108 \\ 6x + 11y - 10z = -74 \\ -2y + 13z = 77 \end{cases}$

54. $\begin{cases} 8x - 11y + 14z = 40 \\ 5x + 15y - 5z = 35 \\ 17x - 9y = -157 \end{cases}$

Find the Mistake *For Exercises 55 and 56, explain the mistake; then find the correct solution.*

55. $\begin{cases} x + 3y = 13 \\ -4x - y = -26 \end{cases}$

$$\left[\begin{array}{cc|c} 1 & 3 & 13 \\ -4 & -1 & -26 \end{array}\right]$$

$$4R_1 + R_2 \longrightarrow \left[\begin{array}{cc|c} 1 & 3 & 13 \\ 0 & 11 & -13 \end{array}\right]$$

$$\frac{1}{11}R_2 \longrightarrow \left[\begin{array}{cc|c} 1 & 3 & 13 \\ 0 & 1 & -\frac{13}{11} \end{array}\right]$$

Solution: $\left(\frac{182}{11}, -\frac{13}{11}\right)$

56. $\begin{cases} x - y + z = 8 \\ 3x - z = -9 \\ 4y + z = -6 \end{cases}$

$$\left[\begin{array}{ccc|c} 1 & -1 & 1 & 8 \\ 3 & 0 & -1 & -9 \\ 4 & 0 & 1 & -6 \end{array}\right]$$

$$-3R_1 + R_2 \longrightarrow \left[\begin{array}{ccc|c} 1 & -1 & 1 & 8 \\ 0 & 3 & -4 & -33 \\ 4 & 0 & 1 & -6 \end{array}\right]$$

$$\begin{array}{r} \\ \frac{1}{3}R_2 \longrightarrow \\ -4R_1 + R_3 \longrightarrow \end{array} \left[\begin{array}{ccc|c} 1 & -1 & 1 & 8 \\ 0 & 1 & -\frac{4}{3} & -11 \\ 0 & 4 & -3 & -38 \end{array}\right]$$

$$-4R_2 + R_3 \longrightarrow \left[\begin{array}{ccc|c} 1 & -1 & 1 & 8 \\ 0 & 1 & -\frac{4}{3} & -11 \\ 0 & 0 & \frac{7}{3} & 6 \end{array}\right]$$

$$\frac{3}{7}R_3 \longrightarrow \left[\begin{array}{ccc|c} 1 & -1 & 1 & 8 \\ 0 & 1 & -\frac{4}{3} & -11 \\ 0 & 0 & 1 & \frac{18}{7} \end{array}\right]$$

Solution: $\left(-\frac{15}{7}, -\frac{53}{7}, \frac{18}{7}\right)$

Objective 3

For Exercises 57–68, translate the problem to a system of equations; then solve using matrices.

57. Brad purchased three grilled chicken sandwiches and two drinks for $12.90, and Angel purchased seven grilled chicken sandwiches and four drinks for $29.30. Find the price of one grilled chicken sandwich and one drink.

58. Sharika purchased three general admission tickets and two student tickets to a college play for $55, and Yo Chen purchased two general admission tickets and four student tickets for $50. Find the cost of one general admission and one student ticket.

59. The two longest rivers in the world, the Nile and the Amazon, have a combined length of 8050 miles. The Nile is 250 miles longer than the Amazon. Find the length of both rivers.

Of Interest

During the new and full moon, a small tidal wave, called a tidal bore, sweeps up the Amazon. The tidal bore can travel 450 miles upstream at speeds in excess of 40 miles per hour, causing waves of 16 feet or more along the riverbank.

60. The longest vehicular tunnel in the world is the Saint Gotthard Tunnel in Switzerland, and the second longest is the Arlberg Tunnel in Austria. The total length of the two tunnels is 18.8 miles. The Saint Gotthard Tunnel is 1.4 miles longer than the Arlberg. Find the length of each tunnel. (*Source: Webster's New World Book of Facts.*)

61. Nikita invested a total of $10,000 in certificates of deposit that pay 5% annually and in a money market account that pays 6% annually. If the total interest earned in one year from the two investments is $536, find the principal that was invested in each.

62. An athlete received a $60,000 signing bonus, which he invested in two funds—a money market fund paying 5% annually and a growth fund paying 7% annually. The total annual interest received from the two investments is $3440. Find the principal invested in each fund.

63. John bought two CDs, four books, and three DVDs for $164, and Tanelle bought five CDs, two books, and two DVDs for $160. If the sum of the cost of one CD and one book equals the cost of one DVD, find the cost of one of each.

64. Angel bought 3 pounds of salmon, 2 pounds of tuna, and 5 pounds of cod for $63. Sara bought 6 pounds of salmon, 3 pounds of tuna, and 5 pounds of cod for $94. The cost of 1 pound of salmon and 1 pound of tuna is the same as the cost of 3 pounds of cod. Find the cost of 1 pound of each.

65. In Green Bay's 31 to 25 victory over Pittsburgh in Super Bowl XLV, Green Bay scored touchdowns (6 points), extra points (1 point), and field goals (3 points). The number of touchdowns equaled the number of extra points. Also, the number of touchdowns was three more than the number of field goals. Find how many of each type of score Green Bay had. (*Source:* www.nfl.com.)

66. In the 2012 NCAA women's basketball championship game, Baylor scored forty-seven times for a total of 80 points using a combination of 3-point field goals, 2-point field goals, and free throws (1 point). If the sum of the number of 3-point and 2-point field goals was eleven more than the number of free throws, find the number of each type.

67. An electrical circuit has three points of connection with different voltage measurements at each of the three connections. An engineer has written the following equations to describe the voltages. Find each voltage.

$$\begin{aligned} 4v_1 - v_2 &= 30 \\ -2v_1 + 5v_2 - v_3 &= 10 \\ -v_2 + 5v_3 &= 4 \end{aligned}$$

68. An engineer has written the following system of equations to describe the forces in pounds acting on a steel structure. Find the forces.

$$\begin{aligned} 6F_1 - F_2 &= 350 \\ 9F_1 - 2F_2 - F_1 &= -100 \\ 3F_2 - F_3 &= 250 \end{aligned}$$

Review Exercises

Exercises 1–6 Expressions

[1.3] *For Exercises 1–6, simplify.*

1. $(-3)(4) - (-4)(5)$

2. $2(2 - 8) + 3[-1 - (-6)] - 4(4 - 6)$

3. $\dfrac{(-5)(-1) - (9)(3)}{(2)(-1) - (3)(3)}$

4. $\dfrac{2\left(\frac{5}{4}\right) - 8\left(\frac{1}{4}\right)}{\left(\frac{1}{2}\right)\left(\frac{5}{4}\right) - \left(\frac{3}{2}\right)\left(\frac{1}{4}\right)}$

5. $\dfrac{(1.1)(-0.2) - (1.7)(-0.5)}{(0.2)(-0.2) - (0.5)(-0.5)}$

6. $\dfrac{1[-45 - (-9)] + 7(-9 - 2) - 2(-27 - 30)}{1(6 - 3) - 2(-9 - 2) - 2[9 - (-4)]}$

4.5 Solving Systems of Linear Inequalities

Objectives

1 Graph the solution set of a system of linear inequalities.

2 Solve applications involving a system of linear inequalities.

Warm-up

For Exercises 1 and 2, graph.

[3.1] 1. $x + 2y = 6$

[3.4] 2. $y < 2x + 1$

Objective 1 Graph the solution set of a system of linear inequalities.

In Section 2.3, we learned to solve linear inequalities in one variable. In Section 3.4, we learned how to graph linear inequalities in two variables. In this section, we develop a graphical approach to solving *systems of linear inequalities* in two variables.

Consider the system of linear inequalities $\begin{cases} x + 2y < 6 \\ 2x - y \geq 2 \end{cases}$. First, let's graph each inequality separately.

Note Recall from Section 3.4 that we use a dashed line with $<$ or $>$ and that points on a dashed line are not in the solution set.

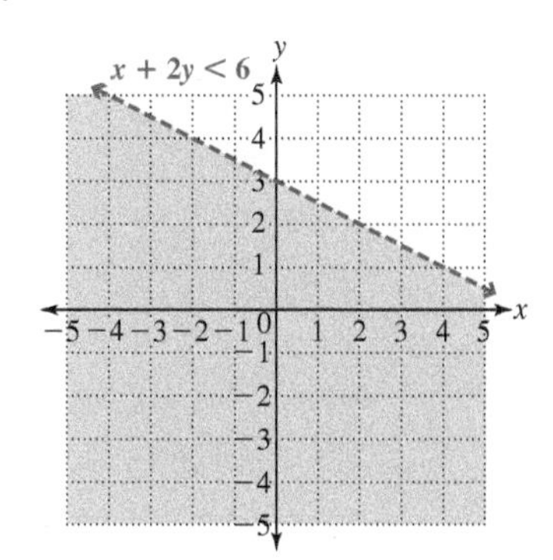

In the graph of $x + 2y < 6$, the shaded region contains all ordered pairs that make $x + 2y < 6$ true.

Note Recall from Section 3.4 that we determine which side of the line to shade by choosing an ordered pair not on the line and checking to see if it makes the inequality true. If it does, we shade on the side of the line containing that ordered pair. If it does not, we shade the other side of the line.

Answers to Warm-up

1.

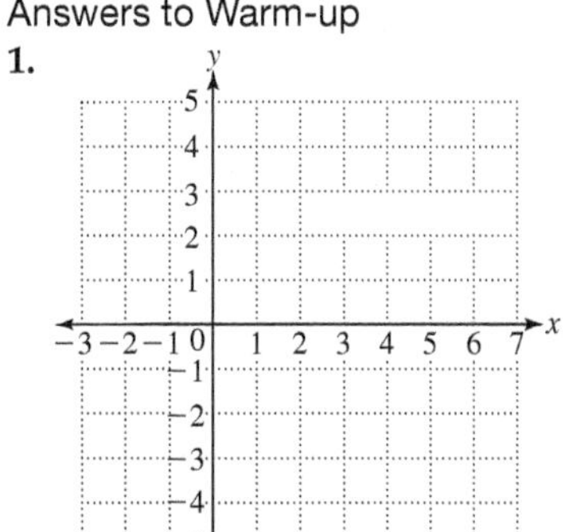

2.

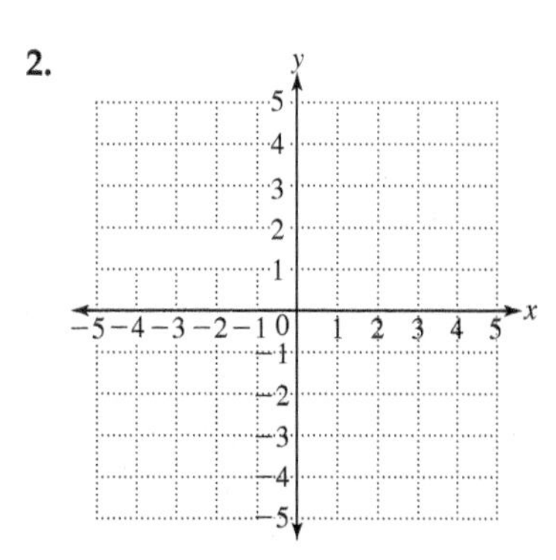

Note Recall from Section 3.4 that we use a solid line with $\le$ or $\ge$ and that points on a solid line are part of the solution set.

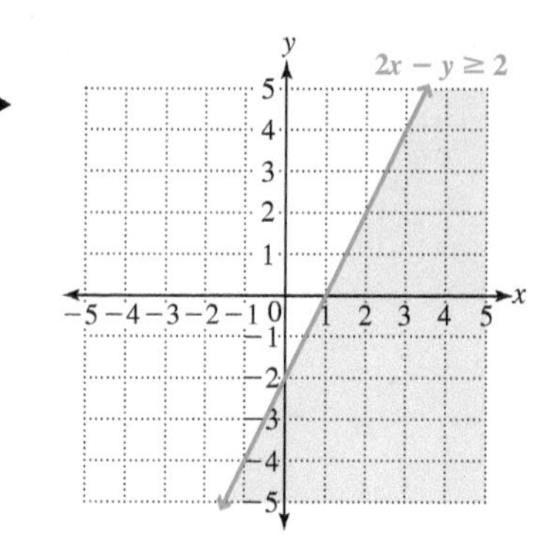

In the graph of $2x - y \ge 2$, ordered pairs in the shaded region and on the line itself make $2x - y \ge 2$ true.

Now we put the two graphs together on the same grid to determine the solution set for the system. A solution for a system of inequalities is an ordered pair that makes every inequality in the system true.

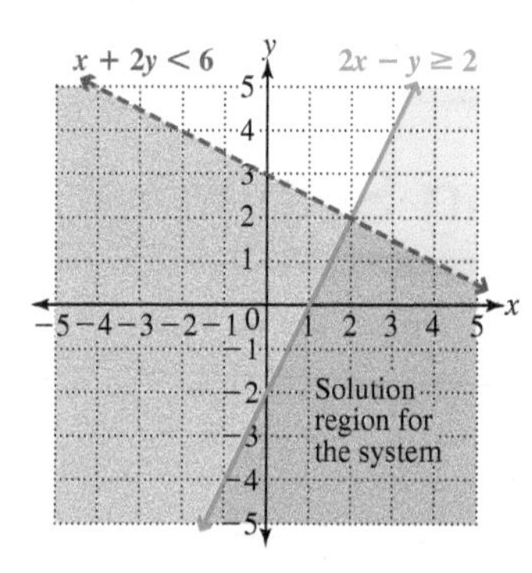

On our graph containing both $x + 2y < 6$ and $2x - y \ge 2$, the region where the shading overlaps contains ordered pairs that make both inequalities true. So all ordered pairs in this region are in the solution set for the system. Also, ordered pairs on the solid line for $2x - y \ge 2$ where it touches the region of overlap are in the solution set for the system, whereas ordered pairs on the dashed line for $x + 2y < 6$ are not. This implies the following procedure.

Procedure Solving a System of Linear Inequalities in Two Variables

To solve a system of linear inequalities in two variables, graph all of the inequalities on the same grid. The solution set for the system contains all ordered pairs in the region where the inequalities' solution sets overlap along with ordered pairs on the portion of any solid line that touches the region of overlap.

To check, we can select a point in the solution region such as $(3, 0)$ and verify that it makes both inequalities true.

First inequality:	**Second inequality:**	
$x + 2y < 6$	$2x - y \ge 2$	
$3 + 2(0) \overset{?}{<} 6$	$2(3) - 0 \overset{?}{\ge} 2$	Replace x with 3 and y with 0 in both inequalities.
$3 < 6$ True.	$6 \ge 2$ True.	

Because $(3, 0)$ makes both inequalities true, it is a solution to the system. Although we selected only one ordered pair in the solution region, remember that *every* ordered pair in that region is a solution.

Connection Remember that for a system of two linear equations, a solution is a point of intersection of the graphs of the two equations. Similarly, for a system of linear inequalities, the solution set is a region of intersection of the graphs of the two inequalities.

Example 1 Graph the solution set for the system of inequalities.

a. $\begin{cases} x + 3y > 6 \\ y < 2x - 1 \end{cases}$

Solution: Graph the inequalities on the same grid. Because both lines are dashed, the solution set for the system contains only those ordered pairs in the region of overlap (the purple shaded region).

Note Ordered pairs on dashed lines are not part of the solution region.

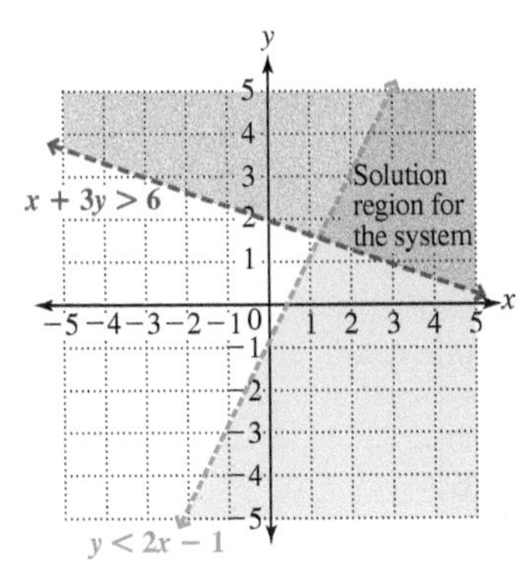

b. $\begin{cases} y < 2 \\ 5x - y \le 4 \end{cases}$

Solution: Graph the inequalities on the same grid. The solution set for this system contains all ordered pairs in the region of overlap (purple shaded region) together with all ordered pairs on the portion of the solid blue line that touches the purple shaded region.

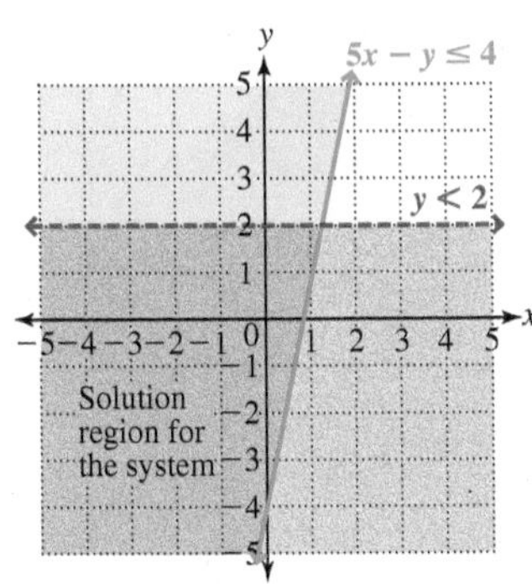

Your Turn 1 Graph the solution set for the system of inequalities.

a. $\begin{cases} x + y > -1 \\ 4x - y < 3 \end{cases}$ **b.** $\begin{cases} x > -3 \\ 3x - 4y \le -8 \end{cases}$

Connection Notice that the lines in Example 2 are parallel. Recall that in a system of linear equations, if the lines are parallel, the system is inconsistent because there is no point of intersection. With linear inequalities, the lines must be parallel *and* the shaded regions must not overlap for the system to be inconsistent.

Inconsistent Systems

Some systems of linear inequalities have no solution. We say that these systems are inconsistent.

Example 2 Graph the solution set for the system of inequalities.

$$\begin{cases} x - 4y \ge 8 \\ y \ge \dfrac{1}{4}x + 3 \end{cases}$$

Solution: Graph the inequalities on the same grid.

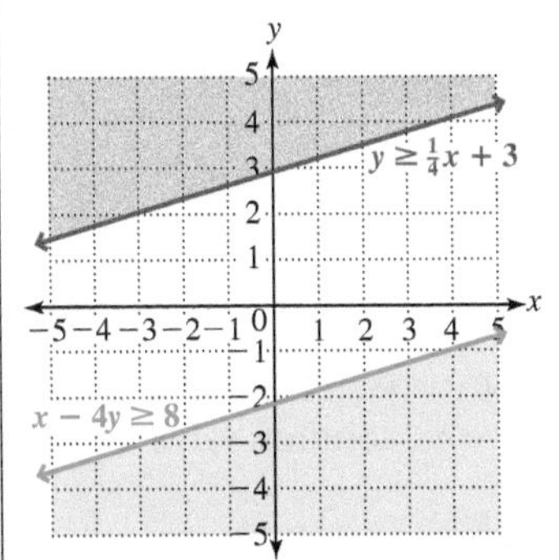

The lines appear to be parallel. We can verify by writing $x - 4y \ge 8$ in slope–intercept form and comparing the slopes and y-intercepts.

$$x - 4y \ge 8$$

$$-4y \ge -x + 8 \quad \text{Subtract } x \text{ from both sides.}$$

$$y \le \frac{1}{4}x - 2 \quad \text{Divide both sides by } -4\text{, which changes the direction of the inequality symbol.}$$

The slopes are equal, and the y-intercepts differ; so the lines are in fact parallel. Because the lines are parallel and the shaded regions do not overlap, there is no solution region for this system. The system is inconsistent.

Answers to Your Turn 1

a.

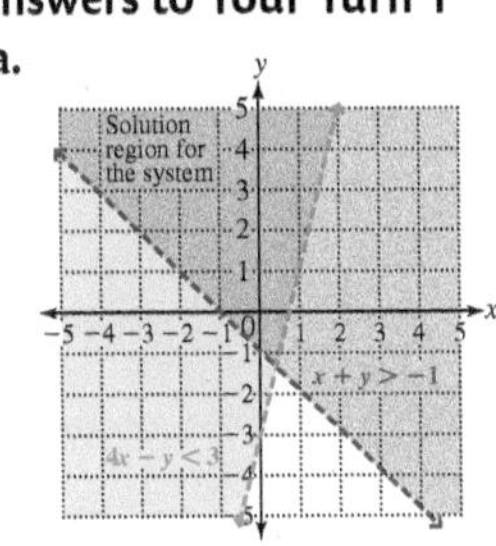

b. Solution region for the system (including the portion of the red line touching this region)

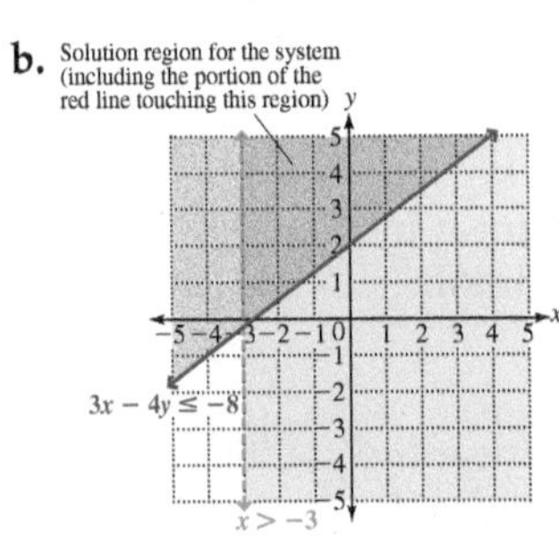

Objective 2 Solve applications involving a system of linear inequalities.

Example 3 A home interiors store stocks two different-size prints from an artist. The manager wants to purchase at least 15 prints for the store. The artist sells the smaller print for \$20 and the larger print for \$30. The manager cannot spend more than \$800 for the prints. Write a system of inequalities that describes the manager's order; then solve the system by graphing.

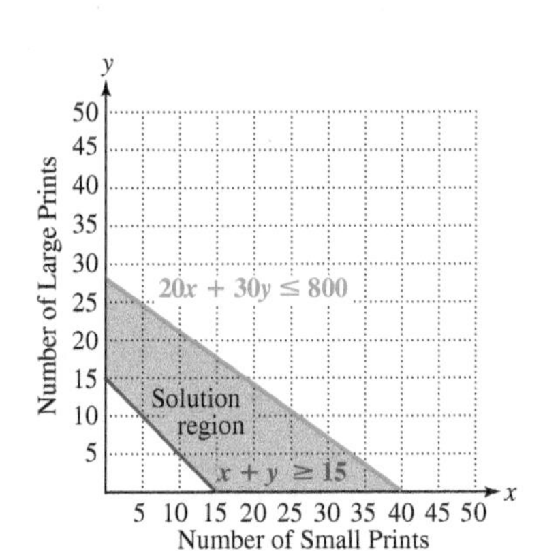

Understand We must translate to a system of inequalities, then solve the system.

Plan and Execute Let x represent the number of small prints and y represent the number of large prints ordered.

Relationship 1: The manager wants to purchase at least 15 prints.

The words *at least* indicate that the combined number of prints is to be greater than or equal to 15; so $x + y \geq 15$.

Relationship 2: The manager cannot spend more than \$800.

Because the small prints cost \$20 each, $20x$ describes the amount spent on small prints. Similarly, $30y$ describes the amount spent on large prints. The total cannot exceed \$800; so $20x + 30y \leq 800$.

$$\text{Our system: } \begin{cases} x + y \geq 15 \\ 20x + 30y \leq 800 \end{cases}$$

Answer (See the graph.) Because the manager cannot order a negative number of prints, the solution set is confined to quadrant I. Any ordered pair in the solution region or on a portion of either line touching the solution region is a solution for the system. However, assuming that only whole prints can be purchased, only ordered pairs of whole numbers in the solution set, such as (10, 15) and (25, 10), are realistic.

Check We will check one ordered pair in the solution region. The ordered pair (10, 15) indicates an order of 10 small prints and 15 large prints so that 25 prints are ordered (which is more than 15) that cost a total of $20(10) + 30(15) = \$650$ (which is less than \$800).

Answer to Your Turn 3

$$\text{System: } \begin{cases} 2l + 2w \leq 200 \\ l \geq w + 10 \end{cases}$$

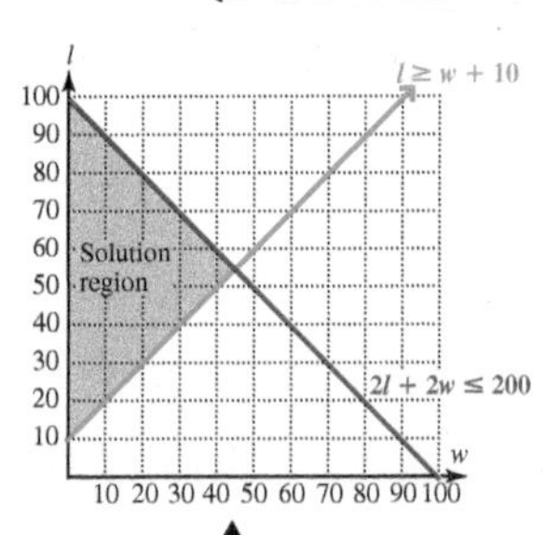

Note Because length and width must be positive numbers, the solution region is confined to quadrant 1.

Your Turn 3 In designing a building with a rectangular base, an architect decides that the perimeter cannot exceed 200 feet. Also, the length will be at least 10 feet more than the width. Write a system of inequalities that describes the design requirements; then solve the system graphically.

4.5 Exercises

For Extra Help MyMathLab®

Note: Exercises marked with a ★ represent challenging exercises.

Objective 1

Prep Exercise 1 Explain how to check a possible solution to a system of inequalities.

Prep Exercise 2 Explain how to determine the region that is the solution set for a system of linear inequalities.

Prep Exercise 3 What circumstances would cause a system of inequalities to have no solution?

Prep Exercise 4 Write a system of linear inequalities whose solution set is the entire first quadrant.

Prep Exercise 5 Write a system of linear inequalities whose solution set is the entire third quadrant.

Prep Exercise 6 Write a system of linear inequalities whose solution set is the entire fourth quadrant.

For Exercises 1–24, graph the solution set for the system of inequalities. See Examples 1 and 2.

1. $\begin{cases} x + y > 4 \\ x - y < 6 \end{cases}$

2. $\begin{cases} x - y > 2 \\ x + y > 4 \end{cases}$

3. $\begin{cases} x + y < -2 \\ x - y > 6 \end{cases}$

4. $\begin{cases} x - y < -5 \\ x + y < 3 \end{cases}$

5. $\begin{cases} 2x + y \geq -1 \\ x - y \leq 5 \end{cases}$

6. $\begin{cases} x - y < -5 \\ 2x - y < -7 \end{cases}$

7. $\begin{cases} x + y < 3 \\ x - 2y \geq 2 \end{cases}$

8. $\begin{cases} x + 3y \leq 6 \\ x - y > 5 \end{cases}$

9. $\begin{cases} 2x + y > 9 \\ y > \dfrac{1}{4}x \end{cases}$

10. $\begin{cases} 3x + 4y \leq -9 \\ y < 3x \end{cases}$

11. $\begin{cases} y < x \\ y > -x + 1 \end{cases}$

12. $\begin{cases} y < x \\ y < 2x - 3 \end{cases}$

13. $\begin{cases} 3x > -4y \\ 3x + 4y \leq -8 \end{cases}$

14. $\begin{cases} 2y - x \geq 6 \\ 2x - 4y \geq 5 \end{cases}$

15. $\begin{cases} y < 3x + 1 \\ 3x - y \leq 4 \end{cases}$

16. $\begin{cases} x - 2y > 3 \\ 2x - 4y \leq 20 \end{cases}$

17. $\begin{cases} x > 2 \\ 3x - 2y > 6 \end{cases}$

18. $\begin{cases} y \leq -1 \\ x + 2y > 3 \end{cases}$

19. $\begin{cases} x \geq -3y \\ y \geq 2x \end{cases}$

20. $\begin{cases} x > 2y \\ x + y > 6 \end{cases}$

21. $\begin{cases} y > 2 \\ x < -1 \end{cases}$

22. $\begin{cases} y \geq -1 \\ x < 2 \end{cases}$

23. $\begin{cases} 2x + y \geq 1 \\ x - y > -1 \\ x > 2 \end{cases}$

24. $\begin{cases} x + y \geq 1 \\ x - y \geq -5 \\ y > -2 \end{cases}$

Find the Mistake *For Exercises 25–28, explain the mistake.*

25. $\begin{cases} x + y > 3 \\ x - y \leq 2 \end{cases}$

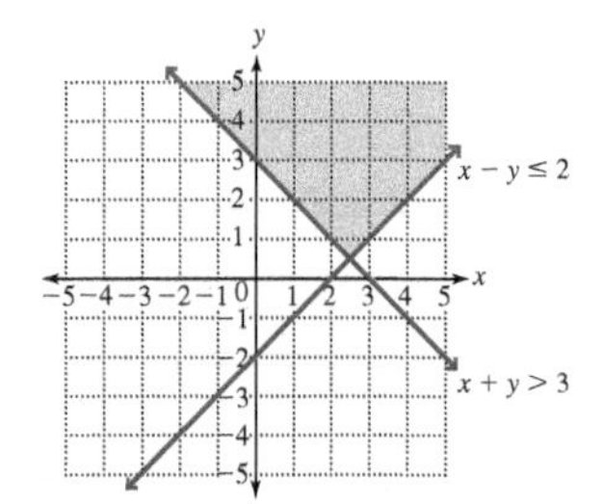

26. $\begin{cases} y < \frac{1}{2}x \\ y \leq -3x + 1 \end{cases}$

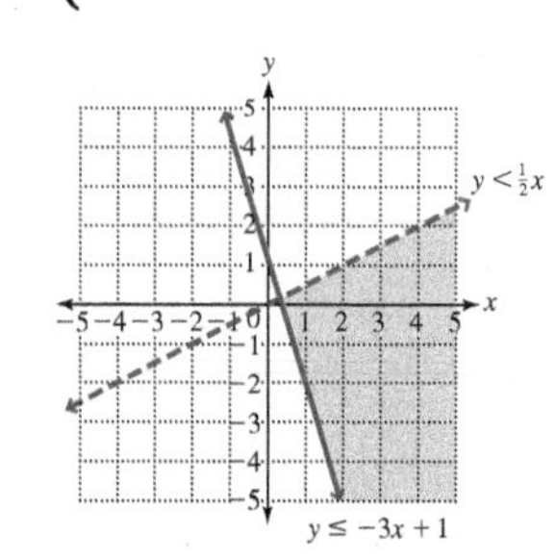

27. $\begin{cases} x < 3 \\ y > 4 \end{cases}$

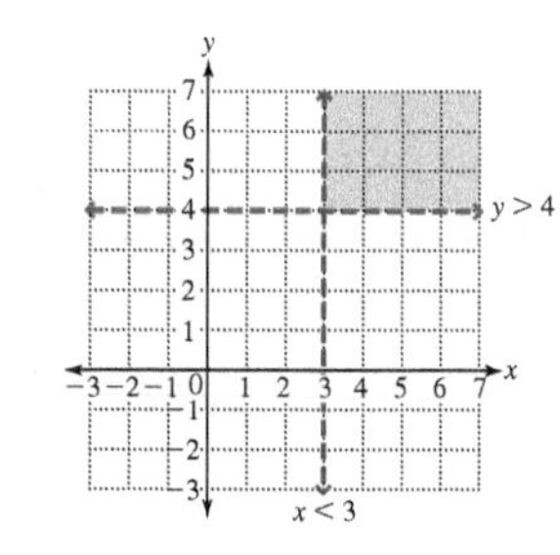

28. 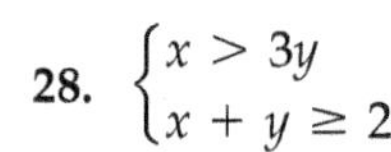

$\begin{cases} x > 3y \\ x + y \geq 2 \end{cases}$

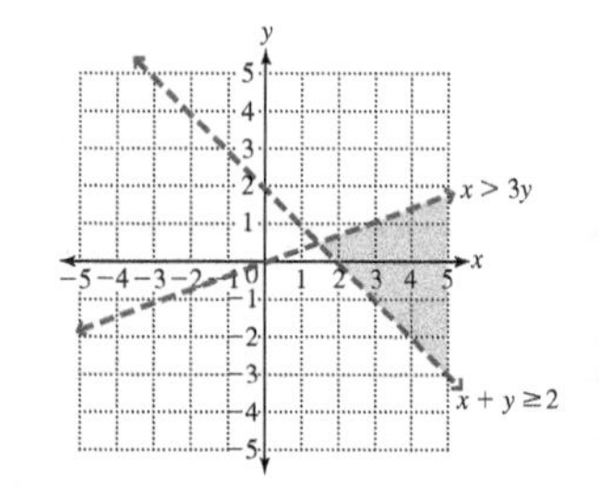

Objective 2

Prep Exercise 7 When translating "at least," which inequality symbol is used?

Prep Exercise 8 When translating "at most," which inequality symbol is used?

For Exercises 29–32, solve. See Example 3.

29. To be admitted to a certain college, prospective students must have a combined verbal and math score of at least 1100 on the SAT. To be admitted to the engineering program at this college, students must also have a math score of at least 500.

a. Write two inequalities describing the requirements to be admitted to the college of engineering.

b. The maximum score on each part of the SAT is 800. Write two inequalities describing these maximum scores.

c. Solve the system of inequalities described by the four inequalities in parts a and b by graphing. Let the horizontal axis be verbal and the vertical axis be math.

d. Give two different combinations of verbal and math scores that satisfy the admission requirements to the college of engineering.

30. An architect is designing a rectangular platform for an auditorium. The client wants the length of the platform to be at least twice the width, and the perimeter must not exceed 200 feet.

a. Write a system of inequalities that describes the specifications for the platform.

b. Solve the system by graphing. Let the horizontal axis be the width and the vertical axis be the length.

c. Give two different combinations of length and width that satisfy the requirements of the client.

★ **31.** A company sells two versions of software: the regular version and the deluxe version. The regular version sells for $15.95, and the deluxe version sells for $20.95. The company gives a bonus to any salesperson who sells at least 100 units and has at least $1700 in total sales.

a. Write a system of inequalities that describes the requirements a salesperson must meet to receive the bonus.

b. Solve the system by graphing. Let the horizontal axis be the regular version and the vertical axis be the deluxe version.

c. Give two different combinations of number of units sold that meet the requirements for receiving the bonus.

★ **32.** An investor is trying to decide how much to invest in two different stocks. She has decided that she will invest at most $5000. The "safe" stock is very stable and is projected to return about 5% over the next year. The other stock is more risky, but it could return about 9% if the company grows as expected. She wants her total dividend at the end of the year to be at least $300.

a. Write a system of inequalities that describes the amount invested in the two stocks and the return on that investment.

b. Solve the system by graphing. Let the horizontal axis be the "safe" stock and the vertical be the "risky" stock.

c. Give two different combinations of investment amounts that return at least $300.

Review Exercises

Exercises 1–4 Expressions

[1.3] *For Exercises 1–4, evaluate.*

1. 3^2

2. -10^2

3. $\sqrt{25}$

4. $\sqrt{\dfrac{16}{49}}$

Exercises 5 and 6 Equations and Inequalities

[3.5] **5.** Given $f(x) = 3x - 5$, find $f(3)$.

[3.5] **6.** Given $f(x) = 3x^2 - 2x + 5$, find $f(-2)$.

Chapter 4 Summary and Review Exercises

Complete each incomplete definition, rule, or procedure; study the key examples; and then work the related exercises.

4.1 Solving Systems of Linear Equations Graphically

Definitions/Rules/Procedures

A **system of equations** is a(n) ______________ equations.

A solution for a system of equations is an ordered set of numbers that ______________________________.

To verify or check a solution to a system of equations:

1. Replace each ____________ in each equation with its corresponding ____________.
2. Verify that each equation is ____________.

Key Example(s)

Determine whether the ordered pair is a solution to the system.

$$\begin{cases} x + 2y = 12 \\ y = -3x + 11 \end{cases}$$

a. $(-2, 7)$

$$\begin{aligned} x + 2y &= 12 & y &= -3x + 11 \\ -2 + 2(7) &\stackrel{?}{=} 12 & 7 &\stackrel{?}{=} -3(-2) + 11 \\ -2 + 14 &\stackrel{?}{=} 12 & 7 &\stackrel{?}{=} 6 + 11 \\ 12 &= 12 & 7 &\neq 17 \end{aligned}$$

$(-2, 7)$ is not a solution to the system.

b. $(2, 5)$

$$\begin{aligned} x + 2y &= 12 & y &= -3x + 11 \\ 2 + 2(5) &\stackrel{?}{=} 12 & 5 &\stackrel{?}{=} -3(2) + 11 \\ 2 + 10 &\stackrel{?}{=} 12 & 5 &\stackrel{?}{=} -6 + 11 \\ 12 &= 12 & 5 &= 5 \end{aligned}$$

$(2, 5)$ is a solution to the system.

Exercises 1 and 2 Equations and Inequalities

[4.1] *For Exercises 1 and 2, determine whether the given ordered pair is a solution to the given system of equations.*

1. $(-2, 7)$; $\begin{cases} x + y = 5 \\ 3x - y = -13 \end{cases}$

2. $(3, 2)$; $\begin{cases} 2x + y = 8 \\ x - y = -1 \end{cases}$

Definitions/Rules/Procedures

To solve a system of linear equations graphically:

1. Graph each equation.
   - **a.** If the lines intersect at a single point, the ____________ of that point form the solution.
   - **b.** If the lines are parallel, there is ____________.
   - **c.** If the lines are identical, there are a(n) ____________ of solutions, which are the coordinates of all points on that line.
2. Check your solution.

Key Example(s)

Solve the system of equations graphically.

$$\begin{cases} 2x + 3y = 8 \\ y = 5x - 3 \end{cases}$$

Graph the two equations; then find the point of intersection.

$2x + 3y = 8$ $y = 5x - 3$ $(1, 2)$

The two lines intersect at the point $(1, 2)$, so $(1, 2)$ is the solution to the system.

Exercises 3–6 Equations and Inequalities

[4.1] *For Exercises 3–6, solve the system graphically.*

3. $\begin{cases} y = 2x + 1 \\ x + y = -8 \end{cases}$

4. $\begin{cases} x - y = 6 \\ 2x + y = 6 \end{cases}$

5. $\begin{cases} y = -3x - 4 \\ 9x + 3y = 12 \end{cases}$

6. $\begin{cases} 5x - 2y = 10 \\ y = \frac{5}{2}x - 5 \end{cases}$

Definitions/Rules/Procedures	Key Example(s)
A **consistent system of equations** is a system of equations that has ________________. An **inconsistent system of equations** is a system of equations that has ________________. **To classify a system of two linear equations in two unknowns,** write the equations in slope–intercept form and compare the slopes and y-intercepts. 1. **Consistent system with independent equations:** The system has a single solution at the point of ____________. The graphs are ____________. They have ____________ slopes. 2. **Consistent system with dependent equations:** The system has a(n) ____________ number of solutions. The graphs are ____________. They have the ____________ slope and ____________ y- intercept. 3. **Inconsistent systems:** The system has ____________ solution. The graphs are ____________ lines. They have the ____________ slope but different ____________.	Determine whether the system is consistent with independent equations, consistent with dependent equations, or inconsistent and discuss the number of solutions for the system. $\begin{cases} x + 4y = 6 \\ 2x + 8y = -8 \end{cases}$ **Solution:** Write the equations in slope–intercept form. $x + 4y = 6$ $\quad$ $2x + 8y = -8$ $4y = -x + 6$ $\quad$ $8y = -2x - 8$ $y = -\frac{1}{4}x + \frac{3}{2}$ $\quad$ $y = -\frac{1}{4}x - 1$ Because the slopes are the same $\left(\text{both } -\frac{1}{4}\right)$ but the y-intercepts are different, this system is inconsistent. This means there is no solution to the system.

Exercises 7–10 Equations and Inequalities

[4.1] *For Exercises 7–10:* *a. Determine whether the system of equations is consistent with independent equations, consistent with dependent equations, or inconsistent.*

b. How many solutions does the system have?

7.

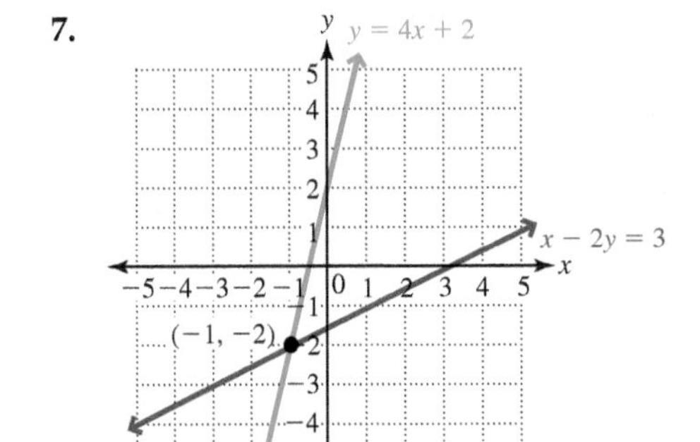

8.

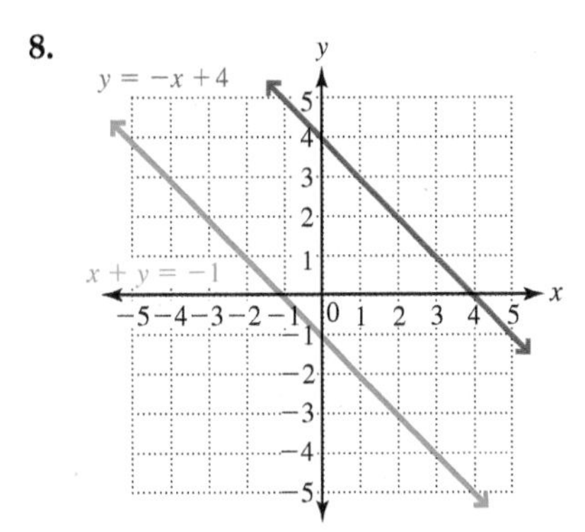

9. $\begin{cases} x - 3y = 10 \\ 2x + 3y = 5 \end{cases}$

10. $\begin{cases} x - 3y = 15 \\ 2x - 6y = 30 \end{cases}$

Definitions/Rules/Procedures	Key Example(s)
To find the solution of a system of two linear equations using the substitution method: 1. Isolate one of the ______ in one of the equations. 2. In the other equation, ______ the expression you found in step 1 for that variable. 3. ______ this new equation. (It will now have only one variable.) 4. Using one of the equations containing both variables, ______ the value you found in step 3 for that variable and solve for the value of the other ______. 5. Check the solution in the original equations.	Solve the system using the substitution method. $\begin{cases} y = 3x - 1 \\ 6x - y = -2 \end{cases}$ Because y is already isolated in the first equation, we substitute $3x - 1$ in place of y in the second equation. $6x - y = -2$ $6x - (3x - 1) = -2$ Substitute $3x - 1$ for y. $6x - 3x + 1 = -2$ Distribute. $3x + 1 = -2$ Combine like terms. $3x = -3$ Subtract 1 from both sides. $x = -1$ Divide both sides by 3. Now solve for the value of y by substituting 1 in place of x in one of the equations. We will use $y = 3x - 1$. $y = 3x - 1$ $y = 3(-1) - 1$ Substitute -1 for x. $y = -3 - 1$ $y = -4$ Solution: $(-1, -4)$

Exercises 11–14 Equations and Inequalities

[4.2] *For Exercises 11–14, solve the system of equations using substitution.*

11. $\begin{cases} y = -3x + 5 \\ x + 2y = -10 \end{cases}$

12. $\begin{cases} 6x + 3y = 11 \\ x - 6y = 4 \end{cases}$

13. $\begin{cases} 2x + 3y = 9 \\ y = -\frac{2}{3}x + 3 \end{cases}$

14. $\begin{cases} 12x - 3y = -6 \\ y - 4x = -2 \end{cases}$

Definitions/Rules/Procedures	Key Example(s)
To solve a system of two linear equations using the elimination method: 1. Write the equations in ______ form ($Ax + By = C$). 2. Use the multiplication principle of equality to eliminate ______ or ______ (optional). 3. If necessary, multiply one or both equations by a number (or numbers) so that they have a pair of terms that are ______.	Solve the system using the elimination method. $\begin{cases} \frac{1}{4}x + \frac{1}{2}y = -1 \\ 4x + y = 5 \end{cases}$ **Solution:** First, we eliminate the fractions in the first equation. $4\left(\frac{1}{4}x + \frac{1}{2}y\right) = 4(-1)$ $x + 2y = -4$

Definitions/Rules/Procedures	Key Example(s)
4. ______ the equations. The result should be an equation in terms of one variable. **5.** ______ the equation from step 4 for the value of that variable. **6.** Using an equation containing both variables, ______ the value you found in step 5 for the corresponding variable and solve for the value of the other variable. **7.** Check your solution in the ______.	Now we will eliminate y by multiplying the second equation by -2, then add the two equations. $x + 2y = -4 \quad\quad x + 2y = -4$ $4x + y = 5 \xrightarrow{\text{Multiply by } -2.} -8x - 2y = -10$ Add equations. $-7x + 0 = -14$ Divide both sides by -7. $-7x = -14$ $x = 2$ Now solve for the value of y by substituting 2 for x in one of the equations. We will use $4x + y = 5$. $4x + y = 5$ $4(2) + y = 5$ Substitute 2 for x. $8 + y = 5$ Multiply. $y = -3$ Subtract 8 from both sides. **Solution:** $(2, -3)$

Exercises 15–18 Equations and Inequalities

[4.3] *For Exercises 15–18, solve the system of equations using elimination.*

15. $\begin{cases} x + y = 7 \\ x - y = -5 \end{cases}$ **16.** $\begin{cases} 3x + 5y = 2 \\ 2x - 3y = 14 \end{cases}$ **17.** $\begin{cases} 0.45x + 0.2y = -2.6 \\ 0.6x - 0.5y = -7.3 \end{cases}$ **18.** $\begin{cases} \frac{2}{3}x - \frac{3}{5}y = \frac{8}{15} \\ x + \frac{1}{4}y = -\frac{3}{2} \end{cases}$

4.2 Solving Systems of Linear Equations in Three Variables

Definitions/Rules/Procedures	Key Example(s)
To solve a system of three linear equations with three unknowns using elimination: **1.** Write each equation in the form ______. **2.** Eliminate one variable from one pair of equations using the ______ method. **3.** If necessary, eliminate the ______ from another pair of equations. **4.** Steps 2 and 3 result in two equations with the same ______. Solve these equations using the elimination method. **5.** To find the third variable, substitute the values of the variables found in step 4 into any of the three ______ that contain the third variable. **6.** Check the ordered triple in ______.	Solve the following system of equations. $\begin{cases} x + 2y - z = -6 & \text{(Equation 1)} \\ 2x - 3y + 4z = 26 & \text{(Equation 2)} \\ -x + 2y - 3z = -18 & \text{(Equation 3)} \end{cases}$ Eliminate x by adding Equation 1 and Equation 3. $x + 2y - z = -6$ $-x + 2y - 3z = -18$ $4y - 4z = -24$ Divide both sides by 4. $y - z = -6$ (Eq. 4) Eliminate x again using Equations 2 and 3. $2x - 3y + 4z = 26 \quad\quad 2x - 3y + 4z = 26$ $-x + 2y - 3z = -18 \rightarrow -2x + 4y - 6z = -36$ Multiply by 2. $y - 2z = -10$ (Eq. 5)

Definitions/Rules/Procedures	Key Example(s)
	Use Equations 4 and 5 to solve for z. $y - z = -6$, $y - 2z = -10$ Multiply by -1. → $y - z = -6$, $-y + 2z = 10$; $z = 4$ Substitute 4 for z in Equation 4 and solve for y. $y - 4 = -6$ $y = -2$ Substitute -2 for y and 4 for z in Equation 1 and solve for x. $x + 2(-2) - 4 = -6$ $x - 4 - 4 = -6$ $x = 2$ The solution is $(2, -2, 4)$.

Exercises 19–22 Equations and Inequalities

[4.2] *For Exercises 19–22, solve the system using the elimination method.*

19. $\begin{cases} x + y + z = -2 \\ 2x + 3y + 4z = -10 \\ 3x - 2y - 3z = 12 \end{cases}$

20. $\begin{cases} x + y + z = 0 \\ 2x - 4y + 3z = -12 \\ 3x - 3y + 4z = 2 \end{cases}$

21. $\begin{cases} 3x + 4y = -3 \\ -2y + 3z = 12 \\ 4x - 3z = 6 \end{cases}$

22. $\begin{cases} 2x + 2y - 3z = 5 \\ x = -3y + 4z \\ z = -x - 2y + 5 \end{cases}$

4.3 Solving Applications Using Systems of Equations

Definitions/Rules/Procedures	Key Example(s)
To solve a problem using a system of equations: **1.** Select a variable to represent each ________. **2.** Write a system of equations. **3.** Solve the system.	A vendor sells small and large boxes of popcorn for \$3 and \$5, respectively. If the vendor sold 177 boxes of popcorn for a total of \$695, how many of each size were sold? Let $x =$ the number of small boxes sold. Let $y =$ the number of large boxes sold. The total number of boxes sold, 177, translates to $x + y = 177$. The total revenue is \$695, which translates to $3x + 5y = 695$. System $\begin{cases} x + y = 177 \\ 3x + 5y = 695 \end{cases}$ To solve the system, we use the elimination method because there are no isolated variables. $x + y = 177$ Multiply by -5. → $-5x - 5y = -885$ $3x + 5y = 695$ → $3x + 5y = 695$ We eliminated y. $-2x + 0 = -190$ Solve for x. $x = 95$ Substitute 95 for x in an equation. $95 + y = 177$ We chose $x + y = 177$. $y = 82$ Isolate y. The vendor sold 95 small and 82 large boxes.

Exercises 23–30 Equations and Inequalities

[4.3] *For Exercises 23–30, solve.*

23. A vendor at a football game sells hotdogs and hamburgers. Hotdogs sell for $4 and hamburgers for $6. If the vendor sold a total of 226 hotdogs and hamburgers for a combined total of $1048, how many hotdogs and how many hamburgers did she sell?

24. Two angles are supplementary. If the measure of the greater of the two angles is 10° less than four times the lesser angle, find the measures of both angles.

25. Linea passes mile marker 2 at 8:15 A.M. running on a trail at 4 miles per hour Tanya, who is cycling at 12 miles per hour, passes the same mile marker at 8:45 A.M. At what time will Tanya catch up to Linea?

26. A kayaker takes 45 minutes to travel 1 mile upstream, whereas it takes him only 15 minutes to travel one mile downstream. Find the kayaker's speed in still water and how much the current changes his speed.

27. Juan has three times as much money invested in stocks as in his savings account. The stocks return 8% on his investment and the savings account returns 2%. If the total return was $624, how much was invested in stocks and in the savings account?

28. How many milliliters of a 20% sulfuric acid solution and a 50% sulfuric acid solution must be mixed together to make 300 milliliters of a 30% sulfuric acid solution.

29. John bought vitamin supplements that cost $8 each, film that cost $4 per roll, and bags of candy that cost $3 per bag. He bought a total of eight items that cost $32. The number of rolls of film was one less than the number of bags of candy. Find the number of each type of item that he bought.

30. At a swim meet, 5 points are awarded for each first-place finish, 3 points for each second-place finish, and 1 point for each third-place finish. Shawnee Mission South High School scored a total of 38 points. The number of first-place finishes was one more than the number of second-place finishes. The number of third-place finishes was three times the number of second-place finishes. Find the number of first-, second-, and third-place finishes for the school.

4.4 Solving Systems of Linear Equations Using Matrices

Definitions/Rules/Procedures

A **matrix** is a(n) ______________ of numbers.

An **augmented matrix** is made up of the ______________ and the ______________ terms of a system. The ______________ terms are separated from the ______________ by a dashed vertical line.

An augmented matrix is in **row echelon form** if its coefficient portion has 1's on the diagonal from ______________ to ______________ and 0s below the 1's.

Row operations:

The solution of a system is not affected by the following row operations in its augmented matrix.

1. Any two rows may be ______________.
2. The elements of any row may be ______________ (or ______________) by any nonzero real number.
3. Any row may be replaced by a row resulting from ______________ the elements of that row (or multiples of that row) to a(n) ______________ of the elements of any other row.

Row operations can be used to solve a system of linear equations by transforming the augmented matrix of the system into an equivalent matrix that is in ______________ form.

Note: Systems of three equations with three unknowns are solved in a similar manner.

Key Example(s)

An example of a matrix is $\begin{bmatrix} -4 & 2 \\ 3 & -1 \end{bmatrix}$.

The augmented matrix for the system

$$\begin{cases} 2x - 4y = 5 \\ x + 6y = -4 \end{cases} \text{ is } \left[\begin{array}{cc:c} 2 & -4 & 5 \\ 1 & 6 & -4 \end{array}\right].$$

An augmented matrix in row echelon form is

$$\left[\begin{array}{cc:c} 1 & 4 & -5 \\ 0 & 1 & 6 \end{array}\right].$$

Solve $\begin{cases} 2x + 3y = -5 \\ x + 2y = -4 \end{cases}$ using the row echelon method.

The augmented matrix is $\left[\begin{array}{cc:c} 2 & 3 & -5 \\ 1 & 2 & -4 \end{array}\right]$.

$$\left[\begin{array}{cc:c} 1 & 2 & -4 \\ 2 & 3 & -5 \end{array}\right]$$ We need row 1, column 1 to be a 1; so interchange the two rows.

$$-2R_1 + R_2 \longrightarrow \left[\begin{array}{cc:c} 1 & 2 & -4 \\ 0 & -1 & 3 \end{array}\right]$$ We need row 2, column 1 to be 0; so multiply row 1 by -2 and add it to row 2.

$$-R_2 \longrightarrow \left[\begin{array}{cc:c} 1 & 2 & -4 \\ 0 & 1 & -3 \end{array}\right]$$ Multiply row 2 by -1 to get row echelon form.

Row 2 means $y = -3$, and row 1 means $x + 2y = -4$.

$x + 2(-3) = -4$ Substitute -3 for y and solve for x.

$x = 2$

The solution is $(2, -3)$.

Exercises 31 and 32 Equations and Inequalities

[4.5] *For Exercises 31 and 32, solve the system using the row echelon method.*

31. $\begin{cases} x - 4y = 8 \\ x + 2y = 2 \end{cases}$

32. $\begin{cases} x + y + z = 2 \\ 2x + y + 2z = 1 \\ 3x + 2y + z = 1 \end{cases}$

4.5 Solving Systems of Linear Inequalities

Definitions/Rules/Procedures

To solve a system of linear inequalities in two variables, graph all of the inequalities on the same grid. The solution set for the system contains all ordered pairs in the region where the inequalities' solutions sets ______________ along with ordered pairs on the portion of any ______________ line that touches the region of overlap.

Key Example(s)

Solve $\begin{cases} x + y > -2 \\ y \le 3x - 1 \end{cases}$.

Solution:

$x + y > -2$

$y \le 3x - 1$

Solution region for the system (including all ordered pairs on the portion of the solid blue line touching this region).

Exercises 33–34 Equations and Inequalities

[4.5] *For Exercises 33 and 34, graph the solution set for the system of inequalities.*

33. $\begin{cases} 2x + y \geq -4 \\ -3x + y \geq -1 \end{cases}$

34. $\begin{cases} -3x + 4y < 12 \\ 2x - y \leq -3 \end{cases}$

[4.5] *Solve using a system of linear inequalities.*

35. A theater has 500 seats. For a particular show, they must make at least \$16,000 to break even. Tickets cost \$40 if purchased in advance and \$50 if purchased on the day of the show.

a. Write a system of inequalities that describes the number of tickets sold in advance, x, versus the number of tickets sold on the day of the show, y.

b. Solve the system by graphing.

c. Give a combination of ticket sales that would exceed \$16,000.

Learning Strategy

The best way to prepare for an exam is study, study, study!

—Ruben C.

Chapter 4 Practice Test

For Extra Help

Step-by-step test solutions are found on the Chapter Test Prep Videos available in MyMathLab® *or on* YouTube.

For Exercises 1 and 2, determine whether the given ordered pair is a solution to the system of equations.

1. $(-1, 3)$; $\begin{cases} 3x + 2y = 3 \\ 4x - y = -7 \end{cases}$

2. $(2, 2, 3)$; $\begin{cases} 2x - 3y + z = 1 \\ -2x + y - 3z = 11 \\ 3x + y + 3z = 14 \end{cases}$

3. Solve by graphing: $\begin{cases} x + 3y = 1 \\ -2x + y = 5 \end{cases}$

For Exercises 4–9, solve the system of equations using substitution or elimination. Note that some systems may be inconsistent or consistent with dependent equations.

4. $\begin{cases} 2x + y = 15 \\ y = 7 - x \end{cases}$

5. $\begin{cases} x - 2y = 1 \\ 3x - 5y = 4 \end{cases}$

6. $\begin{cases} 3x - 2y = -8 \\ 2x + 3y = -14 \end{cases}$

7. $\begin{cases} 4x + 6y = 2 \\ 6x + 9y = 3 \end{cases}$

8. $\begin{cases} x + 2y + z = 2 \\ x + 4y - z = 12 \\ 3x - 3y - 2z = -11 \end{cases}$

9. $\begin{cases} x + 2y - 3z = -9 \\ 3x - y + 2z = -8 \\ 4x - 3y + 3z = -13 \end{cases}$

For Exercises 10 and 11, solve the equations using the row echelon method.

10. $\begin{cases} x + 2y = -6 \\ 3x + 4y = -10 \end{cases}$

11. $\begin{cases} x + y + z = 6 \\ 3x - 2y + 3z = -7 \\ 4x - 2y + z = -12 \end{cases}$

For Exercises 12 and 13, solve the system of equations using the method of your choice.

12. $\begin{cases} 3x + 4y = 14 \\ 2x - 3y = -19 \end{cases}$

13. $\begin{cases} x + y + z = -1 \\ 3x - 2y + 4z = 0 \\ 2x + 5y - z = -11 \end{cases}$

14. Graph the solution set for the system of inequalities.
$\begin{cases} 2x - 3y < 1 \\ x + 2y \le -2 \end{cases}$

For Exercises 15–20, solve.

15. Excedrin surveyed workers in various professions who get headaches at least once a year on the job. Nine more accountants than waiters/waitresses in the survey got a headache. If the combined number of accountants and waiters/waitresses that got a headache was 163, how many people in each of those professions got a headache? (*Source:* the *State* newspaper.)

16. When asked what the Internet most resembled, three times as many people said a library as opposed to a highway. If 240 people were polled, how many people considered the Internet to be a library? (*Source: bLINK* magazine.)

17. A boat traveling with the current took 3 hours to go 30 miles. The same boat went 12 miles in 3 hours against the current. What is the rate of the boat in still water?

18. Janice invested \$12,000 in two funds. One of the funds returned 6% interest, and the other returned 8% interest after one year. If the total interest for the year was \$880, how much did she invest in each fund?

19. Tickets for the senior play at Apopka High School cost \$3 for children, \$5 for students, and \$8 for adults. There were 800 tickets sold for a total of \$4750. The number of adult tickets sold was 50 more than two times the number of children tickets. Find the number of each type of ticket.

20. A landscaper wants to plan a bordered rectangular garden area in a yard. Because she currently has 200 feet of border materials, the perimeter needs to be, at most, 200 feet. She thinks the garden will look best if the length is at least 10 feet more than the width.

    a. Write a system of inequalities to describe the situation.

    b. Solve the system by graphing.

    c. Give a combination of length and width that satisfies the requirements for the garden.

Chapters 1–4 Cumulative Review Exercises

For Exercises 1–3, answer true or false.

[3.2] **1.** The graph of $y = -3x + 5$ is a line with slope of $-3x$.

[1.3] **2.** $-(-4)^3 = -64$

[4.1] **3.** The ordered pair (1, 1) is a solution for the system of equations
$$\begin{cases} x = y \\ 9y - 13 = -4x. \end{cases}$$

For Exercises 4–6, fill in the blank.

[1.4] **4.** Complete using the distributive property: $4(2x + 3) =$ __________.

[2.6] **5.** If $|x| > 5$, then ______ or ______.

[3.3] **6.** The graph of $y = -2x + 5$ is __________ to the graph of $y = \frac{1}{2}x - 3$.

Exercises 7–9 Expressions

For Exercises 7–9, simplify.

[1.3] **7.** $-(6 - 2^2)^2 + 15$

[1.3] **8.** $-|2 - 8| + 3(2)$

[1.4] **9.** $\frac{3}{8}x + 4y + 12 - \frac{1}{2}x - \frac{7}{10}y - 7$

Exercises 10–30 Equations and Inequalities

For Exercises 10–13, solve.

[2.1] **10.** $3.6 - 4(0.7p - 15) = 2.2p - 1.4$

[2.1] **11.** $2(x - 3) + 4x = 6(x - 2)$

[2.3] **12.** $4z - 32 - 5z < 6z + 13 + 2z$

[2.6] **13.** $|4x + 12| + 2 < 10$

[2.1] **14.** $A = P + Prt$ for t

[1.4] **15.** For what value(s) of x is $\frac{x}{x + 2}$ undefined?

[2.2] **16.** A house is situated on a trapezoidal lot as shown below. If the area around the house is to be sodded, find the area to be sodded.

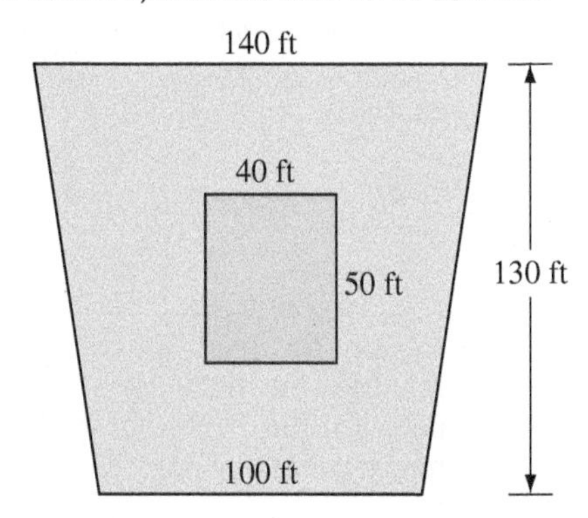

For Exercises 17–20, graph.

[3.1] **17.** $y = 3x + 1$

[3.1] **18.** $4x + 5y = 10$

[3.1] **19.** $x = -4$

[3.4] **20.** $2x - 7y < 14$

[3.3] *For Exercises 21–23, write the equations of the lines. Write the equation in slope–intercept form.*

21. Passing through $(-2, -1)$, $m = -\frac{1}{5}$

22. Passing through $(4, -7)$ and $(-2, -1)$

23. Passing through $(-4, 2)$ and parallel to the graph of $3x + 4y = 4$. Write the equation in standard form.

For Exercises 24–26, solve the system.

[4.1] **24.** $\begin{cases} 4x - 3y = -2 \\ 6x - 7y = 7 \end{cases}$

[4.2] **25.** $\begin{cases} x + y + z = 5 \\ 2x + y - 2z = -5 \\ x - 2y + z = 8 \end{cases}$

[4.4] **26.** $\begin{cases} 4x - 3y < 12 \\ 3x + y > 6 \end{cases}$

For Exercises 27–30, solve.

[2.2] **27.** A concert hall is rectangular in shape and measures 50 feet longer than it is wide. If the perimeter is 420 feet, what are its length and width?

[4.3] **28.** In the construction of a roof frame, a support beam is connected to a horizontal joist, forming two angles. If the greater angle is 15° less than twice the smaller angle, what are the measures of the two angles?

[4.3] **29.** Roxanne received an inheritance of $50,000. She invested part of it at 3% and the remainder at 2.5%. How much did she invest at each rate if the total interest from both investments was $1425 per year?

[4.3] **30.** A paint contractor paid $588 for 24 gallons of paint to paint the inside and outside of a house. If the paint for the inside costs $22 per gallon and the paint for the outside costs $28 per gallon, how many gallons of each did he buy?

CHAPTER

5

Exponents, Polynomials, and Polynomial Functions

Chapter Overview

In this chapter, we learn about a specific group of expressions, called *polynomials*, that we use to build more complicated types of equations and functions. Because polynomials involve exponents we also focus on the rules of exponents. More specifically, the following topics will be explored:

- ▶ Rules of exponents and scientific notation.
- ▶ Add, subtract, multiply, and divide polynomial expressions.
- ▶ Synthetic division and the remainder theorem.

5.1 Exponents and Scientific Notation

Objectives

1. Use the product rule of exponents to simplify expressions.
2. Use the quotient rule of exponents to simplify expressions.
3. Use the power rule of exponents to simplify expressions.
4. Convert between scientific notation and standard form.
5. Simplify products, quotients, and powers of numbers in scientific notation.

Warm-up

[1.3] *For Exercises 1 and 2, write each exponential form as a product, but do not calculate.*

1. $(-2)^4$

2. $\left(\frac{3}{4}\right)^3$

[4.3] **3.** Graph $\begin{cases} 4x - 3y < 12 \\ 3x + y > 6 \end{cases}$

Recall from Section 1.3 that natural-number exponents indicate repeated multiplication of a base number.

Example 1 Evaluate.

a. $(-8)^2$

Solution: $(-8)^2 = (-8)(-8) = 64$

b. $(-5)^3$

Solution: $(-5)^3 = (-5)(-5)(-5) = -125$

c. 3^5

Solution: $3^5 = 3 \cdot 3 \cdot 3 \cdot 3 \cdot 3 = 243$

d. $(-3)^5$

Solution: $(-3)^5 = (-3)(-3)(-3)(-3)(-3) = -243$

e. $(-2)^4$

Solution: $(-2)^4 = (-2)(-2)(-2)(-2) = 16$

f. -2^4

Solution: $-2^4 = -[2 \cdot 2 \cdot 2 \cdot 2] = -16$

Warning Expressions such as $(-2)^4$ and -2^4 in Examples 1(e) and 1(f) mean different things. In $(-2)^4$, the base is -2, whereas in -2^4, the base is 2. So the minus sign in -2^4 means to find the additive inverse of 2^4.

Example 1 suggests the following rules.

Rules Evaluating Exponential Forms

If n is a natural number and a is a real number, then $a^n = \underbrace{a \cdot a \cdot \cdots \cdot a}_{n \text{ factors of } a}$

If $a < 0$ and n is even, then a^n is positive.

If $a < 0$ and n is odd, then a^n is negative.

Objective 1 Use the product rule of exponents to simplify expressions.

So far, we have developed rules only for exponents that are natural numbers (positive integers). However, exponents can be positive, zero, or negative. To show what an exponent of 0 means, we need another rule of exponents called the *product rule.*

Consider $a^2 \cdot a^3$. To find the product, we can expand to individual factors.

$$a^2 \cdot a^3 = a \cdot a \cdot a \cdot a \cdot a = a^5$$

Note The product, a^5, can also be found by adding the exponents:

$$a^2 \cdot a^3 = a^{2+3} = a^5$$

Our example suggests the following product rule of exponents.

Rule Product Rule of Exponents

If a is a real number and m and n are natural numbers, then $a^m \cdot a^n = a^{m+n}$.

Answers to Warm-up

1. $(-2)(-2)(-2)(-2)$

2. $\frac{3}{4} \cdot \frac{3}{4} \cdot \frac{3}{4}$

3.

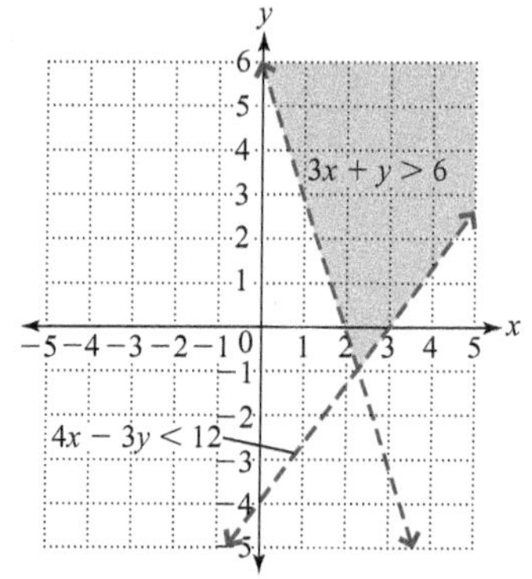

Example 2 Simplify using the product rule of exponents.

a. $3^2 \cdot 3^4$

Solution: $3^2 \cdot 3^4 = 3^{2+4} = 3^6$

b. $n^5 \cdot n^7$

Solution: $n^5 \cdot n^7 = n^{5+7} = n^{12}$

c. $7x \cdot 2x^3$

Solution: $7x \cdot 2x^3 = 7 \cdot 2 \cdot x \cdot x^3$ Use the commutative property to group like bases.

$= 14x^{1+3}$ Multiply the coefficients and use the product rule of exponents.

$= 14x^4$ Simplify the exponent.

d. $(-9m^2n)(5m^3n^7p)$

Solution: $(-9m^2n)(5m^3n^7p)$

$= -9 \cdot 5 \cdot m^2 \cdot m^3 \cdot n \cdot n^7 \cdot p$ Use the commutative property to group like bases.

$= -45m^{2+3}n^{1+7}p$ Multiply the coefficients and use the product rule of exponents.

$= -45m^5n^8p$ Simplify the exponents.

Learning Strategy

If we understand why a formula is invented and why it is meaningful as well as what (physical or intuitive) message it's meant to contain, then we can remember and apply it much more easily and intuitively than if we only know the formula as a string of letters and numbers.

—Zheng Alick Z.

Your Turn 2 Simplify using the product rule of exponents.

a. $5^3 \cdot 5^6$ **b.** $x^2 \cdot x^7$ **c.** $4y^2 \cdot 8y^5$ **d.** $(-6ab^3)(7abc^2)$

We can now use the product rule to show what 0 as an exponent means. Consider $a^0 \cdot a^n$, where $a \neq 0$. Using the product rule of exponents we can simplify.

$$a^0 \cdot a^n = a^{0+n} = a^n$$

Because the product of a^0 and a^n is a^n, we can say that a^0 must be equal to 1.

Rule Zero as an Exponent

If a is a real number and $a \neq 0$, then $a^0 = 1$.

Example 3 Simplify. Assume that variables do not equal 0.

a. $(-32)^0$

Solution: $(-32)^0 = 1$

b. $9n^0$

Solution: $9n^0 = 9(1)$ Replace n^0 with 1.

$= 9$ Multiply.

Note The base of the exponent 0 is n, not $9n$.

Your Turn 3 Simplify. Assume that variables do not equal 0.

a. $(-28)^0$ **b.** y^0 **c.** $5x^0$ **d.** $-4m^0$

Objective 2 Use the quotient rule of exponents to simplify expressions.

Now let's consider dividing exponential forms with the same base, as in $\frac{a^5}{a^3}$, where $a \neq 0$. To determine the rule, we can expand the exponential expressions and divide out the three of the common factors of a.

Answers to Your Turn 2
a. 5^9 **b.** x^9 **c.** $32y^7$ **d.** $-42a^2b^4c^2$

Answers to Your Turn 3
a. 1 **b.** 1 **c.** 5 **d.** −4

$$\frac{a^5}{a^3} = \frac{a \cdot a \cdot a \cdot a \cdot a}{a \cdot a \cdot a} = a \cdot a = a^2$$

Note The quotient, a^2, can also be found by subtracting the exponents.

$$\frac{a^5}{a^3} = a^{5-3} = a^2$$

Our example suggests the following quotient rule of exponents.

Connection We have seen that $a^2 \cdot a^3 = a^{2+3} = a^5$. Because multiplication and division are inverse operations, it makes sense that when dividing the same base, we subtract the exponents.

Rule Quotient Rule for Exponents

If m and n are integers and a is a real number where $a \neq 0$, then $\frac{a^m}{a^n} = a^{m-n}$.

Note We cannot let a equal 0 because if a were replaced with 0, we would have

$$\frac{0^m}{0^n} = \frac{0}{0}, \text{ which is indeterminate.}$$

Note that the divisor's exponent is subtracted from the dividend's exponent. This order is important because subtraction is not a commutative operation.

Connection In Objective 1, we used the product rule of exponents to see that $y^0 = 1$. We can expand the exponential forms and divide out all common factors to affirm this fact.

$$\frac{y^5}{y^5} = \frac{y \cdot y \cdot y \cdot y \cdot y}{y \cdot y \cdot y \cdot y \cdot y} = 1$$

When we divide exponential forms that have the same base and the same exponent, the result is 1.

Example 4 Divide using the quotient rule for exponents. Assume that variables in denominators are not equal to 0.

a. $\frac{x^{11}}{x^5}$

Solution: $\frac{x^{11}}{x^5} = x^{11-5} = x^6$

b. $\frac{y^5}{y^5}$

Solution: $\frac{y^5}{y^5} = y^{5-5} = y^0 = 1$

c. $\frac{36x^8}{4x^2}$

Solution: $\frac{36x^8}{4x^2} = \frac{36}{4} \cdot \frac{x^8}{x^2}$ Separate the coefficients and variables.

$= 9x^{8-2}$ Divide the coefficients and use the quotient rule for the variables.

$= 9x^6$ Simplify.

Note After this example, we no longer show this step.

d. $\frac{18t^2u^8v}{30t^2u}$

Solution: $\frac{18t^2u^8v}{30t^2u} = \frac{18}{30}t^{2-2}u^{8-1}v$ Divide the coefficients and use the quotient rule for the variables.

$= \frac{3}{5}t^0u^7v$ Simplify.

$= \frac{3}{5}u^7v$ $t^0 = 1$

VISUAL

Learning Strategy

If you are a visual learner, look at the fraction line as a subtraction sign between the numerator's exponent and the denominator's exponent.

Your Turn 4 Divide using the quotient rule for exponents. Assume that variables in the denominator are not equal to 0.

a. $\frac{y^{13}}{y^8}$ **b.** $\frac{15t^4}{3t^4}$ **c.** $\frac{-12m^5n^3p}{28mn^3}$

Answers to Your Turn 4

a. y^5 **b.** 5 **c.** $-\frac{3}{7}m^4p$

We can use the quotient rule to deduce what a negative integer exponent means. Suppose we have a division of exponential expressions in which the divisor's exponent is greater than the dividend's exponent, as in $\frac{a^2}{a^5}$. We can use two approaches to simplify this expression.

Using the quotient rule: $\frac{a^2}{a^5} = a^{2-5} = a^{-3}$

Expanding the exponential forms: $\frac{a^2}{a^5} = \frac{a \cdot a}{a \cdot a \cdot a \cdot a \cdot a} = \frac{1}{a \cdot a \cdot a} = \frac{1}{a^3}$

Both results are correct, indicating that $a^{-3} = \frac{1}{a^3}$, which suggests the following rule for negative exponents.

Rule Negative Exponent

If a is a real number, where $a \neq 0$ and n is a natural number, then $a^{-n} = \frac{1}{a^n}$.

Example 5 Evaluate the exponential form.

a. 10^{-3}

Solution: $10^{-3} = \frac{1}{10^3} = \frac{1}{1000}$

Warning Notice that $10^{-3} \neq -1000$.

b. $(-2)^{-4}$

Solution: $(-2)^{-4} = \frac{1}{(-2)^4} = \frac{1}{(-2)(-2)(-2)(-2)} = \frac{1}{16}$

Your Turn 5 Evaluate the exponential form.

a. 4^{-4}

b. $(-3)^{-2}$

What if a negative exponent is in the denominator of a fraction? We use $a^{-n} = \frac{1}{a^n}$ to rewrite the fraction without a negative exponent.

$$\frac{1}{3^{-2}} = \frac{1}{\frac{1}{3^2}} = 1 \cdot \frac{3^2}{1} = 3^2 = 9$$

If we remove the intermediate steps, we see that $\frac{1}{3^{-2}} = 3^2$, which suggests the following rule.

Rule Negative Exponent in the Denominator

If a is a real number, where $a \neq 0$ and n is a natural number, then $\frac{1}{a^{-n}} = a^n$.

Example 6 Evaluate or rewrite with positive exponents only.

a. $\frac{1}{2^{-4}}$

Solution: $\frac{1}{2^{-4}} = 2^4 = 16$

b. $\frac{3}{x^{-6}}$

Solution: $\frac{3}{x^{-6}} = \frac{3}{1} \cdot \frac{1}{x^{-6}} = 3x^6$

Answers to Your Turn 5

a. $\frac{1}{256}$ b. $\frac{1}{9}$

Your Turn 6 Evaluate or rewrite with positive exponents only.

a. $\dfrac{1}{5^{-2}}$

b. $\dfrac{7}{y^{-3}}$

After using the rules of exponents to simplify, if the exponent in the result is negative, we rewrite the exponential form so that the exponent is positive.

Example 7 Simplify and write the result with a positive exponent.

a. $t^3 \cdot t^{-5}$

Solution: $t^3 \cdot t^{-5} = t^{3+(-5)}$ Use the product rule of exponents.

$= t^{-2}$ Simplify the exponent.

$= \dfrac{1}{t^2}$ Write the expression with a positive exponent.

b. $\dfrac{12x^3}{9x^7}$

Connection In Chapter 7, we will name expressions such as $\dfrac{12x^3}{9x^7}$ *rational expressions* and use this same technique to simplify them.

Solution: $\dfrac{12x^3}{9x^7} = \dfrac{12}{9}x^{3-7}$ Use the quotient rule for exponents.

$= \dfrac{4}{3}x^{-4}$ Simplify.

$= \dfrac{4}{3} \cdot \dfrac{1}{x^4}$ Write with a positive exponent.

$= \dfrac{4}{3x^4}$ Simplify.

Your Turn 7 Simplify and write the result with a positive exponent.

a. $x^5 \cdot x^{-6}$

b. $\dfrac{5y^2}{20y^8}$

c. $\dfrac{k^{-4}}{k^3}$

Objective 3 Use the power rule of exponents to simplify expressions.

We now consider raising a power to a power, as in $(a^2)^3$. To simplify this expression, we need to determine how many factors of a the expression indicates. The exponent 3 means to multiply three factors of a^2. Because each a^2 means two factors of a, there are a total of six factors of a.

Note Using the product rule, we see that we add three 2's, which means that we could multiply 2 by 3 to get 6.

$$(a^2)^3 = a^2 \cdot a^2 \cdot a^2 = a^{2+2+2} = a^6$$

Our example suggests the following power rule of exponents.

Rule A Power Raised to a Power

If a is a real number and m and n are integers, then $(a^m)^n = a^{mn}$.

Example 8 Simplify using the power rule and write the result with a positive exponent.

a. $(t^3)^4$

Solution: $(t^3)^4 = t^{3 \cdot 4} = t^{12}$

b. $(x^{-5})^2$

Solution: $(x^{-5})^2 = x^{-5 \cdot 2} = x^{-10} = \dfrac{1}{x^{10}}$

Answers to Your Turn 6

a. 25 b. $7y^3$

Answers to Your Turn 7

a. $\dfrac{1}{x}$ b. $\dfrac{1}{4y^6}$ c. $\dfrac{1}{k^7}$

Your Turn 8 Simplify using the power rule and write the result with a positive exponent.

a. $(x^5)^2$ **b.** $(a^4)^{-2}$

Raising a Product to a Power

Now consider a product raised to a power, as in $(ab)^2$. The exponent 2 indicates two factors of ab.

$$(ab)^2 = ab \cdot ab$$
$$= a \cdot a \cdot b \cdot b \quad \text{Use the commutative property to rearrange like bases.}$$
$$= a^2b^2 \quad \text{Use the product rule to simplify.}$$

Notice that each factor in the parentheses is raised to the power outside the parentheses, which suggests the following rule.

Rule Raising a Product to a Power
If a and b are real numbers and n is an integer, then $(ab)^n = a^nb^n$.

Example 9 Simplify.

a. $(3a^4)^2$

Solution: $(3a^4)^2 = 3^2(a^4)^2$ Use the rule for raising a product to a power.
$= 9a^{4 \cdot 2}$ Use the power rule.
$= 9a^8$ Simplify the exponent.

b. $(-4xy^2z^6)^3$

Solution: $(-4xy^2z^6)^3 = (-4)^3(x)^3(y^2)^3(z^6)^3$ Use the rule for raising a product to a power.
$= -64x^3y^6z^{18}$ Use the power rule.

Your Turn 9 Simplify.

a. $(4y^5)^3$ **b.** $(-2t^5u^2v)^4$

To raise a quotient to a power, we develop a rule similar to raising a product to a power.

$$\left(\frac{a}{b}\right)^4 = \frac{a}{b} \cdot \frac{a}{b} \cdot \frac{a}{b} \cdot \frac{a}{b} = \frac{a^4}{b^4}$$

Notice that both the numerator and denominator are raised to the power outside the parentheses, which suggests the following rule.

Rule Raising a Quotient to a Power

If a and b are real numbers, where $b \neq 0$ and n is an integer, then $\left(\frac{a}{b}\right)^n = \frac{a^n}{b^n}$.

Answers to Your Turn 8
a. x^{10} **b.** $\frac{1}{a^8}$

Answers to Your Turn 9
a. $64y^{15}$ **b.** $16t^{20}u^8v^4$

Example 10 Simplify.

a. $\left(\frac{m}{n}\right)^6$

Solution: $\left(\frac{m}{n}\right)^6 = \frac{m^6}{n^6}$ Use the rule for raising a quotient to a power.

b. $\left(\frac{3}{x^2}\right)^3$

Solution: $\left(\frac{3}{x^2}\right)^3 = \frac{3^3}{(x^2)^3}$ Use the rule for raising a quotient to a power.

$= \frac{27}{x^6}$ Use the power rule.

Your Turn 10 Simplify.

a. $\left(\frac{2}{a}\right)^4$

b. $\left(\frac{r^4}{5}\right)^3$

What if a fraction is raised to a negative exponent? Again, we use $a^{-n} = \frac{1}{a^n}$ to develop a rule.

$$\left(\frac{3}{4}\right)^{-2} = \frac{1}{\left(\frac{3}{4}\right)^2} = \frac{1}{\frac{9}{16}} = \frac{16}{9}$$

If we remove the intermediate steps in our solution, we see that $\left(\frac{3}{4}\right)^{-2} = \frac{16}{9}$. Because $\frac{16}{9} = \left(\frac{4}{3}\right)^2$, we can conclude that $\left(\frac{3}{4}\right)^{-2} = \left(\frac{4}{3}\right)^2$, which suggests the following rule.

Rule Raising a Quotient to a Negative Power

If a and b are real numbers, where $a \neq 0$ and $b \neq 0$ and n is a natural number, then

$$\left(\frac{a}{b}\right)^{-n} = \left(\frac{b}{a}\right)^n.$$

Example 11 Simplify.

a. $\left(\frac{a}{b}\right)^{-3}$

Solution: $\left(\frac{a}{b}\right)^{-3} = \left(\frac{b}{a}\right)^3$ Use the rule for raising a quotient to a negative power.

$= \frac{b^3}{a^3}$ Use the rule for raising a quotient to a power.

b. $\left(\frac{4}{x^3}\right)^{-2}$

Solution: $\left(\frac{4}{x^3}\right)^{-2} = \left(\frac{x^3}{4}\right)^2$ Use the rule for raising a quotient to a negative power.

$= \frac{(x^3)^2}{4^2}$ Use the rule for raising a quotient to a power.

$= \frac{x^6}{16}$ Use the power rule.

Answers to Your Turn 10

a. $\frac{16}{a^4}$ b. $\frac{r^{12}}{125}$

Your Turn 11 Simplify.

a. $\left(\frac{x}{z}\right)^{-4}$

b. $\left(\frac{a^5}{2}\right)^{-3}$

Summary of Rules of Exponents If a and b are real numbers and m and n are integers, then

$a^m \cdot a^n = a^{m+n}$	Product Rule of Exponents
$a^0 = 1, a \neq 0$	Zero as an Exponent
$\frac{a^m}{a^n} = a^{m-n}, a \neq 0$	Quotient Rule for Exponents
$a^{-n} = \frac{1}{a^n}$ and $\frac{1}{a^{-n}} = a^n, a \neq 0$	Rules for Negative Exponents
$(a^m)^n = a^{mn}$	A Power Raised to a Power
$(ab)^n = a^n b^n$	Raising a Product to a Power
$\left(\frac{a}{b}\right)^n = \frac{a^n}{b^n}, b \neq 0$	Raising a Quotient to a Power
$\left(\frac{a}{b}\right)^{-n} = \left(\frac{b}{a}\right)^n, a \neq 0, b \neq 0$	Raising a Quotient to a Negative Power

Note In more advanced math courses, it is proven that these rules of exponents are true when the exponents are real numbers other than integers.

Often more than one rule is required to simplify an expression.

Example 12 Simplify. Write the result with positive exponents only.

a. $(4a^3b^4c)^3(-2a^4b^2c^3)^4$

Solution:

$(4a^3b^4c)^3(-2a^4b^2c^3)^4 = 4^3(a^3)^3(b^4)^3c^3 \cdot (-2)^4(a^4)^4(b^2)^4(c^3)^4$ Use the product raised to a power rule.

$= 64a^9b^{12}c^3 \cdot 16a^{16}b^8c^{12}$ Use the power to a power rule.

$= 1024a^{25}b^{20}c^{15}$ Use the product rule of exponents.

b. $\frac{(x^{-2})^4}{x^3 \cdot x^4}$

Solution: $\frac{(x^{-2})^4}{x^3 \cdot x^4} = \frac{x^{-8}}{x^7}$ Use the power raised to a power rule in the numerator and the product rule of exponents in the denominator.

$= x^{-15}$ Use the quotient rule for exponents.

$= \frac{1}{x^{15}}$ Use the rule for negative exponents.

c. $\frac{(4a^{-3}b^{-2})^{-2}(6a^3b^{-5})^3}{(2a^{-2}b^{-2})^{-4}}$

Solution:

$\frac{(4a^{-3}b^{-2})^{-2}(6a^3b^{-5})^3}{(2a^{-2}b^{-2})^{-4}} = \frac{4^{-2}a^6b^4 \cdot 6^3a^9b^{-15}}{2^{-4}a^8b^8}$ Use the product to a power and the power to a power rules.

$= \frac{2^4 \cdot 6^3a^{15}b^{-11}}{4^2a^8b^8}$ Use the rules for negative exponents and the product rule of exponents.

$= \frac{3456a^7b^{-19}}{16}$ Multiply 2^4 and 6^3 and use the quotient rule for exponents.

$= \frac{216a^7}{b^{19}}$ Divide the coefficients and use the rules for negative exponents.

Answers to Your Turn 11

a. $\frac{z^4}{x^4}$ b. $\frac{8}{a^{15}}$

Your Turn 12 Simplify. Write the result with positive exponents only.

a. $(4x^2y^3z^{-2})^2(-2x^{-3}y^2z^4)^4$ b. $\dfrac{(z^{-1})^5}{z^5 \cdot z^2}$ c. $\dfrac{(2x^{-3}y^2)^4(3xy^{-3})^{-2}}{(3x^{-4}y^{-2})^{-4}}$

Objective 4 Convert between scientific notation and standard form.

Sometimes we use very large or very small numbers, such as when we describe the vast distances in space or the tiny size of cells. For example, the distance from the Sun to the next nearest star, Proxima Centauri, is about 24,700,000,000,000 miles and a single streptococcus bacterium is about 0.00000075 meters in diameter. The large number of zero digits in these numbers makes them tedious to write. **Scientific notation** allows us to write such numbers more concisely.

Definition **Scientific notation:** A number expressed in the form $a \times 10^n$, where a is a decimal number with $1 \le |a| < 10$ and n is an integer.

The number 2.65×10^4 is in scientific notation because $1 \le |2.65| < 10$ and the exponent 4 is an integer. Although 26.5×10^3 names the same number, it is not in scientific notation because $|26.5| > 10$. Similarly, 0.265×10^5 is not in scientific notation because $|0.265| < 1$. We can multiply 2.65×10^4 to determine its value.

Note Multiplying by 10^4 causes the decimal point to move 4 places to the right. ▶

$$2.65 \times 10^4 = 2.65 \times 10{,}000 = 26{,}500$$

Now consider a number in scientific notation with a negative exponent, such as 3.24×10^{-6}. To write this number in standard form, we can rewrite the power of 10 with a positive exponent.

Note Multiplying by 10^{-6} causes the decimal point to move 6 places to the left. ▶

$$3.24 \times 10^{-6} = 3.24 \times \frac{1}{10^6} = \frac{3.24}{1{,}000{,}000} = 0.00000324$$

Our example suggests that the exponent of the power of 10 determines the number of places the decimal point moves.

Procedure **Changing Scientific Notation to Standard Form**

To change a number from scientific notation, $a \times 10^n$, where $1 \le |a| < 10$ and n is an integer, to standard form, if $n > 0$, move the decimal point to the right n places. If $n < 0$, move the decimal point to the left $|n|$ places.

Example 13 Write each number in standard form.

a. -4.35×10^5

Solution: $-4.35 \times 10^5 = -435{,}000$ Because $n > 0$, we move the decimal point right 5 places.

b. 9.4×10^{-8}

Solution: $9.4 \times 10^{-8} = 0.000000094$ Because $n < 0$, we move the decimal point left 8 places.

Your Turn 13 Write each number in standard form.

a. 5.403×10^6 b. 7×10^{-4}

Now let's convert numbers such as 65,000,000 and 0.000045 from standard form to scientific notation. Because scientific notation begins with a decimal number whose absolute value is greater than or equal to 1 but less than 10, our first step is to determine the position of the decimal point.

Answers to Your Turn 12

a. $\dfrac{256y^{14}z^{12}}{x^8}$ b. $\dfrac{1}{z^{12}}$ c. $\dfrac{144y^6}{x^{30}}$

Answers to Your Turn 13

a. 5,403,000 b. 0.0007

65,000,000 ↑

Decimal point goes here because $1 \le |6.5| < 10$.

0.000045 ↑

Decimal point goes here because $1 \le |4.5| < 10$.

Now we need to establish the power of 10.

$65,000,000 = 6.5 \times 10^7$

Because 65,000,000 is greater than 1, the 7 places between the old decimal point position and the new position are expressed as a positive power of 10.

$0.000045 = 4.5 \times 10^{-5}$

Because 0.000045 is less than 1, the 5 places between the old decimal point position and the new position are expressed as a negative power of 10.

We can summarize the process in the following procedure.

Procedure Changing Standard Form to Scientific Notation

To write a number in scientific notation:

1. Locate the new decimal point position, which will be to the right of the first nonzero digit in the number.
2. Determine the power of 10.
   a. If the number's absolute value is greater than 1, the power is the number of digits between the old decimal point position and the new position expressed as a positive power.
   b. If the number's absolute value is less than 1, the power is the number of digits between the old decimal point position and the new position expressed as a negative power.
3. Delete unnecessary 0's.
   a. If the number's absolute value is greater than 1, delete the zeros to the right of the last nonzero digit.
   b. If the number's absolute value is less than 1, delete the zeros to the left of the first nonzero digit.

Example 14 Write the number in scientific notation.

a. $-46,000,000$

Solution: $-46,000,000 = -4.6 \times 10^7$

Move the decimal point here because $1 \le |-4.6| < 10$.

There are 7 places between the new decimal point position and the original position. Because $|-46,000,000| > 1$, the exponent is positive. After placement of the decimal point the zeros to the right of 6 can be deleted.

b. 0.000038

Solution: $0.000038 = 3.8 \times 10^{-5}$

Move the decimal point here because $1 \le |3.8| < 10$.

There are 5 places between the new decimal point position and the original position. Because $|0.000038| < 1$, the exponent is negative. After placement of the decimal point, the zeros to the left of 3 can be deleted.

Your Turn 14 Write the number in scientific notation.

a. $-206,000,000$ **b.** 0.00000052

Answers to Your Turn 14

a. -2.06×10^8 **b.** 5.2×10^{-7}

Objective 5 Simplify products, quotients, and powers of numbers in scientific notation.

We can use the rules of exponents to simplify products, quotients, and powers of numbers in scientific notation.

Example 15 Simplify using scientific notation. Leave the answer in scientific notation.

a. $(480{,}000)(64{,}000{,}000)$

Solution:

$(480{,}000)(64{,}000{,}000) = (4.8 \times 10^5)(6.4 \times 10^7)$ Change to scientific notation.

$= 4.8 \times 6.4 \times 10^5 \times 10^7$ Use the commutative property.

$= 30.72 \times 10^{5+7}$ Multiply the decimal numbers and use the product rule of exponents.

$= 30.72 \times 10^{12}$ Simplify the exponent.

$= 3.072 \times 10^{13}$ Adjust the decimal point position and exponent so that you have the same number in scientific notation.

Note This number is not in scientific notation because $30.72 \geq 10$. Moving the decimal point 1 place to the left gives the proper position. We add 1 to the exponent to account for this additional place.

b. $\dfrac{0.0004368}{840{,}000}$

Solution: $\dfrac{0.0004368}{840{,}000} = \dfrac{4.368 \times 10^{-4}}{8.4 \times 10^5}$ Change to scientific notation.

$= \dfrac{4.368}{8.4} \times \dfrac{10^{-4}}{10^5}$ Separate the decimal numbers and powers.

$= 0.52 \times 10^{-4-5}$ Divide the decimal numbers and use the quotient rule for exponents.

$= 0.52 \times 10^{-9}$ Simplify the exponent.

$= 5.2 \times 10^{-10}$ Adjust the decimal point position and exponent so that you have the same number in scientific notation.

Note 0.52×10^{-9} is not in scientific notation because $0.52 < 1$. Moving the decimal point 1 place to the right gives the proper position. We subtract 1 from the exponent to account for the new position.

c. $(-200{,}000)^3$

Solution: $(-200{,}000)^3 = (-2 \times 10^5)^3$ Change to scientific notation.

$= (-2)^3 \times (10^5)^3$ Use the rule for raising a product to a power.

$= -8 \times 10^{15}$ Simplify $(-2)^3$ and use the power rule of exponents.

Your Turn 15 Simplify using scientific notation. Leave the answer in scientific notation.

a. $(7{,}500{,}000{,}000)(0.000031)$ b. $\dfrac{28{,}800}{64{,}000{,}000{,}000}$ c. $(-4000)^4$

Answers to Your Turn 15
a. 2.325×10^5
b. 4.5×10^{-7}
c. 2.56×10^{14}

5.1 Exercises

For Extra Help MyMathLab®

Note: Exercises marked with a ★ represent challenging exercises.

Objective 1

Prep Exercise 1 Explain how to simplify an expression such as $x^3 \cdot x^4$.

Prep Exercise 2 If a is a real number and $a \neq 0$, then $a^0 =$ ________.

For Exercises 1–12, simplify using the product rule of exponents. See Example 2.

1. $mn^2 \cdot m^3n^3$
2. $x^5y^2 \cdot xy^4$
3. $3^{10} \cdot 3^2$
4. $2^5 \cdot 2^4$
5. $(-3)^5(-3)^3$
6. $(-4)^4(-4)^6$
7. $(-4p^4q^3)(3p^2q^2)$
8. $(2m^3n^2)(-3mn^9)$
9. $(6r^3st^4)(-3r^4s^5t^8)$
10. $(-8a^5b^2c^7)(-2ab^6c^2)$
11. $(1.2u^3t^9)(3.1u^4t^2)$
12. $(3.1x^2y^4z)(-x^8y^2z^7)$

For Exercises 13–20, evaluate or simplify the exponential expression. See Examples 3, 5, and 6.

13. 15^0
14. 6^0
15. $4x^0$
16. $-3y^0$
17. $(3xy^2)^0$
18. $(-6m^3n^2)^0$
19. -5^2
20. -4^3

Find the Mistake *For Exercises 21 and 22, explain the mistake; then find the correct answer.*

21. $-2^4 = (-2)(-2)(-2)(-2) = 16$
22. $(-2)^3 = -6$

Objective 2

Prep Exercise 3 Explain how to simplify an expression such as $\frac{y^7}{y^5}$.

For Exercises 23–34, divide using the quotient rule for exponents. Write the result with positive exponents. Assume that variables in the denominator are not equal to 0. See Examples 4 and 7.

23. $\frac{h^5}{h^2}$
24. $\frac{m^7}{m^4}$
25. $\frac{a^2}{a^7}$
26. $\frac{u^3}{u^8}$
27. $\frac{6x^2y^6}{3xy^3}$
28. $\frac{-12j^5k^9}{3j^3k^6}$
29. $\frac{18r^3s^7t}{-12r^6s^4t}$
30. $\frac{21m^4n^9}{14m^7n^2}$
31. $\frac{15u^{-8}v^3w^4}{-21u^3v^3w^{-2}}$
32. $\frac{18q^{-7}r^9s^3}{81qr^{-7}s^3}$
33. $\frac{8a^{-2}b^3c^8}{2a^3b^{-4}c^5}$
34. $\frac{27x^7y^{-3}z^{-4}}{9x^4y^{-5}z}$

Prep Exercise 4 If a is a real number where $a \neq 0$ and n is a natural number, then $a^{-n} =$ ___________.

Prep Exercise 5 If a is a real number where $a \neq 0$ and n is a natural number, then $\frac{1}{a^{-n}} =$ ___________.

Prep Exercise 6 After simplifying 5^{-2}, is the result positive or negative? Explain.

For Exercises 35–44, evaluate or simplify the exponential expression. See Examples 5 and 6.

35. 2^{-3}
36. 4^{-2}
37. -3^{-4}
38. -5^{-2}
39. $\frac{1}{5^{-3}}$

40. $\frac{1}{8^{-2}}$
41. $\frac{4}{x^{-3}}$
42. $\frac{6}{b^{-5}}$
43. $\frac{1}{4a^{-6}}$
44. $\frac{1}{9c^{-3}}$

For Exercises 45 and 46, explain the mistake; then find the correct answer.

45. $4^{-1} = -4$
46. $\left(\frac{2}{3}\right)^{-2} = -\frac{4}{9}$

Objective 3

Prep Exercise 7 Explain how to simplify an expression such as $(x^3)^4$.

Prep Exercise 8 If a and b are real numbers and n is an integer, then $(ab)^n =$ ___________.

For Exercises 47–56, simplify using the power rule of exponents and the rule for raising a product to a power. See Examples 8 and 9.

47. $(x^3)^4$
48. $(t^4)^2$
49. $(2x^5)^4$
50. $(4n^6)^3$
51. $(-5x^3y)^2$

52. $(-2m^6n^2)^5$
53. $\left(\frac{3}{4}a^2b^4\right)^3$
54. $\left(-\frac{1}{2}ab^3\right)^4$
55. $(-0.3r^2t^4u)^3$
56. $(0.4ab^5c^6)^4$

Prep Exercise 9 If a and b are real numbers and n is an integer, then $\left(\frac{a}{b}\right)^n =$ ___________.

For Exercises 57–70, simplify using the rules for raising a quotient to a power and raising a fraction to a negative power. See Examples 10 and 11.

57. $\left(\frac{c}{d}\right)^4$
58. $\left(\frac{m}{n}\right)^8$
59. $\left(\frac{3}{x}\right)^3$
60. $\left(\frac{2}{c}\right)^4$
61. $\left(-\frac{4}{x^3}\right)^2$

62. $\left(-\frac{2}{x^5}\right)^3$
63. $\left(\frac{a}{b}\right)^{-6}$
64. $\left(\frac{r}{s}\right)^{-8}$
65. $\left(\frac{x^2}{y^3}\right)^{-4}$
66. $\left(\frac{m^4}{n^7}\right)^{-3}$

67. $\left(\frac{4}{x^{-2}}\right)^{-3}$

68. $\left(\frac{5}{y^{-3}}\right)^{-2}$

69. $\left(\frac{3x^2}{y^3}\right)^{-4}$

70. $\left(\frac{m^4}{2n^5}\right)^{-5}$

For Exercises 71–82, simplify. Write the result with positive exponents only. See Example 12.

71. $(4x^2y^3)^2(-2x^3y^4)^3$

72. $(-3w^2v^2)^3(2w^8v^3)^2$

73. $\frac{(3q^4p^2)^2}{(5q^2p^2)^3}$

74. $\frac{(3u^3v^6)^3}{(-4u^6v^2)^2}$

75. $(3h^3t^5)^{-2}(9h^2t^4)^2$

76. $(3p^3q^8)^2(4p^2q^6)^{-3}$

77. $\frac{(9u^3v^2)^{-1}}{(2u^4v^2)^4}$

78. $\frac{(6x^2y^3z)^{-2}}{(3x^6y^2)^3}$

79. $\frac{(2u^2v^3)^{-3}(4u^{-2}v^3)}{(3u^2v^{-3})^{-2}}$

80. $\frac{(-3m^2n^3)^{-2}(m^4n^8)^3}{(-2m^{-2}n^{-3})^{-3}}$

81. $\frac{(-4a^{-2}b^3c^2)^{-1}(2abc)^{-2}}{(3a^5b^2)^3(2a^{-1}b^{-3}c)^{-3}}$

82. $\frac{(9r^6s^2)^{-2}(3r^{-3}s^2)^3}{(6rst^{-2})^{-1}(3r^2s^3t)^{-2}}$

Objective 4

Prep Exercise 10 To write 4.56×10^7 in standard form, move the decimal point ________ places to the ________.

Prep Exercise 11 To write 9.2×10^{-5} in standard form, move the decimal point ________ places to the ________.

For problems 83–94, write the number in standard form. See Example 13.

83. The Andromeda Galaxy is about 2.9×10^6 light-years from Earth.

84. The Large Magellanic Cloud is a small galaxy about 1.79×10^5 light-years from Earth.

85. The Andromeda Galaxy contains at least 2×10^{11} stars.

86. Some scientists estimate that there may be as many as 4×10^{11} stars in our galaxy, the Milky Way.

Andromeda Galaxy

87. According to *Forbes* magazine, the richest person in the world is Carlos Slim Helu, whose net worth is listed as \$$7.3 \times 10^{10}$. (*Source:* "Forbes World's Richest People," 2013.)

88. The second richest person in the world, according to *Forbes* magazine, is Bill Gates whose net worth is listed as \$$6.6 \times 10^{10}$ (*Source:* "Forbes Richest People," 2013.)

89. Blonde hair has a smaller diameter than darker hair and can be as thin as 1.7×10^{-7} m in diameter.

90. The mass of a dust particle is 7.53×10^{-10} kg.

91. The mass of a neutron is about 1.675×10^{-24} g.

92. The mass of a proton is about 1.673×10^{-24} g.

93. The radioactive decay of plutonium-239 emits an alpha particle weighing 6.645×10^{-27} kg.

94. The mass of an electron is about 9.109×10^{-28} g.

Prep Exercise 12 To write 34,700,000 in scientific notation, between which two digits should the decimal be located? What is the exponent of the 10?

Prep Exercise 13 To write 0.00085 in scientific notation, between which two digits should the decimal be located? What is the exponent of the 10?

For Exercises 95–106, write the number in scientific notation. See Example 14.

95. The all-time highest-rated television program is the last episode of *M*A*S*H*, which aired February 28, 1983, and had 50,150,000 households viewing it. (*Source:* Nielson Media Research, January 1961–August 2008.)

96. The universe is estimated to be 16,500,000,000 years old.

97. In May 2013, the U.S. national debt reached (rounded to the nearest billion) \$16,780,000,000,000. (*Source:* Bureau of the Public Debt.)

98. The star Vega is approximately 155,200,000,000,000 miles from Earth.

99. The estimated world population in 2050 is 9,309,000,000.

100. A gram of hydrogen contains 6,000,000,000,000,000,000 protons.

101. Light with a wavelength of 0.00000055 meter is green in color.

102. The size of a plant cell is 0.00001276 meter wide.

103. The size of the HIV virus that causes AIDS is about 0.0000001 meter.

104. The time light takes to travel 1 meter is 0.000000003 second.

105. The fastest computer in the world, as of 2013, is the Titan supercomputer at Oak Ridge National Laboratory. It can perform a single calculation in about 0.00000000000000005 second.

106. The size of an atomic nucleus of the lead atom is 0.0000000000000071 meter.

Of Interest

It would take the fastest home computer 20 years to perform the same calculations the Titan supercomputer can perform in 1 hour.

For Exercises 107 and 108, write the numbers in order from smallest to largest.

★ **107.** 8.3×10^{6}, 1.2×10^{7}, 6×10^{5}, 7.4×10^{6}, 2.4×10^{8}

108. 6.1×10^{-3}, 7.2×10^{-2}, 9.3×10^{-4}, 3.1×10^{-2}, 4.5×10^{-6}

Objective 5

Prep Exercise 14 Why is 42.5×10^6 not in scientific notation?

For Exercises 109–128, simplify using scientific notation. Leave the answers in scientific notation. See Example 15.

109. $(2100)(30{,}000)$

110. $(600{,}000)(21{,}000)$

111. $(-32{,}000)(410{,}000)$

112. $(520{,}000)(-31{,}000)$

113. $(0.00081)(220{,}000)$

114. $(0.0032)(740{,}000{,}000)$

115. $\frac{8{,}400{,}000}{210}$

116. $\frac{84{,}000{,}000{,}000}{4200}$

117. $\frac{930{,}000{,}000{,}000{,}000}{-3{,}000{,}000}$

118. $\frac{-812{,}000{,}000}{400{,}000}$

119. $\frac{0.0057}{0.0000095}$

120. $\frac{0.0000936}{0.45}$

121. $30{,}000^3$

122. 600^2

123. $20{,}000{,}000^3$

124. 4000^3

125. 0.0005^2

126. 0.008^3

★ **127.** $(-0.000004)^{-2}$

★ **128.** $(-0.0002)^{-3}$

For Exercises 129–132, use the following information to solve. The energy in joules of a single photon of light can be determined by $E = hf$, where h is a constant 6.626×10^{-34} joule-second and f is the frequency of the light in hertz. Write each answer in scientific notation with the decimal number rounded to two decimal places.

129. Find the energy of red light with a frequency of 4.2×10^{14} Hz.

130. Find the energy of blue light with a frequency of 6.1×10^{14} Hz.

131. A photon of light is measured to have 1.4×10^{-19} joule of energy. Find its frequency.

132. A photon of yellow-green light is measured to have 3.6×10^{-19} joule of energy. Find its frequency.

Of Interest

The color of light is determined by its frequency. The human eye is capable of distinguishing frequencies of light from about 3.92×10^{14}Hz (boundary of infrared) up to about 7.89×10^{14}Hz (boundary of ultraviolet).

(*Source*: Emiliani, *The Scientific Companion*.)

For Exercises 133 and 134, use the following information. Albert Einstein discovered that if a mass of m kilograms is converted to pure energy (say, in a nuclear reaction), the amount of energy in joules that is released is described by $E = mc^2$, where c represents the constant speed of light, which is approximately 3×10^8 meters per second.

133. Suppose 4.2×10^{-12} kilogram of plutonium is converted to energy in a reactor. How much energy is released?

134. Suppose 4.2×10^{-8} kilogram of uranium is converted to energy in a nuclear reaction. How much energy is released?

Puzzle Problem Upon being asked for her house number, a mathematician responded, "It is a three-digit perfect square between 100 and 400, and if you rotate the number so that it is upside down, that number is a perfect square also." What is her house number?

Review Exercises

Exercises 1–6 Expressions

[1.2] **1.** Explain how to rewrite $19 - (-6)$ as addition.

[1.2] **2.** What property is illustrated by $x + (y + z) = (x + y) + z$?

[1.4] **3.** Use the distributive property to rewrite the expression $-8(n - 4)$.

[1.4] *For Exercises 4–6, simplify by combining like terms.*

4. $7x + 8y - 4x - 12 + 13y - 5$ **5.** $10x^2 - x + 9 - 12x + 3x^2$ **6.** $2(u + 8) - (5u - 3) - 7$

5.2 Polynomials and Polynomial Functions

Objectives

1. Determine the coefficient and degree of a monomial.
2. Determine the degree of a polynomial and write polynomials in descending order of degree.
3. Add polynomials.
4. Subtract polynomials.
5. Classify and graph polynomial functions.
6. Add and subtract polynomial functions.
7. Solve application problems using polynomial functions.

Warm-up

[1.4] *For Exercises 1 and 2, simplify.*

1. $3x^2 - 6x + 1 + 4x^2 + x - 7$ **2.** $-1(6x^2 + 2x - 1)$

[3.5] **3.** Given $f(x) = x^2 - 2x - 3$, find $f(-1)$.

[5.1] **4.** Simplify $\left(\frac{3}{x^3}\right)^{-2}$. Leave the answer with positive exponents only.

In Section 5.1, we learned about exponents so that we could discuss a class of expressions called polynomials. Notice that we are on the expression level of the Algebra Pyramid.

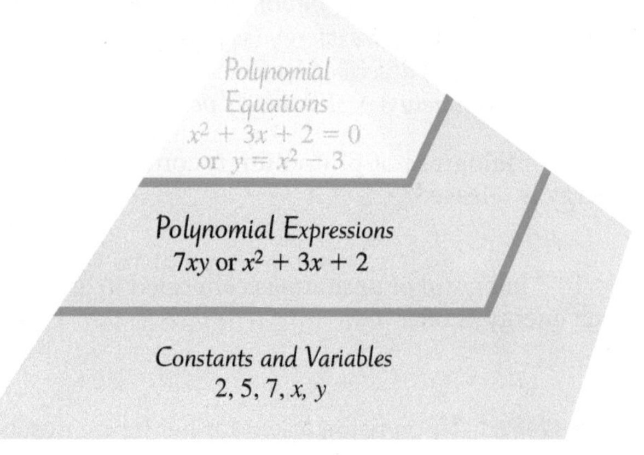

Answers to Warm-up

1. $7x^2 - 5x - 6$
2. $-6x^2 - 2x + 1$
3. 0 **4.** $\frac{x^6}{9}$

Objective 1 Determine the coefficient and degree of a monomial.

We first discuss a special type of term called a **monomial**.

Definition **Monomial:** An expression that is a constant or a product of a constant and variables that are raised to whole number powers.

Examples of monomials: 1.6 $\quad 4x^3 \quad -\frac{3}{4}tu^2$

Examples of expressions that are not monomials: $\frac{x}{y} \quad 6x^{-2} \quad 3x + 2$

It is important to be able to identify the **coefficient** and **degree** of a monomial.

Definitions **Coefficient:** The numerical factor in a monomial.
Degree of a monomial: The sum of the exponents of all variables in the monomial.

Example 1 Identify the coefficient and degree of each monomial.

a. $-6a^7$

Solution: The coefficient is -6, and the degree is 7.

b. $4.3x^3y^7$

Solution: The coefficient is 4.3, and the degree is the sum of the exponents of the variables, which is $3 + 7 = 10$.

c. x

Solution: Because $x = 1x^1$, the coefficient is 1 and the degree is 1.

d. 8

Solution: We can think of 8 as $8x^0$; so the coefficient is 8, and the degree is 0.

Your Turn 1 Identify the coefficient and degree of each monomial.

a. $-8x^4$ **b.** $-0.5x^4y^5$ **c.** t **d.** $\frac{4}{5}$

Objective 2 Determine the degree of a polynomial and write polynomials in descending order of degree.

Now we are ready to formally define a **polynomial**.

Definition **Polynomial:** A monomial or an expression that can be written as a sum of monomials.

Examples of polynomials: $5x \quad 3x + 7 \quad x^2 + 6xy + 3y \quad 4x^3 - x^2 + 2x - 12$

Notice that each variable term in $4x^3 - x^2 + 2x - 12$ has the same variable x; so we say it is a **polynomial in one variable**. A polynomial that has terms with different variables, such as $x^2 + 6xy + 3y$, is called a multivariable polynomial.

Definitions **Polynomial in one variable:** A polynomial in which every variable term has the same variable.
Multivariable polynomial: A polynomial with more than one variable.

Some polynomials have special names. Prefixes indicate the number of terms. For example, a ***monomial***, such as $5x$, has *one* term. A ***binomial***, such as $4x + 8$, has *two* terms. A ***trinomial***, such as $2y^2 + 5y + 8$, has *three* terms. No special names are given to polynomials with more than three terms.

Answers to Your Turn 1

a. coefficient: -8; degree: 4
b. coefficient: -0.5; degree: 9
c. coefficient: 1; degree: 1
d. coefficient: $\frac{4}{5}$; degree: 0

Definitions **Binomial:** A polynomial containing two terms.
Trinomial: A polynomial containing three terms.

By comparing the degrees of the monomials in a polynomial, we can identify the **degree of a polynomial**.

Definition **Degree of a polynomial:** The greatest degree of any of the terms in the polynomial.

Example 2 Identify the degree of each polynomial; then indicate whether the polynomial is a monomial, binomial or trinomial or has no special polynomial name.

a. $5x^6 + x^3 - 10x^2$

Answer: The degree is 6 because it is the greatest degree of any of the terms. The polynomial is a trinomial.

b. $x^5 + 3x^3y^4 - 2xy^3 - y^2 - 6$

Note Remember that the degree of a monomial such as $3x^3y^4$ is the sum of the variables' exponents.

Answer: The degree is 7 because $3x^3y^4$ has the greatest degree of the terms. This polynomial has more than three terms, so it has no special polynomial name.

Your Turn 2 Identify the degree of each polynomial; then indicate whether the polynomial is a monomial, binomial or trinomial or has no special polynomial name.

a. $-8y^9 + 4y$

b. $2x^5 - 9x^6y + x^3y^3 - 15y - 7$

Note Writing multivariable polynomials in descending order of degree will not be required in this course.

To make comparing polynomials easier, mathematicians prefer to write them in *descending order of degree.* For example, $5x^6 + x^3 - 10x^2 + 9$ is in descending order of degree because the first term has the greatest degree, the second term has the next greatest degree, and so on. The first term, $5x^6$, is called the leading term and its coefficient, 5, is the leading coefficient.

Example 3 Write each polynomial in descending order of degree.

a. $x - 3x^3 + 4 + 5x^4$

Solution: $x - 3x^3 + 4 + 5x^4 = 5x^4 - 3x^3 + x + 4$

b. $-6a^2 + 2 + 5a^3 - 7a^5 - 6a$

Solution: $-6a^2 + 2 + 5a^3 - 7a^5 - 6a = -7a^5 + 5a^3 - 6a^2 - 6a + 2$

Your Turn 3 Write each polynomial in descending order of degree.

a. $2n^3 + n^5 - 9n^2 + 8 - 3n^4 + 4n$

b. $-4t^2 + 2t + 16 + 5t^3 - 13t^6$

Objective 3 Add polynomials.

We can perform arithmetic operations with polynomials. First, we consider adding polynomials.

Answers to Your Turn 2
a. degree = 9; binomial
b. degree = 7; no special polynomial name

Answers to Your Turn 3
a. $n^5 - 3n^4 + 2n^3 - 9n^2 + 4n + 8$
b. $-13t^6 + 5t^3 - 4t^2 + 2t + 16$

Example 4 Add. $(3x^2 - 6x + 1) + (4x^2 + x - 7)$

Solution: Removing the parentheses gives a single polynomial that can be simplified by combining like terms.

$(3x^2 - 6x + 1) + (4x^2 + x - 7) = 3x^2 - 6x + 1 + 4x^2 + x - 7$ Remove parentheses.

$= 3x^2 + 4x^2 - 6x + x + 1 - 7$ Collect like terms.

$= 7x^2 - 5x - 6$ Combine like terms.

Note Combining the like terms in order of degree places the resulting polynomial in descending order of degree.

From Example 4, we can summarize how to add polynomials.

Procedure Adding Polynomials

To add polynomials, combine like terms.

Many people prefer to combine the like terms without collecting them or even without removing the parentheses.

Example 5

a. $(t^4 + 9t^3 - 2t - 9) + (t^4 - 4t^3 + 2t - 11)$

Solution: $(t^4 + 9t^3 - 2t - 9) + (t^4 - 4t^3 + 2t - 11)$

$= 2t^4 + 5t^3 + 0 - 20$ Combine like terms.

$= 2t^4 + 5t^3 - 20$

b. $\left(2x^4 + 0.3xy^2 - \frac{3}{4}xy + 5\right) + \left(-6x^4 - 0.5xy^2 + \frac{1}{3}xy + 4\right)$

Solution: $\left(2x^4 + 0.3xy^2 - \frac{3}{4}xy + 5\right) + \left(-6x^4 - 0.5xy^2 + \frac{1}{3}xy + 4\right)$

$= -4x^4 - 0.2xy^2 - \frac{5}{12}xy + 9$ Combine like terms.

Your Turn 5 Add.

a. $(9x^2 - 7x - 5) + (3x^2 + 6x - 3)$

b. $(4n^4 - n^3 + 8n - 15) + (2n^4 - 8n^3 - 8n + 2)$

c. $\left(2.5a^4 - \frac{4}{5}ab^3 + \frac{2}{3}ab - 6\right) + \left(a^4 + \frac{1}{3}ab^3 + ab - 1\right)$

Answers to Your Turn 5

a. $12x^2 - x - 8$

b. $6n^4 - 9n^3 - 13$

c. $3.5a^4 - \frac{7}{15}ab^3 + \frac{5}{3}ab - 7$

Objective 4 Subtract polynomials.

Recall that we can write subtraction as equivalent addition by *adding* the opposite, or *additive inverse*, of the subtrahend. For example, $8 - (-2) = 8 + 2$. We can apply this principle to polynomial subtraction.

The subtraction sign is changed to an addition sign.

$$(9x^2 + 7x + 5) - (6x^2 + 2x + 1) = (9x^2 + 7x + 5) + (-6x^2 - 2x - 1)$$

The subtrahend changes to its additive inverse.

Connection To write the additive inverse of a polynomial, we change the sign of every term. We can also view this process as an application of the distributive property.

$-(6x^2 + 2x + 1)$

$= -1(6x^2 + 2x + 1)$

$= -1 \cdot 6x^2 + (-1) \cdot 2x + (-1) \cdot 1$

$= -6x^2 - 2x - 1$

Our exploration suggests the following procedure for subtracting polynomials.

Procedure Subtracting Polynomials

To subtract polynomials:

1. Write the subtraction statement as an equivalent addition statement.
   a. Change the operation symbol from a subtraction sign to an addition sign.
   b. Change the subtrahend (second polynomial) to its additive inverse. To write the additive inverse, change the sign of each term in the polynomial.
2. Combine like terms.

Example 6 Subtract.

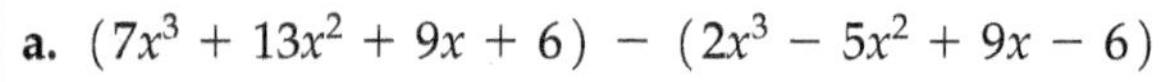

a. $(7x^3 + 13x^2 + 9x + 6) - (2x^3 - 5x^2 + 9x - 6)$

Solution: $(7x^3 + 13x^2 + 9x + 6) - (2x^3 - 5x^2 + 9x - 6)$

Change the minus sign to a plus sign. Change all signs in the subtrahend.

$= (7x^3 + 13x^2 + 9x + 6) + (-2x^3 + 5x^2 - 9x + 6)$

$= 5x^3 + 18^2 + 12$ Combine like terms.

b. $(0.4x^5 - 3x^2y - 0.6xy - y^2) - (x^5 - xy^2 + 1.3xy - 5y^2)$

Solution: $(0.4x^5 - 3x^2y - 0.6xy - y^2) - (x^5 - xy^2 + 1.3xy - 5y^2)$

$= (0.4x^5 - 3x^2y - 0.6xy - y^2) + (-x^5 + xy^2 - 1.3xy + 5y^2)$ Write equivalent addition.

$= -0.6x^5 - 3x^2y + xy^2 - 1.9xy + 4y^2$ Combine like terms.

Learning Strategy (Auditory, Tactile, Visual)

When changing the signs, try using a colored pen so that you can clearly see the sign changes.

Your Turn 6 Subtract.

a. $(6u^4 + u^2 - 12u - 8) - (7u^4 - 5u^2 - u + 4)$

b. $(0.6x^5 - 2xy^2 + y - 1.5) - (1.5x^5 + x^3y^2 - 2xy^2 + 4y - 1.3)$

Objective 5 Classify and graph polynomial functions.

Polynomials can be used to make **polynomial functions**. In this course, we limit ourselves to polynomials in one variable.

Definition Polynomial function: A function of the form $f(x) = ax^m + bx^n + \cdots$ with a finite number of terms, where each coefficient is a real number and each exponent is a whole number.

Examples of polynomial functions: $f(x) = x^2 + 3x + 1$ $\quad g(x) = x^3 - 2$

Polynomial functions are classified by the degree of the polynomial. A degree 0 polynomial is a **constant function**, a degree 1 polynomial is a **linear function**, a degree 2 polynomial is a **quadratic function**, and a degree 3 polynomial is a **cubic function**. For now, we consider only those four classifications.

Definition Constant function: A function of the form $f(x) = c$, where c is a real number.

The graph of a constant function is a horizontal line through $(0, c)$.

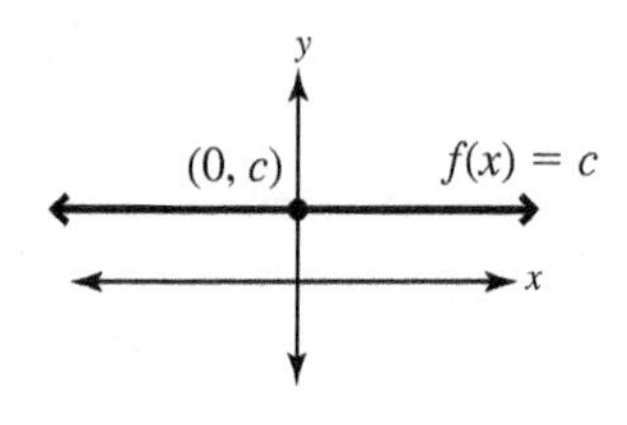

Answers to Your Turn 6

a. $-u^4 + 6u^2 - 11u - 12$

b. $-0.9x^5 - x^3y^2 - 3y - 0.2$

Definition **Linear function:** A function of the form $f(x) = mx + b$, where m and b are real numbers.

Connection Constant functions are special linear functions with $m = 0$.

Graphs of linear functions are lines with slope m and y-intercept $(0, b)$.

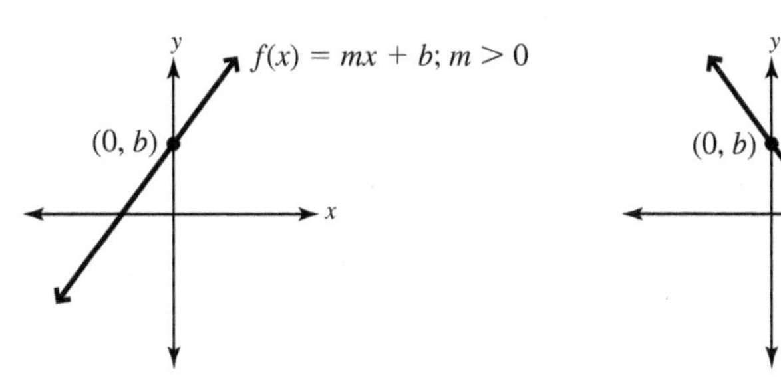

Definition **Quadratic function:** A function of the form $f(x) = ax^2 + bx + c$, where a, b, and c are real numbers and $a \neq 0$.

Graphs of quadratic functions are parabolas with y-intercept at $(0, c)$.

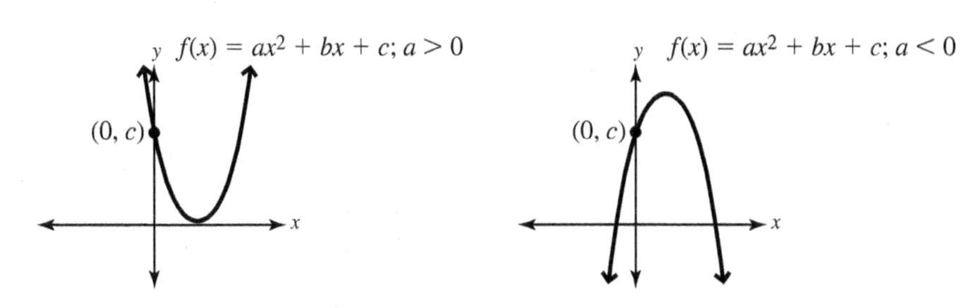

Definition **Cubic function:** A function of the form $f(x) = ax^3 + bx^2 + cx + d$, where a, b, c, and d are real numbers and $a \neq 0$.

Graphs of cubic functions resemble an S-shape with y-intercept at $(0, d)$.

Connection The vertical line test, which we learned in Chapter 3, confirms that all of these are functions.

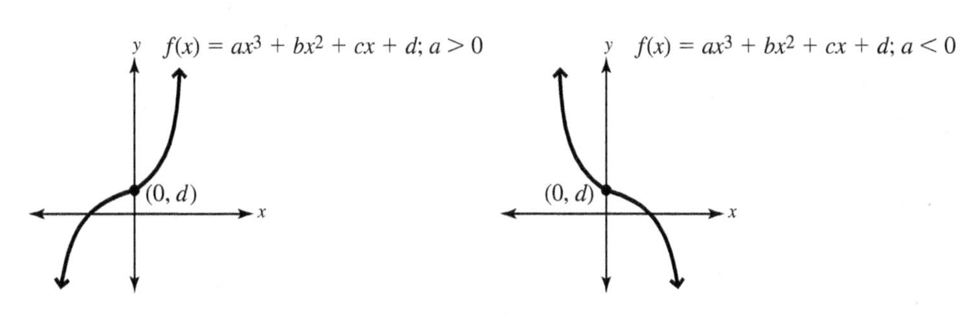

Example 7 Graph and give the domain and range.

a. $f(x) = x^2 - 2x - 3$

Solution: Find enough ordered pairs to generate the graph. Recall that we find an ordered pair by evaluating the function using a chosen value for x. For example, we show the calculation of $f(-1)$.

$$f(-1) = (-1)^2 - 2(-1) - 3 \quad \text{Replace } x \text{ with } -1.$$
$$= 1 + 2 - 3$$
$$= 0$$

The ordered pair is $(-1, 0)$.

Connection Finding $f(x)$ is the same as finding the y-coordinate given an x-coordinate.

Additional ordered pairs are found in a similar way. We list several pairs in the following table.

x	$f(x)$
−2	5
−1	0
0	−3
1	−4
2	−3
3	0
4	5

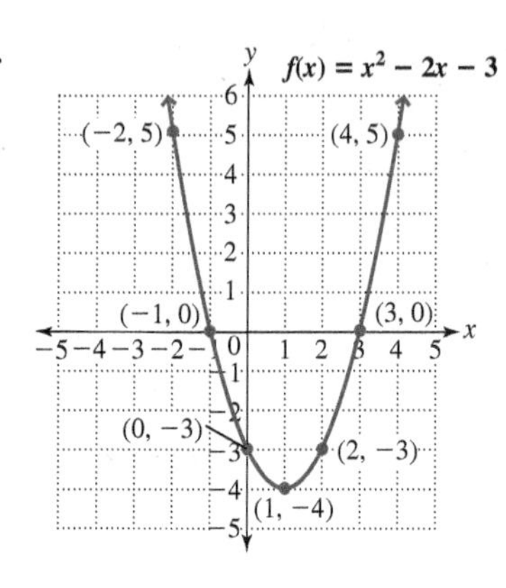

To graph the function, we plot our ordered pairs and then draw a smooth curve. Because this is a quadratic equation, we get a parabola.

Note $\{x \mid -\infty < x < \infty\}$ is the same as $\{x \mid x \text{ is a real number}\}$ or $\mathbb{R}$.

Domain: $\{x \mid -\infty < x < \infty)$ or $(-\infty, \infty)$; Range: $\{y \mid y \geq -4\}$ or $[-4, \infty)$

b. $f(x) = x^3 + 1$

Solution: Find enough ordered pairs to generate the graph. We list several pairs in the following table.

x	$f(x)$
−2	−7
−1	0
0	1
1	2
2	9

To graph the function, we plot our ordered pairs and then draw a smooth curve. Because this is a cubic, we know the graph will resemble an S-shape.

Domain: $\{x \mid -\infty < x < \infty\}$ or $(-\infty, \infty)$;
Range: $\{y \mid -\infty < y < \infty\}$ or $(-\infty, \infty)$

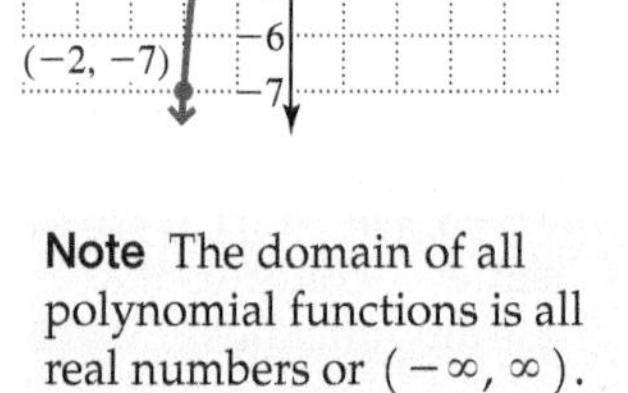

Answers to Your Turn 7

a.

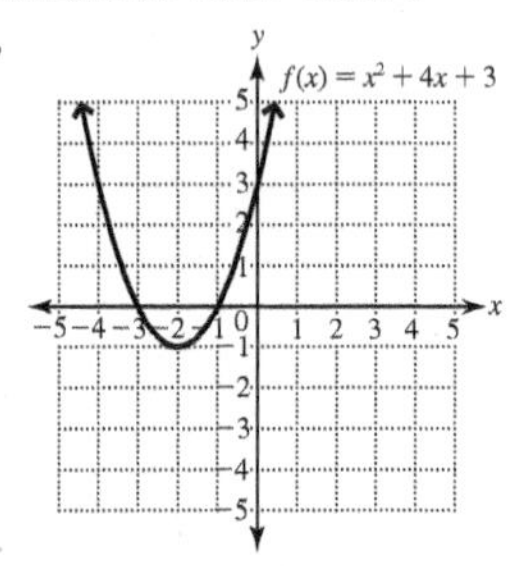

Domain: $\{x \mid -\infty < x < \infty\}$ or $(-\infty, \infty)$
Range: $\{y \mid y \geq -1\}$ or $[-1, \infty)$

b.

Domain: $\{x \mid -\infty < x < \infty\}$ or $(-\infty, \infty)$
Range: $\{y \mid -\infty < y < \infty\}$ or $(-\infty, \infty)$

Your Turn 7 Graph and give the domain and range.

a. $f(x) = x^2 + 4x + 3$ **b.** $f(x) = \frac{1}{2}x^3 - 1$

Note The domain of all polynomial functions is all real numbers or $(-\infty, \infty)$.

Objective 6 Add and subtract polynomial functions.

We can find the sum or difference of two polynomial functions.

Rule Adding or Subtracting Functions

The sum of two functions, $f + g$, is found by $(f + g)(x) = f(x) + g(x)$.
The difference of two functions, $f - g$, is found by $(f - g)(x) = f(x) - g(x)$.

Example 8 Given $f(x) = 2x + 1$ and $g(x) = 7x + 3$, find

a. $f + g$ **b.** $f - g$ **c.** $(f - g)(-4)$

Solution:

a. $(f + g)(x) = (2x + 1) + (7x + 3)$ Use the rule $(f + g)(x) = f(x) + g(x)$.
$= 9x + 4$

b. $(f - g)(x) = (2x + 1) - (7x + 3)$ Use the rule $(f - g)(x) = f(x) - g(x)$.
$= -5x - 2$

c. In part b, we found that $(f - g)(x) = -5x - 2$.
$(f - g)(-4) = -5(-4) - 2$ Replace x with -4 in $(f - g)(x) = -5x - 2$.
$= 20 - 2$ Multiply.
$= 18$ Subtract.

Your Turn 8 Given $h(x) = x^2 - 3x + 5$ and $k(x) = x^2 - 2x - 9$, find

a. $h + k$ **b.** $h - k$ **c.** $(h + k)(-2)$

Objective 7 Solve application problems using polynomial functions.

One common application of polynomial functions is in determining the net profit a business makes given functions for the revenue and the cost.

Example 9 A software company produces an accounting program. The function $R(x) = 0.2x^2 + x + 500$ describes the revenue the company makes from sales, where x represents the number of units sold. The function $C(x) = 10x + 4000$ describes the cost of producing those x units of software.

a. Find a function, $P(x)$, that describes the profit.
b. Find the profit if the company sells 1000 copies of the software.

Solution:

a. Because the net profit is the money left after a business deducts costs, the relationship for finding profit is $P(x) = R(x) - C(x)$.

$P(x) = (0.2x^2 + x + 500) - (10x + 4000)$ Subtract the cost function from the revenue function.

$P(x) = (0.2x^2 + x + 500) + (-10x - 4000)$ Write the equivalent addition.
$P(x) = 0.2x^2 - 9x - 3500$ Combine like terms.

b. Now to find the profit for the sale of 1000 copies, we find $P(1000)$.

$$P(1000) = 0.2(1000)^2 - 9(1000) - 3500 = \$187{,}500$$

Your Turn 9 A small publishing company produces books for counseling teenagers. The function $R(x) = 0.1x^2 + 2x + 250$ describes the revenue a company makes from sales, where x represents the number of books sold. The function $C(x) = 5x + 1200$ describes the cost of producing the books.

a. Find a function, $P(x)$, that describes the profit.
b. Find the profit if the company sells 1000 books.

Answers to Your Turn 8
a. $2x^2 - 5x - 4$ **b.** $-x + 14$
c. 14

Answers to Your Turn 9
a. $P(x) = 0.1x^2 - 3x - 950$
b. \$96,050

5.2 Exercises

For Extra Help MyMathLab®

Note: Exercises marked with a ★ represent challenging exercises.

Objectives 1 and 2

Prep Exercise 1 How do you determine the degree of a monomial?

Prep Exercise 2 Explain the difference between a monomial, a binomial, and a trinomial.

Prep Exercise 3 How do you determine the degree of a polynomial?

For Exercises 1–14, identify the degree of each polynomial; then indicate whether the polynomial is a monomial, binomial, or trinomial or has no special polynomial name. See Examples 1 and 2.

1. $-7uv^3$

2. $-0.4a^3b^2c$

3. $25 - x^2$

4. $x^3 + 4$

5. $5p^3 + 2p^2 - 1$

6. $1.5r^4 - 3r^2 + 8r$

7. $4g^2 - 5g - 11g^3 - 8g^4 + 7$

8. $-7.1k + 2.3k^3 - 8k^2 - 1$

9. $7x + 5x^3 - 19$

10. $-16m^4 + 5m - 7m^2$

11. -7

12. 4

13. $\frac{1}{3}pq - 3p^2q$

14. $3m^2n^2 + 6m^4n$

Objective 3

Prep Exercise 4 How do you add polynomials?

Prep Exercise 5 Given a polynomial in one variable, what does *descending order of degree* mean?

For Exercises 15–24, add and write the resulting polynomial in descending order of degree. See Examples 3–5.

15. $(5x^2 - 3x + 1) + (2x^2 + 7x - 3)$

16. $(3y^2 + 7y - 3) + (4y^2 + 3y + 1)$

17. $(p^4 - 3p^3 + 4p - 1) + (p^4 - 2p^3 - 4p + 7)$

18. $(12r^4 - 5r^2 + 8r - 15) + (7r^4 + 5r^2 + 2r - 9)$

19. $(4u^3 - 6u^2 + u + 11) + (-5u^3 - 3u^2 + u - 5)$

20. $(7p^3 - 9p^2 + 5p - 1) + (-4p^3 - 6p^2 + p + 10)$

21. $\left(\frac{2}{3}u^4 + \frac{3}{4}u^3 - u^2 - u + 3\right) + \left(\frac{2}{3}u^4 - \frac{1}{4}u^3 + 3u^2 + 4u - 1\right)$

22. $\left(\frac{7}{8}w^4 - 5w^3 + 3w^2 - 8\right) + \left(\frac{3}{4}w^4 - \frac{1}{2}w^3 + 4w^2 - 7w + 9\right)$

23. $(3.1t^4 - 2.1t^3 + 7t^2 + 5.8t + 4) + (4.2t^4 + 3.6t^3 - 8t^2 - 3.1t + 3)$

24. $(7.3h^4 - 3.1h^3 - 7.6h^2 + 3.5h + 2.7) + (0.4h^4 - 2.5h^3 + 1.2h^2 - 1.6h - 3.1)$

Objective 4

Prep Exercise 6 When subtracting one polynomial from another, after the operation symbol is changed from a minus sign to a plus sign, what is the next step?

For Exercises 25–34, subtract and write the resulting polynomial in descending order of degree. See Examples 3 and 6.

25. $(7a^3 - a^2 - a) - (8a^3 - 3a + 1)$

26. $(4x^3 - 3x + 4) - (6x^3 - 3x^2 + 5)$

27. $(3m^4 + 2m^3 - m^2 + 1) - (6m^3 - 2m^2 - m + 3)$

28. $(5n^4 + 9n^3 - n^2 - 3) - (6n^3 - 4n^2 + 2n - 6)$

29. $(-7r^3 + 3r^2 - 7r - 4) - (-5r^3 + 2r^2 - 6r - 1)$

30. $(-y^5 + 6y^3 + 7y^2 - y - 3) - (-3y^2 + 7y - 1)$

31. $\left(\frac{4}{5}y^3 - \frac{1}{3}y^2 + 7y - \frac{2}{7}\right) - \left(\frac{1}{5}y^3 + \frac{2}{3}y^2 + \frac{3}{4}y - \frac{1}{7}\right)$

32. $\left(\frac{9}{10}m^5 - \frac{2}{3}m^4 + \frac{3}{5}m^3 - m^2 + \frac{1}{7}m + 8\right) - \left(\frac{11}{10}m^5 + \frac{4}{3}m^4 - \frac{2}{5}m^3 - m^2 + \frac{2}{7}m + 1\right)$

33. $(-4.5w^3 - 5.1w^2 + 2.7w + 4.1) - (3.8w^3 - 1.4w^2 + 3.4w - 2.6)$

34. $(9.1m^4 - 2.5m^3 + 3m^2 + 2.1m - 1) - (-4.2m^4 + 3.2m^3 + 2.1m^2 - 1.1m - 3)$

For Exercises 35–44, perform the indicated operation and write the resulting polynomial in descending order of degree. See Examples 3, 5, and 6.

35. $(-5w^4 - 3w^3 - 8w^2 + w - 14) + (-3w^4 + 6w^3 + w^2 + 12w + 5)$

36. $(6h^9 - 7h^5 - 3h^4 - 2h^3 + 3) + (-h^8 + 6h^6 - 8h^5 + 2h^4 - h^2 - 1)$

37. $(6x^4 + 4x^3 - 3x + 5) - (-2x^3 - 5x^2 + 9x - 8)$

38. $(-9y^5 - y^2 + 13y - 16) - (-4y^3 + 15y^2 - y + 11)$

39. $\left(-\frac{7}{3}g^4 + \frac{4}{5}g^3 + \frac{7}{5}\right) + \left(\frac{4}{3}g^4 - \frac{2}{3}g^2 + \frac{2}{5}\right)$

40. $\left(\frac{8}{5}a^4 - \frac{2}{3}a^3 - \frac{4}{3}\right) + \left(-\frac{3}{5}a^4 + \frac{3}{4}a^2 - \frac{1}{3}\right)$

41. $(4y^2 - 8y + 1) + (5y^2 + 3y + 2) - (6y^2 - 9y + 2)$

42. $(3k^3 - 5k^2 - k - 1) - (2k^3 - 3k^2 - k - 7) + (4k^3 + 4k^2 + k + 7)$

43. $(-6a^3 - 5a^2 + 10) - (9a^3 + 5a^2 - 3a - 1) + (-4a^3 - 2a^2 + 6a - 2)$

44. $-(-r^4 + 3r^2 - 12r - 14) - (3r^4 + 6r^3 + 3r - 8) + (5r^4 - 2r^3 - 3r^2 + 10r - 5)$

★ *For Exercises 45–50, add or subtract. See Examples 5 and 6.*

45. $(8a^2 + 5ab - 2b^2) - (10ab - 6a^2 - 8b^2)$

46. $(3a^2 - 4ab + b^2) - (-2a^2 + 4ab + b^2)$

47. $(7x^3y^4 - 2x^2y^3 - xy + 4) + (-5x^3y^4 + 7x^2y^3 + 8xy - 12)$

48. $(-13a^6 + a^3b^2 + 3ab^3 - 8b^3) + (11a^6 - 3a^2b^3 - 4ab^3 + 10ab^2 + 4b^3)$

49. $(x^2y^2 + 8xy^2 - 12x^2y - 4xy + 7y^2 - 9) - (-3x^2y^2 + 2xy^2 + 4x^2y - xy + 6y^2 + 4)$

50. $(15a^3y^4 + a^2y^2 + 5ay^3 - 7ay + 12y^2 - 11) - (-a^3y^4 + a^2y^2 + 2ay^3 - ay + 9)$

For Exercises 51–58, write an expression for the perimeter in simplest form.

51.

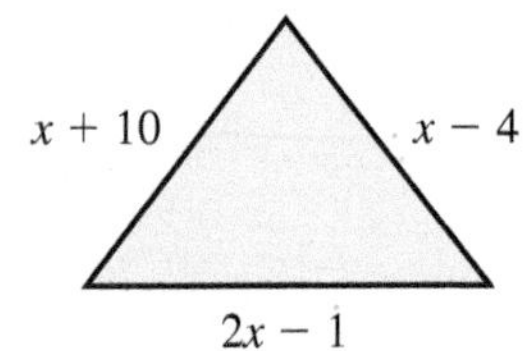

52.

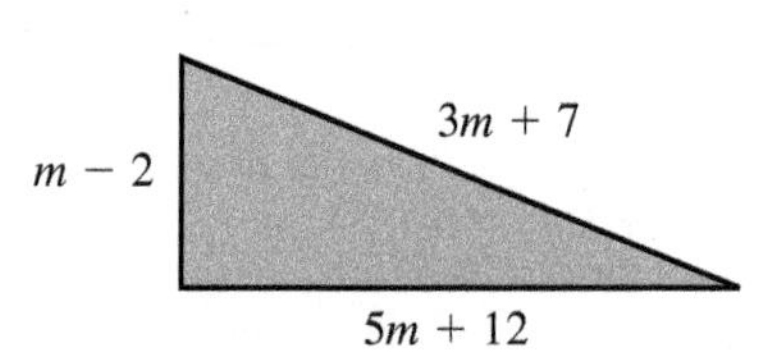

53.

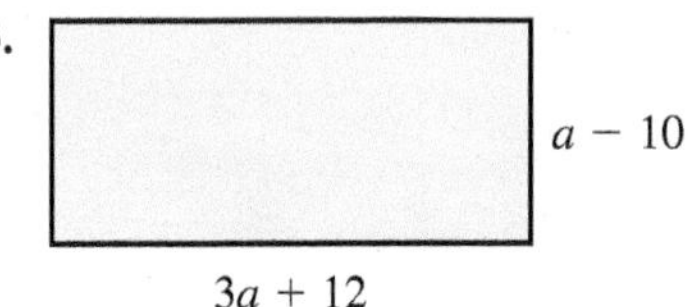

54.

$3x - 2$

$5x + 7$

Objective 5

Prep Exercise 7 Explain how to tell the difference between a constant function, linear function, quadratic function, and cubic function.

For Exercises 55–62, indicate whether the function is a constant function, a linear function, a quadratic function, or a cubic function. See Objective 5.

55. $g(x) = -5$

56. $g(x) = 4$

57. $f(x) = 3x^2 - 2x + 1$

58. $h(x) = x^2 + 4x - 9$

59. $c(x) = 4x - \frac{3}{5}$

60. $h(x) = -3x - 1$

61. $r(x) = -4x^3 + 7x^2 - x + 3$

62. $f(x) = x^3 - 2x^2 + 1$

For Exercises 63–70, graph the function and give the domain and range. See Example 7.

63. $h(x) = -2$

64. $f(x) = 4$

65. $c(x) = -5x + 1$

66. $h(x) = -3x + 4$

67. $g(x) = x^2 + 2x - 3$

68. $g(x) = 3x^2 - 1$

69. $f(x) = x^3 - 2$

70. $c(x) = 2x^3 - 1$

Objective 6

Prep Exercise 8 Given two polynomial functions $f(x)$ and $g(x)$, how do you determine $(f + g)(x)$ and $(f - g)(x)$?

For Exercises 71–78, find: a. $f + g$. *b.* $f - g$. *See Example 8.*

71. $f(x) = -4; g(x) = 3$

72. $f(x) = 7; g(x) = -1$

73. $f(x) = 4x + 3; g(x) = -x - 1$

74. $f(x) = 8x + 3; g(x) = -x + 3$

75. $f(x) = x^2 - 1; g(x) = 3x^2 + 2$

76. $f(x) = 3x^2 + 2; g(x) = 2x^2 - 5x + 9$

77. $f(x) = x^3 + 2x + 5; g(x) = -2x^3 + 2$

78. $f(x) = x^3 - 8; g(x) = -2x^3 + x^2 + 3$

Objective 7

For Exercises 79–82, solve. See Example 9.

79. A pharmaceutical company produces an antibiotic. The function $R(x) = 0.2x^2 + x + 3000$ describes the revenue, in dollars, the company makes from sales, where x represents the number of bottles sold. The function $C(x) = 18x + 2000$ describes the cost of producing the bottles.

a. Find a function, $P(x)$, that describes the profit.

b. Find the profit if the company sells 1000 bottles.

80. A cosmetics company produces a facial cream. The function $R(x) = 0.1x^2 + 2x + 300$ describes the revenue, in dollars, from sales, and x represents the number of jars sold. The function $C(x) = 12x + 275$ describes the cost of producing the jars of cream.

a. Find a function, $P(x)$, that describes the profit.

b. Find the profit if the company sells 500 jars.

81. A small company makes custom chairs. The function $C(x) = 230x + 580$ describes the cost, in dollars, and x represents the number of chairs made. The function $P(x) = 190x + 260$ describes the profit from sales.

a. Find a function, $R(x)$, that describes the revenue.

b. Find the revenue from the sale of 300 chairs.

82. A company makes handwoven area rugs. The function $C(x) = 630x + 320$ describes the cost, in dollars, and x represents the number of rugs made. The function $P(x) = 420x + 260$ describes the profit from sales.

a. Find a function, $R(x)$, that describes the revenue.

b. Find the revenue from the sale of 150 rugs.

For Exercises 83 and 84, use the following information. If we ignore air resistance, the polynomial function $h(x) = -16t^2 + 200$ describes the height, in feet, of a falling object after falling for t seconds from an initial height of 200 feet.

83. What is its height after 0.5 seconds?

84. What is its height after 1.2 seconds?

For Exercises 85 and 86, use the following information. The polynomial function $V(r) = 1.5r^2 - 0.8r + 9.5$ describes the voltage in a circuit, where r represents the resistance in the circuit in ohms.

85. Find the voltage if the resistance is 6 ohms.

86. Find the voltage if the resistance is 8 ohms.

87. The function $F(x) = 0.23x^2 - 3.11x + 34.3$ approximately models the fatality rate of motorcyclists involved in crashes per 100 million vehicle miles traveled each year since 2002. (*Source:* National Highway Traffic Safety Administration, *Traffic Safety Facts*, 2011.)

a. Complete the table of ordered pairs.
b. Graph the function. Note that $x \geq 0$ and $y \geq 0$.
c. Using the model, predict the motorcycle fatality rate in 2020.

	Years since 2002	Fatalities
2002	0	34.3
2004	2	29.0
2006	4	25.54
2008	6	23.92
2010	8	24.14

88. The function $S(x) = -0.045x^2 - 0.2x + 22.4$ approximately models the percentage of high school students who have been smoking each year since 2003. (*Source:* Centers for Disease Control, National Center for Health Statistics, *Health, United States*, 2011.)

a. Complete the table of ordered pairs.
b. Graph the function. Note that $x \geq 0$ and $y \geq 0$.
c. Using the model, predict the percentage of high school students who will smoke in 2020.

	Years since 2003	Students Who Smoked
2003	0	22.4
2005	2	21.82
2007	4	20.88
2009	6	19.58
2011	8	17.92

Review Exercises

Exercises 1 and 2 Expressions

[1.4] *For Exercises 1 and 2, distribute.*

1. $8(3x - 9)$

2. $-6(2y + 7)$

Exercises 3–6 Equations and Inequalities

For Exercises 3 and 4, solve.

[2.1] 3. $5x - 3(x + 2) = 7x - 8$

[2.3] 4. $12x + 9 < 4(2x - 5)$

For Exercises 5 and 6, solve. *a. Write the solution set using set-builder notation.*
b. Write the solution set using interval notation.
c. Graph the solution set.

[2.4] 5. $-11 < 2x - 3 \leq 9$

[2.6] 6. $|2x - 7| \geq 5$

5.3 Multiplying Polynomials

Objectives

1. Multiply a polynomial by a monomial.
2. Multiply two or more polynomials with multiple terms.
3. Determine the product when given special polynomial factors.
4. Multiply polynomial functions.

Warm-up

[1.4] 1. According to the distributive property, $a(b + c) =$ ________.

[5.1] For Exercises 2 and 3, simplify.

2. $(0.2a^2b)(3a^3b)$
3. $-4xy^2z(-5yz^2)$

[5.2] 4. Simplify. $(4x^3 + 3x^2 - 5) - (6x^2 - 4x + 2)$

In Section 5.1, although we had not formally defined monomials, we used the product rule to multiply them. For example, in Example 2(d) of Section 5.1, we multiplied $(-9m^2n)(5m^3n^7p)$.

$$(-9m^2n)(5m^3n^7p) = -45m^{2+3}n^{1+7}p \quad \text{Multiply the coefficients and use the product rule of exponents.}$$

$$= -45m^5n^8p \quad \text{Simplify the exponents.}$$

In this section, we explore how to multiply multiple-term polynomials.

Objective 1 Multiply a polynomial by a monomial.

First, we consider multiplying a polynomial by a monomial, as in $x(x + 4)$. Notice that the distributive property applies.

$$x(x + 4) = x \cdot x + x \cdot 4 \quad \text{Use the distributive property.}$$

$$= x^2 + 4x \quad \text{Simplify.}$$

Note We used the product rule of exponents to simplify $x \cdot x$.

Our example suggests the following procedure.

VISUAL

Learning Strategy

If you are a visual learner, try drawing lines connecting each product.

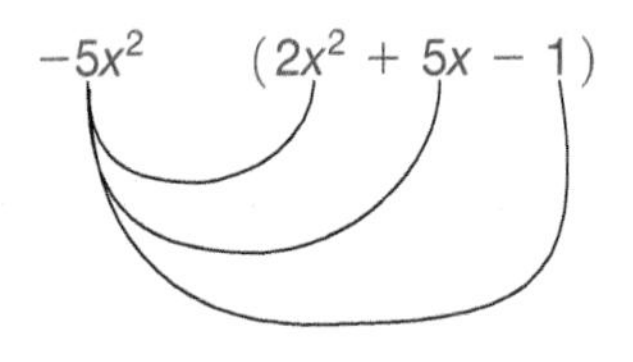

Procedure Multiplying a Polynomial by a Monomial

To multiply a polynomial by a monomial, use the distributive property to multiply each term in the polynomial by the monomial.

Example 1 Multiply.

a. $-5x^2(2x^2 + 5x - 1)$

Solution:

$$-5x^2(2x^2 + 5x - 1) = -5x^2 \cdot 2x^2 - 5x^2 \cdot 5x - 5x^2 \cdot (-1) \quad \text{Distribute } -5x^2.$$

$$= -10x^4 - 25x^3 + 5x^2 \quad \text{Simplify.}$$

b. $0.2a^2b(3a^3b - ab^2 + 0.5ac)$

Solution:

$$0.2a^2b(3a^3b - ab^2 + 0.5ac) = 0.2a^2b \cdot 3a^3b + 0.2a^2b \cdot (-ab^2) + 0.2a^2b \cdot 0.5ac \quad \text{Distribute } 0.2a^2b.$$

$$= 0.6a^5b^2 - 0.2a^3b^3 + 0.1a^3bc \quad \text{Simplify.}$$

Note When multiplying multivariable terms, it is helpful to multiply the coefficients first, then the variables in alphabetical order.

For $0.2a^2b \cdot 3a^3b$, think $0.2 \cdot 3 = 0.6$, $a^2 \cdot a^3 = a^5$, $b \cdot b = b^2$.

Your Turn 1 Multiply.

a. $2x^3(7x^2 - x + 4)$

b. $-4xy^2z(6x^3z - 5yz^2 + 9xz)$

Answers to Your Turn 1

a. $14x^5 - 2x^4 + 8x^3$

b. $-24x^4y^2z^2 + 20xy^3z^3 - 36x^2y^2z^2$

Answers to Warm-up

1. $ab + ac$
2. $0.6a^5b^2$
3. $20xy^3z^3$
4. $4x^3 - 3x^2 + 4x - 7$

Objective 2 Multiply two or more polynomials with multiple terms.

We again use the distributive property to multiply two polynomials with multiple terms, such as $(x + 3)(x + 2)$. Now, however, we distribute the entire $x + 3$ expression.

$$(x + 3)(x + 2) = (x + 3) \cdot x + (x + 3) \cdot 2$$ Distribute $x + 3$.

Note We can skip to this line by multiplying every term in the second polynomial by every term in the first polynomial.

To complete the multiplication, we need to apply the distributive property again.

$= x \cdot x + 3 \cdot x + x \cdot 2 + 3 \cdot 2$ Distribute x in $(x + 3) \cdot x$ and 2 in $(x + 3) \cdot 2$.

$= x^2 + 3x + 2x + 6$ Simplify.

$= x^2 + 5x + 6$ Combine like terms.

Our example suggests the following procedure.

Procedure Multiplying Polynomials

To multiply two polynomials:

1. Multiply each term in the second polynomial by each term in the first polynomial.
2. Combine like terms.

Connection The process of multiplying polynomials is the same as for multiplying numbers. In fact, the same vertical method used to multiply numbers can be used with polynomials. For example, compare $(12)(13)$ with $(x + 2)(x + 3)$.

$$\begin{array}{r} 12 \\ \times 13 \\ \hline 36 \\ +12 \\ \hline 156 \end{array} \qquad \begin{array}{r} x + 2 \\ \times x + 3 \\ \hline 3x + 6 \\ + x^2 + 2x \\ \hline x^2 + 5x + 6 \end{array}$$

Using the distributive property, we multiply each digit in one number by each digit in the other number, then add digits in the like place values. For polynomials, we multiply each term in one polynomial by each term in the other, then combine like terms.

Learning Strategy (Auditory, Tactile, Visual)

As we mentioned earlier, try drawing lines connecting the terms that you multiply so that you can better see the products.

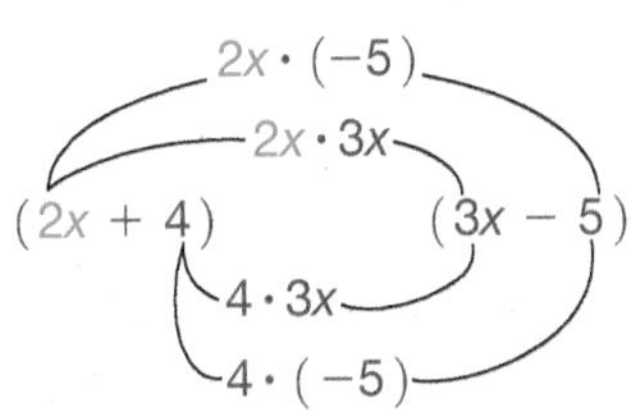

Example 2 Multiply.

a. $(2x + 4)(3x - 5)$

Solution: $(2x + 4)(3x - 5)$

$= 2x \cdot 3x + 2x \cdot (-5) + 4 \cdot 3x + 4 \cdot (-5)$ Multiply each term in $3x - 5$ by each term in $2x + 4$.

$= 6x^2 - 10x + 12x - 20$ Simplify.

$= 6x^2 + 2x - 20$ Combine like terms.

b. $(4m - 3n)(2m - n)$

Solution: $(4m - 3n)(2m - n) = 4m \cdot 2m + 4m \cdot (-n) + (-3n) \cdot 2m + (-3n) \cdot (-n)$ Multiply each term in $2m - n$ by each term in $4m - 3n$.

$= 8m^2 - 4mn - 6mn + 3n^2$ Simplify.

$= 8m^2 - 10mn + 3n^2$ Combine like terms.

c. $(x^2 - 3)(x^2 + 4)$

Solution: $(x^2 - 3)(x^2 + 4) = x^2 \cdot x^2 + x^2 \cdot 4 + (-3) \cdot x^2 + (-3) \cdot 4$ Multiply each term in $x^2 + 4$ by each term in $x^2 - 3$.

$= x^4 + 4x^2 - 3x^2 - 12$ Simplify.

$= x^4 + x^2 - 12$ Combine like terms.

d. $(x^2 + 4)(y + 6)$

Solution: $(x^2 + 4)(y + 6) = x^2 \cdot y + x^2 \cdot 6 + 4 \cdot y + 4 \cdot 6$ Multiply each term in $y + 6$ by each term in $x^2 + 4$.

$= x^2y + 6x^2 + 4y + 24$ Simplify.

Note There are no like terms to combine.

The word *FOIL*, which stands for First Outer Inner Last, is a popular way to remember the process of multiplying two binomials. We use the binomials from Example 2(a) to demonstrate.

Warning FOIL helps keep track of products only when you are multiplying two binomials. When multiplying polynomials with more than two terms, remember to multiply each term in the second polynomial by each term in the first polynomial.

$(2x + 4)(3x - 5)$

First terms: $2x \cdot 3x = 6x^2$

$(2x + 4)(3x - 5)$

Outer terms: $2x \cdot (-5) = -10x$

$(2x + 4)(3x - 5)$

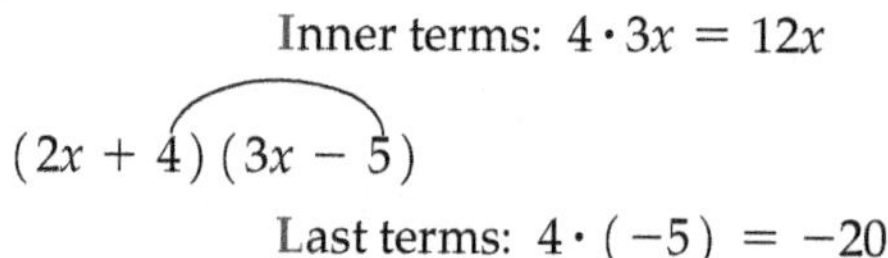

Inner terms: $4 \cdot 3x = 12x$

Last terms: $4 \cdot (-5) = -20$

VISUAL

Learning Strategy

If you are a visual learner, you might note that when drawing lines that connect the first, outside, inside, and last products as follows, the result is a face.

Connection The product of two binomials can be illustrated geometrically. The following rectangle has a length of $x + 7$ and a width of $x + 5$. We can describe its area in two ways:

1. The product of the length and width, $(x + 7)(x + 5)$
2. The sum of the areas of the four internal rectangles, $x^2 + 5x + 7x + 35$

Because these two approaches describe the same area, the two expressions must be equal.

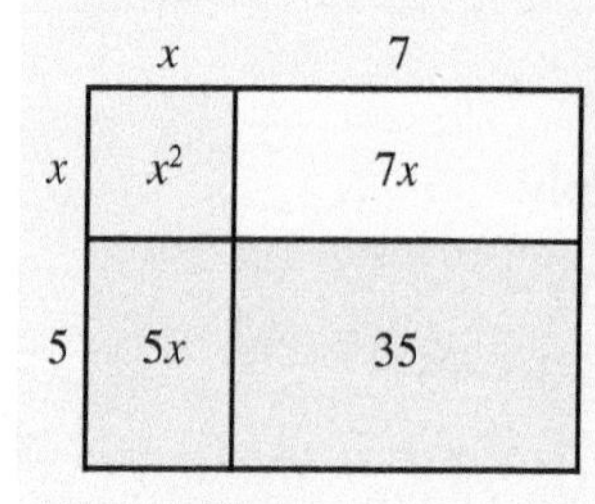

Length $\cdot$ Width = Sum of the areas of the four internal rectangles

$(x + 7)(x + 5) = x^2 + 5x + 7x + 35$

$= x^2 + 12x + 35$ Combine like terms.

Your Turn 2 Multiply.

a. $(x - 9)(x + 3)$ **b.** $(5y - 2)(6y - 7)$ **c.** $(4t^2 - u)(5t^2 + u)$

Answers to Your Turn 2

a. $x^2 - 6x - 27$
b. $30y^2 - 47y + 14$
c. $20t^4 - t^2u - u^2$

Now consider an example of multiplication involving a polynomial with more than two terms, such as a trinomial. Remember that no matter how many terms are in the polynomials, we multiply every term in the second polynomial by every term in the first polynomial.

Example 3 Multiply $(2x + 3)(4x^2 + x - 5)$.

Solution: Multiply each term in $4x^2 + x - 5$ by each term in $2x + 3$.

$$(2x + 3)(4x^2 + x - 5) = 2x \cdot 4x^2 + 2x \cdot x + 2x \cdot (-5) + 3 \cdot 4x^2 + 3 \cdot x + 3 \cdot (-5)$$ Multiply each term in $4x^2 + x - 5$ by each term in $2x + 3$.

$$= 8x^3 + 2x^2 - 10x + 12x^2 + 3x - 15$$ Simplify.

$$= 8x^3 + 14x^2 - 7x - 15$$ Combine like terms.

Warning Notice that FOIL does not make sense with the trinomial in Example 3 because there are too many terms. FOIL handles only the four terms from two binomials.

Your Turn 3 Multiply $(5x - 1)(2x^2 - 9x + 4)$.

Objective 3 Determine the product when given special polynomial factors.

We can use formulas to quickly find the product of some special polynomial factors.

Squaring a Binomial

First, let's look for a pattern that we can use to help us square binomials. There are two cases to consider: $(a + b)^2$ and $(a - b)^2$. Recall that squaring a number or an expression means that we multiply the number or expression by itself.

$$(a + b)^2 = (a + b)(a + b) = a^2 + ab + ab + b^2 = a^2 + 2ab + b^2$$

$$(a - b)^2 = (a - b)(a - b) = a^2 - ab - ab + b^2 = a^2 - 2ab + b^2$$

Rule Squaring a Binomial

If a and b are real numbers, variables, or expressions, then $(a + b)^2 = a^2 + 2ab + b^2$

$(a - b)^2 = a^2 - 2ab + b^2$.

Connection The square of a binomial can also be shown using geometry. In the square shown, the length of each side is $a + b$. We can describe its area in two ways:

1. The length of a side squared, $(a + b)^2$
2. The sum of the areas of the four internal rectangles, $a^2 + ab + ab + b^2$. Because the two approaches describe the same area, the two expressions must be equal.

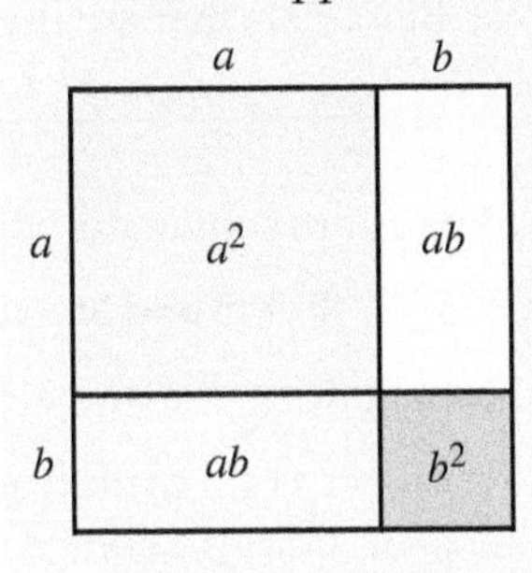

Side squared = Sum of the areas of the four internal rectangles.

$$(a + b)^2 = a^2 + ab + ab + b^2$$
$$= a^2 + 2ab + b^2$$ Combine like terms.

Example 4 Multiply.

a. $(x + 7)^2$

Solution: $(a + b)^2 = a^2 + 2a\ \ b + b^2$

$$(x + 7)^2 = (x)^2 + 2(x)(7) + (7)^2$$ Use $(a + b)^2 = a^2 + 2ab + b^2$, where a is x and b is 7.

$$= x^2 + 14x + 49$$ Simplify.

Warning Notice that $(x + 7)^2$ does *not* equal $x^2 + 7^2$.

Answer to Your Turn 3
$10x^3 - 47x^2 + 29x - 4$

b. $(6r - 5)^2$

Solution:

$(6r - 5)^2 = (6r)^2 - 2(6r)(5) + (5)^2$ Use $(a - b)^2 = a^2 - 2ab + b^2$.

$= 36r^2 - 60r + 25$ Simplify.

c. $(h^3 - 4)^2$

Solution:

$(h^3 - 4)^2 = (h^3)^2 - 2(h^3)(4) + (4)^2$ Use $(a - b)^2 = a^2 - 2ab + b^2$.

$= h^6 - 8h^3 + 16$ Simplify.

d. $(4x + 5y^2)^2$

Solution:

$(4x + 5y^2)^2 = (4x)^2 + 2(4x)(5y^2) + (5y^2)^2$ Use $(a + b)^2 = a^2 + 2ab + b^2$.

$= 16x^2 + 40xy^2 + 25y^4$ Simplify.

e. $[(x + 6) - y]^2$

Solution: If we view $x + 6$ as a and y as b, the expression has the form $(a - b)^2$.

$$\overbrace{(x + 6)}^{a} - \overbrace{y}^{b}]^2 = \overbrace{(x + 6)^2}^{a^2} - 2\overbrace{(x + 6)}^{a}\overbrace{(y)}^{b} + \overbrace{(y)^2}^{b^2}$$

$[(x + 6) - y]^2 = (x + 6)^2 - 2(x + 6)(y) + (y)^2$ Use $(a - b)^2 = a^2 - 2ab + b^2$.

$= x^2 + 2(x)(6) + 6^2 - 2xy - 12y + y^2$ Use $(a + b)^2 = a^2 + 2ab + b^2$ to square $x + 6$.

$= x^2 + 12x + 36 - 2xy - 12y + y^2$ Simplify.

$= x^2 + y^2 - 2xy + 12x - 12y + 36$ Rewrite.

Note When second degree polynomials have more than one variable, they are usually written with the single variable terms of second degree first, then the multivariable term, then the single variable terms of first degree, and finally the constant term.

Your Turn 4 Multiply.

a. $(k - 4)^2$ **b.** $(3y + 1)^2$ **c.** $(x^2 + 5)^2$
d. $(4 - n^3)^2$ **e.** $[5t + (u - 3)]^2$

Multiplying Conjugates

Another special product comes from multiplying **conjugates**.

Definition Conjugates: Binomials that are the sum and difference of the same two terms.

The following binomial pairs are conjugates.

$x + 6$ and $x - 6$ $3t + 5$ and $3t - 5$ $-2c + 9d$ and $-2c - 9d$

Let's multiply general conjugates $a + b$ and $a - b$ to determine the pattern.

$(a + b)(a - b) = a \cdot a + a \cdot (-b) + b \cdot a + b \cdot (-b)$ Use FOIL.

$= a^2 - ab + ab - b^2$ Simplify.

$= a^2 - b^2$ Combine like terms.

Note The like terms $-ab$ and ab are additive inverses, so their sum is 0.

The product, $a^2 - b^2$, is called a *difference of squares*. Our example suggests the following rule.

Rule Multiplying Conjugates

If a and b are real numbers, variables, or expressions, then $(a + b)(a - b) = a^2 - b^2$.

Answers to Your Turn 4

a. $k^2 - 8k + 16$
b. $9y^2 + 6y + 1$
c. $x^4 + 10x^2 + 25$
d. $n^6 - 8n^3 + 16$
e. $25t^2 + u^2 + 10tu - 30t - 6u + 9$

Example 5 Multiply.

a. $(3t + 5)(3t - 5)$

Solution: $(a + b)(a - b) = a^2 - b^2$

$(3t + 5)(3t - 5) = (3t)^2 - (5)^2$ Use $(a + b)(a - b) = a^2 - b^2$, where a is $3t$ and b is 5.

$= 9t^2 - 25$ Simplify.

Note The commutative property tells us that $(t^2 - 5u)(t^2 + 5u)$ is the same as $(t^2 + 5u)(t^2 - 5u)$, so this expression is of the form $(a + b)(a - b)$.

b. $(t^2 - 5u)(t^2 + 5u)$

Solution: $(t^2 - 5u)(t^2 + 5u) = (t^2)^2 - (5u)^2$ Use $(a + b)(a - b) = a^2 - b^2$.

$= t^4 - 25u^2$ Simplify.

c. $[(x + 3) + y][(x + 3) - y]$

Solution: If we view $x + 3$ as a and y as b, the expression has the form $(a + b)(a - b)$.

$$[\underbrace{(x + 3)}_{a} + \underbrace{y}_{b}][\underbrace{(x + 3)}_{a} - \underbrace{y}_{b}] = \underbrace{(x + 3)^2}_{a^2} - \underbrace{y^2}_{b^2}$$

$= x^2 + 2(x)(3) + 3^2 - y^2$ Use $(a + b)^2 = a^2 + 2ab + b^2$ to square $x + 3$.

$= x^2 + 6x + 9 - y^2$ Simplify.

$= x^2 - y^2 + 6x + 9$ Rewrite.

Your Turn 5 Multiply.

a. $(9y - 2)(9y + 2)$

b. $(3h^2 + k^2)(3h^2 - k^2)$

c. $[2m + (n - 4)][2m - (n - 4)]$

Objective 4 Multiply polynomial functions.

We can find the product of two polynomial functions using the following rule.

Rule Multiplying Functions

The product of two functions, $f \cdot g$, is found by $(f \cdot g)(x) = f(x)g(x)$.

Example 6 **a.** If $f(x) = 4x + 7$ and $g(x) = 2x^2 - x - 5$, find $(f \cdot g)(x)$.
b. Find $(f \cdot g)(-2)$.

Solution: **a.** $(f \cdot g)(x) = (4x + 7)(2x^2 - x - 5)$ Use $(f \cdot g)(x) = f(x)g(x)$.

$= 8x^3 - 4x^2 - 20x + 14x^2 - 7x - 35$ Multiply each term in $2x^2 - x - 5$ by each term in $4x + 7$.

$= 8x^3 + 10x^2 - 27x - 35$ Combine like terms.

b. $(f \cdot g)(-2) = 8(-2)^3 + 10(-2)^2 - 27(-2) - 35$ In $(f \cdot g)(x)$ replace x with -2.

$= -64 + 40 + 54 - 35$ Simplify.

$= -5$

Your Turn 6

a. If $f(x) = 3x - 5$ and $g(x) = x^2 - 6x + 2$, find $(f \cdot g)(x)$.

b. Find $(f \cdot g)(-1)$.

Answers to Your Turn 5

a. $81y^2 - 4$

b. $9h^4 - k^4$

c. $4m^2 - n^2 + 8n - 16$

Answers to Your Turn 6

a. $(f \cdot g)(x) = 3x^3 - 23x^2 + 36x - 10$

b. -72

5.3 Exercises

For Extra Help MyMathLab®

Note: Exercises marked with a ★ represent challenging exercises.

Objective 1

Prep Exercise 1 What mathematical property is used to multiply polynomials?

For Exercises 1–20, multiply the polynomial by the monomial. See Example 1.

1. $5x^3(x^2 + 3x - 2)$
2. $4y^5(-3y^3 - 5y^2 + 3)$
3. $-6x^3(2x^2 + 4x - 3)$
4. $-2y^4(y^2 + 7y - 8)$
5. $4n^4(n^3 + 7n^2 - 2n - 3)$
6. $6n^6(n^3 + 4n^2 - n + 3)$
7. $9a^2b^4(3a^4b^2 - 2ab^8)$
8. $3yz^3(2y^4z - 3y^4z^2)$
9. $\frac{1}{4}m^2np^3(2mn^5 - 5m^2n^2 + 8m^2n^3p^2)$
10. $\frac{2}{3}k^2l^3m(2k^3l^2 - 3lm^4 + 6k^2l^7m^2)$
11. $-0.3p^2q^2(8p^3q^7 - 2q + 7p^6)$
12. $0.4a^6c^2(2ac^7 - 3ac^2 - 5a^2c^9)$
13. $-5x^5y^2(4x^2y + 2xy^4 - 5x^8y^5)$
14. $-5a^2b(3ab^3 - 2a^2b^2 + 3ab^5)$
15. $-3r^2s(2r^4s^3 - 6r^2s^2 - 3rs + 3)$
16. $-2x^2y^3(5xy^2 - 3x^2y^5 - 2xy + 4)$
17. $-0.2a^2b^2(2.1a - 6ab + 3a^2b - a^2b^2)$
18. $-0.5h^2k^3(0.6k^3 - 2h^2k^3 - 3hk^4 + 4h^8)$
19. $\frac{1}{3}abc^6(9a^2b^2c^2 - 4a^2bc + 12a)$
20. $\frac{2}{5}xy^2z^3(20x^2y^3 - 5x^2z^2 + 10y)$

Objective 2

Prep Exercise 2 Explain how to multiply two multiple-term polynomials.

Prep Exercise 3 Explain how to multiply two binomials.

For Exercises 21–24, a larger rectangle is formed out of smaller rectangles.

a. *Write an expression in simplest form for the length (along the top).*
b. *Write an expression in simplest form for the width (along the side).*
c. *Write an expression that is the product of the length and width that you found in parts a and b.*
d. *Write an expression in simplest form that is the sum of the areas of each of the smaller rectangles.*
e. *Explain why the expressions in parts c and d are equivalent.*

21.

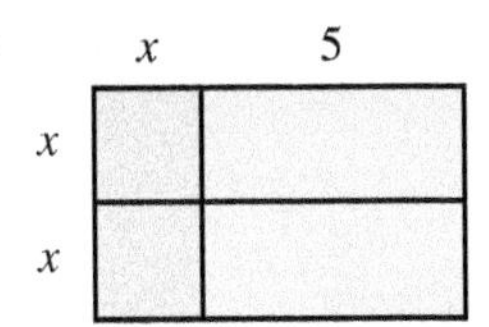

22.

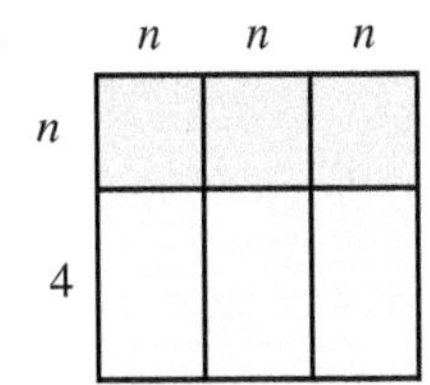

23.

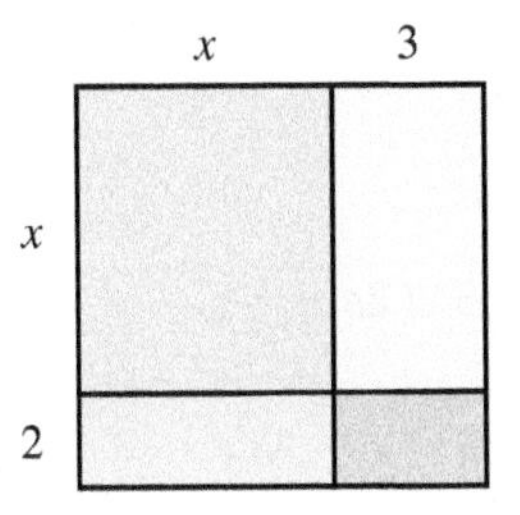

24.

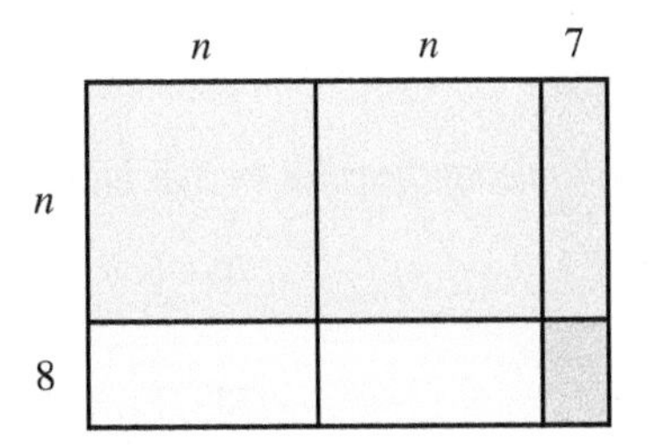

For Exercises 25–36, multiply the binomials. (Use FOIL.) See Example 2.

25. $(2x + 3)(3x + 4)$ **26.** $(3a + 4)(7a + 2)$ **27.** $(3x - 1)(5x + 2)$ **28.** $(4n + 7)(9n - 2)$

29. $(2m - 5n)(3m - 2n)$ **30.** $(2x - 4y)(5x - 3y)$ **31.** $(5m - 3n)(3m + 4n)$ **32.** $(7x + 3y)(3x - 4y)$

33. $(t^2 - 5)(t^2 - 2)$ **34.** $(m^2 - 7)(m^2 - 1)$ **35.** $(a^2 + 6b^2)(a^2 - b^2)$ **36.** $(3x^2 + 4y^2)(x^2 - 5y^2)$

For Exercises 37–48, multiply the polynomials. See Example 3.

37. $(3x - 1)(4x^2 - 2x + 1)$ **38.** $(3x - 1)(2x^2 - 2x + 1)$ **39.** $(7a + 1)(3a^2 - 2a - 2)$

40. $(5x + 1)(3x^2 + 12x + 1)$ **41.** $(7c + 2)(2c^2 - 4c - 3)$ **42.** $(2x - 3)(7x^2 + 3x - 5)$

43. $(4p + 10q)(3p^2 + 2pq + q^2)$ **44.** $(2f + 3g)(f^2 - 5fg - 4g^2)$ **45.** $(2y - 3z)(6y^2 - 2yz + 4z^2)$

46. $(7m - 3n)(2m^2 - 5mn + 3n^2)$ **47.** $(3u^2 - 2u - 1)(3u^2 + 2u + 1)$ **48.** $(2a^2 - 3a + 2)(4a^2 + 2a + 1)$

Objective 3

Prep Exercise 4 If a and b are real numbers, variables, or expressions, then $(a + b)^2 =$ ____________.

Prep Exercise 5 If a and b are real numbers, variables, or expressions, then $(a - b)^2 =$ ____________.

For Exercises 49–60, find the product. See Example 4.

49. $(x + y)^2$ **50.** $(m + n)^2$ **51.** $(4w + 3)^2$

52. $(3k + 2)^2$ **53.** $(4t - 3w)^2$ **54.** $(7a - 2b)^2$

55. $(9 - 5y^2)^2$ **56.** $(6 - 7n^2)^2$ **57.** $[(x + 1) - y]^2$

58. $[(a + 3) - b]^2$ **59.** $[p - (q + 5)]^2$ **60.** $[w - (v + 4)]^2$

Prep Exercise 6 What are conjugates?

For Exercises 61–68, state the conjugate of the given binomial.

61. $x + 8$

62. $y - 4$

63. $3m + 2n$

64. $4q - 9p$

65. $m^2 + n^2$

66. $a^2 - b^2$

67. $-2j - 5k$

68. $-a - 4b$

Prep Exercise 7 If a and b are real numbers, variables, or expressions, then $(a + b)(a - b) =$ ________.

Prep Exercise 8 The product of a conjugate pair is called a ________.

For Exercises 69–78, multiply using the rules for special products. See Example 5.

69. $(2x - 7)(2x + 7)$

70. $(3y + 8)(3y - 8)$

71. $(2q - 5)(2q + 5)$

72. $(3p + 4)(3p - 4)$

73. $(x + 2y)(x - 2y)$

74. $(a + 3b)(a - 3b)$

75. $[(s + 1) + t][(s + 1) - t]$

76. $[(y + 3) + z][(y + 3) - z]$

77. $[3b + (c + 2)][3b - (c + 2)]$

78. $[6k - (x + 1)][6k + (x + 1)]$

For Exercises 79–82, write an expression for the area in simplest form.

79.

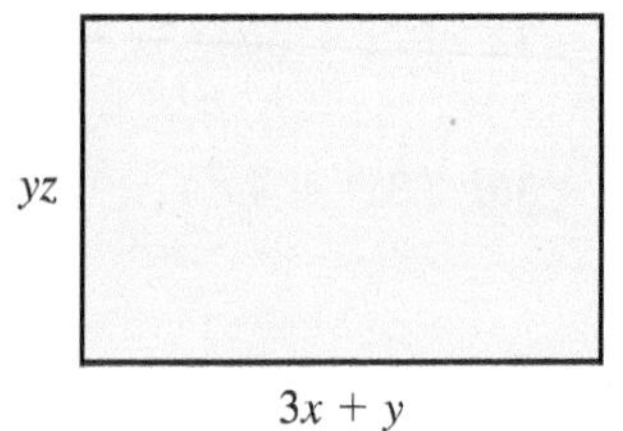

80.

81.

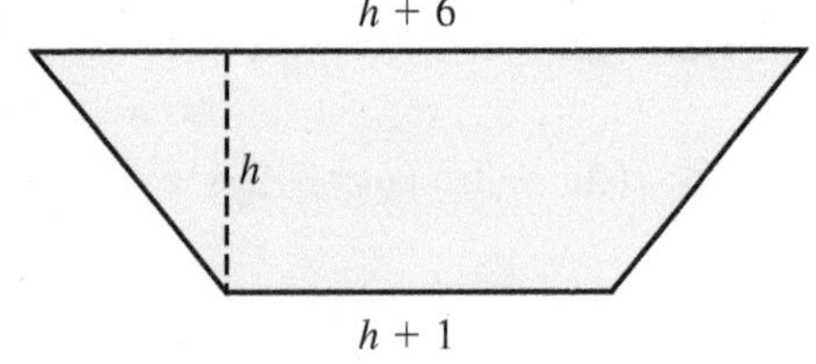

82.

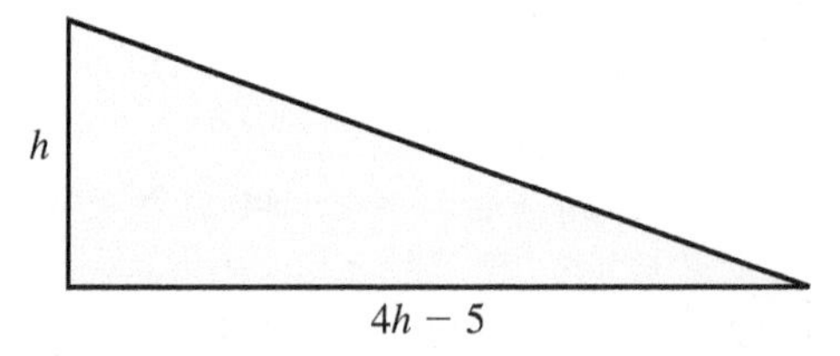

Find the Mistake *For Exercises 83–86, find and explain the mistake; then find the correct product.*

83.
$$(3x - 5)(2x - 9) = 6x^2 - 27x - 10x + 45$$
$$= 6x^2 - 17x + 45$$

84.
$$(t^2 + 3)(t^2 - 1) = t^2 - t^2 + 3t^2 - 3$$
$$= t^2 + 2t^2 - 3$$
$$= 3t^2 - 3$$

85.
$$(2x - 7)(2x + 7) = 4x - 14x + 14x - 49$$
$$= 4x - 49$$

86. $(5y + 8)^2 = 25y^2 + 64$

For Exercises 87–106, find the product. See Examples 1–5.

87. $-4wv(-w^4v^3 - 4x^2wv^7 + 5xw^2v)$

88. $-x^2y(7x^3y^7 - 5x^2y^2 + 1)$

89. $(8r^2 - 3s)(8r^2 + 3s)$

90. $(4b - 5c^2)(4b + 5c^2)$

91. $(2u^2 + 3v^2)^2$

92. $(3r^2 - 7s^2)^2$

93. $(2a^2 - 5b^2)(a^2 - 4b^2)$

94. $(7y^2 + 3x^2)(2y^2 + 5x^2)$

95. $(5q^2 - 3t)(3q^2 + 4t)$

96. $(7k^2 - 2j)(2k^2 + 3j)$

97. $3r^2s^3\left(r^4 - \frac{1}{9}r^3s - \frac{1}{6}r^2s^2 + \frac{1}{3}rs - 1\right)$

98. $4x^4y^2\left(x^3 + \frac{1}{8}x^2y - \frac{1}{16}xy^3 + \frac{1}{4}xy - 3\right)$

99. $-0.1t^2r^3(3.5t^2r^3 - 8t^3r^2 + 2.2t^3r - 2tr)$

100. $-0.2a^3b^2(4.3a^4b^5 - 5ab^4 + 3.2a^2b^3 - 4ab)$

101. $[(3m - 4) + n][(3m - 4) - n]$

102. $[(2c - 5) - d][(2c - 5) + d]$

103. $(x^2 + 3xy + y^2)(4x^2 + 2xy - y^2)$

104. $(x^2 + 6x + 9)(x^2 + 4x + 4)$

105. $[(x^2 - 4) + 3]^2$

106. $[(t^2 + 3) - 5]^2$

Objective 4

For Exercises 107–110, find the products of functions. See Example 6.

107. If $f(x) = 3x - 2$ and $g(x) = 2x - 1$, find
a. $(f \cdot g)(x)$. **b.** $(f \cdot g)(-4)$. **c.** $(f \cdot g)(2)$.

108. If $f(x) = 3x - 4$ and $g(x) = 4x - 1$, find
a. $(f \cdot g)(x)$. **b.** $(f \cdot g)(3)$. **c.** $(f \cdot g)(-1)$.

109. If $f(x) = 3x + 4$ and $g(x) = 4x^2 - 3x + 2$, find
a. $(f \cdot g)(x)$. **b.** $(f \cdot g)(5)$. **c.** $(f \cdot g)(-3)$.

110. If $f(x) = x^2 + 3x - 7$ and $g(x) = 3x + 2$, find
a. $(f \cdot g)(x)$. **b.** $(f \cdot g)(-2)$. **c.** $(f \cdot g)(4)$.

111. If $f(x) = x^2 + 3x - 1$, find $f(n + 1)$.

112. If $f(x) = 3x^2 - x$, find $f(m - 4)$.

113. If $f(x) = 2x^2 - 1$, find $f(t - 6)$.

114. If $f(x) = 3x^2 - 2x - 5$, find $f(2u + 1)$.

★ *For Exercises 115–118, find $f(x + h) - f(x)$.*

115. $f(x) = 2x - 3$

116. $f(x) = 3x + 5$

117. $f(x) = 2x^2 + 3x - 4$

118. $f(x) = 3x^2 - 2x + 5$

★ **119.** A rectangular room has a width that is 2 feet less than the length. Write an expression for the area of the room.

★ **120.** A circular metal plate for a machine has radius r inches. A larger plate with radius $r + 2$ inches is to be used in another part of the machine. Write an expression in simplest form for the sum of the areas of the two circles.

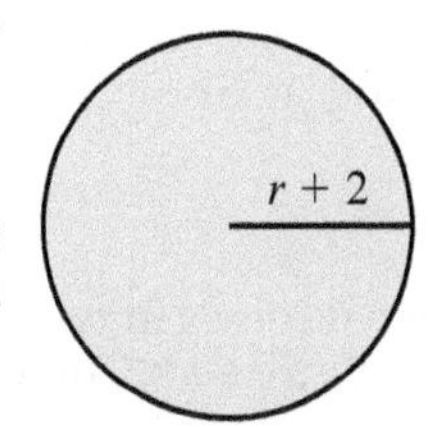

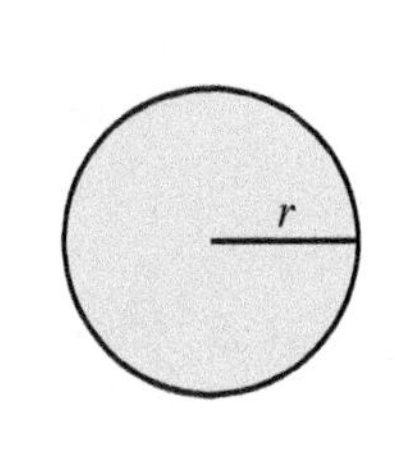

For Exercises 121 and 122, write an expression for the volume in simplest form.

121.

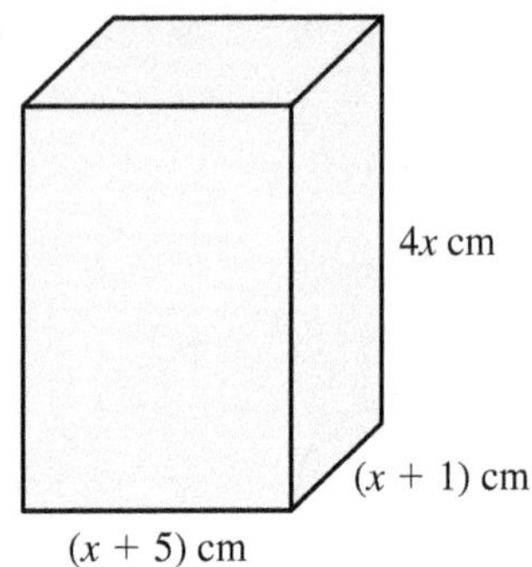

122.

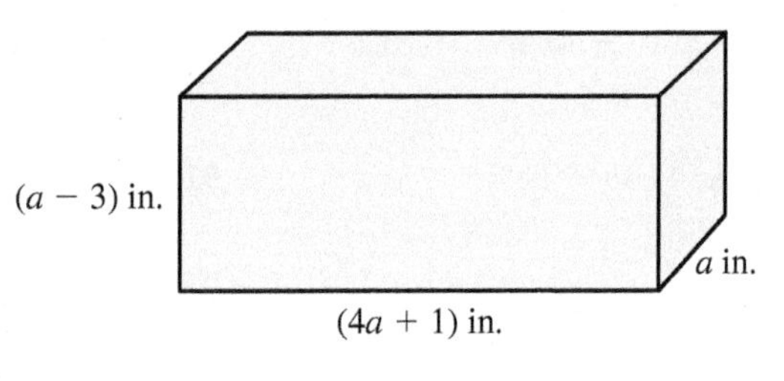

123. A shipping crate's design specifies a length that is triple the width and a height that is 5 feet more than the length. Let w represent the width. Write an expression for the volume in terms of w.

124. A fish tank's design specifies a length that is 8 inches longer than the width. The height of the tank is to be 5 inches longer than the width. Let w represent the width and write an expression for the volume of the tank in terms of w.

★ **125.** A right circular cylinder is to have a height that is 3 inches more than the radius. Write an expression for the volume of the cylinder in terms of the radius.

★ **126.** A cone with a circular base is to have a radius that is 4 centimeters less than the height. Write an expression for the volume of the cone in terms of the height.

Review Exercises

Exercises 1 and 2 Expressions

[5.1] **1.** Multiply $(-4.26 \times 10^6)(2.1 \times 10^5)$. Write the answer in scientific notation.

[5.1] **2.** Divide $\frac{1.5 \times 10^5}{2.4 \times 10^{-4}}$. Write the answer in scientific notation.

Exercises 3–6 Equations and Inequalities

[2.1] **3.** Solve: $\frac{5}{6}x - \frac{1}{8} = \frac{3}{4}x + \frac{1}{2}$

[2.1] **4.** Solve for t in $d = 30t - n$.

[3.3] **5.** Write the equation of a line passing through the point $(-3, 5)$ that is perpendicular to $y = 5x - 2$.

[4.3] **6.** A box of cereal comes in two sizes. The smaller box sells for \$3.50, and the larger box sells for \$4.80. In one week, a grocer sold a total of 84 boxes of that cereal and made \$349.90 in total revenue from the sale. Use a system of equations to find the number of each size box sold.

5.4 Dividing Polynomials

Objectives

1 Divide a polynomial with multiple terms by a monomial.

2 Use long division to divide polynomials.

3 Divide polynomial functions.

Warm-up

[5.1] For Exercises 1–3, simplify. Leave answers with positive exponents only.

1. $\dfrac{32x^5y^3}{8x^2y}$ 2. $\dfrac{42u^2}{6u^2}$ 3. $\dfrac{2xy}{8x^2y}$

[5.3] 4. Simplify. $(6a + b)(2a - 3b)$

In Section 5.1, we learned the quotient rule for exponents. Although we did not say so at that time, we used the quotient rule to divide monomials, as shown here.

$$\frac{28x^5}{4x^2} = \frac{28}{4}x^{5-2} = 7x^3$$

In this section, we use the quotient rule for exponents and divide polynomials with more terms.

Objective 1 Divide a polynomial with multiple terms by a monomial.

First, we consider dividing a polynomial by a monomial, as in $\dfrac{28x^5 + 20x^3}{4x^2}$. Recall that when fractions with a common denominator are added (or subtracted), the numerators are added and the denominator stays the same.

$$\frac{1}{5} + \frac{2}{5} = \frac{1+2}{5}$$

This process can be reversed so that a sum in the numerator of a fraction can be broken into fractions with each addend over the same denominator.

$$\frac{1+2}{5} = \frac{1}{5} + \frac{2}{5}$$

Learning Strategy

When taking notes, develop your own consistent format that is easily readable to you not only during the time you take them, but afterwards when you need to review. It doesn't matter if your notes conform to any particular style, but what is important is that they must speak to you in a consistent way by bearing a structure that you can easily recognize every time you look at them.

—Zeng Alick Z.

Rule Division of a Polynomial by a Monomial

If a, b, and c are real numbers, variables, or expressions with $c \neq 0$, then

$$\frac{a+b}{c} = \frac{a}{c} + \frac{b}{c}.$$

We can apply this rule to $\dfrac{28x^5 + 20x^3}{4x^2}$.

$$\begin{aligned}\frac{28x^5 + 20x^3}{4x^2} &= \frac{28x^5}{4x^2} + \frac{20x^3}{4x^2}\\ &= 7x^{5-2} + 5x^{3-2}\\ &= 7x^3 + 5x\end{aligned}$$

Note We now have a sum of monomial divisions, which we can simplify separately.

This illustration suggests the following procedure.

Procedure Dividing a Polynomial by a Monomial

To divide a polynomial by a monomial, divide each term in the polynomial by the monomial.

Answers to Warm-up

1. $4x^3y^2$ 2. 7 3. $\dfrac{1}{4x}$

4. $12a^2 - 16ab - 3b^2$

Example 1 Divide.

a. $\dfrac{12u^6 - 18u^4 + 42u^2}{6u^2}$

Solution: $\dfrac{12u^6 - 18u^4 + 42u^2}{6u^2} = \dfrac{12u^6}{6u^2} - \dfrac{18u^4}{6u^2} + \dfrac{42u^2}{6u^2}$ Divide each term in the polynomial by the monomial.

$= 2u^4 - 3u^2 + 7$ Divide.

Connection In Section 5.1, we learned that $\dfrac{42u^2}{6u^2} = 7u^{2-2} = 7u^0 = 7 \cdot 1 = 7.$

b. $(32x^5y^3 - 24x^4y^2 - 2xy) \div 8x^2y$

Solution: $(32x^5y^3 - 24x^4y^2 - 2xy) \div 8x^2y$

$= \dfrac{32x^5y^3 - 24x^4y^2 - 2xy}{8x^2y}$

$= \dfrac{32x^5y^3}{8x^2y} - \dfrac{24x^4y^2}{8x^2y} - \dfrac{2xy}{8x^2y}$ Divide each term in the polynomial by the monomial.

$= 4x^3y^2 - 3x^2y - \dfrac{1}{4x}$

Note Using the quotient rule, we have $\dfrac{2xy}{8x^2y} = \dfrac{1}{4}x^{-1} = \dfrac{1}{4x}$.

Warning A common error is to write the answer of Example 1(b) as $\dfrac{4x^3y^2 - 3x^2y - 1}{4x}$.

Your Turn 1 Divide.

a. $\dfrac{54x^5 + 42x^4 - 24x^3}{6x^3}$

b. $(12a^2b^5 - 20a^4b + 6a^2) \div 4a^2b$

Objective 2 Use long division to divide polynomials.

To divide a polynomial by a polynomial, we can use long division. As a reminder of the process of long division, let's divide 157 by 12 using long division.

Divide: $\dfrac{157}{12}$

```
             Quotient
              ↙
             13
Divisor ⟶ 12)157 ⟵ Dividend
            −12
              37
             −36
Remainder ⟶   1
```

Because the answer has a remainder, we write the result as a mixed number, $13\frac{1}{12}$. Notice that we could check the answer by multiplying the divisor by the quotient and then add the remainder. The result should be the dividend.

Quotient · Divisor + Remainder = Dividend

13 · 12 + 1 = 157

Now consider polynomial division, which follows the same long division process.

Example 2 Divide $\dfrac{x^2 + 5x + 7}{x + 2}$.

Solution:

$$x + 2\overline{)x^2 + 5x + 7}\quad \text{quotient: } x$$

Divide the first terms to determine the first term in the quotient: $x^2 \div x = x$.

Next, we multiply the divisor $x + 2$ by the x in the quotient.

multiply

$$\begin{array}{r} x \\ x + 2\overline{)x^2 + 5x + 7} \\ x^2 + 2x \end{array}$$

Answers to Your Turn 1

a. $9x^2 + 7x - 4$

b. $3b^4 - 5a^2 + \dfrac{3}{2b}$

Next, subtract. Recall that to subtract a polynomial, we change the signs of its terms. After combining terms, we bring down the next term in the dividend, which is 7.

$$\begin{array}{r} x \\ x+2\overline{)x^2+5x+7} \\ -(x^2+2x) \\ \hline \end{array} \xrightarrow{\text{Change signs.}} \begin{array}{r} x \\ x+2\overline{)x^2+5x+7} \\ -x^2-2x \quad \downarrow \\ \hline 3x+7 \end{array}$$

Combine like terms and bring down the next term.

Now we repeat the process with $3x + 7$ as the dividend.

multiply

$$\begin{array}{r} x+3 \\ x+2\overline{)x^2+5x+7} \\ -x^2-2x \\ \hline 3x+7 \\ 3x+6 \end{array}$$

Divide the first term of $3x + 7$ by the first term in the divisor, $x + 2$, which gives $3x \div x = 3$. We then multiply the divisor by this 3.

Subtract. As before, we change the signs of the binomial to be subtracted.

$$\begin{array}{r} x+3 \\ x+2\overline{)x^2+5x+7} \\ -x^2-2x \\ \hline 3x+7 \\ -(3x+6) \\ \hline \end{array} \xrightarrow{\text{Change signs.}} \begin{array}{r} x+3 \\ x+2\overline{)x^2+5x+7} \\ -x^2-2x \\ \hline 3x+7 \\ -3x-6 \\ \hline 1 \end{array}$$

Combine like terms.

Note that we have a remainder, 1. Recall that in the numeric version, we wrote the answer as a mixed number, $13\frac{1}{12}$, which means $13 + \frac{1}{12}$ and is in the following form: quotient + $\frac{\text{remainder}}{\text{divisor}}$. With polynomials, we write a similar expression.

$$\underbrace{x+3}_{\text{quotient}} + \frac{1}{x+2} \begin{array}{l} \leftarrow \text{remainder} \\ \leftarrow \text{divisor} \end{array}$$

To check, we multiply the quotient and the divisor, then add the remainder. The result should be the dividend.

$$\begin{aligned} (x+3)(x+2)+1 &= x^2+2x+3x+6+1 \\ &= x^2+5x+7 \end{aligned}$$

It checks.

Procedure Dividing a Polynomial by a Polynomial

To divide a polynomial by a polynomial:

1. Use long division.
2. If there is a remainder, write the result in the form of quotient + $\frac{\text{remainder}}{\text{divisor}}$.

Example 3 Divide $\frac{15x^2 - 26x + 17}{5x - 2}$.

Solution: Begin by dividing the first term in the dividend by the first term in the divisor: $15x^2 \div 5x = 3x$.

$$\begin{array}{r} 3x \\ 5x-2\overline{)15x^2-26x+17} \\ -(15x^2-6x) \\ \hline \end{array} \xrightarrow{\text{Change signs.}} \begin{array}{r} 3x \\ 5x-2\overline{)15x^2-26x+17} \\ -15x^2+6x \\ \hline -20x+17 \end{array}$$

Combine like terms and bring down the next term.

To find the next part of the quotient, we divide $-20x$ by $5x$, which is -4. We then repeat the multiplication and subtraction steps.

$$\begin{array}{r} 3x - 4 \\ 5x - 2\overline{)15x^2 - 26x + 17} \\ \underline{-15x^2 + 6x} \\ -20x + 17 \\ \underline{-(-20x + 8)} \end{array} \xrightarrow{\text{Change signs.}} \begin{array}{r} 3x - 4 \\ 5x - 2\overline{)15x^2 - 26x + 17} \\ \underline{-15x^2 + 6x} \\ -20x + 17 \\ \underline{+20x - 8} \\ 9 \end{array} \quad \text{Combine like terms.}$$

Answer: $3x - 4 + \dfrac{9}{5x - 2}$

Your Turn 3 Divide.

a. $\dfrac{x^2 + 9x + 18}{x + 3}$

b. $\dfrac{12x^2 - 28x + 17}{2x - 3}$

Using a Placeholder in Long Division

In polynomial division, if the dividend, written in descending order of degree, has a missing term, that term is written with a 0 coefficient as a placeholder.

Example 4 Divide $\dfrac{9x^4 - 5 - 7x^2 - 10x}{3x - 1}$.

Solution: In descending order of degree, the dividend is $9x^4 - 7x^2 - 10x - 5$. Because the x^3 term is missing, we write $0x^3$ as a placeholder in the long division.

$$\begin{array}{r} 3x^3 + x^2 - 2x - 4 \\ 3x - 1\overline{)9x^4 + 0x^3 - 7x^2 - 10x - 5} \\ \underline{-9x^4 + 3x^3} \qquad\qquad\qquad\quad \\ 3x^3 - 7x^2 \qquad\qquad \\ \underline{-3x^3 + x^2} \qquad\qquad \\ -6x^2 - 10x \qquad \\ \underline{+6x^2 - 2x} \qquad \\ -12x - 5 \\ \underline{+12x - 4} \\ -9 \end{array}$$

Note For simplicity, we will use this second approach to write negative remainders. ▼

Answer: $3x^3 + x^2 - 2x - 4 + \dfrac{-9}{3x - 1}$ or $3x^3 + x^2 - 2x - 4 - \dfrac{9}{3x - 1}$

Your Turn 4

Divide $\dfrac{-12x + 8x^3 - 8}{2x + 1}$.

Objective 3 Divide polynomial functions.

We can find the quotient of two functions using the following rule.

Rule Dividing Functions

The quotient of two functions, f/g, is found by $(f/g)(x) = \dfrac{f(x)}{g(x)}$, where $g(x) \neq 0$.

Answers to Your Turn 3

a. $x + 6$

b. $6x - 5 + \dfrac{2}{2x - 3}$

Answer to Your Turn 4

$4x^2 - 2x - 5 - \dfrac{3}{2x + 1}$

Example 5 a. If $f(x) = x^3 - x + 7$ and $g(x) = x + 2$, find $(f/g)(x)$.

b. Find $(f/g)(4)$.

Solution: a.

$$\begin{array}{r} x^2 - 2x + 3 \\ x + 2\overline{)x^3 + 0x^2 - x + 7} \\ \underline{-x^3 - 2x^2} \qquad\quad \\ -2x^2 - x \quad \\ \underline{2x^2 + 4x} \quad \\ 3x + 7 \\ \underline{-3x - 6} \\ 1 \end{array}$$

Answer: $(f/g)(x) = x^2 - 2x + 3 + \dfrac{1}{x+2}$

Note $(f/g)(4)$ can also be found using $\dfrac{f(4)}{g(4)}$. ▶

b. $(f/g)(4) = (4)^2 - 2(4) + 3 + \dfrac{1}{4+2}$ In $(f/g)(x)$, replace x with 4.

$= 16 - 8 + 3 + \dfrac{1}{6}$ Simplify.

$= 11\dfrac{1}{6}$, or $\dfrac{67}{6}$

Answers to Your Turn 5

a. $(f/g)(x) = 3x^2 - 8x + 4 - \dfrac{5}{2x+1}$

b. 20

Your Turn 5

a. If $f(x) = 6x^3 - 13x^2 - 1$ and $g(x) = 2x + 1$, find $(f/g)(x)$.

b. Find $(f/g)(-1)$.

5.4 Exercises

For Extra Help MyMathLab®

Objective 1

Prep Exercise 1 Explain how to divide a polynomial by a monomial.

For Exercises 1–14, divide the polynomial by the monomial. See Example 1.

1. $\dfrac{28a^4 - 7a^3 + 21a^2}{7a^2}$

2. $\dfrac{8k^3 - 4k^2 + 2k}{2k}$

3. $(12u^5 - 6u^4 - 15u^3 + 3u^2) \div 3u^2$

4. $(15m^5 - 10m^4 + 20m^3 - 5m^2) \div 5m^2$

5. $\dfrac{24a^5b^4 - 8a^4b^2 + 16a^2b}{8ab}$

6. $\dfrac{12m^4n^2 - 4m^3n + 8m^2n}{4mn}$

7. $\dfrac{36u^3v^4 + 12uv^5 - 15u^2v^2}{3uv^9}$

8. $\dfrac{16h^3k^4 - 25hk - 4h^2k^2}{4hk^6}$

9. $\dfrac{30x^3y^2 - 45xy^3 - xy}{-5xy}$

10. $\dfrac{18a^5b^2 - 24a^4b + 2a^2b}{-6a^2b}$

11. $(12t^6u^5v - 2t^4u^4 - 16t^3u^2 + 3tu^2) \div (-4t^3u^2)$

12. $(40x^7y^4z - 4x^4y^5 - 16x^3y^4 + 5xy^2) \div (-8x^2y^4)$

13. $\dfrac{18a^3b^2c^2 + 12a^2b^3c - 3a^2b}{9a^2bc}$

14. $\dfrac{24m^4n^3p^2 - 9m^2n^3p + 6m^2p}{12m^2np}$

15. The area of the parallelogram shown is described by the monomial $25xy^2$. Find the height.

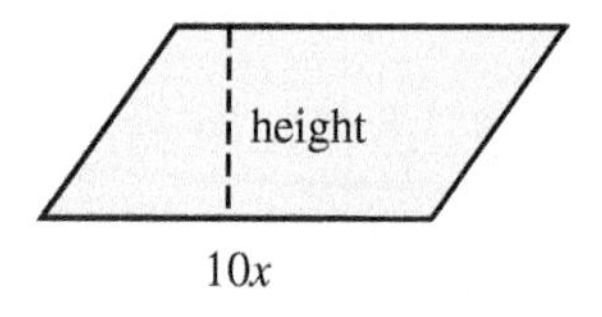

16. The area of the triangle shown is described by the monomial $30m^2n$. Find the base.

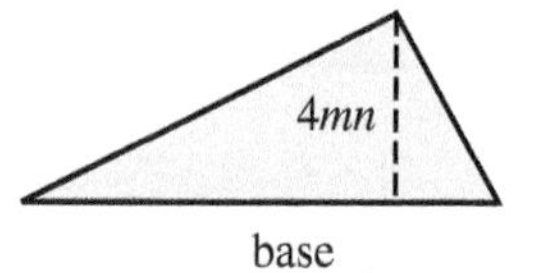

17. A pool company is designing a pool cover in the shape of a rectangle. The area of the cover is described by the polynomial $6x^2 - 12x + 18$, and the length must be $3x$.
   a. Find an expression for the width.

   b. Find the area, length, and width if $x = 5$.

18. The voltage in a circuit is described by the polynomial $48n^3 - 24n^2 + 36n$. The resistance is described by $6n$.
   a. Find an expression for the current (voltage = current × resistance).

   b. Find the voltage, current, and resistance if $n = 2$.

Objective 2

Prep Exercise 2 Explain how to determine the first term in the quotient when using long division to divide $6x^2 + 8x + 7$ by $2x + 1$.

Prep Exercise 3 When dividing polynomials using long division, after determining the first term in the quotient, what is the next step?

Prep Exercise 4 After dividing a polynomial by $6x - 1$, the quotient is $3x^2 - 5x + 2$ and the remainder is 7. Write the final answer expression with the quotient and remainder.

Prep Exercise 5 After dividing a polynomial by $3x + 2$, the quotient is $4x - 8$ and the remainder is -5. Write the final answer expression with the quotient and remainder.

Prep Exercise 6 When setting up the long division for $\dfrac{8x^3 + 9x - 4}{2x - 1}$, why is a 0 placeholder needed?

For Exercises 19–48, use long division to divide the polynomials. See Examples 2–4.

19. $\dfrac{x^2 + 10x + 24}{x + 6}$

20. $\dfrac{c^2 - 11c + 24}{c - 3}$

21. $\dfrac{3p^2 + 4p + 1}{p + 1}$

22. $\dfrac{3m^2 - 17m + 10}{m - 5}$

23. $(5n^2 - 3n - 8) \div (n + 1)$

24. $(3q^2 - 10q + 3) \div (q - 3)$

25. $\dfrac{15x^2 + 2x - 8}{5x + 4}$

26. $\dfrac{4x^2 + 4x - 3}{2x + 3}$

27. $\dfrac{6y^2 + 5y - 4}{3y + 4}$

28. $\dfrac{6r^2 - 13r + 6}{3r - 2}$

29. $\dfrac{2x^3 - 13x^2 + 27x - 18}{2x - 3}$

30. $\dfrac{3m^3 + 10m^2 + 9m + 2}{3m + 1}$

31. $\dfrac{x^3 + 8}{x + 2}$

32. $\dfrac{a^3 + 125}{a + 5}$

33. $\dfrac{y^4 - 16}{y - 2}$

34. $\dfrac{q^4 - 81}{q + 3}$

35. $\dfrac{8z^3 + 125}{2z + 5}$

36. $\dfrac{8b^3 - 27}{2b - 3}$

37. $(2v^3 + 7v^2 + 7v + 2) \div (2v + 1)$

38. $(2w^3 + 8w^2 + 2w - 12) \div (2w + 4)$

39. $\dfrac{-20a^2 + 13a + 21a^3 - 6}{3a - 2}$

40. $\dfrac{14w - 19w^2 + 21w^3 + 24}{3w + 2}$

41. $\dfrac{13q + 12q^3 + 4q^2 + 9}{2q - 1}$

42. $\dfrac{4s^2 - 13s + 9 + 12s^3}{2s - 1}$

43. $\dfrac{3x^4 - 6x^3 + 17x^2 - 24x + 20}{x^2 + 4}$

44. $\dfrac{2x^4 + 7x^3 - 14x^2 - 21x + 24}{x^2 - 3}$

45. $\dfrac{-10c - 8 + 3c^4 - 4c^3 - c^2}{3c + 2}$

46. $\dfrac{2x^4 - 28 - x + x^3 - 19x^2}{2x + 7}$

47. $\dfrac{-u^2 + 5u^3 + 10u + 2}{u + 2}$

48. $\dfrac{6w^4 - 11w^3 - 8 - w^2 + 16w}{2w - 3}$

49. The volume of a rectangular spa is described by $(6t^3 + 30t^2 + 12t - 48)$ cubic feet. The width is to be 6 feet, and the height (depth) is described by $(t + 2)$ feet.
   a. Find an expression for the length of the spa.

   b. Find the length, height, and volume of the spa if $t = 2$.

50. On a blueprint, the specifications for a rectangular room call for the area to be $(3y^4 + 11y^3 + 22y^2 + 23y + 5)$ square feet. It is decided that the length should be described by the binomial $(3y + 5)$ feet.
   a. Find an expression for the width.

   b. Find the length, width, and area of the room if $y = 2$.

Objective 3

For Exercises 51–56, divide the polynomial functions. See Example 5.

51. If $f(x) = 24x^4 - 42x^3 - 30x^2$ and $g(x) = 6x^2$, find
a. $(f/g)(x)$. **b.** $(f/g)(4)$. **c.** $(f/g)(-2)$.

52. If $r(x) = 54x^6 - 36x^4 - 27x^3$ and $t(x) = -9x^3$, find
a. $(r/t)(x)$. **b.** $(r/t)(3)$. **c.** $(r/t)(-1)$.

53. If $h(x) = 6x^3 + 7x^2 - 9x + 2$ and $k(x) = 2x - 1$, find
a. $(h/k)(x)$. **b.** $(h/k)(3)$. **c.** $(h/k)(-1)$.

54. If $f(x) = 18x^3 - 39x^2 + 30x - 8$ and $g(x) = 3x - 2$, find
a. $(f/g)(x)$. **b.** $(f/g)(5)$. **c.** $(f/g)(-4)$.

55. If $n(x) = x^3 - x^2 - 7x + 4$ and $p(x) = x - 3$, find
a. $(n/p)(x)$. **b.** $(n/p)(4)$. **c.** $(n/p)(-2)$.

56. If $c(x) = 2x^4 + 11x^3 - 40x - 20$ and $d(x) = x + 4$, find
a. $(c/d)(x)$. **b.** $(c/d)(2)$. **c.** $(c/d)(-2)$.

Review Exercises

Exercises 1–6 Equations and Inequalities

For Exercises 1 and 2, solve.

[2.1] 1. $9x - (6x + 2) = 5x - 2[4x - (6 - x)]$

[2.3] 2. $7x - 5 \geq 8 - (2x + 1)$

[3.3] 3. Find the equation of a line perpendicular to the graph of $5x - 2y = 10$ that passes through the point $(3, -1)$.

[3.5] 4. Is the relation graphed to the right a function? Explain.

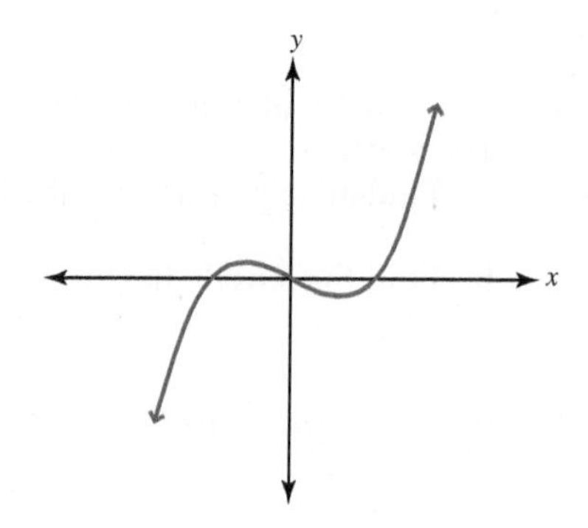

[4.2] 5. Solve the following system using elimination: $\begin{cases} 2x + 3y - z = 7 \\ x + y - 5z = -14 \\ -3x - 5y + z = -12 \end{cases}$

[4.5] 6. Solve the system of inequalities: $\begin{cases} x - y > 3 \\ 2x + y \leq 4 \end{cases}$

5.5 Synthetic Division and the Remainder Theorem

Objectives

1 Use synthetic division to divide a polynomial by a binomial in the form $x - c$.

2 Use the remainder theorem to evaluate polynomials.

Warm-up

[3.5] **1.** If $P(x) = x^5 - 8x^3 - 5x^2 + x + 11$, find $P(3)$.

[5.4] **2.** Divide $\frac{x^2 + 5x + 7}{x + 2}$ using long division.

[5.3] **3.** Simplify $(x + 2)(x + 3) + 1$. How does Exercise 3 relate to Exercise 2?

Objective 1 Use synthetic division to divide a polynomial by a binomial in the form $x - c$.

We saw in Section 5.4 that long division can be a tedious process. When a polynomial is divided by a binomial in the form $x - c$, where c is a constant, we can streamline the process by focusing on just the coefficients in the polynomials using a method called *synthetic division.*

To get a sense of how synthetic division looks, let's look again at the long division for $\frac{x^2 + 5x + 7}{x + 2}$, which was Example 2 in Section 5.4. To the right of that long division, we gradually remove bits so that we are left with the bare essentials for finding the quotient.

Long Division	Variables Removed	Bare Essentials
$x + 3$	$1 + 3$	$1 + 3$
$x + 2\overline{)x^2 + 5x + 7}$	$1 + 2\overline{)1 + 5 + 7}$	$1 - (-2)\overline{)1 + 5 + 7}$
$-x^2 - 2x$	$-1 - 2$	-2
$3x + 7$	$3 + 7$	$+7$
$-3x - 6$	$-3 - 6$	-6
1	1	1

Note In polynomial long division, the terms in the positions that we've highlighted in blue are always additive inverses. Because they always cancel, we can remove them to get the "bare-essentials" version, as seen to the right.

Note We have rewritten $x + 2$ as $x - (-2)$ so that we have the form $x - c$.

We can place these "bare-essential" values in an even more compact form, which is the form for synthetic division.

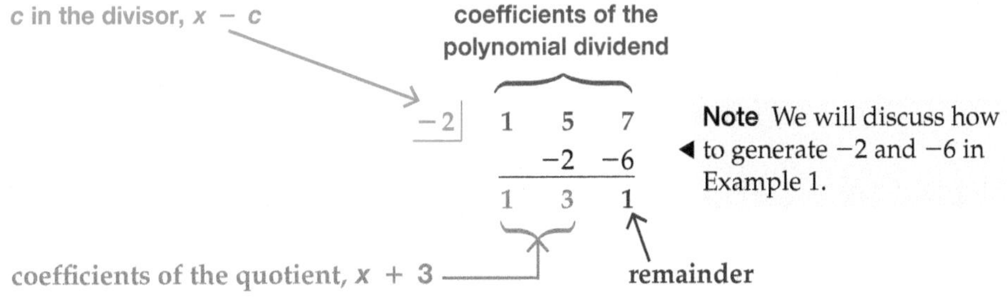

Now let's examine the steps of synthetic division in Example 1.

Answers to Warm-up

1. -4 **2.** $x + 3 + \frac{1}{x + 2}$

3. $x^2 + 5x + 7$, It is the check for Exercise 2.

Example 1 Use synthetic division to divide.

a. $\dfrac{x^2 + 5x + 7}{x + 2}$

Solution: First, we need the divisor to be in the form $x - c$. So we rewrite $x + 2$ as $x - (-2)$, and we see that $c = -2$. We then set up the synthetic division with c and the coefficients of the dividend, as shown.

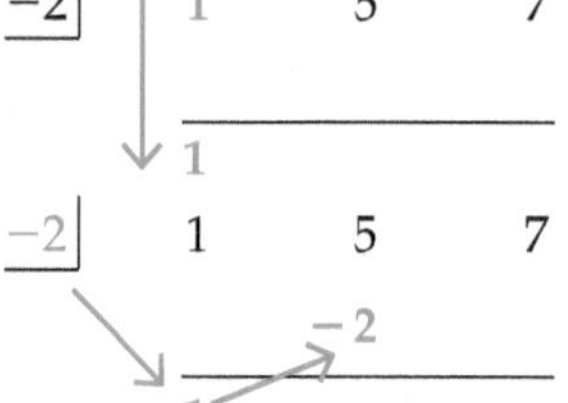

Our first step is to bring down the leading coefficient.

Multiply 1 by −2 and place the result underneath the next coefficient, 5.

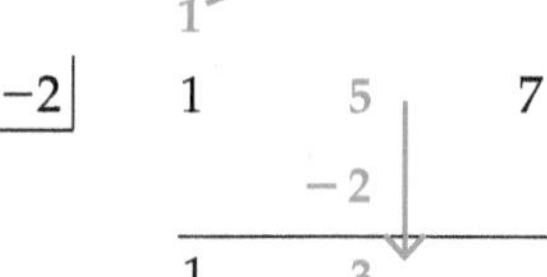

Add: $5 - 2 = 3$.

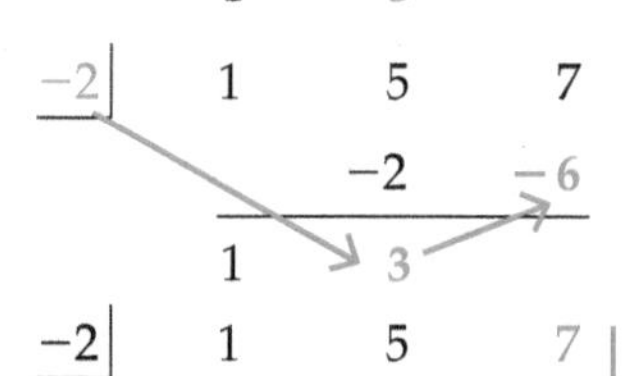

Multiply 3 by −2 and place the result underneath the next coefficient, 7.

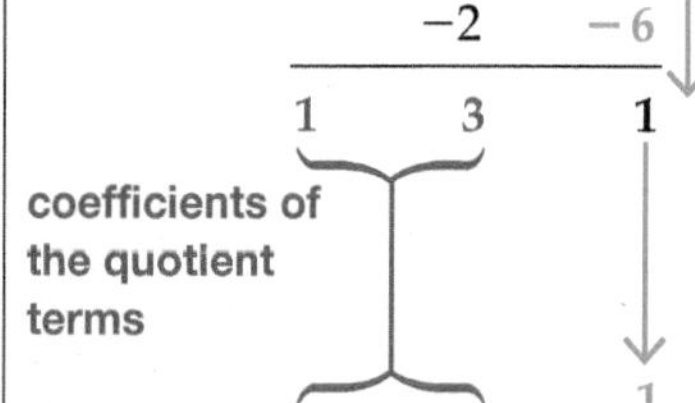

Add: $7 - 6 = 1$, which is the remainder.

coefficients of the quotient terms

remainder

Answer: $1x + 3 + \dfrac{1}{x + 2}$ or $x + 3 + \dfrac{1}{x + 2}$

Note In synthetic division, the degree of the quotient polynomial is always one less than the degree of the dividend polynomial. Because the degree of the dividend $x^2 + 5x + 7$ is 2, the degree of the quotient, $x + 3$, is 1.

b. $(x^4 - 6x^3 + 35x - 17) \div (x - 4)$

Solution: Looking at the divisor, $x - 4$, we see that $c = 4$.

Note Because the polynomial dividend has no x^2 term, we place a 0 in the second degree position as a placeholder.

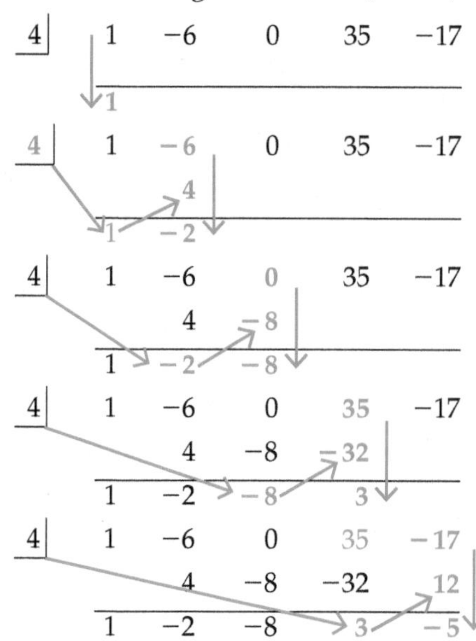

List the coefficients in the dividend and bring down the leading coefficient.

Multiply 1 by 4 and place the result underneath the next coefficient, −6. Then add.

Multiply −2 by 4 and place the result underneath the next coefficient, 0. Then add.

Multiply −8 by 4 and place the result underneath the next coefficient, 35. Then add.

Multiply 3 by 4 and place the result underneath the next coefficient, −17. Then add.

Answer: $x^3 - 2x^2 - 8x + 3 - \dfrac{5}{x - 4}$

Your Turn 1 Use synthetic division to divide.

a. $\dfrac{4x^3 + 6x^2 - 9x + 27}{x + 3}$

b. $(x^4 + x^3 - 26x^2 + 27x - 15) \div (x - 4)$

Objective 2 Use the remainder theorem to evaluate polynomials.

There is a helpful connection between evaluating a polynomial using a value c and dividing that polynomial by $x - c$. In Example 1, we used synthetic division to divide $x^2 + 5x + 7$ by $x + 2$, which means that $c = -2$, and we found the remainder to be 1. If we evaluate the polynomial by substituting -2 for x, we will get that remainder value, 1.

$$(-2)^2 + 5(-2) + 7 = 4 - 10 + 7 = 1$$

This suggests the following rule, which is called the *remainder theorem*. The proof of this theorem is beyond the scope of this text.

Rule The Remainder Theorem

Given a polynomial $P(x)$, the remainder of $\dfrac{P(x)}{x - c}$ is equal to $P(c)$.

Example 2 For $P(x) = x^5 - 8x^3 - 5x^2 + x + 11$, use the remainder theorem to find $P(3)$.

Note Using synthetic division and the remainder theorem avoids the big numbers that we often get when evaluating exponential terms.

Solution: To use the remainder theorem to find $P(3)$, we need to divide $P(x)$ by $x - 3$. We use synthetic division.

$$\begin{array}{r|rrrrrr} 3 & 1 & 0 & -8 & -5 & 1 & 11 \\ & & 3 & 9 & 3 & -6 & -15 \\ \hline & 1 & 3 & 1 & -2 & -5 & -4 \end{array}$$

The remainder is -4, so $P(3) = -4$.

We can verify our answer by finding $P(3)$ using substitution.

$$P(3) = (3)^5 - 8(3)^3 - 5(3)^2 + (3) + 11 = 243 - 216 - 45 + 3 + 11 = -4$$

Answers to Your Turn 1

a. $4x^2 - 6x + 9$

b. $x^3 + 5x^2 - 6x + 3 - \dfrac{3}{x - 4}$

Answer to Your Turn 2

8

Your Turn 2 For $P(x) = x^4 + 4x^3 + 23x - 2$, use the remainder theorem to find $P(-5)$.

5.5 Exercises

For Extra Help MyMathLab®

Note: Exercises marked with a ★ represent challenging exercises.

Objective 1

Prep Exercise 1 What form does the divisor need to have so that synthetic division can be used?

Prep Exercise 2 Suppose you are to use synthetic division to divide $\dfrac{x^3 - 8x^2 - 5x + 9}{x + 6}$. What is the value of c?

Prep Exercise 3 If you were to use synthetic division for $\dfrac{x^4 - 5x^2 - 3x + 11}{x - 2}$, what list of coefficients would you write?

Prep Exercise 4 Explain how to finish the synthetic division shown.

$$\begin{array}{r|rrrr} 2 & 1 & 4 & -8 & -9 \\ & & 2 & 12 & \\ \hline & 1 & 6 & 4 & \end{array}$$

Prep Exercise 5 Where does the remainder appear in synthetic division?

Prep Exercise 6 When using synthetic division, how do you determine the degree of the first term in the quotient expression?

For Exercises 1–6: a. Write the binomial divisor.
b. Write the polynomial dividend.
c. Write the polynomial quotient with its remainder (if there is a remainder). See Objective 1.

1.
$$\begin{array}{r|rrr} -3 & 1 & 7 & 12 \\ & & -3 & -12 \\ \hline & 1 & 4 & 0 \end{array}$$

2.
$$\begin{array}{r|rrr} 9 & 1 & -14 & 45 \\ & & 9 & -45 \\ \hline & 1 & -5 & 0 \end{array}$$

3.
$$\begin{array}{r|rrrr} 3 & 1 & 4 & -25 & 7 \\ & & 3 & 21 & -12 \\ \hline & 1 & 7 & -4 & -5 \end{array}$$

4.
$$\begin{array}{r|rrrr} -2 & 1 & -2 & -6 & 7 \\ & & -2 & 8 & -4 \\ \hline & 1 & -4 & 2 & 3 \end{array}$$

5.
$$\begin{array}{r|rrrrr} 4 & 2 & -8 & -5 & 17 & 10 \\ & & 8 & 0 & -20 & -12 \\ \hline & 2 & 0 & -5 & -3 & -2 \end{array}$$

6.
$$\begin{array}{r|rrrrr} -5 & 3 & 16 & 0 & -27 & -9 \\ & & -15 & -5 & 25 & 10 \\ \hline & 3 & 1 & -5 & -2 & 1 \end{array}$$

For Exercises 7–32, divide using synthetic division. See Example 1.

7. $\dfrac{x^2 - 5x + 6}{x - 2}$

8. $\dfrac{x^2 + 8x - 9}{x - 1}$

9. $\dfrac{2x^2 - x - 5}{x - 4}$

10. $\dfrac{3x^2 - 11x - 7}{x - 5}$

11. $\dfrac{x^3 - x^2 - 5x + 6}{x - 2}$

12. $\dfrac{2x^3 - 6x^2 - 5x + 15}{x - 3}$

13. $\dfrac{3x^3 + 8x^2 + 3x - 2}{x + 2}$

14. $\dfrac{5x^3 + 32x^2 + 52x + 16}{x + 2}$

15. $(3x^3 - x^2 - 22x + 24) \div (x + 3)$

16. $(8x^3 - 10x^2 - x + 3) \div (x - 1)$

17. $\dfrac{2x^3 - 5x^2 - x + 6}{x + 2}$

18. $\dfrac{3x^3 - 5x^2 - 16x + 12}{x - 2}$

19. $(x^3 - 2x^2 + x - 3) \div (x - 2)$

20. $(x^3 + 6x^2 + 2x - 3) \div (x + 6)$

21. $\dfrac{2x^3 + x^2 - 4}{x + 3}$

22. $\dfrac{4x^3 - 2x + 5}{x - 2}$

23. $\dfrac{3x^3 - 4x^2 + 2}{x - 3}$

24. $\dfrac{x^3 + 4x^2 - 7}{x + 3}$

25. $(x^3 - 7x + 6) \div (x - 2)$

26. $(x^3 + 2x + 135) \div (x + 5)$

27. $\dfrac{x^3 + 27}{x - 3}$

28. $\dfrac{x^3 - 8}{x + 2}$

29. $\dfrac{x^4 + 16}{x + 2}$

30. $\dfrac{x^4 + 81}{x - 3}$

★ **31.** $\dfrac{6x^3 + x^2 - 11x - 6}{x + \frac{2}{3}}$

★ **32.** $\dfrac{2x^3 - 13x^2 + 27x - 18}{x - \frac{3}{2}}$

Objective 2

Prep Exercise 7 In your own words, explain what the remainder theorem says.

Prep Exercise 8 What is the advantage of using the remainder theorem as opposed to substitution when evaluating a polynomial?

For Exercises 33–44, use the remainder theorem to find $P(c)$. See Example 2.

33. For $P(x) = 3x^3 - 8x^2 + 8x - 3$, find $P(3)$.

34. For $P(x) = 8x^3 + 2x^2 - 13x + 3$, find $P(1)$.

35. For $P(x) = 6x^3 - 13x^2 + x + 2$, find $P(2)$.

36. For $P(x) = x^3 + x^2 - 2x + 12$, find $P(3)$.

37. For $P(x) = 3x^3 - 11x - 7$, find $P(-3)$.

38. For $P(x) = 2x^3 - 26x - 24$, find $P(-2)$.

39. For $P(x) = x^4 - x^3 - 2x^2 + x - 2$, find $P(1)$.

40. For $P(x) = x^4 + 3x^2 - 15x + 7$, find $P(2)$.

41. For $P(x) = 2x^4 - 6x^2 - 5x + 4$, find $P(-1)$.

42. For $P(x) = 3x^4 + x^3 + 5x - 1$, find $P(-1)$.

43. For $P(x) = x^5 - x^4 + 3x^2 - 3x - 3$, find $P(-2)$.

44. For $P(x) = x^5 + 5x^4 - x - 6$, find $P(-5)$.

45. The area of the parallelogram shown is described by $9x^2 - 10x - 16$.
  a. Find an expression for the length of the base.
  b. Find the base, height, and area of the parallelogram if $x = 12$ inches.

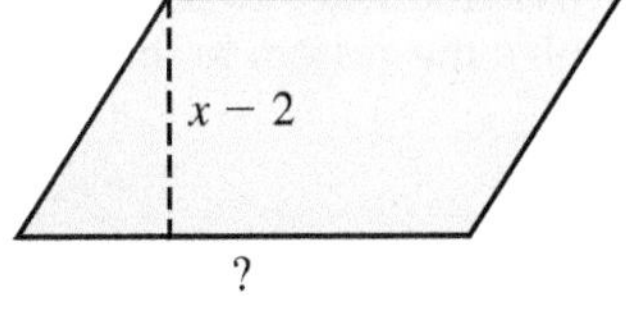

46. The area of the rectangle shown is $12n^2 + 65n - 42$.
  a. Find the length.
  b. Find the length, width, and area of the rectangle if $n = 9$ centimeters.

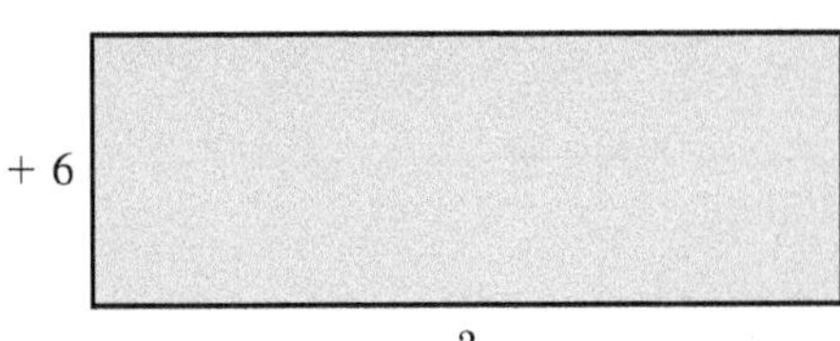

47. The volume of the walk-in freezer shown is described by $2y^3 - 7y^2 - 38y + 88$.
  a. Find an expression for the height of the room.
  b. Find the length, width, height, and volume of the room if $y = 10$ feet.

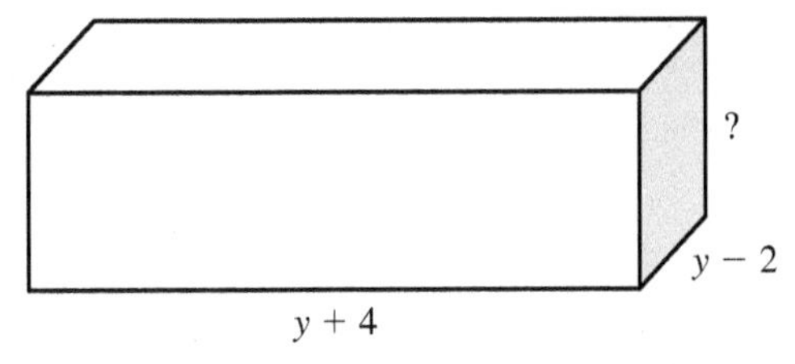

48. The volume of the storage box shown is described by $2x^3 - 168x - 320$.
  a. Find an expression for the length of the box.
  b. Find the length, width, height, and volume of the box if $x = 16$ inches.

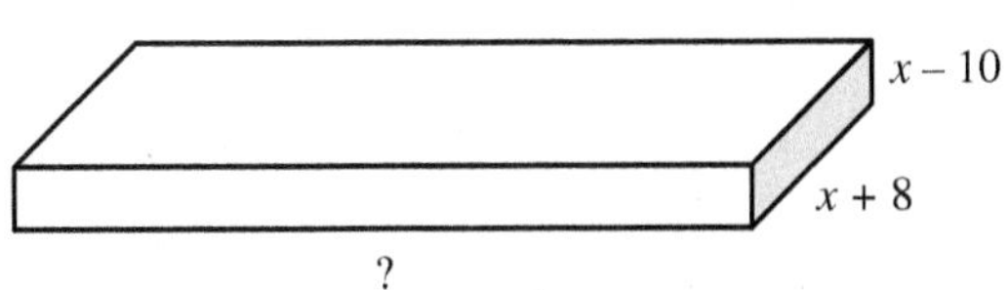

Review Exercises

Exercises 1 and 2 Expressions

[1.1] **1.** Write a set containing the first five prime numbers.

[3.2] **2.** Find the slope of a line passing through $(3, -5)$ and $(-2, 5)$.

Exercises 3–6 Equations and Inequalities

For Exercises 3 and 4, solve. a. Write the solution set in set-builder notation.
b. Write the solution set in interval notation.
c. Graph the solution set.

[2.4] **3.** $2x - 3 < 1$ or $2x - 3 > -5$

[2.6] **4.** $|3x - 1| \leq 8$

[3.4] **5.** Graph $y > 2$.

[4.3] **6.** The perimeter of an enclosure is to be 100 feet, and the length needs to be twice the width. Write a system of equations describing the situation; then solve the system to find the length and width.

Chapter 5 Summary and Review Exercises

Complete each incomplete definition, rule, or procedure; study the key examples; and then work the related exercises.

5.1 Exponents and Scientific Notation

Definitions/Rules/Procedures	Key Example(s)
If n is a natural number and a is a real number, then $a^n =$ ________.	Evaluate. **a.** $(0.2)^3 = (0.2)(0.2)(0.2) = 0.008$ **b.** $(-2)^4 = (-2)(-2)(-2)(-2) = 16$ **c.** $(-2)^5 = (-2)(-2)(-2)(-2)(-2) = -32$
Exponents Summary If a and b are real numbers, m and n are integers, and no denominators are 0, then	Evaluate or simplify.
Product Rule of Exponents: $a^m \cdot a^n =$ ________.	**a.** $x^2 \cdot x^5 = x^{2+5} = x^7$
Zero as an Exponent: $a^0 =$ ________, $a \neq 0$	**b.** $5^0 = 1$
Quotient Rule for Exponents: $\frac{a^m}{a^n} =$ ________.	**c.** $\frac{y^6}{y^2} = y^{6-2} = y^4$
Negative Exponents: $a^{-n} =$ ________ and $\frac{1}{a^{-n}} =$ ________.	**d.** $3^{-4} = \frac{1}{3^4} = \frac{1}{81}$ and $\frac{1}{4^{-3}} = 4^3 = 64$
Raising a Power to a Power: $(a^m)^n =$ ________.	**e.** $(t^3)^4 = t^{3 \cdot 4} = t^{12}$
Raising a Product to a Power: $(ab)^n =$ ________.	**f.** $(5mn^2)^3 = 5^{1\cdot 3}m^{1\cdot 3}n^{2\cdot 3} = 125m^3n^6$
Raising a Quotient to a Power: $\left(\frac{a}{b}\right)^n =$ ________.	**g.** $\left(\frac{2}{5}\right)^3 = \frac{2^3}{5^3} = \frac{8}{125}$
Quotients Raised to a Negative Power: $\left(\frac{a}{b}\right)^{-n} =$ ________.	**h.** $\left(\frac{2}{3}\right)^{-4} = \left(\frac{3}{2}\right)^4 = \frac{81}{16}$

Exercises 1–14 Expressions

[5.1] *For Exercises 1–4, evaluate the exponential expression.*

1. $\left(\frac{2}{3}\right)^{-3}$

2. -4^2

3. 2^{-5}

4. $14x^0$

[5.1] *For Exercises 5 and 6, simplify using the product rule of exponents.*

5. $(4x^2)(5xy^3)$

6. $\left(-\frac{2}{3}m^2n^4\right)(3mn^5)$

[5.1] *For Exercises 7 and 8, simplify using the quotient rule for exponents.*

7. $\frac{6x^4}{2x^9}$

8. $-\frac{7a^3bc^9}{3abc^{15}}$

[5.1] *For Exercises 9 and 10, simplify using the power rule of exponents.*

9. $(3m^5n)^2$

10. $\dfrac{(6m^2n^3p)^{-1}}{(2m^6n^2)^3}$

[5.1] *For Exercises 11–14, use the rules for exponents to simplify.*

11. $(3j^2k^5)^2(0.1jk^3)^3$

12. $\dfrac{(2s^3t)^{-1}}{(3st)^2(s^4t)^8}$

13. $\dfrac{(5m^7n^3)^{-2}}{(3m)^{-1}(3m)^2}$

14. $\left(\dfrac{2a}{b^4}\right)^{-3}$

Definitions/Rules/Procedures	Key Example(s)
To change a number **from scientific notation,** $a \times 10^n$, where $1 \le \lvert a\rvert < 10$ and n is an integer, **to standard form,** if $n > 0$, move the decimal point to the ________ n places. If $n < 0$, move the decimal point to the ________ $\lvert n\rvert$ places.	Write in standard form. **a.** $4.8 \times 10^6 = 4{,}800{,}000$ **b.** $3.1 \times 10^{-4} = 0.00031$
To write a number **in scientific notation,** **1.** Locate the new decimal point position, which will be to the right of the ________ in the number. **2.** Determine the power of 10, **a.** If the number's absolute value is greater than 1, the power is the number of digits between the ________ and the ________ expressed as a positive power. **b.** If the number's absolute value is less than 1, the power is the number of digits between the ________ and the ________ expressed as a negative power. **3.** Delete the unnecessary 0's. **a.** If the number's absolute value is greater than 1. delete the zeros to the ________ of the last nonzero digit. **b.** If the number's absolute value is less than 1, delete the zeros to the ________ of the first nonzero digit.	Write in scientific notation. **a.** $943{,}000 = 9.43 \times 10^5$ **b.** $0.00000082 = 8.2 \times 10^{-7}$

Exercises 15–21 Expressions

[5.1] *For Exercises 15 and 16, write the number in standard form.*

15. The mass of a hydrogen atom is 1.6736×10^{-24} gram.

16. It is speculated that the universe is about 1.65×10^{10} years old.

[5.1] *For Exercises 17 and 18, write in scientific notation.*

17. The mass of a dust particle is 0.000000000753 kilogram.

18. The speed of light is 300,000,000 meters per second.

[5.1] *For Exercises 19 and 20, simplify.*

19. $(5.1 \times 10^4)(-2 \times 10^6)$

20. $\dfrac{8.12 \times 10^{-8}}{2 \times 10^{-5}}$

21. The energy in joules of a single photon of light can be determined by $E = hf$, where h is a constant 6.626×10^{-34} joule-second and f is the frequency of the light in hertz. Find the energy of blue light with a frequency of 6.1×10^{14}Hz.

5.2 Polynomials and Polynomial Functions

Definitions/Rules/Procedures	Key Example(s)
A **monomial** is an expression that is a ________ or a product of a ________ and ________ that are raised to whole number powers.	Examples of monomials are 3, x, $3x$, $3x^3y^2$
The **coefficient** is the ________________ in a monomial.	The numerical coefficient of $3x^3y^2$ is 3 and the numerical coefficient of x^4 is 1.
The **degree of a monomial** is the ________________ of all variables in the monomial.	The degree of $3x$ is 1 and the degree of $3x^3y^2$ is $3 + 2 = 5$
A **polynomial** is a ________ or an expression that can be written as a ________________.	Examples of polynomials are $-5x^2$, $-5x^2 + 2$, and $6y^4 - 4y^2 - 6y - 2$.
A **polynomial in one variable** is a polynomial in which every term has the ________________.	$6y^4 - 4y^2 - 6y - 2$ is a polynomial in y.
A **multivariable polynomial** is a polynomial with ________________.	$3x^2 - 5xy + 2y^2$ is a multivariable polynomial.
A **binomial** is a polynomial containing ________ terms.	$4x + 3$ *and* $3y^3 + 5y$ are binomials.
A **trinomial** is a polynomial containing ________ terms.	$4y^2 + 7y - 10$ is a trinomial.
The **degree** of a polynomial is the ________________ of any of the terms in the polynomial.	The degree of $6y^4 - 4y^2 - 6y - 2$ is 4.

Exercises 22–25 Expressions

[5.2] *For Exercises 22–25, identify the degree of each polynomial; then indicate whether the polynomial is a monomial, binomial, or trinomial or has no special polynomial name.*

22. $-\dfrac{1}{4} + 2c^2 + 8.7c^3 + \dfrac{2}{5}c$

23. $7m^2 + 1$

24. $-8xy^4$

25. $9h^5 + 4h^3 - h^2 + 1$

Definitions/Rules/Procedures	Key Example(s)
To **add polynomials,** combine ______________. To **subtract polynomials;** **1.** Write the subtraction statement as an equivalent ______________ statement. **a.** Change the operation symbol from a ______________ sign to an ______________ sign. **b.** Change the subtrahend (second polynomial) to its ______________. To write the ______________, change the sign of each term in the polynomial **2.** Combine like ______________.	Add. $(5x^3 + 12x^2 - 9x + 1) + (7x^2 - x - 13)$ $= 5x^3 + 12x^2 + 7x^2 - 9x - x + 1 - 13$ $= 5x^3 + 19x^2 - 10x - 12$ Subtract. $(6y^3 - 5y + 18) - (y^3 - 5y + 7)$ $= (6y^3 - 5y + 18) + (-y^3 + 5y - 7)$ $= 6y^3 - y^3 - 5y + 5y + 18 - 7$ $= 5y^3 + 11$

Exercises 26–30 Expressions

[5.2] *For Exercises 26 and 27, perform the indicated operation and write the resulting polynomial in descending order.*

26. $(3c^3 + 2c^2 - c - 1) + (8c^2 + c - 10)$

27. $(y^2 + 3y + 6) - (-5y^2 + 3y - 8)$

[5.2] *For Exercises 28 and 29, add or subtract the polynomials.*

28. $(x^2y^3 - xy^2 + 2x^2y - 4xy + 7y^2 - 9) + (-3x^2y^2 + 2xy^2 + 4x^2y - xy + 6y^2 + 4)$

29. $(4hk - 8k^3) - (5kh + 3k - 2k^3)$

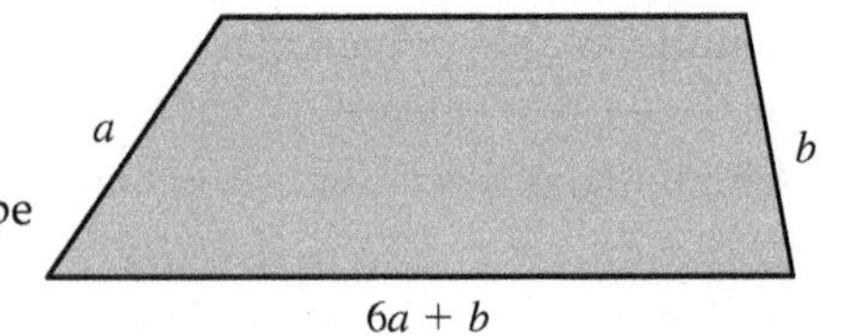

30. a. Write a polynomial in simplest form that describes the perimeter of the shape shown.

b. Find the perimeter if $a = 12$ centimeters and $b = 9$ centimeters.

Definitions/Rules/Procedures	Key Example(s)
The graph of a **constant function** is a ______________ through $(0, c)$. Graphs of **linear functions,** which have the form $f(x) = mx + b$, are lines with slope ______________ and y-intercept ______________. Graphs of **quadratic functions,** which have the form $f(x) = ax^2 + bx + c, a \neq 0$, are ______________ with y-intercept ______________. Graphs of cubic functions, which have the form $f(x) = ax^3 + bx^2 + cx + d, a \neq 0$, resemble an ______________ with y-intercept ______________.	Graph $f(x) = x^2 - 1$.

x	y
–2	3
–1	0
0	–1
1	0
2	3

Exercises 31–38 **Equations and Inequalities**

[5.2] *For Exercises 31–34, indicate whether the function is a constant function, a linear function, a quadratic function, or a cubic function.*

31.

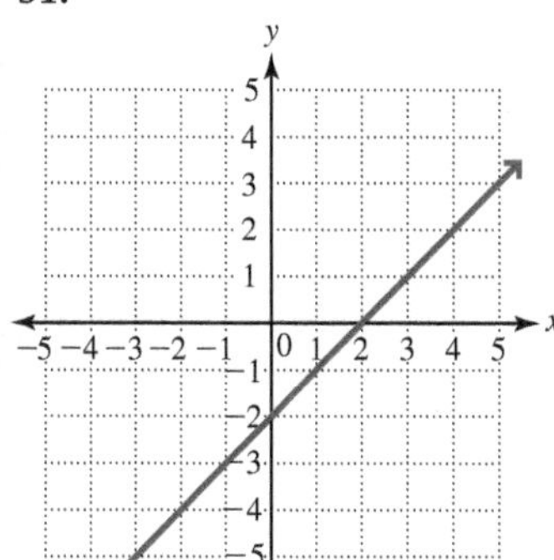

32.

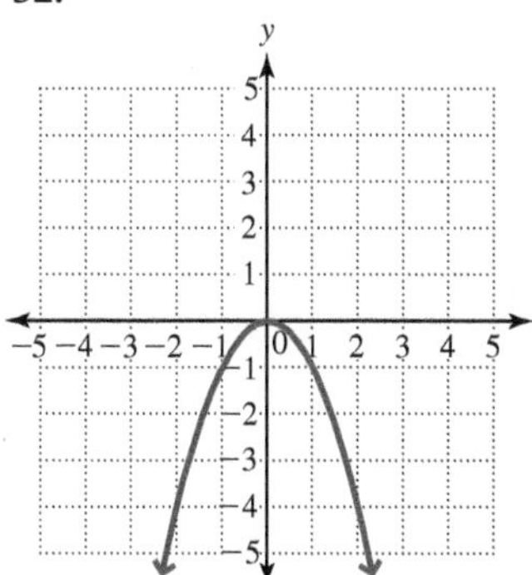

33.

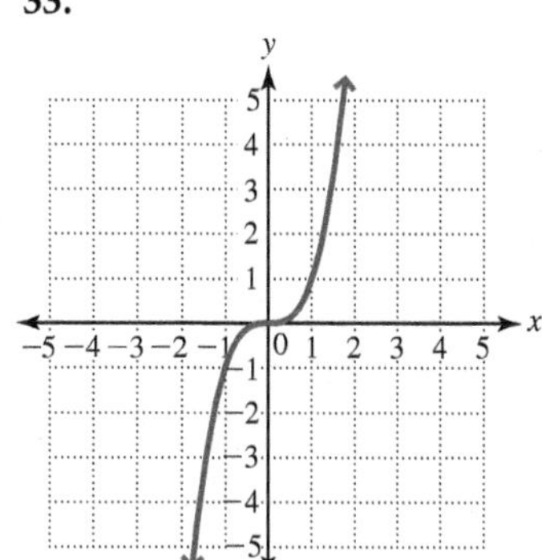

34.

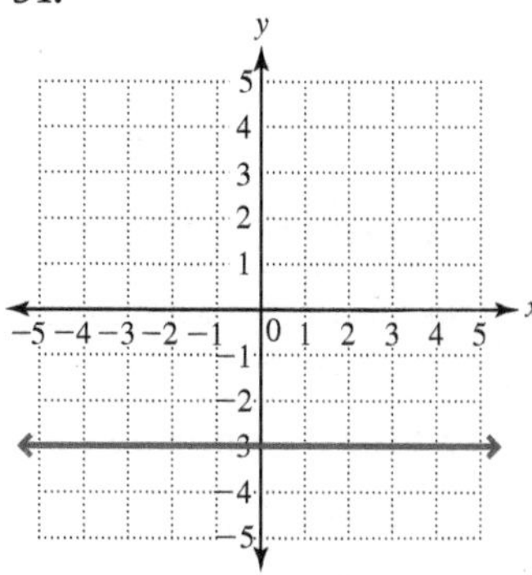

[5.2] *For Exercises 35–38, graph the function and give the domain and range.*

35. $f(x) = x^2 - 3$

36. $g(x) = -\frac{1}{3}x + 2$

37. $h(x) = x^3 - 1$

38. $w(x) = 7$

Definitions/Rules/Procedures	Key Example(s)
The **sum** of two functions, $f + g$, is found by $(f + g)(x) =$ ________________. The **difference** of two functions, $f - g$, is found by $(f - g)(x) =$ ________________.	If $f(x) = 5x^2 - 2x + 4$ and $g(x) = 3x^2 - 8x - 12$, find $(f + g)(x)$ and $(f - g)(x)$. $(f + g)(x) = (5x^2 - 2x + 4) + (3x^2 - 8x - 12)$ $= 8x^2 - 10x - 8$ $(f - g)(x) = (5x^2 - 2x + 4) - (3x^2 - 8x - 12)$ $= (5x^2 - 2x + 4) + (-3x^2 + 8x + 12)$ $= 2x^2 + 6x + 16$

Exercises 39–41 **Equations and Inequalities**

[5.2] **39.** If $f(x) = x^3 + 2x + 5$ and $g(x) = -2x^3 + 2$, find

a. $(f + g)(x)$. **b.** $(f + g)(2)$. **c.** $(f - g)(x)$. **d.** $(f - g)(-4)$.

40. If we ignore air resistance, the polynomial function $h(x) = -16t^2 + 200$ describes the height in feet of a falling object after falling for t seconds from an initial height of 200 feet. What is its height after 3 seconds?

41. The function $R(x) = 0.2x^2 + x + 1000$ describes the revenue that a software company makes from the sale of one of its programs, where x represents the number of copies sold. The function $C(x) = 10x + 2000$ describes the cost of producing the software.

a. Find a function, $P(x)$, that describes the profit.

b. Find the profit if the company sells 200 copies of the program.

5.3 Multiplying Polynomials

Definitions/Rules/Procedures	Key Example(s)
To **multiply a polynomial by a monomial**, use the ___________ property to multiply each term in the polynomial by the monomial.	Multiply. **a.** $5xy(7x^2 + 3x - 6)$ $= 5xy \cdot 7x^2 + 5xy \cdot 3x + 5xy \cdot (-6)$ $= 35x^3y + 15x^2y - 30xy$
To multiply two polynomials, **1.** Multiply each term in the ___________ by each term in the ___________. **2.** Combine like terms.	**b.** $(4x + 7)(2x - 1)$ $= 4x \cdot 2x + 4x \cdot (-1) + 7 \cdot 2x + 7 \cdot (-1)$ $= 8x^2 - 4x + 14x - 7$ $= 8x^2 + 10x - 7$
Conjugates are binomials that are the ___________ and ___________ of the same two terms.	
Special Products: If a and b are real numbers, variables, or expressions, then $(a + b)^2 =$ ___________ $(a - b)^2 =$ ___________ $(a + b)(a - b) =$ ___________	**c.** $(5n + 6)^2 = 25n^2 + 60n + 36$ **d.** $(3x - 4y)^2 = 9x^2 - 24xy + 16y^2$ **e.** $(t^2 + 9)(t^2 - 9) = t^4 - 81$

Exercises 42–55 Expressions

[5.3] *For Exercises 42–53, multiply.*

42. $3x^2y^3(2x^2 - 3x + 1)$

43. $-pq(2p^2 - 3pq + 3q^2)$

44. $(x - 3)(2x + 8)$

45. $(9w - 1)(3w + 2)$

46. $(3r - s)(6r + 7s)$

47. $(x^2 - 1)(x^2 + 2)$

48. $(x - 1)(2x^3 - 3x^2 + 4x + 7)$

49. $(3t^2 - t + 1)(4t^2 + t - 1)$

50. $(3a - 5)(3a + 5)$

51. $(2p - 1)^2$

52. $(8k + 3)^2$

53. $(4h^2 + 7)(4h^2 - 7)$

[5.3] 54. What is the conjugate of $(5m - 2)$?

55. A fish tank's design specifies a length that is 10 inches longer than the width. The height of the tank is to be 5 inches longer than the width. Let w represent the width and write an expression for the volume of the tank in terms of w.

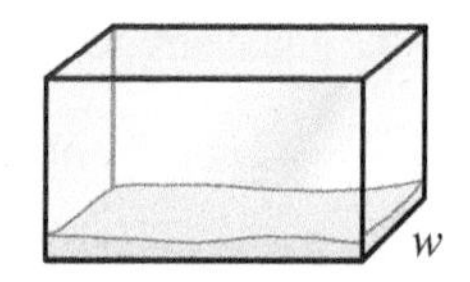

Definitions/Rules/Procedures	Key Example(s)
The product of two functions, $f \cdot g$, is defined by $(f \cdot g)(x) =$ ________.	If $f(x) = x - 7$ and $g(x) = x^2 - x + 1$, find $(f \cdot g)(x)$ and $(f \cdot g)(-2)$. **Solution:** $(f \cdot g)(x) = (x - 7)(x^2 - x + 1)$ $= x^3 - x^2 + x - 7x^2 + 7x - 7$ $= x^3 - 8x^2 + 8x - 7$ $(f \cdot g)(-2) = (-2)^3 - 8(-2)^2 + 8(-2) - 7 = -63$

Exercises 56 and 57 Equations and Inequalities

[5.3] 56. If $f(x) = 2x + 7$ and $g(x) = x^2 + 4x - 1$, find
a. $(f \cdot g)(x)$.
b. $(f \cdot g)(-1)$.

[5.3] 57. If $f(x) = 4x^2 - x$, find $f(x - 4)$.

5.4 Dividing Polynomials

Definitions/Rules/Procedures	Key Example(s)
If a, b, and c are real numbers, variables, or expressions with $c \neq 0$, then $\frac{a + b}{c} =$ ________. To **divide a polynomial by a monomial**, divide each term of the ________ by the ________.	Divide. $\frac{28x^6 + 36x^3 - 8x^2}{4x^2} = \frac{28x^6}{4x^2} + \frac{36x^3}{4x^2} - \frac{8x^2}{4x^2}$ $= 7x^4 + 9x - 2$

Exercises 58 and 59 Expressions

[5.4] *For Exercises 58 and 59, divide the polynomial by the monomial.*

58. $\frac{20m^5 - 5m^4 + 15m^3 - 5m^2}{5m^2}$

59. $\frac{8x^2 - 2x + 3}{2x}$

Definitions/Rules/Procedures	Key Example(s)
To **divide a polynomial by a polynomial,** **1.** Use ________ division. **2.** Is there is a remainder, write the result in the form, ________ + ________.	$\frac{2x^3 - 14x - 5}{x + 2}$ $\begin{array}{r} 2x^2 - 4x - 6 \\ x + 2\overline{)2x^3 + 0x^2 - 14x - 5} \\ \underline{-2x^3 - 4x^2} \qquad\qquad \\ -4x^2 - 14x \qquad \\ \underline{4x^2 + 8x} \qquad \\ -6x - 5 \\ \underline{6x + 12} \\ 7 \end{array}$ **Answer:** $2x^2 - 4x - 6 + \frac{7}{x + 2}$

Definitions/Rules/Procedures	Key Example(s)
The quotient of two functions, f/g, is found by $(f/g)(x) =$ ________.	If $f(x) = 5x^2 - 13x - 10$ and $g(x) = x - 3$, find $(f/g)(x)$ and $(f/g)(4)$. $$\begin{array}{r} 5x + 2 \\ x - 3\overline{)5x^2 - 13x - 10} \\ \underline{-5x^2 + 15x} \\ 2x - 10 \\ \underline{-2x + 6} \\ -4 \end{array}$$ **Answers:** $(f/g)(x) = 5x + 2 - \frac{4}{x - 3}$ $(f/g)(4) = 5(4) + 2 - \frac{4}{4 - 3} = 22 - 4 = 18$

Exercises 60–64 Expressions

[5.4] *For Exercises 60–62, divide using long division.*

60. $\frac{x^2 + 5x + 7}{x + 2}$

61. $(x^2 - 4x + 4) \div (x - 2)$

62. $\frac{4x^2 + 4x - 3}{2x - 1}$

[5.4] 63. If $f(x) = 9x^4 - 7x^2 - 10x - 5$ and $g(x) = 3x - 1$, find $(f/g)(x)$.

64. The voltage across a component in an amplifier circuit is described by the expression $(3y^4 + 11y^3 + 22y^2 + 23y + 5)$ volts. The resistance in the component is described by $(3y + 5)$ ohms.

a. Find an expression that describes the current. (The formula for voltage is voltage = current × resistance.)

b. Find the current, which is measured in amps, if $y = 3$.

5.5 Synthetic Division and the Remainder Theorem

Definitions/Rules/Procedures	Key Example(s)
When a polynomial is divided by a binomial in the form ________, where c is a constant, we can use synthetic division.	Divide using synthetic division. $\frac{2x^3 - 14x - 5}{x + 2}$ $$\begin{array}{r\|rrrr} -2 & 2 & 0 & -14 & -5 \\ & & -4 & 8 & 12 \\ \hline & 2 & -4 & -6 & 7 \end{array}$$ **Answer:** $2x^2 - 4x - 6 + \frac{7}{x + 2}$

Exercises 65–68 Expressions

[5.5] *For Exercises 65–68, divide using synthetic division.*

65. $\frac{5x^2 - 3x + 2}{x - 1}$

66. $\frac{2x^3 - 6x^2 - 5x + 15}{x - 3}$

67. $(4x^3 - 2x + 5) \div (x - 2)$

68. $\dfrac{x^3 - 8}{x - 2}$

Definitions/Rules/Procedures	Key Example(s)
Given a polynomial $P(x)$, the remainder of $\dfrac{P(x)}{x - c}$ is equal to __________.	For $P(x) = 5x^3 - 18x^2 + x - 6$, use the remainder theorem to find $P(4)$. $\begin{array}{r\|rrrr} 4 & 5 & -18 & 1 & -6 \\ & & 20 & 8 & 36 \\ \hline & 5 & 2 & 9 & 30 \end{array}$ **Answer:** $P(4) = 30$

Exercises 69 and 70 Expressions

[5.5] *For Exercises 69 and 70, use the remainder theorem to find $P(c)$.*

69. For $P(x) = 5x^3 - x^2 + 8x + 3$, use the remainder theorem to find $P(1)$.

70. For $P(x) = 7x^3 - 2x + 1$, use the remainder theorem to find $P(-2)$.

Chapter 5 Practice Test

For Extra Help

Step-by-step test solutions are found on the Chapter Test Prep Videos available in MyMathLab® *or on* YouTube.

1. Write 7.2×10^{-3} in standard form.

2. Write 0.00357 in scientific notation.

For Exercises 3–8, simplify.

3. $5x^3(3x^2y)$

4. $(3x^4y)^{-2}$

5. $\dfrac{8u^7v}{2u^4v^5}$

6. $\left(-\dfrac{2}{3}t^5u^2v\right)^4$

7. $(6 \times 10^5)(2.1 \times 10^4)$

8. $\dfrac{8.4 \times 10^{10}}{4.2 \times 10^3}$

For Exercises 9 and 10, indicate whether the function is a constant function, a linear function, a quadratic function, or a cubic function.

9.

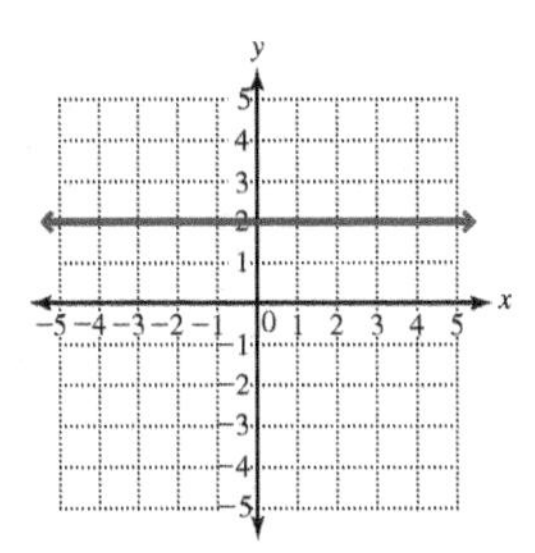

10.

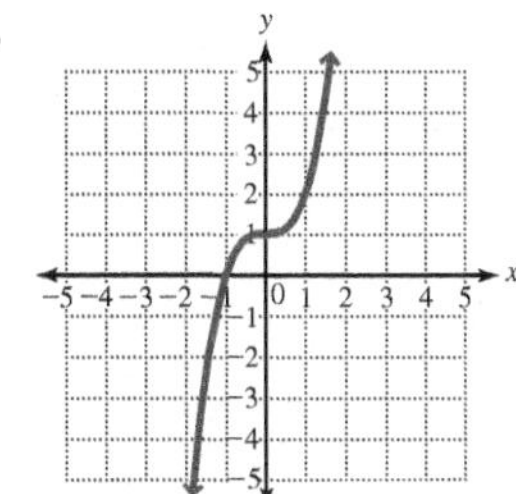

11. Graph $f(x) = x^2 - 1$ and give the domain and range.

For Exercises 12–18 perform the indicated operation.

12. $(a^2 - 4ab + b^2) - (a^2 + 4ab + b^2)$

13. $(6r^5 - 9r^4 - 2r^2 + 8) + (5r^5 + 2r^4 + 7r^3 - 8r^2 - r - 9)$

14. $(3x - 4y)(5x + 2y)$

15. $4m^3n^9(3m^2 - 2mn + 5n^2)$

16. $(7k - 2j)^2$

17. $(2x - 3)(7x^2 + 3x - 5)$

18. $(4h - 3)(4h + 3)$

For Exercises 19 and 20, use $f(x) = 5x - 8$ and $g(x) = 3x + 2$.

19. Find $(f \cdot g)(x)$.

20. Find $(f \cdot g)(2)$.

For Exercises 21 and 22, divide.

21. $\dfrac{8k^3 - 4k^2 + 2k}{2k}$

22. $\dfrac{3m^3 + 10m^2 + 9m + 2}{3m + 1}$

23. Divide $\dfrac{2x^3 + 5x^2 - 8}{x + 3}$ using synthetic division.

24. For $P(x) = x^3 - 5x + 4$, use the remainder theorem to find $P(-3)$.

25. A chemical company produces a respiratory medication. The function $R(x) = 0.1x^2 + x + 5000$ describes the revenue a company makes from sales, where x represents the number of vials sold. The function $C(x) = 15x + 3000$ describes the cost of producing those vials.

a. Find a function, $P(x)$, that describes the profit.

b. Find the profit if the company sells 1000 vials.

Chapters 1–5 Cumulative Review Exercises

For Exercises 1–4, answer true or false.

[1.2] **1.** The identity for multiplication is 0.

[2.3] **2.** If $-3x \geq 18$, then $x \leq -6$.

[3.4] **3.** The boundary line of the graph of $3x + 4y < 15$ is a solid line.

[5.3] **4.** The conjugate of $3x + 5$ is $3x - 5$.

For Exercises 5–8, fill in the blanks.

[2.1] **5.** An equation whose solution set is every real number for which the equation is defined is a(n) ________.

[2.5] **6.** The absolute value equation $|3x + 7| = 5$ is equivalent to the two equations ________ and ________.

[4.1] **7.** A system of equations that has no solution is ________.

[5.5] **8.** When synthetic division is used to divide, the divisor must be in the form ________.

Exercises 9–18 Expressions

[1.4] **9.** Evaluate $\sqrt{(x_2 - x_1)^2 + (y_2 - y_1)^2}$ for $x_2 = 6, x_1 = -2, y_2 = -3$, and $y_1 = 3$.

[1.4] **10.** For what value(s) of the variable is $\dfrac{x - 3}{x + 5}$ undefined?

[1.4] **11.** Translate "negative 5 times the difference of twelve and a number, decreased by 4" to an algebraic expression.

For Exercises 12–14, simplify. Write the result with positive exponents only.

[5.1] **12.** $(-2a^4b^{-3})(4a^{-2}b^{-1})$

[5.1] **13.** $\dfrac{8x^{-2}y^3z^8}{2x^3y^{-4}z^4}$

[5.1] **14.** $(-3x^4y^{-2})^3$

For Exercises 15 and 16, simplify and write the answer in descending order of degree.

[5.2] **15.** $(5x^3 - 7x^2 + 2x - 3) + (-8x^3 + 7x^2 + 3x - 5)$

[5.2] **16.** $(3x^2 + 6x - 2) - (-4x^2 - 7x + 1)$

[5.3] **17.** Multiply $(5x + 2y)(3x - 2y)$.

[5.4] **18.** Divide $\dfrac{6x^3 + 14x^2 - 17x + 10}{3x + 10}$ using long division.

Exercises 19–30 Equations and Inequalities

For Exercises 19–22, solve.

[2.1] **19.** $P = 2\pi r + 2d$ for d

[2.4] **20.** $-6 \leq -4x - 2 \leq 6$

[2.6] **21.** $3|2x - 5| > 9$

[4.1] **22.** $\begin{cases} 5x - 3y = 2 \\ 3x - y = -2 \end{cases}$

For Exercises 23 and 24, graph.

[3.1] 23. $y = \frac{3}{4}x - 2$

[3.4] 24. $2x - 3y > -6$

[3.3] 25. Write the equation of the line containing $(-6, 2)$ that is parallel to the graph of $5x - 3y = 8$.

[3.5] 26. Determine whether the given graph is the graph of a function.

For Exercises 27–30, solve.

[2.2] 27. Tom, Bill, and Sam go to a basketball game. The seat numbers are consecutive integers whose sum is 162. What are the seat numbers?

[2.2] 28. Franz left home in Atlanta, traveling north on I-75 at an average rate of 50 miles per hour. An hour later, his son Hanz also left home in Atlanta, traveling north on I-75 at an average rate of 60 miles per hour. How long will it take Hanz to catch Franz?

[2.2] [4.3] 29. John's Nursery sells azaleas for \$6 each and rosebushes for \$10 each. In one day, the number of rosebushes sold was seven more than twice the number of azaleas sold. If the total amount of money received for the plants was \$798, how many of each type of plant was sold?

[2.2] [4.3] 30. The length of a rectangle is 2 feet more than three times the width. If the perimeter is 84 feet, find the length and width.

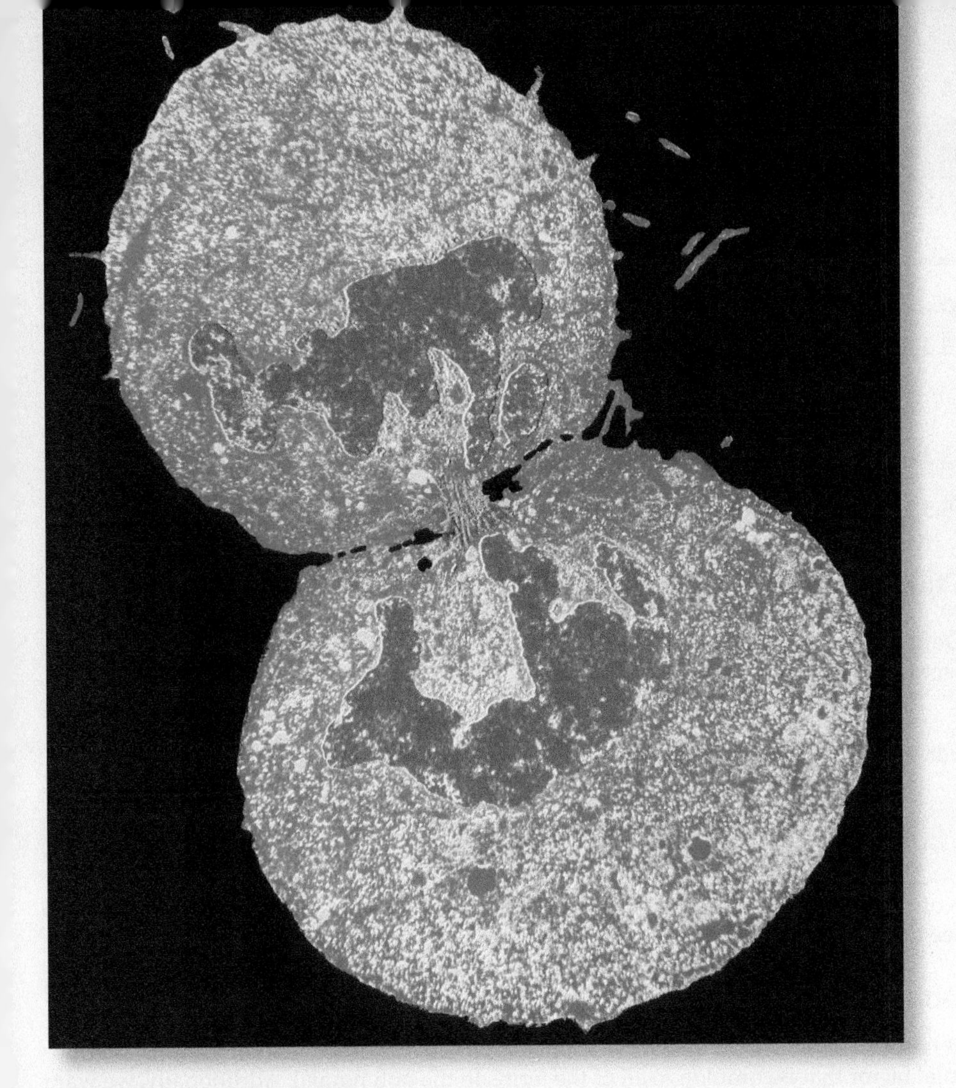

CHAPTER

6 Factoring

Chapter Overview

In this chapter, we learn how to *factor* polynomial expressions. Factoring is rewriting an expression as a product. It is used extensively in simplifying the more complex expressions that we will study in future chapters. The flow of the chapter is as follows:

- Learn various methods for factoring polynomials.
- Solve equations using factoring.
- Graph special polynomial functions.

6.1 Greatest Common Factor and Factoring by Grouping

Objectives

1 Find the greatest common factor of a set of terms.
2 Factor a monomial GCF out of the terms of a polynomial.
3 Factor polynomials by grouping.

Warm-up

[5.4] *For Exercises 1 and 2, find the quotients.*

1. $\dfrac{16x^3y - 40x^2}{8x^2}$

2. $\dfrac{-18x^4y^3 + 9x^2y^2z - 12x^3y}{-3x^2y}$

[5.3] 3. Find the product. $-4y(5xy^2 - 9)$

[5.5] 4. Divide using synthetic division. $\dfrac{x^3 - x^2 - 5x + 6}{x - 2}$

Often in mathematics, we need a number or an expression to be in **factored form.**

Definition **Factored form:** A number or an expression written as a product of factors.

For example, the following polynomials have been rewritten in factored form.

$$2x + 8 = \underbrace{2(x + 4)}_{\text{Factored form}} \qquad x^2 + 5x + 6 = \underbrace{(x + 2)(x + 3)}_{\text{Factored form}}$$

Notice that we can check an expression's factored form by multiplying the factors to see if their product is the original expression. Writing factored form is called *factoring*.

Objective 1 Find the greatest common factor of a set of terms.

When factoring polynomials, the first step is to determine whether there is a monomial factor that is common to all of the terms in the polynomial. Additionally, we want that monomial factor to be the **greatest common factor** of the terms.

Definition **Greatest common factor (GCF) of a set of terms:** A monomial with the greatest coefficient and degree that evenly divides all of the given terms.

For example, the greatest common factor of $12x^2$ and $18x^3$ is $6x^2$ because 6 is the greatest numerical value that evenly divides both 12 and 18 and x^2 is the highest power of x that evenly divides both x^2 and x^3. Notice that x^2 has the *smaller* exponent, which provides a clue as to how to determine the GCF. We can use prime factorizations to help us find GCFs.

AUDITORY

Learning Strategy
If you are an auditory learner, you might try to remember the procedure by thinking about the words in the procedure: To find the **greatest common** factor, use the **common** primes raised to their **smallest** exponent. Notice that the words *greatest* and *smallest* are opposites.

Procedure **Finding the GCF**

To find the GCF of two or more monomials:

1. Write the prime factorization in exponential form for each monomial. Treat variables like prime factors.
2. Write the GCF's factorization by including the prime factors (and variables) common to all of the factorizations, each raised to its smallest exponent in the factorizations.
3. Multiply the factors in the factorization created in step 2.

Note If there are no common prime factors, the GCF is 1.

Answers to Warm-up
1. $2xy - 5$
2. $6x^2y^2 - 3yz + 4x$
3. $-20xy^3 + 36y$
4. $x^2 + x - 3$

Connection A factor tree is one way to find the prime factorization of a number. Following is a factor tree for 24 showing that its prime factorization is $2 \cdot 2 \cdot 2 \cdot 3$, or $2^3 \cdot 3$.

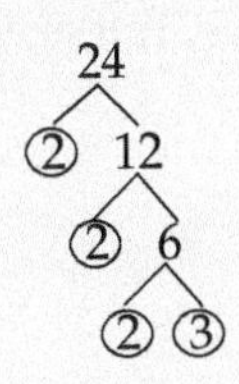

Example 1 Find the GCF of $24x^2y$ and $60x^3$.

Solution: Write the prime factorization of each monomial, treating the variables like prime factors.

$$24x^2y = 2^3 \cdot 3 \cdot x^2 \cdot y$$
$$60x^3 = 2^2 \cdot 3 \cdot 5 \cdot x^3$$

The common prime factors are 2, 3, and x. The smallest exponent for the factor of 2 is 2. The smallest exponent of 3 is 1. The smallest exponent of the x is 2. The GCF is the product of 2^2, 3, and x^2.

$$\text{GCF} = 2^2 \cdot 3 \cdot x^2 = 12x^2$$

Your Turn 1 Find the GCF.

a. $32r^2t$ and $48r^3ts$

b. $35a^2$ and $9b$

Objective 2 Factor a monomial GCF out of the terms of a polynomial.

Earlier we mentioned that $2(x + 4)$ is the factored form of $2x + 8$. Notice that 2 is the GCF of the terms $2x$ and 8. This suggests the following procedure for factoring a monomial GCF out of the terms of a polynomial.

Procedure Factoring a Monomial GCF out of a Polynomial

To factor a monomial GCF out of the terms of a polynomial:

1. Find the GCF of the terms in the polynomial.
2. Rewrite the polynomial as a product of the GCF and the quotient of the polynomial and the GCF.

$$\text{Polynomial} = \text{GCF}\left(\frac{\text{Polynomial}}{\text{GCF}}\right)$$

Example 2 Factor.

a. $18x^2 + 6x$

Solution: The GCF of $18x^2$ and $6x$ is $6x$.

Note 6 is the largest number that divides both 18 and 6 evenly, and x has the smaller exponent of the x^2 and x.

$$18x^2 + 6x = 6x\left(\frac{18x^2 + 6x}{6x}\right)$$ Rewrite using the form $\text{GCF}\left(\frac{\text{polynomial}}{\text{GCF}}\right)$.

$$= 6x\left(\frac{18x^2}{6x} + \frac{6x}{6x}\right)$$ Separate the terms.

$$= 6x(3x + 1)$$ Divide the terms by the GCF.

Check: We can check using the distributive property to see if the product of the two factors is the given polynomial.

$$6x(3x + 1) = 6x \cdot 3x + 6x \cdot 1$$ Distribute $6x$.

$$= 18x^2 + 6x$$ The product is the given polynomial.

b. $16x^3y - 40x^2$

Solution: The GCF of $16x^3y$ and $40x^2$ is $8x^2$.

Note From this point on, when we factor out a monomial GCF, we do not show the division steps.

$$16x^3y - 40x^2 = 8x^2\left(\frac{16x^3y - 40x^2}{8x^2}\right)$$ Rewrite using the form $\text{GCF}\left(\frac{\text{polynomial}}{\text{GCF}}\right)$.

$$= 8x^2\left(\frac{16x^3y}{8x^2} - \frac{40x^2}{8x^2}\right)$$ Separate the terms.

$$= 8x^2(2xy - 5)$$ Divide the terms by the GCF.

Answers to Your Turn 1
a. $16r^2t$ **b.** 1

Your Turn 2 Factor.

a. $12xy - 4x$

b. $15m^3n - 21mn^2 + 27mnp$

Factoring When the First Term Is Negative

Generally, we prefer the first term inside parentheses to be positive. To avoid a negative first term in the parentheses, factor out the negative of the GCF, if needed.

Note Factoring out $4y$ instead of $-4y$ does give an equivalent expression, but the first term in the parentheses is negative.

$$-20xy^3 + 36y = 4y(-5xy^2 + 9)$$

We could have a positive first term by rearranging the terms in the parentheses.

$$= 4y(9 - 5xy^2)$$

Although the first term is positive in $9 - 5xy^2$, the terms are no longer in descending order, which is why factoring out the negative of the GCF is usually the preferred approach.

Example 3 Factor.

a. $-20xy^3 + 36y$

Solution: Because the first term is negative, we factor out the negative of the GCF, which is $-4y$.

$$-20xy^3 + 36y = -4y\left(\frac{-20xy^3 + 36y}{-4y}\right)$$
$$= -4y(5xy^2 - 9)$$

b. $-18x^4y^3 + 9x^2y^2z - 12x^3y$

Solution: We factor out the negative of the GCF, $-3x^2y$.

$$-18x^4y^3 + 9x^2y^2z - 12x^3y = -3x^2y\left(\frac{-18x^4y^3 + 9x^2y^2z - 12x^3y}{-3x^2y}\right)$$
$$= -3x^2y(6x^2y^2 - 3yz + 4x)$$

Your Turn 3 Factor.

a. $-40ab - 35b$

b. $-30t^4u^5 - 24t^3u^4 + 48tu^2v$

Before we discuss additional techniques of factoring, it is important to state that no matter what type of polynomial we are asked to factor, we always first consider whether a monomial GCF (other than 1) can be factored out of the polynomial.

Factoring When the GCF Is a Polynomial

Sometimes when factoring, the GCF is a polynomial.

Example 4 Factor $a(c + 5) + b(c + 5)$.

Solution: Notice that this expression is a sum of two products, $a(c + 5)$ and $b(c + 5)$. Further, note that $(c + 5)$ is the GCF of the two products.

Note The parentheses are filled the same way, by dividing the original expression by the GCF.

$$a(c + 5) + b(c + 5) = (c + 5)\left(\frac{a(c + 5) + b(c + 5)}{c + 5}\right)$$
$$= (c + 5)(a + b)$$

Your Turn 4 Factor $6n(m - 3) - 7(m - 3)$.

Objective 3 Factor polynomials by grouping.

Factoring out a polynomial GCF as we did in Example 4 is an intermediate step in a process called *factoring by grouping*, which is a technique that we try when factoring a four-term polynomial such as $ac + 5a + bc + 5b$. The method is called *grouping* because we group pairs of terms and look for a common factor in each group. We begin by pairing the first two terms as one group and the last two terms as a second group.

$$ac + 5a + bc + 5b = (ac + 5a) + (bc + 5b) \quad \text{Group pairs of terms.}$$

Answers to Your Turn 2
a. $4x(3y - 1)$
b. $3mn(5m^2 - 7n + 9p)$

Answers to Your Turn 3
a. $-5b(8a + 7)$
b. $-6tu^2(5t^3u^3 + 4t^2u^2 - 8v)$

Answer to Your Turn 4
$(m - 3)(6n - 7)$

Notice that the first two terms have a common factor of a and the last two terms have a common factor of b. If we factor the a out of the first two terms and the b out of the last two terms, we have the same expression that we factored in Example 4.

$$= a(c + 5) + b(c + 5) \quad \text{Factor out } a \text{ and } b.$$
$$= (c + 5)(a + b) \quad \text{Factor out } c + 5.$$

Our example suggests the following procedure.

Procedure Factoring by Grouping

To factor a four-term polynomial by grouping:

1. Factor out any monomial GCF (other than 1) that is common to all four terms.
2. Group pairs of terms and factor the GCF out of each pair.
3. If there is a common binomial factor, factor it out.
4. If there is no common binomial factor, interchange the middle two terms and repeat the process. If there is still no common binomial factor, the polynomial cannot be factored by grouping.

Example 5 Factor.

a. $10a + 4ab + 15 + 6b$

Solution: There is no monomial GCF. Because the polynomial has four terms, we try to factor by grouping.

$$10a + 4ab + 15 + 6b = (10a + 4ab) + (15 + 6b) \quad \text{Group pairs of terms.}$$
$$= 2a(5 + 2b) + 3(5 + 2b) \quad \text{Factor out } 2a \text{ from } 10a + 4ab \text{ and 3 from } 15 + 6b.$$
$$= (5 + 2b)(2a + 3) \quad \text{Factor out } 5 + 2b.$$

b. $16xy - 24x + 10y - 15$

Solution: There is no monomial GCF. Because the polynomial has four terms, we try to factor by grouping.

$$16xy - 24x + 10y - 15 = (16xy - 24x) + (10y - 15) \quad \text{Group pairs of terms.}$$
$$= 8x(2y - 3) + 5(2y - 3) \quad \text{Factor } 8x \text{ out of } 16xy - 24x \text{ and 5 out of } 10y - 15.$$
$$= (2y - 3)(8x + 5) \quad \text{Factor out } 2y - 3.$$

c. $24m^2n + 6mn - 60m^2 - 15m$

Solution: A monomial GCF, $3m$, is common to all four terms.

$$24m^2n + 6mn - 60m^2 - 15m = 3m(8mn + 2n - 20m - 5) \quad \text{Factor } 3m \text{ out of all four terms.}$$
$$= 3m[(8mn + 2n) + (-20m - 5)] \quad \text{Group pairs of terms.}$$
$$= 3m[2n(4m + 1) + (-5)(4m + 1)] \quad \text{Factor } 2n \text{ out of } 8mn + 2n \text{ and } -5 \text{ out of } -20m - 5.$$
$$= 3m(4m + 1)(2n - 5) \quad \text{Factor out } 4m + 1.$$

Note Because the first term in $-20m - 5$ is negative, we factored out -5. Also remember that when factoring by grouping, we need a common binomial factor. Factoring out -5 gives us that common binomial factor, $4m + 1$, whereas factoring out 5 does not.

Your Turn 5 Factor.

a. $12x + 9bx + 8 + 6b$ **b.** $3m^2 + 6m + 4mn + 8n$ **c.** $40x^2y - 60x^2 - 8xy + 12x$

Answers to Your Turn 5

a. $(4 + 3b)(3x + 2)$
b. $(m + 2)(3m + 4n)$
c. $4x(2y - 3)(5x - 1)$

6.1 Exercises

For Extra Help MyMathLab®

Note: Exercises marked with a ★ represent challenging exercises.

Objective 1

Prep Exercise 1 Is $4x + 8y$ in factored form? Explain.

Prep Exercise 2 Given a set of monomials, after finding their prime factorizations, explain how to create the GCF's factorization.

For Exercises 1–8, find the GCF. See Example 1.

1. $4x^3y^2, 24x^2y$
2. $6m^4n^9, 15mn^5$
3. $12u^3v^6, 28u^8v^8$
4. $45g^7h^6, 35g^4h^7$
5. $10a^3b^2c^8, 14abc^8, 20a^7b^3c^4$
6. $35u^8v^2w^3, 40u^3v^7w, 25u^6v^3w^4$
7. $5(a + b), 7(a + b)$
8. $8(m - n), 11(m - n)$

Objective 2

Prep Exercise 3 After finding the GCF of the terms in a polynomial, explain how to rewrite the polynomial in factored form.

Prep Exercise 4 When factoring a polynomial whose first term is negative, explain how to avoid having a negative first term in the parentheses of the factored form.

For Exercises 9–28, factor out the GCF. See Examples 2–4.

9. $15c^4d - 20c^2$
10. $12xy^7 - 8y^5$
11. $x^5 - x^3 + x^2$
12. $t^6 - t^4 - t^3$
13. $25xy - 50xz + 100x^2$
14. $60ab + 80ac - 20a^2$
15. $-14u^2v^2 - 7uv^2 + 7uv$
16. $-2x^2y^2 - 8x^3y + 2xy$
17. $9a^7b^3 + 3a^4b^2 - 6a^2b$
18. $12m^6n^8 - 18m^3n^3 + 9m^2n$
19. $3w^3v^4 + 39w^2v + 18wv^2$
20. $15a^6b + 3a^4b^3 + 9ab^2$
21. $18ab^3c - 36a^2b^2c + 24a^5b^2c^8$
22. $24a^2b^5c - 18a^5b^8c + 12ab^3c^2$
23. $-8x^2y + 16xy^2 - 12xy$
24. $-20p^2q - 24pq + 16pq^2$
25. $m(n - 3) + 4(n - 3)$
26. $a(b - 5) + 7(b - 5)$
27. $6(b + 2c) - a(b + 2c)$
28. $5(y - 3z) - x(y - 3z)$

Objective 3

Prep Exercise 5 What types of polynomials are factored by grouping?

Prep Exercise 6 When factoring by grouping, after factoring out the monomial GCF, what is the next step?

For Exercises 29–48, factor by grouping. See Example 5.

29. $ax + ay + bx + by$

30. $xy + by + cx + bc$

31. $u^3 + 3u^2 + 3u + 9$

32. $x^3 + 5x^2 + 2x + 10$

33. $mn + np - 3m - 3p$

34. $am - an - 5m + 5n$

35. $cd + d + c + 1$

36. $pq + p + q + 1$

37. $2a^2 - a + 2a - 1$

38. $d^2 - 3d + d - 3$

39. $3ax + 6ay + 8by + 4bx$

40. $3ac + 6ad + 2bc + 4bd$

41. $h^2 + 8h - hk - 8k$

42. $t^2 + 4t - tu - 4u$

43. $3x^2 + 3y^2 - ax^2 - ay^2$

44. $ax^2 + ay^2 - 5x^2 - 5y^2$

45. $3p^3 - 6p^2q + 2pq^2 - 4q^3$

46. $15c^3 - 5c^2d + 6cd^2 - 2d^3$

47. $2x^3 - 8x^2y - 3xy^2 + 12y^3$

48. $4a^3 - 12a^2b - 3ab^2 + 9b^3$

For Exercises 49–60, factor completely. See Example 5.

49. $2ab + 2bx - 2ac - 2cx$

50. $3am - 3an + 3bm - 3bn$

51. $12xy + 4y + 30x + 10$

52. $30ab + 40a + 15b + 20$

53. $3a^2y - 12a^2 + 9ay - 36a$

54. $5uv^2 - 5v^2 + 15uv - 15v$

55. $3m^3 + 6m^2n - 10m^2 - 20mn$

56. $3ab^2 + 6b^3 - 5ab - 10b^2$

57. $15st^2 - 5t^2 - 30st + 10t$

58. $12ac + 12cx - 3ac^2 - 3c^2x$

59. $5x^3y - 20x^2y + 5xy - 20y$

60. $10a^2b^2 - 10b^3 + 15a^2b - 15b^2$

Find the Mistake *For Exercises 61–64, explain the mistake; then write the correct factored form.*

61. $24x^2y^3 + 36x^3y = 6x^2(4y^3 + 6xy)$

62. $8m^4n - 40m^3 = 8m^3n(m - 5)$

63. $9a^2b^3c - 18a^4b = 9a^2b(bc - 2a^2)$

64. $15h^3k + 12h^3 - 3h^2k = 3h^2k(5h^2 + 4h - 1)$

For Exercises 65–68, use a graphing calculator to verify that the factoring is correct. If it is not, write the correct factored form.

65. $3x^4 - 6x^3 = 3x^3(x - 2)$

66. $-3x^2 - 6x^3 + 9x = -3x(x - 1)(2x + 3)$

67. $4x^3 + 14x^2 + 8x = 2x(x^2 + 7x + 4)$

68. $2x^5 - 6x^3 = 2x^3(2x - 1)$

For Exercises 69 and 70, write an expression for the area of the shaded region; then factor completely.

69.

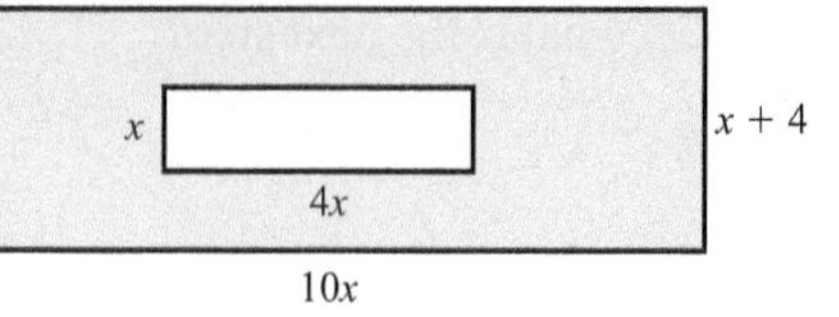

70.

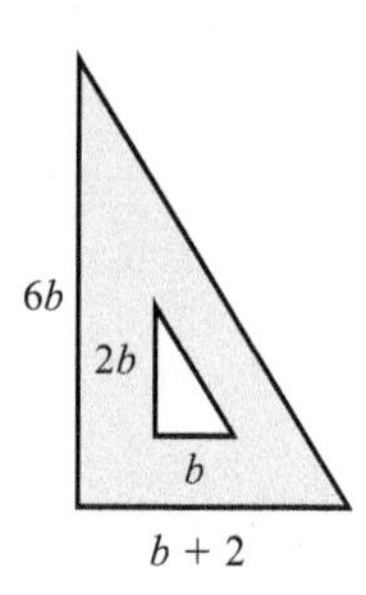

For Exercises 71 and 72: a. *Write an expression for the area of the blue shaded region.*
b. *Write the expression in factored form.*
c. *Using the factored expression, find the area if $x = 3$ feet.*

★ **71.**

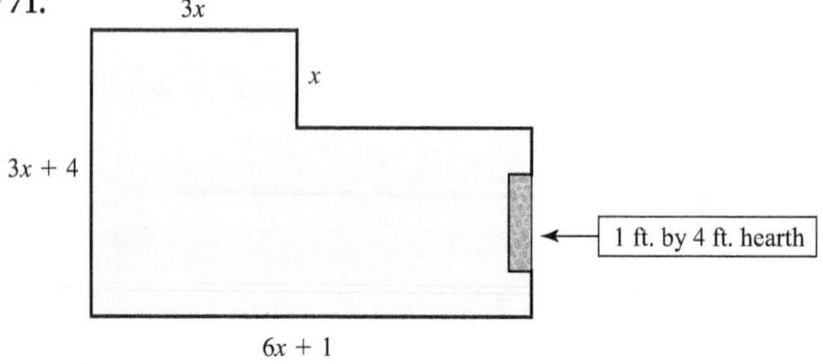

★ **72.**

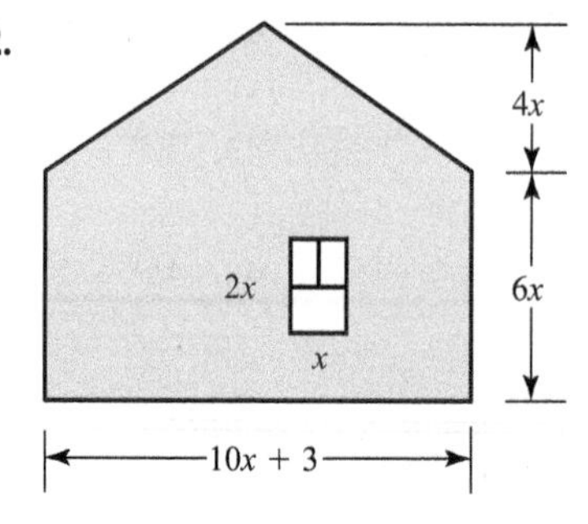

★ **73.** **a.** Write an expression for the volume of the water tower shown. (Treat the base as a cylinder even though the sphere dips into its top slightly).

b. Write the expression in factored form.

c. Find the volume of a tank if $r = 5$ feet. Round to the nearest tenth.

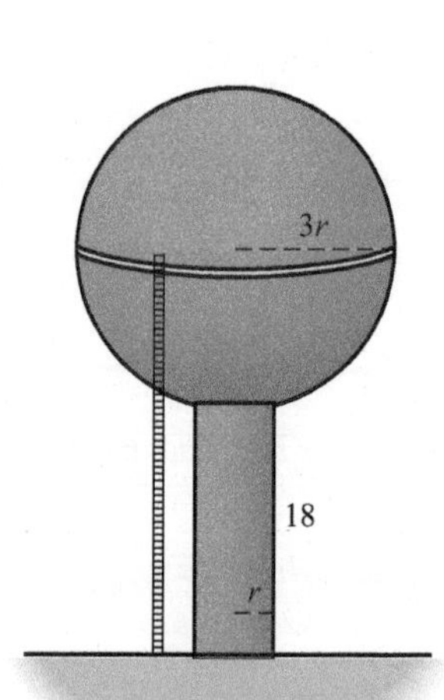

★ **74.** **a.** Write an expression for the volume of the seed spreader shown.

b. Write the expression in factored form.

c. Find the volume of the spreader if $r = 2.5$ feet and $h = 1.5$ feet. Round to the nearest tenth.

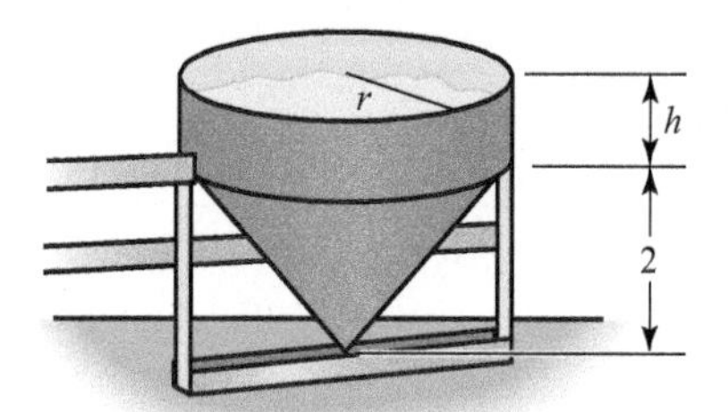

75. The final price after a discount is the *initial price minus the product of the discount rate and the initial price.*
a. Translate the italicized phrase to an expression.
b. Write the expression in factored form.
c. Use the factored expression to find the final price if the initial price is \$54.95 and the discount rate is 40%.

76. The price a store charges after marking up its inventory to make a profit is *the sum of the initial price and the product of the markup rate and the initial price.*
a. Translate the italicized phrase to an expression.
b. Write the expression in factored form.
c. Use the factored expression to find the price the store charges if the initial price is \$24.95 and the markup rate is 30%.

77. The final balance in an account that earns simple interest is *the sum of the principal and the product of the principal, interest rate, and time.*
a. Translate the italicized phrase to an expression.
b. Write the expression in factored form.
c. Use the factored expression to find the final balance in an account containing a principal of \$850 at an interest rate of 3% for half of a year.

78. The distance an object travels after accelerating from an initial velocity is *the product of the initial velocity and the time added to one-half the product of the acceleration and the square of the time.*
a. Translate the italicized phrase to an expression.
b. Write the expression in factored form.
c. Use the factored expression to find the distance, in feet, that a car travels if it accelerates 7 feet per second per second for 3 seconds from an initial velocity of 58 feet per second.

79. The surface area of a cylinder is *the product of 2, π, and the square of the radius, all added to the product of 2, π, the radius, and the height.*
a. Translate the italicized phrase to an expression.
b. Write the expression in factored form.
c. Use the factored expression to find the surface area of a cylinder with a radius of 15 inches and a height of 5 inches. Round to the nearest hundredth.

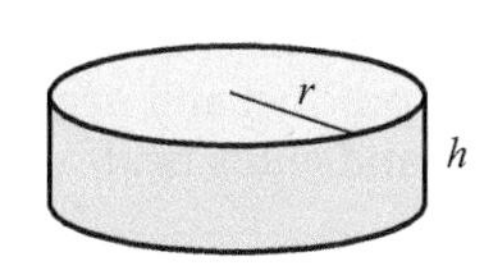

80. The volume of a capsule is *the product of $\frac{4}{3}$, π, and the cube of the radius of the capsule, all added to the product of π, the square of the radius, and the height of the cylindrical portion of the capsule.*
a. Translate the italicized phrase to an expression.
b. Write the expression in factored form.
c. Use the factored expression to find the volume of a capsule with a radius of 5 feet and a height of 8 feet. Round to the nearest hundredth.

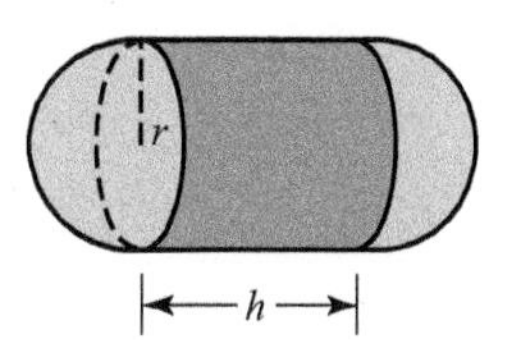

Review Exercises

Exercises 1–6 Expressions

[5.1] **1.** Write -2.45×10^7 in standard form.

[5.1] **2.** Write 0.000092 in scientific notation.

[5.3] For Exercises 3–6, multiply.

3. $(x + 3)(x + 5)$

4. $(x - 6)(x - 4)$

5. $(2x + 7)(3x - 1)$

6. $2x(4x - 5)(x + 3)$

6.2 Factoring Trinomials

Objectives

1. Factor trinomials of the form $x^2 + bx + c$.
2. Factor trinomials of the form $ax^2 + bx + c$, where $a \neq 1$, by trial.
3. Factor trinomials of the form $ax^2 + bx + c$, where $a \neq 1$, by grouping.
4. Factor trinomials of the form $ax^2 + bx + c$ using substitution.

Warm-up

[5.3] *For Exercises 1–3, find the products.*

1. $(x - 3)(x + 5)$ **2.** $(3x - 4)(2x - 1)$ **3.** $5y^2(4y - 1)(2y + 3)$

[6.1] **4.** Factor. $5a + ab + 15 + 3b$

In this section, we learn to factor trinomials such as $x^2 + 5x + 6$ and $3x^2 - x - 10$.

Objective 1 Factor trinomials of the form $x^2 + bx + c$.

First, we consider trinomials of the form $x^2 + bx + c$ in which the coefficient of the first (variable squared) term is 1. If a trinomial in this form is factorable, it will have two binomial factors. For example, $x^2 + 5x + 6 = (x + 2)(x + 3)$. Let's use FOIL to multiply $(x + 2)(x + 3)$ and look for patterns that will help us factor.

$$(x + 2)(x + 3) = \overset{F}{x^2} + \overset{O}{3x} + \overset{I}{2x} + \overset{L}{6}$$
$$= x^2 + \underbrace{5x}_{\text{Sum of 2 and 3}} + \underbrace{6}_{\text{Product of 2 and 3}}$$

Note The product of these two numbers is the last term, and their sum is the coefficient of the middle term.

This suggests the following procedure.

Note If the last term of the trinomial is positive, both signs of the binomial factors will be the sign of the middle term of the trinomial. If the last term of the trinomial is negative, the signs of the binomial factors will be opposites.

Procedure Factoring $x^2 + bx + c$

To factor a trinomial of the form $x^2 + bx + c$:

1. Find two numbers with a product equal to c and a sum equal to b.
2. The factored trinomial will have the form $(x + \square)(x + \square)$, where the second terms are the numbers found in Step 1.

Example 1 Factor.

a. $x^2 - 9x + 20$

Solution: We must find a pair of numbers whose product is 20 and whose sum is −9. Note that if two numbers have a positive product and negative sum, both must be negative. We list the possible products and sums next.

Product	Sum
$(-1)(-20) = 20$	$-1 + (-20) = -21$
$(-4)(-5) = 20$	$-4 + (-5) = -9$
$(-2)(-10) = 20$	$-2 + (-10) = -12$

← This is the correct combination, so −4 and −5 are the second terms in the binomials of the factored form.

Answer: $x^2 - 9x + 20 = (x + (-4))(x + (-5))$
$= (x - 4)(x - 5)$

◀ **Note** We omit this step in future examples.

Check: Multiply the factors to verify that their product is the original trinomial.

$$(x - 4)(x - 5) = x^2 - 5x - 4x + 20 = x^2 - 9x + 20 \quad \text{It checks.}$$

All further checks will be left to the reader.

Answers to Warm-up

1. $x^2 + 2x - 15$
2. $6x^2 - 11x + 4$
3. $40y^4 + 50y^3 - 15y^2$
4. $(5 + b)(a + 3)$

b. $x^2 + 2x - 15$

Solution: We must find a pair of numbers whose product is -15 and whose sum is 2. Because the product is negative, the two numbers must have different signs.

Product	Sum
$(-1)(15) = -15$	$-1 + 15 = 14$
$(-3)(5) = -15$	$-3 + 5 = 2$

← This is the correct combination, so -3 and 5 are the second terms in each binomial of the factored form.

Answer: $x^2 + 2x - 15 = (x - 3)(x + 5)$

Your Turn 1 Factor.

a. $n^2 - 8n + 12$

b. $y^2 + 10y - 24$

Now let's consider cases in which we factor out a monomial GCF first.

Example 2 Factor.

a. $4x^3 - 20x^2 + 24x$

Solution: $4x^3 - 20x^2 + 24x = 4x(x^2 - 5x + 6)$ Factor out the monomial GCF, $4x$.

$= 4x(x - 2)(x - 3)$ Factor $x^2 - 5x + 6$ by finding two numbers, -2 and -3, whose product is 6 and whose sum is -5.

b. $x^4 - 9x^3 + 15x^2$

Solution: $x^4 - 9x^3 + 15x^2 = x^2(x^2 - 9x + 15)$ Factor out the monomial GCF, x^2.

To factor $x^2 - 9x + 15$, we need a pair of numbers whose product is 15 and whose sum is -9.

Note Because the sign of the middle term is negative, $(1)(15)$ and $(3)(5)$ need not be considered.

Product	Sum
$(1)(15) = 15$	$1 + 15 = 16$
$(3)(5) = 15$	$3 + 5 = 8$
$(-1)(-15) = 15$	$-1 + (-15) = -16$
$(-3)(-5) = 15$	$-3 + (-5) = -8$

Note There are no factors of 15 whose sum is -9.

The trinomial $x^2 - 9x + 15$ has no binomial factors with integer terms, so $x^2(x^2 - 9x + 15)$ is the final factored form.

A polynomial such as $x^2 - 9x + 15$ that cannot be factored is like a prime number in that its only factors are 1 and the expression itself. We say such an expression is *prime*.

Your Turn 2 Factor.

a. $3n^3 - 18n^2 + 24n$

b. $y^5 + 5y^4 - 18y^3$

Let's consider some cases of trinomials that contain two variables.

Example 3 Factor $3mn^3 + 12mn^2 - 96mn$.

Solution: $3mn^3 + 12mn^2 - 96mn = 3mn(n^2 + 4n - 32)$ Factor out the monomial GCF, $3mn$.

$= 3mn(n - 4)(n + 8)$ Factor $n^2 + 4n - 32$.

Answers to Your Turn 1

a. $(n - 2)(n - 6)$

b. $(y - 2)(y + 12)$

Answers to Your Turn 2

a. $3n(n - 2)(n - 4)$

b. $y^3(y^2 + 5y - 18)$

In Example 3, we saw that factoring out the monomial GCF left us with a trinomial factor of the form $x^2 + bx + c$. Now let's examine some cases where the trinomial's factors contain both variables.

Example 4 Factor.

a. $x^2 - 8xy + 12y^2$

Solution: This trinomial has the form $x^2 + bx + c$ if we view b as $-8y$ and c as $12y^2$. We must find a pair of terms whose product is $12y^2$ and whose sum is $-8y$. These terms must be $-2y$ and $-6y$.

$$x^2 - 8xy + 12y^2 = (x - 2y)(x - 6y)$$

b. $2t^2u + 4tu^2 - 30u^3$

Solution: This trinomial has a monomial GCF, $2u$.

$$2t^2u + 4tu^2 - 30u^3 = 2u(t^2 + 2tu - 15u^2)$$ Factor out the monomial GCF, $2u$.

In the remaining trinomial factor, we can view the coefficient of t in the middle term as $2u$. We must find a pair of terms whose product is $-15u^2$ and whose sum is $2u$. These terms have to be $-3u$ and $5u$.

$$= 2u(t - 3u)(t + 5u)$$

Your Turn 4 Factor.

a. $x^2 + 9xy + 20y^2$

b. $x^3 + 4x^2y - 21xy^2$

Objective 2 Factor trinomials of the form $ax^2 + bx + c$, where $a \neq 1$, by trial.

Now let's factor trinomials of the form $ax^2 + bx + c$, where $a \neq 1$, such as $5x^2 + 13x + 6$. One way to factor these trinomials is to try various combinations of the factors of the first and last terms in the binomials until we find a combination that produces the middle term, which is the sum of the inner and outer products of the binomial factors.

Example 5 Factor $5x^2 + 13x + 6$.

Solution: Because all of the terms are positive, we know that all of the terms in the binomial factors will be positive.

The *first* terms must multiply to equal $5x^2$. These terms must be $5x$ and x.

$$5x^2 + 13x + 6 = (\square + \square)(\square + \square)$$

The *last* terms must multiply to equal 6. These factors could be 1 and 6, or 2 and 3.

Now try various combinations of $5x$ and x as *first* terms with 1 and 6 or 2 and 3 as *last* terms to see which combination produces the correct middle term, $13x$. Remember that the middle term is the sum of the *inner* and *outer* products.

$(5x + 1)(x + 6) = 5x^2 + 30x + x + 6 = 5x^2 + 31x + 6$
$(5x + 6)(x + 1) = 5x^2 + 5x + 6x + 6 = 5x^2 + 11x + 6$
$(5x + 2)(x + 3) = 5x^2 + 15x + 2x + 6 = 5x^2 + 17x + 6$
} Incorrect combinations

$(5x + 3)(x + 2) = 5x^2 + 10x + 3x + 6 = 5x^2 + 13x + 6$ Correct combination

Answer: $5x^2 + 13x + 6 = (5x + 3)(x + 2)$

Answers to Your Turn 4
a. $(x + 4y)(x + 5y)$
b. $x(x - 3y)(x + 7y)$

VISUAL

Learning Strategy

If you are a visual learner, you may find it helpful to draw lines to connect each product.

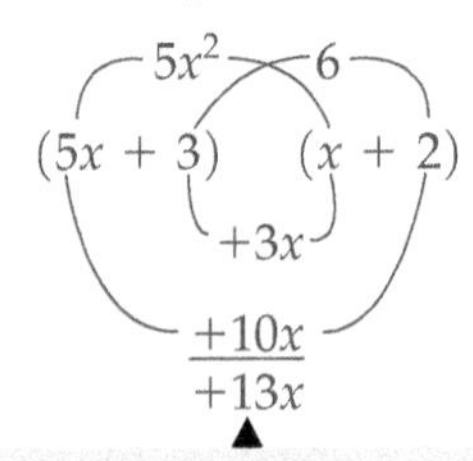

The sum of the *inner* and *outer* products verifies that this is the correct combination.

Our example suggests the following procedure.

Procedure Factoring by Trial

To factor a trinomial of the form $ax^2 + bx + c$, where $a \neq 1$, by trial:

1. Look for a monomial GCF in all of the terms. If there is one, factor it out.
2. Write a pair of *first* terms whose product is ax^2. $(\square + \quad)(\square + \quad)$
3. Write a pair of *last* terms whose product is c. $(\quad + \square)(\quad + \square)$
4. Verify that the sum of the *inner* and *outer* products is bx (the middle term of the trinomial).

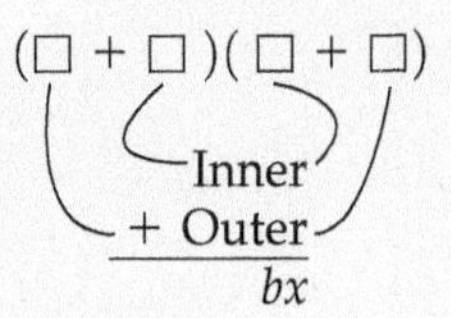

If the sum of the inner and outer products is not bx, try the following:

a. Exchange the last terms of the binomials from step 3; then repeat step 4.
b. For each additional pair of last terms, repeat steps 3 and 4.
c. For each additional pair of first terms, repeat steps 2–4.

Example 6 Factor.

a. $8x^2 - 10x - 7$

Solution: There is no common monomial factor.

The product of the *first* terms must be $8x^2$. These terms could be x and $8x$, or $2x$ and $4x$.

$$8x^2 - 10x - 7 = (\quad + \quad)(\quad - \quad)$$

The product of the *last* terms must be 7, which means they must be 1 and 7. We have already included the appropriate signs.

Now we multiply binomials with various combinations of these first and last terms until we find a combination whose inner and outer products combine to equal $-10x$.

$(x + 1)(8x - 7) = 8x^2 - 7x + 8x - 7 = 8x^2 + x - 7$
$(x + 7)(8x - 1) = 8x^2 - x + 56x - 7 = 8x^2 + 55x - 7$ } Incorrect combinations

$(2x + 1)(4x - 7) = 8x^2 - 14x + 4x - 7 = 8x^2 - 10x - 7$ Correct combination

Note We did not show every possible combination. We did not try $(8x + 7)(x - 1)$ since it gives the same product as $(8x - 7)(x + 1)$ except for the sign of the middle term.

Answer: $8x^2 - 10x - 7 = (2x + 1)(4x - 7)$

b. $6x^2 - 11x + 4$

Solution: There is no common monomial factor. Because 4 is positive and $-11x$ is negative, we know that both last terms in the binomial factors must have minus signs.

The product of the *first* terms must be $6x^2$. These terms could be x and $6x$, or $2x$ and $3x$.

$$6x^2 - 11x + 4 = (\quad - \quad)(\quad - \quad)$$

The product of the *last* terms must be 4. These terms could be 1 and 4, or 2 and 2.

Note If a polynomial does not have a common factor, then no factors of the polynomial can have a common factor. This means that $(6x - 4)(x - 1)$ and $(6x - 2)(x - 2)$ are impossible because $6x - 4$ and $6x - 2$ each have a common factor of 2.

Now find a combination whose inner and outer products combine to equal $-11x$.

$$(6x - 1)(x - 4) = 6x^2 - 24x - x + 4 = 6x^2 - 25x + 4$$
$$(6x - 4)(x - 1) = 6x^2 - 6x - 4x + 4 = 6x^2 - 10x + 4$$
$$(6x - 2)(x - 2) = 6x^2 - 12x - 2x + 4 = 6x^2 - 14x + 4$$
$$(3x - 1)(2x - 4) = 6x^2 - 12x - 2x + 4 = 6x^2 - 14x + 4$$

Incorrect combinations

$$(3x - 4)(2x - 1) = 6x^2 - 3x - 8x + 4 = 6x^2 - 11x + 4$$

Correct combination

Answer: $6x^2 - 11x + 4 = (3x - 4)(2x - 1)$

c. $3x^2 - 5x - 6$

Solution: There is no common monomial factor. Because -6 is negative, the last signs of the binomial factors must be different.

The product of the *first* terms must be $3x^2$. These terms must be $3x$ and x.

$$3x^2 - 5x - 6 = (\quad + \quad)(\quad - \quad)$$

The product of the *last* terms must be -6, which means they must be 1 and -6, -1 and 6, -2 and 3, or 2 and -3.

Note The following combinations cannot be considered because one of the factors in each has a common factor of 3.

$(3x + 6)(x - 1)$
$(3x - 6)(x + 1)$
$(3x + 3)(x - 2)$
$(3x - 3)(x + 2)$

Now find a combination whose inner and outer products combine to equal $-5x$.

$$(3x + 1)(x - 6) = 3x^2 - 18x + x - 6 = 3x^2 - 17x - 6$$
$$(3x - 1)(x + 6) = 3x^2 + 18x - x - 6 = 3x^2 + 17x - 6$$
$$(3x - 2)(x + 3) = 3x^2 + 9x - 2x - 6 = 3x^2 + 7x - 6$$
$$(3x + 2)(x - 3) = 3x^2 - 9x + 2x - 6 = 3x^2 - 7x - 6$$

Incorrect combinations

Because there are no more possibilities, $3x^2 - 5x - 6$ cannot be factored.

Answer: $3x^2 - 5x - 6$ is prime.

d. $15x^2y + 18xy^2 - 24y^3$

Solution: $15x^2y + 18xy^2 - 24y^3 = 3y(5x^2 + 6xy - 8y^2)$ Factor out the monomial GCF, $3y$.

Now we factor $5x^2 + 6xy - 8y^2$. Because $-8y^2$ is negative, we know that the signs of the last terms in the binomial factors must be different.

The product of the *first* terms must be $5x^2$. So they must be x and $5x$.

$$3y(5x^2 + 6xy - 8y^2) = 3y(\quad + \quad)(\quad - \quad)$$

The product of the *last* terms must be $8y^2$. These terms could be y and $8y$, or $2y$ and $4y$.

Now find a combination whose inner and outer products combine to equal $6xy$.

$$3y(x + y)(5x - 8y) = 3y(5x^2 - 8xy + 5xy - 8y^2) = 3y(5x^2 - 3xy - 8y^2)$$
$$3y(x + 8y)(5x - y) = 3y(5x^2 - xy + 40xy - 8y^2) = 3y(5x^2 + 39xy - 8y^2)$$

Incorrect combinations

$$3y(x + 2y)(5x - 4y) = 3y(5x^2 - 4xy + 10xy - 8y^2) = 3y(5x^2 + 6xy - 8y^2)$$

Correct combination

Answer: $15x^2y + 18xy^2 - 24y^3 = 3y(x + 2y)(5x - 4y)$

Your Turn 6 Factor.

a. $6t^2 - 19t + 10$ b. $8n^2 - 10n + 3$ c. $4x^2 - 7x - 6$ d. $6x^3 - 28x^2y - 48xy^2$

Objective 3 Factor trinomials of the form $ax^2 + bx + c$, where $a \neq 1$, by grouping.

Because trial and error can be tedious, an alternative method is to factor by grouping, which we learned in Section 6.1. Recall that we grouped pairs of terms in a four-term polynomial, then factored out the GCF from each pair. Because a trinomial of the form $ax^2 + bx + c$ has only three terms, we split its bx term into two like terms to create a four-term polynomial that we can factor by grouping.

To split the bx term, we use the fact that if $ax^2 + bx + c$ is factorable, b will equal the sum of a pair of factors of the product of a and c. For example, in $5x^2 + 13x + 6$ from Example 5, the product of a and c is $(5)(6) = 30$; so we need to find two factors of 30 whose sum is b, 13. It is helpful to list the factor pairs and their sums in a table.

Factors of $ac = 30$	Sum of Factors of ac
$(1)(30) = 30$	$1 + 30 = 31$
$(2)(15) = 30$	$2 + 15 = 17$
$(3)(10) = 30$	$3 + 10 = 13$
$(5)(6) = 30$	$5 + 6 = 11$

← Notice that 3 and 10 form the only factor pair of 30 whose sum is 13.

Now we can write $13x$ as $3x + 10x$ or $10x + 3x$ and then factor by grouping.

$$\begin{aligned} 5x^2 + 13x + 6 &= 5x^2 + 3x + 10x + 6 \\ &= x(5x + 3) + 2(5x + 3) \\ &= (5x + 3)(x + 2) \end{aligned} \qquad \text{or} \qquad \begin{aligned} &5x^2 + 10x + 3x + 6 \\ &= 5x(x + 2) + 3(x + 2) \\ &= (x + 2)(5x + 3) \end{aligned}$$

Note The order of the like terms does not matter.

Every trinomial factorable by grouping has only one factor pair of ac whose sum is b, which suggests the following procedure.

Procedure Factoring $ax^2 + bx + c$, Where $a \neq 1$, by Grouping

To factor a trinomial of the form $ax^2 + bx + c$, where $a \neq 1$, by grouping:

1. Look for a monomial GCF in all of the terms. If there is one, factor it out.
2. Find two factors of the product ac whose sum is b.
3. Write a four-term polynomial in which bx is written as the sum of two like terms whose coefficients are the two factors you found in step 2.
4. Factor by grouping.

Example 7 Factor by grouping.

a. $6x^2 - 11x + 4$

Solution: For this trinomial, $a = 6$, $b = -11$, and $c = 4$; so $ac = (6)(4) = 24$. We must find two factors of 24 whose sum is -11. Because 4 is positive and -11 is negative, the two factors must be negative.

Factors of $ac = 24$	Sum of Factors of ac
$(-1)(-24) = 24$	$-1 + (-24) = -25$
$(-2)(-12) = 24$	$-2 + (-12) = -14$
$(-3)(-8) = 24$	$-3 + (-8) = -11$
$(-4)(-6) = 24$	$-4 + (-6) = -10$

← Correct

Note You do not need to list all possible combinations as we did here. We listed them to illustrate that only one combination is correct.

Answers to Your Turn 6

a. $(2t - 5)(3t - 2)$
b. $(2n - 1)(4n - 3)$
c. prime
d. $2x(3x + 4y)(x - 6y)$

$6x^2 - 11x + 4 = 6x^2 - 3x - 8x + 4$ Write $-11x$ as $-3x - 8x$.

$= 3x(2x - 1) - 4(2x - 1)$ Factor $3x$ out of $6x^2 - 3x$ and -4 out of $-8x + 4$.

$= (2x - 1)(3x - 4)$ Factor out $2x - 1$.

b. $40y^4 + 50y^3 - 15y^2$

Solution: $40y^4 + 50y^3 - 15y^2 = 5y^2(8y^2 + 10y - 3)$ Factor out the monomial GCF, $5y^2$.

For the trinomial factor, note that $a = 8, b = 10$, and $c = -3$; so $ac = (8)(-3) = -24$. We must find two factors of -24 whose sum is 10. Because -3 is negative and 10 is positive, the two factors will have different signs and the factor with the greater absolute value must be positive.

Factors of $ac = 24$	Sum of Factors of ac	
$(-1)(24) = -24$	$-1 + 24 = 23$	
$(-2)(12) = -24$	$-2 + 12 = 10$	← Correct
$(-3)(8) = -24$	$-3 + 8 = 5$	
$(-4)(6) = -24$	$-4 + 6 = 2$	

$5y^2(8y^2 + 10y - 3) = 5y^2(8y^2 - 2y + 12y - 3)$ Write $10y$ as $-2y + 12y$.

$= 5y^2[2y(4y - 1) + 3(4y - 1)]$ Factor $2y$ out of $8y^2 - 2y$ and 3 out of $12y - 3$.

$= 5y^2(4y - 1)(2y + 3)$ Factor out $4y - 1$.

Your Turn 7 Factor by grouping.

a. $10x^2 - 19x + 6$

b. $36x^4y + 3x^3y - 60x^2y$

Objective 4 Factor trinomials of the form $ax^2 + bx + c$ using substitution.

Some rather complicated-looking polynomials are actually in the form $ax^2 + bx + c$ and can be factored using a method called *substitution.*

Example 8 Factor using substitution.

a. $6(y - 3)^2 + 17(y - 3) + 5$

Solution: If we substitute another variable, such as u, for $y - 3$, we can see that the polynomial is in the form $au^2 + bu + c$.

$$6(y - 3)^2 + 17(y - 3) + 5$$

$$6u^2 + 17u + 5$$ Substitute u for $y - 3$.

Now we can factor $6u^2 + 17u + 5$ by trial or grouping.

$$6u^2 + 17u + 5 = (2u + 5)(3u + 1)$$

Remember that u was substituted for $y - 3$, which means that we must substitute $y - 3$ back for u to have the factored form for our original polynomial.

$= (2u + 5)(3u + 1)$

$= [2(y - 3) + 5][3(y - 3) + 1]$ Substitute $y - 3$ for u.

This resulting factored form can be simplified.

$= [2y - 6 + 5][3y - 9 + 1]$ Distribute 2 and 3.

$= (2y - 1)(3y - 8)$ Combine like terms.

Answers to Your Turn 7

a. $(5x - 2)(2x - 3)$

b. $3x^2y(3x + 4)(4x - 5)$

Learning Strategy

Even if you don't get a homework grade, make sure you do the homework. Teachers assign homework for a reason, often quizzes are based on the homework. If you get something wrong, make sure you go back and see why it is wrong and learn from it.

—Ellyn G.

b. $8x^4 + 2x^2 - 15$

Solution: Notice that the degree of the middle term, 2, is half the degree of the first term, 4, and the last term has no variable at all. If we substitute u for x^2, we can see that the polynomial has the form $ax^2 + bx + c$.

$$8(x^2)^2 + 2x^2 - 15 \quad \text{◀ Note } x^4 = (x^2)^2$$

$$8u^2 + 2u - 15 \quad \text{Substitute } u \text{ for } x^2.$$

Now we can factor $8u^2 + 2u - 15$ by trial or grouping.

$$8u^2 + 2u - 15 = (4u - 5)(2u + 3)$$

To finish, we substitute x^2 for u so that we have the factored form for the original polynomial.

$$= (4u - 5)(2u + 3)$$

$$= (4x^2 - 5)(2x^2 + 3) \quad \text{Substitute } x^2 \text{ for } u.$$

Your Turn 8 Factor using substitution.

a. $24(t + 2)^2 - 22(t + 2) + 3$ **b.** $15n^4 - 19n^2 + 6$

Answers to Your Turn 8
a. $(4t + 5)(6t + 11)$
b. $(5n^2 - 3)(3n^2 - 2)$

6.2 Exercises

For Extra Help MyMathLab®

Note: Exercises marked with a ★ represent challenging exercises.

Objective 1

Prep Exercise 1 To factor $x^2 + 8x + 12$, find two numbers whose product is ________ and whose sum is ________.

Prep Exercise 2 To factor $x^2 + 3x - 10$, find two numbers whose product is ________ and whose sum is ________.

Prep Exercise 3 Complete the factored form by inserting the correct signs.
$x^2 - 5x + 6 = (x$ ______ $2)(x$ ______ $3)$

Prep Exercise 4 Complete the factored form by inserting the correct signs.
$x^2 - x - 6 = (x$ ______ $2)(x$ ______ $3)$

Prep Exercise 5 Complete the factored form by inserting the correct signs.
$x^2 + x - 6 = (x$ ______ $2)(x$ ______ $3)$

For Exercises 1–14, factor completely. See Example 1.

1. $r^2 + 8r + 7$

2. $y^2 + 12y + 11$

3. $w^2 - 2w - 3$

4. $x^2 + 4x - 5$

5. $a^2 + 9a + 18$

6. $y^2 + 9y + 20$

7. $y^2 - 13y + 36$

8. $x^2 - 13x + 30$

9. $m^2 + 2m - 8$

10. $b^2 - 7b - 18$

11. $b^2 - 6b - 40$

12. $n^2 - n - 30$

13. $x^2 - 5x - 18$

14. $x^2 - 3x - 12$

For Exercises 15–22, factor completely. See Examples 2 and 3.

15. $3st^2 + 24st + 21s$

16. $4x^2y + 24xy + 20y$

17. $5y^3 - 65y^2 + 60y$

18. $7n^3 - 35n^2 + 28n$

19. $6au^3 + 6au^2 - 36au$

20. $5k^2lm - 15klm - 50lm$

21. $3x^2y^3 - 12x^2y^2 + 30x^2y$

22. $r^4q^2 - 2r^3q^2 - 20r^2q^2$

For Exercises 23–30, factor the trinomials containing two variables. See Example 4.

23. $p^2 + 11pq + 18q^2$

24. $r^2 + 8rs + 15s^2$

25. $u^2 - 13uv + 42v^2$

26. $a^2 - 10ab + 24b^2$

27. $x^2 - 5xy - 14y^2$

28. $m^2 + 2mn - 15n^2$

29. $a^2 - ab - 42b^2$

30. $h^2 - 4hk - 21k^2$

Objective 2

Prep Exercise 6 To factor $8x^2 + 97x + 12$ by trial, what are all possible factor pairs that could occupy the *first* terms in the binomial factors?

Prep Exercise 7 To factor $8x^2 + 97x + 12$ by trial, what are all possible factor pairs that could occupy the *last* terms in the binomial factors?

Prep Exercise 8 When factoring by trial, how do you know you have the correct combination of terms in the binomial factors?

For Exercises 31–62, factor completely. See Examples 5 and 6.

31. $3a^2 + 10a + 7$

32. $5p^2 - 16p + 3$

33. $2w^2 - 3w - 2$

34. $2y^2 + 5y - 3$

35. $4r^2 - r + 2$

36. $2a^2 - 13a + 1$

37. $4q^2 - 9q + 2$

38. $3c^2 + 13c + 4$

39. $6b^2 + 7b - 3$

40. $8b^2 - 6b - 5$

41. $16m^2 + 24m + 9$

42. $25a^2 + 40a + 16$

43. $4x^2 + 5x - 6$

44. $6p^2 - 7p - 10$

45. $2w^2 + 15wv + 7v^2$

46. $3r^2 + 8rv + 5v^2$

47. $5x^2 - 16xy + 3y^2$

48. $3a^2 - 10ab + 7b^2$

49. $16x^2 - 10xy + y^2$

50. $16h^2 + 10hk + k^2$

51. $6a^2 - 13ab - 10b^2$

52. $8p^2 + 8pq - 9q^2$

53. $3t^2 + 19tu - 14u^2$

54. $3x^2 - xy - 14y^2$

55. $3m^2 - 10mn - 8n^2$

56. $5a^2 + 26a - 24$

57. $12a^2 - 17ab + 6b^2$

58. $6u^2 + 13uv + 6v^2$

59. $22m^3 + 200m^2 + 18m$

60. $8x^2y + 60xy + 28y$

61. $4u^2v + 2uv^2 - 30v^3$

62. $24m^4 - 80m^3 - 64m^2$

Objective 3

Prep Exercise 9 To factor $3x^2 + 11x + 10$ by grouping, find two numbers whose product is __________ and whose sum is __________.

Prep Exercise 10 To factor $5x^2 - 4x - 12$ by grouping, find two numbers whose product is __________ and whose sum is __________.

For Exercises 63–74, factor by grouping. See Example 7.

63. $3y^2 + 16y + 5$

64. $5r^2 - 16r + 3$

65. $6c^2 + 11c + 6$

66. $4b^2 + 15b + 6$

67. $3t^2 - 17t + 10$

68. $3k^2 + 14k + 8$

69. $6x^2 - x - 15$

70. $10u^2 + 11u - 6$

71. $32x^2y + 24xy - 36y$

72. $12m^3 + 10m^2 - 12m$

73. $15a^2b - 25ab^2 - 10b^3$

74. $2x^2y + 6xy^2 - 8y^3$

Objective 4

Prep Exercise 11 To factor $x^4 + 7x^2 + 10$ using substitution, substitute u for ____________.

Prep Exercise 12 To factor $2(x - 3)^2 - 11(x - 3) + 12$ using substitution, substitute u for ____________.

For Exercises 75–86, factor using substitution. See Example 8.

75. $x^4 - x^2 - 2$

76. $m^4 - 2m^2 - 3$

77. $8r^4 + 2r^2 - 3$

78. $6w^4 + 7w^2 - 3$

79. $15x^4 - 11x^2 + 2$

80. $12r^4 - 19r^2 + 5$

81. $y^6 - 16y^3 + 48$

82. $3h^6 + 14h^3 - 5$

83. $7(x + 1)^2 + 8(x + 1) + 1$

84. $2(m - 3)^2 + 7(m - 3) + 6$

85. $3(a + 2)^2 - 10(a + 2) - 8$

86. $8(x - 3)^2 + 6(x - 3) - 9$

Find the Mistake *For Exercises 87–90, identify the mistake in the factored form; then give the correct factored form.*

87. $x^2 - x - 6 = (x - 2)(x + 3)$

88. $x^2 - 5x + 6 = (x - 6)(x + 1)$

89. $4x^2 + 16x + 16 = (2x + 4)(2x + 4)$

90. $2x^2 - 11x + 12 = (2x - 4)(x - 3)$

★ *For Exercises 91–94, find all natural-number values of b that make the trinomial factorable.*

91. $x^2 + bx + 16$

92. $x^2 + bx + 20$

93. $x^2 + bx - 63$

94. $x^2 - bx - 35$

★ *For Exercises 95–98, find a natural number c that makes the trinomial factorable.*

95. $x^2 + 9x + c$

96. $x^2 - 8x + c$

97. $x^2 + x - c$

98. $x^2 + 4x - c$

Puzzle Problem A census taker came to a house where a man lived with his three daughters.

"What are the ages of your three daughters?" asked the census taker.

"The product of their ages is 72, and the sum of their ages is my house number."

"But that clearly is not enough information," insisted the census taker.

"All right," said the man. "The oldest in years loves chocolate milk." The census taker was then able to determine the age of each daughter. Find the ages of the three daughters. (*Source*: Michael A. Stueben, *Discover* magazine; November 1983.)

Review Exercises

Exercises 1–4 Expressions

[5.3] *For Exercises 1–3, multiply.*

1. $(2x + 3)(2x + 3)$ **2.** $(4y - 1)(4y + 1)$ **3.** $(n - 2)(n^2 + 2n + 4)$

[5.4] **4.** Divide using long division: $\dfrac{6x^3 - 5x^2 + 3x - 2}{x - 1}$

Exercises 5 and 6 Equations and Inequalities

For Exercises 5 and 6, solve, then:
a. Write the solution set using set-builder notation.
b. Write the solution set using interval notation.
c. Graph the solution set.

[2.3] **5.** $5(x - 6) + 2x \geq 3x - 4$ [2.6] **6.** $|2x + 7| < 5$

6.3 Factoring Special Products and Factoring Strategies

Objectives

1 Factor perfect square trinomials.
2 Factor a difference of squares.
3 Factor a difference of cubes.
4 Factor a sum of cubes.
5 Use various strategies to factor polynomials.

Warm-up

[5.3] *For Exercises 1–3, find the products.*

1. $(5x - 2y)^2$ **2.** $(3x + 4y)(3x - 4y)$ **3.** $(2x - 3)(4x^2 + 6x + 9)$

[6.2] **4.** Factor. $6a^2 - 11a - 10$

In Section 5.3, we explored special products found by squaring binomials and by multiplying conjugates. In this section, we see how to factor those special products.

Answers to Warm-up
1. $25x^2 - 20xy + 4y^2$
2. $9x^2 - 16y^2$
3. $8x^3 - 27$
4. $(3a + 2)(2a - 5)$

Objective 1 Factor perfect square trinomials.

A perfect square trinomial is the product resulting from squaring a binomial. Recall the following rules for squaring binomials that we developed in Section 5.3.

Note In these perfect square trinomials, the first and last terms are squares and the middle term is twice the product of their square roots.

$$(a + b)^2 = a^2 + 2ab + b^2$$
$$(a - b)^2 = a^2 - 2ab + b^2$$

To use these rules when factoring, we simply reverse them.

Rules Factoring Perfect Square Trinomials

$$a^2 + 2ab + b^2 = (a + b)^2$$
$$a^2 - 2ab + b^2 = (a - b)^2$$

Example 1 Factor.

a. $16x^2 + 40x + 25$

Solution: This trinomial is a perfect square because it has the form $a^2 + 2ab + b^2$, where $a = 4x$ and $b = 5$.

$a^2 = (4x)^2 = 16x^2$ $\quad$ $b^2 = 5^2 = 25$

$$16x^2 + 40x + 25$$

$2ab = 2(4x)(5) = 40x$

Using the rule $a^2 + 2ab + b^2 = (a + b)^2$, where $a = 4x$ and $b = 5$, we have

$$16x^2 + 40x + 25 = (4x + 5)^2.$$

Check: $(4x + 5)^2 = (4x)^2 + 2(4x)(5) + 5^2 = 16x^2 + 40x + 25$

Further checks will be left to the reader.

b. $9y^2 - 42y + 49$

Solution: This trinomial is a perfect square because it has the form $a^2 - 2ab + b^2$, where $a = 3y$ and $b = 7$.

$9y^2 - 42y + 49 = (3y - 7)^2$ Use the rule $a^2 - 2ab + b^2 = (a - b)^2$, where $a = 3y$ and $b = 7$.

c. $25x^2 - 20xy + 4y^2$

Solution: Perfect square trinomials can have two variables. In this case, we have the form $a^2 - 2ab + b^2$, where $a = 5x$ and $b = 2y$.

$25x^2 - 20xy + 4y^2 = (5x - 2y)^2$ Use $a^2 - 2ab + b^2 = (a - b)^2$, where $a = 5x$ and $b = 2y$.

d. $m^3n + 8m^2n + 16mn$

Solution: Remember to factor out any monomial GCF first.

$m^3n + 8m^2n + 16mn = mn(m^2 + 8m + 16)$ Factor out the monomial GCF, mn.

$= mn(m + 4)^2$ Factor the perfect square trinomial using $a^2 + 2ab + b^2 = (a + b)^2$, where $a = m$ and $b = 4$.

Your Turn 1 Factor.

a. $4x^2 + 28x + 49$ **b.** $9n^2 - 48n + 64$

c. $16h^2 + 72hk + 81k^2$ **d.** $2t^3u^2 - 20t^2u^2 + 50tu^2$

Answers to Your Turn 1
a. $(2x + 7)^2$ **b.** $(3n - 8)^2$
c. $(4h + 9k)^2$ **d.** $2tu^2(t - 5)^2$

Objective 2 Factor a difference of squares.

Another special product we considered in Section 5.3 was that of conjugates, which are binomials that differ only in the sign separating the terms, as in $3x + 2$ and $3x - 2$. Note that the product of conjugates is a *difference of squares.*

$$(3x + 2)(3x - 2) = 9x^2 - 4$$

This term is the square of $3x$. This term is the square of 2.

The rule for multiplying conjugates that we developed in Section 5.3 is

$$(a + b)(a - b) = a^2 - b^2.$$

Reversing this rule tells us how to factor a difference of squares.

Warning A *sum* of squares, $a^2 + b^2$, is prime and cannot be factored.

Rule Factoring a Difference of Squares

$$a^2 - b^2 = (a + b)(a - b)$$

Example 2 Factor.

a. $9x^2 - 16y^2$

Solution: This binomial is a difference of squares because $9x^2 - 16y^2 = (3x)^2 - (4y)^2$.

$$a^2 - b^2 = (a + b)(a - b)$$

$9x^2 - 16y^2 = (3x)^2 - (4y)^2 = (3x + 4y)(3x - 4y)$ Use $a^2 - b^2 = (a + b)(a - b)$ with $a = 3x$ and $b = 4y$.

b. $25x^4 - 64$

Solution: This binomial is a difference of squares because $25x^4 - 64 = (5x^2)^2 - (8)^2$.

$25x^4 - 64 = (5x^2)^2 - (8)^2 = (5x^2 + 8)(5x^2 - 8)$ Use $a^2 - b^2 = (a + b)(a - b)$ with $a = 5x^2$ and $b = 8$.

c. $5n^4 - 20n^6$

Solution: This binomial has a monomial GCF, $5n^4$.

$5n^4 - 20n^6 = 5n^4(1 - 4n^2)$ Factor out the monomial GCF, $5n^4$.

$= 5n^4(1 + 2n)(1 - 2n)$ Factor $1 - 4n^2$ using $a^2 - b^2 = (a + b)(a - b)$, where $a = 1$ and $b = 2n$.

d. $x^4 - 81$

Solution: This binomial is a difference of squares because $x^4 - 81 = (x^2)^2 - (9)^2$.

$x^4 - 81 = (x^2)^2 - (9)^2 = (x^2 + 9)(x^2 - 9)$ Use $a^2 - b^2 = (a + b)(a - b)$ with $a = x^2$ and $b = 9$.

Notice that the factor $x^2 - 9$ is another difference of squares, with $a = x$ and $b = 3$; so we can factor further.

Note The factor $x^2 + 9$ is a sum of squares, which is prime. ▶

$$= (x^2 + 9)(x + 3)(x - 3)$$

e. $(x - y)^2 - 9$

Solution: This is a difference of squares because $(x - y)^2 - 9 = (x - y)^2 - 3^2$, which is in the form $a^2 - b^2$ with $a = x - y$ and $b = 3$.

$(x - y)^2 - 9 = (x - y)^2 - 3^2$

$= [(x - y) + 3][(x - y) - 3]$ Use $a^2 - b^2 = (a + b)(a - b)$ with $a = x - y$ and $b = 3$.

$= (x - y + 3)(x - y - 3)$ Simplify.

Your Turn 2 Factor.

a. $x^2 - 36$ **b.** $16h^4 - 49k^6$ **c.** $24y^5 - 54y^3$ **d.** $t^4 - 16$ **e.** $(3m + n)^2 - 25$

Answers to Your Turn 2

a. $(x + 6)(x - 6)$
b. $(4h^2 + 7k^3)(4h^2 - 7k^3)$
c. $6y^3(2y + 3)(2y - 3)$
d. $(t^2 + 4)(t + 2)(t - 2)$
e. $(3m + n + 5)(3m + n - 5)$

Objective 3 Factor a difference of cubes.

Another special form that we can factor is $a^3 - b^3$, which is a *difference of cubes*. A difference of cubes is the product of a binomial in the form $a - b$ and a trinomial in the form $a^2 + ab + b^2$.

$$(a - b)(a^2 + ab + b^2) = a^3 + a^2b + ab^2 - a^2b - ab^2 - b^3 \quad \text{Multiply.}$$
$$= a^3 - b^3 \quad \text{Simplify.}$$

Our multiplication suggests the following rule for factoring a difference of cubes.

Warning The trinomial $a^2 + ab + b^2$ is not a perfect square and cannot be factored. Remember, a perfect square trinomial has the form $a^2 + 2ab + b^2$.

Rule Factoring a Difference of Cubes

$$a^3 - b^3 = (a - b)(a^2 + ab + b^2)$$

Example 3 Factor.

a. $8x^3 - 27$

Solution: This binomial is a difference of cubes because $8x^3 - 27 = (2x)^3 - (3)^3$. To factor, we use the rule $a^3 - b^3 = (a - b)(a^2 + ab + b^2)$ with $a = 2x$ and $b = 3$.

$$a^3 - b^3 = (a - b)(a^2 + a\,b + b^2)$$
$$8x^3 - 27 = (2x)^3 - (3)^3 = (2x - 3)[(2x)^2 + (2x)(3) + (3)^2] \quad \text{Substitute } 2x \text{ for } a \text{ and } 3 \text{ for } b.$$
$$= (2x - 3)(4x^2 + 6x + 9) \quad \text{Simplify.}$$

b. $64m^5 - 125m^2n^3$

Solution: This binomial has a monomial GCF, m^2.

$$64m^5 - 125m^2n^3 = m^2(64m^3 - 125n^3) \quad \text{Factor out the monomial GCF, } m^2.$$
$$= m^2(4m - 5n)[(4m)^2 + (4m)(5n) + (5n)^2] \quad \text{Factor } 64m^3 - 125n^3 \text{ using } a^3 - b^3 = (a - b)(a^2 + ab + b^2) \text{ with } a = 4m \text{ and } b = 5n.$$
$$= m^2(4m - 5n)(16m^2 + 20mn + 25n^2) \quad \text{Simplify.}$$

Note $16m^2 + 20mn + 25n^2$ is prime.

Your Turn 3 Factor.

a. $1 - t^3$

b. $54xy^3 - 128x$

Objective 4 Factor a sum of cubes.

A *sum of cubes* has the form $a^3 + b^3$ and can be factored using a pattern similar to that for the difference of cubes. A sum of cubes is the product of a binomial in the form $a + b$ and a trinomial in the form $a^2 - ab + b^2$.

$$(a + b)(a^2 - ab + b^2) = a^3 - a^2b + ab^2 + a^2b - ab^2 + b^3 \quad \text{Multiply.}$$
$$= a^3 + b^3 \quad \text{Simplify.}$$

Our multiplication suggests the following rule for factoring a sum of cubes.

Warning In the rule for the sum of cubes, the trinomial factor $a^2 - ab + b^2$ cannot be factored.

Rule Factoring a Sum of Cubes

$$a^3 + b^3 = (a + b)(a^2 - ab + b^2)$$

Answers to Your Turn 3

a. $(1 - t)(1 + t + t^2)$

b. $2x(3y - 4)(9y^2 + 12y + 16)$

 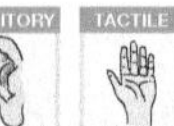 

Learning Strategy

A mnemonic for remembering the signs of the factorization of the sum or difference of cubes is "SOAP."

$a^3 + b^3 = (a + b)(a^2 - ab + b^2)$
$a^3 - b^3 = (a - b)(a^2 + ab + b^2)$

↑ S a m e ↑ O p p o s i t e ↑ A l w a y s P o s i t i v e

Example 4 Factor.

a. $27x^3 + 125$

Solution: This binomial is a sum of cubes because $27x^3 + 125 = (3x)^3 + (5)^3$. To factor, we use $a^3 + b^3 = (a + b)(a^2 - ab + b^2)$ with $a = 3x$ and $b = 5$.

$$a^3 + b^3 = (a + b)(a^2 - a\,b + b^2)$$

$$27x^3 + 125 = (3x)^3 + (5)^3 = (3x + 5)((3x)^2 - (3x)(5) + (5)^2)$$ Substitute $3x$ for a and 5 for b.

$$= (3x + 5)(9x^2 - 15x + 25)$$ Simplify.

b. $64tu^5 + 27t^4u^2$

Solution: This binomial has a monomial GCF, tu^2.

$64tu^5 + 27t^4u^2 = tu^2(64u^3 + 27t^3)$ Factor out the monomial GCF, tu^2.

$= tu^2(4u + 3t)[(4u)^2 - (4u)(3t) + (3t)^2]$ Factor $64u^3 + 27t^3$ using $a^3 + b^3 = (a + b)(a^2 - ab + b^2)$ $a = 4u$ and $b = 3t$.

$= tu^2(4u + 3t)(16u^2 - 12tu + 9t^2)$ Simplify.

c. $(2x + y)^3 + 27$

Solution: This is a sum of cubes because $(2x + y)^3 + 27 = (2x + y)^3 + 3^3$, which is in the form $a^3 + b^3$ with $a = 2x + y$ and $b = 3$.

$(2x + y)^3 + 27 = (2x + y)^3 + 3^3$

$= [(2x + y) + 3][(2x + y)^2 - 3(2x + y) + 3^2]$ Use $a^3 + b^3 = (a + b)(a^2 - ab + b^2)$ with $a = 2x + y$ and $b = 3$.

$= (2x + y + 3)(4x^2 + 4xy + y^2 - 6x - 3y + 9)$ Simplify.

Your Turn 4 Factor.

a. $64 + y^3$ **b.** $81n^5 + 24n^2$ **c.** $(3m - 2n)^3 + 64$

Objective 5 Use various strategies to factor polynomials.

We conclude our exploration of factoring by mixing up the various types of factorable expressions. The challenge will be in determining which of the various factoring techniques is appropriate for the given expression. Use the following outline as a general guide.

Procedure Factoring a Polynomial

To factor a polynomial, factor out any monomial GCF and then consider the number of terms in the polynomial.

I. **Four terms:** Try to factor by grouping.
II. **Three terms:** Determine whether the trinomial is a perfect square.
  A. If the trinomial is a perfect square, consider its form.
    1. If it is in the form $a^2 + 2ab + b^2$, the factored form is $(a + b)^2$.
    2. If it is in the form $a^2 - 2ab + b^2$, the factored form is $(a - b)^2$.
  B. If the trinomial is not a perfect square, consider its form.
    1. If it is in the form $x^2 + bx + c$, find two factors of c whose sum is b and write the factored form as $(x + \text{first number})(x + \text{second number})$.
    2. If it is in the form $ax^2 + bx + c$, where $a \neq 1$, use trial. Or find two factors of ac whose sum is b; write these factors as coefficients of two like terms that, when combined, equal bx; and factor by grouping.

Answers to Your Turn 4

a. $(4 + y)(16 - 4y + y^2)$
b. $3n^2(3n + 2)(9n^2 - 6n + 4)$
c. $(3m - 2n + 4)(9m^2 - 12mn + 4n^2 - 12m + 8n + 16)$

III. **Two terms:** Determine whether the binomial is a difference of squares, sum of cubes, or difference of cubes.
   A. If it is a difference of squares, $a^2 - b^2$, the factors are conjugates and the factored form is $(a + b)(a - b)$. Note that a sum of squares cannot be factored.
   B. If it is a difference of cubes, $a^3 - b^3$, the factored form is $(a - b)(a^2 + ab + b^2)$.
   C. If it is a sum of cubes, $a^3 + b^3$, the factored form is $(a + b)(a^2 - ab + b^2)$.

Note Always check to see if any of the factors can be factored further.

Example 5 Factor completely.

a. $15x^5 + 60x^3$

Solution: Remove the GCF of $15x^3$.

$$15x^5 + 60x^3 = 15x^3(x^2 + 4) \quad \text{Factor out the monomial GCF, } 15x^3.$$

The binominal $x^2 + 4$ is a sum of squares, which cannot be factored using real numbers.

b. $15ac - 5ad - 3bc + bd$

Solution: There is no monomial GCF. Because there are four terms, we factor by grouping.

$$\begin{aligned} 15ac - 5ad - 3bc + bd &= (15ac - 5ad) + (-3bc + bd) && \text{Group pairs of terms.} \\ &= 5a(3c - d) - b(3c - d) && \text{Factor } 5a \text{ out of } 15ac - 5ad \text{ and } -b \text{ out of } -3bc + bd. \\ &= (3c - d)(5a - b) && \text{Factor out } 3c - d. \end{aligned}$$

c. $4x^2 - 20xy + 25y^2$

Solution: There is no monomial GCF. Because there are three terms, we determine whether it is a perfect square trinomial. It is a perfect square trinomial of the form $a^2 - 2ab + b^2 = (a - b)^2$ with $a = 2x$ and $b = 5y$.

$$4x^2 - 20xy + 25y^2 = (2x - 5y)^2$$

d. $12x^2 + 25x - 7$

Solution: This trinomial has no monomial GCF. It is of the form $ax^2 + bx + c$, where $a \neq 1$, so we try using trial.

The product of the *first* terms must be $12x^2$.
The factors could be x and $12x$, $2x$ and $6x$, or $3x$ and $4x$.

$$12x^2 + 25x - 7 = (\quad + \quad)(\quad - \quad)$$

The product of the *last* terms must be 7. We have already included the appropriate signs, so these factors are 1 and 7.

After trying various combinations of the preceding terms, we find the following correct combination.

$$12x^2 + 25x - 7 = (3x + 7)(4x - 1)$$

e. $27m^3 - n^3$

Solution: There is no monomial GCF. Because it is a binomial, we determine whether it is the difference of squares or the sum or difference of cubes. It is the difference of cubes in the form $a^3 - b^3 = (a - b)(a^2 + ab + b^2)$ with $a = 3m$ and $b = n$.

$$27m^3 - n^3 = (3m - n)(9m^2 + 3mn + n^2)$$

f. $36r^2 - 1$

Solution: There is no monomial GCF. Because it is a binomial, we determine whether it is the difference of squares or the sum or difference of cubes. It is the difference of squares with $a = 6r$ and $b = 1$.

$$36r^2 - 1 = (6r + 1)(6r - 1)$$

Your Turn 5 Factor completely.

a. $12m^4n + 27m^2n^3$ **b.** $6xz + 9xw - 8yz - 12yw$ **c.** $25a^2 + 30ab + 9b^2$
d. $12a^2 - 7a - 10$ **e.** $8m^3 + 125n^3$ **f.** $16x^2 - 25$

Often it is necessary to use more than one type of factoring in order to factor an expression completely.

Example 6 Factor completely.

a. $5y^4 - 10y^3 - 75y^2$

Solution: $5y^4 - 10y^3 - 75y^2 = 5y^2(y^2 - 2y - 15)$ Factor out the monomial GCF, $5y^2$.

The trinomial $y^2 - 2y - 15$ is not a perfect square and is in the form $x^2 + bx + c$, so we look for two numbers whose product is -15 and whose sum is -2. Those two numbers are 3 and -5.

$= 5y^2(y + 3)(y - 5)$ Factor $y^2 - 2y - 15$.

b. $8m^3n^2 - 24m^2n^3 + 18mn^4$

Solution: $8m^3n^2 - 24m^2n^3 + 18mn^4 = 2mn^2(4m^2 - 12mn + 9n^2)$ Factor out $2mn^2$.

The trinomial $4m^2 - 12mn + 9n^2$ is a perfect square.

$= 2mn^2(2m - 3n)^2$ Factor $4m^2 - 12mn + 9n^2$ using $a^2 - 2ab + b^2 = (a - b)^2$, where $a = 2m$ and $b = 3n$.

c. $9(3y - 1)^2 - 36y^2$

Solution: $9(3y - 1)^2 - 36y^2 = 9[(3y - 1)^2 - 4y^2]$ Factor out the monomial GCF, 9.

The expression $(3y - 1)^2 - 4y^2$ is a difference of squares, where $a = 3y - 1$ and $b = 2y$.

$= 9[(3y - 1) + 2y)][(3y - 1) - 2y]$ Factor $(3y - 1)^2 - 4y^2$ using $a^2 - b^2 = (a + b)(a - b)$.

$= 9(5y - 1)(y - 1)$ Simplify by combining like terms.

d. $t^5 - 4t^3 - 8t^2 + 32$

Solution: We have a four-term polynomial with no monomial GCF, so we try to factor by grouping.

$t^5 - 4t^3 - 8t^2 + 32 = t^3(t^2 - 4) - 8(t^2 - 4)$ Factor t^3 out of $t^5 - 4t^3$ and -8 out of $-8t^2 + 32$.

$= (t^2 - 4)(t^3 - 8)$ Factor out $t^2 - 4$.

$= (t + 2)(t - 2)(t - 2)(t^2 + 2t + 4)$ Factor using $a^2 - b^2 = (a + b)(a - b)$ and $a^3 - b^3 = (a - b)(a^2 + ab + b^2)$.

$= (t + 2)(t - 2)^2(t^2 + 2t + 4)$ Simplify by writing $(t - 2)(t - 2)$ as $(t - 2)^2$.

Answers to Your Turn 5
a. $3m^2n(4m^2 + 9n^2)$
b. $(2z + 3w)(3x - 4y)$
c. $(5a + 3b)^2$
d. $(3a + 2)(4a - 5)$
e. $(2m + 5n)(4m^2 - 10mn + 25n^2)$
f. $(4x + 5)(4x - 5)$

Note The binomial $t^2 - 4$ is a difference of squares, and $t^3 - 8$ is a difference of cubes; so we can factor these factors further.

e. $h^2 + 2h + 1 - 25k^2$

Solution: $h^2 + 2h + 1 - 25k^2 = (h + 1)^2 - 25k^2$ Factor the perfect square trinomial $h^2 + 2h + 1$.

The expression $(h + 1)^2 - 25k^2$ is a difference of squares, where $a = h + 1$ and $b = 5k$.

$= (h + 1 + 5k)(h + 1 - 5k)$ Factor $(h + 1)^2 - 25k^2$ using $a^2 - b^2 = (a + b)(a - b)$.

Answers to Your Turn 6

a. $2t(5u - 1)(3u - 1)$
b. $x^2y(6 - x)^2$ or $x^2y(x - 6)^2$
c. $6m^3(n + 3)(n - 1)$
d. $(x - 3)(x + 3)^2(x^2 - 3x + 9)$
e. $(2x - 3 + 6y)(2x - 3 - 6y)$

Your Turn 6 Factor completely.

a. $30tu^2 - 16tu + 2t$
b. $36x^2y - 12x^3y + x^4y$
c. $6m^3(n + 1)^2 - 24m^3$
d. $x^5 - 9x^3 + 27x^2 - 243$
e. $4x^2 - 12x + 9 - 36y^2$

6.3 Exercises

For Extra Help MyMathLab®

Note: Exercises marked with a ★ represent challenging exercises.

Objective 1

Prep Exercise 1 Why is $4x^2 + 20x + 25$ a perfect square trinomial?

Prep Exercise 2 Complete the rules for factoring perfect square trinomials.
$a^2 + 2ab + b^2 =$ ______.
$a^2 - 2ab + b^2 =$ ______.

For Exercises 1–12, factor the perfect square. See Example 1.

1. $x^2 + 10x + 25$
2. $y^2 + 8y + 16$
3. $b^2 - 4b + 4$
4. $m^2 - 12m + 36$
5. $25u^2 - 30u + 9$
6. $9m^2 - 24m + 16$
7. $n^2 + 24mn + 144m^2$
8. $w^2 + 10wv + 25v^2$
9. $9q^2 - 30pq + 25p^2$
10. $4r^2 - 20rs + 25s^2$
11. $4p^2 - 28pq + 49q^2$
12. $36a^2 + 60ab + 25b^2$

Objective 2

Prep Exercise 3 Complete the rules for factoring a difference of squares.
$a^2 - b^2 =$ ____________.

Prep Exercise 4 What are the binomial factors of a difference of squares called?

For Exercises 13–26, factor the difference of squares. See Example 2.

13. $a^2 - y^2$
14. $a^2 - b^2$
15. $25x^2 - 4$
16. $16m^2 - 49$
17. $100u^2 - 49v^2$
18. $121x^2 - 144y^2$
19. $9x^2 - 36$
20. $2x^2 - 50y^2$
21. $x^4 - 16$
22. $y^4 - 81$
23. ★ $9(x - 3)^2 - 16$
24. ★ $25(a + 2)^2 - 9$
25. ★ $16z^2 - 9(x - y)^2$
26. ★ $100a^2 - 9(b + c)^2$

Objectives 3 and 4

Prep Exercise 5 Complete the rules for factoring a difference or sum of cubes.
$a^3 - b^3 =$ ________________.
$a^3 + b^3 =$ ________________.

For Exercises 27–44, factor the sum or difference of cubes. See Examples 3 and 4.

27. $m^3 - 27$

28. $y^3 - 125$

29. $125x^3 + 27$

30. $27y^3 + 64$

31. $27x^3 - 8$

32. $8y^3 - 125$

33. $u^3 + 125v^3$

34. $27k^3 + 8m^3$

35. $27m^3 - 125m^6n^3$

36. $1000a^3b^6 - 27b^3$

37. $(u + 3)^3 + 8$

38. $27 + (x - 2)^3$

39. $27 - (a + b)^3$

40. $64 - (x - y)^3$

★**41.** $64x^3 + 27(y + z)^3$

★**42.** $(t + u)^3 + 64$

★**43.** $64d^3 - 27(x + y)^3$

★**44.** $27m^3 - 8(y + b)^3$

For Exercises 45–48, find the natural number b that makes the expression a perfect square trinomial.

45. $16x^2 + bx + 25$

46. $25x^2 + bx + 64$

47. $4x^2 - bx + 81$

48. $36x^2 - bx + 25$

For Exercises 49–52, find the natural number c that completes the perfect square trinomial.

49. $x^2 + 8x + c$

50. $x^2 + 12x + c$

51. $9x^2 - 24x + c$

52. $4x^2 - 28x + c$

Objective 5

Prep Exercise 6 What is always the first step in factoring?

Prep Exercise 7 What method of factoring do you try if the polynomial has four terms?

Prep Exercise 8 What are the three types of factorable binomials?

For Exercises 53–92, factor completely. See Examples 5 and 6.

53. $12m^3n^2 + 20m^2n^4$

54. $24a^4b^3 + 6a^2b$

55. $x^2 + 8x + 15$

56. $p^2 + p - 30$

57. $x^2 - 16$

58. $x^2 - y^2$

59. $12c^2 - 8c - 15$

60. $18c^2 - 9c - 20$

61. $ax - xy - ay + y^2$

62. $2ab - 2ac - 3bd + 3cd$

63. $4x^2 - 28x + 49$

64. $9a^2 + 30a + 25$

65. $25x^2 + 36y^2$

66. $64a^2 + 49b^2$

67. $12a^3b^2c + 3a^2b^2c^2 + 9abc^3$

68. $20x^2y^3z^4 - 12x^5yz^2 + 8x^3y^3z$

69. $b^3 + 125$

70. $y^3 - 8$

71. $15x^2 - 12x - 2$

72. $8x^2 - 14x - 3$

73. $6b^2 + b - 2$

74. $2a^2 - 5a - 12$

75. $ab - 36ab^3$

76. $5m^2 - 45$

77. $16c^2 - 24c + 9$

78. $25x^2 + 20x + 4$

79. $18x^3 - 3x^2 - 36x$

80. $18z^3 - 78z^2 - 180z$

81. $x^4 - 16$

82. $t^4 - 81$

83. $20a^2 - 9a + 20$

84. $6a^2 + 19a - 15$

85. $12u^2 - 84uv + 147v^2$

86. $ax^2 + 4ax + 4a$

87. $36a^4b^3 - 39a^3b^4 - 12a^2b^5$

88. $24x^5y - 2x^4y^2 - 12x^3y^3$

89. $6x^5y + 5x^4y^2 - 12x^3y^3$

90. $8a^4b^2 - 13a^3b^3 - 15a^2b^4$

91. $x^2y + 3x - xy^2 - 3y$

92. $a^2b - 4a - ab^2 + 4b$

★ *For Exercises 93–104, factor completely. These may require more than one factoring rule or additional simplification. See Example 6.*

93. $27 + (x - 1)^3$

94. $2x(4x - 1)^3 - 54x^4$

95. $5m^5 + 10m^3n^2 + 5mn^4$

96. $72x^6y - 24x^3y^3 + 2y^5$

97. $(n - 3)^2 + 6(n - 3) + 9$

98. $(2t + 1)^2 - 10(2t + 1) + 25$

99. $64t^6 - t^3u^3$

100. $(x + y)^3 + (x - y)^3$

101. $2x^3 + 3x^2 - 2xy^2 - 3y^2$

102. $3x^4 - x^3y - 24x + 8y$

103. $36y^2 - (x^2 - 8x + 16)$

104. $24(m - n)^2 - 54n^2$

For Exercises 105 and 106, write a polynomial for the area of the shaded region; then factor completely.

105.

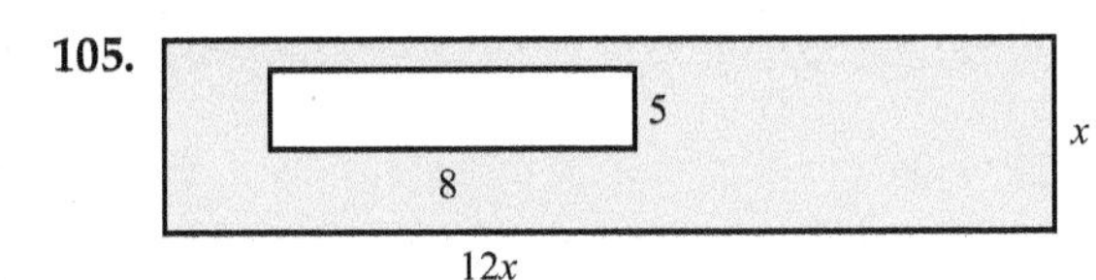

106.

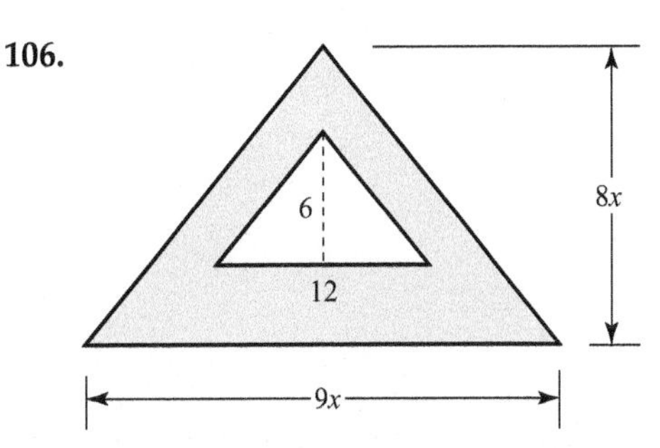

For Exercises 107 and 108, write a polynomial for the volume of the shaded region; then factor completely.

107.

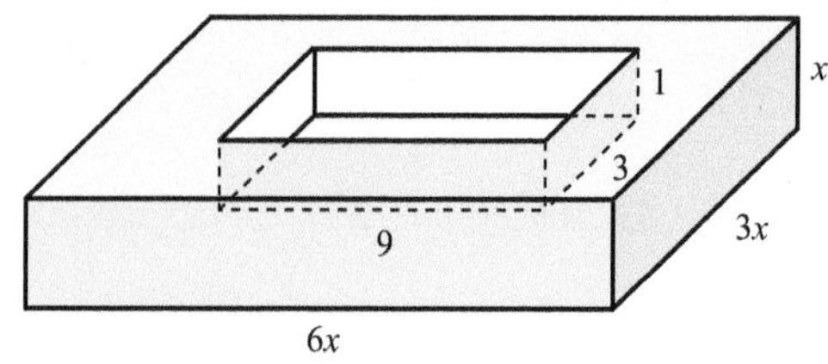

108.

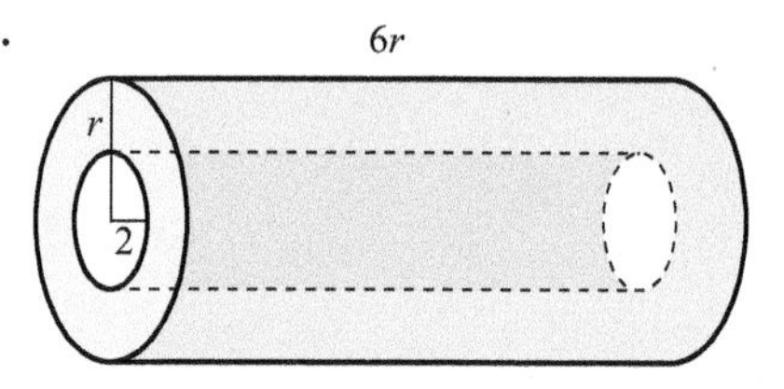

For Exercises 109 and 110, given an expression for the area of each rectangle, find the dimensions. Assume that the dimensions (lengths and widths) are polynomial factors of the area.

109.

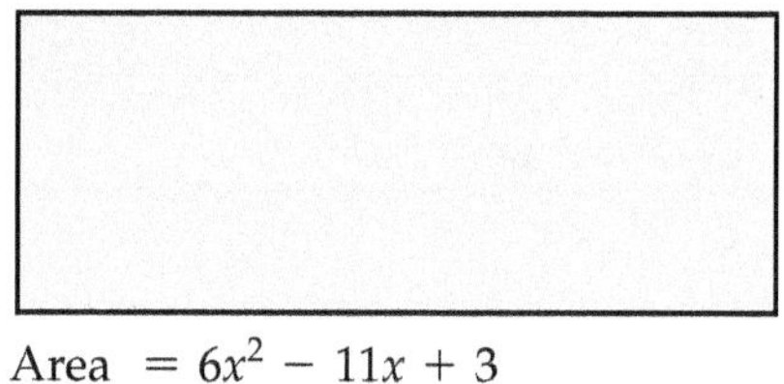

Area $= 6x^2 - 11x + 3$

110.

Area $= 16n^2 - 25$

For Exercises 111 and 112, given an expression for the volume of each box, find the dimensions. Assume that the dimensions (lengths, widths, and heights) are polynomial factors of the volume.

111.

Volume $= 15x^3 + 55x^2 + 30x$

112.

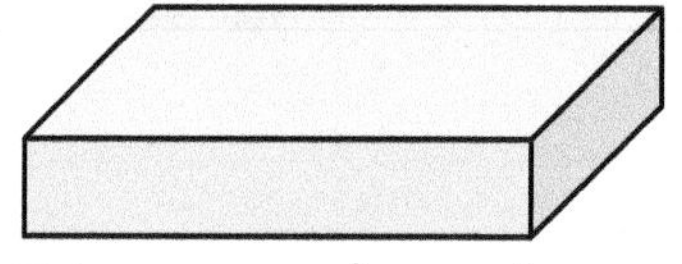

Volume $= 16x^3 + 32x^2 - 240x$

Review Exercises

Exercises 1–6 Equations and Inequalities

For Exercises 1 and 2, solve.

[2.1] 1. $9x - 5(3x + 2) = 12x - (4x - 3)$

[2.5] 2. $|2x - 5| = 7$

[4.3] 3. An investor invests a total of $6000 in three different plans. The total invested in the first two plans is equal to twice the amount invested in the third plan. The annual interest rates of the three plans are 4%, 6%, and 8% If she earns $370 in one year, use a system of equations to find the amount she invested in each plan.

For Exercises 4 and 5, graph.

[3.5] 4. $f(x) = -\frac{3}{4}x - 2$

[3.4] 5. $y > 2x - 3$

[3.3] 6. Write the equation of a line that passes through the points $(3, -2)$ and $(-4, 1)$ in standard form and in slope–intercept form.

6.4 Solving Equations by Factoring

Objectives

1. Use the zero-factor theorem to solve equations by factoring.
2. Find the intercepts of quadratic and cubic functions.
3. Solve problems involving quadratic equations.

Warm-up

[2.1] **1.** Solve $4t + 5 = 0$.

For Exercises 2–4, factor.

[6.2] **2.** $6x^2 - 21x - 12$

[6.1, 6.3] **3.** $t^3 + 3t^2 - 4t - 12$

[6.2] **4.** $-16t^2 + 12t + 40$

Objective 1 Use the zero-factor theorem to solve equations by factoring.

We are now ready to move up the Algebra Pyramid from polynomial expressions to **polynomial equations,** which we solve using factoring.

Definition **Polynomial equation:** An equation that equates two polynomials.

Note Inequalities are also in the top level of the Algebra Pyramid, but we will not be exploring polynomial inequalities at this point.

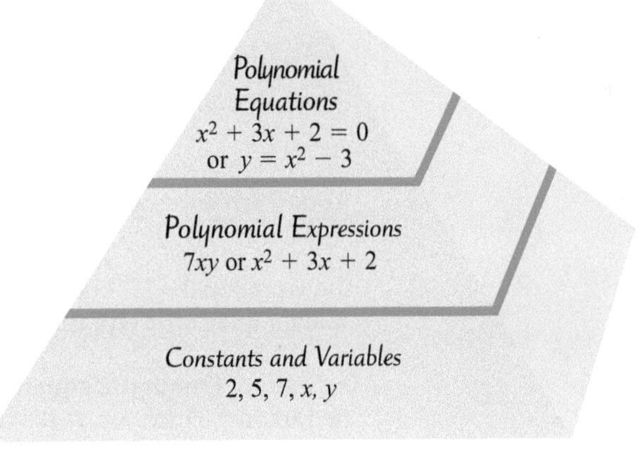

To solve these polynomial equations by factoring, we use the *zero-factor theorem,* which states that if the product of two or more factors is 0, then at least one of the factors is equal to 0.

Rule **Zero-Factor Theorem**

If a and b are real numbers and $ab = 0$, then $a = 0$ or $b = 0$.

Note The zero-factor theorem also holds for three or more factors.

Example 1 Solve $(x + 5)(x - 2) = 0$.

Solution: Because we have a product of two factors, $x + 5$ and $x - 2$, equal to 0, one or both of the factors must equal 0. When using the zero-factor theorem, we set each factor equal to 0 and then solve each of those equations.

$x + 5 = 0$ or $x - 2 = 0$ Use the zero-factor theorem.

$x = -5$ $\quad x = 2$ Solve each equation.

Check: Verify that -5 and 2 satisfy the original equation, $(x + 5)(x - 2) = 0$.

For $x = -5$: $(-5 + 5)(-5 - 2) \stackrel{?}{=} 0$
$(0)(-7) = 0$ True

For $x = 2$: $(2 + 5)(2 - 2) \stackrel{?}{=} 0$
$(7)(0) = 0$ True

-5 and 2 check; therefore, both are solutions.

Your Turn 1 Solve $(n - 6)(n + 1) = 0$.

Answer to Your Turn 1
$n = 6$ and -1

Answers to Warm-up

1. $t = -\frac{5}{4}$
2. $3(2x + 1)(x - 4)$
3. $(t + 3)(t + 2)(t - 2)$
4. $-4(4t + 5)(t - 2)$

To use the zero-factor theorem to solve a polynomial equation, we need the equation to be in **standard form**, which means that one side is equal to 0.

Definition **Polynomial equation in standard form:** $P = 0$, where P is a polynomial in terms of one variable written in descending order of degree.

The polynomial needs to be written in factored form so that we can use the zero-factor theorem.

Note The polynomial $3x^2 - 11x - 4$ has two different factors, $3x + 1$ and $x - 4$, which is why the equation $3x^2 - 11x - 4 = 0$ has two different solutions. In general, the number of solutions a polynomial equation in standard form has depends on the number of different factors the polynomial has.

Example 2 Solve $3x^2 - 11x = 4$.

Solution: First, we need the equation in standard form. Then to use the zero-factor theorem, we factor the polynomial.

$$3x^2 - 11x - 4 = 0 \quad \text{Subtract 4 from both sides to get standard form.}$$
$$(3x + 1)(x - 4) = 0 \quad \text{Factor } 3x^2 - 11x - 4.$$
$$3x + 1 = 0 \quad \text{or} \quad x - 4 = 0 \quad \text{Use the zero-factor theorem.}$$
$$3x = -1 \qquad x = 4 \quad \text{Solve each equation.}$$
$$x = -\frac{1}{3}$$

The solutions are $-\frac{1}{3}$ and 4. We will leave the check to the reader.

Your Turn 2 Solve $12y^2 - 2 = 5y$.

The equation $3x^2 - 11x = 4$ from Example 2 is a special type of polynomial equation called a **quadratic equation in one variable**.

Definition **Quadratic equation in one variable:** An equation that can be written in the form $ax^2 + bx + c = 0$, where a, b, and c, are real numbers and $a \neq 0$.

Connection The equation $(x + 5)(x - 2) = 0$ from Example 1 is a quadratic equation. We can see why if we multiply the factors so that the equation becomes $x^2 + 3x - 10 = 0$, which is clearly a quadratic equation.

Notice that $a \neq 0$ means that in a quadratic equation, the degree of the polynomial is always 2. Also note that if the equation is in the form $ax^2 + bx + c = 0$, it is in standard form. For example, $3x^2 - 12x + 4 = 0$ and $6x^2 - 24 = 0$ are quadratic equations in standard form.

Although we have focused on quadratic equations so far, we can use the zero-factor theorem to solve polynomial equations of degree greater than 2.

Note Recall that the zero- factor theorem works for three or more factors as well. The polynomial $x^3 - 4x^2 - 12x$ has three different factors, so the polynomial equation $x^3 - 4x^2 - 12x = 0$ has three different solutions.

Example 3 Solve $x^3 - 4x^2 = 12x$.

Solution:
$$x^3 - 4x^2 - 12x = 0 \quad \text{Subtract } 12x \text{ from both sides to write in standard form.}$$
$$x(x^2 - 4x - 12) = 0 \quad \text{Factor out the monomial GCF, } x.$$
$$x(x - 6)(x + 2) = 0 \quad \text{Factor } x^2 - 4x - 12.$$
$$x = 0 \quad \text{or} \quad x - 6 = 0 \quad \text{or} \quad x + 2 = 0 \quad \text{Use the zero-factor theorem to solve.}$$
$$x = 6 \qquad x = -2$$

The solutions are 0, 6, and −2. We will leave the check to the reader.

Your Turn 3 Solve $12x^3 + 20x^2 - 8x = 0$.

Answer to Your Turn 2

$y = \frac{2}{3}$ and $-\frac{1}{4}$

Answer to Your Turn 3

$x = 0, \frac{1}{3},$ and -2

The equation $x^3 - 4x^2 = 12x$ from Example 3 is called a **cubic equation in one variable.**

Definition Cubic equation in one variable: An equation that can be written in the form $ax^3 + bx^2 + cx + d = 0$, where a, b, c, and d are real numbers and $a \neq 0$.

Notice in the definition that $a \neq 0$ means that in a cubic equation, the degree-3 term, ax^3, must be present.

Connection In Examples 2 and 3, notice that the degree of the polynomial in each equation corresponds to the number of solutions the equation has. In general, the degree of a polynomial equation indicates the *maximum* number of real solutions the equation could have. The actual number of real solutions depends upon the number of unique factors the polynomial has. Because a degree-2 polynomial can have, at most, two different factors, a quadratic equation can have, at most, two different solutions. A degree-3 polynomial can have, at most, three different factors; so a cubic equation could have, at most, three solutions.

We can now summarize the process of solving polynomial equations using the zero-factor theorem.

Procedure Solving Polynomial Equations Using Factoring

To solve a polynomial equation using factoring:

1. Write the equation in standard form. (Set one side equal to 0 with the other side in descending order of degree when possible.)
2. Write the polynomial in factored form.
3. Use the zero-factor theorem to solve.

Example 4 Solve.

a. $3x(2x - 7) = 12$

Solution:

$6x^2 - 21x = 12$ — Distribute $3x$.

$6x^2 - 21x - 12 = 0$ — Subtract 12 from both sides to get standard form.

$3(2x^2 - 7x - 4) = 0$ — Factor out the monomial GCF, 3.

$3(2x + 1)(x - 4) = 0$ — Factor $2x^2 - 7x - 4$.

$2x + 1 = 0$ or $x - 4 = 0$ — Use the zero-factor theorem.

$2x = -1$ $\quad x = 4$ — Solve each equation.

$x = -\frac{1}{2}$

Note When using the zero-factor theorem, we disregard constant factors such as 3 because they have no variable and, therefore, cannot be 0.

The solutions are $-\frac{1}{2}$ and 4.

b. $\frac{1}{2}(x^2 - 3) + \frac{1}{12}x = \frac{1}{3}(x^2 - 2)$

Solution:

$12\left[\frac{1}{2}(x^2 - 3) + \frac{1}{12}x\right] = 12\left[\frac{1}{3}(x^2 - 2)\right]$ — Multiply both sides by the LCD, 12, to clear the fractions.

$12 \cdot \frac{1}{2}(x^2 - 3) + 12 \cdot \frac{1}{12}x = 12 \cdot \frac{1}{3}(x^2 - 2)$ — Distribute 12.

$6(x^2 - 3) + x = 4(x^2 - 2)$ — Simplify.

$6x^2 - 18 + x = 4x^2 - 8$ — Distribute 6 on the left side and 4 on the right side.

$$2x^2 + x - 10 = 0$$ Subtract $4x^2$ from and add 8 to both sides.

$$(2x + 5)(x - 2) = 0$$ Factor.

$$2x + 5 = 0 \quad \text{or} \quad x - 2 = 0$$ Use the zero-factor theorem.

$$2x = -5 \qquad x = 2$$ Solve each equation.

$$x = -\frac{5}{2}$$

The solutions are $-\frac{5}{2}$ and 2.

c. $9y(y^2 + 1) = 4y(6y - 1) - 3y$

Solution: $9y^3 + 9y = 24y^2 - 4y - 3y$ Distribute $9y$ and $4y$.

$$9y^3 + 9y = 24y^2 - 7y$$ Combine like terms.

$$9y^3 - 24y^2 + 16y = 0$$ Subtract $24y^2$ from and add $7y$ to both sides.

$$y(9y^2 - 24y + 16) = 0$$ Factor out the monomial GCF, y.

$$y(3y - 4)^2 = 0$$ Factor $9y^2 - 24y + 16$, which is a perfect square.

$$y = 0 \quad \text{or} \quad 3y - 4 = 0$$ Use the zero-factor theorem.

$$3y = 4$$ Solve each equation.

$$y = \frac{4}{3}$$

Note Even though this is a third-degree polynomial equation, there are only two solutions because there are only two unique factors.

The solutions are 0 and $\frac{4}{3}$.

d. $t^3 + 3t^2 - 13 = 7t - (3t + 1)$

Solution: $t^3 + 3t^2 - 13 = 7t - 3t - 1$ Distribute.

$$t^3 + 3t^2 - 13 = 4t - 1$$ Combine like terms.

$$t^3 + 3t^2 - 4t - 12 = 0$$ Subtract $4t$ from and add 1 to both sides to get standard form.

$$t^2(t + 3) - 4(t + 3) = 0$$ Factor by grouping.

$$(t + 3)(t^2 - 4) = 0$$ Factor out $t + 3$.

$$(t + 3)(t + 2)(t - 2) = 0$$ Factor $t^2 - 4$, which is a difference of squares.

$$t + 3 = 0 \quad \text{or} \quad t + 2 = 0 \quad \text{or} \quad t - 2 = 0$$ Use the zero-factor theorem.

$$t = -3 \qquad t = -2 \qquad t = 2$$ Solve each equation.

The solutions are -3, -2, and 2.

Your Turn 4 Solve.

a. $3x(3x - 10) = -24$

b. $\frac{1}{4}(x^2 + 8) + 2x = \frac{1}{8}(x - 2)$

c. $4n(n^2 - 2n) = 2n(2n - 3) - 3n$

d. $h^2(2h + 1) - 3 = 18h + 6$

Objective 2 Find the intercepts of quadratic and cubic functions.

In Section 5.2, we graphed some basic polynomial functions. Now we will see a connection between solutions of polynomial equations and the x-intercepts of their corresponding polynomial functions.

Look at the following graphs. Notice that quadratic functions with their parabolic graph could have zero, one, or two x-intercepts. Similarly, cubic functions with their "sideways" S-shaped graph could have one, two, or three x-intercepts.

Answers to Your Turn 4

a. $2, \frac{4}{3}$ b. $-6, -\frac{3}{2}$

c. $0, \frac{3}{2}$ d. $-\frac{1}{2}, 3, -3$

Quadratic Functions

$$f(x) = a^2 + bx + c$$

Note For simplicity, we have shown cases only where $a > 0$. However, the same number of x-intercepts are possible when $a < 0$.

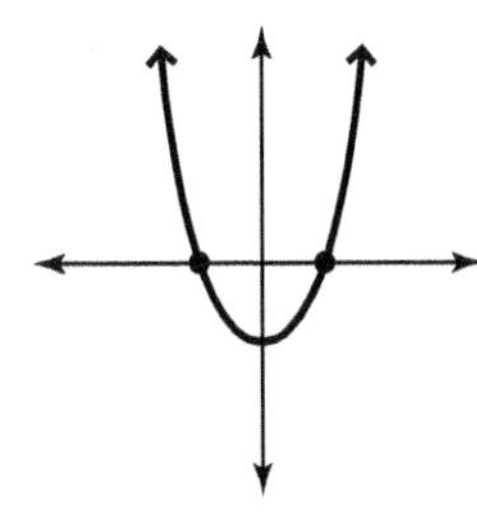

Two x-intercepts

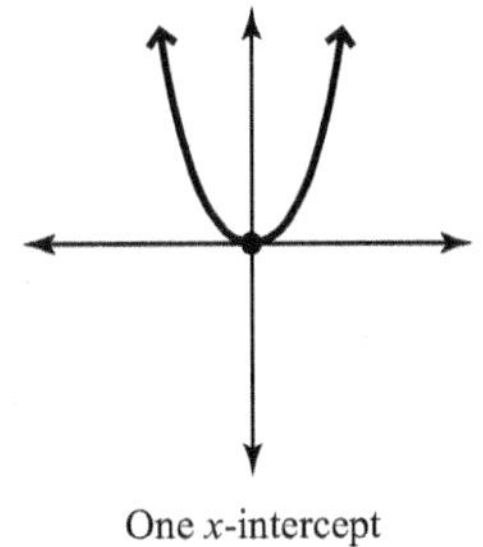

One x-intercept

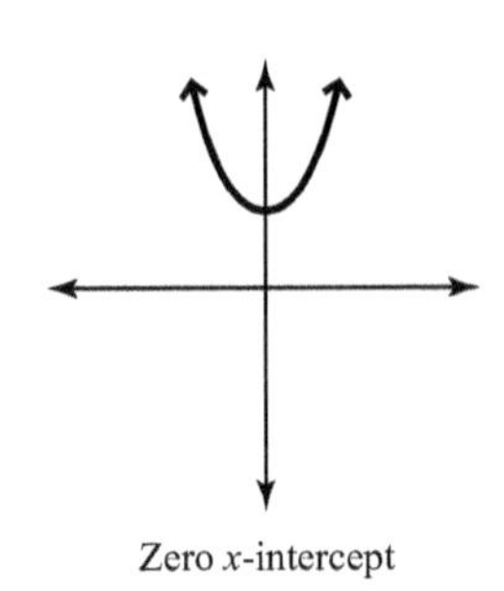

Zero x-intercept

Cubic Functions

$$f(x) = ax^3 + bx^2 + cx + d$$

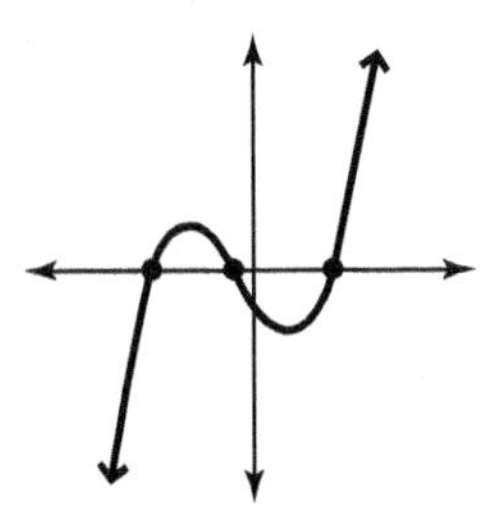

Three x-intercepts

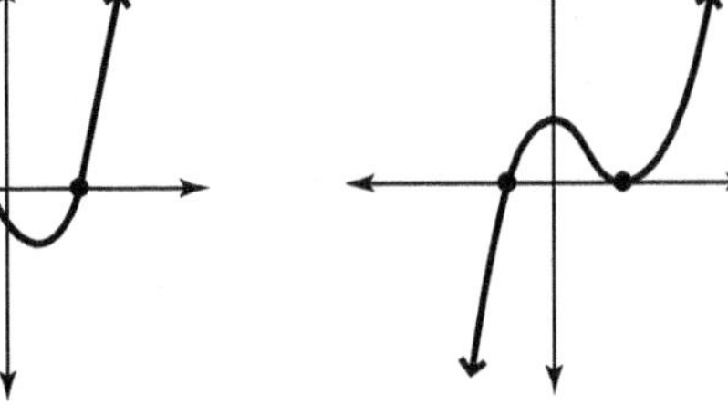

Two x-intercepts

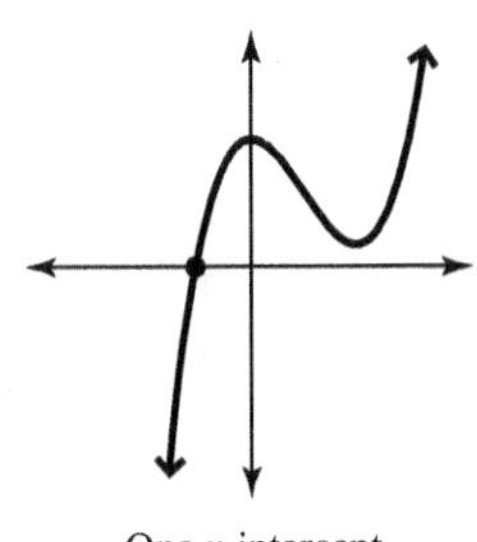

One x-intercept

Recall that to find x-intercepts, we set $f(x) = 0$ and solve for x. Setting a polynomial function equal to 0 creates a polynomial equation. So the real solutions to that equation are the x-coordinates of the function's x-intercepts. Consequently, if the polynomial is factorable, we should be able to find those x-intercepts using the zero-factor theorem.

Note Because x-intercepts occur when $f(x) = 0$, they are often called the **zeros** of a function.

Connection Earlier in this section, we noted that the degree of the polynomial in a polynomial equation indicates the maximum number of solutions the equation can have. We noted that the actual number of solutions depends on the number of distinct factors the polynomial has. Those connections also apply to the number of x-intercepts that a polynomial function has. For example, the polynomial $x^2 + x - 6$ has two factors, $x + 3$ and $x - 2$. So the polynomial equation $x^2 + x - 6 = 0$ has two solutions, -3 and 2. Consequently, the polynomial function $f(x) = x^2 + x - 6$ has two x-intercepts, $(-3, 0)$ and $(-2, 0)$.

Example 5 Find the x-intercept(s) and then sketch the graph.

a. $f(x) = x^2 + x - 6$

Solution: To find the x-intercepts, we let $f(x) = 0$ and then solve for x.

$$x^2 + x - 6 = 0 \quad \text{Set } f(x) = 0.$$

$$(x + 3)(x - 2) = 0 \quad \text{Factor.}$$

$$x + 3 = 0 \quad \text{or} \quad x - 2 = 0 \quad \text{Use the zero-factor theorem.}$$

$$x = -3 \qquad x = 2 \quad \text{Solve each equation.}$$

The x-intercepts are $(-3, 0)$ and $(2, 0)$.

To sketch the graph, we also find the y-intercept by finding $f(0)$.

$$f(0) = (0)^2 + (0) - 6 = -6$$

The y-intercept is $(0, -6)$.

Finding a few more points is helpful:

x	$f(x)$
−4	6
−2	−4
−1	−6
1	−4
3	6

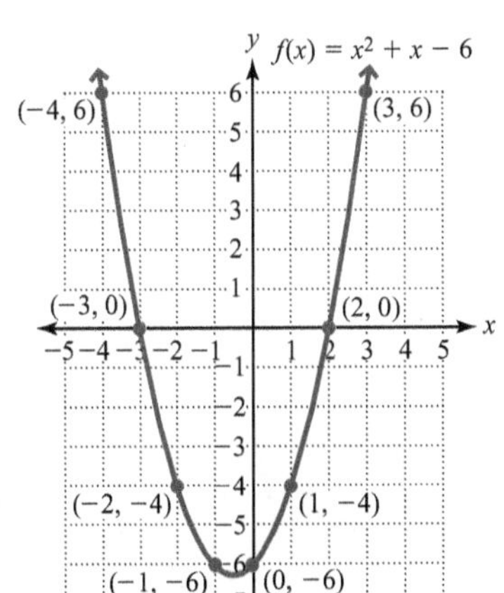

b. $f(x) = (x + 2)(x - 1)(x - 4)$

Solution: $(x + 2)(x - 1)(x - 4) = 0$ Set $f(x) = 0$.

$x + 2 = 0$ or $x - 1 = 0$ or $x - 4 = 0$ Use the zero-factor theorem.

$x = -2$ $\quad x = 1$ $\quad x = 4$ Solve each equation.

The x-intercepts are $(-2, 0)$, $(1, 0)$, and $(4, 0)$.

Multiplying the factors in the function reveals this function to be a cubic function.

$f(x) = (x^2 + x - 2)(x - 4)$ Multiply $x + 2$ and $x - 1$.

$= x^3 - 3x^2 - 6x + 8$ Multiply $x^2 + x - 2$ and $x - 4$.

To sketch the graph, we also find the y-intercept by finding $f(0)$.

$$f(0) = (0)^3 - 3(0)^2 - 6(0) + 8 = 8$$

The y-intercept is $(0, 8)$.

Finding a few more points is helpful.

x	f(x)
−1	10
2	−8
3	−10

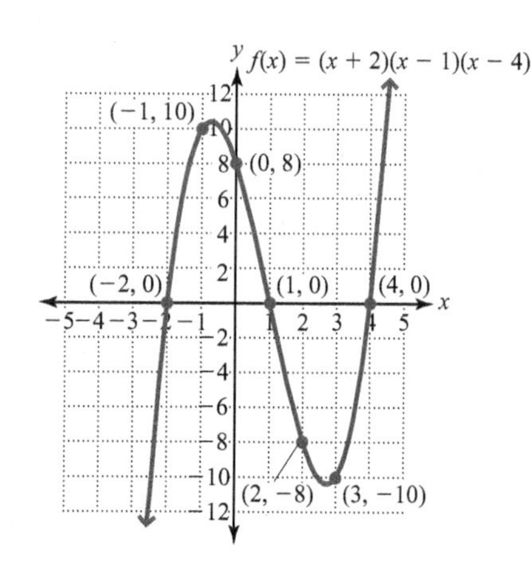

Your Turn 5 Find the x-intercepts and then sketch the graph.

a. $f(x) = x^2 - 4$ **b.** $f(x) = x(x + 2)(x - 3)$

Connection A function's x-intercepts can be found using a graphing calculator. On a TI-83 or TI-84 Plus, for example, use the **ZERO** function. After graphing the function, select **ZERO** from the **CALC** menu. Enter the left and right bounds of the x-intercept that you want to identify; then press ENTER one more time at the Guess prompt to get the coordinates of that intercept.

This feature can be used to find the solutions to a polynomial equation written in standard form. For example, let's solve the equation from Example 3, $x^3 - 4x^2 = 12x$. Written in standard form, the equation is $x^3 - 4x^2 - 12x = 0$. The corresponding polynomial function is $f(x) = x^3 - 4x^2 - 12x$ and its graph is shown to the right. Using the **ZERO** function for each x-intercept gives $x = 0$, $x = 6$, and $x = -2$, which are the solutions we found in Example 3 for the equation $x^3 - 4x^2 = 12x$.

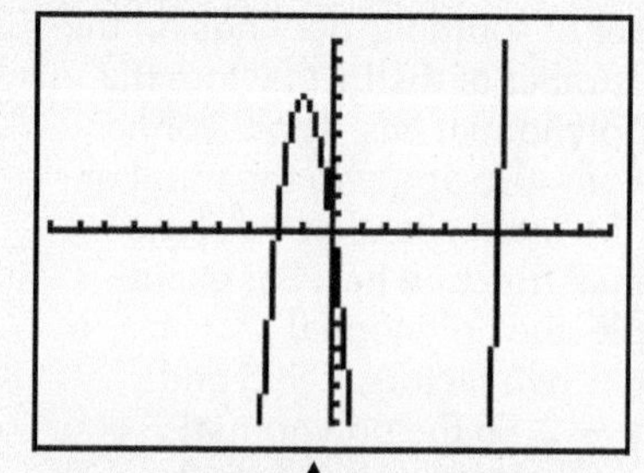

Note In the graph shown, we have our window set with Xmin $= -10$, Xmax $= 10$, Ymin $= -10$, and Ymax $= 10$. To view the missing bend, use Ymin $= -50$.

We can also use these x-intercepts and the zero-factor theorem in reverse to determine the polynomial's factored form. Rewriting $x = 0$, $x = 6$, and $x = -2$, equal to 0 gives $x = 0$, $x - 6 = 0$, and $x + 2 = 0$. Those are the factors in the factored form, so $x^3 - 4x^2 - 12x = x(x - 6)(x + 2)$.

Answers to Your Turn 5

a. $(2, 0), (-2, 0)$

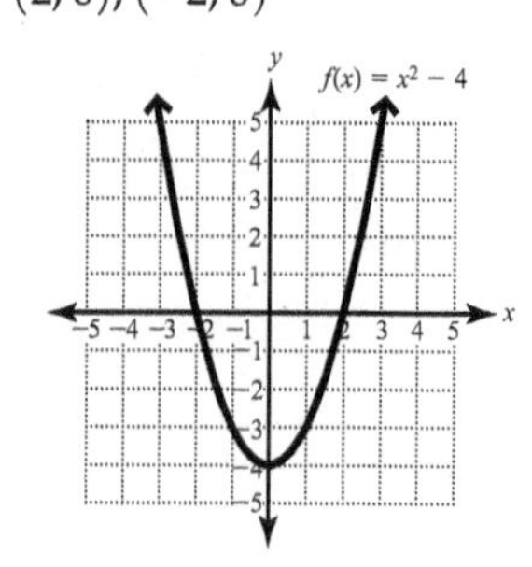

b. $(0, 0), (-2, 0), (3, 0)$

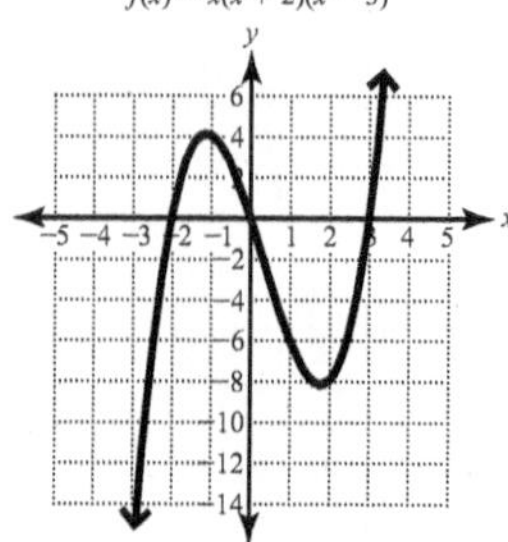

Objective 3 Solve problems involving quadratic equations.

Many applications can be solved using polynomial equations.

Example 6 The equation $h = -16t^2 + v_0t + h_0$ describes the height, h, in feet of an object t seconds after being thrown upward with an initial velocity of v_0 feet per second from an initial height of h_0 feet. Suppose a rock is thrown upward with an initial velocity of 12 feet per second from a 40-foot tower. How many seconds does it take the rock to reach the ground?

Solution: $0 = -16t^2 + 12t + 40$ — In $h = -16t^2 + v_0t + h_0$, substitute 0 for h, 12 for v_0, and 40 for h_0.

$0 = -4(4t^2 - 3t - 10)$ — Factor out the monomial GCF, -4.

$0 = -4(4t + 5)(t - 2)$ — Factor $4t^2 - 3t - 10$.

$4t + 5 = 0$ or $t - 2 = 0$ — Use the zero-factor theorem to solve.

$t = -\frac{5}{4}$ $\quad$ $t = 2$

Note Remember, we disregard the constant factor, -4 since $-4 \neq 0$.

Answer: Our answer must describe the amount of time *after* the rock is thrown; so only the positive value, 2, makes sense. This means that the rock takes 2 seconds to hit the ground.

Your Turn 6 An object is thrown upward with an initial velocity of 16 feet per second and an initial height of 60 feet. How many seconds does the object take to reach the ground?

One of the most well-known theorems in mathematics is the Pythagorean theorem, named after the Greek mathematician Pythagoras. The theorem relates the lengths of the sides of all right triangles. Recall that in a right triangle, the two sides that form the 90° angle are the *legs* and the side directly across from the 90° angle is the *hypotenuse.*

Of Interest

Pythagoras (c. 569–500 B.C.) was a Greek mathematician who lived in the town of Croton in what is now Italy. The theorem that now bears his name was known and used in Egypt and elsewhere before Pythagoras's time. It bears his name, however, because he was the first to prove that the relationship is true for all right triangles. In fact, the Greeks were the first culture to value mathematical proofs as opposed to using mathematics solely to solve problems.

Rule Pythagorean Theorem

If a right triangle has legs of lengths a and b and hypotenuse of length c, then $a^2 + b^2 = c^2$.

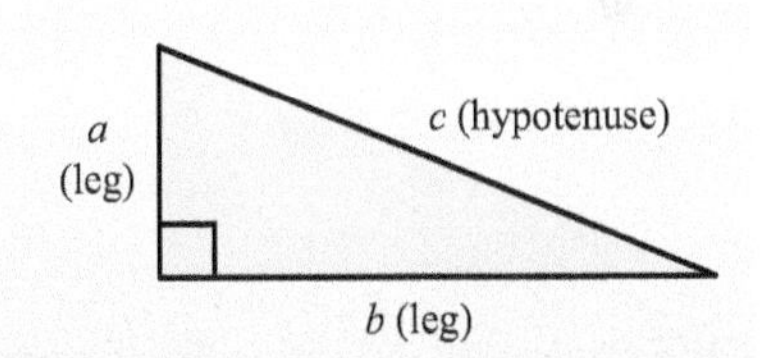

We can use the Pythagorean theorem to find an unknown length in a right triangle if we know the other two lengths.

Example 7 The figure shows a portion of a roof frame to be constructed. Find the length, in feet, of each side.

Solution:

$(x)^2 + (x + 1)^2 = (x + 9)^2$ — In $a^2 + b^2 = c^2$, substitute x for a, $x + 1$ for b, and $x + 9$ for c.

$x^2 + x^2 + 2x + 1 = x^2 + 18x + 81$ — Square the binomials.

$x^2 - 16x - 80 = 0$ — Subtract x^2, $18x$, and 81 from both sides to get standard form.

$(x - 20)(x + 4) = 0$ — Factor.

$x - 20 = 0$ or $x + 4 = 0$ — Use the zero-factor theorem.

$x = 20$ $\quad$ $x = -4$

Answer: Because x describes a length in feet, only the positive solution is sensible; so x must be 20 feet. This means that $x + 1$ is 21 feet and $x + 9$ is 29 feet.

Answer to Your Turn 6

$2\frac{1}{2}$ sec.

Your Turn 7 A sail is to be designed as shown to the right. Find its dimensions in feet.

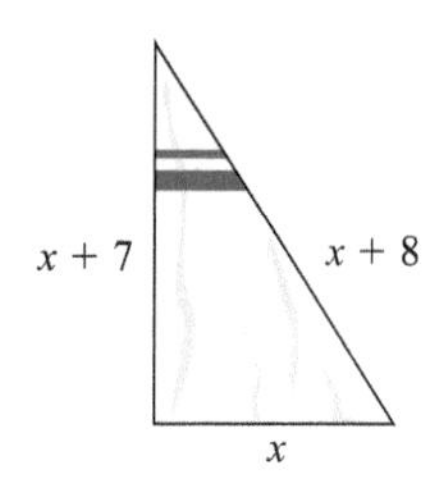

Answer to Your Turn 7
5 ft., 12 ft., 13 ft.

6.4 Exercises

For Extra Help MyMathLab®

Note: Exercises marked with a ★ represent challenging exercises.

Objective 1

Prep Exercise 1 To use the zero-factor theorem the equation must be in standard form, which means that one side of the equation is ________.

Prep Exercise 2 Explain how to apply the zero-factor theorem to solve $(x - 6)(x + 3) = 0$.

For Exercises 1–12, solve. See Example 1.

1. $x(x + 4) = 0$

2. $y(y - 8) = 0$

3. $(x - 3)(x + 2) = 0$

4. $(y + 4)(y + 3) = 0$

5. $(3b - 2)(2b + 5) = 0$

6. $(4a - 3)(2a - 5) = 0$

7. $x(x - 3)(2x + 5) = 0$

8. $a(3a - 4)(a + 5) = 0$

9. $(x + 2)(x - 5)(x + 7) = 0$

10. $(y - 3)(y + 4)(y - 6) = 0$

11. $(2x + 3)(3x - 1)(4x + 7) = 0$

12. $(2a - 5)(3a + 2)(5a - 1) = 0$

For Exercises 13–50, solve. See Examples 2–4.

13. $d^2 - 4d = 0$

14. $r^2 + 6r = 0$

15. $x^2 - 9 = 0$

16. $a^2 - 121 = 0$

17. $m^2 + 14m + 45 = 0$

18. $r^2 - 2r - 3 = 0$

19. $2x^2 + 3x - 2 = 0$

20. $3c^2 - c - 2 = 0$

21. $6a^2 + 13a - 5 = 0$

22. $8d^2 - 14d + 3 = 0$

23. $x^2 + 6x + 9 = 0$

24. $b^2 - 10b + 25 = 0$

25. $6y^2 = 3y$

26. $15v^2 = 5v$

27. $x^2 = 25$

28. $u^2 = 64$

29. $n^2 + 6n = 27$

30. $p^2 - 8p = 20$

31. $p^2 = 3p - 2$

32. $m^2 = 11m - 24$

33. $2v^2 + 5 = -7v$

34. $6k^2 + 1 = -7k$

35. $a(a - 5) = 14$

36. $c(2c - 11) = -5$

37. $4x(x + 7) = -49$

38. $4m(m - 3) = -9$

39. $(x + 1)(x + 2) = 20$

40. $(x + 2)(x + 4) = 35$

41. $x^3 + x^2 - 6x = 0$

42. $h^3 + 2h^2 - 8h = 0$

43. $12x(x + 1) + 3 = 5(2x + 1) + 2$

44. $6 - 4x(2 + x) = 8(1 - x) - 3$

45. $(2v + 1)(v - 2) = -v(v + 2)$

46. $(2x + 1)(2x + 3) = -2x(2x + 1)$

47. $9x(x^2 + x) = 3x(x - 2) + 5x$

48. $4x(x^2 - 3x) = 2x(4x - 8) - 9x$

49. $x^3 + 2x^2 - 15 = 13x - (4x - 3)$

50. $a^3 - 3a^2 - 6a = -16 - (2a - 4)$

★ *For Exercises 51–56, write a polynomial equation in standard form with the given solutions.*

51. $-3, 2$

52. $1, 5$

53. $-\frac{2}{3}, 4$

54. $-2, -\frac{1}{4}$

55. $-1, 0, 3$

56. $-2, 0, 4$

Objective 2

Prep Exercise 3 What is the degree of the polynomial in a quadratic equation?

Prep Exercise 4 What is the degree of the polynomial in a cubic equation?

Prep Exercise 5 Is $f(x) = (x - 1)(x + 4)$ a quadratic or cubic function? How many x-intercepts does it have?

Prep Exercise 6 Is $f(x) = (x + 2)(x - 3)(x - 5)$ a quadratic or cubic function? How many x-intercepts does it have?

Prep Exercise 7 Sketch the graph of a general cubic function with two x-intercepts.

Prep Exercise 8 Sketch the graph of a general quadratic function with one x-intercept.

For Exercises 57–62, match the graph to the function. See Example 5.

57. $f(x) = (x + 3)(x - 1)$

58. $f(x) = x(x + 1)(x - 2)$

59. $f(x) = 2x^2 - 7x + 3$

60. $f(x) = x^2 - 4$

61. $f(x) = x^3 + 2x^2 - 3x$

62. $f(x) = x^3 + x^2 - 6x$

a.

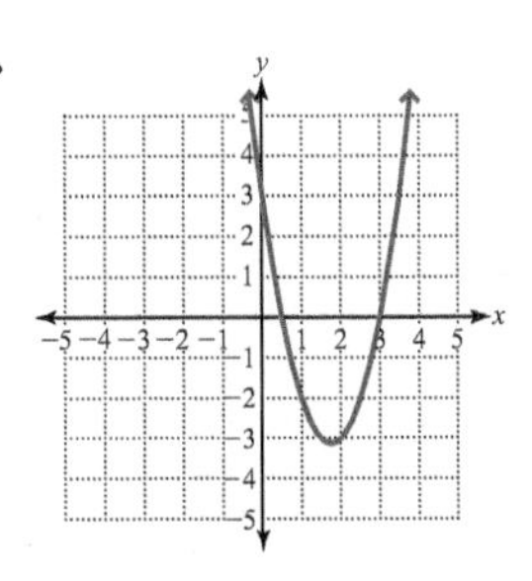

b.

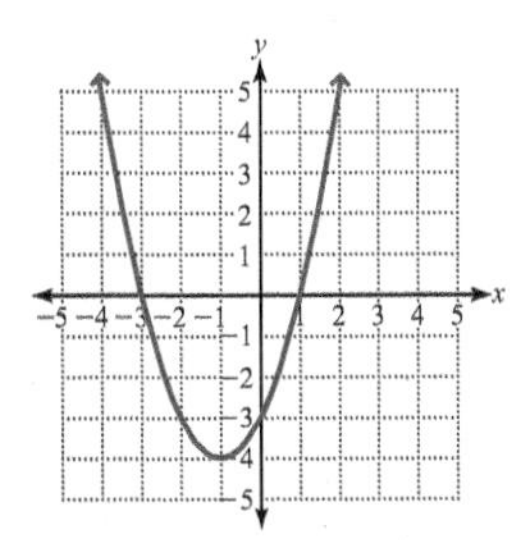

c.

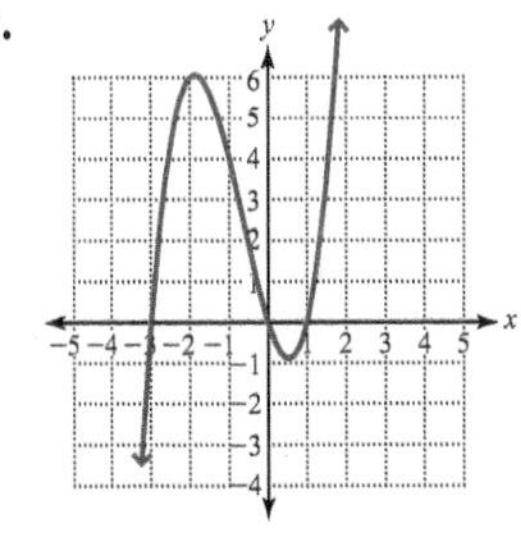

d.

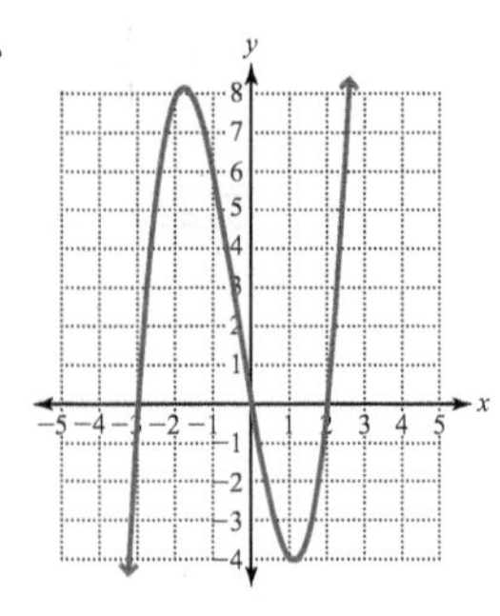

e.

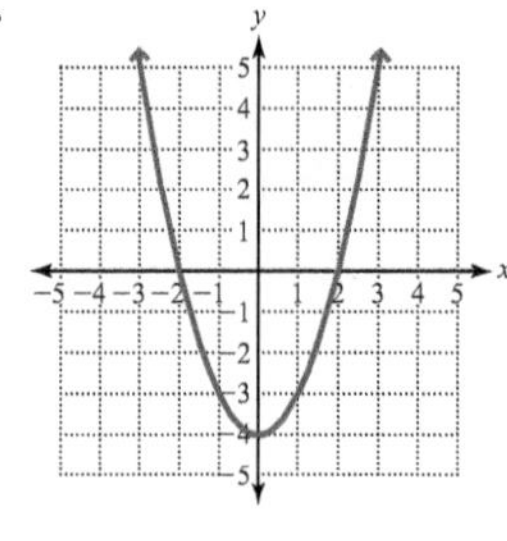

f.

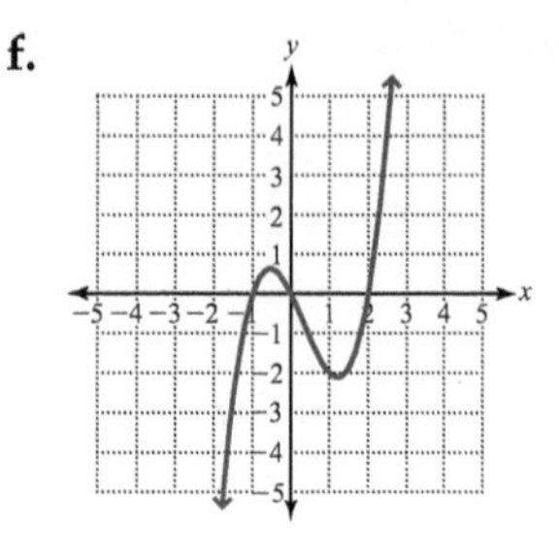

Prep Exercise 9 Explain how to find the x-intercepts of a function.

For Exercises 63–70, find the x-intercepts and then sketch the graph. See Example 5.

63. $f(x) = x^2 - 25$

64. $f(x) = x^2 - 1$

65. $f(x) = x^2 - 6x + 5$

66. $f(x) = x^2 - 9x + 14$

67. $f(x) = x(x + 4)(x - 1)$

68. $f(x) = x(x + 1)(x - 6)$

69. $f(x) = x^3 - x^2 - 6x$

70. $f(x) = x^3 - x^2 - 2x$

For Exercises 71–76, use a graphing calculator to determine the x-intercepts.

71. $f(x) = x^2 - 9x + 18$

72. $f(x) = x^2 + 7x + 12$

73. $f(x) = 6x^2 + 7x - 5$

74. $f(x) = 10x^2 - 27x + 5$

75. $f(x) = x^3 - x^2 - 2x$

76. $f(x) = x^3 - 9x$

For Exercises 77–82, use a graphing calculator to solve the equation. (Hint: Find the x-intercepts of the corresponding polynomial function.)

77. $x^2 = 3x + 40$

78. $28 - 3x = x^2$

79. $8x^2 - 12 = 18x - 7$

80. $20x^2 + 9 = 15 - 7x$

81. $x^3 + x^2 = 12x$

82. $x^3 + 10 = 4x^2 + 7x$

Objective 3

Prep Exercise 10 Complete the Pythagorean theorem. If a right triangle has legs of lengths a and b and hypotenuse of length c, then ________.

For Exercises 83–86, find the length of each side of the right triangle shown. See Example 7.

83.

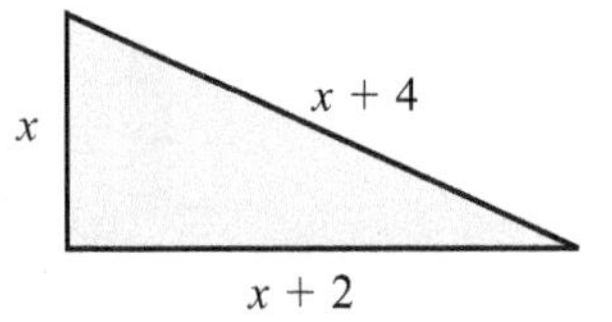

84.

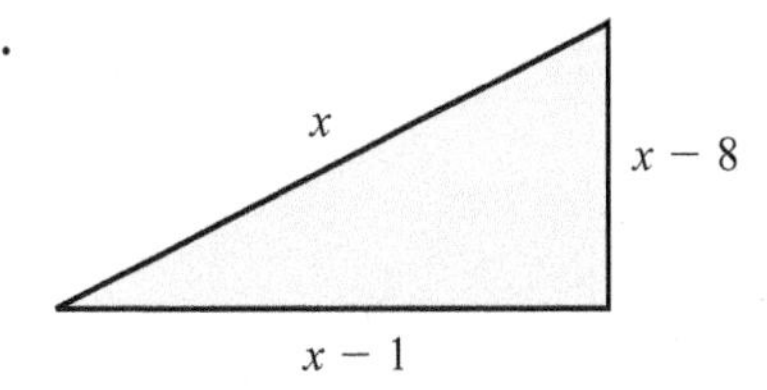

85.

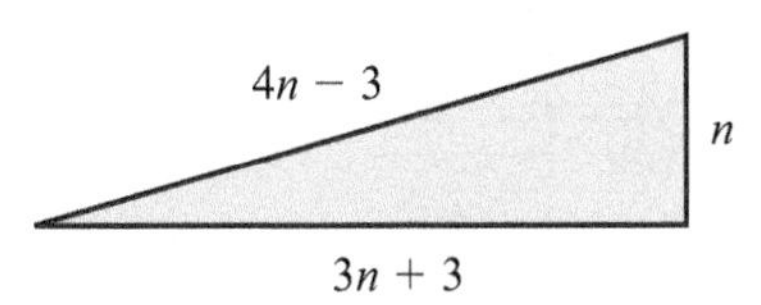

86.

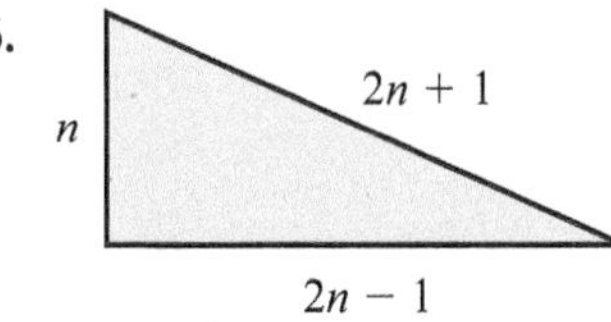

For Exercises 87–98, solve. See Examples 6 and 7.

87. Find a number such that 55 plus the square of the number is the same as 16 times the number.

88. One natural number is four times another. The product of the two numbers is 100. Find both numbers.

89. The sum of the squares of three consecutive positive odd integers is 155.

90. The sum of the squares of three consecutive positive even integers is 200. Find the integers.

91. The length of a small rectangular garden is 4 meters more than the width. If the area is 320 square meters, find the dimensions of the garden.

92. A design on the front of a marketing brochure calls for a triangle with a base that is 6 centimeters less than the height. If the area of the triangle is to be 216 square centimeters, what is the length of the base and the height?

93. The design of a small building calls for a rectangular shape with base dimensions of 22 feet by 28 feet. The architect decides to change the shape of the building to a circle but wants it to have the same base area. If we use $\frac{22}{7}$ to approximate π, what is the radius of the circular building?

94. Find the radius of a circle that has the same area as a rectangle whose length is 14 meters and whose width is 11 meters. Use $\frac{22}{7}$ for π.

95. A steel plate used in a machine is in the shape of a trapezoid and has an area of 85.5 square inches. The dimensions are shown. Note that the length of the base is equal to the height. Calculate the height.

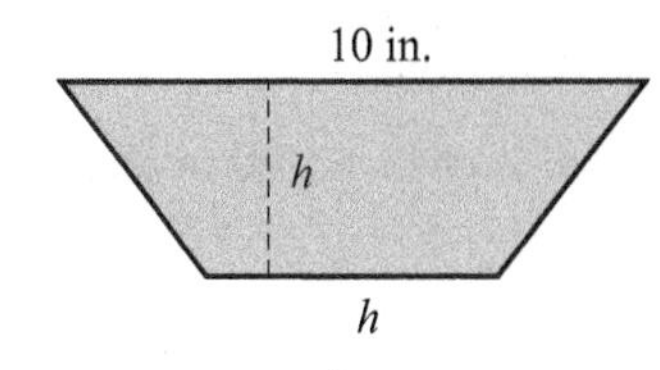

96. The front elevation of one wing of a house is shown. Because of budget constraints, the total area of the front of this wing must be 352 square feet. The height of the triangular portion is 14 feet less than the base. Find the base length.

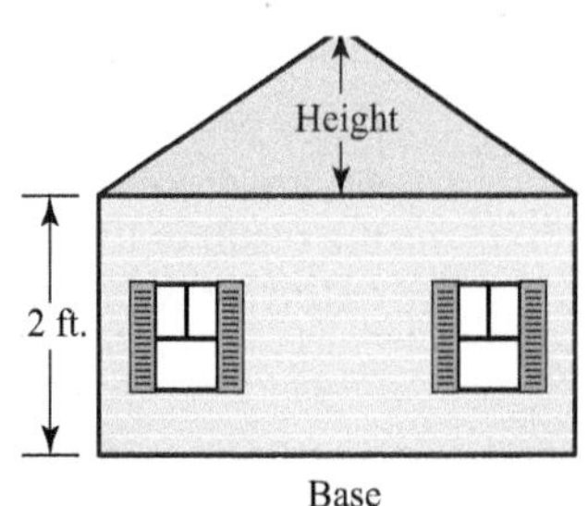

97. A support wire attached to a power pole is to be replaced. The wire needs to be 2 feet longer than the height of the pole. The wire will be staked into the ground 10 feet from the base of the pole. Find the length of the wire and the height at which it is attached to the pole.

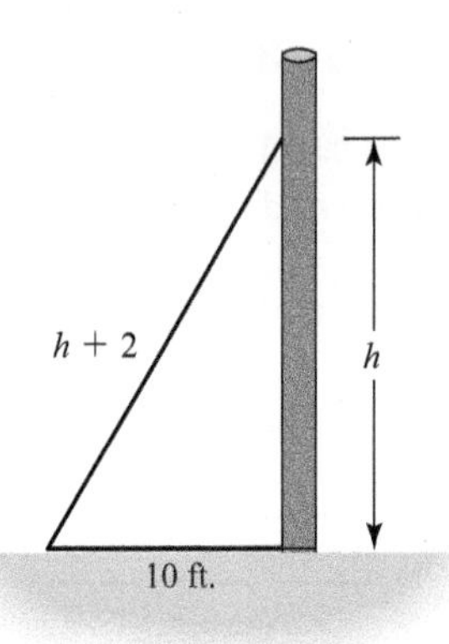

★**98.** Beams in a steel-frame roof structure are configured as shown. The length of the horizontal joist is 2 feet less than the length of the roof support beam. Find the length of the joist and the roof support beam.

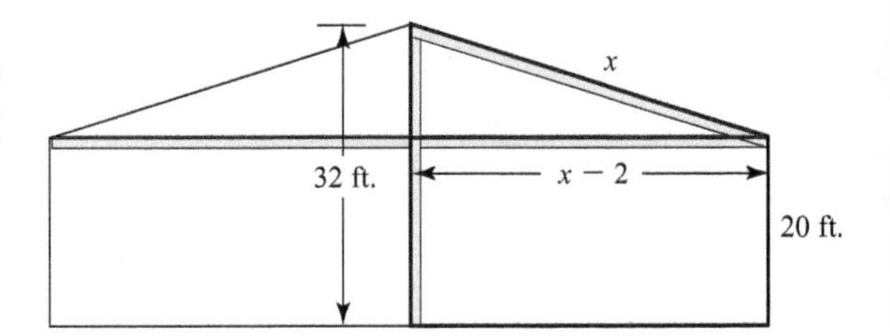

For Exercises 99 and 100, use the formula $h = -16t^2 + v_0t + h_0$, *where h is the final height in feet, t is the time of travel in seconds,* v_0 *is the initial velocity in feet per second, and* h_0 *is the initial height in feet of an object traveling upward. See Example 6.*

★**99.** A ball is thrown upward at 4 feet per second from a building 29 feet high. When will the ball be 9 feet above the ground?

★**100.** A rocket is fired from the ground with an upward velocity of 200 feet per second. How many seconds will the rocket take to return to the ground?

For Exercises 101 and 102, use the formula $B = P\left(1 + \frac{r}{n}\right)^{nt}$, *which is used to calculate the final balance of an investment or a loan after being compounded. Following is a list of what each variable represents:*

B represents the final balance.
P represents the principal, which is the amount invested.
r represents the annual percentage rate.
t represents the time in years in which the principal is compounded.
n represents the number of times the principal is compounded in a year.

101. LaQuita invests $4000 in an account that is compounded annually. If after two years her balance is $4840, what was the interest rate of the account?

102. David invests $1000 in an account that is compounded semiannually (every six months). If after one year his balance is $1210, what is the interest rate of the account?

Puzzle Problem Gary's and Rene's ages are integers, and the sum of Gary's age and the square of Rene's age is 62. However, if you add the square of Gary's age to Rene's age, the result is 176. Find their ages. (*Source:* Henry Ernest Dudney, *The Best of Discover Magazine's Mind Benders, 1984.*)

Review Exercises

Exercises 1 and 2 Expressions

[5.1] **1.** Write 0.0004203 in scientific notation.

[5.4] [5.5] **2.** Divide: $\dfrac{x^3 + x^2 - 10x + 11}{x + 4}$

Exercises 3–6 Equations and Inequalities

For Exercises 3 and 4, graph the solution set.

[2.6] **3.** $|2x - 5| \geq 1$

[4.5] **4.** $\begin{cases} x + y < 5 \\ 2x - y \geq 4 \end{cases}$

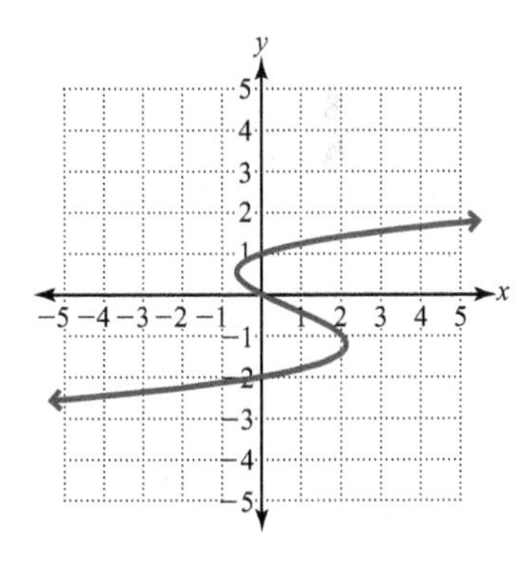

[3.5] **5.** Is the relation graphed to the right a function?

[5.2] **6.** If $f(x) = 5x^3 + 6x - 19$ and $g(x) = 3x^2 - 6x - 8$, find $(f + g)(x)$.

Chapter 6 Summary and Review Exercises

Complete each incomplete definition, rule, or procedure; study the key examples; and then work the related exercises.

6.1 Greatest Common Factor and Factoring by Grouping

Definitions/Rules/Procedures	Key Example(s)
To find the **GCF** of two or more monomials: **1.** Write the ______________ in exponential form for each monomial. Treat variables like prime factors. **2.** Write the GCF's factorization by including the prime factors (and variables) __________ to all of the factorizations, each raised to its __________ exponent in the factorizations. **3.** __________ the factors in the factorization created in Step 2. **Note** If there are no common prime factors, the GCF is 1.	Find the GCF of $54x^3y$ and $180x^2yz$ using prime factorization. $54x^3y = 2 \cdot 3^3 \cdot x^3 \cdot y$ $180x^2yz = 2^2 \cdot 3^2 \cdot 5 \cdot x^2 \cdot y \cdot z$ $\text{GCF} = 2 \cdot 3^2 \cdot x^2 \cdot y = 18x^2y$

Exercises 1–4 Expressions

[6.1] *For Exercises 1–4, find the GCF.*

1. $16x^4y^2, 24x^3y^9$

2. $20mn^3, 15m^4n^2$

3. $2u^4v^3, 3xy^9$

4. $4(x + 1), 6(x + 1)$

Definitions/Rules/Procedures	Key Example(s)
To factor a **monomial GCF** out of the terms of a polynomial: **1.** Find the __________ of the terms of the polynomial. **2.** Write the polynomial as a product of the __________ and the quotient of the __________ and the __________. Polynomial = __________(__________)	Factor $54x^3y - 180x^2yz$. $54x^3y - 180x^2yz = 18x^2y\left(\frac{54x^3y - 180x^2yz}{18x^2y}\right)$ $= 18x^2y(3x - 10z)$

Exercises 5–12 Expressions

[6.1] *For Exercises 5–12, factor out the GCF.*

5. $u^6 - u^4 - u^2$

6. $13d^2 - 26de$

7. $4h^2k^6 - 2h^5k$

8. $12cd - 4c^2d^2 + 10c^4d^4$

9. $16p^6q^3 - 12p^8q^5 + 13p^7q^2$

10. $9w^6v^8 + 6wv^6 - 12w^3v^3$

11. $17(w + 3) - m(w + 3)$

12. $2y(x + 5) + (x + 5)$

Definitions/Rules/Procedures	Key Example(s)
To **factor a four-term polynomial by grouping:** **1.** Factor out any monomial ________ (other than 1) that is ________ to all four terms. **2.** Group pairs of terms and factor the ________ out of each pair. **3.** If there is a common ________ factor, factor it out. **4.** If there is no common ________ factor, ________ the middle two terms and repeat the process. If there is still no common ________ factor, the polynomial cannot be factored by grouping.	Factor by grouping: $12x^3 + 15x^2 - 8x - 10$ $12x^3 + 15x^2 - 8x - 10 = 3x^2(4x + 5) - 2(4x + 5)$ $= (4x + 5)(3x^2 - 2)$

Exercises 13–20 Expressions

[6.1] *For Exercises 13–20, factor completely by grouping.*

13. $3m + mn + 6 + 2n$

14. $a^3 + 3a^2 + 3a + 9$

15. $2x + 2y - ax - ay$

16. $bc^2 + bd^2 - 5c^2 - 5d^2$

17. $x^2y - x^2s - ry + rs$

18. $4k^2 - k + 4k - 1$

19. $8uv^2 - 4v^2 + 20uv - 10v$

20. $5c^2d^2 - 5d^3 + 20c^2d - 20d^2$

6.2 Factoring Trinomials

Definitions/Rules/Procedures	Key Example(s)
To **factor a trinomial** of the form $x^2 + bx + c$: **1.** Find two numbers with a product equal to ________ and a sum equal to ________. **2.** The factored trinomial will have the factored form $(x + \square)(x + \square)$, where the second terms are the numbers found in ________. **Note** The signs of b and c may cause one or both signs in the binomial factors to be minus signs.	Factor $x^2 - 7x + 12$. **Product** — **Sum** $(-1)(-12) = 12$ — $-1 + (-12) = -13$ $(-3)(-4) = 12$ — $-3 + (-4) = -7$ Correct combination $x^2 - 7x + 12 = (x - 3)(x - 4)$
To **factor a trinomial** in the form $ax^2 + bx + c$: where $a \neq 1$, by **trial,** **1.** Look for a monomial ________ in all of the terms. If there is one, factor it out. **2.** Write a pair of *first* terms whose product is ________. **3.** Write a pair of *last* terms whose product is ________. **4.** Verify that the sum of the ________ and ________ products is bx (the middle term of the trinomial). If the sum of the ________ and ________ products is not bx, try the following: **a.** Exchange the ________ terms of the binomials from step 3; then repeat step 4. **b.** For each additional pair of ________ terms, repeat steps 3 and 4. **c.** For each additional pair of ________ terms, repeat steps 2–4.	Factor $36x^3 - 42x^2 - 8x$. First, factor out the monomial GCF, $2x$. $= 2x(18x^2 - 21x - 4)$ Factors of 18 are 1 and 18, 2 and 9, and 3 and 6. Factors of 4 are 1 and 4, and 2 and 2. We try various combinations of those factors in the first and last positions of binomial factors, checking each combination using FOIL. The correct combination is $(3x - 4)(6x + 1)$ because it gives a middle term of $-21x$. $= 2x(3x - 4)(6x + 1)$

Exercises 21–34 Expressions

[6.2] *For Exercises 21–34, factor the trinomials completely.*

21. $a^2 - 10a + 9$ **22.** $m^2 + 20m + 51$ **23.** $y^2 + 2y - 48$

24. $x^2 - 7x - 30$ **25.** $3x^2 - x - 14$ **26.** $16h^2 + 10h + 1$

27. $4u^2 - 2u + 3$ **28.** $6t^2 + t - 15$ **29.** $s^2 - 11st + 10t^2$

30. $6u^2 + 13uv + 6v^2$ **31.** $5m^2 - 16mn + 3n^2$ **32.** $4x^2 + 5xy - 8y^2$

33. $b^4 - 7b^3 - 18b^2$ **34.** $2x^3 - 16x^2 - 40x$

Definitions/Rules/Procedures	Key Example(s)
To **factor a trinomial** in the form $ax^2 + bx + c$, where $a \neq 1$, by **grouping:** **1.** Look for a monomial ________ in all of the terms. If there is one, factor it out. **2.** Find two factors of the product ________ whose sum is ________. **3.** Write a four-term polynomial in which ________ is written as the ________ of the two like terms whose coefficients are the two factors you found in step 2. **4.** Factor by ________.	Factor $15xy^3 + 54xy^2 - 24xy$. $= 3xy(5y^2 + 18y - 8)$ Factor out the monomial GCF, $3xy$. Multiply a and c: $5(-8) = -40$. Now we find two factors of -40 whose sum is 18. Note that 20 and -2 work; so we write the middle term, $18y$, as $20y - 2y$ and then factor by grouping. $= 3xy(5y^2 + 20y - 2y - 8)$ $= 3xy[5y(y + 4) - 2(y + 4)]$ $= 3xy(y + 4)(5y - 2)$

Exercises 35–38 Expressions

[6.2] *For Exercises 35–38, factor completely by grouping.*

35. $2u^2 + 9u + 10$ **36.** $3m^2 - 8m + 4$ **37.** $10u^2 + 7u - 3$ **38.** $6y^2 - 13y - 5$

Definitions/Rules/Procedures	Key Example(s)
If a trinomial is **not** of the **form** $ax^2 + bx + c$ but it resembles that form, we can try ________ to see if it is in fact of the form $ax^2 + bx + c$. If it is, we can factor using trial or grouping.	Factor $6(x - 4)^2 - 19(x - 4) - 7$. This polynomial resembles the form $ax^2 + bx + c$; so we try substitution, substituting u for $x - 4$. $= 6u^2 - 19u - 7$ Substitute u for $x - 4$. $= (3u + 1)(2u - 7)$ Factor. $= [3(x - 4) + 1][2(x - 4) - 7]$ Substitute $x - 4$ for u. $= [3x - 12 + 1][2x - 8 - 7]$ Distribute. $= (3x - 11)(2x - 15)$ Simplify.

Exercises 39–42 Expressions

[6.2] *For Exercises 39–42, factor completely using substitution.*

39. $x^4 + 5x^2 + 4$ **40.** $3c^4 + 13c^2 + 4$ **41.** $2h^6 + 9h^3 + 4$ **42.** $3(k + 1)^2 - 2(k + 1) - 5$

6.3 Factoring Special Products and Factoring Strategies

Definitions/Rules/Procedures	Key Example(s)
Rules for **factoring special products:** Perfect square trinomials: $a^2 + 2ab + b^2 =$ ______ $a^2 - 2ab + b^2 =$ ______ Difference of squares: $a^2 - b^2 =$ ______ Difference of cubes: $a^3 - b^3 =$ ______ Sum of cubes: $a^3 + b^3 =$ ______ **Note** A sum of squares, $a^2 + b^2$, is prime.	Factor. **a.** $25x^2 + 40x + 16 = (5x + 4)^2$ **b.** $49t^2 - 28tu + 4u^2 = (7t - 2u)^2$ **c.** $9y^2 - 25 = (3y + 5)(3y - 5)$ **d.** $8x^3 - 125 = (2x - 5)(4x^2 + 10x + 25)$ **e.** $n^3 + 125 = (n + 5)(n^2 - 5n + 25)$

Exercises 43–60 Expressions

[6.3] *For Exercises 43–48, factor the perfect square completely.*

43. $x^2 + 6x + 9$

44. $y^2 + 12y + 36$

45. $m^2 - 4m + 4$

46. $w^2 - 14w + 49$

47. $9d^2 + 30d + 25$

48. $4c^2 - 28c + 49$

[6.3] *For Exercises 49–56, factor the difference of squares completely.*

49. $h^2 - 9$

50. $p^2 - 64$

51. $9d^2 - 4$

52. $81k^2 - 100$

53. $2w^2 - 50$

54. $4q^2 - 36$

55. $25y^2 - 9z^2$

56. $x^4 - 81$

[6.3] *For Exercises 57–60, factor the sum or difference of cubes completely.*

57. $c^3 - 27$

58. $m^3 + 64$

59. $27b^3 + 8a^3$

60. $64d^3 - 8c^3$

Definitions/Rules/Procedures	Key Example(s)
To **factor a polynomial,** factor out any monomial GCF and then consider the number of terms in the polynomial. **I. Four terms:** Try to factor by ______. **II. Three terms:** Determine whether the trinomial is a ______. **A.** If the trinomial is a ______, consider its form. **1.** If it is in the form $a^2 + 2ab + b^2$, the factored form is ______. **2.** If it is in the form $a^2 - 2ab + b^2$, the factored form is ______. **B.** If the trinomial is not a perfect square, consider its form. **1.** If it is in the form $x^2 + bx + c$, find two factors of ______ whose sum is ______ and write factored form as $(x + \text{first number})(x + \text{second number})$.	Factor completely. **a.** $4y(2x + 1)^2 - 36x^2y$ $= 4y[(2x + 1)^2 - 9x^2]$ Factor out the monomial GCF, $4y$. $= 4y[(2x + 1) + 3x][(2x + 1) - 3x]$ Factor the difference of squares. $= 4y(5x + 1)(1 - x)$ Simplify.

Definitions/Rules/Procedures	Key Example(s)
2. If it is in the form $ax^2 + bx + c$, where $a \neq 1$, use ______. Or find two factors of ______ whose sum is ______; write these factors as coefficients of two like terms that, when combined equal bx; and factor by grouping. **III. Two terms:** Determine whether the binomial is a difference of squares, sum of cubes or difference of cubes. **A.** If it is a difference of squares, $a^2 - b^2$, the factored form is ______. Note that the sum of squares, $a^2 + b^2$, cannot be factored. **B.** If it is a difference of cubes, $a^3 - b^3$, the factored form is ______. **C.** If it is a sum of cubes, $a^3 + b^3$, the factored form is ______. **Note** Always check to see if any of the factors can be factored further.	**b.** $x^5 - 9x^3 + 8x^2 - 72$ $= x^3(x^2 - 9) + 8(x^2 - 9)$ Factor by grouping. $= (x^2 - 9)(x^3 + 8)$ Factor out $x^2 - 9$. $= (x + 3)(x - 3)(x + 2)(x^2 - 2x + 4)$ Factor the difference of squares and the sum of cubes.

Exercises 61–68 Expressions

[6.3] *For Exercises 61–68, factor completely.*

61. $9d^2 - 6d + 1$

62. $3m^2 - 3n^2$

63. $2a^2 - 5a - 12$

64. $x^2 + 9$

65. $3p^4 + 3p^3 - 90p^2$

66. $x^4 - 16$

67. $15a^3b^2c^7 + 3a^2b^4c^2 + 5a^9bc^3$

68. $w^3 - (2 + y)^3$

6.4 Solving Equations by Factoring

Definitions/Rules/Procedures	Key Example(s)
A **polynomial equation** is an equation that equates ______. A **polynomial equation in standard form** is an equation of the form ______ where P is a polynomial in terms of one variable written in ______ order of degree. A **quadratic equation in one variable** is an equation that can be written in the form ______, where a, b, and c are real numbers and $a \neq 0$. A **cubic equation in one variable** is an equation that can be put in the form ______ where a, b, c and d are real numbers and $a \neq 0$.	

Definitions/Rules/Procedures	Key Example(s)
Zero-factor theorem: If a and b are real numbers and $ab = 0$, then ________ or ________. To solve a **polynomial equation using factoring:** 1. Write the equation in standard form. (Set one side equal to ________ with the other side in ________ order of degree when possible.) 2. Write the polynomial in ________ form. 3. Use the ________ theorem to solve.	Solve. **a.** $2x^2 = 4 - 7x$ $2x^2 + 7x - 4 = 0$ Write in standard form. $(2x - 1)(x + 4) = 0$ Factor. $2x - 1 = 0$ or $x + 4 = 0$ Use the zero-factor theorem. $2x = 1$ $x = -4$ Solve the equations. $x = \frac{1}{2}$ **b.** $n^3 + n(n - 1) = 3n^2 + 2(4n - 9)$ $n^3 + n^2 - n = 3n^2 + 8n - 18$ Distribute. $n^3 - 2n^2 - 9n + 18 = 0$ Write in standard form. $n^2(n - 2) - 9(n - 2) = 0$ Factor by grouping. $(n - 2)(n^2 - 9) = 0$ Factor out $n - 2$. $(n - 2)(n + 3)(n - 3) = 0$ Factor the difference of squares. $n - 2 = 0$ or $n + 3 = 0$ or $n - 3 = 0$ Use the zero-factor theorem. $n = 2$ $n = -3$ $n = 3$

Exercises 69–74 Equations and Inequalities

[6.4] *For Exercises 69–74, solve.*

69. $(x + 4)(x - 1) = 0$

70. $x^2 - 64 = 0$

71. $w^2 + 2w - 3 = 0$

72. $6y^2 = 7y - 1$

73. $4x(x + 7) + 9 = -40$

74. $b^3 - 2b^2 + 20 = 2b^2 - (16 - 9b)$

Definitions/Rules/Procedures	Key Example(s)
Given a function $f(x)$, we can find its x-intercepts by solving the equation ________.	Find the x-intercepts of the function $f(x) = x^2 + x - 6$ and then graph the function. $x^2 + x - 6 = 0$ Set $f(x) = 0$. $(x + 3)(x - 2) = 0$ Factor. $x + 3 = 0$ or $x - 2 = 0$ Use the zero-factor theorem. $x = -3$ $x = 2$ Solve each equation. The x-intercepts are $(-3, 0)$ and $(2, 0)$.

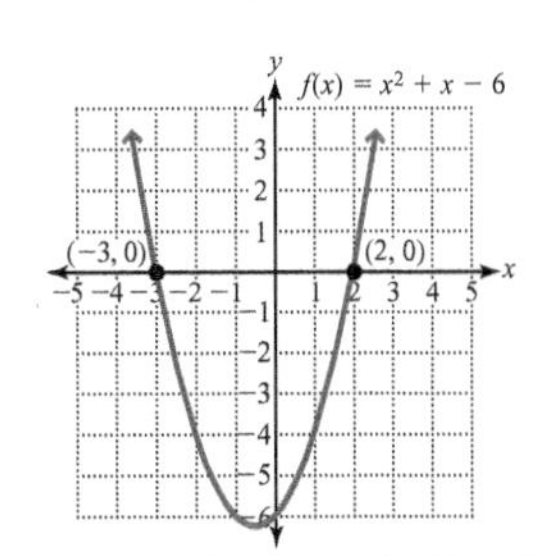

Exercises 75 and 76 Equations and Inequalities

[6.4] *For Exercises 75 and 76, find the x-intercepts and then sketch the graph.*

75. $f(x) = x^2 - 4$

76. $f(x) = x(x + 3)(x - 2)$

Definitions/Rules/Procedures	Key Example(s)
Pythagorean theorem: Given a right triangle where a and b represent the lengths of the legs and c represents the length of the hypotenuse, ______________. 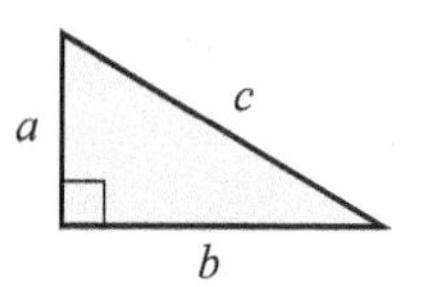	Find the unknown lengths in the accompanying right triangle. 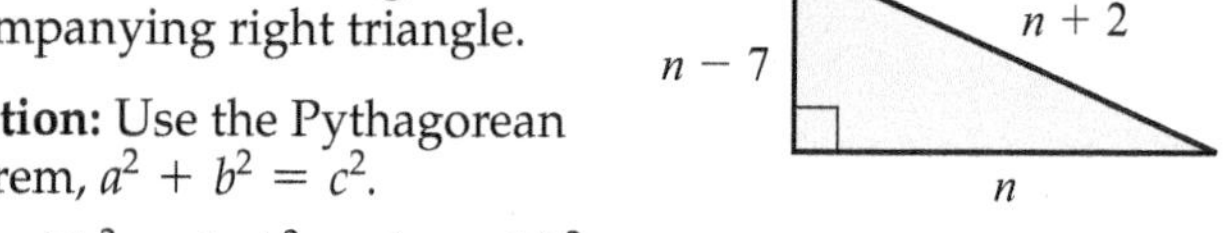 **Solution:** Use the Pythagorean theorem, $a^2 + b^2 = c^2$. $(n - 7)^2 + (n)^2 = (n + 2)^2$ $2n^2 - 14n + 49 = n^2 + 4n + 4$ $n^2 - 18n + 45 = 0$ Write in standard form. $(n - 15)(n - 3) = 0$ Factor. $n - 15 = 0$ or $n - 3 = 0$ Use the zero-factor theorem. $n = 15$ $n = 3$ Solve the equations. **Answer:** If $n = 3$, then $n - 7 = -4$, which makes no sense as a length; so n must be 15. Therefore, $n - 7 = 8$ and $n + 2 = 17$.

Exercises 77–84 Equations and Inequalities

[6.4] *For Exercises 77–84, solve.*

77. The product of two natural numbers is 56. If the second number is 10 more than the first, find the two numbers.

78. Find two consecutive positive integers whose product is 110.

79. Find two consecutive positive even integers whose product is 288.

80. A rectangular room is 3 feet longer than it is wide. If the area of the room is 88 square feet, find the dimensions of the room.

81. A stage is to be constructed in the shape of a trapezoid with a base length that is three times the height. If the area is to be 672 square feet, find the height and length of the base.

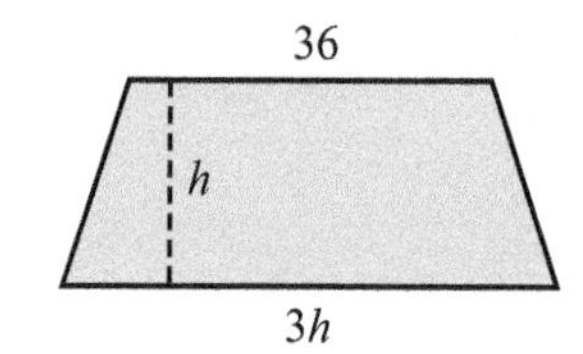

82. Find the length of each side of the right triangle shown.

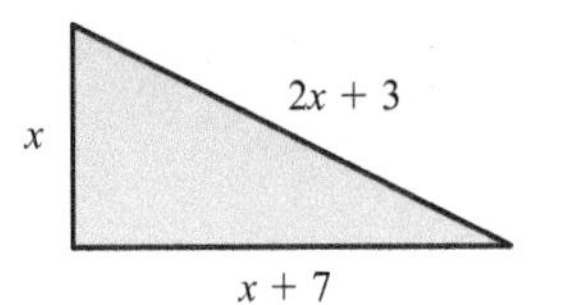

83. A sound designer is trying to find optimal positions for speakers in a room. If she places a speaker on a wall as shown, find the height of the speaker and its distance from her ear.

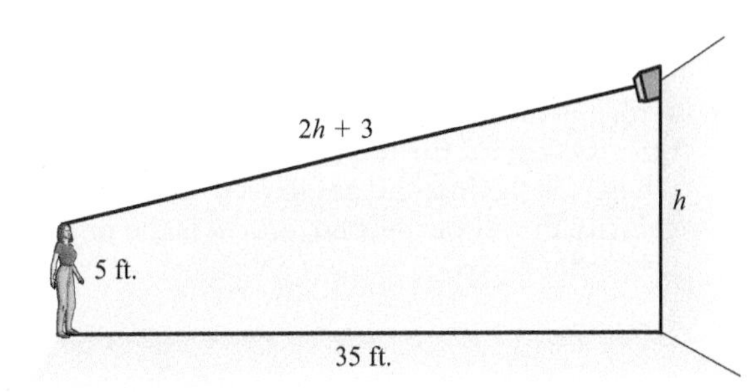

84. A ball is thrown upward at 12 feet per second from a building 60 feet high. Find the time it takes for the ball to be 6 feet above the ground. Use the formula $h = -16t^2 + v_0t + h_0$, where h is the final height in feet, t is the time of travel in seconds, v_0 is the initial velocity in feet per second, and h_0 is the initial height in feet of an object traveling upward.

Learning Strategy

When preparing for an exam, give yourself enough time. Study a day or two before the exam and the morning of the exam. Eat breakfast and be calm.

—Biridianna N

Chapter 6 Practice Test

For Extra Help

Step-by-step test solutions are found on the Chapter Test Prep Videos available in MyMathLab® *or on* YouTube.

1. Find the GCF of $14m^7n^2$, $21m^3n^9$.

Factor completely.

2. $3m + 6m^3 - 9m^6$

3. $3m + 3n - m - n$

4. $9n^2 - 16$

5. $8x^3 - 27$

6. $y^2 + 14y + 49$

7. $q^2 - 2q - 48$

8. $3ab^2 - 30ab + 24a$

9. $5 - 125t^2$

10. $6d^2 + d - 2$

11. $8c^3 + 8d^3$

12. $w^3 + 2w^2 + 3w + 6$

13. $s^4 - 81$

14. $5p^2 - 7pq - 12q^2$

Solve.

15. $8x^2 - 14x + 5 = 0$

16. $x(x + 3) - 6 = 12$

17. $2x^3 + x^2 = 15x$

18. Find the length of each side of the right triangle shown.

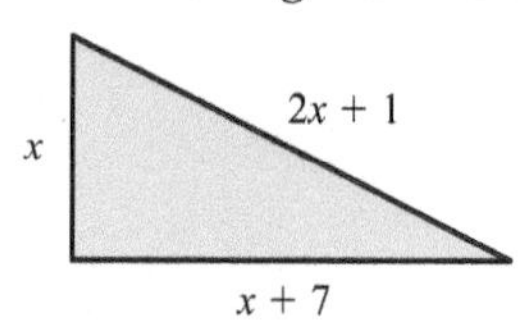

19. A rock is thrown upward at 4 feet per second from a building 56 feet high. Find the time it takes for the rock to hit the ground. Use the formula $h = -16t^2 + v_0t + h_0$, where h is the final height in feet, t is the time of travel in seconds, v_0 is the initial velocity in feet per second, and h_0 is the initial height in feet of an object traveling upward.

Find the x-intercepts and graph the function.

20. $f(x) = x^2 - 9$

Chapters 1–6 Cumulative Review Exercises

For Exercises 1–4, answer true or false.

[3.5] **1.** The graph of $y = 2x - 1$ is the same as the graph of $f(x) = 2x - 1$.

[3.5] **2.** $f(x) = 3x - x^2$ is an example of a linear function.

[5.1] **3.** When a number n is written in scientific notation where $0 < |n| < 1$, the exponent is negative.

[6.3] **4.** The expression $a^2 + b^2$ can be factored to $(a + b)(a + b)$.

For Exercises 5–7, fill in the blank.

[3.3] **5.** The slopes of parallel lines are ______________.

[5.1] **6.** To multiply monomials, (a) multiply ______________, (b) ______________ the exponents of the like bases, and (c) write any unlike variable bases unchanged in the product.

[6.3] **7.** Complete the rule for factoring a sum of cubes.
$a^3 + b^3 =$ ______________

Exercises 8–16 Expressions

For Exercises 8–11, simplify.

[1.3] **8.** $\dfrac{4(3^2 - 10) - 4}{3 - \sqrt{25 - 16}}$

[5.2] **9.** $(2x^3 + 5x^2 - 19x + 14) - (17x^2 + 2x - 6)$

[5.3] **10.** $(2x + 1)(4x^2 - 3x + 2)$

[5.1] **11.** $\left(\dfrac{2}{x}\right)^{-3}$

For Exercises 12–16, factor.

[6.1] **12.** $4m - 16m^2$

[6.2] **13.** $8x^2 + 29x - 12$

[6.3] **14.** $49a^2 + 84a + 36$

[6.3] **15.** $64p^3 + 8$

[6.1] **16.** $3x - 6y + ax - 2ay$

Exercises 17–30 Equations and Inequalities

For Exercises 17–19, solve.

[2.1] **17.** $9x - [14 + 2(x - 3)] = 8x - 2(3x - 1)$

[2.5] **18.** $|2x - 3| - 9 = 16$

[6.4] **19.** $5x^2 + 26x = 24$

For Exercises 20 and 21, solve. *a. Write the solution set using set-builder notation.*
b. Write the solution set using interval notation.
c. Graph the solution set.

[2.3] 20. $\frac{3}{4}x + \frac{2}{3} \leq \frac{5}{6}x + \frac{1}{2}$

[2.6] 21. $2|x - 3| - 2 < 10$

For Exercises 22 and 23, solve.

[4.1] 22. $\begin{cases} 5x - 2y = -1 \\ 3x + 2y = -7 \end{cases}$

[4.2] 23. $\begin{cases} x - 2y + z = 8 \\ x + y + z = 5 \\ 2x + y - 2z = -5 \end{cases}$

[3.5] 24. If $f(x) = 2x^2 + 3x - 5$. find $f(-3)$.
[5.5]

[4.5] 25. Graph the solution set for the system of inequalities. $\begin{cases} x + y < 6 \\ y \geq 2x \end{cases}$

[6.4] 26. For $f(x) = x^2 + x - 6$, find the x- and y-intercepts; then graph.

For Exercises 27–30, solve.

[2.4] 27. A mail-order DVD club offers one bonus DVD if the total of the order is from \$50 to \$80 and two bonus DVDs if the total order is \$80 or more. Lyle decides to buy the lowest-priced DVDs at \$10 each.

a. In what range would the number of DVDs Lyle orders have to be for him to receive one bonus DVD?

b. In what range would the number of DVDs Lyle orders have to be for him to receive two bonus DVDs?

[2.2] 28. Adults aged 20–24 spend 2 minutes longer driving than twice the time of teenagers aged 15–19. If the combined minutes per day that both groups drive is 77 minutes, how many minutes does each group drive? (*Source:* U.S. Department of Transportation.)

[2.2] 29. Two angles are complementary. The larger angle is six less than twice the smaller angle. Find the measure of the two angles.

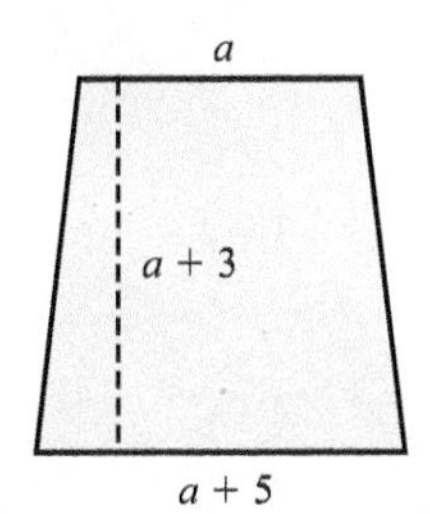

[6.4] 30. A metal lid is in the shape of a trapezoid as shown. If the area of the metal used is to be 60 square inches, find each length represented in the figure.

CHAPTER

7

Rational Expressions and Equations

Chapter Overview

In this chapter, we explore rational expressions, rational equations, and rational functions. The chapter flow will be similar to that of Chapters 5 and 6. We will:

- Define rational expressions and functions.
- Perform arithmetic operations with rational expressions.
- Solve equations containing rational expressions.

7.1 Simplifying, Multiplying, and Dividing Rational Expressions

Objectives

1. Simplify rational expressions.
2. Multiply rational expressions.
3. Divide rational expressions.
4. Evaluate rational functions.
5. Find the domain of a rational function.
6. Graph rational functions.

Warm-up

[6.1–6.3] *For Exercises 1–3, factor.*

1. $4x^2 + 16x$

2. $x^4 - 81$

3. $2b^2 - 13b + 20$

[6.4] 4. Solve. $6b^2 - 11b - 10 = 0$

Objective 1 Simplify rational expressions.

Now that we have explored polynomials, we are ready to learn about **rational expressions.**

Connection Notice that the definition of a rational expression is like the definition of a rational number. Recall that a rational number can be expressed in the form $\frac{a}{b}$, where a and b are integers and $b \neq 0$.

Definition Rational expression: An expression that can be written in the form $\frac{P}{Q}$, where P and Q are polynomials and $Q \neq 0$.

Some rational expressions are

$$\frac{4x^2y^3}{3xy^2}, \quad \frac{3x-5}{x^2-16}, \quad \text{and} \quad \frac{2x^2-3x+6}{x^2-x-6}.$$

Simplifying a rational expression is like simplifying a fraction. To simplify a fraction to lowest terms, we use the rule $\frac{an}{bn} = \frac{a \cdot 1}{b \cdot 1} = \frac{a}{b}$ when $b \neq 0$ and $n \neq 0$. We can rewrite this rule so that it applies to rational expressions.

Rule Fundamental Principle of Rational Expressions

$\frac{PR}{QR} = \frac{P \cdot 1}{Q \cdot 1} = \frac{P}{Q}$, where P, Q, and R are polynomials and Q and R are not 0.

The rule indicates that a factor common to both the numerator and denominator can be divided out of a rational expression. Like rational numbers, rational expressions are in lowest terms if the greatest common factor of the numerator and denominator is 1.

Consider the following comparison between simplifying a fraction and simplifying a similar rational expression.

$$\frac{14}{21} = \frac{2 \cdot 7}{3 \cdot 7} = \frac{2 \cdot 1}{3 \cdot 1} = \frac{2}{3} \qquad \frac{2a}{3a} = \frac{2 \cdot a}{3 \cdot a} = \frac{2 \cdot 1}{3 \cdot 1} = \frac{2}{3}$$

7 is the common factor here.

a is the common factor here.

Note This simplification is true when $a \neq 0$. If $a = 0$, the original expression is indeterminate. We must avoid variable values that make an expression undefined or indeterminate.

Procedure Simplifying Rational Expressions to Lowest Terms

To simplify a rational expression to lowest terms:

1. Factor the numerator and denominator completely.
2. Divide out all common factors in the numerator and denominator.
3. Multiply the remaining factors in the numerator and the remaining factors in the denominator.

Answers to Warm-up

1. $4x(x + 4)$

2. $(x^2 + 9)(x + 3)(x - 3)$

3. $(2b - 5)(b - 4)$

4. $b = -\frac{2}{3}, b = \frac{5}{2}$

Example 1 Simplify.

a. $\dfrac{18x^4}{21x^2}$

Solution: Factor the numerator and denominator completely; then divide out all common factors.

Note To the right of each answer in Example 1, we have excluded those values that cause the original expression to be undefined or indeterminate. In future examples, we will not have these statements because we assume that variables would not be replaced with such values.

Note We will no longer show the step with the 1s and will simply highlight the common factors that are eliminated.

$$\frac{18x^4}{21x^2} = \frac{2\cdot3\cdot3\cdot x\cdot x\cdot x\cdot x}{3\cdot7\cdot x\cdot x}$$

Note The common factors are a single 3 and two x's. These form the GCF that we divide out, which is $3x^2$.

$$= \frac{2\cdot1\cdot3\cdot1\cdot x\cdot x}{1\cdot7\cdot1}$$

$$= \frac{6x^2}{7}$$ Answer if $x \neq 0$

There are different styles for showing the process of dividing out common factors. For example, some people prefer using cancel marks.

$$\frac{18x^4}{21x^2} = \frac{2\cdot\overset{1}{\cancel{3}}\cdot3\cdot\overset{1}{\cancel{x}}\cdot\overset{1}{\cancel{x}}\cdot x\cdot x}{\underset{1}{\cancel{3}}\cdot7\cdot\underset{1}{\cancel{x}}\cdot\underset{1}{\cancel{x}}} = \frac{6x^2}{7}$$

b. $\dfrac{4x^2 + 16x}{3x^2 + 12x}$

Solution:

$$\frac{4x^2 + 16x}{3x^2 + 12x} = \frac{4\cdot x\cdot(x + 4)}{3\cdot x\cdot(x + 4)}$$

Factor the numerator and denominator completely; then divide out the common factors, x and $x + 4$. Recall that $\frac{x}{x} = 1$ for $x \neq 0$ and $\frac{x + 4}{x + 4} = 1$ for $x \neq -4$.

$$= \frac{4}{3}$$ Answer if $x \neq 0, -4$

c. $\dfrac{x^2 - 9}{x^2 - x - 12}$

Solution:

Note $\dfrac{x + 3}{x + 3} = 1$ for $x \neq -3$.

$$\frac{x^2 - 9}{x^2 - x - 12} = \frac{(x + 3)(x - 3)}{(x + 3)(x - 4)}$$

Factor the numerator and denominator completely; then divide out the common factor, $x + 3$.

$$= \frac{x - 3}{x - 4}$$ Answer if $x \neq -3, 4$

d. $\dfrac{x^3 + 27}{3x^2 + 5x - 12}$

Solution:

Note $\dfrac{x + 3}{x + 3} = 1$ for $x \neq -3$.

$$\frac{x^3 + 27}{3x^2 + 5x - 12} = \frac{(x + 3)(x^2 - 3x + 9)}{(x + 3)(3x - 4)}$$

Factor the numerator and denominator completely; then divide out the common factor, $x + 3$.

$$= \frac{x^2 - 3x + 9}{3x - 4}$$ Answer if $x \neq -3, \frac{4}{3}$

e. $\dfrac{2x^2 - 11x + 15}{x^4 - 81}$

Solution:

$$\frac{2x^2 - 11x + 15}{x^4 - 81} = \frac{(2x-5)(x-3)}{(x^2+9)(x^2-9)}$$

$$= \frac{(2x-5)(x-3)}{(x^2+9)(x+3)(x-3)}$$ Factor the numerator and denominator completely; then divide out the common factor, $x - 3$.

Note $\frac{x-3}{x-3} = 1$ for $x \neq 3$.

$$= \frac{2x-5}{(x^2+9)(x+3)}$$ Answer if $x \neq -3, 3$

Warning A common error is to divide out common terms instead of common factors, as in $\frac{6+3}{3} = \frac{6+\cancel{3}}{\cancel{3}} = 6 + 1 = 7$. Using the order of operations, we see that the correct answer is $\frac{6+3}{3} = \frac{9}{3} = 3$. The mistake in the first simplification is that we divided out a *term*, not a *factor*. Remember, terms are separated by + and − signs and factor implies multiplication. Therefore, to simplify a rational expression, you must first factor the numerator and denominator completely and then divide out the common factors. Examples with rational expressions follow.

Correct: $\frac{x^2-6x+8}{x^2+2x-8} = \frac{(x-4)\overset{1}{\cancel{(x-2)}}}{(x+4)\underset{1}{\cancel{(x-2)}}} = \frac{x-4}{x+4}$

$x - 2$ is a factor common to both the numerator and the denominator, so we can divide both the numerator and the denominator by $x - 2$.

Incorrect: $\frac{x^2-6x+8}{x^2+2x-8} = \frac{\cancel{x^2}-6x+\cancel{8}}{\cancel{x^2}+2x-\cancel{8}} = \frac{1-6x+1}{1+2x-1} = \frac{2-\cancel{6}x}{\cancel{2}x} = \frac{1-3x}{x}$

x^2 and 8 are terms common to the numerator and the denominator. Because they are not factors, it is wrong to divide them out.

Learning Strategy

The book offers practice problems as well as the solutions. Do the practice problems and check answers. If wrong, keep practicing until you get the correct solutions.
—Ruben C.

Your Turn 1 Simplify.

a. $\frac{24a^3}{18a^5}$ b. $\frac{3x^2+6x}{5x^2+10x}$ c. $\frac{x^2-4}{x^2+9x+14}$ d. $\frac{x^3-8}{x^2-4}$ e. $\frac{x^4-16}{3x^2+2x-8}$

Answers to Your Turn 1

a. $\frac{4}{3a^2}$ b. $\frac{3}{5}$ c. $\frac{x-2}{x+7}$
d. $\frac{x^2+2x+4}{x+2}$
e. $\frac{(x^2+4)(x-2)}{3x-4}$

Rational expressions containing binomial factors that are additive inverses require an extra step. In Example 2, let's look at binomial factors that are additive inverses.

Example 2 Simplify $\frac{3x^2-7x-6}{9-3x}$.

Solution: $\frac{3x^2-7x-6}{9-3x}$

$$= \frac{(3x+2)(x-3)}{3(3-x)}$$ Factor the numerator and denominator completely.

It appears that there are no common factors. However, $x - 3$ and $3 - x$ are additive inverses because $(x-3) + (3-x) = 0$. We can get common factors by factoring -1 out of either expression. We will use the expression in the numerator.

$$= \frac{(3x+2)(-1)(3-x)}{3(3-x)}$$ Factor -1 out of $x - 3$. $(x-3) = -1(-x+3) = (-1)(3-x)$

Note We could have factored the -1 out of the denominator.

$$\frac{(3x+2)(x-3)}{3(-1)(x-3)} = \frac{3x+2}{-3}$$

$$= \frac{-3x-2}{3} \text{ or } -\frac{3x+2}{3}$$ Divide out the common factor $3 - x$; then multiply the remaining factors.

Example 2 also illustrates the rule of sign placement in a fraction or rational expression. Remember that with a negative fraction (or rational expression), the minus sign can be placed in the numerator or denominator or aligned with the fraction line.

Rule Sign Placement

$$-\frac{P}{Q} = \frac{-P}{Q} = \frac{P}{-Q}, \text{ where } P \text{ and } Q \text{ are polynomials and } Q \neq 0.$$

Warning A rational expression such as $\frac{4 + x}{4 - x}$ looks like it could be simplified by factoring -1 out of $4 + x$ or $4 - x$. Although $4 + x$ and $4 - x$ appear to be additive inverses, they are not because $(4 + x) + (4 - x) = 8$, not 0. Consequently, factoring -1 out of the numerator or denominator as in $\frac{4 + x}{(-1)(-4 + x)}$ does not give common factors. In fact, $\frac{4 + x}{4 - x}$ is in simplest form.

Your Turn 2 Simplify $\frac{x^2 + 3x - 10}{2x - x^2}$.

Objective 2 Multiply rational expressions.

To multiply fractions, we use the rule $\frac{a}{b} \cdot \frac{c}{d} = \frac{ac}{bd}$, where $b \neq 0$ and $d \neq 0$. We can rewrite this rule so that it applies to multiplying rational expressions.

Rule Multiplying Rational Expressions

$$\frac{P}{Q} \cdot \frac{R}{S} = \frac{PR}{QS}, \text{ where } P, Q, R, \text{ and } S \text{ are polynomials and } Q \text{ and } S \neq 0.$$

We use the same procedure for multiplying rational expressions as for multiplying rational numbers.

Connection Multiplying rational expressions is very much like simplifying rational expressions to lowest terms.

Procedure Multiplying Rational Expressions

To multiply rational expressions:

1. Factor each numerator and denominator completely.
2. Divide out factors common to both the numerator and denominator.
3. Multiply numerator by numerator and denominator by denominator.
4. Simplify as needed.

Example 3 Find the product.

a. $\frac{8a^2b}{7a^3b} \cdot \frac{21ab^2}{4a^2b}$

Solution: $\frac{8a^2b}{7a^3b} \cdot \frac{21ab^2}{4a^2b} = \frac{2 \cdot 2 \cdot 2 \cdot a \cdot a \cdot b}{7 \cdot a \cdot a \cdot a \cdot b} \cdot \frac{3 \cdot 7 \cdot a \cdot b \cdot b}{2 \cdot 2 \cdot a \cdot a \cdot b}$ Factor the numerators and denominators completely.

$= \frac{2}{1} \cdot \frac{3 \cdot b}{a \cdot a}$ Divide out the common factors of 2, 7, a, and b.

$= \frac{6b}{a^2}$ Multiply the remaining factors.

Answer to Your Turn 2

$-\frac{x + 5}{x}$

Note Recall that $\frac{a}{a} = 1$ for $a \neq 0$ and $\frac{a-4}{a-4} = 1$ for $a \neq 4$.

b. $-\frac{2a^2}{3a-12} \cdot \frac{5a-20}{30a^3}$

Solution: $-\frac{2a^2}{3a-12} \cdot \frac{5a-20}{30a^3} = -\frac{2 \cdot a \cdot a}{3 \cdot (a-4)} \cdot \frac{5 \cdot (a-4)}{2 \cdot 3 \cdot 5 \cdot a \cdot a \cdot a}$ Factor the numerators and denominators completely.

$= -\frac{1}{3} \cdot \frac{1}{3a}$ Divide out the common factors of 2, 5, a, and $a - 4$.

$= -\frac{1}{9a}$ Multiply the remaining factors.

Note Recall that $\frac{x+2}{x+2} = 1$ for $x \neq -2$, $\frac{x-3}{x-3} = 1$ for $x \neq 3$, and $\frac{x-2}{x-2} = 1$ for $x \neq 2$.

c. $\frac{x^2 - x - 6}{x^2 + 2x - 15} \cdot \frac{x^2 + 4x - 12}{x^2 - 4}$

Solution: $\frac{x^2 - x - 6}{x^2 + 2x - 15} \cdot \frac{x^2 + 4x - 12}{x^2 - 4}$

$= \frac{(x+2)(x-3)}{(x+5)(x-3)} \cdot \frac{(x-2)(x+6)}{(x+2)(x-2)}$ Factor the numerators and denominators completely.

$= \frac{1}{x+5} \cdot \frac{x+6}{1}$ Divide out the common factors of $x + 2$, $x - 3$, and $x - 2$.

$= \frac{x+6}{x+5}$ Multiply the remaining factors.

d. $\frac{3-a}{a^2 + 3a + 2} \cdot \frac{a^2 - 4}{a^2 + 2a - 15}$

Solution: $\frac{3-a}{a^2 + 3a + 2} \cdot \frac{a^2 - 4}{a^2 + 2a - 15}$

$= \frac{3-a}{(a+2)(a+1)} \cdot \frac{(a+2)(a-2)}{(a-3)(a+5)}$ Factor the numerators and denominators completely.

$= \frac{-1(a-3)}{(a+2)(a+1)} \cdot \frac{(a+2)(a-2)}{(a-3)(a+5)}$ Because $3 - a$ and $a - 3$ are additive inverses, factor -1 from $3 - a$ to get $-1(a - 3)$.

$= \frac{-1}{a+1} \cdot \frac{a-2}{a+5}$ Divide out the common factors of $a - 3$ and $a + 2$.

$= \frac{2-a}{(a+1)(a+5)}$ Multiply the remaining factors.

Note We could write the solution as $-\frac{a-2}{(a+1)(a+5)}$, which is often the preferred form.

e. $\frac{x^2 + x - 20}{15x^3} \cdot \frac{30x^2 - 20x}{3x^2 - 14x + 8}$

Solution: $\frac{x^2 + x - 20}{15x^3} \cdot \frac{30x^2 - 20x}{3x^2 - 14x + 8}$

$= \frac{(x+5)(x-4)}{3 \cdot 5 \cdot x \cdot x \cdot x} \cdot \frac{2 \cdot 5 \cdot x \cdot (3x-2)}{(3x-2)(x-4)}$ Factor the numerators and denominators completely.

$= \frac{x+5}{3 \cdot x \cdot x} \cdot \frac{2}{1}$ Divide out the common factors of $x - 4$, $3x - 2$, 5, and x.

$= \frac{2x+10}{3x^2}$ Multiply the remaining factors.

Your Turn 3 Find the product.

a. $\frac{6r^3s}{5r^2s^2} \cdot \frac{15rs^2}{2r^3s^2}$ b. $-\frac{6a^2}{9a+27} \cdot \frac{7a+21}{28a^4}$ c. $\frac{a^2+6a+8}{a^2+5a-6} \cdot \frac{a^2+3a-18}{a^2+a-12}$

d. $\frac{y^2+3y-10}{y^2-25} \cdot \frac{y+4}{2-y}$ e. $\frac{x^2-2x-8}{14} \cdot \frac{16x-40}{2x^2-x-10}$

Answers to Your Turn 3

a. $\frac{9}{rs}$ b. $-\frac{1}{6a^2}$

c. $\frac{a+2}{a-1}$ d. $-\frac{y+4}{y-5}$

e. $\frac{4x-16}{7}$

Objective 3 Divide rational expressions.

To divide fractions, we use the rule $\frac{a}{b} \div \frac{c}{d} = \frac{a}{b} \cdot \frac{d}{c}$, where b, c, and d are not 0. In other words, we multiply by the reciprocal of the divisor. We can rewrite this rule so that it applies to dividing rational expressions.

Procedure Dividing Rational Expressions

To divide rational expressions:

1. Write an equivalent multiplication statement using $\frac{P}{Q} \div \frac{R}{S} = \frac{P}{Q} \cdot \frac{S}{R}$, where P, Q, R, and S are polynomials and Q, R, and $S \neq 0$.
2. Simplify using the procedure for multiplying rational expressions.

Example 4 Find the quotient.

a. $\frac{4x^2 + 8x}{6x^2} \div \frac{5x + 10}{9x^3}$

Solution:

$$\frac{4x^2 + 8x}{6x^2} \div \frac{5x + 10}{9x^3} = \frac{4x^2 + 8x}{6x^2} \cdot \frac{9x^3}{5x + 10}$$ Write an equivalent multiplication statement.

$$= \frac{2 \cdot 2 \cdot x \cdot (x + 2)}{2 \cdot 3 \cdot x \cdot x} \cdot \frac{3 \cdot 3 \cdot x \cdot x \cdot x}{5(x + 2)}$$ Factor the numerators and denominators completely.

$$= \frac{6x^2}{5}$$ Divide out the common factors of 2, 3, x, and $x + 2$ and multiply the remaining factors.

b. $\frac{2x^2 + 7x + 3}{x^2 - 9} \div \frac{2x^2 + 11x + 5}{x^2 - 3x}$

Solution:

$$\frac{2x^2 + 7x + 3}{x^2 - 9} \div \frac{2x^2 + 11x + 5}{x^2 - 3x}$$

$$= \frac{2x^2 + 7x + 3}{x^2 - 9} \cdot \frac{x^2 - 3x}{2x^2 + 11x + 5}$$ Write an equivalent multiplication statement.

$$= \frac{(2x + 1)(x + 3)}{(x + 3)(x - 3)} \cdot \frac{x(x - 3)}{(2x + 1)(x + 5)}$$ Factor the numerators and denominators completely.

$$= \frac{x}{x + 5}$$ Divide out the common factors of $2x + 1$, $x + 3$, and $x - 3$ and multiply the remaining factors.

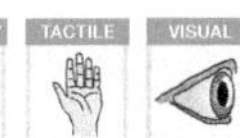

Learning Strategy

When simplifying rational expressions with many factors, write larger and more spread out than you usually write. This reduces clutter, making the common factors easier to see.

c. $\frac{15b^2 + b - 6}{6b^2 - 11b - 10} \div \frac{10b^2 + 4b - 6}{2b^2 - 13b + 20}$

Solution:

$$\frac{15b^2 + b - 6}{6b^2 - 11b - 10} \div \frac{10b^2 + 4b - 6}{2b^2 - 13b + 20}$$

$$= \frac{15b^2 + b - 6}{6b^2 - 11b - 10} \cdot \frac{2b^2 - 13b + 20}{10b^2 + 4b - 6}$$ Write an equivalent multiplication statement.

$$= \frac{(5b - 3)(3b + 2)}{(3b + 2)(2b - 5)} \cdot \frac{(2b - 5)(b - 4)}{2(5b - 3)(b + 1)}$$ Factor the numerators and denominators completely.

$$= \frac{b - 4}{2b + 2}$$ Divide out the common factors of $5b - 3$, $3b + 2$, and $2b - 5$ and multiply the remaining factors.

d. $\dfrac{2x^2 + 5x - 12}{8 - 4x} \div \dfrac{20 + 5x}{6x^3 - 12x^2}$

Solution: $\dfrac{2x^2 + 5x - 12}{8 - 4x} \div \dfrac{20 + 5x}{6x^3 - 12x^2}$

$= \dfrac{2x^2 + 5x - 12}{8 - 4x} \cdot \dfrac{6x^3 - 12x^2}{20 + 5x}$ Write an equivalent multiplication statement.

$= \dfrac{(2x - 3)(x + 4)}{4(2 - x)} \cdot \dfrac{6x^2(x - 2)}{5(4 + x)}$ Factor the numerators and denominators completely.

$= \dfrac{(2x - 3)(x + 4)}{2 \cdot 2 \cdot (2 - x)} \cdot \dfrac{2 \cdot 3x^2(-1)(2 - x)}{5(4 + x)}$ Continue factoring.

Note The binomials $x - 2$ and $2 - x$ are additive inverses, so we factored -1 from $x - 2$. Also, $x + 4 = 4 + x$.

$= \dfrac{2x - 3}{2} \cdot \dfrac{3x^2(-1)}{5}$ Divide out the common factors of 2, $x + 4$, and $2 - x$.

$= -\dfrac{6x^3 - 9x^2}{10}$ Multiply the remaining factors.

Your Turn 4 Find the quotient.

a. $\dfrac{8x^4}{2x^3 + 6x^2} \div \dfrac{6x}{4x + 12}$

b. $\dfrac{2x^2 + 9x + 9}{x^2 - 9} \div \dfrac{2x^2 - 3x - 9}{x^2 - 3x}$

c. $\dfrac{8a^2 - 6a - 9}{2a^2 - 13a + 15} \div \dfrac{12a^2 + a - 6}{3a^2 + 10a - 8}$

d. $\dfrac{8x^2 - 6x - 9}{24x - 4x^2} \div \dfrac{24 + 32x}{6x^3 - 36x^2}$

Objective 4 Evaluate rational functions.

We can now define a **rational function**.

Definition Rational function: A function expressed in terms of rational expressions.

For example, $f(x) = \dfrac{3x + 2}{x^2 - 9}$ is a rational function because $\dfrac{3x + 2}{x^2 - 9}$ is a rational expression. Rational functions are evaluated in the same manner as other functions.

Example 5 Given $f(x) = \dfrac{x^2 + 2x - 8}{x - 3}$, find $f(4)$.

Solution: $f(4) = \dfrac{4^2 + 2(4) - 8}{4 - 3}$ Replace x with 4.

$= \dfrac{16 + 8 - 8}{1}$ Simplify.

$= 16$

Your Turn 5 Given $f(x) = \dfrac{x + 6}{x^2 - 4}$, find the following.

a. $f(3)$ b. $f(-6)$ c. $f(2)$

Answers to Your Turn 4

a. $\dfrac{8x}{3}$ b. $\dfrac{x}{x - 3}$ c. $\dfrac{a + 4}{a - 5}$

d. $-\dfrac{6x^2 - 9x}{16}$

Answers to Your Turn 5

a. $\dfrac{9}{5}$ b. 0 c. undefined

Objective 5 Find the domain of a rational function.

Recall from Section 3.5 that the *domain* of a function is the set of all input values for which the function is defined. With a rational function, we must be careful to identify values that cause the function to be undefined and eliminate those values from its domain. Notice that the function $f(x) = \dfrac{x^2 + 2x - 8}{x - 3}$ is undefined if $x = 3$.

$$f(3) = \frac{3^2 + 2(3) - 8}{3 - 3} = \frac{9 + 6 - 8}{0},$$

which is undefined. Because any real number other than 3 can be used as an input value, the domain of $f(x) = \dfrac{x^2 + 2x - 8}{x - 3}$ is $\{x | x \neq 3\}$. Our example suggests the following procedure.

Procedure Finding the Domain of a Rational Function

To find the domain of a rational function:

1. Write an equation that sets the denominator equal to 0.
2. Solve the equation.
3. Exclude the value(s) found in step 2 from the function's domain.

Example 6 Find the domain of $f(x) = \dfrac{x - 6}{3x^3 + x^2 - 4x}$.

Solution:

$3x^3 + x^2 - 4x = 0$ Set the denominator equal to 0 and then solve.

$x(3x^2 + x - 4) = 0$ Solve by first factoring out the GCF, x.

$x(3x + 4)(x - 1) = 0$ Factor $3x^2 + x - 4$ by trial.

$x = 0$ or $3x + 4 = 0$ or $x - 1 = 0$ Use the zero-factor theorem.

$x = -\dfrac{4}{3}$ $\quad x = 1$ Solve each equation.

Note We do not consider the numerator because the rational expression is undefined or indeterminate only if the denominator is 0.

The function is undefined if x is replaced by 0, $-\frac{4}{3}$, or 1; so the domain is $\left\{x \middle| x \neq 0, -\frac{4}{3}, 1\right\}$.

Your Turn 6 Find the domain of each rational function.

a. $f(x) = \dfrac{2x + 3}{x^2 - 2x - 15}$

b. $f(x) = \dfrac{3x + 4}{2x^3 + 5x^2 + 2x}$

Objective 6 Graph rational functions.

Let's graph the function $f(x) = \dfrac{1}{x - 2}$ by making a table of ordered pairs. Note that 2 is not in its domain; so there will be no point on the graph whose x value is 2. We say that the graph is *discontinuous* at $x = 2$. To complete our table, we use x-values in the domain that are very close to 2.

x	-1	0	1	1.5	1.9	1.99	2.01	2.1	2.5	3	4	5
y	$-\frac{1}{3}$	$-\frac{1}{2}$	-1	-2	-10	-100	100	10	2	1	$\frac{1}{2}$	$\frac{1}{3}$

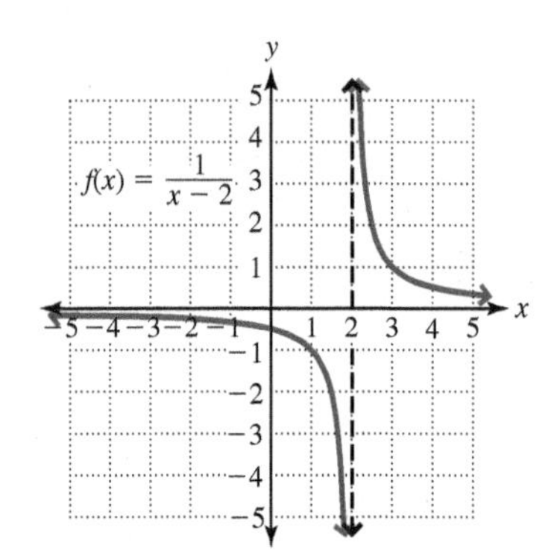

Using the table, we can draw the graph to the left. Note that we indicate the discontinuity at $x = 2$ with a dashed vertical line called a *vertical asymptote.* Also notice that when $x < 2$, the y-values become smaller as x gets closer to 2 and when $x > 2$, the y-values become greater as x gets closer to 2. Graphs of rational functions usually have this type of shape with vertical asymptotes through x-values that make the denominator equal to 0 but do not make the numerator equal to 0.

Answers to Your Turn 6

a. $\{x | x \neq -3, 5\}$

b. $\left\{x \middle| x \neq -\frac{1}{2}, 0, -2\right\}$

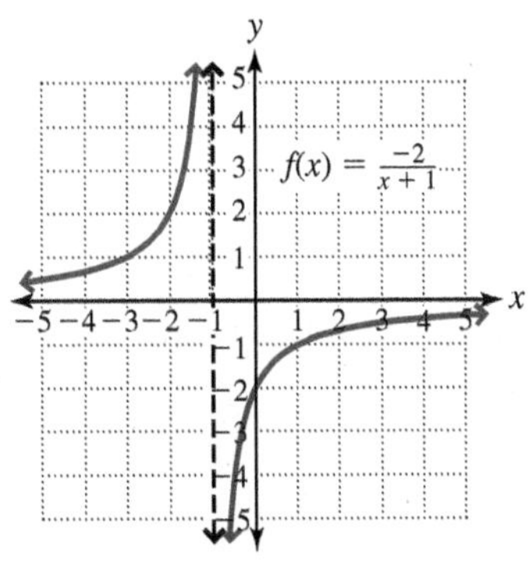

Example 7 Graph $f(x) = \dfrac{-2}{x+1}$.

Solution: Note that $x \neq -1$; so the graph has a vertical asymptote at $x = -1$. We find ordered pairs around that asymptote and then graph, as shown to the right.

x	-4	-3	-2	-1.5	-1.1	-1.01	-0.99	-0.9	-0.5	0	1	2
y	$\frac{2}{3}$	1	2	4	20	200	-200	-20	-4	-2	-1	$-\frac{2}{3}$

Your Turn 7 Graph $f(x) = \dfrac{1}{x-1}$.

Answer to Your Turn 7

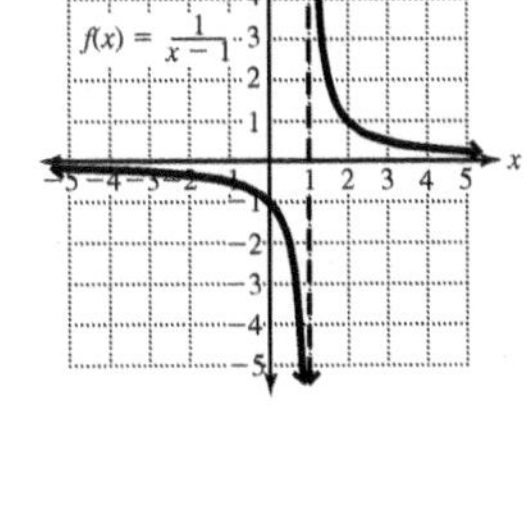

Connection We can check a simplification using a graphing utility by graphing $Y_1 =$ the original expression and $Y_2 =$ the simplified expression and verifying that the graphs are the same. In Example 1(c), we simplified $\dfrac{x^2 - 9}{x^2 - x - 12}$ to $\dfrac{x-3}{x-4}$. Graphing $Y_1 = \dfrac{x^2 - 9}{x^2 - x - 12}$ and $Y_2 = \dfrac{x-3}{x-4}$ in the window $[-10, 10]$ for x and $[-10, 10]$ for y on the TI-83 Plus or TI-84 results in the following graphs.

$Y_1 = \dfrac{x^2 - 9}{x^2 - x - 12}$

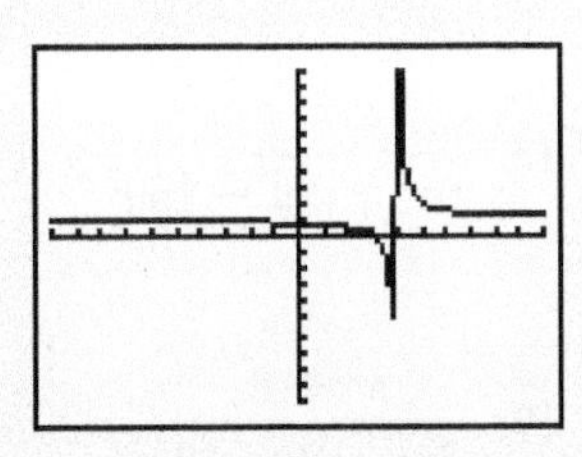

$Y_2 = \dfrac{x-3}{x-4}$

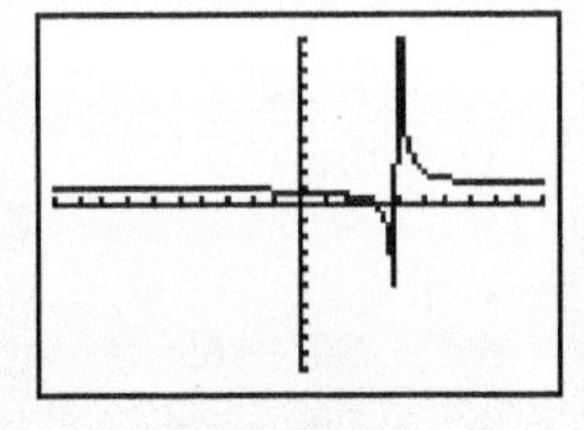

Note that the graphs appear to be the same. However, there is a subtle difference resulting from a difference in the domains of the functions. Factoring the denominator of Y_1, we have $Y_1 = \dfrac{x^2 - 9}{(x+3)(x-4)}$. We can see that its domain is $\{x \mid x \neq -3, 4\}$, whereas the domain of Y_2 is $\{x \mid x \neq 4\}$. The number -3 is excluded from the domain of Y_1 because $x + 3$ is a factor in its denominator. During simplification, $x + 3$ divides out; so -3 is in the domain of Y_2. This means that Y_1 has a tiny " hole" in its graph at $x = -3$, whereas Y_2 does not. The hole in Y_1 cannot be seen with our current window setting. To see the hole in Y_1, clear Y_2 so that only Y_1 is graphed, press ZOOM, select ZDecimal, and press ENTER. The graph of Y_1, now looks like this.

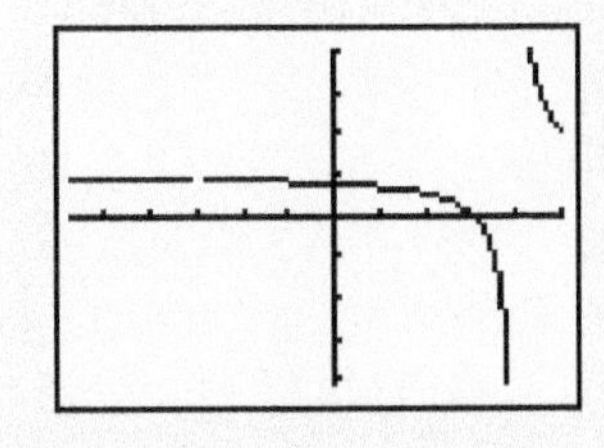

Notice the tiny blank space at $x = -3$. If we use the ZDecimal function on Y_2, we see no such hole because $x = -3$ is in the domain of Y_2. In general, holes occur with factors that divide out, whereas asymptotes occur with factors in a denominator that cannot be divided out.

Note Using ZDecimal results in the window
$Xmin = -4.7,\quad Xmax = 4.7,$
$Ymin = -3.1,\quad Ymax = 3.1.$

7.1 Exercises

For Extra Help MyMathLab®

Objective 1

Prep Exercise 1 A rational expression is an expression that can be written in the form $\frac{P}{Q}$, where P and Q are ______________ and $Q \neq 0$.

Prep Exercise 2 What does it mean for a rational expression to be in lowest terms?

Prep Exercise 3 What is the first step in simplifying a rational expression to lowest terms?

Prep Exercise 4 When simplifying $\frac{x^2 + 4x - 5}{x^2 + 2x - 3}$, why can you not divide out the x^2's?

Prep Exercise 5 What is the difference between terms and factors?

For Exercises 1–30, simplify each rational function. See Examples 1 and 2.

1. $\frac{-28m^3n^5}{16m^7n^6}$
2. $\frac{32c^2d^6}{-24c^4d^7}$
3. $\frac{-32x^4y^3z^2}{24x^2y^2z^2}$
4. $\frac{42w^5x^4y^3}{-28w^2x^3y^3}$
5. $\frac{2x + 12}{3x + 18}$
6. $\frac{3x + 6}{5x + 10}$
7. $\frac{8x^2 + 12x}{20x^2 + 30x}$
8. $\frac{24a^2 - 16a}{36a^2 - 24a}$
9. $\frac{-3x - 15}{4x + 20}$
10. $\frac{-5x - 25}{6x + 30}$
11. $\frac{a^2 + 3a - 10}{a^2 + 2a - 15}$
12. $\frac{c^2 + 3c - 18}{c^2 - c - 6}$
13. $\frac{x^2 - 6x}{x^2 - 8x + 12}$
14. $\frac{x^2 - 2x}{x^2 - 10x + 16}$
15. $\frac{4x^2 - 9y^2}{6x^2 - xy - 15y^2}$
16. $\frac{9x^2 - 25y^2}{6x^2 - 5xy - 25y^2}$
17. $\frac{4a^2 - 4ab - 3b^2}{6a^2 - ab - 2b^2}$
18. $\frac{6a^2 - 5ab - 6b^2}{3a^2 - 7ab - 6b^2}$
19. $\frac{x^3 - 8}{3x^2 - 2x - 8}$
20. $\frac{x^3 + 1}{3x^2 - 4x - 7}$
21. $\frac{x^3 + 27}{3x^2 - 27}$
22. $\frac{4x^3 + 4}{x^2 - 1}$
23. $\frac{xy - 4x + 3y - 12}{xy - 6x + 3y - 18}$
24. $\frac{ab + 5a - 3b - 15}{bc - 2b + 5c - 10}$
25. $\frac{6a^3 + 4a^2 + 9a + 6}{6a^2 + 13a + 6}$
26. $\frac{8a^3 + 6a^2 + 20a + 15}{8a^2 + 10a + 3}$
27. $\frac{2b - a}{a - 2b}$

28. $\dfrac{3m - n}{n - 3m}$

29. $\dfrac{x^2 + 4x - 21}{6 - 2x}$

30. $\dfrac{x^2 - 9x + 20}{20 - 5x}$

Objective 2

Prep Exercise 6 Explain how to multiply rational expressions.

For Exercises 31–56, find the product. See Example 3.

31. $\dfrac{8x^2y^4}{12x^3y^2} \cdot \dfrac{4x^3y^3}{6x^4y^3}$

32. $\dfrac{10a^4b^2}{18a^2b^3} \cdot \dfrac{12a^3b}{15a^2b^4}$

33. $-\dfrac{24a^2bc^3}{16ab^3c^2} \cdot \dfrac{32a^3b^2c^3}{18a^4b^3c^3}$

34. $-\dfrac{30x^3y^2z}{42x^2yz^2} \cdot \dfrac{15x^2y^2z^2}{25x^4yz}$

35. $\dfrac{3x^2y^4}{15x + 10y} \cdot \dfrac{24x + 16y}{12xy^2}$

36. $\dfrac{12p + 16q}{10r^2s} \cdot \dfrac{15r^3s^4}{15p + 20q}$

37. $\dfrac{8x + 12}{18x - 24} \cdot \dfrac{12 - 9x}{10x + 15}$

38. $\dfrac{24x + 16}{30x - 18} \cdot \dfrac{27 - 45x}{30x + 20}$

39. $\dfrac{4a - 16}{3a - 3} \cdot \dfrac{a^2 - 1}{a^2 - 16}$

40. $\dfrac{x^2 - 9}{6x + 18} \cdot \dfrac{5x - 25}{x^2 - 25}$

41. $\dfrac{4x^2y^3}{2x - 6} \cdot \dfrac{12 - 4x}{6x^3y}$

42. $\dfrac{6b^3c^3}{70 - 10b} \cdot \dfrac{5b - 35}{8b^2c^5}$

43. $\dfrac{x^2 - 4}{x^2 - 3x - 10} \cdot \dfrac{x^2 - 8x + 15}{x^2 - 9}$

44. $\dfrac{x^2 + 3x - 10}{x^2 - 16} \cdot \dfrac{x^2 - 9x + 20}{x^2 - 25}$

45. $\dfrac{6x^2 + x - 12}{2x^2 - 5x - 12} \cdot \dfrac{3x^2 - 14x + 8}{9x^2 - 18x + 8}$

46. $\dfrac{2x^2 + x - 15}{4x^2 + 11x - 3} \cdot \dfrac{4x^2 - 9x + 2}{2x^2 - 9x + 10}$

47. $\dfrac{2x^2 - 7x + 6}{8x^2 - 6x - 9} \cdot \dfrac{3x^2 - 13x - 10}{x^2 - 7x + 10}$

48. $\dfrac{8x^2 + 10x - 3}{3x^2 + 4x - 4} \cdot \dfrac{x^2 + 6x + 8}{2x^2 + 11x + 12}$

49. $\dfrac{2x^2 - 11x + 12}{2x^2 + 11x - 21} \cdot \dfrac{3x^2 + 20x - 7}{4 - x}$

50. $\dfrac{3x^2 - 17x + 10}{3x^2 + 7x - 6} \cdot \dfrac{2x^2 + x - 15}{5 - x}$

51. $\dfrac{ac + 3a + 2c + 6}{ad + a + 2d + 2} \cdot \dfrac{ad - 5a + 2d - 10}{bc + 3b - 4c - 12}$

52. $\dfrac{mn + 2m + 4n + 8}{np - 3n + 2p - 6} \cdot \dfrac{pq + 5p - 5q - 25}{mq + 4m + 4q + 16}$

53. $\dfrac{4x^3 - 3x^2 + 4x - 3}{4x^3 - 3x^2 - 8x + 6} \cdot \dfrac{2x^3 + x^2 - 4x - 2}{3x^3 + 2x^2 + 3x + 2}$

54. $\dfrac{3x^3 + 2x^2 + 12x + 8}{2x^3 + 5x^2 + 8x + 20} \cdot \dfrac{2x^3 - 5x^2 - 6x + 15}{3x^3 + 2x^2 - 9x - 6}$

55. $\dfrac{a^3 - b^3}{a^2 + ab + b^2} \cdot \dfrac{2a^2 + ab - b^2}{a^2 - b^2}$

56. $\dfrac{3x^2 - 5xy + 2y^2}{x^2 - y^2} \cdot \dfrac{x^3 + y^3}{x^2 - xy + y^2}$

Objective 3

Prep Exercise 7 What is the first step when dividing rational expressions?

For Exercises 57–78, find the quotient. See Example 4.

57. $\dfrac{8x^7g^3}{15x^2g^4} \div \dfrac{4xg^2}{3x^2g^5}$

58. $\dfrac{10c^5g^5}{12c^2g^4} \div \dfrac{5cg^2}{4c^2g^3}$

59. $\dfrac{4a^3 - 8a^2}{15a} \div \dfrac{3a^2 - 6a}{5a^4}$

60. $\dfrac{6b^2 + 24b}{7b^2} \div \dfrac{7b^3 + 28b^2}{14b^4}$

61. $\dfrac{10x^3 + 15x^2}{18x - 6} \div \dfrac{20x + 30}{9x^2 - 3x}$

62. $\dfrac{12y + 8}{4y - 10} \div \dfrac{48y + 32}{8y^3 - 20y^2}$

63. $\dfrac{3a - 4b}{6a - 9b} \div \dfrac{12b - 9a}{4a - 6b}$

64. $\dfrac{4x - 3y}{12x - 3y} \div \dfrac{6y - 8x}{8x - 2y}$

65. $\dfrac{d^2 - d - 12}{2d^2} \div \dfrac{3d^2 + 13d + 12}{d}$

66. $\dfrac{y^2 + y - 20}{7y^2} \div \dfrac{3y^2 + 19y + 20}{y}$

67. $\dfrac{m^2 - 25n^2}{2m - 12n} \div \dfrac{3m + 15n}{m^2 - 36n^2}$

68. $\dfrac{h^2 - 36j^2}{3h - 21j} \div \dfrac{4h + 24j}{h^2 - 49j^2}$

69. $\dfrac{3x + 24}{4x - 32} \div \dfrac{x^2 + 16x + 64}{x^2 - 16x + 64}$

70. $\dfrac{5w + 15}{3w - 9} \div \dfrac{w^2 + 6w + 9}{w^2 - 6w + 9}$

71. $\dfrac{2x^2 - 7xy + 6y^2}{2x^2 + 7xy + 6y^2} \div \dfrac{4x^2 - 9y^2}{4x^2 + 12xy + 9y^2}$

72. $\dfrac{9y^2 - 4z^2}{3y^2 + 7yz - 6z^2} \div \dfrac{3y^2 - 7yz - 6z^2}{y^2 + 6yz + 9z^2}$

73. $\dfrac{2x^2 - 7x - 4}{3x^2 - 14x + 8} \div \dfrac{2x^2 + 7x + 3}{3x^2 - 8x + 4}$

74. $\dfrac{2x^2 - 9x + 4}{3x^2 - 10x - 8} \div \dfrac{2x^2 + 5x - 3}{3x^2 - 13x - 10}$

75. $\dfrac{27a^3 - 8b^3}{9a^2 - 4b^2} \div \dfrac{2a^2 - 5ab - 12b^2}{6a^2 + 13ab + 6b^2}$

76. $\dfrac{8x^3 - 27y^3}{x^2 + 2xy - 15y^2} \div \dfrac{4x^2 - 9y^2}{2x^2 - 3xy - 9y^2}$

77. $\dfrac{ab + 3a + 2b + 6}{bc + 4b + 3c + 12} \div \dfrac{ac - 3a + 2c - 6}{bc + 4b - 4c - 16}$

78. $\dfrac{xy - 3x + 4y - 12}{xy + 6x - 3y - 18} \div \dfrac{xy + 5x + 4y + 20}{xy + 5x - 3y - 15}$

For Exercises 79–82, perform the indicated operations.

79. $\dfrac{x^2 - 3x - 18}{2x^2 + 13x + 20} \cdot \dfrac{3x^2 + 10x - 8}{4x^2 - 24x} \div \dfrac{3x^2 + 7x - 6}{6x^2 + 11x - 10}$

80. $\dfrac{4x^2 - 31x + 21}{12x^2 - 13x - 4} \cdot \dfrac{8x^2 + 14x + 3}{4x^2 - 19x + 12} \div \dfrac{2x^2 - 11x - 21}{9x^2 - 12x}$

81. $\dfrac{5x^2 - 17x - 12}{12x^2 - 29x + 15} \cdot \left(\dfrac{6x^2 + x - 12}{3x^2 + 14x - 24} \div \dfrac{2x^2 - 5x - 12}{3x^2 + 13x - 30}\right)$

82. $\dfrac{6x^3 - 14x^2}{12x^2 + 4x - 5} \cdot \left(\dfrac{4x^2 + 4x - 3}{9x^2 - 27x + 14} \div \dfrac{6x^2 + 9x}{18x^2 + 3x - 10}\right)$

Find the Mistake *For Exercises 83 and 84, explain the mistake; then work the problem correctly.*

83. $\dfrac{a^2 + b^2}{a^2 - b^2} \cdot \dfrac{a^2 + 2ab - 3b^2}{a^2 + ab - 6b^2} = \dfrac{(a + b)(a + b)}{(a + b)(a - b)} \cdot \dfrac{(a - b)(a + 3b)}{(a + 3b)(a - 2b)} =$
$\dfrac{\cancel{(a + b)}(a + b)}{\cancel{(a + b)}\cancel{(a - b)}} \cdot \dfrac{\cancel{(a - b)}\cancel{(a + 3b)}}{\cancel{(a + 3b)}(a - 2b)} = \dfrac{a + b}{a - 2b}$

84. $\dfrac{a^2 + b^2}{a^2 - b^2} \cdot \dfrac{a^2 + 2ab - 3b^2}{a^2 + ab - 6b^2} = \dfrac{\cancel{a^2} + \cancel{b^2}}{\cancel{a^2} - \cancel{b^2}} \cdot \dfrac{\cancel{a^2} + 2ab - \cancel{3b^2}}{\cancel{a^2} + ab - \cancel{6b^2}} = \dfrac{2ab}{ab - 2}$

Objective 4

Prep Exercise 8 If $f(x) = \dfrac{x^2 + 3x + 2}{x - 5}$, is $f(5)$ defined? Why or why not?

For Exercises 85–92, find the indicated value of the given function. See Example 5.

85. $f(x) = \dfrac{x + 3}{x - 4}$,
   a. $f(3)$
   b. $f(-3)$
   c. $f(4)$

86. $f(x) = \dfrac{x + 5}{x - 2}$,
   a. $f(1)$
   b. $f(-5)$
   c. $f(2)$

87. $f(x) = \dfrac{2x - 3}{3x + 2}$,
   a. $f\left(\dfrac{3}{2}\right)$
   b. $f(-1)$
   c. $f(2)$

88. $f(x) = \dfrac{3x + 5}{4x - 3}$,
   a. $f\left(-\dfrac{5}{3}\right)$
   b. $f(1)$
   c. $f(-2)$

89. $f(x) = \dfrac{3x - 4}{2x^2 - x - 6}$,
   a. $f(3)$
   b. $f\left(\dfrac{4}{3}\right)$
   c. $f(2)$

90. $f(x) = \dfrac{3x^2 + 11x - 4}{4x - 6}$,
   a. $f(-4)$
   b. $f\left(\dfrac{3}{2}\right)$
   c. $f(4)$

91. $f(x) = \dfrac{x^2 + x - 6}{x^2 + x - 12}$,
   a. $f(-3)$
   b. $f(5)$
   c. $f(3)$

92. $f(x) = \dfrac{x^2 - 4x - 5}{x^2 + 5x + 6}$,
   a. $f(5)$
   b. $f(2)$
   c. $f(-2)$

Objective 5

Prep Exercise 9 What numbers must be excluded from the domain of a rational function?

For Exercises 93–104, find the domain of each rational function. See Example 6.

93. $f(x) = \frac{x + 6}{3x + 12}$

94. $f(x) = \frac{2x - 5}{3x + 9}$

95. $f(x) = \frac{3x - 6}{x^2 - 25}$

96. $g(a) = \frac{5a + 15}{a^2 - 64}$

97. $g(c) = \frac{3c - 9}{4c^2 - 81}$

98. $h(n) = \frac{2n^2 - 3n + 6}{25n^2 - 49}$

99. $h(t) = \frac{2t^2 + 4t - 8}{6t^2 + 11t - 10}$

100. $f(a) = \frac{3a + 5}{8a^2 + 6a - 9}$

101. $g(x) = \frac{4x - 8}{x^3 + 4x^2 - 21x}$

102. $h(x) = \frac{3x^2 - 27}{x^3 - 2x^2 - 8x}$

103. $f(x) = \frac{x - 7}{x^3 - 8}$

104. $f(x) = \frac{2x^2 - 8}{x^3 + 64}$

Objective 6

Prep Exercise 10 What feature, represented by a dashed line, is used on the graph of a rational function for values that cause the function to be undefined?

For Exercises 105–108, match each function with its graph. See Objective 6.

105. $f(x) = \frac{1}{x - 3}$

106. $f(x) = \frac{-1}{x + 3}$

107. $f(x) = \frac{-2}{x + 2}$

108. $f(x) = \frac{3}{x - 4}$

a.

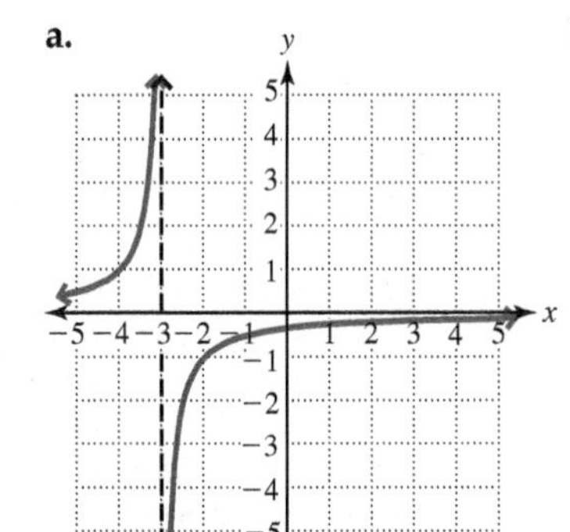

b.

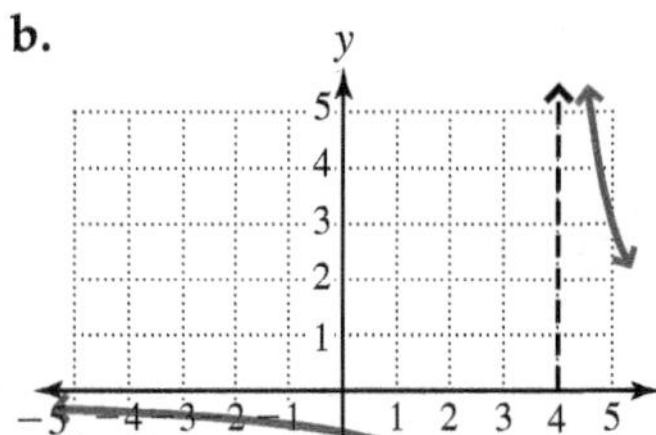

c.

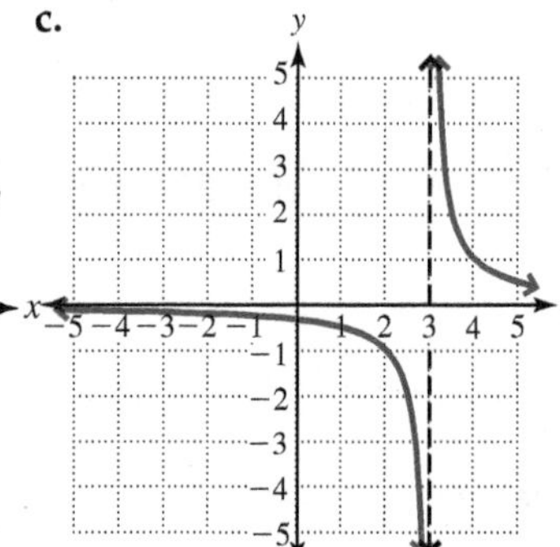

d.

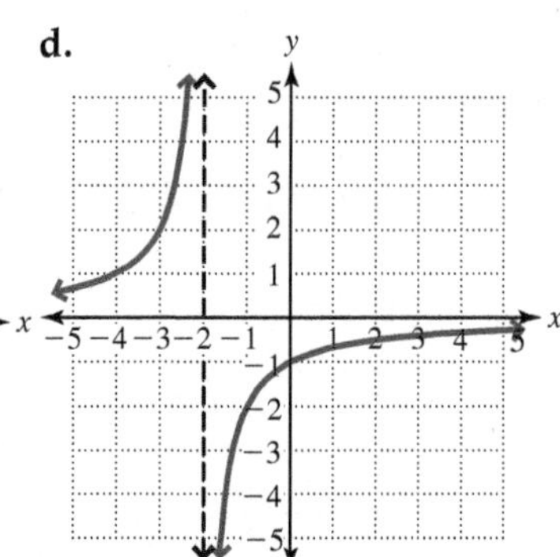

For Exercises 109–112, graph. See Example 7.

109. $f(x) = \frac{1}{x - 2}$

110. $f(x) = \frac{2}{x - 1}$

111. $f(x) = \frac{-2}{x + 1}$

112. $f(x) = \frac{-1}{x - 3}$

For Exercises 113–120, solve.

113. The electrical resistance R in ohms of a wire is given by $R = \frac{5}{d^2}$, where d is the diameter of the wire in millimeters. Find the electrical resistance of a wire whose diameter is 2 millimeters.

114. The weight of an object depends on its distance from the center of the Earth. If a person weighs 180 pounds on the surface of the Earth, this person's weight is given by $w = \frac{2{,}880{,}000{,}000}{d^2}$, where w is the weight in pounds and d is the distance from the center of the Earth in miles. Find the weight of this person 1000 miles above the surface of Earth if the radius of Earth is 4000 miles.

115. If the start-up cost of a small manufacturing business is \$1800 per day and the manufacturing costs are \$15 per unit, the average cost per unit is $C = \frac{15x + 1800}{x}$, where x is the number of units. Find the average cost per unit to produce 200 units.

116. An automobile's velocity starting from rest is $v = \frac{90t}{2t + 18}$, where v is in feet per second and t is in seconds. Find the velocity after 6 seconds.

117. The ordering and transportation cost for the components used in manufacturing a product is given by

$$C = 200\left(\frac{300}{x^2} + \frac{x}{x + 50}\right),$$

where C is the cost in thousands of dollars and x is the size of the order in hundreds. Find the ordering and transportation cost for an order of 5000 components.

118. If 500 bacteria are introduced into a culture, the equation $N = 500\left(1 + \frac{4t}{50 + t^2}\right)$ gives the number, N, of bacteria after t hours. Find the number of bacteria after 5 hours.

119. If a fire truck approaches a stationary observer at a velocity of v miles per hour, the frequency, F, in cycles per second of the truck siren is given by

$$F = \frac{130{,}000}{330 - v}.$$

Find the frequency if the truck is approaching at 70 miles per hour.

120. The concentration (in milliliters per liter) of a medication in the bloodstream t hours after injection into muscle tissue is given by

$$C = \frac{3t}{27 + t^3}, \quad \text{where} \quad t \geq 0.$$

Find the concentration 3 hours after the injection.

Of Interest

When an object in motion produces sound, the sound waves in front of the object are compressed by its motion so that they have a higher frequency (pitch). The waves that trail the object are stretched so that they have a lower frequency. You experience this phenomenon, called the Doppler effect, when a train or car passes by blowing its horn and you hear the pitch change from high to low.

Review Exercises

Exercises 1–6 Expressions

For Exercises 1–4, add or subtract.

[1.2] 1. $\frac{3}{11} + \frac{4}{11}$

[1.2] 2. $\frac{3}{5} - \frac{5}{6}$

[1.4] 3. $\frac{5}{12}y + \frac{7}{30}y$

[1.4] 4. $\frac{11}{15}x + 4 - \frac{2}{15}x - 3$

For Exercises 5 and 6, factor.

[6.1] 5. $6x^2 - 15x$

[6.2] 6. $x^2 + x - 20$

7.2 Adding and Subtracting Rational Expressions

Objectives

1. Add or subtract rational expressions with the same denominator.
2. Find the least common denominator (LCD).
3. Write equivalent rational expressions with the LCD as the denominators.
4. Add or subtract rational expressions with unlike denominators.

Warm-up

For Exercises 1 and 2, simplify.

[5.2] **1.** $(p^2 - 1) - (4p + 4) + (2p^2 + 2p)$

[5.3] **2.** $(2x - 3)(x - 5)$

[6.2] **3.** Factor. $x^2 - 4x - 21$

[7.1] **4.** Simplify to lowest terms. $\frac{x^2 - 9x + 20}{20 - 5x}$

Objective 1 Add or subtract rational expressions with the same denominator.

Recall that when adding or subtracting fractions with the same denominator, we add or subtract the numerators and keep the same denominator. We add and subtract rational expressions the same way.

$$\frac{4a}{13} + \frac{6a}{13} = \frac{4a + 6a}{13} = \frac{10a}{13} \quad \text{or} \quad \frac{b}{2b - 3} - \frac{4}{2b - 3} = \frac{b - 4}{2b - 3}$$

Procedure Adding or Subtracting Rational Expressions with the Same Denominator

To add or subtract rational expressions with the same denominator:

1. Add or subtract the numerators and keep the same denominator.
2. Simplify to lowest terms.

Example 1 Add or subtract.

a. $\frac{3a}{a - 3} - \frac{9}{a - 3}$

Solution:

$$\frac{3a}{a - 3} - \frac{9}{a - 3} = \frac{3a - 9}{a - 3}$$ Subtract the numerators and keep the same denominator.

$$= \frac{3(a - 3)}{a - 3}$$ Factor the numerator.

$$= 3$$ Divide out the common factor, $a - 3$.

b. $\frac{x^2}{x^2 + 3x - 18} + \frac{5x - 6}{x^2 + 3x - 18}$

Solution:

$$\frac{x^2}{x^2 + 3x - 18} + \frac{5x - 6}{x^2 + 3x - 18} = \frac{x^2 + (5x - 6)}{x^2 + 3x - 18}$$ Add the numerators and keep the same denominator.

$$= \frac{x^2 + 5x - 6}{x^2 + 3x - 18}$$ Remove the parentheses.

$$= \frac{(x + 6)(x - 1)}{(x + 6)(x - 3)}$$ Factor the numerator and denominator.

$$= \frac{x - 1}{x - 3}$$ Divide out the common factor, $x + 6$.

Answers to Warm-up

1. $3p^2 - 2p - 5$
2. $2x^2 - 13x + 15$
3. $(x - 7)(x + 3)$
4. $-\frac{x - 5}{5}$ or $\frac{5 - x}{5}$

c. $\dfrac{n^2 + 3n}{n^2 + 4n + 4} - \dfrac{5n + 8}{n^2 + 4n + 4}$

Solution:

$$\frac{n^2 + 3n}{n^2 + 4n + 4} - \frac{5n + 8}{n^2 + 4n + 4} = \frac{(n^2 + 3n) - (5n + 8)}{n^2 + 4n + 4}$$ Subtract the numerators and keep the same denominator.

$$= \frac{(n^2 + 3n) + (-5n - 8)}{n^2 + 4n + 4}$$ Write the equivalent addition.

$$= \frac{n^2 - 2n - 8}{n^2 + 4n + 4}$$ Combine like terms.

$$= \frac{(n - 4)(n + 2)}{(n + 2)(n + 2)}$$ Factor the numerator and denominator.

$$= \frac{n - 4}{n + 2}$$ Divide out the common factor, $n + 2$.

d. $\dfrac{p^2 - 1}{2p^2 - 2} - \dfrac{4p + 4}{2p^2 - 2} + \dfrac{2p^2 + 2p}{2p^2 - 2}$

Solution: $\dfrac{p^2 - 1}{2p^2 - 2} - \dfrac{4p + 4}{2p^2 - 2} + \dfrac{2p^2 + 2p}{2p^2 - 2}$

$$= \frac{(p^2 - 1) - (4p + 4) + (2p^2 + 2p)}{2p^2 - 2}$$ Subtract and add the numerators and keep the same denominator.

$$= \frac{(p^2 - 1) + (-4p - 4) + (2p^2 + 2p)}{2p^2 - 2}$$ Write the equivalent addition.

$$= \frac{3p^2 - 2p - 5}{2p^2 - 2}$$ Combine like terms.

$$= \frac{(3p - 5)(p + 1)}{2(p - 1)(p + 1)}$$ Factor the numerator and denominator.

$$= \frac{3p - 5}{2(p - 1)}$$ Divide out the common factor, $p + 1$.

Your Turn 1 Add or subtract.

a. $\dfrac{4a}{a - 4} - \dfrac{16}{a - 4}$ b. $\dfrac{2x^2 + 3x}{x^2 - 9} + \dfrac{4x + 3}{x^2 - 9}$ c. $\dfrac{2x^2 + 4x - 3}{x^2 - x - 6} - \dfrac{x^2 + 6x + 5}{x^2 - x - 6}$

d. $\dfrac{3m^2 - 8}{m^2 - 5m + 6} + \dfrac{m^2 + 4m - 4}{m^2 - 5m + 6} - \dfrac{2m^2 - 2m + 8}{m^2 - 5m + 6}$

Objective 2 Find the least common denominator (LCD).

To add fractions with unlike denominators, we have to rewrite each fraction as an equivalent fraction with the *least common denominator (LCD)* as its denominator. The least common denominator is the smallest number divisible by all of the denominators. Thus, the LCD contains all of the denominators as factors. Likewise, the first step in adding rational expressions with unlike denominators is to find the least common denominator, which is the polynomial of least degree that has all of the denominators as factors.

Compare finding the LCD of rational numbers with finding the LCD of rational expressions.

$$\frac{5}{48} \text{ and } \frac{7}{27} \qquad \frac{3}{8x^2} \text{ and } \frac{7}{12x^3}$$

Answers to Your Turn 1

a. 4 b. $\dfrac{2x + 1}{x - 3}$

c. $\dfrac{x - 4}{x - 3}$ d. $\dfrac{2m + 10}{m - 3}$

Because the LCD contains all denominators as factors, factor each denominator into prime factors.

$$48 = 2^4 \cdot 3 \qquad 27 = 3^3 \qquad\qquad 8x^2 = 2^3 \cdot x^2 \qquad 12x^3 = 2^2 \cdot 3 \cdot x^3$$

$$\text{LCD} = 2^4 \cdot 3^3 = 432 \qquad\qquad \text{LCD} = 2^3 \cdot 3 \cdot x^3 = 24x^3$$

This suggests the following procedure.

Procedure Finding the LCD

To find the LCD of two or more rational expressions:

1. Find the prime factorization of each denominator.
2. Write a product that contains each unique prime factor the greatest number of times that factor occurs in any factorization. Or, if you prefer to use exponents, write the product that contains each unique prime factor raised to the greatest exponent that occurs on that factor in any factorization.
3. Simplify the product found in step 2.

Example 2 Find the LCD.

a. $\dfrac{5}{6x^3y}$ and $\dfrac{7}{8x^2y^2}$

Solution: First, factor the denominators $6x^3y$ and $8x^2y^2$ by writing their prime factorizations.

$$6x^3y = 2 \cdot 3 \cdot x^3 \cdot y$$
$$8x^2y^2 = 2^3 \cdot x^2 \cdot y^2$$

The unique factors are 2, 3, x, and y. The greatest power of 2 is 3, the greatest power of 3 is 1, the greatest power of x is 3, and the greatest power of y is 2.

$$\text{LCD} = 2^3 \cdot 3 \cdot x^3 \cdot y^2 = 24x^3y^2$$

Warning The LCD is made up of denominators and/or factors of denominators, not terms. A common error is to say that the LCD of $\dfrac{3}{y} + \dfrac{6}{y-2}$ is $y - 2$ instead of $y(y - 2)$.

b. $\dfrac{y}{y-6}$ and $\dfrac{y+3}{y^2+5y}$

Solution: Factor the denominators.

$$y - 6 \text{ is prime} \quad \text{and} \quad y^2 + 5y = y(y + 5)$$

The unique factors are $y - 6$, y, and $y + 5$, and the highest power of each is 1.

$$\text{LCD} = y(y - 6)(y + 5)$$

c. $\dfrac{x+3}{x^2+2x-15}$ and $\dfrac{x^2-4}{x^2-5x+6}$

Solution: Factor the denominators.

$$x^2 + 2x - 15 = (x - 3)(x + 5)$$
$$x^2 - 5x + 6 = (x - 3)(x - 2)$$

The unique factors are $x - 3$, $x + 5$, and $x - 2$, and the highest power of each is 1.

$$\text{LCD} = (x - 3)(x + 5)(x - 2)$$

d. $\dfrac{a+3}{3a^4+15a^3+12a^2}$ and $\dfrac{a^2+3a+2}{2a^2+4a+2}$

Solution: Factor the denominators.

$$3a^4 + 15a^3 + 12a^2 = 3a^2(a + 1)(a + 4)$$
$$2a^2 + 4a + 2 = 2(a + 1)^2$$

The unique factors are 2, 3, a, $a + 1$, and $a + 4$. The highest power of 2, 3, and $a + 4$ is 1, and the highest power of a and $a + 1$ is 2.

$$\text{LCD} = 2 \cdot 3 \cdot a^2 \cdot (a + 1)^2 \cdot (a + 4) = 6a^2(a + 1)^2(a + 4)$$

Your Turn 2 Find the LCD.

a. $\dfrac{7}{15a^2b^4}$ and $\dfrac{11}{12a^3b^4}$

b. $\dfrac{b}{b + 3}$ and $\dfrac{b + 2}{b^2 + 3b}$

c. $\dfrac{y - 6}{y^2 - 7y + 10}$ and $\dfrac{2y - 3}{y^2 + 3y - 10}$

d. $\dfrac{x - 5}{4x^2 + 16x + 16}$ and $\dfrac{3x + 2}{3x^2 - 9x - 30}$

Objective 3 Write equivalent rational expressions with the LCD as the denominators.

Fractions must be written with a common denominator before they can be added or subtracted. The choice of denominators is the LCD.

Note To determine what to multiply a fraction by to convert it to an equivalent fraction with the LCD as its denominator, divide the denominator into the LCD.

Example 3 Write the fractions with the LCD as the denominators.

a. $\dfrac{5}{6x^3y}$ and $\dfrac{7}{8x^2y^2}$

Solution: In Example 2(a), we found that the LCD is $24x^3y^2$. To convert $\dfrac{5}{6x^3y}$ to a fraction whose denominator is $24x^3y^2$, multiply $\dfrac{5}{6x^3y}$ by 1 in the form of $\dfrac{4y}{4y}$, and to convert $\dfrac{7}{8x^2y^2}$ to a fraction whose denominator is $24x^3y^2$, multiply $\dfrac{7}{8x^2y^2}$ by 1 in the form of $\dfrac{3x}{3x}$.

$$\frac{5}{6x^3y} = \frac{5}{6x^3y} \cdot \frac{4y}{4y} = \frac{20y}{24x^3y^2} \quad \text{Multiply } \frac{5}{6x^3y} \text{ by } \frac{4y}{4y}.$$

$$\frac{7}{8x^2y^2} = \frac{7}{8x^2y^2} \cdot \frac{3x}{3x} = \frac{21x}{24x^3y^2} \quad \text{Multiply } \frac{7}{8x^2y^2} \text{ by } \frac{3x}{3x}.$$

b. $\dfrac{x + 3}{x^2 + 2x - 15}$ and $\dfrac{x^2 - 4}{x^2 - 5x + 6}$

Solution: In Example 2(c), we found that the LCD is $(x - 3)(x + 5)(x - 2)$. Factor the denominators. To convert $\dfrac{x + 3}{(x - 3)(x + 5)}$ to a fraction whose denominator is the LCD, multiply by 1 in the form of $\dfrac{x - 2}{x - 2}$, and to convert $\dfrac{x^2 - 4}{(x - 3)(x - 2)}$ to a fraction whose denominator is the LCD, multiply by 1 in the form of $\dfrac{x + 5}{x + 5}$.

$$\frac{x + 3}{x^2 + 2x - 15} = \frac{x + 3}{(x - 3)(x + 5)} \cdot \frac{x - 2}{x - 2} = \frac{x^2 + x - 6}{(x - 3)(x + 5)(x - 2)} \quad \text{Multiply by } \frac{x - 2}{x - 2}.$$

$$\frac{x^2 - 4}{x^2 - 5x + 6} = \frac{x^2 - 4}{(x - 3)(x - 2)} \cdot \frac{x + 5}{x + 5} = \frac{x^3 + 5x^2 - 4x - 20}{(x - 3)(x - 2)(x + 5)} \quad \text{Multiply by } \frac{x + 5}{x + 5}.$$

Your Turn 3 Write the fractions with the LCD as the denominators. (See Your Turn 2.)

a. $\dfrac{7}{15a^2b^4}$ and $\dfrac{11}{12a^3b^4}$

b. $\dfrac{y - 6}{y^2 - 7y + 10}$ and $\dfrac{2y - 3}{y^2 + 3y - 10}$

Answers to Your Turn 2

a. $60a^3b^4$
b. $b(b + 3)$
c. $(y - 2)(y - 5)(y + 5)$
d. $12(x + 2)^2(x - 5)$

Answers to Your Turn 3

a. $\dfrac{28a}{60a^3b^4}, \dfrac{55}{60a^3b^4}$

b. $\dfrac{y^2 - y - 30}{(y - 2)(y - 5)(y + 5)}, \dfrac{2y^2 - 13y + 15}{(y - 2)(y + 5)(y - 5)}$

Objective 4 Add or subtract rational expressions with unlike denominators.

To add or subtract rational expressions with unlike denominators, we use the same process that we use to add or subtract fractions. Compare the following:

$$\frac{2}{3}+\frac{3}{4} \qquad \frac{2}{a+b}+\frac{3}{a-b}$$

$$=\frac{2}{3}\cdot\frac{4}{4}+\frac{3}{4}\cdot\frac{3}{3} \qquad =\frac{2}{a+b}\cdot\frac{a-b}{a-b}+\frac{3}{a-b}\cdot\frac{a+b}{a+b}$$ Rewrite each with the LCD.

$$=\frac{8}{12}+\frac{9}{12} \qquad =\frac{2a-2b}{(a+b)(a-b)}+\frac{3a+3b}{(a+b)(a-b)}$$ Simplify.

$$=\frac{17}{12} \qquad =\frac{5a+b}{(a+b)(a-b)}$$ Add the fractions.

These examples suggest the following procedure.

Procedure Adding or Subtracting Rational Expressions with Different Denominators

To add or subtract rational expressions with different denominators:

1. Find the LCD.
2. Write each rational expression as an equivalent expression with the LCD.
3. Add or subtract the numerators and keep the LCD as the denominator.
4. Simplify.

Example 4 Add or subtract.

a. $\frac{a-3}{3a^2}-\frac{2a-3}{2a}$

Solution: The LCD is $6a^2$, so multiply the numerator and denominator of $\frac{a-3}{3a^2}$ by 2 and the numerator and denominator of $\frac{2a-3}{2a}$ by $3a$.

Note When multiplying by $\frac{2}{2}$ and $\frac{3a}{3a}(a \neq 0)$, we are multiplying by 1, which results in equivalent fractions.

$$\frac{a-3}{3a^2}-\frac{2a-3}{2a}=\frac{(a-3)(2)}{3a^2(2)}-\frac{(2a-3)(3a)}{(2a)(3a)}$$ Write each rational expression with the LCD, $6a^2$.

$$=\frac{2a-6}{6a^2}-\frac{6a^2-9a}{6a^2}$$ Distribute.

$$=\frac{(2a-6)-(6a^2-9a)}{6a^2}$$ Subtract numerators.

$$=\frac{(2a-6)+(-6a^2+9a)}{6a^2}$$ Write the equivalent addition.

$$=\frac{-6a^2+11a-6}{6a^2}$$ Simplify the numerator.

Note You cannot divide out the 6's because they are not factors.

b. $\frac{5}{b-2}+\frac{3}{b+4}$

Solution: Both denominators are prime, so the LCD is $(b-2)(b+4)$.

$$\frac{5}{b-2}+\frac{3}{b+4}=\frac{5(b+4)}{(b-2)(b+4)}+\frac{3(b-2)}{(b+4)(b-2)}$$ Multiply $\frac{5}{b-2}$ by $\frac{b+4}{b+4}$ and $\frac{3}{b+4}$ by $\frac{b-2}{b-2}$.

Note The denominators are not usually multiplied out in case the rational expression can be simplified to lowest terms.

$$= \frac{5b + 20}{(b + 4)(b - 2)} + \frac{3b - 6}{(b + 4)(b - 2)} \quad \text{Distribute.}$$

$$= \frac{(5b + 20) + (3b - 6)}{(b + 4)(b - 2)} \quad \text{Add numerators.}$$

$$= \frac{8b + 14}{(b + 4)(b - 2)} \quad \text{Simplify the numerator.}$$

c. $\dfrac{2x - 3}{3x^2 + 14x + 8} + \dfrac{3x - 1}{3x^2 - 13x - 10}$

Solution: First, find the LCD by factoring the denominators.

$$3x^2 + 14x + 8 = (3x + 2)(x + 4) \quad \text{and} \quad 3x^2 - 13x - 10 = (3x + 2)(x - 5)$$

$$\text{LCD} = (3x + 2)(x + 4)(x - 5)$$

$$\frac{2x - 3}{3x^2 + 14x + 8} + \frac{3x - 1}{3x^2 - 13x - 10}$$

$$= \frac{2x - 3}{(3x + 2)(x + 4)} + \frac{3x - 1}{(3x + 2)(x - 5)} \quad \text{Factor the denominators.}$$

$$= \frac{(2x - 3)(x - 5)}{(3x + 2)(x + 4)(x - 5)} + \frac{(3x - 1)(x + 4)}{(3x + 2)(x - 5)(x + 4)} \quad \text{Write equivalent rational expressions with the LCD, } (3x + 2)(x + 4)(x - 5).$$

Note Again, when multiplying by $\frac{x - 5}{x - 5}$ and $\frac{x + 4}{x + 4}$, we are multiplying by 1 for $x \neq 5$ and $x \neq -4$.

$$= \frac{2x^2 - 13x + 15}{(3x + 2)(x + 4)(x - 5)} + \frac{3x^2 + 11x - 4}{(3x + 2)(x + 4)(x - 5)} \quad \text{Multiply in numerators.}$$

$$= \frac{(2x^2 - 13x + 15) + (3x^2 + 11x - 4)}{(3x + 2)(x + 4)(x - 5)} \quad \text{Add numerators.}$$

$$= \frac{5x^2 - 2x + 11}{(3x + 2)(x + 4)(x - 5)} \quad \text{Simplify the numerator.}$$

d. $\dfrac{x}{x + 3} - \dfrac{21}{x^2 - x - 12}$

Solution: First, find the LCD by factoring $x + 3$ and $x^2 - x - 12$.

$$x + 3 \text{ is prime} \quad \text{and} \quad x^2 - x - 12 = (x - 4)(x + 3)$$

$$\text{LCD} = (x + 3)(x - 4)$$

$$\frac{x}{x + 3} - \frac{21}{x^2 - x - 12} = \frac{x}{x + 3} - \frac{21}{(x - 4)(x + 3)} \quad \text{Factor the denominator.}$$

$$= \frac{x(x - 4)}{(x + 3)(x - 4)} - \frac{21}{(x - 4)(x + 3)} \quad \text{Write equivalent rational expressions with the LCD, } (x - 4)(x + 3)$$

$$= \frac{x^2 - 4x}{(x + 3)(x - 4)} - \frac{21}{(x - 4)(x + 3)} \quad \text{Distribute.}$$

$$= \frac{x^2 - 4x - 21}{(x + 3)(x - 4)} \quad \text{Subtract numerators.}$$

$$= \frac{(x + 3)(x - 7)}{(x + 3)(x - 4)} \quad \text{Factor the numerator.}$$

$$= \frac{x - 7}{x - 4} \quad \text{Divide out the common factor, } x + 3.$$

e. $\dfrac{a+b}{a-b} - \dfrac{2a-3b}{b-a}$

Solution: The expressions $a - b$ and $b - a$ are additive inverses, so we can obtain the LCD by multiplying the numerator and denominator of one rational expression by -1.

$$\frac{a+b}{a-b} - \frac{2a-3b}{b-a} = \frac{a+b}{a-b} - \frac{(2a-3b)(-1)}{(b-a)(-1)}$$ To get the LCD, $a - b$, multiply $\dfrac{2a-3b}{b-a}$ by $\dfrac{-1}{-1}$.

$$= \frac{a+b}{a-b} - \frac{-2a+3b}{a-b}$$ Distribute.

$$= \frac{(a+b)-(-2a+3b)}{a-b}$$ Subtract numerators.

$$= \frac{(a+b)+(2a-3b)}{a-b}$$ Write the equivalent addition.

$$= \frac{3a-2b}{a-b}$$ Simplify the numerator.

Your Turn 4 Add or subtract.

a. $\dfrac{3x+7}{8xy^2} - \dfrac{3x-4}{6x^2y}$

b. $\dfrac{x+6}{4x+16} - \dfrac{x+7}{6x+24}$

c. $\dfrac{x+2}{2x^2+5x-3} + \dfrac{2x-3}{2x^2-5x+2}$

d. $\dfrac{x}{x+5} - \dfrac{35}{x^2+3x-10}$

e. $\dfrac{3x+2y}{x-y} - \dfrac{x-5y}{y-x}$

Answers to Your Turn 4

a. $\dfrac{9x^2+21x-12xy+16y}{24x^2y^2}$

b. $\dfrac{1}{12}$

c. $\dfrac{3x^2+3x-13}{(2x-1)(x+3)(x-2)}$

d. $\dfrac{x-7}{x-2}$

e. $\dfrac{4x-3y}{x-y}$ or $\dfrac{3y-4x}{y-x}$

7.2 Exercises

For Extra Help MyMathLab®

Note: Exercises marked with a ★ represent challenging exercises.

Objective 1

Prep Exercise 1 Explain how to add or subtract rational expressions that have the same denominator.

For Exercises 1–16, add or subtract. Simplify your answers to lowest terms. See Example 1.

1. $\dfrac{3c}{2d} + \dfrac{5c}{2d}$

2. $\dfrac{5y}{4x} + \dfrac{3y}{4x}$

3. $\dfrac{4a+3b}{2a-5b} + \dfrac{3a-4b}{2a-5b}$

4. $\dfrac{4p-3q}{3p-2q} + \dfrac{2p-5q}{3p-2q}$

5. $\dfrac{2x-3y}{x+2y} - \dfrac{4x+2y}{x+2y}$

6. $\dfrac{3c+2d}{2c+d} - \dfrac{5c-3d}{2c+d}$

7. $\dfrac{c^2}{c^2 - 5c + 4} - \dfrac{-2c + 24}{c^2 - 5c + 4}$

8. $\dfrac{a^2}{a^2 + 3a - 10} - \dfrac{-a + 6}{a^2 + 3a - 10}$

9. $\dfrac{y^2 - 2y}{y^2 - 12y + 36} + \dfrac{-5y + 6}{y^2 - 12y + 36}$

10. $\dfrac{r^2 - 3r}{r^2 - 6r + 9} + \dfrac{5r - 15}{r^2 - 6r + 9}$

11. $\dfrac{h^2 - 4h}{h^2 + 10h + 21} - \dfrac{-6h + 3}{h^2 + 10h + 21}$

12. $\dfrac{w^2 + 6}{w^2 + 7w + 10} - \dfrac{2w + 14}{w^2 + 7w + 10}$

13. $\dfrac{2x^2 + x - 3}{x^2 + 6x + 5} + \dfrac{x^2 - 2x + 4}{x^2 + 6x + 5} - \dfrac{2x^2 + x + 4}{x^2 + 6x + 5}$

14. $\dfrac{3x^2 - 2x + 3}{x^2 + x - 12} - \dfrac{x^2 + x - 2}{x^2 + x - 12} - \dfrac{x^2 + 4x - 7}{x^2 + x - 12}$

15. $\dfrac{x^3}{x^2 - xy + y^2} + \dfrac{y^3}{x^2 - xy + y^2}$

16. $\dfrac{a^3}{a^2 - 2a + 4} + \dfrac{8}{a^2 - 2a + 4}$

Objectives 2 and 3

Prep Exercise 2 Explain how to find the LCD of two different denominators.

Prep Exercise 3 Below are the prime factorizations of two denominators, $12x^2y$ and $10x^3$.

$$12x^2y = 2^2 \cdot 3 \cdot x^2 \cdot y$$
$$10x^3 = 2 \cdot 5 \cdot x^3$$

a. List the prime factors that are included in the LCD.

b. Write the prime factorization of the LCD.

Prep Exercise 4 Below are the prime factorizations of two denominators, $x^2 - x - 12$ and $5x^2 - 20x$.

$$x^2 - x - 12 = (x + 3)(x - 4)$$
$$5x^2 - 20x = 5x(x - 4)$$

a. List the prime factors that are included in the LCD.

b. Write the prime factorization of the LCD.

Prep Exercise 5 Suppose the LCD of two rational expressions is $(2x + 3)(x - 4)(3x - 2)$. If the denominator of one of the rational expressions is $6x^2 + 5x - 6$, by what factor would you multiply its numerator and denominator to obtain an equivalent rational expression with the LCD? Explain how you found this factor.

Prep Exercise 6 Suppose the LCD of two rational expressions with unlike denominators is $x - 6$. If the denominator of one of the rational expressions is $6 - x$, by what factor do you multiply its numerator and denominator to obtain an equivalent rational expression with the LCD?

For Exercises 17–32, find the LCD for each pair. Write each fraction pair as equivalent fractions with the LCD as their denominators. See Example 2 and 3.

17. $\dfrac{4}{15a^6b^8}, \dfrac{9}{20a^4b^5}$

18. $\dfrac{11}{12s^6v^3}, \dfrac{13}{18s^4v^5}$

19. $\dfrac{x}{x + 7}, \dfrac{x}{x - 7}$

20. $\dfrac{y}{y - 3}, \dfrac{y}{y - 4}$

21. $\dfrac{7r}{9r+18}, \dfrac{8r}{15r+30}$

22. $\dfrac{11x}{8x-32}, \dfrac{17x}{12x-48}$

23. $\dfrac{3b}{b^2-1}, \dfrac{b}{b^2+3b-4}$

24. $\dfrac{4c}{c^2-9}, \dfrac{2c}{c^2+c-6}$

25. $\dfrac{c-5}{c^2-2c-3}, \dfrac{3+c}{c^2-5c+6}$

26. $\dfrac{w+6}{w^2+6w+5}, \dfrac{2-w}{w^2+w-20}$

27. $\dfrac{n+1}{n^2+8n+16}, \dfrac{n-4}{n^2+5n+4}$

28. $\dfrac{n-3}{n^2-10n+25}, \dfrac{n-1}{n^2-2n-15}$

29. $\dfrac{x+5}{x^3+x^2-6x}, \dfrac{x-3}{x^4+7x^3+12x^2}$

30. $\dfrac{y-2}{y^5-16y^3}, \dfrac{y-1}{y^3+5y^2+4y}$

31. $\dfrac{3}{x}, \dfrac{4}{x^2+4x}, \dfrac{6}{x+4}$

32. $\dfrac{5}{a}, \dfrac{8}{a-5}, \dfrac{-3}{a^2-5a}$

Objective 4

Prep Exercise 7 Suppose you are subtracting two rational expressions. You have found the LCD, and you are at the following point in the solution: $\dfrac{(3a^2-2a+4)-(a^2+5a-1)}{(2a-3)(a+3)}$. What is the next step?

Prep Exercise 8 Two students were adding the same two rational expressions. One got the answer $\dfrac{6}{a-3}$, and the other got the answer $\dfrac{-6}{3-a}$. Explain how it is possible for both students to be correct.

For Exercises 33–78, add or subtract as indicated. Simplify your answers to lowest terms. See Example 4.

33. $\dfrac{5}{8u}-\dfrac{7}{12u}$

34. $\dfrac{6}{15v}-\dfrac{4}{9v}$

35. $\dfrac{3z}{10x}+\dfrac{5z}{4x}$

36. $\dfrac{5w}{6y}+\dfrac{7w}{9y}$

37. $\dfrac{y+6}{4y}+\dfrac{2y-3}{3y}$

38. $\dfrac{4r-2}{5r}+\dfrac{3r-4}{2r}$

39. $\dfrac{m+8}{2m}-\dfrac{m-7}{3m}$

40. $\dfrac{n-9}{7n}-\dfrac{n-2}{4n}$

41. $\dfrac{3p+1}{10p^3q^2}+\dfrac{9p-2}{6p^2q}$

42. $\dfrac{5t-8}{16s^5t^2}+\dfrac{2t-7}{12s^2t^3}$

43. $\dfrac{2a+b}{8a^2b^4}-\dfrac{a-3b}{12a^3b^3}$

44. $\dfrac{3r-2s}{6rs^3}-\dfrac{4r-3s}{8r^2s^2}$

45. $\dfrac{4}{k}-\dfrac{6}{k+2}$

46. $\dfrac{5}{j-5}-\dfrac{9}{j}$

47. $\dfrac{4}{w-3}+\dfrac{5}{w+7}$

48. $\dfrac{3}{c+4}+\dfrac{2}{c-1}$

49. $\dfrac{x-4}{3x+9}-\dfrac{x+5}{6x+18}$

50. $\dfrac{y-3}{2y-8}-\dfrac{y-7}{6y-24}$

51. $\dfrac{2t}{t^2-8t+16}+\dfrac{6}{t-4}$

52. $\dfrac{4s}{s^2+14s+49}+\dfrac{5}{s+7}$

53. $\dfrac{u^2-2u}{u^2-6u+9}+\dfrac{4}{4u-12}$

54. $\dfrac{v^2+4v}{v^2+8v+16}+\dfrac{3}{3v+12}$

55. $\dfrac{x}{x+2}-\dfrac{16}{x^2-4x-12}$

56. $\dfrac{x}{x-5}-\dfrac{45}{x^2-x-20}$

57. $\dfrac{x+1}{x^2-x-6}+\dfrac{2x+8}{x^2+6x+8}$

58. $\dfrac{y+3}{y^2-3y+2}+\dfrac{2y+4}{y^2+y-2}$

59. $\dfrac{2v+5}{v^2-16}-\dfrac{2v+6}{v^2-v-12}$

60. $\dfrac{3t+9}{t^2-9}-\dfrac{3t-2}{t^2-8t+15}$

61. $\dfrac{z-3}{z^2+6z+9}+\dfrac{z+1}{z^2+z-6}$

62. $\dfrac{w-2}{w^2+2w-24}+\dfrac{w-8}{w^2-8w+16}$

63. $\dfrac{x+4}{x^3-36x}-\dfrac{1}{x^2+3x-18}$

64. $\dfrac{3}{x^2+3x-4}-\dfrac{x-2}{x^3-16x}$

65. $\dfrac{3x}{x-2}-\dfrac{4}{x+2}-\dfrac{3}{x^2-4}$

66. $\dfrac{5}{x-4}+\dfrac{2x}{x+4}-\dfrac{2}{x^2-16}$

67. $\dfrac{5}{a^2-4a}+\dfrac{6}{a}+\dfrac{4}{a-4}$

68. $\dfrac{7}{z}-\dfrac{3}{z^2+2z}+\dfrac{4}{z+2}$

69. $\dfrac{3x+1}{x-4}-\dfrac{7}{x}+\dfrac{2}{x^2-4x}$

70. $\dfrac{3}{y}+\dfrac{2y-5}{y+5}-\dfrac{3}{y^2+5y}$

71. $\dfrac{3r}{r+3}-\dfrac{5r}{r-3}+\dfrac{2}{r^2-9}$

72. $\dfrac{2w}{w-1}-\dfrac{3w}{w+1}+\dfrac{5}{w^2-1}$

73. $\dfrac{v}{u-v}-\dfrac{3v}{v-u}$

74. $\dfrac{2x}{x-y}-\dfrac{3x}{y-x}$

75. $\dfrac{2m-n}{m-n}-\dfrac{m-3n}{n-m}$

76. $\dfrac{p-5q}{p-q}-\dfrac{4p-6q}{q-p}$

77. $\dfrac{2a-3b}{3a-b}+\dfrac{3a+2b}{b-3a}$

78. $\dfrac{4x+3y}{2x-5y}+\dfrac{2x-y}{5y-2x}$

Find the Mistake *For Exercises 79–84, explain the mistake; then find the correct sum or difference.*

79. $\dfrac{4}{a}+\dfrac{a}{5}=\dfrac{4+a}{a+5}$

80. $\dfrac{4}{3x}-\dfrac{5}{12x}=\dfrac{1}{3x}-\dfrac{5}{3x}=-\dfrac{4}{3x}$

81. $\dfrac{6v}{2x}+\dfrac{3v}{2x}=\dfrac{9v}{4x}$

82. $\dfrac{4}{w}+\dfrac{6}{w+z}=\dfrac{4(+z)}{w(+z)}+\dfrac{6}{w+z}$

$=\dfrac{4+z+6}{w+z}$

$=\dfrac{10+z}{w+z}$

83. $\dfrac{5c+2}{3c-5}-\dfrac{2c+3}{3c-5}=\dfrac{5c+2-2c+3}{3c-5}$

$=\dfrac{3c+5}{3c-5}$

84. $\dfrac{x-5}{x^2-1}-\dfrac{x+3}{x-1}=\dfrac{x-5-x-3}{x^2-1}=\dfrac{-8}{x^2-1}$

85. If $f(x)=\dfrac{9x+7}{6x}$ and $g(x)=\dfrac{x-3}{6x}$, find each of the following.

a. $(f+g)(x)$ **b.** $(f-g)(x)$ **c.** $(f+g)(4)$ **d.** $(f-g)(-2)$

86. If $f(x)=\dfrac{x^2-2x+4}{x-4}$ and $g(x)=\dfrac{5x-8}{x-4}$, find each of the following.

a. $(f+g)(x)$ **b.** $(f-g)(x)$ **c.** $(f+g)(0)$ **d.** $(f-g)(-5)$

87. If $f(x)=\dfrac{x-14}{x^2-4}$ and $g(x)=\dfrac{x+1}{x-2}$, find each of the following.

a. $(f+g)(x)$ **b.** $(f-g)(x)$ **c.** $(f+g)(1)$ **d.** $(f-g)(-1)$

88. If $f(x)=\dfrac{x+2}{3x-18}$ and $g(x)=\dfrac{2}{3x}$, find each of the following.

a. $(f+g)(x)$ **b.** $(f-g)(x)$ **c.** $(f+g)(-2)$ **d.** $(f-g)(4)$

For Exercises 89–96, solve.

89. The portion of a house that Frank paints after t hours is represented by $\frac{t}{5}$, and the portion that Jose paints is represented by $\frac{t}{3}$. Find the portion of the house painted by Frank and Jose together after t hours.

90. A contractor is building a highway, and the work is divided among three teams. After t months, the portion of the highway completed by team A is represented by $\frac{t}{2}$, the portion by team B is $\frac{t}{4}$, and the portion by team C is $\frac{t}{5}$. Find the portion of the highway constructed by the three teams together after t months.

91. In designing a building, an architect describes the length of a rectangular building as $\frac{3}{x+4}$ and the width as $\frac{1}{x-2}$. Find the perimeter of the building.

92. The thickness of a board is given by $\frac{4}{x+3}$. The board is run through a planer, which removes a layer whose thickness is $\frac{2}{x}$. What is the new thickness of the board?

93. If x represents a number, write and simplify an expression for the sum of three times the number and two times its reciprocal.

94. Write and simplify an expression for the sum of the reciprocals of two consecutive even integers if x represents the smaller integer.

★ **95.** Find the average of $\frac{x}{5}$ and $\frac{x}{3}$.

★ **96.** Write an expression in simplest form for the area of the trapezoid shown.

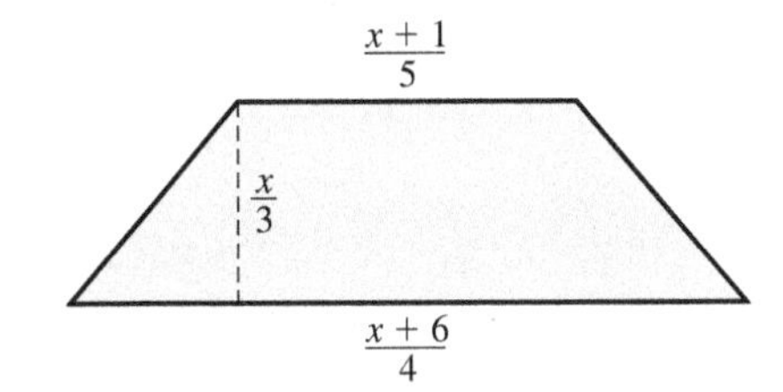

Puzzle Problem A 7-minute hourglass and a 5-minute hourglass can be used together to measure exactly 16 minutes. One rule must be followed. As soon as the top portion of an hourglass is empty, it must immediately be flipped. Explain the steps required to measure 16 minutes.

Review Exercises

Exercises 1–6 Expressions

For Exercises 1–4, perform the indicated operations.

[1.2] **1.** $\frac{1}{3}+\frac{3}{4}$

[1.2] **2.** $\frac{5}{12}\div\frac{3}{8}$

[1.3] **3.** $\left(\frac{2}{3}+\frac{1}{2}\right)\div\left(\frac{3}{4}-\frac{1}{3}\right)$

[1.3] **4.** $12\left(\frac{2}{3}+\frac{5}{6}\right)$

[7.1] *For Exercises 5 and 6, simplify.*

5. $\frac{x^2-9x+8}{x^2-6x-16}$

6. $\frac{3y+2}{27y^3+8}$

7.3 Simplifying Complex Rational Expressions

Objectives

1. Simplify complex rational expressions.
2. Simplify rational expressions with negative exponents.

Warm-up

For Exercises 1 and 2, simplify.

[5.2] 1. $(x^2 + 6x + 9) - (x^2 - 6x + 9)$

[7.1] 2. $\dfrac{x^2 - 9}{x^2} \div \dfrac{x + 3}{x}$

[7.2] 3. Add. $\dfrac{x + 1}{x^2 - x - 6} + \dfrac{2x + 8}{x^2 + 6x + 8}$

Complex rational expressions, or *complex fractions*, have parts that are themselves rational expressions.

Definition Complex rational expression: A rational expression that contains rational expressions in the numerator and/or denominator.

Here are some complex rational expressions.

Note We use a slightly larger fraction line to separate the numerator and denominator of complex rational expressions. ▶

$$\frac{\frac{2}{5}}{15} \qquad \frac{2x}{\frac{4}{2x + 5}} \qquad \frac{1 + \frac{3}{x} - \frac{6}{x^2}}{2 + \frac{3}{x} - \frac{4}{x^2}} \qquad \frac{\frac{x + 2}{x - 4} + \frac{x - 5}{x + 2}}{\frac{x + 1}{x + 2} - \frac{x + 3}{x - 3}}$$

Objective 1 Simplify complex rational expressions.

A complex rational expression is simplified when the numerator and denominator are polynomials with no common factors. There are two common methods for simplifying complex rational expressions.

Procedure Simplifying Complex Rational Expressions

To simplify a complex rational expression:

Method 1

1. If necessary, rewrite the numerator and/or denominator as a single rational expression.
2. Rewrite as a horizontal division problem and simplify.

Method 2

1. Multiply the numerator and denominator of the complex rational expression by the LCD of the fractions in the numerator and denominator.
2. Simplify.

For example, using Method 1, the complex fraction $\dfrac{\frac{2}{3}}{\frac{5}{6}}$ can be expressed as the horizontal division problem $\dfrac{2}{3} \div \dfrac{5}{6}$ and then simplified.

$$\frac{\frac{2}{3}}{\frac{5}{6}} = \frac{2}{3} \div \frac{5}{6} \qquad \text{Write as a horizontal division problem.}$$

$$= \frac{2}{3} \cdot \frac{\overset{2}{\cancel{6}}}{5} \qquad \text{Rewrite as multiplication.}$$

$$= \frac{4}{5} \qquad \text{Simplify.}$$

Answers to Warm-up

1. $12x$ 2. $\dfrac{x - 3}{x}$

3. $\dfrac{3x - 5}{(x - 3)(x + 2)}$

Using Method 2, we multiply the numerator and denominator of the complex fraction by the LCD of the fractions in the numerator and denominator. In this case, the denominators are 3 and 6; so the LCD is 6.

$$\frac{\frac{2}{3}}{\frac{5}{6}} = \frac{\frac{2}{3}\cdot 6}{\frac{5}{6}\cdot 6}$$ Multiply the numerator and denominator by the LCD, 6.

$$= \frac{4}{5}$$ Simplify.

Example 1 Simplify.

a. $\dfrac{\frac{a^2b}{c^2}}{\frac{ab^3}{c^3}}$

Method 1: The numerator and denominator of the complex fraction are already single fractions.

Note Method 1 is usually preferred only when both the numerator and denominator contain monomials; as in Example1(a).

$$\frac{\frac{a^2b}{c^2}}{\frac{ab^3}{c^3}} = \frac{a^2b}{c^2} \div \frac{ab^3}{c^3}$$ Write as a horizontal division problem.

$$= \frac{a^2b}{c^2}\cdot\frac{c^3}{ab^3}$$ Rewrite as a multiplication problem.

$$= \frac{ac}{b^2}$$ Simplify.

Method 2:

Note When we multiply the numerator and denominator by c^3, we are multiplying the fraction by $\frac{c^3}{c^3} = 1$ for $c \neq 0$.

$$\frac{\frac{a^2b}{c^2}}{\frac{ab^3}{c^3}} = \frac{\frac{a^2b}{c^2}\cdot c^3}{\frac{ab^3}{c^3}\cdot c^3}$$ Multiply the numerator and denominator by the LCD of the fractions in the numerator and denominator, c^3. Then divide out the common factors.

$$= \frac{a^2bc}{ab^3}$$ Multiply.

$$= \frac{ac}{b^2}$$ Simplify.

b. $\dfrac{1 - \frac{9}{x^2}}{1 + \frac{3}{x}}$

Method 1: First write the numerator and denominator of the complex fraction as a single fraction, then simplify.

$$\frac{1 - \frac{9}{x^2}}{1 + \frac{3}{x}} = \frac{1\cdot\frac{x^2}{x^2} - \frac{9}{x^2}}{1\cdot\frac{x}{x} + \frac{3}{x}}$$ Write the numerator fractions as equivalent fractions with their LCD, x^2, and write the denominator fractions with their LCD, x.

$$= \frac{\frac{x^2}{x^2} - \frac{9}{x^2}}{\frac{x}{x} + \frac{3}{x}}$$ Simplify.

$$= \frac{\frac{x^2 - 9}{x^2}}{\frac{x + 3}{x}}$$ Subtract in the numerator and add in the denominator.

$$= \frac{x^2 - 9}{x^2} \div \frac{x + 3}{x}$$ Write the complex fraction as a horizontal division problem.

$$= \frac{x^2 - 9}{x^2} \cdot \frac{x}{x + 3}$$ Rewrite as multiplication.

$$= \frac{\cancel{(x + 3)}(x - 3)}{\cancel{x^2}_{x}} \cdot \frac{\cancel{x}}{\cancel{x + 3}}$$ Factor $x^2 - 9$ and divide out the common factors.

$$= \frac{x - 3}{x}$$ Multiply the remaining factors.

Method 2: $$\frac{1 - \frac{9}{x^2}}{1 + \frac{3}{x}} = \frac{\left(1 - \frac{9}{x^2}\right) \cdot x^2}{\left(1 + \frac{3}{x}\right) \cdot x^2}$$ Multiply the numerator and denominator by the LCD of the fractions in the numerator and denominator, x^2.

$$= \frac{1 \cdot x^2 - \frac{9}{x^2} \cdot x^2}{1 \cdot x^2 + \frac{3}{x} \cdot x^2}$$ Distribute x^2.

$$= \frac{x^2 - 9}{x^2 + 3x}$$ Multiply.

$$= \frac{(x + 3)(x - 3)}{x(x + 3)}$$ Factor the numerator and denominator.

$$= \frac{x - 3}{x}$$ Divide out the common factor, $x + 3$.

Your Turn 1 Simplify.

a. $\dfrac{\frac{4cd^3}{3a}}{\frac{2c^4d}{6a^2}}$

b. $\dfrac{1 - \frac{4}{x} - \frac{12}{x^2}}{1 + \frac{6}{x} + \frac{8}{x^2}}$

Method 2 is at least as easy as method 1 and is usually the preferred method. Consequently, the remaining examples of this section are done using method 2 only.

Example 2 Simplify using method 2.

a. $\dfrac{x + \frac{9}{x - 6}}{1 + \frac{3}{x - 6}}$

Answers to Your Turn 1

a. $\dfrac{4ad^2}{c^3}$ b. $\dfrac{x - 6}{x + 4}$

Solution: $\dfrac{x + \dfrac{9}{x-6}}{1 + \dfrac{3}{x-6}} = \dfrac{\left(x + \dfrac{9}{x-6}\right)(x-6)}{\left(1 + \dfrac{3}{x-6}\right)(x-6)}$ Multiply the numerator and denominator by the LCD of the fractions in the numerator and denominator, $x - 6$.

$= \dfrac{x(x-6) + \left(\dfrac{9}{\cancel{x-6}}\right)\cancel{(x-6)}}{1(x-6) + \left(\dfrac{3}{\cancel{x-6}}\right)\cancel{(x-6)}}$ Distribute $x - 6$.

$= \dfrac{x^2 - 6x + 9}{x - 6 + 3}$ Multiply.

$= \dfrac{(x-3)^2}{x-3}$ Factor the numerator and simplify the denominator.

$= x - 3$ Divide out the common factor, $x - 3$.

b. $\dfrac{\dfrac{x+3}{x-3} - \dfrac{x-3}{x+3}}{\dfrac{x+3}{x-3} + \dfrac{x-3}{x+3}}$

Solution: $\dfrac{\left(\dfrac{x+3}{x-3} - \dfrac{x-3}{x+3}\right)(x-3)(x+3)}{\left(\dfrac{x+3}{x-3} + \dfrac{x-3}{x+3}\right)(x-3)(x+3)}$ Multiply the numerator and denominator by the LCD of the fractions in the numerator and denominator, $(x - 3)(x + 3)$.

$= \dfrac{\dfrac{x+3}{\cancel{x-3}}\cdot\cancel{(x-3)}(x+3) - \dfrac{x-3}{\cancel{x+3}}\cdot(x-3)\cancel{(x+3)}}{\dfrac{x+3}{\cancel{x-3}}\cdot\cancel{(x-3)}(x+3) + \dfrac{x-3}{\cancel{x+3}}\cdot(x-3)\cancel{(x+3)}}$ Distribute.

$= \dfrac{(x+3)(x+3) - (x-3)(x-3)}{(x+3)(x+3) + (x-3)(x-3)}$ Simplify.

$= \dfrac{(x^2 + 6x + 9) - (x^2 - 6x + 9)}{(x^2 + 6x + 9) + (x^2 - 6x + 9)}$ Multiply each pair of binomials.

$= \dfrac{(x^2 + 6x + 9) + (-x^2 + 6x - 9)}{(x^2 + 6x + 9) + (x^2 - 6x + 9)}$ Write the equivalent addition in the numerator.

$= \dfrac{12x}{2x^2 + 18}$ Combine like terms.

$= \dfrac{\cancel{2}\cdot 6x}{\cancel{2}(x^2 + 9)}$

$= \dfrac{6x}{x^2 + 9}$ Divide out the common factor, 2.

Your Turn 2 Simplify using method 2.

a. $\dfrac{x + \dfrac{16}{x-8}}{1 + \dfrac{4}{x-8}}$

b. $\dfrac{\dfrac{a-4}{a+4} + \dfrac{a+4}{a-4}}{\dfrac{a-4}{a+4} - \dfrac{a+4}{a-4}}$

Answers to Your Turn 2

a. $x - 4$ **b.** $-\dfrac{a^2 + 16}{8a}$

Objective 2 Simplify rational expressions with negative exponents.

Rational expressions containing terms with negative exponents are complex rational expressions when the terms are rewritten with positive exponents.

Example 3 Simplify $\frac{x^{-1} + y^{-1}}{x^{-2} - y^{-2}}$.

Solution: Using $x^{-n} = \frac{1}{x^n}$, rewrite the expression with positive exponents only.

$$\frac{x^{-1} + y^{-1}}{x^{-2} - y^{-2}} = \frac{\frac{1}{x} + \frac{1}{y}}{\frac{1}{x^2} - \frac{1}{y^2}}$$ Rewrite with positive exponents only.

$$= \frac{\left(\frac{1}{x} + \frac{1}{y}\right) \cdot x^2y^2}{\left(\frac{1}{x^2} - \frac{1}{y^2}\right) \cdot x^2y^2}$$ Multiply numerator and denominator by the LCD, x^2y^2.

$$= \frac{xy^2 + x^2y}{y^2 - x^2}$$ Distribute x^2y^2 and simplify the results.

$$= \frac{xy(y + x)}{(y - x)(y + x)}$$ Factor the numerator and denominator.

$$= \frac{xy}{y - x}$$ Divide out the common factor, $y + x$.

Warning Because x^{-1}, y^{-1}, x^{-2}, and y^{-2} are terms and not factors, they cannot be flip-flopped between the numerator and denominator. That is, the answer is not $\frac{x^2 + y^2}{x - y}$.

Your Turn 3

Simplify $\frac{xy^{-2} - y^{-1}}{x^{-1} - x^{-2}y}$.

Answer to Your Turn 3

$\frac{x^2}{y^2}$

7.3 Exercises

For Extra Help MyMathLab®

Objective 1

Prep Exercise 1 By what would you multiply the numerator and denominator of $\frac{\frac{x+3}{x-2} + 5}{6 - \frac{x+6}{x-3}}$ to simplify using method 2?

Prep Exercise 2 Given the complex rational expression $\frac{\frac{3a+b}{2a-b}}{a-4b}$, is the numerator $3a + b$ or $\frac{3a+b}{2a-b}$? How can you tell?

Prep Exercise 3 Write $\frac{\frac{x+2}{x-4}}{\frac{x-3}{x+5}}$ as a horizontal division problem.

Prep Exercise 4 Write $\left(\frac{c+2d}{2c+3d} - 4\right) \div \left(5 + \frac{3c-d}{c+4d}\right)$ as a complex rational expression.

Prep Exercise 5 Which of the two methods discussed in this section would you use to simplify $\dfrac{x - 3 + \dfrac{x+2}{x-4}}{x + 2 - \dfrac{x-3}{x+4}}$? Why?

Prep Exercise 6 Which of the two methods discussed in this section would you use to simplify $\dfrac{\dfrac{3a^2b^4}{4c^2d^3}}{\dfrac{9ab^2}{8c^4d^2}}$? Why?

For Exercises 1–42, simplify each complex fraction. See Examples 1 and 2.

1. $\dfrac{\dfrac{5}{21}}{\dfrac{3}{14}}$

2. $\dfrac{\dfrac{11}{12}}{\dfrac{3}{8}}$

3. $\dfrac{\dfrac{u}{v^3}}{\dfrac{w}{v^4}}$

4. $\dfrac{\dfrac{r}{s^5}}{\dfrac{t}{s^2}}$

5. $\dfrac{\dfrac{u^7v^2}{w^5}}{\dfrac{u^3v^4}{w}}$

6. $\dfrac{\dfrac{m^6p^5}{n^7}}{\dfrac{m^5p^3}{n^4}}$

7. $\dfrac{\dfrac{15}{8}}{20}$

8. $\dfrac{9}{\dfrac{12}{5}}$

9. $\dfrac{a}{\dfrac{b}{c}}$

10. $\dfrac{\dfrac{m}{n}}{p}$

11. $\dfrac{\dfrac{16}{2x-2}}{\dfrac{8}{x-3}}$

12. $\dfrac{\dfrac{5x+30}{15}}{\dfrac{x-3}{3}}$

13. $\dfrac{\dfrac{x+4}{6}}{\dfrac{2x+8}{18}}$

14. $\dfrac{\dfrac{b-7}{4}}{\dfrac{3b-21}{16}}$

15. $\dfrac{\dfrac{3x^2}{x+2}}{\dfrac{2x}{x+2}}$

16. $\dfrac{\dfrac{5a^2}{a-2}}{\dfrac{2a}{a-2}}$

17. $\dfrac{\dfrac{4}{9} - \dfrac{1}{3}}{\dfrac{7}{12} - \dfrac{5}{18}}$

18. $\dfrac{\dfrac{3}{4} + \dfrac{1}{6}}{\dfrac{5}{8} - \dfrac{5}{12}}$

19. $\dfrac{1 + \dfrac{x}{2}}{1 - \dfrac{x}{2}}$

20. $\dfrac{3 - \dfrac{y}{3}}{5 + \dfrac{y}{3}}$

21. $\dfrac{\dfrac{1}{t} - 1}{\dfrac{1}{t^3} - 1}$

22. $\dfrac{\dfrac{8}{s^3} - 1}{\dfrac{2}{s} - 1}$

23. $\dfrac{1 - \dfrac{2}{v} - \dfrac{3}{v^2}}{1 - \dfrac{4}{v} + \dfrac{3}{v^2}}$

24. $\dfrac{2 + \dfrac{13}{w} + \dfrac{6}{w^2}}{1 + \dfrac{11}{w} + \dfrac{30}{w^2}}$

25. $\dfrac{\dfrac{2}{x} - \dfrac{7}{x^2} - \dfrac{30}{x^3}}{2 + \dfrac{11}{x} + \dfrac{15}{x^2}}$

26. $\dfrac{3 + \dfrac{2}{y} - \dfrac{1}{y^2}}{\dfrac{3}{y} - \dfrac{13}{y^2} + \dfrac{4}{y^3}}$

27. $\dfrac{5 + \dfrac{4}{r+8}}{r+8}$

28. $\dfrac{\dfrac{7}{s-5} + 3}{s-5}$

29. $\dfrac{t+\dfrac{9}{t-7}}{11-\dfrac{3}{t-7}}$

30. $\dfrac{q-\dfrac{2}{q+4}}{5+\dfrac{13}{q+4}}$

31. $\dfrac{\dfrac{x+4}{x^2-9}}{2+\dfrac{1}{x+3}}$

32. $\dfrac{\dfrac{c+3}{c^2-25}}{4+\dfrac{2}{c-5}}$

33. $\dfrac{\dfrac{5}{a+3}-\dfrac{4}{a-3}}{\dfrac{3}{a+3}-\dfrac{2}{a-3}}$

34. $\dfrac{\dfrac{6}{n+4}-\dfrac{5}{n-4}}{\dfrac{2}{n+4}-\dfrac{1}{n-4}}$

35. $\dfrac{\dfrac{r+3}{r-3}-\dfrac{r-3}{r+3}}{\dfrac{r+3}{r-3}+\dfrac{r-3}{r+3}}$

36. $\dfrac{\dfrac{m-7}{m+7}+\dfrac{m+7}{m-7}}{\dfrac{m-7}{m+7}-\dfrac{m+7}{m-7}}$

37. $\dfrac{\dfrac{3x-9}{x-3}-\dfrac{x-3}{x-5}}{\dfrac{x+5}{x-5}+\dfrac{2x-6}{x-3}}$

38. $\dfrac{\dfrac{2x+8}{x+4}+\dfrac{x-4}{x-2}}{\dfrac{3x+12}{x+4}-\dfrac{x+2}{x-2}}$

39. $\dfrac{\dfrac{2}{y^2}-\dfrac{5}{xy}-\dfrac{3}{x^2}}{\dfrac{2}{y^2}+\dfrac{5}{xy}+\dfrac{2}{x^2}}$

40. $\dfrac{\dfrac{3}{y^2}-\dfrac{5}{xy}-\dfrac{2}{x^2}}{\dfrac{3}{y^2}+\dfrac{10}{xy}+\dfrac{3}{x^2}}$

41. $\dfrac{\dfrac{2a}{a+6}+\dfrac{1}{a}}{9a-\dfrac{3}{a+6}}$

42. $\dfrac{8b-\dfrac{7}{b-1}}{\dfrac{4b}{b-1}+\dfrac{21}{b}}$

Objective 2

Prep Exercise 7 Write $\dfrac{3x^{-2}}{4x^{-1}+5}$ with positive exponents only.

Prep Exercise 8 Write $\dfrac{x^{-2}-y^{-1}}{x^{-1}+y}$ with positive exponents only.

For Exercises 43–54, rewrite each as a complex fraction and simplify. See Example 3.

43. $\dfrac{3x^{-1}}{3x^{-1}+1}$

44. $\dfrac{2a^{-1}-1}{2a^{-1}}$

45. $\dfrac{1-9x^{-2}}{1+3x^{-1}}$

46. $\dfrac{1-4a^{-1}}{1-16a^{-2}}$

47. $\dfrac{3x^{-2}+5y^{-1}}{x^{-1}+y^{-1}}$

48. $\dfrac{a^{-2}-b^{-1}}{2a^{-1}+3b^{-2}}$

49. $\dfrac{x^{-2}y^{-2}}{4x^{-1}+3y^{-1}}$

50. $\dfrac{3r^{-1}+4s^{-1}}{r^{-2}s^{-2}}$

51. $\dfrac{36a^{-2} - 25b^{-2}}{6a^{-1} + 5b^{-1}}$

52. $\dfrac{3x^{-1} + 2y^{-1}}{9x^{-2} - 4y^{-2}}$

53. $\dfrac{(4a)^{-1} + 2b^{-2}}{2a^{-1} + b^{-2}}$

54. $\dfrac{3p^{-1} + q^{-2}}{3p^{-1} + (3q)^{-1}}$

Find the Mistake *For Exercises 55–58, example the mistake; then find the correct answer.*

55. $\dfrac{a + \frac{1}{3}}{b + \frac{1}{3}} = \dfrac{3a}{3b} = \dfrac{a}{b}$

56. $\dfrac{\frac{1}{a} - \frac{1}{b}}{a - b} = \dfrac{\frac{a}{1} \cdot \frac{1}{a} - \frac{b}{1} \cdot \frac{1}{b}}{a \cdot a - b \cdot b}$

$= \dfrac{1 - 1}{a^2 - b^2}$

$= \dfrac{0}{a^2 - b^2}$

$= 0$

57. $\dfrac{n + \frac{3}{n}}{\frac{3}{n}} = \dfrac{n + \frac{n}{1} \cdot \frac{3}{n}}{\frac{n}{1} \cdot \frac{3}{n}}$

$= \dfrac{n + 3}{3}$

58. $\dfrac{\frac{1}{m} - \frac{1}{n}}{\frac{1}{m} + \frac{1}{n}} = \dfrac{\frac{mn}{1} \cdot \frac{1}{m} - \frac{mn}{1} \cdot \frac{1}{n}}{\frac{mn}{1} \cdot \frac{1}{m} + \frac{mn}{1} \cdot \frac{1}{n}}$

$= \dfrac{n - m}{n + m}$

$= -1$

For Exercises 59 and 60, average rate can be found using the formula

$$\text{Average rate} = \frac{\text{Total distance}}{\text{Total time}}.$$

59. Jamel went on a short trip to visit a friend who lives 20 miles away. The trip there took $\frac{1}{3}$ of an hour, and the trip back home took $\frac{1}{4}$ of an hour. What was his average rate for the trip?

60. On a recent trip, Ellen traveled the first 30 miles in $\frac{1}{2}$ of an hour. She was in heavy traffic the next 10 miles, which took her $\frac{1}{4}$ of an hour. After the traffic improved, she traveled the next 50 miles in $\frac{2}{3}$ of an hour. What was her average speed for the trip?

For Exercises 61–64, solve.

61. Given the area A, and length, l, of a rectangle, the width can be found using the formula $w = \frac{A}{l}$. Suppose the area of a rectangle is $\frac{3x^2 - 11x - 4}{18}$ square inches and the length is $\frac{x - 4}{2}$ inches. Find the width.

62. If the area, A, and base, b, of a triangle are known, the height can be found using the formula $h = \frac{2A}{b}$. If the area of a triangle is $\frac{10}{4x + 8}$ square inches and the base is $\frac{5}{3x + 6}$ inches, find the height.

63. In electrical circuits, if two resistors with resistance R_1 and R_2 ohms are wired in parallel, the resistance of the circuit is found using the complex rational expression

$$\frac{1}{\frac{1}{R_1} + \frac{1}{R_2}}.$$

a. Simplify this complex rational expression.

b. If a 60-ohm and 40-ohm resistor are wired in parallel, find the resistance of the circuit.

c. If two 40-ohm speakers are wired in parallel, find the resistance of the circuit.

64. Suppose three resistors with resistance R_1, R_2, and R_3 ohms are wired in parallel. The resistance of the circuit is found using the complex rational expression

$$\frac{1}{\frac{1}{R_1} + \frac{1}{R_2} + \frac{1}{R_3}}.$$

a. Simplify this complex rational expression.

b. If a 60-ohm, a 40-ohm, and a 20-ohm resistor are wired in parallel, find the resistance of the circuit.

c. If a 200-ohm, a 400-ohm, and a 600-ohm resistor are wired in parallel, find the resistance of the circuit.

Review Exercises

Exercises 1–6 Equations and Inequalities

For Exercises 1–4, solve.

[2.1] **1.** $3x + 5 = 2$

[2.1] **2.** $4x - 6 = 2x - 14$

[6.4] **3.** $3x^2 + 7x = 0$

[6.4] **4.** $2x^2 - 7x - 15 = 0$

[2.2] **5.** Gretchen walks 2.6 miles in 45 minutes. Find Gretchen's average rate in miles per hour. (Use $d = rt$.)

[2.2] **6.** A northbound car and a southbound car meet on a highway. The northbound car is traveling 40 mph, and the southbound car is traveling 60 mph. How much time elapses from the time they pass each other until they are 20 miles apart?

7.4 Solving Equations Containing Rational Expressions

Objective

1 Solve equations containing rational expressions.

Warm-up

For Exercises 1 and 2, solve.

[2.1] 1. $(x + 1)4 - (x - 3)2 = (x + 2)4$

[6.4] 2. $(x + 2)\cdot x - (x + 4)\cdot 1 = 8$

[7.1] 3. Find the domain of $f(x) = \dfrac{8}{x^2 + 6x + 8}$

[7.3] 4. Simplify. $\dfrac{x + \dfrac{16}{x - 8}}{1 + \dfrac{4}{x - 8}}$

To solve equations containing rational expressions, we multiply both sides of the equation by the LCD. Because all of the denominators divide evenly into the LCD, multiplying both sides of the equation by the LCD eliminates all of the denominators of all of the rational expressions in the equation. We then solve the resulting equation.

Procedure Solving Equations Containing Rational Expressions

To solve an equation containing rational expressions,

1. Eliminate the denominators of the rational expressions by multiplying both sides of the equation by the LCD of all the rational expressions in the equation.
2. Solve the resulting equation using the methods in Chapter 2 (for linear equations) and Chapter 6 (for quadratic equations).
3. Check your solution(s) in the original equation.

Example 1 Solve $\dfrac{5}{x} - \dfrac{3}{4} = \dfrac{1}{2}$.

Note If $x = 0$, then $\dfrac{5}{x}$ is undefined; so the solution cannot be 0. We say more about why it is important to note such values after Example 1.

Solution: To solve, eliminate the rational expressions by multiplying both sides of the equation by the LCD, $4x$.

Note Because of the multiplication property of equality, we can multiply both sides of the equation by the LCD (in this case, $4x$). We do not have to rewrite each term with the LCD.

$4x\left(\dfrac{5}{x} - \dfrac{3}{4}\right) = 4x\left(\dfrac{1}{2}\right)$ Multiply both sides by $4x$.

$4x\left(\dfrac{5}{x}\right) - 4x\left(\dfrac{3}{4}\right) = 2x$ Distribute on the left; multiply on the right.

$20 - 3x = 2x$ Multiply.

$20 = 5x$ Add $3x$ to both sides.

$4 = x$ Divide both sides by 5, We see that 4 is a possible solution.

Check: $\dfrac{5}{4} - \dfrac{3}{4} \stackrel{?}{=} \dfrac{1}{2}$ Substitute 4 for x.

$\dfrac{1}{2} = \dfrac{1}{2}$ The equation is true, so 4 is the solution.

Your Turn 1 Solve $\dfrac{4}{5} - \dfrac{1}{x} = \dfrac{2}{3}$.

Answer to Your Turn 1
$\dfrac{15}{2}$

Answers to Warm-up
1. $x = 1$ **2.** $x = -4, x = 3$
3. $x \neq -4, x \neq -2$ **4.** $x - 4$

Extraneous Solutions

If we multiply both sides of an equation by an expression that contains a variable, we might obtain a solution that, when substituted into the original equation, makes one of its denominators equal to 0 (and therefore makes one of its expressions undefined). We call such an apparent solution an *extraneous solution*, and we discard it.

By inspecting the denominator of each rational expression, you can determine the value(s) that would cause the expression to be undefined before you solve the equation.

Example 2 Solve $\frac{3y}{y-4} = 6 + \frac{12}{y-4}$.

Solution: Notice that if $y = 4$, then $\frac{3y}{y-4}$ and $\frac{12}{y-4}$ are undefined; so 4 cannot be a solution.

$$(y-4)\left(\frac{3y}{y-4}\right) = (y-4)\left(6 + \frac{12}{y-4}\right)$$ Multiply both sides by the LCD, $y - 4$.

$$3y = (y-4)(6) + (y-4)\frac{12}{y-4}$$ Multiply on the left; distribute on the right.

$$3y = 6y - 24 + 12$$ Multiply on the right.

$$-3y = -12$$ Subtract $6y$ from both sides and simplify on the right.

$$y = 4$$ Divide both sides by -3; so 4 is a possible solution.

Check: We already noted that 4 causes expressions in the equation to be undefined; so 4 is extraneous. Because 4 was the only possible solution, this equation has no solution.

Connection Because the equation in Example 2 has no solution, if we were to write its solution set, we would write an empty set, which is denoted by { } or ∅.

Your Turn 2 Solve $\frac{5n}{n-2} - 4 = \frac{10}{n-2}$.

Proportions

If an equation has the form $\frac{a}{b} = \frac{c}{d}$, which is called a *proportion*, multiplying both sides by the LCD, bd, gives the following:

$$\cancel{b}d \cdot \frac{a}{\cancel{b}} = b\cancel{d} \cdot \frac{c}{\cancel{d}}$$

$$ad = bc$$

A faster way to reach that same conclusion is to *cross multiply*.

$$\begin{matrix} ad & & bc \\ \frac{a}{b} & \times & \frac{c}{d} \end{matrix}$$

Warning Cross multiplication can be used *only* when the equation is of the form $\frac{a}{b} = \frac{c}{d}$, a fraction equal to a fraction.

Rule Proportions and Their Cross Products

If $\frac{a}{b} = \frac{c}{d}$, where $b \neq 0$ and $d \neq 0$, then $ad = bc$.

Answer to Your Turn 2
no solution (2 is extraneous)

Connection Cross multiplication is just a shortcut for multiplying both sides by the LCD. Suppose we solved the equation in Example 3 by multiplying both sides by the LCD.

$$(x + 4)(x - 2)\frac{x}{x + 4} = (x + 4)(x - 2)\frac{1}{x - 2}$$

$$(x - 2)x = (x + 4)1$$

This equation is the same equation we had in the first step of the solution.

Example 3 Solve $\frac{x}{x + 4} = \frac{1}{x - 2}$.

Solution: Note that x cannot equal -4 or 2. Because the equation is in the form $\frac{a}{b} = \frac{c}{d}$, we can cross multiply.

$$x(x - 2) = (x + 4)(1) \quad \text{Cross multiply.}$$

$$x^2 - 2x = x + 4 \quad \text{Distribute.}$$

$$x^2 - 3x - 4 = 0 \quad \text{Write the quadratic equation in standard form } (ax^2 + bx + c = 0).$$

$$(x - 4)(x + 1) = 0 \quad \text{Factor.}$$

$$x - 4 = 0 \quad \text{or} \quad x + 1 = 0 \quad \text{Use the zero-factor theorem.}$$

$$x = 4 \quad \text{or} \quad x = -1 \quad \text{Solve each equation; so 4 and } -1 \text{ are possible solutions.}$$

Because $x = 4$ and $x = -1$ are not extraneous solutions, we need only check to see whether our work is correct. We will leave the check to the reader.

Your Turn 3 Solve $\frac{y}{y - 2} = \frac{10}{y + 1}$.

Example 4 Solve.

a. $\frac{x}{x + 4} - \frac{1}{x + 2} = \frac{8}{x^2 + 6x + 8}$

Note Inspecting the denominators, we see that neither -4 nor -2 can be a solution. ▶

Solution: $\frac{x}{x + 4} - \frac{1}{x + 2} = \frac{8}{(x + 4)(x + 2)}$ Factor the denominator $x^2 + 6x + 8$.

$$(x + 4)(x + 2)\left(\frac{x}{x + 4} - \frac{1}{x + 2}\right) = (x + 4)(x + 2)\frac{8}{(x + 4)(x + 2)} \quad \text{Multiply both sides by the LCD, } (x + 4)(x + 2).$$

Note In Example 4, we cannot cross multiply because these equations are not in the form $\frac{a}{b} = \frac{c}{d}$.

$$(x + 4)(x + 2)\frac{x}{x + 4} - (x + 4)(x + 2)\frac{1}{x + 2} = 8 \quad \text{Distribute on the left side and multiply on the right.}$$

$$(x + 2) \cdot x - (x + 4) \cdot 1 = 8 \quad \text{Multiply.}$$

$$x^2 + 2x - x - 4 = 8 \quad \text{Simplify the left side.}$$

$$x^2 + x - 12 = 0 \quad \text{Write the quadratic equation in standard form.}$$

$$(x + 4)(x - 3) = 0 \quad \text{Factor the left side.}$$

$$x + 4 = 0 \quad \text{or} \quad x - 3 = 0 \quad \text{Use the zero-factor theorem.}$$

$$x = -4 \quad \text{or} \quad x = 3 \quad \text{Solve each equation.}$$

Check: We have already noted that -4 cannot be a solution; so 3 is the only possible solution. We will leave the check of $x = 3$ to the reader.

b. $\frac{4}{x^2 - x - 6} - \frac{2}{x^2 + 3x + 2} = \frac{4}{x^2 - 2x - 3}$

Note 3, -2, and -1 cannot be solutions. ▶

Solution: $\frac{4}{(x - 3)(x + 2)} - \frac{2}{(x + 2)(x + 1)} = \frac{4}{(x - 3)(x + 1)}$ Factor the denominators to determine the LCD.

$$(x - 3)(x + 2)(x + 1)\left(\frac{4}{(x - 3)(x + 2)} - \frac{2}{(x + 2)(x + 1)}\right) \quad \text{Multiply both sides by the LCD, } (x - 3)(x + 2)(x + 1).$$

$$= (x - 3)(x + 2)(x + 1)\left(\frac{4}{(x - 3)(x + 1)}\right)$$

Answer to Your Turn 3
4, 5

$$(x + 1)4 - (x - 3)2 = (x + 2)4$$ Distribute and simplify.

$$4x + 4 - 2x + 6 = 4x + 8$$ Simplify.

$$2x + 10 = 4x + 8$$ Combine like terms on the left side.

$$2 = 2x$$ Subtract $2x$ and 8 from both sides.

$$1 = x$$ Divide both sides by 2.

Check: Because 1 does not make any denominator equal to 0, it is not an extraneous solution. We will leave the check of $x = 1$ to the reader.

Your Turn 4 Solve.

a. $\dfrac{12}{x^2 - 4} = \dfrac{3}{x - 2} + 1$

b. $\dfrac{7}{n^2 + 3n - 10} = \dfrac{6}{n^2 + 4n - 5} + \dfrac{3}{n^2 - 3n + 2}$

Warning Make sure you understand the difference between performing operations with rational expressions and solving equations containing rational expressions. For example, an *expression* such as $\dfrac{2}{x} + \dfrac{3}{x + 1}$ can be *evaluated* or *rewritten* (not solved). To rewrite this expression, we add the two rational expressions.

$$\frac{2}{x} + \frac{3}{x + 1} = \frac{2(x + 1)}{x(x + 1)} + \frac{3(x)}{(x + 1)(x)}$$ Rewrite each rational expression with the LCD, $x(x + 1)$.

$$= \frac{2x + 2}{x(x + 1)} + \frac{3x}{x(x + 1)}$$ Multiply each numerator.

$$= \frac{5x + 2}{x(x + 1)}$$ Add the numerators and keep the LCD.

An *equation* such as $\dfrac{2}{x} + \dfrac{3}{x + 1} = \dfrac{17}{12}$ can be *solved*. First, we eliminate the rational expressions by multiplying both sides of the equation by the LCD, $12x(x + 1)$.

$$12x(x + 1)\left(\frac{2}{x} + \frac{3}{x + 1}\right) = 12x(x + 1)\frac{17}{12}$$ Multiply both sides by the LCD.

$$12(x + 1)(2) + 12x(3) = x(x + 1)(17)$$ Simplify.

$$24x + 24 + 36x = 17x^2 + 17x$$ Multiply.

$$0 = 17x^2 - 43x - 24$$ Write in standard form.

$$0 = (17x + 8)(x - 3)$$ Factor.

$$x = -\frac{8}{17} \quad \text{or} \quad x = 3$$ Use the zero-factor theorem.

Answers to Your Turn 4
a. -5 (2 is extraneous)
b. no solution (-5 is extraneous)

7.4 Exercises

For Extra Help MyMathLab®

Objective 1

Prep Exercise 1 Explain the difference between solving an equation involving rational expressions and adding rational expressions. For example, how is solving $\dfrac{y}{3y + 6} + \dfrac{3}{4y + 8} = \dfrac{3}{4}$ different from adding $\dfrac{y}{3y + 6} + \dfrac{3}{4y + 8} + \dfrac{3}{4}$?

Prep Exercise 2 What is an extraneous solution? What can cause extraneous solutions of rational equations to be introduced?

Prep Exercise 3 What are the possible extraneous solutions of the equation in Prep Exercise 6? Why?

Prep Exercise 4 Is $y = -3$ an extraneous solution of $\frac{4y}{y+3} + \frac{12}{y+3} = 3$? Why or why not?

For Exercises 1–6, without solving the equations, find the value(s) of the variable that will result in an extraneous solution.

1. $\frac{3}{x-4} = \frac{1}{2}$
2. $\frac{7}{5} = \frac{4}{y+2}$
3. $\frac{4}{x+6} - \frac{3}{x-2} = \frac{2}{2}$
4. $\frac{6}{a+4} + \frac{2}{a-1} = \frac{3}{4}$
5. $\frac{4x}{x^2-4} + \frac{3}{x^2+5x+6} = \frac{7x}{x^2+x-6}$
6. $\frac{3y}{y^2+2y-8} - \frac{2}{y^2+3y-10} = \frac{6y}{y^2+9y+20}$

Prep Exercise 5 When solving equations that contain rational expressions, why should you multiply both sides of the equation by the LCD?

Prep Exercise 6 To solve the equation $\frac{3}{x^2-1} + \frac{2}{x^2+3x+2} = \frac{6}{x^2+x-2}$, by what expression would you multiply both sides of the equation?

Prep Exercise 7 An equation of the form $\frac{a}{b} = \frac{c}{d}$ is called a ____________.

Prep Exercise 8 What special rule can be used to solve an equation of the form $\frac{a}{b} = \frac{c}{d}$, where $b \neq 0$ and $d \neq 0$?

For Exercises 7–60, solve and check. Identify any extraneous solutions. See Examples 1–4.

7. $\frac{4}{3u} - \frac{1}{2u} = \frac{5}{6}$
8. $\frac{3}{2x} - \frac{4}{6x} = \frac{5}{12}$
9. $\frac{5}{4w} - \frac{3}{5w} = -\frac{13}{40}$
10. $\frac{7}{2z} + \frac{4}{3z} = \frac{29}{18}$
11. $\frac{7}{x-3} = 8 - \frac{1}{x-3}$
12. $\frac{5}{y+2} + 9 = -\frac{4}{y+2}$
13. $\frac{t}{t+4} = 3 - \frac{12}{t+4}$
14. $\frac{8}{z-5} - 2 = \frac{z}{z-5}$
15. $\frac{6a}{a+5} = \frac{3}{a+5} + 3$
16. $\frac{5b}{b-8} - 4 = \frac{3}{b-8}$
17. $\frac{4x}{2x-3} = 3 + \frac{6}{2x-3}$
18. $\frac{9a}{3a-2} + 2 = \frac{6}{3a-2}$
19. $\frac{3x}{x+3} = \frac{6}{x+7}$
20. $\frac{4}{x+1} = \frac{2x}{x+6}$
21. $\frac{m}{m-1} = \frac{3m}{4m-3}$
22. $\frac{5p}{p-5} = \frac{p}{p-1}$
23. $\frac{2y}{3y-6} + \frac{3}{4y-8} = \frac{1}{4}$
24. $\frac{2y}{3y+6} - \frac{3}{4y+8} = \frac{1}{4}$
25. $\frac{a^2}{a+2} - 3 = \frac{a+6}{a+2} - 4$
26. $\frac{3b+4}{b-4} = \frac{b^2}{b-4} + 3$
27. $2x - \frac{12}{x} = 5$
28. $2x - \frac{15}{x} = 7$
29. $\frac{x+1}{x+2} + \frac{5}{2x} = \frac{3x+2}{x+2}$
30. $\frac{x+2}{x+4} - \frac{3}{2x} = \frac{2x-5}{x+4}$
31. $\frac{2x+1}{x+2} - \frac{x+1}{3x-2} = \frac{x}{x+2}$
32. $\frac{2x-2}{x+1} - \frac{x-2}{2x-3} = \frac{x}{2x-3}$
33. $\frac{4}{x+5} - \frac{2}{x-5} = \frac{4x-20}{x^2-25}$

34. $\frac{6}{x+2} - \frac{4}{x-2} = \frac{3x-22}{x^2-4}$

35. $\frac{x}{x-2} - \frac{4}{x-1} = \frac{2}{x^2-3x+2}$

36. $\frac{w}{w+5} - \frac{2}{w-3} = -\frac{16}{w^2+2w-15}$

37. $\frac{6}{t^2+t-12} = \frac{4}{t^2-t-6} + \frac{1}{t^2+6t+8}$

38. $\frac{7}{v^2-6v+5} - \frac{2}{v^2-4v-5} = \frac{3}{v^2-1}$

39. $\frac{5}{x^2-x-6} - \frac{8}{x^2+2x-15} = \frac{3}{x^2+7x+10}$

40. $\frac{8}{a^2+4a-12} - \frac{5}{a^2+a-6} = \frac{2}{a^2+9a+18}$

41. $\frac{p+1}{p^2+8p+15} - \frac{2}{p^2+7p+10} = \frac{1}{p^2+5p+6}$

42. $\frac{w-8}{w^2+2w-8} + \frac{5}{w^2+w-6} = \frac{3}{w^2+7w+12}$

43. $\frac{a-5}{a+5} + \frac{a+15}{a-5} = 2 - \frac{10}{a^2-25}$

44. $\frac{b+7}{b+3} + \frac{b-5}{b-3} = 2 + \frac{6}{b^2-9}$

45. $\frac{3n}{n^2-2n-15} = \frac{2n}{n-5} + \frac{n}{n+3}$

46. $\frac{c^2+4}{c^2+c-2} = \frac{2c}{c+2} - \frac{2c+1}{c-1}$

47. $\frac{x+5}{x-5} - \frac{x-5}{x+5} = \frac{40}{x^2-25}$

48. $\frac{48}{p^2-16} = \frac{p-4}{p+4} - \frac{p+4}{p-4}$

49. $\frac{2}{2x^2-4x-10} = \frac{3}{3x^2+6x-3}$

50. $\frac{3}{3x^2-3x-24} = \frac{5}{5x^2-x-20}$

51. $1 = \frac{3}{a-2} - \frac{12}{a^2-4}$

52. $1 - \frac{2}{u+1} = \frac{4}{u^2-1}$

53. $\frac{w+2}{w^2-5w+6} + \frac{2w}{w^2-w-2} = -\frac{2}{w^2-2w-3}$

54. $\frac{2}{q^2-5q+6} = \frac{q}{q^2-4} + \frac{q-2}{q^2-q-6}$

55. $\frac{2p}{2p-3} = \frac{15-32p^2}{4p^2-9} + \frac{3p}{2p+3}$

56. $\frac{3x^2+11x+4}{9x^2-4} + \frac{2x}{3x+2} = \frac{4x}{3x-2}$

57. $\frac{4x}{x^2+4x} - \frac{2x}{x^2+2x} = \frac{3x-2}{x^2+6x+8}$

58. $\frac{9v}{v^2-4v} - \frac{3v}{2v^2+v} = \frac{v-7}{2v^2-7v-4}$

59. $\frac{6}{x^3+2x^2-x-2} = \frac{6}{x+2} - \frac{3}{x^2-1}$

60. $\frac{1}{y-3} = \frac{6}{y^2-1} + \frac{12}{y^3-3y^2-y+3}$

For Exercises 61–68, equations that you will encounter or have encountered in the exercise sets are given. Solve each one for the indicated variable.

61. $C = \frac{90{,}000p}{100-p}$, for p

62. $v = \frac{50t}{t+5}$, for t

63. $I = \frac{2E}{R+2r}$, for r

64. $I = \frac{E}{R+r}$, for R

65. $\frac{1}{s} + \frac{1}{S} = \frac{1}{f}$, for f

66. $C = \frac{400+3x}{x}$, for x

67. $R = \frac{1}{\frac{1}{R_1} + \frac{1}{R_2}}$, for R_1

68. $F = \frac{f_1 f_2}{f_1 + f_2 - d}$, for d

Find the Mistake *For Exercises 69 and 70, explain the mistake; then find the correct solution(s).*

69. $\dfrac{x}{x-4} = \dfrac{4}{x-4} + 3$

$$(x-4)\frac{x}{x-4} = (x-4)\frac{4}{x-4} + (x-4)\cdot 3$$
$$x = 4 + 3x - 12$$
$$-2x = -8$$
$$x = 4 \text{ So } x = 4 \text{ is the solution.}$$

70. $\dfrac{x}{2} + \dfrac{x}{x+5} = \dfrac{-5}{x+5}$

$$2(x+5)\frac{x}{2} + 2(x+5)\frac{x}{x+5} = 2(x+5)\frac{-5}{x+5}$$
$$x^2 + 5x + 2x = -10$$
$$x^2 + 7x + 10 = 0$$
$$(x+5)(x+2) = 0$$
$$x = -5 \quad \text{or} \quad x = -2$$
So $x = -5$ and $x = -2$ are the solutions.

For Exercises 71–78, solve.

71. Many utility companies use coal to generate electricity, which emits pollutants into the atmosphere. Suppose the cost for removing the pollutants is given by

$$C = \frac{90{,}000\,p}{100 - p},$$

where C is the cost in thousands of dollars and p is the percentage of the pollutants removed. Find the percentage of pollutants removed if the cost is $22,500,000. (*Hint:* First, change $22,500,000 to thousands.)

72. Suppose the velocity of an automobile starting from rest is given by

$$v = \frac{50t}{t + 5},$$

where v is the velocity in miles per hour and t is the time in seconds. Find the number of seconds when the velocity is 25 miles per hour.

73. If a police car with its siren on is approaching a stationary observer, the frequency, F, of the siren is given by

$$F = \frac{132{,}000}{330 - s},$$

where s is the speed of the police car in meters per second. Find the speed of the car if the frequency is 440 cycles per second.

74. For a small business, the per unit cost, C, to order and store x units of a product is given by

$$C = 5x + \frac{2000}{x}.$$

Find the number of units if the cost is $520.

75. If x is the average speed on the outgoing trip and $x - 20$ is the average speed on the return trip, the total time for a trip of 80 miles is given by

$$T = \frac{80}{x} + \frac{80}{x - 20},$$

where T is in hours and x is in miles per hour. Find the average speed of the outgoing trip if the total time for the trip is 6 hours.

76. In optics, the thin lens equation is

$$\frac{1}{s} + \frac{1}{S} = \frac{1}{f},$$

where s is the object distance from the lens, S is the image distance from the lens, and f is the focal length of the lens. If the focal length is 6 centimeters and the image distance is 10 centimeters, find the object distance.

77. The current, I, in amperes in a circuit including an external resistance of R ohms and a cell of electromotive force of E volts and an internal resistance of r ohms is given by

$$I = \frac{E}{R + r}.$$

Find the internal resistance in a circuit whose current is 2 amperes, external resistance is 50 ohms, and electromotive force is 110 volts.

78. If two cells are connected in a series, the current, I, in amperes in a circuit including an external resistance of R ohms and a cell of electromotive force of E volts and an internal resistance of r ohms is given by

$$I = \frac{2E}{R + 2r}.$$

Find the external resistance in a circuit if the current is 11 amperes, the electromotive force is 220 volts, and the internal resistance is 5 ohms.

Puzzle Problem Amy has nine apparently identical bricks. However, one brick weighs slightly less than the others. Using a scale balance and weighing the bricks only twice, she was able to identify the lighter brick. Explain the steps.

Review Exercises

Exercises 1–6 Equations and Inequalities

[2.2, 4.3] *For Exercises 1–6, solve.*

1. Fred has two more nickels than dimes, and the total value of the nickels and dimes is $1.30. How many of each type of coin does he have?

2. The length of a rectangular painting is 2 inches more than the width. If the area of the painting is 80 square inches, find the length and width.

3. One number is 3 more than the other. Find the numbers if the difference of their squares is 39.

4. If the sum of two numbers is 28 and their difference is 4, find the numbers.

5. Jose purchased a saw and a drill. If the drill cost $54 less than the saw and he paid a total of $238 for both, how much did each tool cost?

6. A recreational vehicle leaves New Orleans heading west on I-10 toward Houston at an average rate of 55 miles per hour. Four hours later, a truck also leaves New Orleans on I-10 toward Houston at an average rate of 75 miles per hour. How many hours will it take the truck to catch up with the recreational vehicle?

7.5 Applications with Rational Expressions; Variation

Objectives

1 Use tables to solve problems with two unknowns involving rational expressions.
2 Solve problems involving direct variation.
3 Solve problems involving inverse variation.
4 Solve problems involving joint variation.
5 Solve problems involving combined variation.

Warm-up

[7.4] *For Exercises 1–3, solve.*

1. $\frac{t}{10} + \frac{t}{7} = 1$
2. $\frac{450}{r} = \frac{1200}{r} - \frac{3}{2}$
3. $\frac{20}{x - 5} + \frac{30}{x + 5} = 2$

Objective 1 Use tables to solve problems with two unknowns involving rational expressions.

In Section 4.3, we used tables to organize information involving two unknown amounts. Now let's consider similar problems that lead to equations containing rational expressions.

Answers to Warm-up

1. $t = \frac{70}{17}$ 2. $r = 500$ 3. $x = 0, x = 25$

Problems Involving Work

Tables are helpful in solving problems involving two or more people (or machines) working together to complete a task. In these problems, we are given each person's rate of work and asked to find the time it takes them to complete the task if they work together. For each person involved,

Person's rate of work · Person's time at work = Part of the task completed by that person

Because the people are working together, the sum of their individual parts of the task equals the whole task.

Part completed by one person + Part completed by the other person = Whole task

Learning Strategy

If you don't understand the lecture, don't be afraid to ask questions. If you are too intimidated to ask a question in a large lecture, go to all review sessions and see the professor during office hours. Put a star next to problems that you don't understand to ask later. Sit with a friend in class and ask him/her for help.

—Lauren H.

Example 1 If Mike can paint his room in 10 hours and Susan can paint the same room in 7 hours, how long will it take them to paint the room working together?

Understand Mike paints at a rate of 1 room in 10 hours, or $\frac{1}{10}$ of a room per hour.

Susan paints at a rate of 1 room in 7 hours, or $\frac{1}{7}$ of a room per hour.

People	Rate of Work (rooms per hour)	Time at Work (number of hours)	Part of Task Completed
Mike	$\frac{1}{10}$	t	$\frac{1}{10}t$ or $\frac{t}{10}$ of a room
Susan	$\frac{1}{7}$	t	$\frac{1}{7}t$ or $\frac{t}{7}$ of a room

Note Because they are working together for the same amount of time, we let t represent that amount of time.

Note Multiplying the rate of work and the time at work gives an expression of the part of work completed. For example, if Mike works at a rate of $\frac{1}{10}$ of a room per hour for 4 hours, he can paint $\frac{1}{10} \cdot 4 = \frac{2}{5}$ of a room.

The whole job in this case is 1 room, so we can write an equation that combines their individual expressions for the task completed and set this sum equal to 1 room.

Plan and Execute Part Mike completed + Part Susan completed = 1 room

$$\frac{t}{10} + \frac{t}{7} = 1$$

$$70\left(\frac{t}{10} + \frac{t}{7}\right) = 70(1) \quad \text{Multiply both sides by the LCD, 70.}$$

$$\overset{7}{\cancel{70}} \cdot \frac{t}{\underset{1}{\cancel{10}}} + \overset{10}{\cancel{70}} \cdot \frac{t}{\underset{1}{\cancel{7}}} = 70 \quad \text{Distribute and divide out common factors.}$$

$$7t + 10t = 70 \quad \text{Multiply.}$$

$$17t = 70 \quad \text{Combine like terms.}$$

$$t = \frac{70}{17} \quad \text{Divide both sides by 17.}$$

Answer Working together, it takes Mike and Susan $\frac{70}{17}$, or $4\frac{2}{17}$, hours to paint the room.

Check Mike paints $\frac{1}{10}$ of the room per hour; so if he works alone $\frac{70}{17}$ hours, he paints $\frac{1}{10} \cdot \frac{70}{17} = \frac{7}{17}$ of a room. Susan paints $\frac{1}{7}$ of the room per hour; so in $\frac{70}{17}$ hours, she paints $\frac{1}{7} \cdot \frac{70}{17} = \frac{10}{17}$ of the room. Combining their individual amounts, we see that in $\frac{70}{17}$ hours, they paint $\frac{7}{17} + \frac{10}{17} = \frac{17}{17} = 1$ room.

Your Turn 1 If one pipe can fill a tank in 12 hours and a second pipe can fill a tank in 8 hours, how long will it take to fill the tank with both pipes open?

Motion Problems

Recall that the formula for calculating distance, given the rate of travel and time of travel, is $d = rt$. If we isolate r, we have $r = \frac{d}{t}$. If we isolate t, we have $t = \frac{d}{r}$. These equations suggest that we will encounter rational expressions when describing rate or time.

Example 2 **a.** An airplane flew 450 miles from Miami to Tallahassee in $1\frac{1}{2}$ hours less time than it took to fly 1200 miles from Miami to Newark. If the airplane flew at the same rate on both trips, find the rate of the airplane.

Understand We are to find the rate of the airplane. This situation involves the same rate between each pair of cities, but the distances and times are different. We use a table and the fact that $t = \frac{d}{r}$ to organize the distance, rate, and time of each leg of the trip.

Flight	Distance (miles)	Rate (miles/hour)	Time (hours)
Miami to Tallahassee	450	r	$\frac{450}{r}$
Miami to Newark	1200	r	$\frac{1200}{r}$

Note Because $d = rt$, to describe time, we use $t = \frac{d}{r}$.

We can use the fact that the time from Miami to Tallahassee is $1\frac{1}{2}$, or $\frac{3}{2}$, hours less than the time from Miami to Newark to write the equation.

Plan and Execute Time from Miami to Tallahassee = Time from Miami to Newark $-\frac{3}{2}$

$$\frac{450}{r} = \frac{1200}{r} - \frac{3}{2}$$

Answer to Your Turn 1

$4\frac{4}{5}$ hr.

$$2r\left(\frac{450}{r}\right) = 2r\left(\frac{1200}{r} - \frac{3}{2}\right)$$ Multiply both sides by the LCD, $2r$.

$$2\cancel{r}\cdot\frac{450}{\cancel{r}} = 2\cancel{r}\cdot\frac{1200}{\cancel{r}} - 2\cancel{r}\cdot\frac{3}{\cancel{2}}$$ Distribute and divide out common factors.

$$2\cdot 450 = 2\cdot 1200 - r\cdot 3$$ Simplify.

$$900 = 2400 - 3r$$ Multiply.

$$-1500 = -3r$$ Subtract 2400 from both sides.

$$500 = r$$ Divide both sides by -3

Answer The plane is traveling 500 miles per hour.

Check If the plane flies 450 miles from Miami to Tallahassee at 500 miles per hour, the time required is $\frac{450}{500} = \frac{9}{10}$ of an hour. If the plane flies 1200 miles from Miami to Newark at 500 miles per hour, the required time is $\frac{1200}{500} = \frac{12}{5}$ hours. The time from Miami to Tallahassee is to be $1\frac{1}{2}$ hours less than the time from Miami to Newark and $\frac{12}{5} - \frac{3}{2} = \frac{24}{10} - \frac{15}{10} = \frac{9}{10}$, which is the time from Miami to Tallahassee.

b. A river has a current of 5 miles per hour. If a boat can make a trip of 20 miles upstream and 30 miles downstream in 2 hours, find the rate of the boat in still water.

Understand We are looking for the rate of the boat in still water, so we let this rate be x. When the boat is traveling upstream, the current slows the boat by 5 miles per hour. When the boat is traveling downstream, the current speeds the boat by 5 miles per hour. Again, we use a table to organize our information.

Direction	Distance (miles)	Rate (miles/hour)	Time (hours)
Upstream	20	$x - 5$	$\frac{20}{x - 5}$
Downstream	30	$x + 5$	$\frac{30}{x + 5}$

Note Because x is the rate of the boat in still water, the rate against the current is $x - 5$ and the rate with the current is $x + 5$.

Note Because $d = rt$, to describe time, we use $t = \frac{d}{r}$.

We can write the equation using the fact that the boat can go 20 miles upstream and 30 miles downstream in 2 hours.

Plan and Execute Time upstream + Time downstream = 2 hours

$$\frac{20}{x - 5} + \frac{30}{x + 5} = 2$$

$$(x - 5)(x + 5)\left(\frac{20}{x - 5} + \frac{30}{x + 5}\right) = (x - 5)(x + 5)(2)$$ Multiply both sides by the LCD, $(x - 5)(x + 5)$.

$$\cancel{(x - 5)}(x + 5)\left(\frac{20}{\cancel{x - 5}}\right) + (x - 5)\cancel{(x + 5)}\left(\frac{30}{\cancel{x + 5}}\right) = (x^2 - 25)(2)$$ Distribute.

$$20(x + 5) + 30(x - 5) = 2x^2 - 50$$ Simplify.

$$20x + 100 + 30x - 150 = 2x^2 - 50$$ Distribute.

$$50x - 50 = 2x^2 - 50$$ Combine like terms.

$$0 = 2x^2 - 50x$$ Add 50 to and subtract $50x$ from both sides.

$$0 = 2x(x - 25)$$ Factor.

$2x = 0$ or $x - 25 = 0$ Use the zero-factor theorem.

$x = 0$ or $x = 25$

Answer Because the speed of the boat cannot be 0, the boat travels at 25 miles per hour in still water.

Check If the boat travels at the rate of 25 miles per hour in still water, its speed against the current is $25 - 5 = 20$ miles per hour. To go 20 miles upstream, the boat would take $\frac{20}{20} = 1$ hour. The speed of the boat with the current is $25 + 5 = 30$ miles per hour. To go 30 miles downstream, the boat would take $\frac{30}{30} = 1$ hour. The time going upstream plus the time going downstream is $1 + 1 = 2$, so our solution is correct.

Your Turn 2 A plane flies 1500 miles against the wind in the same amount of time it took to fly 1800 miles with the wind. If the speed of the wind was 50 miles per hour, find the speed of the plane in still air.

Objective 2 Solve problems involving direct variation.

In 2013, the Internal Revenue Service allowed a \$0.565 per mile deduction for each mile driven for business purposes. If d is the amount of the deduction and n is the number of miles driven, then $d = 0.565n$. In the following table, we use that formula to determine the deduction for various values of n.

Number of Miles	Deduction ($d = 0.565n$)
1	0.565
2	1.13
3	1.695
4	2.26
5	2.825

From the table, we see that as the number of miles increases, so does the amount of the deduction. Or, more formally, as values of n increase, so do values of d. In $d = 0.565n$, the two variables, d and n, are said to **vary directly**, or are *directly proportional*, and 0.565 is the constant of variation.

Definition **Direct variation:** Two variables y and x vary directly if $y = kx$. If y varies directly as the nth power of x, then $y = kx^n$, where k is the constant of variation.

In words, direct variation is written as *y varies directly as x* or *y is directly proportional to x*, and these phrases translate to $y = kx$. The expression $y = kx^n$ is translated as *y varies directly as the nth power of x* or *y is directly proportional to the nth power of x*.

In all variation problems in this section, we are given one set of values of the variables, which we use to find the constant of variation, k. We then use this value of k and the equation to find other values of the variable(s).

Procedure Solving Variation Problems

1. Write the equation.
2. Substitute the initial values and find k.
3. Substitute for k in the equation found in step 1.
4. Substitute the additional data into the equation from step 3 and solve for the unknown.

Answer to Your Turn 2
550 mph

Example 3 Solve the direct variations.

a. Suppose y varies directly as x. If $y = 18$ when $x = 5$, find y when $x = 8$.

Solution:

$y = kx$ Translate "y varies directly as x."

$18 = k \cdot 5$ Replace y with 18 and x with 5.

$3.6 = k$ Divide both sides by 5.

In $y = kx$, replacing k with 3.6 gives $y = 3.6x$. Now find y when $x = 8$.

$y = 3.6(8) = 28.8$ In $y = 3.6x$, replace x with 8.

b. Suppose u varies directly as the square of v. If $u = 54$ when $v = 3$, find u when $v = 5$.

Solution:

$u = kv^2$ Translate "u varies directly as the square of v."

$54 = k \cdot 3^2$ Replace u with 54 and v with 3.

$54 = 9k$ Square 3.

$6 = k$ Divide both sides by 9.

In $u = kv^2$, replacing k with 6 gives $u = 6v^2$. Now find u when $v = 5$.

$u = 6(5)^2$ In $u = 6v^2$, replace v with 5

$u = 6(25)$ Square 5.

$u = 150$ Multiply.

Your Turn 3 Suppose m varies directly as the square of n. If $m = 24$ when $n = 2$, find m when $n = -3$.

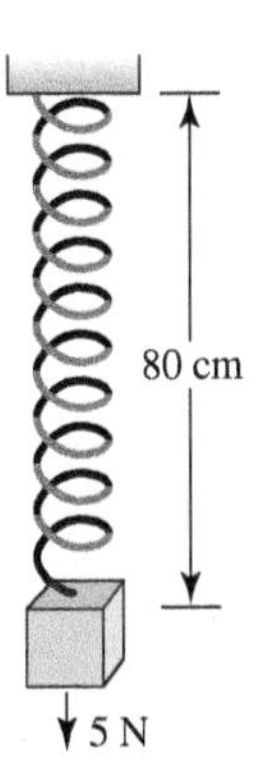

Example 4 In physical science, Hooke's law states that the distance a spring of uniform material and thickness is stretched varies directly with the force applied to the spring. If a force of 5 newtons stretches a spring 80 centimeters, how much force is required to stretch the spring 128 centimeters?

Understand Translating "the distance a spring of uniform material and thickness is stretched varies directly with the force," we write $d = kF$, where d represents distance and F represents the force.

Plan Use $d = kF$, replacing d with 80 centimeters and F with 5 newtons to solve for the value of k. Then use that value in $d = kF$ to solve for the force required to stretch the spring 128 centimeters.

Execute $80 = k \cdot 5$ Replace d with 80 and F with 5.

$16 = k$ Divide both sides by 5.

Replacing k with 16 in $d = kF$, we have $d = 16F$, which we use to solve for F when d is 128 centimeters.

$128 = 16F$ Substitute 128 for d.

$8 = F$ Divide both sides by 16.

Answer To stretch the spring 128 centimeters, a force of 8 newtons is applied.

Check A force of 8 newtons stretches the spring $d = 16(8) = 128$ centimeters.

Your Turn 4 The pressure exerted by a liquid varies directly as the depth beneath the surface. If an object is submerged in seawater to a depth of 3 feet, the pressure is 192 pounds per square foot. Find the pressure per square foot if an object is submerged in seawater to a depth of 8 feet.

Answer to Your Turn 3
54

Answer to Your Turn 4
512 lb./ft.^{2}

Objective 3 Solve problems involving inverse variation.

Suppose a campaign worker must stuff 100 envelopes. If r is the rate at which she can stuff the envelopes and t is the time required, then $t = \frac{100}{r}$. For example, if she can stuff 2 envelopes per minute, it will take her $\frac{100}{2} = 50$ minutes to stuff the envelopes. In the following table, we use $t = \frac{100}{r}$ to see the relationship between the rate at which she can stuff envelopes and the amount of time required to complete the job.

Rate r	Time $\left(t = \frac{100}{r}\right)$
1 per minute	100 minutes
2 per minute	50 minutes
4 per minute	25 minutes

From the table, we see that as the rate at which she stuffs the envelopes increases, the time required decreases. More formally, as values of r increase, values of t decrease. In $t = \frac{100}{r}$, the two variables, t and r, are said to be in **inverse variation**, or are *inversely proportional*, and 100 is the constant of variation.

Definition **Inverse variation:** Two variables y and x vary inversely if $y = \frac{k}{x}$. If $y = \frac{k}{x^n}$, then y varies inversely as the nth power of x, where k is the constant of variation.

In words, inverse variation is written as *y varies inversely as x* or *y is inversely proportional to x*, and these phrases translate to $y = \frac{k}{x}$. Similarly, *y varies inversely as the nth power of x* or *y is inversely proportional to the nth power of x* translates to $y = \frac{k}{x^n}$.

Example 5 Boyle's law states that if the temperature is held constant, the volume of a gas in a closed container is inversely proportional to the pressure applied to it. If a gas has a volume of 4 cubic feet when the pressure is 10 pounds per square foot, find the volume when the pressure is 2.5 pounds per square foot.

Understand Because the volume and pressure vary inversely, we can write $V = \frac{k}{P}$, where V represents volume and p represents pressure.

Plan Use the fact that the volume is 4 cubic feet when the pressure is 10 pounds per square foot to find the value of the constant, k. Then use this value of the constant to find the volume when the pressure is 2.5 pounds per square foot.

Execute

$$4 = \frac{k}{10} \quad \text{Substitute for } V \text{ and } P.$$

$$10 \cdot 4 = 10 \cdot \frac{k}{10} \quad \text{Multiply both sides by 10.}$$

$$40 = k$$

Replacing k with 40 in $V = \frac{k}{P}$, we have $V = \frac{40}{P}$, which we use to solve for V when P is 2.5 pounds per square foot.

$$V = \frac{40}{2.5} \quad \text{Substitute 40 for } P.$$

$$V = 16$$

Answer With a pressure of 2.5 pounds per square foot, the volume is 16 cubic feet.

Your Turn 5 The intensity of a light varies inversely as the square of the distance from the light source. If the intensity is 150 foot-candles (fc) when the distance is 2 meters, find the intensity if the distance is 4 meters.

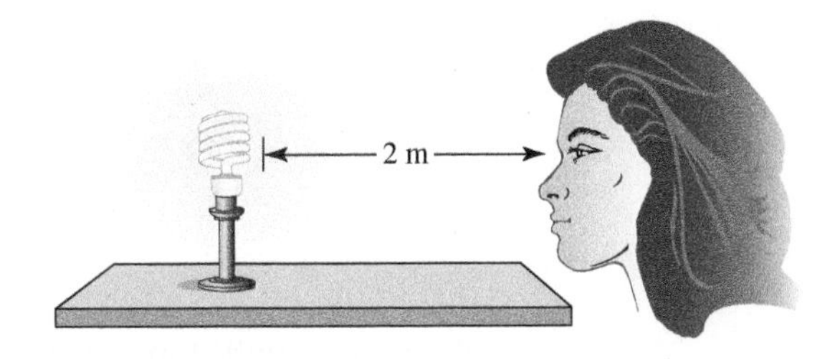

Objective 4 Solve problems involving joint variation.

Often, one quantity will vary as the product of two or more quantities. This is called **joint variation**.

Definition **Joint variation:** If y varies jointly as x and z, then $y = kxz$, where k is the constant of variation.

In words, joint variation is written as *y varies jointly as x and z* or *y is jointly proportional to x and z*, and these phrases translate to $y = kxz$.

Example 6 Suppose y varies jointly with x and z. If $y = 72$ when $x = 4$ and $z = 6$, find y when $x = 2$ and $z = 5$.

Solution:

$y = kxz$ Translate "y varies jointly with x and z."

$72 = k \cdot 4 \cdot 6$ Replace y with 72, x with 4, and z with 6.

$72 = 24k$ Multiply 4 and 6.

$3 = k$ Divide both sides by 24.

In $y = kxz$, replacing k with 3 gives $y = 3xz$. Now find y when $x = 2$ and $z = 5$.

$y = 3 \cdot 2 \cdot 5$ In $y = 3xz$, replace x with 2 and z with 5.

$y = 30$ Multiply.

Your Turn 6 Suppose n varies jointly with p and the square of q. If $n = 90$ when $p = 5$ and $q = 3$, find n when $p = 2$ and $q = 5$.

Objective 5 Solve problems involving combined variation.

Problems involving more than one type of variation are called *combined variation.*

Example 7 Coulomb's law states that the force, F, between two charges q_1 and q_2 in a vacuum varies jointly with the charges and inversely with the square of the distance, d, between them. If the force is 5 dynes when q_1 is 9 electrostatic units and q_2 is 20 electrostatic units and the distance between the charges is 6 centimeters, find the force between the particles if q_1 is 16 electrostatic units, q_2 is 12 electrostatic units, and the distance between the particles is 8 centimeters.

Answer to Your Turn 5
37.5 fc

Answer to Your Turn 6
$n = 100$

Solution: $F = \dfrac{kq_1q_2}{d^2}$ Translate "the force, F, between two charges q_1 and q_2 in a vacuum varies jointly with the charges and inversely with the square of the distance, d, between them."

$5 = \dfrac{k \cdot 9 \cdot 20}{6^2}$ Replace F with 5, q_1 with 9, q_2 with 20, and d with 6.

$5 = \dfrac{180k}{36}$ Multiply.

$5 = 5k$ Simplify.

$1 = k$ Divide both sides by 5.

Replacing k with 1 in $F = \dfrac{kq_1q_2}{d^2}$ gives $F = \dfrac{q_1q_2}{d^2}$. Now find F when $q_1 = 16$, $q_2 = 12$, and $d = 8$.

$F = \dfrac{16 \cdot 12}{8^2}$ In $F = \dfrac{q_1q_2}{d^2}$, replace q_1 with 16, q_2 with 12, and d with 8.

$F = \dfrac{192}{64}$ Multiply.

$F = 3$ Simplify.

Answer: The force is 3 dynes.

Your Turn 7 The maximum height, h, obtained by an object that is launched vertically upward varies directly with the square of the velocity, v, and inversely as the acceleration due to gravity, g. If an object that is launched vertically upward with a velocity of 64 feet per second and acceleration due to gravity of 32 feet per second per second obtains a maximum height of 64 feet, find the maximum height obtained by an object launched vertically upward with a velocity of 128 feet per second and acceleration due to gravity of 32 feet per second per second.

Answer to Your Turn 7
256 ft.

7.5 Exercises

For Extra Help MyMathLab®

Note: Exercises marked with a ★ represent challenging exercises.

Objective 1

Prep Exercise 1 If a person can complete a task in x hours, what part of the task can he complete in 1 hour?

Prep Exercise 2 If a represents the number of hours for one person to complete a task and b represents the number of hours for a second person to complete the same task, represent the part of the task completed in 1 hour while both people work together.

Prep Exercise 3 If a vehicle travels a distance of 100 miles at a rate of r, write an expression for the time, t, it takes the vehicle to travel the 100 miles. Which type of variation is this?

Prep Exercise 4 If Fred can row a canoe x miles per hour in still water and the current in the Wekiva River is 3 miles per hour, write an expression for Fred's rate rowing upstream and an expression for his rate rowing downstream in the river.

For Exercises 1–16, use a table to organize the information; then solve. See Examples 1 and 2.

1. Jason can wash and wax his car in 4 hours. His younger sister can wash and wax the same car in 6 hours. Working together, how fast can they wash and wax the car?

2. Joe can paint the outside of a house in 6 days working alone. His helper, Frank, can paint the outside of the same house in 9 days working alone. How long will it take them to paint the outside of the house working together?

3. Alicia and Geraldine have volunteered to make quilts for a charity auction. If Alicia can make a quilt in 25 days and Geraldine can make a quilt in 35 days, in how many days can they make a quilt working together?

4. It takes Alice 90 minutes to put a futon frame together, and it takes Maya 60 minutes to put the same type of frame together. If they worked together, how long would it take to put a frame together?

5. Working together, it takes two roofers 4 hours to put a new roof on a portable classroom. If the first roofer can do the job by himself in 6 hours, how many hours will it take the second roofer to do the job by himself?

6. Working together, Rita and Tiffany can cut and trim a lawn in 2 hours. If it takes Rita 5 hours to do the lawn by herself, how long will it take Tiffany to do the lawn by herself?

7. With both the cold water and hot water faucets open, it takes 9 minutes to fill a bathtub. The cold water faucet alone takes 15 minutes to fill the tub. How long will it take to fill the tub with just the hot water faucet?

8. The cargo hold of a ship has two loading pipes. Used together, the two pipes can fill the cargo hold in 6 hours. If the larger pipe alone can fill the hold in 8 hours, how many hours will it take the smaller pipe to fill the hold by itself?

9. A bus leaves Valdosta, Georgia, at 10 A.M. traveling north on I-75 at an average rate of 63 miles per hour. At the same time, a car leaves Valdosta traveling south on I-75 at an average rate of 72 miles per hour. At what time will they be 675 miles apart?

10. Sailing in opposite directions, an aircraft carrier and a destroyer leave their base in Hawaii at 5 A.M. If the destroyer sails at 30 miles per hour and the aircraft carrier sails at 20 miles per hour, at what time will the two ships be 300 miles apart?

11. Rapid City is 360 miles from Sioux Falls. At 6 A.M., a freight train leaves Rapid City for Sioux Falls, and at the same time, a passenger train leaves Sioux Falls for Rapid City. The two trains meet at 9 A.M. If the freight train travels $\frac{3}{5}$ of the speed of the passenger train, how fast does the passenger train travel?

12. Houston and Calgary are about 2100 miles apart. At 2 P.M., an airplane leaves Houston for Calgary, flying 250 miles per hour. At the same time, an airplane leaves Calgary for Houston, flying 450 miles per hour. How long will it be before the two airplanes meet?

13. Jack leaves his home in Atlanta, traveling north on I-75 at an average rate of 45 miles per hour. Two hours later, his wife, Frances, leaves home and takes the same route, traveling at an average rate of 60 miles per hour. How long will it take Frances to catch Jack?

14. A ship leaves port traveling 15 miles per hour. Two hours later, a speedboat leaves the same port traveling 40 miles per hour. How long will it take the speedboat to overtake the ship?

15. An airliner flies against the wind from Washington, D.C., to San Francisco in 5.5 hours. It flies back to Washington, D.C., with the wind in 5 hours. If the average speed of the wind is 21 miles per hour, what is the speed of the airliner in still air?

16. A river has a current of 3 miles per hour. A boat goes 40 miles upstream in the same time it goes 50 miles downstream. Find the speed of the boat in still water.

Objective 2

Prep Exercise 5 Translate "m varies directly as n."

Prep Exercise 6 After substituting the initial values into the translated equation, what is the next step?

Prep Exercise 7 If p varies directly as q, then as q decreases, what happens to the value of p? (Assume that $k > 0$).

For Exercises 17–26, solve the direct variations. See Examples 3 and 4.

17. Suppose a varies directly as b. If $a = 4$ when $b = 9$, find a when $b = 27$.

18. Suppose r varies directly as s. If $r = 6$ when $s = 9$, find r when $s = 15$.

19. Suppose y varies directly as the square of x. If $y = 100$ when $x = 5$, find y when $x = 3$.

20. Suppose t varies directly as the square of u. If $t = 45$ when $u = 3$, find t when $u = -2$.

21. Suppose m varies directly as n. If $m = 6$ when $n = 8$, what is the value of n when $m = 9$?

22. Suppose x varies directly as y. If $x = 12$ when $y = 15$, what is the value of y when $x = 4$?

23. The price of salmon at a fish market is constant, so the cost increases with the quantity purchased. Tamika notices that 2.5 pounds cost $16.25. If she plans to buy 6 pounds of salmon, how much will she pay?

24. At a produce stand, the cost of zucchini is constant and sells 3 for $0.89. To the nearest cent, find the cost of 5 zucchini.

25. According to Charles's law, if the pressure is held constant, the volume of a gas varies directly with the temperature measured on the Kelvin scale. If the volume of a gas is 288 cubic centimeters when the temperature is 80 K, find the volume when the temperature is 50 K.

26. According to Ohm's law, the current, I, which is measured in amperes, in an electrical circuit varies directly with the voltage, V. If the current is 0.64 ampere when the voltage is 24 volts, find the current when the voltage is 48 volts.

In Exercises 27–30, use the fact that, ignoring air resistance, the distance a free-falling body falls is directly proportional to the square of the time it has been falling. See Example 4.

27. On Earth, if an object falls 144 feet in 3 seconds, how many seconds will it take the object to fall 400 feet?

28. On Earth, if an object falls 39.2 meters in 2 seconds, how far will it fall in 5 seconds?

29. On the Moon, an object falls 6.48 meters in 2 seconds. How far will it fall in 5 seconds?

30. On the Moon, an object falls 14.58 meters in 3 seconds. How long will it take the object to fall 58.32 meters?

Of Interest

Without air resistance, all objects fall at the same rate. This was demonstrated by astronaut David Scott in 1971 when he dropped a hammer and a feather from the same height on the Moon and they landed on the Moon's surface at the same time.

31. The circumference of a circle varies directly with its diameter. If the circumference is 12.56 feet when the diameter is 4 feet, find the diameter when the circumference is 21.98 feet.

32. The pressure on an object submerged in a liquid varies directly as the depth beneath the surface. If an object is submerged in gasoline to a depth of 6 feet, the pressure is 253.8 pounds per square foot. Find the pressure if the object is submerged 9 feet.

Objective 3

Prep Exercise 8 Translate "t varies inversely as u."

Prep Exercise 9 If m is inversely proportional to n, then as the n quantity increases, what happens to the value of m? (Assume that $k > 0$.)

For Exercises 33–40, solve the inverse variations. See Example 5.

33. Suppose a varies inversely as b. If $a = 3.2$ when $b = 5$, what is b when a is 8?

34. Suppose p varies inversely as q. If p is 8 when q is 3.25, what is q when p is 2?

35. Suppose m varies inversely as n. If n is 6 when m is 11, what is n when m is 8?

36. Suppose x varies inversely as y. If $y = 12$ when $x = 3$, what is y when x is 12?

37. By Charles's law, if the temperature is held constant, the pressure that a gas exerts against the walls of a container is inversely proportional to the volume of the container. A gas is inside a cylinder with a piston at one end that can vary the volume of the cylinder. When the volume of the cylinder is 20 cubic inches, the pressure inside is 40 psi (pounds per square inch). Find the pressure of the gas if the piston compresses the gas to a volume of 16 cubic inches.

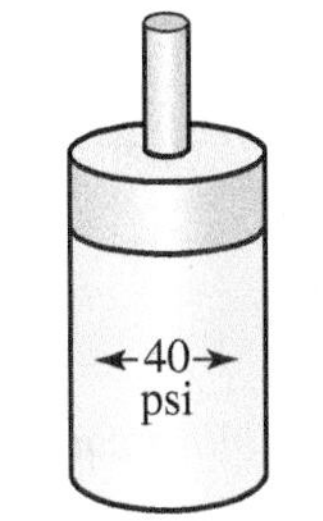

40 psi at 20 cubic inches

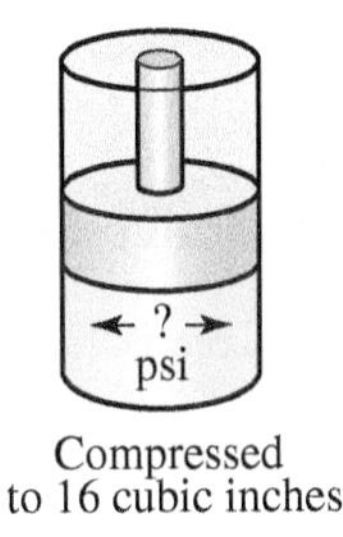

Compressed to 16 cubic inches

38. Find the volume in the cylinder from Exercise 37 if the piston compresses the gas to a pressure of 30 psi.

39. In an electrical conductor, the current, I, which is measured in amperes, varies inversely as the resistance, R, which is measured in ohms. If the current is 10 amperes when the resistance is 15 ohms, find the resistance when the current is 25 amperes.

40. In Exercise 39, find the current when the resistance is 15 ohms.

In Exercises 41 and 42, use the fact that the length of a radio wave varies inversely with its frequency. See Example 5.

41. If the length of a radio wave is 400 meters when the frequency is 900 kilohertz, find the wavelength when the frequency is 600 kilohertz.

42. If the frequency of a radio wave is 800 kilohertz when the wavelength is 200 meters, find the frequency when the radio wavelength is 500 meters.

43. The f-stop setting for a camera lens is inversely proportional to the aperture, which is the size of the opening in the lens. For a particular lens, the f-stop setting is 5.6 when the aperture is 50 mm. Find the aperture when the f-stop is 2.8.

44. The weight of an object varies inversely with the square of the distance from the center of the Earth. At Earth's surface, the weight of the space shuttle is about 4.5 million pounds. At that point, it is 4000 miles from the center of the Earth. How much will the space shuttle weigh when it is in orbit 200 miles above the surface of the Earth?

Objective 4

Prep Exercise 10 Translate "c varies jointly with d and e."

For Exercises 45–52, solve the joint variations. See Example 6.

45. Suppose a varies jointly with b and c. If $a = 96$ when $b = 6$ and $c = 4$, find a when $b = 2$ and $c = 8$.

46. Suppose m varies jointly with p and q. If $m = 70$ when $p = 5$ and $q = 2$, find m when $p = 6$ and $q = 8$.

47. Suppose a varies jointly as the square of b and c. If $a = 96$ when $b = 2$ and $c = 6$, find a when $b = 3$ and $c = 2$.

48. Suppose x varies jointly with y and the square of z. If $x = 40$ when $y = 1$ and $z = 2$, find x when $y = 2$ and $z = 1$.

49. If the width of a rectangular solid is held constant, the volume varies jointly with the length and the height. If the volume is 192 cubic inches when the length is 8 inches and the height is 6 inches, find the volume when the length is 7 inches and the height is 12 inches.

50. For a fixed amount of principal, the simple interest varies jointly with the rate and the time. If the simple interest is \$1000 when the rate is 4% and the time is 5 years, find the simple interest when the rate is 6% and the time is 10 years.

51. The volume of a right circular cylinder varies jointly as the square of the radius and the height. If the volume is 301.6 cubic centimeters when the radius is 4 centimeters and the height is 6 centimeters, find the volume when the radius is 3 centimeters and the height is 6 centimeters.

52. The number of units produced varies jointly with the number of workers and the number of hours worked per worker. If 80 units can be produced by 10 workers who work 40 hours each, how many units can be produced by 15 workers who work 30 hours each?

Objective 5

For Exercises 53–60, solve the combined variations. See Example 7.

53. Suppose y varies directly with x and inversely as z. If $y = 8$ when $x = 4$ and $z = 6$, find y when $x = 5$ and $z = 10$.

54. Suppose m varies directly with n and inversely as p. If $m = 27$ when $n = 6$ and $p = 4$, find m when $n = 9$ and $p = 6$.

55. Suppose y varies jointly with x and z and inversely with n. If $y = 81$ when $x = 4$, $z = 9$, and $n = 8$, find n when $x = 6$, $y = 8$, and $z = 12$.

56. Suppose p varies jointly with q and r and inversely with s. If $p = 15$ when $q = 5$, $r = 3$, and $s = 6$, find p when $q = 5$, $r = 3$, and $s = 9$.

57. The resistance of a wire, R, varies directly with the length and inversely with the square of the diameter. If the resistance is 7.5 ohms when the wire is 6 meters long and the diameter is 0.02 meter, find the resistance in a wire of the same material if the length is 10 meters and the diameter is 0.04 meter.

58. The universal gas law states that the volume of a gas varies directly with the temperature and inversely with the pressure. If the volume of a gas is 1.75 cubic meters when the temperature is 70 K and the pressure is 20 grams per square centimeter, find the volume when the temperature is 80 K and the pressure is 40 grams per square centimeter.

59. Newton's Law of Universal Gravitation states that the force of attraction between two bodies is jointly proportional to their masses and inversely proportional to the square of the distance between them. If the force of attraction between two masses of 4 grams and 6 grams that are 3 centimeters apart is 48 dynes, find the force of attraction between two masses of 2 grams and 12 grams that are 6 centimeters apart.

60. The weight-carrying capacity of a rectangular beam varies jointly with its width and the square of its height and inversely as its length. If a beam is 4 inches wide, 6 inches high, and 10 feet long, it has a carrying capacity of 1400 pounds. Find the carrying capacity of a beam made of the same material if it is 3 inches wide, 5 inches high, and 12 feet long.

★ **61.** The table lists the number of miles a car can drive on the given number of gallons of gas.

a. Plot the ordered pairs of data as points in the coordinate plane. Connect the points to form a graph.

b. Are the number of miles driven and the number of gallons of gas directly proportional or inversely proportional? Explain.

Number of Miles	Number of Gallons
135	6
180	8
225	10
270	12

c. Find the constant of variation. What does it represent?

d. Does the data represent a function? Explain.

★ **62.** The table lists the time required to make a long trip at various average driving speeds.

a. Plot the ordered pairs of data as points in the coordinate plane. Connect the points to form a graph.

b. Are speeds and driving times directly proportional or inversely proportional? Explain.

Speed (in mph)	Driving Time (in hr.)
60	10
50	12
40	15
30	20
20	30

c. Find the constant of variation. What does it represent?

d. Do the data represent a function? Explain.

Review Exercises

Exercises 1–5 Expressions

[1.3] *For Exercises 1 and 2, evaluate.*

1. 2^3

2. -12^2

[5.1] 3. Multiply: $(6a^3b^2)(-5ab^4)$

[5.1] 4. Divide: $\dfrac{-48x^3y^6}{-12x^7y^4}$

[1.4] 5. Combine like terms: $-3(x + 4) - 2(x - 5)$

Exercises 6 Equations and Inequalities

[6.4] 6. Solve: $3x^2 - 13x - 10 = 0$

Chapter 7 Summary and Review Exercises

Complete each incomplete definition, rule, or procedure; study the key examples; and then work the related exercises.

7.1 Simplifying, Multiplying, and Dividing Rational Expressions

Definitions/Rules/Procedures	Key Example(s)
A **rational expression** is an expression that can be written in the form $\frac{P}{Q}$, where P and Q are ________ and $Q \neq 0$.	
To **simplify a rational expression to lowest terms**: 1. ________ the numerator and denominator completely. 2. Divide out all ________ ________ in the numerator and denominator. 3. ________ the remaining factors in the numerator and the remaining factors in the denominator.	Simplify. a. $\frac{12a^3b}{21a^2b^2} = \frac{2\cdot 2\cdot 3\cdot a\cdot a\cdot a\cdot b}{3\cdot 7\cdot a\cdot a\cdot b\cdot b} = \frac{4a}{7b}$ b. $\frac{2x^2-7x-15}{x^2-25} = \frac{(2x+3)(x-5)}{(x+5)(x-5)} = \frac{2x+3}{x+5}$

Exercises 1–12 Expressions

For Exercises 1–12, simplify each rational expression.

1. $\frac{32x^2y^5}{8xy^3}$
2. $\frac{-42p^6q^2}{27p^3q^4}$
3. $-\frac{18m^3n^2}{54m^5n^5}$
4. $\frac{14x+63}{10x+45}$
5. $\frac{4a-20}{7a-35}$
6. $\frac{a^2+4a-21}{a^2+9a+14}$
7. $\frac{2x-3}{6x^2-x-12}$
8. $\frac{25x^2-9}{10x^2+x-3}$
9. $\frac{2c+3d}{8c^2+26cd+21d^2}$
10. $\frac{5-2x}{4x^2-25}$
11. $\frac{8ac-2a+12c-3}{6ac-8a+9c-12}$
12. $\frac{8x^3+27}{2x^2-7x-15}$

Definitions/Rules/Procedures	Key Example(s)
To **multiply rational expressions:** 1. ________ each numerator and denominator completely. 2. Divide out all ________ ________ to both the numerators and denominators. 3. Multiply ________ by ________ and ________ by ________. 4. Simplify as needed.	Multiply. a. $\frac{6m^2}{9n}\cdot\frac{12n^3}{10m^4} = \frac{2\cdot 3\cdot m\cdot m}{3\cdot 3\cdot n}\cdot\frac{2\cdot 2\cdot 3\cdot n\cdot n\cdot n}{2\cdot 5\cdot m\cdot m\cdot m\cdot m}$ $= \frac{4n^2}{5m^2}$ b. $\frac{3x^2-15x}{3x^2-17x+10}\cdot\frac{x^2+5x+6}{6x^2+12x}$ $= \frac{3x(x-5)}{(3x-2)(x-5)}\cdot\frac{(x+3)(x+2)}{2\cdot 3x(x+2)}$ $= \frac{1}{3x-2}\cdot\frac{x+3}{2}$ $= \frac{x+3}{2(3x-2)}$ or $\frac{x+3}{6x-4}$

Exercises 13–22 Expressions

For Exercises 13–22, find each product.

13. $\dfrac{28m^2n^3}{15p^2q^6} \cdot \dfrac{25p^4q^3}{14mn}$

14. $\dfrac{8m - 12n}{32m^3n} \cdot \dfrac{36mn^3}{6m - 9n}$

15. $\dfrac{14x - 21}{30x - 40} \cdot \dfrac{15x - 20}{8x - 12}$

16. $\dfrac{25x^2 - 16}{16x + 24} \cdot \dfrac{12x + 18}{5x - 4}$

17. $\dfrac{2x - 7}{12} \cdot \dfrac{10}{7 - 2x}$

18. $\dfrac{x^2 - 16}{x^2 + 6x + 8} \cdot \dfrac{x^2 - 3x - 10}{x^2 - 25}$

19. $\dfrac{y^2 - 9}{y^2 - 3y - 18} \cdot \dfrac{y^2 - 4y - 12}{y^2 - 6y + 9}$

20. $\dfrac{8x^2 + 2x - 15}{6x^2 + x - 12} \cdot \dfrac{3x^2 - 13x + 12}{3x^2 - 7x - 6}$

21. $\dfrac{ab + 2ad - 3bc - 6cd}{ab - 4ad + 2bc - 8cd} \cdot \dfrac{ab + 5ad + 2bc + 10cd}{ab + 5ad - 3bc - 15cd}$

22. $\dfrac{8x^3 + y^3}{4x^2 - y^2} \cdot \dfrac{6x^2 + 5xy - 4y^2}{4x^2 - 2xy + y^2}$

Definitions/Rules/Procedures	Key Example(s)
To **divide rational expressions:** **1.** Write an equivalent multiplication statement using $\dfrac{P}{Q} \div \dfrac{R}{S} =$ __________, where P, Q, R, and S are polynomials and Q, R, and $S \neq 0$ **2.** Simplify using the procedure for __________ rational expressions	Divide. **a.** $\dfrac{12c^2d}{7cd^3} \div \dfrac{8c^4}{21d^2} = \dfrac{12c^2d}{7cd^3} \cdot \dfrac{21d^2}{8c^4}$ $= \dfrac{2 \cdot 2 \cdot 3 \cdot c \cdot c \cdot d}{7 \cdot c \cdot d \cdot d \cdot d} \cdot \dfrac{3 \cdot 7 \cdot d \cdot d}{2 \cdot 2 \cdot 2 \cdot c \cdot c \cdot c \cdot c}$ $= \dfrac{9}{2c^3}$ **b.** $\dfrac{8a^2 + 6a - 9}{8a^2 + 14a - 15} \div \dfrac{2a^2 + 11a + 12}{4a^2 + 19a + 12}$ $= \dfrac{8a^2 + 6a - 9}{8a^2 + 14a - 15} \cdot \dfrac{4a^2 + 19a + 12}{2a^2 + 11a + 12}$ $= \dfrac{(4a - 3)(2a + 3)}{(4a - 3)(2a + 5)} \cdot \dfrac{(4a + 3)(a + 4)}{(2a + 3)(a + 4)}$ $= \dfrac{4a + 3}{2a + 5}$

Exercises 23–30 Expressions

For Exercises 23–30, find each quotient.

23. $\dfrac{39a^2b^4}{27x^3y^2} \div \dfrac{26a^3b}{28xy^4}$

24. $\dfrac{8y^3 - 20y^2}{9y} \div \dfrac{6y^4 - 15y^3}{10y^2}$

25. $\dfrac{21a + 7b}{16a - 24b} \div \dfrac{42a + 14b}{24a - 36b}$

26. $\dfrac{z^2}{z^2 - 2z - 8} \div \dfrac{9z^3 + 3z^4}{z^2 + 5z + 6}$

27. $\dfrac{4p^2 - 9}{24p + 28} \div \dfrac{10p - 15}{36p^2 - 49}$

28. $\dfrac{16p^2 - 8pq - 3q^2}{8p^2 + 22pq + 5q^2} \div \dfrac{8p^2 - 10pq + 3q^2}{10p^2 + pq - 3q^2}$

29. $\dfrac{x^3 - 8y^3}{x^2 - 36y^2} \div \dfrac{x^2 + 2xy - 8y^2}{x^2 - 2xy - 24y^2}$

30. $\dfrac{xz + 2xw - 3yz - 6yw}{4xz - 2xw + 6yz - 3yw} \div \dfrac{xz - 2xw - 3yz + 6yw}{2xz - 4xw + 3yz - 6yw}$

Definitions/Rules/Procedures	Key Example(s)
A **rational function** is a function expressed in terms of __________ ________________.	
To **evaluate a rational function**, replace the ____________ with the indicated ____________ and simplify.	Given $f(x) = \dfrac{x^2 + 2x + 3}{x - 4}$, find $f(2)$. **Solution:** $f(2) = \dfrac{2^2 + 2 \cdot 2 + 3}{2 - 4}$ $= \dfrac{4 + 4 + 3}{-2}$ $= -\dfrac{11}{2}$
To find the **domain** of a rational function: **1.** Write an equation that sets the ____________ equal to ____________. **2.** Solve the equation. **3.** ____________ the value(s) found in step 2 from the function's domain.	Find the domain of $f(x) = \dfrac{x + 5}{x^2 - 2x - 24}$. **Solution:** $x^2 - 2x - 24 = 0$ Set the denominator equal to 0. $(x + 4)(x - 6) = 0$ Factor. $x + 4 = 0$ or $x - 6 = 0$ Use the zero-factor theorem. $x = -4$ or $x = 6$ Solve each equation. The domain is $\{x \mid x \neq -4, 6\}$.

Exercises 31–34 Expressions

31. Given $f(x) = \dfrac{2x}{5 - x}$, find

a. $f(3)$ b. $f(0)$ c. $f(-1)$

32. Given $g(x) = \dfrac{x + 2}{x^2 - 8x + 15}$, find

a. $g(4)$ b. $g(-2)$ c. $g(5)$

For Exercises 33 and 34, find the domain of the rational function.

33. $f(x) = \dfrac{2x + 4}{3x - 5}$

34. $f(x) = \dfrac{3x - 4}{x^2 + 4x - 12}$

Definitions/Rules/Procedures	Key Example(s)
Graphs of **rational functions** have ______ ______ at values of x that make the denominator equal to 0 but do not make the numerator equal to 0.	Graph $f(x) = \frac{1}{x-3}$. **Solution:** Because the domain of f is $\{x \mid x \neq 3\}$, the graph has an asymptote at $x = 3$.

x	−5	−2	0	2	4	5
y	$-\frac{1}{8}$	$-\frac{1}{5}$	$-\frac{1}{3}$	−1	1	$\frac{1}{2}$

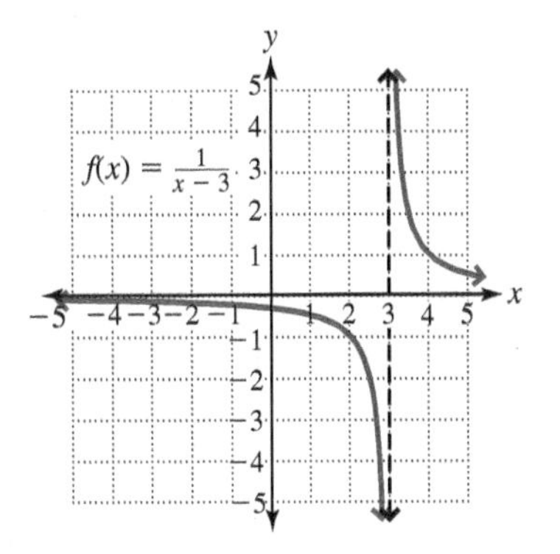

Exercises 35 and 36  Equations and Inequalities

For Exercises 35 and 36, graph.

35. $f(x) = \frac{1}{x-4}$

36. $f(x) = \frac{2}{x+2}$

7.2 Adding and Subtracting Rational Expressions

Definitions/Rules/Procedures	Key Example(s)
To **add or subtract rational expressions** that have the same denominator: **1.** Add or subtract the ______ and keep the same ______. **2.** Simplify to lowest ______.	Add or subtract. Simplify your answers to lowest terms. **a.** $\frac{4a}{a+3} + \frac{12}{a+3} = \frac{4a+12}{a+3}$ Add numerators. $= \frac{4(a+3)}{a+3}$ Factor. $= 4$ Simplify.

Definitions/Rules/Procedures	Key Example(s)
	b. $\frac{y^2+3y}{y^2+4y+4} - \frac{5y+8}{y^2+4y+4}$ $= \frac{(y^2+3y)-(5y+8)}{y^2+4y+4}$ Subtract numerators. $= \frac{(y^2+3y)+(-5y-8)}{y^2+4y+4}$ Write the equivalent addition. $= \frac{y^2-2y-8}{y^2+4y+4}$ Combine like terms. $= \frac{(y-4)(y+2)}{(y+2)(y+2)}$ Factor. $= \frac{y-4}{y+2}$ Simplify.

Exercises 37–40 Expressions

For Exercises 37–40, add or subtract. Simplify your answers to lowest terms.

37. $\frac{5r}{14x} + \frac{3r}{14x}$

38. $\frac{3a+4b}{2a-3b} - \frac{a-2b}{2a-3b}$

39. $\frac{2p^2-p+2}{p^2-16} + \frac{p^2+4p-8}{p^2-16}$

40. $\frac{x^2+2x-5}{x^2-3x-18} - \frac{3x+7}{x^2-3x-18}$

Definitions/Rules/Procedures	Key Example(s)
To find the **LCD** of two or more **rational expressions**: **1.** Find the ____________ ____________ of each denominator. **2.** Write a product that contains each unique prime factor the ____________ number of times that factor occurs in any factorization. Or if you prefer to use exponents, write the product that contains each unique prime factor raised to the ____________ exponent that occurs on that factor in any factorization. **3.** Simplify the product found in step 2.	Find the LCD. **a.** $\frac{7}{8x^3y^2}$ and $\frac{9}{12xy^4}$ $8x^3y^2 = 2^3 \cdot x^3 \cdot y^2$ and $12xy^4 = 2^2 \cdot 3 \cdot x \cdot y^4$ $\text{LCD} = 2^3 \cdot 3 \cdot x^3 \cdot y^4 = 24x^3y^4$ **b.** $\frac{x+6}{x^2+4x+4}$ and $\frac{x^2-2x+4}{3x^2+6x}$ $x^2+4x+4 = (x+2)^2$ and $3x^2+6x = 3x(x+2)$ $\text{LCD} = 3x(x+2)^2$

Exercises 41–48 Expressions

For Exercises 41–48, find the LCD and write each fraction pair as equivalent fractions with the LCD as the denominator.

41. $\frac{8b}{9p^3q^4}$ and $\frac{11c}{12p^2q^5}$

42. $\frac{7}{t-4}$ and $\frac{9}{t+2}$

43. $\dfrac{5u}{8u + 12}$ and $\dfrac{9u}{14u + 21}$

44. $\dfrac{2w}{w^2 - 16}$ and $\dfrac{6w}{w^2 + 5w + 4}$

45. $\dfrac{3a}{a^2 + 6a + 9}$ and $\dfrac{10a}{a^2 - 2a - 15}$

46. $\dfrac{m + 1}{m^2 - 2m - 8}$ and $\dfrac{m - 1}{m^2 - 3m - 4}$

47. $\dfrac{2x - 3}{x^3 + 6x^2 + 8x}$ and $\dfrac{6x + 3}{x^5 + 2x^4 - 8x^3}$

48. $\dfrac{6}{x^3}, \dfrac{4x - 3}{x^2 + 10x + 24}$ and $\dfrac{3x}{4x^3 + 16x^2}$

Definitions/Rules/Procedures	Key Example(s)
To **add or subtract rational expressions** with different denominators: **1.** Find the ___________. **2.** Write each rational expression as an equivalent rational expression with the ___________. **3.** Add or subtract the ___________ and keep the ___________ as the denominator. **4.** Simplify	Add or subtract. **a.** $\dfrac{10x}{21y} + \dfrac{7x}{18y} = \dfrac{10x(6)}{21y(6)} + \dfrac{7x(7)}{18y(7)}$ $= \dfrac{60x}{126y} + \dfrac{49x}{126y}$ $= \dfrac{60x + 49x}{126y}$ $= \dfrac{109x}{126y}$ **b.** $\dfrac{x + 2}{4x - 8} - \dfrac{x + 4}{6x - 12}$ $= \dfrac{x + 2}{4(x - 2)} - \dfrac{x + 4}{6(x - 2)}$ $= \dfrac{3}{3} \cdot \dfrac{x + 2}{4(x - 2)} - \dfrac{2}{2} \cdot \dfrac{x + 4}{6(x - 2)}$ $= \dfrac{3x + 6}{12(x - 2)} - \dfrac{2x + 8}{12(x - 2)}$ $= \dfrac{(3x + 6) - (2x + 8)}{12(x - 2)}$ $= \dfrac{3x + 6 + (-2x - 8)}{12(x - 2)}$ $= \dfrac{x - 2}{12(x - 2)}$ $= \dfrac{1}{12}$

Exercises 49–58 Expressions

For Exercises 49–58, add or subtract as indicated.

49. $\frac{8a}{15x} - \frac{4a}{9x}$

50. $\frac{y-3}{4y} - \frac{y+2}{5y}$

51. $\frac{2t-3}{18t^4u^2} + \frac{5t+1}{12t^3u^5}$

52. $\frac{8}{w-3} - \frac{-3}{w}$

53. $\frac{v+4}{4v+8} - \frac{v-2}{2v-6}$

54. $\frac{t^2-5t}{t^2+8t+16} + \frac{6}{t+4}$

55. $\frac{2w+5}{w^2-25} + \frac{6w}{w^2-3w-10}$

56. $\frac{z+9}{4z^2+33z+35} - \frac{z-12}{z^2+14z+49}$

57. $\frac{3}{a} - \frac{3a+5}{a+3} + \frac{5a-3}{a^2+3a}$

58. $\frac{4x-3y}{3x-y} - \frac{3x-4y}{y-3x}$

7.3 Simplifying Complex Rational Expressions

Definitions/Rules/Procedures	Key Example(s)
To **simplify a complex rational expression**, use one of the following methods. **Method 1: 1.** If necessary, rewrite the numerator and/or denominator as a single ________ ________. **2.** Rewrite as a horizontal ________ problem and simplify.	Simplify $\frac{\frac{4x^2}{x^2-36}}{\frac{2x}{2x^2+9x-18}}$. $\frac{\frac{4x^2}{x^2-36}}{\frac{2x}{2x^2+9x-18}} = \frac{4x^2}{x^2-36} \div \frac{2x}{2x^2+9x-18}$ $= \frac{4x^2}{x^2-36} \cdot \frac{2x^2+9x-18}{2x}$ $= \frac{2 \cdot 2 \cdot x \cdot x}{(x+6)(x-6)} \cdot \frac{(2x-3)(x+6)}{2x}$ $= \frac{2x}{x-6} \cdot \frac{2x-3}{1}$ $= \frac{4x^2-6x}{x-6}$

Definitions/Rules/Procedures	Key Example(s)
Method 2: **1.** Multiply the numerator and denominator of the complex rational expression by the ___________ of the fractions in the numerator and denominator. **2.** Simplify	Simplify $\dfrac{x - \dfrac{8}{x+2}}{1 + \dfrac{2}{x+2}}$. $\dfrac{x - \dfrac{8}{x+2}}{1 + \dfrac{2}{x+2}} = \dfrac{\left(x - \dfrac{8}{x+2}\right)(x+2)}{\left(1 + \dfrac{2}{x+2}\right)(x+2)}$ $= \dfrac{x(x+2) - \dfrac{8}{x+2}\cdot(x+2)}{1(x+2) + \dfrac{2}{x+2}\cdot(x+2)}$ $= \dfrac{x^2 + 2x - 8}{x + 2 + 2}$ $= \dfrac{(x+4)(x-2)}{x+4}$ $= x - 2$

Exercises 59–68 Expressions

For Exercises 59–68, simplify.

59. $\dfrac{\frac{4}{5}}{\frac{3}{10}}$

60. $\dfrac{\frac{4}{3} - \frac{8}{9}}{\frac{5}{6} - \frac{4}{12}}$

61. $\dfrac{\frac{u^4v^2}{w}}{\frac{uv^3}{w^2}}$

62. $\dfrac{\frac{x^6y^3}{t^4}}{\frac{x^2y^2}{t}}$

63. $\dfrac{2w - \frac{w}{4}}{6 - \frac{w}{4}}$

64. $\dfrac{\frac{7}{b-15} - 6}{b - 15}$

65. $\dfrac{1 - \frac{1}{y} - \frac{12}{y}}{1 - \frac{6}{y} + \frac{8}{y}}$

66. $\dfrac{x + \frac{25}{x+10}}{1 - \frac{5}{x+10}}$

67. $\dfrac{\frac{6}{y^2} - \frac{1}{xy} - \frac{12}{x^2}}{\frac{4}{y^2} - \frac{4}{xy} - \frac{3}{x^2}}$

68. $\dfrac{\frac{x+4}{x-4} - \frac{x-4}{x+4}}{\frac{x+4}{x-4} + \frac{x-4}{x+4}}$

7.4 Solving Equations Containing Rational Expressions

Definitions/Rules/Procedures	Key Example(s)
To **solve an equation** containing **rational expressions**: 1. Eliminate the denominators of the rational expressions by __________ both sides of the equation by the __________ of all rational expressions in the equation. 2. Solve the resulting __________ using the methods in Chapter 2 (for linear equations) and 6 (for quadratic equations). 3. Check your solution(s) in the __________ equation.	Solve $\frac{x^2 + 5x - 2}{x + 2} - 3 = \frac{x^2 + 6x}{x + 2} - x$. Note that x cannot be -2 because it causes the denominators to be 0, making those expressions undefined. $(x + 2)\left(\frac{x^2 + 5x - 2}{x + 2} - 3\right) = (x + 2)\left(\frac{x^2 + 6x}{x + 2} - x\right)$ $(x + 2)\cdot\frac{x^2 + 5x - 2}{(x + 2)} - 3(x + 2) = (x + 2)\cdot\frac{x^2 + 6x}{x + 2} - x(x + 2)$ $x^2 + 5x - 2 - 3x - 6 = x^2 + 6x - x^2 - 2x$ $x^2 + 2x - 8 = 4x$ $x^2 - 2x - 8 = 0$ $(x - 4)(x + 2) = 0$ $x - 4 = 0$ or $x + 2 = 0$ $x = 4$ or $x = -2$ Because -2 is extraneous, the only solution is 4. We will leave the check of $x = 4$ to the reader.
If $\frac{a}{b} = \frac{c}{d}$, where $b \neq 0$ and $d \neq 0$, then $ad =$ __________.	Solve $\frac{3}{x + 2} = \frac{6}{3x + 2}$. $3(3x + 2) = 6(x + 2)$ Cross multiply. $9x + 6 = 6x + 12$ $3x = 6$ $x = 2$ We will leave the check to the reader.

Exercises 69–76 Equations and Inequalities

For Exercises 69–76, solve and check. Identify any extraneous solutions.

69. $\frac{5}{4t} - \frac{3}{8t} = \frac{1}{2}$

70. $\frac{12}{m - 2} = 9 + \frac{m}{m - 2}$

71. $\frac{x^2}{x - 4} = \frac{3x + 4}{x - 4} - 3$

72. $\frac{2a - 2}{a + 1} - \frac{a}{2a - 3} = \frac{a - 2}{2a - 3}$

73. $\frac{1}{q + 4} = \frac{q}{3q + 2}$

74. $\frac{5}{v^2 + 5v + 6} - \frac{2}{v^2 - 2v - 8} = \frac{3}{v^2 - 16}$

75. $\frac{3}{x^2 + 5x + 6} = \frac{2}{x + 3} + \frac{x - 3}{x^2 + x - 2}$

76. $\frac{7x}{x^2 - 2x - 8} - \frac{4x}{x^2 - 5x + 4} = \frac{3}{x - 4}$

7.5 Applications with Rational Expressions; Variation

Definitions/Rules/Procedures	Key Example(s)
Applications involving **work** and **motion** frequently lead to equations containing rational expressions.	If Mike can paint his room in 10 hours and Susan can paint the same room in 7 hours, how long would it take them to paint the room working together?

Solution: Let x equal the number of hours to paint the room working together. Set up a table.

	Number of Hours	Part done in 1 hour
Mike	10	$\frac{1}{10}$
Susan	7	$\frac{1}{7}$
Together	x	$\frac{1}{x}$

Since the sum of the parts that each can do in 1 hour equals the part they can do together in 1 hour, the equation is:

$$\frac{1}{10} + \frac{1}{7} = \frac{1}{x}$$

$$70x\left(\frac{1}{10} + \frac{1}{7}\right) = 70x\left(\frac{1}{x}\right)$$ Multiply both sides by the LCD, $70x$.

$$70x \cdot \frac{1}{10} + 70x \cdot \frac{1}{7} = 70x \cdot \frac{1}{x}$$ Distribute and simplify.

$$7x + 10x = 70$$ Add $7x$ and $10x$.

$$17x = 70$$ Divide by sides by 17.

$$x = \frac{70}{17}$$

Therefore, it would take $\frac{70}{17}$ hours to paint the room working together.

Exercises 77–79 Equations and Inequalities

For Exercises 77–79, solve.

77. George can write a chapter for a mathematics textbook in 30 days. Lucille can write a chapter in 45 days. How long would it take them to write a chapter together?

78. The Gulf Stream is an ocean current off the eastern coast of the United States. A Coast Guard cutter can sail 200 miles against the current in the same amount of time it can sail 260 miles with the current. If the current is 3 miles per hour, how fast can the cutter sail in still water?

79. It is 510 road miles from Denver to Salt Lake City. A truck leaves Denver for Salt Lake City at 4 A.M. At the same time, an automobile leaves Salt Lake City for Denver. If the truck travels at 45 miles per hour and the auto travels at 55 miles per hour, at what time will the auto and truck meet? (Give your answer to the nearest minute.)

Definitions/Rules/Procedures	Key Example(s)
If y **varies directly** as x, we translate to an equation, __________, where k is a constant.	The distance a car can travel varies directly with the number of gallons of gas used. On the highway, a Toyota Corolla traveled 280 miles using 8 gallons of gas. How far can the same car travel on 12 gallons of gas? **Solution:** $d = kn$ Translate "the distance a car can travel varies directly as the number of gallons of gas used." $280 = k(8)$ Substitute 280 for d and 8 for n. $\frac{280}{8} = \frac{8k}{8}$ Divide both sides by 8. $35 = k$ In $d = kn$, replacing k with 35 gives $d = 35n$. Now find d when $n = 12$. $d = (35)(12) = 420$ miles

Exercises 80–82 Equations and Inequalities

For Exercises 80–82, solve.

80. Suppose p varies directly as q. If $p = 24$ when $q = 4$, find p when $q = 7$.

81. Suppose s varies directly as the square of t. If $s = 72$ when $t = 3$, find s when $t = 5$.

82. The distance a car can travel varies directly with the amount of gas it carries. On a trip, a Chevy Corvette travels 156 miles using 6 gallons of fuel. How many gallons are required to travel 234 miles?

Definitions/Rules/Procedures	Key Example(s)
If y **varies inversely** as x, we translate to an equation, __________, where k is a constant.	The intensity of light is inversely proportional to the square of the distance from the source. A light meter 6 feet from a lightbulb measures the intensity to be 16 foot-candles. What is the intensity 24 feet from the bulb? **Solution:** $I = \frac{k}{d^2}$ Translate "the intensity of light is inversely proportional to the square of the distance from the source." $16 = \frac{k}{6^2}$ Substitute 16 for I and 6 for d. $36(16) = 36 \cdot \frac{k}{36}$ Multiply both sides by 36. $576 = k$ In $I = \frac{k}{d^2}$, replacing k with 576 gives $I = \frac{576}{d^2}$. Now, find I when $d = 24$. $I = \frac{576}{24^2}$ In $I = \frac{576}{d^2}$, replace d with 24. $I = \frac{576}{576} = 1$ foot-candle

Exercises 83–86 Equations and Inequalities

For Exercises 83–86, solve.

83. Suppose y varies inversely as x. If $y = 8$ when $x = 3$, find x when $y = 4$.

84. Suppose m varies inversely as the square of n. If $m = 16$ when $n = 3$, find m when $n = 4$.

85. Suppose m varies directly with n and inversely with p. If $m = 2$ when $n = 3$ and $p = 9$, find m when $n = 6$ and $p = 4$.

86. If the wavelength of a wave remains constant, the velocity, v, of a wave is inversely proportional to its period, T. In an experiment, waves are created in a fluid so that the period is 7 seconds and the velocity is 4 centimeters per second. If the period is increased to 12 seconds, what is the velocity?

Definitions/Rules/Procedures	Key Example(s)
If y **varies jointly** as x and z, we translate to an equation, __________, where k is a constant.	Suppose m varies jointly as n and p. If $m = 288$ when $n = 3$ and $p = 6$, find p when $m = 240$ and $n = 5$. **Solution:** $m = knp$ Translate "*m* varies jointly with *m* and *p*." $288 = k(3)(6)$ Substitute 3 for *m* and 6 for *p*. $288 = 18k$ Divide both sides by 18. $16 = k$ In $m = knp$, replacing k with 16 gives $m = 16np$. Now find p when $m = 240$ and $n = 5$. $240 = 16(5)(p)$ In $m = 16np$, replace *m* with 240 and *n* with 5. $240 = 80p$ Divide both sides by 80. $3 = p$

Exercises 87–88 Equations and Inequalities

For Exercises 87 and 88, solve.

87. Suppose y varies jointly with x and z. If $y = 40$ when $x = 2$ and $z = 5$, find y when $x = 4$ and $z = 2$.

88. The volume of a right circular cylinder varies jointly with the radius squared and the height. If the volume is 62.8 cubic inches when the radius is 2 inches and the height is 5 inches, find the volume when the radius is 4 inches and the height is 2 inches.

Chapter 7 Practice Test

For Extra Help

Step-by-step test solutions are found on the Chapter Test Prep Videos available in MyMathLab® or on YouTube.

For Exercises 1–4, simplify.

1. $\dfrac{42a^3b^4}{16ab^6}$

2. $\dfrac{6x^2 + 11x - 10}{4x^2 + 4x - 15}$

3. $\dfrac{m^3 - 64n^3}{3m^2 - 14mn + 8n^2}$

4. $\dfrac{2bn - 4bm - 3cn + 6cm}{2b^2 + 7bc - 15c^2}$

For Exercises 5–8, find the products or quotients.

5. $\dfrac{27a^2b^4}{14x^4y} \cdot \dfrac{35xy^3}{18a^5b^2}$

6. $\dfrac{16y^2 - 25}{6y^2 - 17y - 14} \cdot \dfrac{3y^2 + 2y}{8y^2 - 2y - 15}$

7. $\dfrac{6q - 8p}{8p - 2q} \div \dfrac{4p - 3q}{12p - 3q}$

8. $\dfrac{2n^3 - 5n^2 - 6n + 15}{3n^3 + 2n^2 - 9n - 6} \div \dfrac{2n^3 - 5n^2 - 8n + 20}{3n^3 + 2n^2 + 12n + 8}$

9. Given $f(x) = \dfrac{2x - 4}{3x^2 - 7x - 6}$, find the following.

a. $f(-1)$ **b.** $f(2)$ **c.** $f(3)$

10. Find the domain of $f(x) = \dfrac{3x - 4}{2x^2 - x - 10}$.

11. Find the LCD for $\dfrac{3a}{a^2 - 9}$ and $\dfrac{6a}{2a^2 + 13a + 21}$ and write as equivalent fractions with the LCD as the denominators.

For Exercises 12–16, find the sum or difference.

12. $\dfrac{2y^2 + 3}{9y^2} - \dfrac{y - 6}{12y}$

13. $\dfrac{2r^2 - 2r - 5}{r^2 - 16} - \dfrac{-7r + 7}{r^2 - 16}$

14. $\dfrac{t + 3}{t^2 - 10t + 25} - \dfrac{t - 4}{2t^2 - 50}$

15. $\dfrac{3a + 4b}{3a - 5b} + \dfrac{2a - b}{5b - 3a}$

16. $\dfrac{5}{x} - \dfrac{3x - 1}{x^2 + 5x} - \dfrac{2x - 5}{x + 5}$

For Exercises 17–20, simplify.

17. $\dfrac{\dfrac{6a^3b^2}{c^3}}{\dfrac{9a^2b^4}{c^6}}$

18. $\dfrac{6 - \dfrac{1}{x} - \dfrac{15}{x^2}}{4 + \dfrac{4}{x} - \dfrac{3}{x^2}}$

19. $\dfrac{t - \dfrac{14}{t - 5}}{2 - \dfrac{4}{t - 5}}$

20. $\dfrac{9a^{-2} - 4b^{-2}}{3a^{-1} + 2b^{-1}}$

For Exercises 21–24, solve. Identify any extraneous solutions.

21. $\frac{8}{6x} + \frac{5}{2x} = \frac{13}{9}$

22. $\frac{3x}{x-4} = 6 + \frac{12}{x-4}$

23. $\frac{x}{x+1} + \frac{9x}{x^2+3x+2} = \frac{12}{x+2}$

24. $\frac{n}{n+1} + \frac{n+1}{n+2} = \frac{6n+5}{n^2+3n+2}$

25. Working together, it takes Elena and Eduardo 10 days to do an architectural project. If Elena can do the project by herself in 15 days, how long will it take Eduardo to do the project working by himself?

26. An air cargo plane and an airliner take off at the same time from the same airport. They fly in opposite directions. The air cargo plane flies at 420 miles per hour, and the airliner flies at 530 miles per hour. How long is it before the two airplanes are 1900 miles apart?

For Exercises 27–30, solve the variations.

27. Suppose r varies directly with s. If $r = 14$ when $s = 2$, find s when $r = 42$.

28. Suppose p varies jointly with q and r. If $p = 48$ when $q = 4$ and $r = 6$, find p when $q = 3$ and $r = 5$.

29. The intensity of light is inversely proportional to the square of the distance from the source. A light meter 6 feet from a lightbulb measures the intensity to be 16 foot-candles. What is the intensity at 20 feet from the bulb?

30. The distance a car can travel varies directly with the number of gallons the car uses. On the highway, a Toyota Prius traveled 255 miles using 5 gallons of gas. How far can the same car travel on 8 gallons of gas?

Chapters 1–7 Cumulative Review Exercises

For Exercises 1–4, answer true or false.

[1.2] 1. $-4(x + 3) = -4(3 + x)$ is an example of the distributive property.

[1.3] 2. $\sqrt{x^2} = x$ for all values of x.

[3.2] 3. If the slope of a line is $\frac{3}{4}$, the slope of any line perpendicular to that line is $-\frac{4}{3}$.

[4.1] 4. If the graphs of the lines of a system of equations intersect at a point, the coordinates of that point are a solution of the system.

For Exercises 5–7, fill in the blanks.

[7.5] 5. If Fred can paint the inside of a house in 12 hours, he can paint ________ of the house in 1 hour.

[2.5] 6. The equation $|x + 3| = -4$ has ________ solutions.

[5.5] 7. To divide $\dfrac{x^3 - 2x + 5}{x + 2}$ using synthetic division, the divisor is ________.

Exercises 8–18 Expressions

For Exercises 8 and 9, simplify.

[1.3] 8. $\dfrac{2(-4)^2 - 7^2}{5^2 - 3(-3)^2}$

[1.3] 9. $-5 - 2 \cdot 6^2 \div (-36)(-4) + 8$

[1.4] 10. Evaluate $x - y(x^4 - 2z)$ for $x = -2$, $y = -3$, and $z = 6$.

For Exercises 11–16, simplify.

[5.2] 11. $(3x^3 - 2x^2 + 4) - (-4x^3 + 3x^2 - 5x + 6)$

[5.3] 12. $(2a - 3b)(4a^2 + 6ab + 9b^2)$

[5.4] 13. $\dfrac{12x^3y^4 - 16x^4y^3 + 3x^2y^5}{4x^3y^4}$

[7.1] 14. $\dfrac{6x^2 + x - 12}{2x^2 - 5x - 12} \cdot \dfrac{3x^2 - 14x + 8}{9x^2 - 18x + 8}$

[7.2] 15. $\dfrac{2x + 5}{x^2 - 16} - \dfrac{x - 9}{x^2 - x - 12}$

[7.3] 16. $\dfrac{1 - \frac{3}{x} - \frac{4}{x^2}}{\frac{3}{x} - \frac{13}{x^2} + \frac{4}{x^3}}$

For Exercises 17 and 18, factor.

[6.2] 17. $36a^4b^3 - 39a^3b^4 - 12a^2b^5$

[6.3] 18. $108c^3 - 32d^3$

Exercises 19–30 Equations and Inequalities

For Exercises 19–23, solve.

[2.1] **19.** $\frac{3}{4}x + 3 = \frac{2}{3}(x - 6)$

[2.6] **20.** $8 + |3 + 2x| \geq 4$

[4.2] **21.** $\begin{cases} 2x + y - z = 2 \\ x + 3y + 2z = 1 \\ x + y + z = 2 \end{cases}$

[6.4] **22.** $6x^2 - 10 = 11x$

[7.4] **23.** $\frac{3}{x^2 + 2x - 24} + \frac{x - 5}{x^2 - 16} = \frac{x}{x^2 + 10x + 24}$

For Exercises 24 and 25, graph.

[3.1] **24.** $2x - 5y = -10$

[3.4] **25.** $3x - y \leq -6$

[3.3] **26.** Write the equation of the line containing $(-3, -3)$ and $(-1, 1)$. Leave the answer in slope–intercept form.

For Exercises 27–30, solve.

[2.2] **27.** Two angles are supplementary. If one of the angles is 20° more than three times the other angle, what are the angles?

[4.3] **28.** How many pounds of peppermint candy worth $1.80 per pound must be mixed with 15 pounds of butterscotch worth $2.30 a pound to get a mixture worth $2.10 per pound?

[6.4] **29.** The length of a rectangle is 3 feet less than twice the width. If the area is 54 square feet, find the length and width.

[7.5] **30.** If Hal can paint a room in seven hours and Frank can paint the same room in 10 hours, how long would it take them to paint the room working together?

CHAPTER 8

Rational Exponents, Radicals, and Complex Numbers

Chapter Overview

In this chapter, we explore square roots and radicals in more detail. More specifically, we learn the following skills:

- Evaluate and simplify radical expressions.
- Connect radicals with rational exponents.
- Rewrite expressions containing radicals.
- Solve equations containing radicals.
- Explore complex numbers.

8.1 Radical Expressions and Functions

Objectives

1 Find the nth root of a number.

2 Approximate roots using a calculator.

3 Simplify radical expressions.

4 Evaluate radical functions.

5 Find the domain of radical functions.

6 Solve applications involving radical functions.

Warm-up

[1.3] *For Exercises 1 and 2, evaluate.*

1. $\sqrt{49}$

2. $\sqrt{\dfrac{64}{121}}$

[1.4] 3. Evaluate $3x + 5$ for $x = -3$.

[2.3] 4. Solve: $-2x + 6 \geq 0$

In this section through Section 8.5, we focus on the expression portion of our Algebra Pyramid and explore square root and radical expressions.

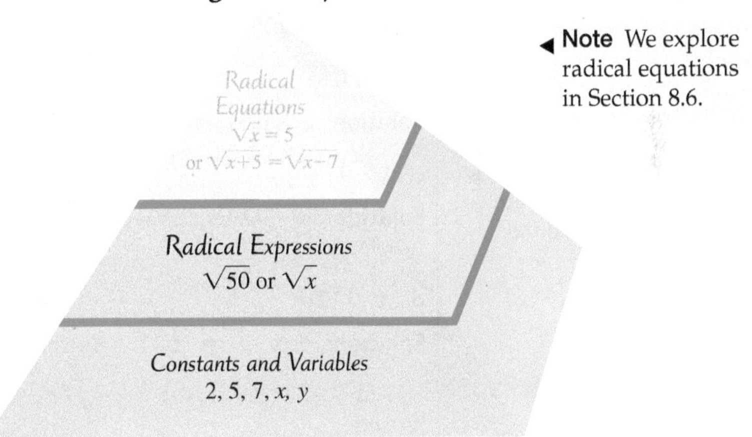

Note We explore radical equations in Section 8.6.

Objective 1 Find the nth root of a number.

In Section 1.3, we learned that a square root of a given number is a number whose square is the given number. We also learned that every positive real number has two square roots, a positive and a negative root. For example, the square roots of 16 are 4 and -4 because $4^2 = 16$ and $(-4)^2 = 16$. We can write the two square roots more compactly as ± 4.

Similarly, a *cube root* of a given number is a number whose cube is the given number. For example, the cube root of 8 is 2 because $2^3 = 8$.

The pattern continues for higher roots. For example, the fourth roots of 81 are ± 3 because $3^4 = 81$ and $(-3)^4 = 81$. Our examples suggest the following definition of ***n*th root**.

Definition ***n*th root:** The number b is an nth root of a number a if $b^n = a$.

Recall from Section 1.3, that the radical sign, $\sqrt{\ }$, denotes the principal (nonnegative) square root of a number. For roots other than square roots, we use the symbol $\sqrt[n]{a}$, read "the nth root of a," where n, the *root index*, indicates which root we are to find. The number a, called the *radicand*, is the number or expression whose root we are to find. The entire expression is called a *radical*, and any expression containing a radical is called a *radical expression.*

Just as $\sqrt{a}$ denotes the principal (nonnegative) square root of a, if n is even, then $\sqrt[n]{a}$ denotes the *principal nth root* of a. For example, $\sqrt[4]{81} = 3$ because 3 is the nonnegative fourth root of 81. Further, for $\sqrt[n]{a}$ to be a real number when n is even, a must be nonnegative because no real number can be raised to an even power to equal a negative number. For example, $\sqrt{-4}$ is not a real number because there is no real number whose square is -4.

Answers to Warm-up

1. 7 **2.** $\dfrac{8}{11}$

3. -4 **4.** $x \leq 3$

If n is odd in $\sqrt[n]{a}$, then a can be positive or negative because a positive number raised to an odd power is positive and a negative number raised to an odd power is negative.

Rule Evaluating *n*th Roots

When evaluating a radical expression $\sqrt[n]{a}$, the sign of a and the index n will determine possible outcomes.

If a is nonnegative, then $\sqrt[n]{a} = b$, where $b \geq 0$ and $b^n = a$.

If a is negative and n is even, then there is no real-number root.

If a is negative and n is odd, then $\sqrt[n]{a} = b$, where b is negative and $b^n = a$.

Example 1 Evaluate each root if possible.

a. $\sqrt{225}$

Solution: $\sqrt{225} = 15$ Because $15^2 = 225$

b. $-\sqrt{0.81}$

Solution: $-\sqrt{0.81} = -0.9$ Because $(-0.9)^2 = 0.81$

Note It can be helpful to think of $-\sqrt{0.81}$ as $-1 \cdot \sqrt{0.81} = -1 \cdot 0.9 = -0.9$.

Note In Section 8.7. we define the square root of a negative number using a set of numbers called the **imaginary numbers**.

c. $\sqrt{-9}$

Solution: $\sqrt{-9}$ is not a real number because there is no real number whose square is -9.

d. $\pm\sqrt{121}$

Solution: $\pm\sqrt{121} = \pm 11$ Because $(11)^2 = 121$ and $(-11)^2 = 121$

e. $\sqrt{\frac{4}{9}}$

Solution: $\sqrt{\frac{4}{9}} = \frac{2}{3}$ Because $\left(\frac{2}{3}\right)^2 = \frac{4}{9}$

f. $\sqrt[3]{8}$

Solution: $\sqrt[3]{8} = 2$ Because $2^3 = 8$

g. $\sqrt[3]{-8}$

Solution: $\sqrt[3]{-8} = -2$ Because $(-2)^3 = -8$

h. $\sqrt[4]{16}$

Solution: $\sqrt[4]{16} = 2$ Because $2^4 = 16$

Your Turn 1 Evaluate each root if possible.

a. $\sqrt{121}$ b. $-\sqrt{4}$ c. $\sqrt{-64}$ d. $\pm\sqrt{0.36}$

e. $\sqrt{\frac{25}{36}}$ f. $\sqrt[3]{64}$ g. $\sqrt[3]{-27}$ h. $-\sqrt[4]{81}$

Objective 2 Approximate roots using a calculator.

Each root we have considered so far has been rational, which means a rational number can express its exact value. But some roots, such as $\sqrt{2}$, are irrational, which means no rational number exists that expresses its exact value. In fact, writing $\sqrt{2}$ with the radical sign is how we express its exact value. However, $\sqrt{2}$ can be approximated using a calculator.

Calculator approximation to nine decimal places: $\sqrt{2} \approx 1.414213562$

Approximation to three decimal places: $\sqrt{2} \approx 1.414$

Approximation to two decimal places: $\sqrt{2} \approx 1.41$

Note Recall that $\approx$ means "approximately equal to."

Answers to Your Turn 1

a. 11 b. -2 c. not a real number d. ± 0.6 e. $\frac{5}{6}$ f. 4 g. -3 h. -3

Example 2 Approximate using a calculator. Round to three decimal places.

a. $\sqrt{12}$

Answer: $\sqrt{12} \approx 3.464$

b. $-\sqrt{38}$

Answer: $-\sqrt{38} \approx -6.164$

c. $\sqrt[3]{45}$

Answer: $\sqrt[3]{45} \approx 3.557$

Your Turn 2 Approximate using a calculator. Round to three decimal places.

a. $\sqrt{19}$ b. $-\sqrt{93}$ c. $\sqrt[3]{63}$

Objective 3 Simplify radical expressions.

The definition of a root can also be used to find roots with variable radicands. Recall that with an even index, the principal root is nonnegative. So at first, we will assume that all variables represent nonnegative values. We will use $(a^m)^n = a^{mn}$ to verify the roots.

Note Remember that with an even index, the principal root is nonnegative. Because we are assuming that the variables are nonnegative, our result accurately indicates the principal square root.

Connection To raise a power to another power using $(a^m)^n = a^{mn}$, we multiply the powers. To find a root of a power, we divide the index into the exponent.

Example 3 Find the root. Assume that variables represent nonnegative values.

a. $\sqrt{x^2}$

Solution: $\sqrt{x^2} = x$ Because $(x)^2 = x^2$

b. $\sqrt{a^6}$

Solution: $\sqrt{a^6} = a^3$ Because $(a^3)^2 = a^6$

c. $\sqrt{16x^8}$

Solution: $\sqrt{16x^8} = 4x^4$ Because $(4x^4)^2 = 16x^8$

d. $\sqrt{\dfrac{25x^8}{49y^2}}$

Solution: $\sqrt{\dfrac{25x^8}{49y^2}} = \dfrac{5x^4}{7y}$ Because $\left(\dfrac{5x^4}{7y}\right)^2 = \dfrac{25x^8}{49y^2}$

e. $\sqrt[3]{y^6}$

Solution: $\sqrt[3]{y^6} = y^2$ Because $(y^2)^3 = y^6$

f. $\sqrt[4]{16x^{12}}$

Solution: $\sqrt[4]{16x^{12}} = 2x^3$ Because $(2x^3)^4 = 16x^{12}$

Connection In parts c and f, notice the similarity between finding the root of a product and raising a product to a power. To raise a product to a power, we raise each factor to the power. To find a root of a product, we find the root of each factor.

Your Turn 3 Find the root. Assume that variables represent nonnegative values.

a. $\sqrt{x^4}$ b. $\sqrt{9x^{10}}$ c. $\sqrt{36a^{12}}$

d. $\sqrt{\dfrac{100x^4}{81y^6}}$ e. $\sqrt[3]{27y^9}$ f. $\sqrt[4]{b^8}$

If the variables can represent *any* real number and the index is even, we must be careful to ensure that the principal root is nonnegative by using absolute value symbols. If the root index is odd, however, we do not need absolute value symbols because the root can be positive or negative depending on the sign of the radicand.

Answers to Your Turn 2

a. 4.359 b. −9.644 c. 3.979

Answers to Your Turn 3

a. x^2 b. $3x^5$ c. $6a^6$ d. $\dfrac{10x^2}{9y^3}$ e. $3y^3$ f. b^2

Note To illustrate why $\sqrt{x^2} = |x|$, suppose we were to evaluate $\sqrt{x^2}$ when $x = -3$. We would have $\sqrt{(-3)^2} = \sqrt{9} = 3$. Notice that the root, 3, is, in fact, the absolute value of -3. If we had incorrectly stated that $\sqrt{x^2} = x$, then $\sqrt{(-3)^2}$ would have had to equal -3, which is not true.

Example 4 Find the root. Assume that variables represent any real number.

a. $\sqrt{x^2}$

Solution: $\sqrt{x^2} = |x|$

b. $\sqrt{25a^6}$

Solution: $\sqrt{25a^6} = 5|a^3|$

c. $\sqrt{(n+1)^2}$

Solution: $\sqrt{(n+1)^2} = |n+1|$

d. $\sqrt{25y^8}$

Solution: $\sqrt{25y^8} = 5y^4$

e. $\sqrt[3]{8n^3}$

Solution: $\sqrt[3]{8n^3} = 2n$

f. $\sqrt[3]{(t-2)^3}$

Solution: $\sqrt[3]{(t-2)^3} = t - 2$

Note In part d, we do not need absolute value because y^4, with its even exponent, is always nonnegative. In parts e and f, we do not use absolute value because the indices are odd.

Your Turn 4 Find the root. Assume that variables represent any real number.

a. $\sqrt{x^4}$ **b.** $\sqrt{9x^{10}}$ **c.** $\sqrt{36a^{12}}$ **d.** $\sqrt{1.21u^6t^2}$

e. $\sqrt{\dfrac{100x^4}{81y^6}}$ **f.** $\sqrt[3]{27y^9}$ **g.** $\sqrt[4]{b^8}$

Objective 4 Evaluate radical functions.

Now that we have learned about radical expressions, let's examine **radical functions**.

Definition **Radical function:** A function containing a radical expression whose radicand has a variable.

Example 5

a. Given $f(x) = \sqrt{3x-2}$, find $f(3)$.

Solution: To find $f(3)$, substitute 3 for x and simplify.

$$f(3) = \sqrt{3(3)-2} = \sqrt{9-2} = \sqrt{7}$$

b. Given $f(x) = \sqrt{2x-6}$, find $f(0)$.

Solution: To find $f(0)$, substitute 0 for x and simplify. $f(0) = \sqrt{2(0)-6} = \sqrt{-6}$, which is not a real number.

Your Turn 5 Given $f(x) = \sqrt{2x+5}$, find each of the following

a. $f(-1)$ **b.** $f(-3)$

Objective 5 Find the domain of radical functions.

In Example 5(b), we found that $f(0)$ did not exist because $\sqrt{-6}$ is not a real number. What does this suggest about the domains of functions involving radicals? Consider the graphs of $f(x) = \sqrt{x}$ and $f(x) = \sqrt[3]{x}$, which we can generate by choosing values for x.

x	$f(x) = \sqrt{x}$
-1	Not real
0	0
1	1
4	2

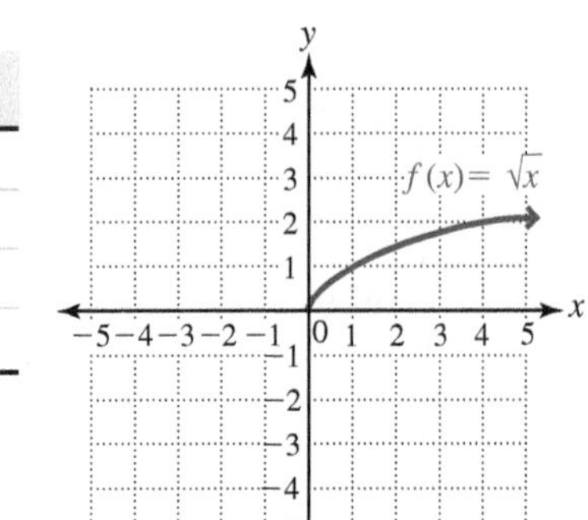

Notice that $\sqrt{x}$ is not a real number when x is negative, which means that the domain for $f(x) = \sqrt{x}$ is $\{x | x \geq 0\}$, or $[0, \infty)$.

Answers to Your Turn 4
a. x^2 **b.** $3|x^5|$ **c.** $6a^6$ **d.** $1.1|u^3t|$ **e.** $\dfrac{10x^2}{9|y^3|}$ **f.** $3y^3$ **g.** b^2

Answers to Your Turn 5
a. $f(-1) = \sqrt{3}$
b. $f(-3)$ is not a real number.

x	$f(x) = \sqrt[3]{x}$
−8	−2
−1	−1
0	0
1	1
8	2

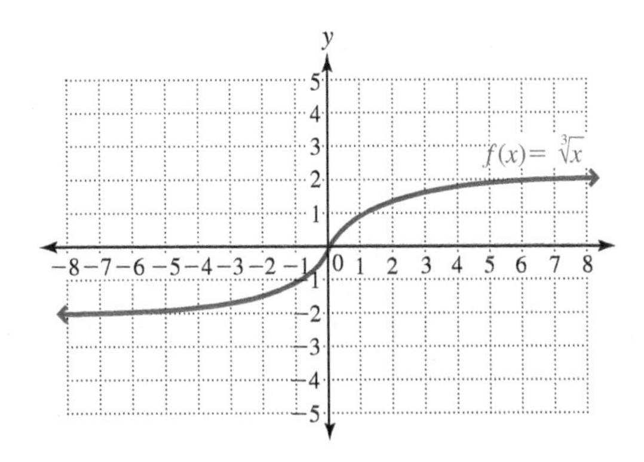

Alternatively, $\sqrt[3]{x}$, with its odd index, is a real number when x is negative; so its domain is all real numbers, or $(-\infty, \infty)$. Our graphs suggest the following conclusion:

Conclusion: The domain of a radical function with an even index must contain values that keep its radicand nonnegative.

Procedure Finding the Domain of a Radical Function

If the root index is *odd*, the domain is $\{x \mid x \text{ is a real number}\}$, or $(-\infty, \infty)$.
If the root index is *even*, the radicand must be nonnegative. Consequently, the domain is the solution set to the inequality "radicand ≥ 0."

Example 6 Find the domain of each of the following.

a. $f(x) = \sqrt{x - 4}$

Solution: Because the index is even, the radicand must be nonnegative.

$x - 4 \geq 0$ Set the radicand ≥ 0.

$x \geq 4$ Add 4 to both sides.

Domain: $\{x \mid x \geq 4\}$, or $[4, \infty)$

b. $f(x) = \sqrt{-2x + 6}$

Solution: $-2x + 6 \geq 0$ Set the radicand ≥ 0.

$-2x \geq -6$ Subtract 6 from both sides.

$x \leq 3$ Divide both sides by -2 and change the direction of the inequality.

Domain: $\{x \mid x \leq 3\}$, or $(-\infty, 3]$

Note Recall that when we divide *both* sides of an inequality by a negative number, we reverse the direction of the inequality.

Your Turn 6 Find the domain of each of the following.

a. $f(x) = \sqrt{2x - 4}$ **b.** $f(x) = \sqrt{-3x - 9}$

Objective 6 Solve applications involving radical functions.

Often, radical functions appear in real-world situations where one variable is a function of another.

Example 7

a. The velocity of a free-falling object is a function of the distance it has fallen. Ignoring air resistance, the velocity of an object, v, in meters per second, can be found after it has fallen h meters by using the formula $v = -\sqrt{19.6h}$. Find the velocity of a stone that has fallen 30 meters after being dropped from a cliff.

Answers to Your Turn 6
a. $\{x \mid x \geq 2\}$, or $[2, \infty)$
b. $\{x \mid x \leq -3\}$, or $(-\infty, -3]$

Solution: $v = -\sqrt{19.6(30)}$ In $v = -\sqrt{19.6h}$, replace h with 30.

$v = -\sqrt{588}$ Multiply within the radical.

$v \approx -24.2$ Evaluate the square root.

Answer: ≈ -24.2 m/sec.

Note: A negative velocity indicates that the object is traveling downward.

b. The period of a pendulum is the amount of time it takes the pendulum to swing from the point of release to the opposite extreme and then back to the point of release. The period is a function of the length. The period, T, measured in seconds, can be found using the formula $T = 2\pi\sqrt{\frac{L}{9.8}}$, where L represents the length of the pendulum in meters. Find the period of a pendulum that is 0.5 meter long.

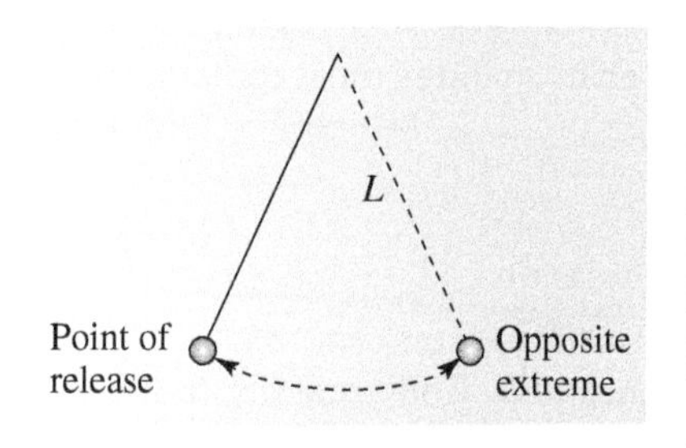

Solution: $T \approx 2(3.14)\sqrt{\frac{0.5}{9.8}}$ In $T = 2\pi\sqrt{\frac{L}{9.8}}$, substitute 3.14 for π and 0.5 for L.

$T \approx 6.28\sqrt{0.051}$ Simplify.

$T \approx 6.28(0.226)$ Approximate the square root.

$T \approx 1.42$ Multiply.

Answer: ≈ 1.42 sec.

Your Turn 7

a. A skydiver jumps from a plane and puts her body into a dive position so that air resistance is minimized. Find her velocity after she falls 100 meters. Use the formula in Example 7(a).

b. Find the period of a pendulum that is 0.2 meter long. Round to the nearest thousandth. Use the formula in Example 7(b).

Answers to Your Turn 7

a. ≈ -44.3 m/sec.
b. ≈ 0.898 sec.

8.1 Exercises

For Extra Help MyMathLab®

Objective 1

Prep Exercise 1 Why are there two square roots for every positive real number?

For Exercises 1–8, find all square roots of each number. See Objective 1.

1. 36
2. 64
3. 121
4. 81
5. 196
6. 400
7. 225
8. 289

Prep Exercise 2 If the index is even, what is the principal root of a number?

Prep Exercise 3 Why is an even root of any negative number not a real number?

For Exercises 9–50, evaluate each root if possible. See Example 1.

9. $\sqrt{25}$
10. $\sqrt{49}$
11. $\sqrt{-64}$
12. $\sqrt{-25}$
13. $-\sqrt{25}$
14. $-\sqrt{100}$
15. $\pm\sqrt{25}$
16. $\pm\sqrt{100}$
17. $\sqrt{1.44}$
18. $\sqrt{1.96}$

19. $\sqrt{-10.64}$ **20.** $\sqrt{-2.25}$ **21.** $-\sqrt{0.0121}$ **22.** $-\sqrt{0.0169}$ **23.** $\sqrt{\frac{49}{81}}$

24. $\sqrt{\frac{64}{169}}$ **25.** $-\sqrt{\frac{144}{169}}$ **26.** $-\sqrt{\frac{25}{4}}$ **27.** $\sqrt[3]{27}$ **28.** $\sqrt[3]{125}$

29. $\sqrt[3]{-64}$ **30.** $\sqrt[3]{-216}$ **31.** $-\sqrt[3]{-216}$ **32.** $-\sqrt[3]{-125}$ **33.** $\sqrt[4]{256}$

34. $\sqrt[4]{81}$ **35.** $\sqrt[4]{-625}$ **36.** $\sqrt[4]{-81}$ **37.** $-\sqrt[4]{16}$ **38.** $-\sqrt[4]{625}$

39. $\sqrt[5]{32}$ **40.** $\sqrt[5]{243}$ **41.** $\sqrt[5]{-243}$ **42.** $\sqrt[5]{-32}$ **43.** $-\sqrt[5]{-32}$

44. $-\sqrt[5]{-3125}$ **45.** $\sqrt[6]{64}$ **46.** $\sqrt[6]{729}$ **47.** $\sqrt[3]{-\frac{8}{27}}$ **48.** $\sqrt[3]{-\frac{64}{125}}$

49. $\sqrt[4]{\frac{16}{81}}$ **50.** $\sqrt[5]{\frac{1}{32}}$

Objective 2

Prep Exercise 4 Why is $\sqrt{16}$ rational whereas $\sqrt{17}$ is irrational?

Prep Exercise 5 How can we express the exact value of a root that is irrational?

For Exercises 51–66, approximate using a calculator. Round to three decimal places. See Example 2.

51. $\sqrt{7}$ **52.** $\sqrt{12}$ **53.** $-\sqrt{11}$ **54.** $-\sqrt{41}$

55. $\sqrt[3]{50}$ **56.** $\sqrt[3]{21}$ **57.** $\sqrt[3]{-53}$ **58.** $\sqrt[3]{-83}$

59. $\sqrt[4]{189}$ **60.** $\sqrt[4]{123}$ **61.** $-\sqrt[4]{85}$ **62.** $-\sqrt[4]{77}$

63. $\sqrt[5]{89}$ **64.** $\sqrt[5]{62}$ **65.** $\sqrt[6]{146}$ **66.** $\sqrt[6]{98}$

Objective 3

Prep Exercise 6 If x is any real number, explain why $\sqrt{x^2} = |x|$ instead of x (with no absolute value).

For Exercises 67–90, find the root. Assume that variables represent nonnegative values. See Example 3.

67. $\sqrt{b^4}$ **68.** $\sqrt{r^8}$ **69.** $\sqrt{16x^2}$ **70.** $\sqrt{81t^2}$ **71.** $\sqrt{100r^8s^6}$

72. $\sqrt{121x^8y^{10}}$ **73.** $\sqrt{0.25a^6b^{12}}$ **74.** $\sqrt{0.81r^4s^{14}}$ **75.** $\sqrt[3]{m^3}$ **76.** $\sqrt[3]{n^6}$

77. $\sqrt[3]{27a^9b^6}$
78. $\sqrt[3]{64u^{12}t^9}$
79. $\sqrt[3]{-64a^3b^{12}}$
80. $\sqrt[3]{-27r^{15}s^3}$
81. $\sqrt[3]{0.008x^{18}}$
82. $\sqrt[3]{0.027r^{12}}$
83. $\sqrt[4]{a^4}$
84. $\sqrt[4]{x^{12}}$
85. $\sqrt[4]{16x^{16}}$
86. $\sqrt[4]{81t^{20}}$
87. $\sqrt[5]{32x^{10}}$
88. $\sqrt[5]{243x^{15}}$
89. $\sqrt[6]{x^{12}y^6}$
90. $\sqrt[7]{s^7t^{21}}$

For Exercises 91–102, find the root. Assume that variables represent any real number. See Example 4.

91. $\sqrt{36m^2}$
92. $\sqrt{9t^6}$
93. $\sqrt{(r-1)^2}$
94. $\sqrt{(k+3)^2}$
95. $\sqrt[4]{256y^{12}}$
96. $\sqrt[4]{16x^{20}}$
97. $\sqrt[3]{27y^3}$
98. $\sqrt[3]{125x^6}$
99. $\sqrt{(y-3)^4}$
100. $\sqrt{(x+2)^4}$
101. $\sqrt[3]{(y-4)^6}$
102. $\sqrt[3]{(n+5)^9}$

Objective 4

Prep Exercise 7 A radical function is a function containing a(n) __________ expression whose __________ has a variable.

For Exercises 103–106, find the indicated value of the function. See Example 5.

103. $f(x) = \sqrt{2x+4}$; find $f(0)$
104. $f(x) = \sqrt{3x+2}$; find $f(3)$
105. $f(x) = \sqrt{4x+3}$; find $f(3)$
106. $f(x) = \sqrt{-2x+3}$; find $f(-2)$

Objective 5

Prep Exercise 8 Explain how to determine the domain of a radical function.

For Exercises 107–110, find the domain. See Example 6.

107. $f(x) = \sqrt{2x-8}$
108. $f(x) = \sqrt{3x+12}$
109. $f(x) = \sqrt{-4x+16}$
110. $f(x) = \sqrt{-2x+6}$

For Exercises 111–114:
- **a.** *Use a graphing calculator to graph the function. See Objective 5.*
- **b.** *Find the domain of the function. See Objective 5 and Example 6.*

111. $f(x) = \sqrt{x-2}$
112. $f(x) = \sqrt{x+3}$

113. $f(x) = \sqrt[3]{x + 1}$

114. $f(x) = \sqrt[4]{x - 3}$

For Exercises 115 and 116, use the following information. Ignoring air resistance, the velocity, v, in meters per second, of an object after falling h meters can be found using the formula $v = -\sqrt{19.6h}$. *Round your answers to the nearest thousandth. See Example 7(a).*

115. Find the velocity of a rock that has been dropped from a cliff after it falls 16 meters.

116. Find the velocity of a ball that has been dropped from a roof after it falls 9 meters.

For Exercises 117 and 118, use the following information. The period, T, of a pendulum in seconds can be found using the formula $T=2\pi\sqrt{\frac{L}{9.8}}$, *where L represents the length of the pendulum in meters. Round your answers to the nearest thousandth. See Example 7(b).*

117. Find the period of a pendulum that is 3 meters long.

118. Find the period of a pendulum that is 6 meters long.

Of Interest

One of the many topics Galileo Galilei studied was pendulums. His interest in them was piqued when he noticed a swinging light fixture in the cathedral of his hometown of Pisa and timed its period using his pulse.

For Exercises 119 and 120, use the following information. The formula $S=\frac{7}{2}\sqrt{2D}$ *can be used to approximate the speed, S, in miles per hour, that a car was traveling prior to braking and skidding a distance, D, in feet, on asphalt. (Source: Harris Technical Services Traffic Accident Reconstructionists.) Round your answers to the nearest thousandth.*

119. Find the speed of a car if the skid distance is 15 feet.

120. Find the speed of a car if the skid distance is 40 feet.

For Exercises 121 and 122, use the following information. Given the lengths a and b of the sides of a right triangle, we can find the hypotenuse, c, using the formula $c=\sqrt{a^2+b^2}$.

121. Three pieces of lumber are to be connected to form a right triangle that will be part of the frame for a roof. If the horizontal piece is to be 12 feet and the vertical piece is to be 5 feet, how long must the connecting piece be?

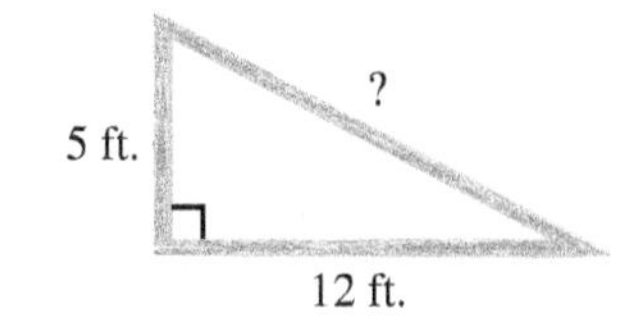

122. A counselor decides to create a ropes course with a zip line. She wants the line to connect from a 40-foot-tall tower to the ground at a point 100 feet from the base of the tower. Assuming that the tower and ground form a right angle, find the length of the zip line. Round your answer to the nearest tenth.

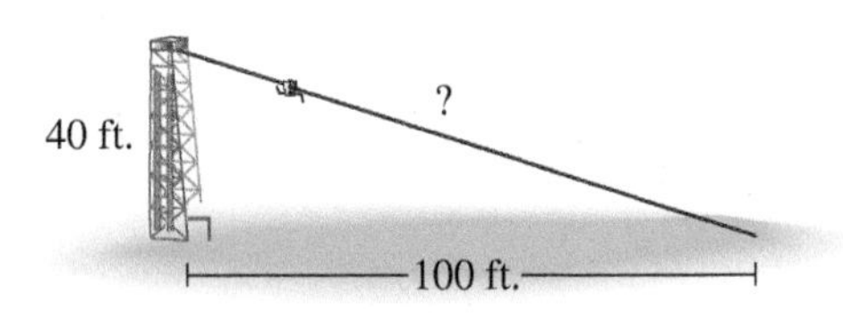

Connection In Section 6.6, we used the Pythagorean theorem, $c^2 = a^2 + b^2$, to find missing side lengths of a right triangle. The formula $c = \sqrt{a^2 + b^2}$ comes from isolating c in that theorem.

For Exercises 123 and 124, use the following information. Two forces, F_1 and F_2, acting on an object at a 90° angle will pull the object with a resultant force, R, at an angle between F_1 and F_2. (See the figure.) The value of the resultant force can be found using the formula $R = \sqrt{F_1^2 + F_2^2}$.

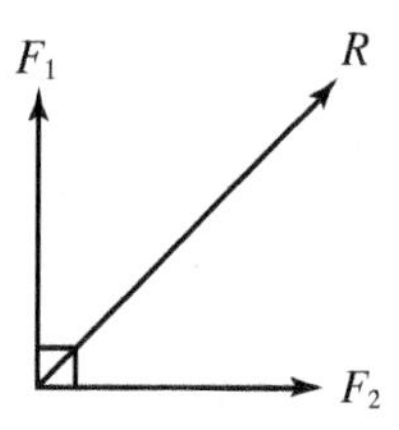

123. Find the resultant force if $F_1 = 9$ N and $F_2 = 12$ N.

◀ **Note** The unit of force in the metric system is the newton (abbreviated N).

124. Find the resultant force if $F_1 = 8$ N and $F_2 = 15$ N.

125. The number of earthquakes worldwide for 2008–2012 with a magnitude of 1.0–1.9 can be approximated using the function $f(x) = 10.5\sqrt{x} + 21$, where x represents the number of years after 2008. (*Hint:* The year 2008 corresponds to $x = 0$.) (*Source:* USGS Earthquake Hazards Program.)

a. Find the approximate number of earthquakes with a magnitude of 1.0–1.9 that occurred in 2010.

b. Find the approximate number of earthquakes with a magnitude of 1.0–1.9 that occurred in 2011.

126. The average mathematics scores for 8th graders in the United States on the International Mathematics and Science Study from 1995 to 2011 can be approximated by the function $f(x) = 4.3\sqrt{x} + 492$, where x represents the number of years after 1995. (*Hint:* The year 1995 corresponds to $x = 0$.) (*Source:* National Center for Education Statistics.)

a. What was the average mathematics score for 8th graders in 2003? Round to the nearest whole number.

b. What was the average mathematics score for 8th graders in 2011? Round to the nearest whole number.

Puzzle Problem A telemarketer calls a house and speaks to the mother of three children. The telemarketer asks the mother, "How old are you?" The woman answers, "41." The marketer then asks, "How old are your children?" The woman replies, "The product of their ages is our house number, 1296, and all of their ages are perfect squares." What are the ages of the children?

Of Interest

In 2011, U.S. 8th graders ranked 9th in mathematics on the International Mathematics and Science Study behind Finland (514), Israel (516), Russian Federation (539), Japan (570), Hong Kong-CHN (586), Chinese Taipei-CHN (609), Singapore (611), and Republic of Korea (613).

Review Exercises

Exercises 1–6 Expressions

[1.3] **1.** Follow the order of operations to simplify.

a. $\sqrt{16 \cdot 9}$

b. $\sqrt{16} \cdot \sqrt{9}$

[5.3] *For Exercises 2–6, multiply.*

2. $x^5 \cdot x^3$

3. $(-9m^3n)(5mn^2)$

4. $4x^2(3x^2 - 5x + 1)$

5. $(7y - 4)(3y + 5)$

6. Write an expression for the area of the figure shown.

$x - 9$

$2x + 1$

8.2 Rational Exponents

Objectives

1. Evaluate rational exponents.
2. Write radicals as expressions raised to rational exponents.
3. Simplify expressions with rational number exponents using the rules of exponents.
4. Use rational exponents to simplify radical expressions.

Warm-up

[8.1] *For Exercises 1 and 2, simplify.*

1. $\sqrt{36x^{10}}$
2. $\sqrt[3]{8x^{12}}$

For Exercises 3 and 4, perform the indicated operation.

[1.2] 3. $-\frac{3}{5}+\frac{3}{4}$

[1.2] 4. $\frac{4}{3}\cdot 3$

Objective 1 Evaluate rational exponents.

Rational Exponents with a Numerator of 1

So far, we have seen only integer exponents. However, expressions can have fractional exponents also, as in $3^{1/2}$. The fractional exponent in $3^{1/2}$ is called a **rational exponent**.

Definition Rational exponent: An exponent that is a rational number.

To discover what rational exponents mean, consider the following:

$$(3^{1/2})^2 = 3^{1/2\cdot 2} = 3^1 = 3 \quad \text{Use } (a^m)^n = a^{mn}\text{, which we learned in Section 5.1.}$$

By the definition of square root, $\sqrt{3}$ is the number whose square is 3; so

$$(\sqrt{3})^2 = 3$$

Our exploration suggests that $3^{1/2} = \sqrt{3}$. Note that the denominator, 2, of the rational exponent is the index of the root. (Remember that square roots have an unwritten index of 2.) This relationship holds for other roots as well.

Note If a is negative and n is odd, the root is negative. If a is negative and n is even, there is no real-number root. Such roots are imaginary and are discussed in Section 8.7.

Rule Rational Exponents with a Numerator of 1

$a^{1/n} = \sqrt[n]{a}$, where n is a natural number other than 1.

Example 1 Rewrite using radicals; then simplify if possible. Assume that all variables represent nonnegative values.

a. $36^{1/2}$

Solution: $36^{1/2} = \sqrt{36} = 6$

b. $81^{1/4}$

Solution: $81^{1/4} = \sqrt[4]{81} = 3$

c. $(-125)^{1/3}$

Solution: $(-125)^{1/3} = \sqrt[3]{-125} = -5$

d. $(-64)^{1/4}$

Solution: $(-64)^{1/4} = \sqrt[4]{-64}$ There is no real-number answer.

Note In $(-64)^{1/4}$, the parentheses group the minus sign with the radicand, whereas in $-36^{1/2}$, the minus sign is not part of the radicand.

e. $-36^{1/2}$

Solution: $-36^{1/2} = -\sqrt{36} = -6$

f. $x^{1/5}$

Solution: $x^{1/5} = \sqrt[5]{x}$

g. $(64x^6)^{1/2}$

Solution: $(64x^6)^{1/2} = \sqrt{64x^6} = 8x^3$

h. $12z^{1/4}$

Solution: $12\sqrt[4]{z}$

Note Only the z is raised to the one-fourth power.

i. $\left(\frac{a^6}{36}\right)^{1/2}$

Solution: $\left(\frac{a^6}{36}\right)^{1/2} = \sqrt{\frac{a^6}{36}} = \frac{a^3}{6}$

Answers to Warm-up

1. $6|x^5|$ 2. $2x^4$ 3. $\frac{3}{20}$ 4. 4

Your Turn 1 Rewrite using radicals; then simplify if possible. Assume that all variables represent positive values.

a. $49^{1/2}$ b. $625^{1/4}$ c. $(-64)^{1/3}$ d. $-81^{1/2}$

e. $y^{1/4}$ f. $(25a^8)^{1/2}$ g. $18n^{1/5}$ h. $\left(\frac{49}{x^{10}}\right)^{1/2}$

Rational Exponents with a Numerator Other Than 1

Let's explore how we can rewrite expressions such as $8^{2/3}$ in which the numerator of the rational exponent is a number other than 1. Note that we could write the fraction $\frac{2}{3}$ as a product.

$$8^{2/3} = 8^{2(1/3)} \quad \text{or} \quad 8^{2/3} = 8^{(1/3)2}$$

Using the rule of exponents $(n^a)^b = n^{ab}$ in reverse, we can write

$$8^{2/3} = 8^{2(1/3)} = (8^2)^{1/3} \quad \text{or} \quad 8^{2/3} = 8^{(1/3)2} = (8^{1/3})^2$$

Applying the rule $a^{1/n} = \sqrt[n]{a}$, we can write

$$8^{2/3} = 8^{2(1/3)} = (8^2)^{1/3} = \sqrt[3]{8^2} \quad \text{or} \quad 8^{2/3} = 8^{(1/3)2} = (8^{1/3})^2 = (\sqrt[3]{8})^2$$

Notice that the denominator of the rational exponent becomes the index of the radical. The numerator of the rational exponent can be written as the exponent of the radicand or as an exponent for the entire radical.

Rule General Rule for Rational Exponents
$a^{m/n} = \sqrt[n]{a^m} = (\sqrt[n]{a})^m$, where $a \geq 0$ and m and n are natural numbers other than 1.

Warning In the expression $-9^{5/2}$, the negative sign tells us to find the opposite of the value of the exponential form.

$$-9^{5/2} = -(9^{5/2}) = -(\sqrt{9})^5$$

Or some people find it helpful to think of $-9^{5/2}$ as $-1 \cdot 9^{5/2}$.

$$-9^{5/2} = -1 \cdot 9^{5/2}$$
$$= -1 \cdot (\sqrt{9})^5$$

Example 2 Rewrite using radicals; then simplify if possible. Assume that all variables represent nonnegative values.

a. $8^{2/3}$

Solution:
$8^{2/3} = \sqrt[3]{8^2}$ Rewrite.
$= \sqrt[3]{64}$ Square the radicand.
$= 4$ Evaluate the root.

or

$8^{2/3} = (\sqrt[3]{8})^2$ Rewrite.
$= (2)^2$ Evaluate the root.
$= 4$ Evaluate the exponential form.

b. $625^{3/4}$

Solution: We could rewrite $625^{3/4}$ as $\sqrt[4]{625^3}$, but calculating 625^3 is tedious without a calculator; so we use the other form of the rule.

$625^{3/4} = (\sqrt[4]{625})^3$ Rewrite.
$= (5)^3$ Evaluate the root.
$= 125$ Evaluate the exponential form.

c. $-9^{5/2}$

Solution: $-9^{5/2} = -(\sqrt{9})^5 = -(3)^5 = -243$

d. $\left(\frac{1}{9}\right)^{5/2}$

Note The simplification is usually easier if we write $a^{m/n}$ as $(\sqrt[n]{a})^m$ if a is a constant and as $\sqrt[n]{a^m}$ if a is a variable or a variable expression.

Solution: $\left(\frac{1}{9}\right)^{5/2} = \left(\sqrt{\frac{1}{9}}\right)^5 = \left(\frac{1}{3}\right)^5 = \frac{1}{243}$

Answers to Your Turn 1
a. $\sqrt{49} = 7$ b. $\sqrt[4]{625} = 5$
c. $\sqrt[3]{-64} = -4$
d. $-\sqrt{81} = -9$
e. $\sqrt[4]{y}$ f. $\sqrt{25a^8} = 5a^4$
g. $18\sqrt[5]{n}$ h. $\sqrt{\frac{49}{x^{10}}} = \frac{7}{x^5}$

e. $x^{2/3}$

Solution: $x^{2/3} = \sqrt[3]{x^2}$ ◄ **Note** Because the base is a variable, we put the exponent beneath the radical sign.

f. $(3x - 5)^{3/5}$

Solution: $(3x - 5)^{3/5} = \sqrt[5]{(3x - 5)^3}$

Your Turn 2 Rewrite using radicals; then simplify if possible. Assume that all variables represent nonnegative values.

a. $32^{3/5}$ **b.** $-49^{3/2}$ **c.** $\left(\frac{1}{16}\right)^{3/2}$ **d.** $y^{5/6}$ **e.** $(3a + 4)^{4/5}$

Negative Rational Exponents

Recall from Section 5.1 that $a^{-b} = \frac{1}{a^b}$. For example, $2^{-3} = \frac{1}{2^3}$. This same rule applies to negative rational exponents.

Rule Negative Rational Exponents

$a^{-m/n} = \frac{1}{a^{m/n}}$, where $a \neq 0$ and m and n are natural numbers with $n \neq 1$.

Example 3 Rewrite using radicals; then simplify if possible.

a. $81^{-1/2}$

Solution: $81^{-1/2} = \frac{1}{81^{1/2}}$ Rewrite the exponential form with a positive exponent by inverting the base and changing the sign of the exponent.

$= \frac{1}{\sqrt{81}}$ Write the rational exponent in radical form.

$= \frac{1}{9}$ Evaluate the square root.

b. $16^{-3/4}$

Solution: $16^{-3/4} = \frac{1}{16^{3/4}}$ Rewrite the exponential form with a positive exponent by inverting the base and changing the sign of the exponent.

$= \frac{1}{(\sqrt[4]{16})^3}$ Write the rational exponent in radical form.

$= \frac{1}{(2)^3}$ Evaluate the radical.

$= \frac{1}{8}$ Simplify the exponential form.

c. $\left(\frac{16}{25}\right)^{-1/2}$

Solution: $\left(\frac{16}{25}\right)^{-1/2} = \frac{1}{\left(\frac{16}{25}\right)^{1/2}}$ Rewrite the exponential form with a positive exponent by inverting the base and changing the sign of the exponent.

$= \frac{1}{\sqrt{\frac{16}{25}}}$ Write the rational exponent in radical form.

Connection Although we used $a^{-m/n} = \frac{1}{a^{m/n}}$ in part c, we could have used $\left(\frac{a}{b}\right)^{-n} = \left(\frac{b}{a}\right)^n$, which we learned in Chapter 5.

Answers to Your Turn 2

a. 8 **b.** -343 **c.** $\frac{1}{64}$

d. $\sqrt[6]{y^5}$ **e.** $\sqrt[5]{(3a + 4)^4}$

Of Interest

John Wallis (1616–1703) was one of the first mathematicians to explain rational and negative exponents. Wallis wrote extensively about physics and mathematics; unfortunately, his work was overshadowed by the work of his countryman Isaac Newton. Newton added to what Wallis began with exponents and roots and popularized the notation that we use today. (*Source:* D. E. Smith, *History of Mathematics*, Dover, 1953.)

$= \dfrac{1}{\frac{4}{5}}$ Evaluate the square root.

$= \dfrac{5}{4}$ Simplify the complex fraction.

d. $(-64)^{-2/3}$

Solution: $(-64)^{-2/3} = \dfrac{1}{(-64)^{2/3}}$ Rewrite the exponential form with a positive exponent.

$= \dfrac{1}{(\sqrt[3]{-64})^2}$ Write the rational exponent in radical form.

$= \dfrac{1}{(-4)^2}$ Evaluate the cube root.

$= \dfrac{1}{16}$ Square -4.

Your Turn 3 Rewrite using radicals; then simplify if possible.

a. $49^{-1/2}$ b. $-27^{-2/3}$ c. $\left(\dfrac{16}{81}\right)^{-3/4}$ d. $(-8)^{-5/3}$

Objective 2 Write radicals as expressions raised to rational exponents.

In upper-level math courses, it is often necessary to write radical expressions in exponential form. To do so, we use the facts that $\sqrt[n]{a^m} = a^{m/n}$ and $(\sqrt[n]{a})^m = a^{m/n}$.

Example 4 Write each of the following in exponential form. Assume that all variables represent positive values.

a. $\sqrt[4]{x^3}$

Solution: $\sqrt[4]{x^3} = x^{3/4}$

b. $\dfrac{1}{\sqrt[3]{x^2}}$

Solution: $\dfrac{1}{\sqrt[3]{x^2}} = \dfrac{1}{x^{2/3}} = x^{-2/3}$

c. $(\sqrt[5]{z})^3$

Solution: $(\sqrt[5]{z})^3 = z^{3/5}$

d. $\sqrt[6]{(3x-5)^5}$

Solution: $\sqrt[6]{(3x-5)^5} = (3x-5)^{5/6}$

Your Turn 4 Write each of the following in exponential form. Assume that all variables represent positive values.

a. $\sqrt[5]{y^2}$ b. $\dfrac{1}{\sqrt[4]{x^3}}$ c. $(\sqrt[7]{a})^3$ d. $\sqrt[3]{(3y-2)^5}$

Objective 3 Simplify expressions with rational number exponents using the rules of exponents.

Rational exponents follow the same rules that we established in Chapter 5 for integer exponents, which we review here.

Answers to Your Turn 3
a. $\dfrac{1}{7}$ b. $-\dfrac{1}{9}$
c. $\dfrac{27}{8}$ d. $-\dfrac{1}{32}$

Answers to Your Turn 4
a. $y^{2/5}$ b. $x^{-3/4}$
c. $a^{3/7}$ d. $(3y-2)^{5/3}$

Rule Rules of Exponents Summary

(Assume that no denominators are 0, that a and b are real numbers, and that m and n are integers.)

Zero as an exponent: $a^0 = 1$, where $a \neq 0$
0^0 is indeterminate.

Negative exponents: $a^{-n} = \frac{1}{a^n}$ and $\frac{1}{a^{-n}} = a^n$ $\left(\frac{a}{b}\right)^{-n} = \left(\frac{b}{a}\right)^n$

Product rule of exponents: $a^m \cdot a^n = a^{m+n}$

Quotient rule for exponents: $\frac{a^m}{a^n} = a^{m-n}$

Raising a power to a power: $(a^m)^n = a^{mn}$

Raising a product to a power: $(ab)^n = a^n b^n$

Raising a quotient to a power: $\left(\frac{a}{b}\right)^n = \frac{a^n}{b^n}$

Example 5 Use the rules of exponents to simplify. Write the answers with positive exponents. Assume that all variables represent positive values.

a. $x^{1/5} \cdot x^{3/5}$

Solution: $x^{1/5} \cdot x^{3/5} = x^{1/5+3/5}$ Use $a^m \cdot a^n = a^{m+n}$.
$= x^{4/5}$ Add the exponents.

b. $(2a^{1/2})(4a^{1/3})$

Solution: $(2a^{1/2})(4a^{1/3}) = 8a^{1/2+1/3}$ Use $a^m \cdot a^n = a^{m+n}$.
$= 8a^{3/6+2/6}$ Rewrite the exponents with a common denominator of 6.
$= 8a^{5/6}$ Add the exponents.

c. $\frac{5^{2/3}}{5^{3/4}}$

Solution: $\frac{5^{2/3}}{5^{3/4}} = 5^{2/3-3/4}$ Use $\frac{a^m}{a^n} = a^{m-n}$.
$= 5^{8/12-9/12}$ Rewrite the exponents with a common denominator of 12.
$= 5^{-1/12}$ Subtract the exponents.
$= \frac{1}{5^{1/12}}$ Rewrite with a positive exponent.

d. $(-4y^{-3/5})(5y^{4/5})$

Solution: $(-4y^{-3/5})(5y^{4/5}) = -20y^{-3/5+4/5}$ Use $a^m \cdot a^n = a^{m+n}$.
$= -20y^{1/5}$ Add the exponents.

e. $(w^{3/4})^2$

Solution: $(w^{3/4})^2 = w^{(3/4)2}$ Use $(a^m)^n = a^{mn}$.
$= w^{3/2}$ Multiply the exponents.

f. $(2a^{2/3}b^{3/5})^3$

Solution: $(2a^{2/3}b^{3/5})^3 = 2^3(a^{2/3})^3(b^{3/5})^3$ Use $(ab)^n = a^n \cdot b^n$.
$= 8a^{(2/3)3}b^{(3/5)3}$ Use $(a^m)^n = a^{mn}$.
$= 8a^2b^{9/5}$ Multiply the exponents.

g. $\dfrac{(3x^{4/3})^3}{x^3}$

Solution: $\dfrac{(3x^{4/3})^3}{x^3} = \dfrac{3^3(x^{4/3})^3}{x^3}$ Use $(ab)^n = a^n \cdot b^n$ in the numerator.

$= \dfrac{27x^{(4/3)3}}{x^3}$ Use $(a^m)^n = a^{mn}$.

$= \dfrac{27x^4}{x^3}$ Simplify.

$= 27x^{4-3}$ Use $\dfrac{a^m}{a^n} = a^{m-n}$.

$= 27x$ Simplify.

Your Turn 5 Use the rules of exponents to simplify. Write the answers with positive exponents. Assume that all variables represent positive values.

a. $(x^{2/3})(x^{1/2})$ b. $\dfrac{y^{2/7}}{y^{5/7}}$ c. $(b^{3/4})^2$ d. $(2x^{3/2}y^{2/3})^4$ e. $\dfrac{(2z^{5/3})^3}{z^2}$

Objective 4 Use rational exponents to simplify radical expressions.

Often, radical expressions can be simplified by rewriting them with rational exponents, simplifying, and then rewriting as a radical expression.

Example 6 Rewrite as a radical with a smaller root index. Assume that all variables represent nonnegative values.

a. $\sqrt[4]{49}$

Solution: $\sqrt[4]{49} = 49^{1/4}$ Rewrite in exponential form.

$= (7^2)^{1/4}$ Write 49 as 7^2.

$= 7^{2 \cdot 1/4}$ Use $(a^m)^n = a^{mn}$.

$= 7^{1/2}$ Simplify.

$= \sqrt{7}$ Write in radical form.

b. $\sqrt[6]{x^4}$

Solution: $\sqrt[6]{x^4} = x^{4/6}$ Rewrite in exponential form.

$= x^{2/3}$ Simplify to lowest terms.

$= \sqrt[3]{x^2}$ Write in radical form.

c. $\sqrt[6]{a^4b^2}$

Solution: $\sqrt[6]{a^4b^2} = (a^4b^2)^{1/6}$ Rewrite in exponential form.

$= (a^4)^{1/6}(b^2)^{1/6}$ Use $(ab)^n = a^nb^n$.

$= a^{4 \cdot 1/6}b^{2 \cdot 1/6}$ Use $(a^m)^n = a^{mn}$.

$= a^{2/3}b^{1/3}$ Simplify.

$= (a^2b)^{1/3}$ Use $(ab)^n = a^nb^n$.

$= \sqrt[3]{a^2b}$ Write in radical form.

Your Turn 6 Rewrite as a radical with a smaller root index. Assume that all variables represent nonnegative values.

a. $\sqrt[4]{36}$ b. $\sqrt[8]{x^6}$ c. $\sqrt[8]{x^6y^2}$

Answers to Your Turn 5

a. $x^{7/6}$ b. $\dfrac{1}{y^{3/7}}$ c. $b^{3/2}$

d. $16x^6y^{8/3}$ e. $8z^3$

Answers to Your Turn 6

a. $\sqrt{6}$ b. $\sqrt[4]{x^3}$ c. $\sqrt[4]{x^3y}$

Writing radicals in exponential form also allows us to multiply and divide radical expressions with different root indices.

Procedure Multiplying and Dividing Radical Expressions with Different Root Indices

To multiply or divide radical expressions with different root indices:

1. Change from radical to exponential form.
2. Multiply or divide using the appropriate rule(s) of exponents.
3. Write the result from step 2 in radical form.

Example 7 Perform the indicated operations. Write the result using a radical. Assume that all variables represent positive values.

a. $\sqrt{x} \cdot \sqrt[3]{x^2}$

Solution:

$\sqrt{x} \cdot \sqrt[3]{x^2} = x^{1/2} \cdot x^{2/3}$ Write in exponential form.

$= x^{1/2+2/3}$ Use $a^m \cdot a^n = a^{m+n}$.

$= x^{3/6+4/6}$ Rewrite the exponents with their LCD.

$= x^{7/6}$ Simplify.

$= \sqrt[6]{x^7}$ Write in radical form.

Note In Section 8.3, we learn to further simplify expressions such as $\sqrt[6]{x^7}$.

b. $\dfrac{\sqrt[4]{x^3}}{\sqrt[3]{x}}$

Solution:

$\dfrac{\sqrt[4]{x^3}}{\sqrt[3]{x}} = \dfrac{x^{3/4}}{x^{1/3}}$ Write in exponential form.

$= x^{3/4-1/3}$ Use $\dfrac{a^m}{a^n} = a^{m-n}$.

$= x^{9/12-4/12}$ Rewrite the exponents with their LCD.

$= x^{5/12}$ Simplify.

$= \sqrt[12]{x^5}$ Write in radical form.

c. $\sqrt{3} \cdot \sqrt[3]{2}$

Solution:

$\sqrt{3} \cdot \sqrt[3]{2} = 3^{1/2} \cdot 2^{1/3}$ Write in exponential form.

$= 3^{3/6} \cdot 2^{2/6}$ Write the exponents with their LCD.

$= (3^3 \cdot 2^2)^{1/6}$ Use $a^n b^n = (ab)^n$.

$= (27 \cdot 4)^{1/6}$ Evaluate 3^3 and 2^2.

$= 108^{1/6}$ Multiply.

$= \sqrt[6]{108}$ Rewrite as a radical.

Your Turn 7 Perform the indicated operations. Write the result using a radical. Assume that all variables represent positive values.

a. $\sqrt[3]{x^2} \cdot \sqrt[4]{x}$ b. $\dfrac{\sqrt[4]{a^3}}{\sqrt[3]{a^2}}$ c. $\sqrt[4]{2} \cdot \sqrt{5}$

Answers to Your Turn 7
a. $\sqrt[12]{x^{11}}$ b. $\sqrt[12]{a}$ c. $\sqrt[4]{50}$

By writing radical expressions using rational exponents, we can also find the root of a root.

Example 8 Write $\sqrt[3]{\sqrt{x}}$ as a single radical. Assume that all variables represent nonnegative values.

Solution: $\sqrt[3]{\sqrt{x}} = (x^{1/2})^{1/3}$ Write in exponential form.

$= x^{(1/2)(1/3)}$ Apply $(a^m)^n = a^{mn}$.

$= x^{1/6}$ Simplify.

$= \sqrt[6]{x}$ Write as a radical.

Your Turn 8 Write $\sqrt[4]{\sqrt[3]{y}}$ as a single radical. Assume that all variables represent nonnegative values.

Answer to Your Turn 8
$\sqrt[12]{y}$

8.2 Exercises

For Extra Help MyMathLab®

Objective 1

Prep Exercise 1 If $4^{1/3}$ were written as a radical expression, what would be the radicand?

Prep Exercise 2 If $4^{3/5}$ were written as a radical expression, what would be the root index?

Prep Exercise 3 If $a^{m/n}$, with n even and m/n in lowest terms, were written as a radical expression, what restrictions would be placed on a? Why?

Prep Exercise 4 If $a^{m/n}$, with n odd, were written as a radical expression, what restrictions would be placed on a? Why?

For Exercises 1–34, rewrite each of the following using radicals; then simplify if possible. Assume that all variables represent nonnegative values. See Examples 1–3.

1. $25^{1/2}$ **2.** $64^{1/2}$ **3.** $-100^{1/2}$ **4.** $-64^{1/2}$

5. $27^{1/3}$ **6.** $216^{1/3}$ **7.** $(-64)^{1/3}$ **8.** $(-125)^{1/3}$

9. $y^{1/4}$ **10.** $w^{1/8}$ **11.** $(144x^8)^{1/2}$ **12.** $(121z^6)^{1/2}$

13. $18r^{1/2}$ **14.** $22a^{1/2}$ **15.** $\left(\frac{x^4}{81}\right)^{1/2}$ **16.** $\left(\frac{n^8}{36}\right)^{1/2}$

17. $64^{2/3}$ **18.** $16^{3/4}$ **19.** $-81^{3/4}$ **20.** $-16^{5/4}$

21. $(-8)^{4/3}$ **22.** $(-27)^{5/3}$ **23.** $16^{-3/4}$ **24.** $8^{-4/3}$

25. $x^{4/5}$ **26.** $m^{5/7}$ **27.** $8n^{2/3}$ **28.** $6a^{5/6}$

29. $(-32)^{-2/5}$ **30.** $(-216)^{-2/3}$ **31.** $\left(\frac{1}{25}\right)^{3/2}$ **32.** $\left(\frac{1}{32}\right)^{3/5}$

33. $(2a + 4)^{5/6}$ **34.** $(5r - 2)^{5/7}$

Objective 2

Prep Exercise 5 When a radical expression is written as an expression with a rational exponent, what number is used as the denominator of the exponent?

For Exercises 35–50, write each of the following in exponential form. Assume that all variables represent positive values. See Example 4.

35. $\sqrt[4]{25}$ **36.** $\sqrt[5]{42}$ **37.** $\sqrt[6]{z^5}$ **38.** $\sqrt[7]{r^5}$

39. $\dfrac{1}{\sqrt[6]{5^5}}$ **40.** $\dfrac{1}{\sqrt[7]{6^2}}$ **41.** $\dfrac{5}{\sqrt[5]{x^4}}$ **42.** $\dfrac{8}{\sqrt[7]{n^9}}$

43. $(\sqrt[3]{5})^7$ **44.** $(\sqrt[5]{8})^6$ **45.** $(\sqrt[7]{x})^2$ **46.** $(\sqrt[3]{m})^8$

47. $\sqrt[4]{(4a-5)^7}$ **48.** $\sqrt[7]{(5w+3)^5}$ **49.** $(\sqrt[5]{2r-5})^8$ **50.** $(\sqrt[5]{3r-6})^9$

Objective 3

Prep Exercise 6 Complete the rules.

$a^m \cdot a^n = a^?$ $\dfrac{a^m}{a^n} = a^?$ $(a^m)^n = a^?$

For Exercises 51–90, use the rules of exponents to simplify. Write the answers with positive exponents. Assume that all variables represent positive values. See Example 5.

51. $x^{1/5} \cdot x^{3/5}$ **52.** $n^{1/7} \cdot n^{5/7}$ **53.** $x^{3/2} \cdot x^{-1/3}$ **54.** $n^{2/3} \cdot n^{-1/2}$

55. $a^{2/3} \cdot a^{3/4}$ **56.** $r^{5/2} \cdot r^{5/3}$ **57.** $(3w^{1/7})(7w^{3/7})$ **58.** $(5p^{1/9})(3p^{4/9})$

59. $(-3a^{2/3})(4a^{3/4})$ **60.** $(8c^{4/5})(-4c^{3/2})$ **61.** $\dfrac{7^{7/3}}{7^{2/3}}$ **62.** $\dfrac{3^{7/9}}{3^{2/9}}$

63. $\dfrac{x^{1/6}}{x^{5/6}}$ **64.** $\dfrac{y^{2/9}}{y^{5/9}}$ **65.** $\dfrac{x^{3/4}}{x^{1/2}}$ **66.** $\dfrac{x^{5/8}}{x^{1/2}}$

67. $\dfrac{r^{3/4}}{r^{2/3}}$ **68.** $\dfrac{m^{3/5}}{m^{1/2}}$ **69.** $\dfrac{x^{-3/7}}{x^{2/7}}$ **70.** $\dfrac{v^{-4/5}}{v^{2/5}}$

71. $\dfrac{a^{3/4}}{a^{-3/2}}$ **72.** $\dfrac{b^{5/6}}{b^{-2/3}}$ **73.** $(5s^{-2/7})(4s^{5/7})$ **74.** $(6u^{8/9})(-6u^{-5/9})$

75. $(-6b^{-5/4})(4b^{3/2})$ **76.** $(-6y^{-5/6})(-7y^{5/3})$ **77.** $(x^{2/3})^3$ **78.** $(r^{3/4})^4$

79. $(a^{5/6})^2$ **80.** $(n^{3/8})^4$ **81.** $(b^{2/3})^{3/5}$ **82.** $(m^{3/2})^{2/5}$

83. $(2x^{2/3}y^{1/2})^6$

84. $(2a^{1/4}b^{3/2})^8$

85. $(8q^{3/2}t^{3/4})^{1/3}$

86. $(16x^{2/3}y^{1/3})^{3/4}$

87. $\frac{(3a^{5/4})^4}{a^2}$

88. $\frac{(5v^{5/2})^2}{v^3}$

89. $\frac{(9z^{7/3})^{1/2}}{z^{5/6}}$

90. $\frac{(36x^{3/4})^{1/2}}{x^{1/8}}$

Objective 4

Prep Exercise 7 Does $\sqrt[4]{100} = \sqrt{10}$? Why or why not?

Prep Exercise 8 Does $\sqrt[3]{3} \cdot \sqrt[5]{2} = \sqrt[15]{1944}$? Why or why not?

For Exercises 91–102, represent each of the following as a radical with a smaller root index. Assume that all variables represent nonnegative values. See Example 6.

91. $\sqrt[4]{4}$

92. $\sqrt[4]{49}$

93. $\sqrt[6]{49}$

94. $\sqrt[6]{100}$

95. $\sqrt[4]{x^2}$

96. $\sqrt[6]{y^3}$

97. $\sqrt[8]{r^6}$

98. $\sqrt[10]{n^6}$

99. $\sqrt[8]{x^6y^2}$

100. $\sqrt[6]{y^2z^4}$

101. $\sqrt[10]{m^4n^6}$

102. $\sqrt[10]{a^2b^8}$

For Exercises 103–114, perform the indicated operations. Write the result using a radical. Assume that all variables represent positive values. See Example 7.

103. $\sqrt[3]{x} \cdot \sqrt{x}$

104. $\sqrt[4]{y} \cdot \sqrt[3]{y^2}$

105. $\sqrt[4]{y^2} \cdot \sqrt[3]{y^2}$

106. $\sqrt[5]{x^4} \cdot \sqrt[3]{x^2}$

107. $\frac{\sqrt[3]{x^4}}{\sqrt[4]{x^2}}$

108. $\frac{\sqrt[3]{y^5}}{\sqrt[4]{y^2}}$

109. $\frac{\sqrt[5]{n^4}}{\sqrt[3]{n^2}}$

110. $\frac{\sqrt[6]{z^4}}{\sqrt{z}}$

111. $\sqrt{5} \cdot \sqrt[3]{3}$

112. $\sqrt[3]{4} \cdot \sqrt{5}$

113. $\sqrt[4]{6} \cdot \sqrt[3]{2}$

114. $\sqrt[4]{4} \cdot \sqrt[3]{2}$

For Exercises 115–118, write each as a single radical. Assume that all variables represent nonnegative values. See Example 8.

115. $\sqrt[3]{\sqrt[3]{x}}$

116. $\sqrt[3]{\sqrt[5]{m}}$

117. $\sqrt{\sqrt[3]{n}}$

118. $\sqrt[4]{\sqrt[5]{z}}$

Review Exercises

Exercises 1–6 Expressions

[1.3] 1. Rewrite $2 \cdot 2 \cdot 2 \cdot 2 \cdot x \cdot x \cdot x \cdot y \cdot y$ using exponents.

[1.3] *For Exercises 2 and 3, simplify using the order of operations agreement.*

2. $\sqrt{16} \cdot \sqrt{9}$

3. $\sqrt[3]{27} \cdot \sqrt[3]{125}$

[5.3] *For Exercises 4 and 5, simplify.*

4. $(2.5 \times 10^6)(3.2 \times 10^5)$

5. $\left(\frac{3}{4}x^3y\right)\left(-\frac{5}{6}xyz^2\right)$

[5.4] 6. Use long division to find the quotient: $\frac{2x^3 - 2x^2 - 19x + 18}{x + 3}$

8.3 Multiplying, Dividing, and Simplifying Radicals

Objectives

1 Multiply radical expressions.
2 Divide radical expressions.
3 Use the product rule to simplify radical expressions.

Warm-up

For Exercises 1 and 2, simplify.

[8.1] 1. $\sqrt{81x^{12}y^4}$

[8.2] 2. $27^{2/3}$

[8.2] *For Exercises 3 and 4, use the rules of exponents to simplify. Write answers with positive exponents. Assume that all variables represent positive values.*

3. $(6x^{3/4})(7x^{4/5})$

4. $\dfrac{(4x^{7/3})^3}{2x^2}$

In this section, we explore some ways to simplify expressions that involve multiplication or division of radicals.

Objective 1 Multiply radical expressions.

Consider the expression $\sqrt{9}\cdot\sqrt{16}$. The usual approach is to find the roots and then multiply those roots. However, we can also multiply the radicands and then find the root of the product.

Find the roots first:
$\sqrt{9}\cdot\sqrt{16} = 3\cdot 4 = 12$

Multiply the radicands first:
$\sqrt{9}\cdot\sqrt{16} = \sqrt{9\cdot 16} = \sqrt{144} = 12$

Both approaches give the same result, suggesting the following rule:

Note To apply this rule, the root indices must be the same.

Rule Product Rule for Radicals
If both $\sqrt[n]{a}$ and $\sqrt[n]{b}$ are real numbers, then $\sqrt[n]{a}\cdot\sqrt[n]{b} = \sqrt[n]{a\cdot b}$.

Example 1 Find the product and simplify. Assume that all variables represent positive values.

a. $\sqrt{3}\cdot\sqrt{27}$

Solution: $\sqrt{3}\cdot\sqrt{27} = \sqrt{3\cdot 27} = \sqrt{81} = 9$

Connection Example 1(a) illustrates that the product of two irrational numbers can be a rational number.

b. $\sqrt{11}\cdot\sqrt{x}$

Solution: $\sqrt{11}\cdot\sqrt{x} = \sqrt{11x}$

c. $\sqrt[3]{4}\cdot\sqrt[3]{2}$

Solution: $\sqrt[3]{4}\cdot\sqrt[3]{2} = \sqrt[3]{4\cdot 2} = \sqrt[3]{8} = 2$

d. $\sqrt[3]{3x}\cdot\sqrt[3]{4x}$

Solution: $\sqrt[3]{3x}\cdot\sqrt[3]{4x} = \sqrt[3]{3x\cdot 4x} = \sqrt[3]{12x^2}$

e. $\sqrt[4]{5}\cdot\sqrt[4]{7x^2}$

Solution: $\sqrt[4]{5}\cdot\sqrt[4]{7x^2} = \sqrt[4]{5\cdot 7x^2} = \sqrt[4]{35x^2}$

f. $\sqrt{\dfrac{5}{x}}\cdot\sqrt{\dfrac{y}{2}}$

Solution: $\sqrt{\dfrac{5}{x}}\cdot\sqrt{\dfrac{y}{2}} = \sqrt{\dfrac{5}{x}\cdot\dfrac{y}{2}} = \sqrt{\dfrac{5y}{2x}}$

Connection Expressions that have a fraction in a radical, such as $\sqrt{\dfrac{5y}{2x}}$, are not considered to be in simplest form. We learn how to simplify them in Section 8.5.

g. $\sqrt{x}\cdot\sqrt{x}$

Solution: $\sqrt{x}\cdot\sqrt{x} = \sqrt{x\cdot x} = \sqrt{x^2} = x$

Answers to Warm-up
1. $9x^6y^2$ 2. 9
3. $42x^{31/20}$ 4. $32x^5$

Your Turn 1 Find the product and simplify. Assume that all variables represent positive values.

a. $\sqrt{2} \cdot \sqrt{32}$ **b.** $\sqrt{5} \cdot \sqrt{a}$ **c.** $\sqrt[3]{5x} \cdot \sqrt[3]{2y}$ **d.** $\sqrt{\frac{2}{a}} \cdot \sqrt{\frac{b}{7}}$ **e.** $\sqrt{a} \cdot \sqrt{a}$

Notice that in Example 1(g), $\sqrt{x} \cdot \sqrt{x} = x$. It is also true that $\sqrt{x} \cdot \sqrt{x} = (\sqrt{x})^2$. Therefore, $(\sqrt{x})^2 = x$. Similarly, $\sqrt[3]{x} \cdot \sqrt[3]{x} \cdot \sqrt[3]{x} = \sqrt[3]{x \cdot x \cdot x} = \sqrt[3]{x^3} = x$. It is also true that $\sqrt[3]{x} \cdot \sqrt[3]{x} \cdot \sqrt[3]{x} = (\sqrt[3]{x})^3$; so $(\sqrt[3]{x})^3 = x$. These examples suggest the following rule:

Rule Raising an *n*th Root to the *n*th Power

For any nonnegative real number a, $(\sqrt[n]{a})^n = a$.

This means that $(\sqrt[3]{4})^3 = 4$, $(\sqrt[5]{19})^5 = 19$, and $(\sqrt[4]{3x^2})^4 = 3x^2$.

Objective 2 Divide radical expressions.

Earlier we developed the product rule for radicals. Now we develop a similar rule for quotients such as $\frac{\sqrt{100}}{\sqrt{25}}$. We can follow the order of operations and divide the roots, or we can divide the radicands and then find the square root of the quotient.

Find the roots first:

$$\frac{\sqrt{100}}{\sqrt{25}} = \frac{10}{5} = 2$$

Divide the radicands first:

$$\frac{\sqrt{100}}{\sqrt{25}} = \sqrt{\frac{100}{25}} = \sqrt{4} = 2$$

Both approaches give the same result, which suggests the following rule:

Rule Quotient Rule for Radicals

If both $\sqrt[n]{a}$ and $\sqrt[n]{b}$ are real numbers, then $\frac{\sqrt[n]{a}}{\sqrt[n]{b}} = \sqrt[n]{\frac{a}{b}}$, where $b \neq 0$.

As with all equations, this rule can be used going from left to right or right to left.

Example 2 Simplify. Assume that variables represent positive values.

a. $\sqrt{\frac{7}{36}}$

Solution: $\sqrt{\frac{7}{36}} = \frac{\sqrt{7}}{\sqrt{36}} = \frac{\sqrt{7}}{6}$

b. $\frac{\sqrt{108}}{\sqrt{3}}$

Solution: $\frac{\sqrt{108}}{\sqrt{3}} = \sqrt{\frac{108}{3}} = \sqrt{36} = 6$

c. $\sqrt[3]{\frac{9}{x^3}}$

Solution: $\sqrt[3]{\frac{9}{x^3}} = \frac{\sqrt[3]{9}}{\sqrt[3]{x^3}} = \frac{\sqrt[3]{9}}{x}$

d. $\frac{\sqrt[3]{15}}{\sqrt[3]{5}}$

Solution: $\frac{\sqrt[3]{15}}{\sqrt[3]{5}} = \sqrt[3]{\frac{15}{5}} = \sqrt[3]{3}$

e. $\sqrt[4]{\frac{y}{81}}$

Solution: $\sqrt[4]{\frac{y}{81}} = \frac{\sqrt[4]{y}}{\sqrt[4]{81}} = \frac{\sqrt[4]{y}}{3}$

Answers to Your Turn 1

a. 8 **b.** $\sqrt{5a}$ **c.** $\sqrt[3]{10xy}$ **d.** $\sqrt{\frac{2b}{7a}}$ **e.** a

Your Turn 2 Simplify. Assume that variables represent positive values.

a. $\sqrt{\dfrac{x}{49}}$ b. $\dfrac{\sqrt{75}}{\sqrt{5}}$ c. $\sqrt[4]{\dfrac{6}{x^4}}$ d. $\dfrac{\sqrt[3]{32}}{\sqrt[3]{4}}$

Objective 3 Use the product rule to simplify radical expressions.

Note In future sections, we explore other conditions that require simplification.

Several conditions exist in which a radical is not considered to be in simplest form. One such condition is when a radicand has a factor that can be written to a power greater than or equal to the index. For example, $\sqrt[3]{81}$ is not in simplest form because the perfect cube 27 is a factor of 81. Our first step in simplifying $\sqrt[3]{81}$ is to rewrite it as $\sqrt[3]{27 \cdot 3}$ so that we can then use the product rule for radicals.

Procedure Simplifying *n*th Roots

To simplify an *n*th root:

1. Write the radicand as a product of the greatest possible perfect *n*th power and a number or an expression that has no perfect *n*th power factors.
2. Use the product rule $\sqrt[n]{ab} = \sqrt[n]{a} \cdot \sqrt[n]{b}$, where a is the perfect *n*th power.
3. Find the *n*th root of the perfect *n*th power radicand.

Connection A list of perfect powers may be helpful.

Perfect squares:

1, 4, 9, 16, 25, 36, 49, 64, 81, 100, . . .

Perfect cubes:

1, 8, 27, 64, 125, 216, . . .

Perfect fourth powers:

1, 16, 81, 256, 625, . . .

Example 3 Simplify.

a. $\sqrt{18}$

Solution: $\sqrt{18} = \sqrt{9 \cdot 2}$ The greatest perfect square factor of 18 is 9, so we write 18 as $9 \cdot 2$.

$= \sqrt{9} \cdot \sqrt{2}$ Use the product rule of roots to separate the factors into two radicals.

$= 3\sqrt{2}$ Simplify the square root of 9.

b. $5\sqrt{72}$

Solution: $5\sqrt{72} = 5 \cdot \sqrt{36 \cdot 2}$ The greatest perfect square factor of 72 is 36, so we write 72 as $36 \cdot 2$.

$= 5 \cdot \sqrt{36} \cdot \sqrt{2}$ Use the product rule of roots to separate the factors into two radicals.

$= 5 \cdot 6 \cdot \sqrt{2}$ Simplify the square root of 36.

$= 30\sqrt{2}$ Multiply $5 \cdot 6$.

Note We can use perfect *n*th power factors other than the greatest perfect *n*th power factor. For example, in simplifying $\sqrt{72}$, instead of $\sqrt{72} = \sqrt{36 \cdot 2} = 6\sqrt{2}$, we could write

$$\sqrt{72} = \sqrt{4 \cdot 18} = 2\sqrt{18} = 2\sqrt{9 \cdot 2} = 2 \cdot 3\sqrt{2} = 6\sqrt{2}.$$

Notice that using the greatest perfect *n*th factor saves steps.

c. $\sqrt[3]{40}$

Solution: $\sqrt[3]{40} = \sqrt[3]{8 \cdot 5}$ The greatest perfect cube factor of 40 is 8.

$= \sqrt[3]{8} \cdot \sqrt[3]{5}$ Use the product rule of roots.

$= 2\sqrt[3]{5}$ Simplify the cube root of 8.

d. $4\sqrt[4]{162}$

Solution: $4\sqrt[4]{162} = 4\sqrt[4]{81 \cdot 2}$ The greatest perfect fourth power factor of 162 is 81.

$= 4\sqrt[4]{81} \cdot \sqrt[4]{2}$ Use the product rule of roots.

$= 4 \cdot 3 \cdot \sqrt[4]{2}$ Simplify the fourth root of 81.

$= 12\sqrt[4]{2}$

Answers to Your Turn 2

a. $\dfrac{\sqrt{x}}{7}$ b. $\sqrt{15}$ c. $\dfrac{\sqrt[4]{6}}{x}$ d. 2

Your Turn 3 Simplify.

a. $\sqrt{150}$ b. $6\sqrt{80}$ c. $\sqrt[3]{108}$ d. $3\sqrt[4]{80}$

Using Prime Factorization

If the greatest perfect nth power of a particular radicand is not obvious, try using the prime factorization of the radicand. Each prime factor that appears twice will have a square root equal to one of the two factors, each prime factor that appears three times will have a cube root equal to one of the three factors, and so on. The remaining factors stay in the radical sign.

Example 4 Simplify the following radicals using prime factorizations.

a. $\sqrt{375}$

Solution: $\sqrt{375} = \sqrt{5 \cdot 5 \cdot 5 \cdot 3}$ Write 375 as the product of its prime factors.

$= 5\sqrt{3 \cdot 5}$ The square root of the pair of 5s is 5.

$= 5\sqrt{15}$ Multiply the prime factors in the radicand.

b. $\sqrt[3]{324}$

Solution: $\sqrt[3]{324} = \sqrt[3]{3 \cdot 3 \cdot 3 \cdot 3 \cdot 2 \cdot 2}$ Write 324 as the product of its prime factors.

$= 3\sqrt[3]{3 \cdot 2 \cdot 2}$ The cube root of the three 3s is 3.

$= 3\sqrt[3]{12}$ Multiply the prime factors in the radicand.

c. $\sqrt[4]{240}$

Solution: $\sqrt[4]{240} = \sqrt[4]{2 \cdot 2 \cdot 2 \cdot 2 \cdot 3 \cdot 5}$ Write 240 as the product of its prime factors.

$= 2\sqrt[4]{3 \cdot 5}$ The fourth root of the four 2s is 2.

$= 2\sqrt[4]{15}$ Multiply the prime factors in the radicand.

Your Turn 4 Simplify the following radicals using prime factorizations.

a. $\sqrt{294}$ b. $\sqrt[3]{324}$ c. $\sqrt[4]{486}$

Simplifying Radicals with Variables

We can use either procedure to simplify radicals whose radicands contain variables. Because $\sqrt[n]{a^m} = a^{m/n}$, we can find an exact root if n divides into m evenly. For example, $\sqrt{x^4} = x^{4/2} = x^2$ and $\sqrt[3]{x^{12}} = x^{12/3} = x^4$. Therefore, to find the nth root, we rewrite the radicand as a product in which one factor has the greatest possible exponent divisible by the index n.

Example 5 Simplify. Assume that variables represent nonnegative values.

a. $\sqrt{x^7}$

Solution: $\sqrt{x^7} = \sqrt{x^6 \cdot x}$ The greatest number smaller than 7 that is divisible by 2 is 6, so write x^7 as $x^6 \cdot x$.

$= \sqrt{x^6} \cdot \sqrt{x}$ Use the product rule of roots.

$= x^3\sqrt{x}$ Simplify $\sqrt{x^6} = x^3$.

b. $3\sqrt{24a^5b^9}$

Solution: $3\sqrt{24a^5b^9} = 3\sqrt{4 \cdot 6 \cdot a^4 \cdot a \cdot b^8 \cdot b}$ Write 24 as $4 \cdot 6$, a^5 as $a^4 \cdot a$, and b^9 as $b^8 \cdot b$.

$= 3\sqrt{4a^4b^8 \cdot 6ab}$ Regroup the factors so that perfect squares are together.

$= 3\sqrt{4a^4b^8} \cdot \sqrt{6ab}$ Use the product rule of roots.

$= 3 \cdot 2a^2b^4 \cdot \sqrt{6ab}$ Simplify $\sqrt{4} = 2$, $\sqrt{a^4} = a^2$, and $\sqrt{b^8} = b^4$.

$= 6a^2b^4\sqrt{6ab}$ Multiply $3 \cdot 2 = 6$.

Connection A list of perfect powers might be helpful.
Perfect squares:

$x^2, x^4, x^6, x^8, \ldots$

Perfect cubes:

$x^3, x^6, x^9, x^{12}, \ldots$

Answers to Your Turn 3
a. $5\sqrt{6}$ b. $24\sqrt{5}$
c. $3\sqrt[3]{4}$ d. $6\sqrt[4]{5}$

Answers to Your Turn 4
a. $7\sqrt{6}$ b. $3\sqrt[3]{12}$ c. $3\sqrt[4]{6}$

c. $y^2\sqrt[3]{y^8}$

Solution: $y^2\sqrt[3]{y^8} = y^2\sqrt[3]{y^6 \cdot y^2}$ The greatest number smaller than 8 that is divisible by 3 is 6, so write y^8 as $y^6 \cdot y^2$.

$= y^2\sqrt[3]{y^6} \cdot \sqrt[3]{y^2}$ Use the product rule of roots.

$= y^2 \cdot y^2 \cdot \sqrt[3]{y^2}$ Simplify $\sqrt[3]{y^6} = y^2$.

$= y^4\sqrt[3]{y^2}$ Multiply $y^2 \cdot y^2 = y^4$.

d. $\sqrt[5]{64x^9y^{12}}$

Solution: $\sqrt[5]{64x^9y^{12}} = \sqrt[5]{32 \cdot 2 \cdot x^5 \cdot x^4 \cdot y^{10} \cdot y^2}$ Write 64 as $32 \cdot 2$, x^9 as $x^5 \cdot x^4$, and y^{12} as $y^{10} \cdot y^2$.

$= \sqrt[5]{32x^5y^{10} \cdot 2x^4y^2}$ Regroup the factors.

$= \sqrt[5]{32x^5y^{10}} \cdot \sqrt[5]{2x^4y^2}$ Use the product rule of roots.

$= 2xy^2\sqrt[5]{2x^4y^2}$ Simplify $\sqrt[5]{32} = 2$, $\sqrt[5]{x^5} = x$, and $\sqrt[5]{y^{10}} = y^2$.

Your Turn 5 Simplify. Assume that variables represent nonnegative values.

a. $\sqrt{n^{11}}$ b. $2\sqrt{45r^7s^3}$ c. $\sqrt[4]{m^{13}}$ d. $\sqrt[3]{a^8b^{10}}$

After using the product or quotient rules, it is often necessary to simplify the results.

Example 6 Find the product or quotient and simplify the results. Assume that variables represent positive values.

a. $\sqrt{3} \cdot \sqrt{6}$

Solution: $\sqrt{3} \cdot \sqrt{6} = \sqrt{18}$ Use the product rule of roots to multiply.

$= \sqrt{9 \cdot 2}$ Write 18 as $9 \cdot 2$.

$= 3\sqrt{2}$ Simplify $\sqrt{9} = 3$.

b. $5\sqrt{3x^3} \cdot 3\sqrt{15x^2}$

Solution: $5\sqrt{3x^3} \cdot 3\sqrt{15x^2} = 5 \cdot 3\sqrt{3x^3} \cdot \sqrt{15x^2}$ Regroup the factors.

$= 15\sqrt{45x^5}$ Multiply.

$= 15\sqrt{9 \cdot 5 \cdot x^4 \cdot x}$ Write 45 as $9 \cdot 5$ and x^5 as $x^4 \cdot x$.

$= 15 \cdot 3x^2\sqrt{5x}$ Simplify $\sqrt{9} = 3$ and $\sqrt{x^4} = x^2$.

$= 45x^2\sqrt{5x}$ Multiply.

c. $\dfrac{\sqrt{288}}{\sqrt{6}}$

Solution: $\dfrac{\sqrt{288}}{\sqrt{6}} = \sqrt{\dfrac{288}{6}}$ Use the quotient rule of roots.

$= \sqrt{48}$ Divide the radicand.

$= \sqrt{16 \cdot 3}$ Write 48 as $16 \cdot 3$.

$= 4\sqrt{3}$ Simplify $\sqrt{16} = 4$.

d. $\dfrac{8\sqrt{756a^8b^5}}{2\sqrt{7a^4b^2}}$

Solution: $\dfrac{8\sqrt{756a^8b^5}}{2\sqrt{7a^4b^2}} = 4\sqrt{\dfrac{756a^8b^5}{7a^4b^2}}$ Divide coefficients and use the quotient rule of radicals.

$= 4\sqrt{108a^4b^3}$ Divide the radicand.

$= 4\sqrt{36a^4b^2 \cdot 3b}$ Rewrite the radicand with a perfect square factor.

Answers to Your Turn 5

a. $n^5\sqrt{n}$ b. $6r^3s\sqrt{5rs}$
c. $m^3\sqrt[4]{m}$ d. $a^2b^3\sqrt[3]{a^2b}$

$= 4 \cdot 6a^2b\sqrt{3b}$ Find the square roots.

$= 24a^2b\sqrt{3b}$ Multiply.

Your Turn 6 Find the product or quotient and simplify the results. Assume that variables represent positive values.

a. $\sqrt{6} \cdot \sqrt{15}$ **b.** $4\sqrt{14x^3} \cdot 2\sqrt{6x^4}$ **c.** $\dfrac{\sqrt{1296}}{\sqrt{12}}$ **d.** $\dfrac{14\sqrt{315x^{11}y^8}}{2\sqrt{5x^6y^5}}$

Answers to Your Turn 6
a. $3\sqrt{10}$ **b.** $16x^3\sqrt{21x}$
c. $6\sqrt{3}$ **d.** $21x^2y\sqrt{7xy}$

8.3 Exercises

For Extra Help MyMathLab®

Objective 1

Prep Exercise 1 For $\sqrt{8} \cdot \sqrt{18}$, explain the difference between using the product rule for radicals and multiplying the approximate roots of 8 and 18.

For Exercises 1–28, find the product and simplify. Assume that variables represent positive values. See Example 1.

1. $\sqrt{2} \cdot \sqrt{32}$
2. $\sqrt{3} \cdot \sqrt{48}$
3. $\sqrt{3x} \cdot \sqrt{27x^5}$
4. $\sqrt{8y^3} \cdot \sqrt{2y}$
5. $\sqrt{6xy^3} \cdot \sqrt{24xy}$
6. $\sqrt{50u^3v^2} \cdot \sqrt{2uv^4}$
7. $\sqrt{5} \cdot \sqrt{13}$
8. $\sqrt{6} \cdot \sqrt{11}$
9. $\sqrt{15} \cdot \sqrt{x}$
10. $\sqrt{17} \cdot \sqrt{y}$
11. $\sqrt[3]{3} \cdot \sqrt[3]{9}$
12. $\sqrt[3]{4} \cdot \sqrt[3]{16}$
13. $\sqrt[3]{5y} \cdot \sqrt[3]{2y}$
14. $\sqrt[3]{6m} \cdot \sqrt[3]{2m}$
15. $\sqrt[4]{3} \cdot \sqrt[4]{7}$
16. $\sqrt[4]{7} \cdot \sqrt[4]{5}$
17. $\sqrt[4]{12w^3} \cdot \sqrt[4]{6w}$
18. $\sqrt[4]{21r^3} \cdot \sqrt[4]{7r}$
19. $\sqrt[4]{3x^2y} \cdot \sqrt[4]{5xy^2}$
20. $\sqrt[4]{2ab^2} \cdot \sqrt[4]{6ab}$
21. $\sqrt[5]{6x^3} \cdot \sqrt[5]{5x^4}$
22. $\sqrt[5]{3m^2} \cdot \sqrt[5]{8m^7}$
23. $\sqrt[6]{4x^2y^3} \cdot \sqrt[6]{2x^3y}$
24. $\sqrt[6]{ab^3} \cdot \sqrt[6]{7a^3b^2}$
25. $\sqrt{\dfrac{7}{2}} \cdot \sqrt{\dfrac{3}{5}}$
26. $\sqrt{\dfrac{5}{2}} \cdot \sqrt{\dfrac{11}{3}}$
27. $\sqrt{\dfrac{6}{x}} \cdot \sqrt{\dfrac{y}{5}}$
28. $\sqrt{\dfrac{a}{3}} \cdot \sqrt{\dfrac{7}{b}}$

Objective 2

Prep Exercise 2 How does the quotient rule avoid approximation in simplifying $\dfrac{\sqrt{125}}{\sqrt{5}}$?

For Exercises 29–44, simplify. Assume that variables represent positive values. See Example 2.

29. $\sqrt{\dfrac{25}{36}}$
30. $\sqrt{\dfrac{49}{81}}$
31. $\sqrt{\dfrac{10}{9}}$
32. $\sqrt{\dfrac{15}{81}}$

33. $\dfrac{\sqrt{180}}{\sqrt{5}}$ **34.** $\dfrac{\sqrt{243}}{\sqrt{3}}$ **35.** $\dfrac{\sqrt{15}}{\sqrt{5}}$ **36.** $\dfrac{\sqrt{21}}{\sqrt{3}}$

37. $\sqrt[3]{\dfrac{4}{w^6}}$ **38.** $\sqrt[3]{\dfrac{7}{v^9}}$ **39.** $\sqrt[3]{\dfrac{5y^2}{27x^9}}$ **40.** $\sqrt[3]{\dfrac{5a}{8r^6}}$

41. $\dfrac{\sqrt[3]{320}}{\sqrt[3]{5}}$ **42.** $\dfrac{\sqrt[3]{162}}{\sqrt[3]{6}}$ **43.** $\sqrt[4]{\dfrac{3u^3}{16x^8}}$ **44.** $\sqrt[4]{\dfrac{3x^2}{81y^4}}$

Objective 3

Prep Exercise 3 Explain why $\sqrt{28}$ is not in simplest form.

Prep Exercise 4 Explain how to simplify a cube root containing a radicand with a perfect cube factor.

Prep Exercise 5 Explain why the expression $3x^2\sqrt[3]{x^5}$ is not in simplest form.

For Exercises 45–80, simplify. Assume that variables represent nonnegative values. See Examples 3–5.

45. $\sqrt{98}$ **46.** $\sqrt{96}$ **47.** $\sqrt{128}$ **48.** $\sqrt{180}$

49. $6\sqrt{80}$ **50.** $4\sqrt{75}$ **51.** $5\sqrt{112}$ **52.** $3\sqrt{147}$

53. $\sqrt{a^7}$ **54.** $\sqrt{d^5}$ **55.** $\sqrt{x^2y^4}$ **56.** $\sqrt{a^6b^2}$

57. $\sqrt{x^6y^8z^{10}}$ **58.** $\sqrt{p^4q^8r^8}$ **59.** $rs^2\sqrt{r^9s^5}$ **60.** $a^2b^3\sqrt{a^{11}b^3}$

61. $3\sqrt{72x^5}$ **62.** $6\sqrt{75d^3}$ **63.** $\sqrt[3]{32}$ **64.** $\sqrt[3]{128}$

65. $\sqrt[3]{x^7}$ **66.** $\sqrt[3]{b^{11}}$ **67.** $\sqrt[3]{x^6y^5}$ **68.** $\sqrt[3]{m^{13}n^9}$

69. $\sqrt[3]{128z^8}$ **70.** $\sqrt[3]{48h^{14}}$ **71.** $2\sqrt[3]{40}$ **72.** $4\sqrt[3]{250}$

73. $\sqrt[4]{80}$ **74.** $\sqrt[4]{162}$ **75.** $3x^2\sqrt[4]{243x^9}$ **76.** $3a^4\sqrt[4]{48a^7}$

77. $\sqrt[5]{486x^{16}}$ **78.** $\sqrt[5]{160n^{18}}$ **79.** $\sqrt[6]{x^8y^{14}z^{11}}$ **80.** $\sqrt[7]{a^{16}b^9c^{12}}$

For Exercises 81–90, find the product and write the answer in simplest form. Assume that variables represent nonnegative values. See Examples 6(a) and 6(b).

81. $\sqrt{3}\cdot\sqrt{21}$ **82.** $\sqrt{5}\cdot\sqrt{15}$ **83.** $5\sqrt{10}\cdot 3\sqrt{6}$ **84.** $2\sqrt{6}\cdot 5\sqrt{21}$

85. $\sqrt{y^3} \cdot \sqrt{y^2}$

86. $\sqrt{m^7} \cdot \sqrt{m^4}$

87. $x\sqrt{x^2y^3} \cdot y^2\sqrt{x^4y^4}$

88. $x\sqrt{x^5y^2} \cdot y\sqrt{xy^3}$

89. $4\sqrt{6c^3} \cdot 3\sqrt{10c^5}$

90. $6\sqrt{15c^2} \cdot 2\sqrt{10c^4}$

For Exercises 91 and 92, write an expression in simplest form for the area of the figure.

91.

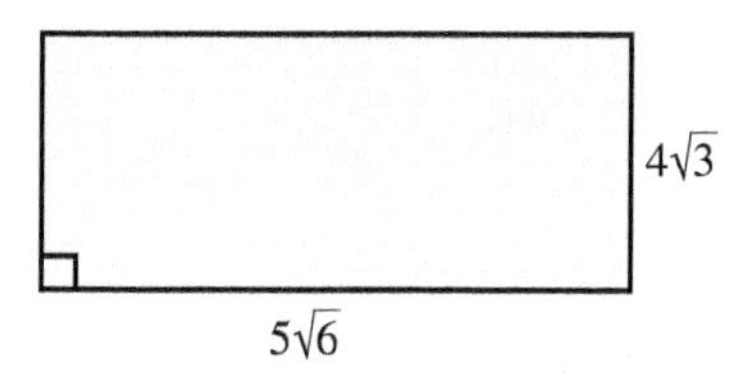

92.

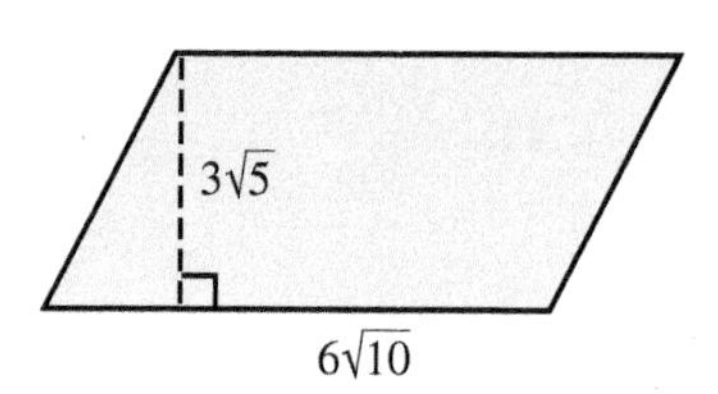

For Exercises 93–104, find the quotient and write the answer in simplest form. Assume that variables represent positive values. See Examples 6(c) and 6(d).

93. $\dfrac{\sqrt{48}}{\sqrt{6}}$

94. $\dfrac{\sqrt{54}}{\sqrt{3}}$

95. $\dfrac{9\sqrt{160}}{3\sqrt{8}}$

96. $\dfrac{10\sqrt{280}}{2\sqrt{10}}$

97. $\dfrac{\sqrt{c^5d^6}}{\sqrt{cd^3}}$

98. $\dfrac{\sqrt{m^6n^5}}{\sqrt{m^3n^3}}$

99. $\dfrac{8\sqrt{45a^5}}{2\sqrt{5a}}$

100. $\dfrac{6\sqrt{48n^7}}{3\sqrt{3n^3}}$

101. $\dfrac{12\sqrt{72c^5}}{4\sqrt{6c^2}}$

102. $\dfrac{15\sqrt{48a^7}}{5\sqrt{2a^2}}$

103. $\dfrac{36\sqrt{96x^6y^{11}}}{4\sqrt{3x^2y^4}}$

104. $\dfrac{54\sqrt{240r^{11}s^{10}}}{9\sqrt{5r^6s^4}}$

For Exercises 105–110, find the product and write the answer in simplest form. Assume that variables represent nonnegative values.

105. $\sqrt{\dfrac{3}{7}} \cdot \sqrt{\dfrac{8}{7}}$

106. $\sqrt{\dfrac{8}{5}} \cdot \sqrt{\dfrac{6}{5}}$

107. $\sqrt{\dfrac{a^3}{2}} \cdot \sqrt{\dfrac{a^5}{2}}$

108. $\sqrt{\dfrac{c^7}{6}} \cdot \sqrt{\dfrac{c^5}{6}}$

109. $\sqrt{\dfrac{3x^5}{2}} \cdot \sqrt{\dfrac{15x^5}{8}}$

110. $\sqrt{\dfrac{5y^3}{3}} \cdot \sqrt{\dfrac{10y^3}{27}}$

For Exercises 111 and 112, write an expression in simplest form for the area of the figure.

111.

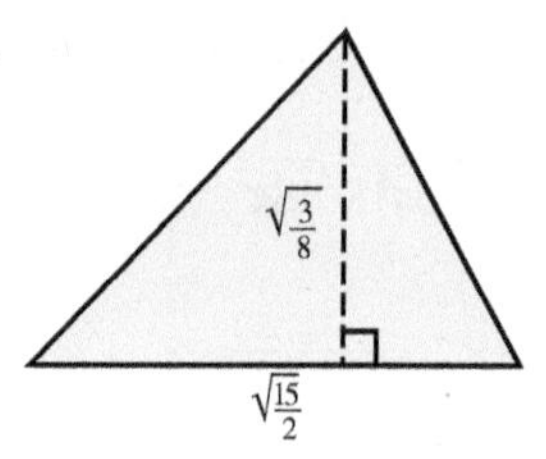

112.

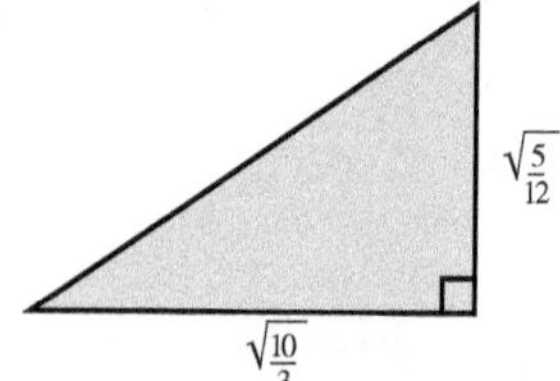

Review Exercises

Exercises 1–6 Expressions

[1.4] *For Exercises 1 and 2, use* $a = 6$ *and* $b = 8$.

1. Does $\sqrt{a^2 + b^2} = \sqrt{a^2} + \sqrt{b^2} = a + b$?

2. Does $\sqrt{(a + b)^2} = a + b$?

[1.4] **3.** Simplify: $6x^2 - 4x - 3x + 2x^2$

[5.3] *For Exercises 4–6, multiply.*

4. $(2a - 3b)(4a + 3b)$

5. $(3m + 5n)(3m - 5n)$

6. $(2x - 3y)^2$

8.4 Adding, Subtracting, and Multiplying Radical Expressions

Objectives

1. Add or subtract like radicals.
2. Use the distributive property in expressions containing radicals.
3. Simplify radical expressions that contain mixed operations.

Warm-up

For Exercises 1–4, simplify.

[8.3] **1.** $\sqrt{3} \cdot \sqrt{6}$

[1.4] **2.** $4x^2 - 5 + 2x^2 + 2$

[5.3] **3.** $2x(3 + 5x)$

[5.3] **4.** $(5 - 3x)(5 + 3x)$

Objective 1 Add or subtract like radicals.

Recall that like terms such as $-8a^2$ and $3a^2$ have identical variables with identical exponents. Similarly, **like radicals** have identical radicands and root indices.

Definition Like radicals: Radical expressions with identical radicands and identical root indices.

The radicals $3\sqrt{2}$ and $5\sqrt{2}$ are like. The radicals $3\sqrt{2}$ and $5\sqrt{3}$ are unlike because their radicands are different. The radicals $3\sqrt{2}$ and $5\sqrt[3]{2}$ are unlike because their root indices are different.

Adding or subtracting like radicals is essentially the same as combining like terms. Remember that the distributive property is at work when we combine like terms because we factor out the common variable, which leaves a sum or difference of the coefficients.

Like terms:

$$3x + 4x = (3 + 4)x$$
$$= 7x$$

Like radicals:

$$3\sqrt{5} + 4\sqrt{5} = (3 + 4)\sqrt{5}$$
$$= 7\sqrt{5}$$

Our example suggests the following procedure for adding like radical expressions:

Procedure Adding Like Radicals

To add or subtract like radicals, add or subtract the coefficients and leave the radical parts the same.

Answers to Warm-up

1. $3\sqrt{2}$
2. $6x^2 - 3$
3. $6x + 10x^2$
4. $25 - 9x^2$

Example 1 Add or subtract. Assume that all variables represent nonnegative values.

a. $7\sqrt{5} + 2\sqrt{5}$

Solution: $7\sqrt{5} + 2\sqrt{5} = (7 + 2)\sqrt{5}$
$= 9\sqrt{5}$

b. $6\sqrt[3]{4} + \sqrt[3]{4}$

Solution: $6\sqrt[3]{4} + \sqrt[3]{4} = (6 + 1)\sqrt[3]{4}$
$= 7\sqrt[3]{4}$

c. $6x\sqrt[4]{3x} - 2x\sqrt[4]{3x}$

Solution: $6x\sqrt[4]{3x} - 2x\sqrt[4]{3x} = (6x - 2x)\sqrt[4]{3x}$
$= 4x\sqrt[4]{3x}$

d. $8\sqrt{3} + 6\sqrt{2} - 5\sqrt{3} + 3\sqrt{2}$

Solution: $8\sqrt{3} + 6\sqrt{2} - 5\sqrt{3} + 3\sqrt{2} = 8\sqrt{3} - 5\sqrt{3} + 6\sqrt{2} + 3\sqrt{2}$ Regroup the terms.
$= (8 - 5)\sqrt{3} + (6 + 3)\sqrt{2}$
$= 3\sqrt{3} + 9\sqrt{2}$

Warning Never add radicands! Consider $\sqrt{9} + \sqrt{16}$. It might be tempting to add the radicands to get $\sqrt{25}$, but $\sqrt{9} + \sqrt{16} \neq \sqrt{25}$. If we calculate each side separately, we see why they are not equivalent.

$$\sqrt{9} + \sqrt{16} \neq \sqrt{25}$$
$$3 + 4 \neq 5$$
$$7 \neq 5$$

Your Turn 1 Add or subtract. Assume that all variables represent nonnegative values.

a. $5\sqrt{6} + 2\sqrt{6}$

b. $5\sqrt[4]{5x^2} + \sqrt[4]{5x^2}$

c. $7y\sqrt[3]{7} - 12y\sqrt[3]{7}$

d. $9x\sqrt{5x} - 2y\sqrt{3y} - 6x\sqrt{5x} + 7y\sqrt{3y}$

In a problem involving addition or subtraction of radicals, if the radicals are not like radicals, it may be possible to simplify one or more of the radicals so that they are like.

Example 2 Add or subtract. Assume that variables represent nonnegative values.

a. $7\sqrt{3} + \sqrt{12}$

Solution: $7\sqrt{3} + \sqrt{12} = 7\sqrt{3} + \sqrt{4 \cdot 3}$ Factor out the perfect square factor, 4.
$= 7\sqrt{3} + \sqrt{4} \cdot \sqrt{3}$ Use the product rule to separate the radicals.
$= 7\sqrt{3} + 2\sqrt{3}$ Simplify.
$= 9\sqrt{3}$ Combine like radicals.

b. $3\sqrt[3]{24} - 2\sqrt[3]{3}$

Solution: $3\sqrt[3]{24} - 2\sqrt[3]{3} = 3\sqrt[3]{8 \cdot 3} - 2\sqrt[3]{3}$ Rewrite 24 as $8 \cdot 3$.
$= 3\sqrt[3]{8} \cdot \sqrt[3]{3} - 2\sqrt[3]{3}$ Use the product rule.
$= 3 \cdot 2\sqrt[3]{3} - 2\sqrt[3]{3}$ Simplify $\sqrt[3]{8}$.
$= 6\sqrt[3]{3} - 2\sqrt[3]{3}$ Multiply.
$= 4\sqrt[3]{3}$ Combine like radicals.

c. $\sqrt{48x^3} + \sqrt{12x^3}$

Solution: $\sqrt{48x^3} + \sqrt{12x^3} = \sqrt{16x^2 \cdot 3x} + \sqrt{4x^2 \cdot 3x}$ Rewrite $48x^3$ as $16x^2 \cdot 3x$ and $12x^3$ as $4x^2 \cdot 3x$.
$= \sqrt{16x^2} \cdot \sqrt{3x} + \sqrt{4x^2} \cdot \sqrt{3x}$ Use the product rule.
$= 4x\sqrt{3x} + 2x\sqrt{3x}$ Find $\sqrt{16x^2}$ and $\sqrt{4x^2}$.
$= 6x\sqrt{3x}$ Combine like radicals.

Answers to Your Turn 1
a. $7\sqrt{6}$ b. $6\sqrt[4]{5x^2}$ c. $-5y\sqrt[3]{7}$
d. $3x\sqrt{5x} + 5y\sqrt{3y}$

d. $3\sqrt[4]{32x^5} + \sqrt[4]{162x^9}$

Solution: $3\sqrt[4]{32x^5} + \sqrt[4]{162x^9}$

$= 3\sqrt[4]{16 \cdot 2 \cdot x^4 \cdot x} + \sqrt[4]{81 \cdot 2 \cdot x^8 \cdot x}$ Write 32 as $16 \cdot 2$, x^5 as $x^4 \cdot x$, 162 as $81 \cdot 2$, and x^9 as $x^8 \cdot x$.

$= 3 \cdot 2x\sqrt[4]{2x} + 3x^2\sqrt[4]{2x}$ Find the roots.

$= 6x\sqrt[4]{2x} + 3x^2\sqrt[4]{2x}$ Multiply.

▶ $= (6x + 3x^2)\sqrt[4]{2x}$ Factor.

Note Although the radicals are like, we cannot combine the coefficients further because they are not like terms.

Your Turn 2 Add or subtract. Assume that variables represent nonnegative values.

a. $4\sqrt{24} - 6\sqrt{54}$ **b.** $4\sqrt{50x^5} - 2\sqrt{18x^5}$

c. $6a\sqrt[3]{54a^4} - 2\sqrt[3]{128a^7}$ **d.** $2\sqrt[3]{81x^7} - 5\sqrt[3]{24x^4}$

Objective 2 Use the distributive property in expressions containing radicals.

Products involving sums and differences of radicals are found in much the same way as products of polynomials. We use the distributive property, multiply binomials using FOIL, and square a binomial (all from Section 5.3).

Example 3 Find the product. Assume that variables represent nonnegative values.

a. $\sqrt{3}(\sqrt{3} + \sqrt{15})$

Solution: $\sqrt{3}(\sqrt{3} + \sqrt{15}) = \sqrt{3} \cdot \sqrt{3} + \sqrt{3} \cdot \sqrt{15}$ Use the distributive property.

$= \sqrt{3 \cdot 3} + \sqrt{3 \cdot 15}$ Use the product rule.

$= \sqrt{9} + \sqrt{45}$ Multiply.

$= 3 + 3\sqrt{5}$ Find $\sqrt{9}$ and simplify $\sqrt{45}$.

b. $2\sqrt{6}(3 + 5\sqrt{5})$

Solution: $2\sqrt{6}(3 + 5\sqrt{5}) = 2\sqrt{6} \cdot 3 + 2\sqrt{6} \cdot 5\sqrt{5}$ Use the distributive property.

$= 2 \cdot 3\sqrt{6} + 2 \cdot 5\sqrt{6 \cdot 5}$ Use the product rule.

$= 6\sqrt{6} + 10\sqrt{30}$ Multiply.

c. $(2 + \sqrt{3})(\sqrt{5} - \sqrt{6})$

Solution: $(2 + \sqrt{3})(\sqrt{5} - \sqrt{6}) = 2\sqrt{5} - 2\sqrt{6} + \sqrt{3} \cdot \sqrt{5} - \sqrt{3} \cdot \sqrt{6}$ Use FOIL.

$= 2\sqrt{5} - 2\sqrt{6} + \sqrt{15} - \sqrt{18}$ Use the product rule

$= 2\sqrt{5} - 2\sqrt{6} + \sqrt{15} - 3\sqrt{2}$ Simplify $\sqrt{18}$.

d. $(3\sqrt{x} + \sqrt{y})(2\sqrt{x} - 5\sqrt{y})$

Solution: $(3\sqrt{x} + \sqrt{y})(2\sqrt{x} - 5\sqrt{y})$

$= 3\sqrt{x} \cdot 2\sqrt{x} - 3\sqrt{x} \cdot 5\sqrt{y} + \sqrt{y} \cdot 2\sqrt{x} - \sqrt{y} \cdot 5\sqrt{y}$ Use FOIL.

$= 6x - 15\sqrt{xy} + 2\sqrt{xy} - 5y$ Use the product rule.

$= 6x - 13\sqrt{xy} - 5y$ Combine like radicals.

e. $(5 + \sqrt{3})^2$

Solution: $(5 + \sqrt{3})^2 = 5^2 + 2 \cdot 5\sqrt{3} + (\sqrt{3})^2$ Use $(a + b)^2 = a^2 + 2ab + b^2$.

$= 25 + 10\sqrt{3} + 3$ Simplify.

$= 28 + 10\sqrt{3}$ Add 25 and 3.

Answers to Your Turn 2

a. $-10\sqrt{6}$ **b.** $14x^2\sqrt{2x}$ **c.** $10a^2\sqrt[3]{2a}$ **d.** $(6x^2 - 10x)\sqrt[3]{3x}$

Answers to Your Turn 3

a. $6\sqrt{11} - 6\sqrt{66}$

b. $2a - 3\sqrt{ab} - 9b$

c. $14 + 6\sqrt{5}$

Your Turn 3 Find the product. Assume that variables represent nonnegative values.

a. $2\sqrt{11}(3 - 3\sqrt{6})$ **b.** $(2\sqrt{a} + 3\sqrt{b})(\sqrt{a} - 3\sqrt{b})$ **c.** $(3 + \sqrt{5})^2$

Radicals in Conjugates

Radical expressions can be conjugates. Like binomial conjugates, conjugates involving radicals differ only in the sign separating the terms. For example, $7 - \sqrt{3}$ and $7 + \sqrt{3}$ are conjugates. Let's explore what happens when we multiply conjugates containing radicals.

Example 4 Find the product.

a. $(2 + \sqrt{3})(2 - \sqrt{3})$

Solution: $(2 + \sqrt{3})(2 - \sqrt{3}) = 2^2 - (\sqrt{3})^2$ Use $(a + b)(a - b) = a^2 - b^2$ (from Section 5.5).

$= 4 - 3$ Simplify.

$= 1$

b. $(\sqrt{5} - 3\sqrt{3})(\sqrt{5} + 3\sqrt{3})$

Solution: $(\sqrt{5} - 3\sqrt{3})(\sqrt{5} + 3\sqrt{3}) = (\sqrt{5})^2 - (3\sqrt{3})^2$ Use $(a - b)(a + b) = a^2 - b^2$.

$= 5 - 9 \cdot 3$ Simplify.

$= 5 - 27$ Multiply.

$= -22$ Subtract.

Note The product of conjugates *always* results in a rational number. This will be useful in Section 8.5 when we rationalize denominators.

Your Turn 4 Find the product.

a. $(4 + \sqrt{10})(4 - \sqrt{10})$

b. $(3\sqrt{5} + 2\sqrt{6})(3\sqrt{5} - 2\sqrt{6})$

Objective 3 Simplify radical expressions that contain mixed operations.

Now let's use the order of operations to simplify radical expressions that have more than one operation.

Example 5 Simplify.

a. $\sqrt{2} \cdot \sqrt{10} + \sqrt{3} \cdot \sqrt{15}$

Solution: $\sqrt{2} \cdot \sqrt{10} + \sqrt{3} \cdot \sqrt{15} = \sqrt{2 \cdot 10} + \sqrt{3 \cdot 15}$ Use the product rule.

$= \sqrt{20} + \sqrt{45}$ Multiply.

$= \sqrt{4 \cdot 5} + \sqrt{9 \cdot 5}$ Rewrite 20 as $4 \cdot 5$ and 45 as $9 \cdot 5$.

$= \sqrt{4} \cdot \sqrt{5} + \sqrt{9} \cdot \sqrt{5}$ Use the product rule.

$= 2\sqrt{5} + 3\sqrt{5}$ Find $\sqrt{4}$ and $\sqrt{9}$.

$= 5\sqrt{5}$ Combine like radicals.

b. $\dfrac{\sqrt{54}}{\sqrt{3}} + \sqrt{32}$

Solution: $\dfrac{\sqrt{54}}{\sqrt{3}} + \sqrt{32} = \sqrt{\dfrac{54}{3}} + \sqrt{16 \cdot 2}$ Use the quotient rule and rewrite 32 as $16 \cdot 2$.

$= \sqrt{18} + \sqrt{16} \cdot \sqrt{2}$ Divide and use the product rule.

$= \sqrt{9 \cdot 2} + 4\sqrt{2}$ Rewrite 18 as $9 \cdot 2$ and find $\sqrt{16}$.

$= \sqrt{9} \cdot \sqrt{2} + 4\sqrt{2}$ Use the product rule.

$= 3\sqrt{2} + 4\sqrt{2}$ Find $\sqrt{9}$.

$= 7\sqrt{2}$ Combine like radicals.

Your Turn 5 Simplify.

a. $2\sqrt{3} \cdot \sqrt{6} + 4\sqrt{7} \cdot \sqrt{14}$

b. $\sqrt{63} + \dfrac{\sqrt{140}}{\sqrt{5}}$

Answers to Your Turn 4
a. 6 **b.** 21

Answers to Your Turn 5
a. $34\sqrt{2}$ **b.** $5\sqrt{7}$

8.4 Exercises

For Extra Help MyMathLab®

Note: Exercises marked with a ★ represent challenging exercises.

Objective 1

Prep Exercise 1 What must be identical in like radicals? What can be different?

Prep Exercise 2 Explain how to add $3\sqrt{2} + 2\sqrt{2}$.

For Exercises 1–14, add or subtract. Assume that variables represent nonnegative values. See Example 1.

1. $9\sqrt{6} - 15\sqrt{6}$

2. $4\sqrt{5} - 13\sqrt{5}$

3. $7\sqrt{a} + 2\sqrt{a}$

4. $5\sqrt{y} + 7\sqrt{y}$

5. $4\sqrt{5} - 2\sqrt{6} + 8\sqrt{5} - 6\sqrt{6}$

6. $9\sqrt{7} - 4\sqrt{11} - 3\sqrt{7} - 5\sqrt{11}$

7. $3a\sqrt{5a} - 4b\sqrt{7b} + 8a\sqrt{5a} + 2b\sqrt{7b}$

8. $12n\sqrt{2n} - 14m\sqrt{5m} - 8n\sqrt{2n} + 18m\sqrt{5m}$

9. $6x\sqrt[3]{9} - 3x\sqrt[3]{9}$

10. $4y\sqrt[3]{3} - y\sqrt[3]{3}$

11. $6x^2\sqrt[4]{5x} - 12x^2\sqrt[4]{5x}$

12. $3y^3\sqrt[4]{8y} - 9y^3\sqrt[4]{8y}$

★ **13.** $3x\sqrt{5x} + 4x\sqrt[3]{5x}$

★ **14.** $4z\sqrt[4]{2z} - 7z\sqrt{2z}$

For Exercises 15–34, add or subtract. Assume that variables represent nonnegative values. See Example 2.

15. $\sqrt{48} - \sqrt{75}$

16. $\sqrt{24} - \sqrt{96}$

17. $\sqrt{80y} - \sqrt{125y}$

18. $\sqrt{27x} + \sqrt{75x}$

19. $\sqrt{80} - 4\sqrt{45}$

20. $\sqrt{20} - 2\sqrt{180}$

21. $3\sqrt{96} - 2\sqrt{54}$

22. $5\sqrt{63} + 2\sqrt{28}$

23. $6\sqrt{48a^3} - 2\sqrt{75a^3}$

24. $4\sqrt{98y^5} - 7\sqrt{128y^5}$

25. $\sqrt{150} - \sqrt{54} + \sqrt{24}$

26. $\sqrt{20} + \sqrt{125} - \sqrt{80}$

27. $2\sqrt{8} - 3\sqrt{48} + 2\sqrt{98} - \sqrt{75}$

28. $3\sqrt{180} - \sqrt{192} - 4\sqrt{80} - \sqrt{48}$

29. $\sqrt[3]{128} + \sqrt[3]{54}$

30. $\sqrt[3]{24} + \sqrt[3]{81}$

31. $4\sqrt[3]{135x^5} - 6x\sqrt[3]{320x^2}$

32. $3a^2\sqrt[3]{500a^4} + 6a\sqrt[3]{108a^7}$

33. $-4\sqrt[4]{32x^9} + 2x\sqrt[4]{162x^5}$

34. $6y\sqrt[4]{243y^6} - 2y^2\sqrt[4]{48y^2}$

Objective 2

Prep Exercise 3 Explain how to multiply $(\sqrt{5} + 3)(\sqrt{5} + 2)$.

For Exercises 35–42 use the distributive property. Assume that variables represent nonnegative values. See Examples 3(a) and 3(b).

35. $\sqrt{2}(3 + \sqrt{2})$

36. $\sqrt{5}(4 - \sqrt{5})$

37. $\sqrt{3}(\sqrt{3} - \sqrt{15})$

38. $\sqrt{6}(\sqrt{6} + \sqrt{3})$

39. $\sqrt{5}(\sqrt{6} + 2\sqrt{10})$

40. $\sqrt{7}(\sqrt{5} - 3\sqrt{14})$

41. $4\sqrt{3x}\,(2\sqrt{3x} - 4\sqrt{6x})$

42. $6\sqrt{2y}\,(3\sqrt{2y} + 2\sqrt{10y})$

For Exercises 43–62, multiply. (Use FOIL.) Assume that variables represent nonnegative values. See Examples 3(c) and 3(d).

43. $(3 + \sqrt{5})(4 - \sqrt{2})$

44. $(2 + \sqrt{5})(6 - \sqrt{3})$

45. $(3 + \sqrt{x})(2 + \sqrt{x})$

46. $(5 - \sqrt{a})(2 + \sqrt{a})$

47. $(2 + 3\sqrt{3})(3 + 5\sqrt{2})$

48. $(7 - 3\sqrt{5})(2 - 2\sqrt{10})$

49. $(\sqrt{3} + \sqrt{5})(\sqrt{5} + \sqrt{7})$

50. $(\sqrt{5} + \sqrt{2})(\sqrt{2} + \sqrt{7})$

51. $(\sqrt{x} + 3\sqrt{y})(\sqrt{x} - 2\sqrt{y})$

52. $(2\sqrt{a} + \sqrt{b})(\sqrt{a} + 4\sqrt{b})$

53. $(4\sqrt{2} + 2\sqrt{5})(3\sqrt{7} - 3\sqrt{3})$

54. $(8\sqrt{2} - 2\sqrt{3})(2\sqrt{5} + 3\sqrt{10})$

55. $(2\sqrt{a} + 3\sqrt{b})(4\sqrt{a} - \sqrt{b})$

56. $(3\sqrt{m} - \sqrt{n})(2\sqrt{m} + 4\sqrt{n})$

57. $(\sqrt[3]{4} + 5)(\sqrt[3]{4} - 8)$

58. $(\sqrt[3]{9} + 5)(\sqrt[3]{9} - 2)$

59. $(\sqrt[3]{9} + \sqrt[3]{4})(\sqrt[3]{3} - \sqrt[3]{2})$

60. $(\sqrt[3]{5} + \sqrt[3]{9})(\sqrt[3]{25} - \sqrt[3]{3})$

★**61.** $(\sqrt[3]{x} + 2)(\sqrt[3]{x^2} - 2\sqrt[3]{x} + 4)$

★**62.** $(\sqrt[3]{r} - 3)(\sqrt[3]{r^2} + 3\sqrt[3]{r} + 9)$

For Exercises 63–70, find the product. See Example 3(e).

63. $(4 + \sqrt{6})^2$

64. $(3 + \sqrt{7})^2$

65. $(4 - \sqrt{2})^2$

66. $(5 - \sqrt{2})^2$

67. $(2 + 2\sqrt{3})^2$

68. $(3 + 2\sqrt{5})^2$

69. $(2\sqrt{3} + 3\sqrt{2})^2$

70. $(4\sqrt{2} - 5\sqrt{6})^2$

For Exercises 71–84, multiply the conjugates. Assume that variables represent nonnegative values. See Example 4.

71. $(4 + \sqrt{3})(4 - \sqrt{3})$

72. $(3 + \sqrt{5})(3 - \sqrt{5})$

73. $(\sqrt{2} + 4)(\sqrt{2} - 4)$

74. $(\sqrt{7} - 3)(\sqrt{7} + 3)$

75. $(6 + \sqrt{x})(6 - \sqrt{x})$

76. $(5 + \sqrt{y})(5 - \sqrt{y})$

77. $(\sqrt{3} + \sqrt{2})(\sqrt{3} - \sqrt{2})$

78. $(\sqrt{5} - \sqrt{3})(\sqrt{5} + \sqrt{3})$

79. $(\sqrt{x} + \sqrt{y})(\sqrt{x} - \sqrt{y})$

80. $(\sqrt{a} - \sqrt{b})(\sqrt{a} + \sqrt{b})$

81. $(4 + 2\sqrt{3})(4 - 2\sqrt{3})$

82. $(7 + 2\sqrt{3})(7 - 2\sqrt{3})$

83. $(3\sqrt{7} + \sqrt{13})(3\sqrt{7} - \sqrt{13})$

84. $(4\sqrt{5} - 3\sqrt{2})(4\sqrt{5} + 3\sqrt{2})$

Objective 3

Prep Exercise 4 Why is $\sqrt{a} + \sqrt{b} \neq \sqrt{a + b}$? Use examples if necessary.

For Exercises 85–92, simplify. See Example 5.

85. $\sqrt{3} \cdot \sqrt{15} + \sqrt{8} \cdot \sqrt{10}$

86. $\sqrt{18} \cdot \sqrt{6} + \sqrt{8} \cdot \sqrt{24}$

87. $3\sqrt{3} \cdot \sqrt{18} - 4\sqrt{18} \cdot \sqrt{12}$

88. $3\sqrt{2} \cdot 2\sqrt{40} - 5\sqrt{12} \cdot \sqrt{15}$

89. $\dfrac{\sqrt{54}}{\sqrt{3}} + \sqrt{72}$

90. $\dfrac{\sqrt{60}}{\sqrt{5}} + \sqrt{48}$

91. $\dfrac{\sqrt{540}}{\sqrt{3}} - 4\sqrt{125}$

92. $\dfrac{\sqrt{288}}{\sqrt{6}} - 6\sqrt{108}$

For Exercises 93 and 94, find the perimeter of the figure in simplest form.

93.

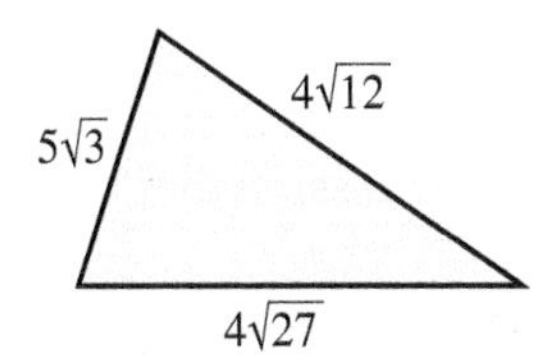

94.

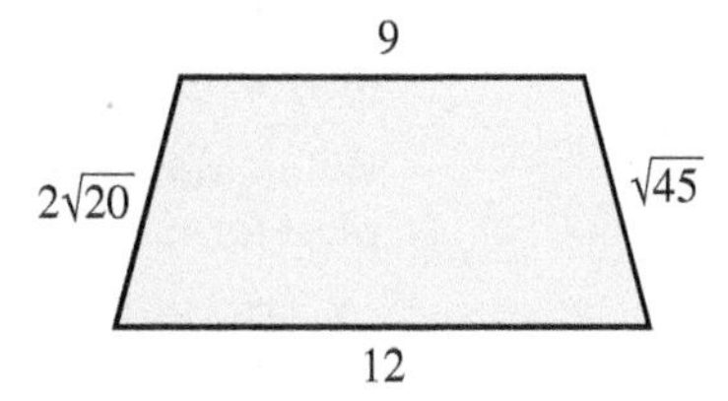

95. Crown molding, which is placed at the top of a wall, is to be installed around the perimeter of the room shown.

a. Write an expression in simplest form for the perimeter of the room.

b. Use a calculator to approximate the perimeter, rounded to the nearest tenth.

c. If crown molding costs $1.89 per foot length, how much will the crown molding cost for this room?

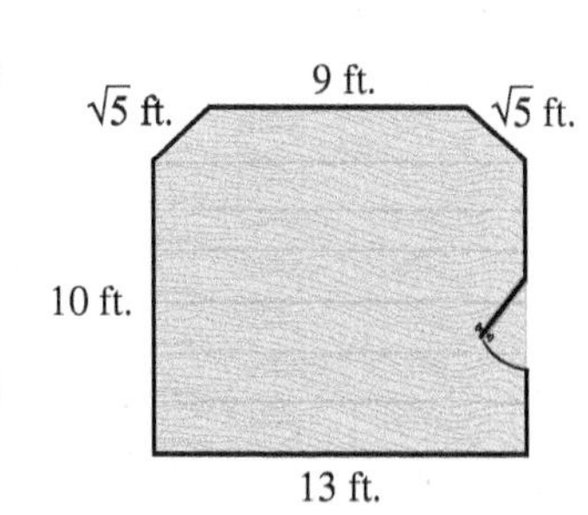

96. A tabletop is to be fitted with veneer strips along the sides.

a. Write an expression in simplest form for the perimeter of the tabletop.

b. Use a calculator to approximate the perimeter, rounded to the nearest tenth.

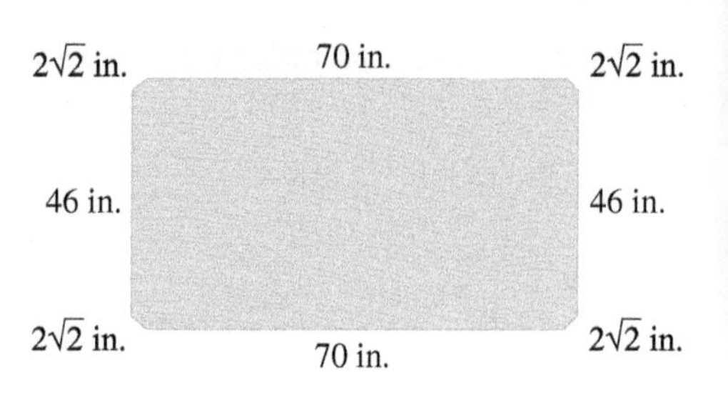

Review Exercises

Exercises 1–4 Expressions

[5.3] **1.** What is the conjugate of $2x - 5$?

[5.3] **2.** Multiply: $(4x + 3)(4x - 3)$

[8.3] **3.** What factor could multiply $\sqrt{8}$ to equal $\sqrt{16}$?

[8.3] **4.** What factor could multiply $\sqrt[3]{2}$ to equal $\sqrt[3]{8}$?

Exercises 5 and 6 Equations and Inequalities

[3.2] **5.** For the equation $5y - 2x = 10$, find the slope and the y-intercept.

[3.1] **6.** Graph: $y = -\frac{1}{3}x + 4$

8.5 Rationalizing Numerators and Denominators of Radical Expressions

Objectives

1. Rationalize denominators.
2. Rationalize denominators that have a sum or difference with a square root term.
3. Rationalize numerators.

Warm-up

[8.3] **1.** Simplify: $\sqrt{3} \cdot \sqrt{3}$
[8.3] **2.** Simplify: $\sqrt[3]{9a^2} \cdot \sqrt[3]{3a}$
[8.3] **3.** By what would you multiply $\sqrt[3]{5}$ to get $\sqrt[3]{125}$?
[8.4] **4.** Simplify: $(\sqrt{6} + \sqrt{2})(\sqrt{6} - \sqrt{2})$

We are now ready to formalize the conditions for a radical expression that is in simplest form. A radical expression is in simplest form if

1. The radicand has no factor raised to a power greater than or equal to the index.
2. There are neither radicals in the denominator of a fraction nor radicands that are fractions.
3. All possible sums, differences, products, and quotients have been found.

In this section, we explore how to simplify expressions that have a radical in the denominator of a fraction, as in $\frac{1}{\sqrt{2}}$.

Objective 1 Rationalize denominators.

If the denominator of a fraction contains a radical, our goal is to *rationalize* the denominator, which means to rewrite the expression so that it has a rational number in the denominator. In general, we multiply the fraction by a well-chosen 1 so that the radical is eliminated. We determine that 1 by finding a factor that multiplies the nth root in the denominator so that its radicand is a perfect nth power.

Answers to Warm-up
1. 3
2. $3a$
3. $\sqrt[3]{25}$
4. 4

Square Root Denominators

In the case of a square root in the denominator, we multiply it by a factor that makes the radicand a perfect square, which allows us to eliminate the square root.

For example, to rationalize $\frac{1}{\sqrt{2}}$, we could multiply by $\frac{\sqrt{2}}{\sqrt{2}}$ because the product's denominator is the square root of a perfect square.

$$\frac{1}{\sqrt{2}} = \frac{1}{\sqrt{2}} \cdot \frac{\sqrt{2}}{\sqrt{2}} = \frac{\sqrt{2}}{\sqrt{4}} = \frac{\sqrt{2}}{2}$$

Note We are not changing the value of $\frac{1}{\sqrt{2}}$ because we are multiplying it by 1 in the form of $\frac{\sqrt{2}}{\sqrt{2}}$.

Any factor that produces a perfect square radicand will work. For example, we could have multiplied $\frac{1}{\sqrt{2}}$ by $\frac{\sqrt{8}}{\sqrt{8}}$.

$$\frac{1}{\sqrt{2}} = \frac{1}{\sqrt{2}} \cdot \frac{\sqrt{8}}{\sqrt{8}} = \frac{\sqrt{8}}{\sqrt{16}} = \frac{\sqrt{4 \cdot 2}}{4} = \frac{2\sqrt{2}}{4} = \frac{\sqrt{2}}{2}$$

Notice, however, that multiplying by $\frac{\sqrt{2}}{\sqrt{2}}$ required fewer steps to simplify than does multiplying by $\frac{\sqrt{8}}{\sqrt{8}}$.

Example 1 Rationalize the denominator. Assume that variables represent positive values.

a. $\frac{2}{\sqrt{5}}$

Solution:

$$\frac{2}{\sqrt{5}} = \frac{2}{\sqrt{5}} \cdot \frac{\sqrt{5}}{\sqrt{5}} \quad \text{Multiply by } \frac{\sqrt{5}}{\sqrt{5}}.$$

$$= \frac{2\sqrt{5}}{\sqrt{25}} \quad \text{Simplify.}$$

$$= \frac{2\sqrt{5}}{5}$$

b. $\sqrt{\frac{5}{8}}$

Solution:

$$\sqrt{\frac{5}{8}} = \frac{\sqrt{5}}{\sqrt{8}} \quad \text{Use the quotient rule of square roots to separate the numerator and denominator into two radicals.}$$

$$= \frac{\sqrt{5}}{\sqrt{8}} \cdot \frac{\sqrt{2}}{\sqrt{2}} \quad \text{Multiply by } \frac{\sqrt{2}}{\sqrt{2}}.$$

$$= \frac{\sqrt{10}}{\sqrt{16}} \quad \text{Multiply.}$$

$$= \frac{\sqrt{10}}{4} \quad \text{Simplify.}$$

Note We chose to multiply by $\frac{\sqrt{2}}{\sqrt{2}}$ because it leads to a smaller perfect square than do other choices such as $\frac{\sqrt{8}}{\sqrt{8}}$. Multiplying by $\frac{\sqrt{8}}{\sqrt{8}}$ produces the same final answer but requires more steps.

Warning Although it may be tempting to do so, we cannot divide out the 4 and 10 because 10 is a radicand, whereas 4 is not. We *never* divide out factors common to a radicand and a number that is not a radicand.

c. $\dfrac{3}{\sqrt{2x}}$

Solution: $\dfrac{3}{\sqrt{2x}} = \dfrac{3}{\sqrt{2x}} \cdot \dfrac{\sqrt{2x}}{\sqrt{2x}}$ Multiply by $\dfrac{\sqrt{2x}}{\sqrt{2x}}$.

$= \dfrac{3\sqrt{2x}}{\sqrt{4x^2}}$ Multiply.

$= \dfrac{3\sqrt{2x}}{2x}$ Simplify.

Your Turn 1 Rationalize the denominator. Assume that variables represent positive values.

a. $\dfrac{1}{\sqrt{7}}$ **b.** $\sqrt{\dfrac{7}{12}}$ **c.** $\dfrac{3}{\sqrt{10x}}$

*n*th-Root Denominators

If the denominator contains a higher-order root such as a cube root, we multiply appropriately to get a perfect cube radicand in the denominator so that we can eliminate the radical. For example, $\dfrac{2}{\sqrt[3]{5}} = \dfrac{2}{\sqrt[3]{5}} \cdot \dfrac{\sqrt[3]{25}}{\sqrt[3]{25}} = \dfrac{2\sqrt[3]{25}}{\sqrt[3]{125}} = \dfrac{2\sqrt[3]{25}}{5}$. We summarize as follows:

Procedure Rationalizing Denominators

To rationalize a denominator containing a single *n*th root, multiply the fraction by a form of 1 so that the product's denominator has a radicand that is a perfect *n*th power.

Example 2 Rationalize the denominator. Assume that variables represent positive values.

a. $\dfrac{3}{\sqrt[3]{2}}$

Solution: $\dfrac{3}{\sqrt[3]{2}} = \dfrac{3}{\sqrt[3]{2}} \cdot \dfrac{\sqrt[3]{4}}{\sqrt[3]{4}}$ Because $\sqrt[3]{2} \cdot \sqrt[3]{4} = \sqrt[3]{8} = 2$, multiply the fraction by $\dfrac{\sqrt[3]{4}}{\sqrt[3]{4}}$.

$= \dfrac{3\sqrt[3]{4}}{\sqrt[3]{8}}$ Multiply.

$= \dfrac{3\sqrt[3]{4}}{2}$ Simplify.

b. $\dfrac{\sqrt[3]{a}}{\sqrt[3]{b}}$

Solution: $\dfrac{\sqrt[3]{a}}{\sqrt[3]{b}} = \dfrac{\sqrt[3]{a}}{\sqrt[3]{b}} \cdot \dfrac{\sqrt[3]{b^2}}{\sqrt[3]{b^2}}$ Because $\sqrt[3]{b} \cdot \sqrt[3]{b^2} = \sqrt[3]{b^3} = b$, multiply the fraction by $\dfrac{\sqrt[3]{b^2}}{\sqrt[3]{b^2}}$.

$= \dfrac{\sqrt[3]{ab^2}}{\sqrt[3]{b^3}}$ Multiply.

$= \dfrac{\sqrt[3]{ab^2}}{b}$ Simplify.

Answers to Your Turn 1

a. $\dfrac{\sqrt{7}}{7}$ **b.** $\dfrac{\sqrt{21}}{6}$ **c.** $\dfrac{3\sqrt{10x}}{10x}$

c. $\sqrt[3]{\frac{5}{9a^2}}$

Solution: $\sqrt[3]{\frac{5}{9a^2}} = \frac{\sqrt[3]{5}}{\sqrt[3]{9a^2}}$ Use the quotient rule to separate the numerator and denominator.

$= \frac{\sqrt[3]{5}}{\sqrt[3]{9a^2}} \cdot \frac{\sqrt[3]{3a}}{\sqrt[3]{3a}}$ Because $\sqrt[3]{9a^2} \cdot \sqrt[3]{3a} = \sqrt[3]{27a^3} = 3a$, multiply the fraction by $\frac{\sqrt[3]{3a}}{\sqrt[3]{3a}}$.

$= \frac{\sqrt[3]{15a}}{\sqrt[3]{27a^3}}$ Multiply.

$= \frac{\sqrt[3]{15a}}{3a}$ Simplify.

d. $\frac{5}{\sqrt[4]{3}}$

Solution: $\frac{5}{\sqrt[4]{3}} = \frac{5}{\sqrt[4]{3}} \cdot \frac{\sqrt[4]{27}}{\sqrt[4]{27}}$ Because $\sqrt[4]{3} \cdot \sqrt[4]{27} = \sqrt[4]{81} = 3$, multiply the fraction by $\frac{\sqrt[4]{27}}{\sqrt[4]{27}}$.

$= \frac{5\sqrt[4]{27}}{\sqrt[4]{81}}$ Multiply.

$= \frac{5\sqrt[4]{27}}{3}$ Simplify.

Your Turn 2 Rationalize the denominator. Assume that variables represent positive values.

a. $\frac{6}{\sqrt[3]{3}}$ b. $\sqrt[3]{\frac{3}{x^2}}$ c. $\frac{4}{\sqrt[3]{4y}}$ d. $\frac{7}{\sqrt[4]{2}}$

Objective 2 Rationalize denominators that have a sum or difference with a square root term.

In Example 4 of Section 8.4, we saw that the product of two conjugates containing square roots does not contain any radicals. Consequently, if the denominator of a fraction contains a sum or difference with a square root term, we can rationalize the denominator by multiplying the fraction by a 1 made up of the conjugate of the denominator. For example, to rationalize $\frac{5}{7 - \sqrt{3}}$, we multiply by $\frac{7 + \sqrt{3}}{7 + \sqrt{3}}$. Because $7 - \sqrt{3}$ and $7 + \sqrt{3}$ are conjugates, their product will not contain any radicals; so the denominator will be rationalized.

$$\frac{5}{7 - \sqrt{3}} = \frac{5}{7 - \sqrt{3}} \cdot \frac{7 + \sqrt{3}}{7 + \sqrt{3}} = \frac{5(7 + \sqrt{3})}{(7)^2 - (\sqrt{3})^2} = \frac{35 + 5\sqrt{3}}{49 - 3} = \frac{35 + 5\sqrt{3}}{46}$$

Procedure Rationalizing a Denominator Containing a Sum or Difference

To rationalize a denominator containing a sum or difference with at least one square root term, multiply the fraction by a form of 1 whose numerator and denominator are the conjugate of the denominator.

Answers to Your Turn 2

a. $2\sqrt[3]{9}$ b. $\frac{\sqrt[3]{3x}}{x}$ c. $\frac{2\sqrt[3]{2y^2}}{y}$ d. $\frac{7\sqrt[4]{8}}{2}$

Example 3 Rationalize the denominator and simplify. Assume that variables represent positive values.

a. $\dfrac{9}{\sqrt{2}+7}$

Solution: $\dfrac{9}{\sqrt{2}+7} = \dfrac{9}{\sqrt{2}+7} \cdot \dfrac{\sqrt{2}-7}{\sqrt{2}-7}$ The conjugate of $\sqrt{2}+7$ is $\sqrt{2}-7$, so we multiply by $\dfrac{\sqrt{2}-7}{\sqrt{2}-7}$.

$= \dfrac{9(\sqrt{2}-7)}{(\sqrt{2})^2-(7)^2}$ Multiply. In the denominator, use the rule $(a+b)(a-b) = a^2 - b^2$.

$= \dfrac{9\sqrt{2}-63}{2-49}$ Simplify.

$= \dfrac{9\sqrt{2}-63}{-47}$ We can simplify the negative denominator by factoring out -1 in the numerator and denominator.

$= \dfrac{-1(63-9\sqrt{2})}{-1(47)}$ After factoring out the -1, the signs of the terms change. Because 63 is now positive, we write it first.

$= \dfrac{63-9\sqrt{2}}{47}$ Divide out the common factor -1.

b. $\dfrac{2\sqrt{3}}{\sqrt{6}-\sqrt{2}}$

Solution: $\dfrac{2\sqrt{3}}{\sqrt{6}-\sqrt{2}} = \dfrac{2\sqrt{3}}{\sqrt{6}-\sqrt{2}} \cdot \dfrac{\sqrt{6}+\sqrt{2}}{\sqrt{6}+\sqrt{2}}$ The conjugate of $\sqrt{6}-\sqrt{2}$ is $\sqrt{6}+\sqrt{2}$, so we multiply by $\dfrac{\sqrt{6}+\sqrt{2}}{\sqrt{6}+\sqrt{2}}$.

$= \dfrac{2\sqrt{3}(\sqrt{6}+\sqrt{2})}{(\sqrt{6})^2-(\sqrt{2})^2}$

$= \dfrac{2\sqrt{18}+2\sqrt{6}}{6-2}$ Multiply in the numerator and evaluate the exponents in the denominator.

$= \dfrac{2\sqrt{9\cdot 2}+2\sqrt{6}}{4}$ Simplify $\sqrt{18}$ by factoring out a perfect square factor in 18.

$= \dfrac{2\cdot 3\sqrt{2}+2\sqrt{6}}{4}$ Simplify $\sqrt{9\cdot 2}$ by finding the square root of 9.

$= \dfrac{2(3\sqrt{2}+\sqrt{6})}{4}$ Factor out the common factor 2 in the numerator.

$= \dfrac{3\sqrt{2}+\sqrt{6}}{2}$ Divide out the common factor 2.

Note We cannot combine $3\sqrt{2}$ and $\sqrt{6}$ because they are not like radicals. ▶

c. $\dfrac{6}{\sqrt{x}-5}$

Solution: $\dfrac{6}{\sqrt{x}-5} = \dfrac{6}{\sqrt{x}-5} \cdot \dfrac{\sqrt{x}+5}{\sqrt{x}+5}$ The conjugate of $\sqrt{x}-5$ is $\sqrt{x}+5$, so we multiply by $\dfrac{\sqrt{x}+5}{\sqrt{x}+5}$.

$= \dfrac{6(\sqrt{x}+5)}{(\sqrt{x})^2-(5)^2}$ Multiply.

$= \dfrac{6\sqrt{x}+30}{x-25}$ Multiply in the numerator and evaluate the exponents in the denominator.

Your Turn 3 Rationalize the denominator and simplify. Assume that variables represent positive values.

a. $\dfrac{9}{\sqrt{5} + 2}$ b. $\dfrac{\sqrt{2}}{\sqrt{5} - \sqrt{3}}$ c. $\dfrac{3}{\sqrt{x} + 4}$

Objective 3 Rationalize numerators.

In later mathematics courses, you may need to rationalize the numerator. We use the same procedure that we use in rationalizing denominators.

Example 4 Rationalize the numerator. Assume that variables represent positive values.

a. $\dfrac{\sqrt{5x}}{4}$

Solution: $\dfrac{\sqrt{5x}}{4} = \dfrac{\sqrt{5x}}{4} \cdot \dfrac{\sqrt{5x}}{\sqrt{5x}}$ To create a perfect square radicand in the numerator, we multiply by $\dfrac{\sqrt{5x}}{\sqrt{5x}}$.

$= \dfrac{\sqrt{25x^2}}{4\sqrt{5x}}$ Multiply.

$= \dfrac{5x}{4\sqrt{5x}}$ Simplify.

b. $\dfrac{3 + \sqrt{2x}}{4}$

Solution: $\dfrac{3 + \sqrt{2x}}{4} = \dfrac{3 + \sqrt{2x}}{4} \cdot \dfrac{3 - \sqrt{2x}}{3 - \sqrt{2x}}$ The conjugate of $3 + \sqrt{2x}$ is $3 - \sqrt{2x}$, so we multiply by $\dfrac{3 - \sqrt{2x}}{3 - \sqrt{2x}}$.

$= \dfrac{3^2 - (\sqrt{2x})^2}{4(3 - \sqrt{2x})}$ Multiply.

$= \dfrac{9 - 2x}{12 - 4\sqrt{2x}}$ Simplify.

Answers to Your Turn 3

a. $9\sqrt{5} - 18$ b. $\dfrac{\sqrt{10} + \sqrt{6}}{2}$

c. $\dfrac{3\sqrt{x} - 12}{x - 16}$

Your Turn 4 Rationalize the numerators. Assume that variables represent positive values.

a. $\dfrac{\sqrt{3a}}{7}$ b. $\dfrac{5 - \sqrt{3a}}{2}$

Answers to Your Turn 4

a. $\dfrac{3a}{7\sqrt{3a}}$ b. $\dfrac{25 - 3a}{10 + 2\sqrt{3a}}$

8.5 Exercises

For Extra Help MyMathLab®

Note: Exercises marked with a ★ represent challenging exercises.

Objective 1

Prep Exercise 1 Explain why each of the following expressions is not in simplest form.

a. $\sqrt{\dfrac{3}{16}}$ b. $\dfrac{5}{\sqrt{3}}$

Prep Exercise 2 Although $\frac{1}{\sqrt{3}}$ and $\frac{\sqrt{3}}{3}$ are equal, explain why $\frac{\sqrt{3}}{3}$ is considered simplest form.

Prep Exercise 3 Explain how to rationalize a denominator that is the square root of a number $\left(\text{for example, } \frac{2}{\sqrt{3}} \text{ or } \frac{\sqrt{5}}{\sqrt{7}}\right)$.

For Exercises 1–24, rationalize the denominator. Assume that variables represent positive values. See Example 1.

1. $\frac{1}{\sqrt{3}}$ **2.** $\frac{1}{\sqrt{7}}$ **3.** $\frac{3}{\sqrt{8}}$ **4.** $\frac{5}{\sqrt{12}}$

5. $\sqrt{\frac{36}{7}}$ **6.** $\sqrt{\frac{64}{3}}$ **7.** $\sqrt{\frac{5}{12}}$ **8.** $\sqrt{\frac{11}{18}}$

9. $\frac{\sqrt{7x^2}}{\sqrt{50}}$ **10.** $\frac{\sqrt{3x^2}}{\sqrt{32}}$ **11.** $\frac{\sqrt{8}}{\sqrt{56}}$ **12.** $\frac{\sqrt{7}}{\sqrt{42}}$

13. $\frac{5}{\sqrt{3a}}$ **14.** $\frac{11}{\sqrt{7b}}$ **15.** $\sqrt{\frac{3m}{11n}}$ **16.** $\sqrt{\frac{5r}{6s}}$

17. $\frac{10}{\sqrt{5x}}$ **18.** $\frac{28}{\sqrt{7a}}$ **19.** $\frac{\sqrt{6x}}{\sqrt{32x}}$ **20.** $\frac{\sqrt{10a}}{\sqrt{18a}}$

21. $\frac{3}{\sqrt{x^3}}$ **22.** $\frac{5}{\sqrt{b^5}}$ **23.** $\frac{8x^2}{\sqrt{2x}}$ **24.** $\frac{18a^4}{\sqrt{6a}}$

Find the Mistake *For Exercises 25 and 26, explain the mistake; then simplify correctly.*

25. $\frac{\sqrt{3}}{\sqrt{2}} = \frac{\sqrt{3}}{\sqrt{2}} \cdot \frac{2}{2} = \frac{2\sqrt{3}}{2}$

26. $\sqrt{\frac{7}{3}} = \frac{\sqrt{7}}{\sqrt{3}} \cdot \frac{\sqrt{3}}{\sqrt{3}} = \frac{\sqrt{21}}{9}$

For Exercises 27–46, rationalize the denominators. Assume that variables represent positive values. See Example 2.

27. $\frac{5}{\sqrt[3]{3}}$ **28.** $\frac{7}{\sqrt[3]{5}}$ **29.** $\sqrt[3]{\frac{5}{2}}$ **30.** $\sqrt[3]{\frac{3}{4}}$

31. $\frac{6}{\sqrt[3]{4}}$ **32.** $\frac{9}{\sqrt[3]{9}}$ **33.** $\frac{m}{\sqrt[3]{n}}$ **34.** $\frac{p}{\sqrt[3]{q}}$

35. $\sqrt[3]{\frac{a}{b^2}}$ **36.** $\sqrt[3]{\frac{m}{n^2}}$ **37.** $\frac{4}{\sqrt[3]{2x}}$ **38.** $\frac{9}{\sqrt[3]{3a}}$

39. $\sqrt[3]{\frac{6}{25a^2}}$ **40.** $\sqrt[3]{\frac{5}{16b^2}}$ **41.** $\frac{5}{\sqrt[4]{4}}$ **42.** $\frac{7}{\sqrt[4]{9}}$

43. $\sqrt[4]{\frac{3}{x^2}}$ **44.** $\sqrt[4]{\frac{5}{y^3}}$ **45.** $\frac{9}{\sqrt[4]{3x^3}}$ **46.** $\frac{12}{\sqrt[4]{2x^2}}$

Objective 2

Prep Exercise 4 Explain how to rationalize a denominator that is a sum or difference with a square root term $\left(\text{for example, } \frac{3}{5 + \sqrt{2}} \text{ or } \frac{2}{\sqrt{x} - \sqrt{y}}\right)$.

For Exercises 47–68, rationalize the denominator and simplify. Assume that variables represent positive values. See Example 3.

47. $\frac{3}{\sqrt{2} + 1}$ **48.** $\frac{3}{\sqrt{5} + 2}$ **49.** $\frac{4}{2 - \sqrt{3}}$ **50.** $\frac{2}{4 - \sqrt{15}}$

51. $\frac{5}{\sqrt{2} + \sqrt{3}}$ **52.** $\frac{7}{\sqrt{6} + \sqrt{7}}$ **53.** $\frac{4}{1 - \sqrt{5}}$ **54.** $\frac{6}{1 - \sqrt{7}}$

55. $\frac{\sqrt{3}}{\sqrt{3} - 1}$ **56.** $\frac{\sqrt{5}}{\sqrt{5} - 1}$ **57.** $\frac{2\sqrt{3}}{\sqrt{3} - 4}$ **58.** $\frac{2\sqrt{5}}{\sqrt{5} - 4}$

59. $\frac{4\sqrt{3}}{\sqrt{7} + \sqrt{2}}$ **60.** $\frac{2\sqrt{2}}{\sqrt{4} + \sqrt{3}}$ **61.** $\frac{8\sqrt{2}}{4\sqrt{2} - \sqrt{6}}$ **62.** $\frac{4\sqrt{3}}{4\sqrt{6} + \sqrt{2}}$

63. $\dfrac{6\sqrt{y}}{\sqrt{y}+1}$ **64.** $\dfrac{4\sqrt{x}}{\sqrt{x}+1}$ **65.** $\dfrac{3\sqrt{t}}{\sqrt{t}+2\sqrt{u}}$ **66.** $\dfrac{2\sqrt{m}}{\sqrt{n}-3\sqrt{m}}$

67. $\dfrac{\sqrt{2y}}{\sqrt{x}-\sqrt{6y}}$ **68.** $\dfrac{\sqrt{14h}}{\sqrt{2h}+\sqrt{k}}$

Objective 3

Prep Exercise 5 Explain how to rationalize the numerator in $\dfrac{2+\sqrt{7x}}{5}$.

For Exercises 69–80, rationalize the numerator. Assume that variables represent positive values. See Example 4.

69. $\dfrac{\sqrt{3}}{2}$ **70.** $\dfrac{\sqrt{7}}{3}$ **71.** $\dfrac{\sqrt{2x}}{5}$ **72.** $\dfrac{\sqrt{7y}}{3}$

73. $\dfrac{\sqrt{8n}}{6}$ **74.** $\dfrac{\sqrt{20t}}{8}$ **75.** $\dfrac{2+\sqrt{3}}{5}$ **76.** $\dfrac{3+\sqrt{2}}{4}$

77. $\dfrac{\sqrt{5x}-6}{9}$ **78.** $\dfrac{\sqrt{2x}+7}{3}$ ★**79.** $\dfrac{5\sqrt{n}+\sqrt{6n}}{2n}$ ★**80.** $\dfrac{4\sqrt{k}-\sqrt{10k}}{5k}$

81. Given $f(x) = \dfrac{5\sqrt{2}}{x}$, find each of the following. Express your answer in simplest form.

a. $f(\sqrt{6})$ **b.** $f(\sqrt{10})$ **c.** $f(\sqrt{22})$

82. Given $g(x) = \dfrac{\sqrt{2}}{x-1}$, find each of the following. Express your answer in simplest form.

a. $g(\sqrt{5})$ **b.** $g(3\sqrt{2})$ **c.** $g(2\sqrt{6})$

83. Graph $f(x) = \dfrac{1}{\sqrt{x}}$; then graph $g(x) = \dfrac{\sqrt{x}}{x}$.

a. What do you notice about the two graphs? What does this indicate about the two functions?

b. Simplify $f(x)$ by rationalizing the denominator. What do you notice?

84. Graph $f(x) = -\dfrac{1}{\sqrt{x}}$; then graph $g(x) = -\dfrac{\sqrt{x}}{x}$.

a. What do you notice about the two graphs? What does this indicate about the two functions?

b. Simplify $f(x)$ by rationalizing the denominator. What do you notice?

85. Previously, we used the formula $T = 2\pi\sqrt{\frac{L}{9.8}}$ to determine the period of a pendulum, where T is the period in seconds and L is the length in meters.

a. Rewrite the formula so that the denominator is rationalized.

b. Rewrite the formula so that the numerator is rationalized.

86. The formula $t = \sqrt{\frac{h}{16}}$ can be used to find the time, t, in seconds for an object to fall a distance of h feet.

a. Rewrite the formula so that the denominator is rationalized.

b. Rewrite the formula so that the numerator is rationalized.

87. The formula $s = \sqrt{\frac{3V}{h}}$ can be used to find the length, s, of each side of the base of a pyramid having a square base, volume V in cubic feet, and height h in feet.

a. Rationalize the denominator in the formula.

b. The volume of the Great Pyramid at Giza in Egypt is approximately 83,068,742 cubic feet, and its height is 449 feet. Find the length of each side of its base.

Of Interest

The Great Pyramid was built for King Khufu from about 2589 to 2566 B.C. The pyramid contains approximately 2,300,000 blocks and weighs about 6.5 million tons, with each block averaging about 2.8 tons.

88. ★ The formula $s = 2\sqrt{\frac{A}{6\sqrt{3}}}$ can be used to find the length, s, of each side of a regular hexagon having an area A.

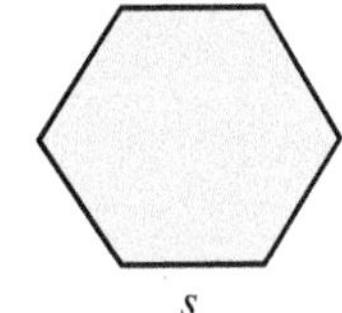

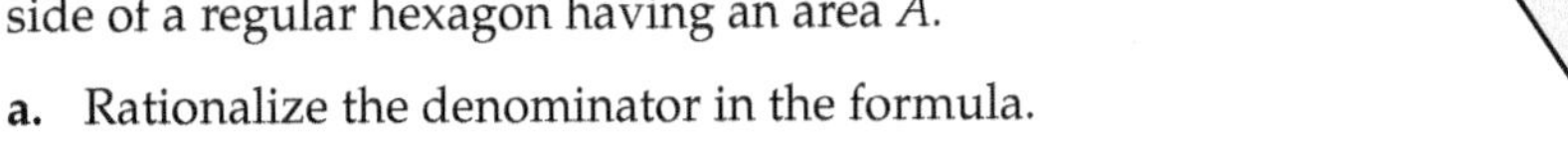

a. Rationalize the denominator in the formula.

b. If A is 100 square meters, write an expression in simplest form for the side length.

c. Use a calculator to approximate the side lengths, rounded to three decimal places.

89. In AC circuits, voltage is often expressed as a *root-mean-square*, or *rms*, value. The formula for calculating the rms voltage, V_{rms}, given the maximum voltage, V_m, value is $V_{rms} = \frac{V_m}{\sqrt{2}}$.

a. Rationalize the denominator in the formula.

b. Given a maximum voltage of 163 V, write an expression for the rms voltage.

c. Use a calculator to approximate the rms voltage, rounded to the nearest tenth.

90. The velocity, in meters per second, of a particle can be determined by the formula $v = \sqrt{\frac{2E}{m}}$, where E represents the kinetic energy, in joules, of the particle and m represents the mass, in kilograms, of the particle.

a. Rationalize the denominator in the formula.

b. A particle with a mass of 1×10^{-6} kilograms has 2.4×10^{7} joules of kinetic energy. Write an expression for its velocity.

c. Use a calculator to approximate the velocity rounded to the nearest tenth.

91. The resistance in a circuit is found to be $\frac{5\sqrt{2}}{3 + \sqrt{6}}$ ohms. Rationalize the denominator.

92. Two charged particles, q_1 and q_2, are separated by a distance of 8 centimeters. The values of the charges are $q_1 = 3 \times 10^{-6}$ coulombs and $q_2 = 1 \times 10^{-6}$ coulombs. Each charged particle exerts an electrical field. At a point between the two particles x centimeters away from q_1, the electric fields cancel each other so that the value of the fields at the point x is 0.

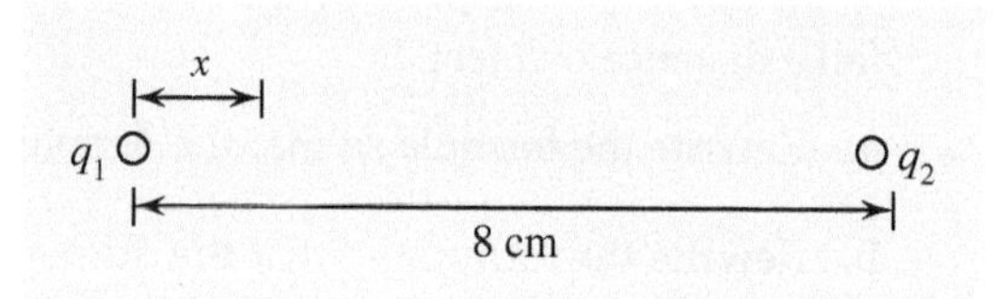

a. Use the formula $x = \frac{l}{1 + \sqrt{\frac{q_2}{q_1}}}$ to find the distance from q_1 at which the electric field is canceled, where l is the distance separating the particles. Write the distance with a rationalized denominator.

b. Use a calculator to approximate the distance, rounded to the nearest tenth.

Review Exercises

Exercises 1 and 2 Expressions

[8.3] **1.** Simplify: $\pm\sqrt{28}$

[6.3] **2.** Factor: $x^2 - 6x + 9$

Exercises 3–6 Equations and Inequalities

[2.1] *For Exercises 3 and 4, solve.*

3. $2x - 3 = 5$

4. $2x - 3 = -5$

[6.4] *For Exercises 5 and 6, solve and check.*

5. $x^2 - 36 = 0$

6. $x^2 - 5x + 6 = 0$

8.6 Radical Equations and Problem Solving

Objective

1 Use the power rule to solve radical equations.

Warm-up

[8.5] **1.** Rationalize the denominator. $\sqrt{\frac{3}{5}}$

[8.3] **2.** Simplify: $(\sqrt{6x} - 1)^2$

For Exercises 2 and 3, solve.

[2.1] **3.** $3x + 4 = 16$

[6.4] **4.** $x^2 + 4x + 4 = 5x + 16$

Answers to Warm-up

1. $\frac{\sqrt{15}}{5}$ **2.** $6x - 1$ **3.** $x = 4$ **4.** $x = -3, 4$

We now explore how to solve **radical equations.**

Definition **Radical equation:** An equation containing at least one radical expression whose radicand has a variable.

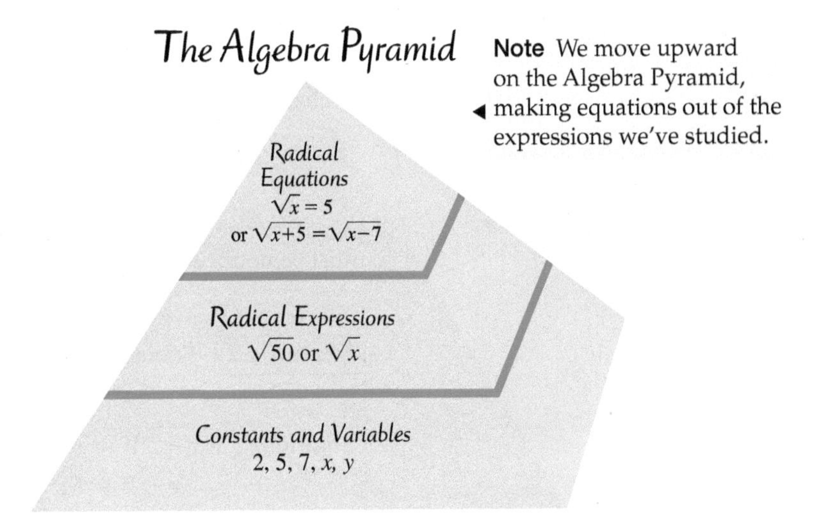

Note We move upward on the Algebra Pyramid, making equations out of the expressions we've studied.

Objective 1 Use the power rule to solve radical equations.

To solve radical equations, we use a principle of equality called the *power rule.*

Rule **Power Rule for Solving Equations**

If both sides of an equation are raised to the same integer power, the resulting equation contains all solutions of the original equation and perhaps some solutions that do not solve the original equation. That is, the solutions of the equation $a = b$ are contained among the solutions of $a^n = b^n$, where n is an integer.

Isolated Radicals

First, we consider equations such as $\sqrt{x} = 9$ in which the radical is isolated. When we use the power rule, we raise both sides of the equation to the same integer power as the root index, then use the principle $(\sqrt[n]{x})^n = x$, which eliminates the radical, leaving its radicand.

Example 1 Solve.

a. $\sqrt{x} = 9$

Solution: $(\sqrt{x})^2 = (9)^2$ Because the root index is 2, we square both sides.

$x = 81$

Check: $\sqrt{81} \stackrel{?}{=} 9$

$9 = 9$ True

b. $\sqrt[3]{y} = -2$

Solution: $(\sqrt[3]{y})^3 = (-2)^3$ Because the root index is 3, we cube both sides.

$y = -8$

Check: $\sqrt[3]{-8} \stackrel{?}{=} -2$

$-2 = -2$ True

Extraneous Solutions

In Section 7.4, we learned that some equations have *extraneous solutions,* which are apparent solutions that do not make the original equation true. Using the power rule can sometimes lead to extraneous solutions, so it is important to check solutions. For example, watch what happens when we use the power rule to solve the equation $\sqrt{x} = -9$.

Warning As we will see in the check, this result is not a solution.

$$(\sqrt{x})^2 = (-9)^2 \quad \text{Square both sides.}$$
$$x = 81$$

By checking 81 in the original equation, we see that it is extraneous.

$$\sqrt{81} \stackrel{?}{=} -9$$
$$9 = -9 \quad \text{This equation is false, so 81 is extraneous.}$$

In fact, $\sqrt{x} = -9$ has no real number solution because if x is a real number, then $\sqrt{x}$ must be nonnegative.

Example 2 Solve.

a. $\sqrt{x - 7} = 8$

Solution: $(\sqrt{x - 7})^2 = (8)^2$ Square both sides.

$x - 7 = 64$ Simplify.

$x = 71$ Add 7 to both sides.

Check: $\sqrt{71 - 7} \stackrel{?}{=} (8)^2$

$\sqrt{64} \stackrel{?}{=} 8$

$8 = 8$ True. The solution is 71.

b. $\sqrt[3]{x - 3} = -1$

Solution: $(\sqrt[3]{x - 3})^3 = (-1)^3$ Cube both sides.

$x - 3 = -1$ Simplify.

$x = 2$ Add 3 to both sides.

Check: $\sqrt[3]{2 - 3} \stackrel{?}{=} -1$

$\sqrt[3]{-1} \stackrel{?}{=} -1$

$-1 = -1$ True. The solution is 2.

Note We can see that this equation has no real-number solution because a principal square root cannot be equal to a negative number. However, we work through the steps to confirm this.

c. $\sqrt{2x + 1} = -3$

Solution: $(\sqrt{2x + 1})^2 = (-3)^2$ Square both sides.

$2x + 1 = 9$ Simplify.

$2x = 8$ Subtract 1 from both sides.

$x = 4$ Divide both sides by 2.

Check: $\sqrt{2(4) + 1} \stackrel{?}{=} -3$

$\sqrt{9} \stackrel{?}{=} -3$

$3 = -3$ False, so 4 is extraneous. This equation has no real-number solution.

Your Turn 2 Solve.

a. $\sqrt{x - 5} = 6$ **b.** $\sqrt[3]{x + 2} = 3$ **c.** $\sqrt{x + 3} = -7$

Answers to Your Turn 2
a. 41 **b.** 25
c. no real-number solution

Radicals on Both Sides of the Equation

As we will see in Example 3, the power rule can be used to solve equations with radicals on both sides of the equal sign.

Example 3 Solve.

a. $\sqrt{6x-1} = \sqrt{x+2}$

Solution: $(\sqrt{6x-1})^2 = (\sqrt{x+2})^2$

$6x - 1 = x + 2$ — Square both sides.

$5x = 3$ — Subtract x from and add 1 to both sides.

$x = \frac{3}{5}$ — Divide both sides by 5.

Check: $\sqrt{6\left(\frac{3}{5}\right) - 1} \stackrel{?}{=} \sqrt{\frac{3}{5} + 2}$

$\sqrt{\frac{18}{5} - \frac{5}{5}} \stackrel{?}{=} \sqrt{\frac{3}{5} + \frac{10}{5}}$

$\sqrt{\frac{13}{5}} = \sqrt{\frac{13}{5}}$ — True. The solution is $\frac{3}{5}$.

b. $\sqrt[3]{5x-2} = \sqrt[3]{3x+2}$

Solution: $(\sqrt[3]{5x-2})^3 = (\sqrt[3]{3x+2})^3$ — Cube both sides.

$5x - 2 = 3x + 2$ — Simplify.

$2x = 4$ — Subtract $3x$ and add 2 to both sides.

$x = 2$ — Divide both sides by 2.

Check: $\sqrt[3]{5(2) - 2} \stackrel{?}{=} \sqrt[3]{3(2) + 2}$

$\sqrt[3]{10 - 2} \stackrel{?}{=} \sqrt[3]{6 + 2}$

$\sqrt[3]{8} \stackrel{?}{=} \sqrt[3]{8}$

$2 = 2$ — True. The solution is 2.

Your Turn 3 Solve.

a. $\sqrt{8x+5} = \sqrt{2x+7}$

b. $\sqrt[3]{6x+9} = \sqrt[3]{10x-3}$

Multiple Solutions

Radical equations may have multiple solutions if, after using the power rule, we are left with a quadratic form.

Example 4 Solve $x + 2 = \sqrt{5x+16}$.

Solution: $(x+2)^2 = (\sqrt{5x+16})^2$ — Square both sides.

$x^2 + 4x + 4 = 5x + 16$ — Use FOIL on the left-hand side.

$x^2 - x - 12 = 0$ — Because the equation is quadratic, we set it equal to 0 by subtracting $5x$ and 16 from both sides.

$(x+3)(x-4) = 0$ — Factor.

$x + 3 = 0$ or $x - 4 = 0$ — Use the zero-factor theorem.

$x = -3$ $\quad$ $x = 4$ — Solve each equation.

Checks:

$-3 + 2 \stackrel{?}{=} \sqrt{5(-3) + 16}$

$-1 \stackrel{?}{=} \sqrt{-15 + 16}$

$-1 \stackrel{?}{=} \sqrt{1}$

$-1 = 1$ — False

$4 + 2 \stackrel{?}{=} \sqrt{5(4) + 16}$

$6 \stackrel{?}{=} \sqrt{20 + 16}$

$6 \stackrel{?}{=} \sqrt{36}$

$6 = 6$ — True. The solution is 4.

Because -3 does not check, it is an extraneous solution. The only solution is 4.

Your Turn 4 Solve $\sqrt{9x+7} = x + 3$.

Answers to Your Turn 3

a. $\frac{1}{3}$ **b.** 3

Answer to Your Turn 4

2 and 1

Radicals Not Isolated

Now we consider radical equations in which the radical term is not isolated. In such equations, we must first isolate the radical term.

Example 5 Solve.

a. $\sqrt{x+1} - 2x = x + 1$

Solution:

$\sqrt{x+1} = 3x + 1$ Add $2x$ to both sides to isolate the radical term.

$(\sqrt{x+1})^2 = (3x+1)^2$ Square both sides.

$x + 1 = 9x^2 + 6x + 1$ Use FOIL on the right-hand side.

$0 = 9x^2 + 5x$ Because the equation is quadratic, we set it equal to 0 by subtracting x and 1 from both sides.

$0 = x(9x + 5)$ Factor.

$x = 0$ or $9x + 5 = 0$ Use the zero-factor theorem.

$9x = -5$

$x = -\frac{5}{9}$

Checks:

$\sqrt{0+1} - 2(0) \stackrel{?}{=} 0 + 1$

$\sqrt{1} - 0 \stackrel{?}{=} 1$

$1 = 1$ True. The solution is 0.

$\sqrt{-\frac{5}{9} + 1} - 2\left(-\frac{5}{9}\right) \stackrel{?}{=} -\frac{5}{9} + 1$

$\sqrt{-\frac{5}{9} + \frac{9}{9}} - 2\left(-\frac{5}{9}\right) \stackrel{?}{=} -\frac{5}{9} + \frac{9}{9}$

$\sqrt{\frac{4}{9}} + \frac{10}{9} \stackrel{?}{=} \frac{4}{9}$

$\frac{2}{3} + \frac{10}{9} \stackrel{?}{=} \frac{4}{9}$

$\frac{6}{9} + \frac{10}{9} \stackrel{?}{=} \frac{4}{9}$

$\frac{16}{9} = \frac{4}{9}$ False

Note This false equation indicates that $-\frac{5}{9}$ is an extraneous solution.

The solution is 0.

b. $\sqrt[4]{3x+4} + 5 = 7$

Solution:

$\sqrt[4]{3x+4} = 2$ Subtract 5 from both sides to isolate the radical term.

$(\sqrt[4]{3x+4})^4 = 2^4$ Raise both sides to the fourth power.

$3x + 4 = 16$ Simplify both sides.

$3x = 12$ Subtract 4 from both sides.

$x = 4$ Divide both sides by 3.

Check:

$\sqrt[4]{3(4)+4} + 5 \stackrel{?}{=} 7$

$\sqrt[4]{12+4} + 5 \stackrel{?}{=} 7$

$\sqrt[4]{16} + 5 \stackrel{?}{=} 7$

$2 + 5 \stackrel{?}{=} 7$

$7 = 7$ True. The solution is 4.

Your Turn 5 Solve.

a. $\sqrt{5x^2 + 6x - 7} + 3x = 5x + 1$ **b.** $\sqrt[4]{3x+6} - 7 = -4$

Answers to Your Turn 5
a. 2 **b.** 25

Using the Power Rule Twice

Some equations may require that we use the power rule twice to eliminate all radicals.

Example 6 Solve $\sqrt{x + 21} = \sqrt{x} + 3$.

Solution: $(\sqrt{x + 21})^2 = (\sqrt{x} + 3)^2$ — Because one of the radicals is isolated, we square both sides.

$x + 21 = (\sqrt{x} + 3)(\sqrt{x} + 3)$ — Simplify.

$x + 21 = x + 3\sqrt{x} + 3\sqrt{x} + 9$ — Use FOIL on the right-hand side.

$x + 21 = x + 6\sqrt{x} + 9$ — Combine like terms.

$12 = 6\sqrt{x}$ — Subtract x and 9 from both sides to isolate the remaining radical expression.

$2 = \sqrt{x}$ — Divide both sides by 6.

$(2)^2 = (\sqrt{x})^2$ — Square both sides.

$4 = x$

Check: $\sqrt{4 + 21} \stackrel{?}{=} \sqrt{4} + 3$

$\sqrt{25} \stackrel{?}{=} 2 + 3$

$5 = 5$ — True. The solution is 4.

Our examples suggest the following procedure.

Procedure Solving Radical Equations

To solve a radical equation:

1. Isolate the radical if necessary. (If there is more than one radical term, isolate one of the radical terms.)
2. Raise both sides of the equation to the same power as the root index of the isolated radical.
3. If all radicals have been eliminated, solve. If a radical term remains, isolate that radical term and raise both sides to the same power as its root index.
4. Check each solution. Any apparent solution that does not check is an extraneous solution.

Answer to Your Turn 6
0 and 4

Your Turn 6 Solve $\sqrt{2x + 1} = \sqrt{x} + 1$.

8.6 Exercises

For Extra Help MyMathLab®

Note: Exercises marked with a ★ represent challenging exercises.

Objective 1

Prep Exercise 1 Explain why we must check all potential solutions to radical equations.

Prep Exercise 2 What is an extraneous solution?

Prep Exercise 3 Explain why there is no real-number solution for the radical equation $\sqrt{x} = -6$.

Prep Exercise 4 Show why $(\sqrt{a})^2 = a$, assuming that $a \geq 0$.

For Exercises 1–22, solve. See Examples 1 and 2.

1. $\sqrt{x} = 2$

2. $\sqrt{y} = 5$

3. $\sqrt{k} = -4$

4. $\sqrt{x} = -1$

5. $\sqrt[3]{y} = 3$ **6.** $\sqrt[3]{m} = 4$ **7.** $\sqrt[3]{z} = -2$ **8.** $\sqrt[3]{p} = -5$

9. $\sqrt{n-1} = 4$ **10.** $\sqrt{x-2} = 3$ **11.** $\sqrt{t+5} = 4$ **12.** $\sqrt{m+8} = 1$

13. $\sqrt{3x-2} = 4$ **14.** $\sqrt{2x+5} = 3$ **15.** $\sqrt{2x+24} = 4$ **16.** $\sqrt{3y+34} = 5$

17. $\sqrt{2n-8} = -3$ **18.** $\sqrt{5x-1} = -6$ **19.** $\sqrt[3]{x-3} = 2$ **20.** $\sqrt[3]{k+2} = 4$

21. $\sqrt[3]{3y-2} = -2$ **22.** $\sqrt[3]{2x+4} = -4$

Prep Exercise 5 Given the radical equation $\sqrt{x+2} + 3x = 4x - 1$, what would be the first step in solving the equation? Why?

For Exercises 23–32 solve. First isolate the radical term. See Example 5.

23. $\sqrt{u-3} - 10 = 1$ **24.** $\sqrt{y+1} - 4 = 2$ **25.** $\sqrt{y-6} + 2 = 9$ **26.** $\sqrt{r-5} + 6 = 10$

27. $\sqrt{6x-5} - 2 = 3$ **28.** $\sqrt{8x+4} - 2 = 4$ **29.** $\sqrt[3]{n+3} - 2 = -4$ **30.** $\sqrt[3]{x-4} + 2 = -3$

31. $\sqrt[4]{x-2} - 2 = -4$ **32.** $\sqrt[4]{m+3} + 2 = 1$

Prep Exercise 6 Give an example of a radical equation that requires you to use the power rule twice in solving the equation. Explain why the principle must be used twice.

For Exercises 33–58, solve. Identify any extraneous solutions. See Examples 3–6.

33. $\sqrt{3x-2} = \sqrt{8-2x}$ **34.** $\sqrt{m+2} = \sqrt{2m-3}$ **35.** $\sqrt{4x-5} = \sqrt{6x+5}$

36. $\sqrt{3x-4} = \sqrt{5x+2}$ **37.** $\sqrt[3]{2r+2} = \sqrt[3]{3r-1}$ **38.** $\sqrt[3]{3h-4} = \sqrt[3]{h+4}$

39. $\sqrt[4]{4x+4} = \sqrt[4]{5x+1}$ **40.** $\sqrt[4]{2x+4} = \sqrt[4]{3x-2}$ **41.** $\sqrt{2x+24} = x + 8$

42. $\sqrt{2x+15} = x + 6$ **43.** $y - 1 = \sqrt{2y-2}$ **44.** $3 + x = \sqrt{7+3x}$

45. $\sqrt{3x+10} - 4 = x$ **46.** $\sqrt{6x+1} - 1 = x$ **47.** $\sqrt{10n+4} - 3n = n + 1$

48. $\sqrt{6x-1} - 6x = 2 - 9x$ **49.** $\sqrt[3]{5x+2} + 2 = 5$ **50.** $\sqrt[3]{4x-1} - 4 = -1$

51. $\sqrt[3]{n^2-2n+5} = 2$ **52.** $\sqrt[3]{y^2+y+7} = 3$ **53.** $1 + \sqrt{x} = \sqrt{2x+1}$

54. $\sqrt{3x + 4} - 2 = \sqrt{x}$

55. $\sqrt{3x + 1} + \sqrt{3x} = 2$

56. $\sqrt{2x + 5} - \sqrt{2x} = 1$

57. $\sqrt{6x + 7} - 2 = \sqrt{2x + 3}$

58. $\sqrt{5x - 1} - 1 = \sqrt{x + 2}$

Find the Mistake *For Exercises 59–62, explain the mistake; then solve correctly.*

59.
$$\begin{aligned} \sqrt{x} &= -9 \\ (\sqrt{x})^2 &= (-9)^2 \\ x &= 81 \end{aligned}$$

60.
$$\begin{aligned} \sqrt{x} &= -2 \\ (\sqrt{x})^2 &= (-2)^2 \\ x &= 4 \end{aligned}$$

61.
$$\begin{aligned} \sqrt{x + 3} &= x - 3 \\ x + 3 &= x^2 - 9 \\ 0 &= x^2 - x - 12 \\ 0 &= (x + 3)(x - 4) \\ x &= -3, 4 \end{aligned}$$

62.
$$\begin{aligned} \sqrt{x + 3} &= 4 \\ (\sqrt{x + 3})^2 &= 4^2 \\ x + 9 &= 16 \\ x + 9 - 9 &= 16 - 9 \\ x &= 7 \end{aligned}$$

For Exercises 63–66, given the period of a pendulum, find the length of the pendulum. Use the formula $T = 2\pi\sqrt{\frac{L}{9.8}}$*, where T is the period in seconds and L is the length in meters.*

63. 2π seconds

64. 6π seconds

65. $\frac{\pi}{2}$ seconds

66. $\frac{\pi}{3}$ seconds

For Exercises 67–70, find the distance an object has fallen. Use the formula $t = \sqrt{\frac{h}{16}}$*, where t is the time in seconds it takes an object to fall a distance of h feet.*

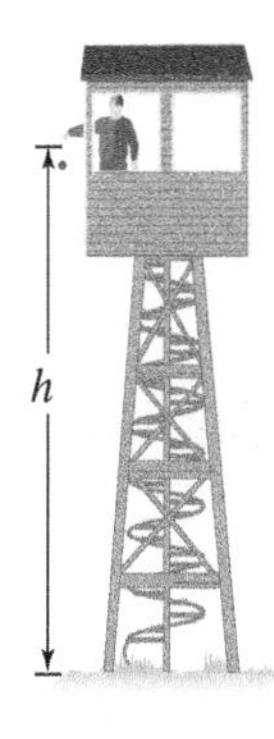

67. 0.3 second

68. 0.5 second

69. $\frac{1}{4}$ second

70. $\frac{1}{8}$ second

For Exercises 71–74, find the skid distance after a car has braked hard at a given speed. Use the formula $S = \frac{7}{2}\sqrt{2D}$*, where D represents the skid distance in feet on asphalt and S represents the speed of the car in miles per hour. Round your answers to hundredths. (Source: Harris Technical Services, Traffic Accident Reconstructionists.)*

71. 30 miles per hour

72. 60 miles per hour

73. 45 miles per hour

74. 75 miles per hour

For Exercises 75–78, use the following information. Two forces, F_1 and F_2, acting on an object at a 90° angle will pull the object with a resultant force, R, at an angle between F_1 and F_2. (See the figure.) The value of the resultant force can be found by the formula $R = \sqrt{F_1^2 + F_2^2}$.

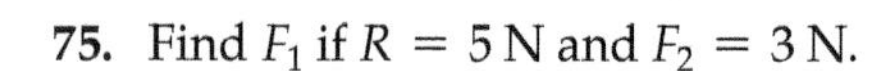

75. Find F_1 if $R = 5$ N and $F_2 = 3$ N.

76. Find F_1 if $R = 10$ N and $F_2 = 8$ N.

★ **77.** Find F_2 if $R = 3\sqrt{5}$ N and $F_1 = 3$ N.

★ **78.** Find F_2 if $R = 2\sqrt{13}$ N and $F_1 = 6$ N.

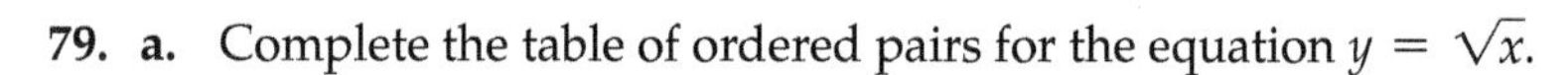

79. a. Complete the table of ordered pairs for the equation $y = \sqrt{x}$.

x	0	1	4	9	16	
y	0	1	2			5

b. Graph the ordered pairs in the coordinate plane; then connect the points to make a curve.

c. Does the graph extend below or to the left of the origin? Explain.

d. Does the curve represent a function? Explain.

80. a. Complete the table of ordered pairs for the equation $y = \sqrt{x} + 2$.

x	0	1	4	9	16	
y	2	3	4			7

b. Graph the ordered pairs in the coordinate plane; then connect the points to make a curve.

c. Does the graph extend below or to the left of the origin? Explain.

d. Does the curve represent a function? Explain.

81. Use a graphing calculator to graph $y = \sqrt{x}$, $y = 2\sqrt{x}$, and $y = 3\sqrt{x}$. Based on your observations, as you increase the size of the coefficient, what happens to the graph?

82. Use a graphing calculator to compare the graph of $y = \sqrt{x}$ to the graph of $y = -\sqrt{x}$. What effect does the negative sign have on the graph?

83. Use a graphing calculator to graph $y = \sqrt{x}$, $y = \sqrt{x} + 2$, and $y = \sqrt{x} - 2$. Based on your observations, what effect does adding or subtracting a constant have on the graph of $y = \sqrt{x}$?

84. Based on your conclusions from Exercises 81–83, without graphing, describe the graph of $y = -2\sqrt{x} + 3$. Use a graphing calculator to graph the equation to confirm your description.

Review Exercises

Exercises 1–6 Expressions

[5.1] *For Exercises 1–3, evaluate.*

1. 3^4

2. $(-0.2)^3$

3. $\left(\frac{2}{5}\right)^{-4}$

[5.1] *For Exercises 4–6, use the rules of exponents to simplify.*

4. $(x^3)(x^5)$

5. $(n^4)^6$

6. $\frac{y^7}{y^3}$

8.7 Complex Numbers

Objectives

1 Write imaginary numbers using i.

2 Perform arithmetic operations with complex numbers.

3 Raise i to powers.

Warm-up

[8.6] **1.** Solve: $\sqrt{7-x} = x - 1$

[8.4] *For Exercises 2–4, simplify.*

2. $(-1 + 13\sqrt{2}) - (8 - 3\sqrt{2})$ **3.** $(5\sqrt{3})(4 - 7\sqrt{3})$

4. $(6 - 5\sqrt{2})(6 + 5\sqrt{2})$

We have said that the square root of a negative number is not a real number. In this section, we learn about the *imaginary number system,* in which square roots of negative numbers are expressed using a notation involving the letter i.

Of Interest

When the idea of finding square roots of negative numbers was first introduced, members of the established mathematics community said that this type of number existed, but only in the imagination of those finding them. From then on, the numbers have been called "imaginary numbers."

Objective 1 Write imaginary numbers using i.

Using the product rule of square roots, any square root of a negative number can be rewritten as a product of a real number and the **imaginary unit**, which we express as i.

Definition **Imaginary unit:** The number represented by i, where $i = \sqrt{-1}$ and $i^2 = -1$.

A number written with the imaginary unit is called an **imaginary number**.

Definition **Imaginary number:** A number that can be expressed in the form $a + bi$, where a and b are real numbers, i is the imaginary unit, and $b \neq 0$.

Note An imaginary number of the form bi is a pure imaginary number.

Example 1 Write each imaginary number as a product of a real number and i.

a. $\sqrt{-9}$

Solution:
$\sqrt{-9} = \sqrt{-1 \cdot 9}$ Factor out -1 in the radicand.
$= \sqrt{-1} \cdot \sqrt{9}$ Use the product rule of square roots.
$= i \cdot 3$ Replace $\sqrt{-1}$ with i.
$= 3i$ Simplify.

b. $\sqrt{-54}$

Solution:
$\sqrt{-54} = \sqrt{-1 \cdot 54}$ Factor out -1 in the radicand.
$= \sqrt{-1} \cdot \sqrt{54}$ Use the product rule of square roots.
$= i\sqrt{9 \cdot 6}$ Write 54 as $9 \cdot 6$.
$= 3i\sqrt{6}$ Simplify.

Note In a product with the imaginary unit, it is customary to write integer factors first, then i, then square root factors.

Answers to Warm-up
1. 3 **2.** $-9 + 16\sqrt{2}$
3. $20\sqrt{3} - 105$ **4.** -14

Example 1 suggests the following procedure:

Procedure Rewriting Imaginary Numbers

To write an imaginary number $\sqrt{-n}$ in terms of the imaginary unit i:

1. Separate the radical into two factors, $\sqrt{-1} \cdot \sqrt{n}$.
2. Replace $\sqrt{-1}$ with i.
3. Simplify $\sqrt{n}$.

Your Turn 1 Write each imaginary number as a product of a real number and i.

a. $\sqrt{-64}$ b. $\sqrt{-18}$ c. $\sqrt{-48}$

Objective 2 Perform arithmetic operations with complex numbers.

We now have two distinct sets of numbers, the set of real numbers and the set of imaginary numbers. There is yet another set of numbers, called the set of **complex numbers**, that contains both real and imaginary numbers.

Definition **Complex number:** A number that can be expressed in the form $a + bi$, where a and b are real numbers and i is the imaginary unit.

Note When the coefficient of i is a radical expression, as in $-2.1 - i\sqrt{5}$ and $4i\sqrt{3}$, we write i to the left of the radical sign so that i does not look like it is part of the radicand.

When written in the form $a + bi$, a complex number is said to be in *standard form*. Following are some examples of complex numbers written in standard form.

$$2 + 3i \qquad 4 - 7i \qquad -2.1 - i\sqrt{5}$$

Note that if $a = 0$, then the complex number is purely an imaginary number, such as the following:

$$-6i \qquad 4i\sqrt{3}$$

If $b = 0$, then the complex number is a real number.

The following Venn diagram shows how all complex numbers are either real numbers or imaginary numbers.

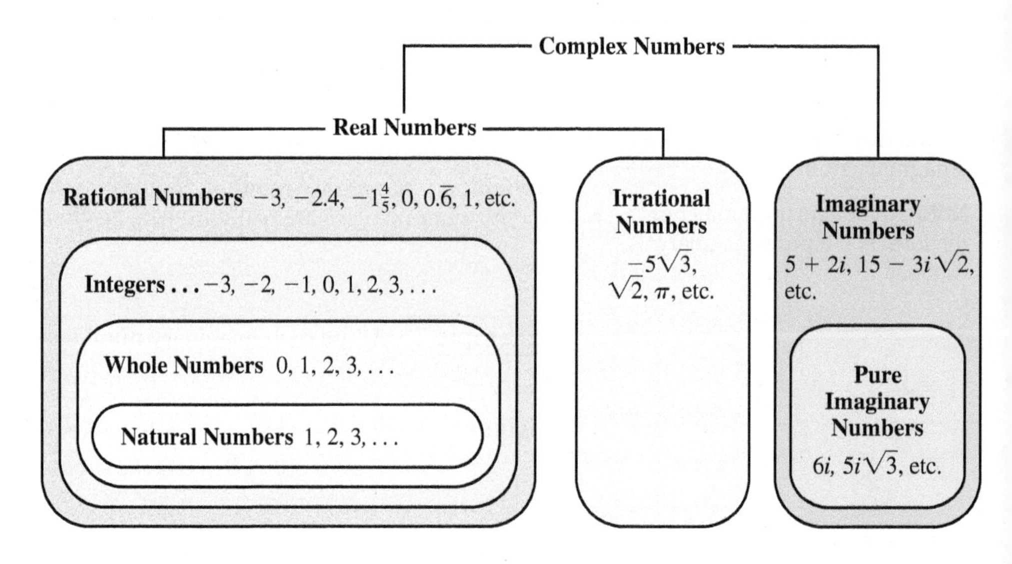

Answers to Your Turn 1
a. $8i$ b. $3i\sqrt{2}$ c. $4i\sqrt{3}$

We can perform arithmetic operations with complex numbers. In general, we treat the complex numbers just like polynomials, where i is like a variable.

Adding and Subtracting Complex Numbers

Example 2 Add or subtract.

a. $(-8 + 7i) + (5 - 19i)$

Solution: We add complex numbers just like we add polynomials—by combining like terms.

$$(-8 + 7i) + (5 - 19i) = -3 - 12i$$

b. $(-1 + 13i) - (8 - 3i)$

Solution: We subtract complex numbers just like we subtract polynomials—by writing an equivalent addition and changing the signs in the second complex number.

$$\begin{aligned}(-1 + 13i) - (8 - 3i) &= (-1 + 13i) + (-8 + 3i)\\ &= -9 + 16i\end{aligned}$$

Your Turn 2 Add or subtract.

a. $(-3 - 5i) + (4 - 6i)$ **b.** $(4 - 7i) - (-3 + 2i)$

Multiplying Complex Numbers

We multiply complex numbers the same way that we multiply monomials and binomials. However, we must be careful when simplifying because these products may contain i^2, which is equal to -1.

Example 3 Multiply.

a. $(9i)(-8i)$

Solution: Multiply the same way that we multiply monomials.

$$\begin{aligned}(9i)(-8i) &= -72i^2 && \text{Multiply.}\\ &= -72(-1) && \text{Replace } i^2 \text{ with } -1.\\ &= 72\end{aligned}$$

b. $(5i)(4 - 7i)$

Solution: Multiply the same way that we multiply a binomial by a monomial.

$$\begin{aligned}(5i)(4 - 7i) &= 20i - 35i^2 && \text{Distribute.}\\ &= 20i - 35(-1) && \text{Replace } i^2 \text{ with } -1.\\ &= 20i + 35\\ &= 35 + 20i && \text{Write in standard form, } a + bi.\end{aligned}$$

c. $(8 - 3i)(2 + i)$

Solution: Multiply the same way that we multiply binomials.

$$\begin{aligned}(8 - 3i)(2 + i) &= 16 + 8i - 6i - 3i^2 && \text{Use FOIL.}\\ &= 16 + 2i - 3(-1) && \text{Combine like terms and replace } i^2 \text{ with } -1.\\ &= 16 + 2i + 3 && \text{Multiply.}\\ &= 19 + 2i && \text{Write in standard form, } a + bi.\end{aligned}$$

d. $(6 - 5i)(6 + 5i)$

Solution: Note that these complex numbers are conjugates.

$$\begin{aligned}(6 - 5i)(6 + 5i) &= 36 + 30i - 30i - 25i^2 && \text{Use FOIL.}\\ &= 36 - 25(-1) && \text{Combine like terms and replace } i^2 \text{ with } -1.\\ &= 36 + 25\\ &= 61\end{aligned}$$

Note The product of these two complex numbers is a real number.

Answers to Your Turn 2
a. $1 - 11i$ **b.** $7 - 9i$

The complex numbers multiplied in Example 3(d) are called **complex conjugates**.

Definition **Complex conjugate:** The complex conjugate of a complex number $a + bi$ is $a - bi$.

Other examples of complex conjugates follow.

$$4 + i \text{ and } 4 - i$$
$$9 - 7i \text{ and } 9 + 7i$$

The product of complex conjugates is always a real number. In general, $(a + bi)(a - bi) = a^2 - b^2i^2 = a^2 - b^2(-1) = a^2 + b^2$, which is a real number.

Your Turn 3 Multiply.

a. $(-4i)(-7i)$ **b.** $(-3i)(6 - i)$ **c.** $(8 - 3i)(5 + 7i)$ **d.** $(7 + 3i)(7 - 3i)$

Connection Recall that we rationalize denominators to clear undesired square root expressions from a denominator. The imaginary unit i represents a square root expression, $\sqrt{-1}$, which is why we rationalize denominators that contain i.

Dividing Complex Numbers

When dividing by a pure imaginary number, we use the fact that $i^2 = -1$, and when dividing by a complex number, we use the fact that the product of complex conjugates is a real number. The process is similar to rationalizing denominators. For example, to rationalize the denominator of $\frac{3}{2\sqrt{5}}$, we multiply by $\frac{\sqrt{5}}{\sqrt{5}}$. Similarly, to divide $\frac{3}{2i}$, we multiply by $\frac{i}{i}$. To rationalize the denominator of $\frac{4}{2 + \sqrt{3}}$, we multiply by $\frac{2 - \sqrt{3}}{2 - \sqrt{3}}$, where $2 - \sqrt{3}$ is the conjugate of $2 + \sqrt{3}$. Similarly, to divide $\frac{4}{2 + i}$, we multiply by $\frac{2 - i}{2 - i}$, where $2 - i$ is the complex conjugate of $2 + i$.

Example 4 Divide. Write in standard form.

a. $\frac{9}{2i}$

Solution: Because $i^2 = -1$, multiplying by $\frac{i}{i}$ eliminates i from the denominator.

$$\frac{9}{2i} = \frac{9}{2i} \cdot \frac{i}{i} \quad \text{Multiply the numerator and denominator by } i.$$
$$= \frac{9i}{2i^2} \quad \text{Multiply.}$$
$$= \frac{9i}{2(-1)} \quad \text{Replace } i^2 \text{ with } -1.$$
$$= \frac{9i}{-2}$$
$$= -\frac{9}{2}i$$

Note We could have thought of $2i$ as $0 + 2i$; so the complex conjugate is $0 - 2i = -2i$ and is multiplied by $\frac{-2i}{-2i}$.

$$\frac{9}{2i} = \frac{9}{2i} \cdot \frac{-2i}{-2i} = \frac{-18i}{-4i^2}$$
$$= \frac{-18i}{-4(-1)} = \frac{-18i}{4}$$
$$= -\frac{9}{2}i$$

b. $\frac{6 + 5i}{7 - 2i}$

Solution: $\frac{6 + 5i}{7 - 2i} = \frac{6 + 5i}{7 - 2i} \cdot \frac{7 + 2i}{7 + 2i}$ Multiply the numerator and denominator by the complex conjugate of the denominator, which is $7 + 2i$.

$$= \frac{42 + 12i + 35i + 10i^2}{49 - 4i^2} \quad \text{Multiply.}$$

Answers to Your Turn 3
a. -28 **b.** $-3 - 18i$
c. $61 + 41i$ **d.** 58

$$= \frac{42 + 47i + 10(-1)}{49 - 4(-1)}$$ Simplify.

$$= \frac{42 + 47i - 10}{49 + 4}$$

$$= \frac{32 + 47i}{53}$$

$$= \frac{32}{53} + \frac{47}{53}i$$ Write in standard form.

Your Turn 4 Divide. Write in standard form.

a. $\frac{8}{3i}$

b. $\frac{6 + i}{4 - 5i}$

Objective 3 Raise *i* to powers.

We have defined i as $\sqrt{-1}$ and have seen that $i^2 = -1$. Raising i to other powers leads to an interesting pattern.

$i^1 = i$ | $i^5 = i^4 \cdot i = 1 \cdot i = i$

$i^2 = -1$ | $i^6 = i^4 \cdot i^2 = 1 \cdot (-1) = -1$

$i^3 = i^2 \cdot i = -1 \cdot i = -i$ | $i^7 = i^4 \cdot i^3 = 1 \cdot (-i) = -i$

$i^4 = (i^2)^2 = (-1)^2 = 1$ | $i^8 = (i^4)^2 = 1^2 = 1$

Note If the exponent on i is an even number not divisible by 4, the result is -1. If the exponent on i is divisible by 4, the result is 1. If the exponent is an odd number that precedes a number divisible by 4, the result is $-i$.

If we continue the pattern, we get the following:

$i^1 = i$	$i^5 = i$	$i^9 = i$
$i^2 = -1$	$i^6 = -1$	$i^{10} = -1$
$i^3 = -i$	$i^7 = -i$	$i^{11} = -i$
$i^4 = 1$	$i^8 = 1$	$i^{12} = 1$

Notice that i to any integer power is i, -1, $-i$, or 1. This pattern allows us to find i to any integer power. We will use the fact that because $i^4 = 1$, $(i^4)^n = 1$ for any integer value of n.

Example 5 Find the powers of i.

a. i^{25}

Solution: $i^{25} = i^{24} \cdot i$ Write i^{25} as $i^{24} \cdot i$ because 24 is the largest multiple of 4 that is smaller than 25.

$= (i^4)^6 \cdot i$ Write i^{24} as $(i^4)^6$ because $i^4 = 1$.

$= 1^6 \cdot i$ Replace i^4 with 1.

$= 1 \cdot i$

$= i$

Note Example 5(b) could have been done as follows:

$$i^{-14} = i^{-12} \cdot i^{-2} = (i^4)^{-3} \cdot i^{-2} = (1)^{-3} \cdot \frac{1}{i^2} = 1 \cdot \frac{1}{-1} = -1$$

b. i^{-14}

Solution: $i^{-14} = \frac{1}{i^{14}}$ Write i^{-14} with a positive exponent.

$= \frac{1}{i^{12} \cdot i^2}$ Write i^{14} as $i^{12} \cdot i^2$ because 12 is the largest multiple of 4 that is smaller than 14.

$= \frac{1}{(i^4)^3 \cdot (-1)}$ Write i^{12} as $(i^4)^3$ and replace i^2 with -1.

$= \frac{1}{1(-1)}$ $(i^4)^3 = 1^3 = 1$

$= -1$

Answers to Your Turn 4

a. $-\frac{8}{3}i$ b. $\frac{19}{41} + \frac{34}{41}i$

Answers to Your Turn 5
a. $-i$ b. -1

Your Turn 5 Find the powers of i.

a. i^{43} b. i^{-18}

8.7 Exercises

For Extra Help MyMathLab®

Objective 1

Prep Exercise 1 What does the imaginary unit i represent?

Prep Exercise 2 Is every real number a complex number? Explain.

Prep Exercise 3 Is every complex number an imaginary number? Explain.

Prep Exercise 4 Explain how to write an imaginary number $\sqrt{-n}$ using the imaginary unit i.

For Exercises 1–16, write the imaginary number using i. See Example 1.

1. $\sqrt{-36}$
2. $\sqrt{-81}$
3. $\sqrt{-5}$
4. $\sqrt{-10}$
5. $\sqrt{-8}$
6. $\sqrt{-12}$
7. $\sqrt{-18}$
8. $\sqrt{-32}$
9. $\sqrt{-27}$
10. $\sqrt{-72}$
11. $\sqrt{-125}$
12. $\sqrt{-80}$
13. $\sqrt{-63}$
14. $\sqrt{-54}$
15. $\sqrt{-245}$
16. $\sqrt{-810}$

Objective 2

Prep Exercise 5 Explain how to add complex numbers.

Prep Exercise 6 Explain how to subtract complex numbers.

For Exercises 17–32, add or subtract. See Example 2.

17. $(9 + 3i) + (-3 + 4i)$
18. $(6 + 4i) + (-2 - 2i)$
19. $(6 + 2i) + (5 - 8i)$
20. $(4 + i) + (7 - 6i)$
21. $(-4 + 6i) - (3 + 5i)$
22. $(8 - 3i) - (-1 - 2i)$
23. $(8 - 5i) - (-3i)$
24. $(19 + 6i) - (-2i)$
25. $(12 + 3i) + (-15 - 13i)$
26. $(14 - 7i) + (-8 - 9i)$
27. $(-5 - 9i) - (-5 - 9i)$
28. $(-4 + 2i) - (-4 + 2i)$

29. $(10 + i) - (2 - 13i) + (6 - 5i)$

30. $(-14 + 2i) - (6 + i) + (19 + 10i)$

31. $(5 - 2i) - (9 - 14i) + (16i)$

32. $(-12i) - (4 - 7i) - (18 + 4i)$

Prep Exercise 7 What is the value of i^2?

For Exercises 33–48, multiply. See Example 3.

33. $(8i)(3i)$

34. $(9i)(5i)$

35. $(-8i)(5i)$

36. $(4i)(-7i)$

37. $2i(6 - 7i)$

38. $6i(9 - i)$

39. $-8i(4 - 9i)$

40. $-7i(5 - 8i)$

41. $(6 + i)(3 - i)$

42. $(5 - 2i)(4 + i)$

43. $(8 + 5i)(5 - 2i)$

44. $(6 - 3i)(7 + 8i)$

45. $(8 + i)(8 - i)$

46. $(4 + 9i)(4 - 9i)$

47. $(3 - 4i)^2$

48. $(5 - 3i)^2$

Prep Exercise 8 Is the expression $\frac{5 - 4i}{3}$ in standard form for a complex number? Explain.

Prep Exercise 9 Explain how to divide $\frac{9 + 5i}{3 - 2i}$.

For Exercises 49–68, divide and write in standard form. See Example 4.

49. $\frac{2}{i}$

50. $\frac{6}{-i}$

51. $\frac{4}{5i}$

52. $\frac{6}{7i}$

53. $\frac{6}{2i}$

54. $\frac{12}{3i}$

55. $\frac{2 + i}{2i}$

56. $\frac{3 + i}{3i}$

57. $\frac{4 + 2i}{4i}$

58. $\frac{6 - 2i}{6i}$

59. $\frac{7}{2 + i}$

60. $\frac{5}{6 + i}$

61. $\frac{2i}{3 - 7i}$

62. $\frac{4i}{5 - 3i}$

63. $\frac{5 - 9i}{1 - i}$

64. $\frac{3 + i}{2 - i}$

65. $\frac{3 + i}{2 + 3i}$

66. $\frac{1 + 3i}{5 + 2i}$

67. $\frac{1 + 6i}{4 + 5i}$

68. $\frac{5 - 6i}{2 - 9i}$

Objective 3

Prep Exercise 10 Complete each rule.

$i = \underline{?}$

$i^2 = \underline{?}$

$i^3 = \underline{?}$

$i^4 = \underline{?}$

For Exercises 69–84, find the powers of i. See Example 5.

69. i^{19} **70.** i^{25} **71.** i^{42} **72.** i^{51}

73. i^{38} **74.** i^{45} **75.** i^{60} **76.** i^{52}

77. i^{-20} **78.** i^{-32} **79.** i^{-30} **80.** i^{-38}

81. i^{-21} **82.** i^{-45} **83.** i^{-35} **84.** i^{-44}

Review Exercises

Exercises 1 and 2 Expressions

[6.3] 1. Factor: $4x^2 - 12x + 9$

[8.3] 2. Simplify: $\pm\sqrt{48}$

Exercises 3–6 Equations and Inequalities

For Exercises 3 and 4, solve.

[2.1] 3. $3x - 2 = 4$

[6.4] 4. $2x^2 - x - 6 = 0$

[3.1] 5. Find the x- and y-intercepts for $2x + 3y = 6$.

[3.5] 6. Graph: $y = x^2 - 3$

Chapter 8 Summary and Review Exercises

Complete each incomplete definition, rule, or procedure; study the key examples; and then work the related exercises.

8.1 Radical Expressions and Functions

Definitions/Rules/Procedures	Key Example(s)
The number b is an nth root of a number a if ________. **If a is nonnegative,** then $\sqrt[n]{a} = b$, where $b \geq 0$ and ________. **If a is negative and n is even,** then there is ________ root. **If a is negative and n is odd,** then $\sqrt[n]{a} = b$, where b is ________ and $b^n = a$.	Find each root if possible. **a.** $\sqrt{36} = 6$ **b.** $\sqrt{-25}$ is not a real number. **c.** $-\sqrt{121} = -11$ **d.** $\sqrt{\frac{49}{100}} = \frac{7}{10}$ **e.** $\sqrt{16x^6} = 4\lvert x^3 \rvert$ **f.** $\sqrt[3]{-27} = -3$ **Note** If we assume that the variables represent nonnegative numbers, the solution for part e is $\sqrt{16x^6} = 4x^3$.

Exercises 1–22 Expressions

[8.1] *For Exercises 1 and 2, find all square roots of the given number.*

1. 121

2. 49

[8.1] *For Exercises 3–8, evaluate the square root.*

3. $\sqrt{169}$

4. $-\sqrt{49}$

5. $\sqrt{-36}$

6. $\sqrt{\frac{1}{25}}$

7. Three pieces of lumber are to be connected to form a right triangle that will be part of the roof frame for a small storage building. If the horizontal piece is to be 4 feet and the vertical piece is to be 3 feet, how long must the connecting piece be?

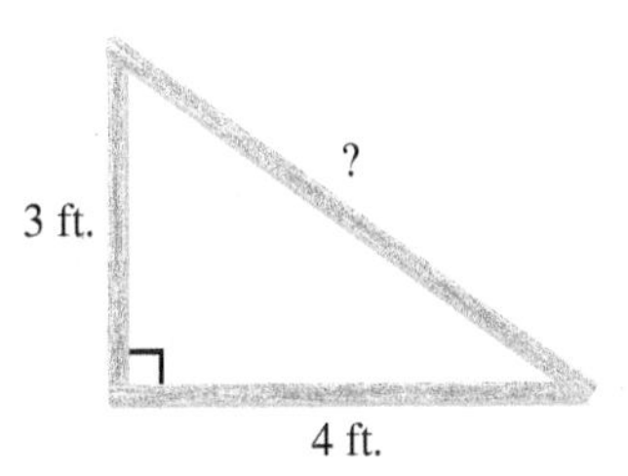

8. In the formula $T = 2\pi\sqrt{\frac{L}{9.8}}$, T is the period of a pendulum in seconds and L is the length of the pendulum in meters. Find the period of a pendulum with a length of 2.45 meters.

[8.1] *For Exercises 9 and 10, use a calculator to approximate each root to the nearest thousandth.*

9. $\sqrt{7}$

10. $\sqrt{90}$

11. The speed of a car can be determined by the length of the skid marks using the formula $S = 2\sqrt{2L}$, where L is the length of the skid mark in feet and S is the speed of the car in miles per hour.

a. Write an expression for the exact speed of the car if the length of the skid is 40 feet.

b. Approximate the speed to the nearest tenth.

12. In the formula $t = \sqrt{\frac{h}{16}}$, t represents the time in seconds it takes an object to fall a distance of h feet.

a. Write an expression in simplest form of the time an object takes to fall 40 feet.

b. Approximate the time to the nearest hundredth.

[8.1] *For Exercises 13–20, simplify. Assume that variables represent nonnegative values.*

13. $\sqrt{49x^8}$ **14.** $\sqrt{144a^6b^{12}}$ **15.** $\sqrt{0.16m^2n^{10}}$ **16.** $\sqrt[3]{x^{15}}$

17. $\sqrt[3]{-64r^9s^3}$ **18.** $\sqrt[4]{81x^{12}}$ **19.** $\sqrt[5]{32x^{15}y^{20}}$ **20.** $\sqrt[7]{x^{14}y^7}$

[8.1] *For Exercises 21 and 22, simplify. Assume that variables represent any real number.*

21. $\sqrt{81x^2}$ **22.** $\sqrt[4]{(x-1)^8}$

Definitions/Rules/Procedures	Key Example(s)
A radical function is a function containing a(n) ____________ expression whose ____________ has a variable.	If $f(x) = \sqrt{4x+3}$, find $f(3)$. Solution: $f(3) = \sqrt{4(3)+3} = \sqrt{12+3} = \sqrt{15}$
To evaluate a radical function, replace the variable with the indicated value and ____________.	
Finding the domain of a radical function If the root index is odd, the domain is $\{x \mid$____________, or (______, ______). If the root index is even, the radicand must be ____________. Consequently, the domain is the solution set to the inequality "radicand ____________."	Find the domain of $f(x) = \sqrt{2x-4}$. Solution: Because $2x - 4$ must be nonnegative, we solve $2x - 4 \geq 0$. $2x - 4 \geq 0$ $2x \geq 4$ $x \geq 2$ Domain: $\{x \mid x \geq 2\}$ or $[2, \infty)$

Exercises 23 and 24 Equations and Inequalities

[8.1] *For Exercises 23 and 24, find the indicated value of the function.*

23. If $f(x) = \sqrt{4x-4}$, find $f(5)$. **24.** If $f(x) = \sqrt[3]{5x+2}$, find $f(-2)$.

Exercises 25 and 26 Expressions

[8.1] *For Exercises 25 and 26, find the domain.*

25. $\sqrt{2x+10}$ **26.** $\sqrt{9-3x}$

8.2 Rational Exponents

Definitions/Rules/Procedures	Key Example(s)
$a^{1/n} =$ ______, where n is a natural number other than 1. **Note:** If a is negative and n is odd, the root is negative. If a is negative and n is even, there is no real-number root.	Evaluate. **a.** $25^{1/2} = \sqrt{25} = 5$ **b.** $(-27)^{1/3} = \sqrt[3]{-27} = -3$ **c.** $(-16)^{1/4} = \sqrt[4]{-16}$, which is not a real number.
$a^{m/n} =$ ______ $=$ ______, where $a \geq 0$ and m and n are natural numbers other than 1.	Simplify. $27^{4/3} = (\sqrt[3]{27})^4 = 3^4 = 81$ Write in exponential form. **a.** $\sqrt[3]{x^2} = x^{2/3}$ **b.** $(\sqrt[6]{x})^5 = x^{5/6}$
$a^{-m/n} =$ ______, where $a \neq 0$ and m and n are natural numbers with $n \neq 1$.	Evaluate. $16^{-3/4} = \frac{1}{16^{3/4}} = \frac{1}{(\sqrt[4]{16})^3} = \frac{1}{2^3} = \frac{1}{8}$ Write $\sqrt[8]{x^6}$ as a radical expression with a smaller root index. Assume variables represent nonnegative values. **Solution:** $\sqrt[8]{x^6} = x^{6/8} = x^{3/4} = \sqrt[4]{x^3}$

Exercises 27–40 Expressions

[8.2] *For Exercises 27–34, rewrite using radicals; then simplify if possible. Assume that variables represent nonnegative values.*

27. $(-64)^{1/3}$ **28.** $(24a^4)^{1/2}$ **29.** $\left(\frac{1}{32}\right)^{3/5}$ **30.** $(5r - 2)^{5/7}$

31. $121^{-1/2}$ **32.** $81^{-3/4}$ **33.** $\left(\frac{16}{49}\right)^{3/2}$ **34.** $81x^{3/4}$

[8.2] *For Exercises 35–40, write in exponential form.*

35. $\sqrt[8]{33}$ **36.** $\frac{8}{\sqrt[7]{n^3}}$ **37.** $(\sqrt[5]{8})^3$ **38.** $(\sqrt[3]{m})^8$ **39.** $(\sqrt[4]{3xw})^3$ **40.** $\sqrt[3]{(a + b)^4}$

Definitions/Rules/Procedures	Key Example(s)
To multiply or divide radical expressions with different root indices: 1. Change from radical to ______ form. 2. Multiply or divide using the appropriate rule(s) of ______. 3. Write the result from step 2 in ______ form.	Simplify. Assume variables represent nonnegative values. **a.** $\sqrt[4]{x} \cdot \sqrt{x} = x^{1/4} \cdot x^{1/2} = x^{1/4+1/2}$ $= x^{1/4+2/4}$ $= x^{3/4} = \sqrt[4]{x^3}$ **b.** $\frac{\sqrt[3]{x^2}}{\sqrt[4]{x}} = \frac{x^{2/3}}{x^{1/4}} = x^{(2/3)-(1/4)}$ $= x^{(8/12)-(3/12)} = x^{5/12} = \sqrt[12]{x^5}$ Write $\sqrt[3]{\sqrt[5]{x}}$ as a single radical. Assume variables represent nonnegative values. **Solution:** $\sqrt[3]{\sqrt[5]{x}} = (x^{1/5})^{1/3} = x^{(1/5)\cdot(1/3)}$ $= x^{1/15} = \sqrt[15]{x}$

Exercises 41–47 Expressions

[8.2] *For Exercises 41–47, use the rules of exponents to simplify. Assume that variables represent positive values.*

41. $x^{2/3} \cdot x^{4/3}$

42. $(4m^{1/4})(8m^{5/4})$

43. $\dfrac{y^{3/5}}{y^{4/5}}$

44. $\dfrac{b^{2/5}}{b^{-3/5}}$

45. $(k^{2/3})^{3/4}$

46. $(2xy^{1/5})^{3/4}$

47. Write $\sqrt[4]{\sqrt[3]{x^2}}$ as a single radical.

8.3 Multiplying, Dividing, and Simplifying Radicals

Definitions/Rules/Procedures	Key Example(s)
Product rule for radicals If both $\sqrt[n]{a}$ and $\sqrt[n]{b}$ are real numbers, then $\sqrt[n]{a} \cdot \sqrt[n]{b} =$ ______.	Simplify. Assume that variables represent nonnegative values. **a.** $\sqrt{2} \cdot \sqrt{50} = \sqrt{2 \cdot 50} = \sqrt{100} = 10$ **b.** $\sqrt{3x} \cdot \sqrt{12x^3} = \sqrt{3x \cdot 12x^3} = \sqrt{36x^4} = 6x^2$ **c.** $\sqrt[3]{5x} \cdot \sqrt[3]{3x} = \sqrt[3]{15x^2}$
Raising an nth root to the nth power For any nonnegative real number a, $(\sqrt[n]{a})^n =$ ______.	Simplify $(\sqrt[4]{x^3})^4$. Assume variables represent nonnegative values. **Solution:** $(\sqrt[4]{x^3})^4 = x^3$.

Exercises 48–57 Expressions

[8.3] *For Exercises 48–57, simplify. Assume that variables represent positive values.*

48. $\sqrt{3} \cdot \sqrt{27}$

49. $\sqrt{5x^5} \cdot \sqrt{20x^3}$

50. $\sqrt[3]{2} \cdot \sqrt[3]{4}$

51. $\sqrt[4]{7} \cdot \sqrt[4]{6}$

52. $\sqrt[5]{3x^2y^3} \cdot \sqrt[5]{5x^2y}$

53. $4\sqrt{6} \cdot 7\sqrt{15}$

54. $\sqrt{x^9} \cdot \sqrt{x^6}$

55. $4\sqrt{10c} \cdot 2\sqrt{6c^4}$

56. $a\sqrt{a^3b^2} \cdot b^2\sqrt{a^5b^3}$

57. $(\sqrt[4]{6ab^2})^4$

Definitions/Rules/Procedures	Key Example(s)
Quotient rule for radicals If both $\sqrt[n]{a}$ and $\sqrt[n]{b}$ are real numbers, then $\dfrac{\sqrt[n]{a}}{\sqrt[n]{b}} =$ ______ where $b \neq 0$.	Simplify. Assume variables represent positive values. **a.** $\dfrac{\sqrt{45}}{\sqrt{5}} = \sqrt{\dfrac{45}{5}} = \sqrt{9} = 3$ **b.** $\dfrac{\sqrt[3]{128x^7}}{\sqrt[3]{2x}} = \sqrt[3]{\dfrac{128x^7}{2x}} = \sqrt[3]{64x^6} = 4x^2$ **c.** $\sqrt[4]{\dfrac{8}{x^4}} = \dfrac{\sqrt[4]{8}}{\sqrt[4]{x^4}} = \dfrac{\sqrt[4]{8}}{x}$

Exercises 58–61 Expressions

[8.3] *For Exercises 58–61, simplify. Assume that variables represent positive values.*

58. $\sqrt{\dfrac{49}{121}}$

59. $\sqrt[3]{-\dfrac{27}{8}}$

60. $\dfrac{9\sqrt{160}}{3\sqrt{8}}$

61. $\dfrac{36\sqrt{96x^6y^{11}}}{4\sqrt{3x^2y^4}}$

Definitions/Rules/Procedures	Key Example(s)
To simplify an nth root: 1. Write the radicand as the product of the greatest possible perfect ________ power and a number or an expression that has no perfect ________ power factors. 2. Use the product rule $\sqrt[n]{ab} = \sqrt[n]{a} \cdot \sqrt[n]{b}$, where a is the ________. 3. Find the nth root of the ________ radicand.	Simplify. Assume that variables represent nonnegative values. **a.** $\sqrt{50} = \sqrt{25 \cdot 2} = \sqrt{25} \cdot \sqrt{2} = 5\sqrt{2}$ **b.** $\sqrt{48x^3} = \sqrt{16x^2 \cdot 3x} = \sqrt{16x^2} \cdot \sqrt{3x} = 4x\sqrt{3x}$ **c.** $\sqrt[4]{48a^9} = \sqrt[4]{16a^8 \cdot 3a} = \sqrt[4]{16a^8} \cdot \sqrt[4]{3a} = 2a^2\sqrt[4]{3a}$ Perform the operation and simplify. $\sqrt{6} \cdot \sqrt{10} = \sqrt{60} = \sqrt{4 \cdot 15} = \sqrt{4} \cdot \sqrt{15} = 2\sqrt{15}$

Exercises 62–65 Expressions

[8.3] *For Exercises 62–65, simplify. Assume that variables represent positive values.*

62. $4b\sqrt{27b^7}$

63. $5\sqrt[3]{108}$

64. $2\sqrt[3]{40x^{10}}$

65. $2x^4\sqrt[4]{162x^7}$

8.4 Adding, Subtracting, and Multiplying Radical Expressions

Definitions/Rules/Procedures	Key Example(s)
Like radicals are radical expressions with identical ________ and identical ________. **To add or subtract** like radicals, add or subtract the coefficients and leave the ________ the same.	Add or subtract. Assume that variables represent nonnegative values. **a.** $2\sqrt{x} - 7\sqrt{x} = (2 - 7)\sqrt{x} = -5\sqrt{x}$ **b.** $4\sqrt[3]{5} + 2\sqrt[3]{5} = (4 + 2)\sqrt[3]{5} = 6\sqrt[3]{5}$ Simplify the radicals; then combine like radicals. $9\sqrt{3} - \sqrt{12} + 6\sqrt{75} = 9\sqrt{3} - \sqrt{4 \cdot 3} + 6\sqrt{25 \cdot 3}$ $= 9\sqrt{3} - 2\sqrt{3} + 6 \cdot 5\sqrt{3}$ $= 9\sqrt{3} - 2\sqrt{3} + 30\sqrt{3}$ $= 37\sqrt{3}$

Exercises 66–73 Expressions

[8.4] *For Exercises 66–73, add or subtract. Assume that variables represent nonnegative values.*

66. $-5\sqrt{n} + 2\sqrt{n}$

67. $3y^3\sqrt[4]{8y} - 9y^3\sqrt[4]{8y}$

68. $\sqrt{45} + \sqrt{20}$

69. $4\sqrt{24} - 6\sqrt{54}$

70. $\sqrt{150} - \sqrt{54} + \sqrt{24}$

71. $4\sqrt{72x^2y} - 2x\sqrt{128y} + 5\sqrt{32x^2y}$

72. $\sqrt[3]{250x^4y^5} + \sqrt[3]{128x^4y^5}$

73. $\sqrt[4]{48} + \sqrt[4]{243}$

Definitions/Rules/Procedures	Key Example(s)
To simplify radical expressions that contain more than one operation, use the ________________ agreement.	Find the product. $(\sqrt{3}-4)(\sqrt{3}+6) = \sqrt{3}\cdot\sqrt{3} + \sqrt{3}\cdot 6 - 4\cdot\sqrt{3} - 4\cdot 6$ Use FOIL. $= \sqrt{9} + 6\sqrt{3} - 4\sqrt{3} - 24$ Multiply. $= 3 + 6\sqrt{3} - 4\sqrt{3} - 24$ Simplify $\sqrt{9}$. $= -21 + 2\sqrt{3}$ Combine like radicals. Simplify. **a.** $\sqrt{6}\cdot\sqrt{8} + \sqrt{5}\cdot\sqrt{15} = \sqrt{48} + \sqrt{75}$ $= \sqrt{16\cdot 3} + \sqrt{25\cdot 3}$ $= 4\sqrt{3} + 5\sqrt{3} = 9\sqrt{3}$ **b.** $\frac{\sqrt{60}}{\sqrt{5}} + \sqrt{48} = \sqrt{\frac{60}{5}} + \sqrt{48}$ $= \sqrt{12} + \sqrt{48}$ $= \sqrt{4\cdot 3} + \sqrt{16\cdot 3}$ $= 2\sqrt{3} + 4\sqrt{3} = 6\sqrt{3}$

Exercises 74–82 Expressions

[8.4] *For Exercises 74–82, find the product. Assume that variables represent nonnegative values.*

74. $\sqrt{5}(\sqrt{3}+\sqrt{2})$

75. $\sqrt[3]{7}(\sqrt[3]{3}+2\sqrt[3]{7})$

76. $3\sqrt{6}(2-3\sqrt{6})$

77. $(\sqrt{2}-\sqrt{3})(\sqrt{5}+\sqrt{7})$

78. $(\sqrt[3]{2}-4)(\sqrt[3]{4}+2)$

79. $(\sqrt[4]{6x}+2)(\sqrt[4]{2x}-1)$

80. $(\sqrt{5a}+\sqrt{3b})(\sqrt{5a}-\sqrt{3b})$

81. $(2\sqrt{3}-\sqrt{5})(2\sqrt{3}+\sqrt{5})$

82. $(\sqrt{2}-\sqrt{5})^2$

8.5 Rationalizing Numerators and Denominators of Radical Expressions

Definitions/Rules/Procedures	Key Example(s)
A radical expression is in simplest form if 1. The radicand has no factor raised to a power greater than or equal to the ____________. 2. There are neither radicals in the denominator of a fraction nor radicands that are ____________. 3. All possible ____________, ____________, ____________, and ____________ have been found. **To rationalize** a denominator containing a single *n*th root, multiply the fraction by a form of 1 so that the product's denominator has a radicand that is a(n) ______________.	Rationalize the denominator. **a.** $\frac{7}{\sqrt{3}} = \frac{7}{\sqrt{3}}\cdot\frac{\sqrt{3}}{\sqrt{3}} = \frac{7\sqrt{3}}{3}$ **b.** $\frac{4}{\sqrt[3]{3}} = \frac{4}{\sqrt[3]{3}}\cdot\frac{\sqrt[3]{9}}{\sqrt[3]{9}} = \frac{4\sqrt[3]{9}}{\sqrt[3]{27}} = \frac{4\sqrt[3]{9}}{3}$

Exercises 83–88 Expressions

[8.5] *For Exercises 83–92, rationalize the denominator and simplify. Assume that variables represent positive values.*

83. $\frac{1}{\sqrt{2}}$

84. $\frac{3}{\sqrt[3]{3}}$

85. $\sqrt{\frac{4}{7}}$

86. $\frac{\sqrt[4]{5x^2}}{\sqrt[4]{2}}$

87. $\sqrt[3]{\frac{17}{3y^2}}$

88. $\frac{\sqrt[4]{9}}{\sqrt[4]{27}}$

Definitions/Rules/Procedures	Key Example(s)
To rationalize a denominator containing a sum or difference with at least one square root term, multiply the fraction by a form of 1 whose numerator and denominator are the ______________ of the denominator.	Rationalize the denominator and simplify. $\frac{6}{4-\sqrt{5}} = \frac{6}{4-\sqrt{5}} \cdot \frac{4+\sqrt{5}}{4+\sqrt{5}} = \frac{6 \cdot 4 + 6 \cdot \sqrt{5}}{16-5} = \frac{24+6\sqrt{5}}{11}$

Exercises 89–92 Expressions

[8.5] *For Exercises 89–92, rationalize the denominator and simplify. Assume that variables represent positive values.*

89. $\frac{4}{\sqrt{2}-\sqrt{3}}$

90. $\frac{1}{4+\sqrt{3}}$

91. $\frac{1}{2-\sqrt{n}}$

92. $\frac{2\sqrt{3}}{3\sqrt{2}-2\sqrt{3}}$

Definitions/Rules/Procedures	Key Example(s)
The numerator of a rational expression is rationalized the same way a(n) ______________ is rationalized.	Rationalize the numerator. **a.** $\frac{\sqrt{6}}{3} = \frac{\sqrt{6}}{3} \cdot \frac{\sqrt{6}}{\sqrt{6}} = \frac{\sqrt{36}}{3\sqrt{6}} = \frac{6}{3\sqrt{6}} = \frac{2}{\sqrt{6}}$ **b.** $\frac{2-\sqrt{3}}{5} = \frac{2-\sqrt{3}}{5} \cdot \frac{2+\sqrt{3}}{2+\sqrt{3}} = \frac{4-3}{5(2+\sqrt{3})} = \frac{1}{10+5\sqrt{3}}$

Exercises 93–96 Expressions

[8.5] *For Exercises 93–96, rationalize the numerator. Assume that variables represent positive values.*

93. $\frac{\sqrt{10}}{6}$

94. $\frac{\sqrt{3x}}{5}$

95. $\frac{2-\sqrt{3}}{8}$

★**96.** $\frac{2\sqrt{t}+\sqrt{3t}}{5t}$

8.6 Radical Equations and Problem Solving

Definitions/Rules/Procedures

A **radical equation** is an equation containing at least one ______________ whose radicand has a(n) ______________.

Power rule for solving equations

If both sides of an equation are raised to the same ______________, the resulting equation contains all solutions of the original equation and perhaps some solutions that do not solve the original equation. That is, the solutions of the equation $a = b$ are contained among the solutions of $a^n = b^n$, where n is an integer.

To solve a radical equation:

1. Isolate the ______________ if necessary. (If there is more than one ______________, isolate one of the ______________.)
2. Raise both sides of the equation to the same power as the ______________ of the isolated radical.
3. If all radicals have been eliminated, solve. If a(n) ______________ remains, isolate it and raise both sides to the same power as its ______________.
4. Check each solution. Any apparent solution that does not check is a(n) ______________ solution.

Key Example(s)

Solve $\sqrt{x-5} = 7$.

$(\sqrt{x-5})^2 = 7^2$ Square both sides.

$x - 5 = 49$ Multiply.

$x = 54$ Add 5 to both sides.

Check: $\sqrt{54-5} = 7$

$\sqrt{49} = 7$ True

Solve $\sqrt{n+14} = n + 2$.

$(\sqrt{n+14})^2 = (n+2)^2$ Square both sides.

$n + 14 = n^2 + 4n + 4$

$0 = n^2 + 3n - 10$ Subtract n and 14 from both sides.

$0 = (n+5)(n-2)$ Factor.

$n + 5 = 0$ or $n - 2 = 0$ Use the zero-factor theorem.

$n = -5$ $\quad n = 2$

Check: $\sqrt{-5+14} = -5 + 2$ $\quad \sqrt{2+14} = 2 + 2$

$\sqrt{9} = -3$ False $\quad \sqrt{16} = 4$ True

The only solution is 2. (−5 is extraneous.)

Solve $3 + \sqrt{t} = \sqrt{t+21}$.

$(3 + \sqrt{t})^2 = (\sqrt{t+21})^2$ Square both sides.

$9 + 6\sqrt{t} + t = t + 21$ Multiply.

$6\sqrt{t} = 12$ Subtract 9 and t from both sides.

$\sqrt{t} = 2$ Divide both sides by 6.

$(\sqrt{t})^2 = 2^2$ Square both sides.

$t = 4$

We will leave the check to the reader.

Exercises 97–111 Equations and Inequalities

[8.6] *For Exercises 97–108, solve. Identify any extraneous solutions.*

97. $\sqrt{x} = 9$

98. $\sqrt{y} = -3$

99. $\sqrt{w-1} = 3$

100. $\sqrt[3]{3x-2} = -2$

101. $\sqrt[4]{x-2} - 3 = -1$

102. $\sqrt{y+1} = \sqrt{2y-4}$

103. $\sqrt{x-6} = x + 2$

104. $\sqrt[3]{3x+10} - 4 = 5$

105. $\sqrt[4]{x+8} = \sqrt[4]{2x+1}$

106. $\sqrt{5n-1} = 4 - 2n$

107. $1 + \sqrt{x} = \sqrt{2x+1}$

108. $\sqrt{3x+1} = 2 - \sqrt{3x}$

109. The speed of a car can be determined by the length of the skid marks using the formula $S = 2\sqrt{2L}$, where L is the length of the skid mark in feet and S is the speed of the car in miles per hour. Find the length of skid marks if the driver brakes hard at a speed of 50 miles per hour.

110. In the formula $T = 2\pi\sqrt{\frac{L}{9.8}}$, T is the period of a pendulum in seconds and L is the length of the pendulum in meters. If the period of the pendulum is $\frac{\pi}{3}$ seconds, find the length to the nearest thousandth.

111. In the formula $t = \sqrt{\frac{h}{16}}$, t represents the time in seconds it takes an object to fall a distance of h feet. Find the distance an object falls in 0.3 second.

8.7 Complex Numbers

Definitions/Rules/Procedures	Key Example(s)
The **imaginary unit** is i, where $i =$ ________ and $i^2 =$ ________.	
An **imaginary number** is a number that can be expressed in the form ________, where a and b are ________, i is the ________, and ________.	
A **complex number** is a number that can be expressed in the form ________, where a and b are ________ and i is the ________.	
The complex conjugate of $a + bi$ is ________.	
To write an imaginary number $\sqrt{-n}$ in terms of the imaginary unit i: **1.** Separate the radical into two factors, ________. **2.** Replace $\sqrt{-1}$ with ________. **3.** Simplify $\sqrt{n}$.	Write using the imaginary unit. **a.** $\sqrt{-36} = \sqrt{-1} \cdot \sqrt{36} = 6i$ **b.** $\sqrt{-32} = \sqrt{-1} \cdot \sqrt{32}$ $= i \cdot 4\sqrt{2} = 4i\sqrt{2}$
To add complex numbers, combine ________.	Add $(5 - 6i) + (9 + 2i)$. $(5 - 6i) + (9 + 2i) = 14 - 4i$
To subtract complex numbers, write the equivalent addition and change the ________ of the second complex number; then combine like terms.	Subtract $(7 - i) - (3 + 5i)$. $(7 - i) - (3 + 5i) = (7 - i) + (-3 - 5i)$ $= 4 - 6i$

Exercises 112–115 Expressions

[8.7] *For Exercises 112 and 113, write the imaginary number using i.*

112. $\sqrt{-9}$

113. $\sqrt{-20}$

[8.7] *For Exercises 114 and 115, add or subtract.*

114. $(3 + 2i) + (5 - 8i)$

115. $(7 - 3i) - (-2 + 4i)$

Definitions/Rules/Procedures	Key Example(s)
To multiply complex numbers, follow the same procedures for multiplying ________ and ________. Remember that $i^2 = -1$.	Multiply $(6 + 5i)(2 - 3i)$. $(6 + 5i)(2 - 3i) = 12 - 18i + 10i - 15i^2$ $= 12 - 8i - 15(-1)$ $= 12 - 8i + 15$ $= 27 - 8i$

Exercises 116–119 Expressions

[8.7] *For Exercises 116–119, multiply and simplify.*

116. $(3i)(4i)$

117. $2i(4 - i)$

118. $(6 + 2i)(4 - i)$

119. $(5 - i)^2$

Definitions/Rules/Procedures	Key Example(s)
To divide by a pure imaginary number, multiply by ______. **To divide complex numbers,** rationalize the denominator using the __________.	Divide and write $\frac{2-3i}{4+5i}$ in standard form. $\frac{2-3i}{4+5i} = \frac{2-3i}{4+5i} \cdot \frac{4-5i}{4-5i}$ $= \frac{8 - 10i - 12i + 15i^2}{16 - 25i^2}$ $= \frac{8 - 22i + 15(-1)}{16 - 25(-1)}$ $= \frac{-7 - 22i}{41}$ $= -\frac{7}{41} - \frac{22}{41}i$

Exercises 120–125 Expressions

[8.7] *For Exercises 120–125, divide and write in standard form.*

120. $\frac{5}{i}$

121. $\frac{3}{-i}$

122. $\frac{4}{3i}$

123. $\frac{7+i}{5i}$

124. $\frac{3}{2+i}$

125. $\frac{5+i}{2-3i}$

Definitions/Rules/Procedures	Key Example(s)
Powers of i: $i^1 =$ ______ $i^2 =$ ______ $i^3 =$ ______ $i^4 =$ ______	Find the powers of i. **a.** $i^{19} = i^{16} \cdot i^3 = (i^4)^4 \cdot (-i)$ $= 1^4(-i) = 1(-i) = -i$ **b.** $i^{-21} = \frac{1}{i^{21}} = \frac{1}{i^{20} \cdot i} = \frac{1}{(i^4)^5 \cdot i} = \frac{1}{1^5 \cdot i}$ $= \frac{1}{i} = \frac{1}{i} \cdot \frac{i}{i} = \frac{i}{i^2}$ $= \frac{i}{-1} = -i$

Exercises 126 and 127 Expressions

[8.7] *For Exercises 126 and 127, find the powers of i.*

126. i^{20}

127. i^{15}

Learning Strategy

During exams, I like to work one or two problems that don't count as much and then move on to the bigger problems. That way, if I run out of time, I will have received the most possible credit.

—Jason J.

Chapter 8 Practice Test

For Extra Help

Step-by-step test solutions are found on the Chapter Test Prep Videos available in MyMathLab® *or on* YouTube.

For Exercises 1 and 2, evaluate the square root.

1. $\sqrt{36}$

2. $\sqrt{-49}$

For Exercises 3–12, simplify. Assume that variables represent positive values.

3. $\sqrt{81x^2y^5}$

4. $\sqrt[3]{54}$

5. $\sqrt[4]{4x} \cdot \sqrt[4]{4x^5}$

6. $-\sqrt[3]{-27r^{15}}$

7. $\dfrac{\sqrt{5}}{\sqrt{45}}$

8. $\dfrac{\sqrt[4]{1}}{\sqrt[4]{81}}$

9. $6\sqrt{7} - \sqrt{7}$

10. $(\sqrt{3} - 1)^2$

11. $x^{2/3} \cdot x^{-4/3}$

12. $(\sqrt[3]{2} - 4)(\sqrt[3]{4} + 2)$

For Exercises 13 and 14, write in exponential form. Assume that variables represent non-negative values.

13. $\sqrt[5]{8x^3}$

14. $\sqrt[3]{(2x + 5)^2}$

For Exercises 15 and 16, rationalize the denominator and simplify. Assume that variables represent positive values.

15. $\dfrac{1}{\sqrt[3]{4}}$

16. $\dfrac{\sqrt{x}}{\sqrt{x} + \sqrt{y}}$

For Exercises 17–19, solve the equation. Identify any extraneous solutions.

17. $\sqrt{3x - 2} = 8$

18. $\sqrt[4]{x + 8} = \sqrt[4]{2x + 1}$

19. $\sqrt{2x + 1} - \sqrt{x + 1} = 2$

For Exercises 20–22, simplify and write the answer in standard form $(a + bi)$.

20. $(2 - i) - (4 + 3i)$

21. $(4 - i)(4 + i)$

22. $\dfrac{2}{4 - 3i}$

23. Write an expression in simplest form for the area of the figure.

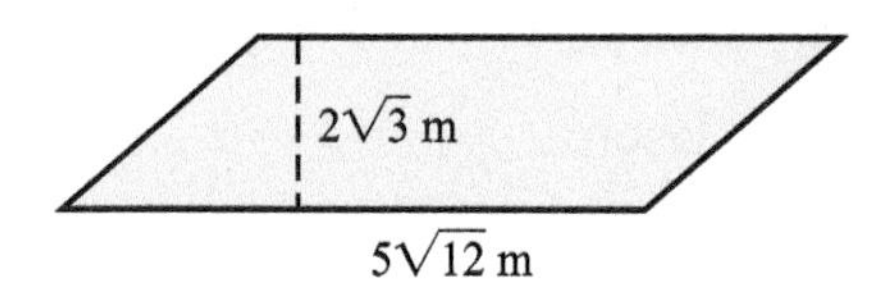

24. The formula $t = \sqrt{\frac{h}{16}}$ describes the amount of time t, in seconds, that an object falls a distance of h feet.

a. Write an expression in simplest form for the exact amount of time an object falls a distance of 12 feet.

b. Find the distance an object falls in 2 seconds.

25. The formula $S = \frac{7}{4}\sqrt{D}$ can be used to approximate the speed, S, in miles per hour, that a car was traveling prior to braking and skidding a distance D, in feet, on ice. (*Source:* Harris Technical Services, Traffic Accident Reconstructionists.)

a. Find the exact speed of a car if the skid length measures 40 feet long.

b. Find the length of the skid marks that a car traveling 30 miles per hour makes if it brakes hard and skids to a halt.

Chapters 1–8 Cumulative Review Exercises

For Exercises 1–4, answer true or false

[2.6] **1.** If $|x| > 3$, then $-3 < x < 3$.

[4.1] **2.** $(2, -3)$ is a solution of $\begin{cases} 4x + 3y = -1 \\ 2x - 5y = -11. \end{cases}$

[8.4] **3.** The conjugate of $4 - 2\sqrt{3}$ is $4 + 2\sqrt{3}$.

[8.7] **4.** 3 is a complex number.

For Exercises 5–7, fill in the blanks.

[3.1] **5.** Given an equation of the form $ax + by = c$, to find the ____________, let $y = 0$.

[4.1] **6.** A system of equations that has no solution is ____________.

[7.2] **7.** The LCD for $\frac{5}{8x^3y^2}$ and $\frac{3}{12x^5y}$ is ____________.

Exercises 8–18 Expressions

[1.3] **8.** Simplify $4 - 6(5 - 3^2) + 12 \div 2 - 6$.

[5.1] **9.** Simplify $\frac{7.44 \times 10^{-2}}{3.1 \times 10^4}$. Write the answer in standard form.

For Exercises 10–17, simplify. Write the answers with positive exponents only.

[5.1] **10.** $\frac{(x^{-2})^3(x^3)^4}{(x^{-3})^3}$

[5.1] **11.** $(3x^2 - 4y)^2$

[5.4] **12.** $\frac{8x^3 - 22x + 8}{2x - 3}$

[7.1] **13.** $\frac{2x^2 + 7x + 3}{x^2 - 9} \div \frac{2x^2 + 11x + 5}{x^2 - 3x}$

[8.4] **14.** $2\sqrt{5a^2} \cdot 3\sqrt{10a^5}$

[8.4] **15.** $3x\sqrt[3]{24x^4} - 4x^2\sqrt[3]{81x}$

[8.2] **16.** $32^{-3/5}$

[8.7] **17.** $(4 + 5i)(2 - 3i)$

[8.5] **18.** Rationalize the denominator of $\frac{2 + \sqrt{2}}{4 - \sqrt{2}}$.

Exercises 19–30 Equations and Inequalities

For Exercises 19–22, solve. Identify any extraneous solutions.

[2.1] **19.** $A = \frac{1}{2}h(B + b)$ for b

[6.4] **20.** $(x + 2)(x + 4) = 63$

[7.4] **21.** $\dfrac{3y}{y-4} - \dfrac{12}{y-4} = 6$

[8.6] **22.** $\sqrt{3x+10} = x + 2$

[3.2] **23.** Find the slope and y-intercept of the graph of $5x - 2y = -15$.

[3.3] **24.** Write the equation of the line containing $(-4, 2)$ that is parallel to the graph of $y = 3x - 5$. Leave the answer in slope–intercept form.

For Exercises 25 and 26, graph.

[3.1] **25.** $y = -\dfrac{5}{3}x + 2$

[4.5] **26.** $\begin{cases} 2x - y > -4 \\ y > -\dfrac{2}{3}x - 2 \end{cases}$

For Exercises 27–30, solve.

[2.2] **27.** Hernando earns $1\dfrac{1}{4}$ times his normal hourly wage for all hours in excess of 40 hours per week. Last week he worked 46 hours and earned $782.80. What is his normal hourly wage?

[7.5] **28.** A river has a current of 5 miles per hour. A boat can make a trip of 20 miles upstream in the same time it can make a trip of 30 miles downstream. Find the rate of the boat in still water.

[7.5] **29.** Suppose y varies jointly as x and the square of z. If $y = 72$ when $x = 4$ and $z = 3$, find y when $x = 3$ and $z = 4$.

[8.6] **30.** The period T, in seconds, of a pendulum whose length is L, in meters, is given by $T = 2\pi\sqrt{\dfrac{L}{9.8}}$. Find the length of a pendulum whose period is 4π seconds.

CHAPTER

9

Quadratic Equations and Functions

Chapter Overview

Previously, we learned to solve quadratic equations by factoring. In this chapter, we learn three additional methods:

- The square root principle.
- Completing the square.
- The quadratic formula.

In addition, we expand our knowledge of graphing quadratic functions, solving inequalities, and working with functions.

9.1 The Square Root Principle and Completing the Square

Objectives

1 Use the square root principle to solve quadratic equations.
2 Solve quadratic equations by completing the square.

Warm-up *For Exercises 1 and 2, simplify.*

[8.3] **1.** $\pm\sqrt{98}$ [8.7] **2.** $\pm\sqrt{-9}$

[6.3] **3.** Write $x^2 - 7x + \frac{49}{4}$ as the square of a binomial.

[2.1] **4.** Solve: $x - \frac{7}{2} = \frac{\sqrt{37}}{2}$

Objective 1 Use the square root principle to solve quadratic equations.

In Section 6.4, we solved quadratic equations such as $x^2 = 25$ by subtracting 25 from both sides, factoring, and then using the zero-factor theorem. Let's recall that process.

$$x^2 - 25 = 0$$

$$(x-5)(x+5) = 0 \quad \text{Factor.}$$

$$x - 5 = 0 \quad \text{or} \quad x + 5 = 0 \quad \text{Use the zero-factor theorem.}$$

$$x = 5 \qquad x = -5 \quad \text{Solve for } x \text{ in each equation.}$$

Another approach to solving $x^2 = 25$ involves square roots. Notice that the solutions to this equation must be numbers that can be squared to equal 25. Those numbers are the square roots of 25, which are 5 and -5. This suggests a new rule called the *square root principle*.

Note The expression $\pm\sqrt{a}$ is read "plus or minus the square root of a."

Rule Square Root Principle
If $x^2 = a$, where a is a real number, then $x = \sqrt{a}$ or $x = -\sqrt{a}$.
It is common to indicate the positive and negative solutions by writing $\pm\sqrt{a}$.

Connection In Section 8.3, we learned how to simplify square roots of numbers that have perfect square factors.

For example, if $x^2 = 64$, then $x = \sqrt{64} = 8$ or $x = -\sqrt{64} = -8$. Or we could simply write $x = \pm\sqrt{64} = \pm 8$.

Solve Equations in the Form $x^2 = a$

The square root principle is especially useful for solving equations in the form $x^2 = a$ when a is not a perfect square.

Example 1 Solve.

a. $x^2 = 40$

Solution: $x^2 = 40$

$$x = \pm\sqrt{40} \quad \text{Use the square root principle.}$$

$$x = \pm\sqrt{4 \cdot 10} \quad \text{Simplify by factoring out a perfect square.}$$

$$x = \pm 2\sqrt{10}$$

Note The $\pm$ symbol means the two solutions are $2\sqrt{10}$ and $-2\sqrt{10}$.

We can check the two solutions using the original equation.

Check: $(2\sqrt{10})^2 \stackrel{?}{=} 40$ Replace x with $2\sqrt{10}$.
$4 \cdot 10 = 40$ True

Check: $(-2\sqrt{10})^2 \stackrel{?}{=} 40$ Replace x with $-2\sqrt{10}$.
$4 \cdot 10 = 40$ True

For the remaining examples, the checks will be left to the reader.

Answers to Warm-up
1. $\pm 7\sqrt{2}$ **2.** $\pm 3i$
3. $\left(x - \frac{7}{2}\right)^2$
4. $x = \frac{7 + \sqrt{37}}{2}$

Note In this chapter, complex number solutions are allowed. If solutions are to be restricted to real numbers only, we say, "Solve over the real numbers."

b. $x^2 = -9$

Solution: $x^2 = -9$

$x = \pm\sqrt{-9}$ Use the square root principle.

$x = \pm 3i$ Write the imaginary number using *i*.

Your Turn 1 Solve.

a. $x^2 = 50$ **b.** $x^2 = -100$

Solve Equations in the Form $ax^2 + b = c$

If an equation is in the form $ax^2 + b = c$, we use the addition and multiplication principles of equality to isolate x^2 and then use the square root principle.

Example 2 Solve.

a. $x^2 + 2 = 100$

Solution: $x^2 + 2 = 100$

$x^2 = 98$ Subtract 2 from both sides to isolate x^2.

$x = \pm\sqrt{98}$ Use the square root principle.

$x = \pm\sqrt{49 \cdot 2}$ Factor out a perfect square.

$x = \pm 7\sqrt{2}$ Simplify using $\sqrt{ab} = \sqrt{a} \cdot \sqrt{b}$.

b. $5x^2 + 2 = 62$

Solution: $5x^2 + 2 = 62$

$5x^2 = 60$ Subtract 2 from both sides.

$x^2 = 12$ Divide both sides by 5.

$x = \pm\sqrt{12}$ Use the square root principle.

$x = \pm\sqrt{4 \cdot 3}$ Factor out a perfect square.

$x = \pm 2\sqrt{3}$ Simplify using $\sqrt{ab} = \sqrt{a} \cdot \sqrt{b}$.

Your Turn 2 Solve.

a. $x^2 - 6 = 42$ **b.** $2x^2 + 11 = 65$

Solve Equations in the Form $(ax + b)^2 = c$

In an equation in the form $(ax + b)^2 = c$, notice that the expression $ax + b$ is squared. We can use the square root principle to eliminate the square by thinking of this form as follows:

$$\text{If } (ax + b)^2 = c, \text{ then } ax + b = \pm\sqrt{c}.$$

Example 3 Solve.

a. $(x + 6)^2 = 49$

Solution: $(x + 6)^2 = 49$

$x + 6 = \pm\sqrt{49}$ Use the square root principle.

$x + 6 = \pm 7$ Simplify.

$x = -6 \pm 7$ Subtract 6 from both sides.

$x = -6 + 7$ or $x = -6 - 7$ Simplify by separating the two solutions.

$x = 1$ $x = -13$

Answers to Your Turn 1
a. $\pm 5\sqrt{2}$ **b.** $\pm 10i$

Answers to Your Turn 2
a. $\pm 4\sqrt{3}$ **b.** $\pm 3\sqrt{3}$

b. $(5x - 1)^2 = 10$

Solution: $(5x - 1)^2 = 10$

$5x - 1 = \pm\sqrt{10}$ Use the square root principle.

$5x = 1 \pm \sqrt{10}$ Add 1 to both sides to isolate 5*x*.

$x = \dfrac{1 \pm \sqrt{10}}{5}$ Divide both sides by 5 to solve for *x*.

c. $(8x + 6)^2 = -32$

Solution: $(8x + 6)^2 = -32$

$8x + 6 = \pm\sqrt{-32}$ Use the square root principle.

$8x = -6 \pm \sqrt{-16 \cdot 2}$ Subtract 6 from both sides and simplify the square root.

$8x = -6 \pm 4i\sqrt{2}$ Rewrite the imaginary number using *i* notation.

$x = \dfrac{-6 \pm 4i\sqrt{2}}{8}$ Divide both sides by 8 to solve for *x*.

$x = -\dfrac{6}{8} \pm \dfrac{4\sqrt{2}}{8}i$ Write the complex number in standard form *a* + *bi*.

$x = -\dfrac{3}{4} \pm \dfrac{\sqrt{2}}{2}i$ Simplify.

Your Turn 3 Solve.

a. $(x - 3)^2 = 25$ **b.** $(2x + 1)^2 = 14$ **c.** $(4x - 5)^2 = -12$

Objective 2 Solve quadratic equations by completing the square.

To make use of the square root principle, we need one side of the equation to be a perfect square and the other side to be a constant, as in $(x + 2)^2 = 5$. But suppose we are given an equation such as $x^2 + 6x = 2$, whose left-hand side is an "incomplete" square. We can use the addition principle of equality to add an appropriate number (9 in this case) to both sides so that the left-hand side becomes a perfect square. We call this process *completing the square.*

$x^2 + 6x + 9 = 2 + 9$ Adding 9 to both sides completes the square on the left side.

We can now factor the left-hand side of the equation, then use the square root principle to finish solving the equation.

$(x + 3)^2 = 11$ Write the left-hand side in factored form.

$x + 3 = \pm\sqrt{11}$ Use the square root principle to eliminate the square.

$x = -3 \pm \sqrt{11}$ Subtract 3 from both sides to isolate *x*.

But how do we determine the number to add to complete the square? That number is the square of half of the coefficient of *x*.

$$x^2 + 6x + 9$$

$$\left(\frac{6}{2}\right)^2 = (3)^2$$

Half of 6 is 3, and 3 squared is 9.

Connection The product of every perfect square in the form $(x + b)^2$ is a trinomial in the form $x^2 + 2bx + b^2$. Notice that half of *x*'s coefficient, 2*b*, is *b*, which is then squared to equal the last term.

Also notice that half of the coefficient of *x* is the constant in the factored form.

$$x^2 + 6x + 9 = (x + 3)^2$$

$$\left(\frac{6}{2}\right)^2 = (3)^2$$

Half of 6 is 3, which is the constant in the factored form.

Answers to Your Turn 3

a. 8, −2

b. $\dfrac{-1 \pm \sqrt{14}}{2}$ **c.** $\dfrac{5}{4} \pm \dfrac{\sqrt{3}}{2}i$

Equations in the Form $x^2 + bx = c$

To solve a quadratic equation by completing the square, we need the equation in the form $x^2 + bx = c$ so that the coefficient of x^2 is 1. Once the equation is in the form $x^2 + bx = c$, we complete the square and then use the square root principle.

Example 4 Solve by completing the square.

a. $x^2 + 12x + 15 = 0$

Solution: We first write the equation in the form $x^2 + bx = c$.

$x^2 + 12x = -15$ Subtract 15 from both sides to get the form $x^2 + bx = c$.

$x^2 + 12x + 36 = -15 + 36$ Complete the square by adding 36 to both sides.

$(x + 6)^2 = 21$ Factor the left side and simplify the right.

$x + 6 = \pm\sqrt{21}$ Use the square root principle.

$x = -6 \pm \sqrt{21}$ Subtract 6 from both sides to isolate x.

Note We found 36 by squaring half of 12.

$$\left(\frac{12}{2}\right)^2 = 6^2 = 36$$

b. $x^2 - 7x + 8 = 5$

Solution: $x^2 - 7x = -3$ Subtract 8 from both sides to get the form $x^2 + bx = c$.

$x^2 - 7x + \frac{49}{4} = -3 + \frac{49}{4}$ Complete the square by adding $\frac{49}{4}$ to both sides.

$\left(x - \frac{7}{2}\right)^2 = \frac{37}{4}$ Factor the left side and simplify the right side.

$x - \frac{7}{2} = \pm\sqrt{\frac{37}{4}}$ Use the square root principle.

$x = \frac{7}{2} \pm \sqrt{\frac{37}{4}}$ Add $\frac{7}{2}$ to both sides to isolate x.

$x = \frac{7}{2} \pm \frac{\sqrt{37}}{2}$ Simplify the square root.

$x = \frac{7 \pm \sqrt{37}}{2}$ Combine the fractions.

Note We found $\frac{49}{4}$ by squaring half of -7.

$$\left(\frac{-7}{2}\right)^2 = \frac{49}{4}$$

Your Turn 4 Solve by completing the square.

a. $x^2 + 8x - 29 = 0$ **b.** $x^2 - 9x - 6 = 5$

Equations in the Form $ax^2 + bx = c$, Where $a \neq 1$

Note Our procedure for completing the square does not work unless the coefficient of the squared term is 1.

So far, our equations have been in the form $x^2 + bx = c$, where the coefficient of the x^2 term is already 1. To solve an equation such as $2x^2 + 12x = 3$, we need to divide both sides of the equation by 2 $\left(\text{or multiply both sides by } \frac{1}{2}\right)$ so that the coefficient of x^2 becomes 1.

$\frac{2x^2 + 12x}{2} = \frac{3}{2}$ Divide both sides by 2.

$x^2 + 6x = \frac{3}{2}$ Simplify.

We can now solve by completing the square.

$x^2 + 6x + 9 = \frac{3}{2} + 9$ Add 9 to both sides to complete the square.

$(x + 3)^2 = \frac{21}{2}$ Factor the left side and simplify the right side.

Answers to Your Turn 4

a. $-4 \pm 3\sqrt{5}$ **b.** $\frac{9 \pm 5\sqrt{5}}{2}$

$$x + 3 = \pm\sqrt{\frac{21}{2}}$$ Use the square root principle.

$$x = -3 \pm \sqrt{\frac{21}{2}}$$ Subtract 3 from both sides to isolate x.

$$x = -3 \pm \frac{\sqrt{21}}{\sqrt{2}} \cdot \frac{\sqrt{2}}{\sqrt{2}}$$ Rationalize the denominator.

$$x = -3 \pm \frac{\sqrt{42}}{2}$$

Note The answer $-3 \pm \frac{\sqrt{42}}{2}$ can also be written as $\frac{-6 \pm \sqrt{42}}{2}$.

We can now write a procedure for solving any quadratic equation by completing the square.

Connection In Section 9.4, we use the process of completing the square to rewrite quadratic functions in a form that allows us to easily determine features of the graph.

Procedure Solving Quadratic Equations by Completing the Square

To solve a quadratic equation by completing the square:

1. Write the equation in the form $x^2 + bx = c$.
2. Complete the square by adding $\left(\frac{b}{2}\right)^2$ to both sides.
3. Write the completed square in factored form and simplify the right side, $c + \left(\frac{b}{2}\right)^2$.
4. Use the square root principle to eliminate the square.
5. Isolate the variable.
6. Simplify as needed.

Example 5 Solve by completing the square.

a. $2x^2 - 8 = 5x$

Solution: $2x^2 - 5x = 8$ Rewrite in the form $ax^2 + bx = c$.

$$\frac{2x^2 - 5x}{2} = \frac{8}{2}$$ Divide both sides by 2.

$$x^2 - \frac{5}{2}x = 4$$ Simplify.

$$x^2 - \frac{5}{2}x + \frac{25}{16} = 4 + \frac{25}{16}$$ Add $\frac{25}{16}$ to both sides to complete the square.

Note To complete the square, square half of $-\frac{5}{2}$.

$$\left(\frac{1}{2} \cdot -\frac{5}{2}\right)^2 = \left(-\frac{5}{4}\right)^2 = \frac{25}{16}$$

$$\left(x - \frac{5}{4}\right)^2 = \frac{89}{16}$$ Factor the left side and simplify the right side.

$$x - \frac{5}{4} = \pm\sqrt{\frac{89}{16}}$$ Use the square root principle.

$$x = \frac{5}{4} \pm \frac{\sqrt{89}}{4}$$ Add $\frac{5}{4}$ to both sides and simplify the square root.

$$x = \frac{5 \pm \sqrt{89}}{4}$$ Combine like fractions.

b. $4x^2 - 8x + 7 = 2$

Solution:

$4x^2 - 8x = -5$ Rewrite in the form $ax^2 + bx = c$.

$\frac{4x^2 - 8x}{4} = -\frac{5}{4}$ Divide both sides by 4.

$x^2 - 2x = -\frac{5}{4}$ Simplify.

$x^2 - 2x + 1 = -\frac{5}{4} + 1$ Add 1 to both sides to complete the square.

$(x - 1)^2 = -\frac{1}{4}$ Factor the left side and simplify the right side.

$x - 1 = \pm\sqrt{-\frac{1}{4}}$ Use the square root principle.

$x = 1 \pm \frac{1}{2}i$ Add 1 to both sides and simplify the square root.

Answers to Your Turn 5

a. $\frac{-3 \pm 2\sqrt{2}}{3}$ or $-1 \pm \frac{2\sqrt{2}}{3}$

b. $4 \pm 3i$

Your Turn 5 Solve by completing the square.

a. $9x^2 + 18x = -1$ **b.** $x^2 - 8x + 11 = -14$

9.1 Exercises

MyMathLab®

Note: Exercises marked with a ★ represent challenging exercises.

Objective 1

Prep Exercise 1 Explain why there are two solutions to an equation in the form $x^2 = a$.

For Exercises 1–10, solve and check. See Example 1.

1. $x^2 = 49$ **2.** $x^2 = 144$ **3.** $y^2 = \frac{4}{25}$ **4.** $t^2 = \frac{9}{49}$ **5.** $n^2 = 1.44$

6. $p^2 = 1.21$ **7.** $z^2 = 45$ **8.** $m^2 = 72$ **9.** $w^2 = -25$ **10.** $c^2 = -49$

Prep Exercise 2 Write a formula for the solutions of $ax^2 - b = c$ by solving for x.

For Exercises 11–30, solve and check. Begin by using the addition or multiplication principles of equality to isolate the squared term. See Example 2.

11. $n^2 - 7 = 42$ **12.** $y^2 - 5 = 59$ **13.** $y^2 - 16 = 65$ **14.** $k^2 + 5 = 30$

15. $4n^2 = 36$ **16.** $5y^2 = 125$ **17.** $25t^2 = 9$ **18.** $16d^2 = 49$

19. $4h^2 = -16$ **20.** $3k^2 = -108$ **21.** $\frac{5}{6}x^2 = \frac{24}{5}$ **22.** $-\frac{2}{3}m^2 = -\frac{27}{32}$

23. $2x^2 + 5 = 21$

24. $4x^2 - 11 = 97$

25. $5y^2 - 7 = -97$

26. $4n^2 + 20 = -76$

27. $\frac{3}{4}y^2 - 5 = 3$

28. $\frac{25}{9}m^2 - 2 = 6$

29. $0.2t^2 - 0.5 = 0.012$

30. $0.5p^2 + 1.28 = 1.6$

Prep Exercise 3 Write a formula for the solutions of $(ax - b)^2 = c$ by solving for x.

For Exercises 31–46, solve and check. Use the square root principle to eliminate the square. See Example 3.

31. $(x + 8)^2 = 49$

32. $(y + 3)^2 = 36$

33. $(5n - 3)^2 = 16$

34. $(6h - 5)^2 = 81$

35. $(m - 8)^2 = -16$

36. $(t - 2)^2 = -4$

37. $(4k - 1)^2 = 40$

38. $(3x - 4)^2 = 80$

39. $(m - 7)^2 = -12$

40. $(t - 5)^2 = -28$

41. $\left(y - \frac{3}{4}\right)^2 = \frac{9}{16}$

42. $\left(x + \frac{4}{9}\right)^2 = \frac{25}{81}$

43. $\left(\frac{5}{9}d - \frac{1}{2}\right)^2 = \frac{1}{36}$

44. $\left(\frac{3}{4}h + \frac{4}{5}\right)^2 = \frac{1}{100}$

45. $(0.4x + 3.8)^2 = 2.56$

46. $(0.8n - 6.8)^2 = 1.96$

Find the Mistake *For Exercises 47–50, explain the mistake; then find the correct solutions.*

47.
$$x^2 - 15 = 34$$
$$x^2 = 49$$
$$x = \sqrt{49}$$
$$x = 7$$

48.
$$x^2 = 20$$
$$x = \sqrt{20}$$
$$x = 2\sqrt{5}$$

49.
$$(x - 5)^2 = -6$$
$$x - 5 = \pm\sqrt{6}$$
$$x = 5 \pm \sqrt{6}$$

50.
$$(x - 1)^2 = -12$$
$$x - 1 = \pm\sqrt{-12}$$
$$x = 1 \pm 2\sqrt{3}$$

For Exercises 51–56, solve; then use a calculator to approximate the irrational solutions rounded to three places.

51. $x^2 = 96$

52. $t^2 = 56$

53. $y^2 - 15 = 5$

54. $x^2 - 13 = 35$

55. $(n - 6)^2 = 15$

56. $(m + 3)^2 = 10$

Objective 2

Prep Exercise 4 Given an expression of the form $x^2 + bx$, what must be added to complete the square?

Prep Exercise 5 If $x^2 + bx + c$ is a perfect square trinomial, what is its factored form?

For Exercises 57–64: **a.** *Add a term to the expression to make it a perfect square.*
b. *Factor the perfect square. See Objective 2.*

57. $x^2 + 14x$

58. $c^2 + 8c$

59. $n^2 - 10n$

60. $a^2 - 12a$

61. $y^2 - 9y$

62. $m^2 - 11m$

63. $s^2 - \frac{2}{3}s$

64. $y^2 - \frac{4}{5}y$

Prep Exercise 6 Given an equation in the form $x^2 + bx = c$, explain how to complete the square.

Prep Exercise 7 Consider the equation $x^2 - 7x + 12 = 0$. Which is a better method for solving the equation, factoring and then using the zero-factor theorem or completing the square? Explain.

Prep Exercise 8 Consider the equation $x^2 + 4x + 5 = 0$. Which is a better method for solving the equation, factoring and then using the zero-factor theorem or completing the square? Explain.

For Exercises 65–76, solve the equation by completing the square. See Example 4.

65. $w^2 + 2w = 15$

66. $p^2 + 8p = 9$

67. $r^2 - 2r + 50 = 0$

68. $c^2 - 6c + 45 = 0$

69. $k^2 = 9k - 18$

70. $a^2 = 6a - 8$

71. $b^2 - 2b - 11 = 5$

72. $x^2 + 2x + 7 = 9$

73. $h^2 - 6h + 3 = -26$

74. $j^2 - 4j + 25 = -3$

75. $u^2 + \frac{1}{2}u = \frac{3}{2}$

76. $y^2 + \frac{1}{3}y = \frac{2}{3}$

For Exercises 77–88, solve the equation by completing the square. Begin by writing the equation in the form $x^2 + bx = c$. See Example 5.

77. $4x^2 + 16x = 9$

78. $4m^2 - 8m = 5$

79. $9x^2 - 18x + 5 = 0$

80. $16x^2 - 32x + 7 = 0$

81. $2n^2 - n - 3 = 0$

82. $2x^2 - 5x - 12 = 0$

83. $4x^2 = -16x + 7$

84. $4x^2 = 24x + 11$

85. $5k^2 + k - 2 = 0$

86. $3s^2 - 4s = 2$

87. $3a^2 - 8 = -4a$

88. $3x^2 - 4 = -8x$

Find the Mistake *For Exercises 89 and 90, explain the mistake. Find the correct solution.*

89.

$$\begin{aligned} 3x^2 + 4x &= 2 \\ 3x^2 + 4x + 4 &= 2 + 4 \\ (3x + 2)^2 &= 6 \\ 3x + 2 &= \pm\sqrt{6} \\ 3x &= -2 \pm \sqrt{6} \\ x &= \frac{-2 \pm \sqrt{6}}{3} \end{aligned}$$

90.

$$\begin{aligned} x^2 + 6x &= 7 \\ x^2 + 6x + 9 &= 7 + 9 \\ (x + 3)^2 &= 16 \\ x + 3 &= \sqrt{16} \\ x + 3 &= 4 \\ x &= -3 + 4 \\ x &= 1 \end{aligned}$$

For Exercises 91–102, solve.

91. A square sheet of metal has an area of 196 square inches. What is the length of each side?

92. A severe thunderstorm warning is issued by the National Weather Service for a square area covering 14,400 square miles. What is the length of each side of the square area?

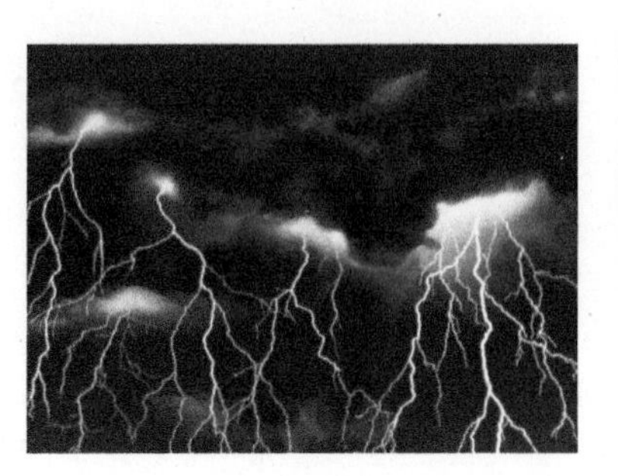

93. A tank is to be made for an aquarium so that the length is 8 feet, the width is twice the height, and the volume is 144 cubic feet. Find the height and width of the tank.

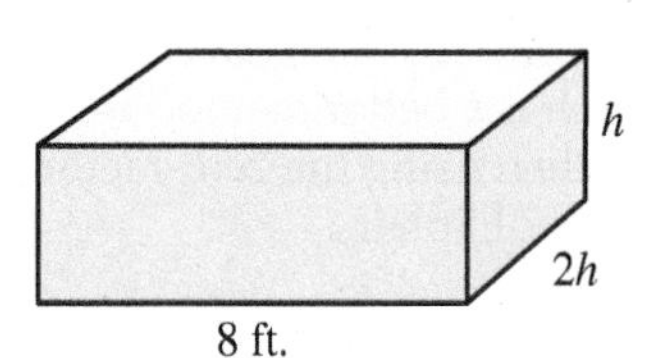

94. The length of a swimming pool is to be three times the width. If the depth is to be a constant 4 feet and the volume is to be 4800 cubic feet, find the length and width.

95. The Arecibo radio telescope is like a giant satellite dish that covers a circular area of approximately $23{,}256.25\pi$ square meters. Find the diameter of the dish.

Of Interest

The Arecibo radio telescope, the world's largest radio telescope, analyzes radiation and signals emitted by objects in space. It was built in a natural depression near Arecibo, Puerto Rico.

96. A field is planted in a circular pattern. A watering device is to be constructed with pipe in a line extending from the center of the field to the edge of the field. The pipe is set on wheels so that it can rotate around the field and cover the entire field with water. If the area of the field is 7225π square feet, how long will the watering pipe be?

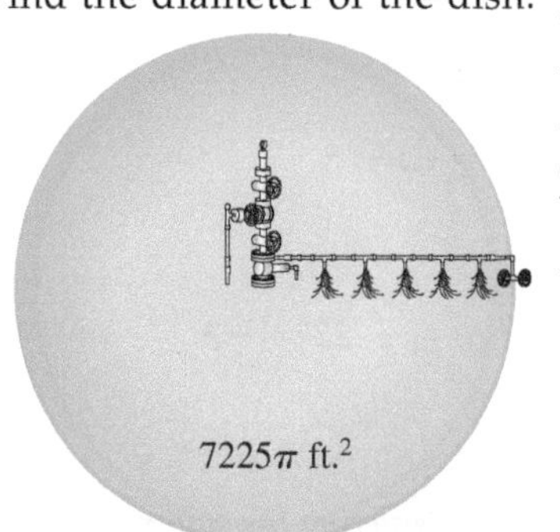

97. The corner of a sheet of plywood has been cut out as shown. If the area of the piece that was removed was 256 square inches, find x.

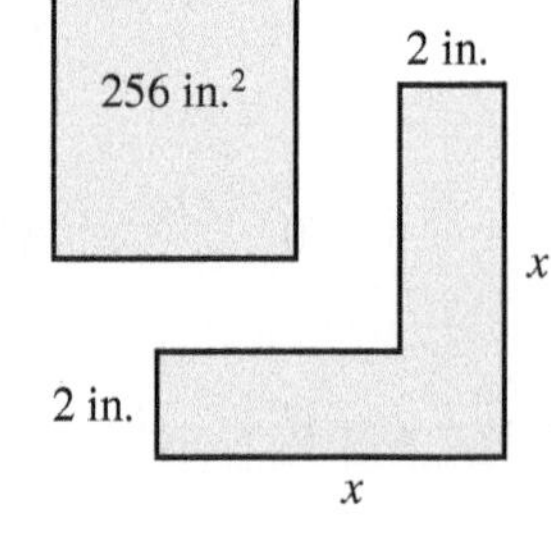

98. The area of the hole in the washer shown is 25π square millimeters. Find r, the radius of the washer.

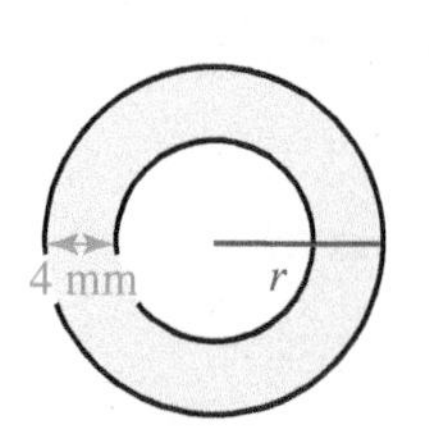

99. An LCD computer monitor measures 20 inches across its diagonal. If the width is 4 inches less than the length, find the dimensions of the monitor.

100. Two identical right triangles are placed together to form the frame of a roof. If the base of each triangle is 7 feet longer than its height and its hypotenuse is 13 feet, what are the dimensions of the base and height?

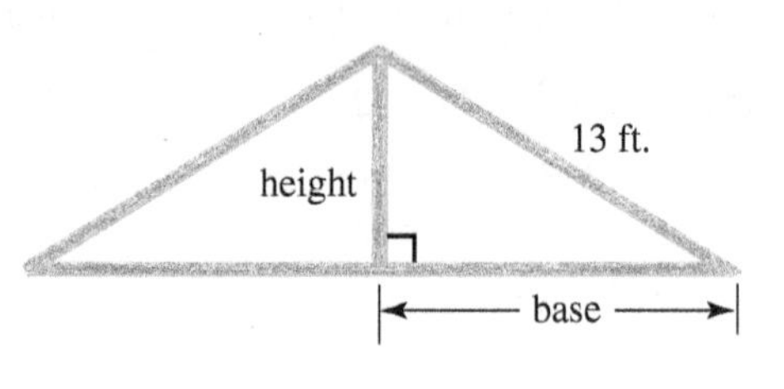

★ **101.** A plastic panel is to have a rectangular hole cut as shown.

a. Find l so that the area remaining after the hole is cut is 1230 square centimeters.

b. Find the length and width of the plastic panel.

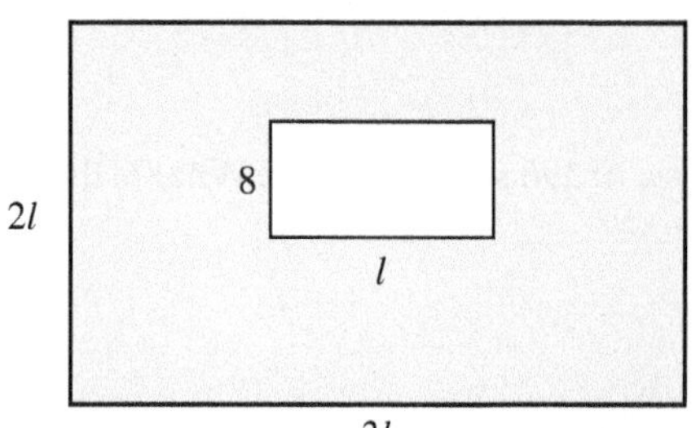

★ **102.** A 6-inch-wide groove with a height of h is to be cut into a wood block as shown.

a. Find h so that the volume remaining in the block after the groove is cut is 360 cubic inches.

b. Find the height and width of the block.

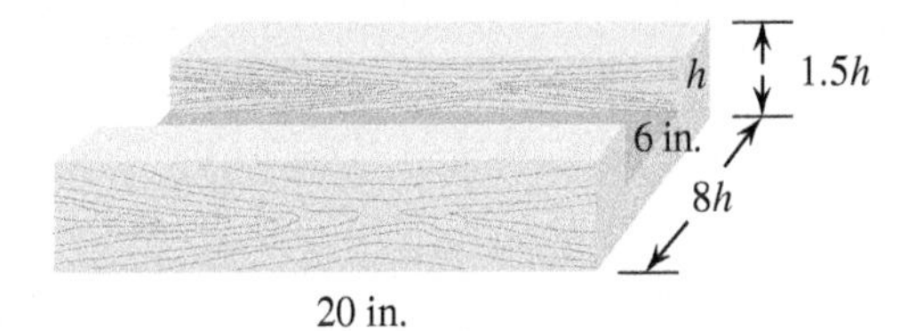

For Exercises 103 and 104, if an object is dropped, the formula $d = 16t^2$ describes the distance d in feet that the object falls in t seconds.

103. Suppose a cover for a ceiling light falls from a 9-foot ceiling. How long does the cover take to hit the floor?

104. A construction worker tosses a scrap piece of lumber from the roof of a house. How long does it take the piece of lumber to reach the ground 25 feet below?

For Exercises 105 and 106, use the following information. In physics, when an object is in motion, it has kinetic energy. The formula $E = \frac{1}{2}mv^2$ is used to calculate the kinetic energy E of an object with a mass m and velocity v. If the mass is measured in kilograms and the velocity is in meters per second, the kinetic energy will be in units called joules (J).

105. Suppose an object with a mass of 50 kilograms has 400 joules of kinetic energy. Find its velocity.

106. In a crash test, a vehicle with a mass of 1200 kilograms is found to have kinetic energy of 117,600 joules just before impact. Find the velocity of the vehicle just before impact.

Puzzle Problem Without using a calculator, which of the following four numbers is a perfect square? (*Hint:* Write a list of smaller perfect squares and look for a pattern.)

9,456,804,219,745,618
2,512,339,789,576,516
7,602,985,471,286,543
4,682,715,204,643,182

Review Exercises

Exercises 1–6 Expressions

[6.3] *For Exercises 1 and 2, factor.*

1. $x^2 - 10x + 25$

2. $x^2 + 6x + 9$

[8.1] *For Exercises 3 and 4, simplify. Assume that variables represent nonnegative values.*

3. $\sqrt{36x^2}$

4. $\sqrt{(2x + 3)^2}$

[1.3, 8.1, 8.7] *For Exercises 5 and 6, simplify.*

5. $\dfrac{-4 + \sqrt{8^2 - 4(2)(5)}}{2(2)}$

6. $\dfrac{-6 - \sqrt{6^2 - 4(5)(2)}}{2(5)}$

9.2 Solving Quadratic Equations Using the Quadratic Formula

Objectives

1 Solve quadratic equations using the quadratic formula.
2 Use the discriminant to determine the number of real solutions that a quadratic equation has.
3 Find the x- and y-intercepts of a quadratic function.
4 Solve applications using the quadratic formula.

Warm-up

[9.1] 1. Solve by completing the square: $3x^2 + 4x - 2 = 0$

[1.4] 2. Evaluate $b^2 - 4ac$ for $a = 0.5$, $b = -0.8$, and $c = 2.5$.

[1.3, 8.3] 3. Simplify: $\dfrac{-8 \pm \sqrt{8^2 - 4(1)(-29)}}{2(1)}$

In this section, we solve quadratic equations using a formula called the *quadratic formula.* Using this formula is much easier than completing the square.

Objective 1 Solve quadratic equations using the quadratic formula.

To derive the quadratic formula, we begin with the general form of the quadratic equation, $ax^2 + bx + c = 0$, and assume that $a > 0$. We follow the procedure for solving a quadratic equation by completing the square.

$$ax^2 + bx + c = 0$$

$$ax^2 + bx = -c \quad \text{Subtract } c \text{ from both sides.}$$

$$x^2 + \frac{b}{a}x = -\frac{c}{a} \quad \text{Divide both sides by } a \text{ so that the coefficient of } x^2 \text{ is 1.}$$

Note To complete the square, square half of $\frac{b}{a}$.

$$\left(\frac{1}{2}\cdot\frac{b}{a}\right)^2 = \left(\frac{b}{2a}\right)^2 = \frac{b^2}{4a^2}$$

$$x^2 + \frac{b}{a}x + \frac{b^2}{4a^2} = -\frac{c}{a} + \frac{b^2}{4a^2} \quad \text{Complete the square.}$$

$$\left(x + \frac{b}{2a}\right)^2 = -\frac{4ac}{4a^2} + \frac{b^2}{4a^2}$$

Factor on the left side. On the right side, write $-\frac{c}{a}$ with the LCD $4a^2$ to add the rational expressions.

$$\left(x + \frac{b}{2a}\right)^2 = \frac{b^2 - 4ac}{4a^2}$$

Add the rational expressions. We rearranged b^2 and $-4ac$ in the numerator.

$$x + \frac{b}{2a} = \pm\sqrt{\frac{b^2 - 4ac}{4a^2}} \quad \text{Use the square root principle to eliminate the square.}$$

$$x + \frac{b}{2a} = \pm\frac{\sqrt{b^2 - 4ac}}{2a} \quad \text{Simplify the square root in the denominator.}$$

$$x = -\frac{b}{2a} \pm \frac{\sqrt{b^2 - 4ac}}{2a} \quad \text{Subtract } \frac{b}{2a} \text{ from both sides to isolate } x.$$

Note The result is the same if $a < 0$.

$$x = \frac{-b \pm \sqrt{b^2 - 4ac}}{2a} \quad \text{Combine the rational expressions.}$$

This final equation is the *quadratic formula.* It can be used to solve any quadratic equation simply by replacing a, b, and c with the corresponding values from the given equation.

Note A quadratic equation must be in the form $ax^2 + bx + c = 0$ so that a, b, and c can be identified for use in the quadratic formula.

Procedure Using the Quadratic Formula

To solve a quadratic equation in the form $ax^2 + bx + c = 0$, where $a \neq 0$, use the quadratic formula:

$$x = \frac{-b \pm \sqrt{b^2 - 4ac}}{2a}$$

Answers to Warm-up

1. $\dfrac{-2 \pm \sqrt{10}}{3}$
2. -4.36 3. $-4 \pm 3\sqrt{5}$

Example 1 Solve.

a. $3x^2 + 10x - 8 = 0$

Solution: This equation is in the form $ax^2 + bx + c = 0$, where $a = 3$, $b = 10$, and $c = -8$. So we can use the quadratic formula, $x = \frac{-b \pm \sqrt{b^2 - 4ac}}{2a}$.

$$x = \frac{-10 \pm \sqrt{10^2 - 4(3)(-8)}}{2(3)}$$ Replace a with 3, b with 10, and c with -8.

$$x = \frac{-10 \pm \sqrt{196}}{6}$$ Simplify in the radical and the denominator.

$$x = \frac{-10 \pm 14}{6}$$ Evaluate $\sqrt{196}$.

$$x = \frac{-10 + 14}{6} \quad \text{or} \quad x = \frac{-10 - 14}{6}$$ Split up the $\pm$ to calculate the two solutions.

$$x = \frac{4}{6} \qquad x = \frac{-24}{6}$$

$$x = \frac{2}{3} \qquad x = -4$$ Simplify to lowest terms.

Connection Because the radicand 196 is a perfect square, the two solutions are rational numbers. As we consider more examples in this section, note how the radicand determines the type of solutions.

b. $x^2 + 8x - 15 = 14$

Solution: $x^2 + 8x - 29 = 0$ Subtract 14 from both sides to get the form $ax^2 + bx + c = 0$.

$$x = \frac{-8 \pm \sqrt{8^2 - 4(1)(-29)}}{2(1)}$$ Use the quadratic formula, replacing a with 1, b with 8, and c with -29.

$$x = \frac{-8 \pm \sqrt{180}}{2}$$ Simplify in the radical and the denominator.

$$x = \frac{-8 \pm \sqrt{36 \cdot 5}}{2}$$ Use the product rule of radicals.

$$x = \frac{-8 \pm 6\sqrt{5}}{2}$$ Simplify the radical.

$$x = \frac{-8}{2} \pm \frac{6\sqrt{5}}{2}$$ Separate into two rational expressions to simplify.

$$x = -4 \pm 3\sqrt{5}$$ Simplify by dividing out the 2.

Connection Because the radicand 180 is not a perfect square, the two solutions are irrational numbers.

Warning Always make sure the radical is simplified *before* trying to simplify the fraction.

c. $9x^2 + 4 = -12x$

Solution: $9x^2 + 12x + 4 = 0$ Add $12x$ to both sides to get the form $ax^2 + bx + c = 0$.

$$x = \frac{-12 \pm \sqrt{12^2 - 4(9)(4)}}{2(9)}$$ Use the quadratic formula, replacing a with 9, b with 12, and c with 4.

$$x = \frac{-12 \pm \sqrt{0}}{18}$$ Simplify in the radical and the denominator.

$$x = -\frac{2}{3}$$ Simplify the radical and simplify the fraction to lowest terms.

Connection Because the radicand is 0, this equation has only one solution.

d. $3x^2 + 9 = -8x + 2$

Solution: $3x^2 + 8x + 7 = 0$ Add $8x$ to and subtract 2 from both sides to get the form $ax^2 + bx + c = 0$.

$$x = \frac{-8 \pm \sqrt{8^2 - 4(3)(7)}}{2(3)}$$ Use the quadratic formula, replacing a with 3, b with 8, and c with 7.

$$x = \frac{-8 \pm \sqrt{-20}}{6}$$ Simplify in the radical and the denominator.

Connection The negative radicand causes the two solutions to be nonreal complex numbers.

Note In standard form, the two complex solutions are $-\frac{4}{3} + \frac{\sqrt{5}}{3}i$ and $-\frac{4}{3} - \frac{\sqrt{5}}{3}i$.

$$x = \frac{-8 \pm 2i\sqrt{5}}{6}$$ Simplify the radical.

$$x = \frac{-4 \pm i\sqrt{5}}{3}$$ Simplify to lowest terms.

Your Turn 1 Solve using the quadratic formula.

a. $4x^2 - 5x - 6 = 0$
b. $4x^2 = 15 - 2x$
c. $x^2 - 8x + 16 = 0$
d. $x^2 + 8 = 2 - 4x$

Choosing a Method for Solving Quadratic Equations

We have learned several methods for solving quadratic equations. The following table summarizes the methods and conditions that make each method the best choice.

Methods for Solving Quadratic Equations

Method	When the Method Is Beneficial
1. Factoring (Section 6.4)	Use when the quadratic equation can be easily factored.
2. Square root principle (Section 9.1)	Use when the quadratic equation can be easily written in the form $ax^2 = c$ or $(ax + b)^2 = c$.
3. Completing the square (Section 9.1)	Keep in mind that this is rarely the best method, but it is important for future topics.
4. Quadratic formula (Section 9.2)	Use when factoring is not easy or is not possible with integer coefficients.

Objective 2 Use the discriminant to determine the number of real solutions that a quadratic equation has.

In the Connection boxes for Example 1, we pointed out how the radicand in the quadratic formula affects the solutions to a given quadratic equation. The expression $b^2 - 4ac$, which is the radicand, is called the **discriminant**.

Definition **Discriminant:** The discriminant is the radicand, $b^2 - 4ac$, in the quadratic formula.

We use the discriminant to determine the number and type of solutions to a quadratic equation.

Procedure **Using the Discriminant**

Given a quadratic equation in the form $ax^2 + bx + c = 0$, where a, b, and c are rational numbers and $a \neq 0$, to determine the number and type of solutions, evaluate the discriminant $b^2 - 4ac$.

If the **discriminant is positive**, the equation has two real-number solutions. The solutions will be rational if the discriminant is a perfect square and irrational otherwise.

If the **discriminant is 0**, the equation has one rational solution.

If the **discriminant is negative**, the equation has two nonreal complex solutions.

Answers to Your Turn 1

a. $2, -\frac{3}{4}$ b. $\frac{-1 \pm \sqrt{61}}{4}$
c. 4 d. $-2 \pm i\sqrt{2}$

Note When the discriminant is 0, the solution is $\frac{-b \pm \sqrt{0}}{2a} = -\frac{b}{2a}$.

Example 2 Use the discriminant to determine the number and type of solutions.

a. $3x^2 - 7x = 8$

Solution: $3x^2 - 7x - 8 = 0$ Write the equation in the form $ax^2 + bx + c = 0$.

$(-7)^2 - 4(3)(-8)$ In $b^2 - 4ac$, replace a with 3, b with -7, and c with -8.

Note Follow the order of operations when evaluating $b^2 - 4ac$ and note that $b^2 \geq 0$ if b is a real number.

$= 49 + 96$

$= 145$

Warning 145 is the *value of the discriminant*, not a solution for the equation $3x^2 - 7x = 8$.

Because the discriminant is positive, this equation has two real-number solutions. Because 145 is not a perfect square, the solutions are irrational.

b. $x^2 = \frac{4}{5}x - \frac{4}{25}$

Solution: $x^2 - \frac{4}{5}x + \frac{4}{25} = 0$ Write the equation in the form $ax^2 + bx + c = 0$.

$25 \cdot x^2 - \frac{25}{1} \cdot \frac{4}{5}x + \frac{25}{1} \cdot \frac{4}{25} = 25 \cdot 0$ Eliminate the fractions by multiplying both sides of the equation by the LCD, 25. This will make the discriminant easier to evaluate.

Note Multiplying both sides of a quadratic equation by a constant does not change the solutions, but does change the value of the discriminant.

$25x^2 - 20x + 4 = 0$

$(-20)^2 - 4(25)(4)$ In $b^2 - 4ac$, replace a with 25, b with -20, and c with 4.

$= 400 - 400$

$= 0$

Note Because the discriminant is 0, the solution is

$$-\frac{b}{2a} = -\frac{(-20)}{2(25)} = \frac{20}{50} = \frac{2}{5},$$

which is a rational number.

Because the discriminant is zero, this equation has only one real solution.

c. $0.5x^2 - 0.8x + 2.5 = 0$

Note We could avoid calculations with decimal numbers by multiplying the original equation through by 10 so that it becomes $5x^2 - 8x + 25 = 0$.

Solution: $(-0.8)^2 - 4(0.5)(2.5)$ In $b^2 - 4ac$, replace a with 0.5, b with -0.8, and c with 2.5.

$= 0.64 - 5$

$= -4.36$

Because the discriminant is negative, there are two nonreal complex solutions.

Your Turn 2 Use the discriminant to determine the number and type of solutions for the equation. If the solutions are real, state whether they are rational or irrational.

a. $5x^2 + 8x - 9 = 0$ **b.** $\frac{5}{2}x^2 + \frac{5}{6} = \frac{4}{3}x$ **c.** $x^2 - 0.12x = -0.0036$

Objective 3 Find the x- and y-intercepts of a quadratic function.

In Chapter 3, we learned that x-intercepts are points where a graph intersects the x-axis. Because an x-intercept is on the x-axis, the y-coordinate of the point is 0. In Sections 5.2 and 6.4, we graphed quadratic functions, which have the form $f(x) = ax^2 + bx + c$ (or $y = ax^2 + bx + c$), and learned that their graphs are parabolas. Notice that when we replace y with 0 to find the x-intercepts in $y = ax^2 + bx + c$, we have the quadratic equation $ax^2 + bx + c = 0$, which has two, one, or no real-number solutions. As a result, quadratic functions in the form $y = ax^2 + bx + c$ have two, one, or no x-intercepts. The following graphs illustrate the possibilities.

Answers to Your Turn 2

a. discriminant $= 244$; two irrational solutions

b. discriminant $= -236$; two nonreal complex solutions

c. discriminant $= 0$; one rational solution

Two x-intercepts: If $0 = ax^2 + bx + c$ has two real-number solutions, which occurs if $b^2 - 4ac > 0$, then the graph of $y = ax^2 + bx + c$ has two x-intercepts and looks like one of the following graphs.

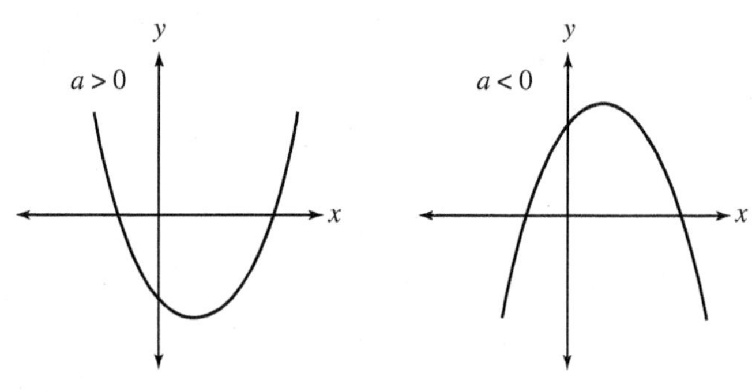

Note In Section 6.4, we noted that if $a > 0$, the parabola opens up and if $a < 0$, the parabola opens down.

One x-intercept: If $0 = ax^2 + bx + c$ has one real-number solution, which occurs if $b^2 - 4ac = 0$, then the graph of $y = ax^2 + bx + c$ has one x-intercept and looks like one of the following graphs.

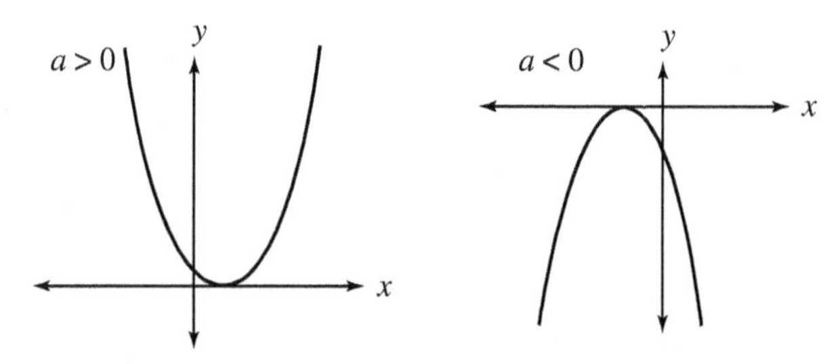

No x-intercepts: If $0 = ax^2 + bx + c$ has no real-number solution, which occurs if $b^2 - 4ac < 0$, then the graph of $y = ax^2 + bx + c$ has no x-intercepts and looks like one of the following graphs.

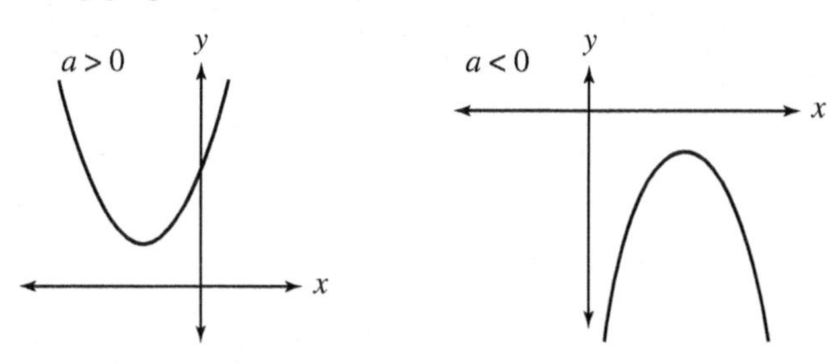

We also have learned that a y-intercept is where a graph intersects the y-axis. To find y-intercepts, we let $x = 0$ and solve for y. Notice that when we replace x with 0 in $y = ax^2 + bx + c$, we have $y = a(0)^2 + b(0) + c = c$; so the y-intercept is $(0, c)$.

Example 3 Find the x- and y-intercepts of $y = x^2 - 2x - 8$; then graph.

Solution: Find the x-intercepts:

$0 = x^2 - 2x - 8$ Replace y with 0.

$0 = (x - 4)(x + 2)$ Factor.

$x - 4 = 0$ or $x + 2 = 0$

$x = 4$ $x = -2$

x-intercepts: $(4, 0)$ and $(-2, 0)$

Connection We could have solved $0 = x^2 - 2x - 8$ using the quadratic formula.

$$x = \frac{-(-2) \pm \sqrt{(-2)^2 - 4(1)(-8)}}{2(1)}$$

$$= \frac{2 \pm \sqrt{36}}{2} = \frac{2 \pm 6}{2}$$

$$= 1 \pm 3$$

$$= 4 \text{ or } -2$$

y-intercept: $(0, -8)$ The y-intercept is $(0, c)$, and c is -8 in this equation.

Note We also could have calculated the y-intercept:

$$y = (0)^2 + 2(0) - 8 = -8$$

Because $a = 1$, which is positive, the graph opens upward. Now we can get a rough sketch of the graph.

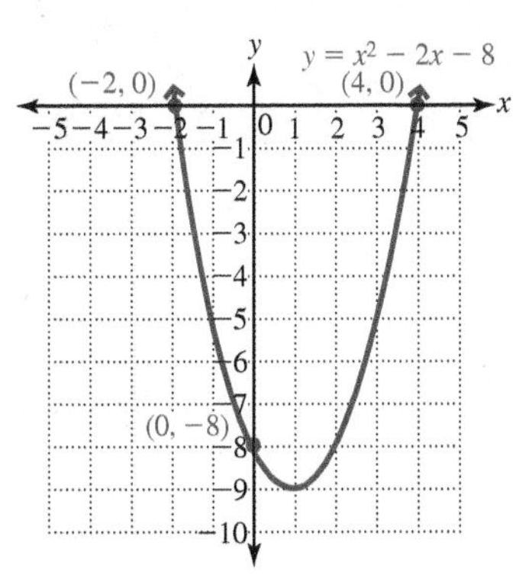

Note Although knowing the intercepts and knowing whether the graph opens up or down is helpful, knowing the exact location of the vertex would improve accuracy. We learn how to find the coordinates of the vertex in Section 9.4.

Your Turn 3 Find the x- and y-intercepts. Verify on a graphing calculator.

a. $y = 3x^2 - 5x + 1$ **b.** $y = x^2 - 8x + 16$ **c.** $y = 2x^2 - 3x + 5$

Objective 4 Solve applications using the quadratic formula.

In physics, the general formula for describing the height of an object after it has been thrown upward is $h = \frac{1}{2}gt^2 + v_0t + h_0$, where g represents the acceleration due to gravity, t is the time in flight, v_0 is the initial velocity, and h_0 is the initial height. For Earth, the acceleration due to gravity is -32.2 ft./sec.^{2} or -9.8 m/sec.^{2}

Example 4 In an extreme games competition, a motorcyclist jumps with an initial velocity of 70 feet per second from a ramp height of 25 feet, landing on a ramp with a height of 15 feet. Find the time the motorcyclist is in the air.

Solution:

$15 = \frac{1}{2}(-32.2)t^2 + 70t + 25$ In $h = \frac{1}{2}gt^2 + v_0t + h_0$, replace with $h = 15$, $v_0 = 70$, and $h_0 = 25$. Because the units are in feet, we use -32.2 ft./sec.^{2} for g.

$15 = -16.1t^2 + 70t + 25$ Multiply.

$0 = -16.1t^2 + 70t + 10$ Subtract 15 from both sides to get the form $ax^2 + bx + c = 0$.

$x = \frac{-70 \pm \sqrt{70^2 - 4(-16.1)(10)}}{2(-16.1)}$ Use the quadratic formula, replacing a with -16.1, b with 70, and c with 10.

$x = \frac{-70 \pm \sqrt{5544}}{-32.2}$ Simplify in the radical and the denominator.

$x \approx -0.138$ or 4.486 Approximate the two irrational solutions.

Answer: Because the time cannot be negative, the motorcycle is in the air approximately 4.486 seconds.

Your Turn 4 A ball is thrown from an initial height of 1.5 meters with an initial velocity of 8 meters per second. Find the time for the ball to land on the ground ($h = 0$). Approximate the time to the nearest thousandth.

Answers to Your Turn 3

a. x-intercepts: $\left(\frac{5 + \sqrt{13}}{6}, 0\right), \left(\frac{5 - \sqrt{13}}{6}, 0\right)$
y-intercept: $(0, 1)$

b. x-intercept: $(4, 0)$
y-intercept: $(0, 16)$

c. no x-intercepts
y-intercept: $(0, 5)$

Answer to Your Turn 4
1.802 sec.

9.2 Exercises

For Extra Help MyMathLab®

Note: Exercises marked with a ★ represent challenging exercises.

Objective 1

Prep Exercise 1 Write the quadratic formula.

Prep Exercise 2 Discuss the advantages and disadvantages of using the quadratic formula to solve quadratic equations.

Prep Exercise 3 Are there quadratic equations that cannot be solved using factoring? Explain.

Prep Exercise 4 Under what conditions will a quadratic equation have only one solution? Modify the quadratic formula to describe this single solution.

For Exercises 1–8, rewrite each quadratic equation in the form $ax^2 + bx + c = 0$; then identify a, b, and c. See Objective 1.

1. $x^2 - 3x + 7 = 0$

2. $x^2 + 6x - 15 = 0$

3. $3x^2 - 9x = 4$

4. $2x^2 - 8x = -11$

5. $x = 1.5x^2 + 0.2$

6. $x = 0.8x^2 + 4.5$

7. $\frac{3}{4}x = -\frac{1}{2}x^2 + 6$

8. $\frac{1}{4}x = \frac{5}{6}x^2 + 6$

For Exercises 9–32, solve using the quadratic formula. See Example 1.

9. $x^2 + 9x + 20 = 0$

10. $x^2 + 8x + 15 = 0$

11. $4x^2 + 5x = 6$

12. $2x^2 - x = 3$

13. $x^2 - 9x = 0$

14. $x^2 - 6x = 0$

15. $x^2 - 8x = -16$

16. $x^2 + 14x = -49$

17. $3x^2 + 4x = 4$

18. $5x^2 - 13x = 6$

19. $x^2 + 2 = 2x$

20. $x^2 + 5 = 4x$

21. $x^2 - x - 1 = 0$

22. $x^2 + 3x - 5 = 0$

23. $3x^2 + 10x + 5 = 0$

24. $3x^2 - x - 3 = 0$

25. $-4x^2 = 5 - 6x$

26. $-5x^2 = 3 - 4x$

27. $18x^2 + 2 = -15x$

28. $10x^2 - 12 = -7x$

29. $3x^2 - 4x = -3$

30. $3x^2 - 6x = -8$

31. $6x^2 - 4 = 3x$

32. $6x^2 - 6 = 13x$

For Exercises 33–44, solve using the quadratic formula. If the solutions are irrational, give an exact answer and an approximate answer rounded to the nearest thousandth. (Hint: You might first clear the fractions or decimals by multiplying both sides by an appropriately chosen number.)

33. $2x^2 + 0.1x = 0.03$

34. $4x^2 + 8.6x - 2.4 = 0$

35. $x^2 + \frac{1}{2}x - 3 = 0$

36. $x^2 + \frac{1}{3}x - \frac{2}{3} = 0$

37. $x^2 - \frac{49}{36} = 0$

38. $x^2 - \frac{64}{25} = 0$

39. $\frac{1}{2}x^2 + \frac{3}{2} = x$

40. $\frac{1}{3}x^2 + 2 = \frac{2}{3}x$

41. $x^2 - 0.5 = -0.06x$

42. $0.5x^2 - 0.4 = 0.25x$

43. $1.2x^2 - 0.6x = -0.5$

44. $2.4x^2 + 4.5 = 6.3x$

Find the Mistake *For Exercises 45–48, explain the mistake; then solve correctly.*

45. Solve $3x^2 - 7x + 1 = 0$ using the quadratic formula.

$$\frac{-7 \pm \sqrt{(-7)^2 - (4)(3)(1)}}{2(3)} = \frac{-7 \pm \sqrt{49 - 12}}{6} = \frac{-7 \pm \sqrt{37}}{6}$$

46. Solve $2x^2 - 6x - 5 = 0$ using the quadratic formula.

$$\frac{6 \pm \sqrt{(-6)^2 - (4)(2)(5)}}{2(2)} = \frac{6 \pm \sqrt{36 - 40}}{4} = \frac{3}{2} \pm \frac{1}{2}i$$

47. Solve $x^2 - 2x + 3 = 0$ using the quadratic formula.

$$\frac{-(-2) \pm \sqrt{(-2)^2 - (4)(1)(3)}}{2(1)} = \frac{2 \pm \sqrt{4 - 12}}{2} = \frac{2 \pm \sqrt{-8}}{2}$$

48. Solve $x^2 - 8 = 0$ using the quadratic formula.

$$\frac{-(-8) \pm \sqrt{(-8)^2 - (4)(1)(0)}}{2(1)} = \frac{8 \pm \sqrt{64 - 0}}{2} = \frac{8 \pm \sqrt{64}}{2} = \frac{8 \pm 8}{2} = 8, 0$$

Objective 2

Prep Exercise 5 What part of the quadratic formula is the discriminant? Write the formula for the discriminant.

For Exercises 49–58, use the discriminant to determine the number and type of solutions for the equation. If the solution(s) are real, state whether they are rational or irrational. See Example 2.

49. $x^2 + 10x = -25$

50. $2x^2 - 8x + 8 = 0$

51. $\frac{1}{4}x^2 - 4x = -4$

52. $\frac{1}{2}x^2 - 5 = -3x$

53. $3x^2 + 10x - 8 = 0$

54. $2x^2 - 3x - 2 = 0$

55. $x^2 - x + 3 = 0$

56. $x^2 + 4x = -5$

57. $x^2 - 6x + 6 = 0$

58. $3x^2 = 13x - 8$

For Exercises 59–68, indicate which of the following methods is the best choice for solving the given equation: factoring, using the square root principle, or using the quadratic formula. Then solve the equation.

59. $x^2 - 81 = 0$

60. $x^2 - 64 = 0$

61. $2x^2 - 4x - 3 = 0$

62. $3x^2 - 6x - 4 = 0$

63. $x^2 + 6x = 0$

64. $x^2 + 14x + 45 = 0$

65. $(x + 7)^2 = 40$

66. $(x - 3)^2 = 48$

67. $x^2 = 8x - 19$

68. $x^2 - 4x = -10$

Objective 3

Prep Exercise 6 If x-intercepts for a quadratic function are imaginary numbers, what does that indicate about its graph?

For Exercises 69–76, find the x- and y-intercepts. See Example 3.

69. $y = x^2 - x - 2$

70. $y = x^2 - 3x + 2$

71. $y = 4x^2 - 12x + 9$

72. $y = 9x^2 + 6x + 1$

73. $y = 2x^2 + 15x - 8$

74. $y = -15x^2 + x + 6$

75. $y = -2x^2 + 3x - 6$

76. $y = -3x^2 - 2x - 5$

Objective 4

For Exercises 77–84, translate to a quadratic equation; then solve using the quadratic formula.

77. A positive integer squared plus five times its consecutive integer is equal to 71. Find the integers.

78. The square of a positive integer minus twice its consecutive integer is equal to 22. Find the integers.

79. A right triangle has side lengths that are three consecutive integers. Use the Pythagorean theorem to find the lengths of those sides. (Remember that the hypotenuse in a right triangle is always the longest side.)

80. A right triangle has side lengths that are consecutive even integers. Use the Pythagorean theorem to find the lengths of those sides. (Remember that the hypotenuse in a right triangle is always the longest side.)

81. The length of a rectangular fence gate is 3.5 feet less than three times the width. Find the length and width of the gate if the area is 34 square feet.

82. A small access door for an attic storage area is designed so that the width is 9.5 feet less than two times the length. Find the length and width if the area is 52 square feet.

83. An architect is experimenting with two different shapes of a room as shown.
    a. Find x so that the rooms have the same area.
    b. Complete the dimensions for the L-shaped room.

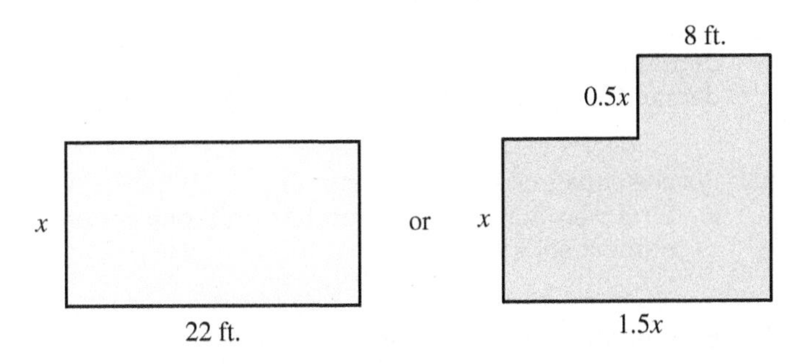

84. A cylinder is to be made so that its volume is equal to that of a sphere with a radius of 9 inches. If a cylinder is to have a height of 4 inches, find its radius.

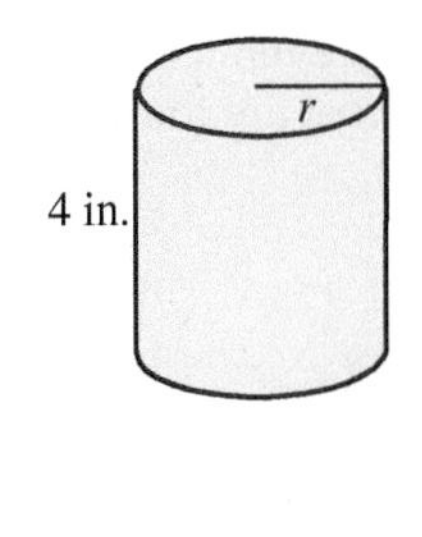

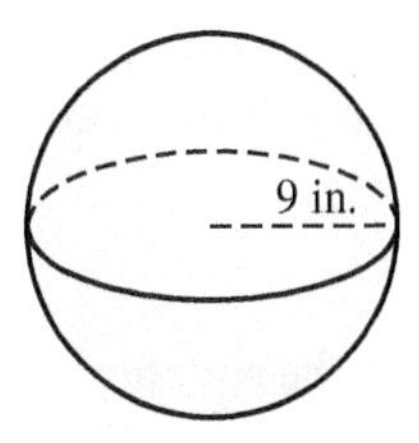

For Exercises 85–88, use the formula $h = \frac{1}{2}gt^2 + v_0t + h_0$, where g represents the acceleration due to gravity, t is the time in flight, v_0 is the initial velocity, and h_0 is the initial height. For Earth, the acceleration due to gravity is -32.2 ft./sec.^{2} or -9.8 m/sec.^{2} Approximate irrational answers to the nearest hundredth. See Example 4.

85. Robbie Knievel jumped the Grand Canyon on a motorcycle. Suppose that on takeoff, his motorcycle had a vertical velocity of 48 feet per second and at the end of the launch ramp his altitude was 485 feet above sea level. If he landed on a ramp that was 460 feet above sea level, how long was he in flight?

86. In the extreme games competition, Brian Deegan successfully performed the first 360-degree flip ever attempted in a competition. Suppose his motorcycle had a vertical velocity of 12.8 meters per second from the end of a ramp that was 8 meters high and he landed on a ramp at a point 4 meters above the ground. How long was he in the air?

87. A platform diver dives from a platform that is 10 meters above the water.
    a. Write the equation that describes her height during the dive. (Assume that her initial velocity is 0.)
    b. Find the time it takes the diver to be 5 meters above the water.
    c. Find the time it takes the diver to enter the water.

88. In a cliff diving championship in Acapulco, a diver dives from a cliff at a height of 70 feet.
    a. Write the equation that describes his height during the dive. (Assume that his initial velocity is 0.)
    b. Find the time it takes the diver to be 50 feet above the water.
    c. Find the time it takes the diver to enter the water.

89. The expression $0.5n^2 + 2.5n$ describes the gross income from the sale of a particular software product, where n is the number of units sold in thousands. The expression $4.5n + 16$ describes the cost of producing the n units. Find the number of units that must be produced and sold for the company to break even. (To *break even* means that the gross income and cost are the same.)

★ **90.** An economist and a marketing manager discover that the expression $2n^2 + 5n$ models the price of a CD based on its demand, where n is the number of units (in millions) the market demands. The expression $-0.5n^2 + 17.1$ describes the price of the CD based on the number of units (also in millions) supplied to the market. Find the number of units that must be demanded and supplied so that the price based on demand is equal to the price based on supply.

★ **91.** For the equation $2x^2 - 5x + c = 0$:

a. Find c so that the equation has only one rational number solution.

b. Find the range of values of c for which the equation has two real-number solutions.

c. Find the range of values of c for which the equation has no real-number solution.

★ **92.** For the equation $4x^2 + 6x + c = 0$:

a. Find c so that the equation has only one rational number solution.

b. Find the range of values of c for which the equation has two real-number solutions.

c. Find the range of values of c for which the equation has no real-number solution.

★ **93.** For the equation $ax^2 + 12x + 8 = 0$:

a. Find a so that the equation has only one rational number solution.

b. Find the range of values of a for which the equation has two real-number solutions.

c. Find the range of values of a for which the equation has no real-number solution.

★ **94.** For the equation $3x^2 + bx + 8 = 0$:

a. Find b so that the equation has only one rational number solution.

b. Find all positive values of b for which the equation has two real-number solutions.

c. Find all positive values of b for which the equation has no real-number solution.

Puzzle Problem In the equation shown, A, B, C, D, and E are five consecutive positive integers where $A < B < C < D < E$. What are the integers?

$$A^2 + B^2 + C^2 = D^2 + E^2$$

Review Exercises

Exercises 1–4 Expressions

[6.2] *For Exercises 1 and 2, factor.*

1. $u^2 - 9u + 14$

2. $3u^2 - 2u - 16$

For Exercises 3 and 4, simplify.

[5.1] **3.** $(x^2)^2$

[8.2] **4.** $(x^{1/3})^2$

Exercises 5 and 6 Equations and Inequalities

For Exercises 5 and 6, solve.

[6.4] **5.** $5u^2 + 13u = 6$

[7.4] **6.** $\frac{7}{3u} = \frac{5}{u} - \frac{1}{u - 5}$

9.3 Solving Equations That Are Quadratic in Form

Objectives

1 Solve equations by rewriting them in quadratic form.
2 Solve equations that are quadratic in form by using substitution.
3 Solve application problems using equations that are quadratic in form.

Warm-up

For Exercises 1–4, solve.

[9.2] **1.** $2x^2 + 5 = 6x$

[7.4] **2.** $\frac{200}{x} = \frac{200}{x+1} + 10$

[8.6] **3.** $\sqrt[3]{x} = -2$

[8.6] **4.** $x - \sqrt{x} - 2 = 0$

Many equations that are not quadratic equations are **quadratic in form** and can be solved using the same methods used to solve quadratic equations.

Definition An equation is **quadratic in form** if it can be rewritten as a quadratic equation $au^2 + bu + c = 0$, where $a \neq 0$ and u is a variable or an expression.

Objective 1 Solve equations by rewriting them in quadratic form.

Equations with Rational Expressions

In Section 7.4, we solved equations containing rational expressions by multiplying both sides of the equation by the LCD. Those rewritten equations are often quadratic. Remember that equations containing rational expressions sometimes have extraneous solutions.

Connection Recall that an apparent solution for an equation with rational expressions is extraneous if it causes one or more of the denominators to equal 0.

Example 1 Solve $\frac{3}{x+1} = 1 - \frac{3}{x(x+1)}$.

Solution:

$x(x+1) \cdot \frac{3}{x+1} = x(x+1)\left(1 - \frac{3}{x(x+1)}\right)$ — Multiply both sides by the LCD $x(x+1)$.

$x(x+1) \cdot \frac{3}{x+1} = x(x+1) \cdot 1 - x(x+1) \cdot \frac{3}{x(x+1)}$ — Distribute $x(x+1)$.

$3x = x^2 + x - 3$ — Simplify both sides.

$0 = x^2 - 2x - 3$ — Subtract $3x$ from both sides to get the quadratic form $ax^2 + bx + c = 0$.

$0 = (x-3)(x+1)$ — Factor.

$x - 3 = 0$ or $x + 1 = 0$ — Use the zero-factor theorem.

$x = 3$ $\quad x = -1$ — Solve each equation.

Checks: $x = 3$

$\frac{3}{3+1} \stackrel{?}{=} 1 - \frac{3}{3(3+1)}$

$\frac{3}{4} \stackrel{?}{=} 1 - \frac{3}{12}$

$\frac{3}{4} = \frac{3}{4}$ True

$x = -1$

$\frac{3}{-1+1} \stackrel{?}{=} 1 - \frac{3}{-1(-1+1)}$

$\frac{3}{0} \stackrel{?}{=} 1 - \frac{3}{0}$

Note The expression $\frac{3}{0}$ is undefined, so -1 is extraneous.

The solution is 3. (-1 is extraneous.)

Your Turn 1 Solve $\frac{x}{x-2} = \frac{6}{x} + \frac{4}{x(x-2)}$.

Answer to Your Turn 1
4 (2 is extraneous.)

Answers to Warm-up

1. $\frac{3}{2} \pm \frac{1}{2}i$
2. $x = -5, x = 4$
3. $x = -8$
4. $x = 4$ (1 is extraneous.)

Equations Containing Radicals

In Section 8.6, we found that after the power rule was used on equations containing radicals, the result was often a quadratic equation. Remember that these radical equations sometimes have extraneous solutions.

Connection Recall that an apparent solution for an equation containing radicals is extraneous if it makes the original equation false.

Example 2 Solve.

a. $\sqrt{x} + x = 6$

Solution: $\sqrt{x} = 6 - x$ — Subtract x from both sides to isolate the radical.

$(\sqrt{x})^2 = (6 - x)^2$ — Square both sides.

$x = 36 - 12x + x^2$ — Simplify both sides.

$0 = x^2 - 13x + 36$ — Subtract x from both sides to write in quadratic form.

$0 = (x - 4)(x - 9)$ — Factor.

$x - 4 = 0$ or $x - 9 = 0$ — Use the zero-factor theorem.

$x = 4 \qquad x = 9$ — Solve each equation.

Checks: $x = 4$ $\qquad$ $x = 9$

$\sqrt{4} + 4 \stackrel{?}{=} 6 \qquad \sqrt{9} + 9 \stackrel{?}{=} 6$

$2 + 4 = 6$ True $\qquad 3 + 9 = 6$ False; so 9 is extraneous.

The solution is 4. (9 is extraneous.)

b. $\sqrt{x - 1} = 2x - 1$

Solution: $(\sqrt{x - 1})^2 = (2x - 1)^2$ — Square both sides of the equation.

$x - 1 = 4x^2 - 4x + 1$ — Simplify both sides.

$0 = 4x^2 - 5x + 2$ — Subtract x from and add 1 to both sides to write in quadratic form.

We cannot factor $4x^2 - 5x + 2$, so we use the quadratic formula.

Note Written in standard form, these solutions are $\frac{5}{8} + \frac{\sqrt{7}}{8}i$ and $\frac{5}{8} - \frac{\sqrt{7}}{8}i$.

$$x = \frac{-(-5) \pm \sqrt{(-5)^2 - 4(4)(2)}}{2(4)} = \frac{5 \pm \sqrt{-7}}{8} = \frac{5 \pm i\sqrt{7}}{8}$$

Check: Because the solutions are complex, we will not check them.

Your Turn 2 Solve.

a. $x - 2\sqrt{x} - 3 = 0$ $\qquad$ **b.** $\sqrt{6x - 11} = 3x - 1$

Objective 2 Solve equations that are quadratic in form by using substitution.

Recall that an equation that is quadratic in form can be written as $au^2 + bu + c = 0$, where $a \neq 0$ and u is an *expression*. We now explore a method for solving these equations where we substitute u for an expression. To use substitution, it is important to note a pattern with the exponents of the terms in a quadratic equation. Notice that the degree of the first term, au^2, is 2; the degree of the middle term, bu, is 1; and the third term is a constant. If a trinomial has one term with an expression raised to the second power, a second term with that same expression raised to the first power, and a third term that is a constant, the equation is quadratic in form and we can use substitution to solve it.

Consider the equation $x^4 - 13x^2 + 36 = 0$. Notice that we can rewrite the equation as $(x^2)^2 - 13(x^2) + 36 = 0$, so it is quadratic in form. By substituting u for each x^2, we have a "friendlier" form of quadratic equation, which we can then solve by factoring, completing the square, or using the quadratic formula.

Answers to Your Turn 2

a. 9 (1 is extraneous.)

b. $\frac{2 \pm 2i\sqrt{2}}{3}$

Note Given an equation containing a trinomial, if the degree of the first term is twice the degree of the middle term and the third term is a constant, try using substitution.

$$(x^2)^2 - 13(x^2) + 36 = 0$$

$$u^2 - 13u + 36 = 0 \quad \text{Substitute } u \text{ for } x^2.$$

$$(u - 9)(u - 4) = 0 \quad \text{Factor.}$$

$$u - 9 = 0 \quad \text{or} \quad u - 4 = 0 \quad \text{Use the zero-factor theorem.}$$

$$u = 9 \qquad u = 4 \quad \text{Solve each equation.}$$

Note that these solutions are for u, not x. We must substitute x^2 back in place of u to finish solving for x.

$$x^2 = 9 \qquad x^2 = 4 \quad \text{Substitute } x^2 \text{ for } u.$$

$$x = \pm\sqrt{9} \qquad x = \pm\sqrt{4} \quad \text{Use the square root principle.}$$

$$x = \pm 3 \qquad x = \pm 2 \quad \text{Simplify.}$$

Note The equation $x^4 - 13x^2 + 36 = 0$ has four solutions: 3, −3, 2, and −2.

Our example suggests the following procedure.

Procedure Using Substitution to Solve Equations That Are Quadratic in Form

To solve equations that are quadratic in form using substitution:

1. Rewrite the equation so that it is in the form $au^2 + bu + c = 0$.
2. Solve the quadratic equation for u.
3. Substitute for u and solve.
4. Check the solutions.

Example 3 Solve.

a. $(x + 2)^2 - 2(x + 2) - 8 = 0$

Solution: If we substitute u for $x + 2$, we have an equation in the form $au^2 + bu + c = 0$.

$$(x + 2)^2 - 2(x + 2) - 8 = 0$$

$$u^2 - 2u - 8 = 0 \quad \text{Substitute } u \text{ for } x + 2.$$

$$(u - 4)(u + 2) = 0 \quad \text{Factor.}$$

$$u - 4 = 0 \quad \text{or} \quad u + 2 = 0 \quad \text{Use the zero-factor theorem.}$$

$$u = 4 \qquad u = -2 \quad \text{Solve each equation for } u.$$

$$x + 2 = 4 \qquad x + 2 = -2 \quad \text{Substitute } x + 2 \text{ for } u.$$

$$x = 2 \qquad x = -4 \quad \text{Solve each equation for } x.$$

Check: Verify that 2 and −4 make $(x + 2)^2 - 2(x + 2) - 8 = 0$ true. We will leave this check to the reader.

b. $x^{2/3} - x^{1/3} - 6 = 0$

Solution: Because $x^{2/3} - x^{1/3} - 6 = 0$ can be written as $(x^{1/3})^2 - x^{1/3} - 6 = 0$, it is quadratic in form. We substitute u for $x^{1/3}$.

$$(x^{1/3})^2 - x^{1/3} - 6 = 0 \quad \text{Rewrite in quadratic form.}$$

$$u^2 - u - 6 = 0 \quad \text{Substitute } u \text{ for } x^{1/3}.$$

$$(u - 3)(u + 2) = 0 \quad \text{Factor.}$$

$$u - 3 = 0 \quad \text{or} \quad u + 2 = 0 \quad \text{Use the zero-factor theorem.}$$

$$u = 3 \qquad u = -2 \quad \text{Solve each equation for } u.$$

$$x^{1/3} = 3 \qquad x^{1/3} = -2 \quad \text{Substitute } x^{1/3} \text{ for } u.$$

$$(x^{1/3})^3 = 3^3 \qquad (x^{1/3})^3 = (-2)^3 \quad \text{Cube both sides of the equations.}$$

$$x = 27 \qquad x = -8 \quad \text{Simplify.}$$

Checks: $x = 27$

$$27^{2/3} - 27^{1/3} - 6 \stackrel{?}{=} 0$$
$$(\sqrt[3]{27})^2 - \sqrt[3]{27} - 6 \stackrel{?}{=} 0$$
$$3^2 - 3 - 6 \stackrel{?}{=} 0$$
$$9 - 3 - 6 \stackrel{?}{=} 0$$
$$0 = 0 \quad \text{True}$$

$x = -8$

$$(-8)^{2/3} - (-8)^{1/3} - 6 \stackrel{?}{=} 0$$
$$(\sqrt[3]{-8})^2 - \sqrt[3]{-8} - 6 \stackrel{?}{=} 0$$
$$(-2)^2 - (-2) - 6 \stackrel{?}{=} 0$$
$$4 + 2 - 6 \stackrel{?}{=} 0$$
$$0 = 0 \quad \text{True}$$

Your Turn 3 Solve.

a. $x^4 - 10x^2 + 9 = 0$ **b.** $(n - 2)^2 + 4(n - 2) - 12 = 0$ **c.** $x^{2/3} - x^{1/3} - 2 = 0$

Objective 3 Solve application problems using equations that are quadratic in form.

Example 4 Solve.

a. The average speed of a car is 10 miles per hour more than the average speed of a bus. The bus takes 1 hour longer than the car to travel 200 miles. Find how long it takes the car to travel 200 miles.

Understand: Both the car and bus travel 200 miles, but it takes the bus 1 hour longer than it takes the car. The rate of the car is 10 miles per hour more than the rate of the bus.

Plan: We use a table to organize the information, then write an equation, which we can solve.

Execute: We let x represent the time for the car to travel 200 miles. Because the bus travels 1 more hour, $x + 1$ describes the time for the bus to travel 200 miles.

Vehicle	d	t	r
bus	200	$x + 1$	$\frac{200}{x+1}$
car	200	x	$\frac{200}{x}$

Note We use $r = \frac{d}{t}$ to describe each rate.

Because the rate of the car is 10 miles per hour faster than the rate of the bus, we can say that (the rate of the car) = (the rate of the bus) + 10.

$$\frac{200}{x} = \frac{200}{x+1} + 10$$

$$x(x+1)\left(\frac{200}{x}\right) = x(x+1)\left(\frac{200}{x+1} + 10\right) \quad \text{Multiply both sides by the LCD } x(x+1).$$

$$(x+1)(200) = x(x+1)\left(\frac{200}{x+1}\right) + x(x+1)(10) \quad \text{Multiply on the left and distribute on the right.}$$

$$200x + 200 = 200x + 10x^2 + 10x \quad \text{Simplify.}$$

$$0 = 10x^2 + 10x - 200 \quad \text{Subtract } 200x \text{ and } 200 \text{ from both sides to get quadratic form.}$$

$$0 = x^2 + x - 20 \quad \text{Divide both sides by 10.}$$

$$0 = (x+5)(x-4) \quad \text{Factor.}$$

$$x + 5 = 0 \quad \text{or} \quad x - 4 = 0 \quad \text{Use the zero-factor theorem.}$$

$$x = -5 \qquad x = 4 \quad \text{Solve each equation.}$$

Answer: Because time cannot be negative, it takes the car 4 hours to travel 200 miles.

Answers to Your Turn 3
a. $\pm 3, \pm 1$ **b.** ± 4 **c.** $8, -1$

Check: The rate of the car is $\frac{200}{x} = \frac{200}{4} = 50$ miles per hour; so in 4 hours, the car travels $50(4) = 200$ miles. The bus travels 200 miles in $4 + 1 = 5$ hours at a rate of $\frac{200}{4 + 1} = \frac{200}{5} = 40$ miles per hour; so in 5 hours, the bus travels $5(40) = 200$ miles. Notice also that the car's rate is 10 miles per hour faster than the bus's.

b. Bobby and Pam manufacture and install blinds. Working together, they can install blinds in every window of an average-sized house in $1\frac{1}{3}$ hours. Working alone, Bobby takes 2 hours longer than Pam to do the same installation. Working alone, how long does each person take to install blinds in an average-sized house?

Understand: We are given the time it takes Bobby and Pam working together and are asked to find how long it takes each person individually. We also know that it takes Bobby 2 hours longer than it takes Pam.

Plan: We use a table to organize the information, write an equation, and then solve.

Execute: We let x represent the number of hours it takes Pam to install blinds working alone. Because Bobby takes 2 more hours, his time working alone is represented by $x + 2$.

Worker	Time to Complete the Job Alone	Rate of Work	Time at Work	Part of Job Completed
Pam	x	$\frac{1}{x}$	$\frac{4}{3}$	$\frac{4}{3x}$
Bobby	$x + 2$	$\frac{1}{x + 2}$	$\frac{4}{3}$	$\frac{4}{3(x + 2)}$

Note Multiplying the rate of work and the time at work gives an expression of the part of the job completed.

The total job in this case is 1 average-sized house, so we can write an equation that combines Pam's and Bobby's individual expressions for work completed and set this sum equal to 1.

$$(\text{Part Pam does}) + (\text{Part Bobby does}) = 1 \text{ (the entire job)}$$

$$\frac{4}{3x} + \frac{4}{3(x + 2)} = 1$$

$$3x(x + 2)\left(\frac{4}{3x} + \frac{4}{3(x + 2)}\right) = 3x(x + 2)(1)$$ Multiply both sides by the LCD $3x(x + 2)$.

$$3x(x + 2)\left(\frac{4}{3x}\right) + 3x(x + 2)\left(\frac{4}{3(x + 2)}\right) = 3x^2 + 6x$$ Distribute on the left and multiply on the right.

$$4(x + 2) + 4x = 3x^2 + 6x$$ Continue simplifying.

$$4x + 8 + 4x = 3x^2 + 6x$$ Distribute.

$$8x + 8 = 3x^2 + 6x$$ Combine like terms.

$$0 = 3x^2 - 2x - 8$$ Subtract $8x$ and 8 from both sides.

$$0 = (3x + 4)(x - 2)$$ Factor.

$$3x + 4 = 0 \quad \text{or} \quad x - 2 = 0$$ Use the zero-factor theorem.

$$3x = -4 \qquad\qquad x = 2$$ Solve each equation.

$$x = -\frac{4}{3}$$

Answer: Because negative time makes no sense in the context of this problem, it takes Pam 2 hours to install blinds working alone and Bobby $x + 2 = 2 + 2 = 4$ hours working alone.

Check: Because Pam can install blinds in 2 hours, she can install them in $\frac{1}{2}$ of an average-sized house in 1 hour. Because Bobby takes 4 hours to do the same work, he can install blinds in $\frac{1}{4}$ of an average-sized house in 1 hour. Together, they can install blinds in $\frac{1}{2} + \frac{1}{4} = \frac{2}{4} + \frac{1}{4} = \frac{3}{4}$ of an average-sized house in 1 hour; so in $1\frac{1}{3}$ hours, they can install blinds in $1\frac{1}{3} \cdot \frac{3}{4} = \frac{4}{3} \cdot \frac{3}{4} = 1$ average-sized house.

Your Turn 4 Solve.

a. The average speed of a passenger train is 25 miles per hour more than the average speed of a car. The time required for the car to travel 300 miles is 2 hours more than the time required for the train to travel the same number of miles. Find the average speed of the car.

b. Terri and Tommy run a flea market. It takes Terri 2 hours longer to put out the merchandise than it does Tommy. If they can put out the merchandise together in $1\frac{7}{8}$ hours, how long does it take each working alone?

Answers to Your Turn 4

a. 50 mph

b. It takes Tommy 3 hr. and Terri 5 hr.

9.3 Exercises

For Extra Help MyMathLab®

Note: Exercises marked with a ★ represent challenging exercises.

Objective 1

Prep Exercise 1 What does it mean to say that an equation is quadratic in form?

Prep Exercise 2 Is $x^{3/4} - x^{1/4} - 6 = 0$ quadratic in form? Why or why not?

For Exercises 1–12, solve the equations with rational expressions. Identify any extraneous solutions. See Example 1.

1. $\frac{60}{x+2} = \frac{60}{x} - 5$

2. $\frac{120}{x+2} = \frac{120}{x} - 5$

3. $\frac{1}{x} + \frac{1}{x+2} = \frac{3}{4}$

4. $\frac{1}{x} + \frac{3}{x-4} = \frac{7}{8}$

5. $\frac{1}{p-4} + \frac{1}{4} = \frac{8}{p^2-16}$

6. $\frac{1}{y-5} - \frac{10}{y^2-25} = -\frac{1}{5}$

7. $\frac{6}{2y+5} = \frac{2}{y+5} + \frac{1}{5}$

8. $\frac{6}{2x+3} = \frac{2}{x-6} + \frac{4}{3}$

9. $1 + 2x^{-1} - 8x^{-2} = 0$

10. $1 + 5x^{-1} + 6x^{-2} = 0$

11. $3 + 13x^{-1} - 10x^{-2} = 0$

12. $3 - 5x^{-1} - 12x^{-2} = 0$

For Exercises 13–24, solve the equations with radical expressions. Identify any extraneous solutions. See Example 2.

13. $x - 8\sqrt{x} + 15 = 0$

14. $x - 3\sqrt{x} + 2 = 0$

15. $2x - 5\sqrt{x} - 7 = 0$

16. $3x + 4\sqrt{x} - 4 = 0$

17. $\sqrt{2a+5} = 3a - 3$

18. $\sqrt{3b+1} = 5b - 3$

19. $\sqrt{2m - 8} - m - 1 = 0$

20. $\sqrt{8x - 9} - x - 4 = 0$

21. $\sqrt{4x + 1} = \sqrt{x + 2} + 1$

22. $\sqrt{4x + 1} = \sqrt{x + 3} + 2$

23. $\sqrt{2x + 1} - \sqrt{3x + 4} = -1$

24. $\sqrt{2x - 1} - \sqrt{4x + 5} = -2$

Objective 2

Prep Exercise 3 To solve $3\left(\frac{x + 2}{3}\right)^2 + 13\left(\frac{x + 2}{3}\right) - 10 = 0$ using substitution, what would your substitution be?

Prep Exercise 4 To solve $x^4 - 7x^2 + 12 = 0$ using substitution, what would your substitution be?

For Exercises 25–46, solve using substitution. Identify any extraneous solutions. See Example 3.

25. $x^4 - 10x^2 + 9 = 0$

26. $x^4 - 13x^2 + 36 = 0$

27. $4x^4 - 13x^2 + 9 = 0$

28. $9x^4 - 37x^2 + 4 = 0$

29. $x^4 + 5x^2 - 36 = 0$

30. $x^4 - 12x^2 - 64 = 0$

31. $(x + 2)^2 + 6(x + 2) + 8 = 0$

32. $(x - 3)^2 + 2(x - 3) - 15 = 0$

33. $2(x + 3)^2 - 9(x + 3) - 5 = 0$

34. $3(x - 1)^2 + 4(x - 1) - 4 = 0$

35. $\left(\frac{x - 1}{2}\right)^2 + 8\left(\frac{x - 1}{2}\right) + 15 = 0$

36. $\left(\frac{x + 2}{3}\right)^2 - 5\left(\frac{x + 2}{3}\right) - 6 = 0$

37. $2\left(\frac{x + 2}{2}\right)^2 + \left(\frac{x + 2}{2}\right) - 3 = 0$

38. $3\left(\frac{x - 3}{3}\right)^2 + 5\left(\frac{x - 3}{3}\right) + 2 = 0$

39. $x^{2/3} - 5x^{1/3} + 6 = 0$

40. $x^{2/3} - 3x^{1/3} + 2 = 0$

41. $2x^{2/3} - 3x^{1/3} - 2 = 0$

42. $3x^{2/3} - 4x^{1/3} - 4 = 0$

43. $x^{1/2} - 5x^{1/4} + 6 = 0$

44. $x^{1/2} - 4x^{1/4} + 3 = 0$

45. $5x^{1/2} + 8x^{1/4} - 4 = 0$

46. $2x^{1/2} - x^{1/4} - 3 = 0$

Objective 3

Prep Exercise 5 If Alisha can clean her house in x hours, what part of her house can she clean in 1 hour?

Prep Exercise 6 If a car travels 400 miles in x hours, how many miles does it travel in 1 hour?

For Exercises 47–58, solve. See Example 4.

47. The average rate of a bus is 15 miles per hour more than the average rate of a truck. The truck takes 1 hour longer than the bus to travel 180 miles. How long does the bus take to travel 180 miles?

48. A charter business has two types of small planes, a jet and a twin propeller plane. The average air speed of the prop plane is 120 miles per hour less than the jet's. The prop plane takes 1 hour longer to travel 720 miles than the jet does. How long does the prop plane take to travel the 720 miles?

★ **49.** The average speed of Wesley Korir, who won the Boston Marathon in 2012, was 3.66 miles per hour faster than the person who finished 1 hour behind him. Find the time Wesley took to run the 26-mile course.

★ **50.** The average speed of Sharon Cherop, who won first place among women in the 2012 Boston Marathon was 2.91 miles per hour faster than the person who finished 1 hour behind her. Find the time Sharon took to run the 26-mile course.

51. Suppose the time required for a bus to travel 360 miles is 3 hours more than the time required for a motorcycle to travel 300 miles. If the average rate of the motorcycle is 15 miles per hour more than the average rate of the bus, find the average rate of the bus.

52. Suppose the time required for a truck to travel 400 miles is 3 hours more than the time required for a car to travel 300 miles. If the average rate of the car is 10 miles per hour more than the average rate of the truck, find the average rate of the truck.

53. After training hard for a year, a novice cyclist discovers that she has increased her average rate by 6 miles per hour and can travel 36 miles in 1 hour less than she did a year ago.
a. What was her old average rate? What is her new average rate?

b. What was her old time to travel 36 miles? What is her new time for 36 miles?

54. A high school track coach determines that his fastest long-distance runner runs 2 miles per hour faster than his slowest runner. Compared to the faster runner, the slower runner takes a half hour longer to run 12 miles.
a. Find the average rates of both runners.

b. Find the time for both runners to run 12 miles.

55. Billy and Jody are commercial fishermen. Working alone, it takes Billy 2 hours longer to run the hoop nets than it takes Jody working alone. Together they can run the hoop nets in $2\frac{2}{5}$ hours. How long does it take each working alone?

56. Using a riding mower, Fran can mow the grass at the campground in 4 hours less time than it takes Donnie using a push mower. Together they can mow the grass in $2\frac{2}{3}$ hours. How long does it take each working alone?

★ **57.** A newspaper has two presses, one of which is older than the other. By itself, the newer press can print all of the copies for a typical day in a half hour less time than the older press takes. When running at the same time, the presses print all of a typical day's copies in 2 hours. How long would it take to print a day's worth if they worked alone?

Of Interest

The world record for completing a marathon was set by Geoffrey Mutai of Kenya in the 2011 Boston Marathon with a time of 2:03:02. The world record for women was set by Paula Radcliffe in the 2003 London Marathon with a time of 2:15:25.

(*Source:* marathonguide.com.)

★ **58.** A school's copy center prints the school newsletter in a half hour using two copiers running at the same time. If the center uses only one copier, the faster of the two copiers takes 6 minutes less time than the other copier to print all of the newsletters. How long does each copier take to print the newsletters working alone?

On a guitar, the frequency of the vibrating string is related to the tension on the string. Suppose two strings of the same diameter and length are placed on an instrument and wound to different tensions. The formula $\frac{F_1^2}{F_2^2} = \frac{T_1}{T_2}$ *describes the relationship between their frequency and tension.*

Of Interest

In addition to tension, the diameter and length of a string also affect its frequency. Increasing the diameter or length of a string under the same tension decreases its frequency.

★ **59.** One string is wound on a guitar to a tension of 50 pounds. A second string of the same diameter is wound to 60 pounds. If the string with the greater tension has a frequency that is 40 vibrations per second greater than the other string, what are the frequencies of the two strings? Round to the nearest ten.

★ **60.** Two strings of the same diameter are wound on a banjo to different tensions: one at 80 pounds and the other at 90 pounds. The frequency of the string under less tension is 20 vibrations per second less than the other string. What are the frequencies of the two strings? Round to the nearest ten.

Review Exercises

Exercises 1 and 2 Expressions

[1.4] 1. Evaluate $-\frac{b}{2a}$ when a is 2 and b is 8.

[1.4] 2. Evaluate $\frac{4ac - b^2}{4a}$ when $a = 2, b = -8$, and $c = 5$.

Exercises 3–6 Equations and Inequalities

[3.1] 3. Find the x- and y-intercepts $2x + 3y = 6$

[3.5] 4. If $f(x) = 2x^2 - 3x + 1$, find $f(-1)$.

[5.2, 6.4] *For Exercises 5 and 6, graph.*

5. $f(x) = x^2 - 3$

6. $f(x) = 3x^2$

9.4 Graphing Quadratic Functions

Objectives

1 Graph quadratic functions of the form $f(x) = ax^2$.
2 Graph quadratic functions of the form $f(x) = ax^2 + k$.
3 Graph quadratic functions of the form $f(x) = a(x - h)^2$.
4 Graph quadratic functions of the form $f(x) = a(x - h)^2 + k$.
5 Graph quadratic functions of the form $f(x) = ax^2 + bx + c$.
6 Solve applications involving parabolas.

Warm-up

[9.3] **1.** Solve: $x^4 - 10x^2 + 9 = 0$

[1.4] **2.** Evaluate $-\frac{b}{2a}$ and $\frac{4ac - b^2}{4a}$ for $a = -0.8$, $b = 2.4$, and $c = 6$.

[9.1] **3.** Solve $8 = x^2 - 2x$ by completing the square.

[3.5] **4.** Given $f(x) = 2x^2 - 8x + 5$, find $f(2)$.

In Sections 5.2 and 6.4 we learned that quadratic functions have the form $f(x) = ax^2 + bx + c$, where a, b, and c are real numbers and $a \neq 0$. By plotting many ordered pairs, we found that the graphs of these functions are parabolas that open up if $a > 0$ and down if $a < 0$. By replacing x with 0, we found that the y-intercept is $(0, c)$.

By replacing y (or $f(x)$) with 0, we found the x-intercepts. We also learned that every parabola will have symmetry about a line called its *axis of symmetry* and the axis of symmetry always passes through a point called the *vertex*, which is the lowest point on a parabola that opens upward or the highest point on a parabola that opens downward. Look at the following graphs.

x	$f(x) = x^2$
-3	9
-2	4
-1	1
0	0
1	1
2	4
3	9

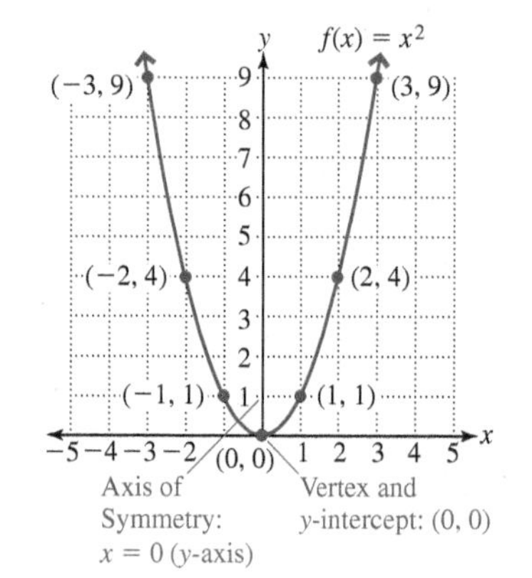

x	$f(x) = -x^2 + 4$
-3	-5
-2	0
-1	3
0	4
1	3
2	0
3	-5

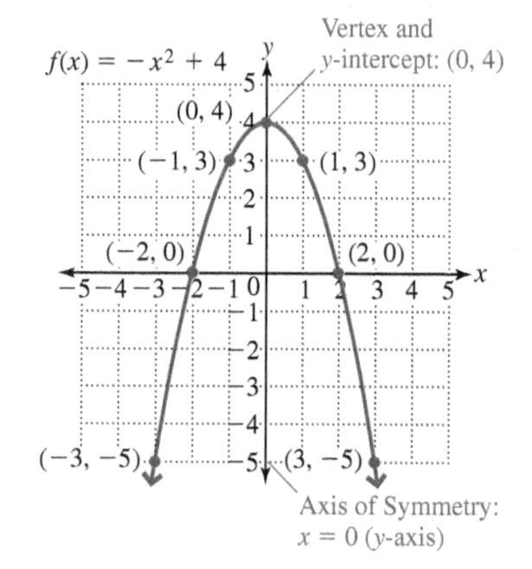

Notice that the graph of $f(x) = x^2$ opens upward because $a = 1$, which is positive, whereas the graph of $f(x) = -x^2 + 4$ opens downward because $a = -1$, which is negative. In this section, we learn more about the graphs of quadratic functions. Later in the section, we revisit the form $f(x) = ax^2 + bx + c$ and learn how to determine a parabola's vertex and axis of symmetry from the values of a and b.

Objective 1 Graph quadratic functions of the form $f(x) = ax^2$.

First, we consider $f(x) = ax^2$ to discover more about how a affects the parabola. We will see that a affects not only whether the parabola opens upward or downward but also how wide the parabola is.

Example 1 Compare the graphs of each function.

a. $f(x) = \frac{1}{4}x^2$, $g(x) = \frac{1}{3}x^2$, $h(x) = \frac{1}{2}x^2$, $k(x) = x^2$, and $m(x) = 2x^2$

Answers to Warm-up
1. $x = \pm 1, \pm 3$
2. $1.5, 7.8$
3. $x = 4, x = -2$
4. -3

Solution: We graph all five functions on the same grid.

Note All of these graphs open upward because a is positive. Also notice that as a increases, the parabolas appear narrower.

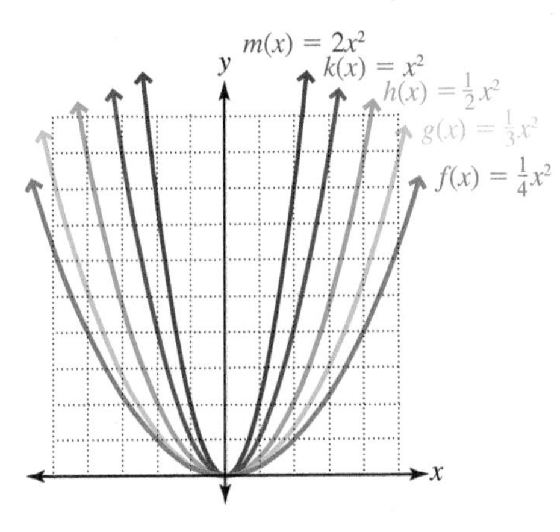

b. $f(x) = -\frac{1}{4}x^2, g(x) = -\frac{1}{3}x^2, h(x) = -\frac{1}{2}x^2, k(x) = -x^2$, and $m(x) = -2x^2$

Note All of these graphs open downward because a is negative. Also notice that as $|a|$ increases, the parabolas appear narrower.

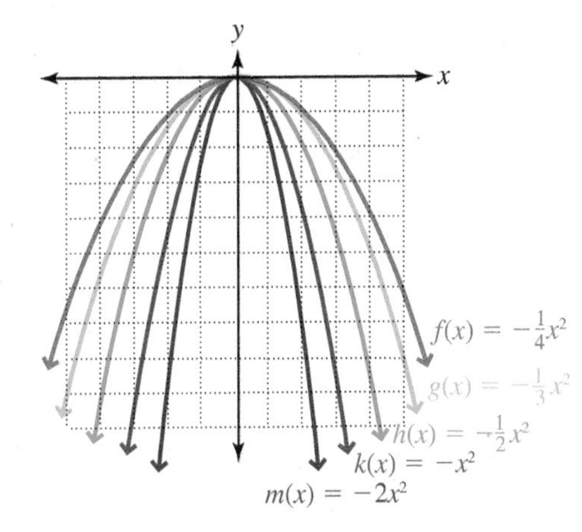

Answers to Your Turn 1

a.

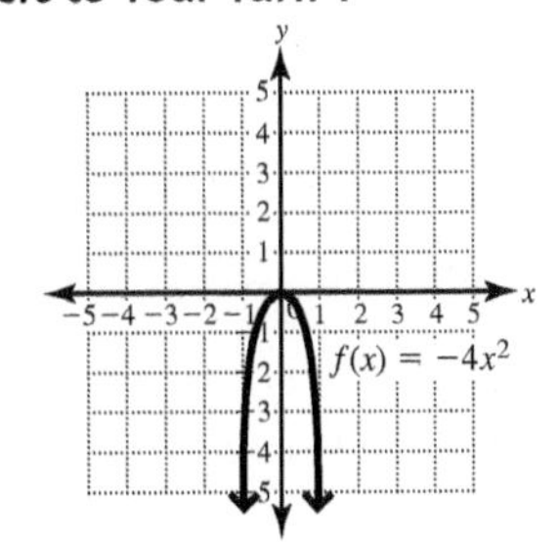

b.

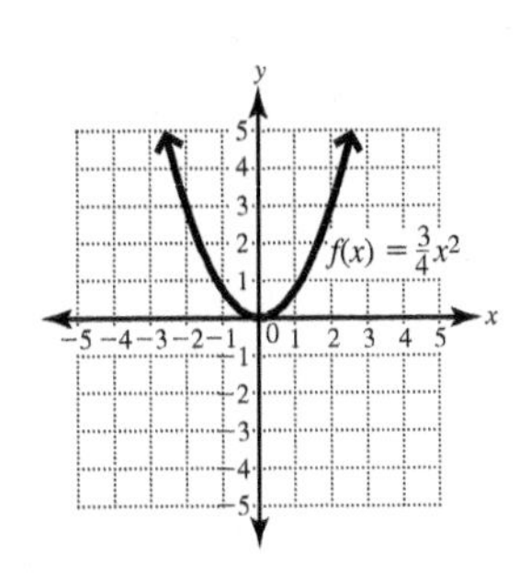

Example 1 suggests the following conclusions about functions of the form $f(x) = ax^2$.

Conclusion: Given a function in the form $f(x) = ax^2$, the axis of symmetry is $x = 0$ (the y-axis) and the vertex is $(0, 0)$. Also, the greater the absolute value of a, the narrower the parabola appears (or the smaller the absolute value of a, the wider the parabola appears).

Your Turn 1 Graph the following functions.

a. $f(x) = -4x^2$

b. $f(x) = \frac{3}{4}x^2$

Objective 2 Graph quadratic functions of the form $f(x) = ax^2 + k$.

We have learned that the vertex of a parabola of the form $f(x) = ax^2$ is at $(0, 0)$ and the axis of symmetry is $x = 0$ (the y-axis). Now let's consider quadratic functions of the form $f(x) = ax^2 + k$, where k is a constant. We will see that for the same value of a, the graphs of $f(x) = ax^2 + k$ and $f(x) = ax^2$ have the same width and shape but the graph of $f(x) = ax^2 + k$ is shifted up or down k units on the y-axis from the origin so that the vertex is at $(0, k)$.

Example 2 Graph the following functions.

a. $g(x) = 2x^2 + 3$ and $h(x) = 2x^2 - 4$

Solution: We graph both functions on the same grid.

Note The vertex of the "basic" function $f(x) = 2x^2$ is the origin $(0, 0)$. The vertex of $g(x) = 2x^2 + 3$ is shifted up 3 and the vertex of $h(x) = 2x^2 - 4$ is shifted down 4 from the origin.

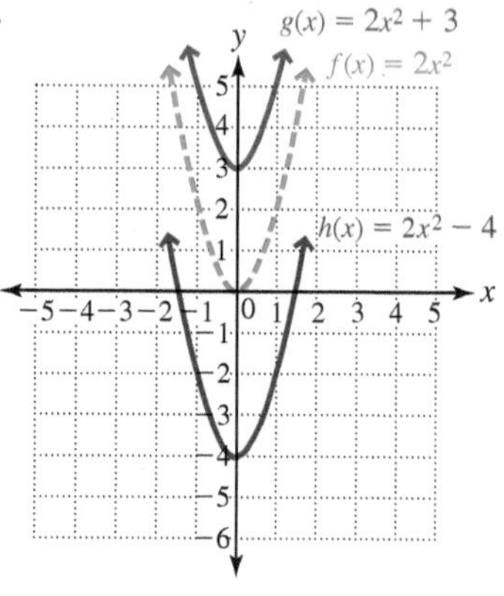

b. $g(x) = -x^2 + 4$ and $h(x) = -x^2 - 2$

Solution:

Note The vertex of the "basic" function $f(x) = -x^2$ is the origin $(0, 0)$. The vertex of $g(x) = -x^2 + 4$ is shifted up 4 and the vertex of $h(x) = -x^2 - 2$ is shifted down 2 from the origin.

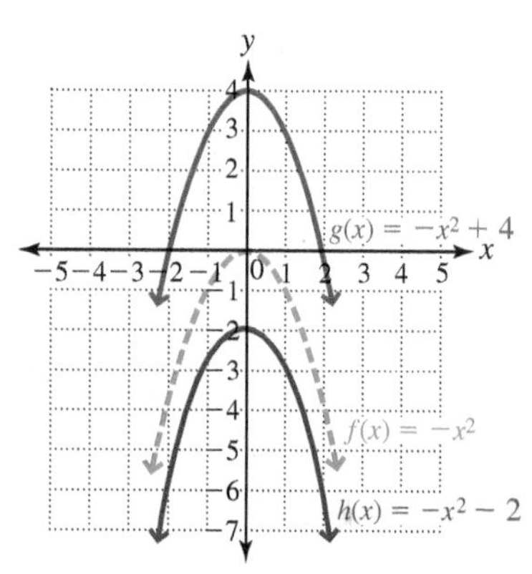

Example 2 suggests the following conclusion about the constant k in functions of the form $f(x) = ax^2 + k$.

Conclusion: Given a function in the form $f(x) = ax^2 + k$, if $k > 0$, then the graph of $f(x) = ax^2$ is shifted k units *up* from the origin. If $k < 0$, then the graph of $f(x) = ax^2$ is shifted $|k|$ units *down* from the origin. The new position of the vertex is $(0, k)$. The axis of symmetry is $x = 0$ (the y-axis).

Your Turn 2 Graph the following functions.

a. $f(x) = 3x^2 - 2$ **b.** $f(x) = -\frac{1}{2}x^2 + 4$

Answers to Your Turn 2

a.

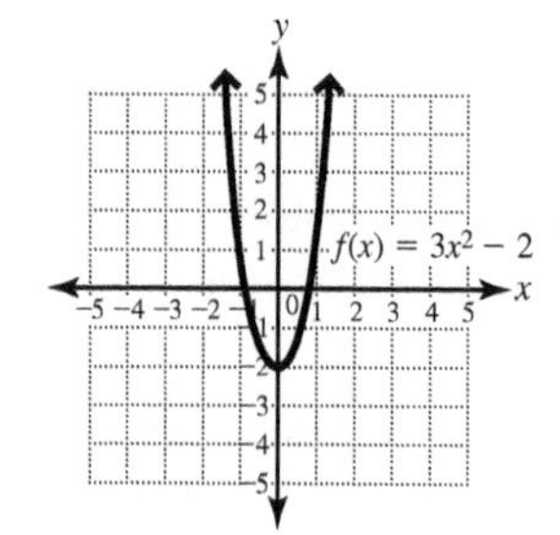

b.

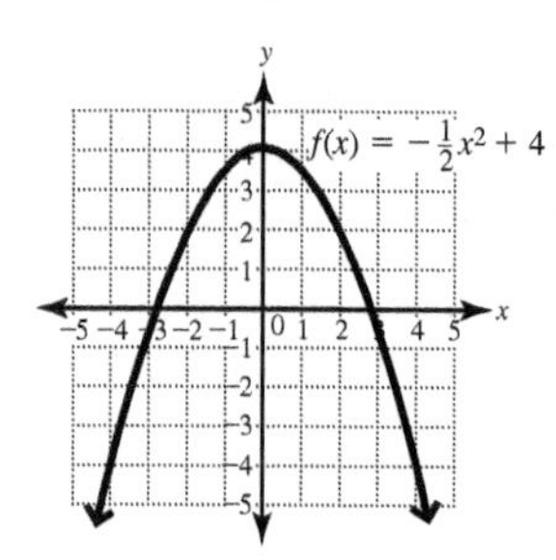

Objective 3 Graph quadratic functions of the form $f(x) = a(x - h)^2$.

We now consider the form $f(x) = a(x - h)^2$, and we will see that the constant h causes the parabola to shift right or left.

Example 3 Graph $m(x) = 2(x - 3)^2$ and $n(x) = 2(x + 4)^2$.

Solution:

Note The vertex of the "basic" function $f(x) = 2x^2$ is $(0, 0)$, and the axis of symmetry is $x = 0$. The graph of $m(x) = 2(x - 3)^2$ has the same shape as $f(x) = 2x^2$, only the vertex and axis of symmetry are shifted **right** 3 units from the origin along the x-axis so that they are $(3, 0)$ and $x = 3$, respectively. Similarly, the vertex and axis of symmetry of $n(x) = 2(x + 4)^2$ are shifted **left** 4 units from the origin along the x-axis so that they are $(-4, 0)$ and $x = -4$, respectively.

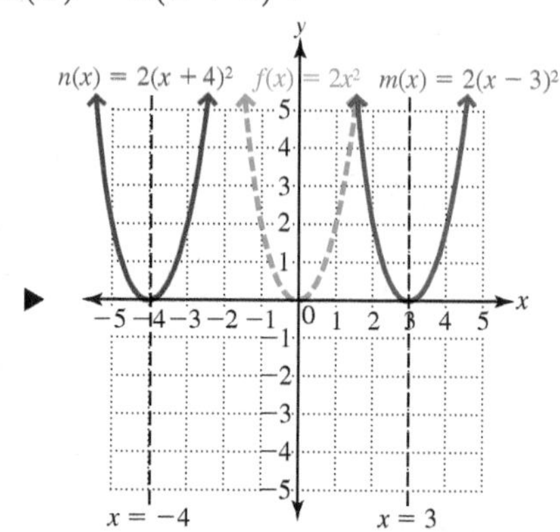

Example 3 suggests the following conclusion about h in $f(x) = a(x - h)^2$.

Conclusion: Given a function in the form $f(x) = a(x - h)^2$, if $h > 0$, then the graph of $f(x) = ax^2$ is shifted h units *right* from the origin. If $h < 0$, then the graph of $f(x) = ax^2$ is shifted $|h|$ units *left* from the origin. The new position of the vertex is $(h, 0)$, and the axis of symmetry is $x = h$.

Your Turn 3 Graph the following functions.

a. $f(x) = -2(x + 1)^2$ **b.** $g(x) = \frac{1}{3}(x - 4)^2$

Objective 4 Graph quadratic functions of the form $f(x) = a(x - h)^2 + k$.

Examples 2 and 3 suggest that the graph of a function of the form $f(x) = a(x - h)^2 + k$ has the same shape as $f(x) = ax^2$ but the vertex is shifted from the origin to (h, k) and the axis of symmetry is shifted from $x = 0$ to $x = h$. We can summarize all we have learned with the following rule.

Rule Parabola with Vertex (h, k)

The graph of a function in the form $f(x) = a(x - h)^2 + k$ is a parabola with vertex at (h, k). The equation of the axis of symmetry is $x = h$. The parabola opens upward if $a > 0$ and downward if $a < 0$. The larger the $|a|$, the narrower the graph.

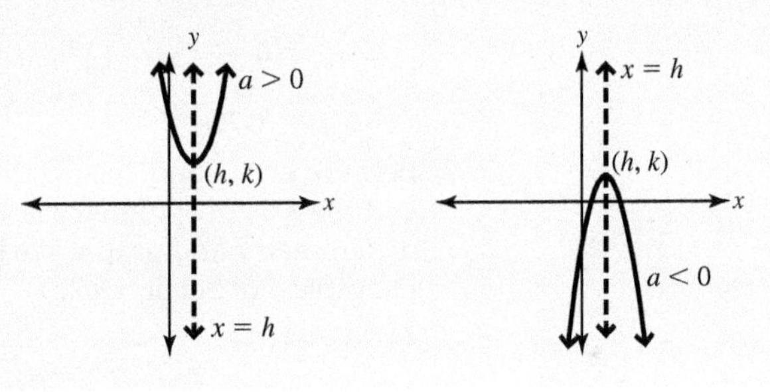

Now let's graph quadratic functions of the form $f(x) = a(x - h)^2 + k$. Recall that the domain of a function is the set of all possible x-values and the range is the set of all possible y-values.

Example 4 Given $f(x) = 2(x - 3)^2 + 1$, determine whether the graph opens upward or downward, find the vertex and axis of symmetry, and draw the graph. Find the domain and range.

Solution: We see that $f(x) = 2(x - 3)^2 + 1$ is in the form $f(x) = a(x - h)^2 + k$, where $a = 2, h = 3$, and $k = 1$. Because a is positive 2, the parabola opens upward. The vertex is at $(3, 1)$, and the axis of symmetry is $x = 3$. To complete the graph, we find a few points on either side of the axis of symmetry.

Note We find these additional points by choosing x-values on either side of the axis of symmetry and using the equation to find the corresponding y-values.

x	y
2	3
1	9
4	3
5	9

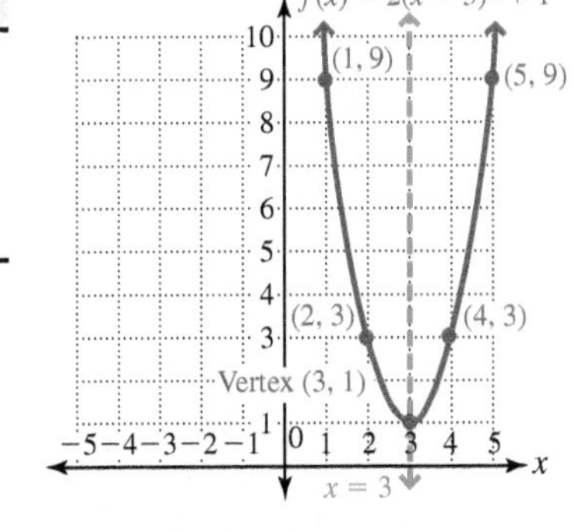

Domain: $\{x|x \text{ is a real number}\}$, or $(-\infty, \infty)$
Range: $\{y|y \geq 1\}$ or $[1, \infty)$

Answers to Your Turn 3

a.

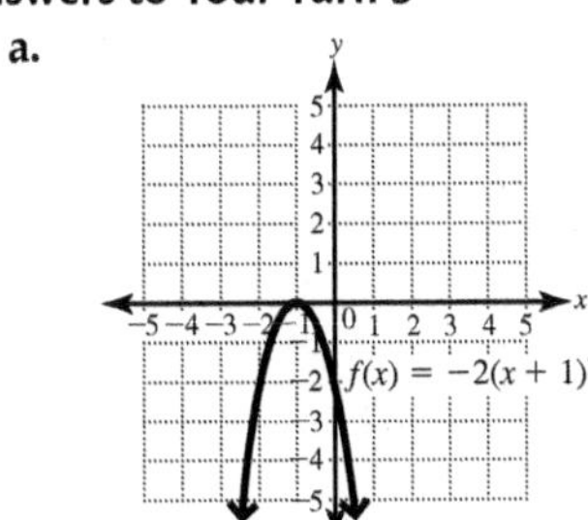

b.

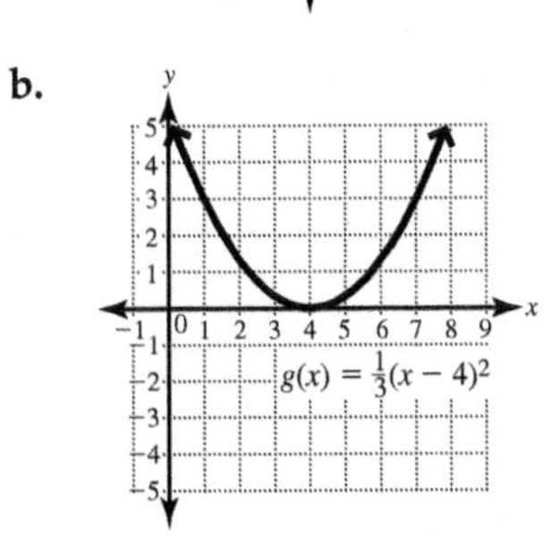

Your Turn 4 Given $f(x) = -\frac{1}{2}(x - 1)^2 + 3$, determine whether the graph opens upward or downward, find the vertex and axis of symmetry, and draw the graph. Find the domain and range.

Objective 5 Graph quadratic functions of the form $f(x) = ax^2 + bx + c$.

The advantage of the form $f(x) = a(x - h)^2 + k$ is that we can "see" the vertex and axis of symmetry and we can "see" whether the parabola opens upward or downward by looking at the equation. If an equation is in the form $f(x) = ax^2 + bx + c$, it can be transformed into $f(x) = a(x - h)^2 + k$ by completing the square.

Example 5 Write the function in the form $f(x) = a(x - h)^2 + k$. Then determine whether the graph opens upward or downward, find the vertex and axis of symmetry, and draw the graph. Find the domain and range.

a. $f(x) = x^2 - 2x - 8$

Solution: $y = x^2 - 2x - 8$ Replace $f(x)$ with y to make manipulations more workable.

$y + 8 = x^2 - 2x$ Add 8 to both sides of the equation to isolate the x terms.

$y + 8 + 1 = x^2 - 2x + 1$ Add 1 to both sides of the equation to complete the square.

$y + 9 = (x - 1)^2$ Factor the right side as a perfect square.

$y = (x - 1)^2 - 9$ Subtract 9 from both sides to solve for y.

Because $a = 1$ and 1 is positive, the graph opens upward. The vertex is at $(1, -9)$, and the axis of symmetry is $x = 1$. Plot a few points on either side of the axis of symmetry and graph. We include the x- and y-intercepts, which we found in Example 3 of Section 9.2.

	x	$f(x)$
x-intercept →	−2	0
	−1	−5
y-intercept →	0	−8
vertex →	1	−9
	2	−8
	3	−5
x-intercept →	4	0

Domain: $\{x \mid x \text{ is a real number}\}$, or $(-\infty, \infty)$

Range: $\{y \mid y \geq -9\}$ or $[-9, \infty)$

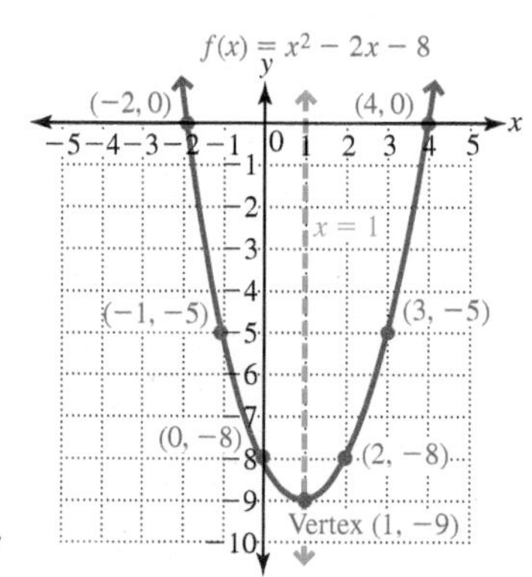

Note The points we found are 1, 2, and 3 units to the left and right of the axis of symmetry.

Note Remember, to find x-intercepts, we replace $f(x)$ (or y) with 0 and solve for x. To find y-intercepts, we replace x with 0 and solve for y.

b. $f(x) = 2x^2 + 12x + 19$

Solution: $y = 2x^2 + 12x + 19$ Replace $f(x)$ with y.

$y - 19 = 2x^2 + 12x$ Subtract 19 from both sides of the equation.

$y - 19 = 2(x^2 + 6x)$ Factor 2 out of the right side of the equation.

$y - 19 + 18 = 2(x^2 + 6x + 9)$ Complete the square inside the parentheses.

Note We added 18 to the left side of the equation because $2 \cdot 9 = 18$ was added to the right side.

$y - 1 = 2(x + 3)^2$ Combine terms on the left and factor on the right.

$y = 2(x + 3)^2 + 1$ Add 1 to both sides.

Because $a = 2$ and 2 is positive, the graph opens upward. The vertex is at $(-3, 1)$, and the axis of symmetry is $x = -3$. Plot a few points on either side of the axis of symmetry and graph.

Answer to Your Turn 4
opens downward; vertex: $(1, 3)$; axis: $x = 1$; domain: $\{x \mid x \text{ is a real number}\}$, or $(-\infty, \infty)$; range: $\{y \mid y \leq 3\}$ or $(-\infty, 3]$

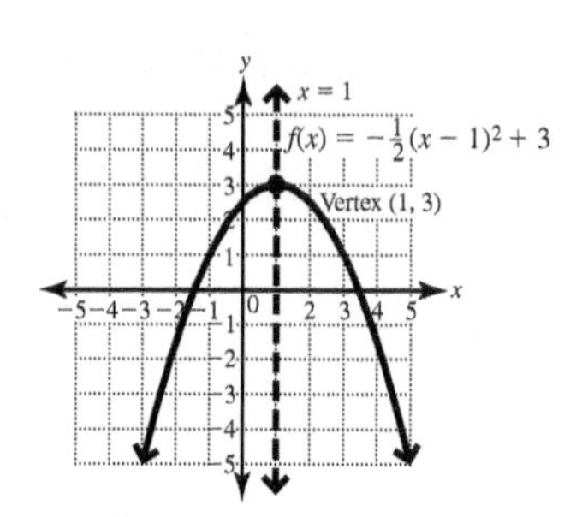

Note The points we found are 1 and 2 units to the right and left of the axis of symmetry.

x	$f(x)$
-2	3
-1	9
-4	3
-5	9

Note that there are no x-intercepts and the y-intercepts is $(0, 19)$.

Domain: $\{x \mid x \text{ is a real number}\}$, or $(-\infty, \infty)$

Range: $\{y \mid y \geq 1\}$ or $[1, \infty)$

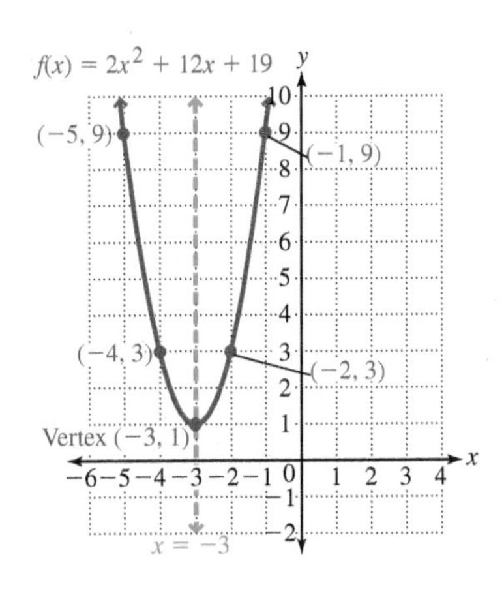

Your Turn 5 Write the function in the form $f(x) = a(x - h)^2 + k$; determine whether the graph opens upward or downward; find the x- and y-intercepts, vertex, and axis of symmetry; and draw the graph. Find the domain and range.

a. $f(x) = x^2 - 2x - 3$ **b.** $f(x) = -2x^2 + 8x - 6$

We also can find the vertex using a formula that we derive by completing the square on the general form $f(x) = ax^2 + bx + c$.

$$y = ax^2 + bx + c \quad \text{Replace } f(x) \text{ with } y.$$

$$y - c = ax^2 + bx \quad \text{Isolate the } x \text{ terms.}$$

$$y - c = a\left(x^2 + \frac{b}{a}x\right) \quad \text{Factor out } a.$$

$$y - c + \frac{b^2}{4a} = a\left(x^2 + \frac{b}{a}x + \frac{b^2}{4a^2}\right) \quad \text{Complete the square.}$$

$$y - \frac{4ac - b^2}{4a} = a\left(x + \frac{b}{2a}\right)^2 \quad \text{Find the LCD of the two terms on the left. Factor on the right.}$$

$$y = a\left(x + \frac{b}{2a}\right)^2 + \frac{4ac - b^2}{4a} \quad \text{Add } \frac{4ac - b^2}{4a} \text{ to both sides.}$$

Note Adding $\left[\frac{1}{2}\left(\frac{b}{a}\right)\right]^2$ or $\frac{b^2}{4a^2}$ completes the square inside the parentheses. Because those parentheses are multiplied by a, we add $a \cdot \frac{b^2}{4a^2} = \frac{b^2}{4a}$ to the left side to keep the equation balanced.

Notice that if we rewrite the last line above as $y = a\left(x - \left(-\frac{b}{2a}\right)\right)^2 + \frac{4ac - b^2}{4a}$, we see that $h = -\frac{b}{2a}$ and $k = \frac{4ac - b^2}{4a}$ so that the vertex is $\left(-\frac{b}{2a}, \frac{4ac - b^2}{4a}\right)$. This also means that the axis of symmetry is $x = -\frac{b}{2a}$. Also, although the y-coordinate of the vertex is $\frac{4ac - b^2}{4a}$, it is usually easier to find the vertex by substituting the x-coordinate into the original function.

Procedure **Finding the Vertex of a Quadratic Function in the Form $f(x) = ax^2 + bx + c$**

Given an equation in the form $f(x) = ax^2 + bx + c$, to determine the vertex:

1. Find the x-coordinate using the formula $x = -\frac{b}{2a}$.
2. Find the y-coordinate by evaluating $f\left(-\frac{b}{2a}\right)$ or $\frac{4ac - b^2}{4a}$.

Answers to Your Turn 5

a. x-intercepts: $(-1, 0)$, $(3, 0)$
y-intercept: $(0, -3)$
$f(x) = (x - 1)^2 - 4$;
opens upward; vertex: $(1, -4)$;
axis: $x = 1$

domain: $\{x \mid x \text{ is a real number}\}$, or $(-\infty, \infty)$; range: $\{y \mid y \geq -4$ or $[-4, \infty)$

b. $f(x) = -2(x - 2)^2 + 2$;
x-intercepts: $(1, 0)$, $(3, 0)$
y-intercepts: $(0, -6)$;
opens downward;
vertex: $(2, 2)$; axis: $x = 2$

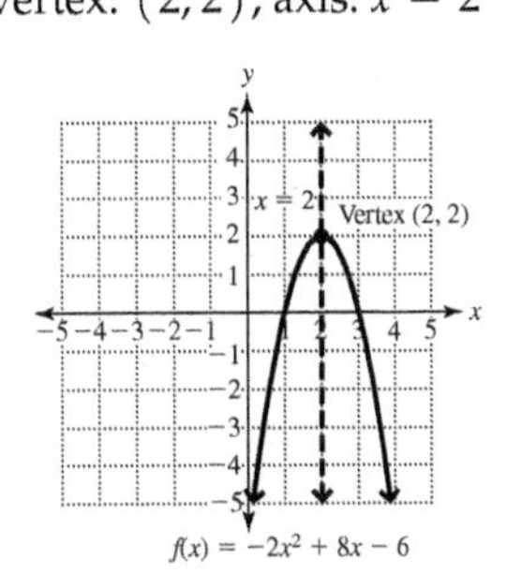

domain $\{x \mid x \text{ is a real number}\}$, or $(-\infty, \infty)$;
range: $\{y \mid y \leq 2\}$ or $(-\infty, 2]$

Example 6 For the function $f(x) = 2x^2 - 8x + 5$, find the coordinates of the vertex.

Note Using $\frac{4ac - b^2}{4a}$ also gives the y-coordinate:

$$\frac{4(2)(5) - (-8)^2}{4(2)} = \frac{-24}{8} = -3$$

However, substituting the x-value, 2, into the original function is easier.

Solution: First find the x-coordinate of the vertex using $-\frac{b}{2a}$.

$$x\text{-coordinate of the vertex: } -\frac{b}{2a} = -\frac{(-8)}{2(2)} = \frac{8}{4} = 2$$

Now find the y-coordinate by evaluating $f(2)$.

$$y\text{-coordinate of the vertex: } f(2) = 2(2)^2 - 8(2) + 5 = 8 - 16 + 5 = -3$$

The vertex is $(2, -3)$.

Your Turn 6 For the equation $f(x) = 3x^2 + 18x - 19$, find the vertex.

Objective 6 Solve applications involving parabolas.

Because the vertex of a parabola is the highest point in parabolas that open downward or the lowest point in parabolas that open upward, we can find a minimum or maximum value in applications involving quadratic functions.

Example 7 A toy rocket is launched straight up with an initial velocity of 40 feet per second. The equation $h = -16t^2 + 40t$ describes the height, h, of the rocket t seconds after being launched.

a. Find the maximum height the rocket reaches.
b. Find the amount of time the rocket is in the air.
c. Draw the graph of the height of the rocket.

Solution:

a. Because the graph of $h = -16t^2 + 40t$ is a parabola that opens down $(a = -16)$, the maximum height occurs at its vertex.

$$t\text{-coordinate of the vertex: } -\frac{b}{2a} = -\frac{40}{2(-16)} = -\frac{40}{-32} = \frac{5}{4} = 1.25 \text{ seconds}$$

To find the h-coordinate of the vertex, we replace t with 1.25 in $h = -16t^2 + 40t$.

$$h = -16(1.25)^2 + 40(1.25) = 25 \text{ feet}$$

The vertex is $(1.25, 25)$, so the maximum height is 25 feet, which occurs 1.25 seconds after the rocket is launched.

Note We could have found the vertex by writing the equation in $y = a(x - h)^2 + k$ form, which would be $h = -16(t - 1.25)^2 + 25$.

b. The time the rocket is in the air is from launch until it returns to the ground. At launch and upon returning to the ground, the rocket's height is 0; so we need to find t when $h = 0$.

$$0 = -16t^2 + 40t \quad \text{Replace } h \text{ with 0.}$$

$$0 = -8t(2t - 5) \quad \text{Factor out a common factor of } -8t.$$

$$-8t = 0 \quad \text{or} \quad 2t - 5 = 0 \quad \text{Use the zero-factor theorem.}$$

$$t = 0 \qquad 2t = 5$$

$$t = 2.5$$

This means that height is 0 when $t = 0$ and when $t = 2.5$ seconds, so the rocket is in the air for 2.5 seconds.

Answer to Your Turn 6
$(-3, -46)$

Note The graph of $h = -16t^2 + 40t$ is **not** the flight path of the rocket. Also, because the situation involves a rocket being launched from the ground (0 feet) and returning to the ground, only the portion of the graph in the first quadrant is realistic.

c. The graph of $h = -16t^2 + 40t$ shows that at 0 seconds, the rocket is at 0 feet, which is when it is launched. Its height increases to a maximum of 25 feet 1.25 seconds after launch. Then the height decreases back to 0 feet 2.5 seconds after launch.

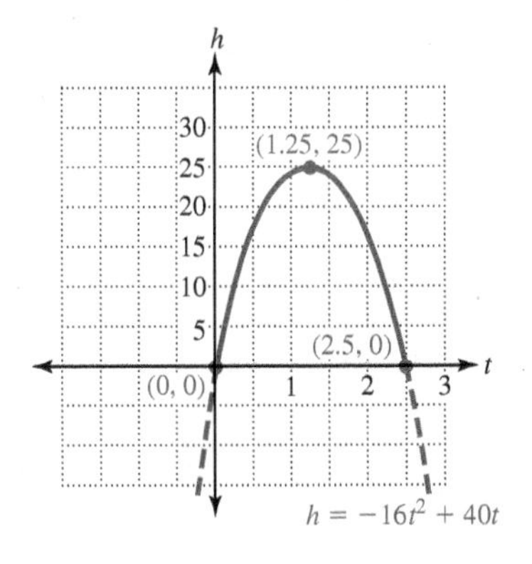

Your Turn 7 A soccer player kicks a ball straight up with an initial velocity of 56 feet per second. The equation $h = -16t^2 + 56t$ describes the height, h, of the ball t seconds after being kicked.

a. After how many seconds is the ball at its maximum height?
b. What is the maximum height the ball reaches?
c. How long is the ball in the air?

If an object is thrown or shot upward and outward, gravity causes its path, or trajectory, to be in the shape of a parabola.

Note x represents the horizontal distance the object travels, and y represents the vertical distance the object travels.

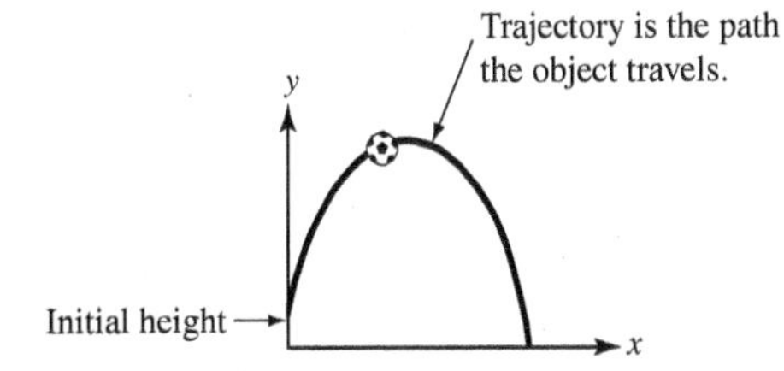

Example 8 The equation $y = -0.8x^2 + 2.4x + 6$ when $x \geq 0$ and $y \geq 0$ describes the trajectory of a ball thrown upward and outward from an initial height of 6 feet.

a. What is the maximum height the ball reaches?
b. How far does the ball travel horizontally?
c. Graph the trajectory of the ball.

Solution:

a. The maximum height will be the y-coordinate of the vertex.

$$x\text{-coordinate of the vertex: } -\frac{b}{2a} = -\frac{(2.4)}{2(-0.8)} = 1.5$$

$$y\text{-coordinate of the vertex: } y = -0.8(1.5)^2 + 2.4(1.5) + 6 = 7.8$$

The vertex is $(1.5, 7.8)$, so the maximum height is 7.8 feet.

b. To find how far the ball travels, we need to find the x-value when it hits the ground, which is where $y = 0$.

$0 = -0.8x^2 + 2.4x + 6$ Substitute 0 for y in the equation.

$x = \dfrac{-(2.4) \pm \sqrt{(2.4)^2 - 4(-0.8)(6)}}{2(-0.8)}$ Substitute $a = -0.8$, $b = 2.4$, and $c = 6$ into the quadratic formula.

$= \dfrac{-2.4 \pm \sqrt{24.96}}{-1.6} \approx -1.62 \text{ or } 4.62$

The negative value does not make sense in the context of this problem, so the distance the ball travels must be approximately 4.62 feet.

Answers to Your Turn 7
a. 1.75 sec. **b.** 49 ft. **c.** 3.5 sec.

c.

Note Because the equation describes the trajectory when $x \geq 0$ and $y \geq 0$, we consider only the part of the graph in the first quadrant.

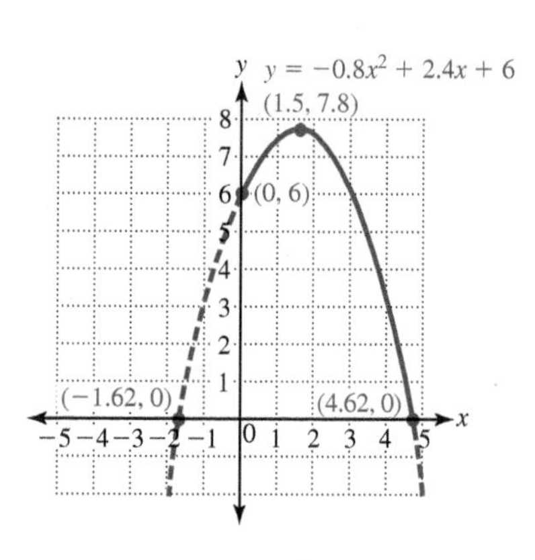

9.4 Exercises

For Extra Help MyMathLab®

Note: Exercises marked with a ★ represent challenging exercises.

Objectives 1–4

Prep Exercise 1 In an equation in the form $f(x) = a(x - h)^2 + k$, what determines whether the parabola opens upward or downward?

Prep Exercise 2 What is the axis of symmetry in a parabola that opens up or down?

Prep Exercise 3 In an equation in the form $f(x) = a(x - h)^2 + k$, what are the coordinates of the vertex? What is the equation of the axis of symmetry?

For Exercises 1–16, find the coordinates of the vertex and write the equation of the axis of symmetry. See Examples 4–6.

1. $f(x) = 5(x - 2)^2 - 3$

2. $h(x) = 2(x - 1)^2 + 4$

3. $g(x) = -2(x + 1)^2 - 5$

4. $k(x) = -(x + 4)^2 - 3$

5. $k(x) = x^2 + 2$

6. $g(x) = x^2 + 5$

7. $h(x) = 3(x + 5)^2$

8. $f(x) = 5(x + 1)^2$

9. $f(x) = -0.5x^2$

10. $k(x) = 2.5x^2$

11. $f(x) = x^2 - 4x + 8$

12. $g(x) = x^2 + 6x + 8$

13. $k(x) = 2x^2 + 16x + 27$

14. $f(x) = 3x^2 - 6x + 1$

15. $g(x) = -3x^2 + 2x + 1$

16. $h(x) = -2x^2 - 3x + 2$

Objective 5

For Exercises 17–30: a. *State whether the parabola opens upward or downward.*
b. *Find the coordinates of the vertex.*
c. *Write the equation of the axis of symmetry.*
d. *Graph. See Examples 1–4.*

17. $h(x) = -3x^2$

18. $f(x) = 2x^2$

19. $k(x) = \frac{1}{4}x^2$

20. $g(x) = -\frac{1}{3}x^2$

21. $f(x) = 4x^2 - 3$

22. $h(x) = -2x^2 + 4$

23. $g(x) = -0.5x^2 + 2$

24. $k(x) = 0.5x^2 - 3$

25. $f(x) = (x - 3)^2 + 2$

26. $h(x) = (x - 1)^2 - 3$

27. $k(x) = -2(x + 1)^2 - 3$

28. $g(x) = -2(x + 3)^2 + 1$

29. $h(x) = \frac{1}{3}(x - 2)^2 - 1$

30. $f(x) = -\frac{1}{4}(x - 3)^2 - 2$

Prep Exercise 4 Suppose the y-intercept for a given quadratic equation is $(0, -5)$. If the vertex is at $(3, -2)$, what can you conclude about the x-intercepts?

Prep Exercise 5 If the solutions for a quadratic equation are imaginary numbers, what does that indicate about the graph of the corresponding quadratic function?

Prep Exercise 6 Given an equation in the form $y = ax^2 + bx + c$, what is the x-coordinate of the vertex?

For Exercises 31–36:
a. Find the x- and y-intercepts.
b. Write the equation in the form $f(x) = a(x - h)^2 + k$.
c. State whether the parabola opens upward or downward.
d. Find the coordinates of the vertex.
e. Write the equation of the axis of symmetry.
f. Graph. See Examples 5 and 6.
g. Find the domain and range.

31. $h(x) = x^2 + 6x + 9$

32. $k(x) = x^2 - 4x + 4$

33. $g(x) = -3x^2 + 6x - 5$

34. $k(x) = -x^2 - 2x + 3$

35. $f(x) = 2x^2 + 6x + 3$

36. $h(x) = -3x^2 - 6x + 4$

Objective 6

For Exercises 37–48, solve. See Examples 7 and 8.

37. A toy rocket is launched with an initial velocity of 45 meters per second. The equation $h = -4.9t^2 + 45t$ describes the height, h, of the rocket in meters t seconds after being launched.

a. After how many seconds does the rocket reach its maximum height?

b. What is the maximum height the rocket reaches?

c. How long is the rocket in the air?

38. A ball is drop-kicked straight up with an initial velocity of 36 feet per second. The equation $h = -16t^2 + 36t$ describes the height, h, of the ball in feet t seconds after being kicked.
a. After how many seconds does the ball reach its maximum height?

b. What is the maximum height the ball reaches?

c. How long is the ball in the air?

39. The equation $y = -0.8x^2 + 3.2x + 6$ models the trajectory of a ball thrown upward and outward from a height of 6 feet. (Assume that $x \geq 0$ and $y \geq 0$.)
a. What is the maximum height the ball reaches?

b. How far does the ball travel horizontally?

c. Graph the trajectory of the ball.

40. The javelin toss is an event in track and field in which participants try to throw a javelin the farthest. Suppose the equation $y = -0.02x^2 + 1.3x + 8$ models the trajectory of one particular throw. (Assume that x and y represent distances in meters and that $x \geq 0$ and $y \geq 0$.)
a. What is the maximum height the javelin reaches?

b. How far does the javelin travel horizontally?

c. Graph the trajectory of the javelin.

41. A record company discovers that the number of CDs sold each week after release follows a parabolic pattern. The function $n(t) = -200t^2 + 4000t$ describes the number, n, of CDs an artist sold each of t weeks after the release of the album.
a. Which week had the greatest number of CDs sold?

b. How many CDs sold that week?

42. The function $n(t) = -3t^2 + 42t$ describes the number, n, of tickets sold for a play each of t days after tickets went on sale.
a. What day had the greatest number of tickets sold?

b. How many tickets sold that day?

★ **43.** A farmer has enough materials to build a fence with a total length of 400 feet. He wants the enclosed space to be rectangular and wants to maximize the area enclosed. Find the length and width so that the area is maximized.

★ **44.** An architect wants the length and width of a rectangular building to be a total of 150 feet long. She also wants to maximize the rectangular base area that the building occupies. Find the length and width so that the base area is maximized.

45. The function $C(n) = n^2 - 110n + 5000$ describes a company's cost, C, of producing n units of its product. Find the number of units the company should produce to minimize its cost.

46. The function $P(t) = 0.001t^2 - 0.24t + 59.90$ roughly models a particular stock's closing price, P, each of t days of trading during one year.
a. After how many days is the price at its lowest?

b. What was the lowest price during that year?

★ **47.** One integer is 12 more than another. If their product is minimized, find the integers and their product.

★ **48.** The greater of two integers minus the smaller integer gives a result of 20. Find the two integers so that their product is minimized. Also find their product.

Puzzle Problem Given the vertex and the x- and y-intercepts of a parabola, reconstruct the equation.

a. Vertex $(-3, -1)$
Intercepts $(-2, 0)$, $(-4, 0)$, $(0, 8)$

b. Vertex $(-1, 0)$
Intercepts $(-1, 0)$, $(0, 1)$

c. Vertex $(-2, 1)$
Intercepts $(0, 5)$

d. Vertex $(1, 0)$
Intercepts $(1, 0)$, $(0, 1)$

Review Exercises

Exercise 1 Expressions

[1.3] **1.** Evaluate: $-|-2 + 3 \cdot 4| - 4^0$

Exercises 2–6 Equations and Inequalities

[6.4] **2.** Solve $x^2 + 2x = 15$ by factoring.

[9.1] **3.** Solve $x^2 = -20$ using the square root principle.

For Exercises 4–6, solve; then graph the solution set.

[2.3] **4.** $4x - 8 \leq 2x + 1$

[2.6] **5.** $|x - 3| \geq 8$

[4.5] **6.** $\begin{cases} x + y > 2 \\ 2x - 3y \leq 6 \end{cases}$

9.5 Solving Nonlinear Inequalities

Objectives

1 Solve quadratic and other inequalities.
2 Solve rational inequalities.

Warm-up

[9.4] **1.** Graph: $y = x^2 - 4x + 1$

[1.4] **2.** Evaluate $\dfrac{x - 4}{x + 2}$ for $x = 5$.

[2.4] **3.** Write $0 \leq x \leq 5$ in interval notation and graph on a number line.

[7.4] **4.** Solve: $\dfrac{x + 5}{x - 1} = 4$

Answers to Warm-up

1.

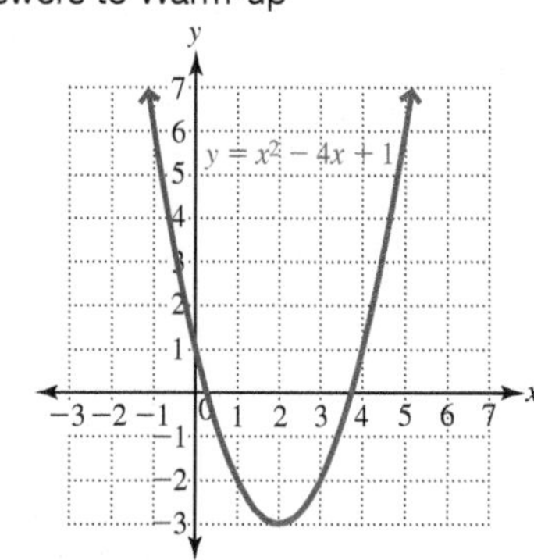

2. $\dfrac{1}{7}$

3. $[0, 5]$;

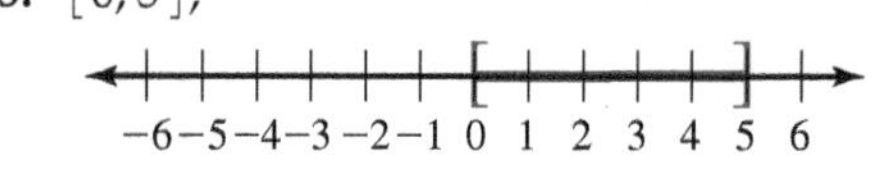

4. $x = 3$

Objective 1 Solve quadratic and other inequalities.

Now that we have solved quadratic equations, we can learn how to solve **quadratic inequalities**.

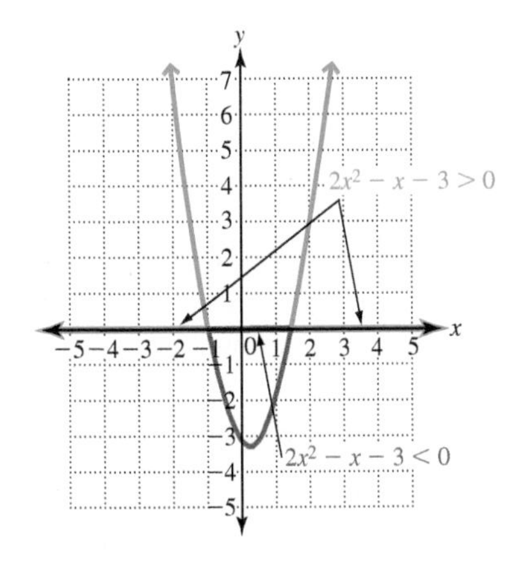

Definition Quadratic inequality: An inequality that can be written in the form $ax^2 + bx + c > 0$ or $ax^2 + bx + c < 0$, where $a \neq 0$.
Note: The symbols $<$ and $>$ can be replaced with $\leq$ and $\geq$.

For example, $2x^2 - x - 3 < 0$ and $2x^2 - x - 3 \geq 0$ are quadratic inequalities. Let's see what we can learn about these inequalities by looking at the graph of the corresponding quadratic function: $y = 2x^2 - x - 3$. Note that by letting $y = 0$, we have the corresponding equation $2x^2 - x - 3 = 0$, and solving this equation gives the x-intercepts $x = -1$ and $x = \frac{3}{2}$. Also notice that those x-intercepts divide the x-axis into three intervals: $(-\infty, -1)$, $\left(-1, \frac{3}{2}\right)$, and $\left(\frac{3}{2}, \infty\right)$.

Notice that the parabola is above the x-axis in intervals $(-\infty, -1)$ and $\left(\frac{3}{2}, \infty\right)$, shown in blue. Evaluating $y = 2x^2 - x - 3$ using any x-value in these intervals produces a y-value that is greater than 0; so these intervals are the solution sets for $2x^2 - x - 3 > 0$.

The parabola is below the x-axis in the interval $\left(-1, \frac{3}{2}\right)$, shown in red. Evaluating $y = 2x^2 - x - 3$ using any x-value in this interval produces a y-value that is less than 0; so this interval is the solution set for $2x^2 - x - 3 < 0$.

The following number lines indicate various solution sets based on the inequality.

$2x^2 - x - 3 > 0$

Solution set: $(-\infty, -1) \cup \left(\frac{3}{2}, \infty\right)$

$2x^2 - x - 3 < 0$

Solution set: $\left(-1, \frac{3}{2}\right)$

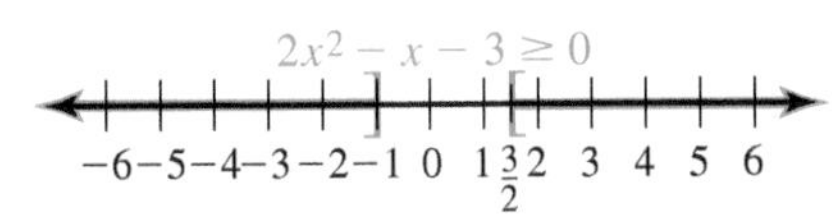

Solution set: $(-\infty, -1] \cup \left[\frac{3}{2}, \infty\right)$

$2x^2 - x - 3 \leq 0$

Solution set: $\left[-1, \frac{3}{2}\right]$

Based on these observations, we use the following procedure to solve quadratic inequalities.

Procedure Solving Quadratic Inequalities

1. Solve the related equation $ax^2 + bx + c = 0$.
2. Plot the solutions of $ax^2 + bx + c = 0$ on a number line. These solutions divide the number line into intervals.
3. Choose a test number from each interval and substitute the number into the inequality. If the test number makes the inequality *true*, then all numbers in that interval will solve the inequality. If the test number makes the inequality *false*, then no numbers in that interval will solve the inequality.
4. State the solution set of the inequality: It is the union of all intervals that solve the inequality. If the inequality symbols are $\leq$ or $\geq$, the values from step 2 are included. If the symbols are $<$ or $>$, they are not solutions.

Example 1 Solve. Write the solution set using interval notation; then graph the solution set on a number line.

a. $x^2 - x < 6$

Solution:

$x^2 - x = 6$	Write the related equation.
$x^2 - x - 6 = 0$	Subtract 6 from both sides to get quadratic form.
$(x - 3)(x + 2) = 0$	Factor.
$x - 3 = 0$ or $x + 2 = 0$	Use the zero-factor theorem.
$x = 3$ $\quad$ $x = -2$	These are the x-intercepts of the graph of $y = x^2 - x - 6$.

Plot −2 and 3 on a number line (x-axis), which divides the number line into three intervals.

−6 −5 −4 −3 −2 −1 0 1 2 3 4 5 6
$(-\infty, -2)$ $\quad$ $(-2, 3)$ $\quad$ $(3, \infty)$

Note −2 and 3 are not part of the solution set because the inequality symbol is <.

Choose a test number from each interval and substitute that value into $x^2 - x < 6$.

For $(-\infty, -2)$, we choose $x = -3$.

$$(-3)^2 - (-3) < 6$$
$$9 + 3 < 6$$
$$12 < 6$$

This is false, so $(-\infty, -2)$ is not in the solution set.

For $(-2, 3)$, we choose $x = 0$.

$$0^2 - 0 < 6$$
$$0 < 6$$

This is true, so $(-2, 3)$ is in the solution set.

For $(3, \infty)$, we choose $x = 4$.

$$4^2 - 4 < 6$$
$$16 - 4 < 6$$
$$12 < 6$$

This is false, so $(3, \infty)$ is not in the solution set.

Because $(-2, 3)$ is the only interval that has solutions to the inequality, it is the solution set. Following is the graph of the solution set.

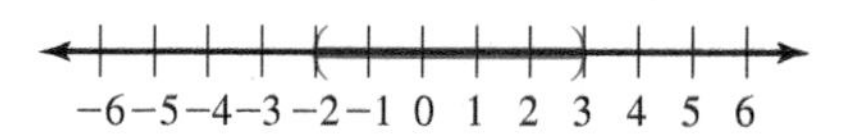

Connection Notice that the solution set for $x^2 - x - 6 < 0$ corresponds to the interval where the graph of $f(x) = x^2 - x - 6$ is below the x-axis and the x-intercepts are the endpoints of the interval and are not included in the solution set.

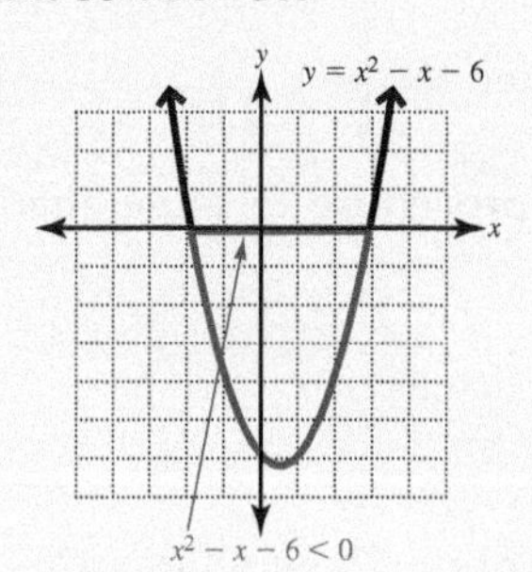

b. $x^2 + 3x \geq 0$

Solution:

$x^2 + 3x = 0$	Write the related equation.
$x(x + 3) = 0$	Factor.
$x = 0$ or $x + 3 = 0$	Use the zero-factor theorem.
$x = -3$	

Plot −3 and 0 on a number line and note the intervals.

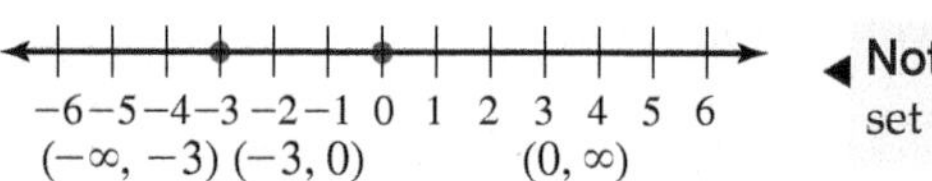

Note −3 and 0 are included in the solution set because the inequality symbol is ≥.

Note Because we have already determined that −3 and 0 are in the solution set, we do not include them in any of our test intervals, which is why we use parentheses for each interval.

Choose a test number from each interval and test in $x^2 + 3x \geq 0$.

For $(-\infty, -3)$, we choose $x = -4$.

$$(-4)^2 + 3(-4) \geq 0$$
$$16 - 12 \geq 0$$
$$4 \geq 0$$

This is true, so $(-\infty, -3)$ is in the solution set.

For $(-3, 0)$, we choose $x = -1$.

$$(-1)^2 + 3(-1) \geq 0$$
$$1 - 3 \geq 0$$
$$-2 \geq 0$$

This is false, so $(-3, 0)$ is not in the solution set.

For $(0, \infty)$, we choose $x = 1$.

$$1^2 + 3(1) \geq 0$$
$$1 + 3 \geq 0$$
$$4 \geq 0$$

This is true, so $(0, \infty)$ is in the solution set.

Connection Notice that the solution set for $x^2 + 3x \ge 0$ corresponds to the intervals where the graph of $f(x) = x^2 + 3x$ is above the x-axis and the x-intercepts are endpoints in the intervals and are included in the solution set.

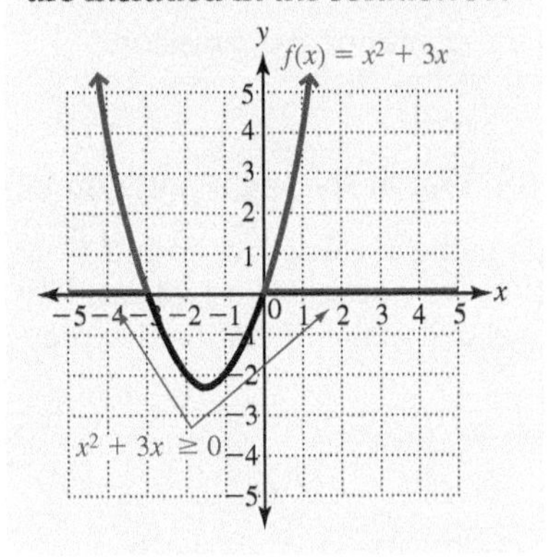

The solution set is $(-\infty, -3] \cup [0, \infty)$, which is graphed next.

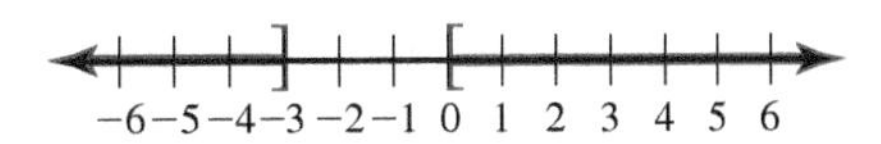

c. $(x - 1)^2 > -2$

Solution: Because $(x - 1)^2$ is always 0 or positive, it is always greater than -2. So every real number is a solution.

Solution set: $\mathbb{R}$, or $(-\infty, \infty)$

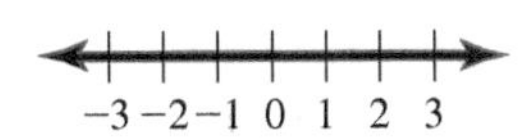

d. $(x - 1)^2 < -2$

Solution: Because $(x - 1)^2$ is never negative, its value can never be less than -2. So there are no real solutions for $(x - 1)^2 < -2$.

Solution set: $\varnothing$

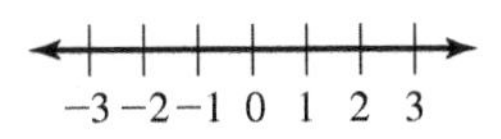

Connection Look at the graph of $f(x) = (x - 1)^2$.

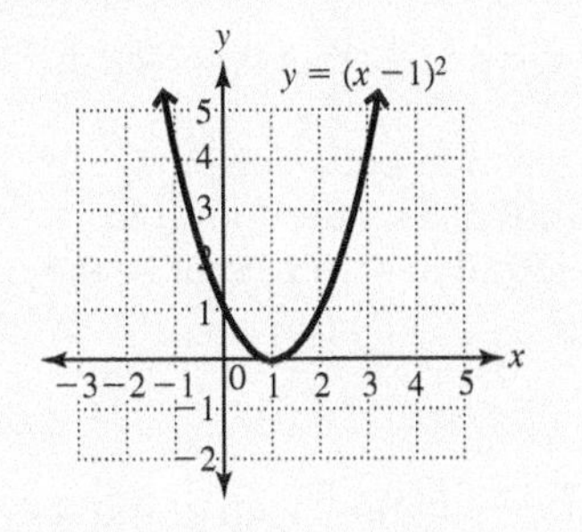

Notice that $f(x)$ is always 0 or positive. So $(x - 1)^2 > -2$ is true for all real x-values and $(x - 1)^2 < -2$ is false for all real x-values.

Your Turn 1 Solve. Write the solution set using interval notation; then graph the solution set on a number line.

a. $x^2 - 2x - 8 \le 0$ b. $x^2 - 4x > 0$ c. $(3x - 8)^2 < -5$

Other Polynomial Inequalities

We use a similar procedure for expressions with more than two factors. In the following example, we condense the procedure to a table.

Example 2 Solve $(x + 4)(x + 1)(x - 3) \le 0$. Write the solution set in interval notation; then graph the solution set on a number line.

Solution: $(x + 4)(x + 1)(x - 3) = 0$ Write the related equation.

$x + 4 = 0$ or $x + 1 = 0$ or $x - 3 = 0$ Set each factor equal to 0.

$x = -4$ $x = -1$ $x = 3$ Solve the equations.

Plot -4, -1, and 3 on a number line and note the intervals.

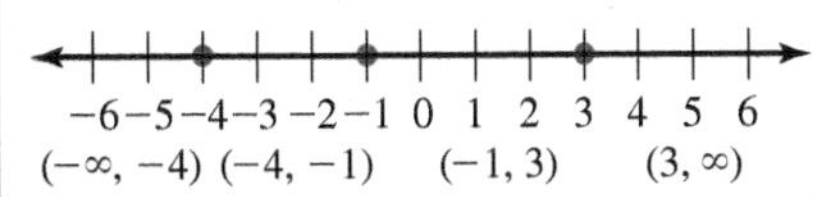

Note -4, -1, and 3 are included in the solution set because the inequality symbol is $\le$. Remember that we test intervals around these values, so they are not included in those intervals.

Interval	$(-\infty, -4)$	$(-4, -1)$	$(-1, 3)$	$(3, \infty)$
Test Number	-5	-2	0	4
Test Results	$-32 \le 0$	$10 \le 0$	$-12 \le 0$	$40 \le 0$
True or False	True	False	True	False

Therefore, the solution set is $(-\infty, -4] \cup [-1, 3]$, which is graphed next.

Note Because the inequality is $\le$, -4, -1, and 3 are included in the solution set.

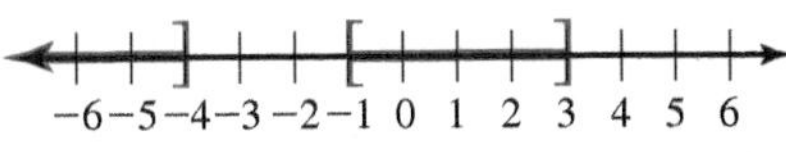

Answers to Your Turn 1

a. $[-2, 4]$

b. $(-\infty, 0) \cup (4, \infty)$

c. $\varnothing$

Your Turn 2 Solve. Write the solution set using interval notation; then graph the solution set on a number line.

$$(x + 3)(x - 2)(x + 5) \le 0$$

Objective 2 Solve rational inequalities.

Now we consider solving **rational inequalities.**

Definition **Rational inequality:** An inequality containing a rational expression.

For example, $\frac{x + 2}{x - 3} > 4$ is a rational inequality. Recall that in solving a polynomial inequality, we divided the number line into intervals using x-values we found from solving its related equation. With rational inequalities, we not only find values that solve the related equation but also must consider values that make the rational expression undefined $(\text{denominator} = 0)$.

Procedure **Solving Rational Inequalities**

1. Find all values that make any denominator equal to 0. These values must be excluded from the solution set.
2. Solve the related equation.
3. Plot the numbers found in steps 1 and 2 on a number line and label the intervals.
4. Choose a test number from each interval and determine whether it solves the inequality.
5. The solution set is the union of all regions whose test number solves the inequality. If the inequality symbol is $\le$ or $\ge$, include the values found in step 2. The solution set never includes the values found in step 1 because they make a denominator equal to 0.

Example 3 Solve. Write the solution set using interval notation; then graph the solution set on a number line.

a. $\frac{x - 4}{x + 2} \ge 0$

Solution: First, we find the values that make the denominator equal to 0.

$$x + 2 = 0$$
$$x = -2$$

Now solve the related equation.

$$\frac{x - 4}{x + 2} = 0 \quad \text{Write the related equation.}$$

$$(x + 2)\frac{x - 4}{x + 2} = (x + 2)(0) \quad \text{Multiply both sides by the LCD, } x + 2.$$

$$x - 4 = 0 \quad \text{Simplify.}$$

$$x = 4$$

Plot -2 and 4 on a number line and label the intervals.

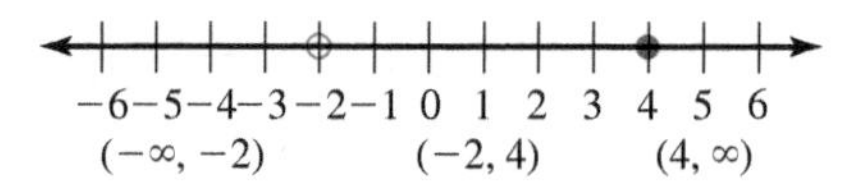

Answer to Your Turn 2
$(-\infty, -5] \cup [-3, 2]$

−7 −6 −5 −4 −3 −2 −1 0 1 2 3

Again, we use a table.

Interval	$(-\infty, -2)$	$(-2, 4)$	$(4, \infty)$
Test Number	-3	0	5
Test Results	$7 \geq 0$	$-2 \geq 0$	$\frac{1}{7} \geq 0$
True or False	True	False	True

Note Remember, because -2 makes $\frac{x-4}{x+2}$ undefined, it is not included in the solution set. Because the inequality is $\geq$, 4 is included in the solution set.

The solution set is $(-\infty, -2) \cup [4, \infty)$ and is graphed next.

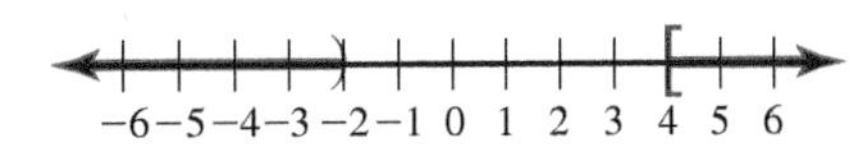

b. $\frac{x+5}{x-1} < 4$

Solution: Find the values that make the denominator equal to 0.

$$x - 1 = 0$$
$$x = 1$$

Solve the related equation.

$$\frac{x+5}{x-1} = 4 \quad \text{Write the related equation.}$$

$$(x-1)\frac{x+5}{x-1} = (x-1)(4) \quad \text{Multiply both sides by the LCD, } x - 1.$$

$$x + 5 = 4x - 4 \quad \text{Simplify.}$$

$$-3x = -9 \quad \text{Subtract } 4x \text{ and 5 from both sides.}$$

$$x = 3$$

Plot 1 and 3 on a number line and label the intervals.

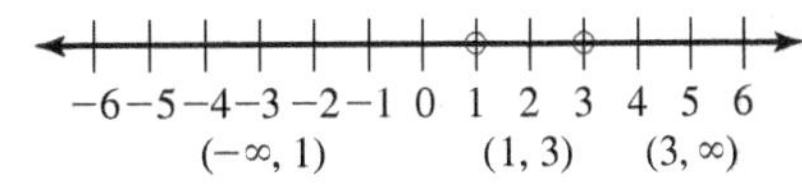

Interval	$(-\infty, 1)$	$(1, 3)$	$(3, \infty)$
Test Number	0	2	4
Test Results	$-5 < 4$	$7 < 4$	$3 < 4$
True or False	True	False	True

Note Because 1 makes $\frac{x+5}{x-1}$ undefined, it is not included in the solution set. Also, 3 is not in the solution set because $\frac{x+5}{x-1} = 4$ when $x = 3$.

The solution set is $(-\infty, 1) \cup (3, \infty)$ and is graphed next.

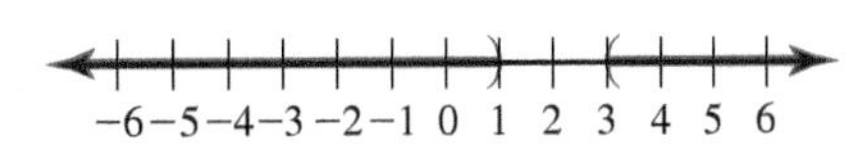

Your Turn 3 Solve. Write the solution set using interval notation; then graph the solution set on a number line.

a. $\frac{x+3}{x-1} < 0$

b. $\frac{x-2}{x+4} \leq 3$

Answers to Your Turn 3

a. $(-3, 1)$

−4 −3 −2 −1 0 1 2

b. $(-\infty, -7] \cup (-4, \infty)$

−9 −8 −7 −6 −5 −4 −3 −2

9.5 Exercises

For Extra Help MyMathLab®

Note: Exercises marked with a ★ represent challenging exercises.

Objective 1

Prep Exercise 1 Explain how to solve $x^2 + 2x - 15 \geq 0$.

Prep Exercise 2 If the graph of $y = ax^2 + bx + c$ intersects the x-axis at -2 and 2, what intervals need to be tested to solve $ax^2 + bx + c > 0$?

Prep Exercise 3 Is it possible to have a quadratic inequality whose solution set is the empty set? Explain.

Prep Exercise 4 Is it possible to have a quadratic inequality whose solution set is one number? Explain.

For Exercises 1–8, the graph of a quadratic function is given. Use the graph to solve each equation and inequality. For solution sets that involve intervals, use interval notation. See Objective 1.

1. a. $x^2 + 6x + 5 = 0$

b. $x^2 + 6x + 5 < 0$

c. $x^2 + 6x + 5 > 0$

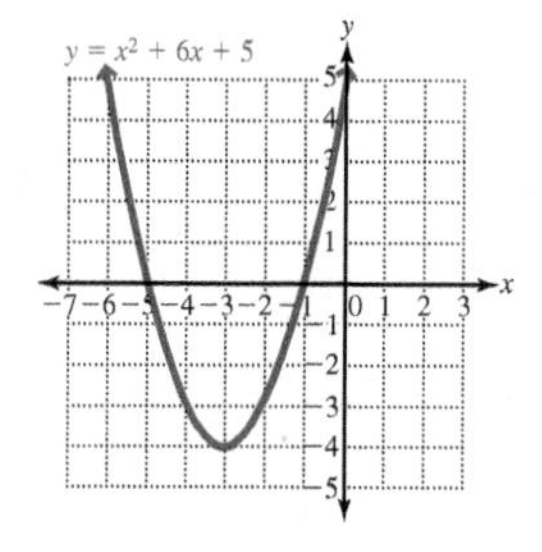

2. a. $x^2 + x - 2 = 0$

b. $x^2 + x - 2 > 0$

c. $x^2 + x - 2 < 0$

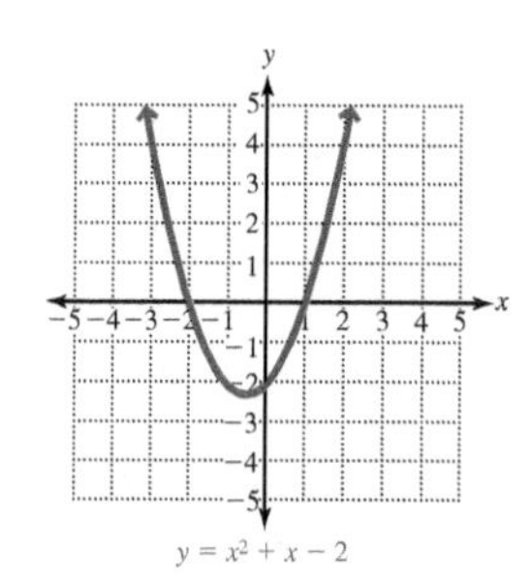

3. a. $-x^2 + 2x + 3 = 0$

b. $-x^2 + 2x + 3 \leq 0$

c. $-x^2 + 2x + 3 \geq 0$

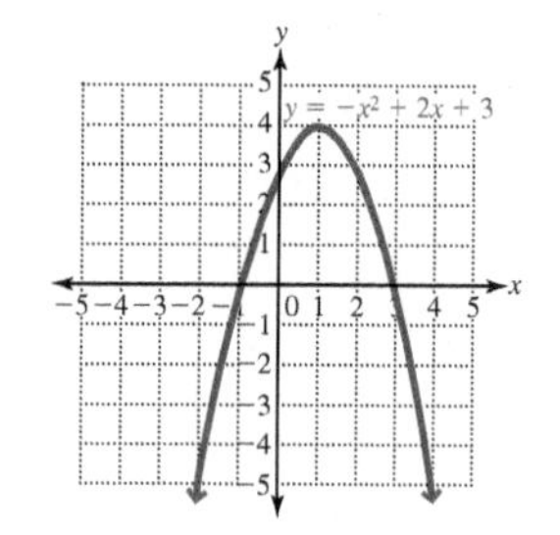

4. a. $-x^2 + 5x = 0$

b. $-x^2 + 5x \leq 0$

c. $-x^2 + 5x \geq 0$

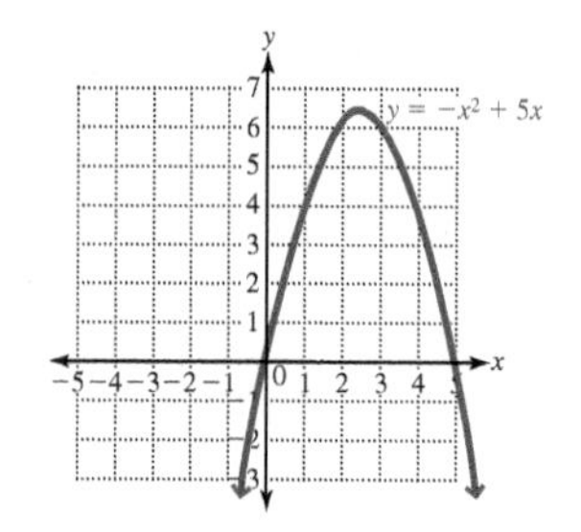

5. a. $x^2 + 4x + 4 = 0$

b. $x^2 + 4x + 4 > 0$

c. $x^2 + 4x + 4 < 0$

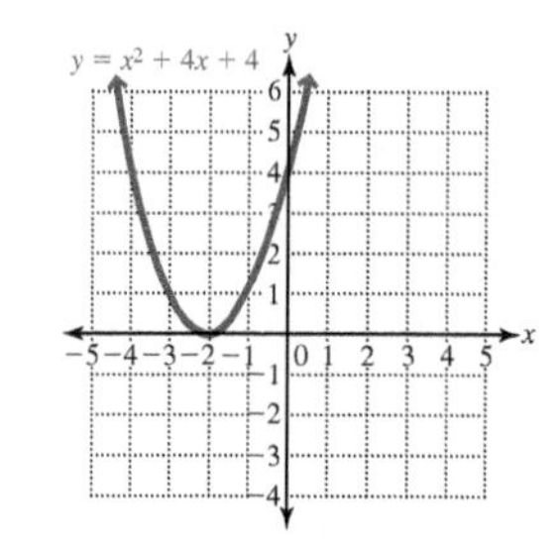

6. a. $x^2 - 6x + 9 = 0$

b. $x^2 - 6x + 9 < 0$

c. $x^2 - 6x + 9 > 0$

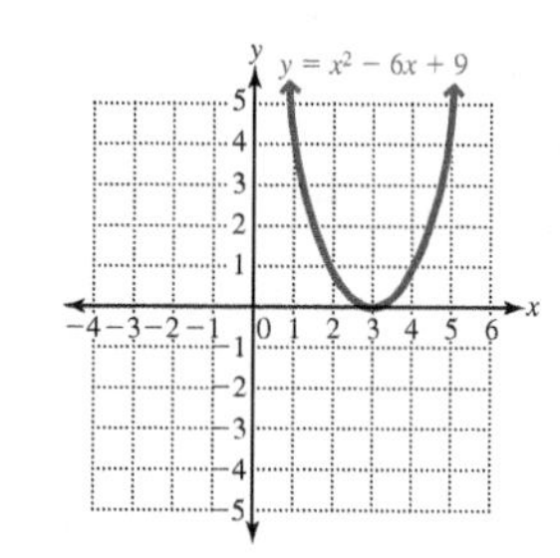

7. **a.** $x^2 + x + 4 = 0$

b. $x^2 + x + 4 \le 0$

c. $x^2 + x + 4 \ge 0$

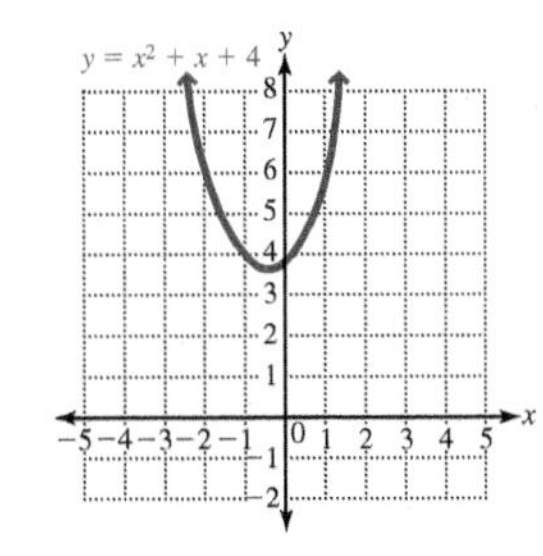

8. **a.** $-x^2 - x - 2 = 0$

b. $-x^2 - x - 2 \le 0$

c. $-x^2 - x - 2 \ge 0$

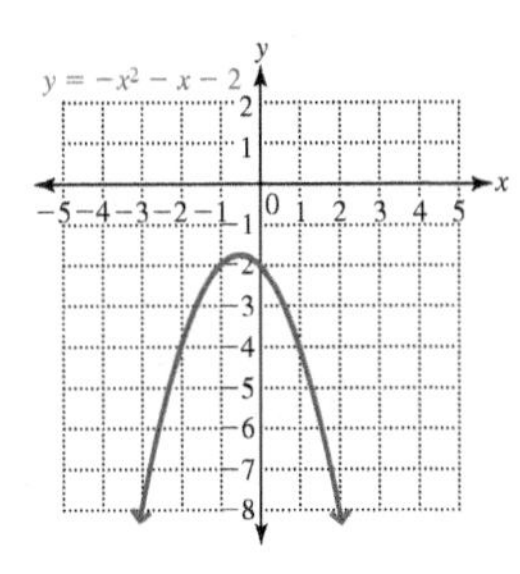

For Exercises 9–38, solve. Write the solution set using interval notation; then graph the solution set on a number line. See Examples 1 and 2.

9. $(x + 4)(x + 2) < 0$

10. $(x + 3)(x + 1) < 0$

11. $(x - 2)(x - 5) > 0$

12. $(x + 4)(x - 3) > 0$

13. $x^2 + 5x + 4 < 0$

14. $x^2 + 6x + 5 < 0$

15. $x^2 - 4x + 3 > 0$

16. $x^2 - 8x + 7 > 0$

17. $b^2 - 6b + 8 \le 0$

18. $c^2 - 9c + 14 \le 0$

19. $y^2 - 5 \ge 4y$

20. $z^2 - 3 \ge 2z$

21. $a^2 - 3a < 10$

22. $b^2 + 4b \le 21$

23. $y^2 + 6y + 9 \ge 0$

24. $x^2 + 10x + 25 \ge 0$

25. $2c^2 - 4c + 7 \le 0$

26. $2a^2 - 3a + 5 \le 0$

27. $4r^2 + 21r + 5 > 0$

28. $4s^2 + 5s - 6 < 0$

29. $x^2 - 5x > 0$

30. $x^2 + 3x > 0$

31. $x^2 \le 6x$

32. $x^2 \le 3x$

33. $(x + 1)^2 \ge -9$

34. $(2x - 1)^2 > -4$

35. $(x - 4)(x + 2)(x + 4) \ge 0$

36. $(x + 5)(x - 3)(x + 1) \ge 0$

37. $(x + 2)(x + 6)(x - 1) < 0$

38. $(x - 3)(x + 3)(x + 1) < 0$

For Exercises 39–44, solve using the quadratic formula.

39. $3c^2 + 4c - 1 < 0$

40. $5a^2 - 10a + 2 < 0$

41. $4r^2 + 8r - 3 \ge 0$

42. $2y^2 - 6y - 5 \ge 0$

43. $-0.2a^2 - 1.6a - 2 \le 0$

44. $-0.1x^2 - 1.2x + 4 > 0$

Objective 2

Prep Exercise 5 The quadratic inequality $(x - 2)(x + 5) < 0$ and the rational inequality $\frac{x-2}{x+5} < 0$ have the same solution sets. Explain how this is possible.

Prep Exercise 6 To solve $\frac{x+2}{(x+5)(x-3)} \ge 0$, what intervals need to be tested?

For Exercises 45–62, solve the rational inequalities. Write the solution set using interval notation; then graph the solution set on a number line. See Example 3.

45. $\frac{a+4}{a-1} > 0$

46. $\frac{m-6}{m+2} \ge 0$

47. $\frac{n+1}{n+5} \le 0$

48. $\frac{b-2}{b+3} < 0$

49. $\frac{6}{x+4} > 0$

50. $\frac{3}{x-3} < 0$

51. $\frac{c}{c+3} < 3$

52. $\frac{m}{m-2} < 2$

53. $\frac{a+5}{a-4} > 4$

54. $\frac{j+4}{j-4} > 5$

55. $\frac{p+3}{p-3} \geq 4$

56. $\frac{c+1}{c-4} \geq 6$

57. $\frac{(k+3)(k-2)}{k-5} \leq 0$

58. $\frac{(m+1)(m-3)}{m+4} \geq 0$

59. $\frac{(2x-1)^2}{x} \geq 0$

60. $\frac{(4x+3)^2}{x} \geq 0$

61. $\frac{x^2-7x+10}{x+1} < 0$

62. $\frac{x^2-2x-8}{x-1} \geq 0$

For Exercises 63–66, solve.

63. If a ball is thrown upward with an initial velocity of 80 feet per second from the top of a building 96 feet high, the height, h, above the ground after t seconds is given by $h = -16t^2 + 80t + 96$, where h is in feet.

a. After how many seconds will the ball hit the ground? (*Hint*: Think about the value of h when the ball is on the ground.)

b. After how many seconds is the ball 192 feet above the ground?

c. Find the interval of time when the ball is more than 192 feet above the ground.

d. Use the answers to parts a, b, and c to find the intervals of time when the ball is less than 192 feet above the ground.

64. If an object is dropped from the top of a cliff that is 256 feet high, the equation giving the height, h, above the ground is $h = 256 - 16t^2$, where h is in feet and t is in seconds.

a. After how many seconds will the object hit the ground?

b. After how many seconds is the object 112 feet above the ground?

c. Find the interval of time when the object is more than 112 feet above the ground.

d. Find the interval of time when the ball is less than 112 feet above the ground.

65. In the parallelogram shown, the height is to be 2 inches less than x. The base is to be 6 inches more than x.

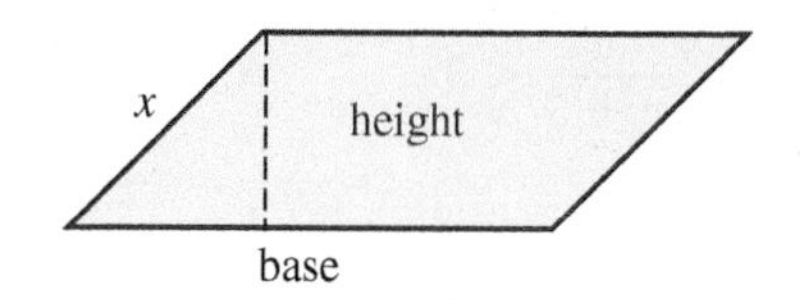

a. Find the range of values for x so that the area of the parallelogram is at least 20 square inches.

b. Find the range of values for the base and the height.

66. An auditorium is to be designed roughly in the shape of a box. The height is set to be 40 feet. It is preferred that the length of the space be 20 feet more than the width.

a. Find the range of values for the width so that the volume of the space is at least 140,000 cubic feet.

b. Find the range of values for the length.

For Exercises 67 and 68, use the formula for the slope of a line: $m = \dfrac{y_2 - y_1}{x_2 - x_1}$.

★ **67.** Suppose a line is to be drawn in the coordinate plane so that it passes through the point at $(2, 5)$. Find the range of values for the second point (x_2, y_2) so that $x_2 = y_2$ and the slope of the line is at most $\frac{1}{2}$.

★ **68.** Suppose a line is to be drawn in the coordinate plane so that it passes through the point at $(-3, 2)$. Find the range of values for the second point (x_2, y_2) so that $x_2 = y_2$ and the slope of the line is at least $\frac{1}{4}$.

Review Exercises

Exercises 1 and 2 Expressions

[9.1] *For Exercises 1 and 2, add a term to the expression to make it a perfect square; then factor the perfect square.*

1. $x^2 - 6x$

2. $x^2 + 5x$

Exercises 3–6 Equations and Inequalities

[8.6] **3.** Solve: $\sqrt{x - 2} = 4$

[9.4] **4.** Does the graph of $f(x) = -x^2 + 2x - 3$ open upward or downward? Why?

[9.4] **5.** Find the x-intercepts for $y = x^2 + 4x - 12$.

[9.4] **6.** What are the coordinates of the vertex of the graph of $f(x) = x^2 - 6x + 5$?

Chapter 9 Summary and Review Exercises

Complete each incomplete definition, rule, or procedure; study the key examples; and then work the related exercises.

9.1 The Square Root Principle and Completing the Square

Definitions/Rules/Procedures	Key Example(s)
Square root principle If $x^2 = a$, where a is a real number, then $x =$ __________ or $x =$ __________. It is common to indicate the positive and negative solutions by writing __________.	Solve. **a.** $x^2 = 36$ **Solution:** $x = \pm\sqrt{36}$ $x = \pm 6$ **b.** $(x - 3)^2 = 12$ **Solution:** $x - 3 = \pm\sqrt{12}$ $x = 3 \pm 2\sqrt{3}$

Exercises 1–11 Equations and Inequalities

[9.1] *For Exercises 1–8, solve and check.*

1. $x^2 = 16$

2. $y^2 = \frac{1}{36}$

3. $k^2 + 2 = 30$

4. $3x^2 = 42$

5. $5h^2 + 24 = 9$

6. $(x + 7)^2 = 25$

7. $(x - 9)^2 = -16$

8. $\left(m + \frac{3}{5}\right)^2 = \frac{16}{25}$

[9.1] *For Exercises 9–11, solve.*

9. A crop circle appeared July 7, 2003, in Windham Hill, England, and covered $22{,}500\pi$ square feet. Find the radius of the circle.

10. Using the formula $E = \frac{1}{2}mv^2$, where E represents the kinetic energy in joules of an object with a mass of m kilograms and a velocity of v meters per second, find the velocity of an object with a mass of 50 kilograms and 400 joules of kinetic energy.

11. A right circular cylinder is to be constructed so that its volume is equal to that of a sphere with a radius of 9 inches. If the cylinder is to have a height of 4 inches, find the radius of the cylinder.

Definitions/Rules/Procedures	Key Example(s)
To solve a quadratic equation by completing the square: **1.** Write the equation in the form __________. **2.** Complete the square by adding __________ to both sides. **3.** Write the completed square in __________ form. **4.** Use the __________ principle to eliminate the square. **5.** Isolate the __________. **6.** Simplify as needed.	Solve by completing the square. $x^2 + 6x - 7 = 8$ $x^2 + 6x = 15$ — Add 7 to both sides. $x^2 + 6x + 9 = 15 + 9$ — Complete the square. $(x + 3)^2 = 24$ — Factor. $x + 3 = \pm\sqrt{24}$ — Use the square root principle. $x = -3 \pm 2\sqrt{6}$ — Subtract 3 from both sides and simplify the square root.

Exercises 12–15 Equations and Inequalities

[9.1] *For Exercises 12–15, solve by completing the square.*

12. $m^2 + 8m = -7$ **13.** $u^2 - 6u - 12 = 100$ **14.** $2b^2 - 6b + 7 = 0$ **15.** $u^2 + \frac{1}{4}u = \frac{3}{4}$

9.2 Solving Quadratic Equations Using the Quadratic Formula

Definitions/Rules/Procedures	Key Example(s)
To solve a quadratic equation in the form $ax^2 + bx + c = 0$, where $a \neq 0$, use the quadratic formula: $x =$ ______________	Solve $3x^2 - 4x + 2 = 0$. **Solution:** In the quadratic formula, replace a with 3, b with -4, and c with 2. $x = \frac{-(-4) \pm \sqrt{(-4)^2 - 4(3)(2)}}{2(3)} = \frac{4 \pm \sqrt{16 - 24}}{6} = \frac{4 \pm \sqrt{-8}}{6} = \frac{4 \pm 2\sqrt{-2}}{6} = \frac{2}{3} \pm \frac{\sqrt{2}}{3}i$

Exercises 16–21 Equations and Inequalities

[9.2] *For Exercises 16–19, solve using the quadratic formula.*

16. $p^2 - 5 = -2p$ **17.** $3x^2 - 2x + 1 = 0$ **18.** $2t^2 + t - 5 = 0$ **19.** $2x^2 + 0.1x - 0.03 = 0$

[9.2] *For Exercises 20 and 21, solve.*

20. The length of a small rectangular shed is 4 feet more than its width. If the area of the base is 285 square feet, what are the dimensions of the base of the shed?

21. A ramp is constructed so that it is a right triangle with a base that is 9 feet longer than its height. If the hypotenuse is 17 feet, find the dimensions of the base and height. Round to the nearest hundredth.

Definitions/Rules/Procedures	Key Example(s)
Given a quadratic equation in the form $ax^2 + bx + c = 0$, where a, b, and c are rational numbers and $a \neq 0$, the discriminant is ______________. If the **discriminant is positive**, the equation has ______________ real-number solutions. The solutions will be rational if the discriminant is a(n) ______________ and irrational otherwise. If the **discriminant is 0**, the equation has ______________ rational solution. If the **discriminant is negative**, the equation has two ______________ solutions.	Use the discriminant to determine the number and type of solutions for $5x^2 - 7x + 8 = 0$. **Solution:** In the discriminant, replace a with 5, b with -7, and c with 8. $(-7)^2 - 4(5)(8) = 49 - 160 = -111$ Because the discriminant is negative, the equation has two nonreal complex solutions.

Exercises 22–25 Equations and Inequalities

[9.2] *For Exercises 22–25, find the discriminant and determine the number and type of solutions for the equation.*

22. $b^2 - 4b - 12 = 0$ **23.** $6z^2 - 7z + 5 = 0$ **24.** $k^2 + 6k + 9 = 0$ **25.** $0.8x^2 + 1.2x + 0.3 = 0$

9.3 Solving Equations That Are Quadratic in Form

Definitions/Rules/Procedures	Key Example(s)
An equation is **quadratic in form** if it can be rewritten as a quadratic equation $au^2 + bu + c = 0$, where $a \neq 0$ and u is a(n) ________ or a(n) ________. If an equation involves rational expressions, multiply both sides by the ________. Be sure to check for ________ solutions.	Solve $\frac{6}{x-2} + \frac{6}{x-1} = 5.$ **Solution:** First multiply both sides by the LCD, $(x-2)(x-1)$. $6(x-1) + 6(x-2) = 5(x-2)(x-1)$ $12x - 18 = 5x^2 - 15x + 10$ Multiply. $0 = 5x^2 - 27x + 28$ Write in $ax^2 + bx + c = 0$ form. $0 = (5x-7)(x-4)$ Factor. $5x - 7 = 0$ or $x - 4 = 0$ Use the zero-factor theorem. $x = \frac{7}{5}$ $x = 4$ Solve each equation. **Check:** Verify that $\frac{7}{5}$ and 4 solve the original equation.

Exercises 26–29 Equations and Inequalities

[9.3] *For Exercises 26–29, solve. Identify any extraneous solutions.*

26. $\frac{1}{y} + \frac{1}{y+3} = \frac{2}{3}$ **27.** $\frac{1}{u} + \frac{1}{u-5} = \frac{10}{u^2 - 25}$ **28.** $6 - 5x^{-1} + x^{-2} = 0$ **29.** $2 - 3x^{-1} - x^{-2} = 0$

Definitions/Rules/Procedures	Key Example(s)
If an equation has radicals, ________ a radical and square both sides. Continue the process as needed until all radicals have been eliminated. Solve the resulting equation and check for ________ solutions.	Solve $\sqrt{4x+4} = 2x - 2.$ $4x + 4 = 4x^2 - 8x + 4$ Square both sides. $0 = 4x^2 - 12x$ Write in $ax^2 + bx + c = 0$ form. $0 = 4x(x-3)$ Factor. $4x = 0$ or $x - 3 = 0$ Use the zero-factor theorem. $x = 0$ $x = 3$ Solve each equation. **Check:** $x = 0$: $\sqrt{4(0)+4} = 2(0) - 2$; $\sqrt{0+4} = 0 - 2$; $2 = -2$ False $x = 3$: $\sqrt{4(3)+4} = 2(3) - 2$; $\sqrt{12+4} = 6 - 2$; $4 = 4$ True $x = 3$ is the only solution. ($x = 0$ is extraneous.)

Exercises 30–33 Equations and Inequalities

[9.3] *For Exercises 30–33, solve. Identify any extraneous solutions.*

30. $14\sqrt{x} + 45 = 0$ **31.** $\sqrt{4m} = 3m - 1$ **32.** $\sqrt{6r + 13} = 2r + 1$ **33.** $\sqrt{21t + 2} + t = 2 + 4t$

Definitions/Rules/Procedures	Key Example(s)
To solve equations that are **quadratic in form** using substitution: **1.** Rewrite the equation so that it is in the form ________________. **2.** Solve the quadratic equation for ____________. **3.** ____________ for u and solve. **4.** Check the solutions.	Solve $(a + 2)^2 + 7(a + 2) + 12 = 0.$ $u^2 + 7u + 12 = 0$ Substitute u for $a + 2$. $(u + 3)(u + 4) = 0$ Factor. $u + 3 = 0$ or $u + 4 = 0$ Use the zero-factor $u = -3$ $u = -4$ theorem. $a + 2 = -3$ or $a + 2 = -4$ Substitute $a + 2$ for u; $a = -5$ $a = -6$ solve for a. **Check:** Verify that -5 and -6 solve the original equation.

Exercises 34–39 Equations and Inequalities

[9.3] *For Exercises 34–39, solve the equations using substitution. Identify any extraneous solutions.*

34. $x^4 - 5x^2 + 6 = 0$

35. $2m^4 - 3m^2 + 1 = 0$

36. $6(x + 5)^2 - 5(x + 5) + 1 = 0$

37. $\left(\frac{x - 1}{3}\right)^2 + 10\left(\frac{x - 1}{3}\right) + 9 = 0$

38. $p^{2/3} - 11p^{1/3} + 24 = 0$

39. $5a^{1/2} + 13a^{1/4} - 6 = 0$

9.4 Graphing Quadratic Functions

Definitions/Rules/Procedures	Key Example(s)
The graph of a function in the form $f(x) = a(x - h)^2 + k$ is a parabola with vertex at ____________. The equation of the axis of symmetry is ____________. The parabola opens upward if ____________ and downward if ____________. The larger the $\lvert a \rvert$, the narrower the graph.	For $f(x) = -3(x + 1)^2 - 2$: **a.** Determine whether the graph opens upward or downward. **b.** Find the vertex. **c.** Write the equation of the axis of symmetry. **d.** Graph. **Answers:** **a.** downward **b.** $(-1, -2)$ **c.** $x = -1$ **d.** 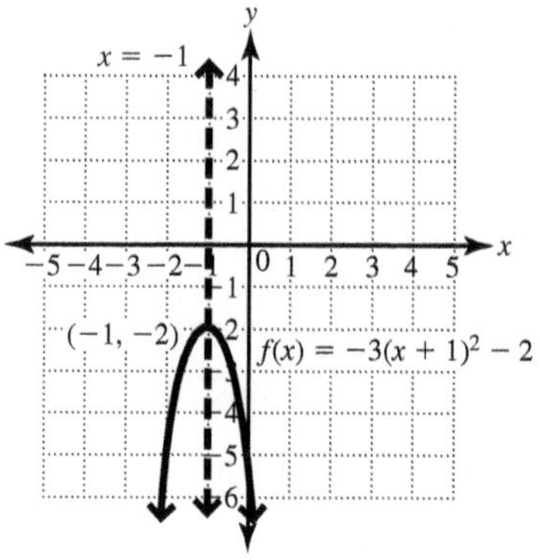

Exercises 40–43 Equations and Inequalities

[9.4] *For Exercises 40–43:* *a. Find the x- and y-intercepts.*
b. State whether the parabola opens upward or downward.
c. Find the coordinates of the vertex.
d. Write the equation of the axis of symmetry.
e. Graph.

40. $f(x) = -2x^2$

41. $g(x) = \frac{1}{2}x^2 + 1$

42. $h(x) = -\frac{1}{3}(x - 2)^2$

43. $k(x) = 4(x + 3)^2 - 2$

Definitions/Rules/Procedures	Key Example(s)
Given an equation in the form $f(x) = ax^2 + bx + c$, to determine the vertex: **1.** Find the x-coordinate using the formula $x =$ ________. **2.** Find the y-coordinate by evaluating $f\left(-\frac{b}{2a}\right)$ or the formula ________.	For $f(x) = 2x^2 - 12x + 19$: **a.** Determine whether the graph opens upward or downward. **b.** Find the vertex. **c.** Write the equation of the axis of symmetry. **d.** Graph. **Answers:** **a.** upward **b.** For the x-coordinate, use $-\frac{b}{2a}$. $x = -\frac{(-12)}{2(2)} = 3 \quad a = 2, b = -12$ For the y-coordinate, evaluate $f(3)$. $f(3) = 2(3)^2 - 12(3) + 19$ $= 18 - 36 + 19 = 1$ Vertex: $(3, 1)$ **c.** $x = 3$ **d.** 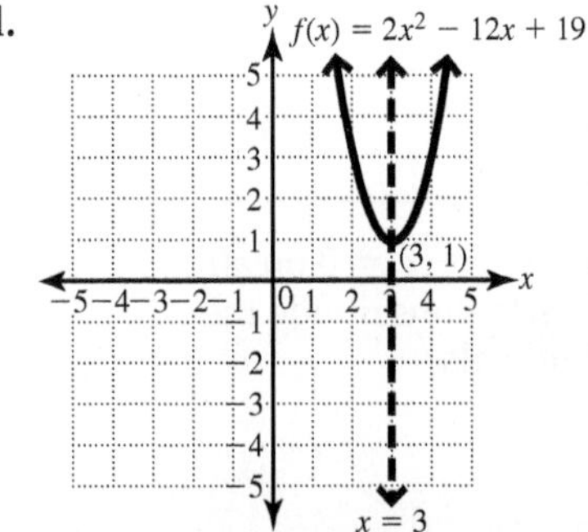 **Note** As an alternative approach, we could have transformed $f(x) = 2x^2 - 12x + 19$ to $f(x) = 2(x - 3)^2 + 1$ by completing the square.

Exercises 44–47 Equations and Inequalities

[9.4] *For Exercises 44 and 45:*
a. *Write the equation in the form $f(x) = a(x - h)^2 + k$.*
b. *Find the x- and y-intercepts.*
c. *State whether the parabola opens upward or downward.*
d. *Find the coordinates of the vertex.*
e. *Write the equation of the axis of symmetry.*
f. *Graph.*
g. *Find the domain and range.*

44. $m(x) = x^2 + 2x - 1$

45. $p(x) = -0.5x^2 + 4x - 6$

[9.4] *For Exercises 46 and 47, solve.*

46. With an initial velocity of 24 feet per second, an acrobat is launched upward from one end of a lever. The function $h = -16t^2 + 24t$ describes the height, h, of the acrobat t seconds after being launched.

a. After how many seconds does the acrobat reach maximum height?

b. What is the maximum height the acrobat reaches?

c. How long is the acrobat in the air?

d. Graph the function.

47. The longest punt on record in the NFL was by Steve O'Neal in a game between the New York Jets and the Denver Broncos on September 21, 1969. The function $y = -0.03x^2 + 2.16x$ models the trajectory of the punt. (Note that x and y are distances in yards.)

a. Find the maximum height the punt reached.

b. Find the horizontal distance the punt traveled.

Of Interest

The yardage you found is the horizontal distance the punt carried in the air. It was actually recorded as a 98-yard punt, which includes the amount of roll after the punt landed and was downed.

9.5 Solving Nonlinear Inequalities

Definitions/Rules/Procedures	Key Example(s)
A **quadratic inequality** is an inequality that can be written in the form ________ or ________, where $a \neq 0$. Note: The symbols $<$ and $>$ can be replaced with $\leq$ and $\geq$.	Solve $x^2 + 2x - 15 \geq 0$. $x^2 + 2x - 15 = 0$ Write the related equation. $(x + 5)(x - 3) = 0$ Factor. $x + 5 = 0$ or $x - 3 = 0$ Use the zero-factor theorem. $x = -5$ $x = 3$ Solve each equation.
Solving quadratic inequalities 1. Solve the related equation ________. 2. Plot the solutions of $ax^2 + bx + c = 0$ on a(n) ________. These solutions divide the ________ into intervals.	Plot -5 and 3 on a number line and label the intervals. −8 −7 −6 −5 −4 −3 −2 −1 0 1 2 3 4 5 6 $(-\infty, -5)$ $(-5, 3)$ $(3, \infty)$

Definitions/Rules/Procedures	Key Example(s)
3. Choose a test number from each interval and substitute the number into the inequality. If the test number makes the inequality *true*, then all numbers in that interval will ______________. If the test number makes the inequality *false*, then ______________ in that interval will solve the inequality.	Choose a number from each interval and test it in the original inequality.
4. State the solution set of the inequality: It is the ______________ of all intervals that solve the inequality. If the inequality symbols are $\le$ or $\ge$, the values from step 2 are included. If the symbols are $<$ or $>$, they are not solutions.	

Interval	$(-\infty, -5)$	$(-5, 3)$	$(3, \infty)$
Test Value	-6	0	4
Result	$9 \ge 0$	$-15 \ge 0$	$9 \ge 0$
True/False	True	False	True

The solution set is $(-\infty, -5] \cup [3, \infty)$.

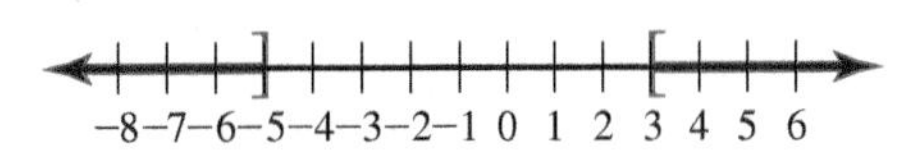

Exercises 48–52  Equations and Inequalities

[9.5] *For Exercises 48–51, solve the inequalities.*

48. $(x + 5)(x - 3) > 0$

49. $n^2 - 6n \le -8$

50. $x^2 + 9x + 14 < 0$

51. $(x + 3)(x - 1)(x - 2) \ge 0$

[9.5] *For Exercise 52, solve.*

52. In a triangle, the height is to be 2 inches less than x. The base is to be 4 inches more than x.

a. Find the range of values for x so that the area of the triangle is at least 56 square inches.

b. Find the range of values for the base and the height.

Definitions/Rules/Procedures	Key Example(s)
A **rational inequality** is an inequality containing a(n) ______________.	Solve $\dfrac{x+5}{x-1} > 4$.
Solving rational inequalities	
1. Find all values that make any denominator equal to ______________. These values must be excluded from the solution set.	Find the value(s) that make any denominator equal to 0. $x - 1 = 0$, $x = 1$
2. Solve the related ______________.	Solve the related equation.

$$\frac{x+5}{x-1} = 4 \qquad \text{Write the related equation.}$$

$$(x-1)\frac{x+5}{x-1} = (x-1)(4) \qquad \text{Multiply both sides by the LCD.}$$

$$x + 5 = 4x - 4 \qquad \text{Simplify.}$$

$$9 = 3x \qquad \text{Subtract } x \text{ from and add 5 to both sides.}$$

$$3 = x \qquad \text{Divide by 3.}$$

Definitions/Rules/Procedures

3. Plot the numbers found in steps 1 and 2 on a(n) ______________ and label the intervals.

4. Choose a(n) ______________ from each interval and determine whether it solves the inequality.

5. The solution set is the ______________ of all regions whose test number solves the inequality. If the inequality symbol is $\le$ or $\ge$, include the values found in step ______________. The solution set never includes the values found in step ______________.

Key Example(s)

Plot 1 and 3 on the number line and label the intervals.

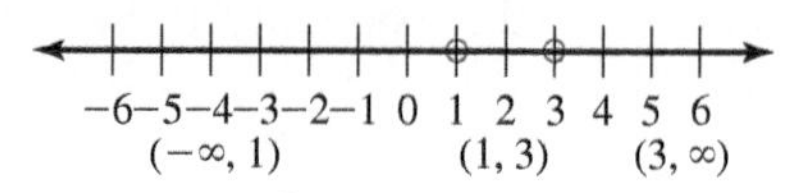

Choose a number from each interval and test it in the original inequality.

Interval	$(\infty, 1)$	$(1, 3)$	$(3, \infty)$
Test Value	0	2	4
Result	$-5 > 4$	$7 > 4$	$3 > 4$
True/False	False	True	False

The solution set is $(1, 3)$.

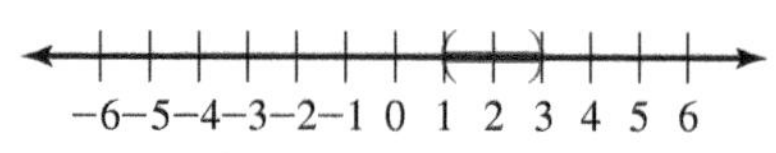

Exercises 53–56 Equations and Inequalities

[9.5] *For Exercises 53–56, solve the rational inequalities.*

53. $\dfrac{a+3}{a-1} \ge 0$

54. $\dfrac{r}{r+2} < 2$

55. $\dfrac{n-3}{n-4} \le 5$

56. $\dfrac{(k+2)(k-3)}{k-5} < 0$

Learning Strategy

Before taking an exam, make sure you're well rested and eat something beforehand. Make sure you do any homework the teacher assigns. Go back and redo the problems you struggled with in the homework and find similar problems to do.

—Ellyn G.

Chapter 9 Practice Test

For Extra Help

Step-by-step test solutions are found on the Chapter Test Prep Videos available in MyMathLab® *or on* YouTube.

For Exercises 1 and 2, use the square root principle to solve and check.

1. $x^2 = 81$

2. $(x-3)^2 = 20$

For Exercises 3 and 4, solve by completing the square.

3. $x^2 - 8x = -4$

4. $3m^2 - 6m = 5$

For Exercises 5 and 6, solve using the quadratic formula.

5. $2x^2 + x - 6 = 0$

6. $x^2 - 8x + 15 = 0$

For Exercises 7–10, solve using any method.

7. $u^2 - 16 = -6u$

8. $4w^2 + 6w + 3 = 0$

9. $4w^2 + 62 + 3 = 0$

10. $2x^2 + 4x = 0$

11. $x^2 + 16 = 0$

12. $3k^2 = -5k$

For Exercises 13–16, solve. Identify any extraneous solutions.

13. $\frac{1}{x+2} + \frac{1}{x} = \frac{5}{12}$

14. $3 - x^{-1} - 2x^{-2} = 0$

15. $9\sqrt{x} + 8 = 0$

16. $\sqrt{x+8} - x = 2$

For Exercises 17 and 18, solve the equations using substitution. Identify any extraneous solutions.

17. $9a^4 + 26a^2 - 3 = 0$

18. $(x+1)^2 + 3(x+1) - 4 = 0$

19. For $f(x) = -x^2 + 6x - 4$:

a. Find the x- and y-intercepts.

b. Write the equation in the form $f(x) = a(x-h)^2 + k$.

c. State whether the parabola opens upward or downward.

d. Find the coordinates of the vertex.

e. Write the equation of the axis of symmetry.

f. Graph.

g. Find the domain and range.

For Exercises 20 and 21: a. Solve the inequality.
b. Graph the solution set on a number line.

20. $(x+1)(x-4) \le 0$

21. $\frac{x+2}{x-1} > 0$

For Exercises 22–25, solve.

22. A ball is thrown downward from a window in a tall building. The distance, d, the ball traveled is given by the equation $d = 16t^2 + 32t$, where t is the time traveled in seconds. How long will it take the ball to fall 180 feet?

23. A rectangular parking space must have an area of 400 square feet. The length is to be 20 feet more than the width. Find the dimensions of the parking space.

24. An archer shoots an arrow in a field. Suppose the equation $y = -0.02x^2 + 1.3x + 8$ models the trajectory of the arrow. (Assume that x and y represent distances in meters and that $x \geq 0$ and $y \geq 0$.)

a. What is the maximum height the arrow reaches?

b. How far does the arrow travel horizontally?

c. Graph the trajectory of the arrow.

25. A zoo is planning to install a new aquarium tank. The tank is to be in the shape of a box with a height of 12 feet. It is preferred that the length be 15 feet more than the width.

a. Find the range of values for the width so that the volume of the space is, at most, 12,000 cubic feet.

b. Find the range of values for the length.

Chapters 1–9 Cumulative Review Exercises

For Exercises 1–3, answer true or false.

[3.1] 1. The point with coordinates $(45, -12)$ is in quadrant II.

[4.1] 2. A system of equations that has no solution is said to be consistent.

[5.2] 3. The degree of the monomial $5x^3y$ is 4.

For Exercises 4–6, fill in the blank.

[5.2] 4. The graph of a polynomial function of the form $f(x) = ax^2 + bx + c$ is a(n) ______________ that opens up if ______________.

[8.3] 5. If $\sqrt[n]{a}$ and $\sqrt[n]{b}$ are both real numbers, then $\sqrt[n]{a} \cdot \sqrt[n]{b} =$ ______________.

[9.1] 6. If $x^2 = a$ and a is a real number, then $x =$ ______________ or $x =$ ______

Exercises 7–15 Expressions

For Exercises 7–9, simplify.

[5.2] 7. $(-3x - 4) - (3x + 2)$

[5.1] 8. $(8n^2)(-7mn^3)$

[5.3] 9. $(x - 3)(4x^2 - 2x + 1)$

[6.3] *For Exercises 10 and 11, factor completely.*

10. $m^3 + 8$

11. $x^2 + 10x + 25$

For Exercises 12–14, simplify.

[7.1] 12. $\dfrac{4u^2 + 4u + 1}{u + 2u^2} \cdot \dfrac{u}{2u^2 - u - 1}$

[7.2] 13. $\dfrac{3}{x - 2} - \dfrac{2}{x + 2}$

[8.2] 14. $x^{3/4} \cdot x^{-1/4}$

[8.5] 15. Rationalize the denominator of $\dfrac{\sqrt{n}}{\sqrt{m} - \sqrt{n}}$.

Exercises 16–30 Equations and Inequalities

For Exercises 16–20, solve.

[2.1] 16. $3n - 5 = 7n + 9$

[2.5] 17. $|3x + 4| + 3 = 7$

[6.4] 18. $2x^2 + 7x = 15$

[7.4] 19. $\dfrac{5}{x - 2} - \dfrac{3}{x} = \dfrac{11}{3x}$

[8.6] 20. $\sqrt{5x - 4} = 9$

[9.2] 21. Solve $x^2 - 6x + 11 = 0$ using the quadratic formula.

For Exercises 22 and 23: ***a.*** *Graph the solution set on a number line.*
b. *Write the solution set in set-builder notation.*
c. *Write the solution set in interval notation.*

[2.4] 22. $-2 < x + 4 < 5$

[9.5] 23. $x^2 + x \leq 12$

[3.1] 24. Graph $f(x) = -2x$.

[3.3] 25. Write the equation of a line in standard form that passes through the point $(-2, 4)$ and is perpendicular to the line $3x - 2y = 6$.

[4.2] 26. Solve the system: $\begin{cases} x + y + z = 5 \\ 2x + y - 2z = -5 \\ x - 2y + z = 8 \end{cases}$

For Exercises 27–30, solve.

[4.3] 27. A trucking company finds that the average speed of experienced drivers is 10 miles per hour more than the average speed of inexperienced drivers. If inexperienced drivers take 1 hour longer than experienced drivers to drive 300 miles, find how long it takes the experienced driver to drive 300 miles.

[4.3] 28. A chemist has a bottle containing 50 ml of 15% saline solution and a bottle of 40% saline solution. She wants a 30% solution. How much of the 40% solution must be added to the 15% solution so that a 30% concentration is created?

[7.5] 29. If the wavelength of a wave remains constant, the velocity, v, of a wave is inversely proportional to its period, T. In an experiment, waves are created in a pool of water so that the period is 8 seconds and the velocity is 6 centimeters per second. If the period is increased to 15 seconds, what is the velocity?

[7.5] 30. The formula $t = \sqrt{\frac{h}{16}}$ describes the amount of time, t, in seconds, that an object falls a distance of h feet.

[8.1] a. Write an expression in simplest form for the exact amount of time an object falls a distance of 24 feet.

[8.6] b. Find the distance an object falls in 3 seconds.

CHAPTER 10

Exponential and Logarithmic Functions

Chapter Overview

Exponential and logarithmic functions are among the most important in mathematics. A few of their applications include the following:

- Population growth
- Electrical circuits
- Sound decibels
- pH
- Richter scale
- Solving of equations

10.1 Composite and Inverse Functions

Objectives

1 Find the composition of two functions.

2 Show that two functions are inverses.

3 Show that a function is one-to-one.

4 Find the inverse of a function.

5 Graph a given function's inverse function.

Warm-up

[5.2, 5.3] **1.** Simplify: $(2x - 3)^2 + 1$

[8.6] **2.** Solve for y: $x = \sqrt{y - 3}$

Objective 1 Find the composition of two functions.

In Sections 5.2–5.4, we learned to perform the basic operations with functions. We now explore a new operation. Recall that a function pairs elements from an "input" set called the *domain* with elements in an "output" set called the *range*. If the output elements of one function are used as input elements in a second function, the resulting function is the **composition** of the two functions. Composition occurs when one quantity depends on a second quantity, which, in turn, depends on a third quantity.

We can visualize the composition of two functions as a machine with two stages, which are the two functions. For example, some coffee machines have two stages. Roasted coffee beans, which are ground in the first stage (first function) are put in the machine. The grounds are then fed into the second stage (second function), which runs hot water over the grounds to produce coffee.

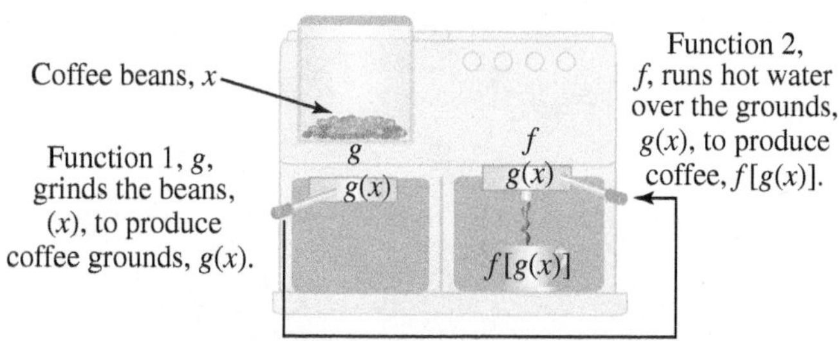

Note The notation $(g \circ f)(x)$ is read "g composed with f of x," or simply "g of f of x."

Warning Do not confuse $(f \circ g)(x)$ with $(f \cdot g)(x)$.

Definition Composition of functions: If f and g are functions, then the composition of f and g is defined as $(f \circ g)(x) = f[g(x)]$ for all x in the domain g for which $g(x)$ is in the domain of f. The composition of g and f is defined as $(g \circ f)(x) = g[f(x)]$ for all x in the domain of f for which $f(x)$ is in the domain of g.

For example, suppose a pebble is dropped in a calm lake, causing ripples to form in concentric circles. If the radius of the outer ripple increases at a rate of 2 feet per second, we can describe the length of the radius as a function of t by using $r(t) = 2t$, where t is in seconds. The area of the circle formed by the outer ripple is also a function described by $A(r) = \pi r^2$. If we substitute the output of $r(t)$, which is $2t$ for r, into $A(r)$, we form the composite function $(A \circ r)(t)$.

$$(A \circ r)(t) = A[r(t)]$$
$$= \pi(2t)^2 \quad \text{Substitute } 2t \text{ for } r \text{ in } A(r) = \pi r^2.$$
$$= 4\pi t^2 \quad \text{Simplify.}$$

The composite function $(A \circ r)(t)$ gives the area of the circle as a function of time.

Example 1 If $f(x) = 2x - 3$ and $g(x) = x^2 + 1$, find the following.

a. $(f \circ g)(2)$

Solution:
$(f \circ g)(2) = f[g(2)]$ Definition of composition
$= f(5)$ $g(2) = 2^2 + 1 = 4 + 1 = 5$
$= 2(5) - 3$ In $f(x)$, replace x with 5.
$= 7$ Simplify.

Note We first find $g(2)$, which is 5, then substitute that result into f.

Answers to Warm-up

1. $4x^2 - 12x + 10$

2. $y = x^2 + 3$

b. $(g \circ f)(2)$

Solution:
$(g \circ f)(2) = g[f(2)]$ Definition of composition
$= g(1)$ $f(2) = 2(2) - 3 = 4 - 3 = 1$
$= 1^2 + 1$ In $g(x)$, replace x with 1.
$= 2$ Simplify.

Note We could have found $(g \circ f)(2)$ from part b by first finding $(g \circ f)(x) = 4x^2 - 12x + 10$, then substituting 2 for x:
$$(g \circ f)(2) = 4(2)^2 - 12(2) + 10 = 16 - 24 + 10 = 2$$

c. $(f \circ g)(x)$

Solution:
$(f \circ g)(x) = f[g(x)]$ Definition of composition
$= f(x^2 + 1)$ Replace $g(x)$ with $x^2 + 1$.
$= 2(x^2 + 1) - 3$ In $f(x)$, replace x with $x^2 + 1$.
$= 2x^2 + 2 - 3$ Multiply.
$= 2x^2 - 1$ Add.

Note We could have found $(f \circ g)(2)$ from part a by first finding $(f \circ g)(x) = 2x^2 - 1$, then substituting 2 for x:
$$(f \circ g)(2) = 2(2)^2 - 1 = 8 - 1 = 7$$

d. $(g \circ f)(x)$

Solution:
$(g \circ f)(x) = g[f(x)]$ Definition of composition
$= g(2x - 3)$ Replace $f(x)$ with $2x - 3$.
$= (2x - 3)^2 + 1$ In $g(x)$, replace x with $2x - 3$.
$= 4x^2 - 12x + 9 + 1$ Multiply.
$= 4x^2 - 12x + 10$ Add.

Generally, $(f \circ g)(x) \neq (g \circ f)(x)$, as in Examples 1(c) and (d), but there are special functions for which $(f \circ g)(x) = (g \circ f)(x)$.

Your Turn 1 If $f(x) = x^2 + 2$ and $g(x) = 3x + 5$, find the following.

a. $f[g(-3)]$ **b.** $g[f(2)]$ **c.** $f[g(x)]$

Objective 2 Show that two functions are inverses.

Operations that undo each other, such as addition and subtraction, are called *inverse operations*. For example, if we begin with a number x and add a second number y, we have $x + y$. Now we subtract y (the number we just added) from that result, and we have $x + y - y = x$, which is the original number.

Likewise, two functions that undo each other under composition are called **inverse functions**.

Definition **Inverse functions:** Two functions f and g are inverses if and only if $(f \circ g)(x) = x$ for all x in the domain of g and $(g \circ f)(x) = x$ for all x in the domain of f.

Loosely speaking, f and g are inverse functions if you evaluate f for a value x in its domain, substitute that result into g and evaluate, and get x again and vice versa.

Procedure **Inverse Functions**

To determine whether two functions f and g are inverses of each other:

1. Show that $f[g(x)] = x$ for all x in the domain of g.
2. Show that $g[f(x)] = x$ for all x in the domain of f.

Answers to Your Turn 1

a. $f[g(-3)] = 18$
b. $g[f(2)] = 23$
c. $f[g(x)] = 9x^2 + 30x + 27$

Example 2 Verify that f and g are inverses.

a. $f(x) = 3x + 2, g(x) = \dfrac{x-2}{3}$

Solution: We need to show that $f[g(x)] = x$ and $g[f(x)] = x$.

$$f[g(x)] = f\left(\frac{x-2}{3}\right) \quad \text{Substitute } \frac{x-2}{3} \text{ for } g(x).$$
$$= 3\left(\frac{x-2}{3}\right) + 2 \quad \text{Substitute } \frac{x-2}{3} \text{ for } x \text{ in } f(x).$$
$$= x - 2 + 2 \quad \text{Multiply.}$$
$$= x \quad \text{So } f[g(x)] = x.$$
$$g[f(x)] = g(3x + 2) \quad \text{Substitute } 3x + 2 \text{ for } f(x).$$
$$= \frac{3x + 2 - 2}{3} \quad \text{Replace } x \text{ with } 3x + 2 \text{ in } g(x).$$
$$= \frac{3x}{3} \quad \text{Add.}$$
$$= x \quad \text{So } g[f(x)] = x.$$

Because $f[g(x)] = x$ and $g[f(x)] = x$, f and g are inverses.

b. $f(x) = x^3 + 5$ and $g(x) = \sqrt[3]{x-5}$

Solution: We need to show that $f[g(x)] = x$ and $g[f(x)] = x$.

$$f[g(x)] = f(\sqrt[3]{x-5}) \quad \text{Replace } g(x) \text{ with } \sqrt[3]{x-5}.$$
$$= (\sqrt[3]{x-5})^3 + 5 \quad \text{Replace } x \text{ with } \sqrt[3]{x-5} \text{ in } f(x).$$
$$= x - 5 + 5 \quad \text{Apply } (\sqrt[n]{x})^n = x.$$
$$= x \quad \text{So } f[g(x)] = x.$$
$$g[f(x)] = g(x^3 + 5) \quad \text{Replace } f(x) \text{ with } x^3 + 5.$$
$$= \sqrt[3]{x^3 + 5 - 5} \quad \text{Replace } x \text{ with } x^3 + 5 \text{ in } g(x).$$
$$= \sqrt[3]{x^3} \quad \text{Add.}$$
$$= x \quad \text{So } g[f(x)] = x.$$

Because $f[g(x)] = x$ and $g[f(x)] = x$, f and g are inverses.

Your Turn 2 Verify that $f(x) = 5x + 6$ and $g(x) = \dfrac{x-6}{5}$ are inverse functions.

Before finding inverse functions, we need to determine what types of functions have inverses. Consider the following:

$f = \{(1,2), (3,4), (5,6), (x,y)\}$; domain $= \{1,3,5,x\}$ and range $= \{2,4,6,y\}$
$g = \{(2,1), (4,3), (6,5), (y,x)\}$; domain $= \{2,4,6,y\}$ and range $= \{1,3,5,x\}$

Note that the ordered pairs in g are the ordered pairs of f with x and y interchanged. Because $(1,2)$ is in f, by definition, $f(1) = 2$. Because $(2,1)$ is in g, $g(2) = 1$. Therefore, $f[g(2)] = f(1) = 2$ and $g[f(1)] = g(2) = 1$. Similarly, $f[g(4)] = f(3) = 4$ and $g[f(3)] = g(4) = 3$. In particular, $f[g(y)] = f(x) = y$ and $g[f(x)] = g(y) = x$ for all real numbers x and y. Consequently, f and g are inverses.

In general, to find the inverse of a function, we reverse the ordered pairs by interchanging x and y. To indicate the inverse of the function f, we use a special notation f^{-1} instead of g. The following figure illustrates inverse functions.

Answer to Your Turn 2

$$f[g(x)] = f\left[\frac{x-6}{5}\right]$$
$$= 5\left(\frac{x-6}{5}\right) + 6$$
$$= x - 6 + 6$$
$$= x$$
$$g[f(x)] = g[5x + 6]$$
$$= \frac{5x + 6 - 6}{5}$$
$$= \frac{5x}{5}$$
$$= x$$

Because $f[g(x)] = x$ and $g[f(x)] = x$, f and g are inverses.

Warning Do not confuse f^{-1} with raising an expression to the negative 1 power. It is usually clear from the context that f^{-1} means an inverse function.

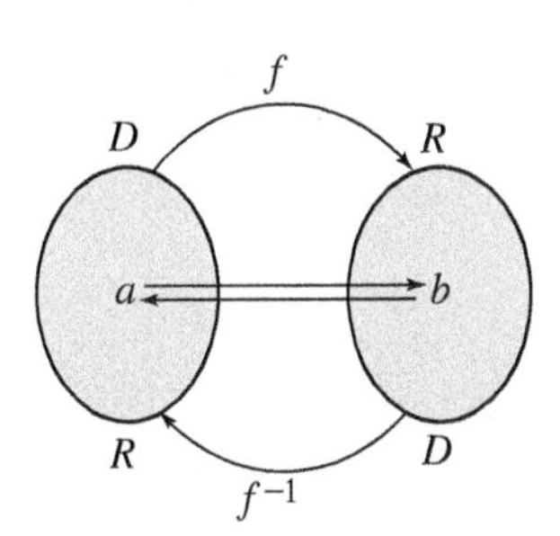

Note Notice that f sends a to b and f^{-1} sends b back to a. Hence, $f^{-1}[f(a)] = a$. Also, f^{-1} sends b to a and f sends a back to b. Hence, $f[f^{-1}(b)] = b$. We also see that the domain of f is the range of f^{-1} and the range of f is the domain of f^{-1}.

Before we can find inverse functions, we need to explore one more concept.

Objective 3 Show that a function is one-to-one.

If we merely reverse the ordered pairs of a function, we do not always get an inverse function. Consider the following sets in which the ordered pairs of B are the ordered pairs of A with x and y interchanged:

$$A = \{(1,2),(2,4),(3,2),(-2,5)\}$$
$$B = \{(2,1),(4,2),(2,3),(5,-2,)\}$$

Note that A represents a function but B does not because 2 in the domain of B is paired with 1 and 3. How did this happen? Because the ordered pairs of B are those of A with x and y interchanged, two ordered pairs of A having the same y-value of 2 $[(1,2)$ and $(3,2)]$ became two ordered pairs with the same x-value in B. Therefore, if the inverse of a function is to be a function, no two ordered pairs of the function may have the same y-value. Such a function is called a **one-to-one function**.

A function is one-to-one if each value in the range corresponds to only one value in the domain. In terms of x and y, two different x-values must result in two different y-values, which suggests the following formal definition.

Definition **One-to-one function:** A function f is one-to-one if for any two numbers a and b in its domain, when $f(a) = f(b), a = b$ and when $a \neq b, f(a) \neq f(b)$.

It is possible to determine whether a function is one-to-one by looking at its graph. If two ordered pairs have the same y-value, the corresponding points lie on the same horizontal line. Consequently, if a function is one-to-one, the graph cannot be intersected by any horizontal line in more than one point.

Rule **Horizontal Line Test for One-to-One Functions**
Given a function's graph, the function is one-to-one if every horizontal line that can intersect the graph does so at one and only one point.

Example 3 Determine whether the following graphs are of one-to-one functions.

a. $f(x) = x^3$

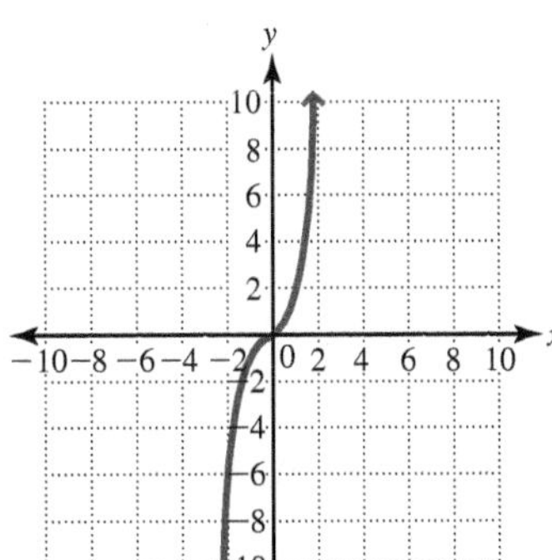

b. $f(x) = x^2 + 2$

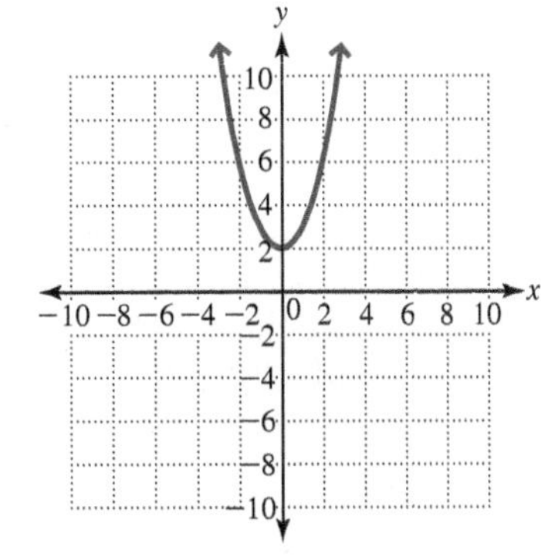

c. $f(x) = \sqrt{36 - x^2}$

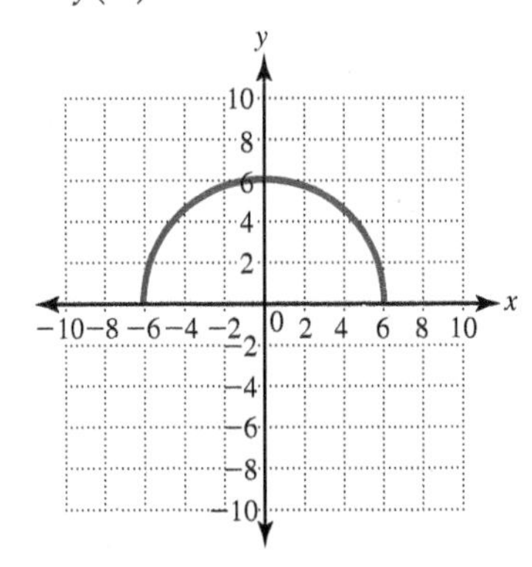

Solution:

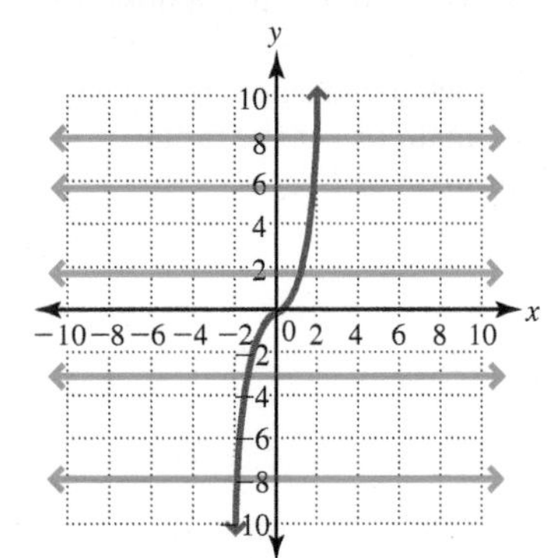

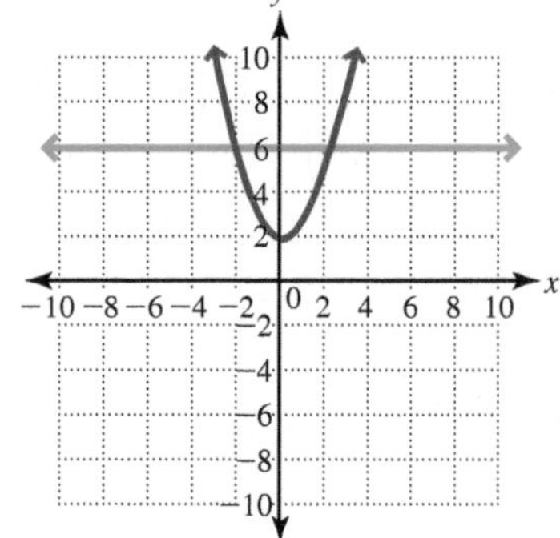

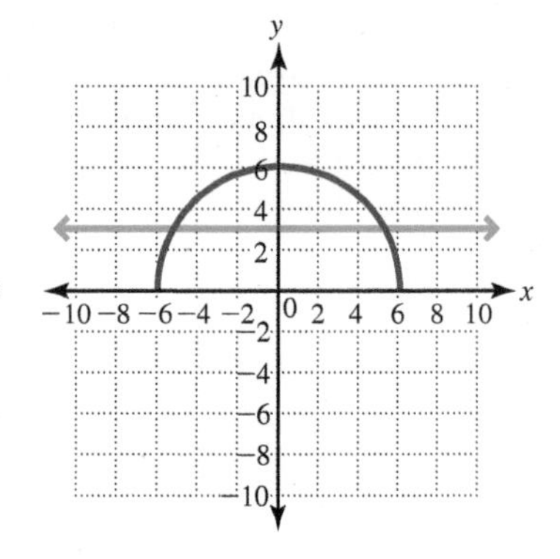

a. Every horizontal line that can intersect this graph does so at one and only one point, so the function is one-to-one.

b. A horizontal line can intersect this graph in more than one point, so the function is not one-to-one.

c. A horizontal line can intersect this graph in more than one point, so the function is not one-to-one.

Your Turn 3 Determine whether the following graphs are of one-to-one functions.

a.

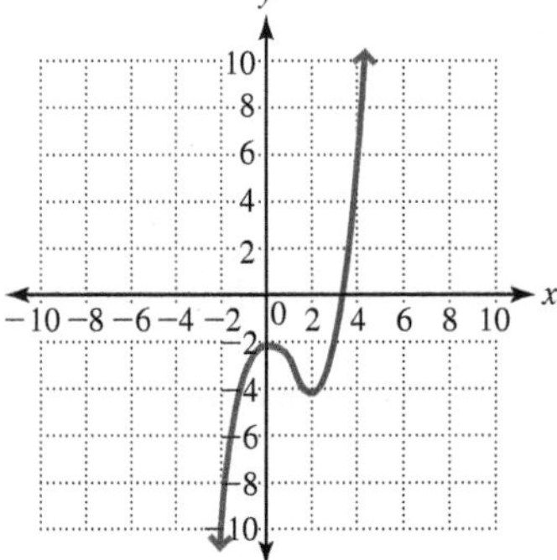

b.

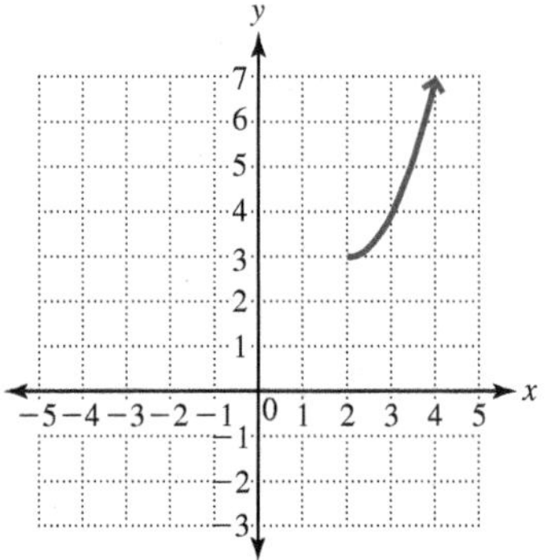

Objective 4 Find the inverse of a function.

Interchanging the ordered pairs of a function that is not one-to-one results in a relation that is not a function. Consequently, we have the following:

Rule Existence of Inverse Functions

A function has an inverse function if and only if the function is one-to-one.

We have already seen that to find the inverse of a function, we need to interchange the x- and y-values of all of the ordered pairs of f. That is, we replace x with y and y with x in the equation defining the function. However, we are not finished. We now have the inverse in the form $f(y) = x$, so we must solve this equation for y to have y as a function of x. The steps are summarized as follows.

Procedure Finding the Inverse Function of a One-to-One Function

1. If necessary, replace $f(x)$ with y.
2. Replace all x's with y's and y's with x's.
3. Solve the equation from step 2 for y.
4. Replace y with $f^{-1}(x)$.

Answers to Your Turn 3

a. no **b.** yes

Example 4 Find $f^{-1}(x)$ for each of the following one-to-one functions.

a. $f(x) = 2x + 4$

The domain and range of f is the set of all real numbers, so the domain and range of f^{-1} is also the set of all real numbers.

Solution:

$y = 2x + 4$ Replace $f(x)$ with y

$x = 2y + 4$ Replace x with y and y with x.

$\frac{x - 4}{2} = y$ Solve for y.

$f^{-1}(x) = \frac{x - 4}{2}$ Replace y with $f^{-1}(x)$.

To verify that we have found the inverse, we need to show that $f[f^{-1}(x)] = x$ and $f^{-1}[f(x)] = x$.

$$f[f^{-1}(x)] = f\left(\frac{x-4}{2}\right) = 2\left(\frac{x-4}{2}\right) + 4 = x - 4 + 4 = x$$

$$f^{-1}[f(x)] = f^{-1}(2x + 4) = \frac{(2x + 4) - 4}{2} = \frac{2x}{2} = x$$

Because $f[f^{-1}(x)] = x$ and $f^{-1}[f(x)] = x$, they are inverses.

b. $f(x) = x^3 + 2$

The domain and range of f is the set of all real numbers, so the domain and range of f^{-1} is also the set of all real numbers.

Solution:

$y = x^3 + 2$ Replace $f(x)$ with y.

$x = y^3 + 2$ Replace x with y and y with x.

$x - 2 = y^3$ Begin solving for y by subtracting 2 from both sides.

$\sqrt[3]{x - 2} = y$ Take the cube root of each side to solve for y.

$f^{-1}(x) = \sqrt[3]{x - 2}$ Replace y with $f^{-1}(x)$.

We can verify that f and f^{-1} are inverses, as in part a.

c. $f(x) = \sqrt{x - 3}$

The domain of f is $[3, \infty)$, and the range is $[0, \infty)$. Therefore, the domain of f^{-1} is $[0, \infty)$ and the range of f^{-1} is $[3, \infty)$.

Solution:

$y = \sqrt{x - 3}$ Replace $f(x)$ with y.

$x = \sqrt{y - 3}$ Replace x with y and y with x.

$x^2 = y - 3$ Begin solving for y by squaring both sides.

$x^2 + 3 = y$ Add 3 to both sides to solve for y.

The domain of $y = x^2 + 3$ is all real numbers, but the domain of f^{-1} is $[0, \infty)$. Therefore, we write $f^{-1}(x) = x^2 + 3, x \geq 0$.

To verify that these are inverses, we need to show that $f[f^{-1}(x)] = x$ and $f^{-1}[f(x)] = x$.

$$f[f^{-1}(x)] = f(x^2 + 3) = \sqrt{(x^2 + 3) - 3} = \sqrt{x^2 + 3 - 3} = \sqrt{x^2}$$

Note Recall $\sqrt{x^2} = |x|$ for x a real number.

Because the domain of f^{-1} is $[0, \infty)$, $\sqrt{x^2} = x$. So $f[f^{-1}(x)] = x$.

$$f^{-1}[f(x)] = f^{-1}(\sqrt{x - 3}) = (\sqrt{x - 3})^2 + 3 = x - 3 + 3 = x$$

Because $f[f^{-1}(x)] = x$ and $f^{-1}[f(x)] = x$, they are inverses.

Your Turn 4 Find $f^{-1}(x)$ for each of the following one-to-one functions.

a. $f(x) = 5x + 2$

b. $f(x) = \frac{x + 3}{x}$

Answers to Your Turn 4

a. $f^{-1}(x) = \frac{x - 2}{5}$

b. $f^{-1}(x) = \frac{3}{x - 1}$

Objective 5 Graph a given function's inverse function.

In the following figure, we have plotted pairs of points whose coordinates are interchanged and have graphed the line $y = x$.

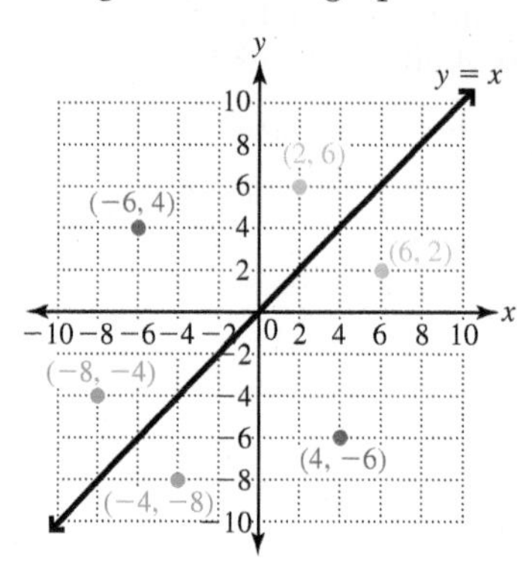

If the graph were folded along the line $y = x$, the points whose coordinates are interchanged would fall on top of each other. These points, therefore, are symmetric with respect to the line $y = x$. It can be shown that the graphs of any two points of the form (a, b) and (b, a) are symmetric with respect to the graph of $y = x$. Similarly, because the ordered pairs of f and f^{-1} have interchanged coordinates, the graphs of f and f^{-1} are symmetric with respect to the line $y = x$.

Learning Strategy

If you are a tactile learner, imagine placing a mirror on the line $y = x$. The graphs of f and f^{-1} are reflections in the line $y = x$.

Rule Graphs of Inverse Functions

The graphs of f and f^{-1} are symmetric with respect to the graph of $y = x$.

Following are the graphs of f and f^{-1} for Example 4 along with the graph of $y = x$.

a. $f(x) = 2x + 4$

$f^{-1}(x) = \dfrac{x-4}{2}$

b. $f(x) = x^3 + 2$

$f^{-1}(x) = \sqrt[3]{x-2}$

c. $f(x) = \sqrt{x-3}$

$f^{-1}(x) = x^2 + 3, \quad x \geq 0$

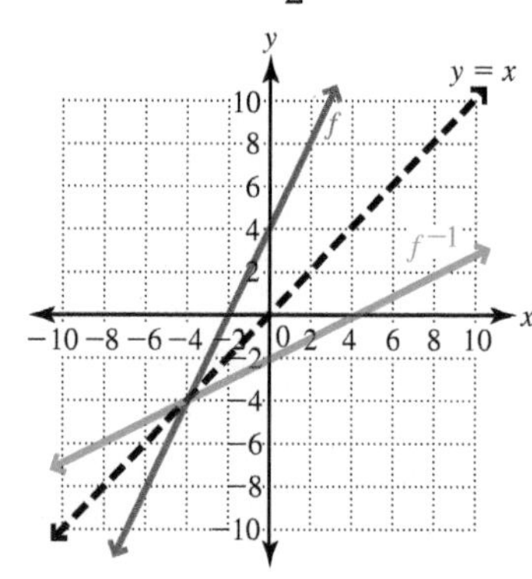

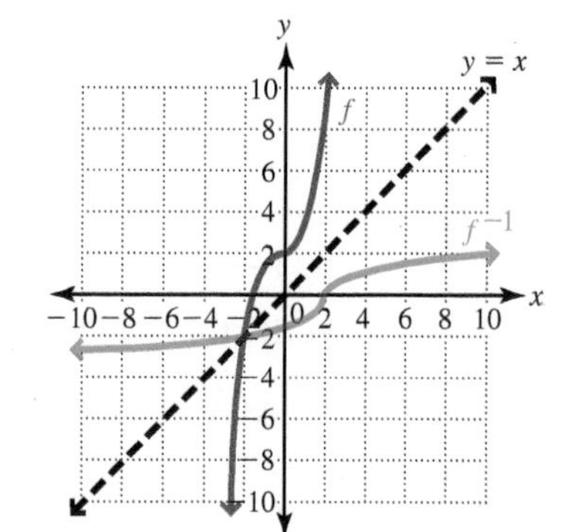

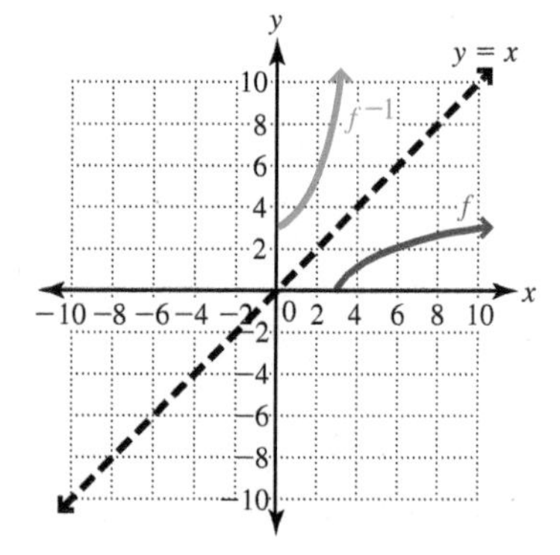

Example 5 Sketch the inverse of the functions whose graphs are shown in parts a and b.

a.

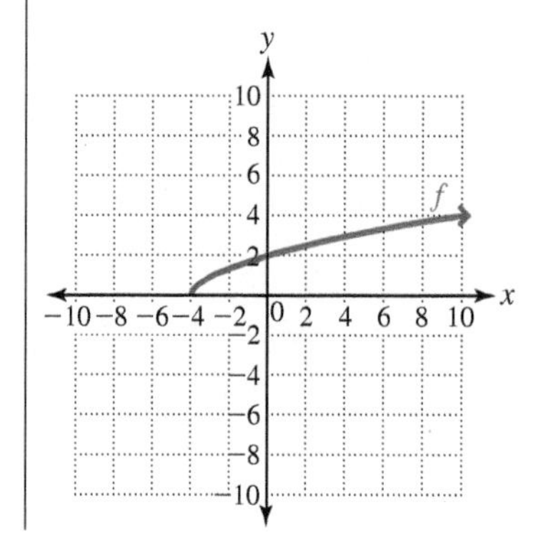

Solution: Draw the line $y = x$ and reflect the graph in the line.

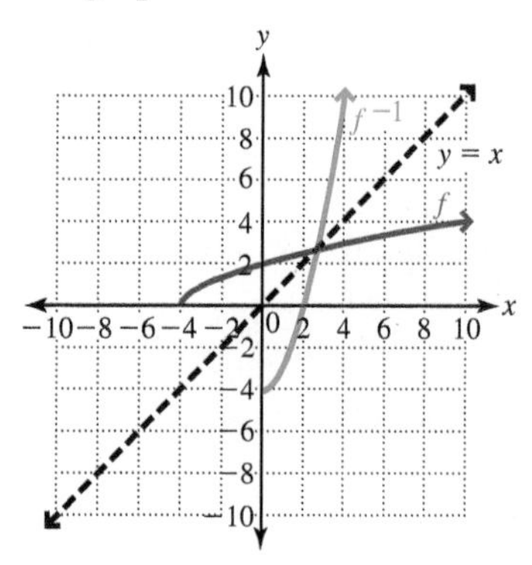

b.

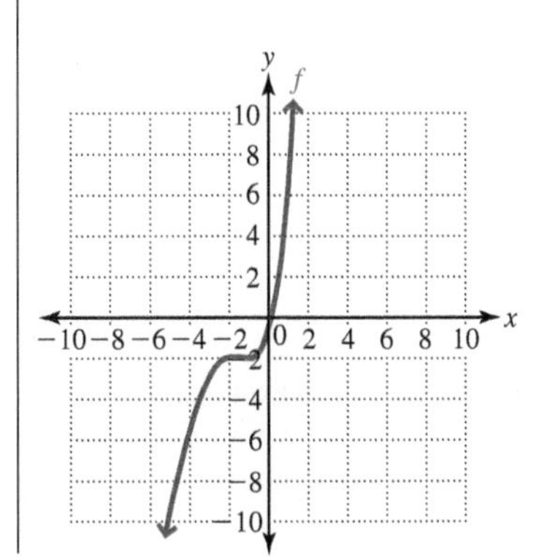

Solution: Draw the line $y = x$ and reflect the graph in the line.

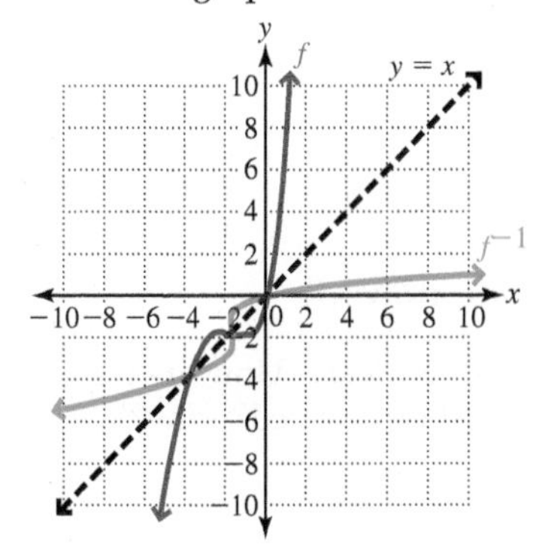

Your Turn 5 Sketch the graph of the inverse of the functions whose graphs are given.

a.

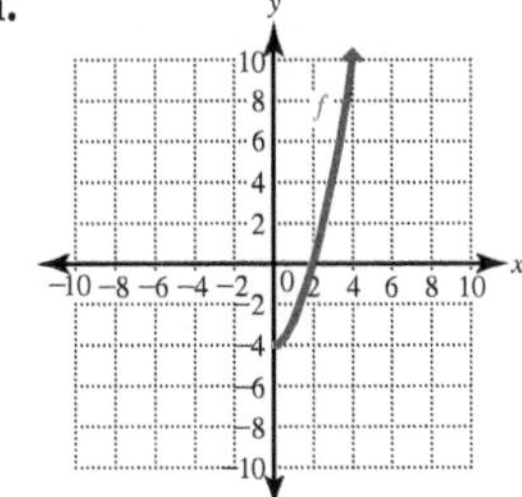

b.

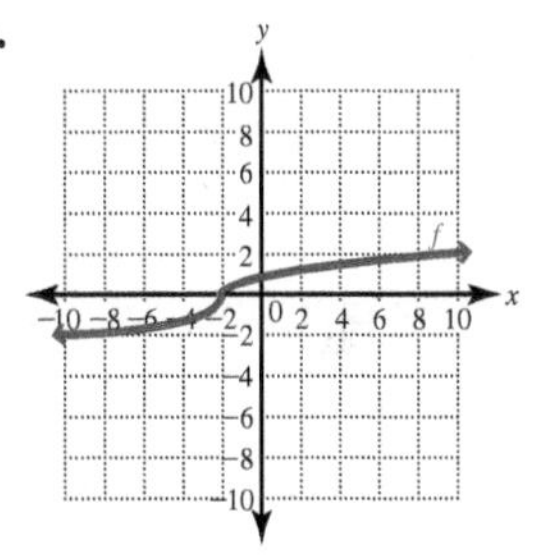

Answers to Your Turn 5

a.

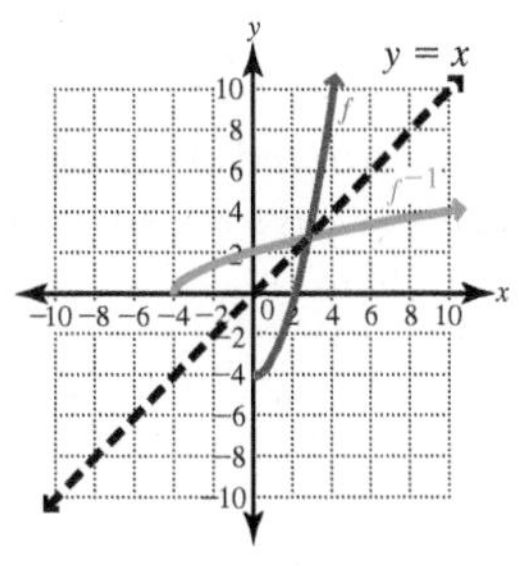

b.

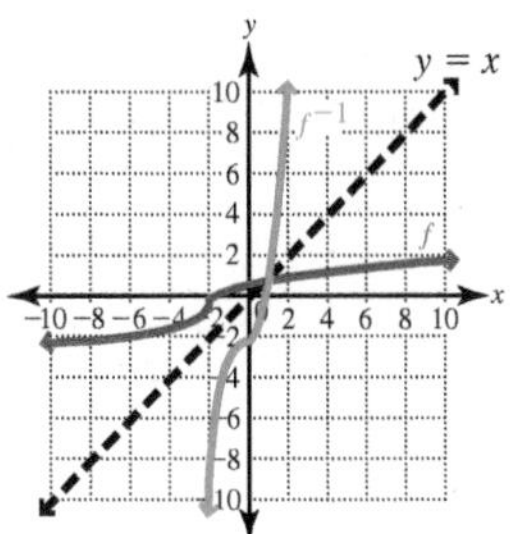

10.1 Exercises

For Extra Help MyMathLab®

Objective 1

Prep Exercise 1 If f and g are functions, then the composition of f and g is defined as $(f \circ g)(x) =$ _______________ for all x in the domain g for which $g(x)$ is in the domain of f.

Prep Exercise 2 If f and g are functions, then the composition of g and f is defined as $(g \circ f)(x) =$ _______________ for all x in the domain f for which $f(x)$ is in the domain of g.

For Exercises 1–12, if $f(x) = 3x + 5$, $g(x) = x^2 + 3$, and $h(x) = \sqrt{x + 1}$, find each composition. See Example 1.

1. $(f \circ g)(0)$

2. $(g \circ f)(0)$

3. $(h \circ f)(1)$

4. $(h \circ g)(3)$

5. $(f \circ g)(-2)$

6. $(g \circ f)(-1)$

7. $(f \circ g)(x)$

8. $(g \circ f)(x)$

9. $(f \circ h)(x)$

10. $(h \circ g)(x)$

11. $(h \circ f)(0)$

12. $(g \circ h)(0)$

For Exercises 13–22, find $(f \circ g)(x)$ and $(g \circ f)(x)$. See Example 1.

13. $f(x) = 2x - 2, g(x) = 3x + 4$

14. $f(x) = 4x + 7, g(x) = 3x + 4$

15. $f(x) = x + 2, g(x) = x^2 + 1$

16. $f(x) = x^2 - 3, g(x) = x + 5$

17. $f(x) = x^2 + 3x - 4; g(x) = 3x$

18. $f(x) = -3x, g(x) = x^2 + 2x + 4$

19. $f(x) = \sqrt{x + 2}, g(x) = 2x - 5$

20. $f(x) = 5x + 2, g(x) = \sqrt{x - 5}$

21. $f(x) = \frac{x + 1}{x}, g(x) = \frac{x - 3}{x}$

22. $f(x) = \frac{x + 4}{x}, g(x) = \frac{2 - x}{x}$

Objective 2

Prep Exercise 3 How are the domains and ranges of inverse functions related? Why?

Prep Exercise 4 If the coordinates of the ordered pairs of a function are interchanged, is the result always the inverse function? Why or why not?

For Exercises 23 and 24, answer each question. See Objective 2.

23. If the domain of f is $[3, \infty)$ and the range is $[0, \infty)$, what are the domain and range of f^{-1}?

24. If f and g are inverse functions and $f(2) = 5$, then $g(5) =$ ______________.

For Exercises 25–28, determine whether the following functions f and g are inverses. See Objective 2.

25. $f = \{(1, 2), (-1, -3), (3, 4), (2, -5)\}, g = \{(2, 1), (-3, -1), (4, 3), (-5, 2)\}$

26. $f = \{(4, -3), (1, 4), (5, 2), (-3, 1), (-1, 3)\}, g = \{(-3, 4), (4, 1), (2, 5), (1, -3), (3, -1)\}$

27. $f = \{(-2, -2), (3, -3), (-4, 4), (-6, -6)\}, g = \{(-2, -2), (-3, 3), (-4, 4), (-6, -6)\}$

28. $f = \{(5, 5), (-4, 4), (2, -2), (-7, -7)\}, g = \{(-5, -5), (4, -4), (-2, 2), (7, 7)\}$

Prep Exercise 5 What must be shown to prove that the two functions f and g are inverses?

For Exercises 29–42, determine whether f and g are inverses by determining whether $(f \circ g)(x) = x$ and $(g \circ f)(x) = x$. See Example 2.

29. $f(x) = x + 5, g(x) = x - 5$

30. $f(x) = x - 8, g(x) = x + 8$

31. $f(x) = 6x, g(x) = \frac{x}{6}$

32. $f(x) = -\frac{x}{3}, g(x) = -3x$

33. $f(x) = 2x - 3, g(x) = \frac{x + 3}{2}$

34. $f(x) = \frac{x - 4}{5}, g(x) = 5x + 4$

35. $f(x) = x^3 - 4, g(x) = \sqrt[3]{x + 4}$

36. $f(x) = x^5 + 3, g(x) = \sqrt[5]{x - 3}$

37. $f(x) = x^2, g(x) = \sqrt{x}$

38. $f(x) = x^4, g(x) = \sqrt[4]{x}$

39. $f(x) = x^2, x \geq 0; g(x) = \sqrt{x}$

40. $f(x) = x^2 + 2, x \geq 0; g(x) = \sqrt{x - 2}$

41. $f(x) = \dfrac{3}{x + 5}, g(x) = \dfrac{3 - 5x}{x}$

42. $f(x) = \dfrac{x}{x + 4}, g(x) = \dfrac{4x}{1 - x}$

For Exercises 43 and 44, answer each question.

43. Is $f(x) = \dfrac{1}{x}$ its own inverse? Explain.

44. Is $f(x) = x$ its own inverse? Explain.

Objective 3

Prep Exercise 6 If given the graph of a function, what kind of test can be performed to determine whether it is one-to-one?

For Exercises 45–48, determine whether the function is one-to-one. See Example 3.

45.

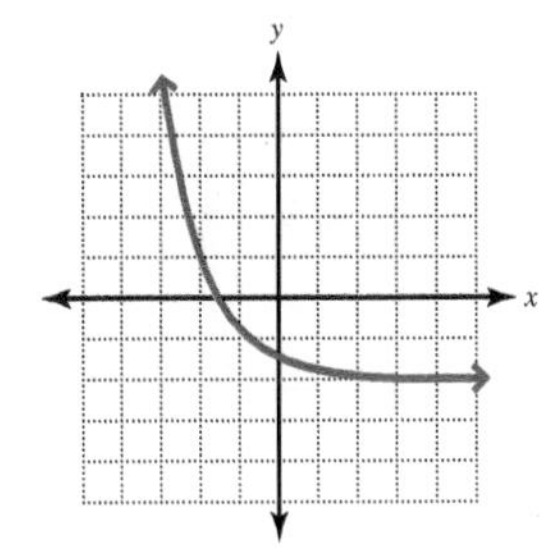

46.

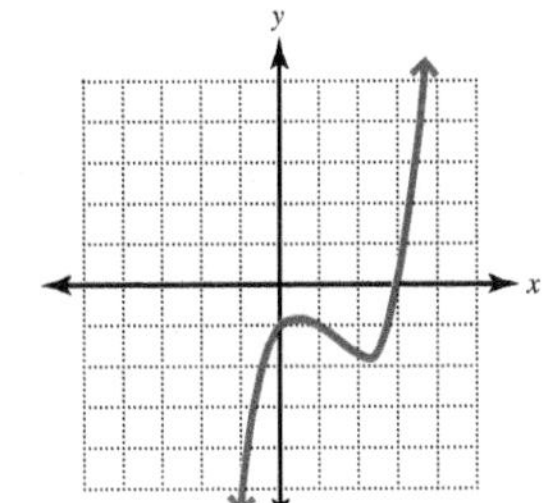

47.

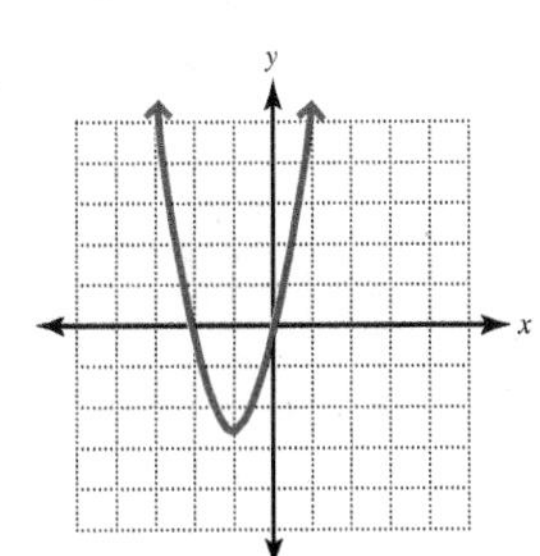

48.

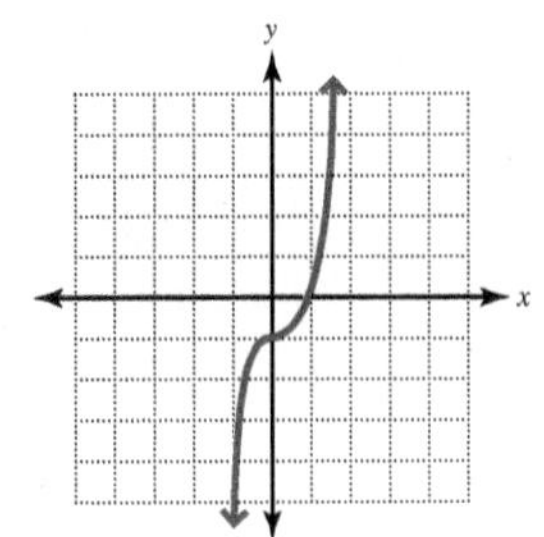

Objective 4

Prep Exercise 7 When finding the inverse of a one-to-one function, after exchanging x and y, what is the next step?

For Exercises 49–70, find $f^{-1}(x)$ for each of the following one-to-one functions f. See Example 4.

49. $f = \{(-3, 2), (-1, -3), (0, 4), (4, 6)\}$

50. $f = \{(-4, 2), (-1, -3), (2, 3), (5, 7)\}$

51. $f = \{(7, -2), (9, 2), (-4, 1), (3, 3)\}$

52. $f = \{(1, 2), (4, 5), (-3, 3), (-9, 0)\}$

53. $f(x) = x + 6$

54. $f(x) = x - 4$

55. $f(x) = 2x + 3$

56. $f(x) = 4x - 5$

57. $f(x) = x^3 - 1$

58. $f(x) = x^3 - 3$

59. $f(x) = \dfrac{2}{x + 2}$

60. $f(x) = \dfrac{-3}{x - 3}$

61. $f(x) = \dfrac{x+2}{x-3}$

62. $f(x) = \dfrac{x-4}{x+2}$

63. $f(x) = \sqrt{x-2}$

64. $f(x) = \sqrt{2x-4}$

65. $f(x) = 2x^3 + 4$

66. $f(x) = 4x^3 - 5$

67. $f(x) = \sqrt[3]{x+2}$

68. $f(x) = \sqrt[3]{x+1}$

69. $f(x) = 2\sqrt[3]{2x+4}$

70. $f(x) = 4\sqrt[3]{2x-6}$

Objective 5

Prep Exercise 8 If two one-to-one functions are inverses, then their graphs are symmetric about what line?

For Exercises 71–82, sketch the graph of the inverse of each of the following functions. See Example 5.

71.

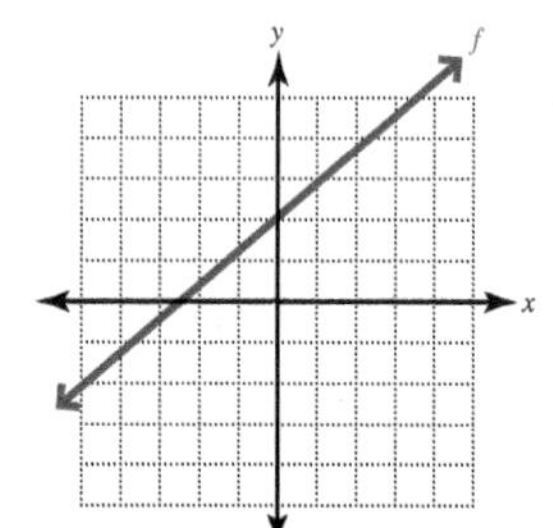

72.

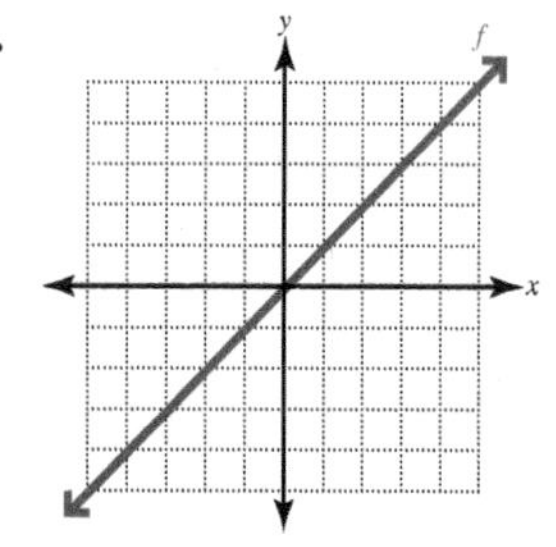

73.

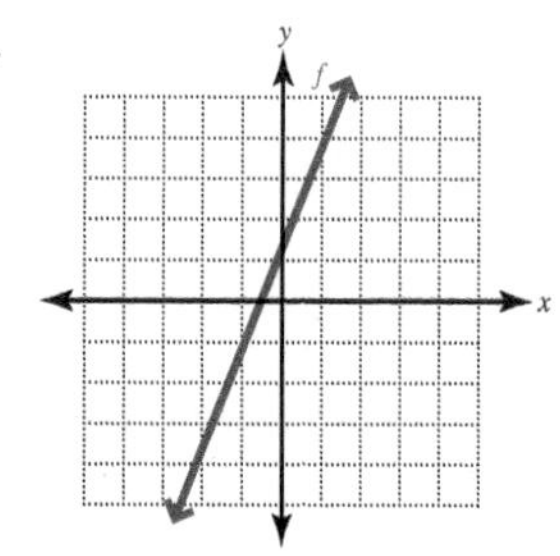

74.

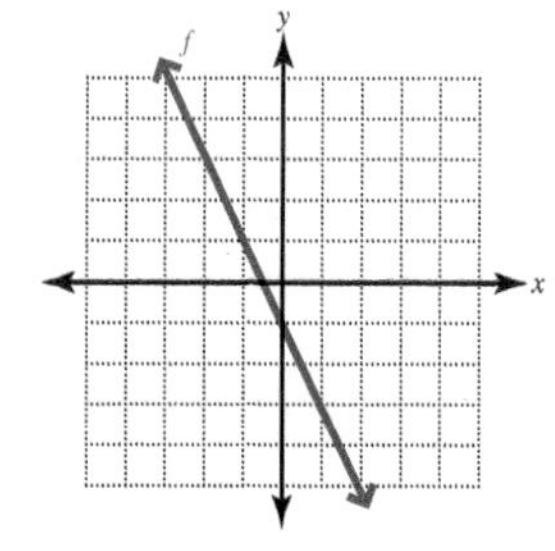

75.

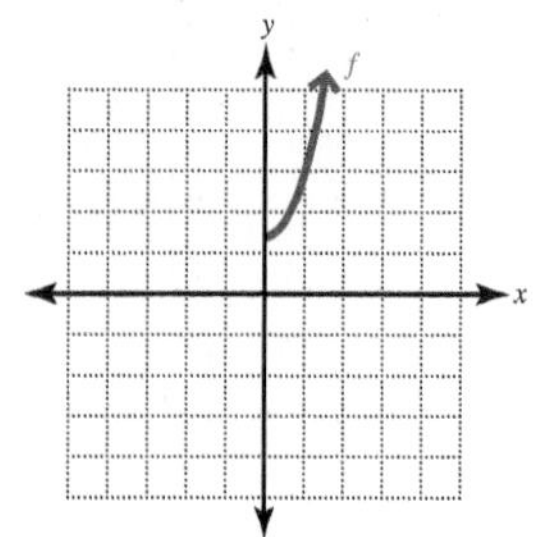

76.

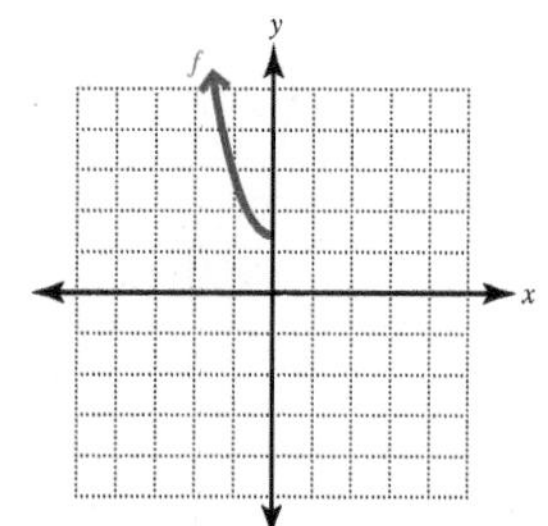

77.

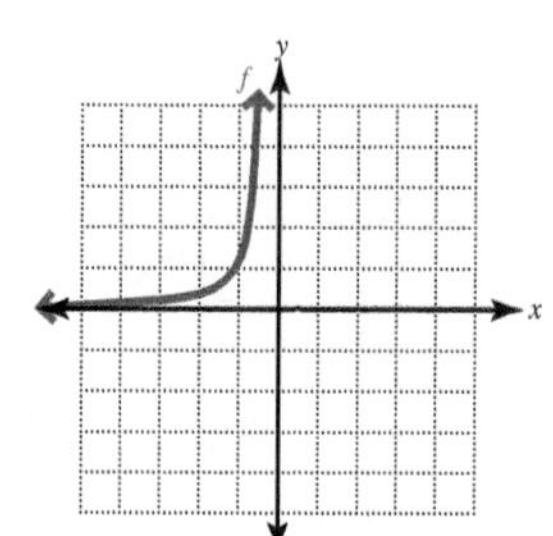

78.

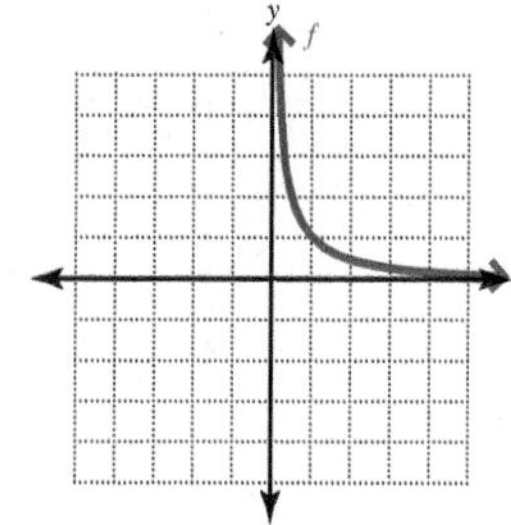

79.

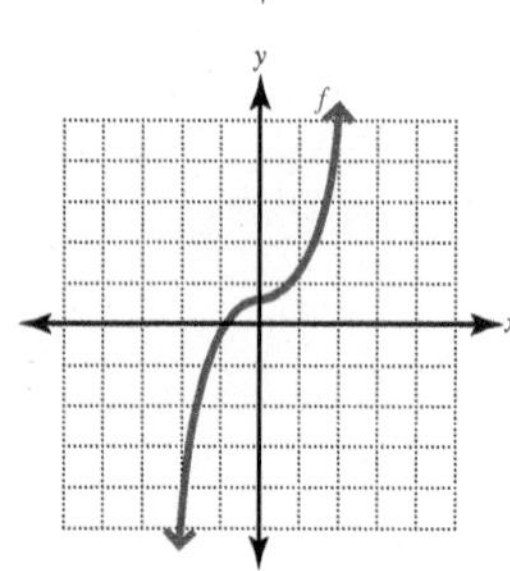

80.

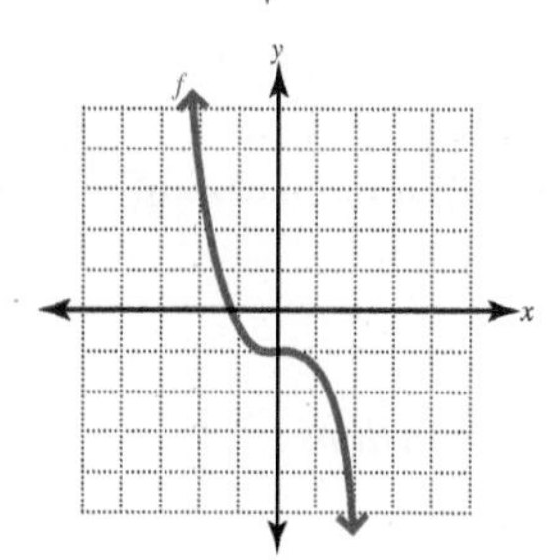

81.

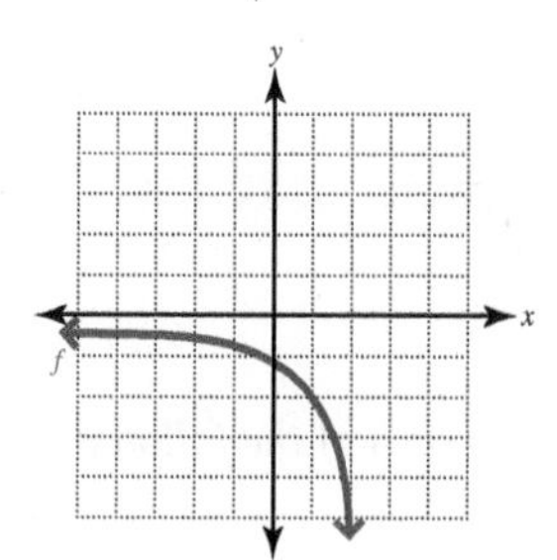

82.

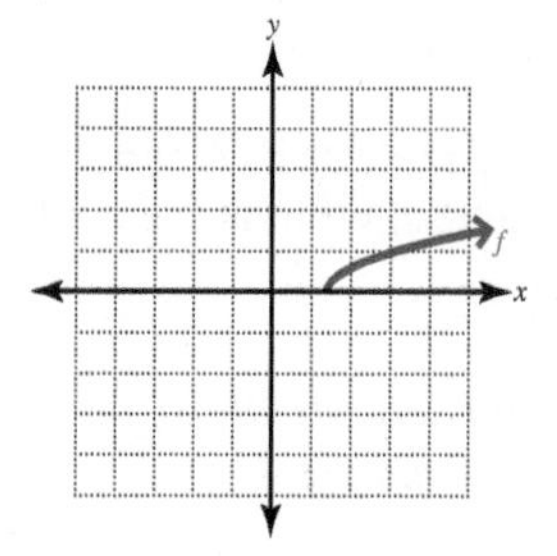

For Exercises 83–86, solve each problem.

83. If a salesperson works for \$100 per week plus 5% commission on sales, the weekly salary is $y = 0.05x + 100$, where y represents the salary and x represents the sales.

  a. Find the inverse function.

  b. What does each variable of the inverse function represent?

  c. Use the inverse function to find the sales for a week in which the salary was \$350.

84. If a Toyota Prius averages 50 miles per hour on a trip, then $y = 50x$, where y represents the number of miles traveled and x represents the number of hours.
a. Find the inverse function.

b. What does each variable of the inverse function represent?

c. Use the inverse function to find the number of hours to travel 210 miles.

85. An office building installed 105 fans, some of which measure 36 inches with the remaining fans measuring 54 inches. If the 36-inch fans cost \$45 each and the 54-inch fans cost \$65 each, the cost of the fans is $y = 45x + 65(105 - x)$, where y represents the cost and x represents the number of 36-inch fans.
a. Find the inverse function.

b. What does each variable of the inverse function represent?

c. Use the inverse function to find the number of 36-inch fans if the total cost was \$5225.

86. A painting contractor purchased 24 gallons of paint for the interior and exterior of a house. If he paid \$18 per gallon for the interior paint and \$22 per gallon for the exterior paint, the amount he paid is $y = 18x + 22(24 - x)$, where y represents the amount paid and x represents the number of gallons of interior paint.
a. Find the inverse function.

b. What does each variable of the inverse function represent?

c. Use the inverse function to find the number of gallons of interior paint if he paid a total of \$488 for the 24 gallons of paint.

For Exercises 87–90, answer the question.

87. If $(-4, 5)$ is an ordered pair on the graph of g, what are the coordinates of an ordered pair on the graph of g^{-1}?

88. If $f(2) = 4$ and $f^{-1}(a) = 2$, find a.

89. A linear function is of the form $f(x) = ax + b, a \neq 0$. Find $f^{-1}(x)$.

90. The square root function is defined by $f(x) = \sqrt{x}$. Find $f^{-1}(x)$. Be careful!

Review Exercises

Exercises 1–5 Expressions

[1.3] 1. Write 32 as 2 to a power.

For Exercises 2–5, evaluate each expression.

[1.3] 2. 2^3

[1.3] 3. $\left(-\frac{1}{3}\right)^3$

[5.1] 4. 4^{-2}

[8.2] 5. $4^{3/2}$

Exercise 6 Equations and Inequalities

[9.4] 6. What are the coordinates of the vertex of $g(x) = (x - 3)^2 + 1$?

10.2 Exponential Functions

Objectives

1. Define and graph exponential functions.
2. Solve equations of the form $b^x = b^y$ for x.
3. Use exponential functions to solve application problems.

Warm-up

[10.1] **1.** Find the inverse function: $f(x) = x^3 - 2$

[6.1] **2.** Write 81 as 3 to a power.

For Exercises 3 and 4, evaluate.

[5.1] **3.** $\left(\frac{1}{4}\right)^{-2}$ [8.2] **4.** $27^{2/3}$

Objective 1 Define and graph exponential functions.

Previously, we defined rational number exponents. For example, we know that $b^3 = b \cdot b \cdot b$, $b^{2/3} = \sqrt[3]{b^2}$, $b^0 = 1$, and $b^{-3} = \frac{1}{b^3}$.

It can be shown that irrational number exponents have meaning as well, and we can approximate expressions such as $2^{\sqrt{3}}$ and 5^{π} by using rational approximations for the exponents. Therefore, the exponential expression b^x has meaning if x is any real number (rational or irrational). So we can define the **exponential function** as follows.

Note The definition of the exponential function has two restrictions on b. If $b = 1$, then $f(x) = b^x = 1^x = 1$, which is a linear function. If $b < 0$, we may get values for which the function is not defined as a real number. For example, if $f(x) = (-4)^x$, then $f\left(\frac{1}{2}\right) = (-4)^{1/2} = \sqrt{-4}$, which is not a real number.

Definition Exponential function: If $b > 0$, $b \neq 1$, and x is any real number, then the exponential function is $f(x) = b^x$.

We graph exponential functions by plotting enough points to determine the graph's shape. We will find that the graph has one typical shape if $b > 1$ and a different typical shape if $0 < b < 1$.

Example 1 Graph.

a. $f(x) = 2^x$ and $g(x) = 4^x$

Solution: Choose some values of x and find the corresponding values of $f(x)$ and $g(x)$ (which are the y-values).

x	−3	−2	−1	0	1	2	3
$f(x)$	$\frac{1}{8}$	$\frac{1}{4}$	$\frac{1}{2}$	1	2	4	8
$g(x)$	$\frac{1}{64}$	$\frac{1}{16}$	$\frac{1}{4}$	1	4	16	64

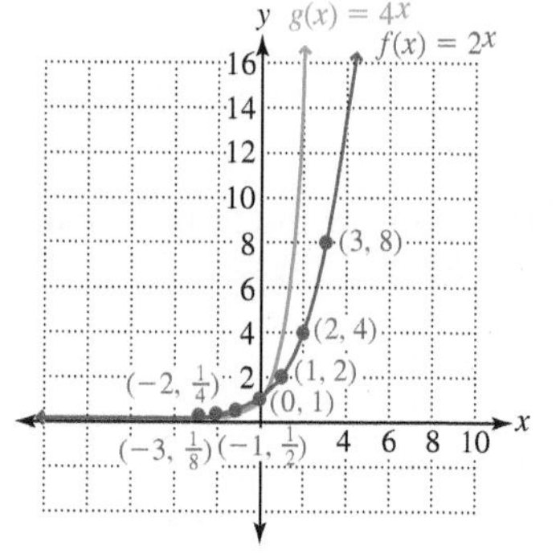

Plotting the ordered pairs for each function gives smooth curves typical of the graphs of $f(x) = b^x$ with $b > 1$. Comparing the graphs of $f(x) = 2^x$ and $g(x) = 4^x$, we can see that the greater the value of b, the steeper the graph.

b. $h(x) = \left(\frac{1}{2}\right)^x$ and $k(x) = \left(\frac{1}{4}\right)^x$

Answers to Warm-up

1. $f^{-1}(x) = \sqrt[3]{x + 2}$
2. 3^4
3. 16
4. 9

Solution: Choose some values of x and find the corresponding values of $h(x)$ and $k(x)$.

x	-3	-2	-1	0	1	2	3
$h(x)$	8	4	2	1	$\frac{1}{2}$	$\frac{1}{4}$	$\frac{1}{8}$
$k(x)$	64	16	4	1	$\frac{1}{4}$	$\frac{1}{16}$	$\frac{1}{64}$

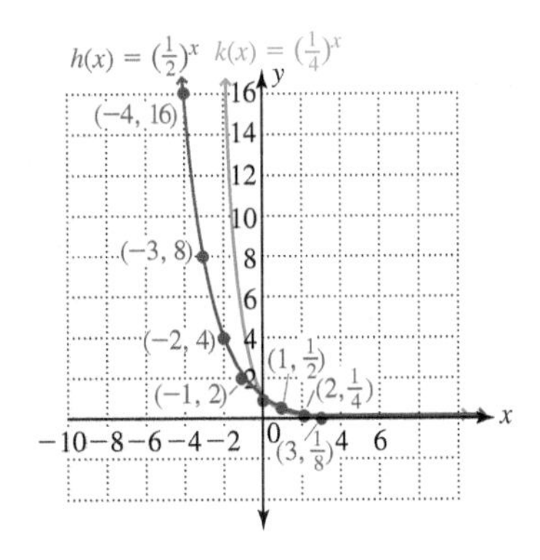

After plotting the ordered pairs for each function, we see graphs that are typical of exponential functions in the form $f(x) = b^x$ with $0 < b < 1$. Comparing the graphs of $h(x) = \left(\frac{1}{2}\right)^x$ and $k(x) = \left(\frac{1}{4}\right)^x$, we see that the smaller the value of b, the steeper the graph.

Note We left the points and coordinate labels off the graphs of $g(x)$ and $k(x)$ to avoid additional clutter. From all of the graphs in Example 1, we see that b^x is never negative.

We summarize the graphs of $f(x) = b^x$ as follows.

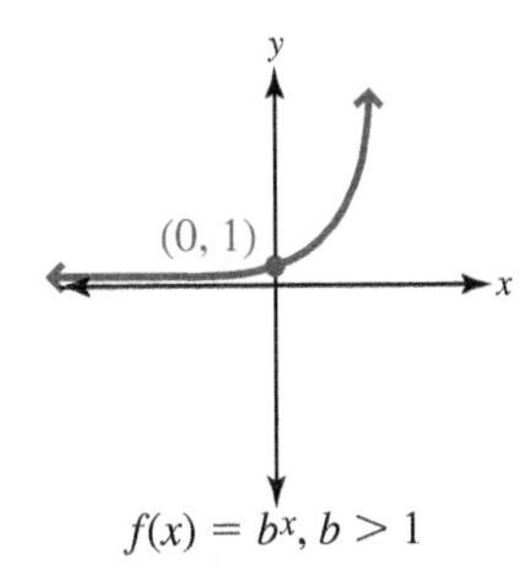

$f(x) = b^x,\ b > 1$

The function is increasing from left to right and is one-to-one. The graph always passes through (0, 1). For negative values of x, the graph approaches the x-axis but never touches it. The larger the values of b, the steeper the graph. The domain is $(-\infty, \infty)$, and the range is $(0, \infty)$.

$f(x) = b^x,\ 0 < b < 1$

The function is decreasing from left to right and is one-to-one. The graph always passes through (0, 1). For positive values of x, the graph approaches the x-axis but never touches it. The smaller the values of x, the steeper the graph. The domain is $(-\infty, \infty)$, and the range is $(0, \infty)$.

Note Any function that increases or decreases along its entire domain is one-to-one because no horizontal line will intersect its graph at more than one point.

More complicated exponential functions are graphed in the same manner.

Example 2 Graph $f(x) = 3^{2x-1}$.

Solution: Find some ordered pairs, plot them, and draw the graph.

x	$y = f(x)$
-1	$3^{-3} = \frac{1}{27}$
0	$3^{-1} = \frac{1}{3}$
1	$3^1 = 3$
2	$3^3 = 27$

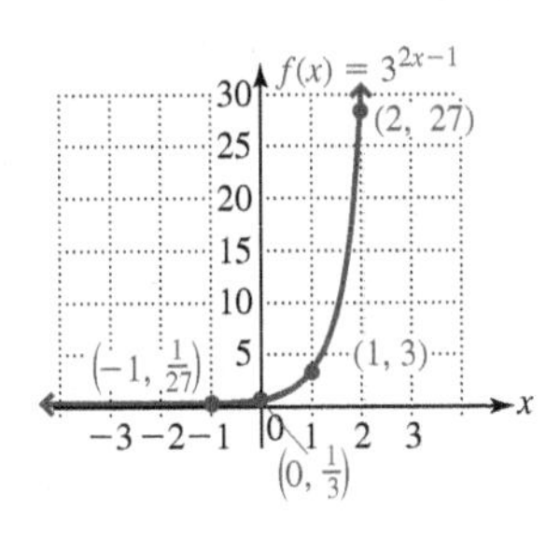

Note $f(x) = 3^{2x-1}$ is not of the form $f(x) = b^x$. Note that it does not pass through $(0, 1)$.

Your Turn 2 Graph $f(x) = 2^{x+1}$.

Answer to Your Turn 2

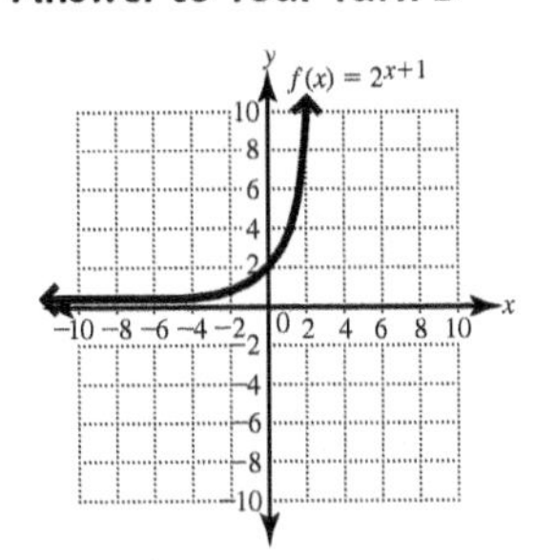

Objective 2 Solve equations of the form $b^x = b^y$ for x.

Earlier we solved equations containing expressions having a variable base and a constant exponent, such as $x^2 = 4$. Now we solve equations that contain expressions having a constant base and a variable exponent, such as $2^x = 16$. We need a new method to solve these equations. Previously, we noted that the exponential function is one-to-one; so for each x-value, there is a unique y-value and vice versa. Consequently, we have the following rule.

Rule One-to-One Property of Exponentials

Given $b > 0$ and $b \neq 1$, if $b^x = b^y$, then $x = y$.

To solve some types of exponential equations, we use the following procedure.

Procedure Solving Exponential Equations

1. If necessary, write both sides of the equation as a power of the same base.
2. If necessary, simplify the exponents.
3. Set the exponents equal to each other.
4. Solve the resulting equation.

Example 3 Solve.

a. $3^x = 81$

Solution: $3^x = 3^4$ Write 81 as 3^4 so that both sides have the same base.

$x = 4$ Set the exponents equal to each other.

Check: $3^x = 81$

$3^4 = 81$ Replace x with 4.

$81 = 81$ True; so $x = 4$ is the solution.

b. $4^x = 32$

Solution: $(2^2)^x = 2^5$ Write 4 as 2^2 and 32 as 2^5 so that both sides have the same base.

$2^{2x} = 2^5$ Simplify $(2^2)^x$ by applying $(a^m)^n = a^{mn}$.

$2x = 5$ Set the exponents equal to each other.

$x = \frac{5}{2}$ Solve for x.

Note We leave the checks for parts b–d to the reader.

c. $8^{x+4} = 4^{2x+3}$

Solution: $(2^3)^{x+4} = (2^2)^{2x+3}$ Write 8 as 2^3 and 4 as 2^2 so that both sides have the same base.

$2^{3x+12} = 2^{4x+6}$ Simplify the exponents by applying $(a^m)^n = a^{mn}$.

$3x + 12 = 4x + 6$ Set the exponents equal to each other.

$6 = x$ Solve for x.

d. $\left(\frac{1}{5}\right)^x = 25$

Solution: $(5^{-1})^x = 5^2$ Write $\frac{1}{5}$ as 5^{-1} and 25 as 5^2 so that both sides have the same base.

$5^{-x} = 5^2$ Simplify $(5^{-1})^x$ by applying $(a^m)^n = a^{am}$.

$-x = 2$ Set the exponents equal to each other.

$x = -2$ Solve for x.

Your Turn 3 Solve.

a. $2^{x+1} = 8$ b. $27^{x-2} = 9^{x}$ c. $\left(\frac{1}{2}\right)^{x} = 16$

Objective 3 Use exponential functions to solve application problems.

Compound Interest Formula

Exponential functions occur in many areas, especially the sciences and business. If P dollars are invested at an annual interest rate of r (written as a decimal) compounded n times per year for t years, the accumulated amount A in the account is given by the formula $A = P\left(1 + \frac{r}{n}\right)^{nt}$.

Example 4 Find the accumulated amount in an account if \$5000 is deposited at 6% compounded quarterly for 10 years. Round the answer to the nearest cent.

Solution: $A = 5000\left(1 + \frac{0.06}{4}\right)^{4(10)}$ In $A = P\left(1 + \frac{r}{n}\right)^{nt}$, substitute 5000 for P, 0.06 for r, 4 for n, and 10 for t.

$A = 5000(1 + 0.015)^{40}$ Simplify.

$A = 5000(1.015)^{40}$

$A = 9070.09$ Evaluate using a calculator and round to the nearest cent (hundredths place).

Answer: After 10 years, the accumulated amount in the account is \$9070.09.

Your Turn 4 If \$3000 is invested at 4% compounded semiannually (twice per year), how much money is in the account at the end of 8 years? Round to the nearest cent.

Half-Life

The *half-life* of a radioactive substance is the amount of time it takes until only half of the original amount of the substance remains. Suppose we begin with 100 grams of a substance that has a half-life of 10 days. After 10 days (one half-life), 50 grams will remain; after 20 days (two half-lives), 25 grams will remain; after 30 days (three half-lives), 12.5 grams will remain; and so on. The formula $A = A_0\left(\frac{1}{2}\right)^{t/h}$ gives the amount remaining, where A_0 is the initial amount, t is the time, and h is the half-life.

Example 5 The isotope ^{45}Ca has a half-life of 165 days. How many grams of a 50-gram sample of ^{45}Ca will remain after 825 days?

Solution: $A = 50\left(\frac{1}{2}\right)^{825/165}$ In $A = A_0\left(\frac{1}{2}\right)^{t/h}$, substitute 50 for A_0, 825 for t, and 165 for h.

$A = 50\left(\frac{1}{2}\right)^{5}$ Simplify.

$A = 1.5625$ Evaluate using a calculator.

Answer: 1.5625 grams of ^{45}Ca will remain after 825 days.

Your Turn 5 The radioactive isotope ^{61}Cr has a half-life of 26 days. How much of a 10-gram sample would remain after 208 days? Give your answer to the nearest thousandth of a gram.

Answers to Your Turn 3
a. 2 b. 6 c. −4

Answer to Your Turn 4
\$4118.36

Answer to Your Turn 5
0.039 g

Aging

Example 6 The number of people in the United States aged 65 and over (in millions) is given in the following table.

Year	Number 65 and over (in millions)
1900	3.1
1910	4.0
1920	4.9
1930	6.7
1940	9.0
1950	12.4
1960	16.7
1970	20.1
1980	25.5
1990	31.4
2000	35.0
2010	40.3

(*Source:* U.S. Bureau of the Census.)

The data can be approximated by the function $y = 3.29(1.025)^x$, where x is the number of years after 1900 and y is the number of people in millions. Use the model to estimate the number of people in the United States aged 65 and over in the year 2020. Round the answer to the nearest tenth of a million.

Solution:

$x = 2020 - 1900 = 120$ Subtract 1900 from 2020 to find the value of x that corresponds to the year 2020.

$y = 3.29(1.025)^{120}$ Substitute 120 for x in $y = 3.29(1.025)^x$.

$y = 63.7$ million Evaluate using a calculator and round to the nearest tenth.

Answer: According to the function, about 63.7 million people will be aged 65 and over in the year 2020.

Connection Following is a graph of the function $y = 3.29(1.025)^x$ with the point corresponding to the solution of Example 6 indicated.

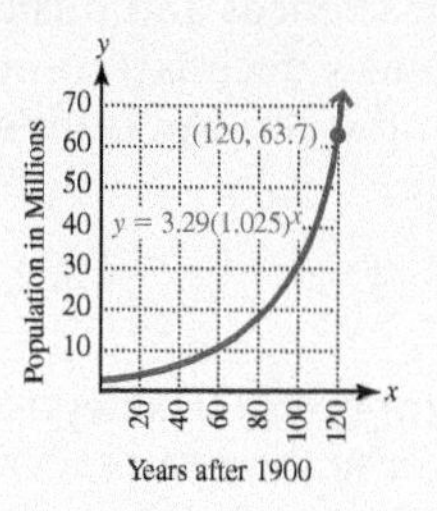

Your Turn 6 The table in the margin shows the number of computers (in millions) in use in the United States for selected years since 1980.

Year	Number of Computers (in millions)
1980	3.1
1985	22.2
1990	51.3
1995	90.2
2000	184
2005	244
2010	306

The data can be approximated by $y = 8.77(1.136)^x$, where x is the number of years after 1980 and y is the number of computers (in millions). Use the model to approximate the number of computers in use in the United States in 2020.

Answer to Your Turn 6
1439.2 million

10.2 Exercises

For Extra Help MyMathLab®

Objective 1

Prep Exercise 1 Is $f(x) = (-2)^x$ an exponential function? Why or why not?

Prep Exercise 2 Find the domain and range of $f(x) = 2^{x-4}$.

Prep Exercise 3 As x gets larger, which graph is steeper: $f(x) = 3^x$ or $g(x) = 1.5^x$? Why?

Prep Exercise 4 Are the graphs of $f(x) = 2^x$ and $g(x) = \left(\frac{1}{2}\right)^x$ symmetric with respect to the y-axis? Explain.

For Exercises 1–18 graph. See Examples 1 and 2.

1. $f(x) = 3^x$

2. $f(x) = 4^x$

3. $f(x) = 4^x - 3$

4. $f(x) = 3^x + 1$

5. $f(x) = \left(\frac{1}{3}\right)^x$

6. $f(x) = \left(\frac{1}{4}\right)^x$

7. $f(x) = \left(\frac{2}{3}\right)^x + 2$

8. $f(x) = \left(\frac{3}{2}\right)^x - 1$

9. $f(x) = -3^x$

10. $f(x) = -2^x$

11. $f(x) = 2^{x-2}$

12. $f(x) = 3^{x+1}$

13. $f(x) = 2^{-x}$

14. $f(x) = 3^{-x}$

15. $f(x) = 2^{2x-3}$

16. $f(x) = 3^{2x+1}$

17. $f(x) = 3^{-x+2}$

18. $f(x) = 2^{-x-1}$

Objective 2

Prep Exercise 5 For $b > 0$ and $b \neq 1$, if $b^x = b^y$, why does $x = y$?

For Exercises 19–36, solve each equation. See Example 3.

19. $2^x = 8$

20. $3^x = 81$

21. $8^x = 32$

22. $27^x = 81$

23. $16^x = 4$

24. $36^x = 216$

25. $6^x = \frac{1}{36}$

26. $3^x = \frac{1}{27}$

27. $\left(\frac{1}{3}\right)^x = 9$

28. $\left(\frac{1}{5}\right)^x = 125$

29. $\left(\frac{2}{3}\right)^x = \frac{8}{27}$

30. $\left(\frac{3}{2}\right)^x = \frac{9}{4}$

31. $\left(\frac{1}{2}\right)^x = 16$

32. $\left(\frac{1}{3}\right)^x = 27$

33. $25^{x+1} = 125$

34. $9^{x+2} = 243$

35. $8^{2x-1} = 32^{x-3}$

36. $4^{2x-1} = 32^{x+2}$

For Exercises 37–40, use a graphing utility.

37. a. Graph $f(x) = 2^x$ and $g(x) = 2^{x+2}$ in the window $[-5, 5]$ for x and $[-1, 10]$ for y.
b. How does the graph of g compare with the graph of f?

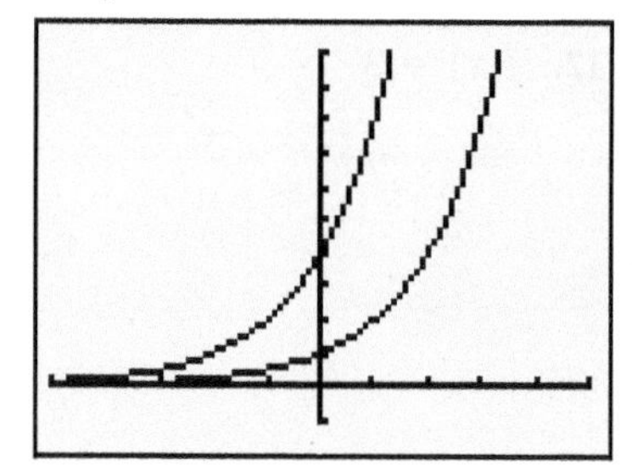

38. a. Graph $f(x) = 3^x$ and $g(x) = 3^x - 3$ in the window $[-3, 3]$ for x and $[-3, 9]$ for y.
b. How does the graph of g compare with the graph of f?

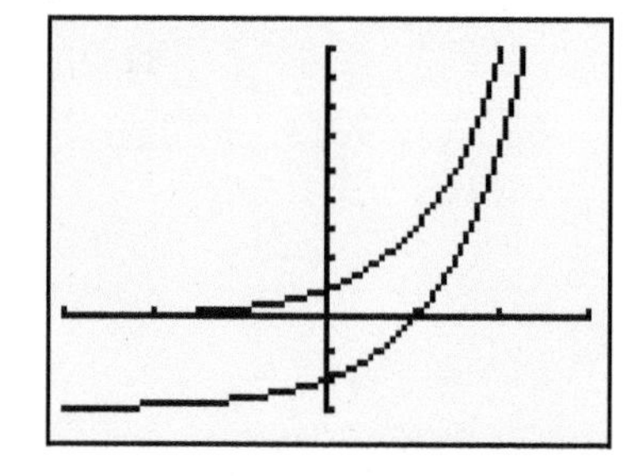

39. a. Graph $f(x) = 2^x$ and $g(x) = 2^{-x}$ in the window $[-5, 5]$ for x and $[-1, 10]$ for y.
b. How does the graph of g compare with the graph of f?

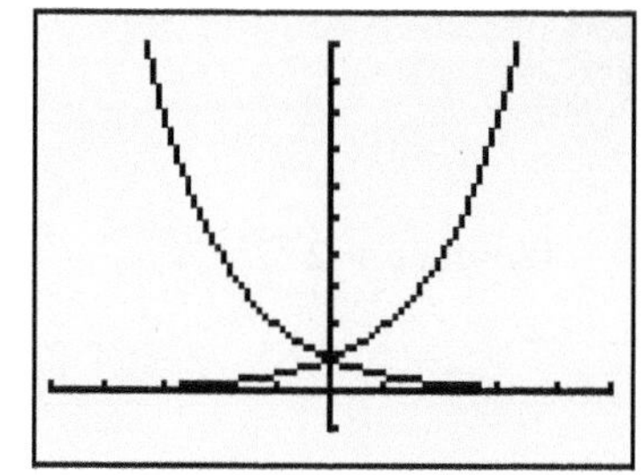

40. a. Graph $f(x) = 2^x$ and $g(x) = -2^x$ in the window $[-5, 5]$ for x and $[-5, 5]$ for y.
b. How does the graph of g compare with the graph of f?

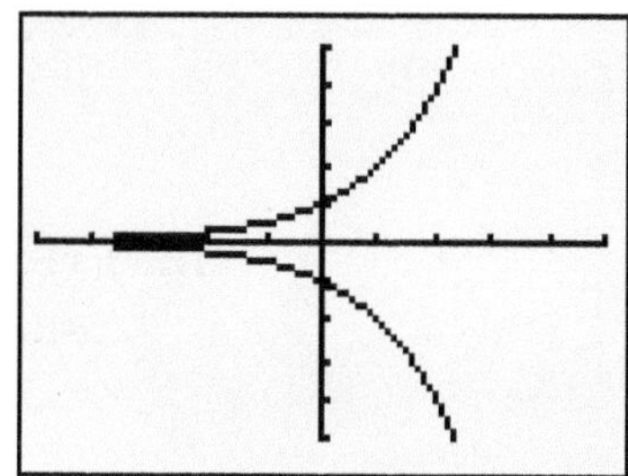

Objective 3

Prep Exercise 6 If we have 10 grams of a substance that has a half-life of 50 days, how many grams will be present after 200 days?

For Exercises 41 and 42, use the following information.

Under ideal conditions, a culture of *E. coli* bacteria doubles in size every 20 minutes. If A_0 is the initial amount and t is the number of minutes passed, the amount present is A, where $A = A_0(2)^{t/20}$.

41. If a culture of *E. coli* begins with 100 cells, how many cells will be present after 120 minutes?

42. If a culture of *E. coli* currently has 500 cells, how many cells were present 90 minutes earlier? (*Hint:* Let $A_0 = 500$, and the time will be negative.)

43. Under ideal conditions, human beings could double their population every 50 years. If A_0 is the initial population and t is the number of years passed, the current population is A, where $A = A_0(2)^{t/50}$. If 6 billion humans were on Earth in 2000, how many humans will there be in 2500 if their growth is uncontrolled?

44. Using the information from Exercise 43, how many human beings were on the Earth in 1900?

For Exercises 45 and 46, use the formula $A = P\left(1 + \frac{r}{n}\right)^{nt}$. *See Example 4.*

45. If \$10,000 is deposited into an account paying 8% interest compounded quarterly, how much will be in the account after 12 years?

46. If \$15,000 is deposited into an account paying 6% interest compounded semiannually, how much will be in the account after nine years?

For Exercises 47 and 48, use the formula $A = A_0\left(\frac{1}{2}\right)^{t/h}$. *See Example 5.*

47. Einsteinium (^{254}ES) has a half-life of 270 days. How much of a 5-gram sample would remain after 2160 days? Give the answer to the nearest thousandth of a gram.

48. Nobelium (^{257}No) has a half-life of 23 seconds. How much of a 100-gram sample would remain after 275 seconds? Give the answer to the nearest thousandth of a gram.

For Exercises 49–54, solve. See Example 6.

49. Since 1960, the U.S. gross domestic product can be approximated by $y = 2632.31(1.033)^x$, where y is the gross domestic product and x is the number of years beginning with 1960. $(x = 1$ represents 1960.$)$ Estimate the U.S. gross domestic product in 2018.

50. Since 1970, the U.S. consumer price index can be approximated by $y = 46.56(1.045)^x$, where y is the consumer price index and x is the number of years beginning with 1970. $(x = 1$ represents 1970.$)$ Estimate the consumer price index in 2019.

51. Chlorine is frequently used to disinfect swimming pools. The concentration should remain between 1.5 and 2.5 parts per million. On a warm, sunny day, 30% of the chlorine can dissipate into the air or combine with other chemicals. If the initial amount of chlorine is 2.5 parts per million, the function $f(x) = 2.5(0.7)^x$ models the amount of chlorine after x days. How much chlorine is in the pool after 2 days?

52. It is estimated that the value of a car depreciates 20% per year for the first five years. If the original price of a car is P, the value, A, of a car after t years is given by $A = P(0.8)^t$. If a car originally cost $25,960, find the value of the car after 3 years to the nearest dollar.

53. Between 1993 and 2012, the number of transistors that can be placed on a single chip has grown significantly, as indicated in the table. The data can be approximated by the function $T(x) = 1.97(1.503)^x$, where x is the number of years after 1993 and $T(x)$ is the number of transistors in millions. If the current trend continues, estimate the number of transistors that could be put on a single chip in 2020. (*Source:* http://en.wikipedia.org.)

Year	Chip	Transistors (millions)
1993	Pentium	3.3
1995	P6	5.5
1997	Pentium II	7.5
1999	Pentium III	9.5
2000	Pentium IV	42
2004	Pentium IV Prescott	125
2006	Core 2 Duo	586
2008	Quad IV	820
2010	8-core Xeon Nehalem-EX	2300
2012	62-core Xeon Phi	5000

54. Dave makes a cup of coffee with cream and places it on the counter to cool. The temperature of the coffee at various times is given in the table shown.

The data can be approximated by the function $T(t) = 146.9(0.989)^t$, where t is the time in seconds and T is the temperature in °F. Estimate the temperature of the coffee after 1 minute.

t (seconds)	T(°F)
0.2	155.8
8.4	133.2
16.6	117.9
24.8	107.9
33	100.7
41.1	94.9
49.3	90.5

Puzzle Problem What is the greatest number that can be written using three numerals?

Review Exercises

Exercises 1–4 Expressions

[5.1] *For Exercises 1–3, evaluate.*

1. 5^3

2. $\left(\frac{1}{3}\right)^{-2}$

3. 4^{-3}

[8.2] **4.** Write $5^{2/3}$ in radical notation.

Exercises 5 and 6 Equations and Inequalities

[10.1] **5.** If $f(x) = 3x + 4$, find $f^{-1}(x)$.

[10.1] **6.** Graph the inverse of the function whose graph is shown.

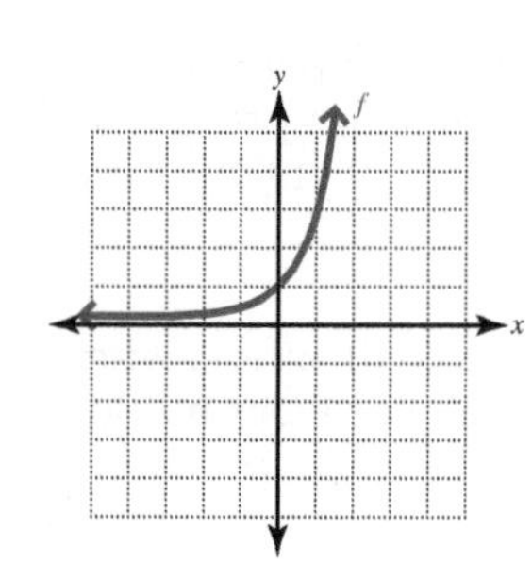

10.3 Logarithmic Functions

Objectives

1. Convert between exponential and logarithmic forms.
2. Solve logarithmic equations by changing to exponential form.
3. Graph logarithmic functions.
4. Solve applications involving logarithms.

Warm-up

[10.2] **1.** Solve: $3^{5x+7} = 9^{x-4}$

[5.1] *For Exercises 2 and 3, evaluate.*

2. $\left(\frac{1}{2}\right)^{-4}$ **3.** $\left(\frac{1}{2}\right)^{0}$

[5.1] **3.** Write $\frac{1}{27}$ as 3 to a power.

Objective 1 Convert between exponential and logarithmic forms.

In Section 10.2, we defined the exponential function as $f(x) = b^x$ with $b > 0$ and $b \neq 1$. Because the exponential function is a one-to-one function, it has an inverse. Following is the graph of $f(x) = 2^x$ and its inverse, which we find by reflecting the graph about the line $y = x$.

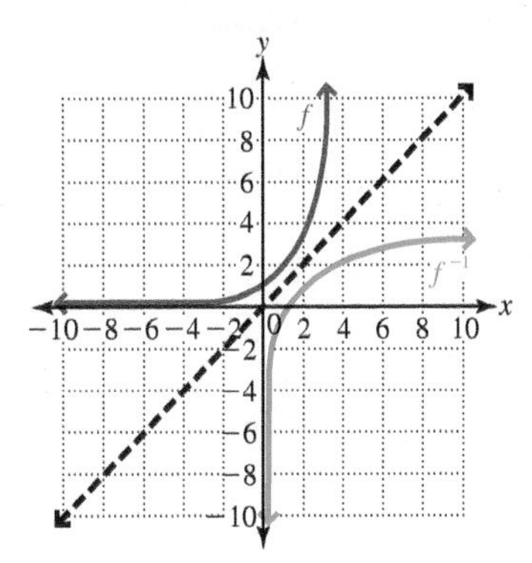

To find the inverse of $f(x) = 2^x$, we use the procedure from Section 10.1.

$$f(x) = 2^x$$

$$y = 2^x \quad \text{Replace } f(x) \text{ with } y.$$

$$x = 2^y \quad \text{Replace } x \text{ with } x \text{ and } x \text{ with } y.$$

The next step is to solve for y, but we haven't learned how to isolate a variable that is an exponent. To rewrite $x = 2^y$, we define a **logarithm**.

Definition Logarithm: If $b > 0$ and $b \neq 1$, then $y = \log_b x$ is equivalent to $x = b^y$.

If we apply this definition to $x = 2^y$ in the preceding equation, we get $y = \log_2 x$. Replacing y with $f^{-1}(x)$, we get $f^{-1}(x) = \log_2 x$; so the inverse function for $f(x) = 2^x$ is $f^{-1}(x) = \log_2 x$. To generalize, exponential functions and *logarithmic functions* are inverses. Consequently, if $y = \log_b x$, the domain is $(0, \infty)$ and the range is $(-\infty, \infty)$ with $b > 0$ and $b \neq 1$.

The expression $\log_b x$ is read "the logarithm base b of x" and *is the exponent to which b must be raised to get x*. Compare the two forms.

The exponent is the logarithm.

$$x = b^y \qquad y = \log_b x$$

The base of the exponent is the base of the logarithm.

Note $y = \log_b x$ means that y is the power to which we raise b to get x.

Answers to Warm-up

1. -5
2. 16
3. 1
4. 3^{-3}

The definition of a logarithm allows us to convert from one form to another. The following table contains pairs of equivalent forms.

Note The logarithmic equations are in the form $\log_b x = y$. The values of x are positive numbers only, and the values of y are both positive and negative.

Logarithmic Form	Exponential Form
$\log_2 8 = 3$	$2^3 = 8$
$\log_{10}\frac{1}{10} = -1$	$10^{-1} = \frac{1}{10}$
$\log_5 1 = 0$	$5^0 = 1$
$\log_{16} 4 = \frac{1}{2}$	$16^{1/2} = 4$
$\log_{1/2} 32 = -5$	$\left(\frac{1}{2}\right)^{-5} = 32$

Example 1 Write in logarithmic form.

a. $3^4 = 81$

Solution: $\log_3 81 = 4$ The base of the exponent is the base of the logarithm, and the exponent is the logarithm.

b. $\left(\frac{1}{2}\right)^{-3} = 8$

Solution: $\log_{1/2} 8 = -3$ The base of the exponent is the base of the logarithm, and the exponent is the logarithm.

c. $9^{1/2} = 3$

Solution: $\log_9 3 = \frac{1}{2}$ The base of the exponent is the base of the logarithm, and the exponent is the logarithm.

Your Turn 1 Write in logarithmic form.

a. $4^3 = 64$ b. $2^{-3} = \frac{1}{8}$ c. $27^{1/3} = 3$

VISUAL

Learning Strategy

If you are a visual learner, try visualizing the following "loop" for rewriting logarithms in exponential form.

equals

$\log_6 36 = 2$

6 raised to 2

Example 2 Write in exponential form.

a. $\log_6 36 = 2$

Solution: $6^2 = 36$ The base of the logarithm is the base of the exponent, and the logarithm is the exponent.

b. $\log_{16} 2 = \frac{1}{4}$

Solution: $16^{1/4} = 2$ The base of the logarithm is the base of the exponent, and the logarithm is the exponent.

c. $\log_{1/2} 16 = -4$

Solution: $\left(\frac{1}{2}\right)^{-4} = 16$ The base of the logarithm is the base of the exponent, and the logarithm is the exponent.

Your Turn 2 Write in exponential form.

a. $\log_4 64 = 3$ b. $\log_8 2 = \frac{1}{3}$ c. $\log_5 \frac{1}{125} = -3$

Answers to Your Turn 1

a. $\log_4 64 = 3$

b. $\log_2 \frac{1}{8} = -3$

c. $\log_{27} 3 = \frac{1}{3}$

Answers to Your Turn 2

a. $4^3 = 64$ b. $8^{1/3} = 2$

c. $5^{-3} = \frac{1}{125}$

Objective 2 Solve logarithmic equations by changing to exponential form.

A logarithmic equation in the form $\log_b x = y$ may have b, x, or y as an unknown.

Procedure Solving Logarithmic Equations

To solve an equation of the form $\log_b x = y$, where b, x, or y is a variable, write the equation in exponential form, $b^y = x$, and then solve for the variable.

Example 3 Solve.

a. $\log_b 16 = 2$

Solution: $b^2 = 16$ Write in exponential form.

$b = \pm 4$ Find the positive and negative square roots of 16.

$b = 4$ The base must be positive, so $b = 4$.

b. $\log_3 \frac{1}{27} = y$

Solution: $3^y = \frac{1}{27}$ Write in exponential form.

$3^y = \frac{1}{3^3}$ Write 27 as 3^3.

$3^y = 3^{-3}$ Write $\frac{1}{3^3}$ as 3^{-3}.

$y = -3$ Set the exponents equal to each other.

c. $\log_{25} x = \frac{1}{2}$

Solution: $25^{1/2} = x$ Write in exponential form.

$5 = x$ Simplify. $25^{1/2} = \sqrt{25} = 5$

d. $\log_{36} \sqrt[4]{6} = y$

Solution: $36^y = \sqrt[4]{6}$

$(6^2)^y = 6^{1/4}$ Write 36 as 6^2 and $\sqrt[4]{6}$ as $6^{1/4}$.

$6^{2y} = 6^{1/4}$ $(6^2)^y = 6^{2y}$

$2y = \frac{1}{4}$ Set the exponents equal to each other.

$y = \frac{1}{8}$ Solve for y.

Your Turn 3 Solve.

a. $\log_b \frac{1}{9} = -2$ b. $\log_5 \frac{1}{25} = y$ c. $\log_{27} x = \frac{1}{3}$

Answers to Your Turn 3
a. 3 b. −2 c. 3

If an equation containing a logarithm is not in the form $\log_b a = c$, try using the addition or multiplication principles of equality to rewrite the equation in that form.

Example 4 Solve $5 - 3\log_2 x = -7$.

Solution: Use the addition principle of equality and multiplication principle of equality to write the equation in the form $\log_b a = c$. We can then change the equation to exponential form to solve for x.

$-3\log_2 x = -12$ Subtract 5 from both sides.

$\log_2 x = 4$ Divide both sides by -3 to isolate the logarithm.

$x = 2^4$ Write in exponential form.

$x = 16$ Simplify.

Your Turn 4 Solve.

a. $9 + \log_3 x = 13$ **b.** $5\log_n 36 = 10$ **c.** $12 - 7\log_4 t = -9$

The definition of logarithm leads to the following two properties.

Rule For any real number b, where $b > 0$ and $b \neq 1$:

1. $\log_b b = 1$.
2. $\log_b 1 = 0$.

Based on the definition of a logarithm, $\log_b b = 1$ because $b^1 = b$ and $\log_b 1 = 0$ because $b^0 = 1$.

Example 5 Find the value.

a. $\log_5 5$ **b.** $\log_e e$ **c.** $\log_{10} 1$

Solution: $\log_5 5 = 1$ **Solution:** $\log_e e = 1$ **Solution:** $\log_{10} 1 = 0$

Your Turn 5 Find the value.

a. $\log_{10} 10$ **b.** $\log_{\sqrt{3}} 1$

Objective 3 Graph logarithmic functions.

To graph logarithmic functions, which have the form $f(x) = \log_b x$, where $b > 0$ and $b \neq 1$, we first change to exponential form so that it will be easier to find ordered pairs.

Procedure Graphing Logarithmic Functions

To graph a function in the form $f(x) = \log_b x$:

1. Replace $f(x)$ with y and write the logarithm in exponential form $x = b^y$.
2. Find ordered pairs that satisfy the equation by assigning values to y and finding x.
3. Plot the ordered pairs and draw a smooth curve through the points.

Example 6 Graph.

a. $f(x) = \log_2 x$

Solution: $y = \log_2 x$ Replace $f(x)$ with y.

$2^y = x$ Write in exponential form.

Choose values for y and find x.

y	0	1	2	3	−1	−2
x	$2^0 = 1$	$2^1 = 2$	$2^2 = 4$	$2^3 = 8$	$2^{-1} = \frac{1}{2}$	$2^{-2} = \frac{1}{4}$

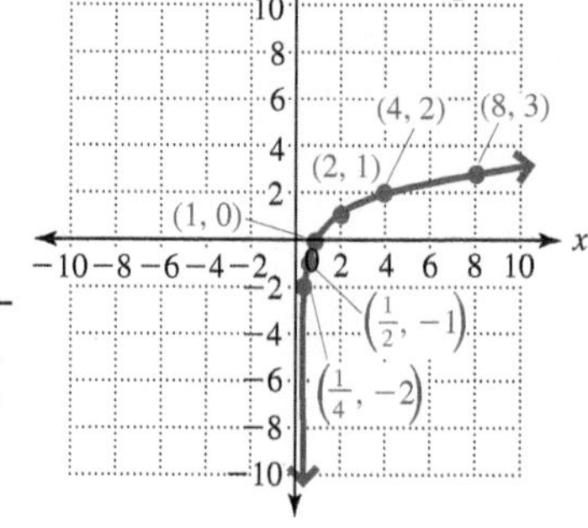

Answers to Your Turn 4
a. 81 **b.** 6 **c.** 64

Answers to Your Turn 5
a. 1 **b.** 0

b. $f(x) = \log_{1/2}x$

Solution: $y = \log_{1/2}x$ Replace $f(x)$ with y.

$x = \left(\frac{1}{2}\right)^y$ Write in exponential form.

Choose values for y and find x.

y	0	1	2	3	−1	−2	−3
x	$\left(\frac{1}{2}\right)^0 = 1$	$\left(\frac{1}{2}\right)^1 = \frac{1}{2}$	$\left(\frac{1}{2}\right)^2 = \frac{1}{4}$	$\left(\frac{1}{2}\right)^3 = \frac{1}{8}$	$\left(\frac{1}{2}\right)^{-1} = 2$	$\left(\frac{1}{2}\right)^{-2} = 4$	$\left(\frac{1}{2}\right)^{-3} = 8$

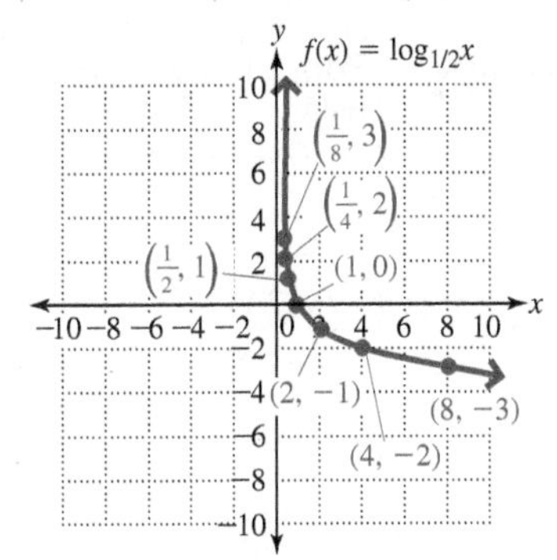

Following is a summary of the key features of the graphs of logarithmic functions.

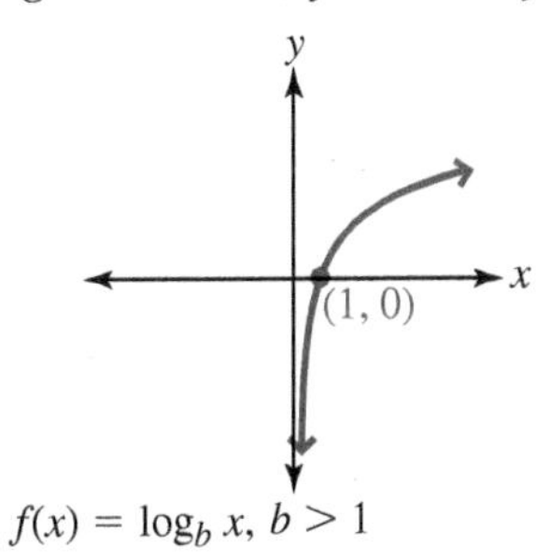

$f(x) = \log_b x,\ b > 1$

The graph passes through $(1, 0)$, approaches the y-axis, and increases. The domain is $(0, \infty)$. The range is $(-\infty, \infty)$.

(1, 0)

$f(x) = \log_b x,\ 0 < b < 1$

The graph passes through $(1, 0)$, approaches the y-axis, and decreases. The domain is $(0, \infty)$. The range is $(-\infty, \infty)$.

Your Turn 6 Graph $f(x) = \log_3 x$.

Objective 4 Solve applications involving logarithms.

Example 7 The function $P = 95 - 30\log_2 x$ models the percent, P, of students that recall the important features of a classroom lecture over time, where x is the number of days that have elapsed since the lecture was given. What percent of the students recall the important features of a lecture 8 days after it was given? (*Source: Psychology in the New Millennium*, 8th Edition, Spencer A. Rathus, Thomson Publishing Company.)

Solution:

$P = 95 - 30\log_2 8$ Substitute 8 for x in $P = 95 - 30\log_2 x$.

$P = 95 - 30(3)$ $\log_2 8 = 3$ because $2^3 = 8$.

$P = 95 - 90$ Multiply.

$P = 5$ Simplify.

Answer: Five percent of the students remember the important features of a lecture 8 days after it is given.

Answer to Your Turn 6

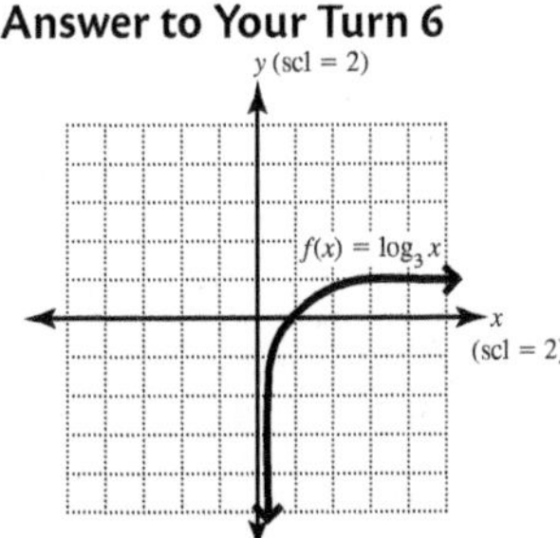

Your Turn 7 Refer to Example 7.

a. Find the percent of students who remember the important features of a lecture 2 days after it was given.

b. When the number of days was decreased by a fourth (from 8 to 2), was the amount retained also decreased by one-fourth?

Answers to Your Turn 7
a. 65% **b.** no

10.3 Exercises

For Extra Help MyMathLab®

Objective 1

Prep Exercise 1 If $f(x) = 3^x$, what is $f^{-1}(x)$? Why?

For Exercises 1–16, write in logarithmic form. See Example 1.

1. $2^5 = 32$

2. $5^4 = 625$

3. $10^3 = 1000$

4. $10^4 = 10{,}000$

5. $e^4 = x$

6. $e^{-2} = z$

7. $5^{-3} = \frac{1}{125}$

8. $7^{-4} = \frac{1}{2401}$

9. $10^{-2} = \frac{1}{100}$

10. $10^{-4} = \frac{1}{10{,}000}$

11. $625^{1/4} = 5$

12. $343^{1/3} = 7$

13. $\left(\frac{1}{4}\right)^2 = \frac{1}{16}$

14. $\left(\frac{3}{4}\right)^3 = \frac{27}{64}$

15. $7^{1/2} = \sqrt{7}$

16. $10^{1/2} = \sqrt{10}$

Prep Exercise 2 Why are logarithms exponents?

Prep Exercise 3 If $f(x) = \log_b x$, what are the restrictions on b?

Prep Exercise 4 If $f(x) = \log_b x$, what are the restrictions on x?

For Exercises 17–32, write in exponential form. See Example 2.

17. $\log_3 81 = 4$

18. $\log_4 64 = 3$

19. $\log_4 \frac{1}{16} = -2$

20. $\log_3 \frac{1}{243} = -5$

21. $\log_{10} 100 = 2$

22. $\log_{10} 1000 = 3$

23. $\log_e a = 5$

24. $\log_e y = -4$

25. $\log_e \frac{1}{e^4} = -4$

26. $\log_e \frac{1}{e^2} = -2$

27. $\log_{1/8} \frac{1}{64} = 2$

28. $\log_{1/4} \frac{1}{256} = 4$

29. $\log_{1/5} 25 = -2$

30. $\log_{1/3} 81 = -4$

31. $\log_7 \sqrt{7} = \frac{1}{2}$

32. $\log_6 \sqrt{6} = \frac{1}{2}$

Objective 2

Prep Exercise 5 In $y = \log_b x$, y is the __________ to which __________ must be raised to get __________.

For Exercises 33–52, solve. See Example 3.

33. $\log_2 x = 5$

34. $\log_3 x = 4$

35. $\log_5 x = -2$

36. $\log_4 x = -2$

37. $\log_3 81 = y$

38. $\log_2 32 = y$

39. $\log_5 \frac{1}{25} = y$

40. $\log_4 \frac{1}{16} = y$

41. $\log_b 1000 = 3$

42. $\log_b 10{,}000 = 4$

43. $\log_m \frac{1}{16} = -4$

44. $\log_n \frac{1}{36} = -2$

45. $\log_{1/2} x = 2$

46. $\log_{1/5} x = 3$

47. $\log_{1/3} h = -5$

48. $\log_{1/6} k = -2$

49. $\log_{1/3} \frac{1}{9} = y$

50. $\log_{1/4} \frac{1}{64} = y$

51. $\log_{1/2} 64 = t$

52. $\log_{1/5} 125 = u$

For Exercises 53–64, solve. See Example 4.

53. $\log_2 x + 4 = 8$

54. $\log_3 x + 2 = 5$

55. $\log_{1/4} h - 2 = 1$

56. $\log_{1/5} k - 4 = -1$

57. $3 \log_b 16 = 12$

58. $2 \log_b 81 = 8$

59. $\frac{1}{3} \log_5 c = 1$

60. $\frac{1}{4} \log_2 d = 2$

61. $3 \log_t 9 + 6 = 12$

62. $2 \log_u 125 + 8 = 14$

63. $\frac{1}{2} \log_4 m - 2 = -3$

64. $\frac{1}{3} \log_5 n - 4 = -5$

Objective 3

Prep Exercise 6 What is the relationship between the graphs of $f(x) = 5^x$ and $g(x) = \log_5 x$? Why?

Exercises 65–68, graph. See Example 6.

65. $f(x) = \log_4 x$

66. $f(x) = \log_5 x$

67. $f(x) = \log_{1/3} x$

68. $f(x) = \log_{1/4} x$

Objective 4

For Exercises 69–72, solve. See Example 7.

69. The percent of adult height attained by a 5- to 15-year-old girl can be approximated by $f(x) = 62 + 35\log_{10}(x - 4)$, where x is the age in years and $f(x)$ is the percent. At age 14, what percent of her adult height has a girl reached?

70. The percent of adult height attained by a 5- to 15-year-old boy can be approximated by $f(x) = 29 + 48.8\log_{10}(x + 1)$, where x is the age in years and $f(x)$ is the percent. At age 9, what percent of his adult height has a boy reached?

71. In Example 7, we learned that the percent of students who recall the important features of a lecture is given by $P = 95 - 30\log_2 x$, where P is the percent and x is the number of days that have elapsed since the lecture was given. After how many days will 35% of the students recall the important features?

72. Using the formula from Exercise 71, find the percent of students who recall the important features of a lecture 1 day after it is given.

73. Why does $\log_b b = 1$ for any value of $b > 0$ and $b \neq 1$?

74. In the definition of a logarithm, $y = \log_b x$, why must $b \neq 1$?

75. Why does $\log_b 1 = 0$?

76. If $f(x) = b^x$, the domain is $(-\infty, \infty)$ and the range is $(0, \infty)$. What are the domain and range of $f(x) = \log_b x$? Why?

Review Exercises

Exercises 1–6 Expressions

[5.1] **1.** Write $\frac{1}{x^6}$ as x to a negative power.

[5.1] *For Exercises 2–5, simplify using the rules of exponents.*

2. $x^4 \cdot x^2$

3. $(x^3)^5$

4. $\frac{x^6}{x^3}$

5. $\frac{(x^3)^2 \cdot x^4}{x^5}$

[8.2] **6.** Write $\sqrt[4]{x^3}$ in exponential form.

10.4 Properties of Logarithms

Objectives

1 Apply the inverse properties of logarithms.
2 Apply the product, quotient, and power properties of logarithms.

Warm-up

[10.3] **1.** Solve: $10 + 7\log_5 x = 31$

[8.2] **2.** Write $\sqrt{\dfrac{a^3}{b}}$ in exponential form.

[5.3] **3.** Multiply: $(x + 2)(x - 3)$

[5.1] **4.** Evaluate: 5^{-2}

Objective 1 Apply the inverse properties of logarithms.

Earlier we developed two properties of logarithms, $\log_b b = 1$ and $\log_b 1 = 0$, which were based on the definition of a logarithm. The fact that $f[f^{-1}(x)] = x$ and $f^{-1}[f(x)] = x$ gives us the following properties of logarithms.

Rule Inverse Properties of Logarithms

For any real numbers b and x, where $b > 0$, $b \neq 1$ and $x > 0$:

1. $b^{\log_b x} = x$.
2. $\log_b b^x = x$.

To prove that $b^{\log_b x} = x$, let $f(x) = b^x$ so that $f^{-1}(x) = \log_b x$.

$f[f^{-1}(x)] = x$ Composition of a function with its inverse is x.

$f[\log_b x] = x$ Replace $f^{-1}(x)$ with $\log_b x$.

$b^{\log_b x} = x$ Replace x with $\log_b x$ in $f(x)$.

To prove that $\log_b b^x = x$, use the definition of a logarithm. $\text{Log}_a b = c$ means that $a^c = b$; so $\log_b b^x = x$ because $b^x = b^x$.

Example 1 Find the value.

a. $3^{\log_3 8}$

Solution: $3^{\log_3 8} = 8$ Use $b^{\log_b x} = x$.

b. $6^{\log_6 x}$

Solution: $6^{\log_6 x} = x$ Use $b^{\log_b x} = x$.

c. $\log_3 3^6$

Solution: $\log_3 3^6 = 6$ Use $\log_b b^x = x$.

d. $\log_5 5^a$

Solution: $\log_5 5^a = a$ Use $\log_b b^x = x$.

Your Turn 1 Find the value.

a. $8^{\log_8 4}$

b. $\log_3 3^{-2}$

Answers to Your Turn 1
a. 4 b. −2

Answers to Warm-up
1. 125
2. $\left(\dfrac{a^3}{b}\right)^{1/2}$
3. $x^2 - x - 6$
4. $\dfrac{1}{25}$

Objective 2 Apply the product, quotient, and power properties of logarithms.

Logarithms were invented to perform operations on very large and very small numbers. With the invention of handheld calculators, they are no longer used for this purpose. However, we still use the properties of logarithms, which are based on the fact that logarithms are exponents.

Warning There is no rule for the logarithms of sums or differences. The $\log_b(x + y) \neq \log_b x + \log_b y$.

Note When the product and quotient rules are used, all of the bases of all of the logarithms **must** be the same.

Rule Further Properties of Logarithms

For real numbers x, y, and b, where $x > 0, y > 0, b > 0$, and $b \neq 1$:

Product Rule of Logarithms: $\log_b xy = \log_b x + \log_b y$
(The logarithm of the product of two numbers is equal to the sum of the logarithms of the numbers.)

Quotient Rule of Logarithms: $\log_b \frac{x}{y} = \log_b x - \log_b y$
(The logarithm of the quotient of two numbers is equal to the difference of the logarithms of the numbers.)

Power Rule of Logarithms: $\log_b x^r = r \log_b x$
(The logarithm of a number raised to a power is equal to the exponent times the logarithm of the number.)

To prove that $\log_b xy = \log_b x + \log_b y$, let $M = \log_b x$ and $N = \log_b y$.

$x = b^M$ and $y = b^N$ — Write each logarithmic equation in exponential form.

$xy = b^M \cdot b^N$ — Multiply the left and right sides of the exponential forms.

$xy = b^{M+N}$ — Add the exponents.

$\log_b xy = M + N$ — Write in logarithmic form.

$\log_b xy = \log_b x + \log_b y$ — Substitute $\log_b x$ for M and $\log_b y$ for N, and the proof is complete.

The proofs of $\log_b \frac{x}{y} = \log_b x - \log_b y$ and $\log_b x^r = r \log_b x$ are similar.

Example 2 Use the product rule of logarithms to write the expression as a sum of logarithms.

a. $\log_{10} xyz$

Solution: $\log_{10} xyz = \log_{10} x + \log_{10} y + \log_{10} z$

b. $\log_b x(x + 3)$

Note $\log_b(x + 3) \neq \log_b x + \log_b 3$.

Solution: $\log_b x(x + 3) = \log_b x + \log_b(x + 3)$

Your Turn 2 Use the product rule of logarithms to write the expression as a sum of logarithms.

a. $\log_a 6x$

b. $\log_2 x(3x + 2)$

Example 3 Use the product rule of logarithms in the form $\log_b x + \log_b y = \log_b xy$ to write the expression as a single logarithm.

a. $\log_3 8 + \log_3 2$

Solution: $\log_3 8 + \log_3 2 = \log_3 8 \cdot 2 = \log_3 16$

b. $\log_8 5 + \log_8 x + \log_8(2x - 3)$

Solution: $\log_8 5 + \log_8 x + \log_8(2x - 3) = \log_8 5x(2x - 3)$
$= \log_8(10x^2 - 15x)$

Your Turn 3 Use the product rule of logarithms to write each of the following as a single logarithm.

a. $\log_6 7 + \log_6 9$

b. $\log_7 2 + \log_7 x + \log_7(2x + 4)$

Answers to Your Turn 2
a. $\log_a 6 + \log_a x$
b. $\log_2 x + \log_2(3x + 2)$

Answers to Your Turn 3
a. $\log_6 63$
b. $\log_7(4x^2 + 8x)$

Example 4 Use the quotient rule of logarithms to write the expression as a difference of logarithms. Leave the answers in simplest form.

a. $\log_5 \frac{5}{11}$

Solution: $\log_5 \frac{5}{11} = \log_5 5 - \log_5 11$

$= 1 - \log_5 11$ Remember, $\log_5 5 = 1$.

b. $\log_4 \frac{x}{x-5}$

Solution: $\log_4 \frac{x}{x-5} = \log_4 x - \log_4(x-5)$ ◀ **Warning** $\log_4(x-5) \neq \log_4 x - \log_4 5$.

Your Turn 4 Use the quotient rule of logarithms to write the expression as a difference of logarithms. Leave your answers in simplest form.

a. $\log_9 \frac{9}{10}$

b. $\log_5 \frac{x}{x+2}$

Example 5 Use the quotient rule of logarithms in the form $\log_b x - \log_b y = \log_b \frac{x}{y}$ to write the expression as a single logarithm.

a. $\log_9 3 - \log_9 x$

Solution: $\log_9 3 - \log_9 x = \log_9 \frac{3}{x}$

b. $\log_{10} x - \log_{10}(x^2 + 4)$

Solution: $\log_{10} x - \log_{10}(x^2 + 4) = \log_{10} \frac{x}{x^2 + 4}$

Your Turn 5 Use the quotient rule of logarithms to write each of the following as a single logarithm.

a. $\log_3 15 - \log_3 5$

b. $\log_4(x^2 + 2) - \log_4(x + 1)$

Example 6 Use the power rule of logarithms to write the expression as a multiple of a logarithm.

a. $\log_4 a^6$

Solution: $\log_4 a^6 = 6 \log_4 a$

b. $\log_b \sqrt[4]{x^3}$

Solution: $\log_b \sqrt[4]{x^3} = \log_b x^{3/4}$ Write $\sqrt[4]{x^3}$ as $x^{3/4}$.

$= \frac{3}{4} \log_b x$

c. $\log_b \frac{1}{x^3}$

Solution: $\log_b \frac{1}{x^3} = \log_b x^{-3}$ Write $\frac{1}{x^3}$ as x^{-3}.

$= -3 \log_b x$ ◀ **Note** $\log_b \frac{1}{x^3}$ could be rewritten as $\log_b 1 - \log_b x^3 = 0 - 3 \log_b x = -3 \log_b x$.

Answers to Your Turn 4

a. $1 - \log_9 10$

b. $\log_5 x - \log_5(x + 2)$

Answers to Your Turn 5

a. $\log_3 3 = 1$

b. $\log_4 \frac{x^2 + 2}{x + 1}$

Your Turn 6 Use the power rule of logarithms to write the expression as a multiple of a logarithm.

a. $\log_a z^4$ **b.** $\log_a \sqrt[3]{x^2}$

Example 7 Use the power rule of logarithms in the form $r \log_b x = \log_b x^r$ to write the expression as a logarithm of a quantity to a power. Leave the answers in simplest form without negative or fractional exponents.

a. $5 \log_4 x$

Solution: $5 \log_4 x = \log_4 x^5$

b. $-2 \log_b 5$

Solution: $-2 \log_b 5 = \log_b 5^{-2}$

$= \log_b \frac{1}{5^2}$ Write 5^{-2} as $\frac{1}{5^2}$.

$= \log_b \frac{1}{25}$ $5^2 = 25$

c. $\frac{2}{3} \log_7 y$

Solution: $\frac{2}{3} \log_7 y = \log_7 y^{2/3}$

$= \log_7 \sqrt[3]{y^2}$ Write $y^{2/3}$ as $\sqrt[3]{y^2}$.

Your Turn 7 Use the power rule of logarithms to write the expression as a logarithm of a quantity to a power.

a. $5 \log_7 x$ **b.** $-3 \log_b z$

Often, it is necessary to use more than one rule to rewrite a logarithmic expression.

Example 8 Write the expression as a sum or difference of multiples of logarithms.

a. $\log_b \frac{z^3}{yz}$

Solution: $\log_b \frac{z^3}{yz} = \log_b z^3 - \log_b yz$ Use the quotient rule.

$= 3 \log_b z - (\log_b y + \log_b z)$ Use the power rule and product rule. Note the use of parentheses.

$= 3 \log_b z - \log_b y - \log_b z$ Remove the parentheses.

b. $\log_3 \sqrt{\frac{a^3}{b}}$

Solution: $\log_3 \sqrt{\frac{a^3}{b}} = \log_3 \left(\frac{a^3}{b}\right)^{1/2}$ Write $\sqrt{\frac{a^3}{b}}$ as $\left(\frac{a^3}{b}\right)^{1/2}$.

$= \frac{1}{2} \log_3 \left(\frac{a^3}{b}\right)$ Use the power rule.

$= \frac{1}{2}(\log_3 a^3 - \log_3 b)$ Use the quotient rule.

$= \frac{1}{2}(3 \log_3 a - \log_3 b)$ Use the power rule.

$= \frac{3}{2} \log_3 a - \frac{1}{2} \log_3 b$ Distribute $\frac{1}{2}$.

Answers to Your Turn 6

a. $4 \log_a z$ **b.** $\frac{2}{3} \log_a x$

Answers to Your Turn 7

a. $\log_7 x^5$ **b.** $\log_b \frac{1}{z^3}$

c. $\log_5 5^2 b^3$

Solution: $\log_5 5^2 b^3 = \log_5 5^2 + \log_5 b^3$ Use the product rule.

$= 2\log_5 5 + 3\log_5 b$ Use the power rule.

$= 2 + 3\log_5 b$ $\log_5 5 = 1$

Your Turn 8 Write the expression as a sum or difference of multiples of logarithms.

a. $\log_a \dfrac{x^2 y^3}{z}$ **b.** $\log_4 \sqrt[4]{\dfrac{x}{y^2}}$ **c.** $\log_8 8^5 m^6$

Example 9 Write the expression as a single logarithm. Leave the answers in simplest form without negative or fractional exponents.

a. $4\log_b 2 - 2\log_b 3$

Solution: $4\log_b 2 - 2\log_b 3 = \log_b 2^4 - \log_b 3^2$ Use the power rule.

$= \log_b \dfrac{2^4}{3^2}$ Use the quotient rule.

$= \log_b \dfrac{16}{9}$ Simplify.

b. $\dfrac{1}{2}(\log_2 5 - \log_2 b)$

Solution: $\dfrac{1}{2}(\log_2 5 - \log_2 b) = \dfrac{1}{2}\log_2 \dfrac{5}{b}$ Use the quotient rule.

$= \log_2\left(\dfrac{5}{b}\right)^{1/2}$ Use the power rule.

$= \log_2 \sqrt{\dfrac{5}{b}}$ Write $\left(\dfrac{5}{b}\right)^{1/2}$ as $\sqrt{\dfrac{5}{b}}$.

c. $\log_a(x+2) + \log_a(x-3)$

Solution: $\log_a(x+2) + \log_a(x-3) = \log_a(x+2)(x-3)$ Apply $\log_b xy = \log_b x + \log_b y$.

$= \log_a(x^2 - x - 6)$ Multiply.

Your Turn 9 Write the expression as a single logarithm. Leave your answers in simplest form without negative or fractional exponents.

a. $2\log_a 4 - 3\log_a x$ **b.** $\dfrac{1}{3}(\log_a x + 2\log_a y)$ **c.** $\log_6 x + \log_6(x-2)$

Answers to Your Turn 8
a. $2\log_a x + 3\log_a y - \log_a z$
b. $\dfrac{1}{4}\log_4 x - \dfrac{1}{2}\log_4 y$
c. $5 + 6\log_8 m$

Answers to Your Turn 9
a. $\log_a \dfrac{16}{x^3}$ **b.** $\log_a \sqrt[3]{xy^2}$
c. $\log_6(x^2 - 2x)$

10.4 Exercises

For Extra Help MyMathLab®

Objective 1

Prep Exercise 1 For any real numbers b and x, where $b > 0$, $b \neq 1$, and $x > 0$, $b^{\log_b x} =$ ____________.

Prep Exercise 2 For any real numbers b and x, where $b > 0$, $b \neq 1$, and $x > 0$, $\log_b b^x =$ ____________.

For Exercises 1–12, find the value. See Example 1.

1. $8^{\log_8 2}$ **2.** $3^{\log_3 7}$ **3.** $a^{\log_a r}$ **4.** $b^{\log_b a}$ **5.** $a^{\log_a 4x}$ **6.** $b^{\log_b 5a}$

7. $\log_3 3^5$ **8.** $\log_7 7^3$ **9.** $\log_e e^y$ **10.** $\log_c c^x$ **11.** $\log_a a^{7x}$ **12.** $\log_b b^{6y}$

Objective 2

Prep Exercise 3 Write the product rule of logarithms: $\log_b xy =$ ________.

Prep Exercise 4 $\log_b(x + y) =$ ________.

For Exercises 13–20, use the product rule to write the expression as a sum of logarithms. See Example 2.

13. $\log_2 5y$ **14.** $\log_3 4z$ **15.** $\log_a pq$ **16.** $\log_b rs$

17. $\log_4 mnp$ **18.** $\log_4 pqr$ **19.** $\log_a x(x - 5)$ **20.** $\log_b x(x + 6)$

For Exercises 21–32, use the product rule to write the expression as a single logarithm. See Example 3.

21. $\log_3 5 + \log_3 8$ **22.** $\log_6 4 + \log_6 7$ **23.** $\log_4 3 + \log_4 9$

24. $\log_9 4 + \log_9 21$ **25.** $\log_a 7 + \log_a m$ **26.** $\log_b 2 + \log_b n$

27. $\log_4 a + \log_4 b$ **28.** $\log_6 r + \log_6 m$ **29.** $\log_a 2 + \log_a x + \log_a(x + 5)$

30. $\log_a 4 + \log_a y + \log_a(y - 5)$ **31.** $\log_4(x + 1) + \log_4(x + 3)$ **32.** $\log_6(x - 3) + \log_6(x + 2)$

Prep Exercise 5 Write the quotient rule of logarithms: $\log_b \frac{x}{y} =$ ________.

For Exercises 33–42, use the quotient rule to write the expression as a difference of logarithms. Leave your answers in simplest form. See Example 4.

33. $\log_2 \frac{7}{9}$ **34.** $\log_4 \frac{5}{6}$ **35.** $\log_a \frac{x}{5}$ **36.** $\log_b \frac{y}{3}$

37. $\log_a \frac{a}{b}$ **38.** $\log_b \frac{a}{b}$ **39.** $\log_a \frac{x}{x - 3}$ **40.** $\log_b \frac{y}{2y + 5}$

41. $\log_4 \frac{2x - 3}{4x + 5}$ **42.** $\log_6 \frac{4x - 1}{3x + 2}$

For Exercises 43–54, use the quotient rule to write the expression as a single logarithm. Leave your answers in simplest form. See Example 5.

43. $\log_6 24 - \log_6 3$ **44.** $\log_7 18 - \log_7 3$ **45.** $\log_2 24 - \log_2 12$ **46.** $\log_3 48 - \log_3 16$

47. $\log_a x - \log_a 3$ **48.** $\log_a r - \log_a 5$ **49.** $\log_4 p - \log_4 q$ **50.** $\log_2 a - \log_2 b$

51. $\log_b x - \log_b(x - 4)$

52. $\log_b y - \log_b(y - 5)$

53. $\log_x(x^2 - x) - \log_x(x - 1)$

54. $\log_a(a^2 + 2a) - \log_a(a + 2)$

Prep Exercise 6 Write the power rule of logarithms: $\log_b x^r =$ ______________.

For Exercises 55–66, use the power rule to write the expression as a multiple of a logarithm. See Example 6.

55. $\log_4 3^6$

56. $\log_3 5^4$

57. $\log_a x^7$

58. $\log_b y^8$

59. $\log_a \sqrt{3}$

60. $\log_a \sqrt{6}$

61. $\log_3 \sqrt[3]{x^2}$

62. $\log_6 \sqrt[4]{y^3}$

63. $\log_a \frac{1}{6^2}$

64. $\log_a \frac{1}{5^4}$

65. $\log_a \frac{1}{y^2}$

66. $\log_a \frac{1}{x^5}$

For Exercises 67–78, use the power rule to write the expression as a logarithm of a quantity to a power. Leave your answers in simplest form without negative or fractional exponents. See Example 7.

67. $4 \log_3 5$

68. $5 \log_3 4$

69. $-3 \log_2 x$

70. $-4 \log_3 y$

71. $\frac{1}{2}\log_7 64$

72. $\frac{1}{3}\log_3 8$

73. $\frac{3}{4}\log_a x$

74. $\frac{5}{6}\log_b y$

75. $\frac{2}{3}\log_a 8$

76. $\frac{3}{4}\log_a 81$

77. $-\frac{1}{2}\log_3 x$

78. $-\frac{1}{3}\log_2 y$

For Exercises 79–90, write the expression as the sum or difference of multiples of logarithms. See Example 8.

79. $\log_a \frac{x^3}{y^4}$

80. $\log_a \frac{x^6}{y^5}$

81. $\log_3 a^4 b^2$

82. $\log_7 m^3 n^5$

83. $\log_a \frac{xy}{z}$

84. $\log_a \frac{pq}{r}$

85. $\log_x \frac{a^2}{bc^3}$

86. $\log_x \frac{c^2}{m^2 n}$

87. $\log_4 \sqrt[4]{\frac{x^3}{y}}$

88. $\log_5 \sqrt{\frac{a^5}{b^3}}$

89. $\log_a \sqrt[3]{\frac{x^2 y}{z^3}}$

90. $\log_3 \sqrt{\frac{c^2 d^3}{m^4}}$

For Exercises 91–104, write the expression as a single logarithm. Leave your answers in simplest form without negative or fractional exponents. See Example 9.

91. $3\log_3 2 - 2\log_3 4$

92. $3\log_4 3 - 2\log_4 9$

93. $4\log_b x + 3\log_b y$

94. $5\log_b a + 4\log_b c$

95. $\frac{1}{2}(\log_a 5 - \log_a 7)$

96. $\frac{1}{2}(\log_4 6 - \log_4 5)$

97. $\frac{2}{3}(\log_a x^2 + \log_a y^3)$

98. $\frac{3}{4}(\log_a m^3 + \log_a n^2)$

99. $\log_b x + \log_b (3x - 2)$

100. $\log_a y + \log_a (2y - 4)$

101. $3\log_a (x - 2) - 4\log_a (x + 1)$

102. $2\log_b (x + 4) - 3\log_b (2x - 1)$

103. $2\log_a x + 4\log_a z - 3\log_a w - 6\log_a u$

104. $5\log_a b + 2\log_a c - 4\log_a d - 3\log_a e$

Review Exercises

Exercises 1 and 2 Expressions

[5.1] *For Exercises 1 and 2, simplify.*

1. $(10^{9.5})(10^{-12})$

2. $\frac{10^{-3}}{10^{-12}}$

Exercises 3–6 Equations and Inequalities

[10.2] 3. If \$10,000 is deposited at 6% compounded quarterly for five years, how much will be in the account?

[10.3] 4. Write $10^{1.6990} = 50$ in logarithmic form.

[10.3] *For Exercises 5 and 6, write in exponential form.*

5. $\log_{10} 45 = 1.6532$

6. $\log_e 0.25 = -1.3863$

10.5 Common and Natural Logarithms

Objectives

1. Define common logarithms and evaluate them using a calculator.
2. Define natural logarithms and evaluate them using a calculator.
3. Solve applications using common logarithms.
4. Solve applications using natural logarithms.

Warm-up

[10.4] **1.** Evaluate: $\log_{10} 10^{9.5}$

[10.4] **2.** Write as a single logarithm: $2\log_n 5 - 3\log_n 2$

[5.1] **3.** Simplify: $\frac{10^{-2.5}}{10^{-12}}$

[10.3] **4.** Solve $6.723 = 3\log_{10} x$ for x. Round to the nearest tenth.

Objective 1 Define common logarithms and evaluate them using a calculator.

Of all possible bases of logarithms, two are most useful. As previously mentioned, logarithms were invented to do computations on very large and very small numbers. Because our system is a base-10 number system, base-10 logarithms were commonly used for this purpose. Consequently, base-10 logarithms are called **common logarithms** and $\log_{10}x$ is written as $\log x$, where the base 10 is understood. Base-10 logarithms are found in engineering, economics, the social sciences, and the natural sciences.

Definition **Common logarithms:** Logarithms with a base of 10. $\text{Log}_{10}x$ is written as $\log x$. Note that $\log 10 = 1$.

Connection Because logarithmic and exponential functions are inverse functions, if $f(x) = 10^x$, then $f^{-1}(x) = \log x$ and if $g(x) = \log x$, then $g^{-1}(x) = 10^x$.

To evaluate common logarithms, we use a calculator. We will round all results to four places.

Example 1 Use a calculator to approximate each common logarithm to four decimal places.

a. $\log 23$

Solution: $\log 23 \approx 1.3617$

Connection Remember that $\log 23 \approx 1.3617$ means that $10^{1.3617} \approx 23$, which provides a way to check. Because we rounded the decimal value, evaluating $10^{1.3617}$ will not give exactly 23.

b. $\log 0.00236$

Solution: $\log 0.00236 \approx -2.6271$

Your Turn 1 Use a calculator to approximate each common logarithm to four decimal places.

a. $\log 436$

b. $\log 0.0724$

Answers to Warm-up

1. 9.5
2. $\log_n\left(\frac{25}{8}\right)$
3. $10^{9.5}$
4. 174.2

Answers to Your Turn 1

a. 2.6395 **b.** −1.1403

Objective 2 Define natural logarithms and evaluate them using a calculator.

The number e is an irrational number whose approximate value is 2.7182818285. It is a universal constant like π. In the natural sciences, compared with base-10 logarithms, base-e logarithms are more prevalent. Because base-e logarithms occur in so many "natural" situations, they are called **natural logarithms**. The notation for $\log_e x$ is $\ln x$, which is read "el en of x."

Definition **Natural logarithms:** Base-e logarithms are called natural logarithms, and $\log_e x$ is written as $\ln x$. Note that $\ln e = 1$.

To find natural logarithms, we use a calculator.

Example 2 Use a calculator to approximate each natural logarithm to four decimal places.

a. $\ln 83$

Solution: $\ln 83 \approx 4.4188$

b. $\ln 0.0055$

Solution: $\ln 0.0055 \approx -5.2030$

Connection Remember that $\ln 83 \approx 4.4188$ means that $e^{4.4188} \approx 83$. Similarly, $\ln 0.0055 = -5.2030$ means that $e^{-5.2030} \approx 0.0055$.

Your Turn 2 Use a calculator to approximate each natural logarithm to four decimal places.

a. $\ln 102$ **b.** $\ln 0.0573$

Objective 3 Solve applications using common logarithms.

Common logarithms can be used to calculate sound intensity and runway length.

Example 3

a. Sound intensity can be measured in watts per unit of area or, more commonly, in decibels. The function $d = 10 \log \frac{I}{I_0}$ is used to calculate sound intensity, where d represents the intensity in decibels, I represents the intensity in watts per unit of area, and I_0 represents the faintest audible sound to the average human ear (which is 10^{-12} watts per square meter). A motorcycle has a sound intensity of about $10^{-2.5}$ watts per square meter. Find the decibel reading for the motorcycle.

Solution: $d = 10 \log \frac{10^{-2.5}}{10^{-12}}$ In $d = 10 \log \frac{I}{I_0}$, substitute $10^{-2.5}$ for I and 10^{-12} for I_0.

$d = 10 \log 10^{9.5}$ Subtract exponents $[-2.5 - (-12) = 9.5]$.

$d = 10(9.5)$ Use $\log_b b^x = x$.

$d = 95$ Simplify.

Answer: The motorcycle has a decibel reading of 95 dB.

Note The abbreviation for decibels is dB.

b. The minimum length of an airport runway needed for a plane to take off is related to the weight of the plane. For some planes, the minimum runway length may be modeled by the function $y = 3 \log x$, where x is the plane's weight in thousands of pounds and y is the length of the runway in thousands of feet. Find the minimum length of a runway needed by a Boeing 737 whose maximum take-off weight is 174,200 pounds.

Answers to Your Turn 2
a. 4.6250 **b.** −2.8595

Solution: $x = \dfrac{174{,}200}{1000} = 174.2$ Divide 174,200 by 1000 to find the value of x corresponding to 174,200 pounds.

$y = 3 \log 174.2$ In $y = 3 \log x$, replace x with 174.2.

$y \approx 6.723$ Evaluate using a calculator.

Answer: Because y is in thousands of feet, the minimum runway length is about $(6.723)(1000) = 6723$ ft.

Your Turn 3

a. The sound intensity of a rock band often exceeds $10^{-0.5}$ watts per square meter. Find the decibel reading of the band.

b. Find the minimum runway length needed for a B-52 Stratofortress whose maximum takeoff weight is 488,000 pounds.

Objective 4 Solve applications using natural logarithms.

If money is deposited into an account and the interest is compounded continuously, then the time t (in years) that it will take an investment of P dollars to grow into A dollars at an interest rate r (written as a decimal) is given by $t = \dfrac{1}{r}\ln\dfrac{A}{P}$.

Example 4 An amount of $5000 is deposited into an account earning 5% annual interest compounded continuously. How many years will it take until the account has reached $10,000?

Solution: $t = \dfrac{1}{0.05}\ln\dfrac{10{,}000}{5000}$ In $t = \dfrac{1}{r}\ln\dfrac{A}{P}$, replace P with 5000, r with 0.05, and A with 10,000.

$t = 20 \ln 2$ Divide.

$t \approx 13.9$ Multiply using a calculator.

Answer: It will take about 13.9 years for $5000 to grow to $10,000 if it is compounded continuously at 5%.

Answers to Your Turn 3
a. 115 dB **b.** ≈ 8065 ft.

Answer to Your Turn 4
≈ 11.5 yr.

Your Turn 4 How long will it take $2000 to grow to $5000 at 8% interest if the interest is compounded continuously?

10.5 Exercises

For Extra Help MyMathLab®

Objectives 1 and 2

Prep Exercise 1 In the common logarithm $\log x$, what is the base?

Prep Exercise 2 In the natural logarithm $\ln x$, what is the base?

Prep Exercise 3 Is $\log x$ positive or negative for $0 < x < 1$? Why?

Prep Exercise 4 Is $\ln x$ positive or negative for $x > 1$? Why?

For Exercises 1–20, use a calculator to approximate each logarithm to four decimal places. See Examples 1 and 2.

1. $\log 64$
2. $\log 27$
3. $\log 0.0067$
4. $\log 0.00087$
5. $\log 435.6$
6. $\log 785.4$
7. $\log(1.5 \times 10^4)$
8. $\log(5.7 \times 10^7)$
9. $\log (1.6 \times 10^{-6})$
10. $\log(7.5 \times 10^{-5})$
11. $\ln 9.34$
12. $\ln 5.33$
13. $\ln 79.2$
14. $\ln 765.4$
15. $\ln 0.034$
16. $\ln 0.0923$
17. $\ln(5.4 \times e^4)$
18. $\ln (2.4 \times e^3)$
19. $\log e$
20. $\ln 10$
21. Use your calculator to find log 0. What happened? Why?
22. Use your calculator to find $\ln(-1)$. What happened? Why?

Prep Exercise 5 If $e^{0.91629} = 2.5$, find ln 2.5.

Prep Exercise 6 Without using a calculator, find the exact value of $\log 10^{\sqrt{5}}$.

For Exercises 23–34, find the exact value of each logarithm using $\log_b b^x = x$.

23. $\log 100$
24. $\log 1000$
25. $\log \dfrac{1}{100}$
26. $\log \dfrac{1}{1000}$
27. $\log \sqrt[3]{10}$
28. $\log \sqrt[4]{10}$
29. $\log 0.001$
30. $\log 0.00001$
31. $\ln e^3$
32. $\ln e^5$
33. $\ln \sqrt{e}$
34. $\ln \sqrt[5]{e}$

Objective 3

For Exercises 35–38, use the formula $d = 10 \log \dfrac{I}{I_{00}}$, where $I_0 = 10^{-12}$ watts/m^2. See Example 3.

35. The sound intensity of a firecracker is 10^{-3} watts per square meter. What is the decibel reading for the firecracker?
36. The sound intensity of a race car is 10^{-1} watts per square meter. What is the decibel reading for the race car?
37. If a noisy office has a decibel reading of 60, what is the sound intensity?
38. If loud thunder has a decibel reading of 80, what is the sound intensity?

For Exercises 39–42, use the following information. In chemistry, the pH of a substance determines whether it is a base $(pH > 7)$ or an acid $(pH < 7)$. To find the pH of a solution, we use the formula $pH = -\log[H_3O^+]$, where $[H_3O^+]$, is the hydronium ion concentration in moles per liter. Note that the pH is unitless.

39. Find the pH of vinegar if $[H_3O^+] = 1.6 \times 10^{-3}$ moles per liter.

40. Find the pH of maple syrup if $[H_3O^+] = 2.3 \times 10^{-7}$ moles per liter.

41. Find the hydronium ion concentration of sauerkraut, which has a pH of 3.5.

42. Find the hydronium ion concentration of blood, which has a pH of 7.4.

Objective 4

For Exercises 43 and 44, use the formula $t = \frac{1}{r}\ln\frac{A}{P}$. *See Example 4.*

43. Find how long it will take $2000 to grow to $5000 at 4% interest if the interest is compounded continuously.

44. Find how long it will take $5000 to grow to $8000 at 5% interest if the interest is compounded continuously.

45. The magnitude of an earthquake is given by the Richter scale, whose formula is $R = \log\frac{I}{I_0}$, where I is the intensity of the earthquake and I_0 is the intensity of a minimal earthquake and is used for comparison purposes. The 1906 San Francisco earthquake had a magnitude of 7.8, and the 1964 Alaska earthquake had a magnitude of 8.4. Compare the intensity of the two earthquakes. (*Hint:* Express the intensity of each in terms of I_0.)

46. Using the formula from Exercise 45, compare the intensity of the 1949 Queen Charlotte Islands earthquake, whose magnitude was 8.1, with the 2004 earth-quake in the Indian Ocean, whose magnitude was estimated at 9.2, that killed more than 225,000 people.

Of Interest

This photograph shows the destruction of San Francisco that was caused by the 1906 earthquake.

47. During an earthquake, energy is released in various forms. The amount of energy radiated from the earthquake as seismic waves is given by $\log E_s = 11.8 + 1.5\,M$, where E_s is measured in ergs and M is the magnitude of the earthquake as given by the Richter scale. Vancouver Island had an earthquake whose magnitude was 7.3. How much energy was released in the form of seismic waves?

48. Using the formula from Exercise 47, find the energy released in the form of seismic waves from the Double Springs Flat earthquake of 1994, whose magnitude was 6.1 on the Richter scale.

49. The purchasing power of $1.00 in 2007 can be approximated by $y = -67.89 + 22.56\ln x$, where y is the number of years beginning with 1970 ($y = 1$ corresponds with 1970) until 2010 and x is the purchasing power, in cents, of $1.00 in 2010. In what year was the purchasing power of $1.00 in 2010 approximately $0.40?

50. The average number of miles per gallon for light trucks can be approximated by $y = 14.27 + 2.98\ln x$, where y is the average number of miles per gallon and x is the number of years beginning with 1975. ($y = 1$ corresponds with 1975.) Find the approximate average number of miles per gallon in 2013.

51. Using the formula from Example 3(b), $y = 3 \log x$, find the minimum runway length needed for a Boeing 717 whose maximum weight is 110,000 pounds at takeoff.

52. Walking speeds in various cities are a function of the population, because as populations increase, so does the pace of life. Average walking speeds can be modeled by the function $W = 0.35 \ln P + 2.74$, where P is the population, in thousands, and W is the walking speed in feet per second. Find the average walking speed in Chicago, whose population was approximately 2,700,000 in 2010.

Review Exercises

Exercises 1 Expressions

[10.4] 1. Write $\log_3(2x + 1) - \log_3(x - 1)$ as a single logarithm.

Exercises 2–6 Equations and Inequalities

[10.3] 2. Write $\log_3(2x + 5) = 2$ in exponential form.

For Exercises 3–6, solve.

[2.1] 3. $3x - (7x + 2) = 12 - 2(x - 4)$

[6.4] 4. $x^2 + 2x = 15$

[7.4] 5. $\frac{5}{x} + \frac{3}{x+1} = \frac{23}{3x}$

[8.6] 6. $\sqrt{5x - 1} = 7$

10.6 Exponential and Logarithmic Equations with Applications

Objectives

1 Solve equations that have variables as exponents.
2 Solve equations containing logarithms.
3 Solve applications involving exponential and logarithmic functions.
4 Use the change-of-base formula.

Warm-up

[10.5] 1. The formula $t = \frac{1}{r} \ln \frac{A}{P}$ can be used to find the time, in years, for a principal amount, P, invested at interest rate, r, to grow to an amount, A. How many years will it take \$2000 invested at 4% interest to grow to \$6000?

[10.4] 2. Write $\log_5 x + \log_5(x + 3)$ as a single logarithm.

[6.4] 3. Solve: $x(x + 3) = 4$

[10.3] 4. Write $\log_2 \frac{5x + 1}{x - 1} = 3$ in exponential form.

Answers to Warm-up

1. ≈ 27.5 yr.
2. $\log_5 x(x + 3)$
3. $x = -4, 1$
4. $\frac{5x + 1}{x - 1} = 2^3$

In Section 10.2, we solved equations that could be put in the form $b^x = b^y$. For example, if $3^x = 81$, then $3^x = 3^4$; so $x = 4$. To use this form, we had to write both sides of the equation as the same base raised to a power. In this section, we solve equations like $3^x = 16$, which is the same as $3^x = 2^4$, where the bases are not the same. To solve this and other exponential and logarithmic equations, we need the following properties.

Rule Properties for Solving Exponential and Logarithmic Equations

For any real numbers b, x, and y, where $b > 0$ and $b \neq 1$:

1. If $b^x = b^y$, then $x = y$.
2. If $x = y$, then $b^x = b^y$.
3. For $x > 0$ and $y > 0$, if $\log_b x = \log_b y$, then $x = y$.
4. For $x > 0$ and $y > 0$, if $x = y$, then $\log_b x = \log_b y$.
5. For $x > 0$, if $\log_b x = y$, then $b^y = x$.

These properties are true because the exponential and logarithmic functions are one-to-one.

Objective 1 Solve equations that have variables as exponents.

To solve equations that have variables as exponents, we use property 4, which says that if two positive numbers are equal, then so are their logarithms.

Example 1 Solve $5^x = 16$.

Solution:

$\log 5^x = \log 16$ Use if $x = y$, then $\log_b x = \log_b y$ (property 4).

$x \log 5 = \log 16$ Use $\log_b x^r = r \log_b x$.

$x = \dfrac{\log 16}{\log 5}$ Divide both sides of the equation by log 5.

The exact solution is $x = \dfrac{\log 16}{\log 5}$. Using a calculator, we find $x \approx 1.7227$ correct to four decimal places.

Check: $5^x = 16$

$5^{1.7227} = 16$ Substitute 1.7727 for x.

$15.9998 \approx 16$ The answer is correct.

Your Turn 1 Solve $6^{2x} = 42$ for x. Round the answer to four decimal places.

In Example 1, we took the common logarithm of both sides, but we could have used natural logarithms (or logarithms of any other base) instead. If one side of the equation contains a power of e, natural logs are preferred so that we can use the fact that $\log_e e^x = x$ or, more simply, $\ln e^x = x$.

Answer to Your Turn 1
1.0430

Example 2 Solve $e^{4x} = 23$.

Note If your calculator automatically puts the left parenthesis, this must be entered as $\ln(23) \div 4$.

Solution: $\ln e^{4x} = \ln 23$ Use if $x = y$; then $\log_b x = \log_b y$.

$4x = \ln 23$ Use $\log_b b^x = x$.

$x = \dfrac{\ln 23}{4}$ Divide both sides by 4.

$x \approx 0.7839$

We will leave the check to the reader.

Your Turn 2 Solve $e^{3x} = 5$ for x. Round the answer to four decimal places.

Objective 2 Solve equations containing logarithms.

Now that we have explored additional properties of logarithms, we can modify the procedure presented in Section 10.3 for solving equations using logarithms.

Procedure Solving Equations Containing Logarithms

To solve equations containing logarithms, use the properties of logarithms to simplify each side of the equation and then use one of the following.

If the simplification results in an equation in the form $\log_b x = \log_b y$, use the fact that $x = y$ and then solve for the variable.

If the simplification results in an equation in the form $\log_b x = y$, write the equation in exponential form, $b^y = x$, and then solve for the variable (as we did in Section 10.3).

Example 3 Solve.

a. $\log_5 x + \log_5(x + 3) = \log_5 4$

Solution: $\log_5 x(x + 3) = \log_5 4$ Use $\log_b xy = \log_b x + \log_b y$ to simplify the left side.

$x(x + 3) = 4$ The equation is in the form $\log_b x = \log_b y$, so $x = y$.

$x^2 + 3x = 4$ Simplify.

$x^2 + 3x - 4 = 0$ Write in $ax^2 + bx + c = 0$ form.

$(x + 4)(x - 1) = 0$ Factor.

$x + 4 = 0$ or $x - 1 = 0$ Set each factor equal to 0.

$x = -4$ or $x = 1$ Solve each equation to find possible solutions.

If -4 is substituted into the original equation, we have $\log_5(-4) + \log_5(-1) = \log_5 4$, but logarithms are defined for positive numbers only. So $x = -4$ is not a solution. A check will show that $x = 1$ is a solution.

b. $\log_3(2x + 5) = 2$

Solution: $3^2 = 2x + 5$ The equation is in the form $\log_b x = y$; so write it in exponential form, $b^y = x$.

$9 = 2x + 5$ Evaluate 3^2.

$4 = 2x$ Subtract 5 from both sides.

$2 = x$ Solve for x.

We will let the reader check that $x = 2$ is a solution.

Answer to Your Turn 2
0.5365

c. $\log_2(5x + 1) - \log_2(x - 1) = 3$

Solution: $\log_2\dfrac{5x + 1}{x - 1} = 3$ Use $\log_b \dfrac{x}{y} = \log_b x - \log_b y$ to simplify the left side.

$\dfrac{5x + 1}{x - 1} = 2^3$ The equation is in the form $\log_b x = y$, so $b^y = x$.

$5x + 1 = 8x - 8$ Multiply both sides by $x - 1$.

$9 = 3x$ Isolate the x term.

$x = 3$ Solve for x.

We will let the reader check that $x = 3$ is a solution.

Your Turn 3 Solve.

a. $\ln x + \ln(x + 2) = \ln 8$ **b.** $\log_3(4x + 2) - \log_3(x - 2) = 2$

Objective 3 Solve applications involving exponential and logarithmic functions.

A wide variety of problems from business and the sciences can be solved using exponential or logarithmic functions. Earlier we used the formula for compound interest, $A = P\left(1 + \dfrac{r}{n}\right)^{nt}$, to find A when given P, r, n, and t. Using logarithms, it is also possible to find t when given A, P, r, and n.

Example 4 How long will it take \$6000 invested at 6% interest compounded quarterly to grow to \$10,000? Round the answer to the nearest tenth of a year.

Solution: $10{,}000 = 6000\left(1 + \dfrac{0.06}{4}\right)^{4t}$ In $A = P\left(1 + \dfrac{r}{n}\right)^{nt}$, replace A with 10,000, P with 6000, r with 0.06, and n with 4.

$\dfrac{5}{3} = (1.015)^{4t}$ Divide both sides by 6000 and simplify inside the parentheses.

$\log\dfrac{5}{3} = \log 1.015^{4t}$ Use if $x = y$; then $\log_b x = \log_b y$.

$\log\dfrac{5}{3} = 4t \log 1.015$ Use $\log_b x^r = r \log_b x$.

$\dfrac{\log\frac{5}{3}}{4 \log 1.015} = t$ Divide both sides by 4 log 1.015.

$8.6 \approx t$ Evaluate using a calculator and round.

Answer: It will take about 8.6 years.

Connection Solving $A = Pe^{rt}$ for t gives the formula $t = \dfrac{1}{r}\ln\dfrac{A}{P}$.

$\dfrac{A}{P} = e^{rt}$ Divide both sides by P to isolate e^{rt}.

$\log_e \dfrac{A}{P} = rt$ Write in log form.

$\dfrac{1}{r}\ln\dfrac{A}{P} = t$ Divide both sides by r, and by definition, $\log_e \dfrac{A}{P}$ is $\ln\dfrac{A}{P}$.

Your Turn 4 How long will it take \$8000 invested at 4% annual interest compounded semiannually to grow to \$12,000? Round to the nearest tenth of a year.

Many banks compound interest continuously. The formula for interest compounded continuously is $A = Pe^{rt}$, where A is the amount in the account, P is the amount deposited, r is the interest rate as a decimal, and t is the time in years.

Example 5 If \$5000 is deposited into an account at 5% interest compounded continuously, how much will be in the account after 9 years?

Solution: $A = 5000e^{0.05(9)}$ In $A = Pe^{rt}$, replace P with 5000, r with 0.05, and t with 9.

$A = 5000e^{0.45}$ Apply $(a^m)^n = a^{mn}$.

$A = \$7841.56$ Evaluate using a calculator.

Answer: There will be \$7841.56 in the account.

Answers to Your Turn 3
a. 2 **b.** 4

Answer to Your Turn 4
10.2 yr.

Example 6 Since the 1980s, the greater Orlando area has been one of the fastest-growing areas in the United States. The following table shows the population of the greater Orlando area for selected years.

Year	Population in Millions
1980	0.805
1985	0.996
1990	1.225
1995	1.428
2000	1.645
2003	1.803
2007	2.033

The data can be approximated by $y = 0.823e^{0.0361t}$, where y is the population in millions and t is the number of years after 1980. (*Source:* U.S. Bureau of the Census.)

a. Assuming that the population continues to grow in the same manner, use the model to estimate the population of Orlando in 2015.

Solution: $t = 2015 - 1980 = 35$ — Subtract 1980 from 2015 to determine the value of t that corresponds to 2015.

$y = 0.823e^{0.0361(35)}$ — In $y = 0.823e^{0.0361t}$, substitute 35 for t.

$y \approx 2.912$ — Evaluate using a calculator.

Answer: Because y is in millions, the estimated population is 2,912,000 in 2015.

b. Find the year in which the population will be 3,500,000.

Solution: $y = \frac{3{,}500{,}000}{1{,}000{,}000} = 3.5$ — Divide 3,500,000 by 1,000,000 to find the value of y that corresponds to 3,500,000.

$3.5 = 0.823e^{0.0361t}$ — In $y = 0.823e^{0.0361t}$, substitute 3.5 for y.

$\frac{3.5}{0.823} = e^{0.0361t}$ — Divide both sides by 0.823.

$\ln\frac{3.5}{0.823} = \ln e^{0.0361t}$ — Use if $x = y$; then $\log_b x = \log_b y$.

$\ln\frac{3.5}{0.823} = 0.0361t$ — Use $\log_b b^x = x$.

$\frac{\ln\frac{3.5}{0.823}}{0.0361} = t$ — Divide both sides by 0.0361.

$40.1 \approx t$ — Calculate.

Note Our manipulations suggest the following formula for calculating t given y:

$$t = \frac{1}{0.0361}\ln\frac{y}{0.823}$$

Answer: Because t represents the number of years after 1980, the population will reach 3,500,000 in $1980 + 40.1 = 2020.1$, which means during 2020.

Your Turn 6

a. The bacteria *Bacillus megaterium* increases at a rate of 4% per minute in a sucrose–salts medium. If 4500 bacteria were present initially, the number, A, present after t minutes is given by the equation $A = 4500e^{0.04t}$. (*Source: Todar's Online Textbook of Bacteriology.*) How many bacteria are present after 26 minutes?

b. After how many minutes will 10,000 bacteria be present?

Answers to Your Turn 6

a. $\approx$12,731 bacteria

b. $\approx$20 min.

Exponential growth and decay can be represented by the equation $A = A_0e^{kt}$, where A is the amount present, A_0 is the initial amount, t is the time, and k is a constant that is determined by the substance. Exponential growth is indicated if $k > 0$; exponential decay, if $k < 0$. Recall that the half-life of a substance is the amount of time until only one-half of the original amount is present.

Plutonium-239 is frequently used as fuel in nuclear reactors to generate electricity. One of the problems with using plutonium is disposing of the radioactive waste, which is extremely dangerous for a very long period of time.

Example 7 A nuclear reactor contains 10 kilograms of radioactive plutonium ^{239}P. Plutonium disintegrates according to the formula $A = A_0e^{-0.0000284t}$.

a. How much will remain after 10,000 years?

Solution: $A = 10e^{-0.0000284(10{,}000)}$ In $A = A_0e^{-0.0000284t}$, replace A_0 with 10 and t with 10,000.

$A = 10e^{-0.284}$ Simplify.

$A \approx 7.53$ Evaluate using a calculator.

Answer: About 7.53 kilograms $\left(\text{or about } \frac{3}{4} \text{ of the original amount}\right)$ will remain after 10,000 years.

b. Find the half-life of ^{239}P.

Solution: $5 = 10e^{-0.0000284t}$ In $A = A_0e^{-0.0000284t}$, replace A with 5 and A_0 with 10.

$\frac{5}{10} = e^{-0.0000284t}$ Divide both sides by 10.

$\ln\frac{5}{10} = \ln e^{-0.0000284t}$ Use if $x = y$; then $\log_b x = \log_b y$.

$\ln\frac{5}{10} = -0.0000284t$ Use $\log_b b^x = x$.

$\frac{\ln\frac{5}{10}}{-0.0000284} = t$ Divide both sides by -0.0000284.

$24{,}406.59 \approx t$ Calculate.

Answer: The half-life of ^{239}P is about 24,400 years.

Note Our manipulations suggest the following formula for calculating t given A and A_0:

$$t = \frac{1}{-0.0000284}\ln\frac{A}{A_0}$$

Your Turn 7 Carbon-14 is a radioactive form of carbon that is present in all living things. Archaeologists and paleontologists frequently use carbon-14 dating in estimating the age of organic fossils. Carbon-14 disintegrates according to the formula $A = A_0e^{-0.000121t}$.

a. If a sample contains 5 grams of carbon-14, how much will be present after 1500 years?
b. What is the half-life of carbon-14?

Objective 4 Use the change-of-base formula.

Sometimes applications involve logarithms other than common or natural logarithms. For example, earlier we were given the formula $P = 95 - 30\log_2 x$, where P is the percent of students who recall the important features of a lecture after x days. To find the percent after 5 days, we need to calculate $\log_2 5$. Most calculators have only base-10 and base-e logarithms; so to calculate $\log_2 5$ using a calculator, we need to write $\log_2 5$

Answers to Your Turn 7
a. $\approx$4.17 g **b.** $\approx$5728 yr.

in terms of common or natural logarithms using the change-of-base formula. To derive the change-of-base formula, we let $y = \log_a x$.

$$a^y = x \quad \text{Write } y = \log_a x \text{ in exponential form.}$$

$$\log_b a^y = \log_b x \quad \text{Take } \log_b \text{ of both sides.}$$

$$y \log_b a = \log_b x \quad \text{Use } \log_b x^r = r \log_b x.$$

$$y = \frac{\log_b x}{\log_b a} \quad \text{Divide both sides by } \log_b a.$$

$$\log_a x = \frac{\log_b x}{\log_b a} \quad \text{Substitute } \log_a x \text{ for } y.$$

Rule Change-of-Base Formula

In general, if $a > 0$, $a \neq 1$, $b > 0$, $b \neq 1$, and $x > 0$, then $\log_a x = \frac{\log_b x}{\log_b a}$.

In terms of common and natural logarithms, $\log_a x = \frac{\log x}{\log a} = \frac{\ln x}{\ln a}$.

Example 8 Use the change-of-base formula to calculate $\log_5 19$. Round the answer to four decimal places.

Note We could have used ln rather than log.

$$\log_5 19 = \frac{\ln 19}{\ln 5} \approx 1.8295$$

Solution: $\log_5 19 = \frac{\log 19}{\log 5} \approx 1.8295$ Use $\log_a x = \frac{\log_b x}{\log_b a}$; then evaluate using a calculator.

Check: $5^{1.8295} = 19.0005 \approx 19$, so the answer is correct.

Your Turn 8 Find $\log_6 25$. Round the answer to four decimal places.

Example 9 Use $P = 95 - 30 \log_2 x$ and the change-of-base formula to find the percent of students who retain the main points of a lecture after 5 days.

Solution: $P = 95 - 30 \log_2 5$ Substitute 5 for x in $P = 95 - 30 \log_2 x$.

$P = 95 - 30 \frac{\ln 5}{\ln 2}$ Use $\log_a x = \frac{\log_b x}{\log_b a}$.

$P \approx 95 - 30(2.3219)$ Evaluate $\frac{\ln 5}{\ln 2}$ using a calculator.

$P \approx 25.34$ Simplify.

About 25% of the students remember the main points of a lecture 5 days later.

Your Turn 9 Use the formula from Example 9 to find the percent of students who retain the main points of a lecture 8 days later.

Answer to Your Turn 8
1.7965

Answer to Your Turn 9
5%

10.6 Exercises

For Extra Help MyMathLab®

Objective 1

Prep Exercise 1 If $b^x = b^y$, then ________.

Prep Exercise 2 When solving $10^{x+2} = 45$, would natural or common logarithms be the better choice? Why?

Prep Exercise 3 What principle is used to solve $100 = (5)^{2n}$?

For Exercises 1–12, solve. Round your answers to four decimal places. See Example 1.

1. $2^x = 9$
2. $3^x = 20$
3. $5^{2x} = 32$
4. $6^{2x} = 48$
5. $5^{x+3} = 10$
6. $6^{x+4} = 38$
7. $8^{x-2} = 6$
8. $5^{x-3} = 12$
9. $4^{x+2} = 5^x$
10. $6^{x+4} = 10^x$
11. $2^{x+1} = 3^{x-2}$
12. $5^{x-3} = 3^{x+1}$

For Exercises 13–20, solve. Round your answers to four decimal places. See Example 2.

13. $e^{3x} = 5$
14. $e^{2x} = 7$
15. $e^{0.03x} = 25$
16. $e^{0.07x} = 32$
17. $e^{-0.022x} = 5$
18. $e^{-0.032x} = 8$
19. $\ln e^{4x} = 24$
20. $\ln e^{5x} = 35$

Objective 2

Prep Exercise 4 If $\log_a m = \log_a n$, then ________.

Prep Exercise 5 What principle is used to solve $\log(x - 3) = \log(3x - 13)$?

Prep Exercise 6 In solving $\log x + \log(x + 2) = \log 15$, we get possible solutions of $x = 3$ and $x = -5$. Why must $x = -5$ be rejected?

For Exercises 21–54, solve. Give exact answers. See Example 3.

21. $\log_4(x + 5) = 2$
22. $\log_3(x + 4) = 2$
23. $\log_4(4x - 8) = 2$
24. $\log_5(3x + 7) = 2$
25. $\log_4 x^2 = 2$
26. $\log_2 x^2 = 6$
27. $\log_6(x^2 + 5x) = 2$
28. $\log_4(x^2 + 6x) = 2$
29. $\log(4x - 3) = \log(3x + 4)$
30. $\log(4x + 1) = \log(2x + 7)$
31. $\ln(3x + 4) = \ln(x - 6)$
32. $\ln(5x + 6) = \ln(3x - 8)$
33. $\log_9(x^2 + 4x) = \log_9 12$
34. $\log_8(x^2 + x) = \log_8 30$
35. $\log_4 x + \log_4 8 = 2$
36. $\log_6 x + \log_6 4 = 2$
37. $\log_2 x - \log_2 5 = 1$
38. $\log_5 x - \log_5 3 = 2$
39. $\log_3 x + \log_3(x + 6) = 3$
40. $\log_2 x + \log_2(x - 3) = 2$

41. $\log_3(2x + 15) + \log_3 x = 3$

42. $\log_2(3x - 2) + \log_2 x = 4$

43. $\log_2(7x + 3) - \log_2(2x - 3) = 3$

44. $\log_3(3x + 3) - \log_3(x - 3) = 2$

45. $\log_2(3x + 8) - \log_2(x + 1) = 2$

46. $\log_3(2x + 1) - \log_3(x - 1) = 1$

47. $\log_8 2x + \log_8 6 = \log_8 10$

48. $\log_5 3x + \log_5 2 = \log_5 4$

49. $\ln x + \ln(2x - 1) = \ln 10$

50. $\ln x + \ln(3x - 5) = \ln 12$

51. $\log x - \log(x - 5) = \log 6$

52. $\log x - \log(x - 2) = \log 3$

53. $\log_6(3x + 4) - \log_6(x - 2) = \log_6 8$

54. $\log_7(5x + 2) - \log_7(x - 2) = \log_7 9$

Objective 3

For Exercises 55–74, solve. See Examples 4–7.

55. How long will it take \$5000 invested at 5% compounded quarterly to grow to \$8000? Round your answer to the nearest tenth of a year.

56. How long will it take \$7000 invested at 6% compounded monthly to grow to \$12,000? Round your answer to the nearest tenth of a year.

57. Assume that \$8000 is deposited into an account at 6% annual interest compounded continuously.
a. How much money will be in the account after 15 years?
b. How long will it take the \$8000 to grow to \$14,000?

58. Assume that \$4000 is deposited into an account at 5% annual interest compounded continuously.
a. How much money will be in the account after 10 years?
b. How long will it take the \$4000 to grow to \$10,000?

59. The probability that a person will have an accident while driving at a given blood alcohol level is approximated by $P(b) = e^{21.5b}$, where b is the blood alcohol level $(0 \le b \le 0.4)$ and P is the percent probability of having an accident.
a. What is the probability of an accident if the blood alcohol level is 0.08, which is legally drunk in many states?
b. Estimate the blood alcohol level when the probability of an accident is 50%.

60. Atmospheric pressure (in pounds per square inch, psi) is a function of the altitude above sea level and can be modeled by $P(a) = 14.7e^{-0.21a}$, where P is the pressure and a is the altitude above sea level in miles.
a. Find the atmospheric pressure at the peak of Mount McKinley, Alaska, which is 3.85 miles above sea level.
b. If the atmospheric pressure at the peak of Mount Everest in Nepal is 4.68 pounds per square inch, find the height of Mount Everest.

61. The population of a mosquito colony increases at a rate of 4% per day. If the initial number of mosquitoes is 500, the number present, A, at the end of t days is given by $A = 500e^{0.04t}$.
a. How many mosquitoes are present after 2 weeks?
b. Find the number of days until 10,000 mosquitoes are present.

62. A cake is removed from the oven at a temperature of 210°F and is left to cool on a counter where the room temperature is 70°F. The cake cools according to the function $T = 70 + 140e^{-0.0231t}$, where T is the temperature of the cake and t is in minutes.
a. What is the temperature of the cake after 40 minutes?
b. After how many minutes will the cake's temperature be 75°F?

63. The world population in 2010 was about 6.8 billion and is increasing at a rate of 1.1% per year. The world population after 2010 can be approximated by the equation $A = A_0e^{0.011t}$, where A_0 is the world population in 2010 and t is the number of years after 2010.
   a. Assuming that the growth rate follows the same trend, find the world population in 2020.
   b. In what year will the world population reach 7 billion?

64. In 2012, Africa had a population of 1072 million and a natural growth rate of 2.3% (2.1 times the world's growth rate). The population growth can be approximated by $A = A_0e^{0.023t}$, where A_0 is the population in millions in 2012 and t is the number of years after 2012.
   a. Excluding immigration, what will the population of Africa be in 2025?
   b. Excluding immigration, in what year will Africa's population reach 1.2 billion?

65. The barometric pressure x miles from the eye of a hurricane is approximated by the function $P = 0.48 \ln(x + 1)$, where P is inches of mercury. (*Source:* A. Miller and R. Anthes, *Meteorology*, 4th Edition, Merrill Publishing.)
   a. Find the barometric pressure 50 miles from the center of the hurricane.
   b. Find the distance from the center where the pressure is 1.5 inches of mercury.

66. The first two-year college, Joliet Junior College in Chicago, was founded in 1901. The number of two-year colleges in the United States grew rapidly, especially during the 1960s, but growth has tapered off. The total number of two-year colleges in the United States since 1960 can be approximated by the function $y = 175.6 \ln x + 513$, where x is the number of years after 1960 and y is the number of two-year colleges. Use the model to estimate the number of two-year colleges in the United States in 2015. (*Source:* American Association of Two-Year Colleges.)

67. The percent, $f(x)$, of adult height attained by a boy who is x years old is modeled by $f(x) = 29 + 48.8 \log(x + 1)$, where $5 \le x \le 15$.
   a. At age 10, about what percent of his adult height has a boy reached?
   b. At what age will a boy attain 75% of his adult height?

68. The annual depreciation rate r of a car purchased for P dollars and worth A dollars after t years can be found by the formula $\log(1 - r) = \frac{1}{t}\log\frac{A}{P}$. Find the depreciation rate of a car that is purchased for \$22,500 and is sold 3 years later for \$10,000.

69. The following table shows the purchasing power of \$1.00 in 1970 for subsequent years. For example, in 1980, \$2.12 had the same purchasing power as \$1.00 in 1970.

Year	Purchasing Power ($)
1970	1.00
1975	1.39
1980	2.12
1985	2.77
1990	3.37
1995	3.92
2000	4.43
2005	5.03
2010	5.61

(*Source:* www.measuringworth.com.)

The data can be approximated by the exponential function $y = 1.244(1.043)^x$, where y is the purchasing power of \$1.00 and x is the number of years after 1970. Assuming that the trend continues, find the approximate number of dollars in 2020 that it will take to have the same purchasing power as \$1.00 had in 1970.

70. The following table shows the U.S. public debt, in trillions of dollars, for selected years.

Year	Public Debt (in trillions)
1990	3.233
1992	4.065
1994	4.693
1996	5.224
1998	5.526
2000	5.674
2002	6.228
2004	7.379
2006	8.507
2008	10.025
2010	13.562

(Source: U.S. Department of the Treasury)

The data can be approximated by the exponential function $y = 3.385(1.063)^x$, where y is the public debt (in trillions) and x is the number of years after 1990. Assuming that the trend continues, what will be the approximate public debt in 2020?

71. Since 1985, the amount spent on recreation in the United States can be approximated by the exponential equation $y = 78.62(1.092)^x$, where y is the amount spent in billions of dollars and x is the number of years after 1985. Assuming that the trend continues, what is the approximate amount that will be spent on recreation in 2018. (*Source:* U.S. Department of Commerce.)

72. Since 1980, the median income for males in the United States can be approximated by the exponential equation $y = 13192.9(1.0369)^x$, where y is the median income and x is the number of years after 1980. Assuming that the trend continues, find the approximate median income for a male in the United States in 2018. (*Source: U.S. Department of Commerce*)

Objective 4

For Exercises 73–80, use the change-of-base formula to find the logarithms. Round your answers to four decimal places. See Example 8.

73. $\log_4 12$ **74.** $\log_5 23$ **75.** $\log_8 3$ **76.** $\log_9 5$

77. $\log_{1/2} 5$ **78.** $\log_{1/3} 4$ **79.** $\log_{1/4} \frac{3}{5}$ **80.** $\log_{1/3} \frac{4}{7}$

For Exercises 81 and 82, use $P = 95 - 30 \log_2 x$ and the change-of-base formula to find the percentage, P, of the lecture retained after the given number of days. Round your answer to the nearest whole percent. See Example 9.

81. 3 days **82.** 7 days

Review Exercises

Exercises 1–6 Expressions

For Exercises 1 and 2, simplify.

[5.2] 1. $(3x + 2) - (2x - 1)$ **[8.4] 2.** $(3\sqrt{5} + 2)(2\sqrt{5} - 1)$

For Exercises 3–6, use $f(x) = x^2 + 4$ and $g(x) = 2x - 1$.

[3.5] 3. Find $f(-5)$. **[5.2] 4.** Find $(f + g)(x)$.

[10.1] 5. Find $(f \circ g)(x)$. **[10.1] 6.** Find $(f \circ g)(2)$.

Chapter 10 Summary and Review Exercises

Complete each incomplete definition, rule, or procedure; study the key examples; and then work the related exercises.

10.1 Composite and Inverse Functions

Definitions/Rules/Procedures	Key Example(s)
Composition of Functions $(f \circ g)(x) =$ ________ for all x in the domain ________ for which $g(x)$ is in the domain of ________. $(g \circ f)(x) =$ ________ for all x in the domain of ____ for which $f(x)$ is in the domain of ________.	If $f(x) = x^2 - 3$ and $g(x) = 3x + 4$, find (a) $(f \circ g)(x)$ and (b) $(g \circ f)(x)$. **a.** $(f \circ g)(x) = f[g(x)]$ $= f(3x + 4)$ Substitute $3x + 4$ for $g(x)$. $= (3x + 4)^2 - 3$ Replace x with $3x + 4$ in $f(x)$. $= 9x^2 + 24x + 13$ Simplify. **b.** $(g \circ f)(x) = g[f(x)]$ $= g(x^2 - 3)$ Substitute $x^2 - 3$ for $f(x)$. $= 3(x^2 - 3) + 4$ Replace x with $x^2 - 3$ in $g(x)$. $= 3x^2 - 5$ Simplify.

Exercises 1–8 Equations and Inequalities

[10.1] *For Exercises 1–4, find each composition if* $f(x) = 3x + 4$ *and* $g(x) = x^2 - 2$.

1. $(f \circ g)(3)$

2. $(g \circ f)(3)$

3. $f[g(0)]$

4. $g[f(0)]$

[10.1] *For Exercises 5–8, find* $(f \circ g)(x)$ *and* $(g \circ f)(x)$.

5. $f(x) = 3x - 6, g(x) = 2x + 3$

6. $f(x) = x^2 + 4, g(x) = 3x - 7$

7. $f(x) = \sqrt{x - 3}, g(x) = 2x - 1$

8. $f(x) = \dfrac{x + 3}{x}, g(x) = \dfrac{x - 4}{x}$

Definitions/Rules/Procedures	Key Example(s)
To determine whether two functions f and g are **inverses** of each other: **1.** Show that $f[g(x)] =$ ________ for all x in the domain of ______. **2.** Show that $g[f(x)] =$ ________ for all x in the domain of ______.	Show that $f(x) = 5x - 6$ and $g(x) = \frac{x + 6}{5}$ are inverse functions. $f[g(x)] = f\left(\frac{x + 6}{5}\right)$ $= 5\left(\frac{x + 6}{5}\right) - 6$ $= x + 6 - 6 = x$ $g[f(x)] = g(5x - 6)$ $= \frac{5x - 6 + 6}{5}$ $= \frac{5x}{5} = x$ Because $f[g(x)] = g[f(x)] = x$, f and g are inverse functions.

Exercises 9–12 Equations and Inequalities

[10.1] *For Exercises 9–12, determine whether f and g are inverse functions.*

9. $f(x) = 3x + 2, g(x) = \frac{x - 2}{3}$

10. $f(x) = x^3 + 6, g(x) = \sqrt[3]{x - 6}$

11. $f(x) = x^2 - 3, x \geq 0; g(x) = \sqrt{x + 2}$

12. $f(x) = \frac{x}{x + 4}, g(x) = \frac{-4x}{x + 1}$

Definitions/Rules/Procedures	Key Example(s)
A function f is **one-to-one** if for any two numbers a and b in its domain, when $f(a) = f(b)$, ________ and when $a \neq b$, ________.	
Given a function's graph, the function is one-to-one if every horizontal line that can intersect the graph does so at ________________.	Determine whether the functions whose graphs follow are one-to-one.
A function has an inverse function if and only if the function is ____________.	
The graphs of f and f^{-1} are symmetric with respect to the graph of ________.	

a.

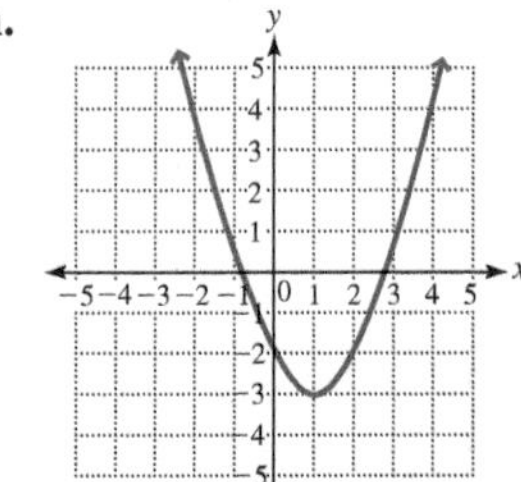

Solution: A horizontal line can intersect this graph in more than one point, so the function is not one-to-one.

b.

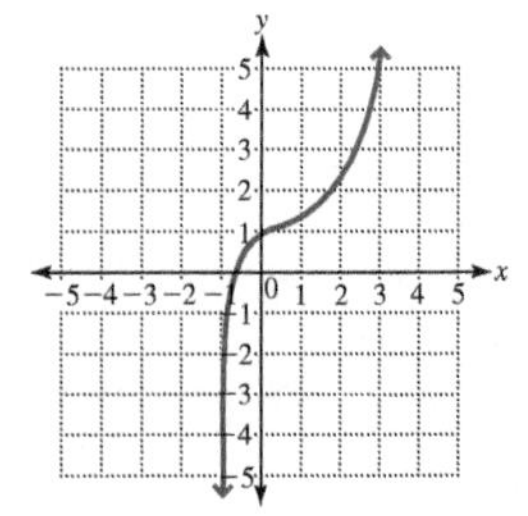

Solution: Every horizontal line that can intersect this graph does so at one and only one point, so the function is one-to-one.

Sketch the inverse of the function whose graph follows.

Solution: Draw the line $y = x$ and reflect the graph in the line.

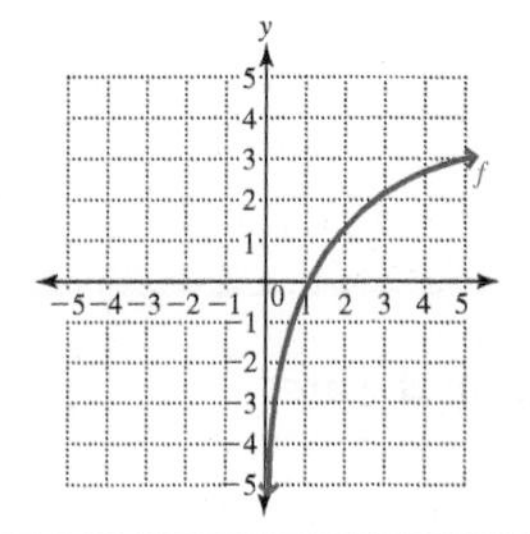

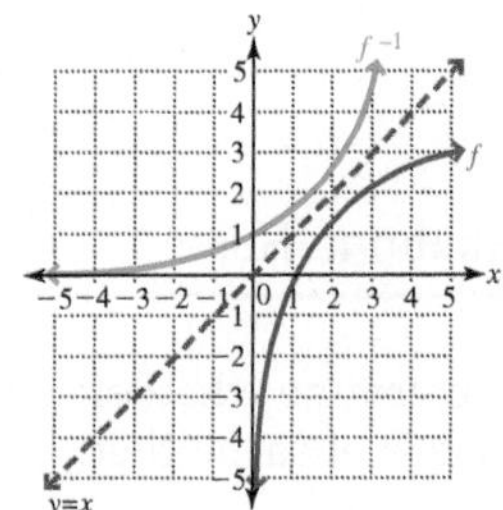

Exercises 13 and 14 Equations and Inequalities

[10.1] *For Exercises 13 and 14, determine whether each is the graph of a one-to-one function. If the function is one-to-one, sketch the graph of the inverse function.*

13.

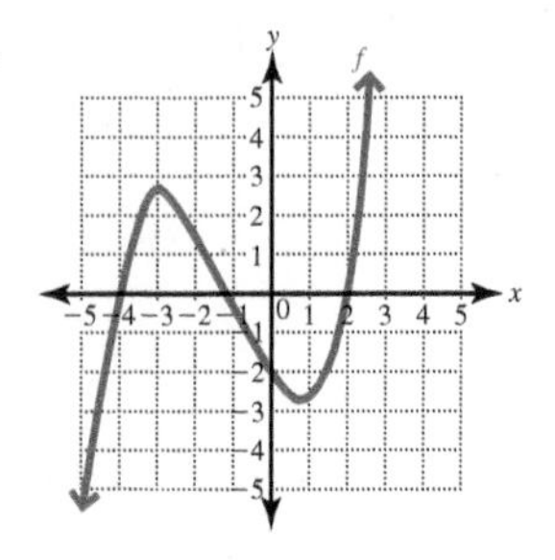

14.

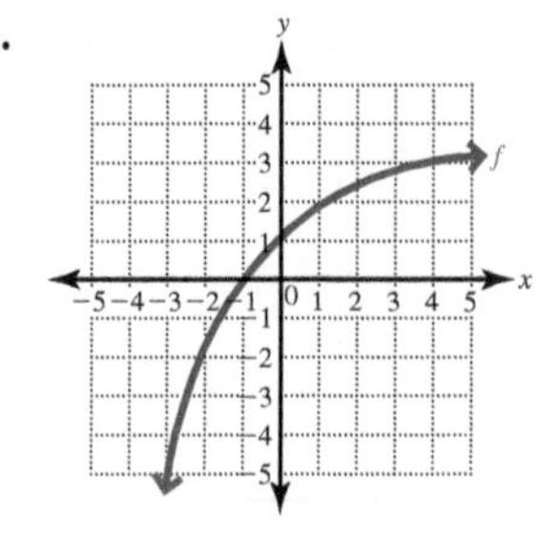

Definitions/Rules/Procedures	Key Example(s)
Finding the **inverse** of a one-to-one function 1. If necessary, replace $f(x)$ with ______. 2. Replace all x's with ______ and all y's with ______. 3. Solve the equation from step 2 for ______. 4. Replace y with ______.	If $f(x) = 3x - 7$, find $f^{-1}(x)$. **Solution:** $f(x) = 3x - 7$ $y = 3x - 7$ Replace $f(x)$ with y. $x = 3y - 7$ Interchange x and y. $\frac{x+7}{3} = y$ Solve for y. $f^{-1}(x) = \frac{x+7}{3}$ Replace y with $f^{-1}(x)$. Verify by showing that $f[f^{-1}(x)] = x$ and $f^{-1}[f(x)] = x$.

Exercises 15–20 Equations and Inequalities

[10.1] *For Exercises 15–18, find $f^{-1}(x)$ for each of the following one-to-one functions.*

15. $f(x) = 5x + 4$

16. $f(x) = x^3 + 6$

17. $f(x) = \frac{4}{x+5}$

18. $f(x) = \sqrt[3]{3x + 2}$

19. If $(4, -6)$ is an ordered pair on the graph of f, what ordered pair is on the graph of f^{-1}?

20. Fill in the blanks. If $f(a) = b$, then $f^{-1}(\_\_\_) = \_\_\_$.

10.2 Exponential Functions

Definitions/Rules/Procedures	Key Example(s)
Graphs of **exponential functions**	Graph $f(x) = 2^x$. Choose some values for x and find the corresponding values of $f(x)$, which are the y-values.

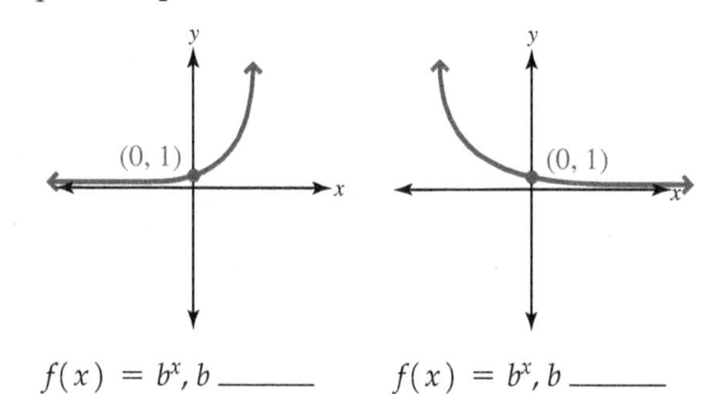

$f(x) = b^x, b$ ______ $f(x) = b^x, b$ ______

x	−2	−1	0	1	2
f(x)	$\frac{1}{4}$	$\frac{1}{2}$	1	2	4

Plot the points and draw the graph.

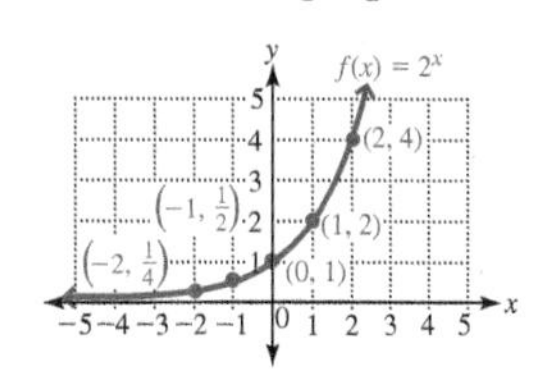

Exercises 21–24 Equations and Inequalities

[10.2] *For Exercises 21–24, graph.*

21. $f(x) = 3^x$

22. $f(x) = \left(\frac{1}{2}\right)^x$

23. $f(x) = 2^{x-3}$

24. $f(x) = 3^{-x+2}$

Definitions/Rules/Procedures	Key Example(s)
Given $b > 0$ and $b \neq 1$, if $b^x = b^y$, then ______.	Solve $8^{x-3} = 16^{2x}$.
Solving **exponential equations** 1. If necessary, write both sides of the equation as a power of the ______. 2. If necessary, simplify the ______. 3. Set the ______ equal to each other. 4. Solve the resulting equation.	**Solution:** $8^{x-3} = 16^{2x}$ $(2^3)^{x-3} = (2^4)^{2x}$ Rewrite 8 and 16 as powers of 2. $2^{3x-9} = 2^{8x}$ Multiply exponents. $3x - 9 = 8x$ Set exponents equal. $-\frac{9}{5} = x$ Solve for *x*.

Exercises 25–36 Equations and Inequalities

[10.2] *For Exercises 25–32, solve each equation.*

25. $5^x = 625$

26. $16^x = 64$

27. $6^x = \frac{1}{36}$

28. $\left(\frac{3}{4}\right)^x = \frac{16}{9}$

29. $4^{x-1} = 64$

30. $5^{x+2} = 25^x$

31. $\left(\frac{1}{3}\right)^{-x} = 27$

32. $8^{3x-2} = 16^{4x}$

33. The median doubling time for a malignant tumor is about 100 days. If there are 500 cells initially, then the number of cells, A, after t days is given by $A = A_0 2^{t/100}$. How many cells are present after one year?

34. If $25,000 is deposited into an account paying 6% interest compounded monthly, how much will be in the account after eight years?

35. The radioactive isotope ^{82}R has a half-life of 107 days. How much of a 50-gram sample remains after 300 days? $\left(\text{Use } A = A_0\left(\frac{1}{2}\right)^{t/h}.\right)$

36. Since 1990, the hourly minimum wage can be approximated by the exponential equation $y = 4.0144(1.028)^x$, where y is the hourly minimum wage and x is the number of years after 1990. Find the approximate minimum wage in 2020.

10.3 Logarithmic Functions

Definitions/Rules/Procedures	Key Example(s)
If $b > 0$ and $b \neq 1$, then $y = \log_b x$ is equivalent to ____________.	Write in logarithmic form. **a.** $3^4 = 81$ **Solution:** $\log_3 81 = 4$ **b.** $\left(\frac{1}{4}\right)^{-3} = 64$ **Solution:** $\log_{1/4} 64 = -3$ Write in exponential form. **a.** $\log_5 125 = 3$ **Solution:** $5^3 = 125$ **b.** $\log_3 \frac{1}{27} = -3$ **Solution:** $3^{-3} = \frac{1}{27}$

Exercises 37–44 Expressions

[10.3] *For Exercises 37–40, write in logarithmic form.*

37. $7^3 = 343$

38. $4^{-3} = \frac{1}{64}$

39. $\left(\frac{3}{2}\right)^4 = \frac{81}{16}$

40. $11^{1/3} = \sqrt[3]{11}$

[10.3] *For Exercises 41–44, write in exponential form.*

41. $\log_9 81 = 2$

42. $\log_{1/5} 125 = -3$

43. $\log_a 16 = 4$

44. $\log_e c = b$

Definitions/Rules/Procedures	Key Example(s)
To solve an equation of the form $\log_b x = y$, where b, x, or y is a variable, write the equation in exponential form, ____________, and then solve for the variable.	Solve. **a.** $\log_2 \frac{1}{16} = x$ **Solution:** $2^x = \frac{1}{16}$ Change to exponential form. $2^x = \frac{1}{2^4}$ $16 = 2^4$ $2^x = 2^{-4}$ $\frac{1}{2^4} = 2^{-4}$ $x = -4$ **b.** $\log_x 49 = 2$ **Solution:** $x^2 = 49$ Change to exponential form. $x = \pm 7$ The square roots of 49 are ± 7. $x = 7$ The base must be positive.
For any real number b, where $b > 0$ and $b \neq 1$: **1.** $\log_b b =$ ____________. **2.** $\log_b 1 =$ ____________.	Find the following logarithms. **a.** $\log_8 8$ Solution: $\log_8 8 = 1$ **b.** $\log_5 1$ Solution: $\log_5 1 = 0$

Exercises 45–52 Equations and Inequalities

[10.3] *For Exercises 45–52, solve.*

45. $\log_3 x = -4$

46. $\log_{1/2} x = -2$

47. $\log_2 32 = x$

48. $\log_{1/4} 16 = x$

49. $\log_x 81 = 4$

50. $\log_x \frac{1}{1000} = 3$

51. $\log_{3/4} \frac{3}{4} = x$

52. $\log_{81} 1 = x$

Definitions/Rules/Procedures	Key Example(s)

To **graph** a function in the form $f(x) = \log_b x$:

1. Replace $f(x)$ with __________ and write the logarithm in exponential form __________.
2. Find ordered pairs that satisfy the equation by assigning values to __________ and finding __________.
3. Plot the ordered pairs and draw a smooth curve through the points.

Graph $y = \log_3 x$.

Solution: Write as $3^y = x$ and assign values to y.

x	y
$\frac{1}{9}$	-2
$\frac{1}{3}$	-1
1	0
3	1
9	2

Exercises 53–56 Equations and Inequalities

[10.3] *For Exercises 53 and 54, graph.*

53. $f(x) = \log_4 x$

54. $f(x) = \log_{1/3} x$

55. If $f(x) = \log_b x$, the domain is $(0, \infty)$, and the range is $(-\infty, \infty)$, what are the domain and range of $g(x) = b^x$? Why?

56. The formula for the number of decibels in a sound is $d = 10 \log \frac{I}{I_0}$. Find the decibel reading of a sound whose intensity is $I = 1000\, I_0$.

10.4 Properties of Logarithms

Definitions/Rules/Procedures	Key Example(s)

For any real numbers b and x, where $b > 0$, $b \neq 1$, and $x > 0$:

1. $b^{\log_b x} =$ __________
2. $\log_b b^x =$ __________

Find the value of each.

a. $6^{\log_6 5}$

Solution: $6^{\log_6 5} = 5$

b. $\log_8 8^a$

Solution: $\log_8 8^a = a$

Definitions/Rules/Procedures	Key Example(s)
For real numbers x, y, and b, where $x > 0$, $y > 0$, $b > 0$, and $b \neq 1$: Product Rule: $\log_b xy =$ ______________.	Use the product rule. **a.** Write $\log_4 4x$ as the sum of logarithms. **Solution:** $\log_4 4x = \log_4 4 + \log_4 x$ $= 1 + \log_4 x$ $\log_4 4 = 1$ **b.** Write $\log_7 4 + \log_7 2$ as a single logarithm. **Solution:** $\log_7 4 + \log_7 2 = \log_7 4 \cdot 2$ $= \log_7 8$
Quotient Rule: $\log_b \frac{x}{y} =$ ______________.	Use the quotient rule. **a.** Write $\log_a \frac{w}{7}$ as the difference of logarithms. **Solution:** $\log_a \frac{w}{7} = \log_a w - \log_a 7$ **b.** Write $\log_7(x + 5) - \log_7(x - 3)$ as a single logarithm. **Solution:** $\log_7(x + 5) - \log_7(x - 3) = \log_7 \frac{x + 5}{x - 3}$
Power Rule: $\log_b x^r =$ ______________.	Use the power rule. **a.** Write $\log_b \sqrt[5]{x^3}$ as a multiple of a logarithm. **Solution:** $\log_b \sqrt[5]{x^3} = \log_b x^{3/5} = \frac{3}{5}\log_b x$ **b.** Write $-3 \log_2 y$ as a logarithm of a quantity to a power. **Solution:** $-3 \log_2 y = \log_2 y^{-3}$ $= \log_2 \frac{1}{y^3}$ Write $\log_a \frac{x^2 y}{z^4}$ as the sum or difference of multiples of logarithms. **Solution:** $\log_a \frac{x^2 y}{z^4} = \log_a x^2 y - \log_a z^4$ Quotient rule $= \log_a x^2 + \log_a y - \log_a z^4$ Product rule $= 2 \log_a x + \log_a y - 4 \log_a z$ Power rule

Definitions/Rules/Procedures	Key Example(s)
	Write $\frac{1}{4}(2\log_5 x - 3\log_5 y)$ as a single logarithm. **Solution:** $\frac{1}{4}(2\log_5 x - 3\log_5 y)$ $= \frac{1}{4}(\log_5 x^2 - \log_5 y^3)$ Power rule $= \frac{1}{4}\log_5 \frac{x^2}{y^3}$ Quotient rule $= \log_5\left(\frac{x^2}{y^3}\right)^{1/4}$ Power rule $= \log_5 \sqrt[4]{\frac{x^2}{y^3}}$

Exercises 57–80 Expressions

[10.4] *For Exercises 57 and 58, find the value.*

57. $3^{\log_3 8}$

58. $\log_4 4^6$

[10.4] *For Exercises 59 and 60, use the product rule to write the expression as a sum of logarithms.*

59. $\log_6 6x$

60. $\log_4 x(2x - 5)$

[10.4] *For Exercises 61 and 62, use the product rule to write the expression as a single logarithm.*

61. $\log_3 4 + \log_3 8$

62. $\log_5 3 + \log_5 x + \log_5(x - 2)$

[10.4] *For Exercises 63 and 64, use the quotient rule to write the expression as a difference of logarithms.*

63. $\log_b \frac{x}{5}$

64. $\log_a \frac{3x - 2}{4x + 3}$

[10.4] *For Exercises 65 and 66, use the quotient rule to write the expression as a single logarithm.*

65. $\log_8 32 - \log_8 16$

66. $\log_2(x + 5) - \log_2(2x - 3)$

[10.4] *For Exercises 67–70, use the power rule to write the expression as a multiple of a logarithm.*

67. $\log_3 7^4$

68. $\log_a \sqrt[3]{x}$

69. $\log_4 \frac{1}{a^4}$

70. $\log_a \sqrt[5]{a^4}$

[10.4] *For Exercises 71 and 72, use the power rule to write the expression as the logarithm of a quantity to a power. Simplify the answer if possible.*

71. $4\log_a x$

72. $\frac{3}{5}\log_a y$

[10.4] *For Exercises 73–76, write the expression as the sum or differences of multiples of logarithms.*

73. $\log_a x^2y^3$

74. $\log_a \frac{c^4}{d^3}$

75. $\log_a \frac{x^2y^3}{z^4}$

76. $\log_a \sqrt{\frac{a^3}{b^4}}$

[10.4] *For Exercises 77–80, write the expression as a single logarithm.*

77. $3\log_x y + 5\log_x z$

78. $3\log_a 4 - 2\log_a 3$

79. $\frac{1}{4}(2\log_a x + 3\log_a y)$

80. $4\log_a(x + 5) + 2\log_a(x - 3)$

10.5 Common and Natural Logarithms

Definitions/Rules/Procedures	Key Example(s)
Base-10 logarithms are called __________ logarithms, and $\log_{10}x$ is written as __________.	Evaluate using a calculator and round to four decimal places.
Common logarithms can be evaluated using the __________ key on a calculator.	**a.** log 356 **Solution:** $\log 356 \approx 2.5514$
Base-*e* logarithms are called __________ logarithms, and $\log_e x$ is written as __________.	**b.** log 0.0059 **Solution:** $\log 0.0059 \approx -2.2291$
Natural logarithms can be evaluated using the __________ key on a calculator.	**c.** ln 72 **Solution:** $\ln 72 \approx 4.2767$ **d.** ln 0.097 **Solution:** $\ln 0.097 \approx -2.3330$

Exercises 81–86 Expressions

[10.5] *For Exercises 81–84, use a calculator to approximate each logarithm to four decimal places.*

81. log 326

82. log 0.0035

83. ln 0.043

84. ln 92

[10.5] *For Exercises 85 and 86, find the exact value without using a calculator.*

85. log 0.00001

86. $\ln \sqrt[4]{e}$

Exercises 87–90 Equations and Inequalities

87. The sound intensity of a clap of thunder was $10^{-3.5}$ watts per square meter. What was the decibel reading? $\left(\text{Use } d = 10\log\frac{I}{I_0}, \text{ where } I_0 = 10^{-12}\text{watts/m}^2.\right)$

88. Using $\text{pH} = -\log[H_3O^+]$ (the hydronium ion concentration), find the pH of an apple whose $[H_3O^+]$ is 0.001259.

89. How long will it take \$6000 to grow to \$10,000 if it is invested at 3% annual interest compounded continuously? (Use $A = Pe^{rt}$.)

90. Using $R = \log\frac{I}{I_0}$ for Richter scale readings, compare the intensity of an earthquake whose Richter scale reading was 7.8 with one whose reading was 6.8. What do you notice? (*Hint:* Solve for I in terms of I_0.)

10.6 Exponential and Logarithmic Equations with Applications

Definitions/Rules/Procedures

For any real numbers b, x, and y, where $b > 0$ and $b \neq 1$:

1. If $b^x = b^y$, then ____________.
2. If $x = y$, then $b^x =$ ____________.
3. For $x > 0$ and $y > 0$, if $\log_b x = \log_b y$, then ____________.
4. For $x > 0$ and $y > 0$, if $x = y$, then $\log_b x =$ ____________.
5. For $x > 0$, if $\log_b x = y$, then ____________ $= x$.

Key Example(s)

Solve $3^x = 4$.

Solution:

$$\log 3^x = \log 4 \quad \text{Use property 4.}$$
$$x \log 3 = \log 4 \quad \text{Power rule}$$
$$x = \frac{\log 4}{\log 3} \quad \text{Divide by log 3.}$$
$$x \approx 1.2619 \quad \text{Evaluate.}$$

Solve $e^{3x} = 12$.

Solution: Because the base is e, take the natural logarithm of both sides.

$$\ln e^{3x} = \ln 12 \quad \text{Use property 4.}$$
$$3x = \ln 12 \quad \text{Use } \log_b b^x = x.$$
$$x = \frac{\ln 12}{3} \approx 0.8283 \quad \text{Divide by 3 and approximate.}$$

How long will it take \$12,000 invested at 4% compounded quarterly to grow to \$15,000?

Solution: Use $A = P\left(1 + \frac{r}{n}\right)^{nt}$ with $A = 15{,}000$, $P = 12{,}000$, $r = 4\% = 0.04$, and $n = 4$. Find t.

$$15{,}000 = 12{,}000\left(1 + \frac{0.04}{4}\right)^{4t} \quad \text{Substitute.}$$
$$\frac{5}{4} = (1.01)^{4t} \quad \text{Divide by 12,000.}$$
$$\log\frac{5}{4} = \log 1.01^{4t} \quad \text{Use property 4.}$$
$$\log\frac{5}{4} = 4t \log 1.01 \quad \text{Power rule}$$
$$\frac{\log\frac{5}{4}}{4 \log 1.01} = t \quad \text{Divide by 4 log 1.01.}$$
$$5.61 \approx t \quad \text{Evaluate.}$$

It will take about 5.6 years.

If \$3000 is invested in an account paying 4.5% compounded continuously, how much will be in the account after 10 years?

Solution: Use $A = Pe^{rt}$ with $P = 3000$, $r = 0.045$, and $t = 10$.

$$A = Pe^{rt}$$
$$A = 3000e^{(0.045)(10)} \quad \text{Substitute.}$$
$$A = 3000e^{0.45} \quad 10(0.045) = 0.45$$
$$A \approx 4704.94 \quad \text{Evaluate.}$$

There will be \$4704.94 in the account.

Definitions/Rules/Procedures	Key Example(s)
Exponential growth and decay can be represented by the equation $A =$ __________, where A is the amount present, A_0 is the __________ amount, t is the time, and k is a constant that is determined by the substance. If $k > 0$, there is exponential __________, and if $k < 0$, there is exponential __________. The half-life of a substance is the amount of time until only one-half of the original amount is present.	The element bismuth has an isotope, ^{200}Bi, that disintegrates according to the formula $A = A_0e^{-0.0198t}$, where t is in minutes. Find the half-life of ^{200}Bi. **Solution:** After one half-life, $\frac{1}{2}A_0$ of the original A_0 grams remains. $A = A_0e^{-0.0198t}$ $\frac{1}{2}A_0 = A_0e^{-0.0198t}$ Substitute $\frac{1}{2}A_0$ for A. $\frac{1}{2} = e^{-0.0198t}$ Divide by A_0. $\ln\frac{1}{2} = \ln e^{-0.0198t}$ Use if $x = y$; then $\log_b x = \log_b y$. $\ln\frac{1}{2} = -0.0198t$ Use $\log_b b^x = x$. $\frac{\ln\frac{1}{2}}{-0.0198} = t$ Divide by -0.0198. $35 \approx t$ Evaluate. The half-life is approximately 35 minutes.

Exercises 91–100 Equations and Inequalities

[10.6] *For Exercises 91–96, solve. Round your answers to four decimal places.*

91. $9^x = 32$

92. $3^{5x} = 19$

93. $6^{2x-1} = 22$

94. $4^{2x-3} = 5^{x+1}$

95. $e^{4x} = 11$

96. $e^{-0.003x} = 5$

97. How long will it take \$15,000 invested at 3% compounded monthly to grow to \$18,000? Round your answer to the nearest tenth of a year. $\left(\text{Use } A = P\left(1 + \frac{r}{n}\right)^{nt}.\right)$

98. Assume that \$7000 is deposited at 7% annual interest compounded continuously

a. How much will be in the account after eight years? (Use $A = Pe^{rt}$.)

b. How long will it take until \$12,000 is in the account?

99. The population of an ant colony is 400 and is increasing at the rate of 3% per month. Use $A = A_0e^{0.03t}$ to answer the following questions.

a. How many ants will be in the colony after one year?

b. After how many months will 800 ants be in the colony?

100. Bacteria reproduce by cell division, and the amount of time required for the cells to divide is called the generation time, G. The generation time is found using the equation $G = \dfrac{t}{3.3 \log \frac{b}{B}}$, where B is the number of bacteria at the beginning of the time interval, b is the number of bacteria at the end of the time interval, and t is the time interval in minutes. Find the generation time of a bacteria population that increases from 1000 to 1,000,000 cells in 4 hours. (*Source: Todar's Online Textbook of Bacteriology.*)

Definitions/Rules/Procedures	Key Example(s)
To solve **equations containing logarithms**, use the properties of logarithms to simplify each side of the equation and then use one of the following. If the simplification results in an equation in the form $\log_b x = \log_b y$, use the fact that ________ and then solve for the variable.	Solve $\log_3 x + \log_3(x - 3) = \log_3 10$. **Solution:** $\log_3 x(x - 3) = \log_3 10$ Product rule $x(x - 3) = 10$ Use property 3. $x^2 - 3x - 10 = 0$ Multiply; then subtract 10. $(x - 5)(x + 2) = 0$ Factor. $x - 5 = 0$ or $x + 2 = 0$ Use the zero-factor theorem. $x = 5$ or $x = -2$ Solve for x We must reject -2 because it results in $\log_3(-2)$ and $\log_3(-5)$ in the original equation, which do not exist.
If the simplification results in an equation in the form $\log_b x = y$, write the equation in exponential form, ________, and then solve for the variable.	Solve $\log_2(2x + 2) - \log_2(x - 4) = 2$. **Solution:** $\log_2 \frac{2x + 2}{x - 4} = 2$ Quotient rule $\frac{2x + 2}{x - 4} = 2^2$ Use property 5 $\frac{2x + 2}{x - 4} = 4$ $2^2 = 4$ $2x + 2 = 4x - 16$ Multiply by $x - 4$. $18 = 2x$ Isolate the x term. $9 = x$ Solve for x.

Exercises 101–108 Equations and Inequalities

[10.6] *For Exercises 101–108, solve. Give exact answers.*

101. $\log_4(x + 8) = 2$

102. $\log_2(x^2 + 2x) = 3$

103. $\log(3x - 8) = \log(x - 2)$

104. $\log 5 + \log x = 2$

105. $\log_3 x - \log_3 4 = 2$

106. $\log_2 x + \log_2(x - 6) = 4$

107. $\log_4 x + \log_4(x + 2) = \log_4 8$

108. $\log_3(5x + 2) - \log_3(x - 2) = 2$

Definitions/Rules/Procedures	Key Example(s)
Change-of-Base Formula In general, if $a > 0, a \neq 1, b > 0, b \neq 1$, and $x > 0$, then $\log_a x = \frac{\log_b \_\_}{\log_b \_\_}$. In terms of common and natural logarithms, $\log_a x = \frac{\log \_\_}{\log \_\_} = \frac{\ln \_\_}{\ln \_\_}$.	Find $\log_5 16$. **Solution:** Use $\log_a x = \frac{\log x}{\log a}$ or $\frac{\ln x}{\ln a}$; then evaluate using a calculator. $\log_5 16 = \frac{\log 16}{\log 5} \approx 1.7227$ or $\log_5 16 = \frac{\ln 16}{\ln 5} \approx 1.7227$

Exercises 109 and 110 Expressions

[10.6] *For Exercises 109 and 110, use the change-of-base formula to approximate each logarithm to four decimal places.*

109. $\log_4 15$

110. $\log_{1/2} 6$

Chapter 10 Practice Test

For Extra Help

Step-by-step test solutions are found on the Chapter Test Prep Videos available in MyMathLab® *or on* YouTube.

1. If $f(x) = x^2 - 6$ and $g(x) = 3x - 5$, find $f[g(x)]$.

2. Use $f(x) = 4x - 3$ to answer parts a–c.

a. Find $f^{-1}(x)$

b. Verify that $f[f^{-1}(x)] = x$ and $f^{-1}[f(x)] = x$.

c. What is the relationship between the graphs of $f(x)$ and $f^{-1}(x)$?

3. The graph of a function f is shown to the right. Graph f^{-1}.

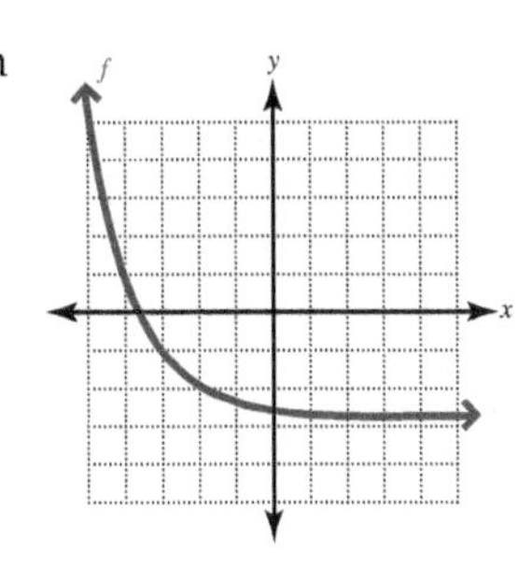

4. Graph $f(x) = 2^{x-1}$.

5. Solve $32^{x-2} = 8^{2x}$.

6. A sum of $20,000 is invested at 5% compounded quarterly.

a. Find the amount in the account after 15 years.

b. Find the number of years until $30,000 is in the account.

7. The isotope ^{98}Nb has a half-life of 30 minutes. Use $A = A_0\left(\frac{1}{2}\right)^{t/n}$.

a. How much of a 100-gram sample remains after 3.6 hours?

b. How long will it take until only 20 grams are remaining?

8. The number of people in the United States aged 65 and over has increased rapidly since 1900 and can be approximated by $y = 3.17(1.026)^x$, where y is the number in millions aged 65 and older and x is the number of years after 1900. (*Source:* U.S. Bureau of the Census.)

a. Find the number aged 65 and older in 1960 and in 2018.

b. In what year were 31.4 million people aged 65 and older?

9. Write $\log_{1/3}81 = -4$ in exponential form.

For Exercises 10 and 11, solve. Give exact answers.

10. $\log_6 \frac{1}{216} = x$

11. $\log_x 625 = 4$

12. Graph $f(x) = \log_2 x$.

For Exercises 13 and 14, write as the sum or difference of multiples of logarithms.

13. $\log_b \frac{x^4y^2}{z}$

14. $\log_b \sqrt[4]{\frac{x^5}{y^7}}$.

15. Write $\frac{3}{4}(2\log_b x + 3\log_b y)$ as a single logarithm.

16. The isotope ^{119}Sn disintegrates according to the function $A = A_0 e^{-0.0028t}$, where t is the time in days.

a. How much of a 300-gram sample remains after 500 days?

b. What is the half-life of ^{119}Sn?

For Exercises 17–19, solve. If necessary, use a calculator to approximate to four decimal places.

17. $6^{x-3} = 19$

18. $\log_3 x + \log_3(x + 6) = 3$

19. $\log(4x + 2) - \log(3x - 2) = \log 2$

20. The number of generations, n, of a population of bacteria cells is given by $n = 3.3 \log \frac{b}{B}$, where B is the number of bacteria cells at the beginning of a time interval and b is the number at the end of the time interval. (*Source: Todar's Online Textbook of Bacteriology.*)

a. Find the number of generations in a bacteria population that increases from 100 to 10,000,000 cells.

b. Find the number of bacteria at the end of the time interval if the number at the beginning is 100 and there are 9.9 generations.

Chapters 1–10 Cumulative Review Exercises

For Exercises 1–3, answer true or false.

[10.3] **1.** If $f(x) = 3^x$, then $f^{-1}(x) = \log_3 x$.

[8.2] **2.** The radical expression $\sqrt[4]{x^3}$ can be written exponentially as $x^{3/4}$.

[5.2] **3.** The range of $f(x) = x^2 + 4$ is $(-\infty, 4]$.

For Exercises 4–6, fill in the blank.

[5.1] **4.** When 0.0000000135 is written in scientific notation, the exponent of 10 is ________.

[6.4] **5.** The solutions of $2x^2 - 9x + 10 = 0$ are the ________ of the graph of $f(x) = 2x^2 - 9x + 10$.

[2.5] **6.** If $|x - 3| = 4$, then ________ or ________.

Exercises 7–16 Expressions

For Exercises 7–14, simplify. Write your answers with positive exponents only.

[1.3] **7.** $6 - 8 \div 2 \cdot 3^2 - 4(3 - 4 \cdot 2^3)$

[5.1] **8.** $\dfrac{(4a^{-4})^3(2a^2)^{-3}}{(2a^{-2})^3}$

[5.1] **9.** $\dfrac{21p^2q^4 - 14p^5q^3 + 6p^3q^7}{7p^3q^5}$

[7.2] **10.** $\dfrac{2x + 3}{x^2 + 6x + 9} - \dfrac{x + 4}{x + 3}$

[8.3] **11.** $4\sqrt{6c^3} \cdot 3\sqrt{10c^5}$

[8.7] **12.** $(4 + 3i)(2 - 4i)$

[10.4] **13.** $\log_b b^{2x}$

[7.1] **14.** $\dfrac{d^2 - d - 12}{2d^2} \div \dfrac{3d^2 + 13d + 12}{d}$

For Exercises 15 and 16, factor completely.

[6.3] **15.** $x^4 - 9x^2 - 4x^2y^2 + 36y^2$

[6.2] **16.** $24x^4y - 30x^3y^2 + 9x^2y^3$

Exercises 17–30 Equations and Inequalities

For Exercises 17–25, solve. Identify any extraneous solutions.

[2.5] **17.** $|2x - 3| - 5 = -4$

[2.6] **18.** $2|6x - 10| > -8$

[4.1] **19.** $\begin{cases} 4x - 5y = -22 \\ 3x + 2y = -5 \end{cases}$

[9.3] **20.** $(2x - 3)^2 - 2(2x - 3) = 8$

[7.4] 21. $\frac{x}{x-2} - \frac{4}{x-1} = \frac{2}{x^2 - 3x + 2}$

[8.6] 22. $\sqrt{x+4} - \sqrt{x-4} = 4$

[10.2] 23. $4^{x+2} = 8$

[10.3] 24. $\frac{1}{2}\log_2 x = 2$

[10.6] 25. $\log(3x - 5) + \log x = \log 12$

[3.2] 26. For $2x - 5y = 10$:

a. find the slope.

b. find the intercepts.

c. draw the graph.

For Exercises 27–30, solve.

[6.4] 27. A frame is in the shape of a right triangle as shown. Find the length of each side.

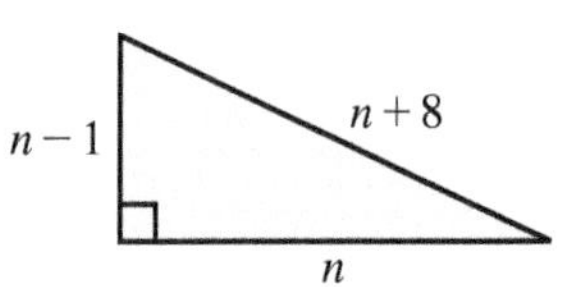

[10.2] 28. If $15,000 is deposited into an account paying 4% compounded monthly, how much will be in the account after three years?

[7.5] 29. Ethan and Pam operate a landscaping company. Ethan can mow, edge, and trim hedges on a quarter-acre lot in 2 hours. Pam takes 1.5 hours to do the same work. How much time would it take them working together?

[7.5] 30. A car travels at an average speed that is 10 miles per hour more than the average speed of a bus. If the bus takes 1 hour longer to travel 300 miles, find how long it takes the car to travel 300 miles.

CHAPTER

11

Conic Sections

Chapter Overview

In this chapter, we expand the graphing of second-degree equations to include the conic sections:

- ▸ Parabolas
- ▸ Circles
- ▸ Ellipses
- ▸ Hyperbolas

We also solve systems of equations with lines and conics and graph systems of inequalities that involve conics.

11.1 Parabolas and Circles

Objectives

1. Graph parabolas of the form $x = a(y - k)^2 + h$.
2. Find the distance and midpoint between two points.
3. Graph circles of the form $(x - h)^2 + (y - k)^2 = r^2$.
4. Find the equation of a circle with a given center and radius.
5. Graph circles of the form $x^2 + y^2 + dx + ey + f = 0$.

Warm-up

[9.4] 1. What are the coordinates of the vertex and the equation of the axis of symmetry of the graph of $y = 2(x - 3)^2 + 1$?

[9.1] *For Exercises 2 and 3, complete the square and write the resulting trinomial as the square of a binomial.*

2. $x^2 - 6x$ **3.** $y^2 + 8y$

[1.4] 4. Evaluate the expression $\sqrt{(x_2 - x_1)^2 + (y_2 - y_1)^2}$ for $x_2 = 3, x_1 = -1, y_2 = -1$, and $y_1 = 2$.

The intersection of a plane with a cone will be a circle, an ellipse, a parabola, or a hyperbola. For that reason, these curves are called **conic sections** or **conics**.

Definition Conic section: A curve in a plane that is the result of intersecting the plane with a cone—more specifically, a circle, an ellipse, a parabola, or a hyperbola.

Circle

Ellipse

Parabola

Hyperbola

Recall from Section 9.4 that we graphed parabolas in the form $y = a(x - h)^2 + k$. The graph opened upward if $a > 0$, opened downward if $a < 0$, had a vertex at (h, k), and had $x = h$ as the axis of symmetry.

Example 1 For $y = 2(x - 3)^2 + 1$, determine whether the graph opens upward or downward, find the vertex and axis of symmetry, and draw the graph.

Solution: The graph opens upward because $a = 2$ and 2 is positive. We compare the equation with the form $y = a(x - h)^2 + k$ and observe that the vertex is at the point with coordinates $(3, 1)$ and the axis of symmetry is $x = 3$. Plot a few points on either side of the axis of symmetry by letting x have values on either side of 3 and finding y.

x	y
2	3
1	9
4	3
5	9

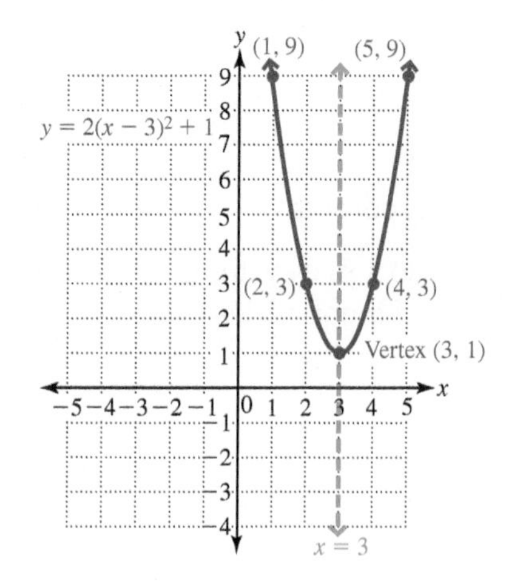

Your Turn 1 For $y = -2(x - 1)^2 + 3$, determine whether the graph opens upward or downward, find the vertex and axis of symmetry, and draw the graph.

Objective 1 Graph parabolas of the form $x = a(y - k)^2 + h$.

If we interchange x and y in the equations of parabolas that open upward and downward, we get the equations of parabolas that open to the left or right. To keep the vertex at (h, k), we also interchange h and k.

Answer to Your Turn 1
opens downward; vertex: $(1, 3)$; axis of symmetry: $x = 1$

Answers to Warm-up
1. $(3, 1), x = 3$
2. $x^2 - 6x + 9, (x - 3)^2$
3. $y^2 + 8y + 16, (y + 4)^2$
4. 5

Rule Equations of Parabolas Opening Left or Right

The graph of an equation in the form $x = a(y - k)^2 + h$ is a parabola with vertex at (h, k). The parabola opens to the right if $a > 0$ and to the left if $a < 0$. The equation of the axis of symmetry is $y = k$.

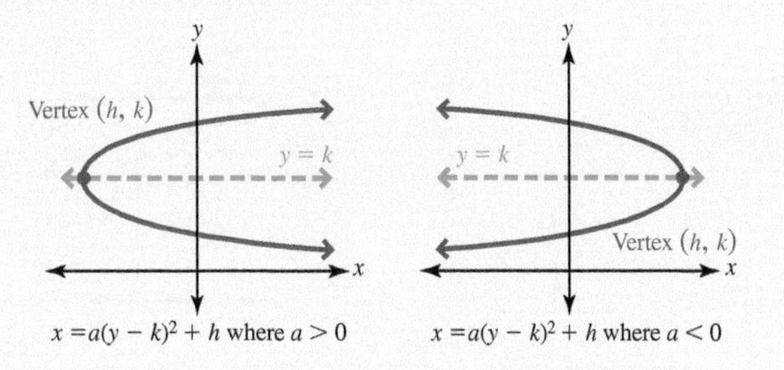

Note Parabolas that open to the left or right are not functions.

Example 2 For each equation, determine whether the graph opens left or right, find the vertex and axis of symmetry, and draw the graph.

a. $x = -2(y + 3)^2 - 2$

Solution: This parabola opens to the left because $a = -2$, which is negative. Rewrite the equation as $x = -2(y-(-3))^2 - 2$. Comparing this equation with $x = a(y - k)^2 + h$, we see that $h = -2$ and $k = -3$. The vertex is at the point with coordinates $(-2, -3)$, and the axis of symmetry is $y = -3$. To graph, plot a few points on either side of the axis of symmetry by letting y equal values on either side of -3 and finding x.

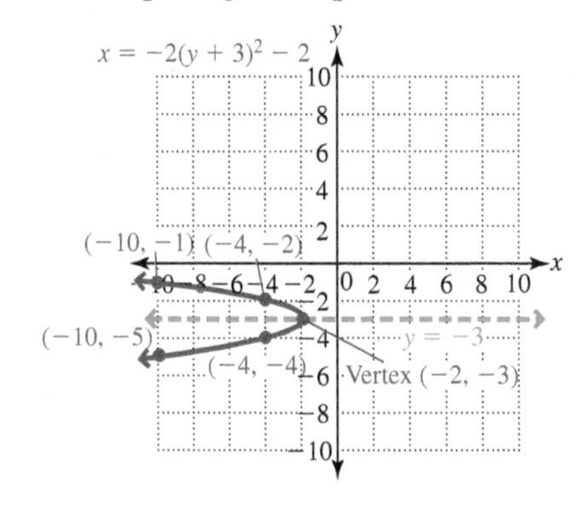

Note In choosing y-values, we went one unit above the axis of symmetry, then two above, then one below, then two below, etc.

x	y
−4	−2
−10	−1
−4	−4
−10	−5

b. $x = 3y^2 + 12y + 8$

Solution: This parabola opens to the right because $a = 3$, which is positive. To find the vertex and axis of symmetry, we need to write the equation in the form $x = a(y - k)^2 + h$.

$x = 3y^2 + 12y + 8$ Original equation

$x - 8 = 3y^2 + 12y$ Subtract 8 from both sides.

$x - 8 = 3(y^2 + 4y)$ Factor out the common factor, 3.

$x - 8 + 12 = 3(y^2 + 4y + 4)$ Complete the square. Note that we added $3 \cdot 4 = 12$ to both sides of the equation.

$x + 4 = 3(y + 2)^2$ Simplify the left side and factor the right side.

$x = 3(y + 2)^2 - 4$ Subtract 4 from both sides of the equation.

The vertex is at the point with coordinates $(-4, -2)$, and the axis of symmetry is $y = -2$.

To complete the graph, let y equal values on either side of -2 and find x.

x	y
-1	-1
8	0
-1	-3
8	-4

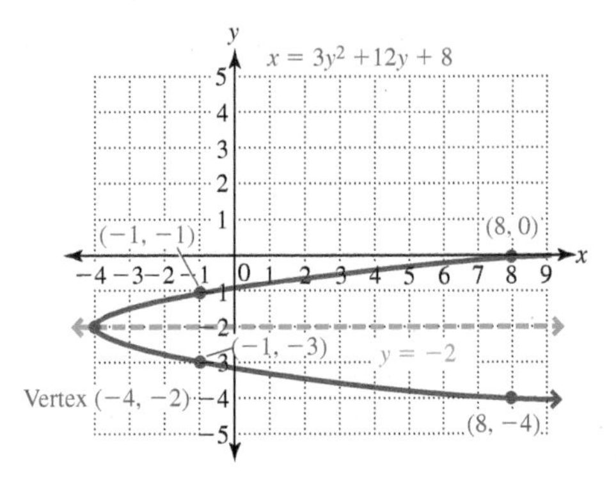

Your Turn 2 For each equation, determine whether the graph opens left or right, find the vertex and axis of symmetry, and draw the graph.

a. $x = -(y - 2)^2 + 1$ **b.** $x = y^2 + 4y + 3$

Objective 2 Find the distance and midpoint between two points.

To derive the other conic's general equations, we need to be able to find the distance between any two points in the coordinate plane. Consider the two points (x_1, y_1) and (x_2, y_2) shown in the following graph.

Note If y_2 were 6 and y_1 were 2, the distance between the points would be $6 - 2 = 4$. Therefore, to calculate the length of the vertical leg, we calculate $y_2 - y_1$.

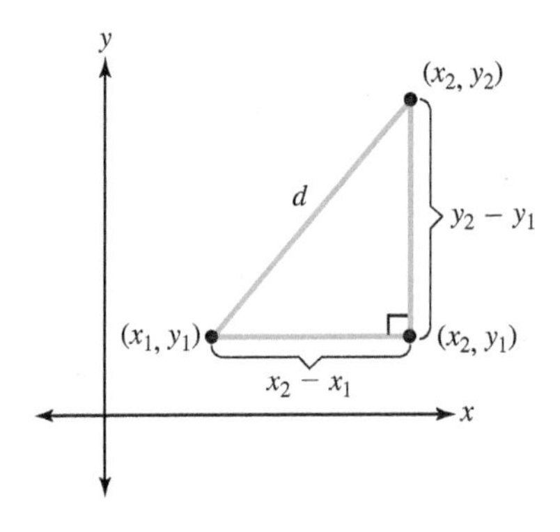

Connection Because the distance between the two points is measured along a line segment that is the hypotenuse of a right triangle, we can use the Pythagorean theorem (see below) to find the distance.

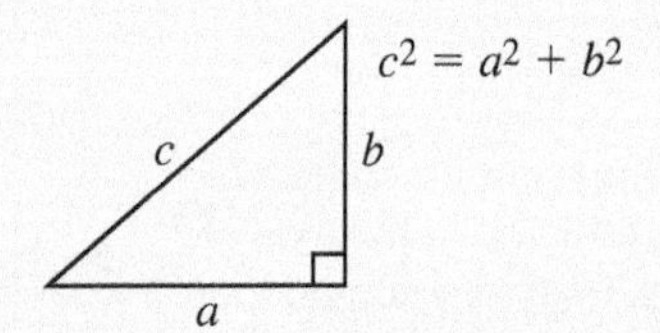

The lengths of the legs of the triangle are $x_2 - x_1$ and $y_2 - y_1$, as illustrated.

Note If x_2 were 4 and x_1 were 1, the distance between the points would be $4 - 1 = 3$. Therefore, to calculate the length of the horizontal leg, we calculate $x_2 - x_1$.

Answers to Your Turn 2

a. opens left; vertex: $(1, 2)$; axis of symmetry: $y = 2$

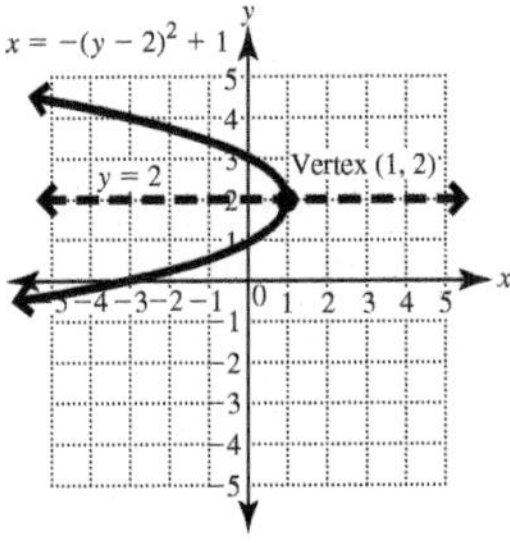

b. opens right; vertex: $(-1, -2)$; axis of symmetry: $y = -2$

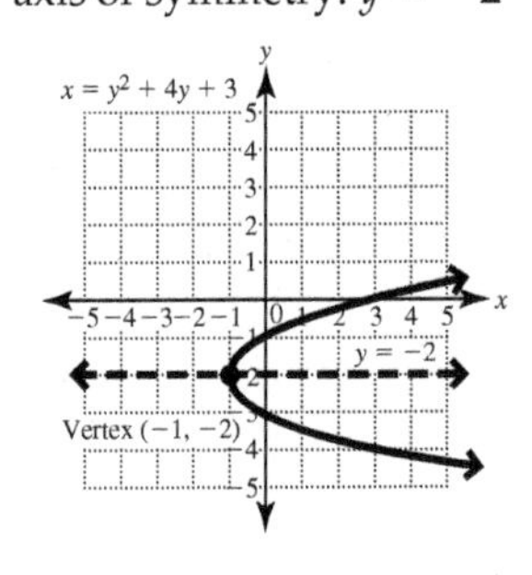

Now we can use the Pythagorean theorem, replacing a with $x_2 - x_1$, b with $y_2 - y_1$, and c with d.

$$d^2 = (x_2 - x_1)^2 + (y_2 - y_1)^2$$

$$d = \pm\sqrt{(x_2 - x_1)^2 + (y_2 - y_1)^2}$$ Use the square root principle to isolate d.

Because d is a distance, it must be positive; so we use only the positive value.

Rule Distance Formula

The distance, d, between two points with coordinates (x_1, y_1) and (x_2, y_2) can be found using the formula

$$d = \sqrt{(x_2 - x_1)^2 + (y_2 - y_1)^2}.$$

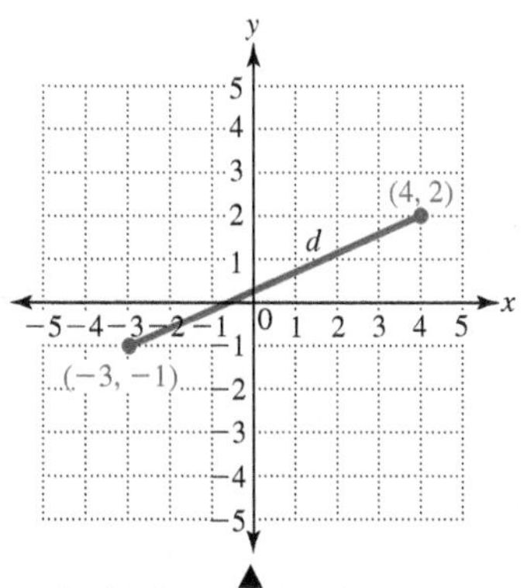

Note The distance formula holds no matter what quadrants the points are in.

Example 3 Find the distance between $(4, 2)$ and $(-3, -1)$. If the distance is an irrational number, also give a decimal approximation rounded to three places.

Solution:

$$d = \sqrt{(x_2 - x_1)^2 + (y_2 - y_1)^2} \quad \text{Use the distance formula.}$$

$$d = \sqrt{(-3 - 4)^2 + (-1 - 2)^2} \quad \text{Let } (4, 2) = (x_1, y_1) \text{ and } (-3, -1) = (x_2, y_2).$$

$$d = \sqrt{(-7)^2 + (-3)^2}$$

$$d = \sqrt{49 + 9}$$

$$d = \sqrt{58}$$

$$d \approx 7.616$$

Note It doesn't matter which ordered pair is (x_1, y_1) and which is (x_2, y_2). Consider Example 3 again with $(-3, -1)$ as (x_1, y_1) and $(4, 2)$ as (x_2, y_2):

$$d = \sqrt{(4 - (-3))^2 + (2 - (-1))^2}$$
$$= \sqrt{(7)^2 + (3)^2} = \sqrt{49 + 9}$$
$$= \sqrt{58} \approx 7.616$$

Your Turn 3 Determine the distance between the given points. If the distance is an irrational number, also give a decimal approximation rounded to three places.

a. $(8, 2)$ and $(3, -4)$ **b.** $(6, -5)$ and $(0, -1)$

On a line segment, the *midpoint* is located equally distant from each of the endpoints. Equivalently, the distance from either endpoint to the midpoint is half the length of the line segment.

Note The coordinates of the midpoint of a line segment are the average of the x-coordinates and the average of the y-coordinates of the endpoints.

Rule **Midpoint Formula**

If the coordinates of the endpoints of a line segment are (x_1, y_1) and (x_2, y_2), the coordinates of the midpoint are

$$\left(\frac{x_1 + x_2}{2}, \frac{y_1 + y_2}{2}\right).$$

Example 4 Find the midpoint of the line segment whose endpoints are $(-3, 2)$ and $(5, 6)$.

Solution:

$$\text{Midpoint} = \left(\frac{x_1 + x_2}{2}, \frac{y_1 + y_2}{2}\right) \quad \text{Use the midpoint formula.}$$

$$= \left(\frac{-3 + 5}{2}, \frac{2 + 6}{2}\right) \quad \text{Let } (-3, 2) = (x_1, y_1) \text{ and } (5, 6) = (x_2, y_2).$$

$$= \left(\frac{2}{2}, \frac{8}{2}\right)$$

$$= (1, 4)$$

Your Turn 4 Find the midpoint of the line segments whose endpoints are given.

a. $(4, -5), (-6, -1)$ **b.** $(3, 1), (-2, -7)$

Answers to Your Turn 3
a. $\sqrt{61} \approx 7.810$
b. $2\sqrt{13} \approx 7.211$

Answers to Your Turn 4
a. $(-1, -3)$ **b.** $\left(\frac{1}{2}, -3\right)$

Objective 3 Graph circles of the form $(x - h)^2 + (y - k)^2 = r^2$.

The second conic section that we consider is the **circle** with **radius** r.

Definitions **Circle:** A set of points in a plane that are equally distant from a central point. The central point is the center. **Radius:** The distance from the center of a circle to any point on the circle.

If the center of a circle is (h, k) and the radius is r, we can use the distance formula to derive the equation of the circle. If (x, y) is any point on the circle, the distance between (x, y) and (h, k) must be the radius, r.

$$\sqrt{(x_2 - x_1)^2 + (y_2 - y_1)^2} = d$$

$$\sqrt{(x - h)^2 + (y - k)^2} = r \quad \text{Substitute } (x, y) \text{ for } (x_2, y_2),\ (h, k) \text{ for } (x_1, y_1) \text{ and } r \text{ for } d.$$

$$(x - h)^2 + (y - k)^2 = r^2 \quad \text{Square both sides.}$$

Rule Standard Form of the Equation of a Circle

The equation of a circle with center at (h, k) and radius r is $(x - h)^2 + (y - k)^2 = r^2$.

Note If the center of a circle is at the origin, then $(h, k) = (0, 0)$ and the equation of the circle becomes $(x - 0)^2 + (y - 0)^2 = r^2$, which simplifies to $x^2 + y^2 = r^2$.

Example 5 Find the center and radius of each circle and draw the graph.

a. $(x - 3)^2 + (y + 2)^2 = 36$

Solution: $(x - 3)^2 + (y - (-2))^2 = 6^2$ Write in the form $(x - h)^2 + (y - k)^2 = r^2$.

Because $h = 3$ and $k = -2$, the center is $(3, -2)$. Because $36 = 6^2$, the radius is 6.

Note To find the radius, we evaluate the square root of 36. Because the radius is a distance, we give only the principal square root.

$$r = \sqrt{36} = 6$$

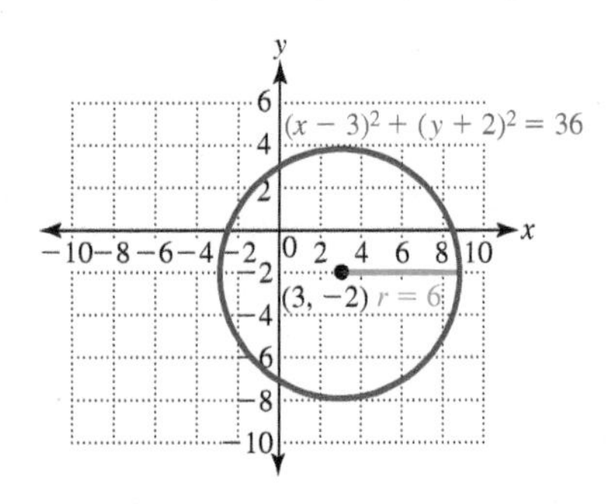

b. $(x + 4)^2 + (y + 1)^2 = 28$

Solution: $(x - (-4))^2 + (y - (-1))^2 = (\sqrt{28})^2$ Write in the form $(x - h)^2 + (y - k)^2 = r^2$.

Because $h = -4$ and $k = -1$, the center is $(-4, -1)$. For this radius, $r = \sqrt{28} = 2\sqrt{7} \approx 5.292$.

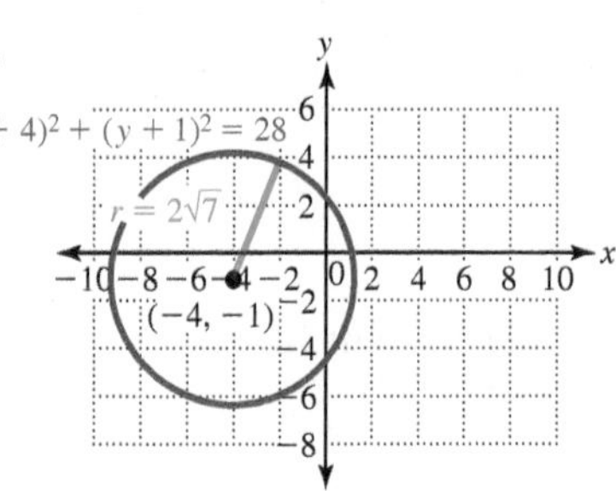

Your Turn 5 Find the center and radius of each circle and draw the graph.

a. $(x - 3)^2 + (y + 5)^2 = 4$ **b.** $(x + 1)^2 + (y - 1)^2 = 18$

Answers to Your Turn 5

a. center: $(3, -5)$; radius: 2

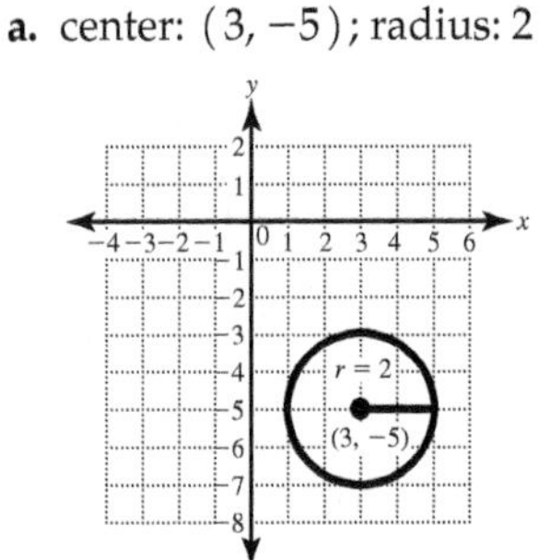

b. center: $(-1, 1)$; radius: $3\sqrt{2}$

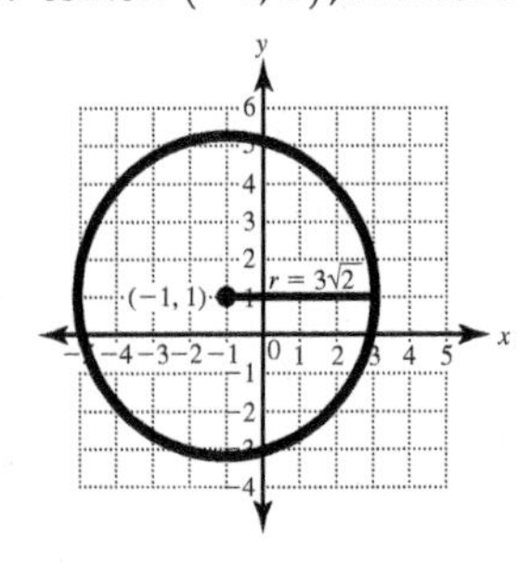

Objective 4 Find the equation of a circle with a given center and radius.

We can also use the standard form of a circle to write equations of circles.

VISUAL

Learning Strategy

Note how the signs of the numbers in the parentheses of the general equation are opposite the signs of the center coordinates.

$(x + 4)^2 + (y - 2)^2 = 64$

Center: $(-4, 2)$

Example 6 Write the equation of each circle in standard form.

a. Center: $(-4, 2)$; radius: 8

Solution: Because the center is at $(-4, 2)$, $h = -4$ and $k = 2$. Also $r = 8$.

$$(x - h)^2 + (y - k)^2 = r^2 \quad \text{Standard form of a circle.}$$
$$(x - (-4))^2 + (y - 2)^2 = 8^2 \quad \text{Substitute for } h, k, \text{ and } r.$$
$$(x + 4)^2 + (y - 2)^2 = 64 \quad \text{Simplify.}$$

b. Center: $(0, 0)$; radius: 2

Solution: Because the center is at $(0, 0)$, the standard form of the equation is $x^2 + y^2 = r^2$.

$$x^2 + y^2 = 2^2 \quad \text{Substitute for } r.$$
$$x^2 + y^2 = 4 \quad \text{Simplify.}$$

Your Turn 6 Write the equation of each circle in standard form.

a. Center: $(4, -2)$; radius: 5

b. Center: $(0, 0)$; radius: 9

Objective 5 Graph circles of the form $x^2 + y^2 + dx + ey + f = 0$.

If the equation of a circle is not given in standard form, we complete the square to write the equation in the form $(x - h)^2 + (y - k)^2 = r^2$.

Example 7 Find the center and radius of the circle whose equation is $x^2 + y^2 - 6x + 8y + 9 = 0$ and draw the graph.

Solution:

$$x^2 + y^2 - 6x + 8y = -9 \quad \text{Subtract 9 from both sides to isolate the variable terms.}$$
$$(x^2 - 6x) + (y^2 + 8y) = -9 \quad \text{Group } x \text{ and } y \text{ terms.}$$
$$(x^2 - 6x + 9) + (y^2 + 8y + 16) = -9 + 9 + 16 \quad \text{Complete the square in } x \text{ and } y \text{ by adding 9 and 16 to both sides of the equation.}$$
$$(x - 3)^2 + (y + 4)^2 = 16 \quad \text{Factor and simplify.}$$

The center is at $(3, -4)$, and the radius is 4.

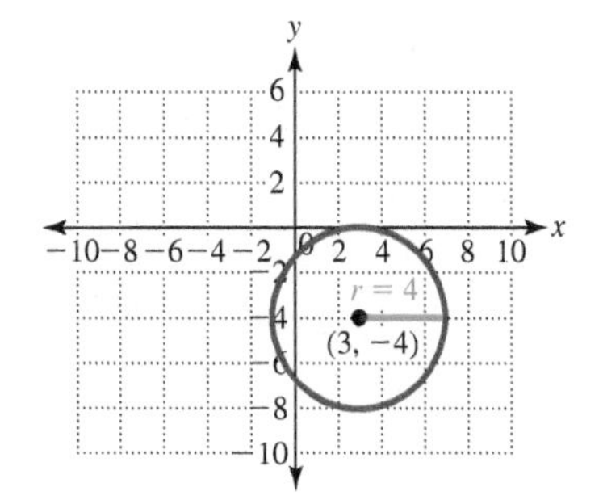

Your Turn 7 Find the center and radius of the circle whose equation is $x^2 + y^2 + 2x - 8y + 8 = 0$ and draw the graph.

Answers to Your Turn 6

a. $(x - 4)^2 + (y + 2)^2 = 25$

b. $x^2 + y^2 = 81$

Answers to Your Turn 7

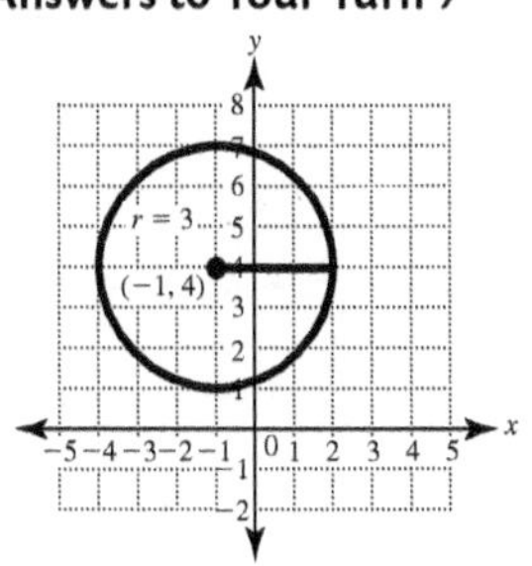

center: $(-1, 4)$; radius: 3

Note Some calculators automatically insert the left parenthesis when the root function is used; others do not.

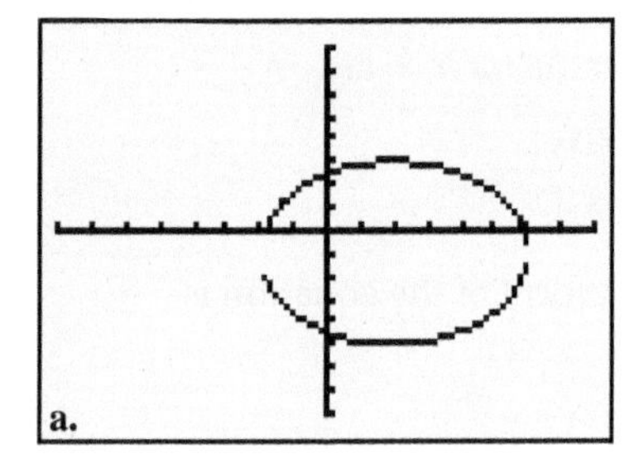

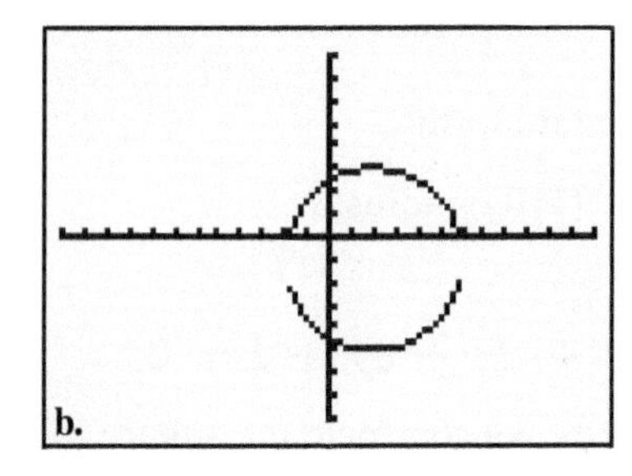

Example 8 Graph $(x-2)^2+(y+1)^2=16$ using a graphing calculator.

Solution: The graph of a circle fails the vertical line test, so the equation of a circle is not a function. To graph equations that are not functions, we can solve for y and graph the two resulting functions on the same screen.

$$(x-2)^2+(y+1)^2=16$$

$$(y+1)^2=16-(x-2)^2 \qquad \text{Subtract } (x-2)^2 \text{ from both sides.}$$

$$y+1=\pm\sqrt{16-(x-2)^2} \qquad \text{Apply the square root principle.}$$

$$y=-1\pm\sqrt{16-(x-2)^2} \qquad \text{Subtract 1 from both sides.}$$

This equation defines two functions. On the graphing calculator, define $Y_1=-1+\sqrt{(16-(x-2)^2)}$ and $Y_2=-1-\sqrt{(16-(x-2)^2)}$ and graph both in the window $[-8,8]$ for x and $[-8,8]$ for y. The resulting graph is labeled "*a*"; note that the graph does not look like the graph of a circle. To make the graph look like a circle, select ZSquare from the ZOOM menu. This results in the graph labeled "*b*."

11.1 Exercises For Extra Help MyMathLab®

Note: Exercises marked with a ★ represent challenging exercises.

Objective 1

Prep Exercise 1 What are the four conic sections?

Prep Exercise 2 In what direction does the parabola defined by $y=2(x-1)^2-3$ open? What is the vertex?

Prep Exercise 3 In what direction does the parabola defined by $x=-2(y+3)^2-4$ open? What is the vertex?

Prep Exercise 4 Is $x^2+2x-3+y=0$ the equation of a circle or a parabola? Explain.

For Exercises 1–22, find the direction the parabola opens, the coordinates of the vertex, and the equation of the axis of symmetry and draw the graph. See Examples 1 and 2.

1. $y=(x-1)^2+2$

2. $y=(x-2)^2+3$

3. $y=-x^2-2x+3$

4. $y=-x^2+4x-1$

5. $x = (y + 2)^2 - 2$

6. $x = (y + 3)^2 + 2$

7. $x = -(y - 1)^2 + 3$

8. $x = -(y - 2)^2 - 1$

9. $x = 2(y + 2)^2 - 4$

10. $x = 3(y + 3)^2 - 4$

11. $x = -3(y + 2)^2 - 5$

12. $x = -2(y - 4)^2 + 1$

13. $x = y^2 + 4y + 3$

14. $x = y^2 - 2y - 3$

15. $x = -y^2 + 6y - 5$

16. $x = -y^2 - 4y - 3$

17. $x = 2y^2 + 8y + 3$

18. $x = 2y^2 - 4y + 1$

19. $x = 3y^2 - 6y + 1$

20. $x = 3y^2 - 12y + 9$

21. $x = -2y^2 + 4y + 5$

22. $x = -3y^2 - 6y - 2$

For Exercises 23–26, match the equation with the correct graph.

23. $x = (y + 3)^2 - 2$

24. $y = (x + 3)^2 - 2$

25. $y = 2x^2 - 8x + 5$

26. $x = 2y^2 - 8y + 5$

a.

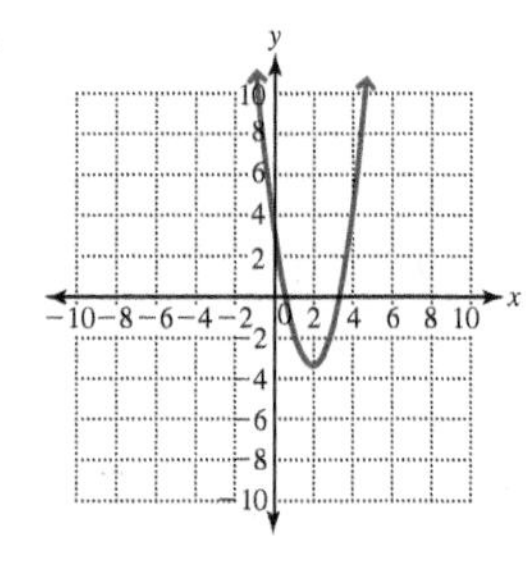

b.

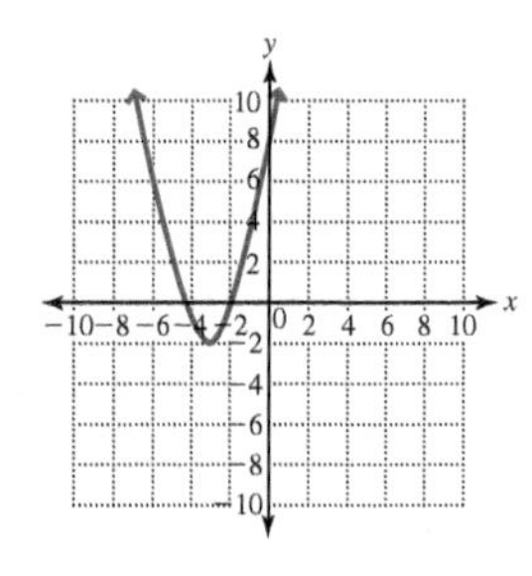

c.

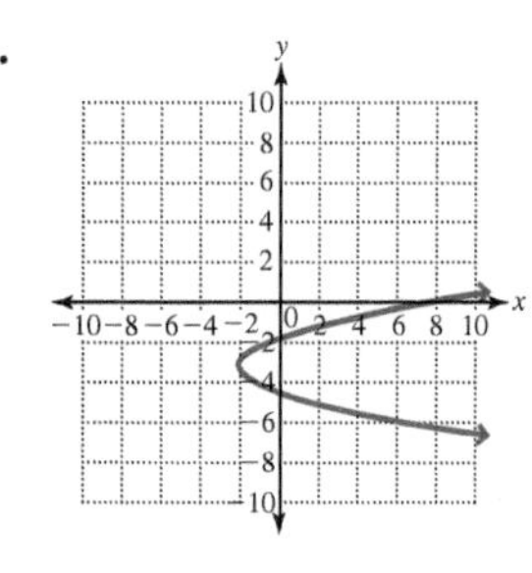

d.

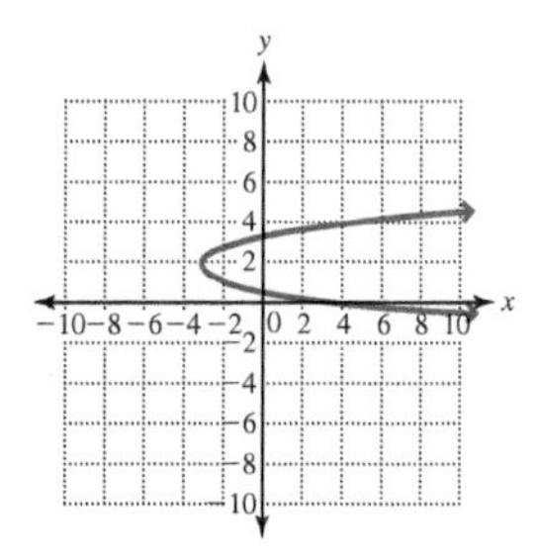

Objective 2

Prep Exercise 5 What formula is used to find the distance between two points (x_1, y_1) and (x_2, y_2)?

Prep Exercise 6 What formula is used to find the midpoint along a segment with endpoints at (x_1, y_1) and (x_2, y_2)?

For Exercises 27–38, find the distance and midpoint between the two points. See Examples 3 and 4.

27. $(-4, 2)$ and $(-1, 6)$

28. $(5, -1)$ and $(1, 2)$

29. $(-8, -4)$ and $(-3, 8)$

30. $(-3, 2)$ and $(3, -6)$

31. $(-8, -10)$ and $(4, -5)$

32. $(4, -6)$ and $(10, 2)$

33. $(2, 4)$ and $(4, 8)$

34. $(-3, 2)$ and $(1, -4)$

35. $(-5, 2)$ and $(3, -2)$

36. $(3, -4)$ and $(7, 2)$

37. $(6, -2)$ and $(1, -5)$

38. $(3, -6)$ and $(-2, 1)$

Objectives 3 and 5

Prep Exercise 7 The center of the circle defined by $(x - h)^2 + (y - k)^2 = r^2$ is at __________, and the radius is __________.

Prep Exercise 8 If the equation of a circle is not given in standard form, we __________ to write the equation in the form $(x - h)^2 + (y - k)^2 = r^2$.

For Exercises 39–42, the coordinates of the center of a circle and a point on the circle are given. Find the radius of the circle.

39. Center: $(4, 2)$; point on the circle: $(8, -1)$

40. Center: $(-4, 6)$; point on the circle: $(2, -2)$

41. Center: $(2, -6)$; point on the circle: $(10, -1)$

42. Center: $(3, -4)$; point on the circle: $(6, 8)$

For Exercises 43–58, find the center and radius and draw the graph. See Examples 5 and 7.

43. $(x - 2)^2 + (y - 1)^2 = 4$

44. $(x - 1)^2 + (y - 3)^2 = 25$

45. $(x + 3)^2 + (y + 2)^2 = 81$

46. $(x + 5)^2 + (y + 4)^2 = 36$

47. $(x - 5)^2 + (y + 3)^2 = 49$

48. $(x + 4)^2 + (y - 2)^2 = 16$

49. $(x - 1)^2 + (y + 1)^2 = 12$

50. $(x + 1)^2 + (y - 3)^2 = 18$

51. $(x + 4)^2 + (y + 2)^2 = 32$

52. $(x - 6)^2 + (y + 5)^2 = 8$

53. $x^2 + y^2 + 8x - 6y + 16 = 0$

54. $x^2 + y^2 - 2x - 6y - 39 = 0$

55. $x^2 + y^2 + 10x - 4y - 35 = 0$

56. $x^2 + y^2 + 12x + 10y + 60 = 0$

57. $x^2 + y^2 + 14x - 4y + 49 = 0$

58. $x^2 + y^2 + 8x - 10y + 16 = 0$

For Exercises 59–62, match the equation with the correct graph.

59. $(x - 2)^2 + (y + 3)^2 = 25$

60. $(x + 2)^2 + (y - 3)^2 = 25$

61. $x^2 + y^2 + 8x + 2y - 8 = 0$

62. $x^2 + y^2 + 2x - 8y - 8 = 0$

a.

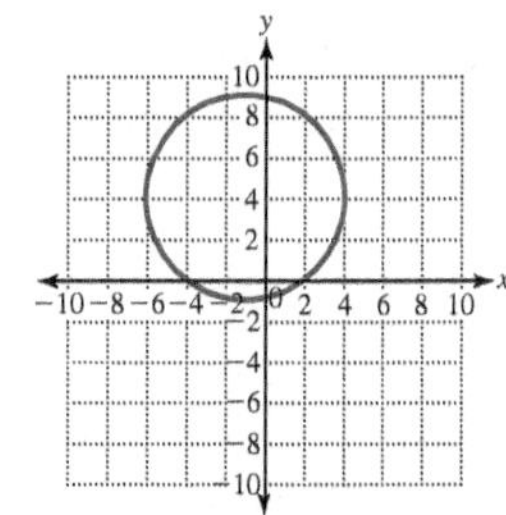

b.

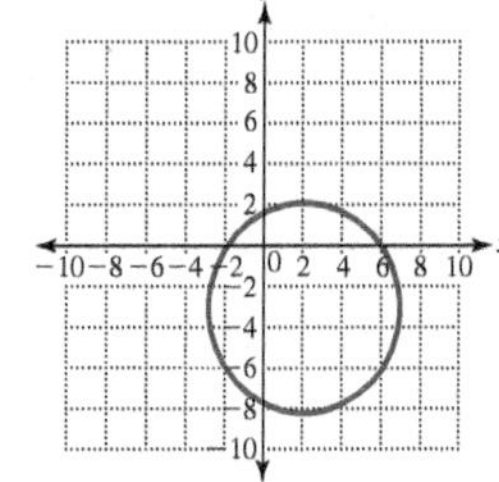

c.

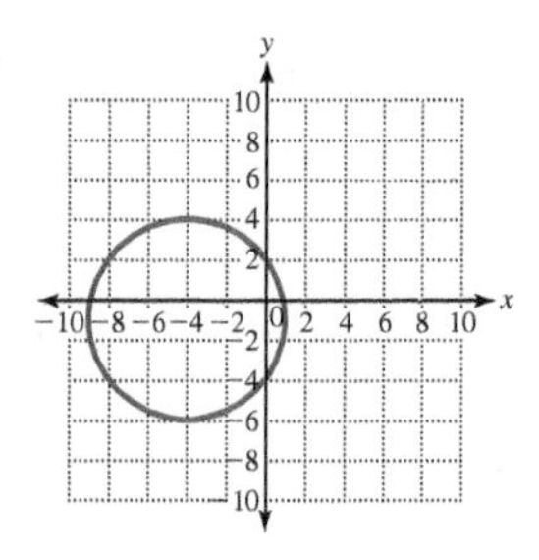

d.

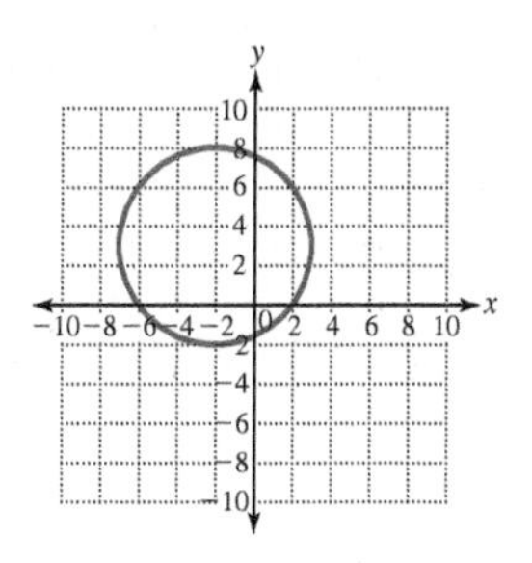

For Exercises 63–66, graph using a graphing calculator. See Example 8.

63. $x^2 + y^2 = 49$

64. $x^2 + y^2 = 36$

65. $(x - 2)^2 + (y + 3)^2 = 25$

66. $(x + 4)^2 + (y - 1)^2 = 9$

Objective 4

For Exercises 67–74, the center and radius of a circle are given. Write the equation of the circle in standard form. See Example 6.

67. Center: $(4, 2)$; radius: 4

68. Center: $(3, 2)$; radius: 11

69. Center: $(-4, -3)$; radius: 5

70. Center: $(-6, -5)$; radius: 2

71. Center: $(6, -2)$; radius: $\sqrt{14}$

72. Center: $(-3, -3)$; radius: $\sqrt{26}$

73. Center: $(-5, 2)$; radius: $3\sqrt{5}$

74. Center: $(6, 2)$; radius: $2\sqrt{6}$

For Exercises 75–78, the center of a circle and a point on the circle are given. Write the equation of the circle in standard form.

75. Center: $(2, 4)$; point on the circle: $(5, 8)$

76. Center: $(-4, 3)$; point on the circle: $(4, 9)$

77. Center: $(2, 4)$; point on the circle: $(7, 16)$

78. Center: $(-6, 8)$; point on the circle: $(-12, 0)$

79. Write the equation of the set of all points that are a distance of 8 units from $(2, -5)$.

80. Write the equation of the set of all points that are a distance of 10 units from $(-3, -7)$.

The diameter of a circle is a line segment whose endpoints lie on the circle and contains the center of the circle. Thus, the center of a circle is the midpoint of a diameter and the diameter is twice the radius.

★ *For Exercises 81–84, the coordinates of the endpoints of a diameter are given. Find the equation of the circle.*

81. $(4, -2), (-2, 6)$

82. $(6, -6), (-2, 0)$

83. $(6, -2), (-2, 8)$

84. $(-8, 6), (-2, 4)$

85. If a rock is thrown vertically upward from the top of a building 112 feet high with an initial velocity of 96 feet per second, the height, h, above ground level after t seconds is given by $h = -16t^2 + 96t + 112$, where h is in feet and t is in seconds.
  a. What is the maximum height the rock will reach?
  b. How many seconds will the rock take to reach its maximum height?
  c. How many seconds will the rock take to hit the ground?

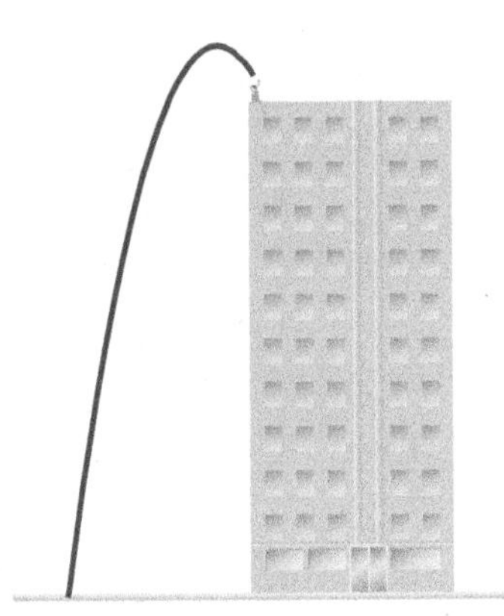

86. The path of a shell fired from ground level is in the shape of the parabola $y = 4x - x^2$, where x and y are given in kilometers.
  a. How high does the shell go?
  b. How far has the shell traveled horizontally when it reaches its maximum height?
  c. How far from its firing point does the shell land?

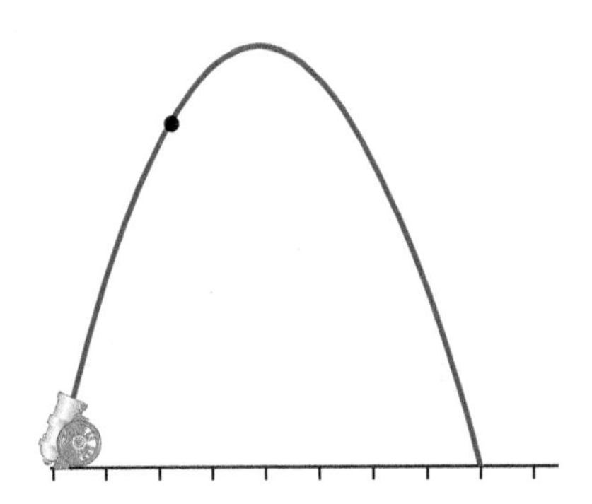

★ 87. Arches in the shapes of parabolas are often used in construction. Find the equation of a parabolic arc that is 18 feet high at its highest point and 30 feet wide at the base, as illustrated in the following figure. Place the origin at the midpoint of the bridges.

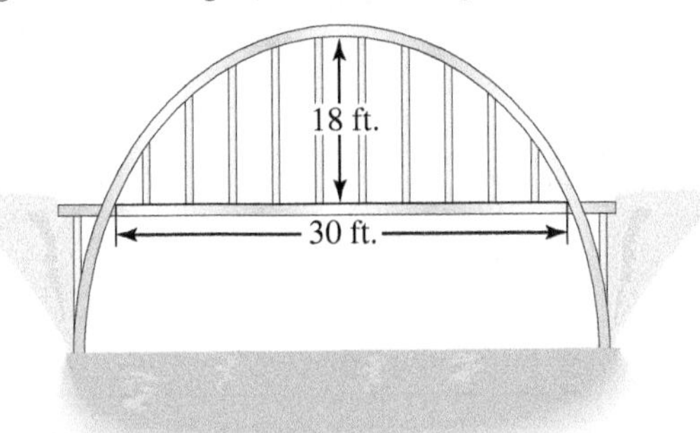

★ 88. The cross sections of satellite dishes are in the shape of parabolas. Find the equation of a dish that is 6 feet across and 1 foot deep if the vertex is at the origin and the parabola is opening upward.

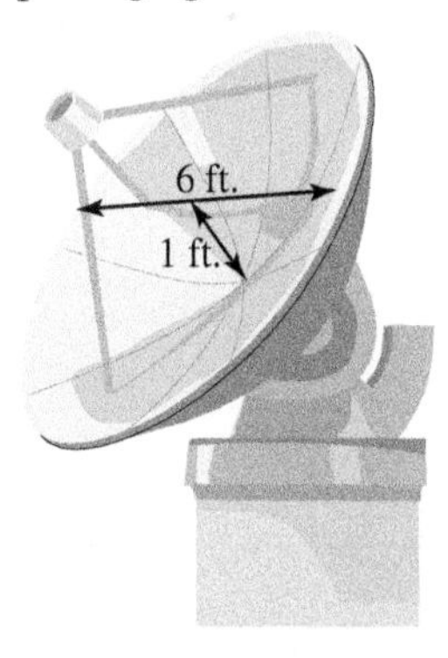

89. The percent of deaths by age per million miles driven can be approximated by the equation $y = 0.0038x^2 - 0.3475x + 8.316$, where x is the age and y is the percent. Find the percent of deaths per million miles driven for drivers 17 years old. (*Source:* Highway Traffic Safety Administration.)

90. The number of drivers involved in fatal accidents for a given blood alcohol content (BAC) can be approximated by $y = -8862.5x^2 + 26622.6x + 332$, where x is the BAC and y is the number of drivers involved in fatal accidents. Find the number of drivers involved in fatal accidents who had a BAC of 0.20. (*Source:* Highway Traffic Safety Administration.)

91. Bill and Don are fishing in the Gulf of Mexico in separate boats that are equipped with radios with a range of 20 miles. If we put Bill's radio at the origin of a coordinate system, what is the equation of all possible locations of Don's boat where the radios would be at their maximum range?

92. A toy plane is attached to a string pinned to the ceiling so that the plane flies in a circle. If the string is 4 feet long, write an equation that describes the path of the plane if the pin is at the origin.

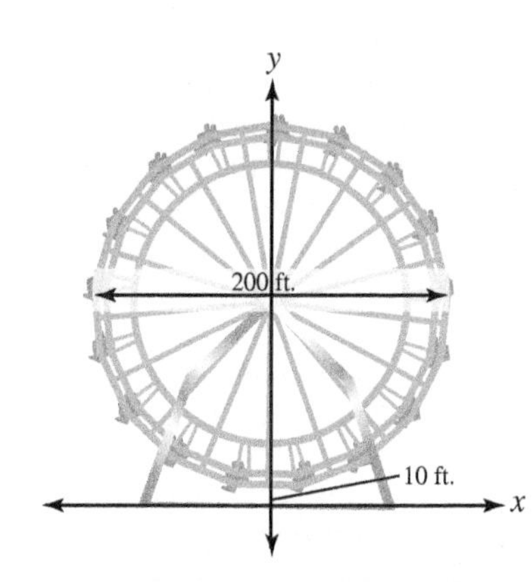

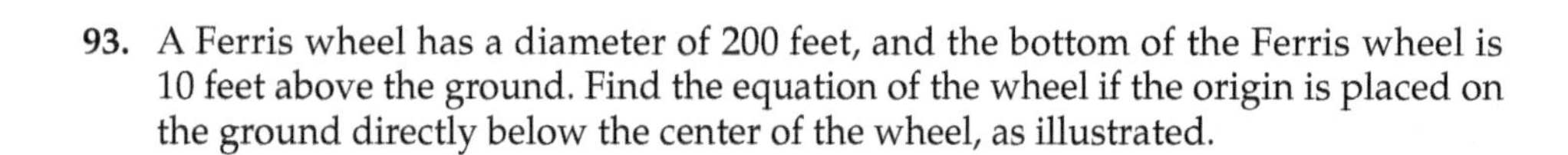
93. A Ferris wheel has a diameter of 200 feet, and the bottom of the Ferris wheel is 10 feet above the ground. Find the equation of the wheel if the origin is placed on the ground directly below the center of the wheel, as illustrated.

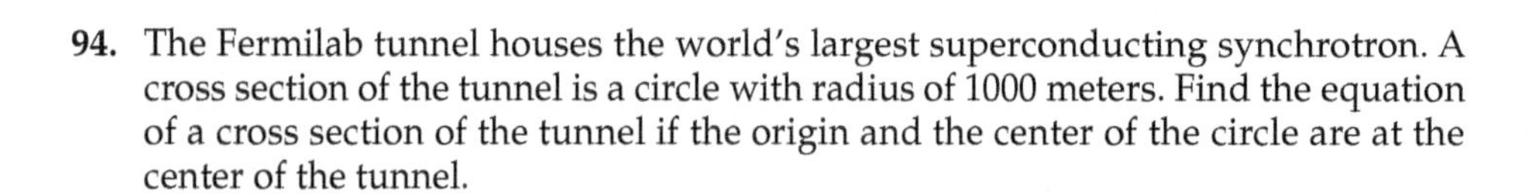
94. The Fermilab tunnel houses the world's largest superconducting synchrotron. A cross section of the tunnel is a circle with radius of 1000 meters. Find the equation of a cross section of the tunnel if the origin and the center of the circle are at the center of the tunnel.

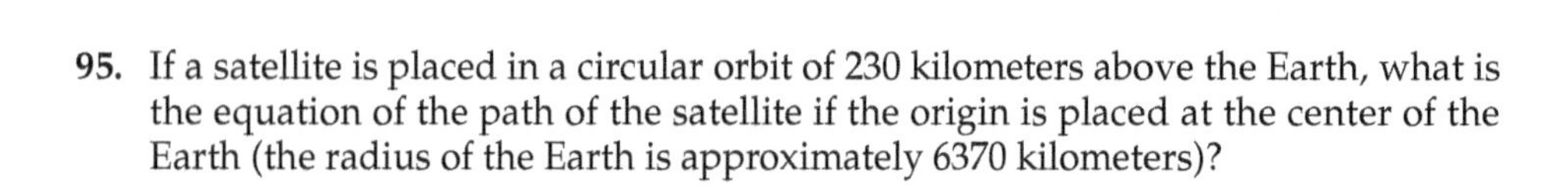
95. If a satellite is placed in a circular orbit of 230 kilometers above the Earth, what is the equation of the path of the satellite if the origin is placed at the center of the Earth (the radius of the Earth is approximately 6370 kilometers)?

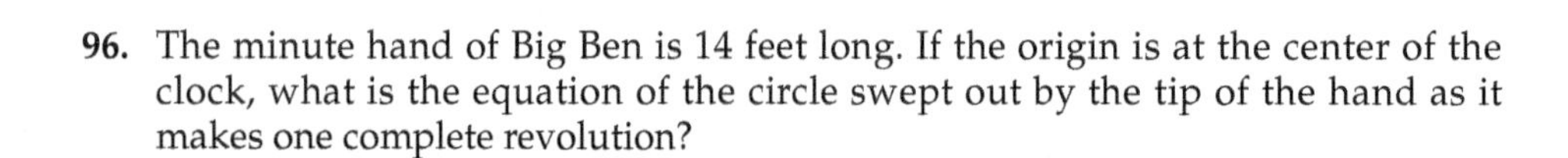
96. The minute hand of Big Ben is 14 feet long. If the origin is at the center of the clock, what is the equation of the circle swept out by the tip of the hand as it makes one complete revolution?

Puzzle Problem Using only a pencil, you can draw both a rough circle and its center without the point of the pencil losing contact with the paper, resulting in a picture like the one shown. Explain how.

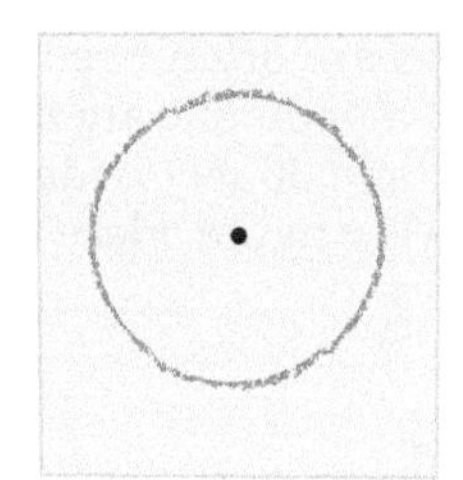

Review Exercises

Exercises 1 and 2 Expressions

[3.1] **1.** Following is a coordinate system with the point $(2, 3)$ plotted. Plot the points that are 4 units to the left and right of $(2, 3)$ and give their coordinates.

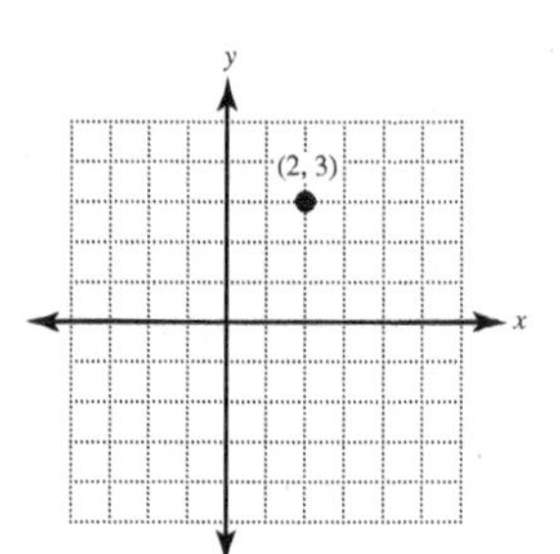

[6.3] **2.** Factor: $25x^2 - 9y^2$

Exercises 3–6 Equations and Inequalities

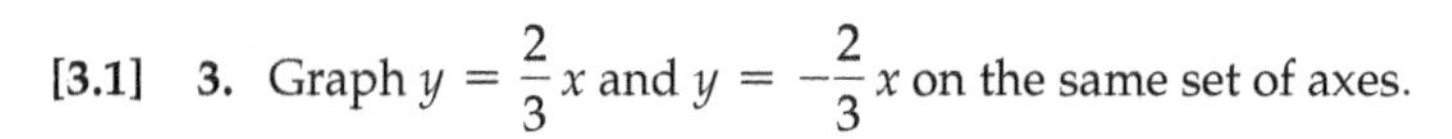
[3.1] **3.** Graph $y = \frac{2}{3}x$ and $y = -\frac{2}{3}x$ on the same set of axes.

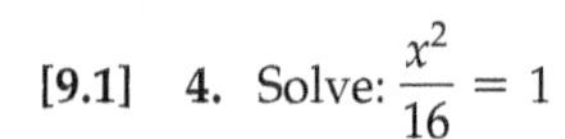
[9.1] **4.** Solve: $\frac{x^2}{16} = 1$

[3.1, 9.1] *For Exercises 5 and 6, find the x- and y- intercepts.*

5. $9x^2 + 16y^2 = 144$

6. $25x^2 + 9y^2 = 225.$

11.2 Ellipses and Hyperbolas

Objectives

1 Graph ellipses.
2 Graph hyperbolas.

Warm-up

[11.1] **1.** Find the center and radius of the circle whose equation is $x^2 + y^2 - 4x + 10y + 13 = 0$.

[3.1] **2.** Find the coordinates of the points that are 6 units to the right and left of the point whose coordinates are $(2, -3)$.

[3.2] **3.** Find the coordinates of a point that is b units above the point whose coordinates are (h, k).

[9.1] **4.** Solve the equation $\frac{y^2}{9} = 1$.

Objective 1 Graph ellipses.

Suppose you drive two nails in a board and tie a string to the two nails. Then you take a pencil, pull the string taut, and draw a figure around the two nails.

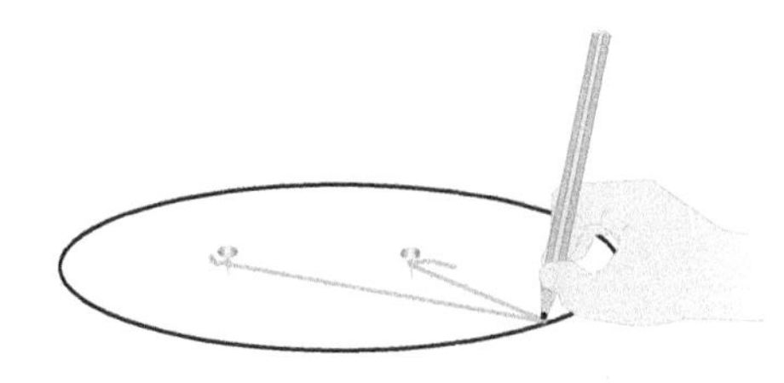

The figure you've drawn is called an **ellipse.** The locations of the two nails are the *focal points.*

Definition Ellipse: The set of all points the sum of whose distances from two fixed points is constant.

Ellipses occur in many situations. The orbits of the planets about the Sun are elliptical with the Sun at one focal point. The orbits of satellites about the Earth are also elliptical. The cams of compound bows are elliptical, which allows a decrease in the amount of effort required to hold the bow at full draw.

In the definition of an ellipse, the two fixed points are the *foci* (plural of *focus*) and the point halfway between the foci is the *center.* The figure at right shows the graph of an ellipse with foci at $(c, 0)$ and $(-c, 0)$, x-intercepts at $(a, 0)$ and $(-a, 0)$, and y-intercepts at $(0, b)$ and $(0, -b)$. Consequently, the center is at the origin, $(0, 0)$. It can be shown that $c^2 = a^2 - b^2$ for ellipses in which $a > b$.

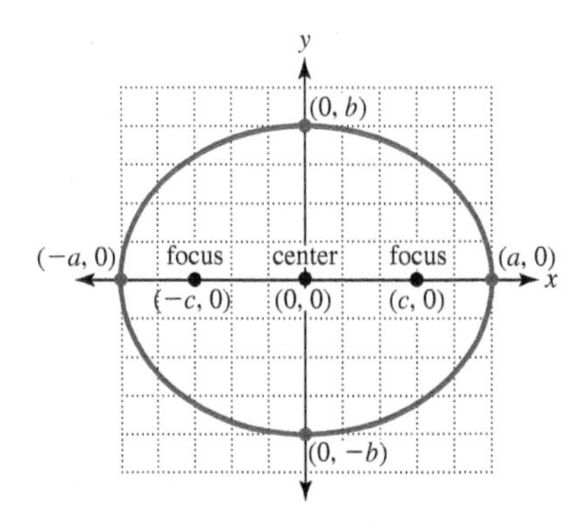

Using the distance formula, it can be shown that an ellipse with these characteristics has the following equation.

Rule Equation of an Ellipse Centered at (0, 0)

The equation of an ellipse with center $(0, 0)$, x-intercepts $(a, 0)$ and $(-a, 0)$, and y-intercepts $(0, b)$ and $(0, -b)$ is

$$\frac{x^2}{a^2} + \frac{y^2}{b^2} = 1.$$

Answers to Warm-up

1. $(2, -5), r = 4$
2. $(8, -3), (-4, -3)$
3. $(h, k + b)$
4. $y = \pm 3$

Example 1 Graph each ellipse and label the x- and y-intercepts.

a. $\dfrac{x^2}{16} + \dfrac{y^2}{9} = 1$

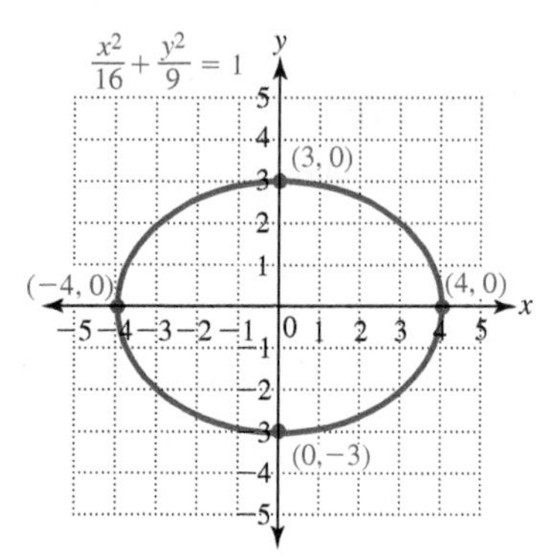

Solution: The equation can be rewritten as $\dfrac{x^2}{4^2} + \dfrac{y^2}{3^2} = 1$, so $a = 4$ and $b = 3$. Consequently, the x-intercepts are $(4, 0)$ and $(-4, 0)$ and the y-intercepts are $(0, 3)$ and $(0, -3)$.

b. $25x^2 + 9y^2 = 225$

Solution: We first need to write the equation in standard form: $\dfrac{x^2}{a^2} + \dfrac{y^2}{b^2} = 1$.

$$\frac{25x^2}{225} + \frac{9y^2}{225} = \frac{225}{225} \quad \text{Divide both sides by 225.}$$

$$\frac{x^2}{9} + \frac{y^2}{25} = 1 \quad \text{Simplify both sides.}$$

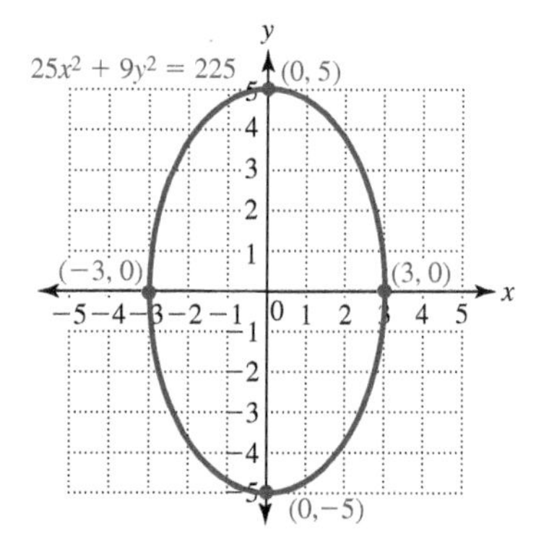

We now see that this is an equation of an ellipse with $a = 3$ and $b = 5$, so the x-intercepts are $(3, 0)$ and $(-3, 0)$ and the y-intercepts are $(0, 5)$ and $(0, -5)$.

Your Turn 1 Graph the ellipse $\dfrac{x^2}{36} + \dfrac{y^2}{9} = 1$. Label the x- and y-intercepts.

If the center of an ellipse is not at the origin, it can be shown that the equation has the following form.

Rule General Equation for an Ellipse

The equation of an ellipse with center (h, k) is $\dfrac{(x-h)^2}{a^2} + \dfrac{(y-k)^2}{b^2} = 1$. The ellipse passes through two points that are a units to the right and left of the center and two points that are b units above and below the center.

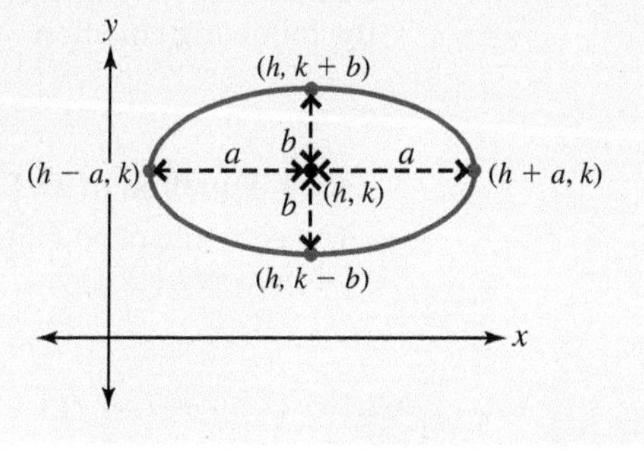

Answer to Your Turn 1

$\dfrac{x^2}{36} + \dfrac{y^2}{9} = 1$

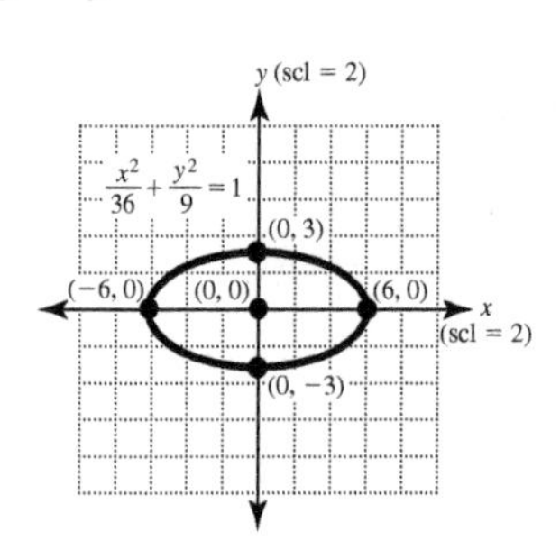

Example 2 Graph the ellipse. Label the center and the points directly above, below, to the left of, and to the right of the center. $\frac{(x-2)^2}{36} + \frac{(y+3)^2}{16} = 1$

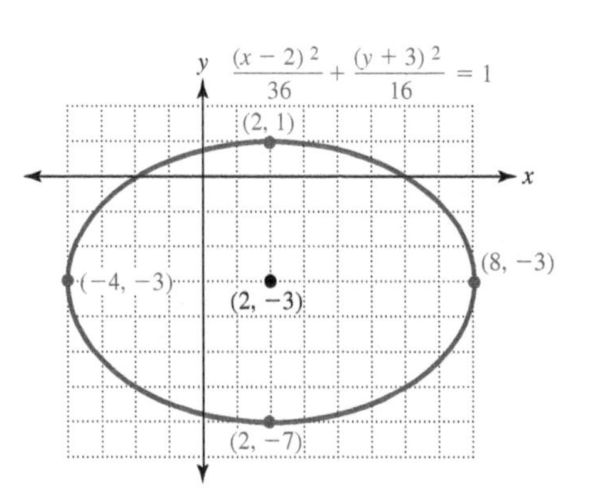

Solution: Because $h = 2$ and $k = -3$, the center of the ellipse is $(2, -3)$. Also, we see that $a = 6$, which means that the ellipse passes through two points 6 units to the right and left of $(2, -3)$. These points are $(8, -3)$ and $(-4, -3)$. Because $b = 4$, the ellipse passes through two points 4 units above and below the center. These points are $(2, 1)$ and $(2, -7)$.

Your Turn 2 Graph the ellipse $\frac{(x+1)^2}{25} + \frac{(y-2)^2}{36} = 1$. Label the center and the points directly above, below, to the right of, and to the left of the center.

Example 3 Graph the ellipse $\frac{x^2}{9} + \frac{y^2}{16} = 1$ using a graphing calculator.

Solution: Because an ellipse is not a function, we solve for y to get two functions that we can input.

$$\frac{y^2}{16} = 1 - \frac{x^2}{9}$$ Subtract $\frac{x^2}{9}$ from both sides.

$$y^2 = 16\left(1 - \frac{x^2}{9}\right)$$ Multiply both sides by 16.

$$y = \pm 4\sqrt{\left(1 - \frac{x^2}{9}\right)}$$ Apply the square root principle.

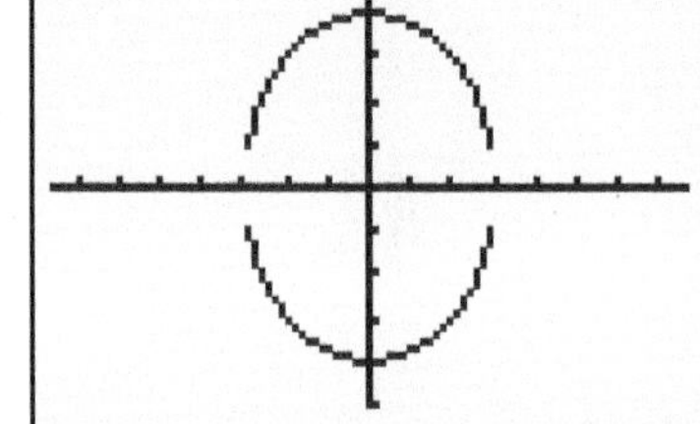

Define $Y_1 = 4\sqrt{\left(1 - \frac{x^2}{9}\right)}$ and $Y_2 = -4\sqrt{\left(1 - \frac{x^2}{9}\right)}$. Graph in a window $[-5, 5]$ for x and $[-5, 5]$ for y and square the window.

Answer to Your Turn 2

$$\frac{(x+1)^2}{25} + \frac{(y-2)^2}{36} = 1$$

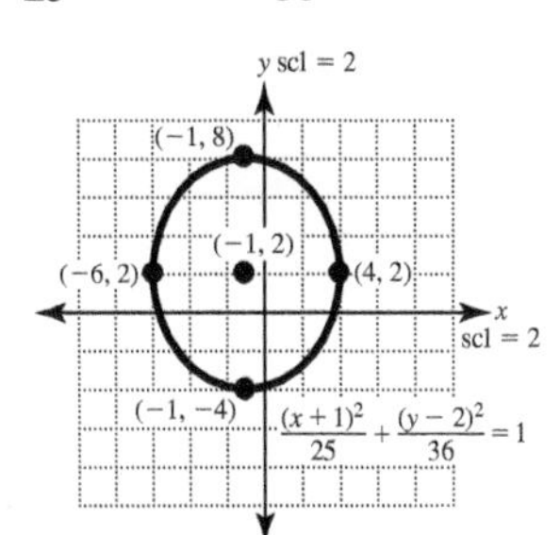

Objective 2 Graph hyperbolas.

The last conic is the **hyperbola**. Applications of hyperbolas include the LORAN navigation system and the orbits of some comets.

Definition Hyperbola: The set of all points the difference of whose distances from two fixed points remains constant.

As with the ellipse, the two fixed points are the *foci* and the point halfway between the foci is the *center*.

 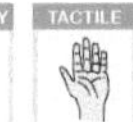 

Learning Strategy

Noting which variable term is positive in the standard form equation of a hyperbola can help you remember how to orient the graph. If the x^2 term is positive, the hyperbola's intercepts are on the x-axis. If the y^2 term is positive, the hyperbola's intercepts are on the y-axis.

Rule Equations of Hyperbolas in Standard Form

The equation of a hyperbola with center $(0, 0)$, x-intercepts $(a, 0)$ and $(-a, 0)$, and no y-intercepts is

$$\frac{x^2}{a^2} - \frac{y^2}{b^2} = 1.$$

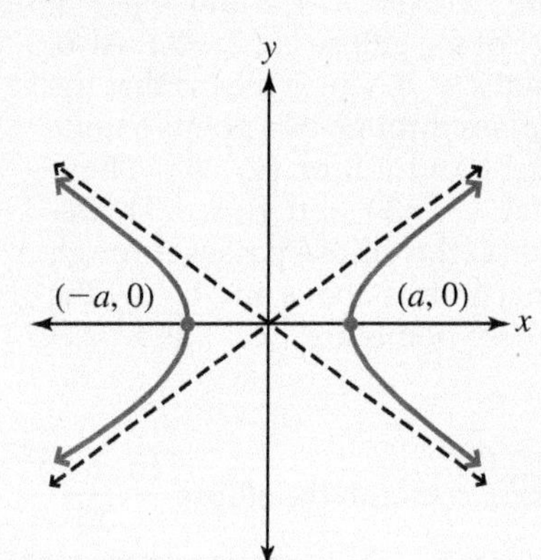

The equation of a hyperbola with center $(0, 0)$, y-intercepts $(0, b)$ and $(0, -b)$, and no x-intercepts is

$$\frac{y^2}{b^2} - \frac{x^2}{a^2} = 1.$$

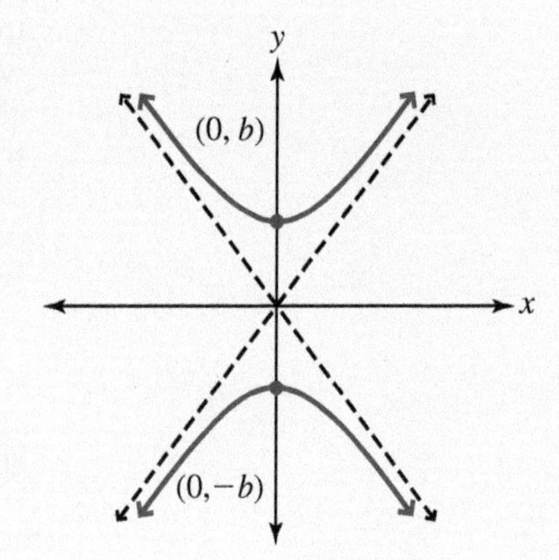

The dashed lines that intersect at the center of a hyperbola are *asymptotes*. They are not a part of the graph, but are used as an aid in graphing. An asymptote is a line that the graph approaches but does not cross as the graph goes away from the origin. The rectangle whose vertices are (a, b), $(-a, b)$, $(a, -b)$, and $(-a, -b)$ is called the *fundamental rectangle,* and the asymptotes are the extended diagonals of the fundamental rectangle, as shown in the following illustration.

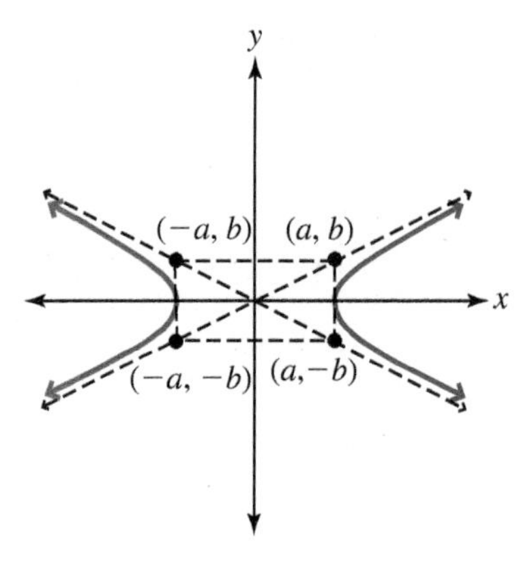

Procedure Graphing a Hyperbola in Standard Form

1. Find the intercepts. If the x^2-term is positive, the x-intercepts are $(a, 0)$ and $(-a, 0)$ and there are no y-intercepts. If the y^2-term is positive, the y-intercepts are $(0, b)$ and $(0, -b)$ and there are no x-intercepts.
2. Draw the fundamental rectangle. The vertices are (a, b), $(-a, b)$, $(a, -b)$, and $(-a, -b)$.
3. Draw the asymptotes, which are the extended diagonals of the fundamental rectangle.
4. Draw the graph so that each branch passes through an intercept and approaches the asymptotes the farther the branches are from the origin.

Example 4 Graph each hyperbola. Also show the fundamental rectangle with its corner points labeled, the asymptotes, and the intercepts.

a. $\dfrac{x^2}{16} - \dfrac{y^2}{9} = 1$

Solution:

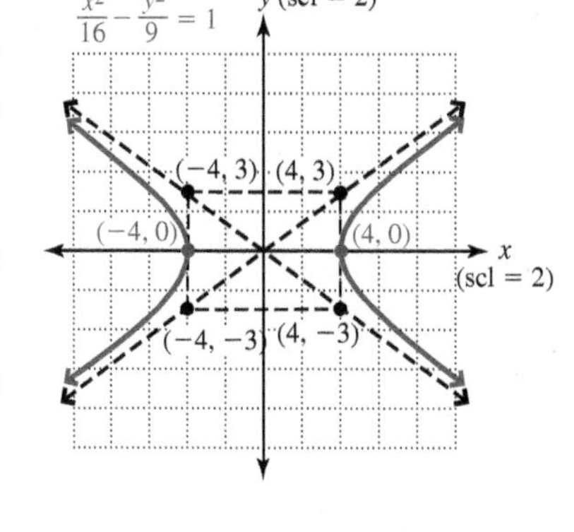

1. This equation can be written as $\dfrac{x^2}{4^2} - \dfrac{y^2}{3^2} = 1$, so $a = 4$ and $b = 3$. Because the x^2-term is positive, the graph has x-intercepts at $(4, 0)$ and $(-4, 0)$.
2. The fundamental rectangle has vertices at $(4, 3)$, $(-4, 3)$, $(4, -3)$, and $(-4, -3)$.

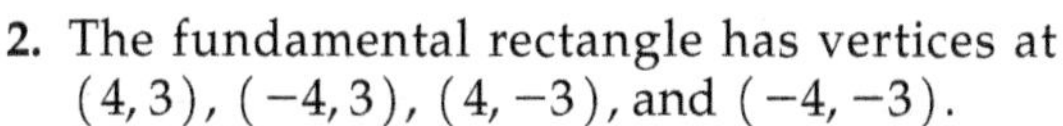

3. Draw the asymptotes.
4. Sketch the graph so that it passes through the x-intercepts and approaches the asymptotes.

b. $\dfrac{y^2}{4} - \dfrac{x^2}{9} = 1$

Solution:

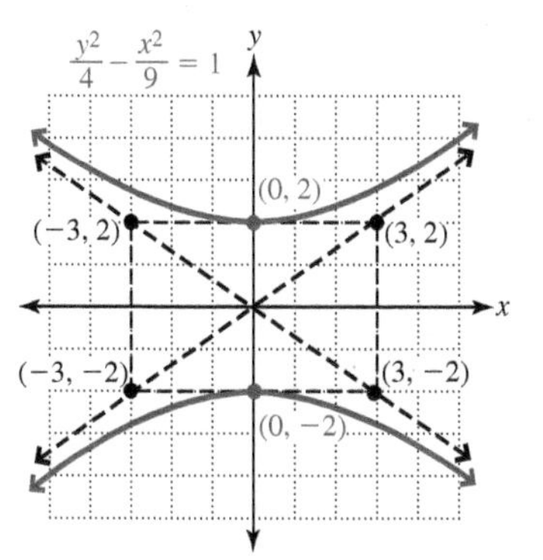

1. This equation can be written as $\dfrac{y^2}{2^2} - \dfrac{x^2}{3^2} = 1$, so $b = 2$ and $a = 3$. Because the y^2-term is positive, the graph has y-intercepts at $(0, 2)$ and $(0, -2)$.
2. The fundamental rectangle has vertices at $(3, 2)$, $(-3, 2)$, $(3, -2)$, and $(-3, -2)$.
3. Draw the asymptotes.
4. Sketch the graph so that it passes through the y-intercepts and approaches the asymptotes.

Your Turn 4 Graph the hyperbola. Also show the fundamental rectangle with its corner points labeled, the asymptotes, and the intercepts.

$$\frac{x^2}{25} - \frac{y^2}{16} = 1$$

Answer to Your Turn 4

$\dfrac{x^2}{25} - \dfrac{y^2}{16} = 1$

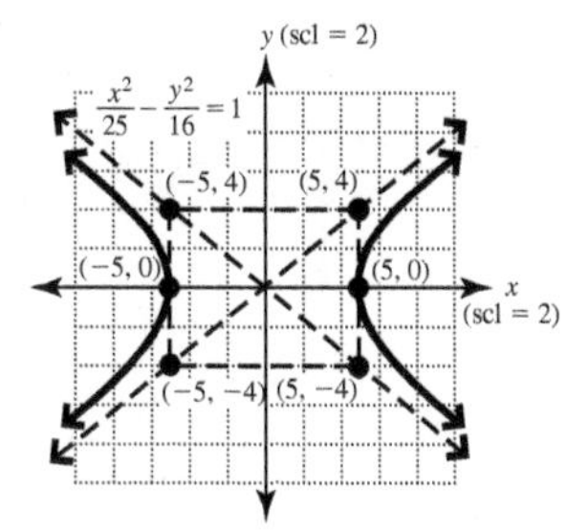

Following is a summary of the conic sections.

Standard Forms		
Parabola	$y = a(x - h)^2 + k, a > 0$	$y = a(x - h)^2 + k, a < 0$
	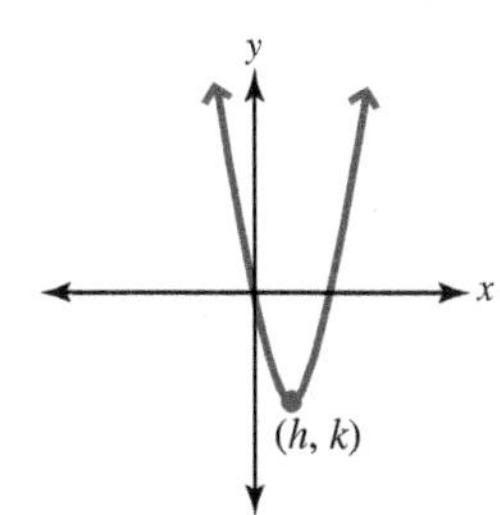	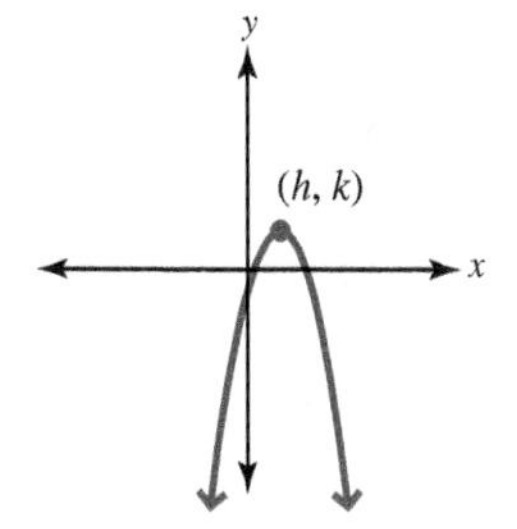

(Continued)

Standard Forms

	$x = a(y-k)^2 + h, a > 0$	$x = a(y-k)^2 + h, a < 0$

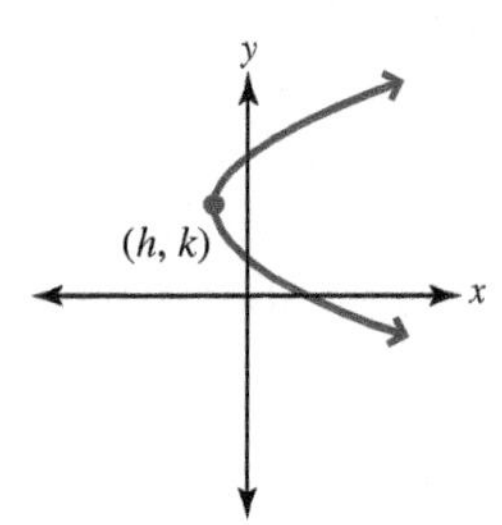

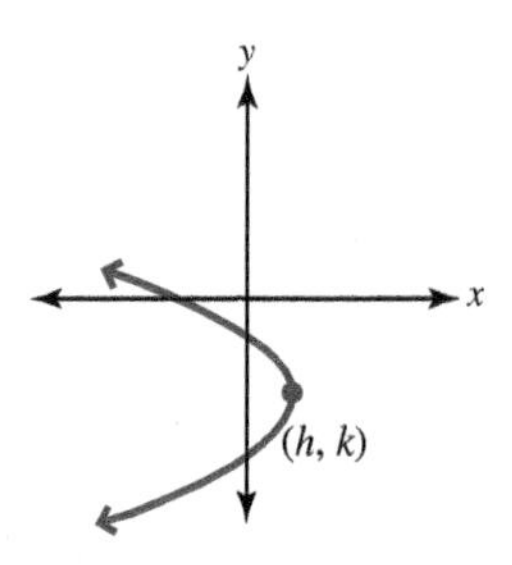

Circle $\quad x^2 + y^2 = r^2 \quad (x-h)^2+(y-k)^2 = r^2$

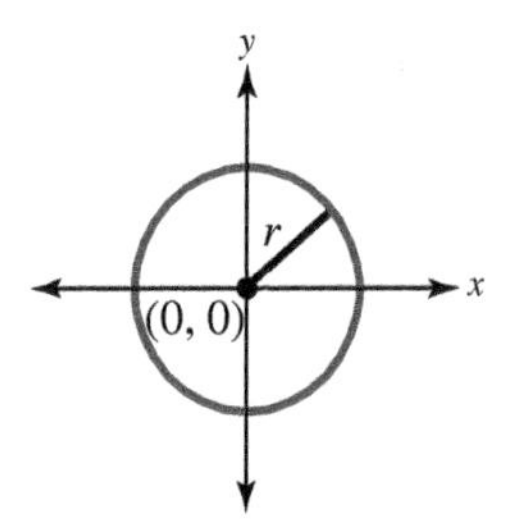

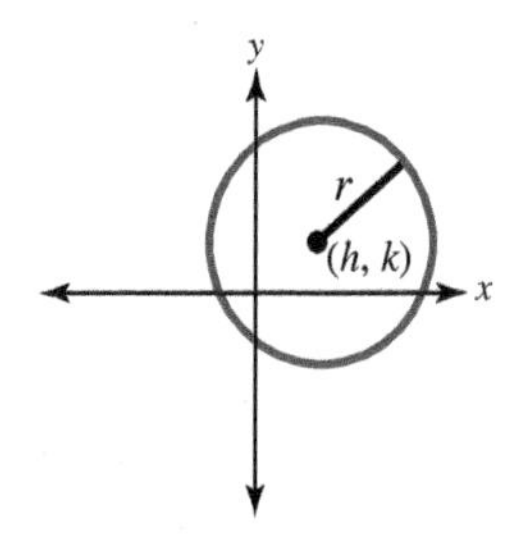

Ellipse $\quad \dfrac{x^2}{a^2} + \dfrac{y^2}{b^2} = 1 \quad \dfrac{(x-h)^2}{a^2} + \dfrac{(y-k)^2}{b^2} = 1$

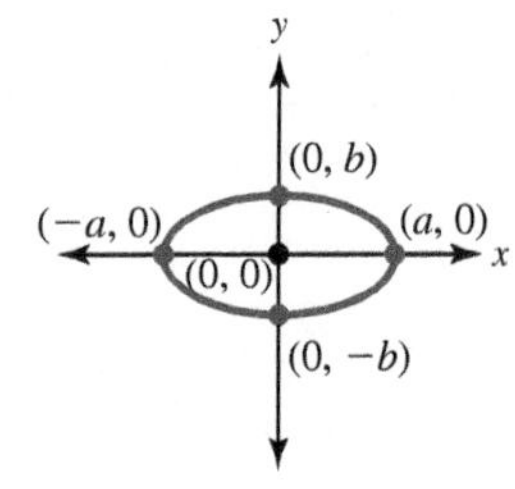

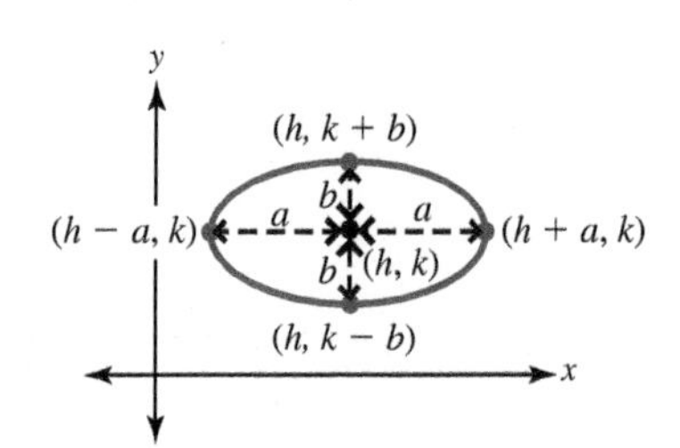

Hyperbola $\quad \dfrac{x^2}{a^2} - \dfrac{y^2}{b^2} = 1$

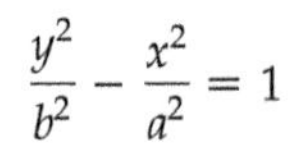
$\dfrac{y^2}{b^2} - \dfrac{x^2}{a^2} = 1$

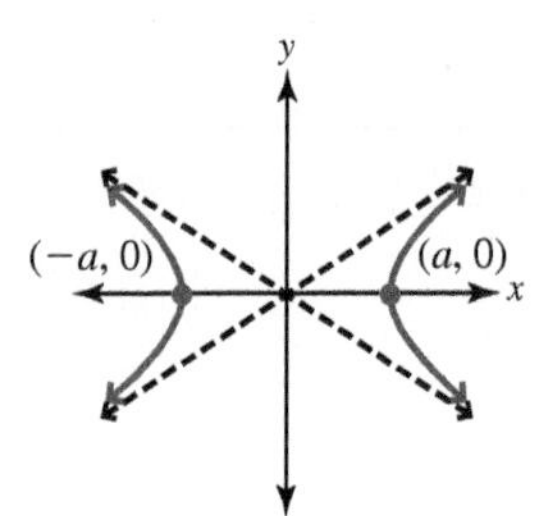

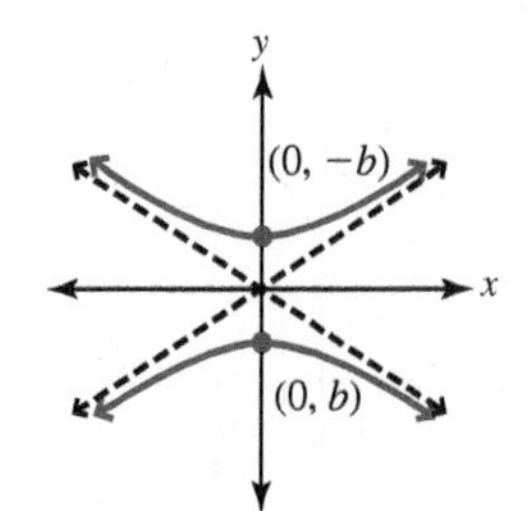

11.2 Exercises

For Extra Help MyMathLab®

Objective 1

Prep Exercise 1 What is the definition of an ellipse?

Prep Exercise 2 In the definitions of the ellipse and hyperbola, two fixed points are mentioned. What are these two points called?

Prep Exercise 3 Answer the following questions about the ellipse defined by $\frac{x^2}{9} + \frac{y^2}{25} = 1$.

a. What is the center?

b. How many units to the left and right of center do you place a point?

c. How many units above and below center do you place a point?

For Exercises 1–8, graph each ellipse. Label the x- and y-intercepts. See Example 1.

1. $\frac{x^2}{81} + \frac{y^2}{64} = 1$

2. $\frac{x^2}{25} + \frac{y^2}{4} = 1$

3. $\frac{x^2}{36} + y^2 = 1$

4. $\frac{x^2}{9} + \frac{y^2}{16} = 1$

5. $4x^2 + 9y^2 = 36$

6. $25x^2 + 4y^2 = 100$

7. $36x^2 + 4y^2 = 144$

8. $x^2 + 9y^2 = 36$

For Exercises 9–12, graph each ellipse. Label the center and the points directly above, below, to the left of, and to the right of the center. See Example 2.

9. $\frac{(x-1)^2}{49} + \frac{(y+3)^2}{25} = 1$

10. $\frac{(x+3)^2}{25} + \frac{(y-2)^2}{64} = 1$

11. $\frac{(x-4)^2}{4} + \frac{(y+3)^2}{36} = 1$

12. $\frac{(x+5)^2}{9} + \frac{(y+3)^2}{25} = 1$

For Exercises 13–16, graph using a graphing calculator. See Example 3.

13. $\frac{x^2}{25} + \frac{y^2}{4} = 1$

14. $\frac{y^2}{12} + \frac{x^2}{8} = 1$

15. $\frac{(y-2)^2}{36} + \frac{(x+4)^2}{25} = 1$

16. $\frac{(x+2)^2}{4} + \frac{(y+1)^2}{16} = 1$

17. If the x-intercepts of an ellipse are $(-3, 0)$ and $(3, 0)$ and the y-intercepts are $(0, 4)$ and $(0, -4)$, what is the equation of the ellipse?

18. If the x-intercepts of an ellipse are $(-6, 0)$ and $(6, 0)$ and the y-intercepts are $(0, 3)$ and $(0, -3)$, what is the equation of the ellipse?

Objective 2

Prep Exercise 4 How do you determine whether a hyperbola opens up and down or left and right?

Prep Exercise 5 Answer the following questions about the hyperbola defined by $\frac{x^2}{9} - \frac{y^2}{25} = 1$.

a. What is the center?

b. What are the x-intercepts?

c. What are the y-intercepts?

d. What are the vertices of the fundamental rectangle?

Prep Exercise 6 Answer the following questions about the hyperbola defined by $\frac{y^2}{16} - \frac{x^2}{81} = 1$.

a. What is the center?

b. What are x-intercepts?

c. What are the y-intercepts?

d. What are the vertices of the fundamental rectangle?

For Exercises 19–26, graph each hyperbola. Also show the fundamental rectangle with its corner points labeled, the asymptotes, and the intercepts. See Example 4.

19. $\frac{x^2}{9} - \frac{y^2}{4} = 1$

20. $\frac{x^2}{16} - \frac{y^2}{25} = 1$

21. $\frac{y^2}{36} - \frac{x^2}{9} = 1$

22. $\frac{y^2}{4} - \frac{x^2}{25} = 1$

23. $9x^2 - y^2 = 36$

24. $x^2 - 4y^2 = 16$

25. $9y^2 - 25x^2 = 225$

26. $16y^2 - 4x^2 = 64$

For Exercises 27 and 28, graph using a graphing calculator. (Hint: Solve the equation for y to find two functions.)

27. $\dfrac{x^2}{36} - \dfrac{y^2}{4} = 1$

28. $\dfrac{y^2}{9} - \dfrac{x^2}{25} = 1$

29. If a hyperbola opens left and right and the vertices of the fundamental rectangle are $(3, 2)$, $(3, -2)$, $(-3, 2)$, and $(-3, -2)$, what is the equation of the hyperbola?

30. If a hyperbola opens up and down and the vertices of the fundamental rectangle are $(5, 3)$, $(5, -3)$, $(-5, 3)$, and $(-5, -3)$, what is the equation of the hyperbola?

For Exercises 31–34, match the equation with the graph.

31. $\dfrac{x^2}{9} + \dfrac{y^2}{25} = 1$

32. $\dfrac{x^2}{9} - \dfrac{y^2}{25} = 1$

33. $\dfrac{x^2}{25} + \dfrac{y^2}{9} = 1$

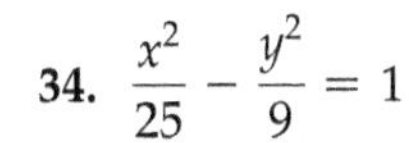

34. $\dfrac{x^2}{25} - \dfrac{y^2}{9} = 1$

a.

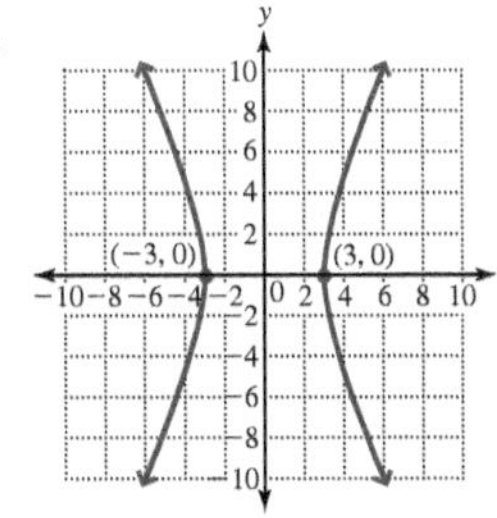

b.

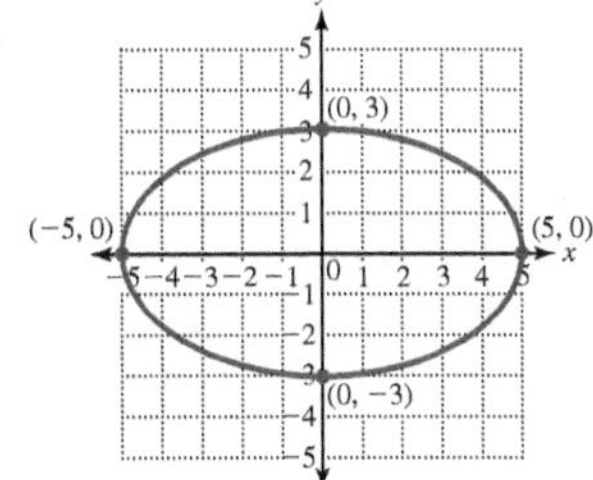

c.

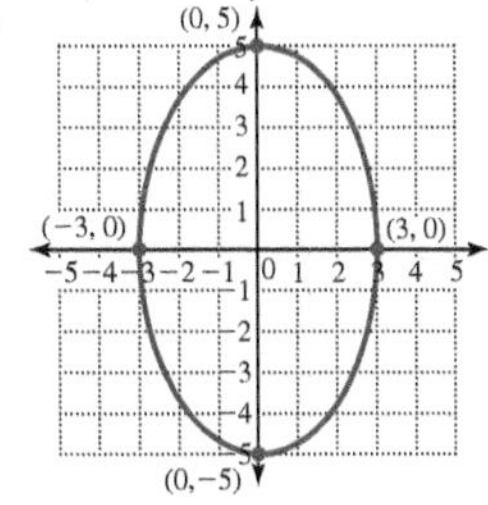

d.

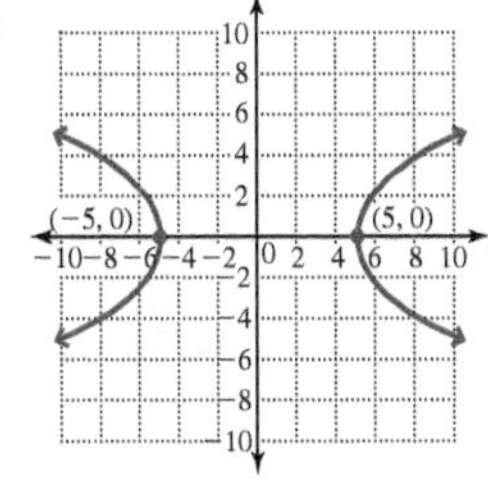

Prep Exercise 7 Which conic has an equation that has x^2 or y^2 but not both?

Prep Exercise 8 How can you tell the equation of an ellipse from the equation of a hyperbola?

For Exercises 35–38, determine whether the graph of the equation is a circle, a parabola, an ellipse, or a hyperbola. Do not draw the graph.

35. $9x^2 + 16y^2 = 144$

36. $16y^2 - 9x^2 = 144$

37. $x^2 + y^2 - 6x + 8y - 75 = 0$

38. $2x^2 - 12x + 23 - y = 0$

For Exercises 39–52, indicate whether the graph of the given equation is a circle, a parabola, an ellipse, or a hyperbola; then draw the graph.

39. $(x-2)^2+(y+2)^2=49$ **40.** $x^2+y^2=81$ **41.** $y=2(x+1)^2+3$ **42.** $x=(y+3)^2-4$

43. $\frac{x^2}{36}+\frac{y^2}{81}=1$ **44.** $\frac{x^2}{36}+\frac{y^2}{16}=1$ **45.** $\frac{x^2}{4}-\frac{y^2}{25}=1$ **46.** $\frac{y^2}{49}-\frac{x^2}{4}=1$

47. $y=2x^2+8x+6$ **48.** $x=y^2+6$ **49.** $\frac{x^2}{16}-\frac{y^2}{16}=1$ **50.** $\frac{y^2}{25}-\frac{x^2}{25}=1$

51. $\frac{(x+3)^2}{9}+\frac{(y+2)^2}{36}=1$ **52.** $\frac{(x-1)^2}{4}+(y+3)^2=1$

For Exercises 53–60, solve.

53. A bridge over a waterway has an arch in the form of half an ellipse. The equation of the ellipse is $400x^2+256y^2=102{,}400$.

a. A sailboat, the top of whose mast is 18 feet above the water, is approaching the arch. Will the mast clear the bridge? Why or why not?

b. How wide is the base of the arch?

54. A highway passes beneath an overpass that is in the shape of half an ellipse. The overpass is 15 feet high at the center and 40 feet wide at the base.

a. What is the equation of the ellipse?

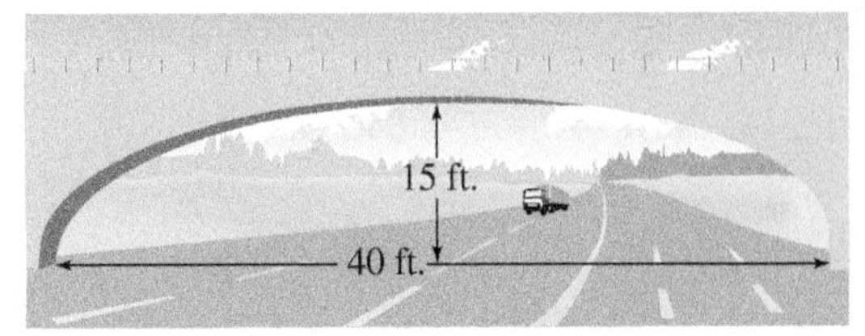

b. A truck that is 10 feet wide carrying a load that is 14 feet above the level of the road is approaching the bridge. If the truck goes down the middle of the road, will the load clear the bridge? Why or why not?

55. The comet Epoch has an orbit in the shape of an ellipse with the Sun at one of the foci. The equation is approximately $\frac{x^2}{3.6^2} + \frac{y^2}{2.88^2} = 1$, where x and y are in astronomical units. (An astronomical unit is 93,000,000 miles.) Sketch the graph of the comet Epoch. (*Source: Orbital Motion*, A. E. Roy, Institute of Physics Publishing, London.)

56. The dwarf planet Pluto has an orbit in the shape of an ellipse with the Sun at one of the foci. The equation is approximately $\frac{x^2}{39.4^2} + \frac{y^2}{38.2^2} = 1$, where x and y are measured in astronomical units. Sketch the graph of the orbit of Pluto. (*Hint:* Make each unit on the x- and y-axes equal to 10.) (*Source: Orbital Motion*, A. E. Roy, Institute of Physics Publishing, London.)

57. Compound bows have elliptical cams that decrease the amount of effort required to hold the bow at full draw. If a cam on a bow is 4 inches from top to bottom and 3 inches across, what is the equation of the ellipse if the origin is at the center of the cam?

58. One of the most popular exercise machines is an elliptical trainer in which your foot moves in an elliptical path. On a typical machine, the length of the stride is about 19 inches and the height varies from 3 to 5 inches depending on the settings. Write the equation for the path that your foot takes if the total length of the stride is 19 inches and the total height is 4 inches. (*Source*: Precor National Headquarters.)

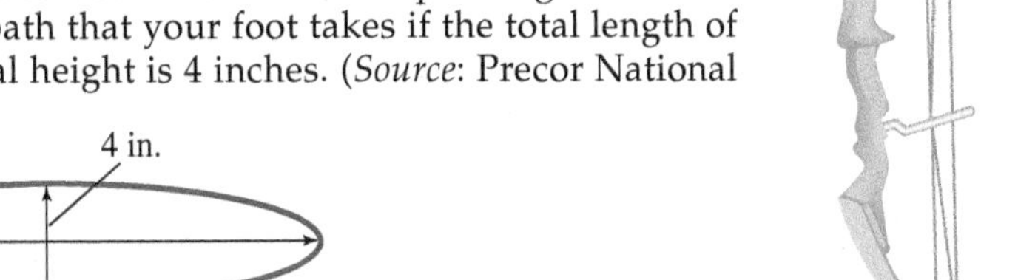

59. If a source of light or sound is placed at one focal point of an elliptic reflector, the light or sound is reflected through the other focal point. This principle is used in a lithotripter that uses sound waves to crush kidney stones by placing a source of sound at one focal point and the kidney stone at the other. If the elliptic reflector is based on the ellipse $\frac{x^2}{25} + \frac{y^2}{16} = 1$, how many units from the center should the kidney stone be placed? (*Hint:* $c^2 = a^2 - b^2$.)

60. The same principle used in Exercise 59 is also used in whispering rooms. The room is in the shape of an elliptic reflector where one person speaks at one focal point and the other person places his or her ear at the other focal point and can hear sounds as faint as a whisper. If a whispering room is based on the ellipse $\frac{x^2}{225} + \frac{y^2}{144} = 1$, how many units from the center of the ellipse would the two people stand? (*Hint:* $c^2 = a^2 - b^2$.)

Puzzle Problem Suppose an ellipse is drawn using the method described on p. 847. Assuming that the ellipse is centered at the origin, if the nails are 8 inches apart and the string is 10 inches in length, what are the x- and y-intercepts?

Review Exercises

Exercises 1–6 Equations and Inequalities

[2.1] **1.** Solve $3x^2 + y = 6$ for y.

[4.1] **2.** Solve the following system using the graphical method. $\begin{cases} x + y = 3 \\ 2x + y = 4 \end{cases}$

[4.1] **3.** Solve the following system using the substitution method. $\begin{cases} 2x + y = 1 \\ 3x + 4y = -6 \end{cases}$

[4.1] **4.** Solve the following system using the elimination method. $\begin{cases} 2x + 3y = 6 \\ 3x - 4y = -25 \end{cases}$

For Exercises 5 and 6, solve.

[6.4] **5.** $3x^2 + 10x - 8 = 0$

[9.1] **6.** $4x^2 = 36$

11.3 Nonlinear Systems of Equations

Objectives

1 Solve nonlinear systems of equations using substitution.

2 Solve nonlinear systems of equations using elimination.

Warm-up

[11.2] **1.** Graph: $9x^2 + 16y^2 = 144$

[2.2] **2.** Solve $4x - y = 1$ for y.

[4.1] **3.** Solve using the substitution method: $\begin{cases} x - 3y = 10 \\ 4x + 5y = 6 \end{cases}$

[4.1] **4.** Solve using the elimination method: $\begin{cases} 5x + 2y = 13 \\ 2x - 5y = 11 \end{cases}$

We solved systems of linear equations in Chapter 4. Now we solve **nonlinear systems** of equations.

Definition **Nonlinear system of equations:** A system of equations that contains at least one nonlinear equation.

The types of equations in a nonlinear system determine the number of solutions that are possible for the system. For example, a system containing a quadratic equation and a linear equation can have 0, 1, or 2 points of intersection and therefore 0, 1, or 2 solutions, respectively, as shown by the following figures.

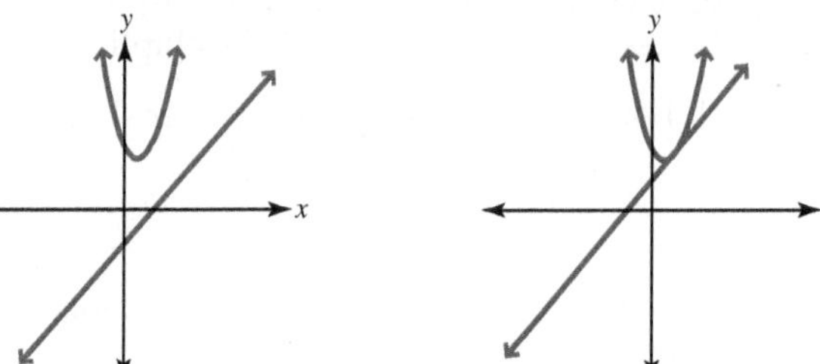

No points of intersection: no solutions

One point of intersection: one solution

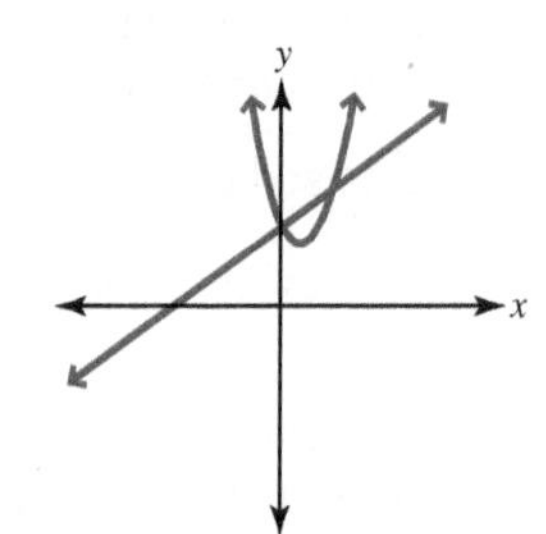

Two points of intersection: two solutions

Connection In Chapter 4, we learned that the number of solutions for a system of linear equations depends on the relative positions of the graphs. We see a similar relationship between the graphs and the number of solutions here.

Answers to Warm-up

1.

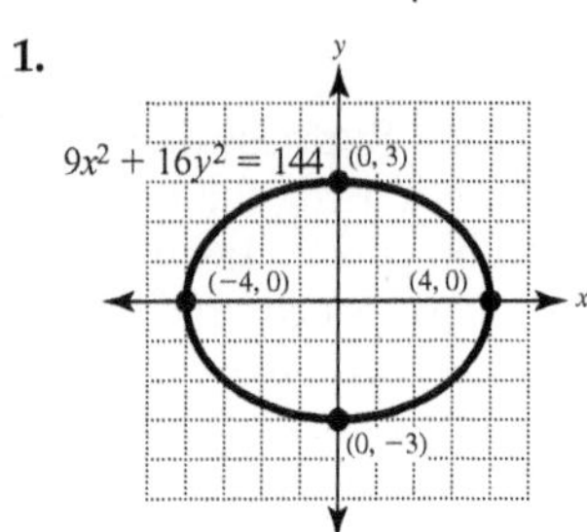

2. $y = 4x - 1$ **3.** $(4, -2)$ **4.** $(3, -1)$

Objective 1 Solve nonlinear systems of equations using substitution.

When solving nonlinear systems, if one equation is linear (or one of the variables in an equation is isolated), the substitution method is usually preferred.

Example 1 Solve using the substitution method.

a. $\begin{cases} y = 2(x+2)^2 - 2 \\ 2x + y = -2 \end{cases}$

Note The graphs are a parabola and a line; so there will be 0, 1, or 2 solutions.

Solution: It is easier to solve the linear equation for one of its variables and substitute into the nonlinear equation. Because the coefficient of y is 1, we solve $2x + y = -2$ for y and get $y = -2x - 2$.

$-2x - 2 = 2(x+2)^2 - 2$ — Substitute $-2x - 2$ for y in $y = 2(x+2)^2 - 2$.

$-2x - 2 = 2(x^2 + 4x + 4) - 2$ — To solve this quadratic equation, we need to write it in the form $ax^2 + bx + c = 0$. First, we square $x + 2$.

$-2x - 2 = 2x^2 + 8x + 6$ — Distribute 2; then combine like terms.

$0 = 2x^2 + 10x + 8$ — Add $2x$ and 2 to both sides.

$0 = x^2 + 5x + 4$ — Divide both sides by 2.

$0 = (x+1)(x+4)$ — Factor.

$0 = x + 1$ or $0 = x + 4$ — Use the zero-factor theorem.

$-1 = x$ or $-4 = x$ — Solve each equation.

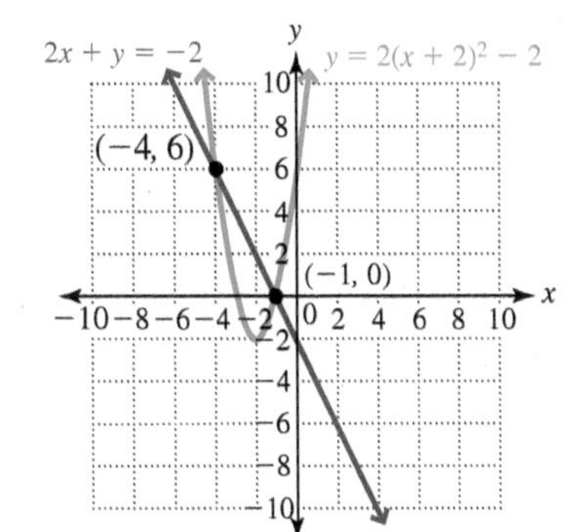

To find y, substitute -1 and -4 for x in $y = -2x - 2$.

$y = -2(-1) - 2$ $\qquad$ $y = -2(-4) - 2$

$y = 2 - 2$ $\qquad$ $y = 8 - 2$

$y = 0$ $\qquad$ $y = 6$

The solutions are $(-1, 0)$ and $(-4, 6)$, as verified by the graph to the left.

b. $\begin{cases} y = \sqrt{x+2} \\ x^2 + y^2 = 8 \end{cases}$

Solution: Because $y = \sqrt{x+2}$ is already solved for y, substitute $\sqrt{x+2}$ for y in $x^2 + y^2 = 8$.

$x^2 + (\sqrt{x+2})^2 = 8$ — Substitute $\sqrt{x+2}$ for y in $x^2 + y^2 = 8$.

$x^2 + x + 2 = 8$ — Square $\sqrt{x+2}$.

$x^2 + x - 6 = 0$ — Because the equation is now quadratic, we subtract 8 from both sides to get the form $ax^2 + bx + c = 0$.

$(x+3)(x-2) = 0$ — Factor.

$x + 3 = 0$ or $x - 2 = 0$ — Use the zero-factor theorem.

$x = -3$ or $x = 2$ — Solve each equation.

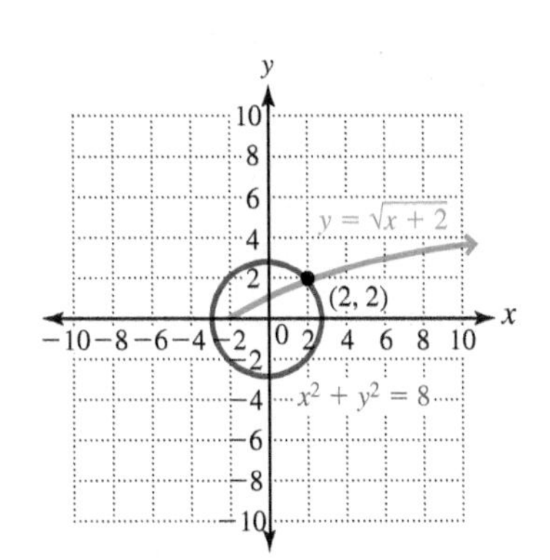

To find y, substitute -3 and 2 for x in $y = \sqrt{x+2}$.

$y = \sqrt{-3+2}$ $\qquad$ $y = \sqrt{2+2}$

$y = \sqrt{-1}$ $\qquad$ $y = \sqrt{4}$

$y = 2$

Because $\sqrt{-1}$ is imaginary, the only real solution is $(2, 2)$, as verified by the graph to the left.

Note A system containing a root function and a circle, as in Example 1 (b), can have 0, 1, or 2 solutions, as shown.

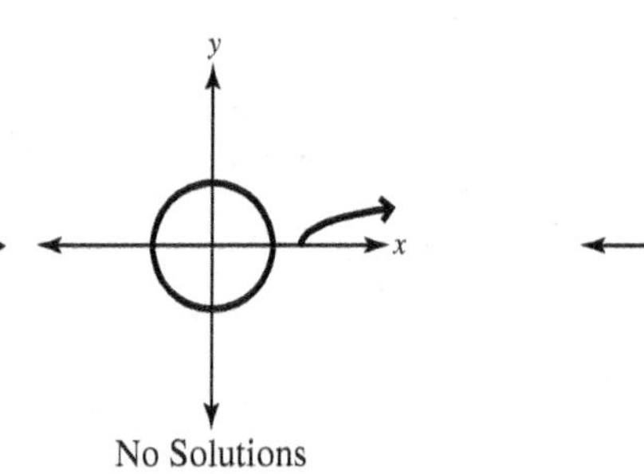

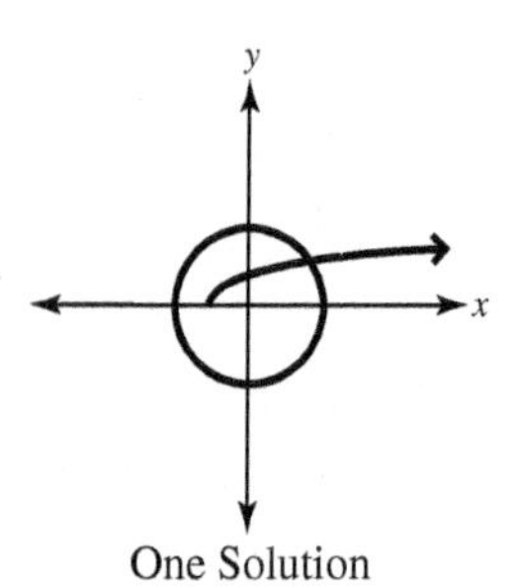

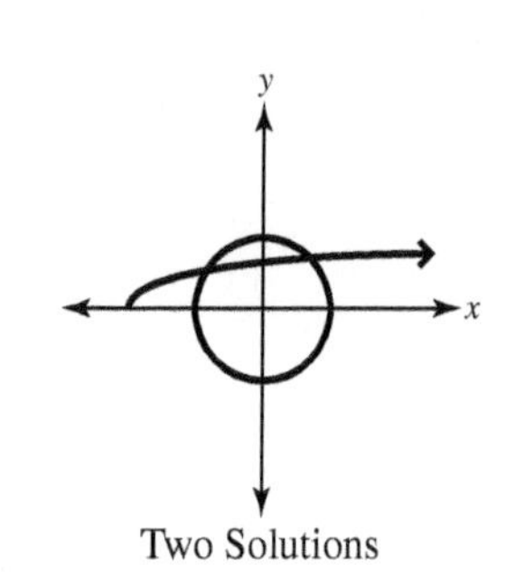

Your Turn 1 Solve using the substitution method.

a. $\begin{cases} 2x - y = 1 \\ x^2 + y^2 = 1 \end{cases}$ b. $\begin{cases} xy = 3 \\ 4x - y = 1 \end{cases}$

Objective 2 Solve nonlinear systems of equations using elimination.

If neither equation contains a radical expression or both equations contain the same powers of the variables, the elimination method can be used.

Example 2 Solve using the elimination method: $\begin{cases} 9x^2 - 4y^2 = 20 \\ x^2 + y^2 = 8 \end{cases}$

Solution: $\begin{cases} 9x^2 - 4y^2 = 20 & \text{(Equation 1)} \\ x^2 + y^2 = 8 & \text{(Equation 2)} \end{cases}$

To eliminate y^2, multiply Equation 2 by 4 and add the equations.

$$9x^2 - 4y^2 = 20 \qquad\qquad 9x^2 - 4y^2 = 20$$

$$x^2 + y^2 = 8 \xrightarrow{\text{Multiply by 4.}} \underline{4x^2 + 4y^2 = 32} \quad \text{Add the equations.}$$

$$13x^2 = 52$$

$$x^2 = 4 \quad \text{Divide both sides by 13.}$$

$$x = \pm 2 \quad \text{Find the square roots of 4.}$$

To find y, substitute 2 and -2 for x in one of the original equations. We will use $x^2 + y^2 = 8$.

Substitute $x = 2$.

$$2^2 + y^2 = 8$$
$$4 + y^2 = 8$$
$$y^2 = 4$$
$$y = \pm 2$$

Therefore, $(2, 2)$ and $(2, -2)$ are solutions.

Substitute $x = -2$.

$$(-2)^2 + y^2 = 8$$
$$4 + y^2 = 8$$
$$y^2 = 4$$
$$y = \pm 2$$

Therefore, $(-2, 2)$ and $(-2, -2)$ are solutions.

This system has four solutions: $(2, 2)$, $(2, -2)$, $(-2, 2)$, and $(-2, -2)$, as verified by the graph to the left.

Answers to Your Turn 1

a. $\left(\frac{4}{5}, \frac{3}{5}\right)$, $(0, -1)$

b. $\left(-\frac{3}{4}, -4\right)$, $(1, 3)$

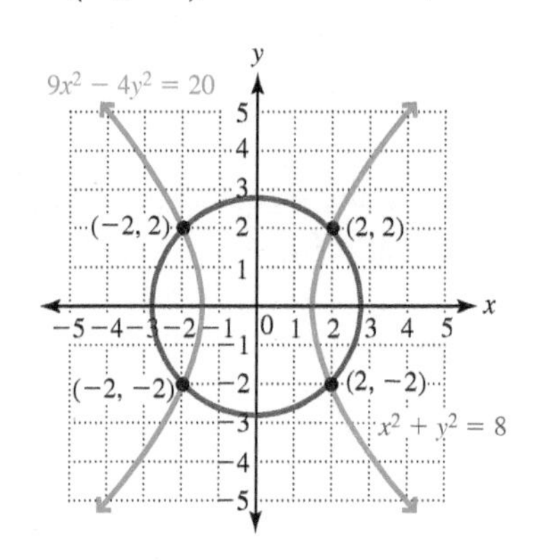

Note If the graphs are a hyperbola and a circle both centered at the origin as in Example 2, there will be 0, 2, or 4 solutions, as shown. ▶

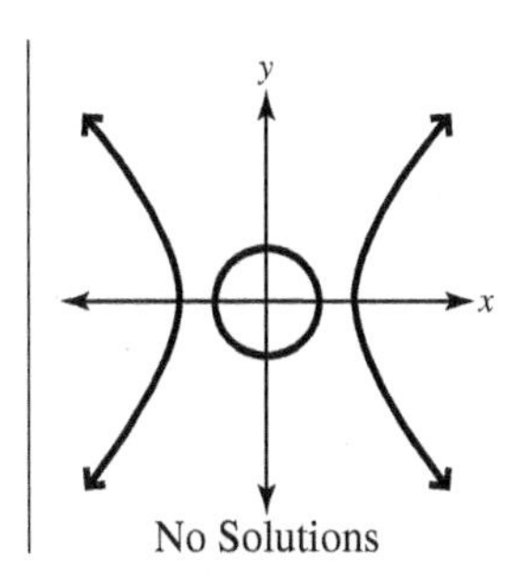

No Solutions

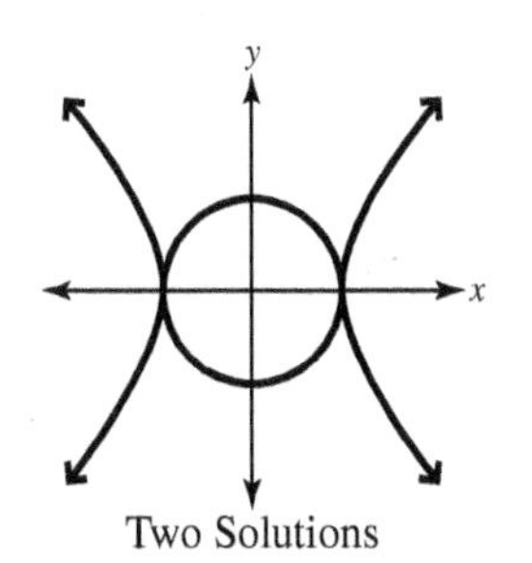

Two Solutions

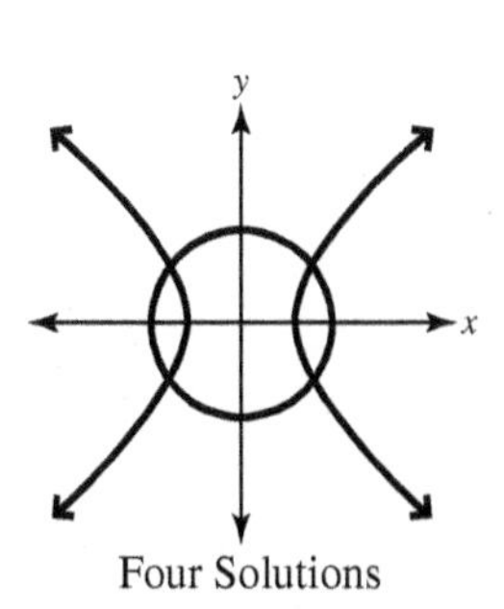

Four Solutions

Answers to Your Turn 2
$(3, 2), (3, -2), (-3, 2), (-3, -2)$

Your Turn 2 Solve using the elimination method: $\begin{cases} 4x^2 + 9y^2 = 72 \\ x^2 + y^2 = 13 \end{cases}$

11.3 Exercises

For Extra Help MyMathLab®

Objectives 1 and 2

Prep Exercise 1 **a.** How many real solutions are possible for a system of two equations whose graphs are a circle and a parabola? **b.** Draw a figure illustrating such a system with two solutions.

Prep Exercise 2 **a.** How many real solutions are possible for a system of two equations whose graphs are a line and a hyperbola? **b.** Draw a figure illustrating such a system with one solution.

Prep Exercise 3 Which method (substitution or elimination) would you use to solve the system $\begin{cases} 2x + y = 8 \\ 4x^2 + 3y^2 = 24 \end{cases}$? Why? (Do not attempt to solve the system.)

Prep Exercise 4 Which method (substitution or elimination) would you use to solve the system $\begin{cases} 3x^2 - 4y^2 = 12 \\ 2x^2 + 3y^2 = 24 \end{cases}$? Why? (Do not attempt to solve the system.)

Prep Exercise 5 Without solving the system, what is the number of possible solutions of the system $\begin{cases} 3x - y = 6 \\ 4x^2 + 9y^2 = 36 \end{cases}$?

Prep Exercise 6 Is it possible for a system of two equations whose graphs are an ellipse and a hyperbola (both centered at the origin) to have three solutions? Why or why not?

For Exercises 1–36, solve. See Examples 1 and 2.

1. $\begin{cases} x^2 + 2y = 1 \\ 2x + y = 2 \end{cases}$

2. $\begin{cases} y = 2x^2 \\ 2x + y = 4 \end{cases}$

3. $\begin{cases} y = x^2 + 4x + 4 \\ 3x - y = -6 \end{cases}$

4. $\begin{cases} y = 6x - x^2 \\ 2x - y = -3 \end{cases}$

5. $\begin{cases} x^2 + y^2 = 25 \\ x - y = -1 \end{cases}$

6. $\begin{cases} x^2 + y^2 = 25 \\ x - 7y = -25 \end{cases}$

7. $\begin{cases} x^2 + y^2 = 10 \\ 3x + y = 6 \end{cases}$

8. $\begin{cases} x^2 + y^2 = 13 \\ 2x + y = 7 \end{cases}$

9. $\begin{cases} x^2 + 2y^2 = 4 \\ x + y = 5 \end{cases}$

10. $\begin{cases} 3x^2 + y^2 = 9 \\ 2x + y = 11 \end{cases}$

11. $\begin{cases} y = 2x^2 + 3 \\ 2x - y = -3 \end{cases}$

12. $\begin{cases} y = -3x^2 - 2 \\ 3x + y = -2 \end{cases}$

13. $\begin{cases} y = (x - 3)^2 + 2 \\ y = -(x - 2)^2 + 3 \end{cases}$

14. $\begin{cases} y = 2(x + 4)^2 + 2 \\ y = -2(x + 3)^2 + 4 \end{cases}$

15. $\begin{cases} x^2 + y^2 = 20 \\ x^2 - y^2 = 12 \end{cases}$

16. $\begin{cases} x^2 + y^2 = 48 \\ x^2 - y^2 = 24 \end{cases}$

17. $\begin{cases} 4x^2 + 9y^2 = 72 \\ x^2 + y^2 = 13 \end{cases}$

18. $\begin{cases} 9x^2 + 4y^2 = 145 \\ x^2 + y^2 = 25 \end{cases}$

19. $\begin{cases} 4x^2 - y^2 = 15 \\ x^2 + y^2 = 5 \end{cases}$

20. $\begin{cases} 9x^2 - 4y^2 = 32 \\ x^2 + y^2 = 5 \end{cases}$

21. $\begin{cases} x^2 + 3y^2 = 36 \\ x = y^2 - 6 \end{cases}$

22. $\begin{cases} 4x^2 + y^2 = 16 \\ y = x^2 - 4 \end{cases}$

23. $\begin{cases} 9x^2 + 4y^2 = 36 \\ 4x^2 - 9y^2 = 36 \end{cases}$

24. $\begin{cases} 9x^2 - 16y^2 = 144 \\ 4x^2 + 9y^2 = 36 \end{cases}$

25. $\begin{cases} y = x^2 \\ x^2 + y^2 = 20 \end{cases}$

26. $\begin{cases} 16x^2 + y^2 = 128 \\ y = 2x^2 \end{cases}$

27. $\begin{cases} 25x^2 - 16y^2 = 400 \\ x^2 + 4y^2 = 16 \end{cases}$

28. $\begin{cases} 9y^2 - 25x^2 = 225 \\ 4y^2 + 25x^2 = 100 \end{cases}$

29. $\begin{cases} xy = 4 \\ 2x^2 - y^2 = 4 \end{cases}$

30. $\begin{cases} xy = 2 \\ 4x^2 + y^2 = 8 \end{cases}$

31. $\begin{cases} y = x^2 - 2x - 3 \\ y = -x^2 + 6x + 7 \end{cases}$

32. $\begin{cases} y = \dfrac{1}{3}x^2 - 2 \\ 3x^2 + 9y^2 = 36 \end{cases}$

33. $\begin{cases} 4x^2 + 5y^2 = 36 \\ 4x^2 - 3y^2 = 4 \end{cases}$

34. $\begin{cases} 4x^2 + 7y^2 = 64 \\ 4x^2 - 3y^2 = 24 \end{cases}$

35. $\begin{cases} x = -y^2 + 2 \\ x^2 - 5y^2 = 4 \end{cases}$

36. $\begin{cases} x = y^2 + 2 \\ 9x^2 - 45y^2 = 36 \end{cases}$

37. Create a system of two equations whose graphs are a circle and a line for which there is no solution. Include the graphs.

38. Create a system of two equations whose graphs are a circle and a hyperbola for which there are exactly two solutions. Include the graphs.

39. The sum of the squares of two integers is 34, and the difference of their squares is 16. Find the integers. There is more than one solution.

40. The difference of the squares of two integers is 32, and their product is 12. Find the integers. There is more than one solution.

41. A computer keyboard has an area of 144 square inches and a perimeter of 52 inches. Find the length and width.

42. A rectangular living room has an area of 48 square meters and a perimeter of 28 meters. Find the dimensions.

43. If p is in dollars and x is in hundreds of units, the demand function for a certain style of chair is given by $p = -3x^2 + 120$ and the supply function is given by $p = 11x + 28$. The *market equilibrium* occurs when the number produced is equal to the number demanded. Find the number of chairs and the price per chair when market equilibrium is reached.

44. If y is in dollars and x is the number of cell phones manufactured (in thousands), a cell phone manufacturer has determined that the cost y to manufacture x cell phones is given by $y = 5x^2 + 30x + 50$ and the revenue from the sales is given by $y = 13x^2$. The *break-even point* is the point (x, y) for which the cost equals the revenue. Solve the system to find the number of units necessary to break even.

For Exercises 45–48, use a graphing calculator to verify the results of the exercise given.

45. Exercise 11

46. Exercise 12

47. Exercise 13

48. Exercise 14

Review Exercises

Exercises 1–6 Equations and Inequalities

[3.4] *For Exercises 1 and 2, determine whether the ordered pair is a solution for* $x + 3y \le 6$.

1. $(0, 0)$

2. $(3, 4)$

[3.4] *For Exercises 3–6, graph.*

3. $y \ge 3$

4. $y < 2x + 3$

5. $2x + 3y < -6$

6. $x \ge -2$

11.4 Nonlinear Inequalities and Systems of Inequalities

Objectives

1 Graph nonlinear inequalities.

2 Graph the solution set of a system of nonlinear inequalities.

Warm-up

[11.3] 1. Solve: $\begin{cases} x^2 + y^2 = 16 \\ x - y = 4 \end{cases}$

[4.5] 2. Graph the solution set of the system of inequalities: $\begin{cases} 5x - 3y > 15 \\ 2x + y < 6 \end{cases}$

Objective 1 Graph nonlinear inequalities.

In Section 3.4, we graphed linear inequalities like $x + 2y > 6$ by graphing the line corresponding to $x + 2y = 6$ and then shading the appropriate region on one side of that boundary line. Recall that we used a dashed line for $<$ and $>$ and a solid line for $\leq$ and $\geq$. We determined which side to shade by using a test point on one side of the boundary.

We use a similar procedure to graph nonlinear inequalities such as $x^2 + y^2 \leq 25$. The boundary is the graph of $x^2 + y^2 = 25$, which is a circle with center at $(0, 0)$ and radius of 5. The solution set of $x^2 + y^2 \leq 25$ contains all ordered pairs on the circle (boundary) along with all ordered pairs inside it or all ordered pairs outside it. To determine which of those two regions is correct, we choose an ordered pair from one of the regions to test in the inequality. Let's test $(0, 0)$.

$0^2 + 0^2 \leq 25$ Substitute (0, 0) into $x^2 + y^2 \leq 25$.

$0 \leq 25$ Simplify. The inequality is true.

Because $0 \leq 25$ is true, all ordered pairs in the region containing $(0, 0)$ are in the solution set; so we shade that region.

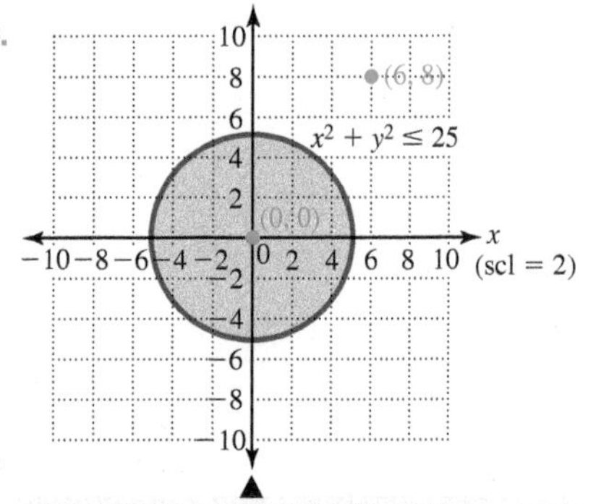

Ordered pairs in the region outside the circle are not in the solution set because they do not solve the inequality. To illustrate, let's test $(6, 8)$.

$6^2 + 8^2 \leq 25$ Substitute (6, 8) into $x^2 + y^2 \leq 25$.

$100 \leq 25$ Simplify. The inequality is false.

Because $100 \leq 25$ is false, the region containing $(6, 8)$ is not in the solution set; so we do not shade that region.

Note If the inequality had been $x^2 + y^2 < 25$, we would have drawn a dashed circle.

Our example suggests the following procedure for graphing nonlinear inequalities.

Procedure Graphing Nonlinear Inequalities

1. Graph the related equation (the boundary curve). If the inequality symbol is $\leq$ or $\geq$, draw the graph as a solid curve. If the inequality symbol is $>$ or $<$, draw a dashed curve.
2. The graph divides the coordinate plane into at least two regions. Test an ordered pair from each region by substituting it into the inequality. If the ordered pair satisfies the inequality, shade the region containing that ordered pair.

Answers to Warm-up

1. $(4, 0), (0, -4)$

2.

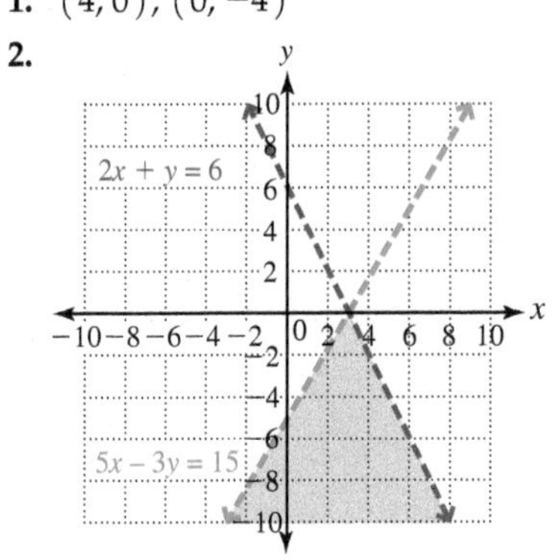

Example 1 Graph the inequality.

a. $y > 2(x - 1)^2 + 3$

Solution: We first graph the related equation, $y = 2(x - 1)^2 + 3$.

Note The parabola is dashed because the inequality is $>$. Also notice that the graph divides the coordinate plane into two regions: region 1 *above* the boundary and region 2 *below*.

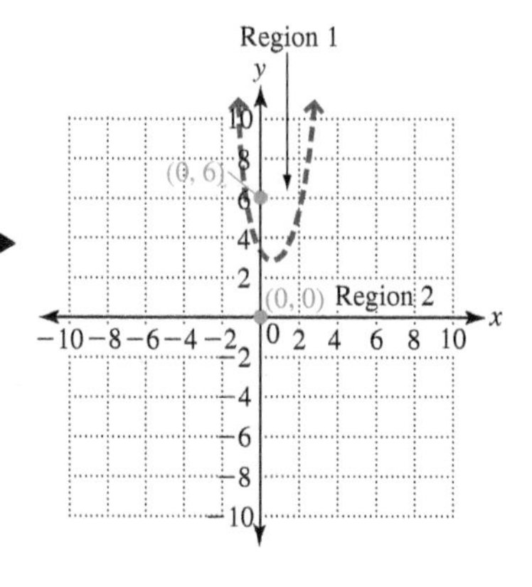

We now test an ordered pair from each region.

Note To make computations easier, choose ordered pairs with 0 for at least one of the coordinates.

Region 1: We choose $(0, 6)$.

$6 > 2(0 - 1)^2 + 3$ Substitute.

$6 > 2(-1)^2 + 3$ Simplify.

$6 > 5$

True; so region 1 is in the solution set.

Region 2: We choose $(0, 0)$.

$0 > 2(0 - 1)^2 + 3$ Substitute.

$0 > 2(-1)^2 + 3$ Simplify.

$0 > 5$

False; so region 2 is not in the solution set.

Because ordered pairs only in region 1 solve the inequality, we shade only that region. The solution set contains all ordered pairs in region 1. Ordered pairs on the parabola are not in the solution set.

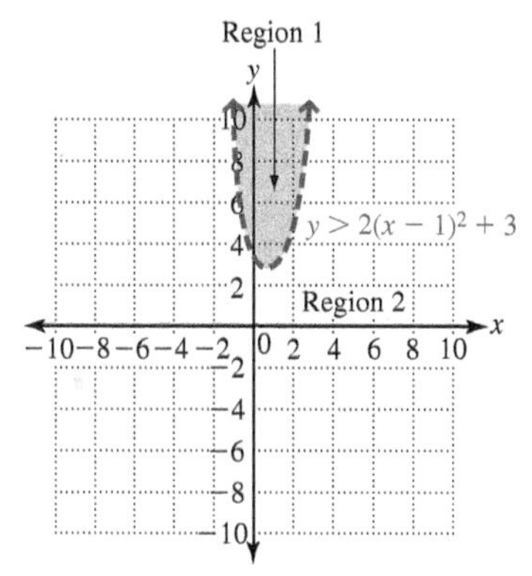

b. $\dfrac{x^2}{9} - \dfrac{y^2}{25} \leq 1$

Solution: Graph the related equation, $\dfrac{x^2}{9} - \dfrac{y^2}{25} = 1$.

Note The hyperbola is solid because the inequality is $\leq$. Also notice that the hyperbola divides the coordinate plane into three regions.

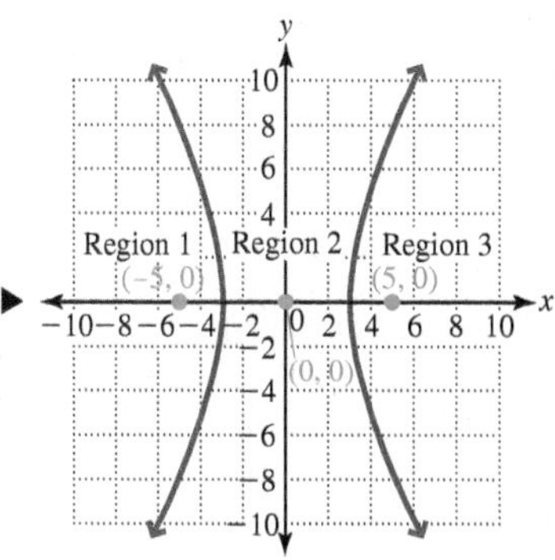

We test an ordered pair from each region.

Region 1: We choose $(-5, 0)$.

$\dfrac{(-5)^2}{9} - \dfrac{0^2}{25} \leq 1$ Substitute.

$\dfrac{25}{9} \leq 1$ Simplify.

False; so region 1 is not in the solution set.

Region 2: We choose $(0, 0)$.

$\dfrac{0^2}{9} - \dfrac{0^2}{25} \leq 1$ Substitute.

$0 \leq 1$ Simplify.

True; so region 2 is in the solution set.

Region 3: We choose $(5, 0)$.

$$\frac{(5)^2}{9} - \frac{0^2}{25} \le 1 \quad \text{Substitute.}$$

$$\frac{25}{9} \le 1 \quad \text{Simplify.}$$

False; so region 3 is not in the solution set.

Because region 2 is the only region containing ordered pairs that solve the inequality, we shade only that region. The solution set contains all ordered pairs on the hyperbola along with all ordered pairs in region 2.

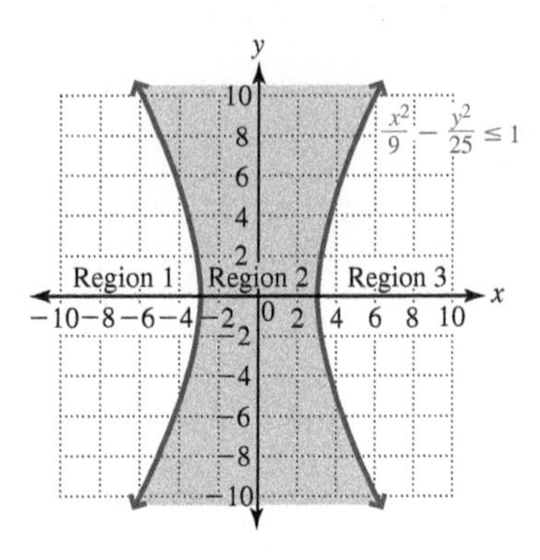

Your Turn 1 Graph the inequality.

a. $y \ge (x + 1)^2 - 3$ b. $\frac{x^2}{25} + \frac{y^2}{9} < 1$

Objective 2 Graph the solution set of a system of nonlinear inequalities.

In Section 4.5, we solved systems of linear inequalities by graphing each inequality on the same grid. The solution set of the system is the intersection of the solution sets of the individual inequalities. We use a similar procedure for systems of nonlinear inequalities.

Example 2 Graph the solution set of the system of inequalities.

a. $\begin{cases} y \ge x^2 - 4 \\ 2x - y < 2 \end{cases}$

Solution: We begin by graphing $y \ge x^2 - 4$ and $2x - y < 2$.

Note The boundary graph, $y = x^2 - 4$, is a parabola opening up with vertex at $(0, -4)$. The test point $(0, 0)$ gives the true statement $0 \ge -4$, so we shade the region containing $(0, 0)$.

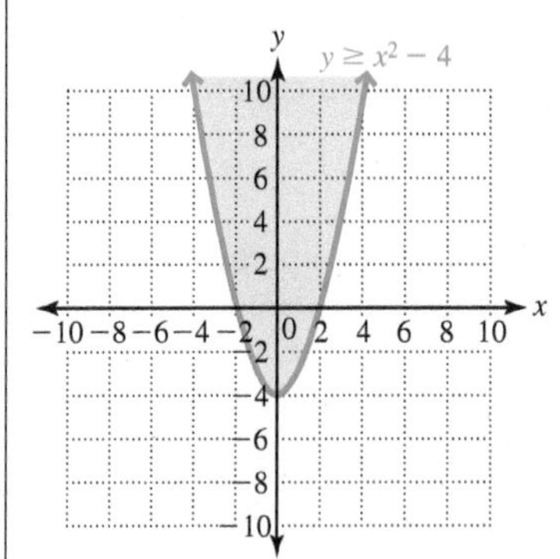

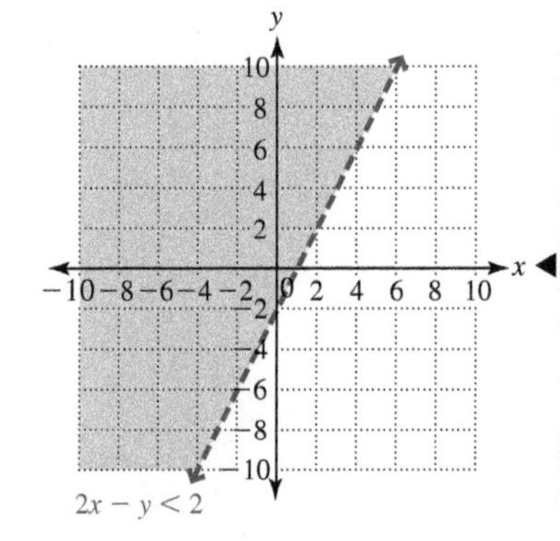

Note The boundary graph, $2x - y = 2$, is a dashed line. The test point $(0, 0)$ gives a true statement $0 < 2$, so we shade the region containing $(0, 0)$.

If we place both graphs on the same grid, their intersection (purple shading) is the solution region for the system. In addition to ordered pairs in the purple shaded region, the solution set also contains all ordered pairs on the portions of the parabola that touch the purple shaded region.

Note Remember, ordered pairs on dashed lines or curves are not in the solution set for a system of inequalities.

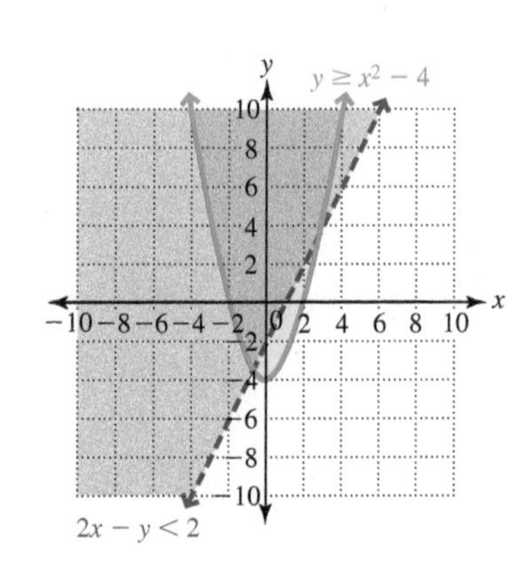

Answers to Your Turn 1

a.

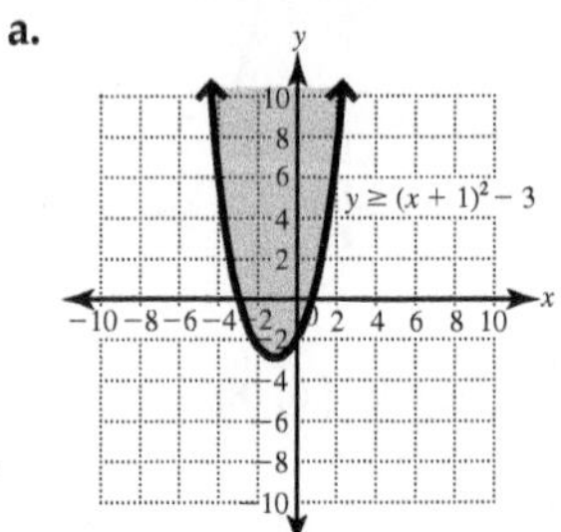

b.

b. $\begin{cases} x^2 + y^2 \le 49 \\ \dfrac{x^2}{16} - \dfrac{y^2}{9} < 1 \\ y \ge 2x + 2 \end{cases}$

Solution: Graph each inequality on the same coordinate system.

The graph of $x^2 + y^2 \le 49$ is a circle and its interior.

The graph of $\dfrac{x^2}{16} - \dfrac{y^2}{9} < 1$ is the region between the branches of the hyperbola with the curve dashed. The graph of $y \ge 2x + 2$ is a solid line and the region above the line. The solution set for the system contains all ordered pairs in the region where the three graphs overlap (purple shaded region) together with all ordered pairs on the portion of the circle and the line that touches the purple shaded region.

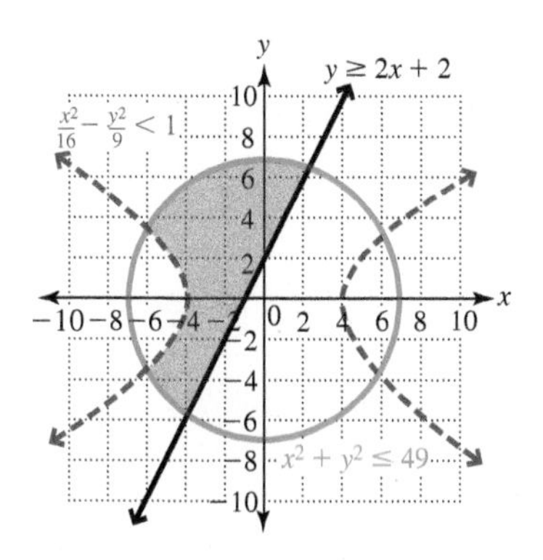

Answers to Your Turn 2

a.

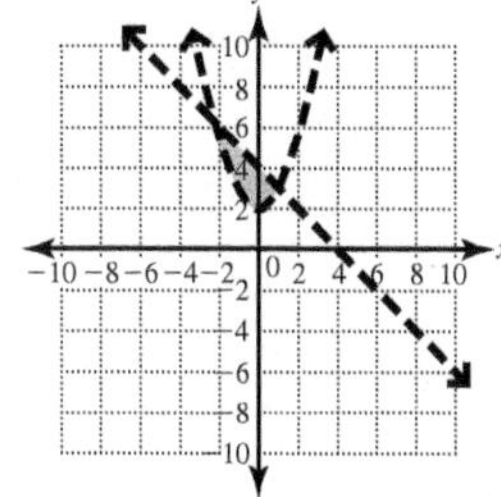

b.

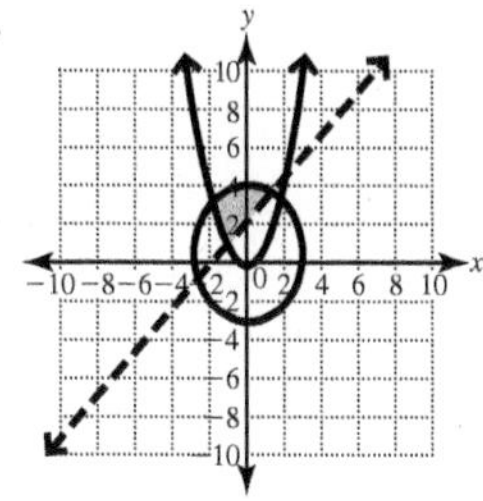

Your Turn 2 Graph the solution set of the system of inequalities.

a. $\begin{cases} y > x^2 + 2 \\ x + y < 4 \end{cases}$

b. $\begin{cases} \dfrac{x^2}{9} + \dfrac{y^2}{16} \le 1 \\ y \ge x^2 \\ -x + y > 2 \end{cases}$

11.4 Exercises

For Extra Help MyMathLab®

Objective 1

Prep Exercise 1 What is the boundary curve for the inequality $9x^2 + 4y^2 > 36$? Is the graph drawn solid or dashed?

Prep Exercise 2 What is the boundary curve for the inequality $y \ge x^2 + 3$? Is the curve drawn solid or dashed?

Prep Exercise 3 Describe the procedure used to graph $\dfrac{x^2}{9} - \dfrac{y^2}{16} > 1$.

For Exercises 1–20, graph the inequality. See Example 1.

1. $x^2 + y^2 \le 9$

2. $x^2 + y^2 \ge 4$

3. $y > x^2$

4. $y < -x^2$

5. $y > 2(x + 2)^2 - 3$

6. $y < (x - 1)^2 + 2$

7. $\frac{x^2}{16} + \frac{y^2}{9} \leq 1$

8. $\frac{x^2}{4} + \frac{y^2}{9} \geq 1$

9. $\frac{x^2}{25} - \frac{y^2}{4} > 1$

10. $\frac{x^2}{36} - \frac{y^2}{16} < 1$

11. $x^2 + y^2 < 16$

12. $x^2 + y^2 > 25$

13. $y \geq -2(x - 3)^2 + 3$

14. $y < -(x + 3)^2 - 2$

15. $\frac{x^2}{16} + \frac{y^2}{4} \leq 1$

16. $\frac{x^2}{49} + \frac{y^2}{25} > 1$

17. $\frac{y^2}{25} - \frac{x^2}{4} \leq 1$

18. $\frac{y^2}{36} - \frac{x^2}{25} > 1$

19. $y < x^2 + 4x - 5$

20. $y > x^2 - 6x - 6$

Objective 2

Prep Exercise 4 For the graph of the two inequalities in a system of inequalities, explain how to determine the solution region.

Prep Exercise 5 Why does the system $\begin{cases} x^2 + y^2 < 1 \\ x^2 + y^2 > 4 \end{cases}$ have no solution?

For Exercises 21–46, graph the solution set of each system of inequalities. See Example 2.

21. $\begin{cases} y < -x^2 \\ 2x - y < 4 \end{cases}$

22. $\begin{cases} y \geq x^2 \\ x + y \leq 3 \end{cases}$

23. $\begin{cases} 3x + 2y \geq -6 \\ x^2 + y^2 \leq 25 \end{cases}$

24. $\begin{cases} x - 2y < -4 \\ x^2 + y^2 < 16 \end{cases}$

25. $\begin{cases} x^2 + y^2 > 4 \\ x^2 + y^2 > 9 \end{cases}$

26. $\begin{cases} x^2 + y^2 \geq 9 \\ x^2 + y^2 \leq 25 \end{cases}$

27. $\begin{cases} y > x^2 + 1 \\ 2x + y < 3 \end{cases}$

28. $\begin{cases} 3x - y \leq 2 \\ y \leq -x^2 + 2 \end{cases}$

29. $\begin{cases} y < -x^2 + 3 \\ y > x^2 - 2 \end{cases}$

30. $\begin{cases} y > -x^2 + 2 \\ y < x^2 + 5 \end{cases}$

31. $\begin{cases} \dfrac{x^2}{25} + \dfrac{y^2}{9} \leq 1 \\ x^2 + y^2 \geq 4 \end{cases}$

32. $\begin{cases} \dfrac{x^2}{9} + \dfrac{y^2}{4} \leq 1 \\ x^2 + y^2 \leq 4 \end{cases}$

33. $\begin{cases} \dfrac{x^2}{25} + \dfrac{y^2}{9} < 1 \\ y > x^2 + 1 \end{cases}$

34. $\begin{cases} \dfrac{x^2}{9} + \dfrac{y^2}{4} < 1 \\ y < -x^2 + 2 \end{cases}$

35. $\begin{cases} \dfrac{x^2}{9} - \dfrac{y^2}{4} \leq 1 \\ \dfrac{x^2}{25} + \dfrac{y^2}{9} \leq 1 \end{cases}$

36. $\begin{cases} \dfrac{x^2}{16} - \dfrac{y^2}{9} \geq 1 \\ \dfrac{x^2}{36} + \dfrac{y^2}{16} \leq 1 \end{cases}$

37. $\begin{cases} \dfrac{x^2}{4} - \dfrac{y^2}{4} > 1 \\ y > 2 \end{cases}$

38. $\begin{cases} \dfrac{x^2}{9} - \dfrac{y^2}{9} > 1 \\ x > 3 \end{cases}$

39. $\begin{cases} 3x + 2y \leq 6 \\ x - y > -3 \\ x + 6y \geq 2 \end{cases}$

40. $\begin{cases} 2x + 3y < 6 \\ x - 2y > -4 \\ x + 5y \geq -4 \end{cases}$

41. $\begin{cases} \dfrac{x^2}{16} + \dfrac{y^2}{4} \le 1 \\ x^2 + y^2 \le 9 \\ y \le x \end{cases}$

42. $\begin{cases} \dfrac{x^2}{9} + \dfrac{y^2}{16} < 1 \\ x^2 + y^2 < 9 \\ y > x + 1 \end{cases}$

43. $\begin{cases} \dfrac{x^2}{49} - \dfrac{y^2}{16} \le 1 \\ \dfrac{x^2}{64} + \dfrac{y^2}{36} \le 1 \\ 2x - y \le -3 \end{cases}$

44. $\begin{cases} \dfrac{x^2}{9} + \dfrac{y^2}{49} \le 1 \\ y \ge x^2 + 3 \\ x + y \le 6 \end{cases}$

45. $\begin{cases} y < 2x^2 + 8 \\ 2x + y > 3 \\ x - y < 4 \end{cases}$

46. $\begin{cases} y < x^2 + 2 \\ 2x + y < 4 \\ 2x - y > -5 \end{cases}$

Review Exercises

Exercises 1–5 Expressions

[1.3] **1.** Simplify: $(3 \cdot 1 + 2) + (3 \cdot 2 + 2) + (3 \cdot 3 + 2) + (3 \cdot 4 + 2)$

[1.4] **2.** Evaluate $\frac{n}{2}(a_1 + a_n)$, when $n = 12$, $a_1 = -8$, and $a_n = 60$.

[1.4] **3.** Evaluate $a_1 + (n - 1)d$, when $a_1 = -12$, $n = 25$, and $d = -3$.

[1.4] *For Exercises 4 and 5, evaluate the expression for $n = 1, 2, 3,$ and 4.*

4. $2n^2 - 3$

5. $\dfrac{(-1)^n}{3n - 2}$

Exercise 6 Equations and Inequalities

[4.3] **6.** The sum of three consecutive odd integers is 207. Find the integers.

Chapter 11 Summary and Review Exercises

Complete each incomplete definition, rule, or procedure; study the key examples; and then work the related exercises.

11.1 Parabolas and Circles

Definitions/Rules/Procedures	Key Example(s)
A **conic section** is a curve in a plane that is the result of intersecting the plane with a(n) ________—more specifically, a circle, an ellipse, a parabola, or a(n) ________.	For $x = y^2 - 4y - 5$, determine whether the graph opens left or right, find the vertex and axis of symmetry, and draw the graph.
The graph of an equation in the form $x = a(y - k)^2 + h$ is a parabola with vertex at ________. The parabola opens to the right if ________ and to the left if ________. The equation of the axis of symmetry is ________.	**Solution:** Because $a = 1$ and $1 > 0$, the graph opens to the right. To find the vertex and axis of symmetry, complete the square.

$$x = y^2 - 4y - 5$$

$$x + 5 = y^2 - 4y \quad \text{Add 5 to both sides.}$$

$$x + 5 + 4 = y^2 - 4y + 4 \quad \text{Add 4 to both sides.}$$

$$x + 9 = (y - 2)^2 \quad \text{Simplify.}$$

$$x = (y - 2)^2 - 9 \quad \text{Subtract 9 from both sides.}$$

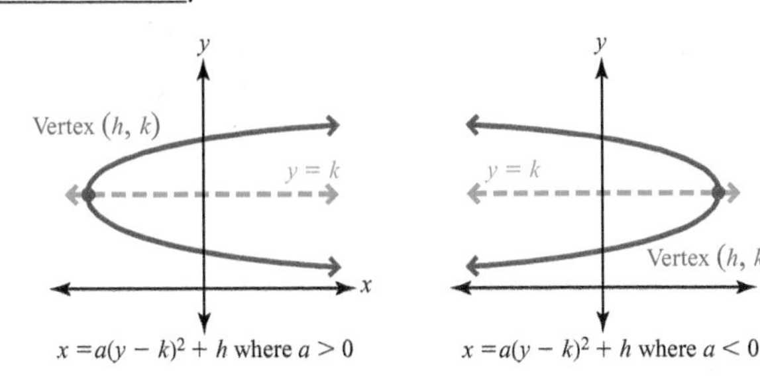

The vertex is $(-9, 2)$, and the equation of the axis of symmetry is $y = 2$.

If necessary, complete the square to put the equation in the proper form.

To graph, choose values of y near the axis of symmetry and plot the graph.

x	y
−5	0
−8	1
−8	3
−5	4

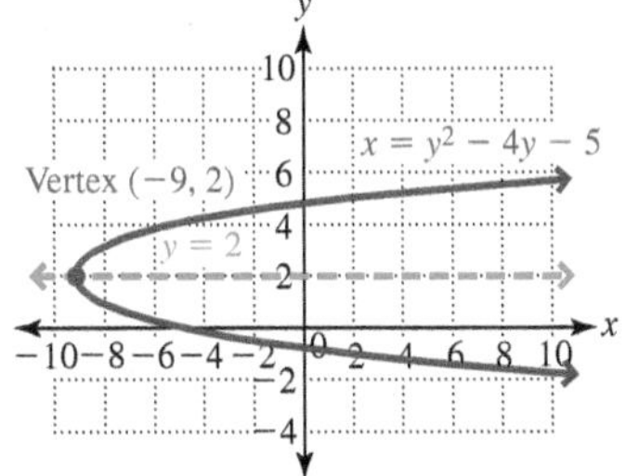

Exercises 1–9 Equations and Inequalities

[11.1] *For Exercises 1–4, find the direction that the parabola opens, the coordinates of the vertex, and the equation of the axis of symmetry. Draw the graph.*

1. $y = 2(x - 3)^2 - 5$

2. $y = -2(x + 2)^2 + 3$

3. $x = -(y - 2)^2 + 4$

4. $x = 2(y + 3)^2 - 2$

[11.1] *For Exercises 5–8, find the direction that the parabola opens, the coordinates of the vertex, and the equation of the axis of symmetry. Draw the graph.*

5. $x = y^2 + 6y + 8$ **6.** $x = 2y^2 - 8y - 6$ **7.** $x = -y^2 - 2y + 3$ **8.** $x = -3y^2 - 12y - 9$

9. If a heavy object is thrown vertically upward with an initial velocity of 32 feet per second from the top of a building 128 feet high, its height above the ground after t seconds is given by $h = -16t^2 + 32t + 128$, where h is the height in feet and t is the time in seconds.

a. What is the maximum height the object will reach?

b. How many seconds will the object take to reach its maximum height?

c. How many seconds will the object take to strike the ground?

Definitions/Rules/Procedures	Key Example(s)
The **distance**, d, between two points with coordinates (x_1, y_1) and (x_2, y_2) can be found using the formula $d =$ ______________________.	Find the distance and midpoint between $(-4, 3)$ and $(2, -1)$. **Solution:** Let $(x_1, y_1) = (-4, 3)$ and $(x_2, y_2) = (2, -1)$. $d = \sqrt{(x_2 - x_1)^2 + (y_2 - y_1)^2}$ $d = \sqrt{(2 - (-4))^2 + (-1 - 3)^2}$ Substitute. $d = \sqrt{52}$ Simplify. $d = \sqrt{4 \cdot 13}$ $d = 2\sqrt{13}$
If the coordinates of the endpoints of a line segment are (x_1, y_1) and (x_2, y_2), the coordinates of the midpoint are ______________________.	Midpoint $= \left(\frac{x_1 + x_2}{2}, \frac{y_1 + y_2}{2}\right)$ Midpoint $= \left(\frac{-4 + 2}{2}, \frac{3 + (-1)}{2}\right)$ Substitute. Midpoint $= \left(\frac{-2}{2}, \frac{2}{2}\right)$ Simplify. Midpoint $= (-1, 1)$

Exercises 10 and 11 Expressions

[11.1] *For Exercises 10 and 11, find the distance and midpoint between the two points.*

10. $(-1, -2)$ and $(-5, 1)$ **11.** $(2, -5)$ and $(-2, 3)$

Definitions/Rules/Procedures	Key Example(s)
A **circle** is a set of points in a plane that are ______ from a central point. The central point is the ______.	Find the center and radius of the circle whose equation is $(x-2)^2+(y+4)^2=25$ and draw the graph.
The radius of a circle is the distance from the ______ to ______.	**Solution:** Rewrite the equation as $(x-2)^2+(y-(-4))^2=5^2$. Because $h=2$ and $k=-4$, the center is $(2,-4)$ and $r=5$.
The **equation of a circle** with center (h,k) and radius r is ______.	

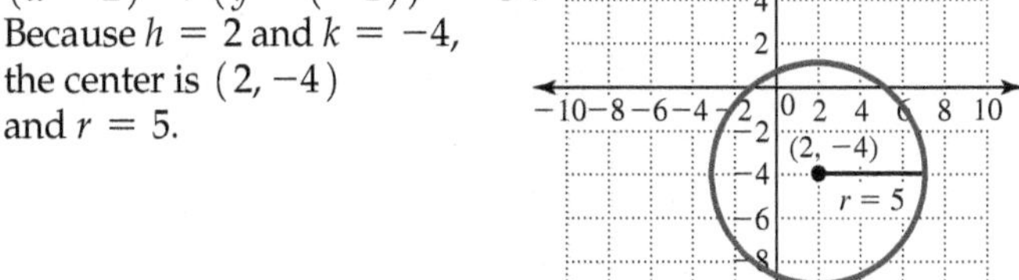

Write the equation of the circle whose center is $(5,-6)$ and whose radius is 8.

Solution: Substitute for h, k, and r in

$$(x-h)^2+(y-k)^2=r^2$$
$$(x-5)^2+(y-(-6))^2=8^2$$
$$(x-5)^2+(y+6)^2=64 \quad \text{Simplify.}$$

To graph circles of the form $x^2+y^2+dx+ey+f=0$, ______ in x and y to write the equation in the form $(x-h)^2+(y-k)^2=r^2$.

Find the center and radius of the circle whose equation is $x^2+y^2+8x-2y+8=0$ and draw the graph.

Solution: Complete the square in x and y.

$$x^2+8x+y^2-2y=-8 \quad \text{Group } x \text{ and } y \text{ terms; subtract 8.}$$
$$x^2+8x+16+y^2-2y+1=-8+16+1 \quad \text{Complete the square.}$$
$$(x+4)^2+(y-1)^2=9 \quad \text{Factor and simplify.}$$

The center is $(-4,1)$, and the radius is 3.

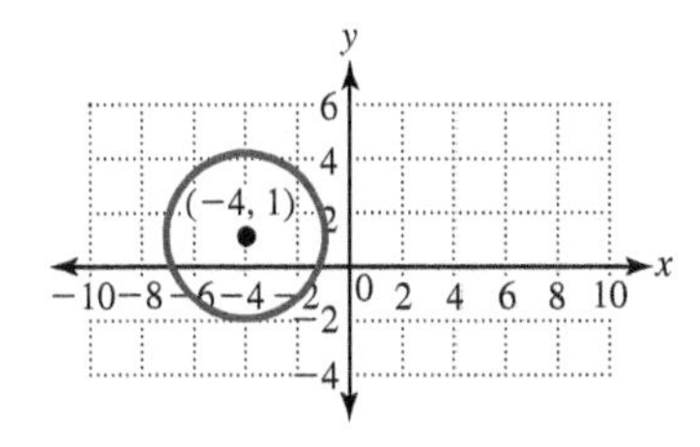

Exercises 12–20  Equations and Inequalities

[11.1] **12.** The center of a circle is at $(6,-4)$, and the circle passes through $(-2,2)$. What is the radius of the circle?

[11.1] *For Exercises 13–16, find the center and radius and draw the graph.*

13. $(x-3)^2+(y+2)^2=25$

14. $(x+5)^2+(y-1)^2=4$

15. $x^2 + y^2 - 4x + 8y + 11 = 0$

16. $x^2 + y^2 + 10x + 2y + 22 = 0$

[11.1] *For Exercises 17 and 18, the center and radius of a circle are given. Write the equation of each circle in standard form.*

17. Center: $(6, -8), r = 9$

18. Center: $(-3, -5), r = 10$

19. To lay out the border of a circular flower bed, sticks are tied to each end of a rope that is 20 feet long. One person holds one of the sticks stationary while another person uses the other stick to trace out a circular path while keeping the rope taut. If the center of the circle is at the stationary stick, what is the equation of the circle?

20. The center of a circle is at $(-6, 8)$, and the circle passes through $(3, -4)$. What is the equation of the circle?

11.2 Ellipses and Hyperbolas

Definitions/Rules/Procedures	Key Example(s)
An **ellipse** is the set of all points the ____________ of whose distances from ____________ is constant. The equation of an ellipse with center $(0, 0)$, x-intercepts $(a, 0)$ and $(-a, 0)$, and y-intercepts $(0, b)$ and $(0, -b)$ is ____________.	Graph $\frac{x^2}{49} + \frac{y^2}{36} = 1$. **Solution:** The equation can be rewritten as $\frac{x^2}{7^2} + \frac{y^2}{6^2} = 1$, so $a = 7$ and $b = 6$. Consequently, the x-intercepts are $(7, 0)$ and $(-7, 0)$ and the y-intercepts are $(0, 6)$ and $(0, -6)$. 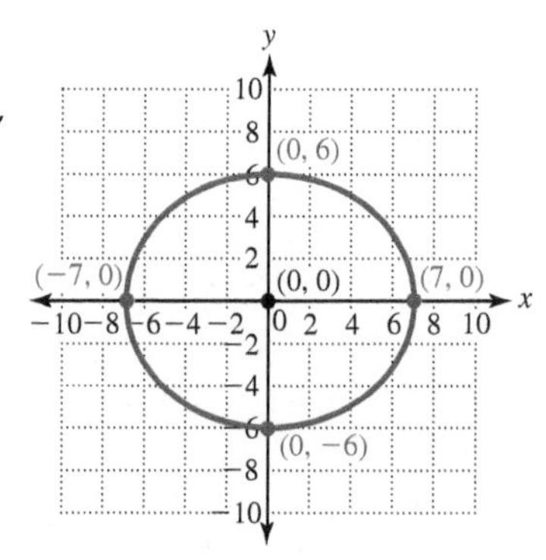
The equation of an ellipse with center (h, k) is ____________. The ellipse passes through two points that are ____________ units to the left and right of the center and two points that are ____________ units above and below the center. 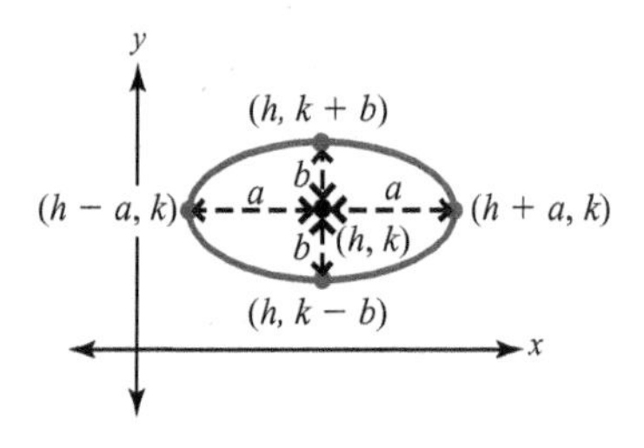 	Graph $\frac{(x+2)^2}{9} + \frac{(y-1)^2}{25} = 1$. **Solution:** The equation can be rewritten as $\frac{(x-(-2))^2}{3^2} + \frac{(y-1)^2}{5^2} = 1$; so $h = -2, k = 1, a = 3$, and $b = 5$. The center is $(-2, 1)$. To find other points on the ellipse, go 3 units right and left of the center and 5 units up and down from the center. 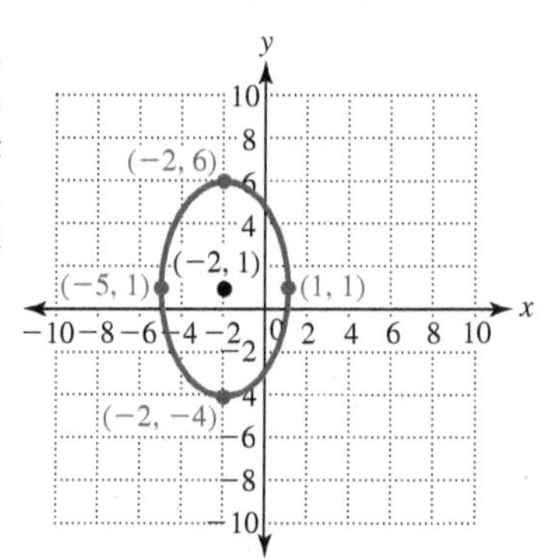

Exercises 21–28 Equations and Inequalities

[11.2] *For Exercises 21–26, graph each ellipse. Label the center and the points directly above, below, to the left of and to the right of the center.*

21. $\frac{x^2}{49} + \frac{y^2}{25} = 1$

22. $\frac{x^2}{9} + \frac{y^2}{25} = 1$

23. $4x^2 + 9y^2 = 36$

24. $25x^2 + 9y^2 = 225$

25. $\frac{(x+2)^2}{16} + \frac{(y-3)^2}{4} = 1$

26. $\frac{(x-1)^2}{9} + \frac{(y+4)^2}{25} = 1$

27. A bridge over a canal is in the shape of half an ellipse. If the highest point of the bridge is 15 feet above the water and the base of the bridge is 50 feet across, what is the equation of the ellipse, half of which forms the bridge?

28. The cam of a compound bow is elliptical and is 4 inches long and 3.25 inches wide, as shown in the figure. What is the equation of the ellipse forming the shape of the cam?

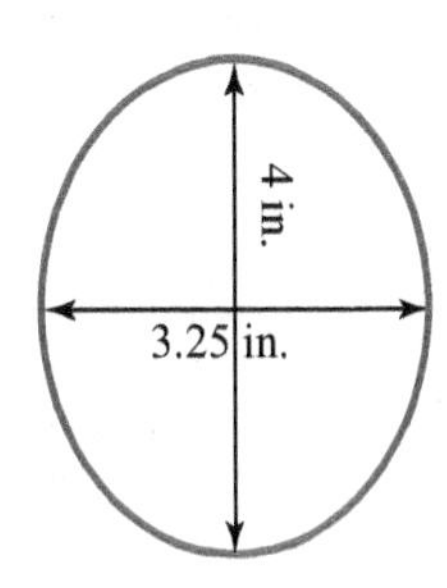

Definitions/Rules/Procedures	Key Example(s)
A **hyperbola** is the set of all points the __________ of whose distances from __________ remains constant.	
The equation of a hyperbola with center $(0, 0)$, x-intercepts $(a, 0)$ and $(-a, 0)$, and no y-intercepts is __________.	

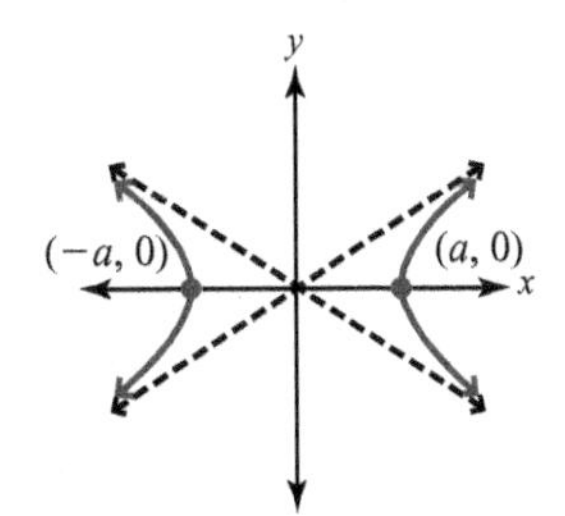

Definitions/Rules/Procedures	Key Example(s)
The equation of a hyperbola with center $(0,0)$, y-intercepts $(0,b)$ and $(0,-b)$, and no x-intercepts is ______________. 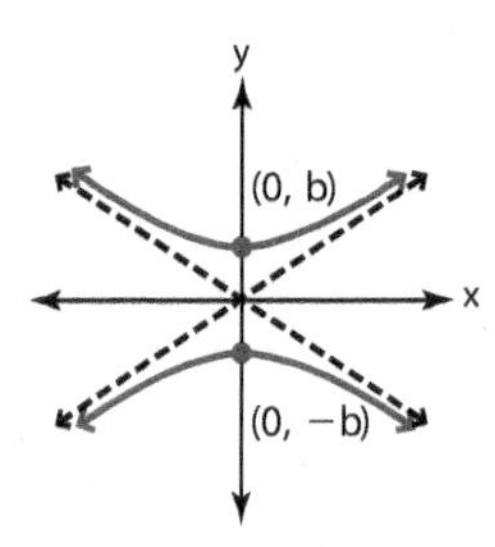	
To graph a hyperbola 1. Find the intercepts. If the x^2-term is positive, the x-intercepts are ______________ and ______________ and there are no ______________-intercepts. If the y^2-term is positive, the y-intercepts are ______________ and ______________ and there are no ______________-intercepts. 2. Draw the fundamental rectangle. The vertices are ______________________________. 3. Draw the asymptotes, which are the extended diagonals of the ______________________________. 4. Draw the graph so that each branch passes through a(n) ______________ and approaches the ______________ the farther the branches are from the origin.	Graph $\frac{x^2}{36} - \frac{y^2}{16} = 1$. **Solution:** The equation can be rewritten as $\frac{x^2}{6^2} - \frac{y^2}{4^2} = 1$, so $a = 6$ and $b = 4$. Because the x^2-term is positive, the x-intercepts are $(6,0)$ and $(-6,0)$ and there are no y-intercepts. The fundamental rectangle has vertices $(6,4)$, $(-6,4)$, $(6,-4)$, and $(-6,-4)$. 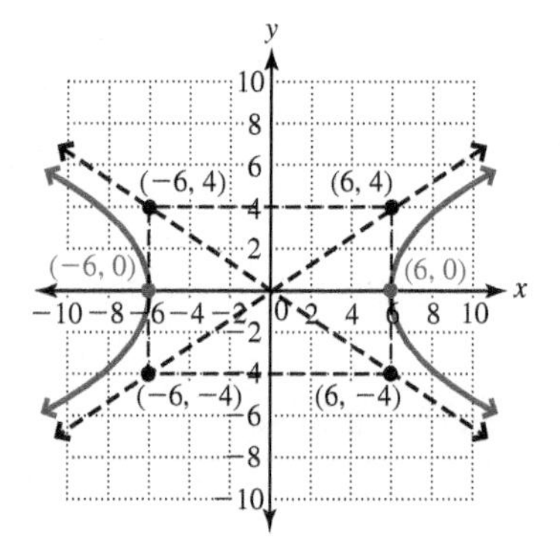

Exercises 29–32 Equations and Inequalities

[11.2] *For Exercises 29–32, graph each hyperbola. Also show the fundamental rectangle with its corner points labeled, the asymptotes, and the intercepts.*

29. $\frac{x^2}{25} - \frac{y^2}{16} = 1$ **30.** $\frac{x^2}{36} - \frac{y^2}{9} = 1$ **31.** $y^2 - 9x^2 = 36$ **32.** $25y^2 - 9x^2 = 225$

11.3 Nonlinear Systems of Equations

Definitions/Rules/Procedures	Key Example(s)
A **nonlinear system of equations** is a system of equations that contains at least one ______ equation. If a nonlinear system has a linear equation, the ______ method is usually preferred. Solve the linear equation for one of its variables and ______ for that variable into the nonlinear equation.	Solve the system $\begin{cases} y = (x-3)^2 + 2 \\ 2x - y = -4 \end{cases}$ by substitution. (**Note:** This is a parabola and a line.) **Solution:** $y = 2x + 4$ — Solve $2x - y = -4$ for y. $2x + 4 = (x-3)^2 + 2$ — Substitute $2x + 4$ for y. $2x + 4 = x^2 - 6x + 11$ — Simplify $0 = x^2 - 8x + 7$ — Write in form $ax^2 + bx + c = 0$. $0 = (x-7)(x-1)$ — Factor. $x - 7 = 0$ or $x - 1 = 0$ — Use the zero-factor theorem. $x = 7$ or $x = 1$ — Solve each equation. To find y, substitute 7 and 1 into $y = 2x + 4$. $y = 2(7) + 4$ $y = 2(1) + 4$ $y = 18$ $y = 6$ Solutions: $(7, 18), (1, 6)$
If neither equation is linear, the ______ method is often preferred.	Solve the system $\begin{cases} 4x^2 + 5y^2 = 36 & \text{(Equation 1)} \\ x^2 + y^2 = 8 & \text{(Equation 2)} \end{cases}$ by elimination. (**Note:** This is an ellipse and a circle.) **Solution:** To eliminate x^2, multiply equation 2 by -4 and add the equations. $4x^2 + 5y^2 = 36$ $4x^2 + 5y^2 = 36$ $x^2 + y^2 = 8 \longrightarrow -4x^2 - 4y^2 = -32$ (Multiply by -4.) $y^2 = 4$ $y = \pm 2$ Substitute 2 and -2 for y into either equation to find x. Let's use $x^2 + y^2 = 8$. $x^2 + 2^2 = 8$ $x^2 + (-2)^2 = 8$ $x^2 + 4 = 8$ $x^2 + 4 = 8$ $x^2 = 4$ $x^2 = 4$ $x = \pm 2$ $x = \pm 2$ Solutions: $(2, 2), (2, -2), (-2, 2)$, and $(-2, -2)$

Exercises 33–44 Equations and Inequalities

[11.3] *For Exercises 33–40, solve.*

33. $\begin{cases} y = 2x^2 - 3 \\ 2x - y = -1 \end{cases}$

34. $\begin{cases} x^2 + y^2 = 17 \\ x - y = 3 \end{cases}$

35. $\begin{cases} 25x^2 + 3y^2 = 100 \\ 2x - y = -3 \end{cases}$

36. $\begin{cases} x^2 + y^2 = 64 \\ x^2 - y^2 = 64 \end{cases}$

37. $\begin{cases} x^2 + y^2 = 25 \\ 25y^2 - 16x^2 = 256 \end{cases}$

38. $\begin{cases} x^2 + 5y^2 = 36 \\ 4x^2 - 7y^2 = 36 \end{cases}$

39. $\begin{cases} y = x^2 - 2 \\ 4x^2 + 5y^2 = 36 \end{cases}$

40. $\begin{cases} y = x^2 - 1 \\ 4y^2 - 5x^2 = 16 \end{cases}$

41. The sum of the squares of two integers is 89, and the difference of their squares is 39. Find the integers.

42. A rectangular rug has a perimeter of 36 feet and an area of 80 square feet. Find the dimensions of the rug.

[11.3] *For Exercises 43 and 44, use a graphing calculator to verify the results of the exercise given.*

43. Exercise 33

44. Exercise 39

11.4 Nonlinear Inequalities and Systems of Inequalities

Definitions/Rules/Procedures

Graphing **nonlinear inequalities**

1. Graph the related __________ (the boundary curve).

   If the inequality symbol is $\le$ or $\ge$, draw the graph as a(n) __________ curve. If the inequality symbol is $<$ or $>$, draw a(n) __________ curve.

2. The graph divides the coordinate plane into at least two regions. Test a(n) __________ from each region by substituting it into the inequality. If the ordered pair satisfies the inequality, shade ____________________.

Key Example(s)

Graph $y \ge (x - 2)^2 - 1$.

Solution: The related equation is a parabola with vertex $(2, -1)$ that opens upward. Because the inequality is $\ge$, all ordered pairs on the parabola are in the solution set; so we draw it with a solid curve.

We choose $(0, 0)$ as a test point.

$0 \ge (0 - 2)^2 - 1$

$0 \ge 3$, which is false; so $(0, 0)$ is not in the solution set.

Choose $(2, 0)$ as a test point.

$0 \ge (2 - 2)^2 - 1$

$0 \ge -1$, which is true; so $(2, 0)$ is in the solution set. Shade the region that contains $(2, 0)$.

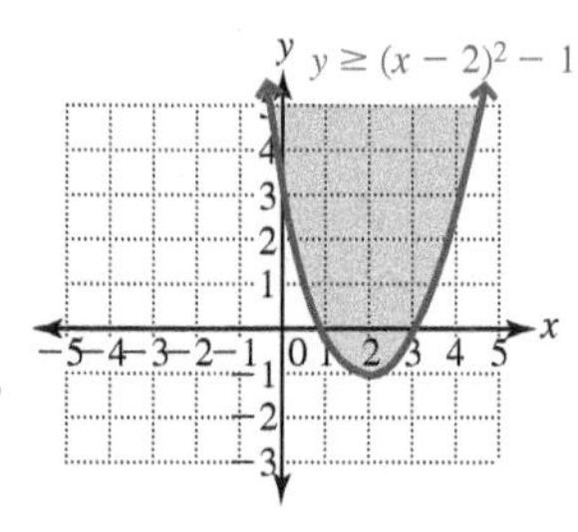

Exercises 45–48 Equations and Inequalities

[11.4] *For Exercises 45–48, graph the inequality.*

45. $x^2 + y^2 \le 64$

46. $y < 2(x - 3)^2 + 4$

47. $\dfrac{y^2}{25} + \dfrac{x^2}{49} > 1$

48. $\dfrac{x^2}{25} - \dfrac{y^2}{36} < 1$

Definitions/Rules/Procedures	Key Example(s)
To graph a **system of nonlinear inequalities**, graph the solution set of each inequality on the same grid. The solution set is the intersection of ______________________.	Graph the solution set of $\begin{cases} \dfrac{x^2}{4} + \dfrac{y^2}{16} \le 1 \\ x^2 + y^2 \ge 9 \end{cases}$. **Solution:** The graph of $\dfrac{x^2}{4} + \dfrac{y^2}{16} \le 1$ is an ellipse and all points inside the ellipse. The graph of $x^2 + y^2 \ge 9$ is a circle and all points outside the circle. So the solution set of the system is the set of all points inside the ellipse and outside the circle.

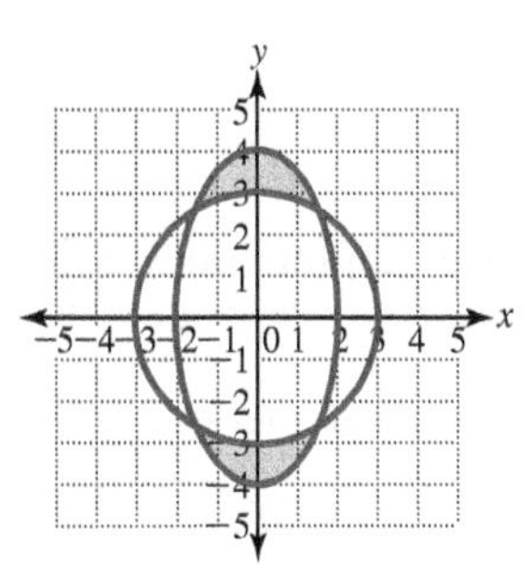

Exercises 49–54 Equations and Inequalities

[11.4] *For Exercises 49–54, graph the solution set of the system of inequalities.*

49. $\begin{cases} y \ge x^2 - 3 \\ 2x + y < 2 \end{cases}$

50. $\begin{cases} x + 2y < 4 \\ x^2 + y^2 \le 25 \end{cases}$

51. $\begin{cases} y > x^2 - 2 \\ y < -x^2 + 1 \end{cases}$

52. $\begin{cases} \dfrac{x^2}{9} + \dfrac{y^2}{25} \le 1 \\ x^2 + y^2 \ge 9 \end{cases}$

53. $\begin{cases} \dfrac{y^2}{4} - \dfrac{x^2}{9} \le 1 \\ \dfrac{x^2}{9} + \dfrac{y^2}{25} \le 1 \end{cases}$

54. $\begin{cases} \dfrac{y^2}{4} + \dfrac{x^2}{16} \le 1 \\ x^2 + y^2 \le 9 \\ y \le x + 1 \end{cases}$

Chapter 11 Practice Test

For Extra Help

Step-by-step test solutions are found on the Chapter Test Prep Videos available in MyMathLab® or on YouTube.

For Exercises 1–3, find the direction the parabola opens, the coordinates of the vertex, and the equation of the axis of symmetry. Draw the graph.

1. $y = 2(x + 1)^2 - 4$

2. $x = -2(y - 3)^2 + 1$

3. $x = y^2 + 4y - 3$

4. Find the distance and midpoint between the points $(2, -3)$ and $(6, -5)$.

For Exercises 5 and 6, find the center and radius and draw the graph.

5. $(x + 4)^2 + (y - 3)^2 = 36$

6. $x^2 + y^2 + 4x - 10y + 20 = 0$

7. Write the equation of the circle with center $(2, -4)$ that passes through the point $(-4, 4)$.

For Exercises 8–11, graph the equation and label relevant points. If the graph is a hyperbola, show the fundamental rectangle with its corner points labeled, the asymptotes, and the intercepts.

8. $16x^2 + 36y^2 = 576$

9. $\dfrac{(x + 2)^2}{4} + \dfrac{(y - 1)^2}{25} = 1$

10. $\dfrac{x^2}{49} - \dfrac{y^2}{25} = 1$

11. $\dfrac{y^2}{9} - \dfrac{x^2}{16} = 1$

For Exercises 12–15, solve.

12. $\begin{cases} y = (x + 1)^2 + 2 \\ 2x + y = 8 \end{cases}$

13. $\begin{cases} 3x - y = 4 \\ x^2 + y^2 = 34 \end{cases}$

14. $\begin{cases} x^2 + y^2 = 13 \\ 3x^2 + 4y^2 = 48 \end{cases}$

15. $\begin{cases} x^2 - 2y^2 = 1 \\ 4x^2 + 7y^2 = 64 \end{cases}$

For Exercises 16 and 17, graph.

16. $y \leq -2(x + 3)^2 + 2$

17. $\dfrac{x^2}{4} + \dfrac{y^2}{9} > 1$

18. Graph the solution set: $\begin{cases} y \geq x^2 - 4 \\ \dfrac{x^2}{9} + \dfrac{y^2}{16} \leq 1 \end{cases}$

19. An arch is in the shape of a parabola as shown in the figure. If we place the origin, O, as indicated, find the height of the arch and the distance across the base if the equation is $y = -\dfrac{1}{2}x^2 + 18$.

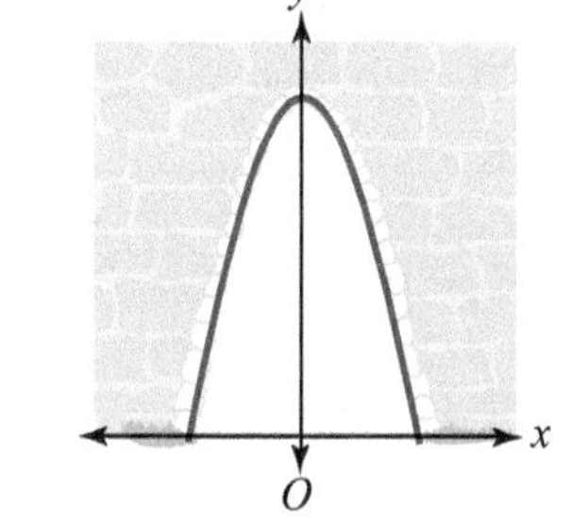

20. The path of a person's foot using an elliptical trainer on a particular setting is an ellipse. If the length of the stride is 19 inches and the height of the stride is 2 inches, what is the equation of the path of the foot?

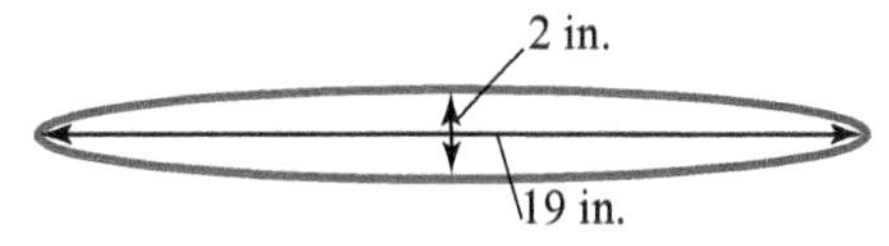

Chapters 1–11 Cumulative Review Exercises

For Exercises 1–3, answer true or false.

[5.1] **1.** A positive base raised to a negative exponent simplifies to a negative number.

[7.4] **2.** If $\frac{a}{b} = \frac{c}{d}$, then $ad = bc$ where $b \neq 0$ and $d \neq 0$.

[10.3] **3.** Logarithms are exponents.

For Exercises 4–6, fill in the blanks.

[4.4] **4.** If the row echelon form of a matrix for a system of equations is $\left[\begin{array}{cc|c} 1 & -3 & 7 \\ 0 & 1 & -5 \end{array}\right]$, then $x =$ ________________ and $y =$ ________________.

[8.7] **5.** $\sqrt{-1} =$ ______.

[11.1] **6.** The formula for finding the distance between two points (x_1, y_1) and (x_2, y_2) is ________________________.

Exercises 7–11 Expressions

For Exercises 7–9, simplify.

[1.3] **7.** $6 - (-3) + 2^{-4}$

[5.1] **8.** $(5m^3n^{-2})^{-3}$

[8.7] **9.** $\sqrt{-20}$

For Exercises 10 and 11, factor completely.

[6.1] **10.** $ax + bx + ay + by$

[6.3] **11.** $k^4 - 81$

Exercises 12–30 Equations and Inequalities

For Exercises 12–17, solve.

[2.5] **12.** $|2x + 1| = |x - 3|$

[8.6] **13.** $\sqrt{x - 3} = 7$

[9.2] **14.** $4x^2 - 2x + 1 = 0$

[9.3] **15.** $9 + 24x^{-1} + 16x^{-2} = 0$

[10.2] **16.** $9^x = 27$

[10.3] **17.** $\log_3(x + 1) - \log_3 x = 2$

[9.5] **18.** Solve the inequality $5m^2 - 3m < 0$; then:

a. Graph the solution set.

b. Write the solution set in set-builder notation.

c. Write the solution set in interval notation.

[5.2, 5.3, 5.4, 10.1] 19. Given $f(x) = 2x + 1$ and $g(x) = x^2 - 1$, find the following.

a. $(f + g)(x)$ b. $(f - g)(x)$ c. $(f \cdot g)(x)$

d. $(f/g)(x)$ e. $(f \circ g)(x)$ f. $(g \circ f)(x)$

[10.3] 20. Write $5^3 = 125$ in logarithmic form.

[10.4] 21. Write as a sum or difference of multiples of logarithms: $\log_5 x^5 y$

[11.1] 22. Write the equation of a circle in standard form with a center at $(2, 4)$ passing through the point $(7, 16)$.

For Exercises 23–25, graph.

[4.5] 23. $\begin{cases} y < -x + 2 \\ y \geq x - 4 \end{cases}$

[10.3] 24. $f(x) = \log_3 x$

[11.2] 25. $\dfrac{x^2}{4} + \dfrac{y^2}{9} = 1$

For Exercises 26–30, solve.

[2.1] 26. The volume of a cylinder can be found using the formula $V = \pi r^2 h$.

a. Solve the formula for h.

b. Suppose the volume of a cylinder is 15π cubic inches. Find its height.

[4.3] 27. During 2008, shrimp overtook tuna as the top-selling seafood in the United States. It is estimated that each person consumed a combined total of 6.3 pounds of these two seafoods and that each person ate $\frac{1}{2}$ pound more shrimp than tuna. How many pounds of each were consumed?

[4.3] 28. A total of $5000 is invested in three funds. The money market fund pays 5%, the income fund pays 6%, and the growth fund pays 3%. The total annual interest from the three accounts is $255, and the amount in the growth fund is $500 less than the amount invested in the money market account. Find the amount invested in each of the funds.

[10.5] **29.** The magnitude of an earthquake was defined in 1935 by Charles Richter as $M = \log\frac{I}{I_0}$. We let I be the intensity of the earthquake measured by the amplitude of a seismograph reading taken 100 km from the epicenter of the earthquake, and we let I_0 be the intensity of a "standard earthquake" whose amplitude is 10^{-4} cm. On January 26, 2001, an earthquake in Rann of Kutch, Gujarat, was measured with an intensity of $10^{2.9}$. What was the magnitude of this earthquake?

[11.2] **30.** A bridge over a canal is in the shape of half an ellipse. If the highest point of the bridge is 20 feet above the water and the base of the bridge is 150 feet across, what is the equation of the ellipse, half of which forms the bridge?

Appendix A

Arithmetic Sequences and Series

Objectives

1 Find the terms of a sequence when given the general term.

2 Define and write arithmetic sequences, find their common difference, and find a particular term.

3 Define and write series, find partial sums, and use summation notation.

4 Write arithmetic series and find their sums.

The word **sequence** is used in mathematics much as it is in everyday life. For example, the classes you attend on a given day occur in a particular order, or sequence. As such, a sequence is an ordered list. Suppose we have a bacteria colony that has an initial population of 10,000 and increases at a rate of 10% each day.

On the second day, we have
$10{,}000 + 0.10(10{,}000) = 10{,}000 + 1000 = 11{,}000$.
On the third day, we have $11{,}000 + 0.10(11{,}000) = 11{,}000 + 1100 = 12{,}100$.
On the fourth day, we have $12{,}100 + 0.10(12{,}100) = 12{,}100 + 1210 = 13{,}310$.
On the fifth day, we have $13{,}310 + 0.10(13{,}310) = 13{,}310 + 1331 = 14{,}641$.

The number of bacteria present after each day forms a sequence that we can summarize as follows.

Days	1	2	3	4	5
Number of Bacteria	10,000	11,000	12,100	13,310	14,641

Based on this, we have the following definition.

Definition **Sequence:** A function list whose domain is $1, 2, 3, \ldots, n$.

The *domain* of our bacteria example is the numbers of the days $\{1, 2, 3, 4, 5\}$. Each number in the *range* of a sequence is called a *term.* So the terms of our bacteria sequence are the numbers of bacteria: 10,000, 11,000, 12,100, 13,310, 14,641. Sequences can be finite or infinite depending on the number of terms. Our bacteria example is a **finite sequence** because it has a finite number of terms. The sequence 2, 4, 6, 8, 10, ... is an **infinite sequence** because it has an infinite number of terms.

Definitions **Finite sequence:** A function with a domain that is the set of natural numbers from 1 to n.
Infinite sequence: A function with a domain that is the set of natural numbers.

Objective 1 Find the terms of a sequence when given the general term.

Because a sequence is a function, we could describe sequences using functional notation. Instead, we use a different notation that emphasizes the fact that the domain is a subset of the natural numbers. We think of the terms of a sequence as $a_1, a_2, a_3, \ldots, a_n$, where the subscript gives the number of the term. Thus, a_n is the nth or *general term* of the sequence and we represent a sequence by giving a formula for a_n. Consider the following sequence.

Term	a_1	a_2	a_3	a_4	a_5	a_n
Term of Sequence	1	4	9	16	25	n^2

We represent this sequence by writing $a_n = n^2$, which means $a_1 = 1^2$, $a_2 = 2^2$, and so on.

Example 1 Find the first three terms of the following sequences and the 25th term.

a. $a_n = 3n - 1$

Solution: We let $n = 1, 2, 3$, and 25 and evaluate.

$$a_1 = 3(1) - 1 = 3 - 1 = 2 \quad \text{Let } n = 1 \text{ and evaluate.}$$
$$a_2 = 3(2) - 1 = 6 - 1 = 5 \quad \text{Let } n = 2 \text{ and evaluate.}$$
$$a_3 = 3(3) - 1 = 9 - 1 = 8 \quad \text{Let } n = 3 \text{ and evaluate.}$$
$$a_{25} = 3(25) - 1 = 75 - 1 = 74 \quad \text{Let } n = 25 \text{ and evaluate.}$$

The first three terms of the sequence are 2, 5, and 8. The 25th term is 74.

b. $a_n = \dfrac{(-1)^n}{n^2 + 1}$

Solution: We let $n = 1, 2, 3$, and 25 and evaluate.

$$a_1 = \frac{(-1)^1}{1^2 + 1} = \frac{-1}{1 + 1} = -\frac{1}{2} \quad \text{Let } n = 1 \text{ and evaluate.}$$
$$a_2 = \frac{(-1)^2}{2^2 + 1} = \frac{1}{4 + 1} = \frac{1}{5} \quad \text{Let } n = 2 \text{ and evaluate.}$$
$$a_3 = \frac{(-1)^3}{3^2 + 1} = \frac{-1}{9 + 1} = -\frac{1}{10} \quad \text{Let } n = 3 \text{ and evaluate.}$$
$$a_{25} = \frac{(-1)^{25}}{25^2 + 1} = \frac{-1}{625 + 1} = -\frac{1}{626} \quad \text{Let } n = 25 \text{ and evaluate.}$$

The first three terms of the sequence are $-\frac{1}{2}, \frac{1}{5}$, and $-\frac{1}{10}$. The 25th term is $-\frac{1}{626}$.

Objective 2 Define and write arithmetic sequences, find their common difference, and find a particular term.

In the sequence $-3, 1, 5, 9, 13, \ldots$, notice that each term after the first is found by adding 4 to the previous term. This is an example of an **arithmetic sequence** or *arithmetic progression*. Any two successive terms of an arithmetic sequence differ by the same amount, which is called the **common difference** and is denoted by d. To find d, choose any term (except the first) and subtract the previous term.

Definition **Arithmetic sequence:** A sequence in which each term after the first is found by adding the same number to the previous term.

Common difference of an arithmetic sequence: The value of d is found by $d = a_n - a_{n-1}$, where a_n is any term in the sequence (except the first) and a_{n-1} is the previous term.

Example 2 Write the first four terms of the following arithmetic sequences.

a. The first term is 2, and the common difference is 5.

Solution: Begin with 2 and find each successive term by adding 5 to the previous term.

$$a_1 = 2, a_2 = 2 + 5 = 7, a_3 = 7 + 5 = 12, a_4 = 12 + 5 = 17$$

The first four terms of the sequence are 2, 7, 12, 17.

b. $a_1 = -1, d = -2$

Solution: Begin with -1 and find each successive term by adding -2 to the previous term.

$$a_1 = -1, a_2 = -1 - 2 = -3, a_3 = -3 - 2 = -5, a_4 = -5 - 2 = -7$$

The first four terms of the sequence are $-1, -3, -5, -7$.

Example 3 Find the common difference, d, for the following arithmetic sequences.

a. $-4, -1, 2, 5, 8, \ldots$

Solution: Pick any term (except the first) and subtract the term before it.

$$d = -1 - (-4) = -1 + 4 = 3$$

Or we could use $d = 5 - 2 = 3$ and so on.

b. $8, 6, 4, 2, 0, \ldots$

Solution: Pick any term (except the first) and subtract the term before it.

$$d = 6 - 8 = -2$$

Or we could use $d = 2 - 4 = -2$ and so on.

If the first term of an arithmetic sequence is a_1 and the common difference is d, then the arithmetic sequence can be written as $a_1, a_1 + d, a_1 + 2d, a_1 + 3d, a_1 + 4d, \ldots$. Note that the coefficient of d is one less than the number of the term; so we have the following rule.

Rule ***n*th Term of an Arithmetic Sequence**

The formula for finding the nth term of an arithmetic sequence is $a_n = a_1 + (n - 1)d$, where a_1 is the first term and d is the common difference.

Example 4 Find the 23rd term and an expression for the nth term of an arithmetic sequence in which $a_1 = -8$ and $d = 3$.

Solution: $a_n = a_1 + (n - 1)d$

$a_{23} = -8 + (23 - 1)(3)$ Substitute **23** for n, -8 for a_1, and **3** for d.

$a_{23} = 58$ Evaluate.

nth term: $a_n = -8 + (n - 1)3$ Substitute for a_1 and d.

$a_n = -11 + 3n$ Simplify.

If we know the first term and one other term, we can use the formula for the nth term of an arithmetic sequence to find the common difference. Consequently, we can find the sequence.

Example 5 The first term of an arithmetic sequence is 1, and the 20th term (a_{20}) is 58. Find the common difference and the first four terms of the sequence.

Solution: We use the fact that we know a_{20} and the formula for the nth term to find d.

$a_n = a_1 + (n - 1)d$

$a_{20} = 1 + (20 - 1)d$ Substitute **1** for a_1 and **20** for n.

$58 = 1 + 19d$ Substitute **58** for a_{20} and solve.

$3 = d$

The first four terms are $1,\ 1 + 3 = 4,\ 4 + 3 = 7,\ 7 + 3 = 10$.

Objective 3 Define and write series, find partial sums, and use summation notation.

When the terms of a sequence are added, the sum is called a **series**.

Definition **Series:** The sum of the terms of a sequence.

Given a sequence $a_1, a_2, a_3, \ldots, a_n$, then $a_1 + a_2 + a_3 + \cdots + a_n$ is the corresponding series. Finite series correspond to finite sequences, and infinite series correspond to infinite sequences. If an expression for the nth term is known, we can write the series in *summation notation* using the capital Greek letter *sigma* (Σ) as follows.

$$a_1 + a_2 + a_3 + \cdots + a_n = \sum_{i=1}^{n} a_i$$

The i is called the *index of summation*; 1 is the *lower limit* of i, and n is the *upper limit* of i. If the upper limit is a natural number, the series is finite, and if the upper limit is ∞, the series is infinite. To find a finite series from summation notation, replace the index of summation (usually i) with its lower limit and evaluate that term, replace i with the lower limit plus 1 and evaluate that term, and so on, until you reach the upper limit. This gives the series. To find the sum, add the terms of the series.

Example 6

a. Write the terms of $\sum_{i=1}^{5}(3i + 2)$ and find the sum of the series.

Solution: Replace i with 1, 2, 3, 4, and 5. Then find the sum.

$$\begin{aligned}\sum_{i=1}^{5}(3i + 2) &= (3\cdot 1 + 2) + (3\cdot 2 + 2) + (3\cdot 3 + 2) + (3\cdot 4 + 2) + (3\cdot 5 + 2)\\ &= 5 + 8 + 11 + 14 + 17\\ &= 55\end{aligned}$$

b. Write the first five terms of the infinite series $\sum_{i=1}^{\infty} 2i^2$.

Solution: Replace i with 1, 2, 3, 4, and 5. Then evaluate.

$$\begin{aligned}\sum_{i=1}^{\infty} 2i^2 &= 2\cdot 1^2 + 2\cdot 2^2 + 2\cdot 3^2 + 2\cdot 4^2 + 2\cdot 5^2 + \cdots\\ &= 2 + 8 + 18 + 32 + 50 + \cdots\end{aligned}$$

Warning Do not confuse the use of i in sigma notation, as in $\sum_{i=1}^{5}(3i + 2)$, with its use as the imaginary unit, as in $7 \pm 5i$. In sigma notation, i is a variable representing natural numbers, whereas in the imaginary unit, i represents $\sqrt{-1}$.

Objective 4 Write arithmetic series and find their sums.

When the terms of an arithmetic sequence are added, it is called an **arithmetic series**.

Definition **Arithmetic series:** The sum of the terms of an arithmetic sequence.

Consequently, an arithmetic series has the form $a_1 + (a_1 + d) + (a_1 + 2d) + (a_1 + 3d) + \cdots + [a_1 + (n - 1)d]$.

Adding a finite number of terms of an infinite series gives a *partial sum*. The symbol S_n is used to indicate the sum of the first n terms. For example, S_5 means add the first five terms. Let's derive a formula for S_n for an arithmetic series as follows.

$$S_n = a_1 + (a_1 + d) + (a_1 + 2d) + (a_1 + 3d) + \cdots + a_n$$

We also need to include the terms between $(a_1 + 3d)$ and a_n. So we write another version of S_n beginning with the last term, a_n, and subtracting the common difference, d, from the previous term.

$$S_n = a_n + (a_n - d) + (a_n - 2d) + (a_n - 3d) + \cdots + a_1$$

To describe the entire sum, we add our two versions of S_n.

$$\begin{array}{rl} S_n = a_1 & + (a_1 + d) + (a_1 + 2d) + (a_1 + 3d) + \cdots + a_n \\ +S_n = a_n & + (a_n - d) + (a_n - 2d) + (a_n - 3d) + \cdots + a_1 \\ \hline 2S_n = (a_1 + a_n) & + (a_1 + a_n) + (a_1 + a_n) + (a_1 + a_n) + \cdots + (a_1 + a_n) \end{array}$$

Because S_n has n terms, there are n terms of $(a_1 + a_n)$; so

$$2S_n = n(a_1 + a_n), \quad \text{or} \quad S_n = \frac{n}{2}(a_1 + a_n)$$

Rule Partial Sum, S_n, of an Arithmetic Series

The sum of the first n terms of an arithmetic series, S_n, called the nth partial sum, is given by

$$S_n = \frac{n}{2}(a_1 + a_n),$$

where n is the number of terms, a_1 is the first term, and a_n is the nth term.

Example 7 Find the sum of the first 20 terms (S_{20}) of the arithmetic series $-6 - 2 + 2 + 6 + \cdots$.

Solution: To find the S_{20}, we first need to find the 20th term. Because $d = 4$,

$$a_{20} = -6 + (20 - 1)(4) = 70$$

$$S_n = \frac{n}{2}(a_1 + a_n)$$

$$S_{20} = \frac{20}{2}(-6 + 70) \quad \text{Substitute 20 for } n, -6 \text{ for } a_1, \text{ and 70 for } a_n.$$

$$S_{20} = 10(64) = 640 \quad \text{Evaluate.}$$

Appendix A Exercises

For Extra Help MyMathLab®

Objective 1

Prep Exercise 1 What is a sequence?

Prep Exercise 2 A sequence with an unlimited number of terms is called a(n) ________ sequence.

Prep Exercise 3 What is an arithmetic sequence?

For Exercises 1–8, write the first four terms of the sequence and the indicated term. See Example 1.

1. $a_n = 2n + 1$, 20th term

2. $a_n = 3n - 4$, 18th term

3. $a_n = n^2 + 2$, 15th term

4. $a_n = n^2 - 3$, 12th term

5. $a_n = \frac{n}{n + 2}$, 22nd term

6. $a_n = \frac{2n}{n + 3}$, 10th term

7. $a_n = \frac{(-1)^n}{n^2 + 1}$, 15th term

8. $a_n = \frac{(-1)^n}{n^2 - 3}$, 26th term

Objective 2

Prep Exercise 4 Explain how to find the common difference of an arithmetic sequence.

For Exercises 9–14, find the common difference, d, for each arithmetic sequence. See Example 3.

9. 2, 7, 12, 17, . . .

10. 3, 11, 19, 27, . . .

11. 25, 22, 19, 16, . . .

12. 42, 36, 30, 24, . . .

13. −12, −5, 2, 9, . . .

14. −24, −27, −30, −33, . . .

Prep Exercise 5 What is the formula for finding the nth term of an arithmetic sequence?

For Exercises 15–20, find the indicated term and an expression for the nth term of the given arithmetic sequence. See Example 4.

15. a_{14} if $a_1 = 14$ and $d = 4$

16. a_{24} if $a_1 = 16$ and $d = 6$

17. a_{28} if $a_1 = -8$ and $d = -3$

18. a_{30} if $a_1 = -7$ and $d = -6$

19. a_{34} of $-5, -1, 3, 7, \ldots$

20. a_{21} of $8, 15, 22, 29, \ldots$

For Exercises 21–26, write the first four terms of the arithmetic sequence with the given characteristics. See Examples 2 and 5.

21. $a_1 = -6, d = 7$

22. $a_1 = -2, d = 6$

23. $a_1 = -5, d = -3$

24. $a_1 = -13, d = -5$

25. $a_1 = 7, a_{18} = 75$

26. $a_1 = 6, a_{22} = 153$

27. Find the first term of an arithmetic sequence if $a_{45} = 143$ and $d = 3$.

28. Find the first term of an arithmetic sequence if $a_{39} = 181$ and $d = 6$.

29. Find the common difference of an arithmetic sequence if the first term is −110 and the 29th term is 2.

30. Find the common difference of an arithmetic sequence if the first term is 78 and the 15th term is −76.

For Exercises 31–34, write the first four terms of the arithmetic sequence with the given d and a_n.

31. $d = 7, a_8 = 41$

32. $d = 4, a_{11} = 27$

33. $d = -3, a_7 = 9$

34. $d = -6, a_{10} = -36$

Objective 3

Prep Exercise 6 What is a series?

For Exercises 35–42, write the series and find the sum. See Example 6.

35. $\sum_{i=1}^{6} i^2$

36. $\sum_{i=1}^{3} 3i^2$

37. $\sum_{i=1}^{4} (2i - 5)$

38. $\sum_{i=1}^{5}(4i+1)$

39. $\sum_{i=1}^{3}(3i^2-4)$

40. $\sum_{i=1}^{4}(-2i^2+5)$

41. $\sum_{i=3}^{6}(4i-3)$

42. $\sum_{i=2}^{5}(4i-2)$

Objective 4

Prep Exercise 7 What does S_n represent?

Prep Exercise 8 What is the formula for finding S_n of an arithmetic series?

For Exercises 43–50, find the given S_n for the arithmetic series. See Example 7.

43. If $a_1 = 10$ and $d = 4$, find S_{25}.

44. If $a_1 = -8$ and $d = 2$, find S_{30}.

45. If $a_1 = 12$ and $d = -3$, find S_{22}.

46. If $a_1 = 18$ and $d = -2$, find S_{30}.

47. $3 + 9 + 15 + 21 + \cdots$. Find S_{15}.

48. $5 + 9 + 13 + 17 + \cdots$. Find S_{25}.

49. $54 + 46 + 38 + 30 + \cdots$. Find S_{12}.

50. $44 + 39 + 34 + 29 + \cdots$. Find S_{18}.

For Exercises 51–58, answer each question.

51. Find the sum of the first 100 natural numbers.

52. Find the sum of the even integers 2 through 200.

53. A concert hall has 60 seats in the first row, 64 in the second, 68 in the third, and so on.
a. How many seats are in the 22nd row?
b. How many seats are in the concert hall if there are 35 rows?

54. Johanna bought a car for \$24,000 with a special no-interest loan and made monthly payments of \$400. Using 24,000 as the first term, write the first five terms of an arithmetic sequence that gives the amount she still owes at the end of each month. How much will she owe at the end of the 25th month?

55. Fence posts are arranged in a triangular stack with 25 on the bottom row, 24 on the next row, 23 on the next row, and so on, until a single post is on the top.
a. How many posts are on the 10th row from the bottom?
b. How many posts are in the stack?

56. Carlos is doing sit-ups to strengthen his abdominal muscles. He did 25 the first night and plans to add 1 each night.
a. Write the first five terms of an arithmetic sequence that gives the number of sit-ups he does each night.
b. Find the number he does on the 30th night.
c. Find the total number of sit-ups he has done after 30 nights.

57. Tanisha takes a job that pays \$28,000 the first year and gives a raise of \$1500 per year.
a. Write the first five terms of an arithmetic sequence that gives her salary at the end of each year.
b. What will her salary be for the 10th year?
c. What are her total earnings for her first 10 years?

58. You have a choice of two jobs. Job A has a starting salary of \$25,000 with raises of \$900 per year, and job B has a starting salary of \$28,000 with raises of \$600 per year.
a. Which job will pay the most during the 10th year?
b. Which job will pay the biggest total amount during the first 15 years?

Appendix B

Geometric Sequences and Series

Objectives

1. Write a geometric sequence and find its common ratio and a specified term.
2. Find partial sums of geometric series.
3. Find the sums of infinite geometric series.
4. Solve applications using geometric series.

Objective 1 Write a geometric sequence and find its common ratio and a specified term.

In Appendix A Arithmetic Sequences and Series, we generated an arithmetic sequence by adding the same number to each term to get the next term. In this section, we multiply each term by the same number to generate a **geometric sequence**. The number we multiply by is called the **common ratio** and is denoted as r. We can find r by dividing any term (except the first) by the term before it.

Definition **Geometric sequence:** A sequence in which every term after the first is found by multiplying the previous term by the same number, called the **common ratio**, r, where $r = \frac{a_n}{a_{n-1}}$.

Note In geometric sequences, the first term is usually denoted as a rather than a_1.

If the first term is a and the common ratio is r, a geometric sequence has the form $a, ar, ar^2, ar^3, ar^4, \ldots, ar^{n-1}$. Thus, the general term of a geometric series is $a_n = ar^{n-1}$.

We notice that the exponent of r is one less than the number of the term, so we have the following rule.

Rule ***n*th Term of a Geometric Sequence**
The formula for finding the nth term of a geometric sequence is $a_n = ar^{n-1}$, where a is the first term and r is the common ratio.

Example 1 A geometric sequence has a first term of 3 and a common ratio of 4.

a. Write the first five terms.

Solution: To find each term, multiply the term before it by 4. So

$a_1 = 3, a_2 = 3(4) = 12, a_3 = 12(4) = 48, a_4 = 48(4) = 192, a_5 = 192(4) = 768.$

The first five terms of the sequence are 3, 12, 48, 192, 768.

b. Find the 10th term.

Solution:
$a_n = ar^{n-1}$ Formula for a_n
$a_{10} = 3(4)^{10-1}$ Substitute **10** for n, **3** for a, and **4** for r.
$a_{10} = 786{,}432$ Evaluate.

c. Find the general term.

Solution:
$a_n = ar^{n-1}$ Formula for a_n
$a_n = 3(4)^{n-1}$ Substitute for a and r.

Example 2 Given the geometric sequence $32, -16, 8, -4, \ldots$, find the common ratio r.

Solution: To find r, divide any term (except the first) by the term before it.

$$r = \frac{-16}{32} = -\frac{1}{2} \quad \text{or} \quad r = \frac{8}{-16} = -\frac{1}{2} \text{ and so on.}$$

Geometric sequences often occur in populations and other applications.

Example 3 The population of rabbits in a large pen increases at a rate of 12% per month. If there are currently 50 rabbits, find the population after 15 months.

Solution: Let P_0 be the initial number of rabbits, P_1 the number after 1 month, P_2 the number after 2 months, and so on. The number of rabbits at the end of each month is the number at the beginning of the month plus an increase of 12% of that number. So the number at the end of each month can be found as follows.

$$P_1 = P_0 + 0.12P_0 = 1.12P_0$$
$$P_2 = P_1 + 0.12P_1 = 1.12P_1 = 1.12(1.12P_0) = (1.12)^2P_0$$
$$P_3 = P_2 + 0.12P_2 = 1.12P_2 = 1.12[(1.12)^2P_0] = (1.12)^3P_0$$

The number of rabbits at the end of each month can be represented by the sequence $P_0, P_1, P_2, P_3, \ldots P_n = P_0, 1.12P_0, (1.12)^2P_0, (1.12)^3P_0, \ldots, (1.12)^{n-1}P_0$, which is a geometric sequence whose first term is P_0 and $r = 1.12$. Consequently, the nth term is $P_0(1.12)^{n-1}$. The number of rabbits at the end of 15 months is the 16th term of the sequence; so

$$P_{16} = P_0(1.12)^{16-1}$$
$$P_{16} = 50(1.12)^{15}$$
$$P_{16} = 273.68$$

Answer: There are about 274 rabbits at the end of 15 months.

Note It can be shown using this procedure that if a population is growing at $p\%$ per unit time and the initial population is P_0, then the population at the end of each unit of time forms a geometric sequence whose first term is P_0 and whose common ratio is $(1 + p)$, where p is $p\%$ written as a decimal. So the geometric sequence is $P_0, P_0(1 + p), P_0(1 + p)^2, P_0(1 + p)^3, \ldots, P_0(1 + p)^{n-1}$. The population after n time periods is the $(n + 1)$st term of the sequence, which is $P_0(1 + p)^n$.

Objective 2 Find partial sums of geometric series.

If we add the terms of an arithmetic sequence, we get an arithmetic series. Likewise, if we add the terms of a geometric sequence, we get a **geometric series**.

Definition **Geometric series:** The sum of the terms of a geometric sequence.

A geometric series is of the form $a + ar + ar^2 + \cdots + ar^{n-1}$ if the series is finite and $a + ar + ar^2 + \cdots$ if the series is infinite.

In the geometric series $3 + 6 + 12 + 24 + \cdots$, we have $a = 3$ and $r = 2$. In $243 - 81 + 27 - 9 + \cdots$, we have $a = 243$ and $r = -\frac{1}{3}$. As with arithmetic series, we can find a formula for partial sum, S_n. Begin with

$$S_n = a + ar + ar^2 + \cdots + ar^{n-1}$$
$$-rS_n = \quad - ar - ar^2 - \cdots - ar^{n-1} - ar^n \qquad \text{Multiply both sides of the equation by } -r.$$
$$S_n - rS_n = a - ar^n \qquad \text{Add the equations.}$$
$$S_n(1 - r) = a(1 - r^n) \qquad \text{Factor both sides.}$$
$$S_n = \frac{a(1 - r^n)}{1 - r} \qquad \text{Divide both sides by } 1 - r.$$

Rule **Partial Sum, S_n, of a Geometric Series**

The sum of the first n terms of a geometric series, S_n, called the nth partial sum, is given by

$$S_n = \frac{a(1 - r^n)}{1 - r},$$

where n is the number of terms, a is the first term, and r is the common ratio $(r \neq 1)$.

Example 4 Find the sum of the first 10 terms of the geometric series: $1 - 3 + 9 - 27 + \cdots$

Solution: We first find r: $r = \frac{-3}{1} = -3$.

$$S_n = \frac{a(1 - r^n)}{1 - r} \quad \text{Formula for } S_n$$

$$S_{10} = \frac{1(1 - (-3)^{10})}{1 - (-3)} \quad \text{Substitute 10 for } n\text{, 1 for } a\text{, and } -3 \text{ for } r.$$

$$S_{10} = -14{,}762 \quad \text{Evaluate.}$$

Objective 3 Find the sums of infinite geometric series.

If the common ratio satisfies $|r| > 1$, the partial sums become infinitely large as n becomes infinitely large. However, if $|r| < 1$, the partial sums approach a value as n becomes infinitely large. This value is called the *limit* of the partial sums and is the sum of the infinite series.

For example, for the series $2 + 1 + \frac{1}{2} + \frac{1}{4} + \cdots$, we have

$$S_5 = \frac{2\left(1 - \left(\frac{1}{2}\right)^5\right)}{1 - \frac{1}{2}} = 3.875$$

$$S_{10} = \frac{2\left(1 - \left(\frac{1}{2}\right)^{10}\right)}{1 - \frac{1}{2}} \approx 3.9960938$$

$$S_{15} = \frac{2\left(1 - \left(\frac{1}{2}\right)^{15}\right)}{1 - \frac{1}{2}} \approx 3.9998779$$

Notice that the greater the value of n, the closer the sum is to 4. The partial sums approach 4 because as n gets larger, $\left(\frac{1}{2}\right)^n$ approaches 0. In the preceding example, $\left(\frac{1}{2}\right)^5 = 0.03125$, $\left(\frac{1}{2}\right)^{10} \approx 0.000977$, and $\left(\frac{1}{2}\right)^{15} \approx 0.0000305$. In general, if $|r| < 1$, then r^n approaches 0 as n becomes large. Consequently, if $|r| < 1$, the formula $S_n = \frac{a(1 - r^n)}{1 - r}$ becomes $S_\infty = \frac{a(1 - 0)}{1 - r} = \frac{a}{1 - r}$ as n becomes infinitely large. We denote the sum as S_∞ rather than S_n.

Rule Sum of an Infinite Geometric Series

If $|r| < 1$, the sum of an infinite geometric series, S_∞, is given by the formula

$$S_\infty = \frac{a}{1 - r},$$

where a is the first term and r is the common ratio. If $|r| \geq 1$, S_∞ does not exist.

Example 5 Find the sum of the infinite geometric series:

$$2 + 1 + \frac{1}{2} + \frac{1}{4} + \cdots$$

Solution: We know that $a = 2$ and $r = \frac{1}{2}$. Because $|r| < 1$, the sum exists.

$$S_\infty = \frac{a}{1 - r}$$

$$S_\infty = \frac{2}{1 - \frac{1}{2}} \quad \text{Substitute 2 for } a \text{ and } \tfrac{1}{2} \text{ for } r.$$

$$S_\infty = 4 \quad \text{Evaluate.}$$

Infinite geometric series also provides us with a method of changing a repeating decimal into a fraction.

Example 6 Write as a fraction: $0.\overline{37}$

Solution: First write $0.\overline{37}$ as an infinite geometric series as follows:

$$0.\overline{37} = 0.37 + 0.0037 + 0.000037 + \cdots$$

$$0.\overline{37} = \frac{37}{100} + \frac{37}{10{,}000} + \frac{37}{1{,}000{,}000} + \cdots$$

The last line is an infinite geometric series with $a = \frac{37}{100}$ and $r = \frac{1}{100}$. Because $|r| < 1$, the sum of this series exists.

$$S_\infty = \frac{a}{1 - r}$$

$$S_\infty = \frac{\frac{37}{100}}{1 - \frac{1}{100}} \quad \text{Substitute } \tfrac{37}{100} \text{ for } a \text{ and } \tfrac{1}{100} \text{ for } r.$$

$$S_\infty = \frac{\frac{37}{100}}{\frac{99}{100}} \quad \text{Simplify.}$$

$$S_\infty = \frac{37}{99}$$

Answer: $0.\overline{37} = \frac{37}{99}$

Objective 4 Solve applications using geometric series.

Example 7 To save for their child's college education, the McBride family put $1000 into a savings account the first year. Each year thereafter they deposited 10% more than the previous year.

a. Write the first five terms of the geometric sequence that gives the amount of money deposited in the account each year.

Solution: From the note next to Example 3, this is a geometric series in which $a = 1000$ and $r = (1 + 0.10) = 1.1$.

$$a_1 = 1000, a_2 = 1000(1.1) = 1100, a_3 = 1000(1.1)^2 = 1210,$$
$$a_4 = 1000(1.10)^3 = 1331, a_5 = 1000(1.10)^4 = 1464.10$$

The first five terms of the sequence are 1000, 1100, 1210, 1331, 1464.10.

b. How much money will the McBrides have deposited in the account at the end of 15 years?

Solution: The series $1000 + 1100 + 1210 + 1331 + 1464.10 + \cdots$ is geometric with $a = 1000$ and $r = 1.1$.

$$S_n = \frac{a(1 - r^n)}{1 - r}$$

$$S_{15} = \frac{1000(1 - 1.1^{15})}{1 - 1.1}$$ Substitute **15** for *n*, **1000** for *a*, and **1.1** for *r*.

$$S_{15} = 31{,}772.48$$ Evaluate.

Answer: They will have deposited \$31,772.48 in the account after 15 years.

Appendix B Exercises

For Extra Help MyMathLab®

Objective 1

Prep Exercise 1 What is a geometric sequence?

Prep Exercise 2 Explain how to find the common ratio of a geometric sequence.

Prep Exercise 3 What is the formula for finding the nth term of a geometry sequence?

For Exercises 1–8:
a. Find the common ratio, r, for the given geometric sequence. See Example 2.
b. Find the indicated term. See Example 1.
c. Find an expression for the general term, a_n. See Example 1.

1. $1, 3, 9, 27, \ldots;$ 9th term

2. $7, 14, 28, 56, \ldots;$ 11th term

3. $-2, 4, -8, 16, \ldots;$ 15th term

4. $-8, 24, -72, 216, \ldots;$ 9th term

5. $243, 81, 27, 9, \ldots;$ 10th term

6. $8, 4, 2, 1, \ldots;$ 8th term

7. $128, -32, 8, -2, \ldots;$ 7th term

8. $125, -25, 5, -1, \ldots;$ 9th term

For Exercises 9–12:
a. Write the first five terms of the geometric sequence, satisfying the given conditions. See Example 1.
b. Find the indicated term. See Example 1.

9. $a = -4, r = -3;$ 8th term

10. $a = -6, r = -2;$ 10th term

11. $a = 243, r = \frac{1}{3};$ 7th term

12. $a = 256, r = -\frac{1}{2};$ 10th term

For Exercises 13–18: *a. Use the formula for the nth term to find r.*
b. Write the first four terms of the geometric sequence.

13. $a = 1, r > 0, a_5 = 16$

14. $a = 3, a_4 = -24$

15. $a = -6, r > 0, a_5 = -96$

16. $a = -4, a_4 = -32$

17. $a = 128, a_4 = 2$

18. $a = -243, a_6 = 1$

19. Find the first term of a geometric sequence in which $r = 3$ and the 5th term is -486.

20. Find the first term of a geometric sequence in which $r = -\frac{1}{2}$ and the 5th term is $-\frac{1}{8}$.

Objective 2

Prep Exercise 4 What is a geometric series?

Prep Exercise 5 What is the formula for finding the sum of the first n terms of a geometric series?

For Exercises 21–26, find the sum of the first n terms of each geometric series for the given value of n. See Example 4.

21. $3 + 9 + 27 + 81 + \cdots, n = 11$

22. $-1 - 2 - 4 - 8 - \cdots, n = 9$

23. $32 + 16 + 8 + 4 + \cdots, n = 9$

24. $81 + 27 + 9 + 3 + \cdots, n = 7$

25. $128 - 32 + 8 - 2 + \cdots, n = 9$

26. $625 - 125 + 25 - 5 + \cdots, n = 8$

Objective 3

Prep Exercise 6 What must be true about the common ratio, r, in order to use the formula $S_\infty = \frac{a}{1 - r}$ to find the sum of an infinite geometric series?

For Exercises 27–32, find the sum of the infinite geometric series if possible. If it is not possible, explain why. See Example 5.

27. $27 + 9 + 3 + \cdots$

28. $8 + 4 + 2 + \cdots$

29. $15 - 9 + \frac{27}{5} - \cdots$

30. $15 - 10 + \frac{20}{3} - \cdots$

31. $9 + 12 + 16 + \cdots$

32. $16 + 20 + 25 + \cdots$

For Exercises 33–36, write each repeating decimal as a fraction. See Example 6.

33. $0.\overline{4}$

34. $0.\overline{7}$

35. $0.\overline{17}$

36. $0.\overline{25}$

For Exercises 37 and 38, answer each question.

37. If $a_n = 500(1.04)^n$,
a. Find the first five terms of the sequence.

b. Find the 10th term of the sequence.

38. If $a_n = 350(1.06)^n$,
a. Find the first five terms of the sequence.

b. Find the 8th term of the sequence.

Objective 4

For Exercises 39–44, solve. See Examples 3 and 7.

39. A population of mink is increasing at a rate of 8% per month. The current mink population is 100.
a. Using 100 as the first term, find the first four terms of the geometric sequence that gives the number of mink at the beginning of each month.

b. Find the number of mink present at the beginning of the 8th month.

c. Find the expression for the general term, a_n.

40. The generation time (the time required for the number present to double) for a particular bacteria is 1 hour. Suppose initially one bacteria was present.
a. Using 1 as the first term, write the first five terms of the geometric sequence giving the number of bacteria present after each hour.

b. Find an expression for the number present after the nth hour.

c. How many bacteria are present after 1 day?

41. Suppose you took a job for a month (20 working days) that paid \$0.01 the first day and your salary doubled each day.
a. Write the first five terms of the geometric sequence that gives your salary each day.

b. Find an expression for the amount earned on the nth day.

c. How much would you earn on the 20th day?

d. What are your total earnings for the month?

42. Damarys deposits \$200 in the bank. Each month thereafter she deposits 5% more than the month before. She does this for 1 year.
a. Find an expression for the amount she deposits in the nth month.

b. Write the first four terms of the geometric sequence that gives the amount of her deposit each month.

c. How much did she deposit in the 10th month?

d. How much does she deposit for the year?

43. A new boat costs \$20,000 and depreciates by 7% each year. What will the boat be worth in 8 years?

44. The isotope ${}_{15}P^{33}$ has a half-life of 25 days. A sample has 400 grams.
a. Find the first five terms of the geometric sequence that gives the amount present at the end of each half-life.

b. Find an expression for the amount present after the nth half-life.

c. Find the amount present after the 10th half-life.

Appendix C

The Binomial Theorem

Objectives

1. Expand a binomial using Pascal's triangle.
2. Evaluate factorial notation and binomial coefficients.
3. Expand a binomial using the binomial theorem.
4. Find a particular term of a binomial expansion.

In this section, we will learn to raise binomials to natural-number powers.

Objective 1 Expand a binomial using Pascal's triangle.

We begin by writing out $(a + b)^n$, where n is a natural number, and look for patterns. These products are called *binomial expansions.*

$$(a + b)^0 = 1$$
$$(a + b)^1 = a + b$$
$$(a + b)^2 = a^2 + 2ab + b^2$$
$$(a + b)^3 = a^3 + 3a^2b + 3ab^2 + b^3$$
$$(a + b)^4 = a^4 + 4a^3b + 6a^2b^2 + 4ab^3 + b^4$$
$$(a + b)^5 = a^5 + 5a^4b + 10a^3b^2 + 10a^2b^3 + 5ab^4 + b^5$$

Conclusions Several patterns can be observed from the preceding expansions.

1. The 1st term in the expansion, a, is raised to the same power as the binomial, and the power of a decreases by 1 in each successive term. Note that the last term contains a^0.
2. The exponent of b is 0 in the 1st term and increases by 1 on each successive term.
3. The sum of the exponents of the variables of each term equals the exponent of the binomial.
4. The number of terms in the expansion is one more than the exponent of the binomial.

Now consider the coefficients of the terms in these expansions.

Coefficients of Expansions

$(a + b)^0$	1
$(a + b)^1$	1 1
$(a + b)^2$	1 2 1
$(a + b)^3$	1 3 3 1
$(a + b)^4$	1 4 6 4 1
$(a + b)^5$	1 5 10 10 5 1

If we arrange the coefficients of each expansion in a triangular array, we see an interesting pattern. Each row begins and ends with 1. Each number inside a row is the sum of the two numbers in the row above it. For example, each 10 in the bottom row comes from adding the 4 and 6 directly above.

This triangular array of numbers is called *Pascal's triangle* in honor of the French mathematician Blaise Pascal. When these observations are used, the expansion of $(a + b)^6$ has seven terms and the variable portions of the terms are $a^6, a^5b, a^4b^2, a^3b^3, a^2b^4, ab^5, b^6$. By continuing the pattern in Pascal's triangle, we can find the coefficients for each of those terms.

$(a + b)^5$	1 5 10 10 5 1
$(a + b)^6$	1 6 15 20 15 6 1

Using the coefficients from the last line of Pascal's triangle and the variables previously listed gives us

$$(a + b)^6 = a^6 + 6a^5b + 15a^4b^2 + 20a^3b^3 + 15a^2b^4 + 6ab^5 + b^6$$

Objective 2 Evaluate factorial notation and binomial coefficients.

Although Pascal's triangle is easy to use, it isn't practical, especially for binomials raised to large powers. Consequently, another method called the *binomial theorem* is often used. Before introducing the binomial theorem, we need **factorial notation.**

Definition Factorial notation: For any natural number n, the symbol $n!$ (read "n factorial") means $n(n-1)(n-2)\ldots 3\cdot 2\cdot 1$.

$0!$ is defined to be 1, so $0! = 1$.

Example 1 Evaluate the following factorials.

a. $5!$

Solution: $5! = 5\cdot 4\cdot 3\cdot 2\cdot 1 = 120$

b. $7!$

Solution: $7! = 7\cdot 6\cdot 5\cdot 4\cdot 3\cdot 2\cdot 1 = 5040$

Sometimes we may not write all of the factors of a factorial. In such cases, the last desired factor is written as a factorial. Below are some alternative ways to write 7!

$$7! = 7\cdot 6\cdot 5\cdot 4\cdot 3\cdot 2\cdot 1 = 7\cdot 6! \quad \text{or} \quad 7! = 7\cdot 6\cdot 5! \quad \text{or} \quad 7! = 7\cdot 6\cdot 5\cdot 4!$$

The coefficients of a binomial expansion can be expressed in terms of factorials using a special notation called the **binomial coefficient**. We will see how the binomial coefficient is used later.

Definition Binomial coefficient: A number written as $\binom{n}{r}$ and defined as $\dfrac{n!}{r!(n-r)!}$.

Example 2 Evaluate the following binomial coefficients.

a. $\binom{6}{2}$

Solution:

$\binom{6}{2} = \dfrac{6!}{2!(6-2)!}$ Substitute 6 for n and 2 for r in $\dfrac{n!}{r!(n-r)!}$.

$= \dfrac{6\cdot 5\cdot 4\cdot 3\cdot 2\cdot 1}{(2\cdot 1)(4\cdot 3\cdot 2\cdot 1)}$ Expand the factorials.

$= 15$ Simplify.

Note The factorials could have been evaluated as follows:

$$\frac{6!}{2!\cdot 4!} = \frac{6\cdot 5\cdot \cancel{4!}}{2\cdot 1\cdot \cancel{4!}} = \frac{6\cdot 5}{2} = \frac{30}{2} = 15$$

b. $\binom{8}{5}$

Solution:

$\binom{8}{5} = \dfrac{8!}{5!(8-5)!}$ Substitute 8 for n and 5 for r in $\dfrac{n!}{r!(n-r)!}$.

$= \dfrac{8!}{5!\cdot 3!}$ Simplify.

$= \dfrac{8\cdot 7\cdot 6\cdot \cancel{5!}}{\cancel{5!}\cdot 3\cdot 2\cdot 1}$ Rewrite $8!$ as $8\cdot 7\cdot 6\cdot 5!$ and simplify.

$= 56$ Evaluate.

Using the formula for evaluating binomial coefficients, we can prove two special cases: $\binom{n}{0} = 1$ and $\binom{n}{n} = 1$.

Objective 3 Expand a binomial using the binomial theorem.

Let's look at the expansion of $(a+b)^6$ and make another observation.

$$(a+b)^6 = a^6 + 6a^5b + 15a^4b^2 + 20a^3b^3 + 15a^2b^4 + 6ab^5 + b^6$$

Look at the 3rd term of the expansion, $15a^4b^2$. The coefficient is 15, and from Example 2(a), we see that $\binom{6}{2} = 15$. The coefficient of the 4th term is 20, and $\binom{6}{3} = 20$. In both cases, the n value of the binomial coefficient, $\binom{n}{r}$, is the exponent of the binomial and the r value is the exponent of b. This observation leads to the **binomial theorem**.

Definition Binomial theorem: For any positive integer n,

$$(a+b)^n = \binom{n}{0}a^n + \binom{n}{1}a^{n-1}b + \binom{n}{2}a^{n-2}b^2 + \binom{n}{3}a^{n-3}b^3 + \cdots + \binom{n}{n}b^n.$$

Example 3 Expand each of the following binomials using the binomial theorem.

a. $(a+b)^6$

Solution:

$$(a+b)^6 = \binom{6}{0}a^6 + \binom{6}{1}a^5b + \binom{6}{2}a^4b^2 + \binom{6}{3}a^3b^3 + \binom{6}{4}a^2b^4 + \binom{6}{5}ab^5 + \binom{6}{6}b^6$$

$$= \frac{6!}{0!6!}a^6 + \frac{6!}{1!5!}a^5b + \frac{6!}{2!4!}a^4b^2 + \frac{6!}{3!3!}a^3b^3 + \frac{6!}{4!2!}a^2b^4 + \frac{6!}{5!1!}ab^5 + \frac{6!}{6!0!}b^6$$

$$= a^6 + 6a^5b + 15a^4b^2 + 20a^3b^3 + 15a^2b^4 + 6ab^5 + b^6$$

Note This is the same result we got earlier using Pascal's triangle. Also note that

$$\binom{6}{3} = \frac{6!}{3!3!} = \frac{6\cdot5\cdot4\cdot3!}{3!3!} = \frac{6\cdot5\cdot4\cdot\cancel{3!}}{3\cdot2\cdot1\cdot\cancel{3!}} = \frac{\cancel{6}\cdot5\cdot4}{\cancel{3}\cdot\cancel{2}\cdot1} = 20, \text{ etc.}$$

b. $(a-3b)^5$

Solution: Write $(a-3b)^5$ as $[a+(-3b)]^5$

$$[a+(-3b)]^5 = \binom{5}{0}a^5 + \binom{5}{1}a^4(-3b) + \binom{5}{2}a^3(-3b)^2 + \binom{5}{3}a^2(-3b)^3 + \binom{5}{4}a(-3b)^4 + \binom{5}{5}(-3b)^5$$

$$= a^5 + 5a^4(-3b) + 10a^3(9b^2) + 10a^2(-27b^3) + 5a(81b^4) + (-243b^5)$$

$$= a^5 - 15a^4b + 90a^3b^2 - 270a^2b^3 + 405ab^4 - 243b^5$$

Note $$\binom{5}{2} = \frac{5!}{2!3!} = \frac{5\cdot4\cdot3!}{2\cdot1\cdot3!} = \frac{5\cdot4\cdot\cancel{3!}}{2\cdot1\cdot\cancel{3!}} = \frac{5\cdot4}{2} = \frac{20}{2} = 10, \text{ etc.}$$

Objective 4 Find a particular term of a binomial expansion.

Sometimes it is necessary to find only a specific term of a binomial expansion without writing out the entire expansion. Look again at the binomial expansion. Note that the 3rd term (which we will call the $(2+1)$st term) is $\binom{n}{2}a^{n-2}b^2$ and the 4th term (which we will call the $(3+1)$st term) is $\binom{n}{3}a^{n-3}b^3$. Similarly, the $(m+1)$st term is $\binom{n}{m}a^{n-m}b^m$. These observations lead to the following.

Rule Finding the $(m+1)$st Term of a Binomial Expansion

The $(m+1)$st term of the expansion $(a+b)^n$ is $\binom{n}{m}a^{n-m}b^m$.

Example 4 Find the indicated term of each of the following binomial expansions.

a. $(a + b)^{11}$, 7th term

Solution: Use the formula for the $(m + 1)$st term with $n = 11$ and $m = 6$ (to find the 7th term).

$$\binom{n}{m}a^{n-m}b^{m} = \binom{11}{6}a^{11-6}b^{6} = 462a^5b^6$$

b. $(2x - 5y)^8$, 4th term

Solution: Write $(2x - 5y)^8$ as $[2x + (-5y)]^8$. Use the formula for the $(m + 1)$st term with $n = 8$ and $m = 3$ (to find the 4th term), $a = 2x$, and $b = -5y$.

$$\binom{n}{m}a^{n-m}b^{m} = \binom{8}{3}(2x)^{8-3}(-5y)^3 = 56(32x^5)(-125y^3) = -224{,}000x^5y^3$$

Appendix C Exercises

For Extra Help MyMathLab®

Objectives 1 and 2

Prep Exercise 1 How many terms are in the expansion of $(5a + b)^{12}$?

Prep Exercise 2 What is the sum of the exponents on x and y for any term in the expansion of $(x + y)^9$?

Prep Exercise 3 How is the symbol 8! read?

Prep Exercise 4 What does 8! mean? What is its value?

For Exercises 1–12, evaluate each expression. See Example 1.

1. $4!$

2. $7!$

3. $(4!)(3!)$

4. $(3!)(2!)$

5. $(6!)(5!)$

6. $(4!)(7!)$

7. $\dfrac{8!}{10!}$

8. $\dfrac{7!}{9!}$

9. $\dfrac{10!}{9!}$

10. $\dfrac{12!}{11!}$

11. $\dfrac{8!}{6!(8-6)!}$

12. $\dfrac{10!}{6!(10-6)!}$

Prep Exercise 5 What is the formula for evaluating the binomial coefficient $\binom{n}{r}$?

For Exercises 13–20, evaluate each binomial coefficient. See Example 2.

13. $\binom{7}{3}$

14. $\binom{5}{2}$

15. $\binom{10}{4}$

16. $\binom{6}{5}$

17. $\binom{7}{7}$

18. $\binom{4}{4}$

19. $\binom{8}{0}$

20. $\binom{9}{0}$

Objective 3

Prep Exercise 6 Using the binomial theorem to expand $(x + 3)^6$, what is the coefficient of the 4th term?

Prep Exercise 7 What is the exponent of b in the 6th term of the expansion of $(a + b)^{13}$? What is the exponent of a in that term?

For Exercises 21–32, use the binomial theorem to expand each of the following. See Example 3.

21. $(a + b)^5$

22. $(a + b)^7$

23. $(x - y)^4$

24. $(x - y)^3$

25. $(2a + b)^3$

26. $(a + 2b)^3$

27. $(x - 2y)^5$

28. $(x - 3y)^4$

29. $(2m + 3n)^6$

30. $(3c + 2d)^5$

31. $(3x - 4y)^4$

32. $(4a - b)^7$

Objective 4

Prep Exercise 8 What is the formula for finding the $(m + 1)$st term of the expansion $(a + b)^n$?

For Exercises 33–40, find the indicated term of each binomial expansion. See Example 4.

33. $(x + y)^8$, 5th term

34. $(a + b)^9$, 4th term

35. $(a - b)^{10}$, 3rd term

36. $(m - n)^7$, 2nd term

37. $(4x + y)^9$, 6th term

38. $(3a + b)^{11}$, 7th term

39. $(3m - 2n)^7$, 4th term

40. $(5x - 3y)^{12}$, 5th term

Appendix D

Solving Systems of Linear Equations Using Cramer's Rule

Objectives

1. Evaluate determinants of 2×2 matrices.
2. Evaluate determinants of 3×3 matrices.
3. Solve systems of equations using Cramer's Rule.

Objective 1 Evaluate determinants of 2×2 matrices.

Systems of linear equations can be solved using a special type of matrix called a **square matrix**.

Definition Square matrix: A matrix with an equal number of rows and columns.

For example, $\begin{bmatrix} 2 & 3 \\ 1 & -4 \end{bmatrix}$ and $\begin{bmatrix} 1 & -3 & 5 \\ -7 & 2 & 9 \\ 0 & 4 & -6 \end{bmatrix}$ are square matrices.

Every square matrix has a *determinant*. We write the determinant of a matrix, A, as $\det(A)$ or $|A|$. The method used to find the determinant of a matrix depends on its size.

Rule Determinant of a 2×2 Matrix

$$\text{If } A = \begin{bmatrix} a_1 & b_1 \\ a_2 & b_2 \end{bmatrix}, \text{ then } \det(A) = \begin{vmatrix} a_1 & b_1 \\ a_2 & b_2 \end{vmatrix} = a_1b_2 - a_2b_1.$$

Warning Be careful to note the difference between [ ] and | |. The notation [ ] denotes a matrix, whereas | | denotes the determinant of a matrix. Also, from the context, we know that | | does not mean absolute value.

Notice that the determinant contains diagonal products, as illustrated by the following diagram.

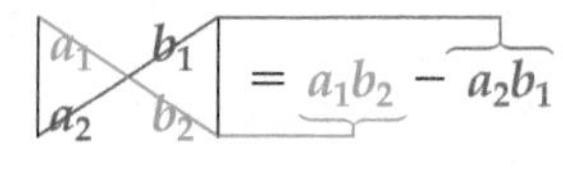

Note Because subtraction is not commutative, make sure you note the order of the two products.

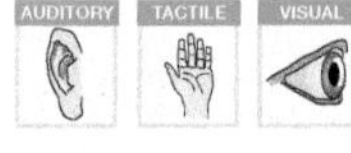

Learning Strategies

The products in the determinant are arranged with the downward cross product subtracting the upward cross product. An easy way to remember this order is that you have to "fall down before you can get up."

Example 1 Find the determinants of the following matrices.

a. $A = \begin{bmatrix} 3 & -2 \\ 2 & 4 \end{bmatrix}$

Solution: $\det(A) = \begin{vmatrix} 3 & -2 \\ 2 & 4 \end{vmatrix} = (3)(4) - (2)(-2) = 12 + 4 = 16$

b. $B = \begin{bmatrix} -3 & 5 \\ 4 & -2 \end{bmatrix}$

Solution: $\det(B) = (-3)(-2) - (4)(5) = 6 - 20 = -14$

c. $M = \begin{bmatrix} 1 & 3 \\ 3 & 9 \end{bmatrix}$

Solution: $\det(M) = (1)(9) - (3)(3) = 9 - 9 = 0$

Note An alternative notation for $\det(A)$ is $|A|$.

Your Turn 1 Find the determinant of $\begin{bmatrix} 1 & -3 \\ -4 & 2 \end{bmatrix}$.

Answer to Your Turn 1
-10

Objective 2 Evaluate determinants of 3 × 3 matrices.

There are various methods of evaluating the determinant of a 3 × 3 matrix. One of the most common methods is *expanding by minors*. Each element of a square matrix has a number called the **minor** for that element.

Definition **Minor of an element of a matrix:** The determinant of the remaining matrix when the row and column in which the element is located are ignored.

Example 2 Find the minor of 2 in $\begin{bmatrix} 2 & -3 & -6 \\ -1 & 5 & -2 \\ 3 & -4 & 1 \end{bmatrix}$.

Solution: To find the minor of 2, we ignore its row and column (shown in blue) and evaluate the determinant of the remaining matrix (shown in red).

$$\begin{bmatrix} 2 & -3 & -6 \\ -1 & 5 & -2 \\ 3 & -4 & 1 \end{bmatrix}$$

$$\begin{vmatrix} 5 & -2 \\ -4 & 1 \end{vmatrix} = (5)(1) - (-4)(-2) = 5 - 8 = -3$$

Your Turn 2 Find the minor of 6 in $\begin{bmatrix} 1 & -3 & 6 \\ -2 & 2 & 0 \\ 4 & -1 & 5 \end{bmatrix}$.

To evaluate the determinant of a 3 × 3 matrix, we will expand by minors along the first column.

Note We can expand by minors along *any* row or column to find the determinant of a 3 × 3 matrix. For simplicity, we have chosen to show expanding by minors only along the first column.

Rule **Evaluating the Determinant of a 3 × 3 Matrix**

$$\begin{vmatrix} a_1 & b_1 & c_1 \\ a_2 & b_2 & c_2 \\ a_3 & b_3 & c_3 \end{vmatrix} = a_1\begin{pmatrix}\text{minor} \\ \text{of } a_1\end{pmatrix} - a_2\begin{pmatrix}\text{minor} \\ \text{of } a_2\end{pmatrix} + a_3\begin{pmatrix}\text{minor} \\ \text{of } a_3\end{pmatrix}$$

$$= a_1\begin{vmatrix} b_2 & c_2 \\ b_3 & c_3 \end{vmatrix} - a_2\begin{vmatrix} b_1 & c_1 \\ b_3 & c_3 \end{vmatrix} + a_3\begin{vmatrix} b_1 & c_1 \\ b_2 & c_2 \end{vmatrix}$$

Warning Notice that the second term of the expansion has a negative sign.

Example 3 Find the determinant of $\begin{bmatrix} 2 & -3 & -4 \\ -1 & 2 & -2 \\ 3 & -4 & 1 \end{bmatrix}$.

Solution: Using the rule for expanding by minors along the first column, we have the following:

$$\begin{vmatrix} 2 & -3 & -4 \\ -1 & 2 & -2 \\ 3 & -4 & 1 \end{vmatrix} = 2\begin{pmatrix}\text{minor} \\ \text{of } 2\end{pmatrix} - (-1)\begin{pmatrix}\text{minor} \\ \text{of } -1\end{pmatrix} + 3\begin{pmatrix}\text{minor} \\ \text{of } 3\end{pmatrix}$$

$$= 2\begin{vmatrix} 2 & -2 \\ -4 & 1 \end{vmatrix} - (-1)\begin{vmatrix} -3 & -4 \\ -4 & 1 \end{vmatrix} + 3\begin{vmatrix} -3 & -4 \\ 2 & -2 \end{vmatrix}$$

$$= 2(2 - 8) + 1(-3 - 16) + 3(6 + 8)$$

$$= -12 + (-19) + 42$$

$$= 11$$

Answer to Your Turn 2

$$\begin{vmatrix} -2 & 2 \\ 4 & -1 \end{vmatrix} = -6$$

Your Turn 3 Find the determinant of $\begin{bmatrix} 3 & -2 & 4 \\ -2 & 3 & 1 \\ 2 & -4 & 2 \end{bmatrix}$.

Objective 3 Solve systems of equations using Cramer's Rule.

Now we can use **Cramer's Rule**, which uses determinants to solve systems of equations. To derive Cramer's Rule, we solve a general system of equations using the elimination method. We show the derivation for a system of two equations in two unknowns.

Of Interest

Cramer's Rule is named after Gabriel Cramer (1704–1752), who was chair of the mathematics department at Geneva, Switzerland. In one of his books, he gave an example that required finding an equation of degree two whose graph passed through five given points. The solution led to a system of five linear equations in five unknowns. So readers could solve the system, Cramer referred them to an appendix, which explained what we now call Cramer's Rule.

$$\begin{cases} a_1x + b_1y = c_1 & \text{(Equation 1)} \\ a_2x + b_2y = c_2 & \text{(Equation 2)} \end{cases}$$

We eliminate y by multiplying Equation 1 by b_2 and Equation 2 by $-b_1$.

$$a_1b_2x + b_1b_2y = b_2c_1 \quad \text{Multiply Equation 1 by } b_2.$$

$$-a_2b_1x - b_1b_2y = -b_1c_2 \quad \text{Multiply Equation 2 by } -b_1.$$

$$a_1b_2x - a_2b_1x = b_2c_1 - b_1c_2 \quad \text{Add the equations.}$$

$$(a_1b_2 - a_2b_1)x = b_2c_1 - b_1c_2 \quad \text{Factor out } x \text{ from the left side.}$$

$$x = \frac{b_2c_1 - b_1c_2}{a_1b_2 - a_2b_1} \quad \text{Divide by } a_1b_2 - a_2b_1.$$

Notice that the numerator is $\begin{vmatrix} c_1 & b_1 \\ c_2 & b_2 \end{vmatrix}$ and the denominator is $\begin{vmatrix} a_1 & b_1 \\ a_2 & b_2 \end{vmatrix}$; so $x = \dfrac{\begin{vmatrix} c_1 & b_1 \\ c_2 & b_2 \end{vmatrix}}{\begin{vmatrix} a_1 & b_1 \\ a_2 & b_2 \end{vmatrix}}$.

If we repeat the same process and solve for y, we get $y = \dfrac{\begin{vmatrix} a_1 & c_1 \\ a_2 & c_2 \end{vmatrix}}{\begin{vmatrix} a_1 & b_1 \\ a_2 & b_2 \end{vmatrix}}$.

A similar approach is used to derive the rule for a system of three equations in three unknowns.

Rule Cramer's Rule

The solution to the system of linear equations $\begin{cases} a_1x + b_1y = c_1 \\ a_2x + b_2y = c_2 \end{cases}$ is

$$x = \frac{\begin{vmatrix} c_1 & b_1 \\ c_2 & b_2 \end{vmatrix}}{\begin{vmatrix} a_1 & b_1 \\ a_2 & b_2 \end{vmatrix}} = \frac{D_x}{D} \text{ and } y = \frac{\begin{vmatrix} a_1 & c_1 \\ a_2 & c_2 \end{vmatrix}}{\begin{vmatrix} a_1 & b_1 \\ a_2 & b_2 \end{vmatrix}} = \frac{D_y}{D}.$$

The solution to the system of linear equations $\begin{cases} a_1x + b_1y + c_1z = d_1 \\ a_2x + b_2y + c_2z = d_2 \\ a_3x + b_3y + c_3z = d_3 \end{cases}$ is

$$x = \frac{\begin{vmatrix} d_1 & b_1 & c_1 \\ d_2 & b_2 & c_2 \\ d_3 & b_3 & c_3 \end{vmatrix}}{\begin{vmatrix} a_1 & b_1 & c_1 \\ a_2 & b_2 & c_2 \\ a_3 & b_3 & c_3 \end{vmatrix}} = \frac{D_x}{D},\ y = \frac{\begin{vmatrix} a_1 & d_1 & c_1 \\ a_2 & d_2 & c_2 \\ a_3 & d_3 & c_3 \end{vmatrix}}{\begin{vmatrix} a_1 & b_1 & c_1 \\ a_2 & b_2 & c_2 \\ a_3 & b_3 & c_3 \end{vmatrix}} = \frac{D_y}{D},\ \text{and } z = \frac{\begin{vmatrix} a_1 & b_1 & d_1 \\ a_2 & b_2 & d_2 \\ a_3 & b_3 & d_3 \end{vmatrix}}{\begin{vmatrix} a_1 & b_1 & c_1 \\ a_2 & b_2 & c_2 \\ a_3 & b_3 & c_3 \end{vmatrix}} = \frac{D_z}{D}.$$

Note Each denominator, D, is the determinant of a matrix containing only the coefficients in the system. To find D_x, we replace the column of x-coefficients in the coefficient matrix with the constants from the system. To find D_y, we replace the column of y-coefficients in the coefficient matrix with the constant terms. We do likewise to find D_z.

Answer to Your Turn 3
26

Example 4 Use Cramer's Rule to solve $\begin{cases} 2x + 3y = -5 \\ 3x - y = 9 \end{cases}$.

Solution: First, we find D, D_x, and D_y.

Note These are the coefficients of the system. ▶ $D = \begin{vmatrix} 2 & 3 \\ 3 & -1 \end{vmatrix} = (2)(-1) - (3)(3) = -2 - 9 = -11$

Note Replace the x-coefficients with the constants. ▶ $D_x = \begin{vmatrix} -5 & 3 \\ 9 & -1 \end{vmatrix} = (-5)(-1) - (9)(3) = 5 - 27 = -22$

Note Replace the y-coefficients with the constants. ▶ $D_y = \begin{vmatrix} 2 & -5 \\ 3 & 9 \end{vmatrix} = (2)(9) - (3)(-5) = 18 + 15 = 33$

Note If $D = 0$ and D_x and $D_y \neq 0$, the system is inconsistent (no solution). If D, D_x, and $D_y = 0$, the system is dependent (infinite number of solutions).

Now we can find x and y.

$$x = \frac{D_x}{D} = \frac{-22}{-11} = 2 \qquad y = \frac{D_y}{D} = \frac{33}{-11} = -3$$

The solution is $(2, -3)$, which we can check by verifying that it satisfies both equations in the system. We will leave the check to the reader.

Your Turn 4 Use Cramer's Rule to solve $\begin{cases} 3x - 2y = -16 \\ x + 3y = 2 \end{cases}$.

Example 5 Use Cramer's Rule to solve $\begin{cases} x + 2y - 2x = -7 \\ 3x - 2y + z = 15 \\ 2x + 3y - 3z = -9 \end{cases}$.

Note We find the determinant of a 3×3 matrix by expanding by minors along the first column. ▶

Solution: We need to find D, D_x, D_y, and D_z.

$$D = \begin{vmatrix} 1 & 2 & -2 \\ 3 & -2 & 1 \\ 2 & 3 & -3 \end{vmatrix} = (1)\begin{vmatrix} -2 & 1 \\ 3 & -3 \end{vmatrix} - (3)\begin{vmatrix} 2 & -2 \\ 3 & -3 \end{vmatrix} + (2)\begin{vmatrix} 2 & -2 \\ -2 & 1 \end{vmatrix}$$

$$= 1(6 - 3) - 3(-6 + 6) + 2(2 - 4)$$
$$= 3 - 0 + (-4)$$
$$= -1$$

$$D_x = \begin{vmatrix} -7 & 2 & -2 \\ 15 & -2 & 1 \\ -9 & 3 & -3 \end{vmatrix} = (-7)\begin{vmatrix} -2 & 1 \\ 3 & -3 \end{vmatrix} - (15)\begin{vmatrix} 2 & -2 \\ 3 & -3 \end{vmatrix} + (-9)\begin{vmatrix} 2 & -2 \\ -2 & 1 \end{vmatrix}$$

$$= -7(6 - 3) - 15(-6 + 6) + (-9)(2 - 4)$$
$$= -21 - 0 + 18$$
$$= -3$$

$$D_y = \begin{vmatrix} 1 & -7 & -2 \\ 3 & 15 & 1 \\ 2 & -9 & -3 \end{vmatrix} = (1)\begin{vmatrix} 15 & 1 \\ -9 & -3 \end{vmatrix} - (3)\begin{vmatrix} -7 & -2 \\ -9 & -3 \end{vmatrix} + (2)\begin{vmatrix} -7 & -2 \\ 15 & 1 \end{vmatrix}$$

$$= 1(-45 + 9) - 3(21 - 18) + 2(-7 + 30)$$
$$= -36 - 9 + 46$$
$$= 1$$

$$D_z = \begin{vmatrix} 1 & 2 & -7 \\ 3 & -2 & 15 \\ 2 & 3 & -9 \end{vmatrix} = (1)\begin{vmatrix} -2 & 15 \\ 3 & -9 \end{vmatrix} - (3)\begin{vmatrix} 2 & -7 \\ 3 & -9 \end{vmatrix} + (2)\begin{vmatrix} 2 & -7 \\ -2 & 15 \end{vmatrix}$$

$$= 1(18 - 45) - 3(-18 + 21) + 2(30 - 14)$$
$$= -27 - 9 + 32$$
$$= -4$$

Answer to Your Turn 4
$(-4, 2)$

Note After finding the values of two of the variables, you could find the value of the third variable by substituting these values back into any one of the original equations.

$$x = \frac{D_x}{D} = \frac{-3}{-1} = 3, \quad y = \frac{D_y}{D} = \frac{1}{-1} = -1, \quad z = \frac{D_z}{D} = \frac{-4}{-1} = 4$$

The solution is $(3, -1, 4)$. We will leave the check to the reader.

Your Turn 5 Use Cramer's Rule to solve $\begin{cases} 2x + 3y - 2z = 11 \\ 3x + y + 4z = -8 \\ x - 3y - 2z = 4 \end{cases}$.

Answer to Your Turn 5
$(1, 1, -3)$

Appendix D Exercises

For Extra Help MyMathLab®

Note: Exercises marked with a ★ represent challenging exercises.

Objectives 1 and 2

Prep Exercise 1 What is a square matrix?

Prep Exercise 2 What is the formula for evaluating $\det(A) = \begin{vmatrix} a_1 & b_1 \\ a_2 & b_2 \end{vmatrix}$?

Prep Exercise 3 Is it possible to find the determinant of $\begin{bmatrix} 1 & 2 & 5 \\ 3 & 6 & 2 \end{bmatrix}$? Why or why not?

Prep Exercise 4 Explain the difference between a matrix and a determinant.

Prep Exercise 5 How do you find the minor of an element of a 3×3 matrix?

For Exercises 1–30, find the determinant. See Examples 1–3.

1. $\begin{bmatrix} 3 & 2 \\ 1 & 5 \end{bmatrix}$

2. $\begin{bmatrix} 4 & 1 \\ 3 & 7 \end{bmatrix}$

3. $\begin{bmatrix} -3 & 5 \\ 2 & 4 \end{bmatrix}$

4. $\begin{bmatrix} 5 & 4 \\ 2 & -6 \end{bmatrix}$

5. $\begin{bmatrix} 3 & -6 \\ 2 & 4 \end{bmatrix}$

6. $\begin{bmatrix} 2 & 8 \\ -3 & 2 \end{bmatrix}$

7. $\begin{bmatrix} -3 & 4 \\ -2 & 5 \end{bmatrix}$

8. $\begin{bmatrix} -3 & -5 \\ 4 & 6 \end{bmatrix}$

9. $\begin{bmatrix} -2 & -3 \\ -4 & 5 \end{bmatrix}$

10. $\begin{bmatrix} -6 & -2 \\ 3 & -4 \end{bmatrix}$

11. $\begin{bmatrix} 0 & 3 \\ -5 & 7 \end{bmatrix}$

12. $\begin{bmatrix} -5 & 0 \\ -4 & 2 \end{bmatrix}$

13. $\begin{bmatrix} 1 & 2 & 1 \\ 3 & 1 & 4 \\ 2 & 3 & 2 \end{bmatrix}$

14. $\begin{bmatrix} 3 & 1 & 4 \\ 2 & 2 & 3 \\ 1 & 4 & 3 \end{bmatrix}$

15. $\begin{bmatrix} -1 & 2 & 0 \\ -3 & 2 & 4 \\ -4 & 2 & 3 \end{bmatrix}$

16. $\begin{bmatrix} -2 & 0 & -1 \\ 3 & -2 & 4 \\ -3 & 2 & 1 \end{bmatrix}$

17. $\begin{bmatrix} 2 & 1 & -3 \\ 0 & -3 & 2 \\ 4 & 1 & -3 \end{bmatrix}$

18. $\begin{bmatrix} 3 & -2 & 4 \\ 3 & 0 & 2 \\ -4 & -2 & 2 \end{bmatrix}$

19. $\begin{bmatrix} 0 & 4 & -2 \\ 3 & 2 & 0 \\ -1 & 4 & 3 \end{bmatrix}$

20. $\begin{bmatrix} 3 & -5 & 0 \\ 2 & -4 & 1 \\ -2 & 0 & -3 \end{bmatrix}$

21. $\begin{bmatrix} 0.3 & -0.5 \\ 1.3 & -0.6 \end{bmatrix}$

22. $\begin{bmatrix} -0.4 & 1.6 \\ -4.7 & 3.1 \end{bmatrix}$

23. $\begin{bmatrix} -0.4 & 0.7 & -1.2 \\ 3.1 & 1.5 & -3.2 \\ 1.6 & -2.2 & -1.5 \end{bmatrix}$

24. $\begin{bmatrix} 1.7 & -3.2 & 4.1 \\ 5.3 & -6.2 & -1.1 \\ -1.3 & 2.3 & -4.5 \end{bmatrix}$

25. $\begin{bmatrix} \frac{1}{2} & -\frac{1}{3} \\ \frac{2}{5} & \frac{3}{5} \end{bmatrix}$

26. $\begin{bmatrix} -\frac{3}{4} & -\frac{3}{5} \\ \frac{3}{2} & \frac{2}{5} \end{bmatrix}$

★ **27.** $\begin{bmatrix} \frac{1}{2} & -\frac{3}{4} & \frac{2}{5} \\ \frac{1}{3} & \frac{1}{5} & -\frac{3}{2} \\ -\frac{3}{4} & \frac{1}{2} & \frac{3}{5} \end{bmatrix}$

★ **28.** $\begin{bmatrix} -\frac{1}{4} & -\frac{3}{2} & \frac{4}{3} \\ \frac{1}{5} & -\frac{5}{4} & \frac{1}{2} \\ -\frac{5}{3} & \frac{1}{4} & -\frac{4}{5} \end{bmatrix}$

★ **29.** $\begin{bmatrix} x & y & 1 \\ 2 & -1 & 3 \\ -2 & 0 & 1 \end{bmatrix}$

★ **30.** $\begin{bmatrix} x & y & 1 \\ -3 & -2 & 4 \\ 3 & -2 & 2 \end{bmatrix}$

Objective 3

Prep Exercise 6 How do you find D_y when solving a system of equations using Cramer's Rule?

For Exercises 31–44, solve using Cramer's Rule. See Example 4.

31. $\begin{cases} x + y = -5 \\ x - 2y = -2 \end{cases}$

32. $\begin{cases} x + 3y = 1 \\ x + y = -3 \end{cases}$

33. $\begin{cases} 2x - 3y = -6 \\ x - y = -1 \end{cases}$

34. $\begin{cases} 2x + 5y = -7 \\ x - y = -7 \end{cases}$

35. $\begin{cases} -x + 2y = -12 \\ 2x - 3y = 20 \end{cases}$

36. $\begin{cases} -x + 2y = -9 \\ 4x + 5y = -3 \end{cases}$

37. $\begin{cases} 4x - 6y = 7 \\ 6x - 9y = 8 \end{cases}$

38. $\begin{cases} 6x - 9y = 17 \\ 8x - 12y = 7 \end{cases}$

39. $\begin{cases} 8x - 3y = 10 \\ 4x + 3y = 14 \end{cases}$

40. $\begin{cases} 2x - y = 10 \\ 4x - 5y = 32 \end{cases}$

41. $\begin{cases} \frac{1}{2}x - \frac{1}{4}y = 0 \\ \frac{3}{4}x + \frac{5}{2}y = \frac{23}{2} \end{cases}$

42. $\begin{cases} \frac{2}{3}x + \frac{1}{4}y = \frac{1}{2} \\ \frac{3}{4}x + \frac{4}{3}y = -\frac{23}{4} \end{cases}$

43. $\begin{cases} 0.2x + 0.5y = 3.4 \\ 0.7x - 0.3y = -0.4 \end{cases}$

44. $\begin{cases} 1.2x - 0.6y = -2.4 \\ 3.1x + 1.3y = -11.9 \end{cases}$

For Exercises 45–56, solve using Cramer's Rule. See Example 5.

45. $\begin{cases} x + y + z = 6 \\ 2x - 4y + 2z = 6 \\ 3x + 2y + z = 11 \end{cases}$

46. $\begin{cases} 2x + y - 3z = -1 \\ x + 2y - 2z = -3 \\ -3x - 4y + z = -3 \end{cases}$

47. $\begin{cases} 3x + y - z = -4 \\ 2x - y + 2z = -7 \\ x - 3y + z = -6 \end{cases}$

48. $\begin{cases} x - y + 3z = -10 \\ 5x + 4y - z = -7 \\ 2x + y - z = -4 \end{cases}$

49. $\begin{cases} 4x + 2y + 3z = 9 \\ 2x - 4y - z = 7 \\ 3x - 2z = 4 \end{cases}$

50. $\begin{cases} 2x - y = -1 \\ 5x - 3y + 2z = 0 \\ 3x + 2y - 3z = -8 \end{cases}$

51. $\begin{cases} 3x + 2y = -12 \\ 3y + 10z = -16 \\ 6x - 2z = 3 \end{cases}$

52. $\begin{cases} 4x + 2z = 7 \\ 8x - 2y = -7 \\ 10y - 2z = -5 \end{cases}$

53. $\begin{cases} \frac{1}{2}x + \frac{1}{3}y + \frac{3}{4}z = \frac{25}{12} \\ \frac{1}{3}x + \frac{2}{9}y + \frac{1}{2}z = \frac{25}{18} \\ \frac{3}{4}x + \frac{1}{4}y - \frac{2}{3}z = -\frac{7}{4} \end{cases}$

54. $\begin{cases} \frac{3}{4}x - \frac{1}{2}y + \frac{1}{2}z = \frac{1}{4} \\ \frac{7}{6}x + \frac{10}{9}y - \frac{2}{5}z = -\frac{623}{90} \\ \frac{7}{4}x + \frac{5}{3}y - \frac{3}{5}z = -\frac{623}{60} \end{cases}$

55. $\begin{cases} 0.3x + 0.4y - 0.6z = 2.6 \\ 0.5x - 0.2y + 0.7z = -0.8 \\ 1.4x + 1.3y - 2.2z = 9.8 \end{cases}$

56. $\begin{cases} 1.2x + 2.1y - 0.5z = 7.3 \\ 3.2x - 2.4y + 1.3z = 6.1 \\ 2.5x + 1.3y - 1.7z = 8.4 \end{cases}$

★ *For Exercises 57–60, find x.*

57. $\begin{vmatrix} 9 & x \\ -6 & 5 \end{vmatrix} = 21$

58. $\begin{vmatrix} 12 & 2 \\ x & -3 \end{vmatrix} = -22$

59. $\begin{vmatrix} 2 & -1 & 0 \\ 0 & x & 1 \\ -2 & 0 & 4 \end{vmatrix} = -38$

60. $\begin{vmatrix} 1 & -2 & 4 \\ -1 & 0 & 2 \\ 0 & x & 5 \end{vmatrix} = -28$

★ *For Exercises 61–66, use the following information. Suppose a triangle has vertices of (x_1, y_1), (x_2, y_2), and (x_3, y_3), as shown in the graph to the right. The area of the triangle is given by* $A = \frac{1}{2}\left|\det\begin{bmatrix} x_1 & y_1 & 1 \\ x_2 & y_2 & 1 \\ x_3 & y_3 & 1 \end{bmatrix}\right|$. *For example, the area of the triangle whose vertices are $(1, 2)$, $(3, -2)$, and $(-2, 5)$ is* $A = \frac{1}{2}\left|\det\begin{bmatrix} 1 & 2 & 1 \\ 3 & -2 & 1 \\ -2 & 5 & 1 \end{bmatrix}\right|$.

Note This notation indicates half of the absolute value of the determinant.

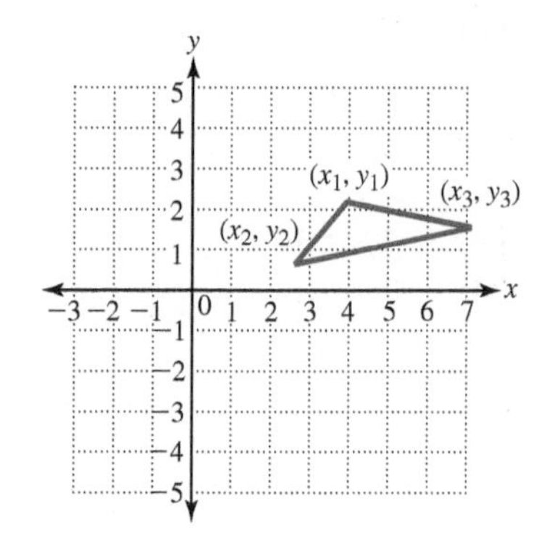

Expanding about the first column, we have

$$A = \frac{1}{2}\left|1\det\begin{bmatrix} -2 & 1 \\ 5 & 1 \end{bmatrix} - 3\det\begin{bmatrix} 2 & 1 \\ 5 & 1 \end{bmatrix} - 2\det\begin{bmatrix} 2 & 1 \\ -2 & 1 \end{bmatrix}\right|$$

$$= \frac{1}{2}|(-2-5) - 3(2-5) - 2(2+2)|$$

$$= \frac{1}{2}|-7+9-8| = \frac{1}{2}|-6| = 3 \text{ square units.}$$

Find the area of the triangles with vertices at the given points.

61. $(2, 4)\ (4, 0)\ (6, 5)$

62. $(0, 2)\ (3, -2)\ (5, 5)$

63. $(-3, 1)\ (2, -3)\ (4, 4)$

64. $(-3, -2)\ (-1, 4)\ (3, -4)$

65. $(-4, -1)\ (1, 3)\ (3, -3)$

66. $(-3, -3)\ (2, 2)\ (4, -1)$

For Exercises 67–72, translate the problem to a system of equations; then solve using Cramer's Rule.

67. The two heaviest known meteorites to be found on Earth's surface are the Hoba West, which was found in Namibia, and the Ahnighito, which was found in Greenland. The total weight of the two meteorites is 90 tons. The Hoba West is twice as heavy as the Ahnighito. Find the weight of each. (*Source: Webster's New World Book of Facts*)

68. The two largest expenses for the average American family are federal taxes and housing, including household expenses. Together these two items total 43.3% of the average family's income. The amount spent on taxes is 12.1% more than the amount spent on housing. Find the percent spent on each. (*Source: Numbers: How Many, How Long, How Far, How Much*)

69. A restaurant makes a soup that includes garbanzo and black turtle beans. The manager purchased 10 pounds of beans at a cost of $8.80. If the garbanzo beans cost $1.00 per pound and the black turtle beans cost $0.70 per pound, how many pounds of each did he purchase?

70. The perimeter of a triangle is 21 inches, and two sides are of equal length. The length of the third side is 3 inches less than the length of the two equal sides. Find the length of each side of the triangle.

71. Coinage bronze is made up of zinc, tin, and copper. The percent of tin is 4 times the percent of zinc. The percent of copper is 19 times the sum of the percents of zinc and tin. Find the percent of zinc, tin, and copper in coinage bronze. (*Source: Webster's New World Book of Facts*)

72. In winning the 2009 men's NCAA basketball championship, North Carolina scored a total of 56 times in their 89 to 72 victory over Michigan State. The sum of the number of 3-point field goals and 2-point field goals was equal to the number of free throws (1 point each). How many of each did North Carolina score? (*Source:* espn.go.com)

Photo Credits

Cover

Papajka/Shutterstock

FM

p. xvii: Pearson Education, Inc.

Chapter 1

p. 1: Romakoma/Shutterstock; p. 42: Classic Image/Alamy

Chapter 2

p. 58: Dmac/Alamy; p. 94: Fotolia; p. 95: Gerard Brown/Dorling Kindersley, Ltd.

Chapter 3

p. 120: Versh/Shutterstock; p. 121: Library of Congress Prints and Photographs Division [LC-USZ62-61365]; p. 142: Mark Karrass/Corbis

Chapter 4

p. 197: Alex Wong/Getty Images; p. 225: Fuse/Thinkstock; p. 226: David Grossman/Alamy; p. 228: Chuck Savage/Corbis; p. 230: Jorg Hackemann/Fotolia; p. 232: Jared C. Tilton/Asp Inc./Cal Sport Media/Newscom; p. 243: Getty Images; p. 244: Bill Shettle/CSM/Landov

Chapter 5

p. 263: Oliver Furrer/Alamy; p. 277: Photodisc/Getty Images; p. 278: Reuters/Landov

Chapter 6

p. 331: G. Murti/Science Source/Photo Researchers, Inc.

Chapter 7

p. 386: Ryan/Fotolia

Chapter 8

p. 460: Carlos Caetano/Shutterstock; p. 474: Bettmann/Corbis

Chapter 9

p. 537: Rudi/Fotolia; p. 546: Don Farrall/Getty Images; p. 553: ZUMA Press, Inc/Alamy; p. 557 (b): Richard Eyre/ iStockphoto; p. 557 (t): Frank Micelotta/Getty Images; p. 566 (b): Justin Lane/EPA/ Newscom; p. 566 (t): Charles Krupa/AP Images

Chapter 10

p. 603: David J. Green/Alamy; p. 615 (b): Kim Steele/Getty Images; p. 615 (t): Paul Sancya/AP Images; p. 621: Reed Kaestner/Corbis; p. 642: Antony Nettle/Alamy; p. 645: Library of Congress Prints and Photographs Division[LC-D4-19236]; p. 654: ZUMA Press/Alamy; p. 655: NASA

Chapter 11

p. 673: Andre Jenny/Alamy; p. 686: Andrew Ward/ Getty Images; p. 696: Claudio Zaccherini/Shutterstock

Appendix A

p. APP-27: Jonathan Blair/Corbis

Collaborative Exercises

Section 1.2 Windchill

Since wind helps exposed portions of your body to lose heat, how cold you feel outside depends not only on the temperature of the air, but also on how windy it is. For the last half of the 20th century, the accompanying table was used to determine a combination of air temperature (in °F) and wind speed called **windchill.** *For example, with a wind speed of 15 mph and an air temperature of 30°F, the windchill is 9°F. That is, the effect on loss of body heat would be similar to 9°F in still air. Use the table to answer the questions that follow.*

Air Temperature (degrees Fahrenheit)

Wind Speed (mph)	45	40	35	30	25	20	15	10	5	0	−5	−10	−15	−20	−25	−30
5	43	37	32	27	22	16	11	6	0	−5	−10	−15	−21	−26	−31	−36
10	34	28	22	16	10	3	−3	−9	−15	−21	−27	−34	−40	−46	−52	−58
15	29	22	15	9	2	−5	−12	−18	−25	−32	−38	−45	−52	−59	−65	−72
20	25	18	11	4	−3	−11	−18	−25	−32	−39	−46	−53	−60	−68	−75	−82
25	22	15	8	0	−7	−15	−22	−30	−37	−44	−52	−59	−67	−74	−82	−89
30	20	13	5	−3	−10	−18	−25	−33	−41	−48	−56	−64	−71	−79	−87	−94
35	19	11	3	−5	−12	−20	−28	−36	−44	−51	−59	−67	−75	−83	−90	−98
40	18	10	2	−6	−14	−22	−30	−38	−46	−53	−61	−69	−77	−85	−93	−101
45	17	9	1	−7	−15	−23	−31	−39	−47	−55	−63	−71	−79	−87	−95	−103
50	17	9	1	−7	−15	−23	−31	−40	−48	−56	−64	−72	−80	−88	−96	−104

Source: National Center for Atmospheric Research

1. For an air temperature of 15°F and a wind speed of 5 mph, what is the windchill?

2. What other temperature/wind speed combinations give the same windchill as in Number 1?

3. For an air temperature of 45°F, verify that the expression **20 − 43** shows the change in windchill as the wind speed *increases* from 5 to 30 mph. Simplify the expression **20 − 43** and comment on the sign of your answer.

4. For an air temperature of 25°F, write and simplify a mathematical expression for the change in windchill if the wind speed *decreases* from 35 mph to 15 mph. Comment on the sign of your answer.

5. For an air temperature of 10°F, what is the difference in windchill when the wind speed decreases from 35 mph to 10 mph?

6. Water freezes at 32°F. If the air temperature is 45°F, estimate how fast the wind must be blowing so that the windchill is 32°F.

7. Examine the row in the table for a wind speed of 5 mph. For an air temperature of 45°F, the windchill temperature is 43°F, or 2 degrees less than the air temperature.

a. Is this difference the same for all air temperatures at a wind speed of 5 mph?

b. How about for other wind speeds?

c. Describe the trend in the difference between air temperature and windchill as you move from left to right across any row in the table. Interpret your answer in terms of how wind affects how cold a person feels?

Section 2.3 Optical Illusion or Confusion?

An optical shop at a local mall advertises the sale shown to the right. The total cost for a pair of glasses is the sum of the costs of the frame and lenses.

1. Let F represent the regular price of a frame; L, the regular price for lenses. Write an expression that describes the total cost of a pair of glasses at the regular price.

2. Does the expression $F + 0.60L$ give the cost of the glasses during the advertised 40% off sale? Why or why not.

3. The regular price for Anna's lenses is $90. Anna chooses frames listed at $120. How much did Anna save by buying her glasses during the sale?

4. Write and solve an inequality to determine the price of the most expensive frame Anna can choose during the sale if she wants to keep the cost of her glasses to at most $125.

5. Every day the optical shop gives a discount of 25% on frames and lenses to seniors. Pat has chosen a $140 frame. She wears bifocals, so her lenses are $260. The shop will apply only one of the discounts. Which would be better for Pat, the 25% senior discount or the advertised 40% off sale?

6. The optical shop has a complete series of economy frames for $60. Write and solve an inequality showing for which lens prices the 40% off sale would be more economical than the senior discount if an economy frame is used.

7. College students are eligible for a 15% discount on glasses that cost over $100. Using F to represent the regular price of a frame and L for the regular price of lenses, write an expression that describes the total cost of a pair of glasses with the student discount.

8. If a student selects $80 frames, write and solve an inequality showing for which lens prices the 40% off sale would be better than the everyday student discount.

Section 3.2 Table Math

Tables, mental math, and visual techniques can serve as tools for understanding and calculating slope and finding ordered pairs for less complex problems.

1. Consider the ordered pairs in the table. Suppose we know that the points form a straight line, and we want to find the slope of the line through the points. The lines outside the table show that as x increases by 2 from 0 to 2, y increases by 7 from 2 to 9. Does this pattern hold for all ordered pairs in the table?

+2 (x: 0 → 2) +7 (y: 2 → 9)

x	y
0	2
2	9
4	16
6	23

2. a. Use the slope formula $m = \dfrac{y_2 - y_1}{x_2 - x_1}$ to find the slope of the line formed by the ordered pairs in the table.

   b. Discuss the connection between the slope you calculated and the pattern shown in Question 1.

3 a. Given the same table as in Question 1, find the two missing values a and b.

   b. Write the ratio of b to a in simplest form. What is this value?

   c. What does your calculation in part b suggest about how to find the slope?

$+a$ (x: 0 → 6) $+b$ (y: 2 → 23)

x	y
0	2
2	9
4	16
6	23

4. Note that the y-intercept is given in the preceding table. Write the equation of the line (in slope–intercept form) that fits the ordered pairs in the table.

5. Use the ratio method and the following table to find the slope of the line through the points $(1, 7)$ and $(4, -4)$. Verify your answer with the slope formula.

x	y
1	7
4	−4

6. In graphing $y = -\frac{3}{4}x + 5$, we already know that the y-intercept is $(0, 5)$ and that the slope is $-\frac{3}{4}$.

We can use a table to find a second point (and even a third checkpoint). Find those points.

x	y
0	5

(+4 applied to x; −3 applied to y)

Section 4.5 Maximizing the Profit

Linear programming is an area of mathematics that solves problems like the one described below. One of the fundamental elements of a linear programming model is the constraint. A constraint is simply an inequality that describes some limited resource required in the problem. For example, suppose you make two types of bicycles, style A and style B. Also suppose that style A requires 40 minutes of welding for each bicycle and style B requires 25 minutes of welding. If you have only 600 minutes of welding time available, the welding-time constraint would be $40A + 25B \leq 600$, *where A represents the number of style A produced and B represents the number of style B produced.*

An aspiring artists' group is preparing for a fund-raising sale. The artists design and construct two types of antique reproduction tables: a semicircular foyer table and a side table. The materials cost \$100 for each foyer table and \$200 for each side table. Both types of table require 2 hours of cutting and carving. The foyer table requires 3.5 hours to assemble, sand, and finish, and the side table requires 1.75 hours to assemble, sand, and finish. The group has \$3300 to purchase all materials; 40 hours available for cutting and carving; and 63 hours available for assembling, sanding, and finishing. For each foyer table they sell, their profit will be \$350. For each side table, their profit will be \$215. How many of each table should they make to maximize their profit?

1. For the artists' fund-raising problem, there are three basic constraints: (1) cost of materials, (2) cutting and carving, and (3) assembling/sanding/finishing. For each of these constraints, write an inequality similar to the one in the bicycle example. Let x represent the number of foyer tables and y the number of side tables.

2. Because x and y represent numbers of tables, they cannot represent negative values. Write inequalities for these two additional constraints.

3. Now graph the system of inequalities described by these five constraints.

4. The region that is the solution of the system of inequalities is called the *feasible region*, and it includes all points that satisfy the system. The goal is to find the optimal solution, which, according to linear programming, is one of the corner points of the feasible region. There should be five corner points: one at the origin, one on each axis, and two more in quadrant I. Determine the coordinates of each corner point. Note that a corner point lying on an axis is the x- or y-intercept for the line passing through that point. Because two lines intersect to form a corner point that is not on an axis, make a system out of their two equations; then solve the system to find their point of intersection.

5. Next, develop an algebraic expression describing the profit gained for selling x of the foyer tables and y of the side tables. This is called the *objective function*.

6. Using the objective function, test each of the points found in step 4 to see which one yields the maximum profit.

7. Using your solution, answer the following questions.
   a. How many of each type of table should the group produce?
   b. How many hours will be spent in each phase of production?
   c. How much money will the group need to purchase the materials?
   d. What amount of profit will the group receive?

Section 5.4 Demo CD

A band decides to produce a demo CD. The band rents recording equipment for $200 and purchases the CDs for $0.50 each.

1. Write a function $C(x)$ that represents the total cost to produce x CDs.
2. Find the cost to produce 50 CDs.
3. Divide your answer to Problem 2 by 50. The result will be the average cost to produce each CD if 50 CDs are made.
4. Find the average cost of producing 100 CDs.
5. Write a function that can be used to calculate the average cost of producing x CDs.
6. As x gets larger, what happens to the average cost?

Section 6.4 The Sandbox

The backyard of a daycare center has a 13-foot-by-18-foot rectangular area that would be perfect for a large sandbox. Enough sand to cover a 150-square-foot play area to a sufficient depth has been purchased. The center wants the sandbox to be rectangular and have a uniform wooden border around the play area for seating. How wide should the border be?

1. The figure to the right is a rough sketch of the desired setup. If x represents the width of the border, write expressions for the length and width of the inside of the box where the sand is placed.

18 ft.
13 ft.
x

2. Write an expression for the area that will be filled with sand.
3. The daycare center has sufficient sand to cover 150 square feet of play area. Write an equation using your area expression from question 2 to ensure that the play area is 150 square feet.
4. Solve your equation from question 3. Which of the two solutions is reasonable in this situation? Why?
5. What are the dimensions of the area to be filled with sand?

Section 7.3 Average Rates

1. A person drives for 10 minutes at 50 miles per hour and then drives 15 minutes at 60 miles per hour. What is her average rate?
2. A person drives 15 miles at 50 miles per hour and then travels another 20 miles at 40 miles per hour. What is his average rate?
3. A person travels a certain distance at 60 miles per hour and then travels that same distance at 70 miles per hour. What is her average rate?

Section 8.6 Building Time

You are constructing a grandfather clock that operates using a pendulum. The period of the pendulum is the time required to complete one full swing. The formula giving the relationship between the length, L (in meters), and the period, T (in seconds), is $T = 2\pi\sqrt{\frac{L}{9.8}}$.

1. Find the length of the pendulum if the period is 1 second.
2. Suppose you want the pendulum to complete one period in 2 seconds. Would you need to increase or decrease the length of the pendulum found in Exercise 1? Use the same formula to determine the length of the pendulum so that the period is 2 seconds.
3. Based on the results of Exercises 1 and 2, what can you conclude about the required length of the pendulum as the period increases?

Section 9.4 Arch Span

The Gateway Arch, located in St. Louis, Missouri, was built from 1963 to 1965 and is the nation's tallest memorial. Although the arch is not a parabola, the equation $h(x) = -0.0063492063x^2 + 630$ *can be used to approximate the height of the arch, where x represents the distance from the axis of symmetry to the arch and h(x) represents its height above the ground.*

1. Using the equation, find the maximum height of the structure.
2. The span of the arch at a given height is the horizontal distance between the two opposing points on the arch at that height. Find the span of the arch at ground level. (*Hint:* Think of ground level as the x-axis. The height is 0 along the x-axis.)

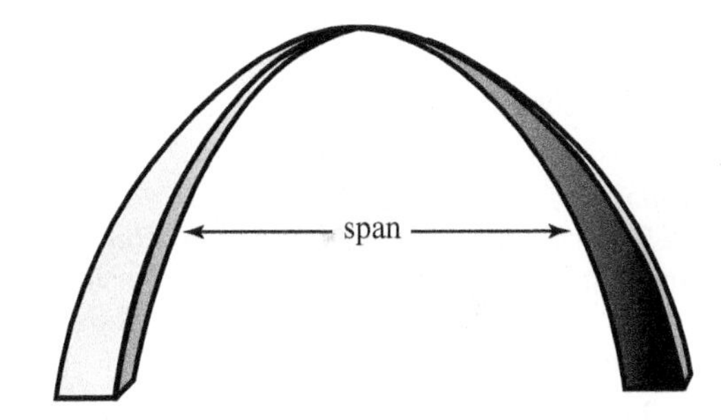

3. How does the span at ground level relate to the maximum height of the arch?
4. The arch has foundations 60 feet below ground level. What is the span of the arch between its foundations?

Section 10.5 Exploring Graphs of Logarithms

Using a graphing calculator, draw the graph of $f(x) = \log x$ *and* $g(x) = \log(10x)$ *in the window* $[0, 100]$ *for x and* $[-2, 4]$ *for y.*

1. Press the TRACE key. Using up or down arrow keys, move from one graph to the other and observe the y-values of several points with the same x-values. What do you observe?
2. How is the graph of $g(x) = \log(10x)$ related to the graph of $f(x) = \log x$? Why?
3. In general, how is the graph of $g(x) = \log(kx)$ for $k > 0$ related to the graph of $f(x) = \log x$?
4. Repeat step 1 using $f(x) = \log x$ and $g(x) = \log \frac{x}{10}$.

5. How is the graph of $g(x) = \log \frac{x}{10}$ related to the graph of $f(x) = \log x$? Why?

6. In general, how is the graph of $g(x) = \log \frac{x}{k}$ for $k > 0$ related to the graph of $f(x) = \log x$?

Section 11.2 The Elliptical Tablecloth

Recall that an ellipse can be drawn by fixing the ends of a string to the foci and then tracing out the ellipse. By considering the following figures, we can determine an expression for the length of the string and a relationship between a, b, and c.

1. Use the figure to write a formula for the length of the string.

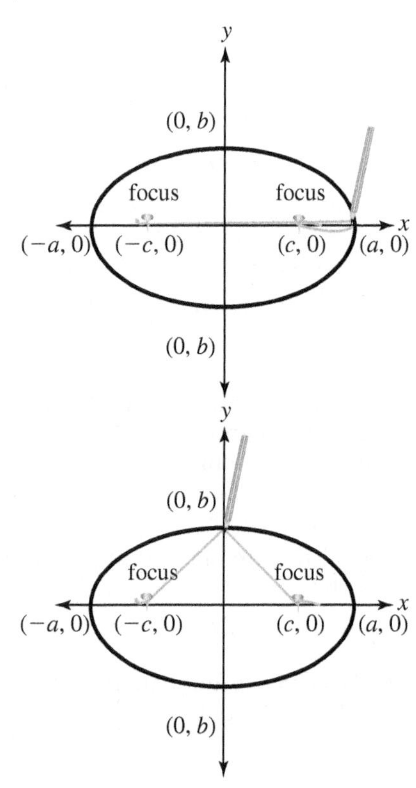

2. In the figure at right, notice that the string forms an isosceles triangle and the y-axis splits that triangle into two identical right triangles.
   a. Find the length of the hypotenuse of each of those right triangles.
   b. What expression describes the length of the string?
   c. Use the Pythagorean theorem to write a formula relating a, b, and c.
   d. Solve the formula for c.

Suppose we are to make an elliptical tablecloth for an elliptical table that is 76 inches long and 58 inches wide. The tablecloth is to drape 6 inches over the edge of the table all the way around, and we need an additional inch for the hem. We have a large rectangular piece of cloth to make the tablecloth. To trace the ellipse on the cloth, we need to know the string length and the location of the foci.

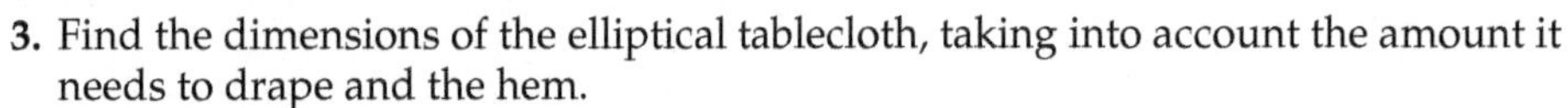

3. Find the dimensions of the elliptical tablecloth, taking into account the amount it needs to drape and the hem.

4. How long must the string be so that the ellipse can be traced?

5. How far from the center are the foci located?

Answers

Chapter 1

Exercise Set 1.1

Prep Exercises **1.** A constant is a symbol that does not vary in value, whereas a variable is a symbol that varies in value. **2.** calculation **3.** equal sign **4.** $<, >, \leq, \geq, \neq$. **5.** Equations and Inequalities; Expressions; Constants and Variables. **6.** A collection of objects. **7.** *braces*, { } **8.** A finite set has a finite (countable) number of elements; an infinite set does not. **9.** $\{\varnothing\}$ is not empty because the set contains one element, the symbol $\varnothing$. **10.** If every element of a set B is an element of a set A, then B is a subset of A. **11.** Rational numbers can be expressed as a ratio of two integers; irrational numbers cannot. **12.** rational, irrational **13.** positive, 0 **14.** right

Exercises **1.** {Saturday, Sunday} **3.** {January, June, July} **5.** {North Carolina, North Dakota} **7.** $\{0, 1, 2, 3, 4\}$ **9.** $\{3, 6, 9, \ldots\}$ **11.** $\{-1, 0, 1\}$ **13.** { } **15.** $\{x \mid x$ is an integer$\}$ **17.** $\{x \mid x$ is a letter of the English alphabet$\}$ **19.** $\{x \mid x$ is a day of the week$\}$ **21.** $\{x \mid x$ is a natural-number multiple of 5$\}$ **23.** **25.** **27.** **29.**

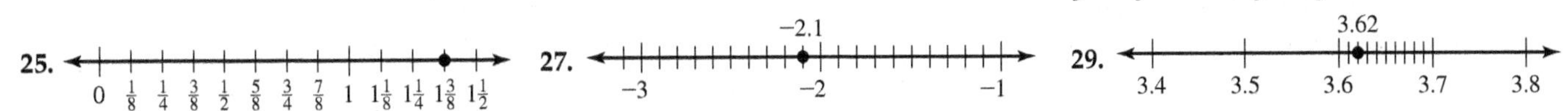

31. false **33.** true **35.** true **37.** true **39.** false **41.** true **43.** true **45.** true **47.** true **49.** false **51.** true **53.** true **55.** true **57.** 2.6 **59.** $1\frac{2}{5}$ **61.** 1 **63.** 8.75 **65.** $>$ **67.** $>$ **69.** $>$ **71.** $=$ **73.** $=$ **75.** $<$ **77.** $-0.6, -0.44, 0, |-0.02|, 0.4, \left|1\frac{2}{3}\right|, 3\frac{1}{4}$ **79.** $-12.6, -9.6, 1, |-1.3|, \left|-2\frac{3}{4}\right|, 2.9$ **81.** {2005, 2006, 2008} **83.** {2006, 2008} **85.** {Internet Explorer, Firefox, Chrome} **87.** {Internet Explorer, Firefox} **89.** {Europe, Asia, North America} **91.** {South America, North America, Asia, Europe}

Exercise Set 1.2

Prep Exercises **1.** The sum of 0 and a number is itself; the product of 1 and a number is itself. **2.** Their sum is 0. **3.** Two numbers that have a product of 1. **4.** The order of the addends is changed with the commutative property of addition, whereas the grouping is changed with the associative property of addition. **5.** To add two numbers that have the same sign, add their absolute values and keep the same sign. **6.** To add two numbers that have different signs, subtract the smaller absolute value from the greater absolute value and keep the sign of the number with the greater absolute value. **7.** n **8.** To write a subtraction statement as an equivalent addition statement, change the operation symbol from a minus sign to a plus sign and change the subtrahend to its additive inverse. **9.** When multiplying or dividing two numbers with different signs, the product or quotient is negative. **10.** positive; negative

Exercises **1.** Additive inverse. **3.** Additive identity. **5.** Multiplicative identity. **7.** Multiplicative inverse. **9.** Multiplicative identity. **11.** Additive inverse. **13.** $-8, \frac{1}{8}$ **15.** $7, -\frac{1}{7}$ **17.** $\frac{5}{8}, -\frac{8}{5}$ **19.** $-0.3, \frac{10}{3}$ **21.** Commutative property of addition. **23.** Distributive property. **25.** Associative property of multiplication. **27.** Associative property of addition. **29.** Commutative property of addition. **31.** Commutative property of multiplication. **33.** -7 **35.** -40 **37.** -6 **39.** -5 **41.** $-\frac{7}{12}$ **43.** $-\frac{19}{24}$ **45.** 6.52 **47.** -7.38 **49.** 7 **51.** -2.7 **53.** -12 **55.** $-\frac{3}{4}$ **57.** 5 **59.** 12 **61.** -4 **63.** 5 **65.** $\frac{13}{10}$ **67.** 0 **69.** 2.55 **71.** -1.6 **73.** -12 **75.** 2 **77.** $-\frac{1}{2}$ **79.** 5 **81.** $-\frac{3}{8}$ **83.** -40 **85.** -6 **87.** 0.54 **89.** -60 **91.** -90 **93.** 2280.83 **95.** $-\$814.66$ **97.** -1281.5 N **99.** \$4.632 million **101. a.** Earth: -177.1 lb.; Moon: -30.25 lb.; Mars: -67.65 lb. **b.** Earth **103.** -51.2 V

Review Exercises: **1.** {Washington, Adams, Jefferson, Madison} **2.** No because an element, 6, is in B but not in A. **3.** infinite **4.** -6 can be written as $-\frac{6}{1}$. **5.** 25 **6.** $=$

Exercises Set 1.3

Prep Exercises **1.** To evaluate an exponential form raised to a natural number exponent, write the base as a factor the number of times indicated by the exponent and then multiply. **2.** Positive. It represents a product of an even number of negative factors. **3.** Negative. This is the additive inverse of 8^6, which is positive. **4.** A number's square is a product resulting from multiplying the number by itself. A number's square root is a number whose square is the given number. **5.** Two, a positive root and a negative root. **6.** None **7.** The principal (nonnegative) square root. **8.** Because a negative number raised to an odd power is negative. **9.** Work from the inside out: parentheses, then brackets, and then braces. **10.** Simplify the numerator and denominator separately; then divide the results.

Exercises **1.** Base = −4; exponent = 3; negative four to the third power, or negative four cubed. **3.** Base = 1; exponent = 7; the additive inverse of one raised to the seventh power. **5.** 625 **7.** 81 **9.** −64 **11.** −36 **13.** 1 **15.** $\frac{9}{64}$ **17.** $-\frac{125}{216}$ **19.** 0.064 **21.** −9.261 **23.** ±15 **25.** ±16 **27.** $\pm\frac{2}{3}$ **29.** No real-number roots exist. **31.** 5 **33.** 2 **35.** 0.6 **37.** −3 **39.** $\frac{2}{3}$ **41.** Not a real number. **43.** 2 **45.** 5 **47.** −55 **49.** 16 **51.** 8 **53.** 20 **55.** 21 **57.** 26.8 **59.** −3.96 **61.** 18 **63.** $-\frac{253}{300}$ **65.** −4 **67.** undefined **69.** The associative property of multiplication was used to multiply $3 \cdot 3$ instead of multiplying $1 \cdot 3$ from left to right. **71.** The distributive property was applied instead of adding $-1 + 36$ in the parentheses. **73.** Mistake: Multiplied before divided. Correct: −2 **75.** Mistake: Found the square root of the addends 16 and 9 instead of their sum. Correct: 20 **77. a.** 72.9 min.; **b.** No, her average time is greater than 72 minutes. **79.** 2362.5 **81.** 1379.186 **83.** 2.64 **85.** $35.5 + 0.10(658 - 500) + 0.12(45) = \$56.70$ **87.** $0.35(814) + 54.50 + 3(89.90) + 112.45 = \$721.55$ **89.** $[2(8.95) + 2(10.95) + 6.95 + 2(1.45)] \div 5 = \$9.93$ **91.** $1200 + 349(3)(12) + 0.15(52{,}000 - 45{,}000) = \$14{,}814$ **93.** 4096 **95.** 128 **97.** 270,000

Review Exercises: **1.** $\{0, 1, 2, 3, 4, 5, 6, 7, 8\}$ **2.** 73 **3.** −60 **4.** −19 **5.** It is an expression because it has no equal sign. **6.** Commutative property of multiplication.

Exercise Set 1.4

Prep Exercises **1.** Addition is commutative. **2.** *From* and *less than.* **3.** To evaluate a variable expression, (1) replace each variable with its corresponding given value and (2) simplify the resulting numerical expression. **4.** $ab + ac$ **5.** Like terms are variable terms that have the same variable(s) raised to the same exponents, or constant terms. **6.** To combine like terms, add or subtract the coefficients and keep the variables and their exponents the same.

Exercises **1.** $5n$ **3.** $3n - 2$ **5.** $5 - p$ **7.** $\frac{n^4}{8}$ **9.** $2n - 20$ **11.** x^4y^2 **13.** $\frac{p}{q} - \frac{1}{2}$ **15.** $m - 3(n + 5)$ **17.** $(4 - t)^5$ **19.** $\frac{6}{7}x + 7$ **21.** $(m - n) - (x + y)$ **23.** Mistake: incorrect order.; Correct: $y - 6$. **25.** Mistake: *sum* means use parentheses.; Correct: $4(r + 7)$. **27.** $5w$ **29.** $2 - 3w$ **31.** $2r$ **33.** $17 - n$ **35.** $\left(t + \frac{1}{2}\right)$ hr. **37.** πd **39.** $\frac{1}{2}h(a + b)$ **41.** $\frac{1}{3}\pi r^2 h$ **43.** $\frac{1}{2}mv^2$ **45.** $\frac{Mm}{d^2}$ **47.** $\sqrt{1 - \frac{v^2}{c^2}}$ **49.** −7 **51.** $-\frac{11}{3}$ **53.** 58 **55.** 10 **57. a.** 35.6 **b.** $\frac{16}{3}$ **59. a.** −3 **b.** $\frac{1}{4}$ **61.** 0 **63.** 6 **65.** −5, 1 **67.** $-\frac{1}{4}$ **69.** $27x - 45$ **71.** $-5m - 10$ **73.** $\frac{1}{12}x - 9$ **75.** $-6.3x - 5.04$ **77.** $-11x$ **79.** $-\frac{2}{7}b^2$ **81.** $7x - 8y - 12$ **83.** $-1.3x - 0.4$ **85.** $2.2h^2 + \frac{8}{3}h + 7$ **87.** $14n - 16$ **89.** $-5a + 7b - 12$ **91. a.** $14 + (6x - 8x)$ **b.** $14 - 2x$ **c.** 20

Review Exercises: **1.** $\{x \mid x \text{ is an integer and } x \geq -2\}$ **2.** −35 **3.** −84 **4.** −49 **5.** Commutative property of addition. **6.** Distributive property.

Chapter 1 Summary and Review Exercises

1.1 Definitions variable; constant; constant, variable, constants, variables; set equal to each other; $\neq, <, >, \leq, \geq$; set $\{1, 2, 3, \ldots\}$; $\{0, 1, 2, 3, \ldots\}$; $\{\ldots, -3, -2, -1, 0, 1, 2, 3, \ldots\}$; subset
1.2 $\frac{a}{b}$; integers; irrational number; 1, the number itself; 0; sum, 0; product, 1; 1.4 term; numerical factor; like terms.

1.1 Definitions/Rules/Procedures elements, members Exercises **1.** {Alaska, Hawaii} **2.** $\{\ldots, -3, -1, 1, 3, 5, \ldots\}$ **3.** $\{5, 10, 15, \ldots\}$ **4.** {s, i, m, p, l, f, y} **5.** $\{x \mid x \text{ is a natural-number multiple of } 3\}$ **6.** $\{x \mid x \text{ is a whole number}\}$ **7.** $\{x \mid x \text{ is a prime number}\}$ **8.** $\{x \mid x \text{ is a day of the week}\}$ **9.** true **10.** false **11.** false **12.** true Definitions/Rules/Procedures $n, -n$; right Exercises **13.** = **14.** > **15.** = **16.** >

1.2 Definitions/Rules/Procedures a; $b + a$; $a + (b + c)$; 0; a; ba; $a(bc)$; $ab + ac$ Exercises **17.** Additive inverse. **18.** Multiplicative inverse. **19.** Additive identity. **20.** Multiplicative identity. **21.** Distributive property. **22.** Associative property of multiplication. **23.** Commutative property of addition. **24.** Commutative property of multiplication. **25.** Associative property of addition. **26.** Distributive property. Definitions/Rules/Procedures absolute values; subtract, greater absolute value; n; minus sign, plus sign, additive inverse Exercises **27.** −1 **28.** 5 **29.** −17 **30.** −7 **31.** −2 **32.** −10 **33.** 17 **34.** −7 Definitions/Rules/Procedures positive; negative; positive, negative; division, multiplication, multiplicative inverse; positive; negative; 0; undefined; indeterminate Exercises **35.** −8 **36.** 15 **37.** −56 **38.** −5 **39.** 2 **40.** −2 **41.** −6 **42.** 25 **[2] 43.** −$220.44

1.3 Definitions/Rules/Procedures b, n; positive; negative Exercises **44.** Base = 2; exponent = 7; the additive inverse of two raised to the seventh power; −128. **45.** Base = −1; exponent = 4; negative one raisedto the fourth power; 1. **46.** −9 **47.** −8 **48.** 16 **49.** $-\frac{8}{125}$ Definitions/Rules/Procedures positive, negative; 0; no; principal (nonnegative); $\frac{\sqrt{a}}{\sqrt{b}}$; index

Exercises **50.** 11 **51.** 3 **52.** 3 **53.** 1 Definitions/Rules/Procedures Exponents/roots; Multiplication/division; Addition/subtraction Exercises **54.** −58 **55.** −61 **56.** −61 **57.** 22 **58.** 625 **59.** −2.6 **60.** 2.83 **61.** 59 + 0.15(37); $64.55 **62.** 12,500,000 codes

1.4 Definitions/Rules/Procedures variables, constants, key words Exercises **63.** $14 - 8n$ **64.** $2(n - 2)$ **65.** $n + \frac{1}{3}(n - 4)$ **66.** $\frac{m}{\sqrt{n}}$ **67.** $\frac{1}{2}(n - 8) - 16$ **68.** $(n + 5) + 20$ **69.** $2w$ **70.** $\left(\frac{1}{3} + t\right)$ hr. Definitions/Rules/Procedures variable; Simplify; denominator equal to 0 Exercises **71.** 9 **72.** 0 **73.** 16 **74.** −21 **75.** 3 **76.** $\frac{1}{2}$ Definitions/Rules/Procedures $ab + ac$; coefficients, variables, exponents Exercises **77.** $-10x - 2$ **78.** $8a + 12b - 16$ **79.** $-x^2 - 2x$ **80.** $m^5 - 2mn^2 + 2mn$ **81.** $-7a^2 + 3ab^2 - 5ab + 3a - 8$ **82.** $3r - 10$

Chapter 1 Practice Test

1. 8.1 [1.1] **2.** $-\frac{11}{4}$ [1.1] **3.** 13 [1.3] **4.** 5 [1.3] **5.** $\frac{1}{2}$ [1.3] **6.** Commutative property of addition. [1.2] **7.** Associative property of multiplication. [1.2] **8.** 8 [1.2] **9.** $\frac{11}{12}$ [1.2] **10.** −7.5 [1.2] **11.** 25 [1.3] **12.** $-\frac{4}{25}$ [1.2] **13.** 2 [1.3] **14.** 4 [1.3] **15.** −10 [1.3] **16.** −6 [1.3] **17.** 18 [1.3] **18.** 0 [1.3] **19.** −8 [1.3] **20.** −$389.50 [1.3] **21.** 5.19 million [1.3] **22.** −138 [1.4] **23.** 2 [1.4] **24.** $-21x - 35$ [1.4] **25.** $-\frac{23}{5}x + \frac{27}{4}y + 2.7$ [1.4]

Chapter 2

Exercise Set 2.1

Prep Exercises **1.** A solution for an equation is a number that makes the equation true when it replaces the variable in the equation. **2.** A solution set is a set containing all of the solutions for a given equation. **3.** Multiply both sides by the LCD. **4.** 1000 **5.** After combining like terms, both sides of the equation are identical. **6.** The solution set for an identity contains every real number for which the equation is defined. **7.** After combining like terms, the variable terms are identical but the constant terms are not. **8.** { } or ∅ **9.** Divide both sides of the equation by r. **10.** Subtract b from both sides.

Exercises **1.** 6 **3.** −1 **5.** 1 **7.** 3 **9.** 1 **11.** 1 **13.** 13 **15.** 4 **17.** 5 **19.** −6.5 **21.** 10 **23.** 3 **25.** 7 **27.** −3 **29.** $\frac{32}{7}$ **31.** 30 **33.** 5 **35.** 0.2 **37.** 3 **39.** No solution. **41.** All real numbers. **43.** Mistake: The distributive property was not used correctly. Correct: $\frac{7}{5}$ **45.** Mistake: Subtracted before distributing into the parentheses. Correct: 3 **47.** Mistake: Did not distribute the − with the 2. Correct: 7 **49.** $C = R - P$ **51.** $b = \frac{A}{h}$ **53.** $p = \frac{A}{2\pi w}$ **55.** $\theta = \frac{2A}{r^2}$ **57.** $M = \frac{Fd^2}{km}$ **59.** $s = \frac{A}{\pi(R + r)}$ **61.** $l = \frac{P - 2w}{2}$ **63.** $y = \frac{6 - 3x}{2}$ **65.** $C = \frac{5}{9}(F - 32)$ **67.** $a = \frac{2(x - vt)}{t^2}$ **69.** Mistake: Subtracted lw instead of dividing by lw. Correct: $h = \frac{V}{lw}$ **71.** Mistake: Subtracted the coefficient of l instead of dividing by the coefficient. Correct: $l = \frac{P - 2w}{2}$

Review Exercises: **1.** {1, 3, 5, 7, 9, 11, 13} **2.** 3314 **3.** $7n - 9$ **4.** $-3(n + 8)$ **5.** $-2x - 6y + 5$ **6.** $-54m + 24$

Exercise Set 2.2

Prep Exercises **1.** (1) Understand the problem. (2) Devise a plan. (3) Execute the plan. (4) Check results. **2.** Answers will vary. Some possible answers: Draw a picture. Make a table. Underline key words. Search for a related example. **3.** is equal to, is, yields, is the same as, produces, results in **4.** The amount of time is the same for both vehicles.

Exercises **1.** −78.5°C **3.** 20 in. **5.** 12 cm **7.** $w = 18$ ft., $l = 24$ ft. **9.** 180 ft.^{2} **11.** $x = 35$ ft., $x + 10 = 45$ ft. **13.** 75 **15.** $3000 **17.** 4.2 slugs **19.** 460.5 kg **21.** 15 m/sec. **23.** 12.5 ft./sec.^{2} **25.** 8 **27.** −9 **29.** −6 **31.** 6 **33.** $108.54 **35.** $999.93 **37.** 37,500 units **39.** 2 hr. **41.** 1.75 hr.

Review Exercises: **1.** $-5\frac{3}{8}, -\frac{1}{6}, 0.02, 4.5\%, \sqrt{48}, |-15.8|$ **2.** 2 **3.** = **4.** > **5.** −6 **6.** $-\frac{15}{8}$

Exercise Set 2.3

Prep Exercises **1.** Any value that makes the inequality true. **2.** For $x < a$, the value of a is not included in the solution set. For $x \le a$, the value of a is included in the solution set. **3.** The set of all values x such that x is greater than or equal to 2. This means that every real number greater than 2 and including 2 is in the solution set for the variable x. **4.** All real numbers less than or equal to 4 are part of the solution set. **5.** All real numbers greater than −2, but not including −2, are in the solution set. **6.** The graph shows that every real number greater than −1 but not including −1 is in the solution set. **7.** The graph shows that 3 and every real number less than 3 is in the solution set. **8.** The direction of an inequality symbol changes when both sides of the inequality are multiplied or divided by a negative number. **9.** ≤ **10.** ≥

Exercises **1. a.** $\{x|x \geq 5\}$ **b.** $[5, \infty)$ **c.** −2 −1 0 1 2 3 4 5 6 7 8 9 10 **3. a.** $\{q|q < -1\}$ **b.** $(-\infty, -1)$ **c.** −6 −5 −4 −3 −2 −1 0 1 2 3 4 5 6 **5. a.** $\left\{p|p < \frac{1}{5}\right\}$ **b.** $\left(-\infty, \frac{1}{5}\right)$ **c.** $\frac{1}{5}$ −1 0 1 2 3 **7. a.** $\{r|r \leq 1.9\}$ **b.** $(-\infty, 1.9]$ **c.** 1.9 0 1 2 **9. a.** $\{r|r < -6\}$ **b.** $(-\infty, -6)$ **c.** −10 −9 −8 −7 −6 −5 −4 −3 −2 −1 0 1 2 **11. a.** $\{y|y \geq 4\}$ **b.** $[4, \infty)$ **c.** −2 −1 0 1 2 3 4 5 6 7 8 9 10 **13. a.** $\{p|p < 4\}$ **b.** $(-\infty, 4)$ **c.** −2 −1 0 1 2 3 4 5 6 7 8 9 10 **15. a.** $\left\{x|x < -\frac{23}{5}\right\}$ **b.** $\left(-\infty, -\frac{23}{5}\right)$ **c.** $-\frac{23}{5}$ −6 −5 −4 −3 **17. a.** $\{a|a < -5\}$ **b.** $(-\infty, -5)$ **c.** −10 −9 −8 −7 −6 −5 −4 −3 −2 −1 0 1 2 **19. a.** $\{x|x < -9\}$ **b.** $(-9, \infty)$ **c.** −11 −10 −9 −8 −7 −6 **21. a.** $\{k|k < 4\}$ **b.** $(-\infty, 4)$ **c.** −3 −2 −1 0 1 2 3 4 5 6 7 8 9 **23. a.** $\{w|w \leq 5\}$ **b.** $(-\infty, 5]$ **c.** −3 −2 −1 0 1 2 3 4 5 6 7 8 9 **25. a.** $\{y|y < 8\}$ **b.** $(-\infty, 8)$ **c.** −3 −2 −1 0 1 2 3 4 5 6 7 8 9 **27. a.** $\left\{x|x < -\frac{11}{6}\right\}$ **b.** $\left(-\infty, -\frac{11}{6}\right)$ **c.** $-\frac{11}{6}$ −3 −2 −1 0 **29. a.** $\left\{m|m \leq \frac{9}{4}\right\}$ **b.** $\left(-\infty, \frac{9}{4}\right]$ **c.** $\frac{9}{4}$ 0 1 2 3 4 **31. a.** $\{x|x \geq -10\}$ **b.** $[-10, \infty)$ **c.** −12 −11 −10 −9 −8 −7 −6 −5 −4 −3 −2 −1 0 **33. a.** $\{z|z > 4\}$ **b.** $(4, \infty)$ **c.** −4 −3 −2 −1 0 1 2 3 4 5 6 7 8 **35. a.** $\{x|x \text{ is a real number}\}$ **b.** $(-\infty, \infty)$ **c.** −4 −3 −2 −1 0 1 2 3 4 **37. a.** $\{\}$ or $\varnothing$ **b.** No interval notation. **c.** −4 −3 −2 −1 0 1 2 3 4 **39.** $\frac{3}{4}x < -6$; $x < -8$ **41.** $5x - 1 > 14$; $x > 3$ **43.** $1 - 4x \leq 25$; $x \geq -6$ **45.** $6 + 2(x - 5) \leq 12$; $x \leq 8$ **47.** 94 or higher. **49.** 80 or less. **51.** 13 ft. or less. **53.** $\approx$ 23.9 in. or less. **55.** At least 60 mph. **57.** 4000 or more lamps. **59. a.** $t < 1948.244$°F **b.** $t \geq 1948.244$°F **61.** 2.5 amps or less.

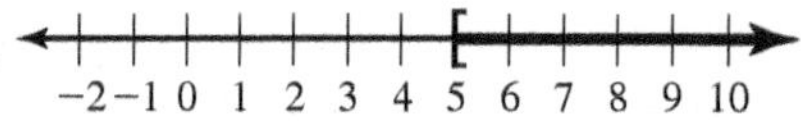

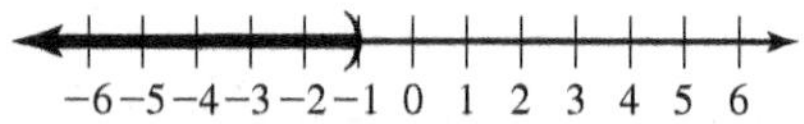

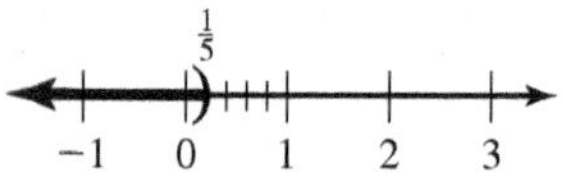

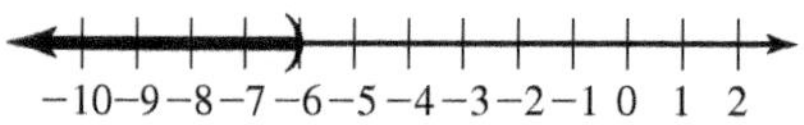

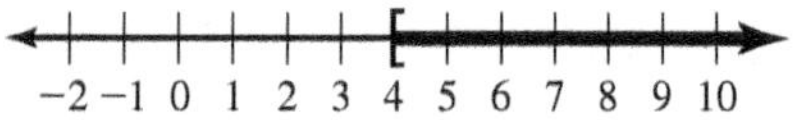

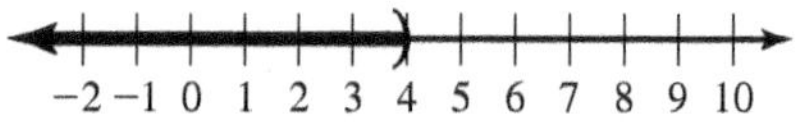

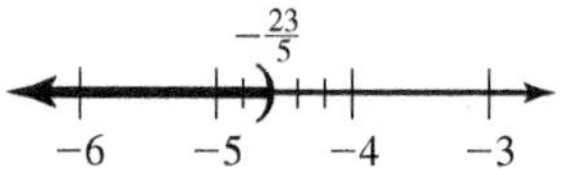

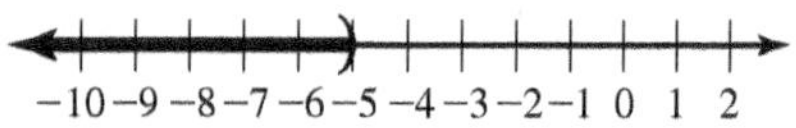

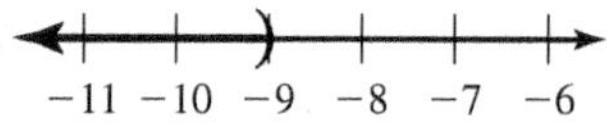

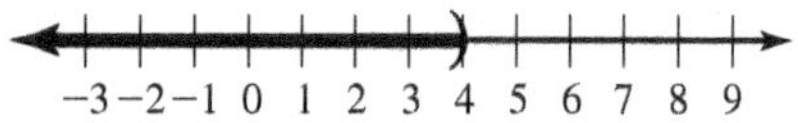

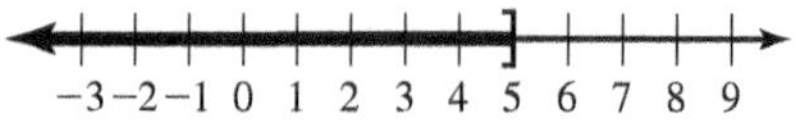

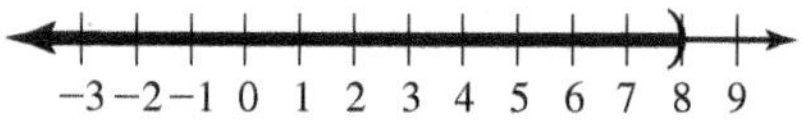

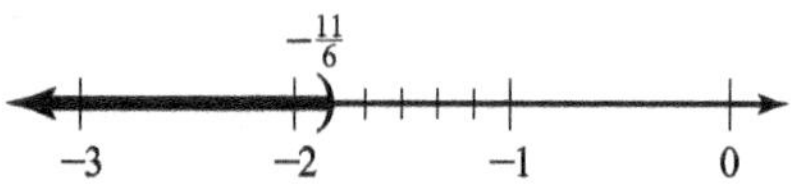

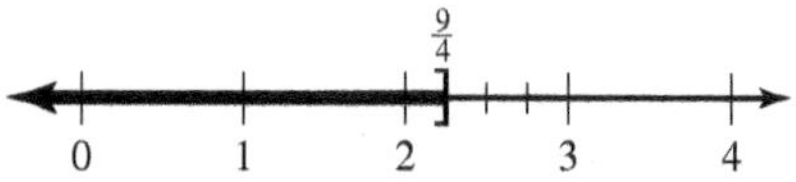

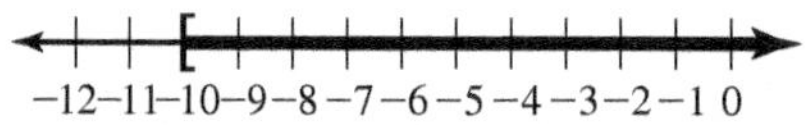

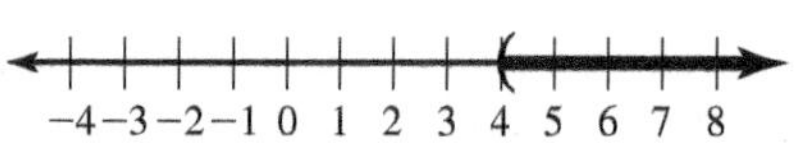

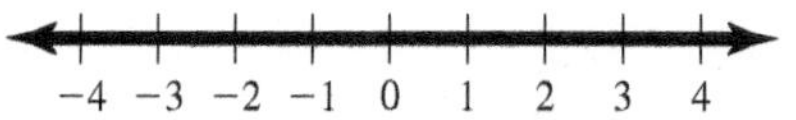

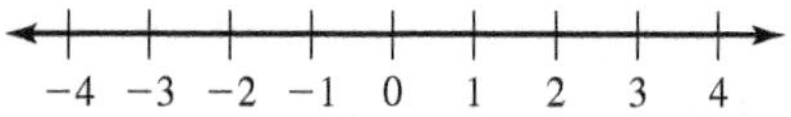

Review Exercises: **1.** yes **2.** Commutative property of addition. **3.** 10 **4.** 45 **5.** −2 **6.** $\frac{32}{7}$

Exercise Set 2.4

Prep Exercises **1.** Two inequalities joined by either *and* or *or* **2.** *and* and *or* **3.** For two sets A and B, the intersection of A and B, symbolized by $A \cap B$, is a set containing only elements that are in both A and B. **4.** Graph the region of overlap of the two inequalities. **5.** For two sets A and B, the union of A and B, symbolized by $A \cup B$, is a set containing each element in either A or B. **6.** Graph the region included in either of the two inequalities.

Exercises **1.** $\{1, 3, 5\}$ **3.** $\{7, 8\}$ **5.** $-4 < x < 5$ **7.** $0 < y \leq 2$ **9.** $-7 < w < 3$ **11.** $0 \leq u \leq 2$ **13.** −4 −3 −2 −1 0 1 2 3 4 5 6 7 8 **15.** −6 −5 −4 −3 −2 −1 0 1 2 3 4 5 6 **17.** −1 0 1 2 3 4 5 6 7 8 9 10 11 **19.** −6 −5 −4 −3 −2 −1 0 1 2 3 4 5 6 **21. a.** −6 −5 −4 −3 −2 −1 0 1 2 3 4 5 6 **b.** $\{x|-3 < x < -1\}$ **c.** $(-3, -1)$ **23. a.** −5 −4 −3 −2 −1 0 1 2 3 4 5 6 7 **b.** $\{x|3 < x \leq 6\}$ **c.** $(3, 6]$ **25. a.** −6 −5 −4 −3 −2 −1 0 1 2 3 4 5 6 **b.** $\{\ \}$ or $\varnothing$ **c.** no interval notation **27. a.** −6 −5 −4 −3 −2 −1 0 1 2 3 4 5 6 **b.** $\{x|-5 \leq x < 3\}$ **c.** $[-5, 3)$ **29. a.** −6 −5 −4 −3 −2 −1 0 1 2 3 4 5 6 **b.** $\{\ \}$ or $\varnothing$ **c.** no interval notation **31. a.** −7 −6 −5 −4 −3 −2 −1 0 1 2 3 4 5 **b.** $\{x|-7 < x < -3\}$ **c.** $(-7, -3)$ **33. a.** −6 −5 −4 −3 −2 −1 0 1 2 3 4 5 6 **b.** $\{x|-1 \leq x \leq 2\}$ **c.** $[-1, 2]$

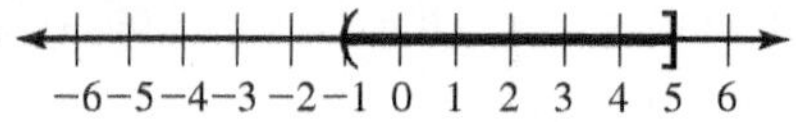

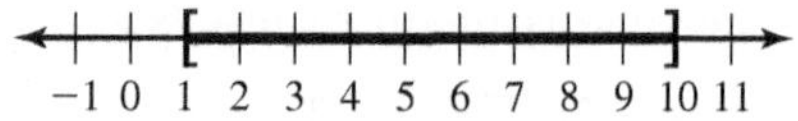

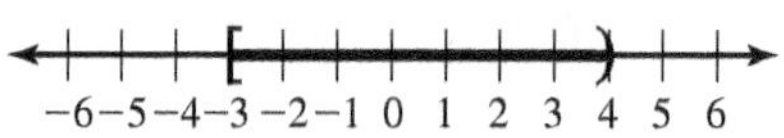

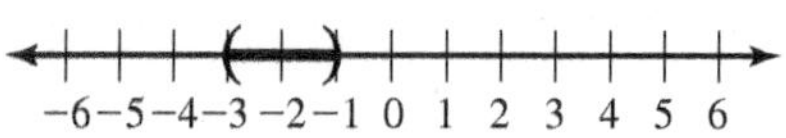

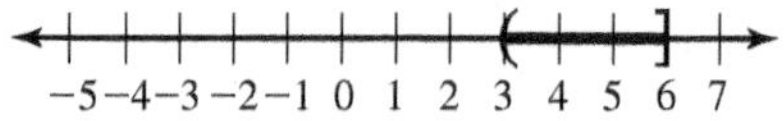

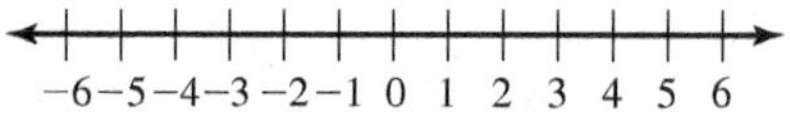

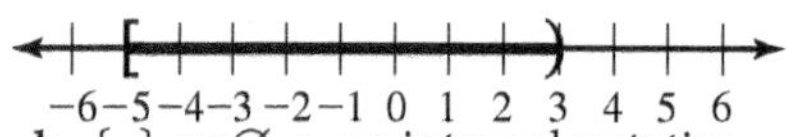

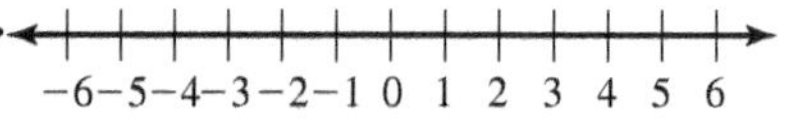

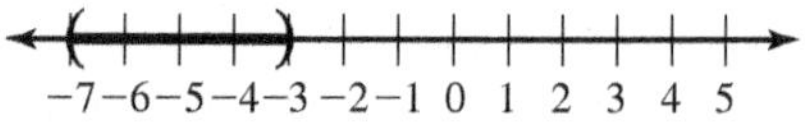

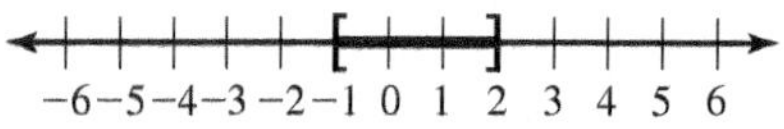

35. a.  $-\frac{2}{3}$ −6 −5 −4 −3 −2 −1 0 1 2 3 4 5 6 **b.** $\left\{x \middle| -\frac{2}{3} \le x < 2\right\}$ **c.** $\left[-\frac{2}{3}, 2\right)$ **37. a.** −6 −5 −4 −3 −2 −1 0 1 2 3 4 5 6 **b.** $\{x \mid 0 \le x < 3\}$ **c.** $[0, 3)$ **39. a.** −6 −5 −4 −3 −2 −1 0 1 2 3 4 5 6 **b.** $\{x \mid 0 \le x \le 3\}$ **c.** $[0, 3]$ **41.** $\{a, c, d, g, o, t\}$

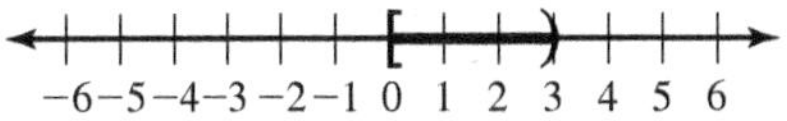

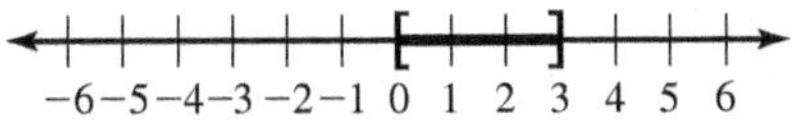

43. $\{w, x, y, z\}$ **45.** 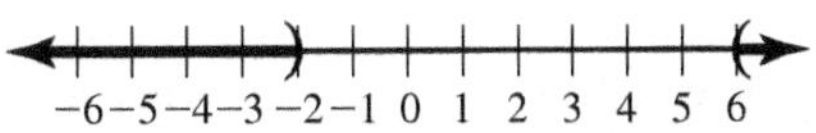 −6 −5 −4 −3 −2 −1 0 1 2 3 4 5 6 **47.** 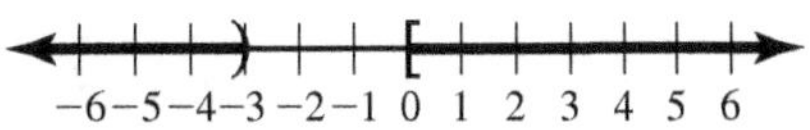 −6 −5 −4 −3 −2 −1 0 1 2 3 4 5 6

49. 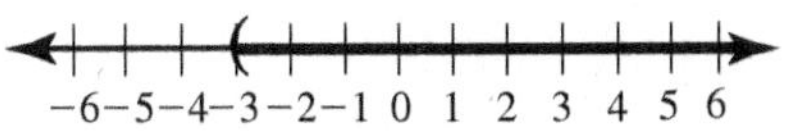 −6 −5 −4 −3 −2 −1 0 1 2 3 4 5 6 **51.** 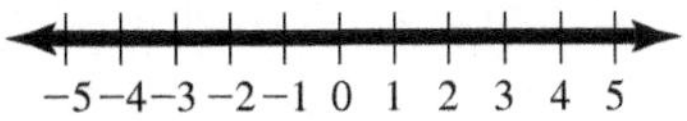 −5 −4 −3 −2 −1 0 1 2 3 4 5

53. a. 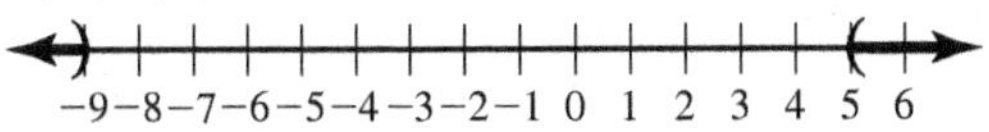 −9 −8 −7 −6 −5 −4 −3 −2 −1 0 1 2 3 4 5 6 **b.** $\{y \mid y < -9 \text{ or } y > 5\}$ **c.** $(-\infty, -9) \cup (5, \infty)$

55. a. 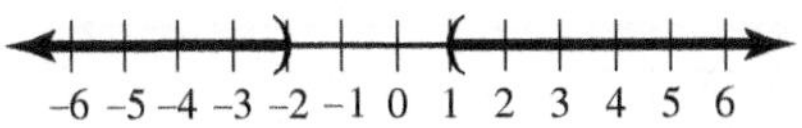 −6 −5 −4 −3 −2 −1 0 1 2 3 4 5 6 **b.** $\{r \mid r < -2 \text{ or } r > 1\}$ **c.** $(-\infty, -2) \cup (1, \infty)$

57. a. 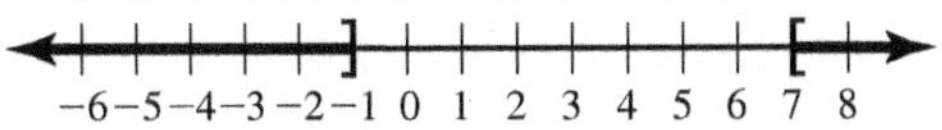 −6 −5 −4 −3 −2 −1 0 1 2 3 4 5 6 7 8 **b.** $\{w \mid w \le -1 \text{ or } w \ge 7\}$ **c.** $(-\infty, -1] \cup [7, \infty)$

59. a.  −6 −5 −4 −3 −2 −1 0 1 2 3 4 5 6 **b.** $\{k \mid k \le 2 \text{ or } k \ge 3\}$ **c.** $(-\infty, 2] \cup [3, \infty)$

61. a. 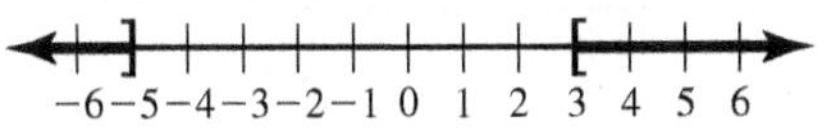 −6 −5 −4 −3 −2 −1 0 1 2 3 4 5 6 **b.** $\{x \mid x \le -5 \text{ or } x \ge 3]$ **c.** $(-\infty, -5] \cup [3, \infty)$

63. a. 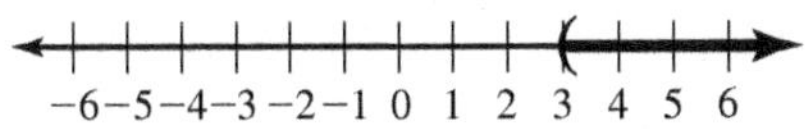 −6 −5 −4 −3 −2 −1 0 1 2 3 4 5 6 **b.** $\{c \mid c > 3\}$ **c.** $(3, \infty)$ **65. a.** 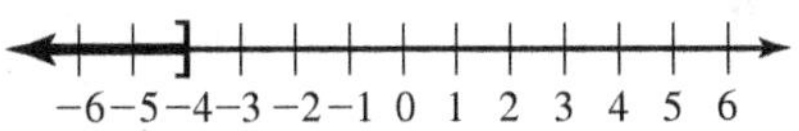 −6 −5 −4 −3 −2 −1 0 1 2 3 4 5 6 **b.** $\{x \mid x \le -4\}$ **c.** $(-\infty, -4]$ **67. a.** 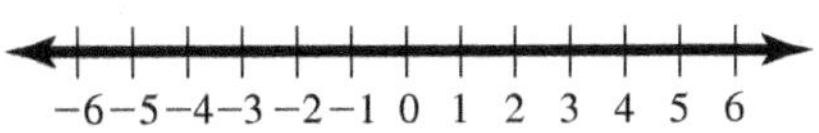 −6 −5 −4 −3 −2 −1 0 1 2 3 4 5 6 **b.** $\{x \mid x \text{ is a real number}\}$, or $\mathbb{R}$ **c.** $(-\infty, \infty)$ **69. a.** 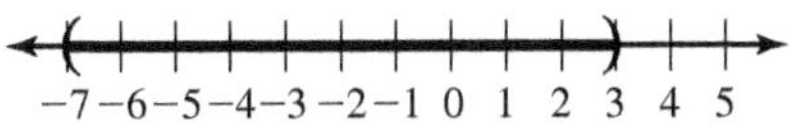 −7 −6 −5 −4 −3 −2 −1 0 1 2 3 4 5 **b.** $\{x \mid -7 < x < 3\}$ **c.** $(-7, 3)$

71. a. 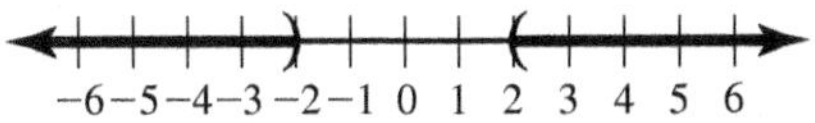 −6 −5 −4 −3 −2 −1 0 1 2 3 4 5 6 **b.** $\{x \mid x < -2 \text{ or } x > 2\}$ **c.** $(-\infty, -2) \cup (2, \infty)$

73. a. 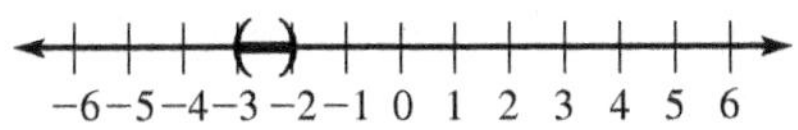 −6 −5 −4 −3 −2 −1 0 1 2 3 4 5 6 **b.** $\{x \mid -3 < x < -2\}$ **c.** $(-3, -2)$ **75. a.** −6 −5 −4 −3 −2 −1 0 1 2 3 4 5 6 **b.** $\{x \mid 3 \le x \le 6\}$ **c.** $[3, 6]$ **77. a.** −6 −5 −4 −3 −2 −1 0 1 2 3 4 5 6 **b.** $\{x \mid 1 < x < 3\}$ **c.** $(1, 3)$

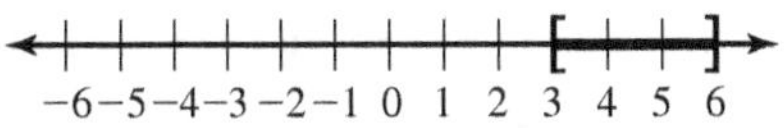

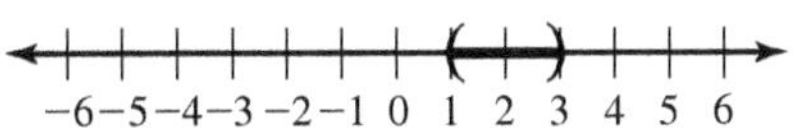

79. a. 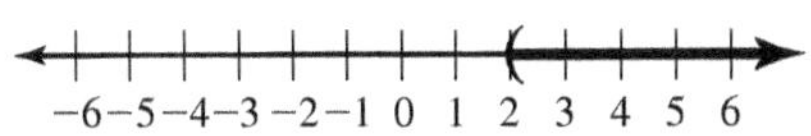 −6 −5 −4 −3 −2 −1 0 1 2 3 4 5 6 **b.** $\{x \mid x > 2\}$ **c.** $(2, \infty)$ **81. a.** 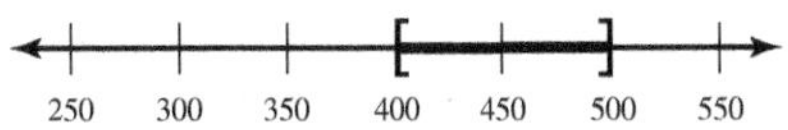 250 300 350 400 450 500 550 **b.** $\{x \mid 400 \le x \le 500\}$ **c.** $[400, 500]$ **83. a.** 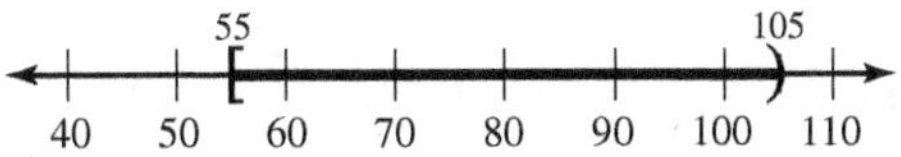 55 105 40 50 60 70 80 90 100 110 **b.** $\{x \mid 55 \le x < 105\}$ **c.** $[55, 105)$

85. a. 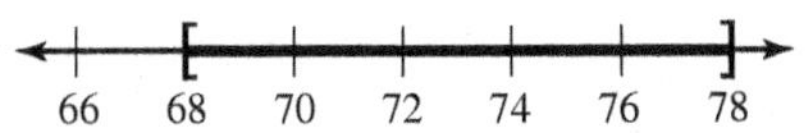 66 68 70 72 74 76 78 **b.** $\{x \mid 68° \le x \le 78°\}$ **c.** $[68, 78]$ **87. a.** 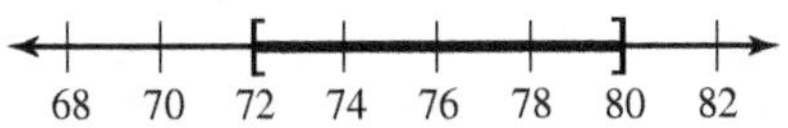 68 70 72 74 76 78 80 82 **b.** $\{x \mid 72° \le x \le 80°\}$ **c.** $[72, 80]$ **89. a.** 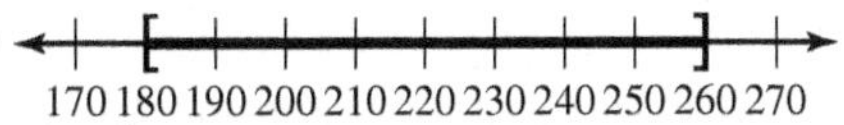 170 180 190 200 210 220 230 240 250 260 270 **b.** $\{x \mid 180 \text{ ft.} \le x \le 260 \text{ ft.}\}$ **c.** $[180, 260]$

Review Exercises: **1.** No, the absolute value of zero is zero and zero is neither negative nor positive. **2.** -11 **3.** -16 **4.** 6 **5.** $\frac{1}{5}$ **6.** -2

Exercise Set 2.5

Prep Exercises **1.** The distance a number is from zero. **2.** We are to find numbers that are 5 units from 0. **3.** If $|n| = a$, where n is a variable or an expression and $a \ge 0$, then $n = a$ or $n = -a$. **4.** There is no solution. **5.** Separate the absolute value equation into two equations: $ax + b = cx + d$ and $ax + b = -(cx + d)$. **6.** After the absolute value equation is separated into two equations, one with the two expressions equal and the other with the two expressions opposites, one of the equations leads to a contradiction.

Exercises **1.** $-2, 2$ **3.** No solution **5.** $-11, 5$ **7.** $2, 3$ **9.** $1, \frac{7}{5}$ **11.** $-\frac{2}{3}, \frac{10}{3}$ **13.** No solution **15.** $\frac{3}{4}$ **17.** $-4, 4$ **19.** $1, -3$ **21.** $12, -4$ **23.** $-\frac{3}{5}, 1$ **25.** $-4, 1$ **27.** $-2, 6$ **29.** $-\frac{9}{2}, \frac{15}{2}$ **31.** $0, 20$ **33.** $-2, 4$ **35.** $\frac{1}{3}, 7$ **37.** $-5, -\frac{3}{5}$ **39.** All real numbers **41.** 1 **43.** $3, -13$ **45.** $-6, 10$ **47.** $\frac{5}{6}, \frac{11}{6}$ **49.** $-\frac{17}{4}, \frac{11}{4}$

Review Exercises: **1.** $x > -7$ **2.** $x \geq -3$ **3.** $n - 2 < 5; n < 7$ **4.** $(-2, 0]$ **5.**

−6 −5 −4 −3 −2 −1 0 1 2 3 4 5 6

6. $-2 < x < \frac{14}{3}$

Exercise Set 2.6

Prep Exercises **1.** $x \geq -a$ and $x \leq a$ or $-a \leq x \leq a$ **2.** Shade between the values of $-a$ and a. **3.** When $a < 0$ **4.** $x \leq -a$ or $x \geq a$ **5.** Shade to the left of $-a$ and to the right of a. **6.** When $a < 0$

Exercises **1. a.**

−6 −5 −4 −3 −2 −1 0 1 2 3 4 5 6

b. $\{x \mid -5 < x < 5\}$ **c.** $(-5, 5)$

3. a.

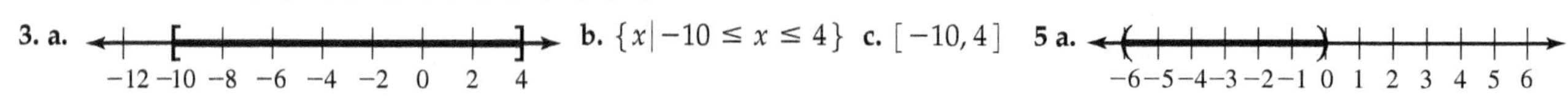

b. $\{x \mid -10 \leq x \leq 4\}$ **c.** $[-10, 4]$ **5 a.** (see number line above) **b.** $\{s \mid -6 < s < 0\}$ **c.** $(-6, 0)$

7 a.

−3 −2 −1 0 1 2 3 4 5 6 7 8

b. $\{m \mid -2 < m < 7\}$ **c.** $(-2, 7)$

9 a.

$\frac{4}{3}$

0 1 2 3

b. $\left\{k \middle| \frac{4}{3} \leq k \leq 2\right\}$ **c.** $\left[\frac{4}{3}, 2\right]$

11 a.

−6 −5 −4 −3 −2 −1 0 1 2 3 4 5 6

b. { } or ∅ **c.** no interval notation

13 a.

−6 −5 −4 −3 −2 −1 0 1 2 3 4 5 6

b. $\{w \mid 0 < w < 6\}$ **c.** $(0, 6)$

15. a.

−12 −10 −8 −6 −4 −2 0 2 4 6 8 10 12

b. $\{c \mid c < -12 \text{ or } c > 12\}$ **c.** $(-\infty, -12) \cup (12, \infty)$

17. a.

−9 −8 −7 −6 −5 −4 −3 −2 −1 0 1 2 3 4 5 6

b. $\{y \mid y \leq -9 \text{ or } y \geq 5\}$ **c.** $(-\infty, -9] \cup [5, \infty)$

19. a.

−4 −2 0 2 4 6 8 10 12 14 16 18 20

b. $\{p \mid p < -2 \text{ or } p > 14\}$ **c.** $(-\infty, -2) \cup (14, \infty)$

21. a.

−6 −5 −4 −3 −2 −1 0 1 2 3 4

b. $\{x \mid x \leq -6 \text{ or } x \geq 2\}$ **c.** $(-\infty, -6] \cup [2, \infty)$

23. a.

$-\frac{5}{2}$

−4 −3 −2 −1 0 1 2

b. $\left\{n \middle| n < -\frac{5}{2} \text{ or } n > 0\right\}$ **c.** $\left(-\infty, -\frac{5}{2}\right) \cup (0, \infty)$

25. a.

−6 −5 −4 −3 −2 −1 0 1 2 3 4 5 6

b. $\{v \mid v \leq -1 \text{ or } v \geq 1\}$ **c.** $(-\infty, -1] \cup [1, \infty)$

27. a.

−6 −5 −4 −3 −2 −1 0 1 2 3 4 5 6

b. $\{y \mid y < -3 \text{ or } y > -1\}$ **c.** $(-\infty, -3) \cup (-1, \infty)$

29. a.

−6 −5 −4 −3 −2 −1 0 1 2 3 4 5 6

b. $\{m \mid m < -5 \text{ or } m > 1\}$ **c.** $(-\infty, -5) \cup (1, \infty)$

31. a.

−6 −5 −4 −3 −2 −1 0 1 2 3 4 5 6

b. $\{x \mid 0 < x < 4\}$ **c.** $(0, 4)$

33. a.

−6 −5 −4 −3 −2 −1 0 1 2 3 4 5 6

b. $\{r \mid r \text{ is a real number}\}$ **c.** $(-\infty, \infty)$

35 a.

−6 −5 −4 −3 −2 −1 0 1 2 3 4 5 6

b. $\{x \mid -4 < x < -2\}$ **c.** $(-4, -2)$

37. a.

−6 −5 −4 −3 −2 −1 0 1 2 3 4 5 6

b. $\{x \mid 0 < x < 1\}$ **c.** $(0, 1)$

39 a.

−14 −12 −10 −8 −6 −4 −2 0 2 4 6 8 10

b. ∅ **c.** no interval notation

41 a.

$\frac{14}{3}$

−6 −5 −4 −3 −2 −1 0 1 2 3 4 5 6

b. $\left\{k \middle| -2 \leq k \leq \frac{14}{3}\right\}$ **c.** $\left[-2, \frac{14}{3}\right]$

43. a.

0 4 8 12 16 20 24 28

b. $\{x \mid x < 4 \text{ or } x > 20\}$ **c.** $(-\infty, 4) \cup (20, \infty)$

45. a. 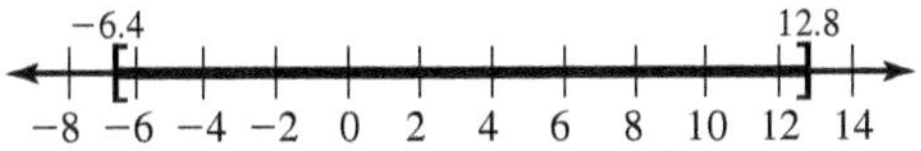

b. $\{y \mid -6.4 \le y \le 12.8\}$ **c.** $[-6.4, 12.8]$

47. a. 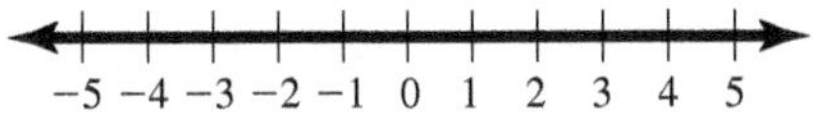

b. $\{p \mid p$ is a real number$\}$ **c.** $(-\infty, \infty)$ **49.** $|x| < 3$ **51.** $|x| \ge 4$

53. $|x + 1| > 1$ **55.** $|x - 3| \le 2$ **57.** $|x| >$ any negative number

Review Exercises: **1.** true **2.** true **3.** -2 **4.** $\frac{15}{2}$ **5.** $y = \frac{C - Ax}{B}$ **6.** 20

Chapter 2 Summary and Review Exercises

2.1 Definitions/Rules/Procedures $a + c = b + c$; $ac = bc$; LCD, LCD, 10; addition, multiplication; constants Exercises **1.** 7 **2.** 4 **3.** 0 **4.** -21 **5.** $\frac{5}{4}$ **6.** $-\frac{3}{4}$ **7.** -1 **8.** All real numbers. **9.** 6 **10.** No solution. **11.** 5 **12.** 8 **13.** $t = \frac{I}{Pr}$ **14.** $w = \frac{P - 2l}{2}$ **15.** $h = \frac{2A}{b}$ **16.** $a = P - b - c$

2.2 Definitions/Rules/Procedures Understand; Plan; Execute; Answer; Check; variable(s), constants, key words

Exercises **17.** -320.8°F **18.** \$4200 **19.** 12 **20.** 4 **21.** \$64.20 **22.** 34,200 units **23.** 3 hr.

2.3 Definitions/Rules/Procedures bracket, solid circle, left, right, parenthesis, open circle, left, right; right, left; <, positive, negative; parentheses, LCD; addition, multiplication, reverse Exercises **24. a.** $\{n \mid n \le 4\}$ **b.** $(-\infty, 4]$ **c.** 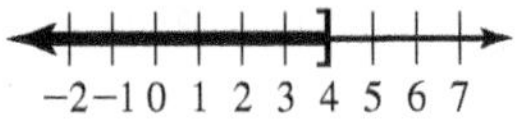

25. a. $\{x \mid x > -2\}$ **b.** $(-2, \infty)$ **c.** 

26. a. $\{m \mid m < -3\}$ **b.** $(-\infty, -3)$ **c.** 

27. a. $\{h \mid h \le -5\}$ **b.** $(-\infty, -5]$ **c.** 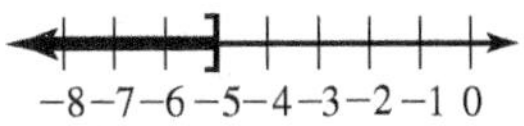

28. a. $\left\{t \mid t \le \frac{18}{5}\right\}$ **b.** $\left(-\infty, \frac{18}{5}\right]$ **c.** 

29. a. $\{u \mid u > 2\}$ **b.** $(2, \infty)$ **c.** 

30. 25 ft. or less.

2.4 Definitions/Rules/Procedures *and, or*; both *A* and *B*; *A* or *B* Exercises **31.** Intersection: $\{1, 5, 9\}$ Union: $\{1, 5, 7, 9\}$ **32.** Intersection: $\{\ \}$ or $\varnothing$ Union: $\{1, 2, 3, 4, 5, 6, 7\}$

Definitions/Rules/Procedures inequality; intersection Exercises **33. a.** 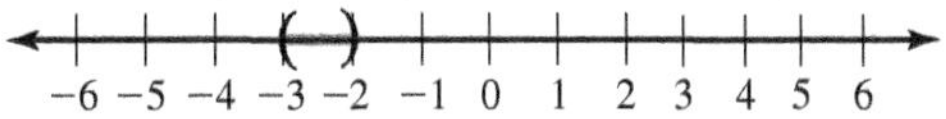

b. $\{x \mid -3 < x < -2\}$ **c.** $(-3, -2)$ **34. a.** 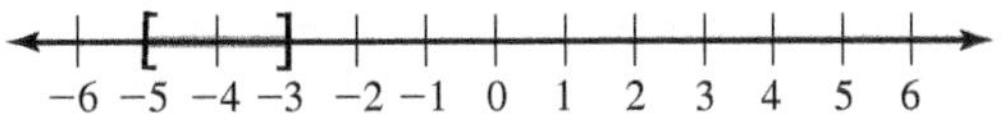

b. $\{x \mid -5 \le x \le -3\}$ **c.** $[-5, -3]$ **35. a.** 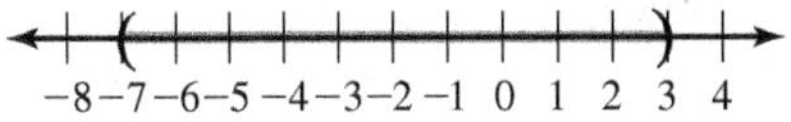

b. $\{x \mid -7 < x < 3\}$ **c.** $(-7, 3)$

36. a. 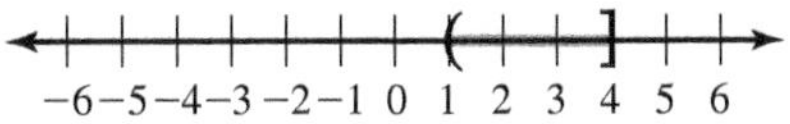

b. $\{x \mid 1 < x \le 4\}$ **c.** $(1, 4]$ **37. a.** 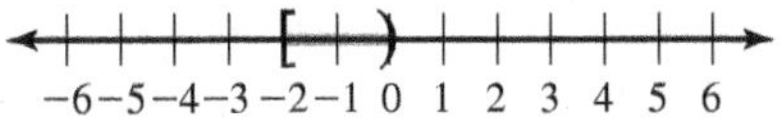

b. $\{x \mid -2 \le x < 0\}$ **c.** $[-2, 0)$ **38. a.** 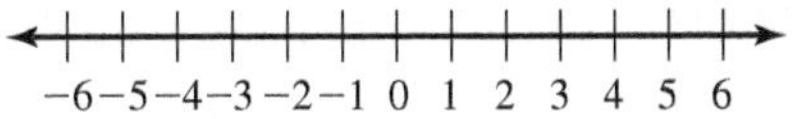

b. $\{\ \}$ or $\varnothing$ **c.** no interval notation

39. a. 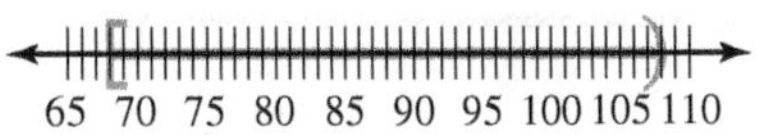

b. $\{x \mid 68 \le x < 108\}$ **c.** $[68, 108)$ **40. a.** 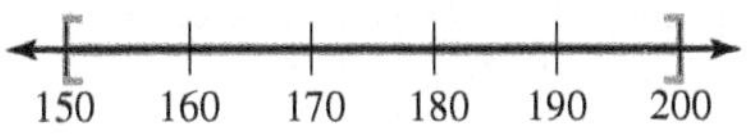

b. $\{x \mid 150 \le x \le 200\}$ **c.** $[150, 200]$ Definitions/Rules/Procedures inequality; union

Exercises **41. a.** 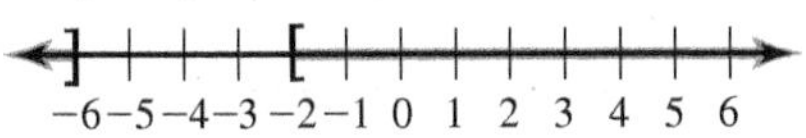

b. $\{w \mid w \le -6$ or $w \ge -2\}$ **c.** $(-\infty, -6] \cup [-2, \infty)$

42. a. 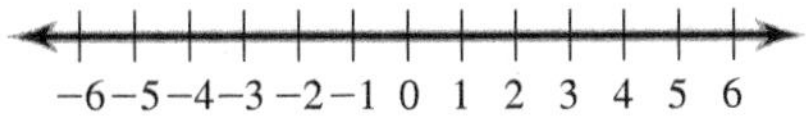

b. $\{x \mid x$ is a real number$\}$ **c.** $(-\infty, \infty)$

43. a. 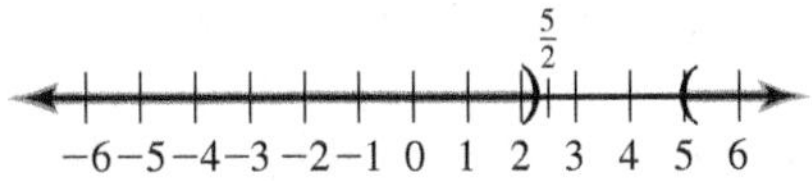

b. $\left\{m \mid m < \frac{5}{2} \text{ or } m > 5\right\}$ **c.** $\left(-\infty, \frac{5}{2}\right) \cup (5, \infty)$

44. a. 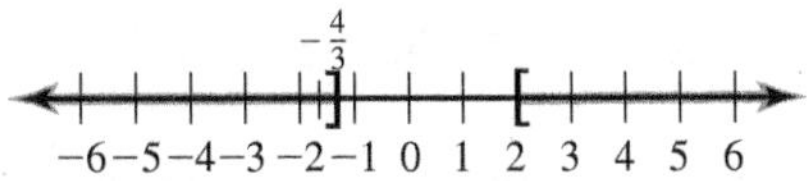

b. $\left\{x \mid x \le -\frac{4}{3} \text{ or } x \ge 2\right\}$ **c.** $\left(-\infty, -\frac{4}{3}\right] \cup [2, \infty)$

45. a. 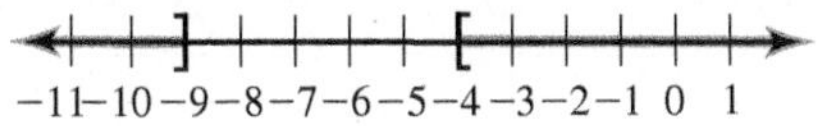

b. $\{x \mid x \le -9$ or $x \ge -4\}$ **c.** $(-\infty, -9] \cup [-4, \infty)$

46. a. −6 −5 −4 −3 −2 −1 0 1 2 3 4 5 6 **b.** $\{w \mid w \le -1 \text{ or } w \ge 1\}$ **c.** $(-\infty, -1] \cup [1, \infty)$

2.5 Definitions/Rules/Procedures $a, -a$; $|ax + b| = c$; $ax + b = c$, $ax + b = -c$ Exercises **47.** $-4, 4$ **48.** $-3, 11$ **49.** $-1, 2$ **50.** no solution **51.** $-7, 15$ **52.** $-3, 3$ **53.** $0, \frac{8}{3}$ **54.** $-1, 11$ Definitions/Rules/Procedures $cx + d$; $-(cx + d)$ Exercises **55.** $0, 2$ **56.** $x = -3, -\frac{1}{5}$ **57.** $-2, \frac{4}{5}$ **58.** $1, \frac{8}{3}$

2.6 Definitions/Rules/Procedures $x > -a, x < a, x \ge -a; x \le a$ Exercises **59. a.** −6 −5 −4 −3 −2 −1 0 1 2 3 4 5 6 **b.** $\{x \mid -5 < x < 5\}$ **c.** $(-5, 5)$ **60. a.** −6 −5 −4 −3 −2 −1 0 1 2 3 4 5 6 **b.** $\{m \mid -5 < m < -1\}$ **c.** $(-5, -1)$

61. a. −6 −5 −4 −3 −2 −1 0 1 2 3 4 5 6 **b.** $\{\ \}$ or $\varnothing$ **c.** no interval notation **62. a.** −6 −5 −4 −3 −2 −1 0 1 2 3 4 5 6 **b.** $\{m \mid -6 \le m \le 0\}$ **c.** $[-6, 0]$ Definitions/Rules/Procedures $x < -a, x > a; x \le -a, x \ge a$ Exercises **63. a.** −6 −5 −4 −3 −2 −1 0 1 2 3 4 5 6 **b.** $\{p \mid p \le -4 \text{ or } p \ge 4\}$ **c.** $(-\infty, -4] \cup [4, \infty]$

64. a. −6 −5 −4 −3 −2 −1 0 1 2 3 4 5 6 7 8 9 10 11 **b.** $\{x \mid x < -4 \text{ or } x > 10\}$ **c.** $(-\infty, -4) \cup (10, \infty)$

65. a. −6 −5 −4 −3 −2 −1 0 1 2 3 4 5 6 **b.** $\{b \mid b < -1 \text{ or } b > 1\}$ **c.** $(-\infty, -1) \cup (1, \infty)$

66. a. −2 −1 0 1 2 3 4 5 6 7 8 9 10 11 **b.** $\{t \mid t < 0 \text{ or } t > 10\}$ **c.** $(-\infty, 0) \cup (10, \infty)$

67. a. $-\frac{7}{2}$ $\frac{13}{2}$ −6 −5 −4 −3 −2 −1 0 1 2 3 4 5 6 7 8 **b.** $\left\{k \mid k \le -\frac{7}{2} \text{ or } k \ge \frac{13}{2}\right\}$ **c.** $\left(-\infty, -\frac{7}{2}\right] \cup \left[\frac{13}{2}, \infty\right)$

68. a. −6 −5 −4 −3 −2 −1 0 1 2 3 4 5 6 **b.** $\{x \mid x \text{ is a real number}\}$ **c.** $(-\infty, \infty)$

Chapter 2 Practice Test

1. $A \cap B = \{h, o, e\}$; $A \cup B = \{h, o, m, u, s, e\}$ [2.4] **2.** 4 [2.1] **3.** 30 [2.1] **4.** $-8, 2$ [2.5] **5.** $-3, 6$ [2.5] **6.** $-8, \frac{2}{3}$ [2.5] **7.** No solution. [2.5] **8.** $b = \frac{2A}{h} - B$ [2.1] **9. a.** −8 −7 −6 −5 −4 −3 −2 −1 0 1 2 3 4 **b.** $\{x \mid -7 < x \le 3\}$ **c.** $(-7, 3]$ [2.4]

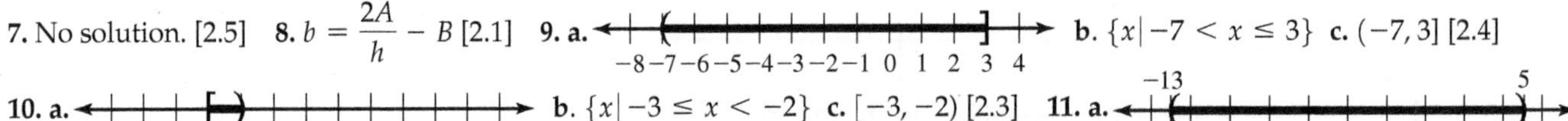

10. a. −6 −5 −4 −3 −2 −1 0 1 2 3 4 5 6 **b.** $\{x \mid -3 \le x < -2\}$ **c.** $[-3, -2)$ [2.3] **11. a.** −13 5 −14 −12 −10 −8 −6 −4 −2 0 2 4 6 **b.** $\{x \mid -13 < x < 5\}$ **c.** $(-13, 5)$ [2.6] **12. a.** −6 −5 −4 −3 −2 −1 0 1 2 3 4 5 6 **b.** $\{x \mid x < -1 \text{ or } x > 3\}$ **c.** $(-\infty, -1) \cup (3, \infty)$ [2.6] **13. a.** −9 −8 −7 −6 −5 −4 −3 −2 −1 0 1 2 3 **b.** $\{x \mid -7 < x < -1\}$ **c.** $(-7, -1)$ [2.6] **14. a.** −6 −5 −4 −3 −2 −1 0 1 2 3 4 5 6 **b.** $\{\ \}$ or $\varnothing$ **c.** no interval notation [2.6] **15. a.** −6 −5 −4 −3 −2 −1 0 1 2 3 4 5 6 **b.** $\{t \mid t \text{ is a real number}\}$ **c.** $(-\infty, \infty)$ [2.6] **16. a.** $-\frac{1}{3}$ −2 −1 0 1 2 3 4 5 6 **b.** $\left\{x \,\middle|\, -\frac{1}{3} \le x \le 3\right\}$ **c.** $\left[-\frac{1}{3}, 3\right]$ [2.6]

17. 12.5 in. [2.2] **18.** \$44.80 [2.2] **19.** 0.4 hr. [2.2] **20.** $5 < x < 6.25$ books; he would have to order six books. [2.4]

Chapters 1–2 Cumulative Review Exercises

1. false **2.** false **3.** false **4.** identity **5.** real number **6.** reverse the inequality symbol **7.** $-\frac{5}{9}$ **8.** -5 **9.** -7 **10.** 80 **11.** -81 **12.** -2 **13.** -24 **14.** 12 **15.** $-\frac{37}{8}$ **16.** $-6x^2 + 12x - 9$ **17.** -4 **18.** $8d^2 + d - 13$ **19.** $8y - 8$ **20.** $12 - 4(5 - x) = 4x - 8$ **21.** -1 **22.** 44 **23.** 2 **24. a.** $\{u \mid u \le -4\}$ **b.** $(-\infty, -4]$ **c.** −7 −6 −5 −4 −3 −2 **25. a.** $\{x \mid 0 \le x \le 5\}$ **b.** $[0, 5]$ **c.** −2 −1 0 1 2 3 4 5 6 7 **26. a.** $\{y \mid -1 \le y < 1\}$ **b.** $[-1, 1)$ **c.** −3 −2 −1 0 1 2 3 **27.** 1, 5 **28.** 4 **29.** 290 **30.** 3 hr.

Chapter 3

Exercise Set 3.1

Prep Exercises **1.** Beginning at the origin, move to the left 4 units along the x-axis, then up 3 units. **2.** $(+,+), (-,+), (-,-), (+,-)$ **3.** Replace the variables in the equation with the corresponding coordinates from the ordered pair. If the resulting equation is true, the ordered pair is a solution. **4. 1.** Replace one of the variables with a chosen number (any number). **2.** Solve the equation for the other variable. **5.** The graph of an equation represents the equation's solution set. **6.** At least two solutions must be found to graph the line. However, it is wise to find three solutions, using the third solution as a check. **7.** Plot the solutions as points in the rectangular coordinate system; then draw a line through the points to form a straight line. Put arrowheads on both ends of the line. **8.** To find an x-intercept, **1.** Replace y with 0 in the given equation. **2.** Solve for x.; To find a y-intercept, **1.** Replace x with 0 in the given equation. **2.** Solve for y.

Exercises **1.** A: $(2, 4)$; B: $(-2, 1)$; C: $(0, -3)$; D: $(4, -5)$ **3. a.** I; **b.** IV; **c.** III; **d.** y-axis

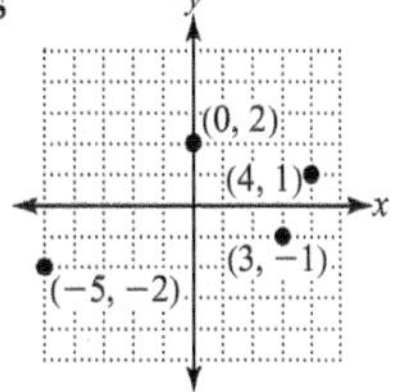

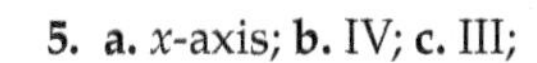
5. a. x-axis; **b.** IV; **c.** III;

d. I

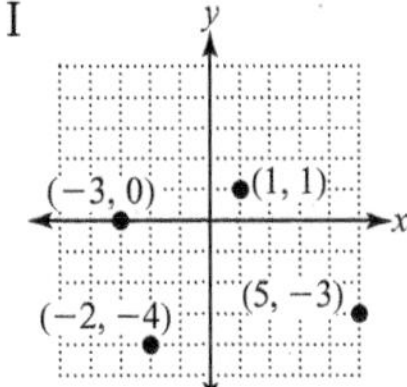

7. a.

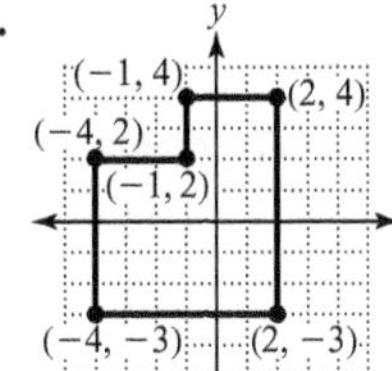

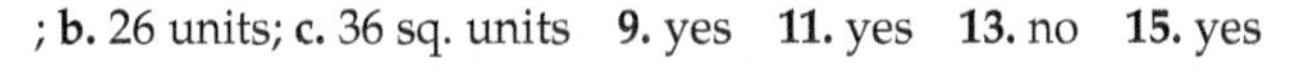
; **b.** 26 units; **c.** 36 sq. units **9.** yes **11.** yes **13.** no **15.** yes

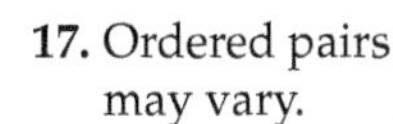
17. Ordered pairs may vary.

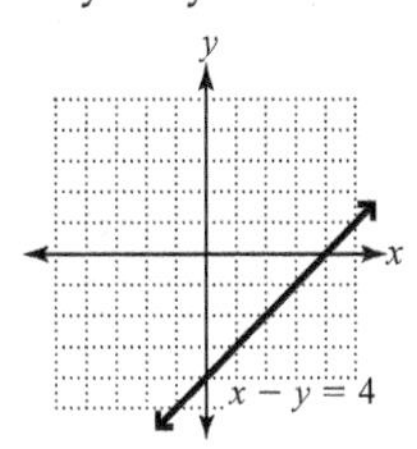

19. Ordered pairs may vary.

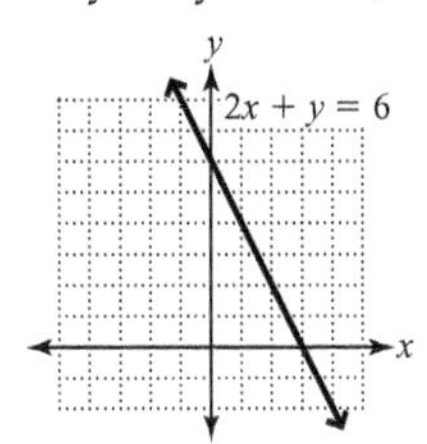

21. Ordered pairs may vary.

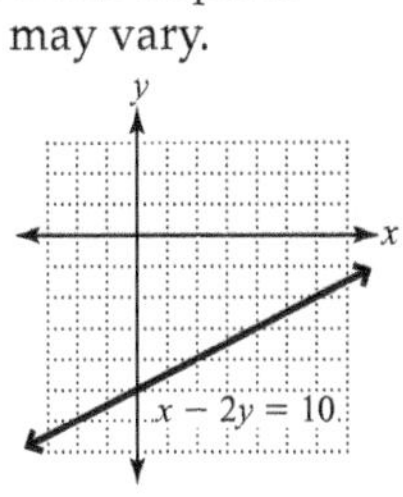

23. Ordered pairs may vary.

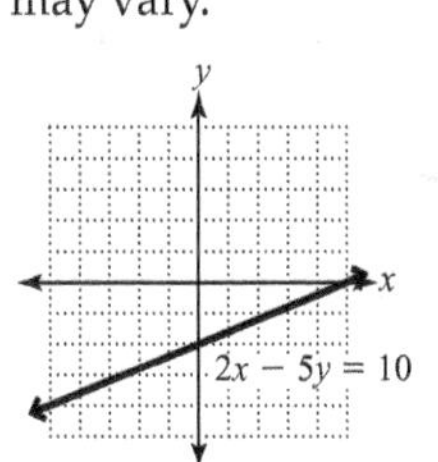

25. Ordered pairs may vary.

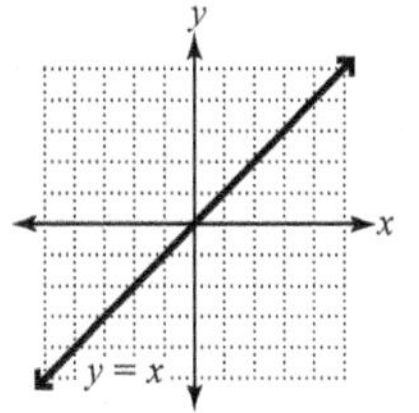

27. Ordered pairs may vary.

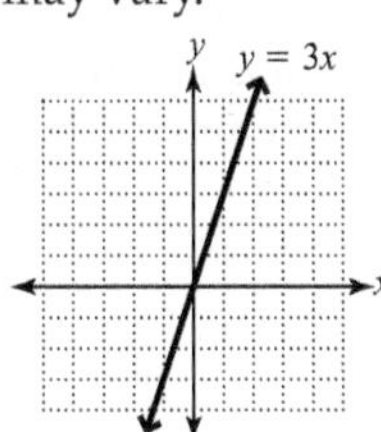

29. Ordered pairs may vary.

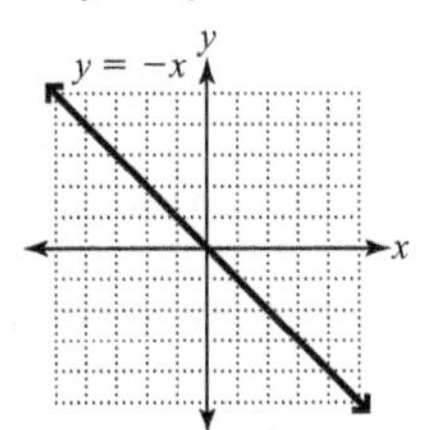

31. Ordered pairs may vary.

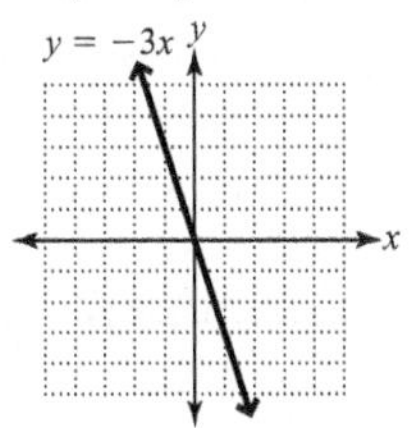

33. Ordered pairs may vary.

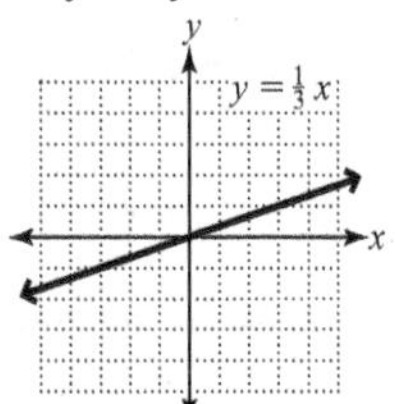

35. Ordered pairs may vary.

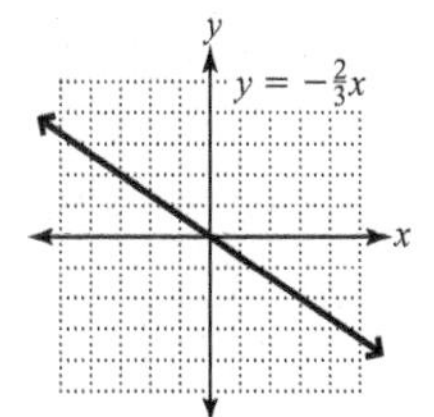

37. Ordered pairs may vary.

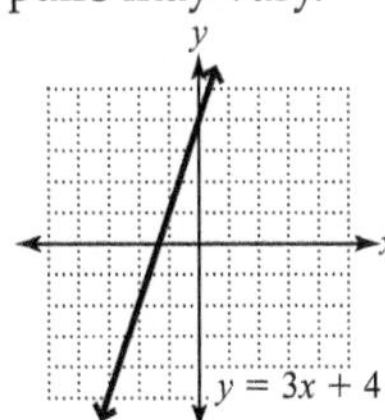

39. Ordered pairs may vary.

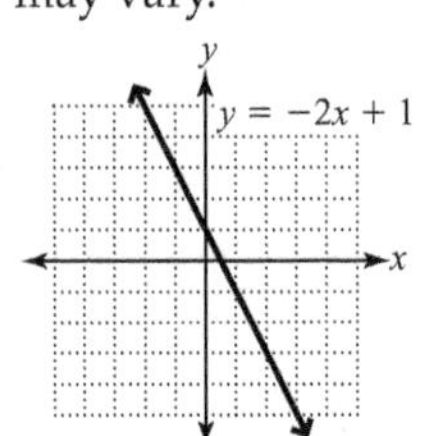

41. Ordered pairs may vary.

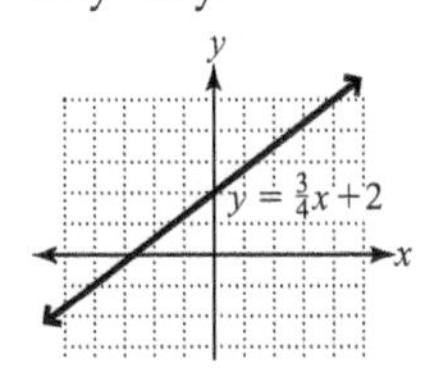

43. Ordered pairs may vary.

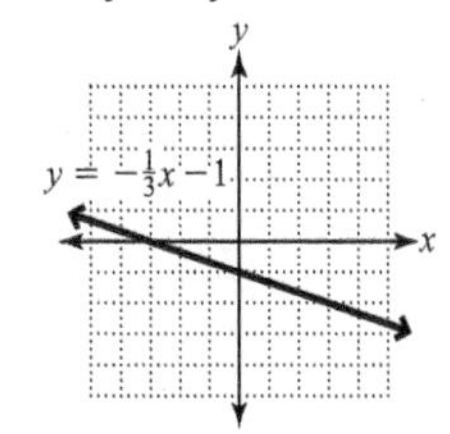

45. Ordered pairs may vary.

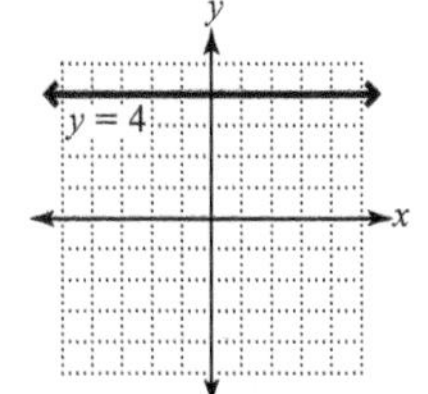

47. Ordered pairs may vary.

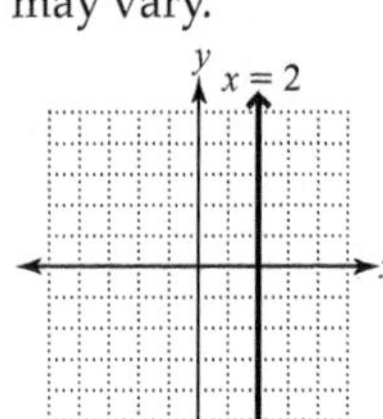

49. $(2, 0)$ $(0, 4)$

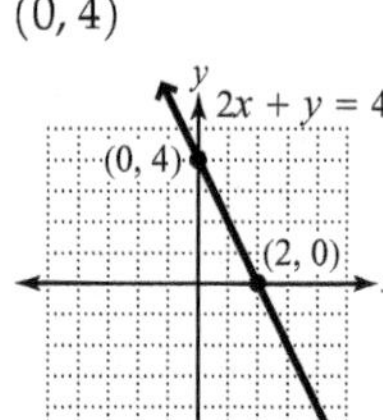

51. $(2, 0)$ $(0, 3)$

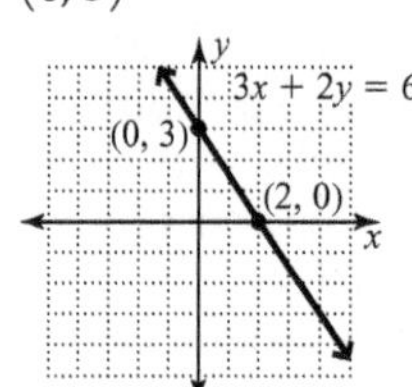

53. $(-3, 0)$ $(0, 2)$

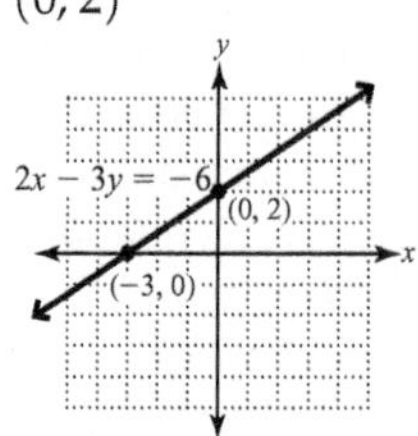

55. $(3, 0)$ $(0, -3)$

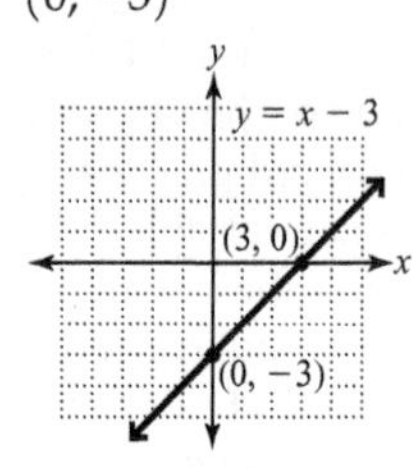

57. $(0, 0)$

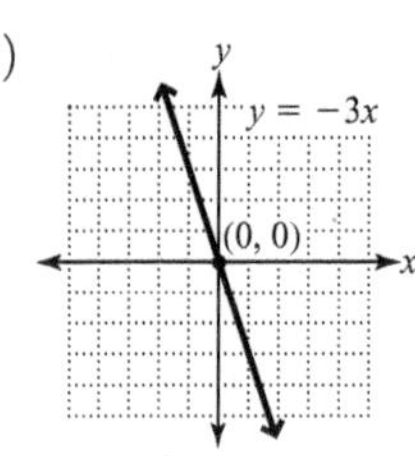

59. $\left(\frac{4}{3}, 0\right)(0, 1)$

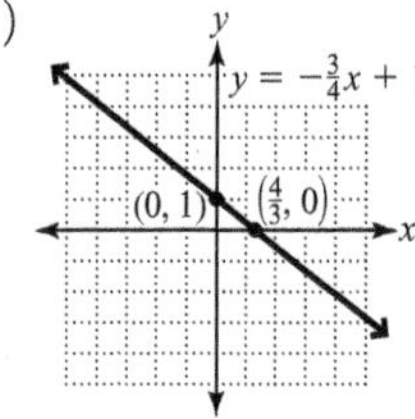

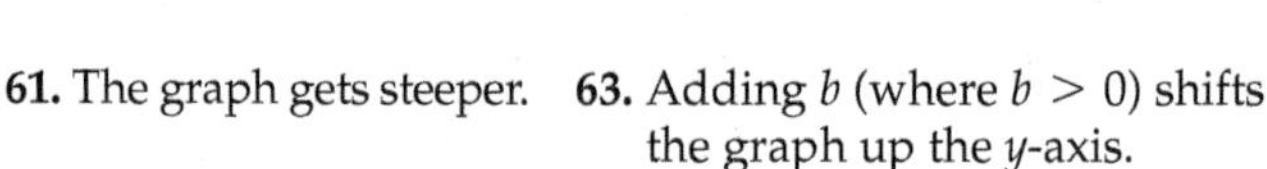

61. The graph gets steeper. **63.** Adding b (where $b > 0$) shifts the graph up the y-axis.

65. a. $(-3, 1), (0, 1), (-2, -3), (1, -3)$ **b.** $(1, 2), (4, 2), (2, -2), (5, -2)$ **c.** $[(x + 4), (y + 1)]$ **67. a.** \$95 **b.** 6 hr.

c.

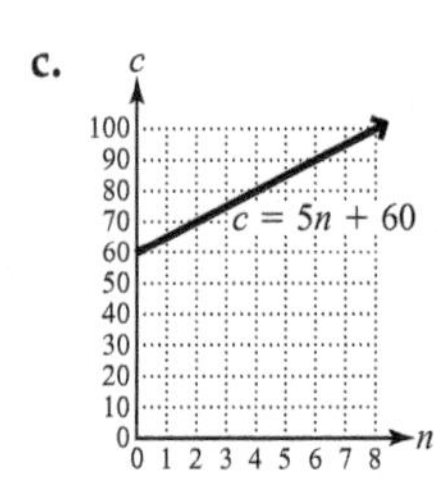

69. a. \$47.50 **b.** 35 min. **c.**

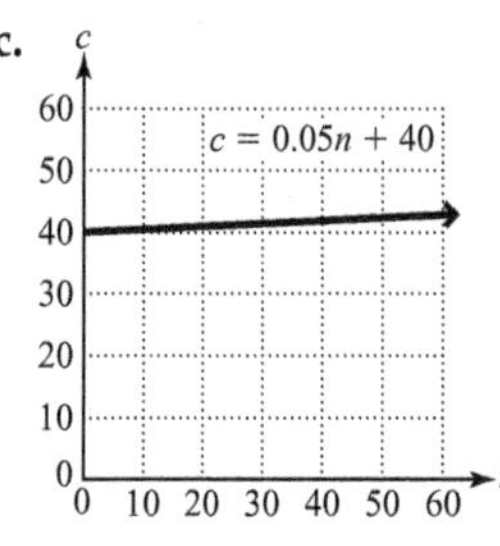

71. a. \$8 **b.** 400 copies **c.**

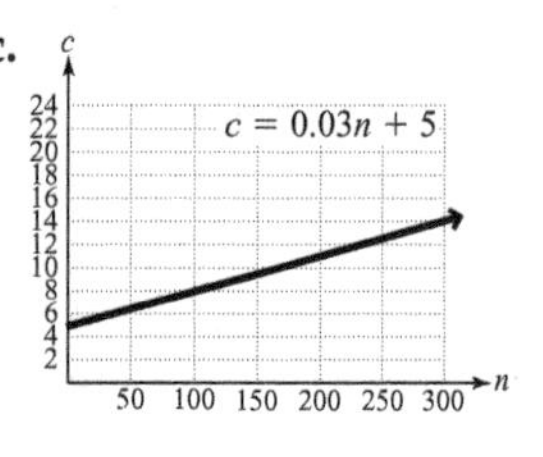

Review Exercises: **1.** Commutative property of addition. **2.** $-\frac{6}{5}$ **3.** -2 **4.** $-\frac{2}{3}$ **5.** $\frac{19}{7}$ **6.** $\{x \mid x > 6\}$; $(6, \infty)$;

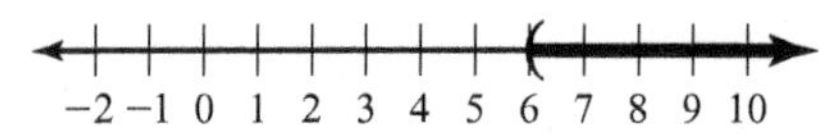

Exercise Set 3.2

Prep Exercises **1.** Slope is the incline of a line. **2.** If $m > 0$, the graph is a line that slants upward from left to right. **3.** If $m < 0$, the graph is a line that slants downward from left to right. **4.** Choose any two points on the line and write a ratio of the vertical rise to the horizontal run between those two points. **5.** The y-intercept is the point where the graph intersects the y-axis. **6.** It is a horizontal line because all the y-coordinates are the same. **7.** It is a vertical line because all the x-coordinates are the same. **8.** Plot $(0, -2)$, then rise 3 and run 4 to find another point. **9.** $m = \frac{y_2 - y_1}{x_2 - x_1}$ **10.** It does not matter. Although choosing (x_1, y_1) and (x_2, y_2) differently changes the signs of the numerator and denominator, after simplifying, the quotient is the same.

Exercises **1.** $y = 3x + 1$ **3.** $y = x - 4$ **5.** upward **7.** downward **9.** $m = \frac{2}{3}$; $(0, 1)$ **11.** Undefined slope; no y-intercept. **13.** e **15.** g **17.** h **19.** d

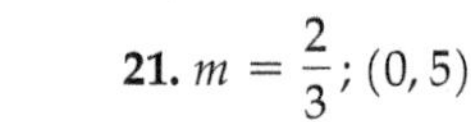

21. $m = \frac{2}{3}$; $(0, 5)$

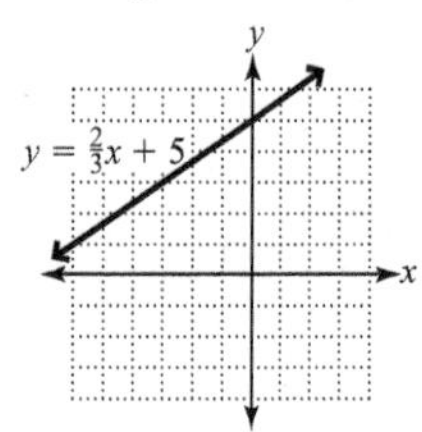

23. $m = -\frac{3}{4}$; $(0, -2)$

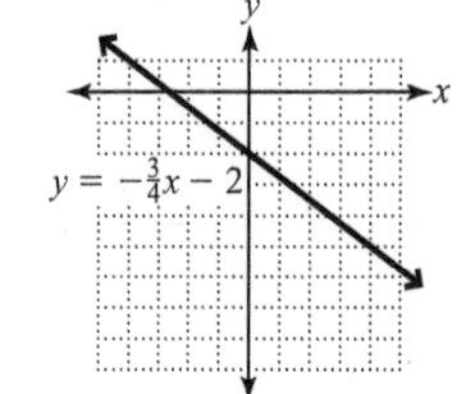

25. $m = 1$; $(0, 4)$

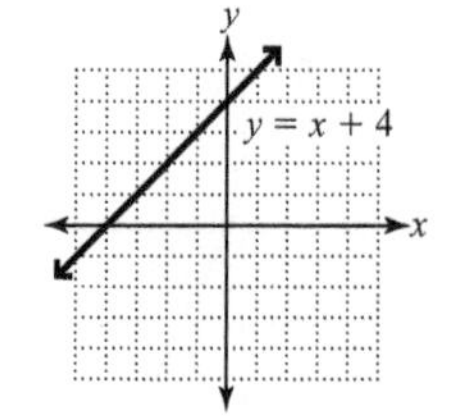

27. $m = -5$; $\left(0, \frac{2}{3}\right)$

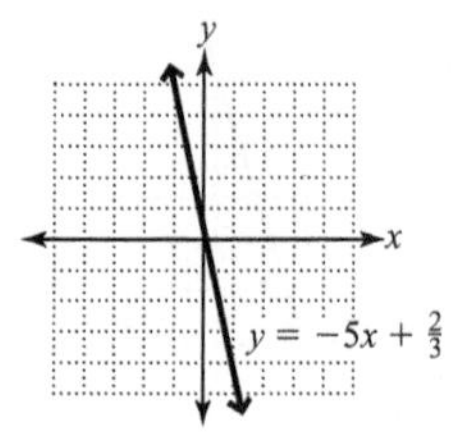

29. $m = -\frac{2}{3}$; $(0, 2)$

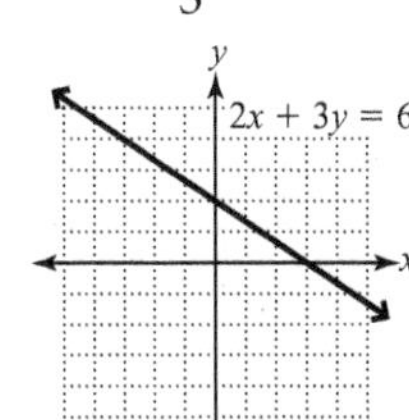

31. $m = \frac{1}{2}$; $\left(0, \frac{7}{2}\right)$

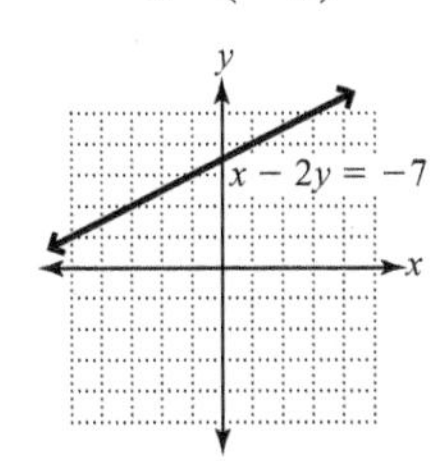

33. $m = \frac{2}{7}$; $\left(0, -\frac{8}{7}\right)$

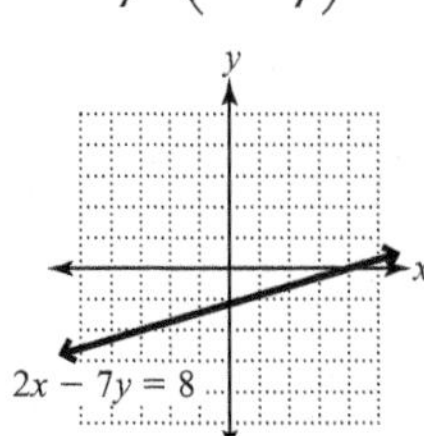

35. $m = 1$; $(0, 0)$

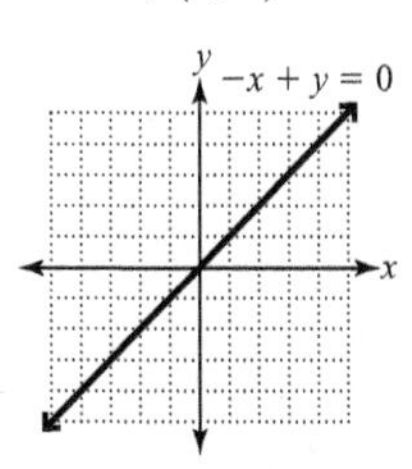

37. -2 **39.** $-\frac{9}{5}$ **41.** $-\frac{3}{2}$ **43.** undefined **45.** 0 **47.** $-\frac{4}{5}$ **49. a.**

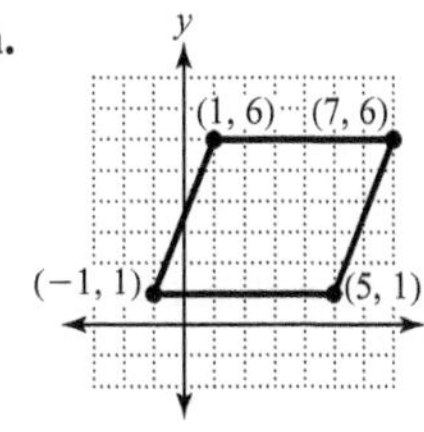

b. Left and right sides: $m = \frac{5}{2}$; top and bottom sides: $m = 0$. **c.** Slopes of parallel sides are equal. **51.** $\frac{1}{12}$ **53.** 0.25 **55. a.** C: (50, 15); D: (100, 80) **b.** $\frac{13}{10}$ **c.** The coaster climbs 13 ft. vertically for every 10 ft. that it moves horizontally. **57. a.**

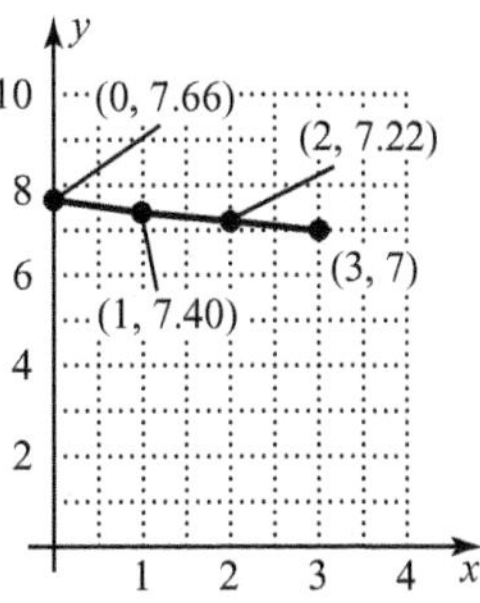

b. -0.22 **c.** \$6.78

59. a.

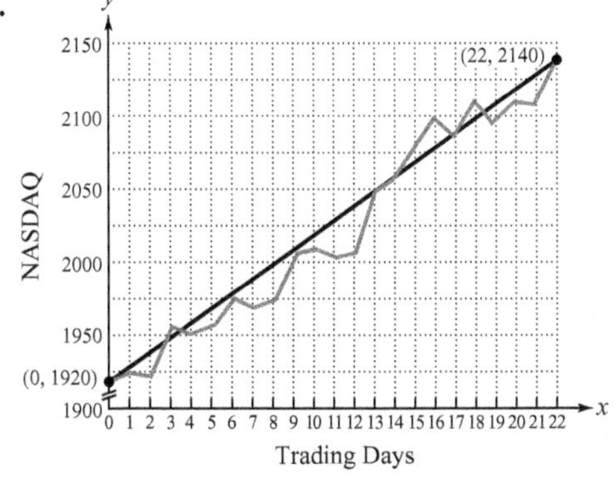

b. 10 **c.** 2150 **61. a.**

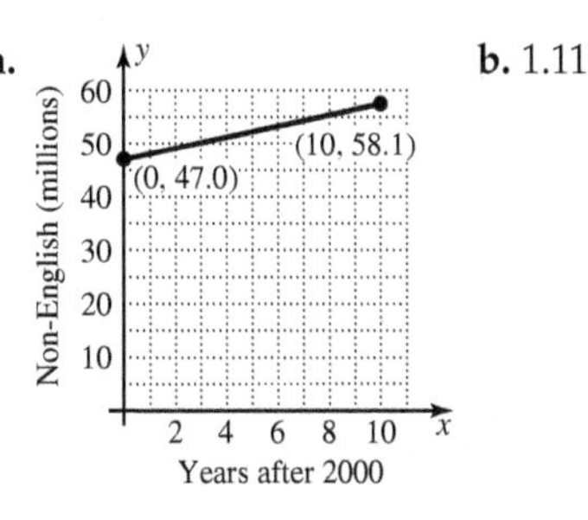

b. 1.11

63. a.

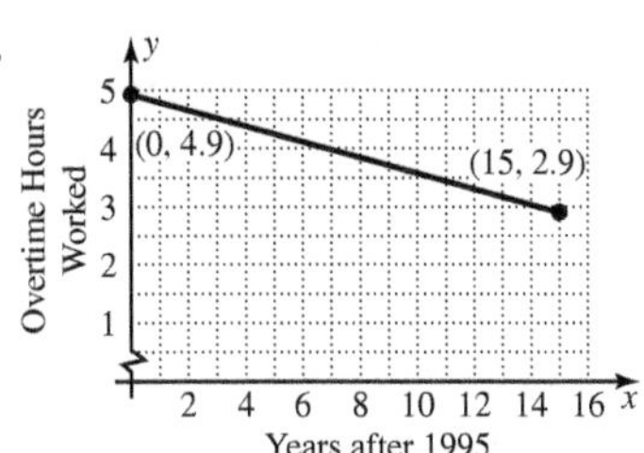

b. $-0.1\overline{3}$

Review Exercises: **1.** 16 **2.** 61 **3.** $15x - 27$ **4.** $-15x + 10$ **5.** $h = \frac{V}{lw}$ **6.** $y = \frac{C - Ax}{B}$

Exercise Set 3.3

Prep Exercises **1.** m represents the slope, b represents the y-coordinate of the y-intercept. **2.** Find the slope using $m = \frac{y_2 - y_1}{x_2 - x_1}$. **3.** $y - y_1 = m(x - x_1)$ **4.** Solve the equation for y. **5.** Multiply both sides by the LCD. **6.** Multiply or divide both sides by -1. **7.** Their slopes are equal. **8.** $-\frac{b}{a}$

Exercises **1.** $y = -4x + 3$ **3.** $y = \frac{3}{5}x - 2$ **5.** $y = -0.2x - 1.5$ **7.** $m = \frac{1}{2}, (0, 2), y = \frac{1}{2}x + 2$ **9.** $m = -2, (0, -4), y = -2x - 4$ **11.** $y = \frac{2}{3}x + 2$ **13.** $y = -\frac{3}{2}x - 6$ **15.** $y = 2x - 5$ **17.** $y = -2x - 10$ **19.** $y = -3x$ **21.** $y = \frac{2}{5}x - 2$ **23.** $y = \frac{4}{5}x - \frac{2}{5}$ **25.** $y = -\frac{3}{2}x + \frac{3}{2}$ **27. a.** $y = -2x + 7$ **b.** $2x + y = 7$ **29. a.** $y = x - 5$ **b.** $x - y = 5$ **31. a.** $y = \frac{2}{3}x - 2$ **b.** $2x - 3y = 6$ **33. a.** $y = \frac{6}{5}x - 6$ **b.** $6x - 5y = -30$ **35. a.** $y = -\frac{1}{13}x + \frac{30}{13}$ **b.** $x + 13y = 30$ **37. a.** $y = -\frac{7}{3}x - \frac{8}{3}$ **b.** $7x + 3y = -8$ **39.** parallel **41.** perpendicular **43.** neither **45.** parallel **47.** perpendicular **49.** perpendicular **51. a.** $y = -5x + 4$ **b.** $5x + y = 4$ **53. a.** $y = 4x + 18$ **b.** $4x - y = -18$ **55. a.** $y = \frac{2}{3}x - 6$ **b.** $2x - 3y = 18$ **57. a.** $y = -\frac{2}{3}x - \frac{1}{3}$ **b.** $2x + 3y = -1$ **59. a.** $y = -\frac{2}{5}x - \frac{41}{5}$ **b.** $2x + 5y = -41$ **61. a.** $y = -3x + 5$ **b.** $3x + y = 5$ **63. a.** $y = \frac{5}{2}x - 3$ **b.** $5x - 2y = 6$ **65. a.** $y = \frac{1}{3}x - \frac{25}{3}$ **b.** $x - 3y = 25$ **67. a.** $y = 4x + 5$ **b.** $4x - y = -5$ **69. a.** $y = -\frac{3}{2}x + \frac{11}{2}$ **b.** $3x + 2y = 11$ **71.** $y = -4$ **73.** $y = 0$ **75. a.** 2.625 **b.** $p = 2.625t + 282.4$ **c.** 334.9 million (or 334,900,000) people **77. a.** -47.75 **b.** $b = -47.75t + 908$ **c.** 144 thousand (or 144,000) barrels **79. a.**

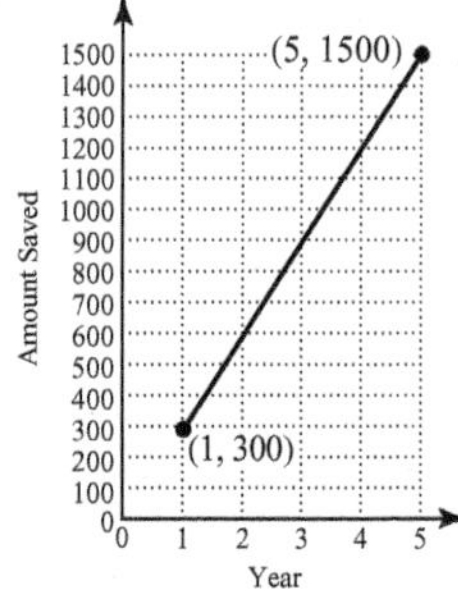

b. 300 **c.** $s = 300t$ **d.** \$3000

81. a.

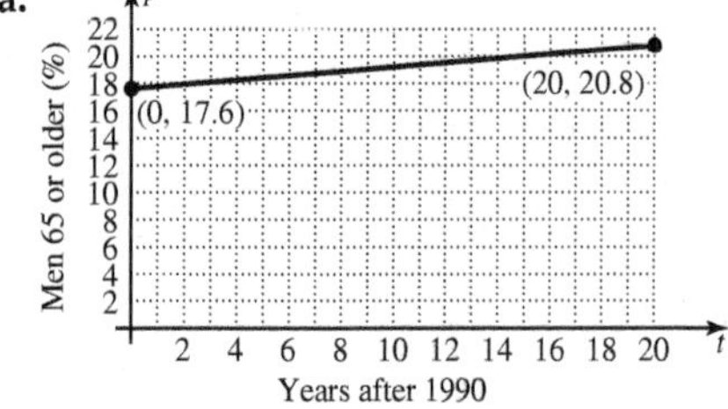

b. 0.16 **c.** $p = 0.16t + 17.6$ **d.** 23.2%

Review Exercises: **1.** $<$ **2.** -2 **3.** $\frac{2}{15}$ **4.** $\{x \mid x \le -5\}; (-\infty, -5]$

$-8\ -7\ -6\ -5\ -4\ -3\ -2\ -1\ 0\ 1\ 2\ 3\ 4$

5. $\{x \mid x \le 2\}; (-\infty, 2]$

$-6\ -5\ -4\ -3\ -2\ -1\ 0\ 1\ 2\ 3\ 4\ 5\ 6$

6. $w \ge 40$ ft.

Exercise Set 3.4

Prep Exercises **1.** $(3, -1)$ is a solution because $3 + (-1) < 4$ is true. **2.** Graph the related equation (the boundary line). **3.** Replace the inequality with an $=$ symbol. **4.** $<, >$ **5.** $\le, \ge$ **6.** Choose an ordered pair on one side of the boundary line. If the ordered pair satisfies the inequality, shade the region that contains it. If the ordered pair does not satisfy the inequality, shade the region on the other side of the boundary line.

Exercises **1.** no **3.** yes **5.** no **7.** yes **9.**

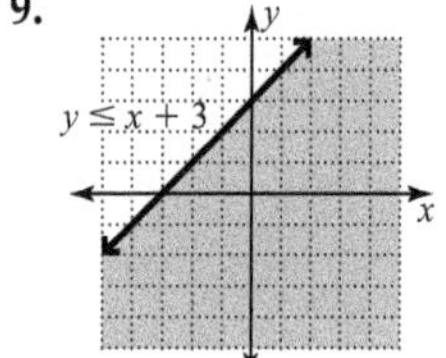

11. $y \ge -4x + 6$

13.

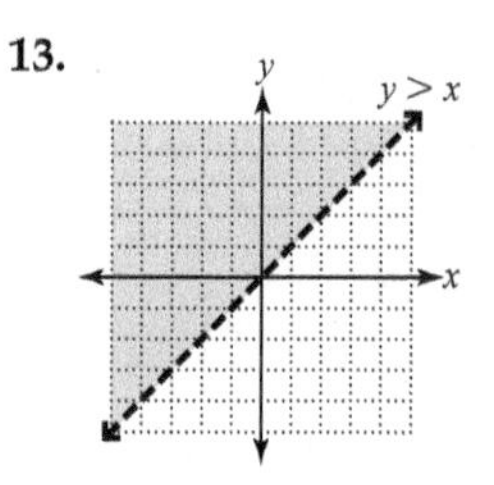

15. $y > -3x$

17.

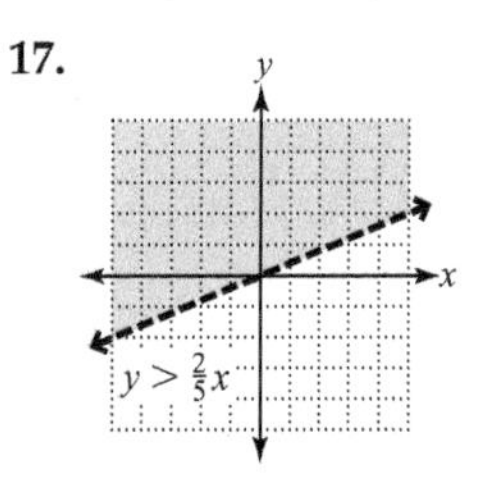

19.

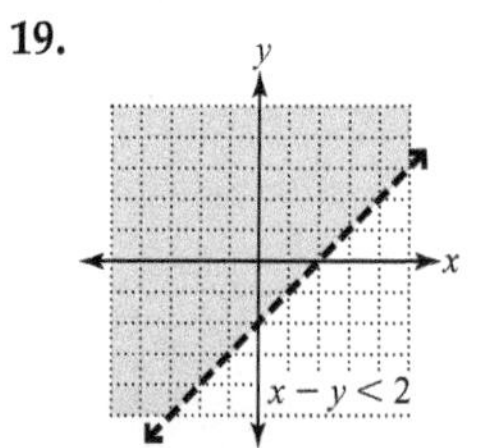

21.

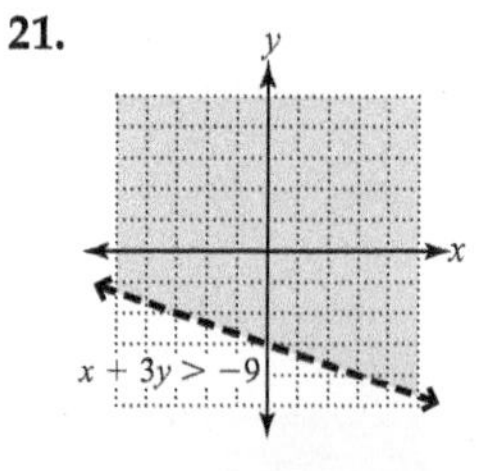

23.

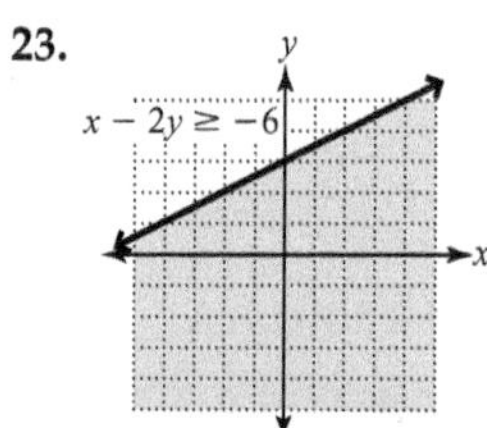

25.

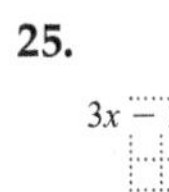
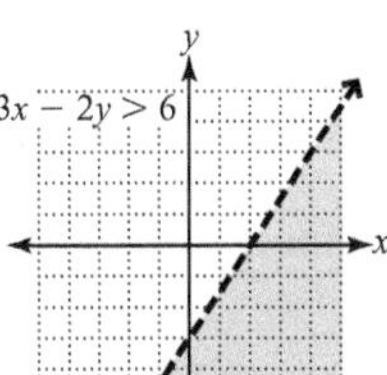

27.

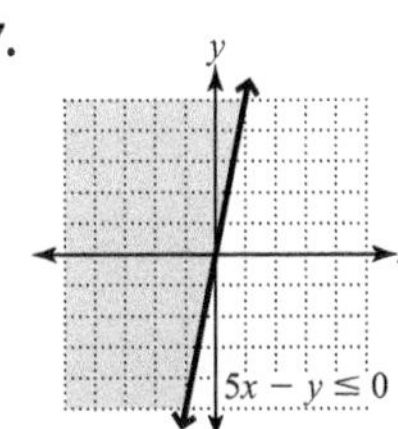

29.

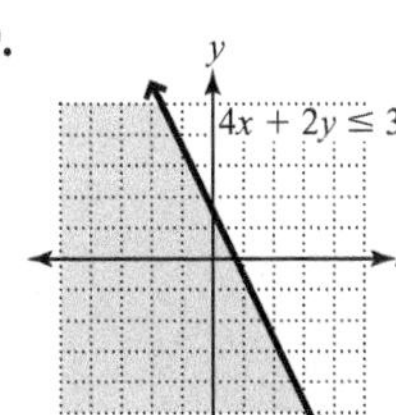

31.

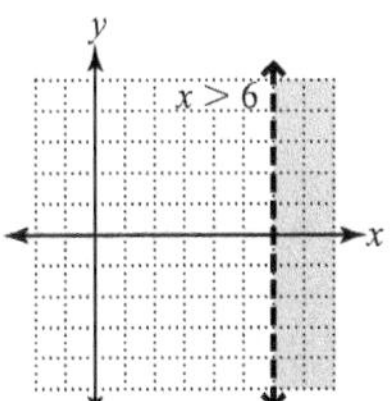

33.

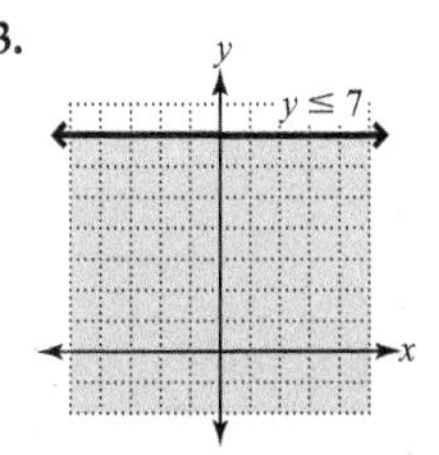

35. a. x represents the number of board games produced, and y represents the number of video games produced.
b.

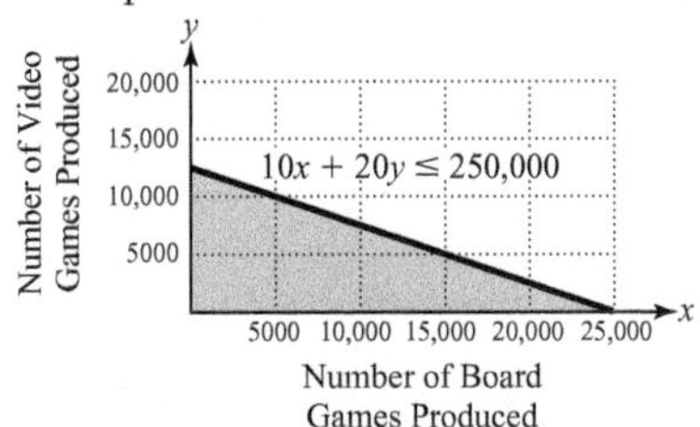

c. All combinations that cost exactly \$250,000 to produce.
d. All combinations that cost less than \$250,000 to produce. **e.** Answers may vary. Two examples are (0, 12,500) and (25,000, 0). **f.** Answers may vary. Two examples are (0, 500) and (10,000, 5000). **g.** No, fractions of a game are not produced.

37. a. $2l + 2w \le 200$ **b.**

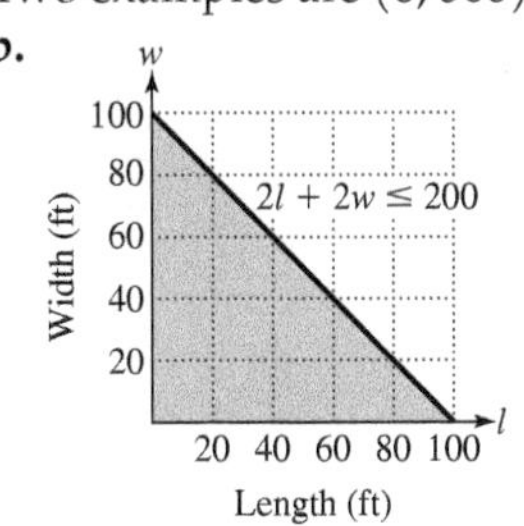

c. Combinations of length and width that make the perimeter exactly 200 ft.
d. All combinations that make the perimeter less than 200 ft.
e. Answers may vary. Two examples are (20, 80) and (60, 40).
f. Answers may vary. Two examples are (20, 40) and (80, 10).

39. a. $12x + 15y \ge 18{,}000$ **b.**

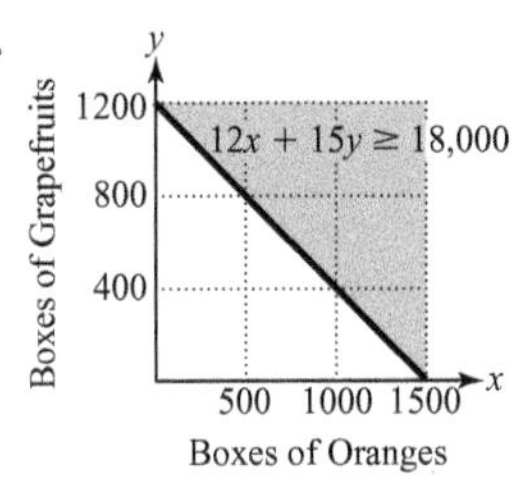

c. All sales combinations that raise exactly \$18,000.
d. All sales combinations that raise more than \$18,000.
e. Answers may vary. Two examples are (500, 800) and (1000, 400).
f. Answers may vary. Two examples are (500, 1000) and (800, 1200).

Review Exercises: **1.** $\{-2, 0, 3, 4\}$ **2.** $\{1, 2, 5, 7\}$ **3.** 13 **4.** 3 **5.** -8 **6.** $(2, 0), \left(0, -\frac{8}{5}\right)$

Exercise Set 3.5

Prep Exercises **1.** A set containing initial values of a relation; its input values; the first coordinates in ordered pairs. **2.** A set containing all values that are paired with the domain in a relation; its output values; the second coordinates in ordered pairs. **3.** A function is a relation in which every value in the domain is assigned to exactly one value in the range. **4.** If each vertical line intersects the graph in at most one point, the relation is a function. If any vertical line intersects the graph at two or more different points, the relation is not a function. **5.** Replace $f(x)$, $f(a)$ in the function with a and simplify. **6.** Yes, its graph will pass the vertical line test. **7.** The graph of every absolute value function is a V shape. Find many ordered pairs, plot them, and then draw a V-shape through those points. **8.** The graph of every quadratic function is a parabola.

Exercises **1.** Domain: {Sarah Lawrence College, New York University, Harvey Mudd College, Columbia College at Columbia University, Wesleyan University}; Range: {\$61,236, \$59,337, \$58,913, \$58,742, \$58,502}; It is a function. **3.** Domain: $\{1, 2, 3, 4, 5, 6\}$; Range: {San Francisco 49ers, Dallas Cowboys, Pittsburgh Steelers, Green Bay Packers, New England Patriots, Oakland/LA Raiders, Washington Redskins, New York Giants, Miami Dolphins, Denver Broncos, Indianapolis/Baltimore Colts, Baltimore Ravens, Chicago Bears, New York Jets, Tampa Bay Buccaneers, Kansas City Chiefs, St. Louis/LA Rams; New Orleans Saints}; It is not a function. **5.** Domain: {1%, 2%, 5%, 6%, 8%, 35%, 42%}; range: {Windows XP, Windows 7, Windows Vista, Mac OS X, iOS, Android, Linux, Other}; It is not a function. **7.** Domain: $\{1, 2, 3, 4, 5\}$; Range: {Thriller, Back in Black, Dark Side of the Moon, Bat out of Hell, Their Greatest Hits 1971–1975}; It is a function. **9.** Domain: $\{-2, 6, 5, -1\}$; Range: $\{4, -2, 1, -6\}$; It is a function. **11.** Domain: $\{-6, 2, -5\}$; Range: $\{2, -3, 7, 4\}$; It is not a function. **13.** Domain: $\{2, -4, 5, 3\}$; Range: $\{-4, 3, -1\}$; It is a function. **15.** Domain: {2000, 2001, 2002, 2003, 2004, 2005, 2006, 2007, 2008, 2009, 2010}; Range: {5920, 5915, 5534, 5574, 5764, 5734, 5840, 5657, 5214, 4551, 4690}; It is a function. **17.** Domain: $\mathbb{R}$ or $(-\infty, \infty)$; Range: $\{y \mid y \le 4\}$ or $(-\infty, 4]$; It is a function. **19.** Domain: $\{x \mid x \ge 0\}$ or $[0, \infty)$; Range: $\mathbb{R}$ or $(-\infty, \infty)$; It is not a function. **21.** Domain: $\{x \mid -4 \le x \le 5\}$ or $[-4, 5]$; Range: $\{-3, -2\}$; It is a function. **23.** Domain: $\{x \mid -4 \le x \le 0\}$ or $[-4, 0]$;

Range: $\{y \mid -3 \le y \le 1\}$ or $[-3, 1]$; It is not a function. **25. a.** -9 **b.** -11 **c.** -7 **d.** $-2a - 11$ **27. a.** 7 **b.** 8 **c.** 10 **d.** $2a^2 - a + 7$ **29. a.** 2 **b.** not real **c.** $\sqrt{5}$ **d.** $\sqrt{3-t}$ **31. a.** 1 **b.** -1 **c.** $\frac{3}{5}$ **d.** $\frac{2}{5}r + 1$ **33. a.** -3 **b.** 6.46 **c.** ≈ -3.96 **d.** $a^2 - 2.1a - 3$ **35. a.** 0 **b.** $\sqrt{21}$ **c.** not real **d.** $\sqrt{n^2 - 4n}$ **37. a.** 1 **b.** $\frac{7}{3}$ **c.** 4 **d.** 13 **39. a.** -2 **b.** undefined **c.** $-\frac{5}{6}$ **d.** 0 **41. a.** 0 **b.** undefined **c.** $\frac{2}{3}$ **d.** $\frac{m}{m^2 - 1}$ **43. a.** -2 **b.** 2 **c.** undefined **d.** not real **45. a.** 2 **b.** 0 **c.** -1 **47. a.** -4 **b.** 0 **c.** -4 **49. a.** -3 **b.** 1 **c.** 2

51.

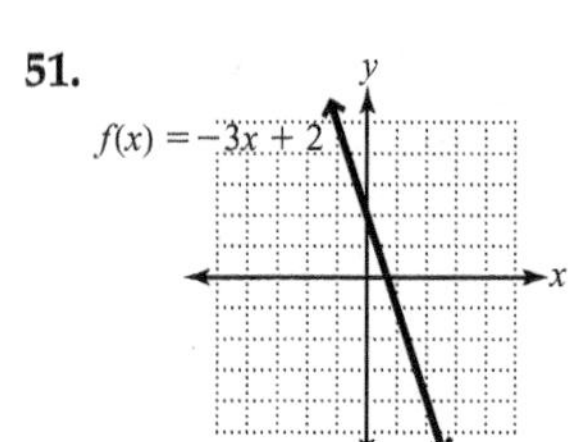

Domain: $\mathbb{R}$ or $(-\infty, \infty)$; Range: $\mathbb{R}$ or $(-\infty, \infty)$

53.

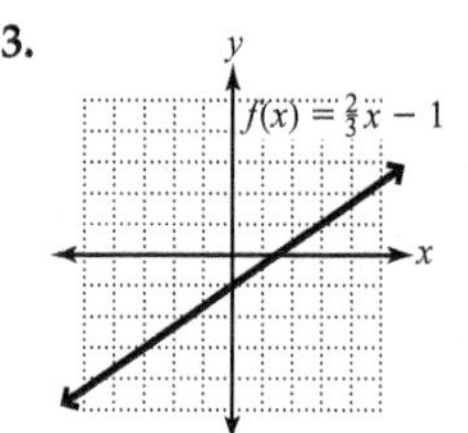

Domain: $\mathbb{R}$ or $(-\infty, \infty)$; Range: $\mathbb{R}$ or $(-\infty, \infty)$

55.

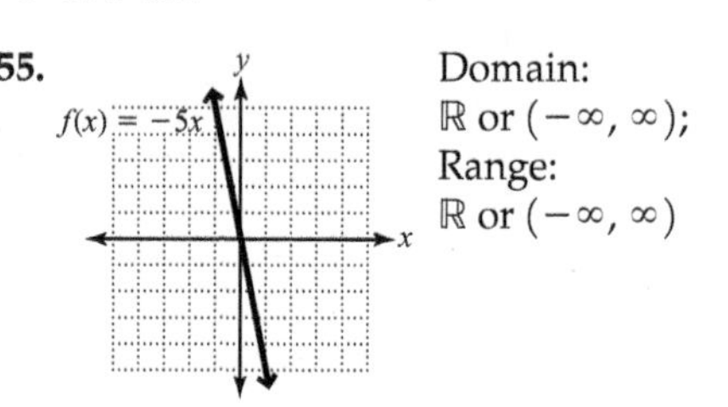

Domain: $\mathbb{R}$ or $(-\infty, \infty)$; Range: $\mathbb{R}$ or $(-\infty, \infty)$

57.

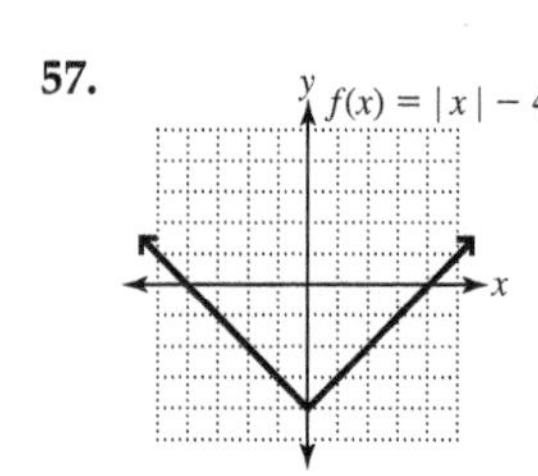

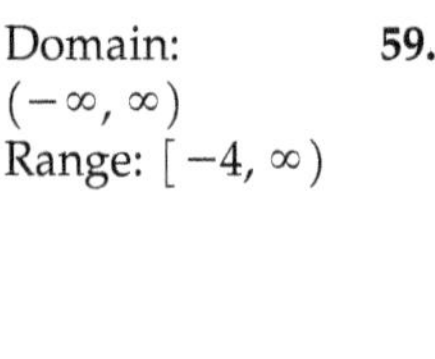

Domain: $(-\infty, \infty)$ Range: $[-4, \infty)$

59.

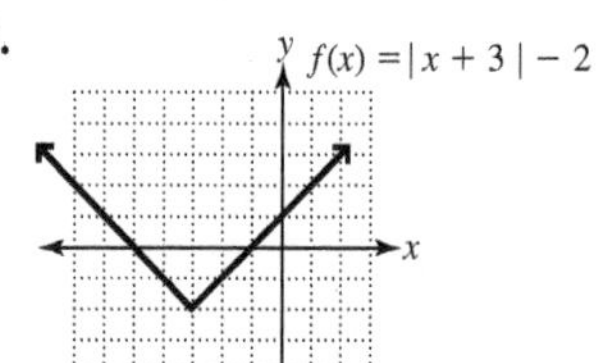

Domain: $(-\infty, \infty)$ Range: $[-2, \infty)$

61.

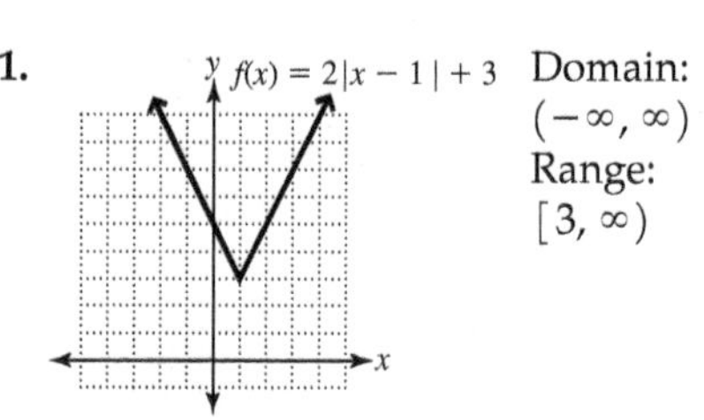

Domain: $(-\infty, \infty)$ Range: $[3, \infty)$

63.

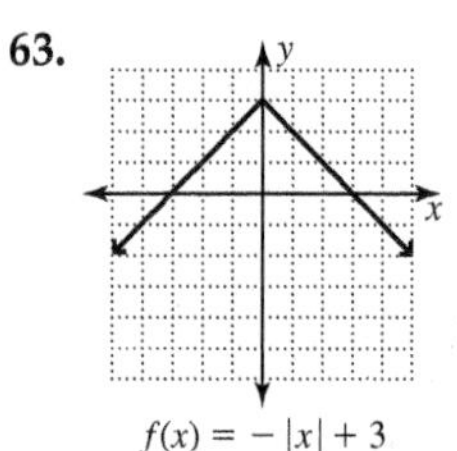

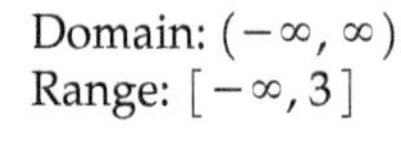

Domain: $(-\infty, \infty)$ Range: $[-\infty, 3]$

65.

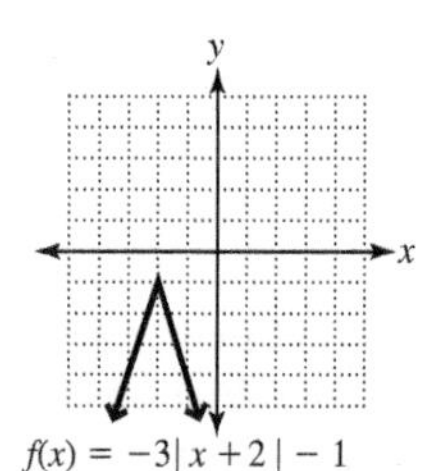

Domain: $(-\infty, \infty)$ Range: $[-\infty, -1]$

67.

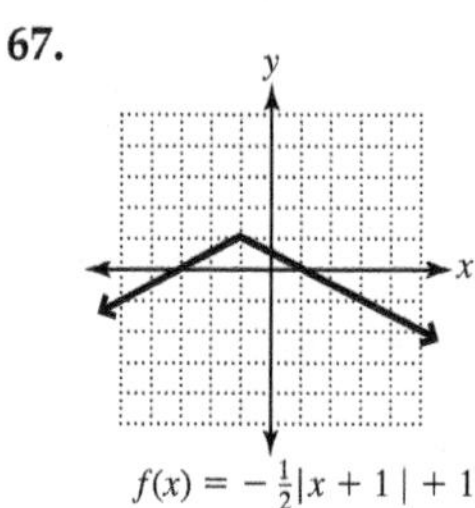

Domain: $(-\infty, \infty)$ Range: $[-\infty, 1]$

69.

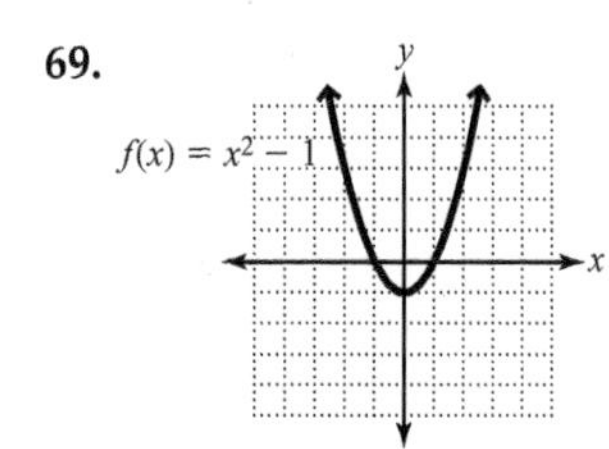

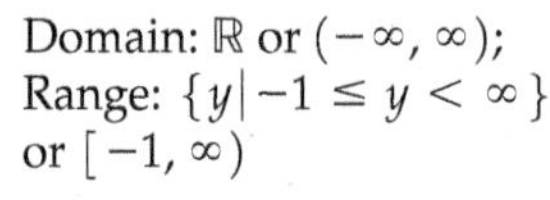

Domain: $\mathbb{R}$ or $(-\infty, \infty)$; Range: $\{y \mid -1 \le y < \infty\}$ or $[-1, \infty)$

71.

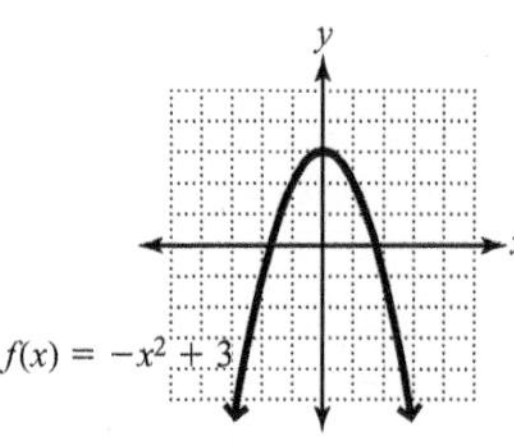

Domain: $\mathbb{R}$ or $(-\infty, \infty)$; Range: $\{y \mid -\infty < y \le 3\}$ or $(-\infty, 3]$

73. a.

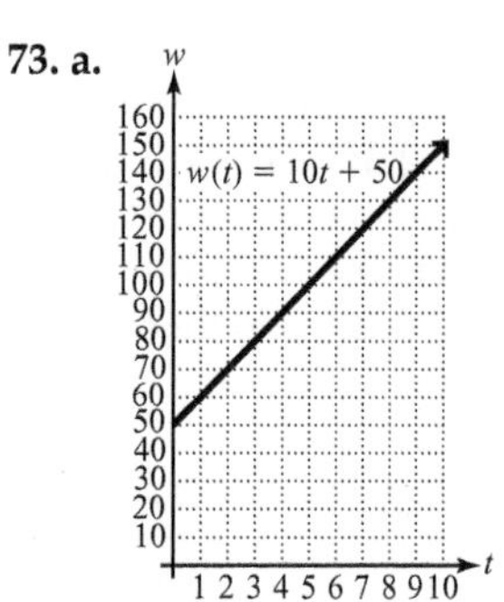

b. \$125 **75. a.** $C(V) = 1.225V + 6$ **b.**

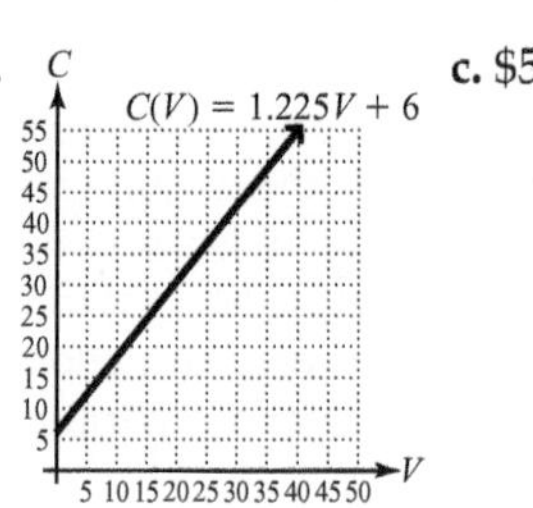

c. \$55

77. a.

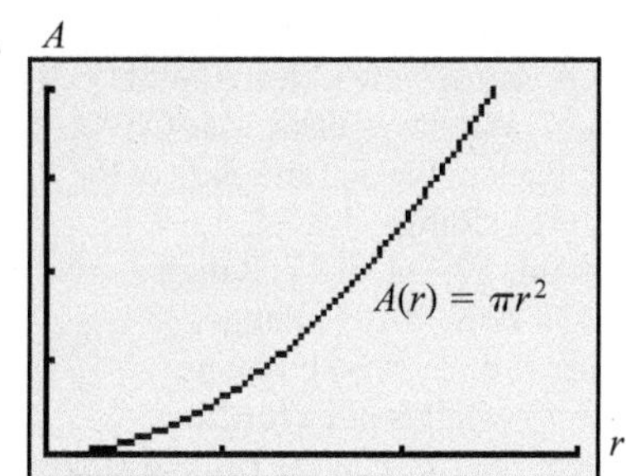

b. no **c.** The radius of the circle is 1.5 units. **d.** 7.07 sq. units

79. a.

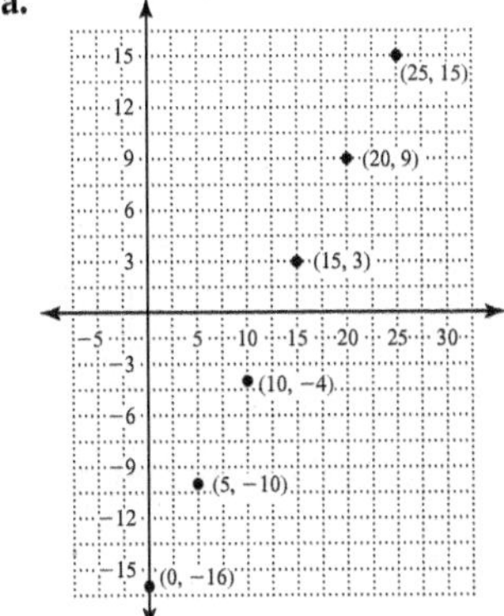

b. yes, 1.24 **c.** $w(a) = 1.24a - 16$ **d.** 33.6°

Review Exercises: **1.** $-2x + 6y - 2$ **2.** -3 **3.** $\{x \mid x \le 2\}, (-\infty, 2]$,

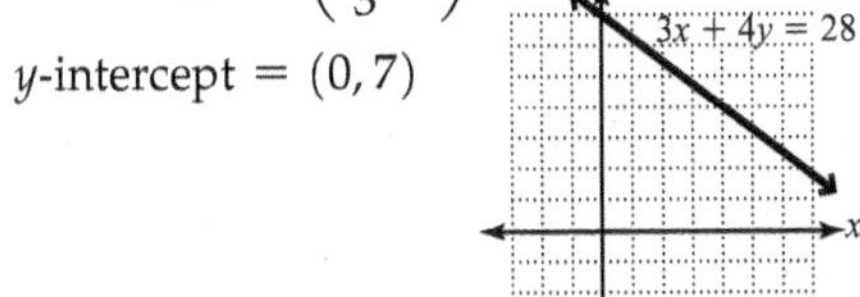

4. 36, 38 **5.** length: 30 in.; width: 24 in. **6.** 12 large, 28 small

Chapter 3 Summary and Review Exercises

3.1 Definitions/Rules/Procedures $Ax + By = C$; number; variable Exercises **1.** no **2.** yes **3.** no **4.** yes

Definitions/Rules/Procedures two; Plot; points, arrowheads; x-axis; y-axis; y; x; x; y; $(0, b)$

Exercises **5.** x-intercept $= (4, 0)$
y-intercept $= (0, -4)$

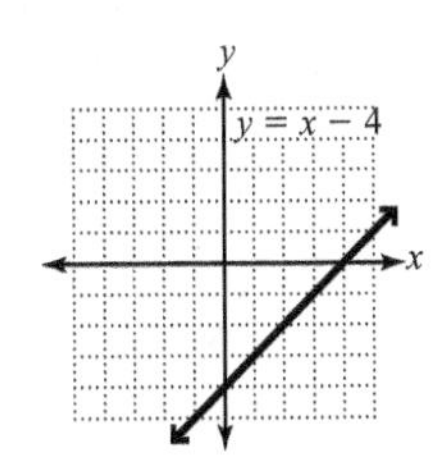

6. x-intercept $= (0, 0)$
y-intercept $= (0, 0)$

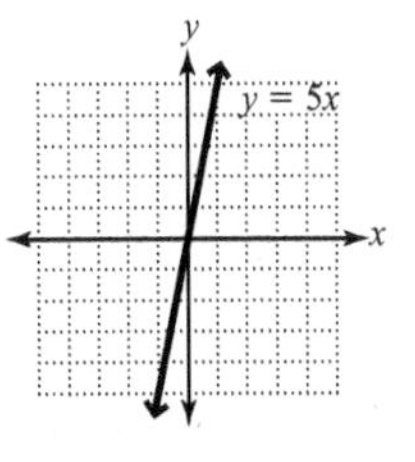

7. x-intercept $= \left(\frac{9}{2}, 0\right)$
y-intercept $= (0, -3)$

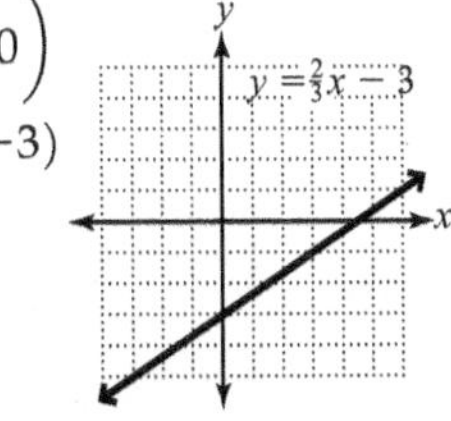

8. x-intercept $= \left(\frac{7}{2}, 0\right)$
y-intercept $= (0, 1)$

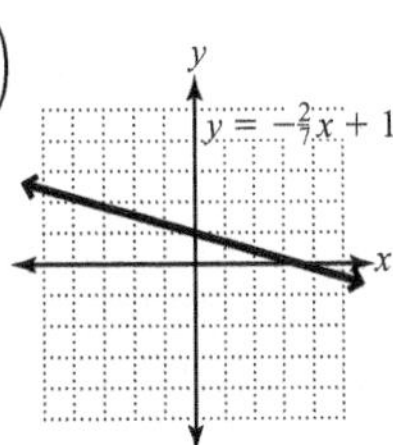

9. x-intercept $= (3, 0)$
y-intercept $= (0, -2)$

10. x-intercept $= \left(\frac{28}{3}, 0\right)$
y-intercept $= (0, 7)$

Definitions/Rules/Procedures horizontal, x, $(0, c)$; vertical, y, $(c, 0)$

Exercises **11.** x-intercept $(3, 0)$, y-intercept none

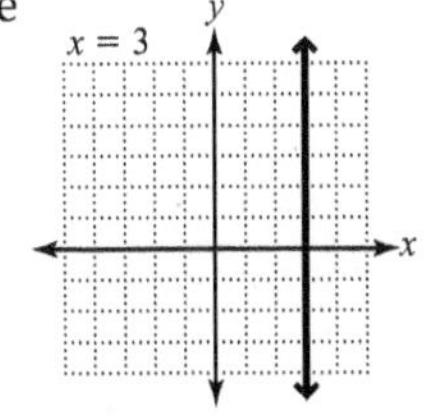

12. x-intercept none, y-intercept $(0, -4)$

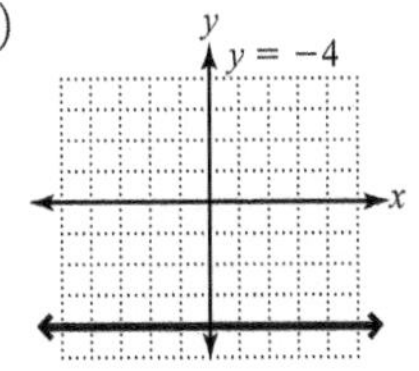

3.2 Definitions/Rules/Procedures vertical change (change in y), horizontal change (change in x); m, $(0, b)$; upward; downward

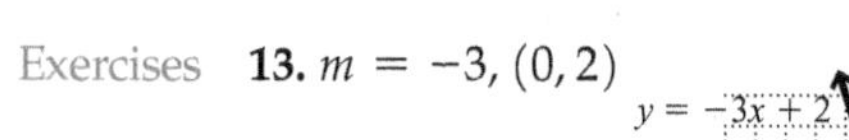

Exercises **13.** $m = -3, (0, 2)$

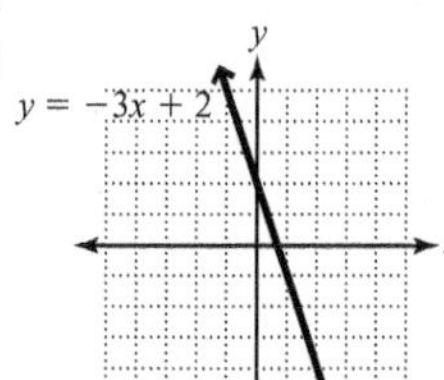

14. $m = \frac{5}{2}, (0, 3)$

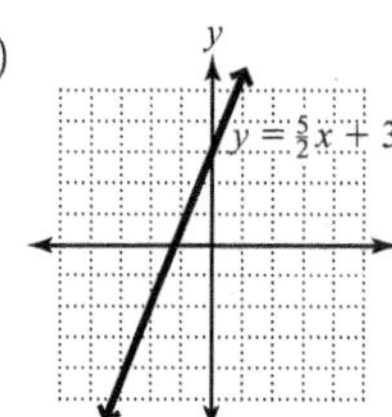

15. $m = -\frac{2}{5}, (0, 3)$

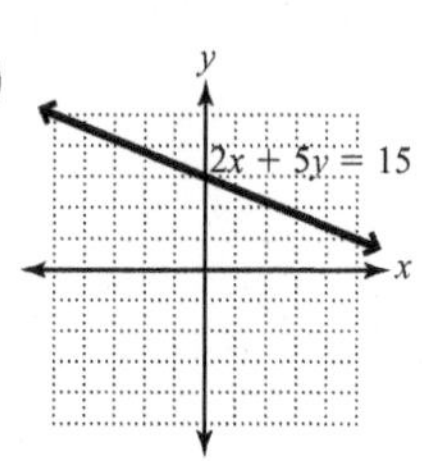

16. $m = 4, (0, -3)$

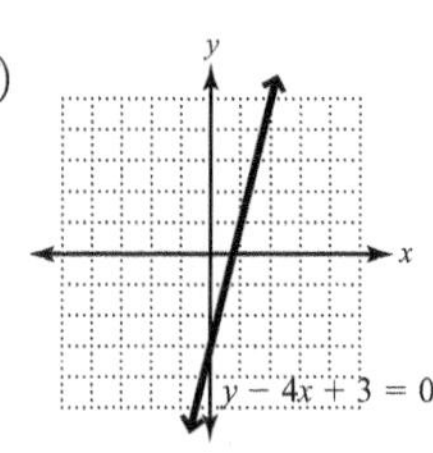

Definitions/Rules/Procedures $\frac{y_2 - y_1}{x_2 - x_1}$; horizontal, 0, $y = c$; vertical, undefined, $x = c$

Exercises **17.** -3 **18.** undefined **19.** 1 **20.** 0 **21.** $\frac{5}{6}$ **22. a.** -0.3 **b.** $y = -0.3x + 12.2$ **c.** 7.7%

3.3 Definitions/Rules/Procedures $y = mx + b$, $\frac{y_2 - y_1}{x_2 - x_1}$, $y = mx + b$ Exercises **23.** $y = 2x - 4$ **24.** $m = -\frac{2}{5}x + 4$ **25.** $y = -0.3x - 1$ **26.** $y = -3x$ **27.** $m = \frac{2}{5}, (0, 2), y = \frac{2}{5}x + 2$ **28.** $m = -2, (0, -1), y = -2x - 1$

Definitions/Rules/Procedures $y - y_1 = m(x - x_1)$, $\frac{y_2 - y_1}{x_2 - x_1}$, $y - y_1 = m(x - x_1)$; $Ax + By = C$ Exercises **29.** $y = -x + 10$ **30.** $y = 6.2x - 23.6$ **31.** $y = \frac{2}{3}x - \frac{8}{3}$ **32.** $y = -3x - 8$ **33. a.** $y = 3x - 6$ **b.** $3x - y = 6$ **34. a.** $y = \frac{7}{2}x + 9$ **b.** $7x - 2y = -18$ **35. a.** $y = \frac{1}{6}x - \frac{8}{3}$ **b.** $x - 6y = 16$ **36. a.** $y = -x + 6$ **b.** $x + y = 6$ Definitions/Rules/Procedures equal, different, $-\frac{b}{a}$

Exercises **37.** parallel **38.** neither **39.** perpendicular **40.** perpendicular **41.** $y = -\frac{3}{5}x - 2$ **42.** $y = \frac{1}{5}x + \frac{23}{5}$ **43.** $y = \frac{3}{2}x - \frac{1}{2}$ **44.** $y = -\frac{1}{3}x - 6$

3.4 Definitions/Rules/Procedures $Ax + By > C, <, \leq, \geq$. (Note: The order symbols may be in any order); equation, equation, solid, dashed; that contains it, on the other side of the boundary line Exercises **45.** no **46.** yes **47.**

y < −2x + 5

48.

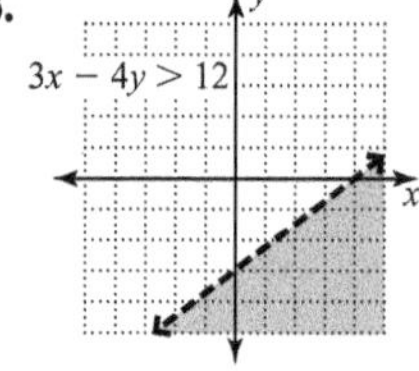

49.

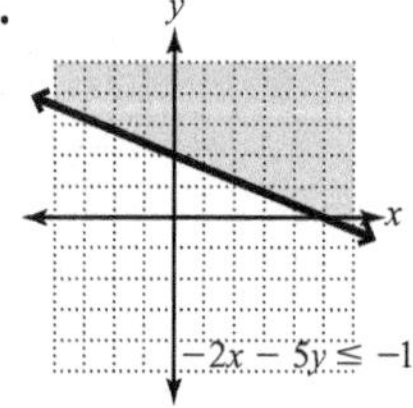

50.

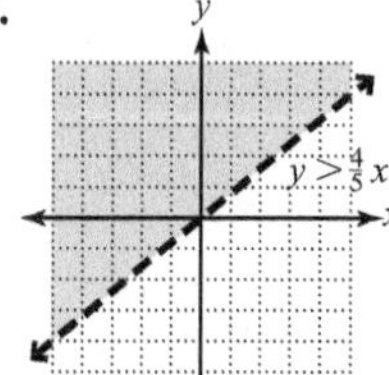

51.

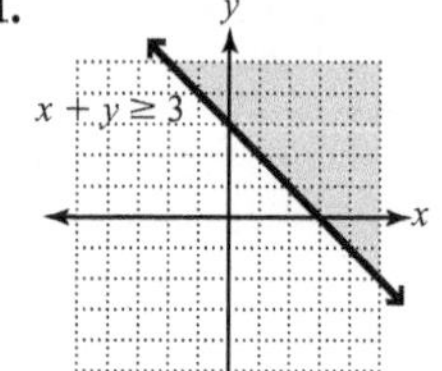

52.

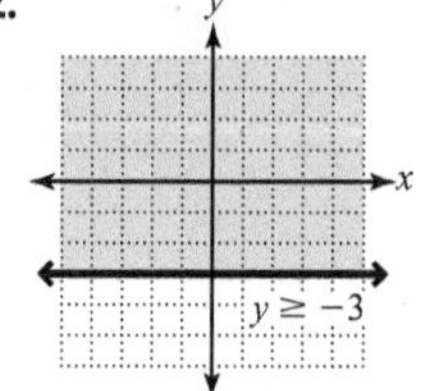

53. a. $35x + 50y \geq 70{,}000$ **b.**

35x + 50y ≥ 70,000; Armchairs; Basic Chairs

c. All combinations of chair sales that make the company break even

($70,000 revenue). **d.** All combinations of chair sales that make the company a profit (more then $70,000 revenue). **e.** Answers may vary. Two examples are (1000, 700) and (2000, 0). **f.** Answers may vary. Two examples are (2000, 500) and (2000, 1000).

3.5 Definitions/Rules/Procedures set of ordered pairs; one value in the range; input, first coordinate (x-coordinate); output, second coordinate (y-coordinate); vertical, vertical, only one point, more than once Exercises **54.** Domain: {McKinley, Logan, Pico de Orizaba, St. Elias, Popocatépetl}; Range: {20,320, 19,551, 18,555, 18,008, 17,930}; it is a function. **55.** Domain: $\{21, 23, 32, 35\}$; Range: {California, Indiana, New York, Ohio, Pennsylvania}; it is not a function. **56.** Domain: $\{x \mid -4 \le x \le 5\}$ or $[-4, 5]$; Range: $\{-2, 2, 3\}$; it is a function. **57.** Domain: $\{x \mid -3 \le x \le 3\}$ or $[-3, 3]$; Range: $\{y \mid 0 \le y \le 3\}$ or $[0, 3]$ it is a function. **58.** Domain: $\mathbb{R}$ or $(-\infty, \infty)$; Range: $\mathbb{R}$ or $(-\infty, \infty)$; it is a function. **59.** Domain: $\{x \mid x \le 3\}$ or $(-\infty, 3]$; Range: $\mathbb{R}$ $(-\infty, \infty)$; it is not a function. Definitions/Rules/Procedures x, a Exercises **60. a.** 0 **b.** -4 **c.** 5 **d.** $n^2 - 4$ **61. a.** $-\frac{1}{7}$ **b.** 0 **c.** undefined

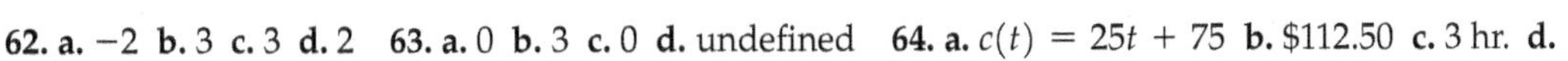

62. a. -2 **b.** 3 **c.** 3 **d.** 2 **63. a.** 0 **b.** 3 **c.** 0 **d.** undefined **64. a.** $c(t) = 25t + 75$ **b.** $112.50 **c.** 3 hr. **d.**

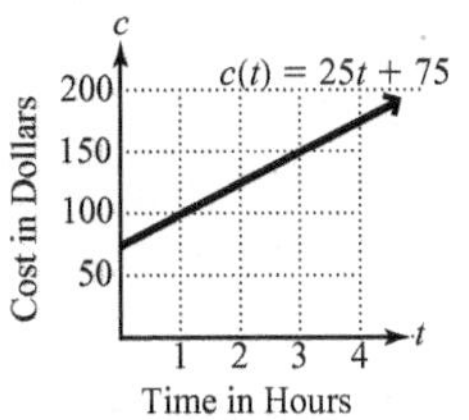

Definitions/Rules/Procedures $mx + b$; line; $a|x - h| + k$; V, V; $ax^2 + bx + c$; parabola

Exercises **65.** $f(x) = -\frac{2}{3}x + 4$

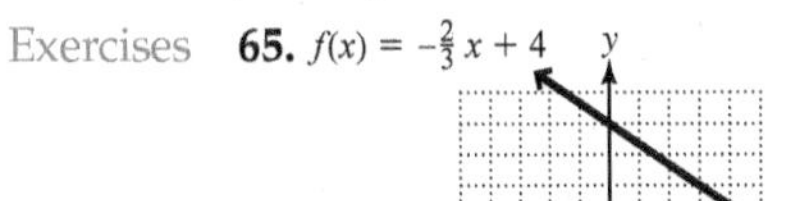

Domain: $\mathbb{R}$
Range: $\mathbb{R}$

66. $f(x) = |x - 2|$

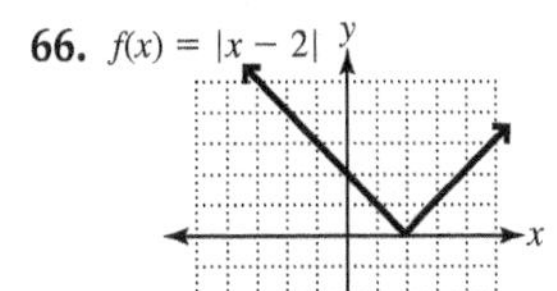

Domain: $\mathbb{R}$
Range: $[0, \infty)$

67. $f(x) = -2|x + 1| + 3$

Domain: $\mathbb{R}$
Range: $(-\infty, 3]$

68. $f(x) = x^2 - 3$

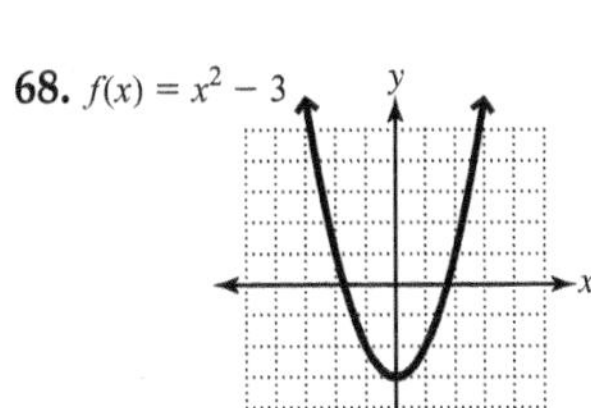

Domain: $\mathbb{R}$
Range: $[-3, \infty)$

Chapter 3 Practice Test

1. A: $(0, 5)$; B: $(3, 3)$; C $(-2, -3)$; D: $(4, -2)$ [3.1] **2.** IV [3.1] **3.** no [3.1] **4.** $m = -\frac{4}{3}, (0, 5)$ [3.2]

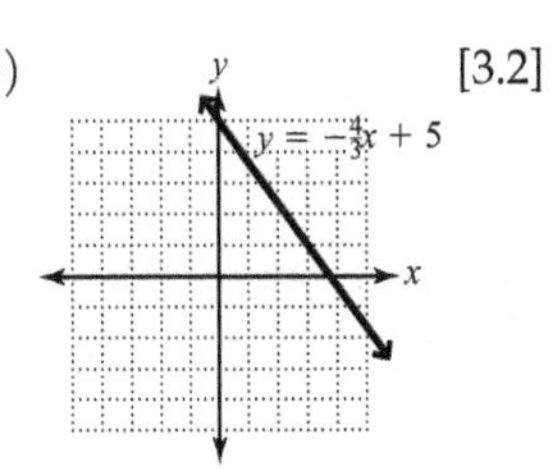

5. $m = \frac{1}{2}, (0, 4)$ [3.2] **6.** 0 [3.2] **7.** $\frac{8}{7}$ [3.2] **8.** $y = \frac{2}{7}x + 5$ [3.3] **9.** $y = \frac{1}{3}x + \frac{2}{3}$ [3.3] **10.** $5x - 4y = -11$ [3. 3]

11. neither [3.3] **12.**

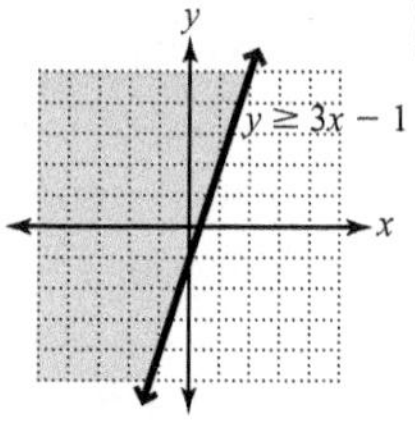

[3.4] **13.**

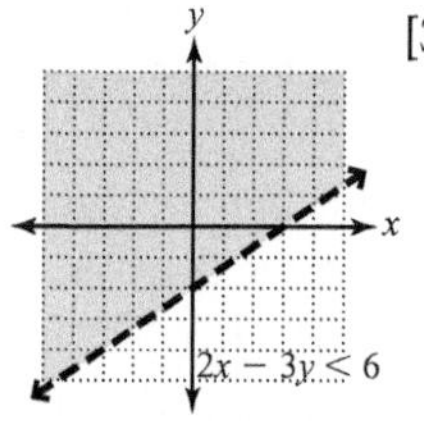

[3.4] **14.** Domain: {Spanish, Chinese, Tagalog, Vietnamese, French}; Range: {37.6, 2.9, 1.6, 1.4, 1.3}; It is a function. [3.5] **15.** Domain: $\{x \mid 0 \leq x \leq 3\}$ or [0, 3]; Range: $\{y \mid -3 \leq y \leq 3\}$ or [−3, 3]; It is not a function. [3.5] **16. a.** 1 **b.** 11 **c.** $2t^2 - 7$ [3.5] **17. a.** −1 **b.** 2 **c.** undefined [3.5]
18. a. Domain: $\{\,|x| - 3 \leq x \leq 3\}$ or [−3, 3]; range: [−2, 1, 3]. [3.5] **b.** 3 [3.5]

19. Domain: ℝ, Range: $[-2, \infty)$

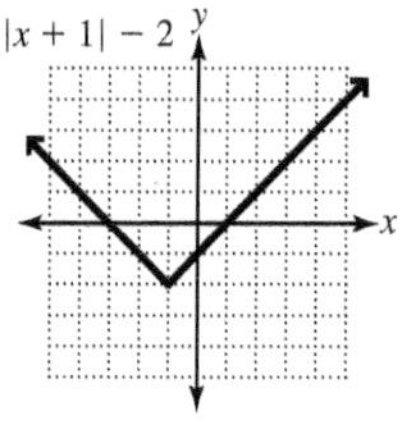

20. Domain: ℝ, Range: $(-\infty, 3]$

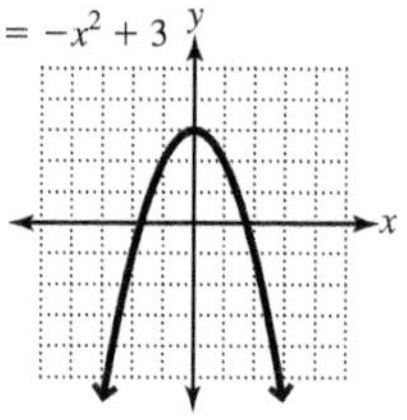

21. a. $2l + 2w \leq 1000$ [3.4] **b.**

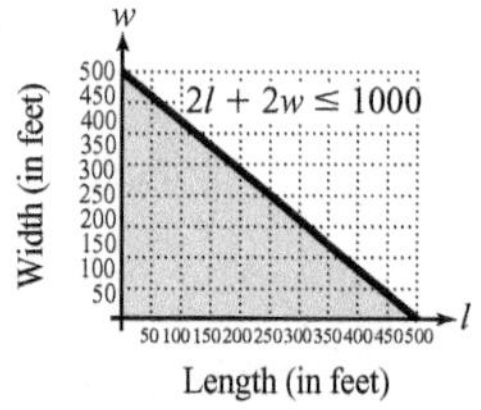

c. Answers may vary. Two possible answers are 300 ft. by 200 ft. and 200 ft. by 200 ft. [3.4]

22. a. $c(w) = 0.45w + 3$ [3.5] **b.**

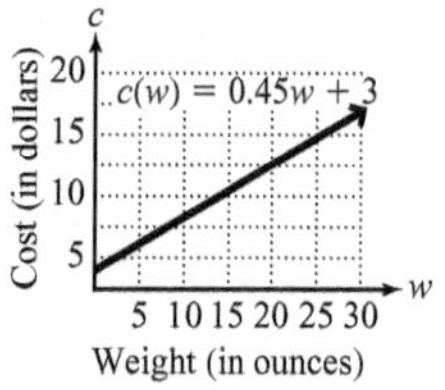

c. $17.40 [3.5]

Chapters 1–3 Cumulative Review Exercises

1. false **2.** true **3.** true **4.** 10 **5.** principal or nonnegative **6.** We perform operations in the following order: a. Within grouping symbols beginning with the innermost: parentheses (), brackets [], braces { }, absolute value | |, above and/or below fraction bars, and radicals √. b. Exponents/roots from left to right, in order as they occur. c. Multiplication/division from left to right, in order as they occur. d. Addition/subtraction from left to right, in order as they occur. **7.** $A \cap B = \varnothing, A \cup B = \{w, e, l, o, v, m, a, t, h\}$
8. 9 **9.** 3 **10.** −135 **11.** undefined **12. a.** $(-4n + 8) - 5n$ **b.** $-9n + 8$ **c.** −28 **13.** multiplicative inverse **14.** distributive property **15.** −29 **16.** $9, -\frac{7}{3}$ **17. a.** $\left\{x \mid x \leq \frac{5}{6}\right\}$ **b.** $\left(-\infty, \frac{5}{6}\right]$ **c.**

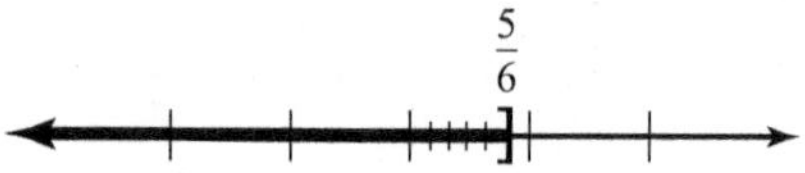

18. a. $\{x \mid x < -4 \text{ or } x > 2\}$
b. $(-\infty, -4) \cup (2, \infty)$ **c.**

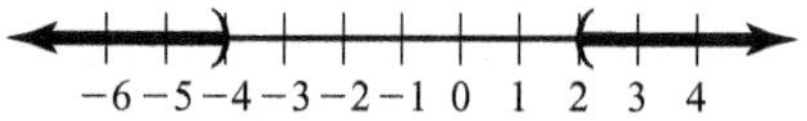

19. $t = \frac{d}{r}$ **20.** 0 **21.** neither **22.** $y = 2x + 6$

23.

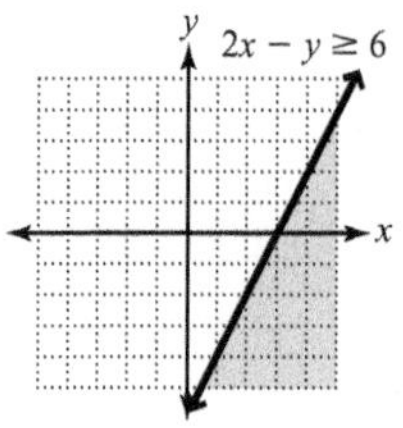

24. 5 **25.**

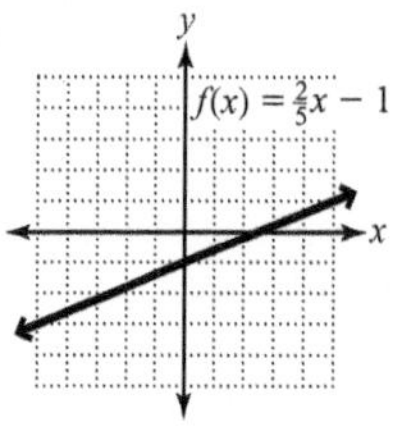

26. 3.6 **27.** 60°, 120° **28.** 4 deluxe, 8 standard **29.** 25 ml **30.** 1000 min. or less

Chapter 4

Exercises Set 4.1

Prep Exercises **1.** Replace each variable in each equation with its corresponding value. Verify that each equation is true. **2.** The lines intersect at one point (the solution). **3.** The lines are parallel. **4.** The lines are identical. **5.** When using the graphing method, if either coordinate in the solution is a fraction or decimal number, we may have to guess the value. Substitution requires no guessing. **6. a.** y in the second equation **b.** $x + 2$ **7.** The resulting equation will no longer contain variables and is a false equation. **8.** The resulting equation will no longer contain variables and is a true equation. **9.** The elimination method is advantageous over graphing when the solution contains fractions or decimal numbers. It is advantageous over substitution when no coefficients are 1. **10.** Eliminate y because $2y$ and $-2y$ are additive inverses. **11.** The resulting equation will no longer contain variables and is a false equation. **12.** The resulting equation will no longer contain variables and is a true equation.

Exercises **1.** yes **3.** no **5.** yes **7.** no **9.** $\begin{cases} x + y = 5 \\ x - y = 3 \end{cases}$ **11.** $\begin{cases} 2l + 2w = 50 \\ w = l - 2 \end{cases}$

13.

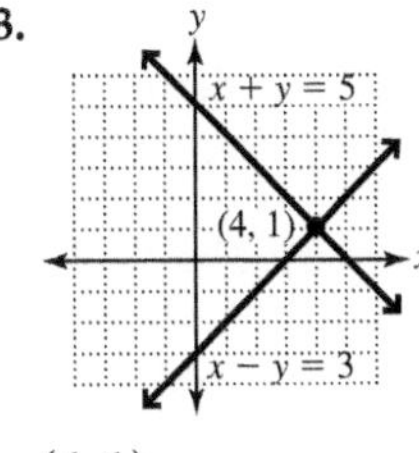

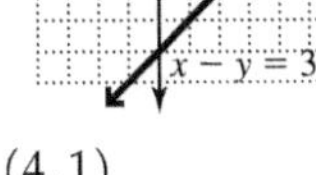

$(4, 1)$

15.

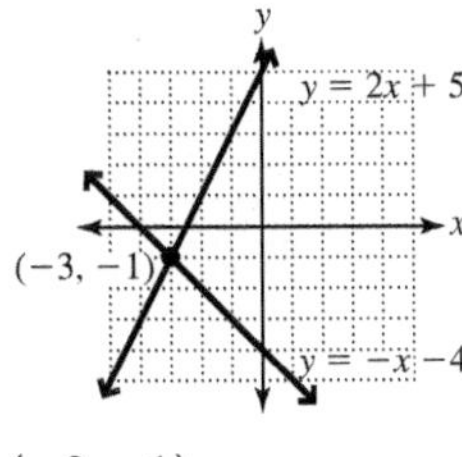

$(-3, -1)$

17.

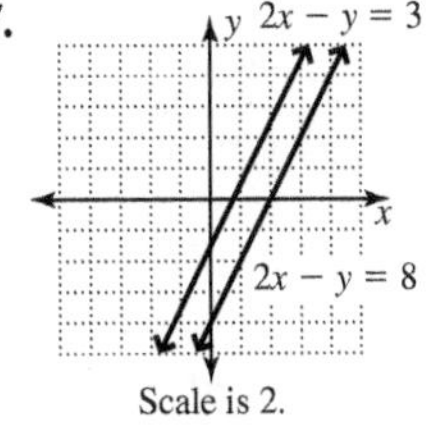

Scale is 2.

No solution.

19.

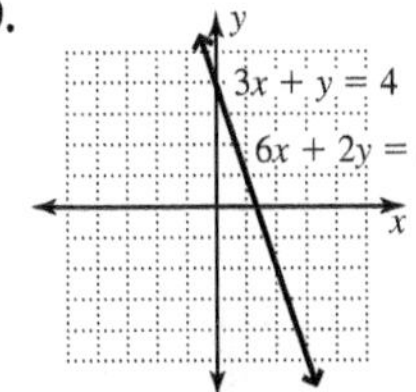

All ordered pairs that solve $3x + y = 4$.

21.

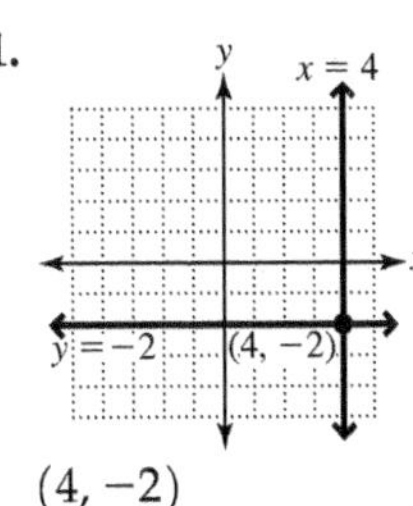

$(4, -2)$

23. a. consistent with independent equations **b.** one solution **25. a.** inconsistent **b.** no solution **27. a.** consistent with dependent equations **b.** infinite number of solutions **29.** Consistent with independent equations. **31.** Consistent with dependent equations. **33.** Inconsistent **35.** $(2, 4)$ **37.** $(-2, 3)$ **39.** $\left(-1, \frac{3}{4}\right)$ **41.** $(1, 2)$ **43.** $(-3, 1)$ **45.** $(2, -2)$ **47.** No solution. **49.** All ordered pairs that solve $x - 2y = 6$. **51.** Mistake: Did not distribute properly. Correct: $(3, 2)$ **53.** $(1, 0)$ **55.** $(3, 1)$ **57.** $(3, -4)$ **59.** $\left(1, -\frac{2}{3}\right)$ **61.** $(1, -2)$ **63.** $\left(\frac{34}{31}, \frac{10}{31}\right)$ **65.** $(-4, 2)$ **67.** $(4, 1)$ **69.** No solution. **71.** All ordered pairs that solve $x + 2y = 4$. **73.** Mistake: Subtracted 7 from 8 instead of adding. Correct: $\left(\frac{15}{2}, \frac{1}{2}\right)$ **75. a.** 5000 **b.** \$4000 **c.** $u > 5000$ **77. a.** $\begin{cases} c = 45 \\ c = 0.10n + 30 \end{cases}$

b.

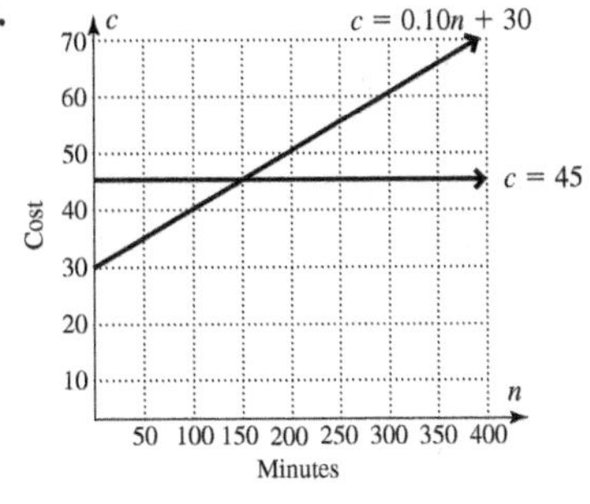

c. 150 min. **d.** plan 1 **e.** plan 2

Review Exercises: **1.** $20x + 28y - 4z$ **2.** $x - 3y$ **3.** $y = -\frac{1}{3}x + 2$ **4.** $x = 6y + 30$ **5.** $x = 2$ **6.** $y = 0$

Exercise Set 4.2

Prep Exercises **1.** It doesn't matter which equations are chosen as long as the second pair is different from the first pair. **2.** We eliminate the same variable from two pairs of equations to get two equations with the same two variables. **3.** Answers may vary. A good choice is to eliminate z using Equations 1 and 2 and 1 and 3. Multiplying Equation 1 by -3 and then adding this result to Equations 2 and 3 takes very few steps. **4.** After adding two equations, if the resulting equation has no variables and is false, the system has no solution. **5.** Two parallel planes intersecting a third plane **6.** It will be on the line of intersection of two planes but not on the third plane.

Exercises **1.** yes **3.** no **5.** yes **7.** $(2, -1, 4)$ **9.** $(-1, 0, 3)$ **11.** $(-1, 1, 3)$ **13.** No solution (inconsistent) **15.** Infinite number of solutions (dependent equations) **17.** $(2, -3, 1)$ **19.** $\left(-2, -\frac{1}{2}, 1\right)$ **21.** $(2, -1, 1)$ **23.** $(-3, 7, -4)$ **25.** $(2, -4, 1)$ **27.** $(4, -4, 2)$

Review Exercises: **1.** $-6x + 12y + 3z - 27$ **2.** $y = -4$ **3.** -1 **4.** $(0, 1)$ **5.** $m = -2$ **6.**

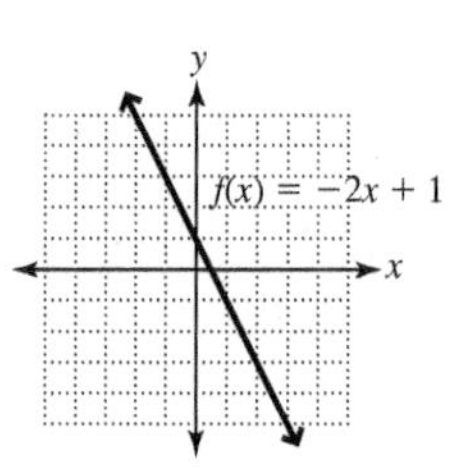

Exercises Set 4.3

Prep Exercises **1. a.** $x + y = 90$ **b.** $x + y = 180$ **2.** $l = 2w + 10$ **3. a.** $5.50x$, $8.00y$ **b.** $x + y = 45$ **c.** $5.50x + 8.00y = 285$ **4.** $y = x + 4$ **5. a.** $75x$, $60y$ **b.** 0.5 hr. **c.** $y = x + 0.5$ or $x = y - 0.5$ **d.** They will have traveled equal distance.; $75x = 60y$ **6.** With the wind: $x + y$; against the wind: $x - y$. **7. a.** $0.30\,x$, $0.10(500)$, $0.15y$ **b.** $y = x + 500$ **c.** $0.30x + 0.10(500) = 0.15y$ **8.** three **9. a.** $y = x + 1$ **b.** $z = y + 1$ **c.** $z = x + 2$ **10.** $l = (s + m) - 0.2$

Exercises **1.** 25°, 65° **3.** 36°, 144° **5.** 38° **7.** 42.5°, 137.5° **9.** Width and length: 55.125 ft. **11.** 28 small, 14 large **13.** The Twilight Saga: Breaking Dawn Part 1: \$712 million; The Twilight Saga: Breaking Dawn Part 2: \$830 million. **15.** 6.7 million **17.** Mathematician: \$99,380; math teacher: \$62,050. **19.** 85.7% **21.** 10 18-inch wreaths; 36 22-inch wreaths **23.** Union: 364,000; Confederacy: 134,000. **25.** 0.2 hr. **27.** 10:00 P.M. **29.** Boat: 14 mph; current: 2 mph. **31.** Plane: 600 mph; wind: 50 mph. **33.** 5 ml of 5%, 5 ml of 20% **35.** 16 oz. **37.** 10 gal. **39.** \$14,000 at 9%, \$3500 at 5% **41.** 6, 2, 8 **43.** 12, 13, 14 **45.** 105°, 35°, 40° **47.** Burger: \$2.50; fries: \$1.50; drink: \$1.00. **49.** 150 child, 250 student, 100 adult **51.** Three 3-point field goals, one 2-point field goal, three free throws. **53.** Zinc: 20 lb.; tin: 60 lb.; copper: 920 lb. **55.** Ham: 3 lb.; turkey: 5 lb.; beef: 2 lb. **57.** \$2000 at 4%, \$2500 at 6%, \$3500 at 7%. **59.** Bicycling: 420; walking: 300; climbing stairs: 600. **61.** $a = -16$, $v_0 = 100$, $h_0 = 150$; $h = -16t^2 + 100t + 150$

Review Exercises **1.** $-6x + 12y + 3z - 27$ **2.** $y = -4$ **3.** -1 **4.** $(0, 1)$ **5.** $m = -2$ **6.**

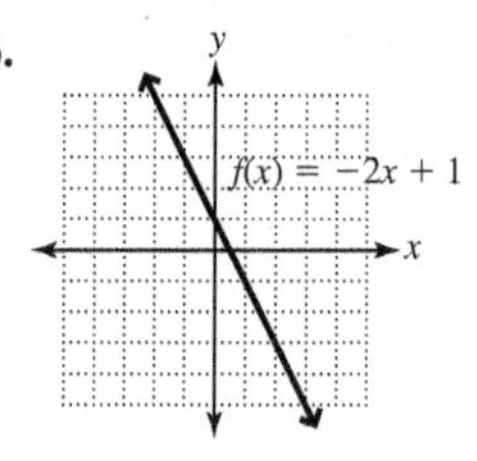

Exercise Set 4.4

Prep Exercises **1.** 4 rows, 2 columns **2.** Arrange each equation so that all variables are in the same order on the left side of the equal sign and the constant term is on the right side of the equal sign. Omit all of the variables and the equal signs. Place a vertical dashed line where the equal signs were so that it is between the coefficients of the variables and the constants. **3.** The dashed line corresponds to the equal signs in the equations. **4.** The rules are the same for both processes. With matrices, we use only the coefficients and the constants. With the elimination method, we also use the variables and the equal sign. **5.** A matrix is in row echelon form when the coefficient portion of the augmented matrix has 1's on the diagonal from upper left to lower right and 0s below the 1s. **6.** The bottom equation represents the value of the last variable. Substitute the known value into the equation above to determine the value of the next variable. Continue until all values for the variables are found.

Exercises **1.** $\left[\begin{array}{cc|c} 14 & 7 & 6 \\ 7 & 6 & 8 \end{array}\right]$ **3.** $\left[\begin{array}{cc|c} 7 & -6 & 1 \\ 0 & -2 & 5 \end{array}\right]$ **5.** $\left[\begin{array}{ccc|c} 1 & -3 & 1 & 4 \\ 2 & -4 & 2 & -4 \\ 6 & -2 & 5 & -4 \end{array}\right]$ **7.** $\left[\begin{array}{ccc|c} 4 & 6 & -2 & -1 \\ 8 & 3 & 0 & -12 \\ 0 & -1 & 2 & 4 \end{array}\right]$ **9.** $(4, 2)$ **11.** $(-6, -5, 1)$

13. $\left[\begin{array}{cc|c} 1 & 3 & -1 \\ 0 & 11 & 4 \end{array}\right]$ **15.** $\left[\begin{array}{ccc|c} 1 & -2 & 4 & 6 \\ 0 & 2 & -1 & -5 \\ 0 & 0 & -2 & 17 \end{array}\right]$ **17.** $\left[\begin{array}{cc|c} 1 & 2 & -2.5 \\ -1 & 3 & 2 \end{array}\right]$ **19.** Replace R_2 with $3R_1 + R_2$. **21.** Replace R_3 with $-2R_2 + R_3$.

23. $(2, -3)$ **25.** $(1, 2)$ **27.** $(-3, -2)$ **29.** $(-2, 0)$ **31.** $(4, 6)$ **33.** $(4, -4)$ **35.** $(-1, 1, 3)$ **37.** $(2, 1, -2)$ **39.** $(2, -3, 1)$ **41.** $(-1, -1, -2)$ **43.** $(-3, 1, 5)$ **45.** $(8, -2, -4)$ **47.** $(3, -4)$ **49.** $(-6, 8)$ **51.** $(4, -3, 5)$ **53.** $(7, -6, 5)$ **55.** Mistake: In the second step, $4R_1 + R_2$ is calculated incorrectly. Correct: The correct calculation is $\left[\begin{array}{cc|c} 1 & 3 & 13 \\ 0 & 11 & 26 \end{array}\right]$. The solution is $(65/11, 26/11)$. **57.** Chicken sandwich: \$3.50; drink: \$1.20 **59.** Nile: 4150 mi.; Amazon: 3900 mi. **61.** \$6400 in CD, \$3600 in money market **63.** CD: \$16; book: \$12, DVD: \$28 **65.** 4 touchdowns, 4 extra points, 1 field goal **67.** $v_1 = 9$, $v_2 = 6$, $v_3 = 2$

Review Exercises: **1.** 8 **2.** 11 **3.** 2 **4.** 2 **5.** 3 **6.** -1

Exercise Set 4.5

Prep Exercises **1.** Select a point in the solution region and verify that it makes both inequalities true. **2.** After graphing each inequality, determine the region of overlap. **3.** Parallel lines with shaded regions that do not overlap **4.** $\begin{cases} x > 0 \\ y > 0 \end{cases}$ **5.** $\begin{cases} x < 0 \\ y < 0 \end{cases}$ **6.** $\begin{cases} x > 0 \\ y < 0 \end{cases}$ **7.** $\geq$ **8.** $\leq$

Exercises

1.

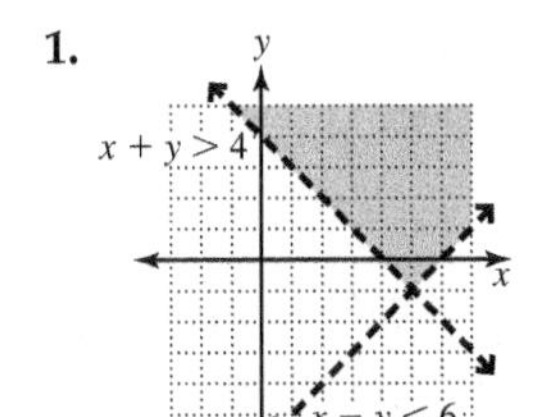

3.

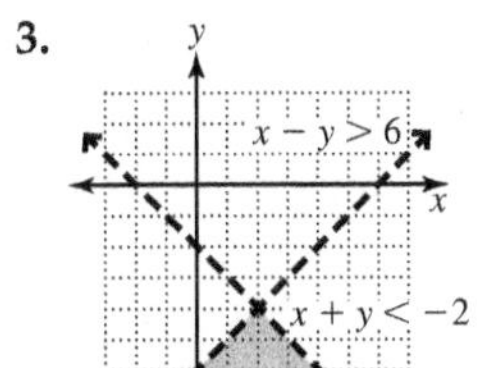

5.

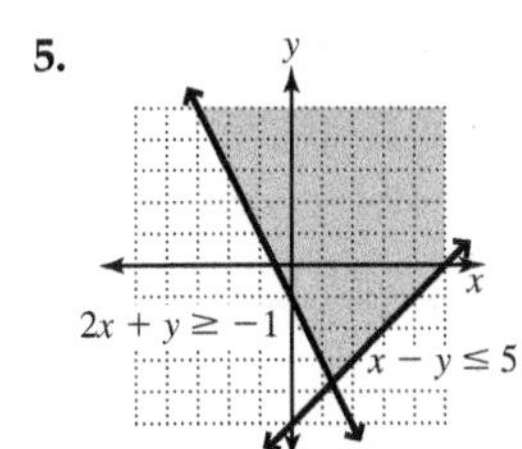

7.

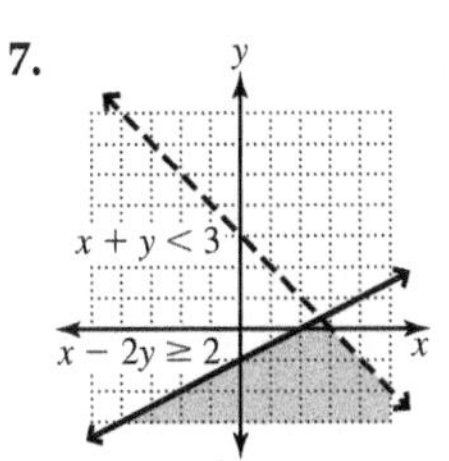

9.

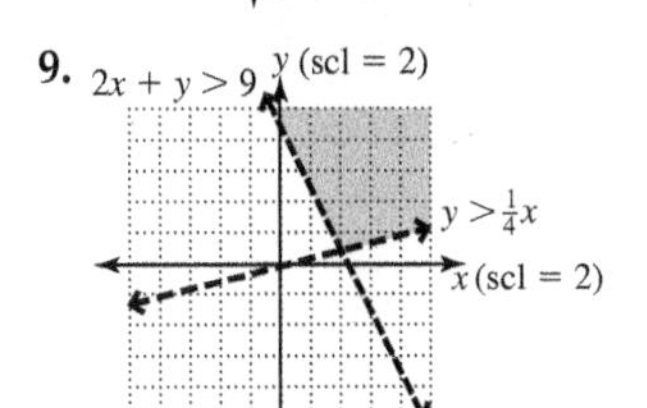

11.

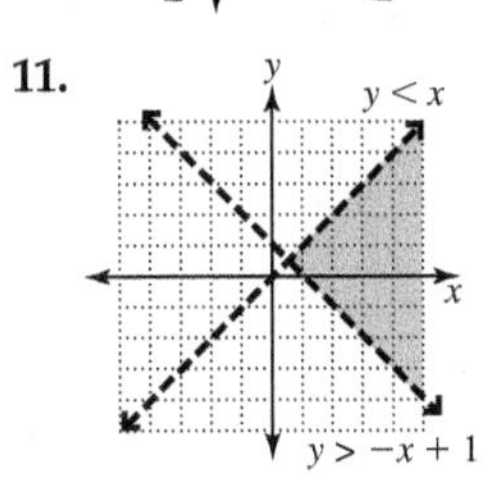

13.

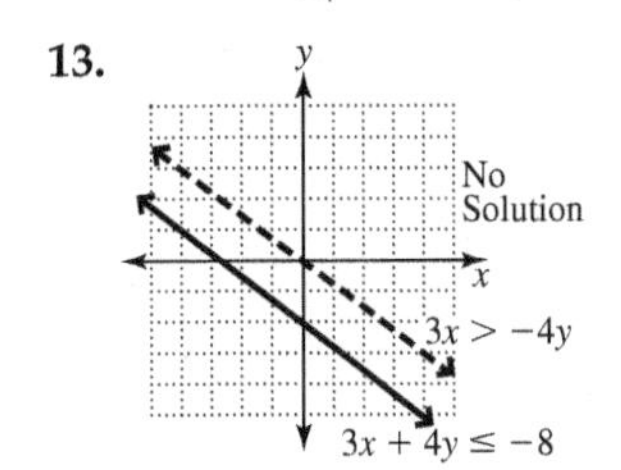

15.

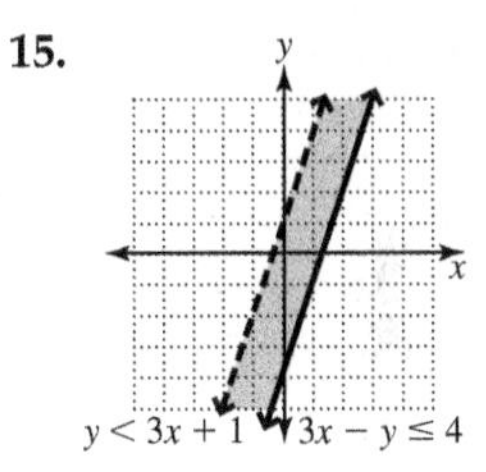

17.

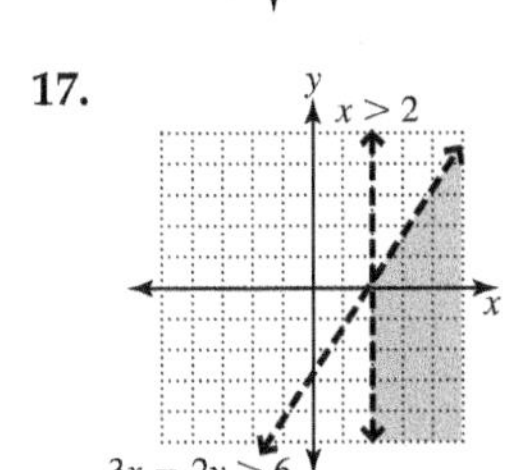

19.

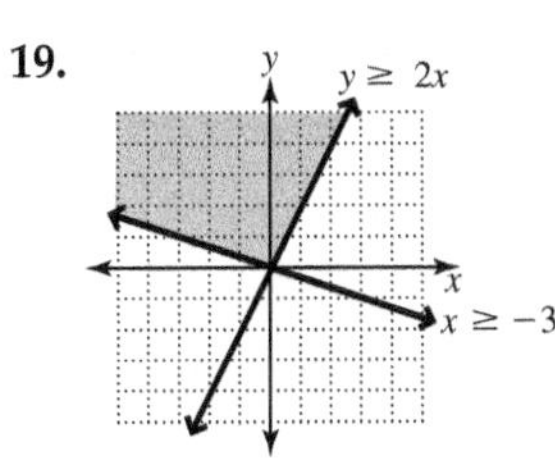

21.

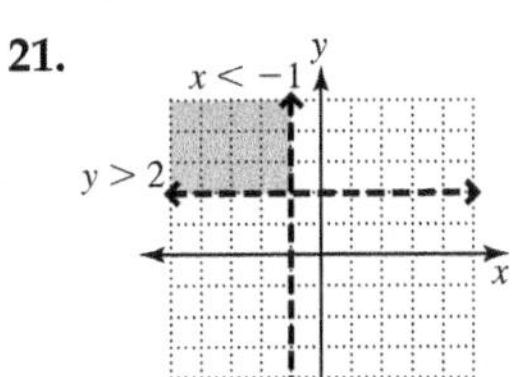

23.

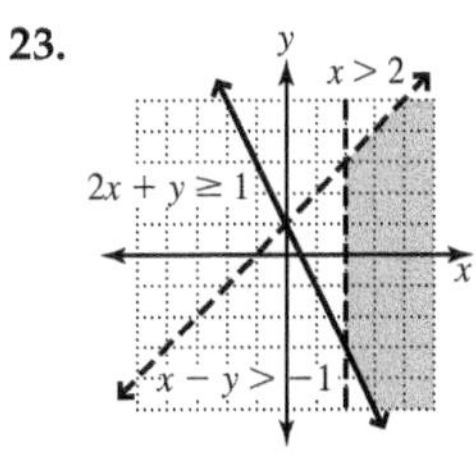

25. $x + y = 3$ should be a dashed line. **27.** The wrong area is shaded. It should be the region containing the point $(1, 5)$.
29. a. $V + M \geq 1100 \quad M \geq 500$ **b.** $V \leq 800 \quad M \leq 800$ **c.**

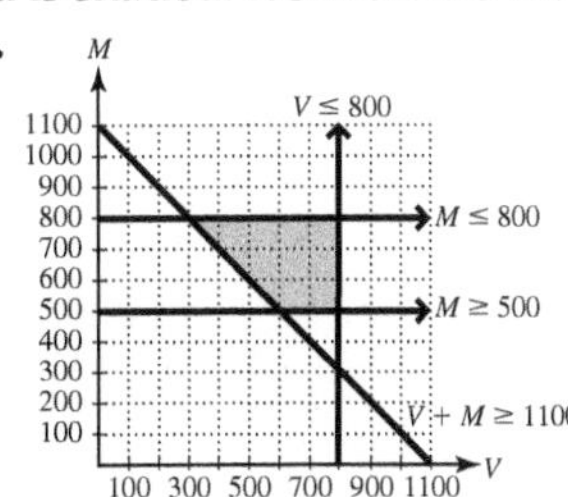

d. Answers will vary. Some possible (V, M) pairs are $(500, 700)$, $(600, 500)$, and $(700, 600)$.

31. a.

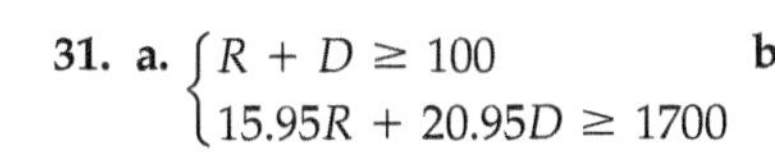

$\begin{cases} R + D \geq 100 \\ 15.95R + 20.95D \geq 1700 \end{cases}$

b.

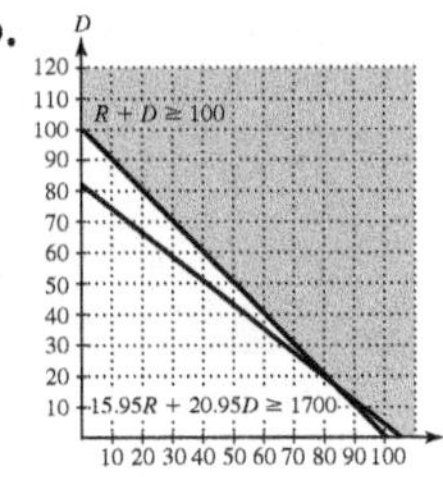

c. Answers will vary. Some possible (R, D) pairs are $(30, 90)$, $(40, 80)$, and $(50, 60)$.

Review Exercises: **1.** 9 **2.** -100 **3.** 5 **4.** $\frac{4}{7}$ **5.** 4 **6.** 21

Chapter 4 Summary and Review Exercises

4.1 Definitions/Rules/Procedures group of two or more; makes all of the equations in the system true; variable, value; true
Exercises **1.** yes **2.** no Definitions/Rules/Procedures coordinates; no solution; infinite number Exercises **3.** $(-3, -5)$
4. $(4, -2)$ **5.** No solution **6.** All ordered pairs that solve $5x - 2y = 10$. Definitions/Rules/Procedures at least one solution; no solution; intersection, different, different; infinite, identical, same, same; no, parallel, same, y-intercepts Exercises **7. a.** consistent with independent equations **b.** one solution **8. a.** inconsistent **b.** no solution **9. a.** consistent with independent equations **b.** one solution **10. a.** consistent with dependent equations **b.** infinite number of solutions Definitions/Rules/Procedures variables; substitute; Solve; substitute, variable Exercises **11.** $(4, -7)$ **12.** $\left(2, -\frac{1}{3}\right)$ **13.** All ordered pairs that solve $2x + 3y = 9$

14. No solution Definitions/Rules/Procedures standard; fractions, decimals; additive inverses; Add; Solve; substitute; original equation Exercises **15.** $(1, 6)$ **16.** $(4, -2)$ **17.** $(-8, 5)$ **18.** $(-1, -2)$

4.2 Definitions/Rules/Procedures $Ax + By + Cz = D$; elimination; same variable; two variables; original equations; all three original equations Exercises **19.** $(1, 0, -3)$ **20.** No solution **21.** $(3, -3, 2)$ **22.** $(4, 0, 1)$

4.3 Definitions/Rules/Procedures unknown Exercises **23.** 154 hotdogs, 72 hamburgers **24.** 142°, 38° **25.** 9:00 A.M. **26.** still water: $2\frac{2}{3}$ mph, current: $1\frac{1}{3}$ mph **27.** stocks: \$7200, savings: \$2400 **28.** 200 ml of 20% solution with 100 ml of 50% solution **29.** 1 vitamin supplement, 3 rolls of film, 4 bags of candy **30.** 4 first place, 3 second place, 9 third place

4.4 Definitions/Rules/Procedures rectangular array; coefficients, constant, constant, coefficients; upper left, lower right; interchanged; multiplied, divided; adding, multiple; row echelon Exercises **31.** $(4, -1)$ **32.** $(-2, 3, 1)$

4.5 Definitions/Rules/Procedures overlap, solid Exercises

33.

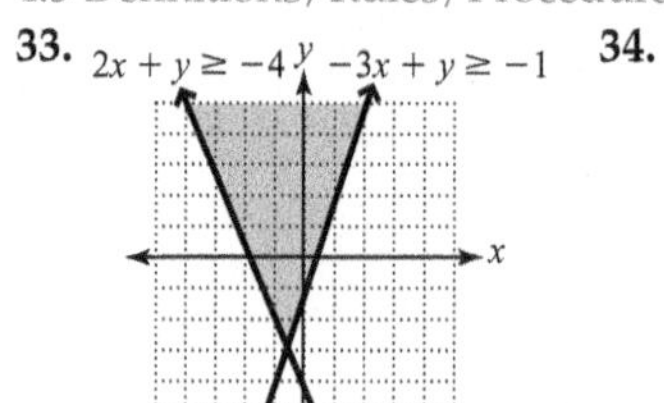

34.

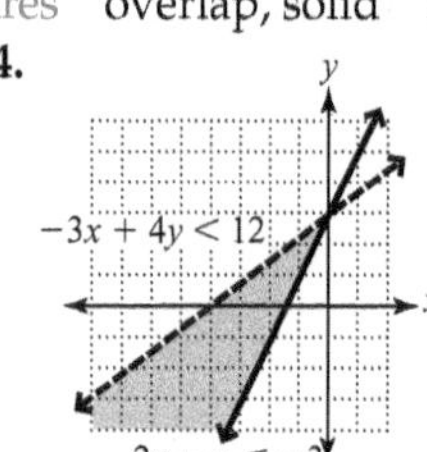

35. a.

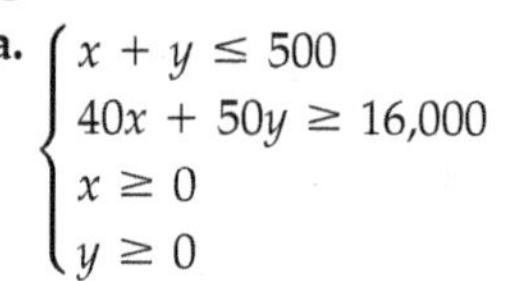

$$\begin{cases} x + y \le 500 \\ 40x + 50y \ge 16{,}000 \\ x \ge 0 \\ y \ge 0 \end{cases}$$

b.

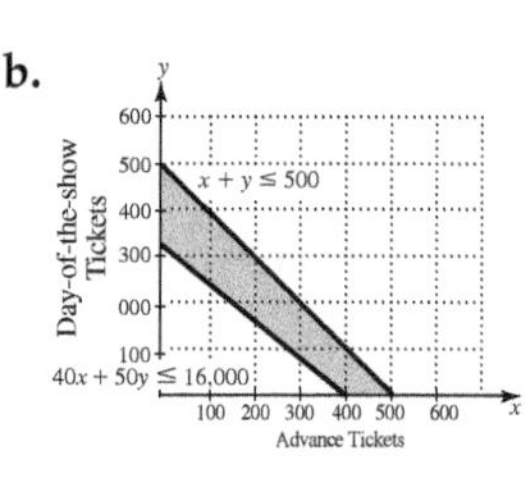

c. Answers may vary. One example is $(450, 50)$.

Chapter 4 Practice Test

1. yes [4.1] **2.** no [4.4] **3.** $(-2, 1)$ [4.1] **4.** $(8, -1)$ [4.2] **5.** $(3, 1)$ [4.2] **6.** $(-4, -2)$ [4.3] **7.** All ordered pairs that solve $4x + 6y = 2$. [4.3] **8.** $(-2, 3, -2)$ [4.4] **9.** $(-4, 2, 3)$ [4.4] **10.** $(2, -4)$ [4.5] **11.** $(-1, 5, 2)$ [4.5] **12.** $(-2, 5)$ [4.6] **13.** $(-4, 0, 3)$ [4.4] **14.**

y
$x + 2y \le -2$
x
$2x - 3y < 1$

[4.6] **15.** 86 accountants, 77 waiters/waitresses [4.2] **16.** 180 [4.2] **17.** 7 mph [4.3] **18.** \$4000 at 6%, \$8000 at 8% [4.3] **19.** Children: 150; student: 300; adult: 350. [4.3]

20. a.

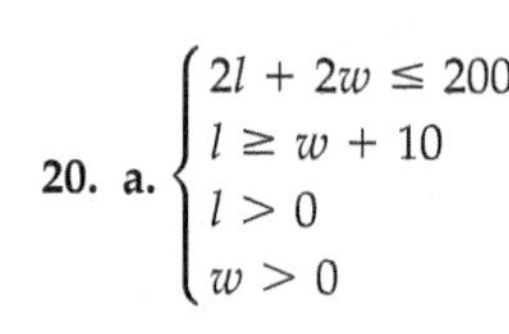

$$\begin{cases} 2l + 2w \le 200 \\ l \ge w + 10 \\ l > 0 \\ w > 0 \end{cases}$$

b.

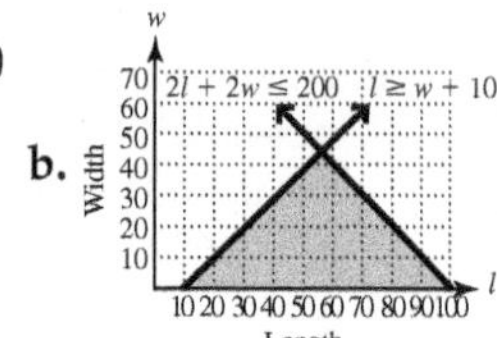

c. Answers may vary. One example is $(60, 20)$. [4.6]

Chapters 1–4 Cumulative Review Exercises

1. false **2.** false **3.** true **4.** $8x + 12$ **5.** $\frac{c}{d}$ **6.** perpendicular **7.** 11 **8.** 0 **9.** $-\frac{1}{8}x + \frac{33}{10}y + 5$ **10.** $p = 13$ **11.** contradiction (no solution) **12.** $z > -5$ **13.** $x = 5$ **14.** $t = \frac{A - P}{Pr}$ **15.** $x = -2$ **16.** 13,600 ft². **17.**

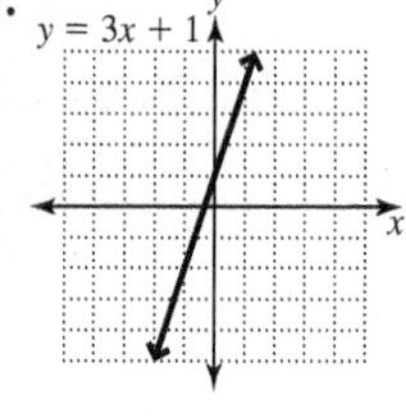

18.

y
x
$4x + 5y = 10$

19.

y
x
$x = -4$

20.

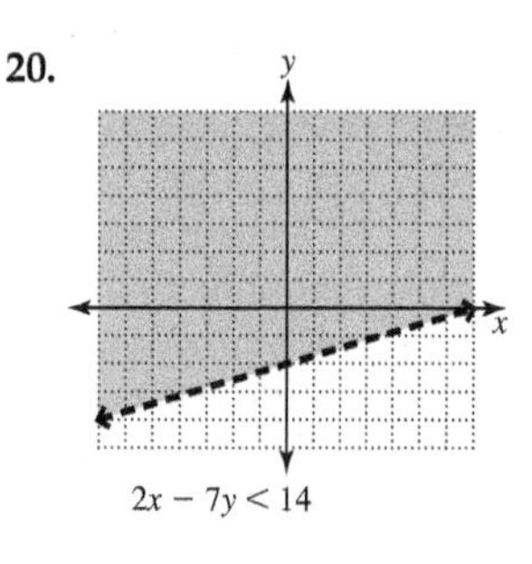

21. $y = -\frac{1}{5}x - \frac{7}{5}$ **22.** $y = -x - 3$ **23.** $3x + 4y = -4$ **24.** $(-3.5, -4)$ **25.** $(2, -1, 4)$ **26.**

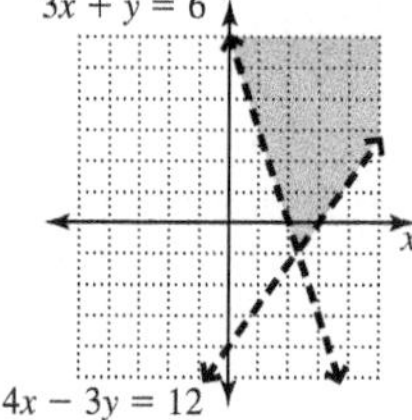

27. $L = 130$ ft. $W = 80$ ft.

28. 115°; 65° **29.** \$35,000 at 3% and \$15,000 at 2.5% **30.** 10 gallons of outside, 14 gallons of inside

Chapter 5

Exercises Set 5.1

Prep Exercises **1.** Keep the same base and add the exponents. **2.** 1 **3.** Keep the same base and subtract the exponents. **4.** $\frac{1}{a^n}$ **5.** a^n **6.** The result is positive because $5^{-2} = \frac{1}{5^2}$, which is positive. **7.** Keep the same base and multiply the exponents. **8.** a^nb^n **9.** $\frac{a^n}{b^n}$ **10.** 7; right **11.** 5; left **12.** Place the decimal point between the 3 and 4.; 7 **13.** Place the decimal point between the 8 and 5.; −4 **14.** Because 42.5 is greater than 10.

Exercises **1.** m^4n^5 **3.** 3^{12} **5.** $(-3)^8$ or 3^8 **7.** $-12p^6q^5$ **9.** $-18r^7s^6t^{12}$ **11.** $3.72u^7t^{11}$ **13.** 1 **15.** 4 **17.** 1 **19.** −25 **21.** Mistake: Assigned the minus sign to the base. Correct: $-(2\cdot 2\cdot 2\cdot 2) = -16$ **23.** h^3 **25.** $\frac{1}{a^5}$ **27.** $2xy^3$ **29.** $-\frac{3s^3}{2r^3}$ **31.** $-\frac{5w^6}{7u^{11}}$ **33.** $\frac{4b^7c^3}{a^5}$ **35.** $\frac{1}{8}$ **37.** $-\frac{1}{81}$ **39.** 125 **41.** $4x^3$ **43.** $\frac{a^6}{4}$ **45.** Mistake: Multiplied 4 by −1. Correct: $\frac{1}{4}$ **47.** x^{12} **49.** $16x^{20}$ **51.** $25x^6y^2$ **53.** $\frac{27}{64}a^6b^{12}$ **55.** $-0.027r^6t^{12}u^3$ **57.** $\frac{c^4}{d^4}$ **59.** $\frac{27}{x^3}$ **61.** $\frac{16}{x^6}$ **63.** $\frac{b^6}{a^6}$ **65.** $\frac{y^{12}}{x^8}$ **67.** $\frac{1}{64x^6}$ **69.** $\frac{y^{12}}{81x^8}$ **71.** $-128x^{13}y^{18}$ **73.** $\frac{9q^2}{125p^2}$ **75.** $\frac{9}{h^2t^2}$ **77.** $\frac{1}{144u^{19}v^{10}}$ **79.** $\frac{9}{2u^4v^{12}}$ **81.** $-\frac{1}{54a^{18}b^{20}c}$ **83.** 2,900,000 **85.** 200,000,000,000 **87.** 73,000,000,000 **89.** 0.00000017 **91.** 0.000000000000000000000001675 **93.** 0.000000000000000000000000006645 **95.** 5.015×10^7 **97.** 1.678×10^{13} **99.** 9.309×10^9 **101.** 5.5×10^{-7} **103.** 1×10^{-7} **105.** 5×10^{-17} **107.** 6×10^5, 7.4×10^6, 8.3×10^6, 1.2×10^7, 2.4×10^8 **109.** 6.3×10^7 **111.** -1.312×10^{10} **113.** 1.782×10^2 **115.** 4×10^4 **117.** -3.1×10^8 **119.** 6×10^2 **121.** 2.7×10^{13} **123.** 8×10^{21} **125.** 2.5×10^{-7} **127.** 6.25×10^{10} **129.** 2.78×10^{-19} J **131.** 2.11×10^{14} Hz **133.** 3.78×10^5 J

Review Exercises: **1.** Change the operation from subtraction to addition and the subtrahend, −6, to its additive inverse, 6. **2.** Associative property of addition. **3.** $-8n + 32$ **4.** $3x + 21y - 17$ **5.** $13x^2 - 13x + 9$ **6.** $-3u + 12$

Exercises Set 5.2

Prep Exercises **1.** Add the exponents of all of the variables in the monomial. **2.** A monomial has one term, a binomial has two terms, and a trinomial has three terms. **3.** The degree of a polynomial is the greatest degree of any of the terms in the polynomial. **4.** Combine like terms. **5.** Descending order of degree means that the terms of the polynomial are written in order from the greatest degree to the least degree. **6.** Change the subtrahend (second polynomial) to its additive inverse. To get the additive inverse, change the sign of each term in the polynomial. **7.** Look at the degree. The degree of a constant function is 0, a linear function is 1, a quadratic function is 2, and a cubic function is 3. **8.** Use $(f + g)(x) = f(x) + g(x)$ and $(f - g)(x) = f(x) - g(x)$.

Exercises **1.** $d = 4$; monomial **3.** $d = 2$; binomial **5.** $d = 3$; trinomial **7.** $d = 4$; no special polynomial name **9.** $d = 3$; trinomial **11.** $d = 0$; monomial **13.** $d = 3$; binomial **15.** $7x^2 + 4x - 2$ **17.** $2p^4 - 5p^3 + 6$ **19.** $-u^3 - 9u^2 + 2u + 6$ **21.** $\frac{4}{3}u^4 + \frac{1}{2}u^3 + 2u^2 + 3u + 2$ **23.** $7.3t^4 + 1.5t^3 - t^2 + 2.7t + 7$ **25.** $-a^3 - a^2 + 2a - 1$ **27.** $3m^4 - 4m^3 + m^2 + m - 2$ **29.** $-2r^3 + r^2 - r - 3$ **31.** $\frac{3}{5}y^3 - y^2 + \frac{25}{4}y - \frac{1}{7}$ **33.** $-8.3w^3 - 3.7w^2 - 0.7w + 6.7$ **35.** $-8w^4 + 3w^3 - 7w^2 + 13w - 9$ **37.** $6x^4 + 6x^3 + 5x^2 - 12x + 13$ **39.** $-g^4 + \frac{4}{5}g^3 - \frac{2}{3}g^2 + \frac{9}{5}$ **41.** $3y^2 + 4y + 1$ **43.** $-19a^3 - 12a^2 + 9a + 9$ **45.** $14a^2 - 5ab + 6b^2$ **47.** $2x^3y^4 + 5x^2y^3 + 7xy - 8$ **49.** $4x^2y^2 + 6xy^2 - 16x^2y - 3xy + y^2 - 13$ **51.** $4x + 5$ **53.** $8a + 4$ **55.** constant **57.** quadratic **59.** linear **61.** cubic

63.

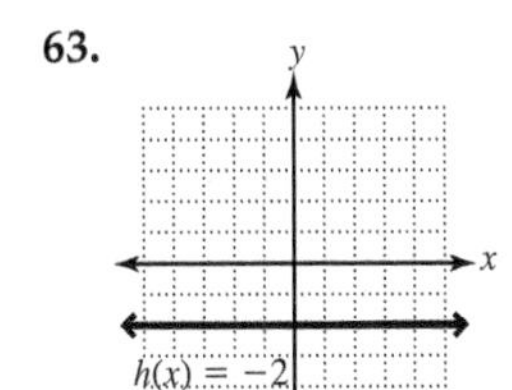

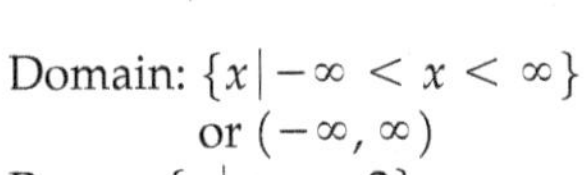

Domain: $\{x \mid -\infty < x < \infty\}$ or $(-\infty, \infty)$
Range: $\{y \mid y = -2\}$ or $[-2, -2]$

65.

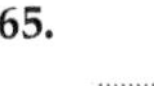

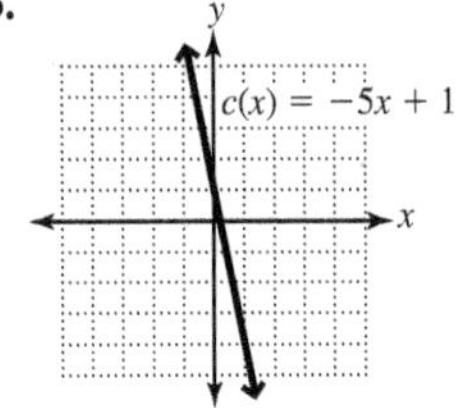

Domain: $\{x \mid -\infty < x < \infty\}$ or $(-\infty, \infty)$
Range: $\{y \mid -\infty < y < \infty\}$ or $(-\infty, \infty)$

67.

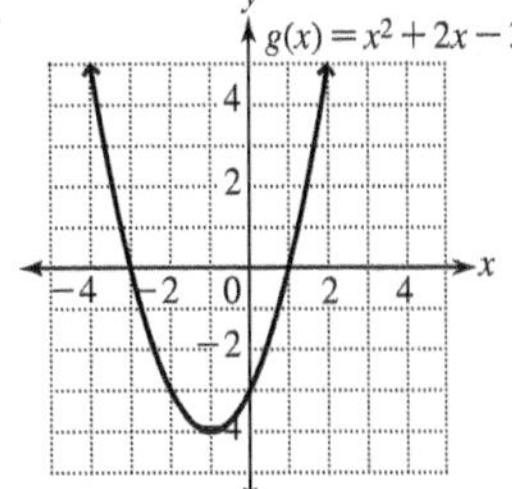

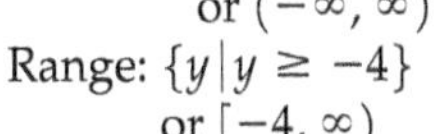

Domain: $\{x \mid -\infty < x < \infty\}$ or $(-\infty, \infty)$
Range: $\{y \mid y \geq -4\}$ or $[-4, \infty)$

69.

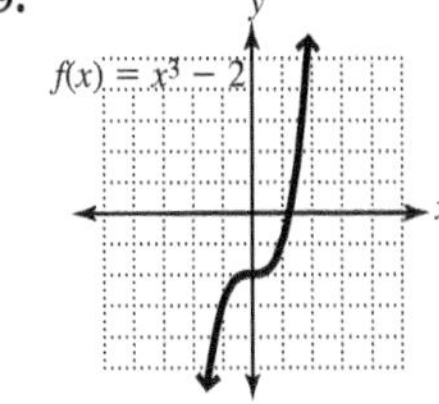

Domain: $\{x \mid -\infty < x < \infty\}$ or $(-\infty, \infty)$
Range: $\{y \mid -\infty < y < \infty\}$ or $(-\infty, \infty)$

71. a. −1 **b.** −7 **73. a.** $3x + 2$ **b.** $5x + 4$ **75. a.** $4x^2 + 1$ **b.** $-2x^2 - 3$ **77. a.** $-x^3 + 2x + 7$ **b.** $3x^3 + 2x + 3$ **79. a.** $P(x) = 0.2x^2 - 17x + 1000$ **b.** \$184,000 **81. a.** $R(x) = 420x + 840$ **b.** \$126,840 **83.** 196 ft. **85.** 58.7 V

87. b.

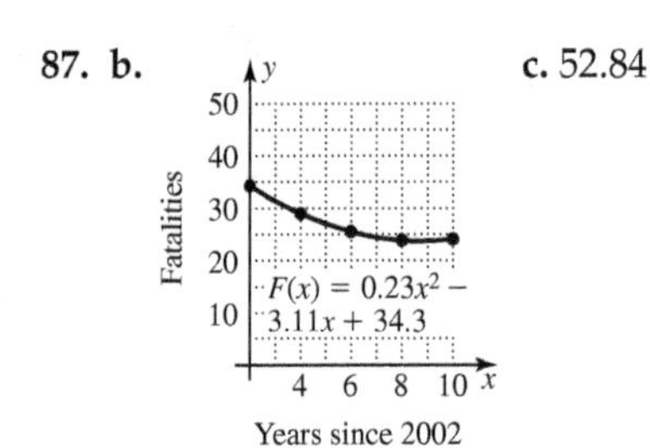

c. 52.84

Review Exercises: **1.** $24x - 72$ **2.** $-12y - 42$ **3.** $\frac{2}{5}$ **4.** $x < -\frac{29}{4}$ **5. a.** $\{x \mid -4 < x \leq 6\}$ **b.** $(-4, 6]$

c.

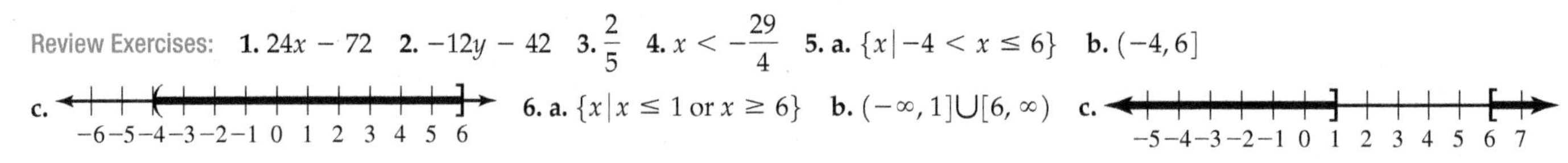

6. a. $\{x \mid x \leq 1 \text{ or } x \geq 6\}$ **b.** $(-\infty, 1] \cup [6, \infty)$ **c.**

Exercises Set 5.3

Prep Exercises **1.** The distributive property. **2.** 1. Multiply every term in the second polynomial by every term in the first polynomial. 2. Combine like terms. **3.** Multiply both terms in the second binomial by both terms in the first binomial (FOIL); then combine like terms. **4.** $a^2 + 2ab + b^2$ **5.** $a^2 - 2ab + b^2$ **6.** Binomials that differ only in the sign separating the terms. **7.** $a^2 - b^2$ **8.** difference of squares.

Exercises **1.** $5x^5 + 15x^4 - 10x^3$ **3.** $-12x^5 - 24x^4 + 18x^3$ **5.** $4n^7 + 28n^6 - 8n^5 - 12n^4$ **7.** $27a^6b^6 - 18a^3b^{12}$ **9.** $\frac{1}{2}m^3n^6p^3 - \frac{5}{4}m^4n^3p^3 + 2m^4n^4p^5$ **11.** $-2.4p^5q^9 + 0.6p^2q^3 - 2.1p^8q^2$ **13.** $-20x^7y^3 - 10x^6y^6 + 25x^{13}y^7$ **15.** $-6r^6s^4 + 18r^4s^3 + 9r^3s^2 - 9r^2s$ **17.** $-0.42a^3b^2 + 1.2a^3b^3 - 0.6a^4b^3 + 0.2a^4b^4$ **19.** $3a^3b^3c^8 - \frac{4}{3}a^3b^2c^7 + 4a^2bc^6$ **21. a.** $x + 5$ **b.** $2x$ **c.** $(x + 5)(2x)$ **d.** $2x^2 + 10x$ **e.** They are equivalent because both describe the area of the figure. **23. a.** $x + 3$ **b.** $x + 2$ **c.** $(x + 3)(x + 2)$ **d.** $x^2 + 5x + 6$ **e.** They are equivalent because both describe the area of the figure. **25.** $6x^2 + 17x + 12$ **27.** $15x^2 + x - 2$ **29.** $6m^2 - 19mn + 10n^2$ **31.** $15m^2 + 11mn - 12n^2$ **33.** $t^4 - 7t^2 + 10$ **35.** $a^4 + 5a^2b^2 - 6b^4$ **37.** $12x^3 - 10x^2 + 5x - 1$ **39.** $21a^3 - 11a^2 - 16a - 2$ **41.** $14c^3 - 24c^2 - 29c - 6$ **43.** $12p^3 + 38p^2q + 24pq^2 + 10q^3$ **45.** $12y^3 - 22y^2z + 14yz^2 - 12z^3$ **47.** $9u^4 - 4u^2 - 4u - 1$ **49.** $x^2 + 2xy + y^2$ **51.** $16w^2 + 24w + 9$ **53.** $16t^2 - 24tw + 9w^2$ **55.** $25y^4 - 90y^2 + 81$ **57.** $x^2 + y^2 - 2xy + 2x - 2y + 1$ **59.** $p^2 + q^2 - 2pq - 10p + 10q + 25$ **61.** $x - 8$ **63.** $3m - 2n$ **65.** $m^2 - n^2$ **67.** $-2j + 5k$ **69.** $4x^2 - 49$ **71.** $4q^2 - 25$ **73.** $x^2 - 4y^2$ **75.** $s^2 - t^2 + 2s + 1$ **77.** $9b^2 - c^2 - 4c - 4$ **79.** $3xyz + y^2z$ **81.** $h^2 + \frac{7}{2}h$ **83.** Mistake: Combined like terms incorrectly. Correct: $6x^2 - 37x + 45$ **85.** Mistake: Multiplied $2x$ by $2x$ incorrectly. Correct: $4x^2 - 49$ **87.** $4v^4w^5 + 16v^8w^2x^2 - 20v^2w^3x$ **89.** $64r^4 - 9s^2$ **91.** $4u^4 + 12u^2v^2 + 9v^4$ **93.** $2a^4 - 13a^2b^2 + 20b^4$ **95.** $15q^4 + 11tq^2 - 12t^2$ **97.** $3r^6s^3 - \frac{1}{3}r^5s^4 - \frac{1}{2}r^4s^5 + r^3s^4 - 3r^2s^3$ **99.** $-0.35t^4r^6 + 0.8t^5r^5 - 0.22t^5r^4 + 0.2t^3r^4$ **101.** $9m^2 - n^2 - 24m + 16$ **103.** $4x^4 + 14x^3y + 9x^2y^2 - xy^3 - y^4$ **105.** $x^4 - 2x^2 + 1$ **107. a.** $6x^2 - 7x + 2$ **b.** 126 **c.** 12 **109. a.** $12x^3 + 7x^2 - 6x + 8$ **b.** 1653 **c.** −235 **111.** $n^2 + 5n + 3$ **113.** $2t^2 - 24t + 71$ **115.** $2h$ **117.** $4xh + 2h^2 + 3h$ **119.** $(l^2 - 2l)$ ft². **121.** $(4x^3 + 24x^2 + 20x)$ cm³ **123.** $(9w^3 + 15w^2)$ ft³. **125.** $(\pi r^3 + 3\pi r^2)$ in³.

Review Exercises: **1.** -8.946×10^{11} **2.** 6.25×10^8 **3.** $\frac{15}{2}$ **4.** $t = \frac{d + n}{30}$ **5.** $y = -\frac{1}{5}x + \frac{22}{5}$ **6.** 41 small boxes, 43 large boxes.

Exercises Set 5.4

Prep Exercises **1.** To divide a polynomial by a monomial, divide each term in the polynomial by the monomial. **2.** Divide $6x^2$ by $2x$. **3.** Multiply the divisor by the quotient. **4.** $3x^2 - 5x + 2 + \frac{7}{6x - 1}$ **5.** $4x - 8 + \frac{-5}{3x + 2}$ or $4x - 8 - \frac{5}{3x + 2}$ **6.** A 0 placeholder is needed because the x^2 term is missing in the dividend.

Exercises **1.** $4a^2 - a + 3$ **3.** $4u^3 - 2u^2 - 5u + 1$ **5.** $3a^4b^3 - a^3b + 2a$ **7.** $\frac{12u^2}{v^5} + \frac{4}{v^4} - \frac{5u}{v^7}$ **9.** $-6x^2y + 9y^2 + \frac{1}{5}$ **11.** $-3t^3u^3v + \frac{1}{2}tu^2 + 4 - \frac{3}{4t^2}$ **13.** $2abc + \frac{4}{3}b^2 - \frac{1}{3c}$ **15.** $\frac{5y^2}{2}$ **17. a.** $2x - 4 + \frac{6}{x}$ **b.** Area: 108; length: 15; width: 7.2. **19.** $x + 4$ **21.** $3p + 1$ **23.** $5n - 8$ **25.** $3x - 2$ **27.** $2y - 1$ **29.** $x^2 - 5x + 6$ **31.** $x^2 - 2x + 4$ **33.** $y^3 + 2y^2 + 4y + 8$ **35.** $4z^2 - 10z + 25$ **37.** $v^2 + 3v + 2$ **39.** $7a^2 - 2a + 3$ **41.** $6q^2 + 5q + 9 + \frac{18}{2q - 1}$ **43.** $3x^2 - 6x + 5$ **45.** $c^3 - 2c^2 + c - 4$ **47.** $5u^2 - 11u + 32 - \frac{62}{u + 2}$ **49. a.** $(t^2 + 3t - 4)$ ft. **b.** $h = 4$ ft., $l = 6$ ft., $V = 144$ ft³. **51. a.** $4x^2 - 7x - 5$ **b.** 31 **c.** 25 **53. a.** $3x^2 + 5x - 2$ **b.** 40 **c.** −4 **55. a.** $x^2 + 2x - 1 + \frac{1}{x - 3}$ **b.** 24 **c.** $-\frac{6}{5}$

Review Exercises: **1.** $\frac{7}{4}$ **2.** $x \ge \frac{4}{3}$ **3.** $2x + 5y = 1$ or $y = -\frac{2}{5}x + \frac{1}{5}$ **4.** Yes, because every value in the domain corresponds to only one value in the range. (It passes the vertical line test.) **5.** $(7, -1, 4)$ **6.**

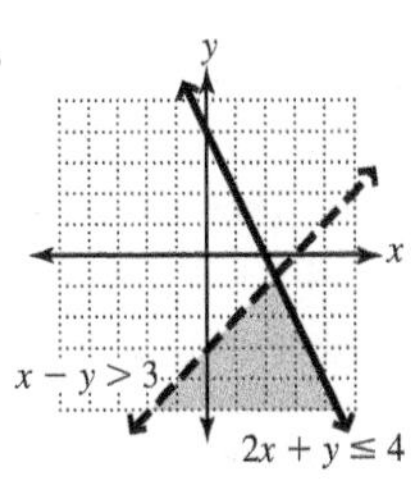

Exercises Set 5.5

Prep Exercises **1.** $x - c$ **2.** -6 **3.** $1, 0, -5, -3, 11$ **4.** Multiply 4 by 2. Put the resulting 8 under the -9. Add $-9 + 8 = -1$, which is the remainder. **5.** It is the last number in the last row. **6.** The degree of the quotient polynomial is 1 less than the degree of the dividend polynomial. **7.** Given a polynomial $P(x)$, the remainder of $\frac{P(x)}{x - c}$ is equal to $P(c)$. **8.** Using synthetic division and the remainder theorem avoids the big numbers that often result from evaluating exponential forms.

Exercises **1. a.** $x + 3$ **b.** $x^2 + 7x + 12$ **c.** $x + 4$ **3. a.** $x - 3$ **b.** $x^3 + 4x^2 - 25x + 7$ **c.** $x^2 + 7x - 4 - \frac{5}{x - 3}$ **5. a.** $x - 4$ **b.** $2x^4 - 8x^3 - 5x^2 + 17x + 10$ **c.** $2x^3 - 5x - 3 - \frac{2}{x - 4}$ **7.** $x - 3$ **9.** $2x + 7 + \frac{23}{x - 4}$ **11.** $x^2 + x - 3$ **13.** $3x^2 + 2x - 1$ **15.** $3x^2 - 10x + 8$ **17.** $2x^2 - 9x + 17 - \frac{28}{x + 2}$ **19.** $x^2 + 1 - \frac{1}{x - 2}$ **21.** $2x^2 - 5x + 15 - \frac{49}{x + 3}$ **23.** $3x^2 + 5x + 15 + \frac{47}{x - 3}$ **25.** $x^2 + 2x - 3$ **27.** $x^2 + 3x + 9 + \frac{54}{x - 3}$ **29.** $x^3 - 2x^2 + 4x - 8 + \frac{32}{x + 2}$ **31.** $6x^2 - 3x - 9$ **33.** 30 **35.** 0 **37.** -55 **39.** -3 **41.** 5 **43.** -33 **45. a.** $9x + 8$ **b.** Base: 116 in.; height: 10 in.; area: 1160 in.^{2}. **47. a.** $2y - 11$ **b.** Length: 14 ft.; width: 8 ft.; height: 9 ft.; volume: 1008 ft.^{3}.

Review Exercises: **1.** $\{2, 3, 5, 7, 11\}$ **2.** -2 **3. a.** $\{x \mid x \text{ is a real number}\}$ **b.** $(-\infty, \infty)$ **c.**

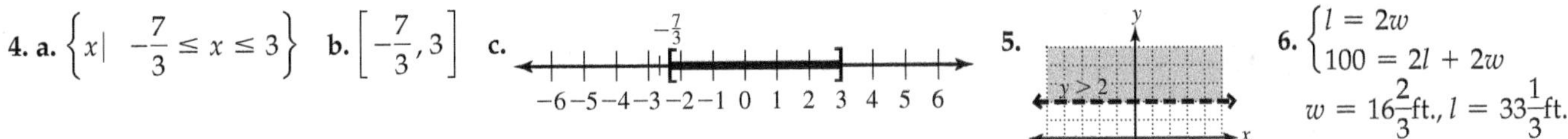

4. a. $\left\{x \mid -\frac{7}{3} \le x \le 3\right\}$ **b.** $\left[-\frac{7}{3}, 3\right]$ **c.**

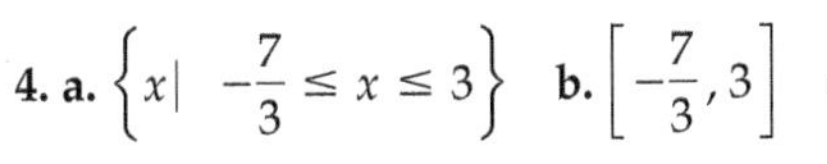

5.

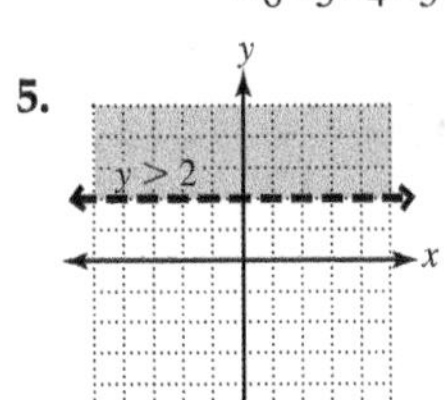

6. $\begin{cases} l = 2w \\ 100 = 2l + 2w \end{cases}$ $w = 16\frac{2}{3}$ft., $l = 33\frac{1}{3}$ft.

Chapter 5 Summary and Review Exercises

5.1 Definitions/Rules/Procedures $\underbrace{a \cdot a \cdot \cdots \cdot a}_{n \text{ factors of } a}$; a^{m+n}; 1; a^{m-n}; $\frac{1}{a^n}$, a^n; a^{mn}; $a^n b^n$; $\frac{a^n}{b^n}$; $\left(\frac{b}{a}\right)^n$ Exercises **1.** $\frac{27}{8}$ **2.** -16 **3.** $\frac{1}{32}$ **4.** 14 **5.** $20x^3y^3$ **6.** $-2m^3n^9$ **7.** $\frac{3}{x^5}$ **8.** $-\frac{7a^2}{3c^6}$ **9.** $9m^{10}n^2$ **10.** $\frac{1}{48m^{20}n^9p}$ **11.** $0.009j^7k^{19}$ **12.** $\frac{1}{18s^{37}t^{11}}$ **13.** $\frac{1}{75m^{15}n^6}$ **14.** $\frac{b^{12}}{8a^3}$

Definitions/Rules/Procedures right, left; first nonzero digit; old decimal point position, new position; old decimal point position, new position; right; left Exercises **15.** 0.000000000000000000000000016736 **16.** 16,500,000,000 **17.** 7.53×10^{-10} **18.** 3×10^8 **19.** -1.02×10^{11} **20.** 4.06×10^{-3} **21.** 4.04186×10^{-19}J

5.2 Definitions/Rules/Procedures constant, constant, variables; numerical factor; sum of the exponents; monomial, sum of monomials; same variable; more than one variable; two; three; greatest degree Exercises **22.** $d = 3$; no special polynomial name **23.** $d = 2$; binomial **24.** $d = 5$; monomial **25.** $d = 5$; no special polynomial name Definitions/Rules/Procedures like terms; addition; subtraction, addition; additive inverse, additive inverse; terms Exercises **26.** $3c^3 + 10c^2 - 11$ **27.** $6y^2 + 14$ **28.** $x^2y^3 - 3x^2y^2 + xy^2 + 6x^2y - 5xy + 13y^2 - 5$ **29.** $-hk - 3k - 6k^3$ **30. a.** $12a + b$ **b.** 153 cm Definitions/Rules/Procedures horizontal line; m, $(0, b)$; parabolas, $(0, c)$; S-shape, $(0, d)$ Exercises **31.** linear **32.** quadratic **33.** cubic **34.** constant

35.

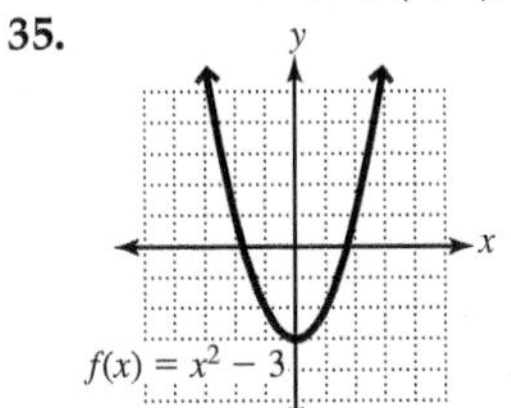

Domain: $\{x \mid -\infty < x < \infty\}$ or $(-\infty, \infty)$
Range: $\{y \mid y \ge -3\}$ or $[-3, \infty)$

36.

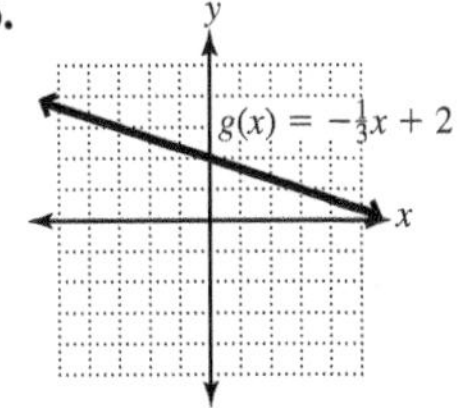

Domain: $\{x \mid -\infty < x < \infty\}$ or $(-\infty, \infty)$
Range: $\{y \mid -\infty < y < \infty\}$ or $(-\infty, \infty)$

37.

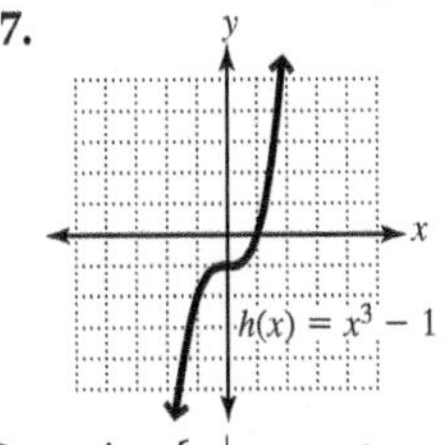

Domain: $\{x \mid -\infty < x < \infty\}$ or $(-\infty, \infty)$
Range: $\{y \mid -\infty < y < \infty\}$ or $(-\infty, \infty)$

38.

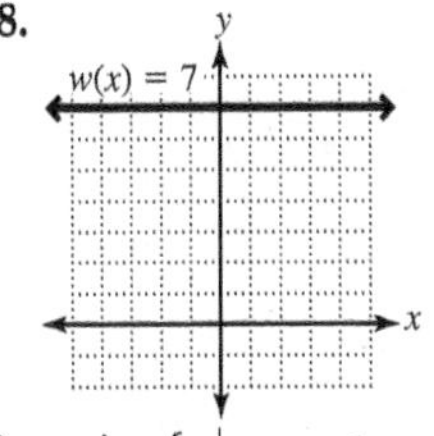

Domain: $\{x \mid -\infty < x < \infty\}$ or $(-\infty, \infty)$
Range: $\{y \mid y = 7\}$ or $[7, 7]$

Definitions/Rules/Procedures $f(x) + g(x)$; $f(x) - g(x)$ Exercises **39. a.** $-x^3 + 2x + 7$ **b.** 3 **c.** $3x^3 + 2x + 3$ **d.** -197 **40.** 56 ft. **41. a.** $P(x) = 0.2x^2 - 9x - 1000$ **b.** $5200

5.3 Definitions/Rules/Procedures distributive; second polynomial, first polynomial; sum, difference; $a^2 + 2ab + b^2$, $a^2 - 2ab + b^2$, $a^2 - b^2$ Exercises **42.** $6x^4y^3 - 9x^3y^3 + 3x^2y^3$ **43.** $-2p^3q + 3p^2q^2 - 3pq^3$ **44.** $2x^2 + 2x - 24$ **45.** $27w^2 + 15w - 2$ **46.** $18r^2 + 15rs - 7s^2$ **47.** $x^4 + x^2 - 2$ **48.** $2x^4 - 5x^3 + 7x^2 + 3x - 7$ **49.** $12t^4 - t^3 + 2t - 1$ **50.** $9a^2 - 25$ **51.** $4p^2 - 4p + 1$ **52.** $64k^2 + 48k + 9$ **53.** $16h^4 - 49$ **54.** $5m + 2$ **55.** $(w^3 + 15w^2 + 50w)$ in.^{3}
Definitions/Rules/Procedures $f(x) \cdot g(x)$ Exercises **56. a.** $2x^3 + 15x^2 + 26x - 7$ **b.** -20 **57.** $4x^2 - 33x + 68$

5.4 Definitions/Rules/Procedures $\frac{a}{c} + \frac{b}{c}$; polynomial, monomial Exercises **58.** $4m^3 - m^2 + 3m - 1$ **59.** $4x - 1 + \frac{3}{2x}$

Definitions/Rules/Procedures long; quotient, $\frac{\text{remainder}}{\text{divisor}}$; $\frac{f(x)}{g(x)}$ Exercises **60.** $x + 3 + \frac{1}{x + 2}$ **61.** $x - 2$ **62.** $2x + 3$ **63.** $3x^3 + x^2 - 2x - 4 - \frac{9}{3x - 1}$ **64. a.** $(y^3 + 2y^2 + 4y + 1)$ amps **b.** 58 amps

5.5 Definitions/Rules/Procedures $x - c$ Exercises **65.** $5x + 2 + \frac{4}{x - 1}$ **66.** $2x^2 - 5$ **67.** $4x^2 + 8x + 14 + \frac{33}{x - 2}$ **68.** $x^2 + 2x + 4$ Definitions/Rules/Procedures $P(c)$ Exercises **69.** 15 **70.** -51

Chapter 5 Practice Test

1. 0.0072 [5.1] **2.** 3.57×10^{-3} [5.1] **3.** $15x^5y$ [5.1] **4.** $\frac{1}{9x^8y^2}$ [5.1] **5.** $\frac{4u^3}{v^4}$ [5.1] **6.** $\frac{16}{81}t^{20}u^8v^4$ [5.1] **7.** 1.26×10^{10} [5.1] **8.** 2×10^7 [5.1] **9.** constant [5.2] **10.** cubic [5.2] **11.**

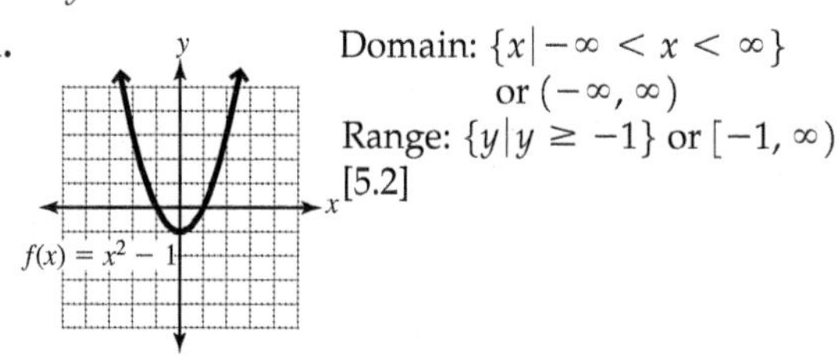

12. $-8ab$ [5.2] **13.** $11r^5 - 7r^4 + 7r^3 - 10r^2 - r - 1$ [5.2] **14.** $15x^2 - 14xy - 8y^2$ [5.3] **15.** $12m^5n^9 - 8m^4n^{10} + 20m^3n^{11}$ [5.3] **16.** $49k^2 - 28jk + 4j^2$ [5.3] **17.** $14x^3 - 15x^2 - 19x + 15$ [5.3] **18.** $16h^2 - 9$ [5.3] **19.** $15x^2 - 14x - 16$ [5.3] **20.** 16 [5.3]

21. $4k^2 - 2k + 1$ [5.4] **22.** $m^2 + 3m + 2$ [5.4] **23.** $2x^2 - x + 3 - \frac{17}{x + 3}$ [5.5] **24.** -8 [5.5] **25. a.** $P(x) = 0.1x^2 - 14x + 2000$ **b.** $88,000 [5.2]

Chapters 1–5 Cumulative Review Exercises

1. false **2.** true **3.** false **4.** true **5.** identity **6.** $3x + 7 = 5$; $3x + 7 = -5$. **7.** inconsistent

8. $x - c$ **9.** 10 **10.** -5 **11.** $-5(12 - x) - 4$ **12.** $-\frac{8a^2}{b^4}$ **13.** $\frac{4y^7z^4}{x^5}$ **14.** $-\frac{27x^{12}}{y^6}$ **15.** $-3x^3 + 5x - 8$

16. $7x^2 + 13x - 3$ **17.** $15x^2 - 4xy - 4y^2$ **18.** $2x^2 - 2x + 1$ **19.** $\frac{P - 2\pi r}{2}$ **20.** $-2 \le x \le 1$ or $[-2, 1]$

21. $\{x \mid x < 1 \text{ or } x > 4\}$ or $(-\infty, 1) \cup (4, \infty)$ **22.** $(-2, -4)$ **23.**

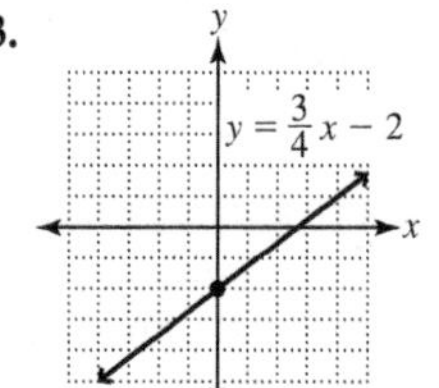

24.

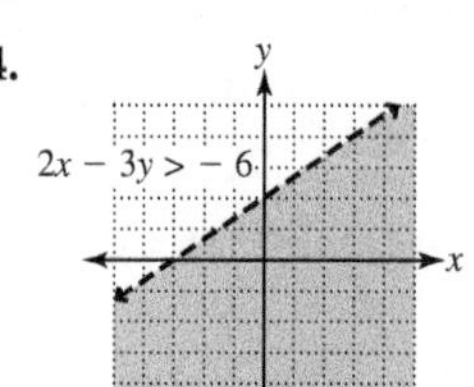

25. $5x - 3y = -36$

26.

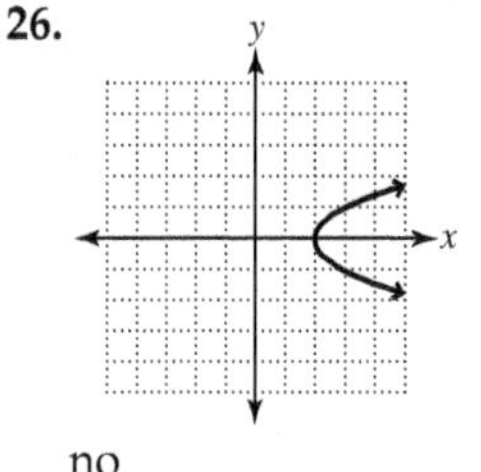

no

27. 53, 54, 55 **28.** 5 hr. **29.** Azalea: 28; rose: 63. **30.** Length: 32 ft.; width: 10 ft.

Chapter 6

Exercise Set 6.1

Prep Exercises **1.** No, the GCF of 4 has not been factored out. **2.** Include only those prime factors (and variables) common to all of the factorizations, each raised to its smallest exponent. **3.** Rewrite the polynomial as a product of the GCF and the quotient of the polynomial and the GCF. Polynomial $= \text{GCF}\left(\frac{\text{Polynomial}}{\text{GCF}}\right)$ **4.** Factor the negative of the GCF. **5.** Four-term polynomials. **6.** Group pairs of terms and factor the GCF out of each pair.

Exercises **1.** $4x^2y$ **3.** $4u^3v^6$ **5.** $2abc^4$ **7.** $a + b$ **9.** $5c^2(3c^2d - 4)$ **11.** $x^2(x^3 - x + 1)$ **13.** $25x(y - 2z + 4x)$ **15.** $-7uv(2uv + v - 1)$ **17.** $3a^2b(3a^5b^2 + a^2b - 2)$ **19.** $3wv(w^2v^3 + 13w + 6v)$ **21.** $6ab^2c(3b - 6a + 4a^4c^7)$ **23.** $-4xy(2x - 4y + 3)$ **25.** $(n - 3)(m + 4)$ **27.** $(b + 2c)(6 - a)$ **29.** $(x + y)(a + b)$ **31.** $(u + 3)(u^2 + 3)$ **33.** $(m + p)(n - 3)$ **35.** $(c + 1)(d + 1)$ **37.** $(2a - 1)(a + 1)$ **39.** $(x + 2y)(3a + 4b)$ **41.** $(h + 8)(h - k)$ **43.** $(x^2 + y^2)(3 - a)$ **45.** $(p - 2q)(3p^2 + 2q^2)$ **47.** $(x - 4y)(2x^2 - 3y^2)$ **49.** $2(b - c)(a + x)$ **51.** $2(3x + 1)(2y + 5)$ **53.** $3a(a + 3)(y - 4)$ **55.** $m(3m - 10)(m + 2n)$ **57.** $5t(t - 2)(3s - 1)$ **59.** $5y(x^2 + 1)(x - 4)$ **61.** Mistake: Did not factor out the GCF, which is $12x^2y$. Correct: $12x^2y(2y^2 + 3x)$ **63.** Mistake: Incorrect power of b in the parentheses. Correct: $9a^2b(b^2c - 2a^2)$ **65.** It is correct. **67.** Correct form: $2x(2x^2 + 7x + 4)$ **69.** $6x^2 + 40x$, $2x(3x + 20)$ **71. a.** $15x^2 + 26x$ **b.** $x(15x + 26)$ **c.** 213 ft.^{2} **73. a.** $36\pi r^3 + 18\pi r^2$ **b.** $18\pi r^2(2r + 1)$ **c.** $\approx$15,550.9 ft.^{3} **75. a.** $p - rp$ **b.** $p(1 - r)$ **c.** \$32.97 **77. a.** $p + prt$ **b.** $p(1 + rt)$ **c.** \$862.75 **79. a.** $2\pi r^2 + 2\pi rh$ **b.** $2\pi r(r + h)$ **c.** $\approx$1884.96 in.^{2}

Review Exercises: **1.** $-24{,}500{,}000$ **2.** 9.2×10^{-5} **3.** $x^2 + 8x + 15$ **4.** $x^2 - 10x + 24$ **5.** $6x^2 + 19x - 7$ **6.** $8x^3 + 14x^2 - 30x$

Exercise Set 6.2

Prep Exercises **1.** 12; 8 **2.** -10; 3 **3.** $-$; $-$ **4.** $+$; $-$ **5.** $-$; $+$ **6.** x and $8x$, $2x$ and $4x$ **7.** 1 and 12, 2 and 6, 3 and 4 **8.** The sum of the *inner* and *outer* products is the middle term. **9.** 30; 11 **10.** -60; -4 **11.** x^2 **12.** $x - 3$

Exercises **1.** $(r + 7)(r + 1)$ **3.** $(w + 1)(w - 3)$ **5.** $(a + 6)(a + 3)$ **7.** $(y - 9)(y - 4)$ **9.** $(m + 4)(m - 2)$ **11.** $(b - 10)(b + 4)$ **13.** prime **15.** $3s(t + 7)(t + 1)$ **17.** $5y(y - 12)(y - 1)$ **19.** $6au(u + 3)(u - 2)$ **21.** $3x^2y(y^2 - 4y + 10)$ **23.** $(p + 9q)(p + 2q)$ **25.** $(u - 7v)(u - 6v)$ **27.** $(x - 7y)(x + 2y)$ **29.** $(a + 6b)(a - 7b)$ **31.** $(3a + 7)(a + 1)$ **33.** $(2w + 1)(w - 2)$ **35.** prime **37.** $(4q - 1)(q - 2)$ **39.** $(3b - 1)(2b + 3)$ **41.** $(4m + 3)^2$ **43.** $(4x - 3)(x + 2)$ **45.** $(2w + v)(w + 7v)$ **47.** $(5x - y)(x - 3y)$ **49.** $(2x - y)(8x - y)$ **51.** prime **53.** $(3t - 2u)(t + 7u)$ **55.** $(3m + 2n)(m - 4n)$ **57.** $(3a - 2b)(4a - 3b)$ **59.** $2m(11m + 1)(m + 9)$ **61.** $2v(2u - 5v)(u + 3v)$ **63.** $(3y + 1)(y + 5)$ **65.** prime **67.** $(3t - 2)(t - 5)$ **69.** $(3x - 5)(2x + 3)$ **71.** $4y(4x - 3)(2x + 3)$ **73.** $5b(3a + b)(a - 2b)$ **75.** $(x^2 + 1)(x^2 - 2)$ **77.** $(4r^2 + 3)(2r^2 - 1)$ **79.** $(5x^2 - 2)(3x^2 - 1)$ **81.** $(y^3 - 12)(y^3 - 4)$ **83.** $(7x + 8)(x + 2)$ **85.** $(3a + 8)(a - 2)$ **87.** Mistake: The signs are incorrect. Correct: $(x + 2)(x - 3)$ **89.** Mistake: The GCF monomial was not factored out. Correct: $4(x + 2)^2$ **91.** 8, 10, 17 **93.** 2, 18, 62 **95.** Answers may vary: (8, 14, 18, and 20 are possible answers.) **97.** Answers may vary: (2, 6, 12, and 20 are a few possible answers.)

Review Exercises: **1.** $4x^2 + 12x + 9$ **2.** $16y^2 - 1$ **3.** $n^3 - 8$ **4.** $6x^2 + x + 4 + \frac{2}{x - 1}$ **5. a.** $\left\{x \middle| x \geq \frac{13}{2}\right\}$ **b.** $\left[\frac{13}{2}, \infty\right)$ **c.** 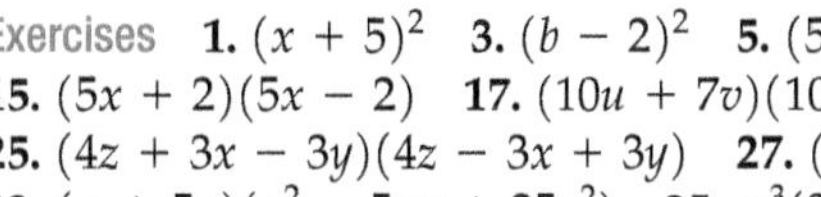 **6. a.** $\{x \mid -6 < x < -1\}$ **b.** $(-6, -1)$ **c.** 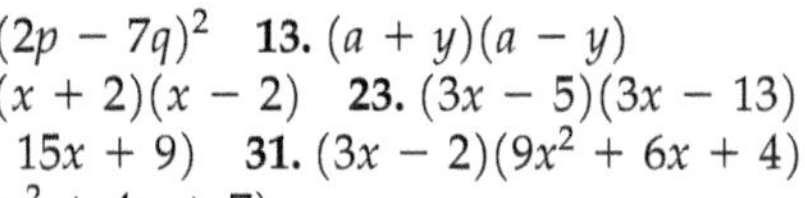

Exercise Set 6.3

Prep Exercises **1.** The first term is the square of $2x$, the last term is the square of 5, and the middle term is twice the produce of those two terms: $2(2x)(5) = 20x$. **2.** $(a + b)^2$; $(a - b)^2$ **3.** $(a + b)(a - b)$ **4.** conjugates **5.** $(a - b)(a^2 + ab + b^2)$; $(a + b)(a^2 - ab + b^2)$ **6.** Factor out any monomial GCF. **7.** grouping **8.** Difference of squares, difference of cubes, sum of cubes.

Exercises **1.** $(x + 5)^2$ **3.** $(b - 2)^2$ **5.** $(5u - 3)^2$ **7.** $(n + 12m)^2$ **9.** $(3q - 5p)^2$ **11.** $(2p - 7q)^2$ **13.** $(a + y)(a - y)$ **15.** $(5x + 2)(5x - 2)$ **17.** $(10u + 7v)(10u - 7v)$ **19.** $9(x + 2)(x - 2)$ **21.** $(x^2 + 4)(x + 2)(x - 2)$ **23.** $(3x - 5)(3x - 13)$ **25.** $(4z + 3x - 3y)(4z - 3x + 3y)$ **27.** $(m - 3)(m^2 + 3m + 9)$ **29.** $(5x + 3)(25x^2 - 15x + 9)$ **31.** $(3x - 2)(9x^2 + 6x + 4)$ **33.** $(u + 5v)(u^2 - 5uv + 25v^2)$ **35.** $m^3(3 - 5mn)(9 + 15mn + 25m^2n^2)$ **37.** $(u + 5)(u^2 + 4u + 7)$ **39.** $(3 - a - b)(9 + 3a + 3b + a^2 + 2ab + b^2)$ **41.** $(4x + 3y + 3z)(16x^2 - 12xy - 12xz + 9y^2 + 18yz + 9z^2)$ **43.** $(4d - 3x - 3y)(16d^2 + 12dx + 12dy + 9x^2 + 18xy + 9y^2)$ **45.** 40 **47.** 36 **49.** 16 **51.** 16 **53.** $4m^2n^2(3m + 5n^2)$ **55.** $(x + 5)(x + 3)$ **57.** $(x + 4)(x - 4)$ **59.** $(2c - 3)(6c + 5)$ **61.** $(a - y)(x - y)$ **63.** $(2x - 7)^2$ **65.** prime **67.** $3abc(4a^2b + abc + 3c^2)$ **69.** $(b + 5)(b^2 - 5b + 25)$ **71.** prime **73.** $(3b + 2)(2b - 1)$ **75.** $ab(1 + 6b)(1 - 6b)$ **77.** $(4c - 3)^2$ **79.** $3x(2x - 3)(3x + 4)$ **81.** $(x^2 + 4)(x + 2)(x - 2)$ **83.** prime **85.** $3(2u - 7v)^2$ **87.** $3a^2b^3(4a + b)(3a - 4b)$ **89.** $x^3y(6x^2 + 5xy - 12y^2)$ **91.** $(xy + 3)(x - y)$ **93.** $(x + 2)(x^2 - 5x + 13)$ **95.** $5m(m^2 + n^2)^2$ **97.** n^2 **99.** $t^3(4t - u)(16t^2 + 4tu + u^2)$ **101.** $(x - y)(x + y)(2x + 3)$ **103.** $(6y + x - 4)(6y - x + 4)$ **105.** $12x^2 - 40$, $4(3x^2 - 10)$ **107.** $18x^3 - 27$, $9(2x^3 - 3)$ **109.** $3x - 1$, $2x - 3$ **111.** $3x + 2$, $x + 3$, $5x$

Review Exercises: **1.** $-\frac{13}{14}$ **2.** $-1, 6$ **3.** \$1500 at 4%, \$2500 at 6%, \$2000 at 8%

4.

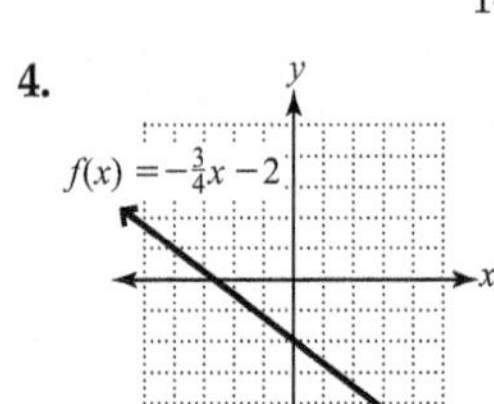

5.

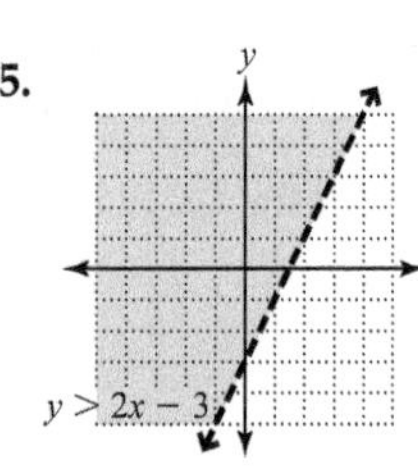

6. $3x + 7y = -5, y = -\frac{3}{7}x - \frac{5}{7}$

Exercise Set 6.4

Prep Exercises **1.** 0 **2.** Set each factor equal to 0. That is, solve $x - 6 = 0$ and $x + 3 = 0$. **3.** 2 **4.** 3 **5.** quadratic; two **6.** cubic; three **7.** (Answers may vary.)

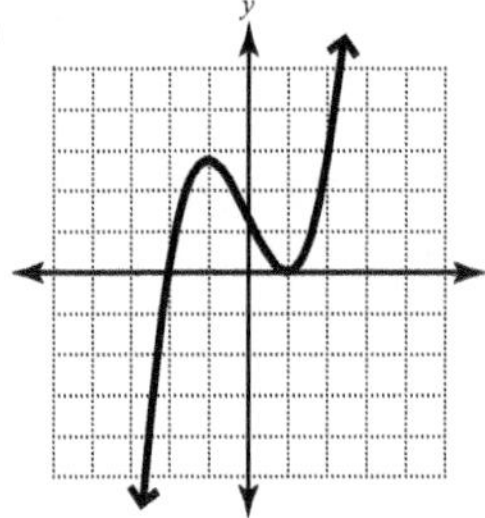

8. (Answers may vary.)

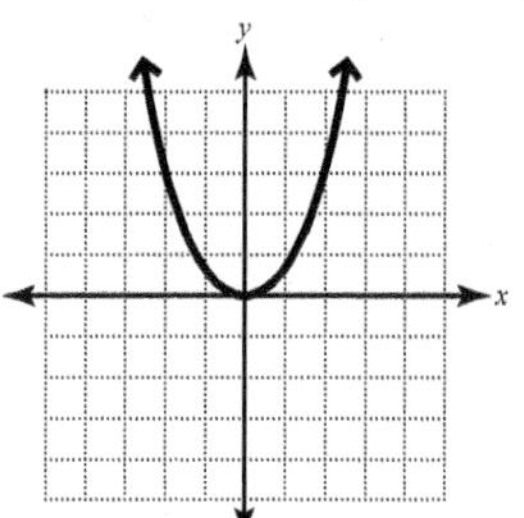

9. Solve $f(x) = 0$. **10.** $a^2 + b^2 = c^2$

Exercises **1.** $-4, 0$ **3.** $-2, 3$ **5.** $-\frac{5}{2}, \frac{2}{3}$ **7.** $-\frac{5}{2}, 0, 3$ **9.** $-7, -2, 5$ **11.** $-\frac{7}{4}, -\frac{3}{2}, \frac{1}{3}$ **13.** $0, 4$ **15.** $-3, 3$ **17.** $-9, -5$ **19.** $-2, \frac{1}{2}$ **21.** $-\frac{5}{2}, \frac{1}{3}$ **23.** -3 **25.** $0, \frac{1}{2}$ **27.** $-5, 5$ **29.** $-9, 3$ **31.** $1, 2$ **33.** $-\frac{5}{2}, -1$ **35.** $-2, 7$ **37.** $-\frac{7}{2}$ **39.** $-6, 3$ **41.** $-3, 0, 2$ **43.** $-\frac{2}{3}, \frac{1}{2}$ **45.** $-\frac{2}{3}, 1$ **47.** $0, -\frac{1}{3}$ **49.** $3, -3, -2$ **51.** $x^2 + x - 6 = 0$ **53.** $3x^2 - 10x - 8 = 0$ **55.** $x^3 - 2x^2 - 3x = 0$ **57.** b **59.** a **61.** c

63. $(-5, 0), (5, 0)$

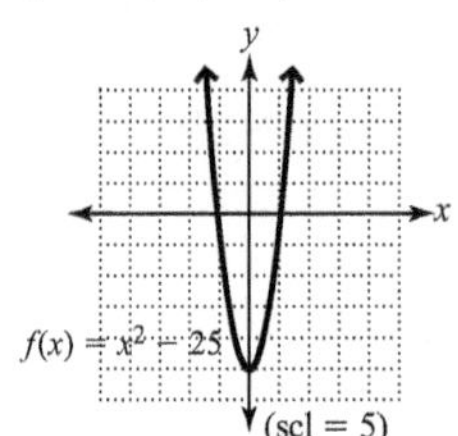

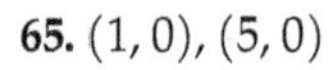

65. $(1, 0), (5, 0)$

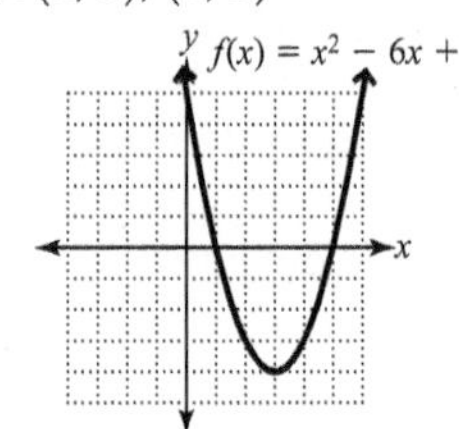

67. $(-4, 0), (0, 0), (1, 0)$

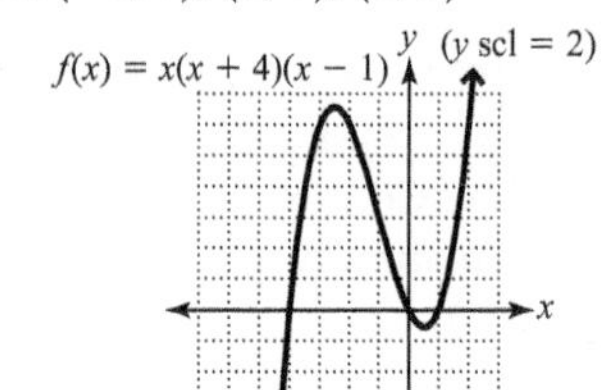

69. $(-2, 0), (0, 0), (3, 0)$

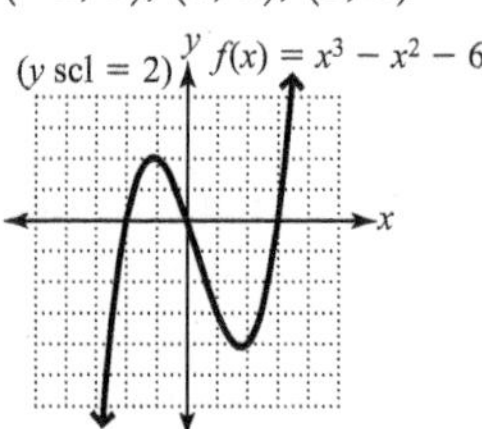

71. $(3, 0), (6, 0)$ **73.** $\left(-\frac{5}{3}, 0\right), \left(\frac{1}{2}, 0\right)$ **75.** $(-1, 0), (0, 0), (2, 0)$ **77.** $-5, 8$ **79.** $-0.25, 2.5$ **81.** $-4, 0, 3$ **83.** $6, 8, 10$ **85.** $7, 24, 25$ **87.** $5, 11$ **89.** $5, 7, 9$ **91.** 16 m by 20 m **93.** 14 ft. **95.** 9 in. **97.** Height: 24 ft.; wire length: 26 ft. **99.** 1.25 sec. **101.** 10%

Review Exercises: **1.** 4.203×10^{-4} **2.** $x^2 - 3x + 2 + \frac{3}{x + 4}$ **3.**

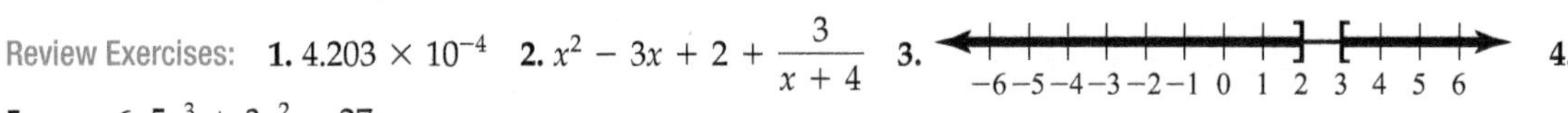

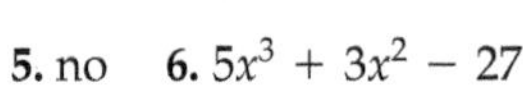

4.

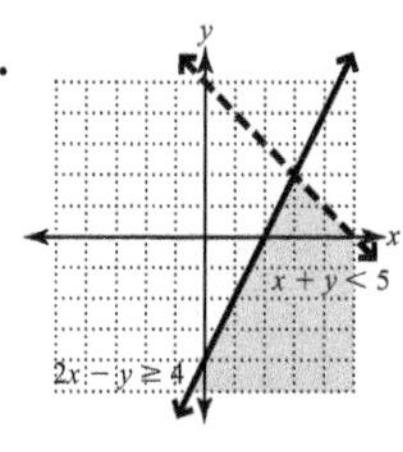

5. no **6.** $5x^3 + 3x^2 - 27$

Chapter 6 Summary and Review Exercises

6.1 Definitions/Rules/Procedures prime factorization; common, smallest; Multiply Exercises **1.** $8x^3y^2$ **2.** $5mn^2$ **3.** 1 **4.** $2(x + 1)$ Definitions/Rules/Procedures GCF; GCF, polynomial, GCF; GCF, $\frac{\text{polynomial}}{\text{GCF}}$ Exercises **5.** $u^2(u^4 - u^2 - 1)$ **6.** $13d(d - 2e)$ **7.** $2h^2k(2k^5 - h^3)$ **8.** $2cd(6 - 2cd + 5c^3d^3)$ **9.** $p^6q^2(16q - 12p^2q^3 + 13p)$ **10.** $3wv^3(3w^5v^5 + 2v^3 - 4w^2)$ **11.** $(w + 3)(17 - m)$ **12.** $(x + 5)(2y + 1)$ Definitions/Rules/Procedures GCF, common; GCF; binomial; binomial, interchange, binomial Exercises **13.** $(n + 3)(m + 2)$ **14.** $(a + 3)(a^2 + 3)$ **15.** $(x + y)(2 - a)$ **16.** $(c^2 + d^2)(b - 5)$ **17.** $(y - s)(x^2 - r)$ **18.** $(4k - 1)(k + 1)$ **19.** $2v(2u - 1)(2v + 5)$ **20.** $5d(c^2 - d)(d + 4)$

6.2 Definitions/Rules/Procedures c, b; step 1; GCF; ax^2; c; outer, inner, outer, inner; last; last; first Exercises **21.** $(a-9)(a-1)$ **22.** $(m+3)(m+17)$ **23.** $(y+8)(y-6)$ **24.** $(x-10)(x+3)$ **25.** $(3x-7)(x+2)$ **26.** $(8h+1)(2h+1)$ **27.** prime **28.** $(3t+5)(2t-3)$ **29.** $(s-10t)(s-t)$ **30.** $(3u+2v)(2u+3v)$ **31.** $(5m-n)(m-3n)$ **32.** prime **33.** $b^2(b-9)(b+2)$ **34.** $2x(x-10)(x+2)$ Definitions/Rules/Procedures GCF, ac, b; bx, sum; grouping Exercises **35.** $(2u+5)(u+2)$ **36.** $(3m-2)(m-2)$ **37.** $(u+1)(10u-3)$ **38.** $(3y+1)(2y-5)$ Definitions/Rules/Procedures substitution Exercises **39.** $(x^2+4)(x^2+1)$ **40.** $(3c^2+1)(c^2+4)$ **41.** $(2h^3+1)(h^3+4)$ **42.** $(3k-2)(k+2)$

6.3 Definitions/Rules/Procedures $(a+b)^2, (a-b)^2$; $(a+b)(a-b)$; $(a-b)(a^2+ab+b^2)$; $(a+b)(a^2-ab+b^2)$ Exercises **43.** $(x+3)^2$ **44.** $(y+6)^2$ **45.** $(m-2)^2$ **46.** $(w-7)^2$ **47.** $(3d+5)^2$ **48.** $(2c-7)^2$ **49.** $(h+3)(h-3)$ **50.** $(p+8)(p-8)$ **51.** $(3d+2)(3d-2)$ **52.** $(9k+10)(9k-10)$ **53.** $2(w+5)(w-5)$ **54.** $4(q+3)(q-3)$ **55.** $(5y+3z)(5y-3z)$ **56.** $(x^2+9)(x+3)(x-3)$ **57.** $(c-3)(c^2+3c+9)$ **58.** $(m+4)(m^2-4m+16)$ **59.** $(3b+2a)(9b^2-6ab+4a^2)$ **60.** $8(2d-c)(4d^2+2cd+c^2)$ Definitions/Rules/Procedures grouping; perfect square; perfect square; $(a+b)^2$; $(a-b)^2$; c, b; trial, ac, b; $(a+b)(a-b)$; $(a-b)(a^2+ab+b^2)$; $(a+b)(a^2-ab+b^2)$ Exercises **61.** $(3d-1)^2$ **62.** $3(m+n)(m-n)$ **63.** $(2a+3)(a-4)$ **64.** prime **65.** $3p^2(p+6)(p-5)$ **66.** $(x^2+4)(x+2)(x-2)$ **67.** $a^2bc^2(15abc^5+3b^3+5a^7c)$ **68.** $(w-2-y)(w^2+2w+wy+4+4y+y^2)$

6.4 Definitions/Rules/Procedures two polynomials; $P=0$, descending; $ax^2+bx+c=0$; $ax^3+bx^2+cx+d=0$; $a=0, b=0$; 0, descending; factored; zero-factor Exercises **69.** $-4, 1$ **70.** $-8, 8$ **71.** $1, -3$ **72.** $\frac{1}{6}, 1$ **73.** $-\frac{7}{2}$ **74.** $-3, 3, 4$ Definitions/Rules/Procedures $f(x)=0$ Exercises **75.** $(-2,0), (2,0)$

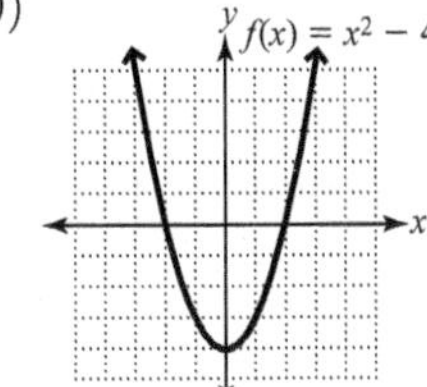

76. $(-3,0), (0,0), (2,0)$

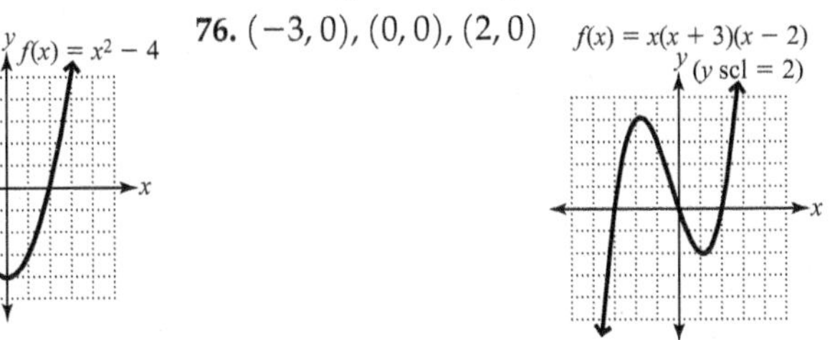

Definitions/Rules/Procedures $a^2+b^2=c^2$ Exercises **77.** 4, 14 **78.** 10, 11 **79.** 16, 18 **80.** 8 ft. by 11 ft. **81.** Height: 16 ft.; base: 48 ft. **82.** 5, 12, 13 **83.** Height: 17 ft.; distance from ear: 37 ft. **84.** $2\frac{1}{4}$ sec.

Chapter 6 Practice Test

1. $7m^3n^2$ [6.1] **2.** $3m(1+2m^2-3m^5)$ [6.1] **3.** $2(m+n)$ [6.1] **4.** $(3n+4)(3n-4)$ [6.3] **5.** $(2x-3)(4x^2+6x+9)$ [6.3] **6.** $(y+7)^2$ [6.3] **7.** $(q-8)(q+6)$ [6.2] **8.** $3a(b^2-10b+8)$ [6.1] **9.** $5(1+5t)(1-5t)$ [6.3] **10.** $(3d+2)(2d-1)$ [6.2] **11.** $8(c+d)(c^2-cd+d^2)$ [6.3] **12.** $(w+2)(w^2+3)$ [6.1] **13.** $(s^2+9)(s+3)(s-3)$ [6.3] **14.** $(5p-12q)(p+q)$ [6.2] **15.** $\frac{1}{2}, \frac{5}{4}$ [6.4] **16.** $-6, 3$ [6.4] **17.** $-3, 0, \frac{5}{2}$ [6.4] **18.** 8, 15, 17 [6.4] **19.** 2 sec. [6.4] **20.** $(-3,0), (3,0)$ [6.4]

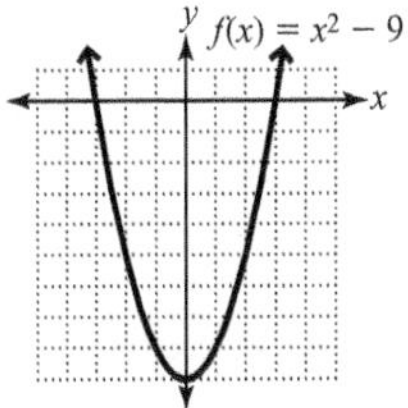

Chapters 1–6 Cumulative Review Exercises

1. true **2.** false **3.** true **4.** false **5.** the same **6.** coefficients, add **7.** $(a+b)(a^2-ab+b^2)$ **8.** undefined **9.** $2x^3-12x^2-21x+20$ **10.** $8x^3-2x^2+x+2$ **11.** $\frac{x^3}{8}$ **12.** $4m(1-4m)$ **13.** $(8x-3)(x+4)$ **14.** $(7a+6)^2$ **15.** $8(2p+1)(4p^2-2p+1)$ **16.** $(x-2y)(3+a)$ **17.** 2 **18.** $-11, 14$ **19.** $\frac{4}{5}, -6$ **20. a.** $\{x \mid x \geq 2\}$ **b.** $[2, \infty)$ **c.**

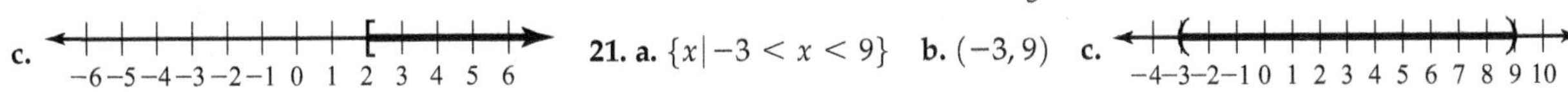

21. a. $\{x \mid -3 < x < 9\}$ **b.** $(-3, 9)$ **c.**

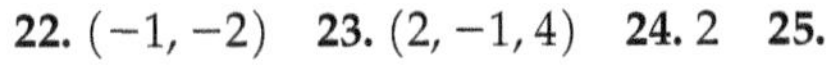

22. $(-1,-2)$ **23.** $(2,-1,4)$ **24.** 2 **25.**

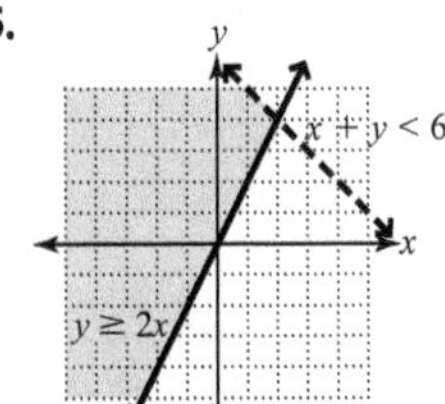

26. $(-3,0), (2,0), (0,-6)$

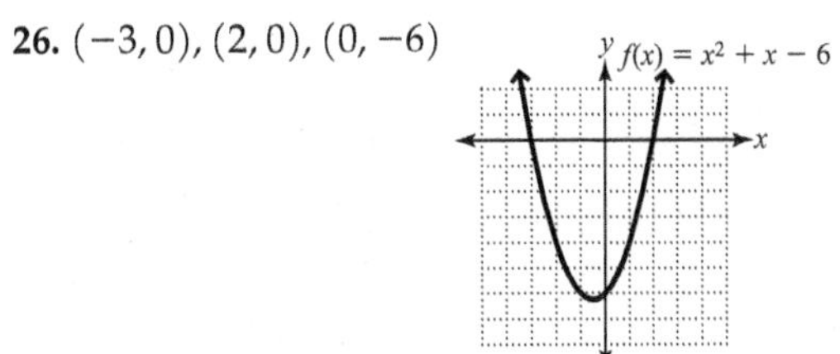

27. a. $5 \leq n < 8$ **b.** $n \geq 8$ **28.** Teens: 25 min.; adults: 52 min. **29.** 32°, 58° **30.** $a = 5$ in., $a+3 = 8$ in., $a+5 = 10$ in.

Chapter 7

Exercise Set 7.1

Prep Exercises **1.** polynomials **2.** The greatest common factor of the numerator and denominator is 1. **3.** Factor the numerator and denominator completely. **4.** They are not factors of the numerator and denominator. **5.** Terms are added or subtracted; factors are multiplied. **6.** Factor each numerator and denominator completely, divide out common factors, multiply the numerators and multiply the denominators, and simplify as needed. **7.** Write an equivalent multiplication statement using $\frac{P}{Q} \div \frac{R}{S} = \frac{P}{Q} \cdot \frac{S}{R}$, where P, Q, R, and S are polynomials and Q, R, and $S \neq 0$. **8.** No, because $f(5)$ yields 0 in the denominator, $x = 5$ is not in the domain of the function. **9.** Numbers that cause the denominator to be 0. **10.** vertical asymptote

Exercises **1.** $-\frac{7}{4m^4n}$ **3.** $-\frac{4x^2y}{3}$ **5.** $\frac{2}{3}$ **7.** $\frac{2}{5}$ **9.** $-\frac{3}{4}$ **11.** $\frac{a-2}{a-3}$ **13.** $\frac{x}{x-2}$ **15.** $\frac{2x-3y}{3x-5y}$ **17.** $\frac{2a-3b}{3a-2b}$ **19.** $\frac{x^2+2x+4}{3x+4}$ **21.** $\frac{x^2-3x+9}{3(x-3)}$ **23.** $\frac{y-4}{y-6}$ **25.** $\frac{2a^2+3}{2a+3}$ **27.** -1 **29.** $-\frac{x+7}{2}$ **31.** $\frac{4y^2}{9x^2}$ **33.** $-\frac{8c}{3b^3}$ **35.** $\frac{2xy^2}{5}$ **37.** $-\frac{2}{5}$ **39.** $\frac{4(a+1)}{3(a+4)}$ **41.** $-\frac{4y^2}{3x}$ **43.** $\frac{x-2}{x+3}$ **45.** 1 **47.** $\frac{3x+2}{4x+3}$ **49.** $-(3x-1)$ **51.** $\frac{(a+2)(d-5)}{(d+1)(b-4)}$ **53.** $\frac{2x+1}{3x+2}$ **55.** $2a-b$ **57.** $\frac{2x^6g^2}{5}$ **59.** $\frac{4a^4}{9}$ **61.** $\frac{x^3}{4}$ **63.** $-\frac{2}{9}$ **65.** $\frac{d-4}{2d(3d+4)}$ **67.** $\frac{(m-5n)(m+6n)}{6}$ **69.** $\frac{3(x-8)}{4(x+8)}$ **71.** $\frac{x-2y}{x+2y}$ **73.** $\frac{x-2}{x+3}$ **75.** $\frac{9a^2+6ab+4b^2}{a-4b}$ **77.** $\frac{b-4}{c-3}$ **79.** $\frac{3x-2}{4x}$ **81.** $\frac{5x+3}{4x-3}$ **83.** Mistake: a^2+b^2 does not factor. Correct: $\frac{a^2+b^2}{(a+b)(a-2b)}$ **85. a.** -6 **b.** 0 **c.** undefined **87. a.** 0 **b.** 5 **c.** $\frac{1}{8}$ **89. a.** $\frac{5}{9}$ **b.** 0 **c.** undefined **91. a.** 0 **b.** $\frac{4}{3}$ **c.** undefined **93.** $\{x \mid x \neq -4\}$ **95.** $\{x \mid x \neq 5, -5\}$ **97.** $\left\{c \mid c \neq \frac{9}{2}, -\frac{9}{2}\right\}$ **99.** $\left\{t \mid t \neq -\frac{5}{2}, \frac{2}{3}\right\}$ **101.** $\{x \mid x \neq 0, -7, 3\}$ **103.** $\{x \mid x \neq 2\}$ **105.** c **107.** d **109.**

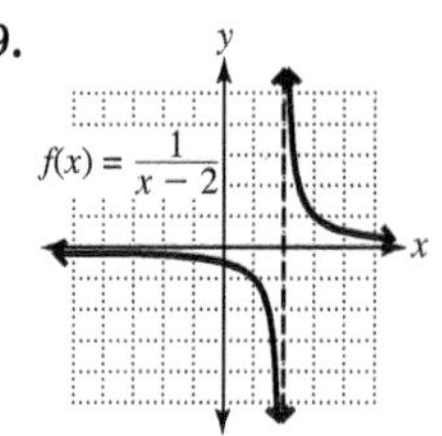

111.

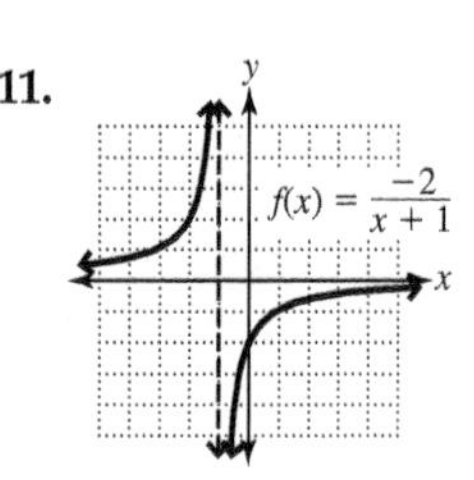

113. $\frac{5}{4}\Omega$ **115.** \$24 **117.** \$124,000 **119.** 500 cycles/sec.

Review Exercises: **1.** $\frac{7}{11}$ **2.** $-\frac{7}{30}$ **3.** $\frac{13}{20}y$ **4.** $\frac{3}{5}x+1$ **5.** $3x(2x-5)$ **6.** $(x+5)(x-4)$

Exercise Set 7.2

Prep Exercises **1.** Add or subtract the numerators and keep the same denominator. Then simplify to lowest terms. **2.** Factor each denominator. Write a product that contains each unique prime factor the greatest number of times that factor occurs in any factorization. Simplify the product. **3. a.** 2, 3, 5, x, and y **b.** $2^2 \cdot 3 \cdot 5 \cdot x^3 \cdot y$ **4. a.** 5, x, $x+3$, and $x-4$ **b.** $5x(x+3)(x-4)$ **5.** Factoring $6x^2+5x-6 = (3x-2)(2x+3)$ shows that it is necessary to multiply by $(x-4)$ to obtain the LCD. **6.** -1 **7.** Subtract the polynomials in the numerator. **8.** Both are correct because $\frac{-6}{3-a} = \frac{6}{-(3-a)} = \frac{6}{a-3}$.

Exercises **1.** $\frac{4c}{d}$ **3.** $\frac{7a-b}{2a-5b}$ **5.** $\frac{-2x-5y}{x+2y}$ **7.** $\frac{c+6}{c-1}$ **9.** $\frac{y-1}{y-6}$ **11.** $\frac{h-1}{h+7}$ **13.** $\frac{x-3}{x+5}$ **15.** $x+y$ **17.** $60a^6b^8$; $\frac{16}{60a^6b^8}, \frac{27a^2b^3}{60a^6b^8}$ **19.** $(x+7)(x-5)$; $\frac{x^2-5x}{(x+7)(x-5)}, \frac{x^2+7x}{(x+7)(x-5)}$ **21.** $45(r+2)$; $\frac{35r}{45(r+2)}, \frac{24r}{45(r+2)}$ **23.** $(b+1)(b-1)(b+4)$; $\frac{3b^2+12b}{(b+1)(b-1)(b+4)}, \frac{b^2+b}{(b+1)(b-1)(b+4)}$ **25.** $(c-3)(c+1)(c-2)$; $\frac{c^2-7c+10}{(c-3)(c+1)(c-2)}, \frac{c^2+4c+3}{(c-3)(c+1)(c-2)}$ **27.** $(n+4)^2(n+1)$; $\frac{n^2+2n+1}{(n+4)^2(n+1)}, \frac{n^2-16}{(n+4)^2(n+1)}$ **29.** $x^2(x+3)(x-2)(x+4)$; $\frac{x^3+9x^2+20x}{x^2(x+3)(x-2)(x+4)}, \frac{x^2-5x+6}{x^2(x+3)(x-2)(x+4)}$ **31.** $x(x+4)$; $\frac{3x+12}{x(x+4)}, \frac{4}{x(x+4)}, \frac{6x}{x(x+4)}$ **33.** $\frac{1}{24u}$ **35.** $\frac{31z}{20x}$ **37.** $\frac{11y+6}{12y}$ **39.** $\frac{m+38}{6m}$ **41.** $\frac{9p+3+45p^2q-10pq}{30p^3q^2}$ **43.** $\frac{6a^2+ab+6b^2}{24a^3b^4}$ **45.** $\frac{8-2k}{k(k+2)}$ **47.** $\frac{9w+13}{(w-3)(w+7)}$ **49.** $\frac{x-13}{6(x+3)}$ **51.** $\frac{8t-24}{(t-4)^2}$ **53.** $\frac{u^2-u-3}{(u-3)^2}$ **55.** $\frac{x-8}{x-6}$ **57.** $\frac{3x-5}{(x-3)(x+2)}$ **59.** $\frac{-3}{(v-4)(v+4)}$ **61.** $\frac{2z^2-z+9}{(z+3)^2(z-2)}$ **63.** $\frac{7x-12}{x(x+6)(x-6)(x-3)}$

65. $\frac{3x^2 + 2x + 5}{x^2 - 4}$ **67.** $\frac{10a - 19}{a(a - 4)}$ **69.** $\frac{3x^2 - 6x + 30}{x(x - 4)}$ **71.** $\frac{-2r^2 - 24r + 2}{r^2 - 9}$ **73.** $\frac{4v}{u - v}$ **75.** $\frac{3m - 4n}{m - n}$ **77.** $\frac{-a - 5b}{3a - b}$

79. Mistake: Added denominators instead of finding the LCD. Correct: $\frac{a^2 + 20}{5a}$ **81.** Mistake: Added denominators. Correct: $\frac{9v}{2x}$

83. Mistake: Did not distribute the subtraction sign to both terms in $2c + 3$. Correct: $\frac{3c - 1}{3c - 5}$ **85. a.** $\frac{5x + 2}{3x}$ **b.** $\frac{4x + 5}{3x}$ **c.** $\frac{11}{6}$ **d.** $\frac{1}{2}$ **87. a.** $\frac{x + 6}{x + 2}$ **b.** $-\frac{x^2 + 2x + 16}{x^2 - 4}$ or $\frac{x^2 + 2x + 16}{4 - x^2}$ **c.** $\frac{7}{3}$ **d.** 5 **89.** $\frac{8t}{15}$ **91.** $\frac{8x - 4}{(x + 4)(x - 2)}$ **93.** $\frac{3x^2 + 2}{x}$ **95.** $\frac{4x}{15}$

Review Exercises: **1.** $\frac{13}{12}$ **2.** $\frac{10}{9}$ **3.** $\frac{14}{5}$ **4.** 18 **5.** $\frac{x - 1}{x + 2}$ **6.** $\frac{1}{9y^2 - 6y + 4}$

Exercise Set 7.3

Prep Exercises **1.** $(x - 2)(x - 3)$ **2.** $3a + b$; the fraction line separating the numerator and denominator of a complex rational expression is slightly larger than the fraction line(s) in the numerator and/or denominator. **3.** $\frac{x + 2}{x - 4} \div \frac{x - 3}{x + 5}$ **4.** $\frac{\frac{c + 2d}{2c + 3d} - 4}{5 + \frac{3c - d}{c + 4d}}$ **5.** Method 2 because the numerator and denominator are not monomials. **6.** Method 1 because the numerator and denominator are monomials and the fraction can easily be written as division. **7.** $\frac{\frac{3}{x^2}}{\frac{4}{x} + 5}$ **8.** $\frac{\frac{1}{x^2} - \frac{1}{y}}{\frac{1}{x} + y}$

Exercises **1.** $\frac{10}{9}$ **3.** $\frac{uv}{w}$ **5.** $\frac{u^4}{w^4v^2}$ **7.** $\frac{3}{32}$ **9.** $\frac{ac}{b}$ **11.** $\frac{x - 3}{x - 1}$ **13.** $\frac{3}{2}$ **15.** $\frac{3x}{2}$ **17.** $\frac{4}{11}$ **19.** $\frac{2 + x}{2 - x}$ **21.** $\frac{t^2}{1 + t + t^2}$ **23.** $\frac{v + 1}{v - 1}$ **25.** $\frac{x - 6}{x(x + 3)}$ **27.** $\frac{5r + 44}{(r + 8)^2}$ **29.** $\frac{t^2 - 7t + 9}{11t - 80}$ **31.** $\frac{x + 4}{2x^2 + x - 21}$ **33.** $\frac{a - 27}{a - 15}$ **35.** $\frac{6r}{r^2 + 9}$ **37.** $\frac{2(x - 6)}{3x - 5}$ **39.** $\frac{x - 3y}{x + 2y}$ **41.** $\frac{2a^2 + a + 6}{9a^3 + 54a^2 - 3a}$ **43.** $\frac{\frac{3}{x}}{\frac{3}{x} + 1} = \frac{3}{3 + x}$ **45.** $\frac{1 - \frac{9}{x^2}}{1 + \frac{3}{x}} = \frac{x - 3}{x}$ **47.** $\frac{\frac{3}{x^2} + \frac{5}{y}}{\frac{1}{x} + \frac{1}{y}} = \frac{3y + 5x^2}{xy + x^2}$ **49.** $\frac{\frac{1}{x^2y^2}}{\frac{4}{x} + \frac{3}{y}} = \frac{1}{4xy^2 + 3x^2y}$ **51.** $\frac{\frac{36}{a^2} - \frac{25}{b^2}}{\frac{6}{a} + \frac{5}{b}} = \frac{6b - 5a}{ab}$ **53.** $\frac{\frac{1}{4a} + \frac{2}{b^2}}{\frac{2}{a} + \frac{1}{b^2}} = \frac{b^2 + 8a}{8b^2 + 4a}$ **55.** Mistake: The +1 was omitted in the multiplication of $\frac{1}{3} \cdot 3$. Correct: $\frac{3a + 1}{3b + 1}$ **57.** Mistake: Did not multiply n by n. Correct: $\frac{n^2 + 3}{3}$ **59.** ≈ 68.6 mph **61.** $\frac{3x + 1}{9}$ in. **63. a.** $\frac{R_1R_2}{R_2 + R_1}$ **b.** 24 Ω **c.** 20 Ω

Review Exercises: **1.** −1 **2.** −4 **3.** $0, -\frac{7}{3}$ **4.** $5, -\frac{3}{2}$ **5.** $\frac{52}{15}$ or $3.4\overline{6}$ mph **6.** 0.2 hr. or 12 min.

Exercise Set 7.4

Prep Exercises **1.** Adding rational expressions yields another rational expression. Solving an equation involving a rational expression yields a numerical value for the variable. **2.** An extraneous solution is an apparent solution that does not satisfy the original equation. It may be caused by multiplying both sides of the equation by a variable expression. **3.** Possible extraneous solutions are 1, −2, and −1. These are the values of x that cause expressions in the equation to be undefined. **4.** Yes. If both sides of the equation are multiplied by $y + 3$, the apparent solution is $y = -3$, but $y + 3 = 0$ if $y = -3$. **5.** Multiplying both sides of the equation by the LCD eliminates the fractions. **6.** Multiply by the LCD, $(x + 1)(x - 1)(x + 2)$. **7.** proportion **8.** Cross multiply; that is use $ad = bc$.

Exercises **1.** 4 **3.** 2, −6 **5.** 2, −2, −3 **7.** 1 **9.** −2 **11.** 4 **13.** 0 **15.** 6 **17.** No solution $\left(\frac{3}{2}\text{ is extraneous}\right)$. **19.** −6, 1 **21.** 0 **23.** −3 **25.** 2 (−2 is extraneous). **27.** $-\frac{3}{2}, 4$ **29.** $2, -\frac{5}{4}$ **31.** 2, −1 **33.** No solution (−5 is extraneous). **35.** 3 (2 is extraneous). **37.** 1 **39.** No solution(3 is extraneous). **41.** 3 (−3 is extraneous). **43.** −16 **45.** $0, \frac{2}{3}$ **47.** 2 **49.** −1 **51.** 1 (2 is extraneous). **53.** $-\frac{2}{3}, 1$ **55.** $\frac{1}{2}, -1$ **57.** 2 (0 is extraneous). **59.** $-\frac{3}{2}, 2$ **61.** $p = \frac{100C}{90{,}000 + C}$ **63.** $r = \frac{2E - IR}{2I}$ **65.** $f = \frac{sS}{S + s}$ **67.** $R_1 = \frac{RR_2}{R_2 - R}$ **69.** 4 is extraneous. There is no solution. **71.** 20% **73.** 30 m/sec. **75.** 40 mph **77.** 5 Ω

Review Exercises **1.** 8 dimes, 10 nickels **2.** length 10 in., width 8 in. **3.** 5 and 8 **4.** 16 and 12 **5.** The saw cost \$146, and the drill cost \$92. **6.** 11 hr.

Exercise Set 7.5

Prep Exercises **1.** $\frac{1}{x}$ **2.** $\frac{1}{a}+\frac{1}{b}$ **3.** $\frac{100}{r}$, inverse **4.** Upstream: $x-3$ mph; downstream: $x+3$ mph **5.** $m=kn$ **6.** Find the value of k. **7.** p decreases. **8.** $t=\frac{k}{u}$ **9.** m decreases. **10.** $c=kde$

Exercises **1.** $2\frac{2}{5}$ hr. **3.** $14\frac{7}{12}$ days **5.** 12 hr. **7.** 22.5 min. **9.** 3 P.M. **11.** 75 mph **13.** 6 hr. **15.** 441 mph **17.** 12 **19.** 36 **21.** 12 **23.** \$39 **25.** 180 cm^3 **27.** 5 sec. **29.** 40.5 m **31.** 7 ft. **33.** 2 **35.** 8.25 **37.** 50 psi **39.** 6 Ω **41.** 600 m **43.** 100 mm **45.** 64 **47.** 72 **49.** 336 in.^{3} **51.** 169.65 cm^3 **53.** 6 **55.** 162 **57.** 3.125 Ω **59.** 12 dyn

61. a.

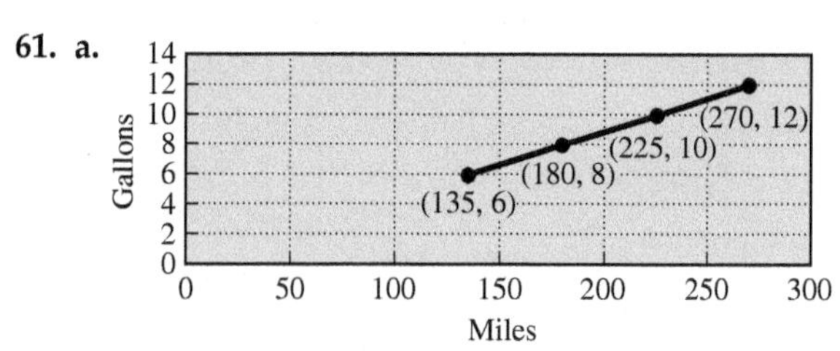

b. They are directly proportional. As the number of miles increases, so does the number of gallons. **c.** $k=22.5$, which represents the miles per gallon the car gets. **d.** Yes, the data represent a function because for any given number of miles, there will be one quantity of gasoline (assuming that the miles per gallon stays constant).

62. a.

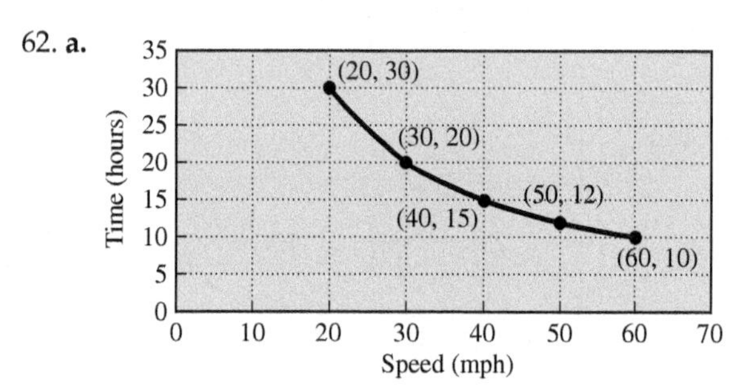

b. They are inversely proportional because as speed increases, time decreases. **c.** $k=600$, which represents the total miles driven. **d.** Yes, the data represent a function because for any given speed, there is only one time required to drive the constant 600 miles.

Review Exercises: **1.** 8 **2.** -144 **3.** $-30a^4b^6$ **4.** $\frac{4y^2}{x^4}$ **5.** $-5x-2$ **6.** $-\frac{2}{3}, 5$

Chapter 7 Summary and Review Exercises

7.1 Definitions/Rules/Procedures polynomials; Factor; common factors; Multiply Exercises **1.** $4xy^2$ **2.** $-\frac{14p^3}{9q^2}$ **3.** $-\frac{1}{3m^2n^3}$ **4.** $\frac{7}{5}$ **5.** $\frac{4}{7}$ **6.** $\frac{a-3}{a+2}$ **7.** $\frac{1}{3x+4}$ **8.** $\frac{5x-3}{2x-1}$ **9.** $\frac{1}{4c+7d}$ **10.** $-\frac{1}{2x+5}$ **11.** $\frac{4c-1}{3c-4}$ **12.** $\frac{4x^2-6x+9}{x-5}$ Definitions/Rules/Procedures Factor; factors common; numerator, numerator, denominator, denominator Exercises **13.** $\frac{10mn^2p^2}{3q^3}$ **14.** $\frac{3n^2}{2m^2}$ **15.** $\frac{7}{8}$ **16.** $\frac{3(5x+4)}{4}$ **17.** $-\frac{5}{6}$ **18.** $\frac{x-4}{x+5}$ **19.** $\frac{y+2}{y-3}$ **20.** $\frac{4x-5}{3x+2}$ **21.** $\frac{b+2d}{b-4d}$ **22.** $3x+4y$ Definitions/Rules/Procedures $\frac{P}{Q}\cdot\frac{S}{R}$ Exercises **23.** $\frac{14y^2b^3}{9x^2a}$ **24.** $\frac{40}{27}$ **25.** $\frac{3}{4}$ **26.** $\frac{1}{3z(z-4)}$ **27.** $\frac{(2p+3)(6p-7)}{20}$ **28.** $\frac{5p+3q}{2p+5q}$ **29.** $\frac{x^2+2xy+4y^2}{x+6y}$ **30.** $\frac{z+2w}{2z-w}$

Definitions/Rules/Procedures rational expressions; variable, value; denominator, 0; Exclude Exercises **31. a.** 3 **b.** 0 **c.** $-\frac{1}{3}$ **32. a.** -6 **b.** 0 **c.** undefined **33.** $\left\{x \mid x \neq \frac{5}{3}\right\}$ **34.** $\{x \mid x \neq -6, 2\}$ Definitions/Rules/Procedures vertical asymptotes

Exercises **35.** **36.**

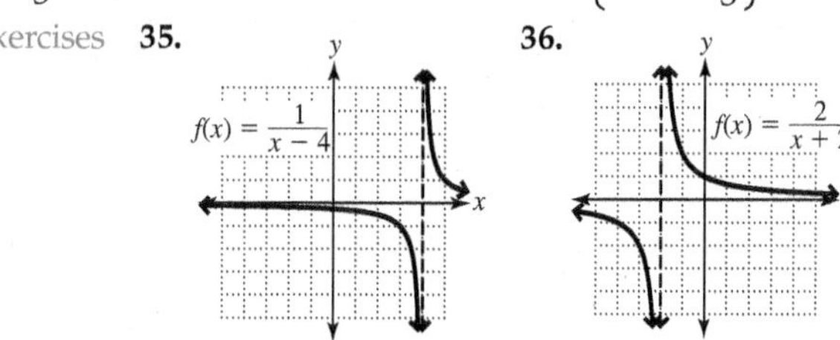

7.2 Definitions/Rules/Procedures numerators, denominator; terms Exercises **37.** $\frac{4r}{7x}$ **38.** $\frac{2a+6b}{2a-3b}$ **39.** $\frac{3p^2+3p-6}{p^2-16}$ **40.** $\frac{x-4}{x-6}$

Definitions/Rules/Procedures prime; factorization, greatest, greatest Exercises **41.** $36p^3q^5; \frac{32bq}{36p^3q^5}, \frac{33cp}{36p^3q^5}$
42. $(t-4)(t+2); \frac{7t+14}{(t-4)(t+2)}, \frac{9t-36}{(t-4)(t+2)}$ **43.** $28(2u+3); \frac{35u}{28(2u+3)}, \frac{36u}{28(2u+3)}$ **44.** $(w+4)(w-4)(w+1);$
$\frac{2w^2+2w}{(w+4)(w-4)(w+1)}, \frac{6w^2-24w}{(w+4)(w-4)(w+1)}$ **45.** $(a+3)^2(a-5); \frac{3a^2-15a}{(a+3)^2(a-5)}, \frac{10a^2+30a}{(a+3)^2(a-5)}$
46. $(m-4)(m+2)(m+1); \frac{m^2+2m+1}{(m-4)(m+2)(m+1)}, \frac{m^2+m-2}{(m-4)(m+2)(m+1)}$ **47.** $x^3(x+4)(x+2)(x-2);$
$\frac{2x^4-7x^3+6x^2}{x^3(x+4)(x+2)(x-2)}, \frac{6x^2+15x+6}{x^3(x+4)(x+2)(x-2)}$ **48.** $4x^3(x+6)(x+4); \frac{24x^2+240x+576}{4x^3(x+6)(x+4)}, \frac{16x^4-12x^3}{4x^3(x+6)(x+4)},$
$\frac{3x^3+18x^2}{4x^3(x+6)(x+4)}$ Definitions/Rules/Procedures LCD; LCD; numerators, LCD Exercises **49.** $\frac{4a}{45x}$ **50.** $\frac{y-23}{20y}$
51. $\frac{4tu^3-6u^3+15t^2+3t}{36t^4u^5}$ **52.** $\frac{11w-9}{w(w-3)}$ **53.** $\frac{-v^2+v-4}{4(v+2)(v-3)}$ **54.** $\frac{t^2+t+24}{(t+4)^2}$ **55.** $\frac{8w^2+39w+10}{(w+5)(w-5)(w+2)}$
56. $\frac{-3z^2+59z+123}{(4z+5)(z+7)^2}$ **57.** $\frac{-3a^2+3a+6}{a(a+3)}$ **58.** $\frac{7x-7y}{3x-y}$

7.3 Definitions/Rules/Procedures rational expression; division; LCD Exercises **59.** $\frac{8}{3}$ **60.** $\frac{8}{9}$ **61.** $\frac{u^3w}{v}$ **62.** $\frac{x^4y}{t^3}$ **63.** $\frac{7w}{24-w}$
64. $\frac{97-6b}{(b-15)^2}$ **65.** $\frac{y-13}{y+2}$ **66.** $x+5$ **67.** $\frac{3x+4y}{2x+y}$ **68.** $\frac{8x}{x^2+16}$

7.4 Definitions/Rules/Procedures multiplying, LCD; equation; original; *bc* Exercises **69.** $\frac{7}{4}$ **70.** 3 **71.** -4 (4 is extraneous)
72. 1, 4 **73.** 1, -2 **74.** $-\frac{122}{29}$ **75.** 2, $-\frac{5}{3}$ **76.** $\frac{1}{3}$

7.5 Definitions/Rules/Procedures Exercises **77.** 18 days **78.** 23 mph **79.** 9:06 A.M. Definitions/Rules/Procedures $y = kx$ Exercises **80.** 42 **81.** 200 Definitions/Rules/Procedures $y = \frac{k}{x}$ Exercises **82.** 9 gal. **83.** 6 **84.** 9 **85.** 9
86. $\frac{7}{3}$ cm/sec. Definitions/Rules/Procedures $y = kxz$ Exercises **87.** 32 **88.** 100.48 in.^{3}

Chapter 7 Practice Test

1. $\frac{21a^2}{8b^2}$ [7.1] **2.** $\frac{3x-2}{2x-3}$ [7.1] **3.** $\frac{m^2+4mn+16n^2}{3m-2n}$ [7.1] **4.** $\frac{n-2m}{b+5c}$ [7.1] **5.** $\frac{15b^2y^2}{4a^3x^3}$ [7.1] **6.** $\frac{y(4y-5)}{(2y-7)(2y-3)}$ [7.1] **7.** -3 [7.1]
8. $\frac{n^2+4}{n^2-4}$ [7.1] **9. a.** $-\frac{3}{2}$ **b.** 0 **c.** undefined [7.1] **10.** $\left\{x \mid x \neq -2, \frac{5}{2}\right\}$ [7.1] **11.** $(a+3)(a-3)(2a+7); \frac{6a^2+21a}{(a+3)(a-3)(2a+7)},$
$\frac{6a^2-18a}{(a+3)(a-3)(2a+7)}$ [7.2] **12.** $\frac{5y^2+18y+12}{36y^2}$ [7.2] **13.** $\frac{2r-3}{r-4}$ [7.2] **14.** $\frac{t^2+25t+10}{2(t+5)(t-5)^2}$ [7.2] **15.** $\frac{a+5b}{3a-5b}$ [7.2]
16. $\frac{-2x^2+7x+26}{x(x+5)}$ [7.2] **17.** $\frac{2ac^3}{3b^2}$ [7.3] **18.** $\frac{3x-5}{2x-1}$ [7.3] **19.** $\frac{t+2}{2}$ [7.3] **20.** $\frac{3b-2a}{ab}$ [7.3] **21.** $\frac{69}{26}$ [7.4] **22.** No solution. (4 is extraneous) [7.4] **23.** -3, 4 [7.4] **24.** 2 (-1 is extraneous) [7.4] **25.** 30 days [7.5] **26.** 2 hr. [7.5] **27.** 6 [7.5] **28.** 30 [7.5] **29.** 1.44 fc [7.5] **30.** 408 mi. [7.5]

Chapters 1 and 7 Cumulative Review Exercises

1. false **2.** false **3.** true **4.** true **5.** $\frac{1}{12}$ **6.** no **7.** -2. **8.** $\frac{17}{2}$ **9.** -5 **10.** 10 **11.** $7x^3-5x^2+5x-2$ **12.** $8a^3-27b^3$
13. $3-\frac{4x}{y}+\frac{3y}{4x}$ **14.** 1 **15.** $\frac{x^2+16x+51}{(x+4)(x-4)(x+3)}$ **16.** $\frac{x(x+1)}{3x-1}$ **17.** $3a^2b^3(4a+b)(3a-4b)$ **18.** $4(3c-2d)(9c^2+6cd+4d^2)$
19. -84 **20.** $\{x \mid x \text{ is a real number}\}$ or $(-\infty, \infty)$ **21.** $(2, -1, 1)$ **22.** $\frac{5}{2}, -\frac{2}{3}$ **23.** $\frac{9}{4}$ **24.**

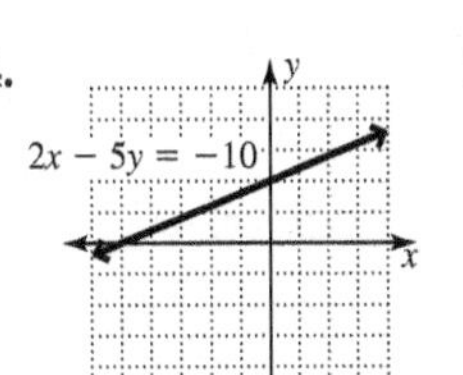

25.

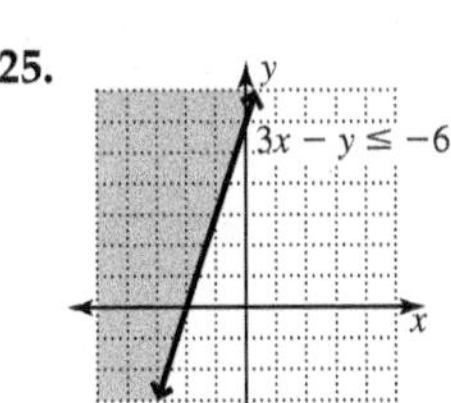

26. $y = 2x + 3$ **27.** 40°, 140° **28.** 10 lb. **29.** Length: 9 ft.; width: 6 ft. **30.** $\frac{70}{17}$ hr.

Chapter 8

Exercise Set 8.1

Prep Exercises **1.** Squaring a number or its additive inverse results in the same positive number. **2.** The nonnegative root of a number **3.** You cannot raise a number to an even power and get a negative value. **4.** $\sqrt{16}$ is rational because its exact value is a rational number, 4. $\sqrt{17}$ is irrational because its exact value cannot be expressed using a rational number. **5.** With the radical symbol **6.** $\sqrt{x^2}$ means the nonnegative square root of x^2 even if x is negative. **7.** radical; radicand **8.** Set the radicand ≥ 0. The solution set is the domain.

Exercises **1.** ± 6 **3.** ± 11 **5.** ± 14 **7.** ± 15 **9.** 5 **11.** Not a real number **13.** -5 **15.** ± 5 **17.** 1.2 **19.** Not a real number **21.** -0.11 **23.** $\frac{7}{9}$ **25.** $-\frac{12}{13}$ **27.** 3 **29.** -4 **31.** 6 **33.** 4 **35.** Not a real number **37.** -2 **39.** 2 **41.** -3 **43.** 2 **45.** 2 **47.** $-\frac{2}{3}$ **49.** $\frac{2}{3}$ **51.** 2.646 **53.** -3.317 **55.** 3.684 **57.** -3.756 **59.** 3.708 **61.** -3.036 **63.** 2.454 **65.** 2.295 **67.** b^2 **69.** $4x$ **71.** $10r^4s^3$ **73.** $0.5a^3b^6$ **75.** m **77.** $3a^3b^2$ **79.** $-4ab^4$ **81.** $0.2x^6$ **83.** a **85.** $2x^4$ **87.** $2x^2$ **89.** x^2y **91.** $6|m|$ **93.** $|r-1|$ **95.** $4|y^3|$ **97.** $3y$ **99.** $(y-3)^2$ **101.** $(y-4)^2$ **103.** 2 **105.** $\sqrt{15}$ **107.** $\{x|x \geq 4\}$, or $[4, \infty)$ **109.** $\{x|x \leq 4\}$, or $(-\infty, 4]$

111. a.

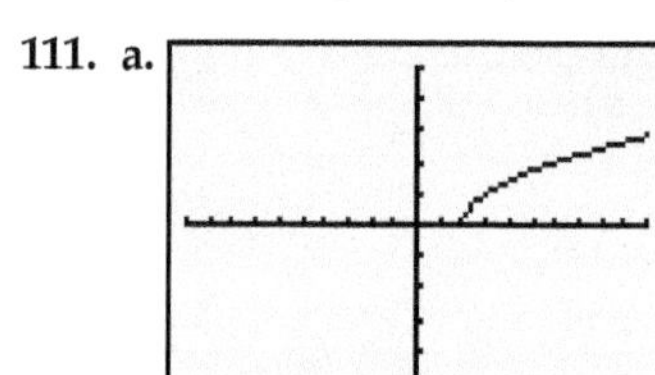

b. $\{x|x \geq 2\}$ or $[2, \infty)$ **113. a.**

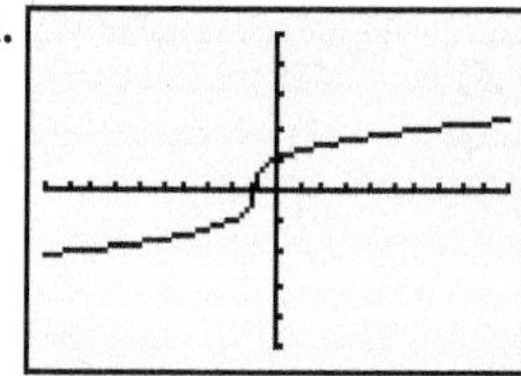

b. $\mathbb{R}$, or $(-\infty, \infty)$

115. ≈ -17.709 m/sec. **117.** ≈ 3.476 sec. **119.** ≈ 19.170 mph **121.** 13 ft. **123.** 15 N **125. a.** ≈ 36 **b.** ≈ 39

Review Exercises: **1. a.** 12 **b.** 12 **2.** x^8 **3.** $-45m^4n^3$ **4.** $12x^4 - 20x^3 + 4x^2$ **5.** $21y^2 + 23y - 20$ **6.** $2x^2 - 17x - 9$

Exercise Set 8.2

Prep Exercises **1.** 4 **2.** 5 **3.** a would be nonnegative because with an even index, the radicand must be nonnegative. **4.** No restrictions would be placed on a because with an odd index, the radicand can be positive or negative. **5.** The index **6.** $m+n$; $m-n$; mn **7.** Yes, because $100^{1/4} = (10^2)^{1/4} = 10^{1/2}$ **8.** Yes, because $3^{1/3} \cdot 2^{1/5} = 3^{5/15} \cdot 2^{3/15} = (3^5 \cdot 2^3)^{1/15} = 1944^{1/15}$

Exercises **1.** $\sqrt{25} = 5$ **3.** $-\sqrt{100} = -10$ **5.** $\sqrt[3]{27} = 3$ **7.** $\sqrt[3]{-64} = -4$ **9.** $\sqrt[4]{y}$ **11.** $\sqrt{144x^8} = 12x^4$ **13.** $18\sqrt{r}$ **15.** $\sqrt{\frac{x^4}{81}} = \frac{x^2}{9}$ **17.** $(\sqrt[3]{64})^2 = 16$ **19.** $-(\sqrt[4]{81})^3 = -27$ **21.** $(\sqrt[3]{-8})^4 = 16$ **23.** $\frac{1}{(\sqrt[4]{16})^3} = \frac{1}{8}$ **25.** $\sqrt[5]{x^4}$ **27.** $8\sqrt[3]{n^2}$ **29.** $\frac{1}{(\sqrt[5]{-32})^2} = \frac{1}{4}$ **31.** $\left(\sqrt{\frac{1}{25}}\right)^3 = \frac{1}{125}$ **33.** $\sqrt[6]{(2a+4)^5}$ **35.** $25^{1/4}$ **37.** $z^{5/6}$ **39.** $5^{-5/6}$ **41.** $5x^{-4/5}$ **43.** $5^{7/3}$ **45.** $x^{2/7}$ **47.** $(4a-5)^{7/4}$ **49.** $(2r-5)^{8/5}$ **51.** $x^{4/5}$ **53.** $x^{7/6}$ **55.** $a^{17/12}$ **57.** $21w^{4/7}$ **59.** $-12a^{17/12}$ **61.** $7^{5/3}$ **63.** $\frac{1}{x^{2/3}}$ **65.** $x^{1/4}$ **67.** $r^{1/12}$ **69.** $\frac{1}{x^{5/7}}$ **71.** $a^{9/4}$ **73.** $20s^{3/7}$ **75.** $-24b^{1/4}$ **77.** x^2 **79.** $a^{5/3}$ **81.** $b^{2/5}$ **83.** $64x^4y^3$ **85.** $2q^{1/2}t^{1/4}$ **87.** $81a^3$ **89.** $3z^{1/3}$ **91.** $\sqrt{2}$ **93.** $\sqrt[3]{7}$ **95.** $\sqrt{x}$ **97.** $\sqrt[4]{r^3}$ **99.** $\sqrt[4]{x^3y}$ **101.** $\sqrt[5]{m^2n^3}$ **103.** $\sqrt[6]{x^5}$ **105.** $\sqrt[6]{y^7}$ **107.** $\sqrt[6]{x^5}$ **109.** $\sqrt[15]{n^2}$ **111.** $\sqrt[6]{1125}$ **113.** $\sqrt[12]{3456}$ **115.** $\sqrt[9]{x}$ **117.** $\sqrt[6]{n}$

Review Exercises: **1.** $2^4x^3y^2$ **2.** 12 **3.** 15 **4.** 8×10^{11} **5.** $-\frac{5}{8}x^4y^2z^2$ **6.** $2x^2 - 8x + 5 + \frac{3}{x+3}$

Exercise Set 8.3

Prep Exercises **1.** Multiplying the approximate roots gives $\sqrt{8} \cdot \sqrt{18} \approx 2.828 \cdot 4.243 = 11.999204$, which is tedious and inexact. Using the product rule for radicals gives $\sqrt{8} \cdot \sqrt{18} = \sqrt{8 \cdot 18} = \sqrt{144} = 12$, which is fast and exact. **2.** The quotient rule transforms the expression to the square root of a perfect square: $\sqrt{\frac{125}{5}} = \sqrt{25} = 5$. **3.** The radicand, 28, has the perfect square 4 as a factor. So $\sqrt{28} = 2\sqrt{7}$. **4.** Rewrite the expression as a product of two radicals, the first containing the perfect cube and the second containing no perfect cubes. Then simplify the first radical. For example, $\sqrt[3]{54x^8}$ can be rewritten as $\sqrt[3]{27x^6} \cdot \sqrt[3]{2x^2}$, then simplified to $3x^2\sqrt[3]{2x^2}$. **5.** The radicand, x^5, has the perfect cube x^3 as a factor.

Exercises **1.** 8 **3.** $9x^3$ **5.** $12xy^2$ **7.** $\sqrt{65}$ **9.** $\sqrt{15x}$ **11.** 3 **13.** $\sqrt[3]{10y^2}$ **15.** $\sqrt[4]{21}$ **17.** $w\sqrt[4]{72}$ **19.** $\sqrt[4]{15x^3y^3}$ **21.** $x\sqrt[5]{30x^2}$ **23.** $\sqrt[6]{8x^5y^4}$ **25.** $\sqrt{\frac{21}{10}}$ **27.** $\sqrt{\frac{6y}{5x}}$ **29.** $\frac{5}{6}$ **31.** $\frac{\sqrt{10}}{3}$ **33.** 6 **35.** $\sqrt{3}$ **37.** $\frac{\sqrt[3]{4}}{w^2}$ **39.** $\frac{\sqrt[3]{5y^2}}{3x^3}$ **41.** 4 **43.** $\frac{\sqrt[4]{3u^3}}{2x^2}$ **45.** $7\sqrt{2}$ **47.** $8\sqrt{2}$ **49.** $24\sqrt{5}$ **51.** $20\sqrt{7}$ **53.** $a^3\sqrt{a}$ **55.** xy^2 **57.** $x^3y^4z^5$ **59.** $r^5s^4\sqrt{rs}$ **61.** $18x^2\sqrt{2x}$ **63.** $2\sqrt[3]{4}$ **65.** $x^2\sqrt[3]{x}$ **67.** $x^2y\sqrt[3]{y^2}$ **69.** $4z^2\sqrt[3]{2z^2}$ **71.** $4\sqrt[3]{5}$ **73.** $2\sqrt[4]{5}$ **75.** $9x^4\sqrt[4]{3x}$ **77.** $3x^3\sqrt[5]{2x}$ **79.** $xy^2z\sqrt[6]{x^2y^2z^5}$ **81.** $3\sqrt{7}$ **83.** $30\sqrt{15}$ **85.** $y^2\sqrt{y}$ **87.** $x^4y^5\sqrt{y}$

89. $24c^4\sqrt{15}$ **91.** $60\sqrt{2}$ **93.** $2\sqrt{2}$ **95.** $6\sqrt{5}$ **97.** $c^2d\sqrt{d}$ **99.** $12a^2$ **101.** $6c\sqrt{3c}$ **103.** $36x^2y^3\sqrt{2y}$ **105.** $\frac{2\sqrt{6}}{7}$ **107.** $\frac{a^4}{2}$ **109.** $\frac{3x^5\sqrt{5}}{4}$ **111.** $\frac{3}{8}\sqrt{5}$

Review Exercises: **1.** no **2.** yes **3.** $8x^2 - 7x$ **4.** $8a^2 - 6ab - 9b^2$ **5.** $9m^2 - 25n^2$ **6.** $4x^2 - 12xy + 9y^2$

Exercise Set 8.4

Prep Exercises **1.** Like radicals have the same index and the same radicand, but their coefficients may be different. **2.** Add the coefficients, 3 + 2, and keep the radical, $\sqrt{2}$, the same. **3.** Multiply each term in the second expression by each term in the first (that is, use FOIL) **4.** $\sqrt{a} + \sqrt{b} \neq \sqrt{a+b}$ because the order of operations is different. For $\sqrt{a} + \sqrt{b}$, the radicals must be evaluated first and then the addition. For $\sqrt{a+b}$, the addition must be evaluated first and then the radical. Example: $\sqrt{36} + \sqrt{64} \neq \sqrt{36 + 64}$ because $6 + 8 \neq 10$.

Exercises **1.** $-6\sqrt{6}$ **3.** $9\sqrt{a}$ **5.** $12\sqrt{5} - 8\sqrt{6}$ **7.** $11a\sqrt{5a} - 2b\sqrt{7b}$ **9.** $3x\sqrt[3]{9}$ **11.** $-6x^2\sqrt[4]{5x}$ **13.** Cannot combine because the radicals are not like **15.** $-\sqrt{3}$ **17.** $-\sqrt{5y}$ **19.** $-8\sqrt{5}$ **21.** $6\sqrt{6}$ **23.** $14a\sqrt{3a}$ **25.** $4\sqrt{6}$ **27.** $18\sqrt{2} - 17\sqrt{3}$ **29.** $7\sqrt[3]{2}$ **31.** $-12x\sqrt[3]{5x^2}$ **33.** $-2x^2\sqrt[4]{2x}$ **35.** $3\sqrt{2} + 2$ **37.** $3 - 3\sqrt{5}$ **39.** $\sqrt{30} + 10\sqrt{2}$ **41.** $24x - 48x\sqrt{2}$ **43.** $12 - 3\sqrt{2} + 4\sqrt{5} - \sqrt{10}$ **45.** $6 + 5\sqrt{x} + x$ **47.** $6 + 10\sqrt{2} + 9\sqrt{3} + 15\sqrt{6}$ **49.** $\sqrt{15} + \sqrt{21} + 5 + \sqrt{35}$ **51.** $x + \sqrt{xy} - 6y$ **53.** $12\sqrt{14} - 12\sqrt{6} + 6\sqrt{35} - 6\sqrt{15}$ **55.** $8a + 10\sqrt{ab} - 3b$ **57.** $2\sqrt[3]{2} - 3\sqrt[3]{4} - 40$ **59.** $1 - \sqrt[3]{18} + \sqrt[3]{12}$ **61.** $x + 8$ **63.** $22 + 8\sqrt{6}$ **65.** $18 - 8\sqrt{2}$ **67.** $16 + 8\sqrt{3}$ **69.** $30 + 12\sqrt{6}$ **71.** 13 **73.** −14 **75.** $36 - x$ **77.** 1 **79.** $x - y$ **81.** 4 **83.** 50 **85.** $7\sqrt{5}$ **87.** $-15\sqrt{6}$ **89.** $9\sqrt{2}$ **91.** $-14\sqrt{5}$ **93.** $25\sqrt{3}$ **95. a.** $42 + 2\sqrt{5}$ ft. **b.** 46.5 ft. **c.** $87.89

Review Exercises: **1.** $2x + 5$ **2.** $16x^2 - 9$ **3.** $\sqrt{2}$ **4.** $\sqrt[3]{4}$ **5.** Slope is $\frac{2}{5}$; y-intercept is $(0, 2)$. **6.**

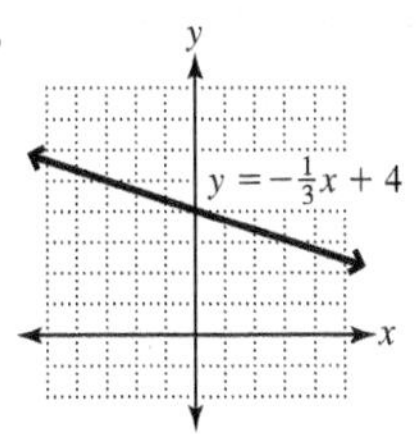

Exercise Set 8.5

Prep Exercises **1. a.** $\sqrt{16}$ is rational. **b.** A radical, $\sqrt{3}$, is in the denominator. **2.** The denominator in $\frac{\sqrt{3}}{3}$ is a rational number, whereas it is irrational in $\frac{1}{\sqrt{3}}$. **3.** Multiply the fraction by a 1 so that the product's denominator has a radicand that is a perfect square. **4.** Multiply the fraction by a 1 whose numerator and denominator are the conjugate of the denominator. **5.** Multiply the numerator and denominator by $2 - \sqrt{7x}$.

Exercises **1.** $\frac{\sqrt{3}}{3}$ **3.** $\frac{3\sqrt{2}}{4}$ **5.** $\frac{6\sqrt{7}}{7}$ **7.** $\frac{\sqrt{15}}{6}$ **9.** $\frac{x\sqrt{14}}{10}$ **11.** $\frac{\sqrt{7}}{7}$ **13.** $\frac{5\sqrt{3a}}{3a}$ **15.** $\frac{\sqrt{33mn}}{11n}$ **17.** $\frac{2\sqrt{5x}}{x}$ **19.** $\frac{\sqrt{3}}{4}$ **21.** $\frac{3\sqrt{x}}{x^2}$ **23.** $4x\sqrt{2x}$ **25.** Mistake: The product of $\sqrt{2}$ and 2 is not 2. Correct: $\frac{\sqrt{6}}{2}$ **27.** $\frac{5\sqrt[3]{9}}{3}$ **29.** $\frac{\sqrt[3]{20}}{2}$ **31.** $3\sqrt[3]{2}$ **33.** $\frac{m\sqrt[3]{n^2}}{n}$ **35.** $\frac{\sqrt[3]{ab}}{b}$ **37.** $\frac{2\sqrt[3]{4x^2}}{x}$ **39.** $\frac{\sqrt[3]{30a}}{5a}$ **41.** $\frac{5\sqrt[4]{4}}{2}$ or $\frac{5\sqrt{2}}{2}$ **43.** $\frac{\sqrt[4]{3x^2}}{x}$ **45.** $\frac{3\sqrt[4]{27x}}{x}$ **47.** $3\sqrt{2} - 3$ **49.** $8 + 4\sqrt{3}$ **51.** $5\sqrt{3} - 5\sqrt{2}$ **53.** $-1 - \sqrt{5}$ **55.** $\frac{3 + \sqrt{3}}{2}$ **57.** $\frac{-6 - 8\sqrt{3}}{13}$ **59.** $\frac{4\sqrt{21} - 4\sqrt{6}}{5}$ **61.** $\frac{32 + 8\sqrt{3}}{13}$ **63.** $\frac{6y - 6\sqrt{y}}{y - 1}$ **65.** $\frac{3t - 6\sqrt{tu}}{t - 4u}$ **67.** $\frac{\sqrt{2xy} + 2y\sqrt{3}}{x - 6y}$ **69.** $\frac{3}{2\sqrt{3}}$ **71.** $\frac{2x}{5\sqrt{2x}}$ **73.** $\frac{2n}{3\sqrt{2n}}$ **75.** $\frac{1}{10 - 5\sqrt{3}}$ **77.** $\frac{5x - 36}{9\sqrt{5x} + 54}$ **79.** $\frac{19}{10\sqrt{n} - 2\sqrt{6n}}$ **81. a.** $\frac{5\sqrt{3}}{3}$ **b.** $\sqrt{5}$ **c.** $\frac{5\sqrt{11}}{11}$ **83. a.** The graphs are identical. The functions are identical. **b.** $f(x) = g(x)$ **85. a.** $T = \pi\sqrt{9.8L}/4.9$ **b.** $T = 2\pi L/\sqrt{9.8L}$ **87. a.** $s = \sqrt{3Vh}/h$ **b.** 745 ft. **89. a.** $V_{rms} = \frac{\sqrt{2}}{2}V_m$ or $\frac{\sqrt{2}V_m}{2}$ **b.** $V_{rms} = \frac{163\sqrt{2}}{2}$ **c.** ≈ 115.3 **91.** $\frac{15\sqrt{2} - 10\sqrt{3}}{3}\Omega$

Review Exercises: **1.** $\pm 2\sqrt{7}$ **2.** $(x - 3)^2$ **3.** 4 **4.** −1 **5.** −6, 6 **6.** 2, 3

Exercise Set 8.6

Prep Exercises **1.** Some of the solutions may be extraneous. **2.** A solution that does not satisfy the original equation **3.** The principal square root of a number cannot equal a negative. **4.** $(\sqrt{a})^2 = \sqrt{a} \cdot \sqrt{a} = \sqrt{a^2} = a$ **5.** Subtract $3x$ from both sides to isolate the radical. This allows use of the power rule to eliminate the radical. **6.** Answers may vary. One example is $\sqrt{x} + 4 = \sqrt{2x + 1}$. The radicals must be totally eliminated.

Exercises **1.** 4 **3.** No real-number solution **5.** 27 **7.** −8 **9.** 17 **11.** 11 **13.** 6 **15.** −4 **17.** No real-number solution **19.** 11 **21.** −2 **23.** 124 **25.** 55 **27.** 5 **29.** −11 **31.** No real-number solution **33.** 2 **35.** No real-number solution (−5 is an extraneous solution) **37.** 3 **39.** 3 **41.** −4 (−10 is an extraneous solution) **43.** 1, 3 **45.** −3, −2 **47.** $\frac{1}{2}\left(-\frac{3}{8}\text{ is an extraneous solution}\right)$

49. 5 **51.** −1, 3 **53.** 0, 4 **55.** $\frac{3}{16}$ **57.** 3 (−1 is an extraneous solution) **59.** Mistake: You cannot take the principal square root of a number and get a negative. Correct: No real-number solution **61.** Mistake: The binomial $x - 3$ was not squared correctly. Correct: $x = 6$ with $x = 1$ an extraneous solution **63.** 9.8 m **65.** 0.6125 m **67.** 1.44 ft. **69.** 1 ft. **71.** 36.73 ft. **73.** 82.65 ft. **75.** 4 N **77.** 6 N **79. a.** 25; 3; 4 **b.**

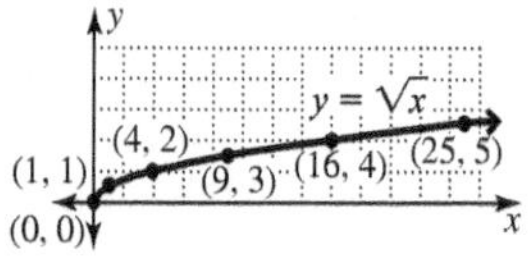

c. No, the x-values must be 0 or positive because real square roots exist only when $x \ge 0$. The y-values must be 0 or positive because, by definition, the principal square root is 0 or positive. **d.** Yes, because it passes the vertical line test.

81. The graph becomes steeper from left to right. **83.** The graph rises or lowers according to the value of the constant.

Review Exercises: **1.** 81 **2.** −0.008 **3.** 39.0625 or $\frac{625}{16}$ **4.** x^8 **5.** n^{24} **6.** y^4

Exercise Set 8.7

Prep Exercises **1.** $\sqrt{-1}$ **2.** Yes, every real number can be expressed as $a + 0i$. **3.** No, every real number is a complex number that is not imaginary. For example, 2 is complex but not imaginary. **4.** Separate into two radicals, $\sqrt{-1} \cdot \sqrt{n}$. Replace $\sqrt{-1}$ with i and simplify $\sqrt{n}$. **5.** We add complex numbers just like we add polynomials—by combining like terms. **6.** We subtract complex numbers just like we subtract polynomials—by writing an equivalent addition and changing the signs in the second complex number. **7.** −1 **8.** No, because it is not in the form $a + bi$. **9.** Multiply the numerator and denominator by $3 + 2i$. **10.** $i, -1, -i, 1$

Exercises **1.** $6i$ **3.** $i\sqrt{5}$ **5.** $2i\sqrt{2}$ **7.** $3i\sqrt{2}$ **9.** $3i\sqrt{3}$ **11.** $5i\sqrt{5}$ **13.** $3i\sqrt{7}$ **15.** $7i\sqrt{5}$ **17.** $6 + 7i$ **19.** $11 - 6i$ **21.** $-7 + i$ **23.** $8 - 2i$ **25.** $-3 - 10i$ **27.** 0 **29.** $14 + 9i$ **31.** $-4 + 28i$ **33.** −24 **35.** 40 **37.** $14 + 12i$ **39.** $-72 - 32i$ **41.** $19 - 3i$ **43.** $50 + 9i$ **45.** 65 **47.** $-7 - 24i$ **49.** $-2i$ **51.** $-\frac{4i}{5}$ **53.** $-3i$ **55.** $\frac{1}{2} - i$ **57.** $\frac{1}{2} - i$ **59.** $\frac{14}{5} - \frac{7}{5}i$ **61.** $-\frac{7}{29} + \frac{3}{29}i$ **63.** $7 - 2i$ **65.** $\frac{9}{13} - \frac{7}{13}i$ **67.** $\frac{34}{41} + \frac{19}{41}i$ **69.** $-i$ **71.** −1 **73.** −1 **75.** 1 **77.** 1 **79.** −1 **81.** $-i$ **83.** i

Review Exercises: **1.** $(2x - 3)^2$ **2.** $\pm 4\sqrt{3}$ **3.** 2 **4.** $-\frac{3}{2}, 2$ **5.** (0, 2), (3, 0) **6.**

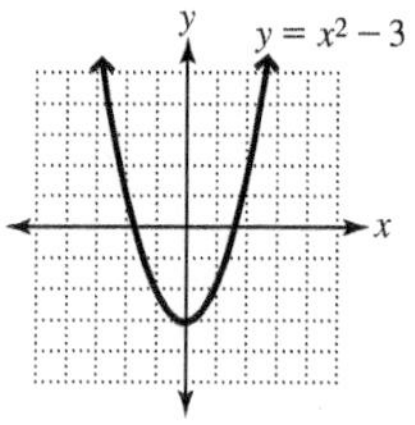

Chapter 8 Summary and Review Exercises

8.1 Definitions/Rules/Procedures $b^n = a$; $b^n = a$; no real-number; negative Exercises **1.** ±11 **2.** ±7 **3.** 13 **4.** −7 **5.** Not a real number **6.** $\frac{1}{5}$ **7.** 5 ft. **8.** π sec. **9.** 2.646 **10.** 9.487 **11. a.** $8\sqrt{5}$ mph **b.** 17.9 mph **12. a.** $\frac{\sqrt{10}}{2}$ sec. **b.** 1.58 sec. **13.** $7x^4$ **14.** $12a^3b^6$ **15.** $0.4mn^5$ **16.** x^5 **17.** $-4r^3s$ **18.** $3x^3$ **19.** $2x^3y^4$ **20.** x^2y **21.** $9|x|$ **22.** $(x - 1)^2$ Definitions/Rules/Procedures radical; radicand; simplify; x is a real number; $-\infty$; ∞; nonnegative; ≥0 Exercises **23.** 4 **24.** −2 **25.** $x \ge -5$ **26.** $x \le 3$

8.2 Definitions/Rules/Procedures $\sqrt[n]{a}$; $\sqrt[n]{a^m}$; $(\sqrt[n]{a})^m$; $\frac{1}{a^{m/n}}$ Exercises **27.** −4 **28.** $2a^2\sqrt{6}$ **29.** $\frac{1}{8}$ **30.** $\sqrt[7]{(5r - 2)^5}$ **31.** $\frac{1}{11}$ **32.** $\frac{1}{27}$ **33.** $\frac{64}{343}$ **34.** $81\sqrt[4]{x^3}$ **35.** $33^{1/8}$ **36.** $8n^{-3/7}$ **37.** $8^{3/5}$ **38.** $m^{8/3}$ **39.** $(3xw)^{3/4}$ **40.** $(a + b)^{4/3}$ Definitions/Rules/Procedures exponential; exponents; radical Exercises **41.** x^2 **42.** $32m^{3/2}$ **43.** $\frac{1}{y^{1/5}}$ **44.** b **45.** $k^{1/2}$ **46.** $2^{3/4}x^{3/4}y^{3/20}$ **47.** $\sqrt[6]{x}$

8.3 Definitions/Rules/Procedures $\sqrt[n]{a \cdot b}$; a Exercises **48.** 9 **49.** $10x^4$ **50.** 2 **51.** $\sqrt[4]{42}$ **52.** $\sqrt[5]{15x^4y^4}$ **53.** $84\sqrt{10}$ **54.** $x^7\sqrt{x}$ **55.** $16c^2\sqrt{15c}$ **56.** $a^5b^4\sqrt{b}$ **57.** $6ab^2$ Definitions/Rules/Procedures $\frac{\sqrt[n]{a}}{\sqrt[n]{b}}$ Exercises **58.** $\frac{7}{11}$ **59.** $-\frac{3}{2}$ **60.** $6\sqrt{5}$ **61.** $36x^2y^3\sqrt{2y}$

Definitions/Rules/Procedures nth; nth; perfect nth power; perfect nth power Exercises **62.** $12b^4\sqrt{3b}$ **63.** $15\sqrt[3]{4}$ **64.** $4x^3\sqrt[3]{5x}$ **65.** $6x^5\sqrt[4]{2x^3}$

8.4 Definitions/Rules/Procedures radicands; root indices; radical parts Exercises **66.** $-3\sqrt{n}$ **67.** $-6y^3\sqrt[4]{8y}$ **68.** $5\sqrt{5}$ **69.** $-10\sqrt{6}$ **70.** $4\sqrt{6}$ **71.** $28x\sqrt{2y}$ **72.** $9xy\sqrt[3]{2xy^2}$ **73.** $5\sqrt[4]{3}$ Definitions/Rules/Procedures order of operations Exercises **74.** $\sqrt{15} + \sqrt{10}$ **75.** $\sqrt[3]{21} + 2\sqrt[3]{49}$ **76.** $6\sqrt{6} - 54$ **77.** $\sqrt{10} + \sqrt{14} - \sqrt{15} - \sqrt{21}$ **78.** $-6 + 2\sqrt[3]{2} - 4\sqrt[3]{4}$ **79.** $\sqrt[4]{12x^2} - \sqrt[4]{6x} + 2\sqrt[4]{2x} - 2$ **80.** $5a - 3b$ **81.** 7 **82.** $7 - 2\sqrt{10}$

8.5 Definitions/Rules/Procedures index; fractions; sums; differences; products; quotients; perfect nth power Exercises **83.** $\frac{\sqrt{2}}{2}$ **84.** $\sqrt[3]{9}$ **85.** $\frac{2\sqrt{7}}{7}$ **86.** $\frac{\sqrt[4]{40x^2}}{2}$ **87.** $\frac{\sqrt[3]{153y}}{3y}$ **88.** $\frac{\sqrt[4]{27}}{3}$ Definitions/Rules/Procedures conjugate Exercises **89.** $-4\sqrt{2} - 4\sqrt{3}$

90. $\frac{4-\sqrt{3}}{13}$ **91.** $\frac{2+\sqrt{n}}{4-n}$ **92.** $\sqrt{6}+2$ Definitions/Rules/Procedures denominator Exercises **93.** $\frac{5}{3\sqrt{10}}$ **94.** $\frac{3x}{5\sqrt{3x}}$ **95.** $\frac{1}{16+8\sqrt{3}}$ **96.** $\frac{1}{10\sqrt{t}-5\sqrt{3t}}$

8.6 Definitions/Rules/Procedures radical expression; variable; integer power; radical; radical term; radical terms; root index; radical term; root index; extraneous Exercises **97.** 81 **98.** No real-number solution **99.** 10 **100.** −2 **101.** 18 **102.** 5 **103.** No real-number solution **104.** $\frac{719}{3}$ **105.** 7 **106.** 1 $\left(\frac{17}{4}\text{ is extraneous.}\right)$ **107.** 0, 4 **108.** $\frac{3}{16}$ **109.** 312.5 ft. **110.** 0.272 sec. **111.** 1.44 ft.

8.7 Definitions/Rules/Procedures $\sqrt{-1}$; −1; $a+bi$; real numbers; imaginary unit; $b \neq 0$; $a+bi$; real numbers; imaginary unit; $a-bi$; $\sqrt{-1}\cdot\sqrt{n}$; i; like terms; signs Exercises **112.** $3i$ **113.** $2i\sqrt{5}$ **114.** $8-6i$ **115.** $9-7i$ Definitions/Rules/Procedures monomials; binomials Exercises **116.** −12 **117.** $2+8i$ **118.** $26+2i$ **119.** $24-10i$ Definitions/Rules/Procedures $\frac{i}{i}$; complex conjugate Exercises **120.** $-5i$ **121.** $3i$ **122.** $-\frac{4i}{3}$ **123.** $\frac{1}{5}-\frac{7}{5}i$ **124.** $\frac{6}{5}-\frac{3}{5}i$ **125.** $\frac{7}{13}+\frac{17}{13}i$ Definitions/Rules/Procedures i; −1; $-i$; 1 Exercises **126.** 1 **127.** $-i$

Chapter 8 Practice Test

1. 6 [8.1] **2.** $7i$ [8.1] **3.** $9xy^2\sqrt{y}$ [8.1] **4.** $3\sqrt[3]{2}$ [8.1] **5.** $2x\sqrt[4]{x^2}$, or $2x\sqrt{x}$ [8.3] **6.** $3r^5$ [8.1] **7.** $\frac{1}{3}$ [8.3] **8.** $\frac{1}{3}$ [8.3] **9.** $5\sqrt{7}$ [8.4] **10.** $4-2\sqrt{3}$ [8.4] **11.** $\frac{1}{x^{2/3}}$ [8.2] **12.** $-6+2\sqrt[3]{2}-4\sqrt[3]{4}$ [8.4] **13.** $8^{1/5}x^{3/5}$ [8.2] **14.** $(2x+5)^{2/3}$ [8.2] **15.** $\frac{\sqrt[3]{2}}{2}$ [8.5] **16.** $\frac{x-\sqrt{xy}}{x-y}$ [8.5] **17.** 22 [8.6] **18.** 7 [8.6] **19.** 24 (0 is an extraneous solution.) [8.6] **20.** $-2-4i$ [8.7] **21.** $17+0i$ or 17 [8.7] **22.** $\frac{8}{25}+\frac{6}{25}i$ [8.7] **23.** 60 m² [8.3] **24. a.** $\frac{\sqrt{3}}{2}$ [8.6] **b.** 64 ft. [8.6] **25. a.** $3.5\sqrt{10}$ mph [8.6] **b.** 293.88 ft or 14,400/49 ft. [8.6]

Chapters 1–8 Cumulative Review Exercises

1. false **2.** false **3.** true **4.** true **5.** x-intercept **6.** inconsistent **7.** $24x^5y^2$ **8.** 28 **9.** 0.0000024 **10.** x^{15} **11.** $9x^4-24x^2y+16y^2$ **12.** $4x^2+6x-2+\frac{2}{2x-3}$ **13.** $\frac{x}{x+5}$ **14.** $30a^3\sqrt{2a}$ **15.** $-6x^2\sqrt[3]{3x}$ **16.** $\frac{1}{8}$ **17.** $23-2i$ **18.** $\frac{5+3\sqrt{2}}{7}$ **19.** $b=\frac{2A-Bh}{h}$ or $\frac{2A}{h}-B$ **20.** 5, −11 **21.** No solution (4 is an extraneous solution.) **22.** 2 (−3 is an extraneous solution.) **23.** slope $=\frac{5}{2}$, y-intercept $=\frac{15}{2}$ **24.** $y=3x+14$ **25.**

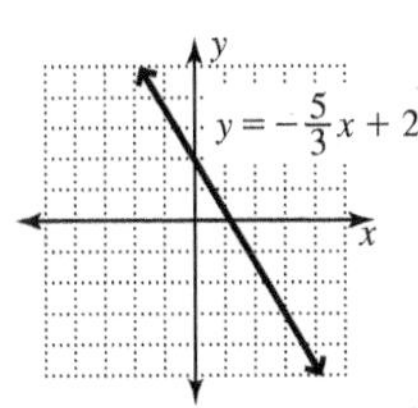

26.

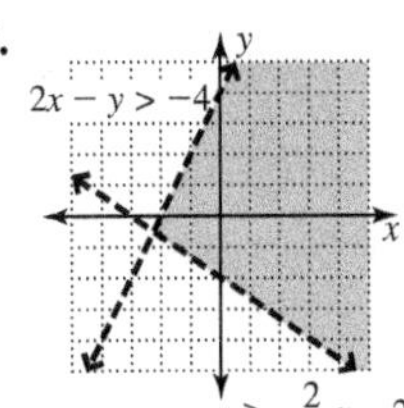

27. \$16.48 **28.** 25 mph **29.** 96 **30.** 39.2 m

Chapter 9

Exercise Set 9.1

Prep Exercises **1.** $x^2=a$ has two solutions because squaring $\sqrt{a}$ and $-\sqrt{a}$ gives a for every real number a. **2.** $x=\pm\sqrt{\frac{b+c}{a}}$ **3.** $x=\frac{b\pm\sqrt{c}}{a}$ **4.** $\left(\frac{b}{2}\right)^2$ **5.** $\left(x+\frac{b}{2}\right)^2$ **6.** Add $\left(\frac{b}{2}\right)^2$ to both sides of the equation. **7.** Because $x^2-7x+12$ is easy to factor, factoring is a better method as it requires fewer steps than completing the square. **8.** Because x^2+4x+5 cannot be factored, completing the square is the only option at this point.

Exercises **1.** ±7 **3.** $\pm\frac{2}{5}$ **5.** ±1.2 **7.** $\pm3\sqrt{5}$ **9.** $\pm5i$ **11.** ±7 **13.** ±9 **15.** ±3 **17.** $\pm\frac{3}{5}$ **19.** $\pm2i$ **21.** $\pm\frac{12}{5}$ **23.** $\pm2\sqrt{2}$ **25.** $\pm3i\sqrt{2}$ **27.** $\pm\frac{4\sqrt{6}}{3}$ **29.** ±1.6 **31.** −15, −1 **33.** $-\frac{1}{5}, \frac{7}{5}$ **35.** $8\pm4i$ **37.** $\frac{1\pm2\sqrt{10}}{4}$ **39.** $7\pm2i\sqrt{3}$ **41.** $0, \frac{3}{2}$ **43.** $\frac{3}{5}, \frac{6}{5}$ **45.** −13.5, −5.5 **47.** Mistake: Gave only the positive solution; Correct: ±7 **49.** Mistake: Changed −6 to 6; Correct: $5\pm i\sqrt{6}$ **51.** $\pm4\sqrt{6}\approx\pm9.798$ **53.** $\pm2\sqrt{5}\approx\pm4.472$ **55.** $6\pm\sqrt{15}\approx9.873, 2.127$ **57. a.** $x^2+14x+49$ **b.** $(x+7)^2$ **59. a.** $n^2-10n+25$ **b.** $(n-5)^2$ **61. a.** $y^2-9y+\frac{81}{4}$ **b.** $\left(y-\frac{9}{2}\right)^2$ **63. a.** $s^2-\frac{2}{3}s+\frac{1}{9}$ **b.** $\left(s-\frac{1}{3}\right)^2$ **65.** −5, 3

67. $1 \pm 7i$ **69.** 3, 6 **71.** $1 \pm \sqrt{17}$ **73.** $3 \pm 2i\sqrt{5}$ **75.** $-\frac{3}{2}, 1$ **77.** $-\frac{9}{2}, \frac{1}{2}$ **79.** $\frac{1}{3}, \frac{5}{3}$ **81.** $-1, \frac{3}{2}$ **83.** $-2 \pm \frac{\sqrt{23}}{2}$ **85.** $\frac{-1 \pm \sqrt{41}}{10}$ **87.** $\frac{-2 \pm 2\sqrt{7}}{3}$ **89.** Did not divide by 3 so that x^2 has a coefficient of 1; then wrote an incorrect factored form; Correct: $\frac{-2 \pm \sqrt{10}}{3}$ **91.** 14 in. **93.** Height: 3 ft.; width: 6 ft. **95.** 305 m **97.** 18 in. **99.** 16 ft. by 12 ft. **101. a.** 15 cm **b.** Length: 45 cm; width: 30 cm **103.** $\frac{3}{4}$ sec. **105.** 4 m/sec.

Review Exercises: **1.** $(x-5)^2$ **2.** $(x+3)^2$ **3.** $6x$ **4.** $2x+3$ **5.** $\frac{-2+\sqrt{6}}{2}$ or $-1+\frac{\sqrt{6}}{2}$ **6.** $\frac{-3-i}{5}$ or $-\frac{3}{5}-\frac{1}{5}i$

Exercise Set 9.2

Prep Exercises **1.** $x = \frac{-b \pm \sqrt{b^2-4ac}}{2a}$ **2.** Answers may vary. One advantage is that it can be used to solve any quadratic equation. One disadvantage is that the steps can be tedious. **3.** Yes, if they have irrational or nonreal complex solutions **4.** When the discriminant is zero. The quadratic formula becomes $x = -\frac{b}{2a}$. **5.** The discriminant is the radicand in the quadratic formula; $b^2 - 4ac$ **6.** The graph does not intersect the x-axis.

Exercises **1.** $a = 1, b = -3, c = 7$ **3.** $a = 3, b = -9, c = -4$ **5.** $a = 1.5, b = -1, c = 0.2$ or $a = -1.5, b = 1, c = -0.2$ **7.** $a = -\frac{1}{2}, b = -\frac{3}{4}, c = 6$ or $a = \frac{1}{2}, b = \frac{3}{4}, c = -6$ **9.** $-5, -4$ **11.** $-2, \frac{3}{4}$ **13.** 0, 9 **15.** 4 **17.** $-2, \frac{2}{3}$ **19.** $1 \pm i$ **21.** $\frac{1 \pm \sqrt{5}}{2}$ **23.** $\frac{-5 \pm \sqrt{10}}{3}$ **25.** $\frac{3 \pm i\sqrt{11}}{4}$ **27.** $-\frac{1}{6}, -\frac{2}{3}$ **29.** $\frac{2 \pm i\sqrt{5}}{3}$ **31.** $\frac{3 \pm \sqrt{105}}{12}$ **33.** $-0.15, 0.1$ **35.** $-2, \frac{3}{2}$ **37.** $\pm\frac{7}{6}$ **39.** $1 \pm i\sqrt{2}$ **41.** $\frac{-3 \pm \sqrt{5009}}{100} \approx -0.738, 0.678$ **43.** $\frac{3 \pm i\sqrt{51}}{12}$ **45.** Mistake: Did not evaluate $-b$. Correct: $\frac{7 \pm \sqrt{37}}{6}$ **47.** Mistake: The result was not completely simplified. Correct: $1 \pm i\sqrt{2}$ **49.** One rational **51.** Two irrational **53.** Two rational **55.** Two nonreal complex **57.** Two irrational **59.** Square root principle or factoring; ± 9 **61.** Quadratic formula; $\frac{2 \pm \sqrt{10}}{2}$ **63.** Factoring; 0, -6 **65.** Square root principle; $-7 \pm 2\sqrt{10}$ **67.** Quadratic formula; $4 \pm i\sqrt{3}$ **69.** $(2, 0), (-1, 0), (0, -2)$ **71.** $\left(\frac{3}{2}, 0\right), (0, 9)$ **73.** $\left(\frac{1}{2}, 0\right), (-8, 0), (0, -8)$ **75.** No x-intercepts, $(0, -6)$ **77.** $x^2 + 5(x+1) = 71; 6, 7$ **79.** $x^2 + (x+1)^2 = (x+2)^2; 3, 4, 5$ **81.** $w(3w - 3.5) = 34$; width: 4 ft.; length: 8.5 ft. **83. a.** $22x = 1.5x^2 + 4x$; 12 ft. **b.** 18 ft., 12 ft., 10 ft., 6 ft., 8 ft., 18 ft. (from bottom clockwise) **85.** ≈ 3.43 sec. **87. a.** $h = -4.9t^2 + 10$ **b.** ≈ 1.01 sec. **c.** ≈ 1.43 sec. **89.** 8000 units **91. a.** $\frac{25}{8}$ **b.** $c < \frac{25}{8}$ **c.** $c > \frac{25}{8}$ **93. a.** $\frac{9}{2}$ **b.** $a < \frac{9}{2}$ **c.** $a > \frac{9}{2}$

Review Exercises: **1.** $(u-7)(u-2)$ **2.** $(3u-8)(u+2)$ **3.** x^4 **4.** $x^{2/3}$ **5.** $\frac{2}{5}, -3$ **6.** 8

Exercise Set 9.3

Prep Exercises **1.** It can be rewritten as a quadratic equation. **2.** No, $x^{3/4} = (x^{1/4})^3$ shows a cubic relationship. **3.** $u = \frac{x+2}{3}$ **4.** $u = x^2$ **5.** $\frac{1}{x}$ **6.** $\frac{400}{x}$

Exercises **1.** $-6, 4$ **3.** $-\frac{4}{3}, 2$ **5.** -8 (4 is extraneous.) **7.** $-7.5, 5$ **9.** $-4, 2$ **11.** $-5, \frac{2}{3}$ **13.** 9, 25 **15.** $\frac{49}{4}$ (1 is extraneous.) **17.** 2 $\left(\frac{2}{9}\text{ is extraneous.}\right)$ **19.** $\pm 3i$ **21.** 2 $\left(-\frac{2}{9}\text{ is extraneous.}\right)$ **23.** 0, 4 **25.** $\pm 3, \pm 1$ **27.** $\pm 1.5, \pm 1$ **29.** $\pm 2, \pm 3i$ **31.** $-6, -4$ **33.** $-3.5, 2$ **35.** $-9, -5$ **37.** $-5, 0$ **39.** 8, 27 **41.** $-\frac{1}{8}, 8$ **43.** 81, 16 **45.** $\frac{16}{625}$ (16 is extraneous.) **47.** 3 hr. **49.** ≈ 2.21 hr. **51.** 45 mph **53. a.** Old rate = 12 mph; new rate = 18 mph **b.** Old time = 3 hr.; new time = 2 hr. **55.** Jody: 4 hr.; Billy: 6 hr. **57.** Older: ≈ 4.27 hr.; newer: ≈ 3.77 hr. **59.** 420 vps, 460 vps

Review Exercises: **1.** -2 **2.** -3 **3.** $(3, 0), (0, 2)$ **4.** 6 **5.**

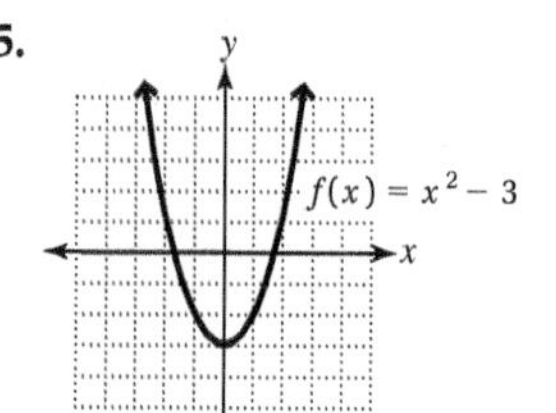

6.

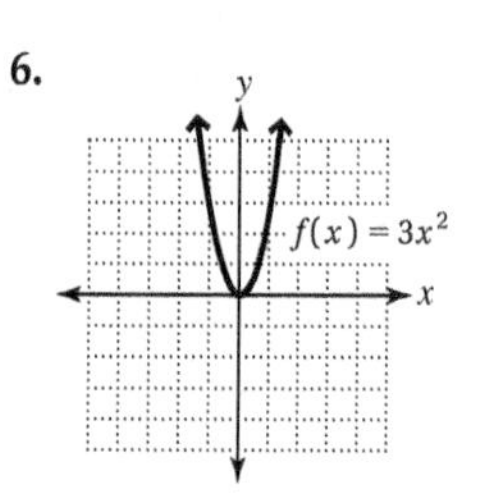

Exercise Set 9.4

Prep Exercises **1.** The sign of a **2.** The axis of symmetry is the vertical line through the vertex. **3.** (h, k); $x = h$ **4.** There are no x-intercepts. **5.** There are no x-intercepts. **6.** $x = -\dfrac{b}{2a}$

Exercises **1.** V: $(2, -3)$; axis: $x = 2$ **3.** V: $(-1, -5)$; axis: $x = -1$ **5.** V: $(0, 2)$; axis: $x = 0$ **7.** V: $(-5, 0)$; axis: $x = -5$ **9.** V: $(0, 0)$; axis: $x = 0$ **11.** V: $(2, 4)$; axis: $x = 2$ **13.** V: $(-4, -5)$; axis: $x = -4$ **15.** V: $\left(\dfrac{1}{3}, \dfrac{4}{3}\right)$; axis: $x = \dfrac{1}{3}$ **17. a.** downward

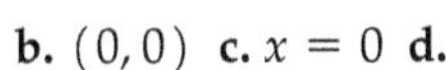
b. $(0, 0)$ **c.** $x = 0$ **d.**

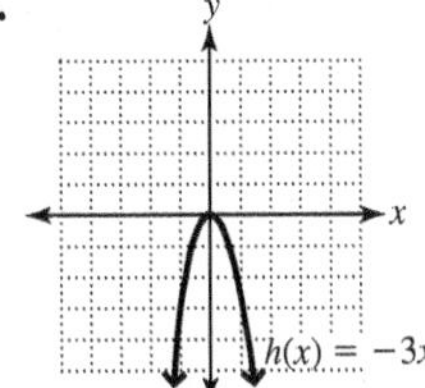

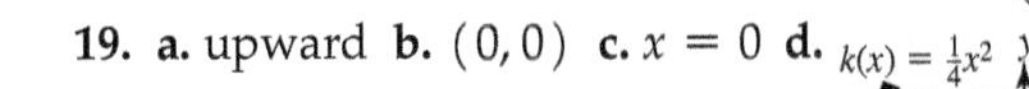
19. a. upward **b.** $(0, 0)$ **c.** $x = 0$ **d.**

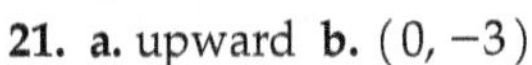
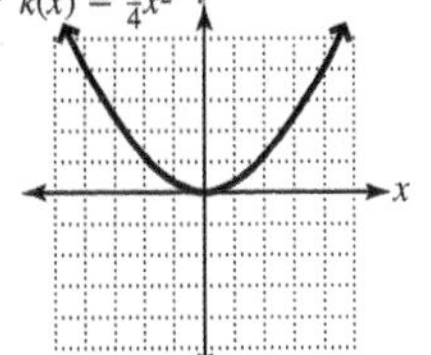

21. a. upward **b.** $(0, -3)$

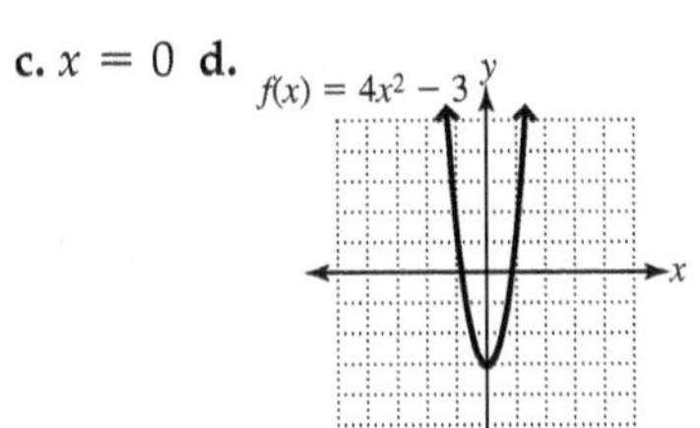
c. $x = 0$ **d.**

23. a. downward **b.** $(0, 2)$ **c.** $x = 0$ **d.**

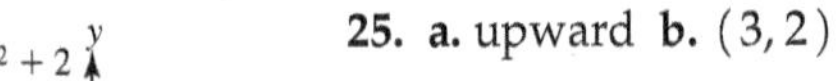
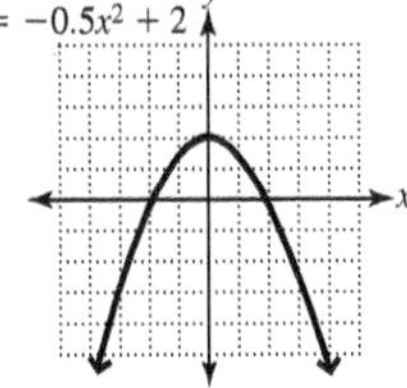

25. a. upward **b.** $(3, 2)$

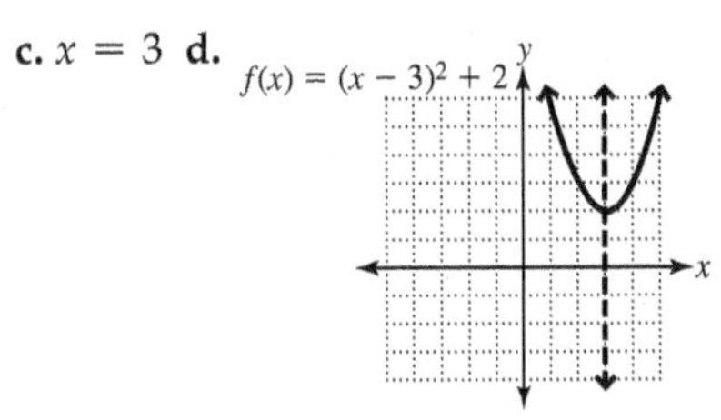
c. $x = 3$ **d.**

27. a. downward **b.** $(-1, -3)$ **c.** $x = -1$ **d.**

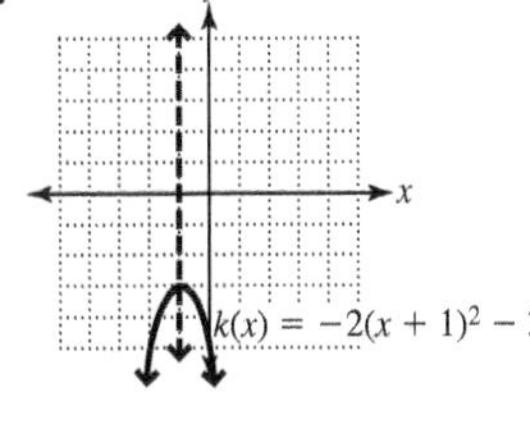

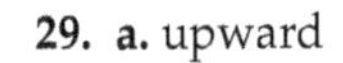
29. a. upward

b. $(2, -1)$ **c.** $x = 2$ **d.**

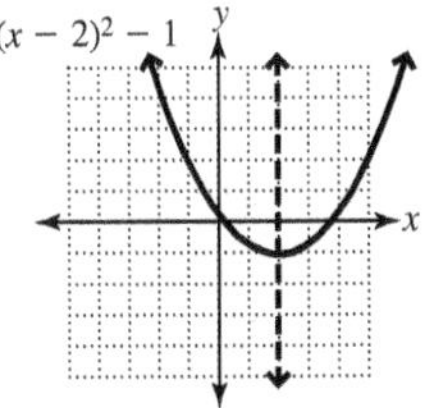

31. a. $(-3, 0)$, $(0, 9)$ **b.** $h(x) = (x + 3)^2$ **c.** upward **d.** $(-3, 0)$ **e.** $x = -3$

f.

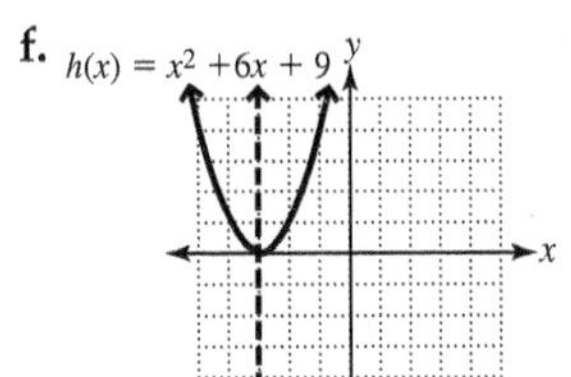

g. domain: $\{x \mid x \text{ is a real number}\}$, or $(-\infty, \infty)$; range: $\{y \mid y \geq 0\}$ or $[0, \infty)$

33. a. no x-intercepts, $(0, -5)$ **b.** $g(x) = -3(x - 1)^2 - 2$ **c.** downward **d.** $(1, -2)$ **e.** $x = 1$ **f.**

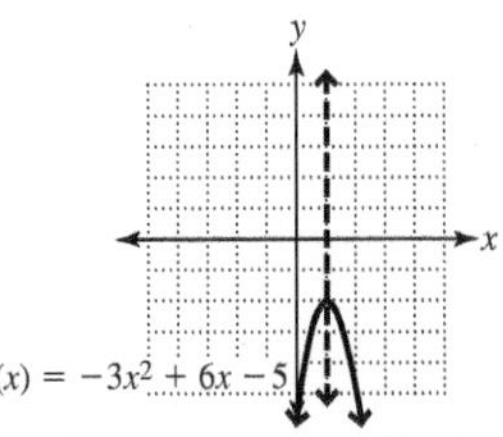

g. domain: $\{x \mid x \text{ is a real number}\}$, or $(-\infty, \infty)$; range: $\{y \mid y \leq -2\}$ or $(-\infty, -2]$ **35. a.** $\left(\dfrac{-3 + \sqrt{3}}{2}, 0\right)$, $\left(\dfrac{-3 - \sqrt{3}}{2}, 0\right)$, $(0, 3)$

b. $f(x) = 2\left(x + \frac{3}{2}\right)^2 - \frac{3}{2}$ **c.** upward **d.** $\left(-\frac{3}{2}, -\frac{3}{2}\right)$ **e.** $x = -\frac{3}{2}$ **f.**

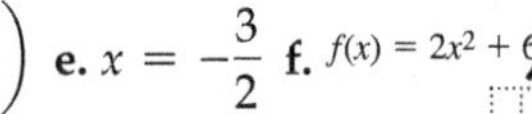

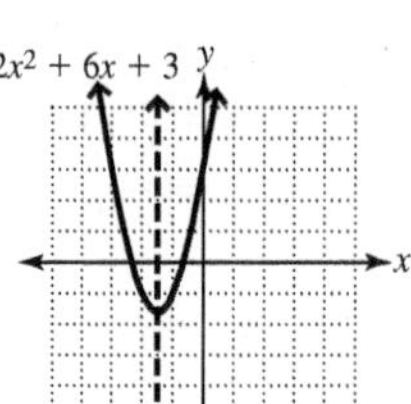

g. domain: $\{x \mid x \text{ is a real number}\}$, or $(-\infty, \infty)$; range: $\left\{y \mid y \geq -\frac{3}{2}\right\}$ or $\left[-\frac{3}{2}, \infty\right)$ **37. a.** 4.59 sec. **b.** 103.32 m **c.** 9.18 sec. **39. a.** 9.2 ft. **b.** ≈5.39 ft.

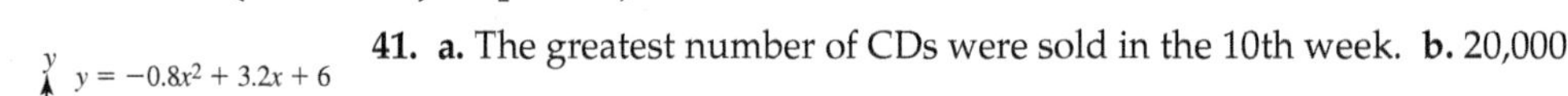

c.

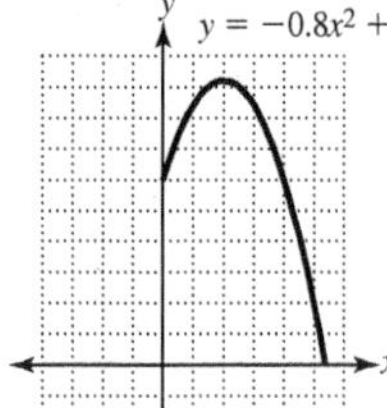

41. a. The greatest number of CDs were sold in the 10th week. **b.** 20,000

43. The area is maximized if the length and width are 100 ft. **45.** 55 units **47.** −6, 6; the product is −36.

Review Exercises: **1.** −11 **2.** −5, 3 **3.** $\pm 2i\sqrt{5}$ **4.** $x \leq 4.5$

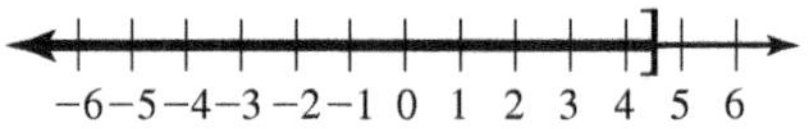

5. $x \leq -5$ or $x \geq 11$

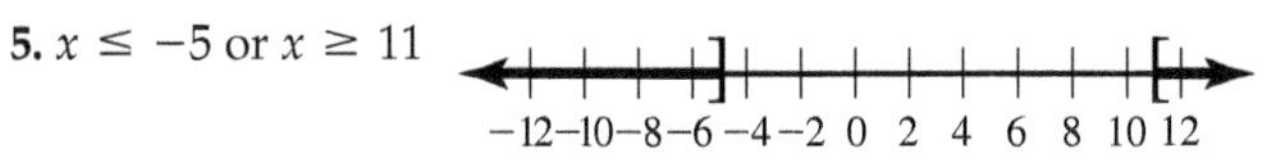

6.

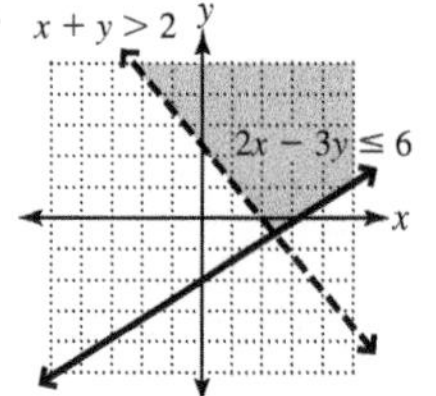

Exercise Set 9.5

Prep Exercises **1.** Solve the related equation and then determine which of the three intervals satisfy the inequality by testing a number in each interval. If a test number satisfies the inequality, then all points in that interval satisfy the inequality.
2. $(-\infty, -2), (-2, 2), (2, \infty)$ **3.** Yes, for example, $x^2 + 2 \leq 0$ has an empty solution set because $x^2 + 2$ is always positive. **4.** Yes, for example, $(x + 3)^2 \leq 0$ has only $x = -3$ as a solution. **5.** In both cases, the intervals to check are $(-\infty, -5), (-5, 2)$, and $(2, \infty)$, and in both cases, only x-values in $(-5, 2)$ satisfy the original inequalities.
6. $(-\infty, -5), (-5, -2), (-2, 3), (3, \infty)$

Exercises **1. a.** $x = -5, -1$ **b.** $(-5, -1)$ **c.** $(-\infty, -5) \cup (-1, \infty)$ **3. a.** $x = -1, 3$ **b.** $(-\infty, -1] \cup [3, \infty)$ **c.** $[-1, 3]$ **5. a.** $x = -2$ **b.** $(-\infty, -2) \cup (-2, \infty)$ **c.** ∅ **7. a.** ∅ **b.** ∅ **c.** ℝ, or $(-\infty, \infty)$ **9.** $(-4, -2)$

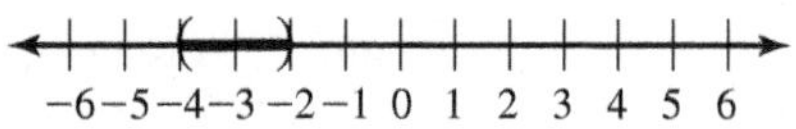

11. $(-\infty, 2) \cup (5, \infty)$ 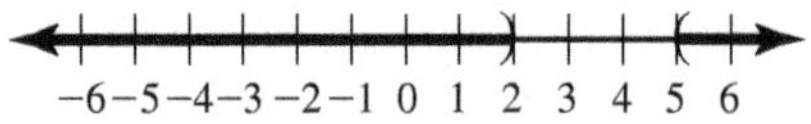 **13.** $(-4, -1)$

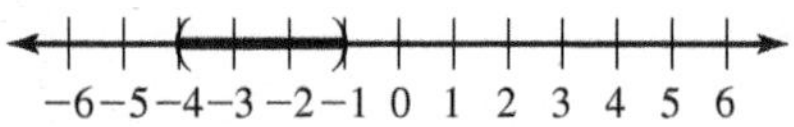

15. $(-\infty, 1) \cup (3, \infty)$ 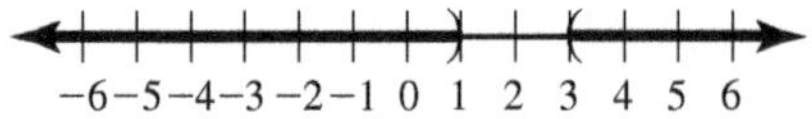 **17.** $[2, 4]$

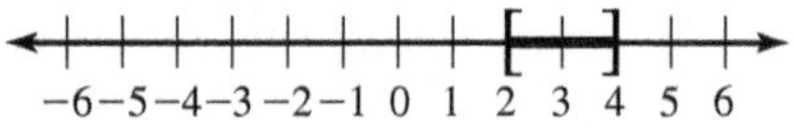

19. $(-\infty, -1] \cup [5, \infty)$ **21.** $(-2, 5)$

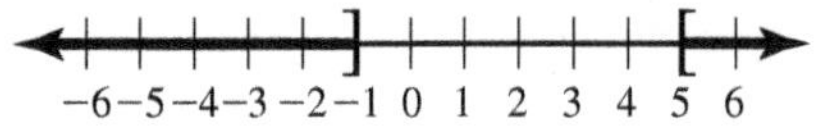

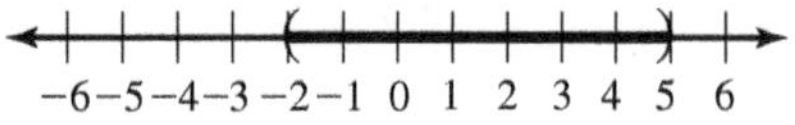

23. $(-\infty, \infty)$, or ℝ 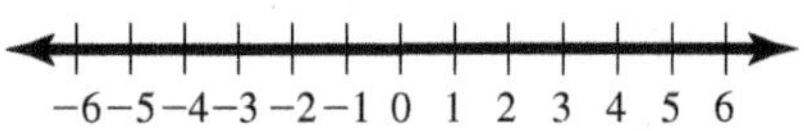 **25.** No solution, or ∅

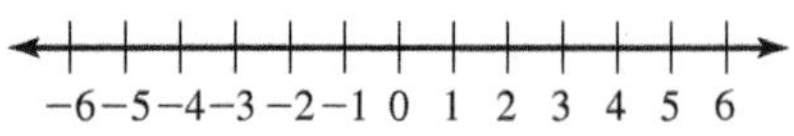

27. $(-\infty, -5) \cup (-0.25, \infty)$

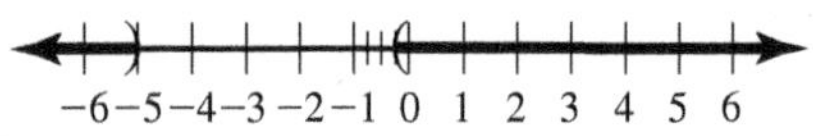

29. $(-\infty, 0) \cup (5, \infty)$ 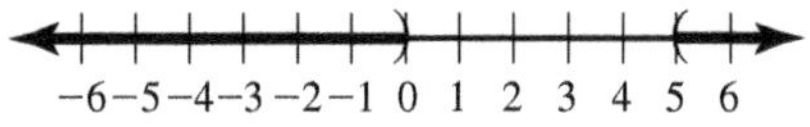 **31.** $[0, 6]$

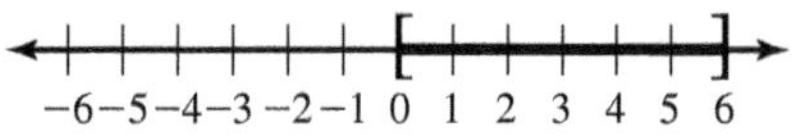

33. $(-\infty, \infty)$, or ℝ 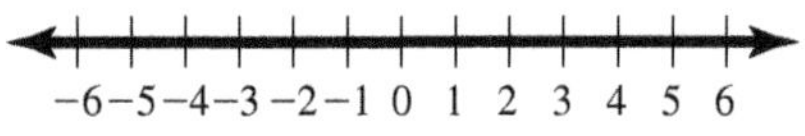 **35.** $[-4, -2] \cup [4, \infty)$

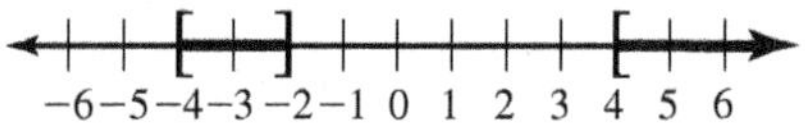

37. $(-\infty, -6) \cup (-2, 1)$ **39.** $\left(\frac{-2 - \sqrt{7}}{3}, \frac{-2 + \sqrt{7}}{3}\right)$

41. $\left(-\infty, \frac{-2 - \sqrt{7}}{2}\right] \cup \left[\frac{-2 + \sqrt{7}}{2}, \infty\right)$ **43.** $(-\infty, -4 - \sqrt{6}] \cup [-4 + \sqrt{6}, \infty)$

45. $(-\infty, -4) \cup (1, \infty)$ **47.** $(-5, -1]$

49. $(-4, \infty)$ **51.** $\left(-\infty, -\frac{9}{2}\right) \cup (-3, \infty)$

53. $(4, 7)$ **55.** $(3, 5]$

57. $(-\infty, -3] \cup [2, 5)$ **59.** $(0, \infty)$

61. $(-\infty, -1) \cup (2, 5)$ **63. a.** 6 sec. **b.** At 2 sec. and again at 3 sec. **c.** $(2, 3)$ sec. **d.** $[0, 2) \cup (3, 6]$ sec. **65. a.** $x \geq 4$ in. **b.** Base ≥ 10 in.; height ≥ 2 in. **67.** $2 < x \leq 8$

Review Exercises: **1.** 9, $(x - 3)^2$ **2.** $\frac{25}{4}, \left(x + \frac{5}{2}\right)^2$ **3.** 18 **4.** Downward, because the coefficient of x^2 is negative **5.** $(-6, 0), (2, 0)$ **6.** $(3, -4)$

Exercise Set 9.6

Prep Exercises **1.** $(p + q)(x) = p(x) + q(x)$ **2.** $(p - q)(x) = p(x) - q(x)$ **3.** Use FOIL to multiply the binomials. **4.** Divide each term in the trinomial by the monomial.

Exercises **1.** $5x - 1; -x + 1$ **3.** $3x - 8; -x - 2$ **5.** $x^2 - 3x + 4; x^2 - 5x + 10$ **7.** $3x^2 - 5x + 2; x^2 - 5x - 8$ **9.** $-2x^2 + 2x + 7; -8x^2 + 6x + 9$ **11.** $2x^2 - 10x$ **13.** $x^2 - x - 2$ **15.** $3x^2 + 8x + 4$ **17.** $x^4 - x^3 - 11x^2 + 25x - 14$ **19.** $6x^4 - 11x^3 + 4x^2 + 2x - 1$ **21.** $x - 3, x \neq 0$ **23.** $2x - 1 + \frac{2}{x}, x \neq 0$ **25.** $x - 5, x \neq 9$ **27.** $x + 1, x \neq \frac{5}{2}$ **29.** $2x - 15 + \frac{68}{x + 5}, x \neq -5$ **31. a.** $7x + 2$ **b.** $3x - 4$ **c.** $10x^2 + 13x - 3$ **d.** $\frac{5x - 1}{2x + 3}, x \neq \frac{-3}{2}$ **33. a.** $2x^2 - 2x + 4$ **b.** $2x^2 - 4x + 6$ **c.** $2x^3 - 5x^2 + 8x - 5$ **d.** $2x - 1 + \frac{4}{x - 1}, x \neq 1$ **35. a.** -12 **b.** 11 **c.** -6 **d.** 2 **37. a.** 16 **b.** 54 **c.** 7 **d.** -8

39. a. $4x + 6$
b.

$f(x)$	$g(x)$	$(f + g)(x)$
−11	1	−10
−5	3	−2
1	5	6
7	7	14
13	9	22

c.

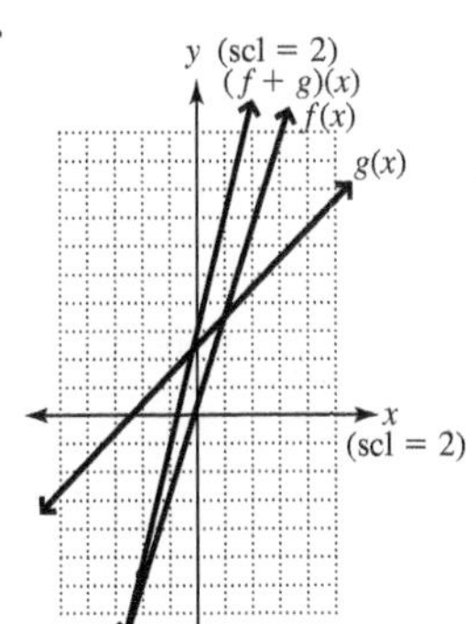

d. The output values for $f(x)$ added to the output values of $g(x)$ are equal to the output values of $(f + g)(x)$.

41. a. $-x - 14$
b.

$h(x)$	$k(x)$	$(h - k)(x)$
−17	−7	−10
−13	−1	−12
−9	5	−14
−5	11	−16
−1	17	−18

c.

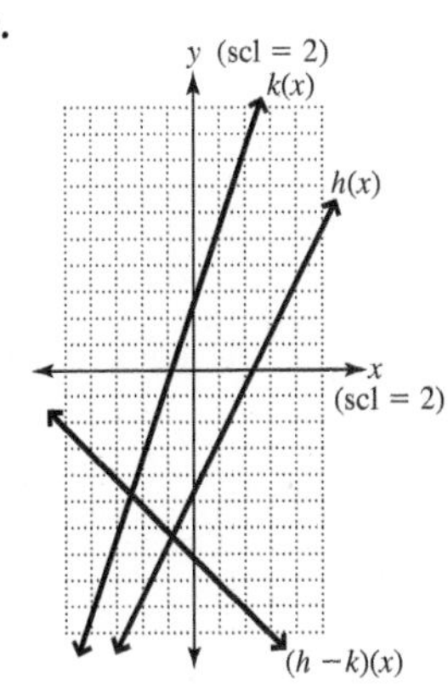

d. The output values for $k(x)$ subtracted from the output values of $h(x)$ are equal to the output values of $(h - k)(x)$.

43. a. $t(x) = 4x + 4$ **b.** \$804 **c.**

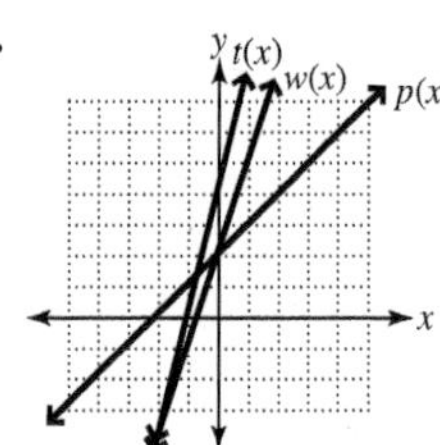

45. a. $p(x) = x^2 - 4x - 12$ **b.** \$9588 profit **c.**

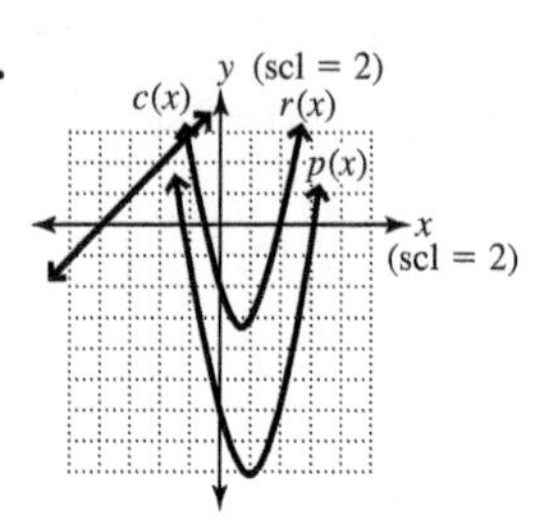

d. 6 units **47. a.** $A(x) = 6x^2 + 4x$ **b.** 66 ft.^{2} **49. a.** $h(x) = 4x + 1$ **b.** 17 cm

Review Exercises: **1.** 5 **2.** $9a^2 - 12a + 5$ **3.** a **4.** $x = \frac{y-4}{2}$ **5.** $x = \sqrt[3]{y-2}$

Chapter 9 Summary and Review Exercises

9.1 Definitions/Rules/Procedures $\sqrt{a}; -\sqrt{a}; \pm\sqrt{a}$ Exercises **1.** ± 4 **2.** $\pm\frac{1}{6}$ **3.** $\pm 2\sqrt{7}$ **4.** $\pm\sqrt{14}$ **5.** $\pm i\sqrt{3}$ **6.** $-12, -2$ **7.** $9 \pm 4i$ **8.** $-\frac{7}{5}, \frac{1}{5}$ **9.** 150 ft. **10.** 4 m/sec. **11.** $9\sqrt{3}$ in. Definitions/Rules/Procedures $x^2 + bx = c; \left(\frac{b}{2}\right)^2$; factored; square root; variable Exercises **12.** $-7, -1$ **13.** $-8, 14$ **14.** $\frac{3 \pm i\sqrt{5}}{2}$ **15.** $-1, \frac{3}{4}$

9.2 Definitions/Rules/Procedures $\frac{-b \pm \sqrt{b^2 - 4ac}}{2a}$ Exercises **16.** $-1 \pm \sqrt{6}$ **17.** $\frac{1 \pm i\sqrt{2}}{3}$ **18.** $\frac{-1 \pm \sqrt{41}}{4}$ **19.** $-\frac{3}{20}, \frac{1}{10}$ **20.** Width: 15 ft.; length: 19 ft. **21.** Height: 6.65 ft.; base: 15.65 ft. Definitions/Rules/Procedures $b^2 - 4ac$; two; perfect square; one; nonreal complex Exercises **22.** $D = 64$; two rational **23.** $D = -71$; two nonreal complex **24.** $D = 0$; one rational **25.** $D = 0.48$; two irrational

9.3 Definitions/Rules/Procedures variable; expression; LCD; extraneous Exercises **26.** $\pm\frac{3\sqrt{2}}{2}$ **27.** $-\frac{5}{2}$ (5 is extraneous.) **28.** $\frac{1}{3}, \frac{1}{2}$ **29.** $\frac{3 \pm \sqrt{17}}{4}$ Definitions/Rules/Procedures isolate; extraneous Exercises **30.** No solution **31.** $1\left(\frac{1}{9}\text{ is extraneous.}\right)$ **32.** $2\left(-\frac{3}{2}\text{ is extraneous.}\right)$ **33.** $\frac{1}{3}, \frac{2}{3}$ Definitions/Rules/Procedures $au^2 + bu + c = 0$; u; Substitute Exercises **34.** $\pm\sqrt{2}, \pm\sqrt{3}$ **35.** $\pm 1, \pm\frac{\sqrt{2}}{2}$ **36.** $-\frac{9}{2}, -\frac{14}{3}$ **37.** $-26, -2$ **38.** 27, 512 **39.** $\frac{16}{625}$ (81 is extraneous.)

9.4 Definitions/Rules/Procedures (h, k); $x = h$; $a > 0$; $a < 0$ Exercises **40. a.** $(0, 0)$ **b.** downward **c.** $(0, 0)$ **d.** $x = 0$ **e.**

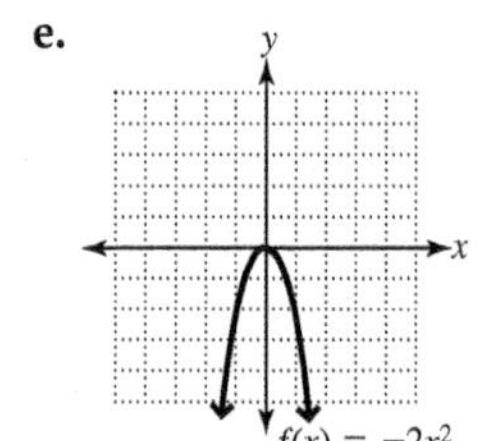

41. a. no x-intercepts, $(0, 1)$ **b.** upward **c.** $(0, 1)$ **d.** $x = 0$ **e.**

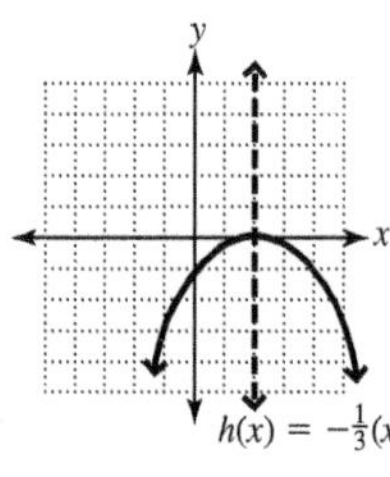

42. a. $(2, 0), \left(0, -\frac{4}{3}\right)$ **b.** downward **c.** $(2, 0)$ **d.** $x = 2$ **e.**

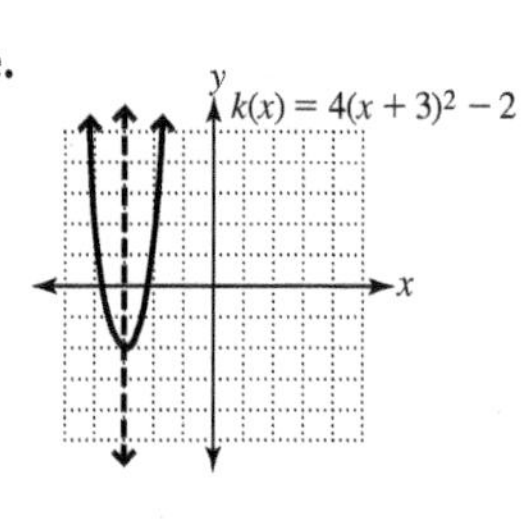

43. a. $\left(-3 + \frac{\sqrt{2}}{2}, 0\right), \left(-3 - \frac{\sqrt{2}}{2}, 0\right)$, $(0, 34)$ **b.** upward **c.** $(-3, -2)$ **d.** $x = -3$ **e.**

Definitions/Rules/Procedures $-\frac{b}{2a}; \frac{4ac - b^2}{4a}$

Exercises **44. a.** $m(x) = (x+1)^2 - 2$ **b.** $(-1+\sqrt{2}, 0), (-1-\sqrt{2}, 0), (0, -1)$ **c.** upward **d.** $(-1, -2)$ **e.** $x = -1$
f.

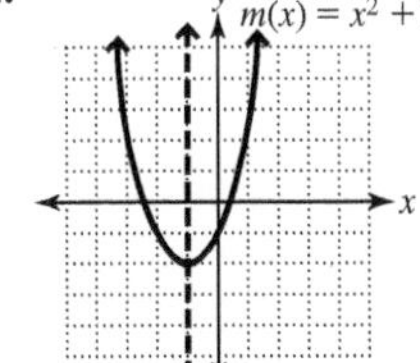

g. domain: $\{x \mid x \text{ is a real number}\}$ or $(-\infty, \infty)$; range: $\{y \mid y \geq -2\}$ or $[-2, \infty)$

45. a. $p(x) = -0.5(x-4)^2 + 2$ **b.** $(2, 0), (6, 0), (0, -6)$ **c.** downward **d.** $(4, 2)$ **e.** $x = 4$ **f.**

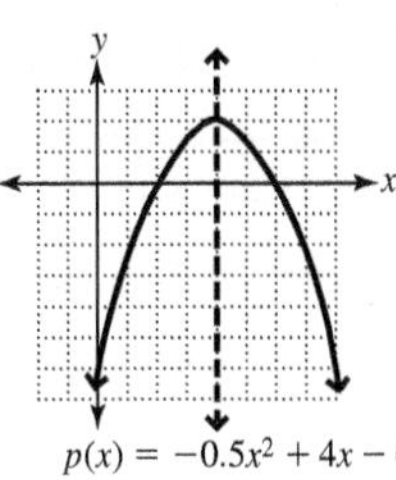

g. domain: $\{x \mid x \text{ is a real number}\}$ or $(-\infty, \infty)$; range: $\{y \mid y \leq 2\}$ or $(-\infty, 2]$ **46. a.** 0.75 sec. **b.** 9 ft. **c.** 1.5 sec.
d.

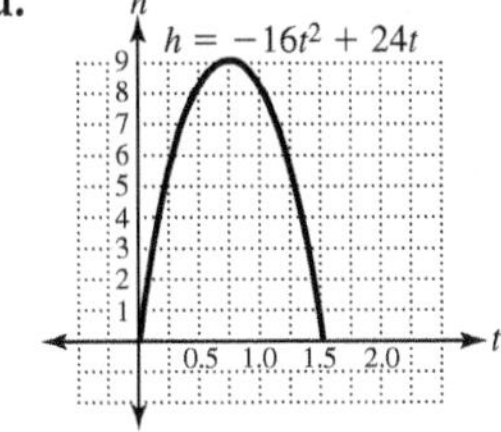

47. a. 38.88 yd. **b.** 72 yd.

9.5 Definitions/Rules/Procedures $ax^2 + bx + c > 0$; $ax^2 + bx + c < 0$; $ax^2 + bx + c = 0$; number line; number line; solve the inequality; no number; union Exercises **48.** $(-\infty, -5) \cup (3, \infty)$ **49.** $[2, 4]$ **50.** $(-7, -2)$ **51.** $[-3, 1] \cup [2, \infty)$ **52. a.** $x \geq 10$ in.
b. Height $\geq$ 8 in., base $\geq$ 14 in. Definitions/Rules/Procedures rational expression; 0; equation; number line; test number; union; 2; 1
Exercises **53.** $(-\infty, -3] \cup (1, \infty)$ **54.** $(-\infty, -4) \cup (-2, \infty)$ **55.** $(-\infty, 4) \cup \left[\frac{17}{4}, \infty\right)$ **56.** $(-\infty, -2) \cup (3, 5)$

9.6 Definitions/Rules/Procedures $f(x) + g(x)$; $f(x) - g(x)$; $f(x) \cdot g(x)$; $\dfrac{f(x)}{g(x)}$ Exercises **57. a.** $x + 10$ **b.** $-7x + 8$
c. $-12x^2 + 33x + 9$ **58. a.** $7x$ **b.** $-9x - 4$ **c.** $-8x^2 - 18x - 4$ **59. a.** $3x^2 + 7$ **b.** $3x^2 - 2x + 3$ **c.** $3x^3 + 5x^2 + 3x + 10$
60. a. $x^2 + x$ **b.** $x^2 - 5x - 2$ **c.** $3x^3 - 5x^2 - 5x - 1$ **61.** $2x^2 - 5x + 4, x \neq 0$ **62.** $x^2 - 6x + 3 + \dfrac{-2}{2x-5}, x \neq \dfrac{5}{2}$
63. a. $c(x) = 4x + 4$ **b.** \$904 **c.**

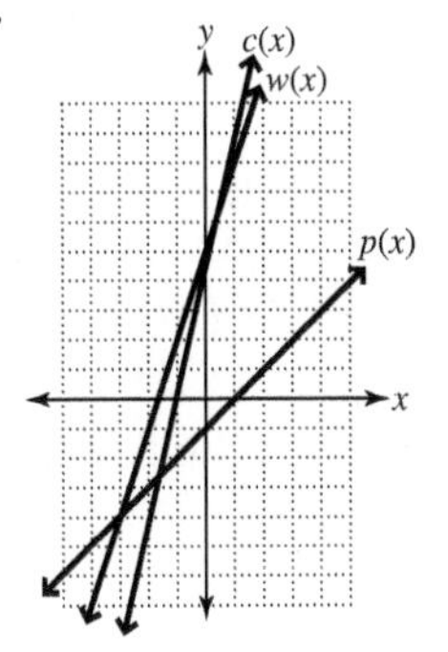

64. a. $h(x) = 2x - 1$ **b.** 17 in.

Chapter 9 Practice Test

1. ± 9 [9.1] **2.** $3 \pm 2\sqrt{5}$ [9.1] **3.** $4 \pm 2\sqrt{3}$ [9.1] **4.** $\dfrac{3 \pm 2\sqrt{6}}{3}$ [9.1] **5.** $-2, \dfrac{3}{2}$ [9.2] **6.** 3, 5 [9.2] **7.** $-8, 2$ [9.1, 9.2]
8. $\dfrac{-3 \pm i\sqrt{3}}{4}$ [9.1, 9.2] **9.** $\pm 4i$ [9.1, 9.2] **10.** $0, -\dfrac{5}{3}$ [9.1, 9.2] **11.** $-\dfrac{6}{5}, 4$ [9.3] **12.** $-\dfrac{2}{3}, 1$ [9.3] **13.** No solution [9.3]
14. 1 (-4 is extraneous.) [9.3] **15.** $\pm\dfrac{1}{3}, \pm i\sqrt{3}$ [9.3] **16.** $-5, 0$ [9.3] **17. a.** $(3 + \sqrt{5}, 0)\ (3 - \sqrt{5}, 0), (0, -4)$

b. $f(x) = -(x-3)^2 + 5$ **c.** downward **d.** $(3,5)$ **e.** $x = 3$ **f.**

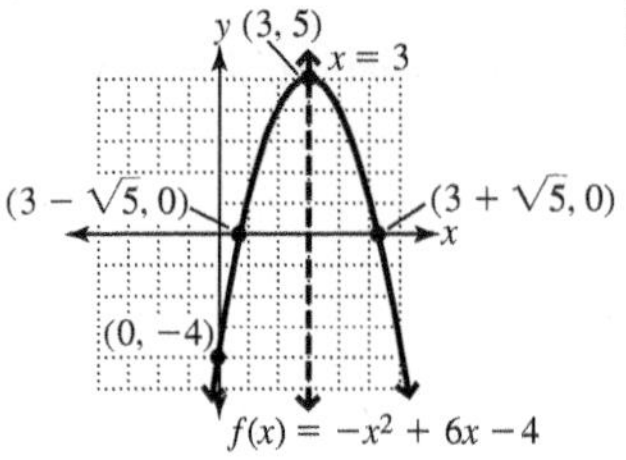

[10.4]

g. domain: $\{x \mid x \text{ is a real number}\}$, or $(-\infty, \infty)$; range: $\{y \mid y \le 5\}$ or $(-\infty, 5]$
18. $[-1, 4]$

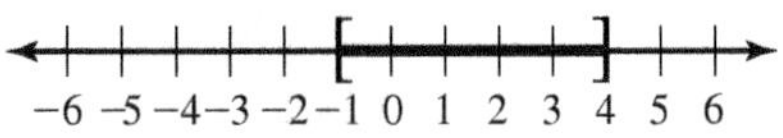

[10.5] **19.** $(-\infty, -2) \cup (1, \infty)$

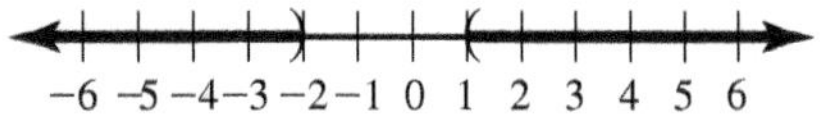

[10.5]

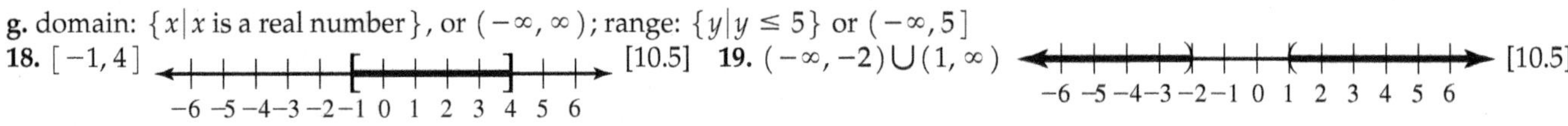

20. a. $4x^2 - 2$ **b.** $-2x^2 + 2$ **c.** $3x^4 - 2x^2$ [10.6] **21.** $5x^2 + 9x - 7$ [10.6] **22.** 2.5 sec. [10.2] **23. a.** 29.125 m **b.** 70.66 m
c.

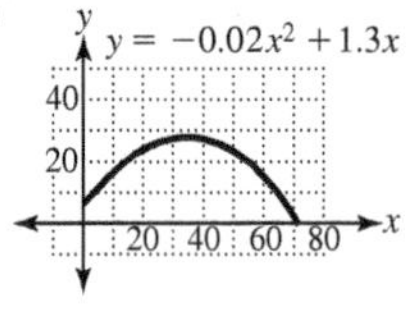

[10.4] **24. a.** $0 < w \le 25$ ft. **b.** $15 < l \le 40$ ft. [10.5] **25. a.** $h(x) = 9x + 2$ **b.** 137 cm [10.6]

Chapters 1–9 Cumulative Review Exercises

1. false **2.** false **3.** true **4.** parabola; $a > 0$ **5.** $\sqrt[n]{ab}$ **6.** $\sqrt{a}; -\sqrt{a}$ **7.** $-6x - 6$ **8.** $-56mn^5$ **9.** $4x^3 - 14x^2 + 7x - 3$ **10.** $(m + 2)(m^2 - 2m + 4)$ **11.** $(x + 5)^2$ **12.** $\dfrac{1}{u - 1}$ **13.** $\dfrac{x + 10}{(x - 2)(x + 2)}$ **14.** $x^{1/2}$ **15.** $\dfrac{\sqrt{mn} + n}{m - n}$ **16.** $-\dfrac{7}{2}$ **17.** $-\dfrac{8}{3}, 0$ **18.** $\dfrac{3}{2}, -5$ **19.** 8 **20.** 17 **21.** $3 \pm i\sqrt{2}$ **22. a.**

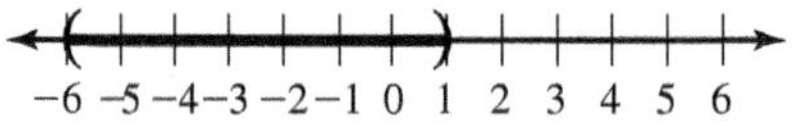

b. $\{x \mid -6 < x < 1\}$ **c.** $(-6, 1)$

23. a.

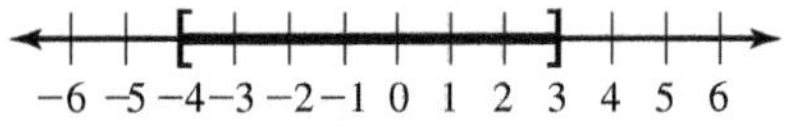

b. $\{x \mid -4 \le x \le 3\}$ **c.** $[-4, 3]$ **24.**

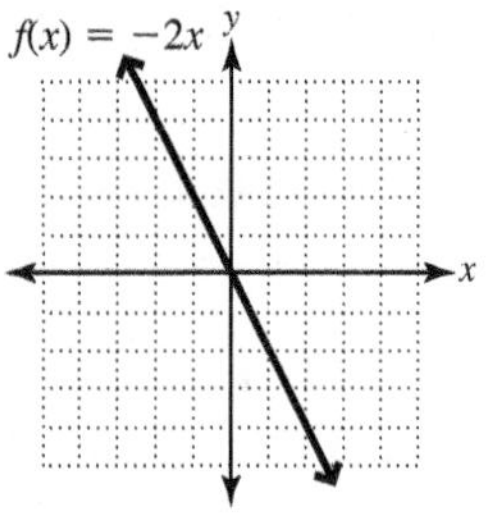

25. $2x + 3y = 8$

26. $(2, -1, 4)$ **27.** 5 hr. **28.** 75 ml **29.** 3.2 cm/sec. **30. a.** $\dfrac{\sqrt{6}}{2}$ sec. **b.** 144 ft.

Chapter 10

Exercise Set 10.1

Prep Exercises **1.** $f[g(x)]$ **2.** $g[f(x)]$ **3.** The domain of a function is the range of its inverse; the range of a function is the domain of its inverse. This occurs because the ordered pairs of f and f^{-1} have each x and y interchanged. **4.** No, the function must be one-to-one. Otherwise, the inverse is not a function. **5.** To prove that functions f and g are inverses, it must be shown that $f[g(x)] = x$ and $g[f(x)] = x$. **6.** Horizontal line test **7.** Solve for y. **8.** $y = x$

Exercises **1.** 14 **3.** 3 **5.** 26 **7.** $3x^2 + 14$ **9.** $3\sqrt{x+1} + 5$ **11.** $\sqrt{6}$ **13.** $(f \circ g)(x) = 6x + 6$; $(g \circ f)(x) = 6x - 2$ **15.** $(f \circ g)(x) = x^2 + 3$; $(g \circ f)(x) = x^2 + 4x + 5$ **17.** $(f \circ g)(x) = 9x^2 + 9x - 4$; $(g \circ f)(x) = 3x^2 + 9x - 12$ **19.** $(f \circ g)(x) = \sqrt{2x - 3}$; $(g \circ f)(x) = 2\sqrt{x + 2} - 5$ **21.** $(f \circ g)(x) = \dfrac{2x - 3}{x - 3}$; $(g \circ f)(x) = \dfrac{1 - 2x}{x + 1}$ **23.** Domain is $[0, \infty)$, and range is $[3, \infty)$. **25.** yes **27.** no **29.** yes **31.** yes **33.** yes **35.** yes **37.** no **39.** yes **41.** yes **43.** Yes, because $(f \circ g)(x) = (g \circ f)(x) = x$ **45.** yes **47.** no **49.** $f^{-1} = \{(2, -3), (-3, -1), (4, 0), (6, 4)\}$ **51.** $f^{-1} = \{(-2, 7), (2, 9), (1, -4), (3, 3)\}$ **53.** $f^{-1}(x) = x - 6$ **55.** $f^{-1}(x) = \dfrac{x - 3}{2}$ **57.** $f^{-1}(x) = \sqrt[3]{x + 1}$ **59.** $f^{-1}(x) = \dfrac{2 - 2x}{x}$ **61.** $f^{-1}(x) = \dfrac{3x + 2}{x - 1}$ **63.** $f^{-1}(x) = x^2 + 2, x \ge 0$ **65.** $f^{-1}(x) = \sqrt[3]{\dfrac{x - 4}{2}}$ **67.** $f^{-1}(x) = x^3 - 2$ **69.** $f^{-1}(x) = \dfrac{x^3}{16} - 2$

71.

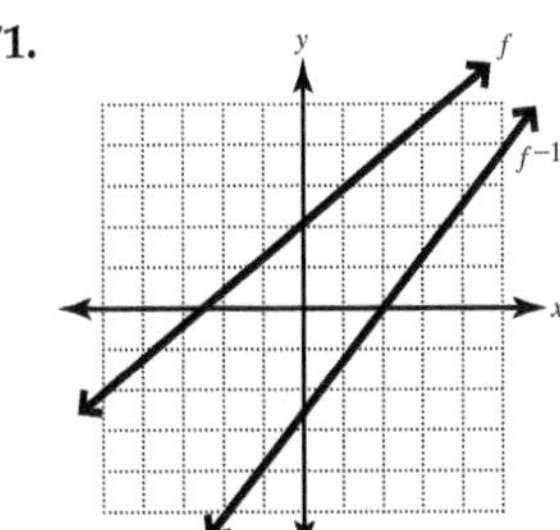

73.

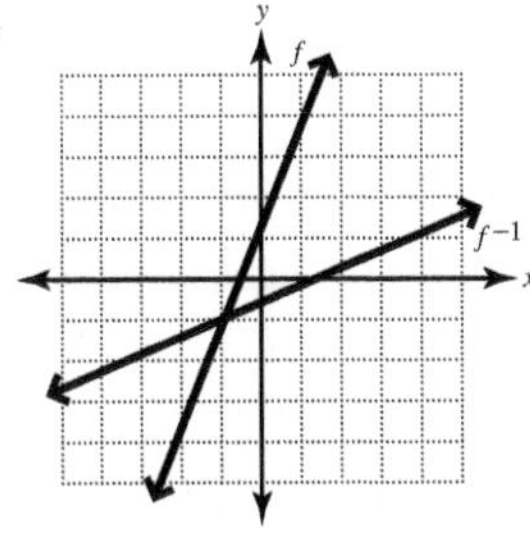

75.

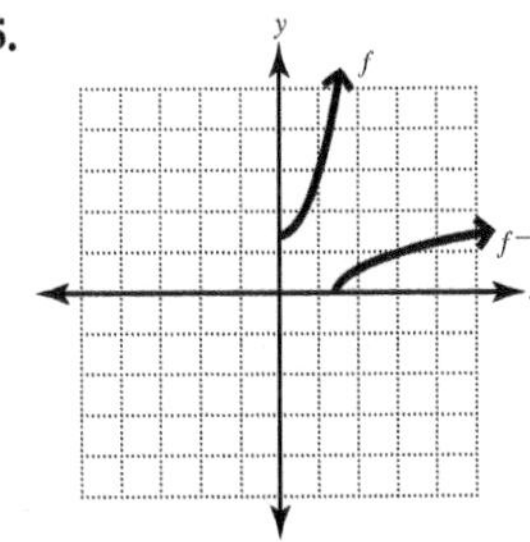

77.

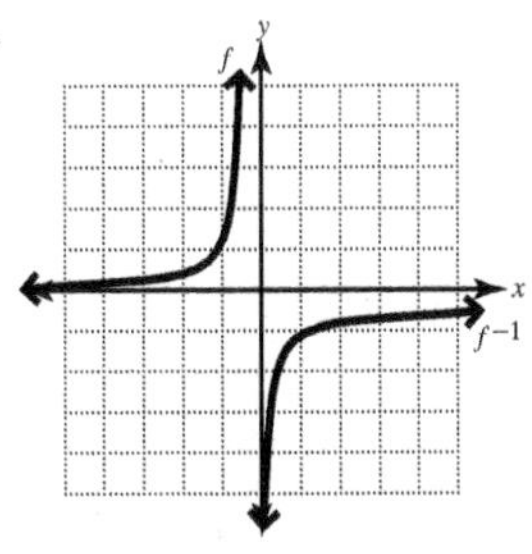

79.

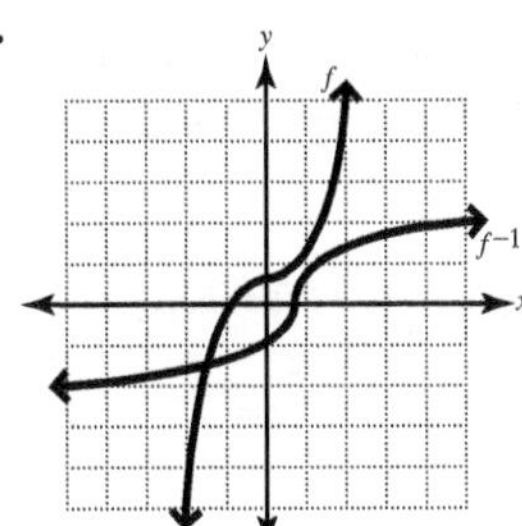

81.

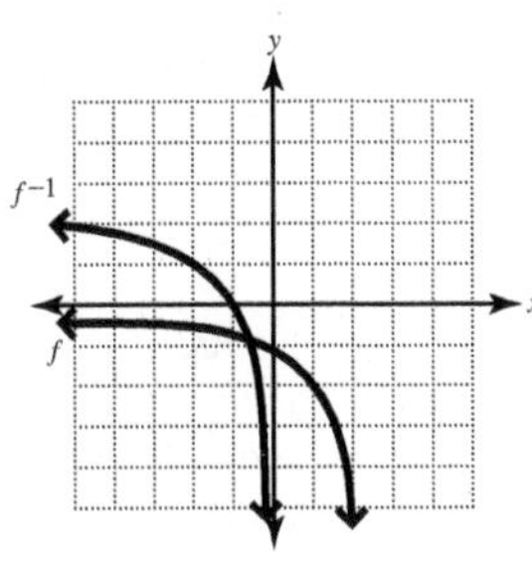

83. a. $y = \dfrac{x - 100}{0.05}$ **b.** x represents the salary; y represents the sales. **c.** \$5000 **85. a.** $y = \dfrac{6825 - x}{20}$ **b.** x represents the cost; y represents the number of 36-in. fans. **c.** 80 **87.** $(5, -4)$ **89.** $f^{-1}(x) = \dfrac{x - b}{a}$

Review Exercises: **1.** $32 = 2^5$ **2.** 8 **3.** $-\dfrac{1}{27}$ **4.** $\dfrac{1}{16}$ **5.** 8 **6.** (3, 1)

Exercise Set 10.2

Prep Exercises **1.** No, because the base is negative **2.** The domain is all real numbers; the range is $(0, \infty)$. **3.** $f(x) = 3^x$ is the steeper graph because it has the greater base. **4.** Yes, the graphs are symmetric. The point (1, 2) on $f(x)$ corresponds to the point $(-1, 2)$ on $g(x)$, and so on. **5.** Because the exponential function is one-to-one **6.** 0.625 g

Exercises **1.**

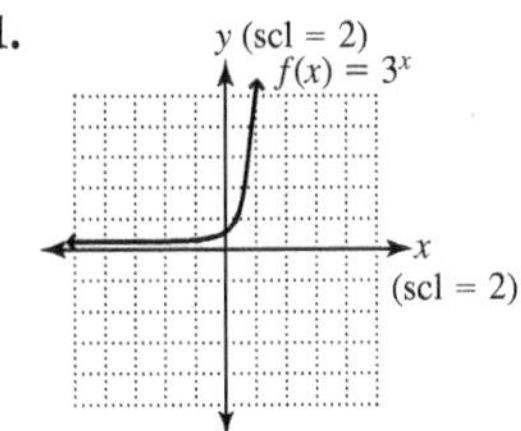

3.

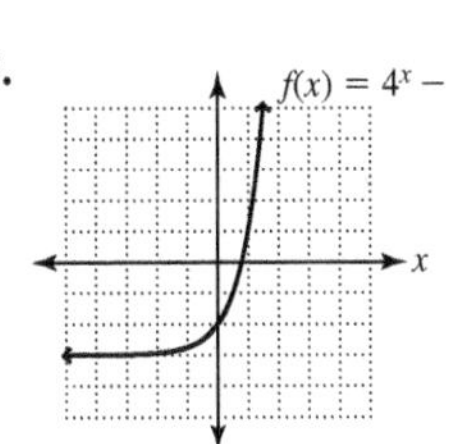

5.

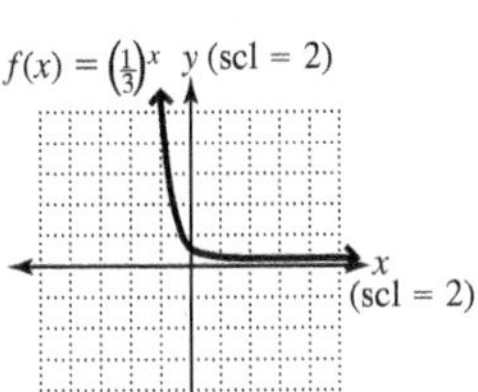

7.

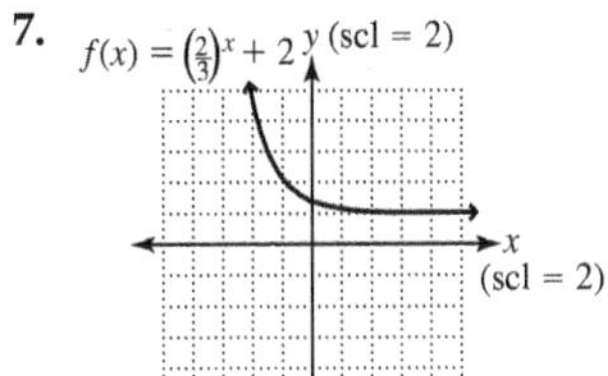

9.

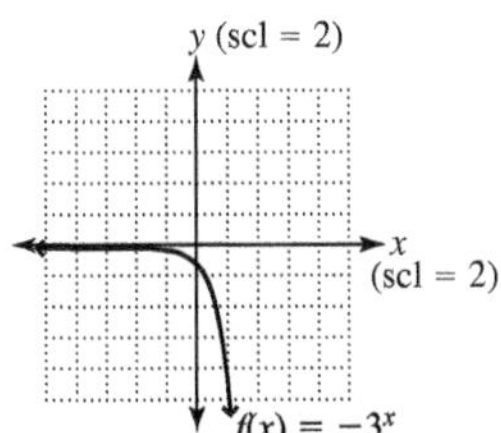

11.

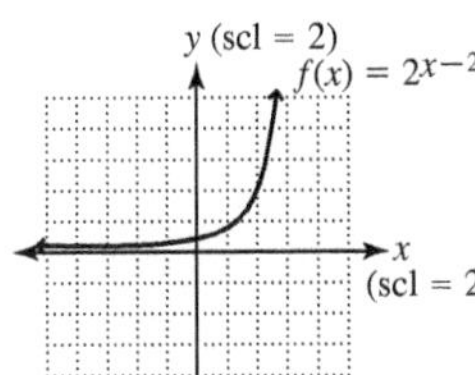
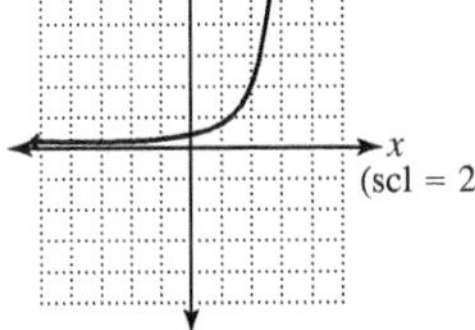

13.

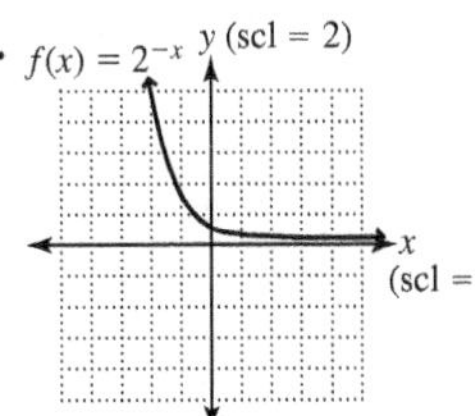
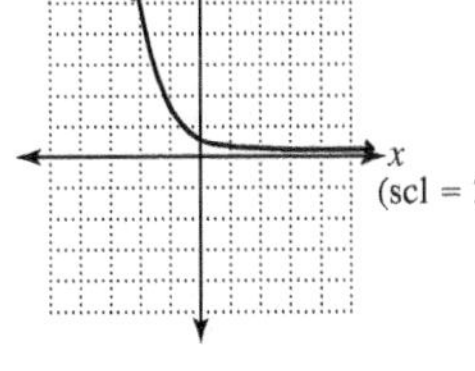

15.

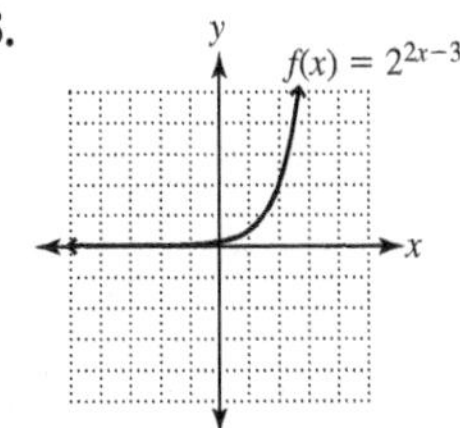

17.

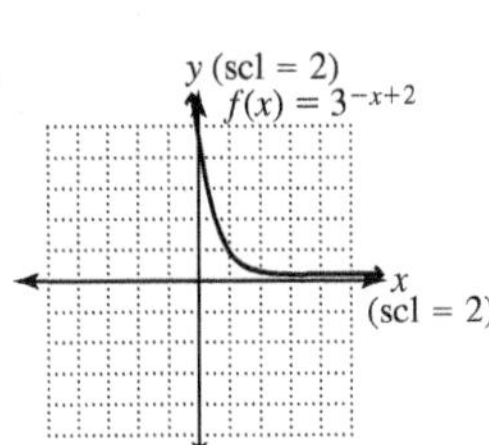

19. 3 **21.** $\dfrac{5}{3}$ **23.** $\dfrac{1}{2}$ **25.** −2 **27.** −2 **29.** 3 **31.** −4 **33.** $\dfrac{1}{2}$ **35.** −12 **37. a.**

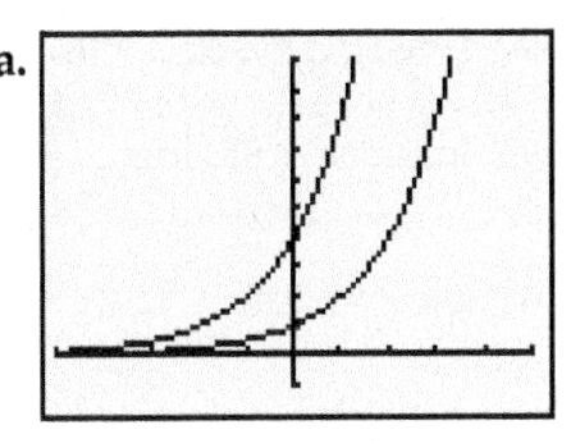

b. The graph of g is the graph of f shifted 2 units to the left. **39. a.**

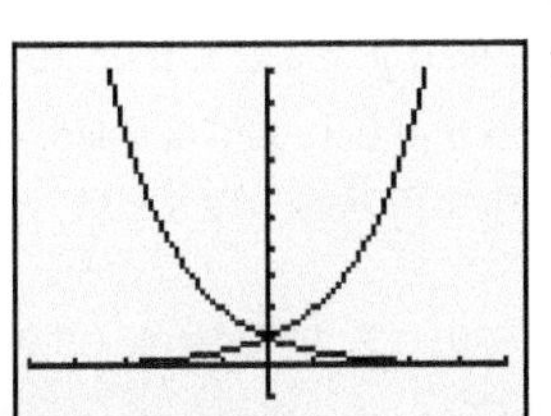

b. The graph of g is the graph of f reflected about the y-axis. **41.** 6400 cells **43.** 6144 billion people **45.** \$25,870.70 **47.** 0.020 g **49.** ≈17,875.41 **51.** 1.225 parts per million **53.** ≈118,130 million transistors

Review Exercises: **1.** 125 **2.** 9 **3.** $\frac{1}{64}$ **4.** $\sqrt[3]{5^2}$ or $(\sqrt[3]{5})^2$ **5.** $f^{-1}(x) = \frac{x-4}{3}$ **6.**

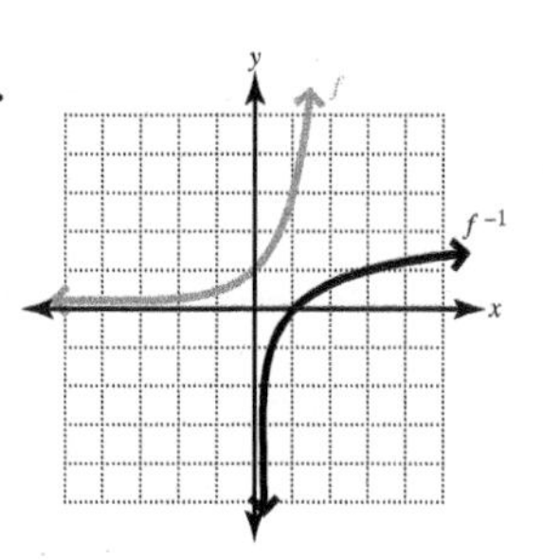

Exercise Set 10.3

Prep Exercises **1.** $f^{-1}(x) = \log_3 x$ because logarithmic and exponential functions are inverses. **2.** Logarithms are exponents because logarithms are inverses of exponential functions. **3.** $b > 0, b \neq 1$ **4.** $x > 0$ **5.** power, b, x **6.** The graphs are symmetric about the line $y = x$ because they are inverses.

Exercises **1.** $\log_2 32 = 5$ **3.** $\log_{10} 1000 = 3$ **5.** $\log_e x = 4$ **7.** $\log_5 \frac{1}{125} = -3$ **9.** $\log_{10} \frac{1}{100} = -2$ **11.** $\log_{625} 5 = \frac{1}{4}$ **13.** $\log_{1/4} \frac{1}{16} = 2$ **15.** $\log_7 \sqrt{7} = \frac{1}{2}$ **17.** $3^4 = 81$ **19.** $4^{-2} = \frac{1}{16}$ **21.** $10^2 = 100$ **23.** $e^5 = a$ **25.** $e^{-4} = \frac{1}{e^4}$ **27.** $\left(\frac{1}{8}\right)^2 = \frac{1}{64}$ **29.** $\left(\frac{1}{5}\right)^{-2} = 25$ **31.** $7^{1/2} = \sqrt{7}$ **33.** 32 **35.** $\frac{1}{25}$ **37.** 4 **39.** -2 **41.** 10 **43.** 2 **45.** $\frac{1}{4}$ **47.** 243 **49.** 2 **51.** -6 **53.** 16 **55.** $\frac{1}{64}$ **57.** 2 **59.** 125 **61.** 3 **63.** $\frac{1}{16}$ **65.**

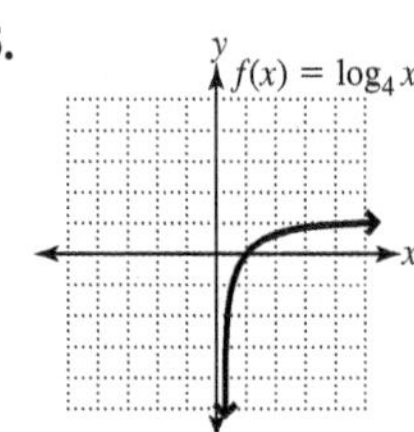

67. $f(x) = \log_{1/3} x$

69. 97% **71.** After 4 days. **73.** $\log_b b = 1$ because $b^1 = b$. **75.** $\log_b 1 = 0$ because $b^0 = 1$.

Review Exercises: **1.** x^{-6} **2.** x^6 **3.** x^{15} **4.** x^3 **5.** x^5 **6.** $x^{3/4}$

Exercise Set 10.4

Prep Exercises **1.** x **2.** x **3.** $\log_b x + \log_b y$ **4.** $\log_b (x + y)$; there is no rule for the logarithm of a sum. **5.** $\log_b x - \log_b y$ **6.** $r \log_b x$

Exercises **1.** 2 **3.** r **5.** $4x$ **7.** 5 **9.** y **11.** $7x$ **13.** $\log_2 5 + \log_2 y$ **15.** $\log_a p + \log_a q$ **17.** $\log_4 m + \log_4 n + \log_4 p$ **19.** $\log_a x + \log_a (x - 5)$ **21.** $\log_3 40$ **23.** $\log_4 27$ **25.** $\log_a 7m$ **27.** $\log_4 ab$ **29.** $\log_a (2x^2 + 10x)$ **31.** $\log_4 (x^2 + 4x + 3)$ **33.** $\log_2 7 - \log_2 9$ **35.** $\log_a x - \log_a 5$ **37.** $1 - \log_a b$ **39.** $\log_a x - \log_a (x - 3)$ **41.** $\log_4 (2x - 3) - \log_4 (4x + 5)$ **43.** $\log_6 8$ **45.** 1 **47.** $\log_a \frac{x}{3}$ **49.** $\log_4 \frac{p}{q}$ **51.** $\log_b \frac{x}{x-4}$ **53.** 1 **55.** $6 \log_4 3$ **57.** $7 \log_a x$ **59.** $\frac{1}{2} \log_a 3$ **61.** $\frac{2}{3} \log_3 x$ **63.** $-2 \log_a 6$ **65.** $-2 \log_a y$ **67.** $\log_3 5^4$ **69.** $\log_2 \frac{1}{x^3}$ **71.** $\log_7 8$ **73.** $\log_a \sqrt[4]{x^3}$ **75.** $\log_a 4$ **77.** $\log_3 \frac{1}{\sqrt{x}}$ **79.** $3 \log_a x - 4 \log_a y$ **81.** $4 \log_3 a + 2 \log_3 b$ **83.** $\log_a x + \log_a y - \log_a z$ **85.** $2 \log_x a - \log_x b - 3 \log_x c$ **87.** $\frac{3}{4} \log_4 x - \frac{1}{4} \log_4 y$ **89.** $\frac{2}{3} \log_a x + \frac{1}{3} \log_a y - \log_a z$ **91.** $\log_3 \frac{1}{2}$ **93.** $\log_b x^4 y^3$ **95.** $\log_a \sqrt{\frac{5}{7}}$ **97.** $\log_a \sqrt[5]{(x^2y^3)^2}$ **99.** $\log_b (3x^2 - 2x)$ **101.** $\log_a \frac{(x-2)^3}{(x+1)^4}$ **103.** $\log_a \frac{x^2 z^4}{w^3 u^6}$

Review Exercises: **1.** $10^{-2.5}$ **2.** 10^9 **3.** \$13,468.55 **4.** $\log_{10} 50 = 1.6990$ **5.** $10^{1.6532} = 45$ **6.** $e^{-1.3863} = 0.25$

Exercise Set 10.5

Prep Exercises **1.** 10 **2.** e **3.** Negative. If $y = \log x$, then $10^y = x$. If $0 < x < 1$, then y must be negative so that $x = \frac{1}{10^y}$ where $y > 0$, which results in values of x such that $0 < x < 1$. Also, any positive value of y results in a value of $x > 1$. Also look at the graph. **4.** Positive. If $y = \ln x$, then $e^y = x$. If $x > 1$, then $e^y > 1$, which is true if $y > 0$. **5.** 0.91629 **6.** $\sqrt{5}$

Exercises **1.** 1.8062 **3.** -2.1739 **5.** 2.6391 **7.** 4.1761 **9.** -5.7959 **11.** 2.2343 **13.** 4.3720 **15.** -3.3814 **17.** 5.6864 **19.** 0.4343 **21.** Error results because the domain of $\log_a x$ is $(0, \infty)$, so log 0 is undefined. **23.** 2 **25.** -2 **27.** $\frac{1}{3}$ **29.** -3 **31.** 3 **33.** $\frac{1}{2}$ **35.** 90 dB **37.** 10^{-6} watts/m^2 **39.** ≈ 2.796 **41.** $10^{-3.5}$ moles/L **43.** ≈ 22.9 **45.** The 1964 Alaska earthquake was about four times as severe as the 1906 San Francisco earthquake. **47.** $10^{22.75}$ ergs **49.** 1986 **51.** ≈ 6124 ft.

Review Exercises: **1.** $\log_3 \frac{2x+1}{x-1}$ **2.** $3^2 = 2x + 5$ **3.** -11 **4.** $-5, 3$ **5.** 8 **6.** 10

Exercise Set 10.6

Prep Exercises **1.** $x = y$ **2.** Common logs because the base is 10 **3.** If $x = y$, then $\log_b x = \log_b y$. **4.** $m = n$ **5.** If $\log_b x = \log_b y$, then $x = y$. **6.** The solution $x = -5$ is rejected because substituting into the equation gives $\log(-5)$ and $\log(-3)$, which are both undefined.

Exercises **1.** 3.1699 **3.** 1.0767 **5.** −1.5693 **7.** 2.8617 **9.** 12.4251 **11.** 7.1285 **13.** 0.5365 **15.** 107.2959 **17.** −73.1563 **19.** 6 **21.** 11 **23.** 6 **25.** ±4 **27.** 4, −9 **29.** 7 **31.** No solution **33.** −6, 2 **35.** 2 **37.** 10 **39.** 3 **41.** $\frac{3}{2}$ **43.** 3 **45.** 4 **47.** $\frac{5}{6}$ **49.** $\frac{5}{2}$ **51.** 6 **53.** 4 **55.** 9.5 yr. **57. a.** \$19,676.82 **b.** ≈9.3 yr. **59. a.** ≈5.58% **b.** ≈0.18 **61. a.** ≈875 mosquitoes **b.** ≈75 days **63. a.** ≈7.59 billion **b.** ≈2.6 yr. after 2010, in 2012 **65. a.** ≈1.89 in. of mercury **b.** ≈21.8 mi. **67. a.** ≈79.8% **b.** ≈7.76, so about age 8 **69.** ≈\$10.21 **71.** ≈\$1435.1 billion **73.** 1.7925 **75.** 0.5283 **77.** −2.3219 **79.** 0.3685 **81.** 47%

Review Exercises: **1.** $x + 3$ **2.** $28 + \sqrt{5}$ **3.** 29 **4.** $x^2 + 2x + 3$ **5.** $4x^2 - 4x + 5$ **6.** 13

Chapter 10 Summary and Review Exercises

10.1 Definitions/Rules/Procedures $f[g(x)]$; g; f; $g[f(x)]$; f; g Exercises **1.** 25 **2.** 167 **3.** −2 **4.** 14 **5.** $(f \circ g)(x) = 6x + 3$, $(g \circ f)(x) = 6x - 9$ **6.** $(f \circ g)(x) = 9x^2 - 42x + 53$, $(g \circ f)(x) = 3x^2 + 5$ **7.** $(f \circ g)(x) = \sqrt{2x - 4}$, $(g \circ f)(x) = 2\sqrt{x - 3} - 1$ **8.** $(f \circ g)(x) = \frac{4x - 4}{x - 4}$, $(g \circ f)(x) = \frac{3 - 3x}{x + 3}$ Definitions/Rules/Procedures x; g; x; f Exercises **9.** yes **10.** yes **11.** no **12.** no Definitions/Rules/Procedures $a = b$; $f(a) \neq f(b)$; one and only one point; one-to-one; $y = x$ Exercises **13.** no **14.** yes Definitions/Rules/Procedures y; y's; x's; y; $f^{-1}(x)$ Exercises **15.** $f^{-1}(x) = \frac{x - 4}{5}$ **16.** $f^{-1}(x) = \sqrt[3]{x - 6}$ **17.** $f^{-1}(x) = \frac{4 - 5x}{x}$ **18.** $f^{-1}(x) = \frac{x^3 - 2}{3}$ **19.** $(-6, 4)$ **20.** $f^{-1}(b) = a$

10.2 Definitions/Rules/Procedures > 1; < 1 Exercises **21.**

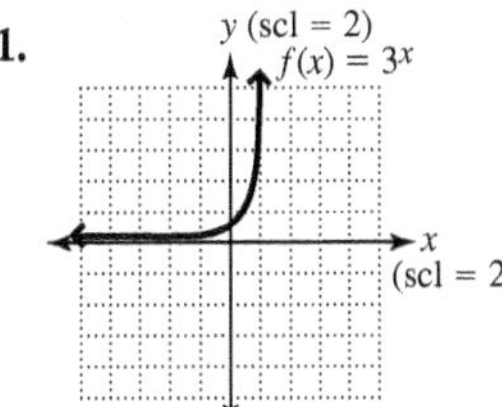

22.

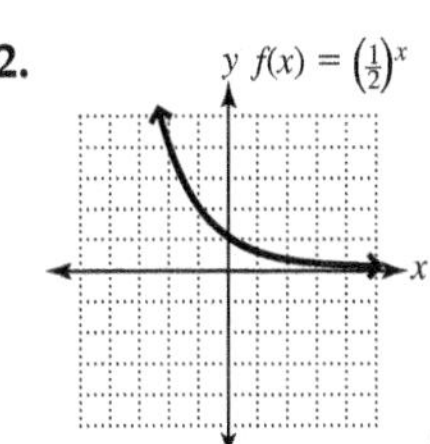

23.

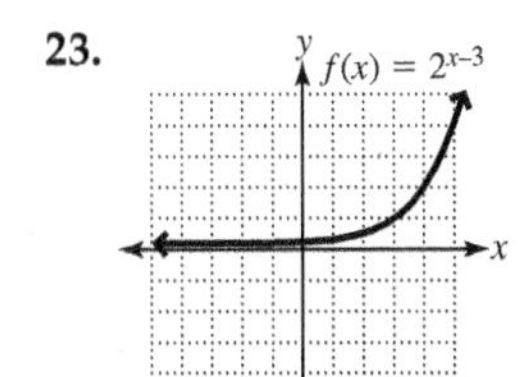

24.

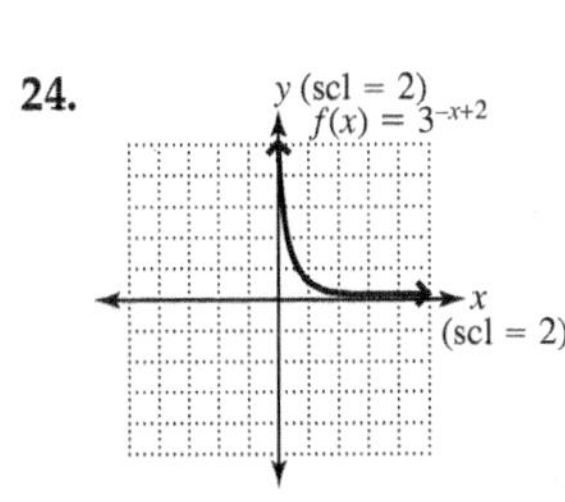

Definitions/Rules/Procedures $x = y$; same base; exponents; exponents

Exercises **25.** 4 **26.** $\frac{3}{2}$ **27.** −2 **28.** −2 **29.** 4 **30.** 2 **31.** 3 **32.** $-\frac{6}{7}$ **33.** ≈6277 cells **34.** \$40,353.57 **35.** ≈7.16 g **36.** ≈\$9.19

10.3 Definitions/Rules/Procedures $x = b^y$ Exercises **37.** $\log_7 343 = 3$ **38.** $\log_4 \frac{1}{64} = -3$ **39.** $\log_{3/2} \frac{81}{16} = 4$ **40.** $\log_{11} \sqrt[3]{11} = \frac{1}{3}$ **41.** $9^2 = 81$ **42.** $\left(\frac{1}{5}\right)^{-3} = 125$ **43.** $a^4 = 16$ **44.** $e^b = c$ Definitions/Rules/Procedures $b^x = y$; 1; 0 Exercises **45.** $\frac{1}{81}$ **46.** 4 **47.** 5 **48.** −2 **49.** 3 **50.** $\frac{1}{10}$ **51.** 1 **52.** 0 Definitions/Rules/Procedures y; $x = b^y$; y; x Exercises **53.**

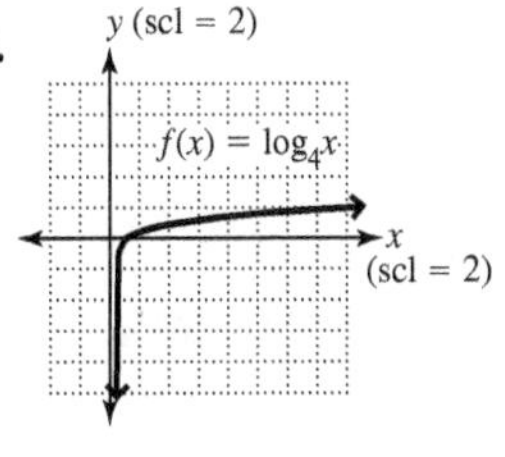

54.

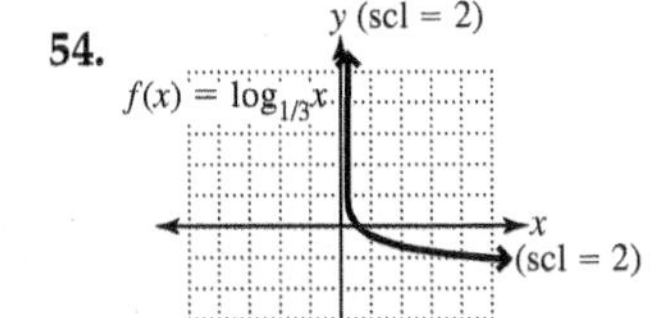

55. Domain is $(-\infty, \infty)$ and range is $(0, \infty)$ because $f(x) = \log_b x$ and $g(x) = b^x$ are inverses. **56.** 30 dB

10.4 Definitions/Rules/Procedures x; x; $\log_b x + \log_b y$; $\log_b x - \log_b y$; $r\log_b x$ Exercises **57.** 8 **58.** 6 **59.** $1 + \log_6 x$ **60.** $\log_4 x + \log_4(2x - 5)$ **61.** $\log_3(32)$ **62.** $\log_5(3x^2 - 6x)$ **63.** $\log_b x - \log_b 5$ **64.** $\log_a(3x - 2) - \log_a(4x + 3)$ **65.** $\log_8 2$ **66.** $\log_2 \frac{x + 5}{2x - 3}$ **67.** $4 \log_3 7$ **68.** $\frac{1}{3}\log_a x$ **69.** $-4 \log_4 a$ **70.** $\frac{4}{5}$ **71.** $\log_a x^4$ **72.** $\log_a \sqrt[5]{y^3}$ **73.** $2 \log_a x + 3 \log_a y$ **74.** $4 \log_a c - 3 \log_a d$ **75.** $2 \log_a x + 3 \log_a y - 4 \log_a z$ **76.** $\frac{3}{2} - 2 \log_a b$ **77.** $\log_x y^3 z^5$ **78.** $\log_a \frac{4^3}{3^2}$ **79.** $\log_a \sqrt[4]{x^2 y^3}$ **80.** $\log_a(x + 5)^4(x - 3)^2$

10.5 Definitions/Rules/Procedures common; $\log x$; LOG; natural; $\ln x$; ln Exercises **81.** 2.5132 **82.** −2.4559 **83.** −3.1466 **84.** 4.5218 **85.** −5 **86.** $\frac{1}{4}$ **87.** 85 dB **88.** ≈2.9 **89.** ≈17 yr. **90.** The 7.8 earthquake was 10 times as severe.

10.6 Definitions/Rules/Procedures $x = y$; b^y; $x = y$; $\log_b y$; b^y; $A_0 e^{kt}$; initial; growth; decay Exercises **91.** 1.5773 **92.** 0.5360 **93.** 1.3626 **94.** 4.9592 **95.** 0.5995 **96.** −536.4793 **97.** 6.1 yr. **98. a.** $12,254.71 **b.** ≈7.7 yr. **99. a.** ≈573 ants **b.** ≈23 months **100.** ≈24 min. Definitions/Rules/Procedures $x = y$; $b^y = x$ Exercises **101.** 8 **102.** −4, 2 **103.** 3 **104.** 20 **105.** 36 **106.** 8 **107.** 2 **108.** 5 Definitions/Rules/Procedures x; a; x; a; x; a Exercises **109.** 1.9534 **110.** −2.5850

Chapter 10 Practice Test

1. $f[g(x)] = 9x^2 - 30x + 19$ [10.1] **2. a.** $f^{-1}(x) = \frac{x + 3}{4}$ [10.1] **b.** $f[f^{-1}(x)] = f\left[\frac{x + 3}{4}\right] = 4\left(\frac{x + 3}{4}\right) - 3 = x + 3 - 3 = x$; $f^1[f(x)] = f^{-1}(4x - 3) = \frac{4x - 3 + 3}{4} = \frac{4x}{4} = x$ [10.1] **c.** The graphs are symmetric about the graph of $y = x$. [10.1]

3.

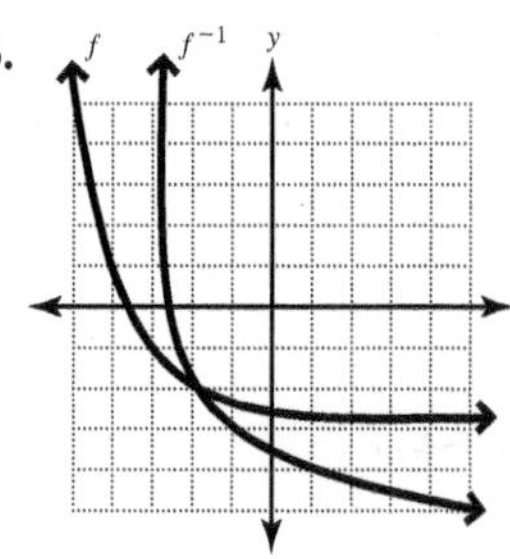

[10.1] **4.**

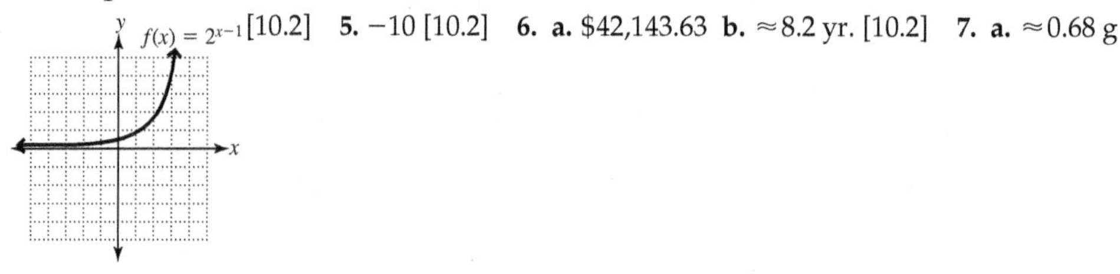

[10.2] **5.** −10 [10.2] **6. a.** $42,143.63 **b.** ≈8.2 yr. [10.2] **7. a.** ≈0.68 g **b.** ≈1.16 hr. [10.2] **8. a.** ≈14.8 million in 1960, ≈65.5 million in 2018 **b.** 1989 [10.2] **9.** $\left(\frac{1}{3}\right)^{-4} = 81$ [10.3] **10.** −3 [11.3] **11.** 5 [10.3] **12.**

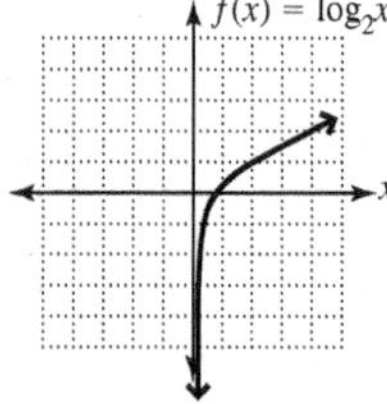

[10.4] **13.** $4 \log_b x + 2 \log_b y - \log_b z$ [10.4] **14.** $\frac{5}{4}\log_b x - \frac{7}{4}\log_b y$ [10.4] **15.** $\log_b \sqrt[4]{x^6 y^9}$ [10.4]

16. a. ≈74 g **b.** ≈247.6 days [10.5] **17.** 4.6433 [10.5] **18.** 3 [10.5] **19.** 3 [10.5] **20. a.** 16.5 generations **b.** 100,000 [10.6]

Chapters 1–10 Cumulative Review Exercises

1. true **2.** true **3.** false **4.** −8. **5.** x-intercept **6.** $x - 3 = 4$; $x - 3 = -4$. **7.** 86 **8.** $\frac{1}{a^{12}}$ **9.** $\frac{3}{pq} - \frac{2p^2}{q^2} + \frac{6q^2}{7}$ **10.** $\frac{-x^2 - 5x - 9}{(x + 3)^2}$ **11.** $24c^4\sqrt{15}$ **12.** $20 - 10i$ **13.** $2x$ **14.** $\frac{d - 4}{2d(3d + 4)}$ **15.** $(x + 2y)(x - 2y)(x + 3)(x - 3)$ **16.** $3x^2y(4x - 3y)(2x - y)$ **17.** 1, 2 **18.** All real numbers. **19.** $(-3, 2)$ **20.** $\frac{1}{2}, \frac{7}{2}$ **21.** 3, (2 is extraneous.) **22.** No solution (5 is extraneous.) **23.** $-\frac{1}{2}$ **24.** 16 **25.** 3 **26. a.** $\frac{2}{5}$ **b.** $(5, 0), (0, -2)$ **c.**

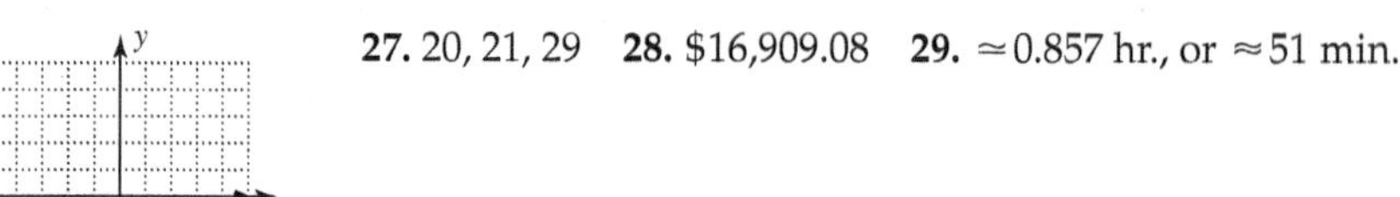

27. 20, 21, 29 **28.** $16,909.08 **29.** ≈0.857 hr., or ≈51 min. **30.** 5 hr.

Chapter 11

Exercise Set 11.1

Prep Exercises **1.** Parabola, circle, ellipse, and hyperbola **2.** Up; $(1, -3)$ **3.** Left; $(-4, -3)$ **4.** It is a parabola because only one variable is squared. **5.** $d = \sqrt{(x_2 - x_1)^2 + (y_2 - y_1)^2}$ **6.** $\left(\frac{x_1 + x_2}{2}, \frac{y_1 + y_2}{2}\right)$ **7.** $(h, k); r$ **8.** complete the square

Exercises **1.** Opens upward; vertex: $(1, 2)$; axis of symmetry: $x = 1$

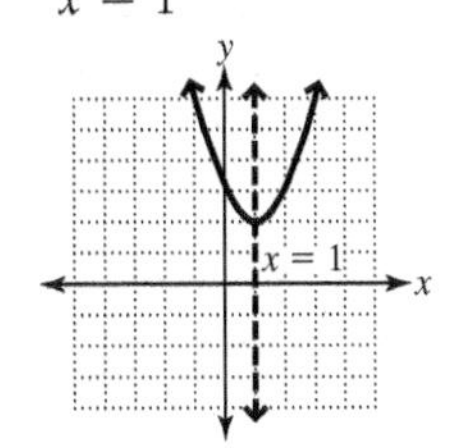

3. Opens downward; vertex: $(-1, 4)$; axis of symmetry: $x = -1$

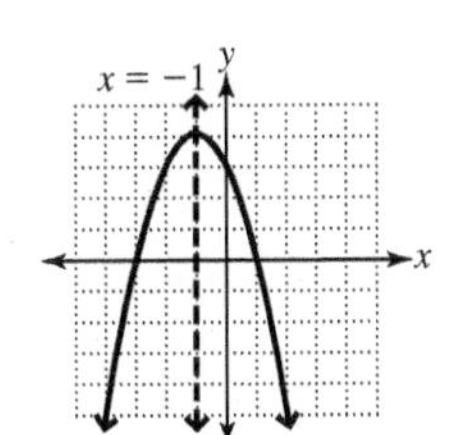

5. Opens right; vertex: $(-2, -2)$; axis of symmetry: $y = -2$

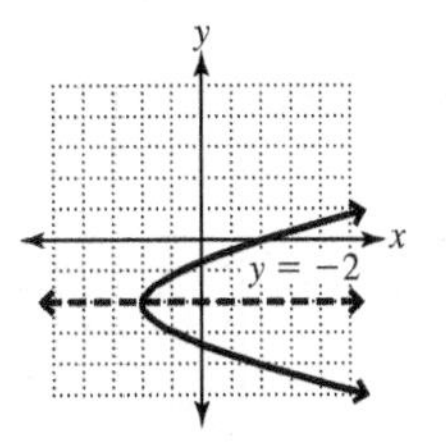

7. Opens left; vertex: $(3, 1)$; axis of symmetry: $y = 1$

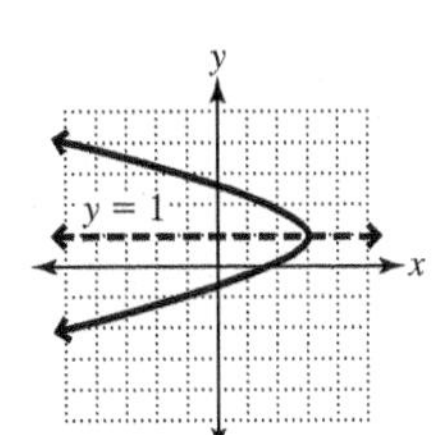

9. Opens right; vertex: $(-4, -2)$; axis of symmetry: $y = -2$

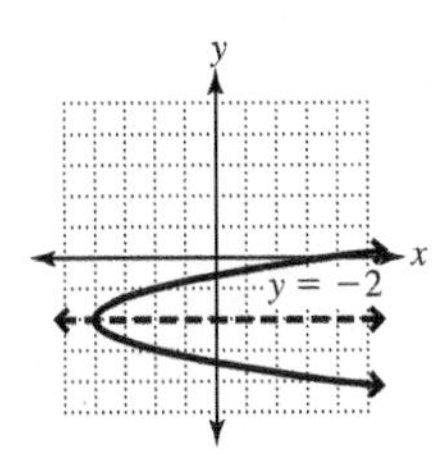

11. Opens left; vertex: $(-5, -2)$; axis of symmetry: $y = -2$

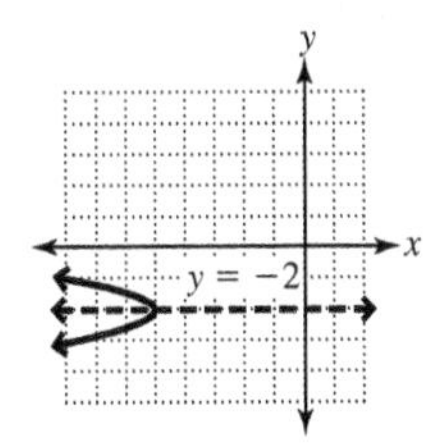

13. Opens right; vertex: $(-1, -2)$; axis of symmetry: $y = -2$

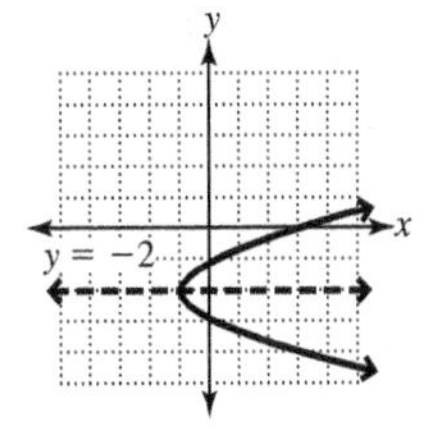

15. Opens left; vertex: $(4, 3)$; axis of symmetry: $y = 3$

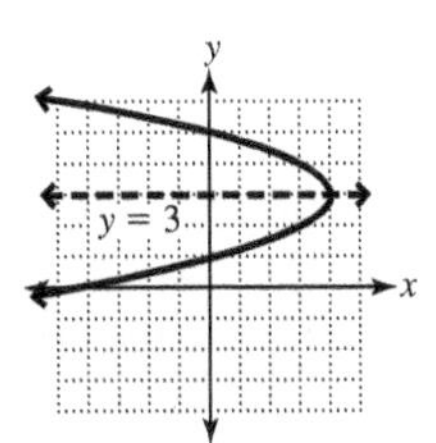

17. Opens right; vertex: $(-5, -2)$; axis of symmetry: $y = -2$

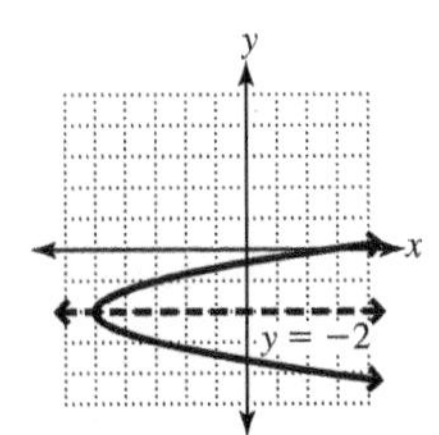

19. Opens right; vertex: $(-2, 1)$; axis of symmetry: $y = 1$

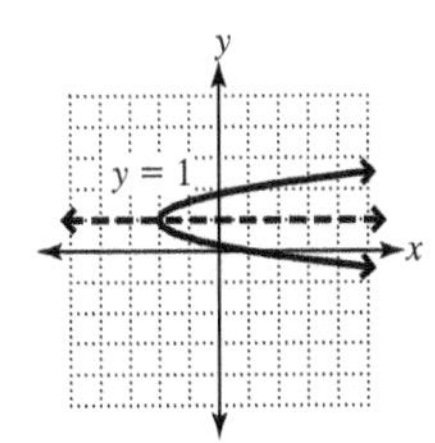

21. Opens left; vertex: $(7, 1)$; axis of symmetry: $y = 1$

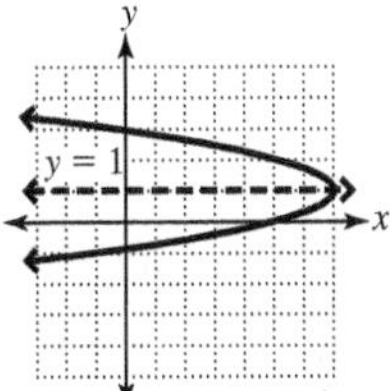

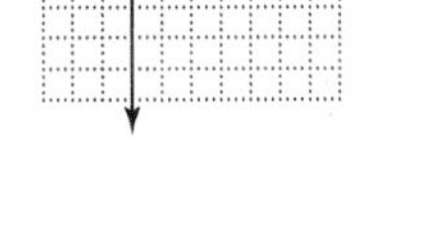

23. c **25.** a **27.** $5, \left(-\frac{5}{2}, 4\right)$ **29.** $13, \left(-\frac{11}{2}, 2\right)$ **31.** $13, \left(-2, -\frac{15}{2}\right)$ **33.** $2\sqrt{5}, (3, 6)$ **35.** $4\sqrt{5}, (-1, 0)$ **37.** $\sqrt{34}, \left(\frac{7}{2}, -\frac{7}{2}\right)$ **39.** 5

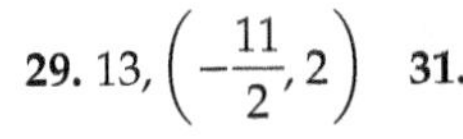

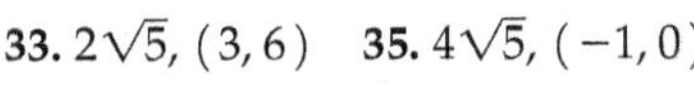

41. $\sqrt{89}$ **43.** Center: $(2, 1)$; radius: 2

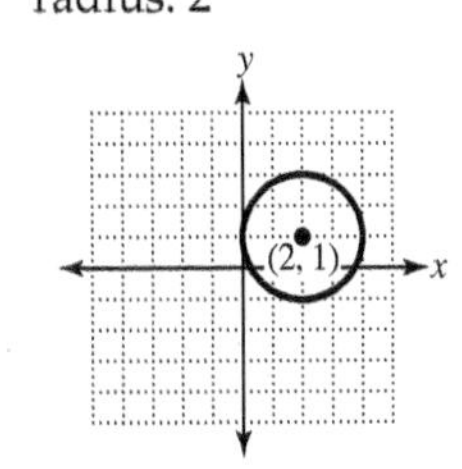

45. Center: $(-3, -2)$; radius: 9

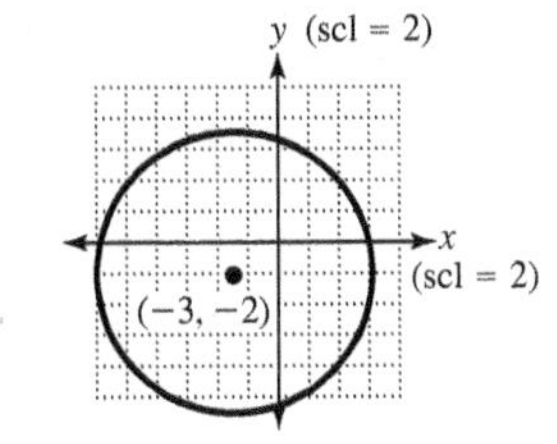

47. Center: $(5, -3)$; radius: 7

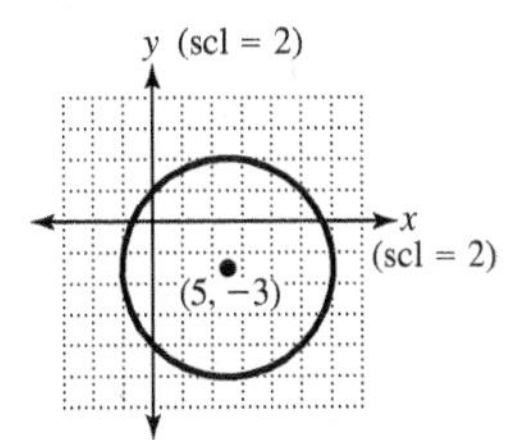

49. Center: $(1, -1)$; radius: $2\sqrt{3}$

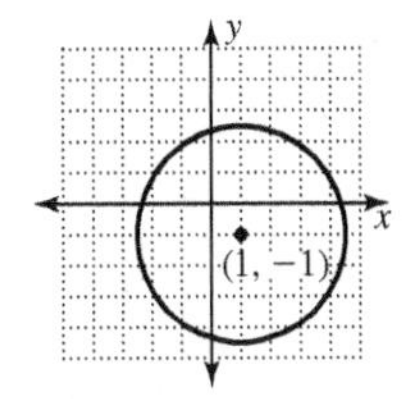

51. Center: $(-4,-2)$; radius: $4\sqrt{2}$

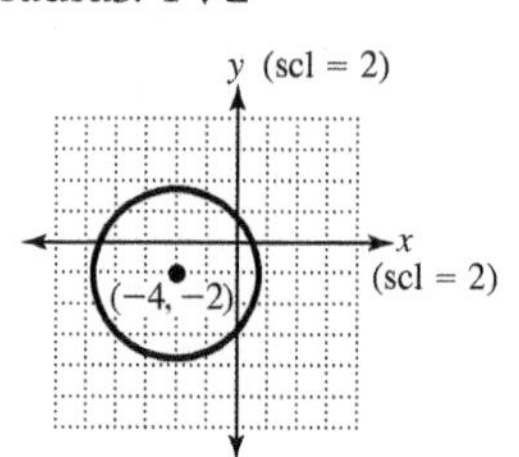

53. Center: $(-4,3)$; radius: 3

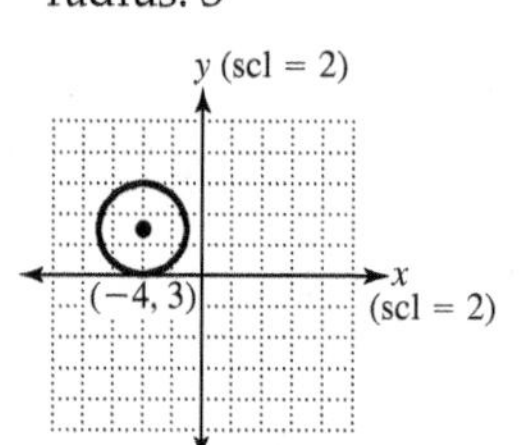

55. Center: $(-5,2)$; radius: 8

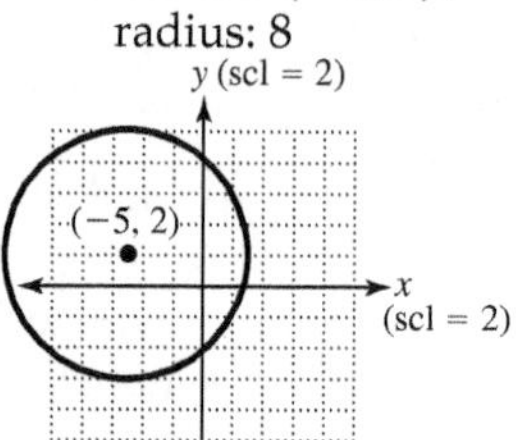

57. Center: $(-7,2)$; radius: 2

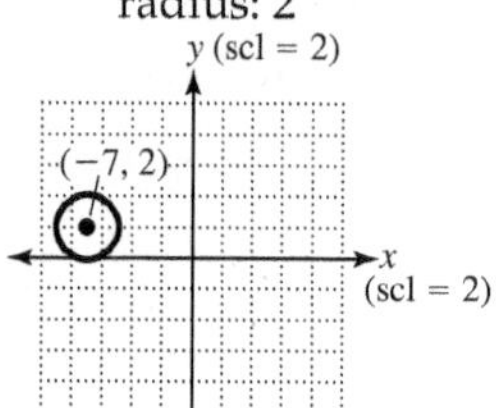

59. b **61.** c **63.**

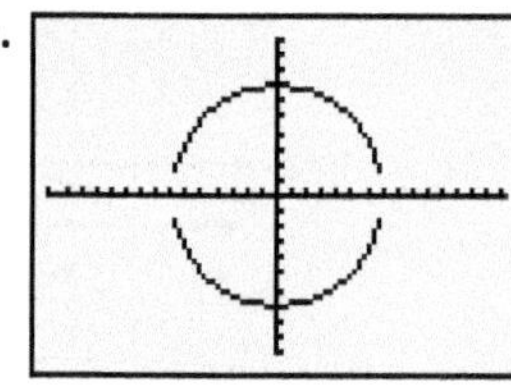

65.

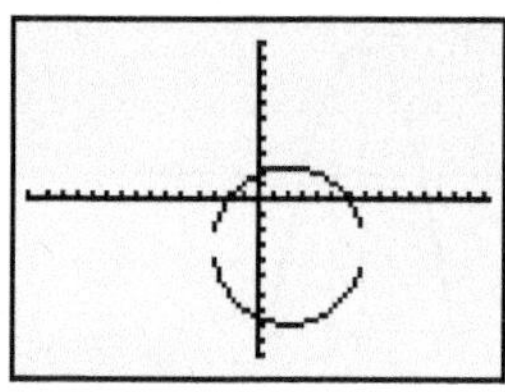

67. $(x-4)^2+(y-2)^2=16$

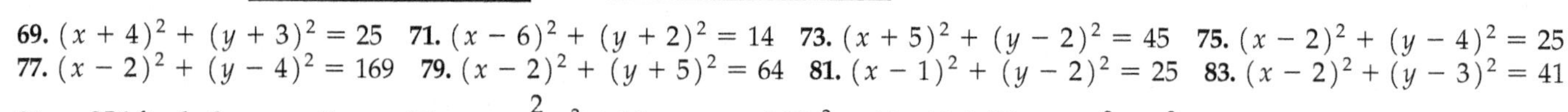

69. $(x+4)^2+(y+3)^2=25$ **71.** $(x-6)^2+(y+2)^2=14$ **73.** $(x+5)^2+(y-2)^2=45$ **75.** $(x-2)^2+(y-4)^2=25$ **77.** $(x-2)^2+(y-4)^2=169$ **79.** $(x-2)^2+(y+5)^2=64$ **81.** $(x-1)^2+(y-2)^2=25$ **83.** $(x-2)^2+(y-3)^2=41$

85. a. 256 ft. **b.** 3 sec. **c.** 7 sec. **87.** $y=-\frac{2}{25}x^2+18$ or $y=-0.08x^2+18$ **89.** 3.5% **91.** $x^2+y^2=400$ **93.** $x^2+(y-110)^2=10{,}000$ **95.** $x^2+y^2=43{,}560{,}000$

Review Exercises: **1.**

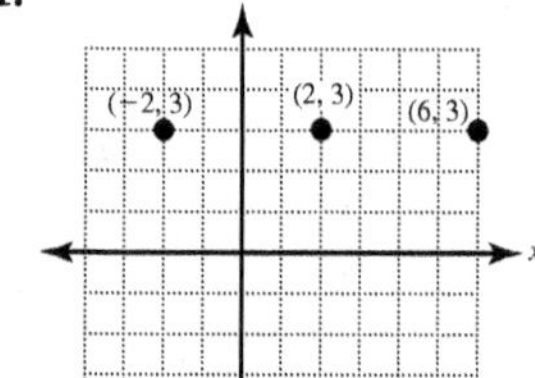

2. $(5x+3y)(5x-3y)$ **3.**

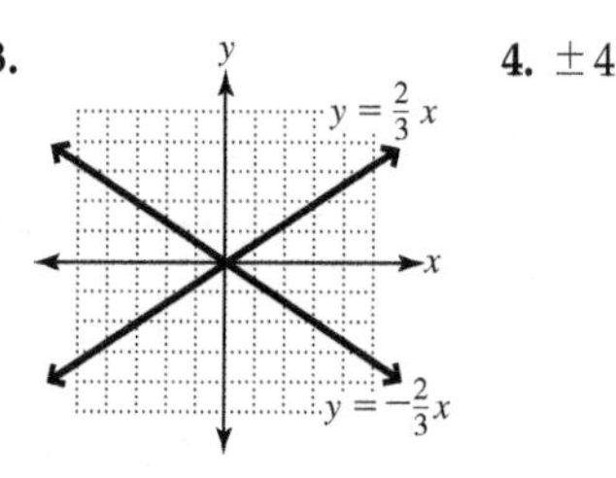

4. ± 4

5. $(4,0),(-4,0),(0,3),(0,-3)$ **6.** $(3,0),(-3,0),(0,5),(0,-5)$

Exercise Set 11.2

Prep Exercises **1.** An ellipse is the set of all points the sum of whose distances from two fixed points is constant. **2.** The fixed points are the foci. **3. a.** $(0,0)$ **b.** 3 units left and right **c.** 5 units up and down **4.** If the x^2-term is positive, the graph has x-intercepts and opens left and right. If the y^2-term is positive, the graph has y-intercepts and opens up and down. **5. a.** $(0,0)$ **b.** $(-3,0)$ and $(3,0)$ **c.** It has no y-intercepts. **d.** $(3,5),(-3,5),(3,-5),(-3,-5)$ **6. a.** $(0,0)$ **b.** It has no x-intercepts. **c.** $(0,-4)$ and $(0,4)$ **d.** $(9,4),(-9,4),(9,-4),(-9,-4)$ **7.** The parabola has only one squared term. **8.** The equation of an ellipse is a sum; the equation of a hyperbola is a difference.

Exercises **1.**

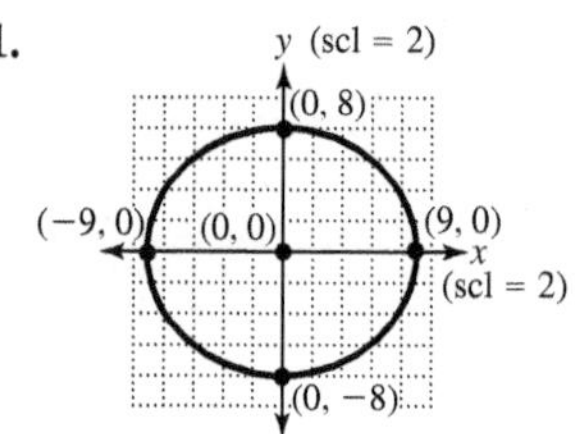

3.

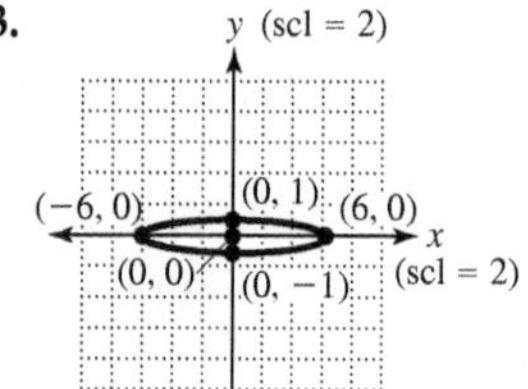

5.

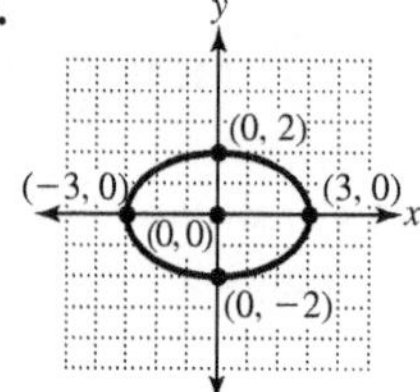

7.

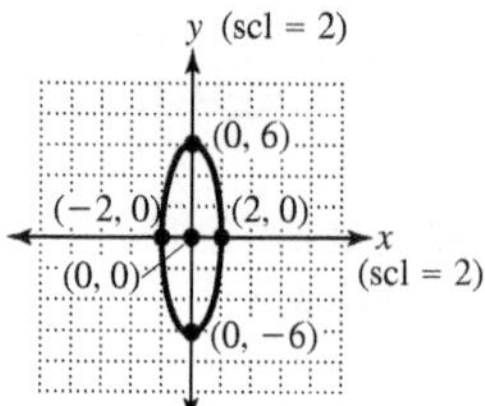

9.

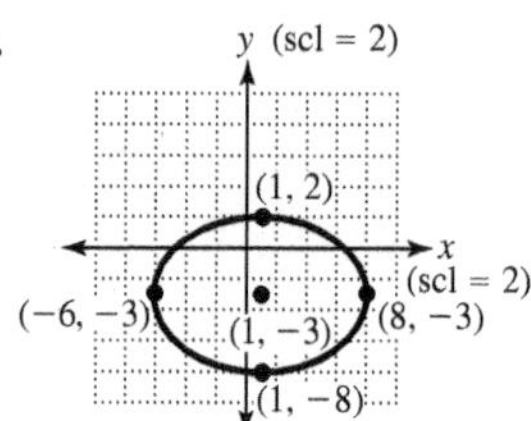

11.

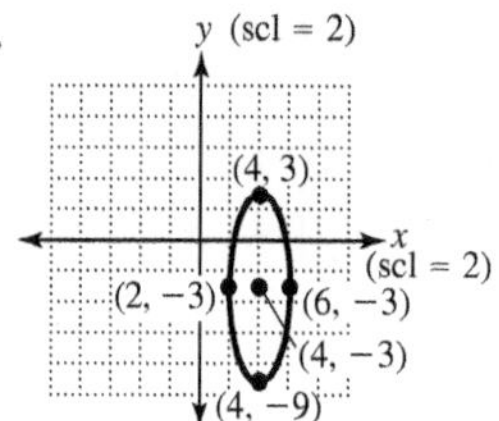

13.

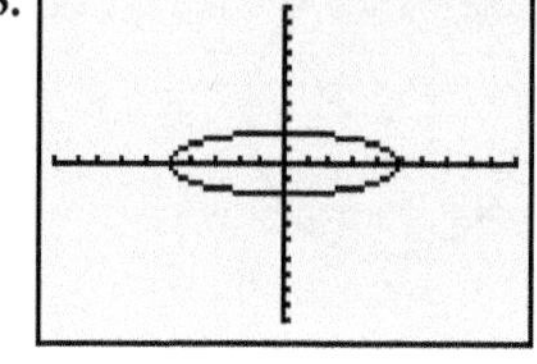

15.

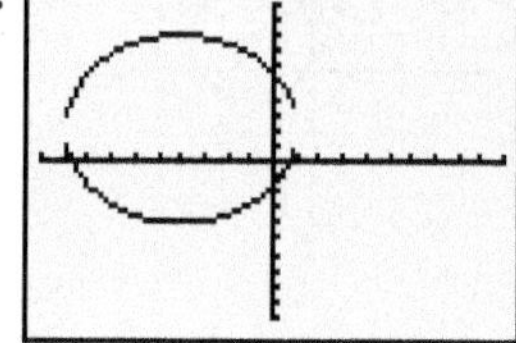

17. $\frac{x^2}{9} + \frac{y^2}{16} = 1$ **19.**

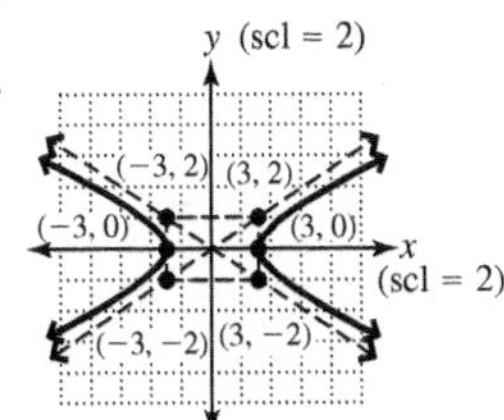

21.

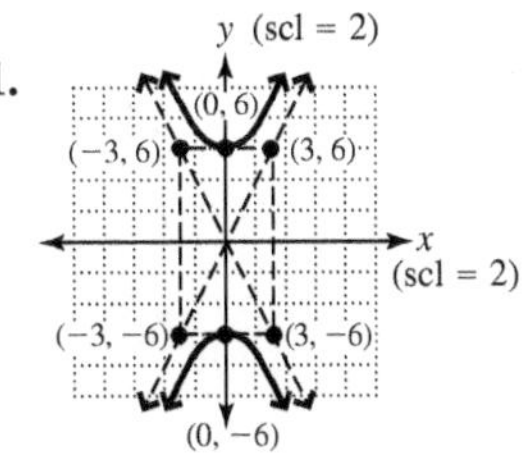

23.

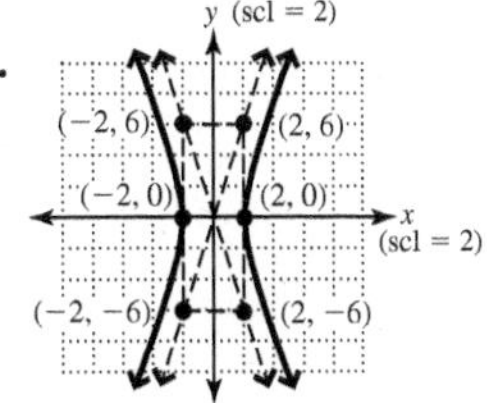

25.

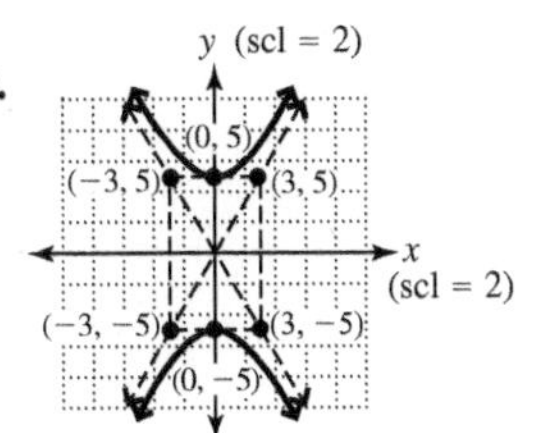

27.

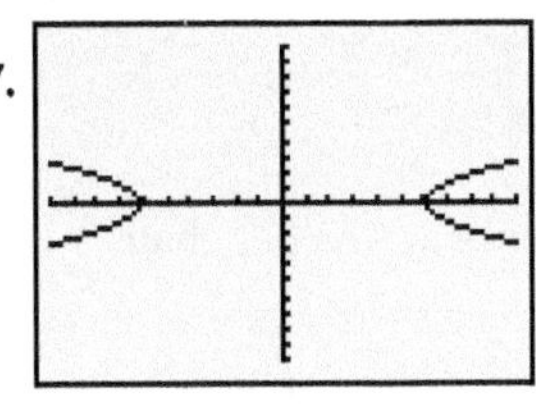

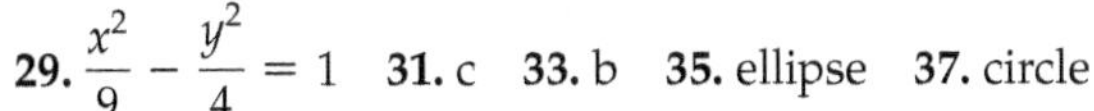

29. $\frac{x^2}{9} - \frac{y^2}{4} = 1$ **31.** c **33.** b **35.** ellipse **37.** circle

39. circle

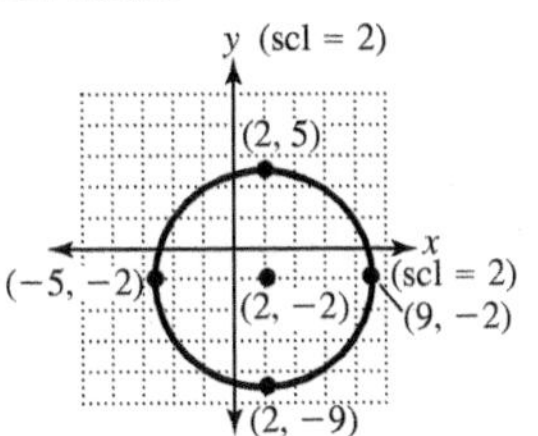

41. parabola

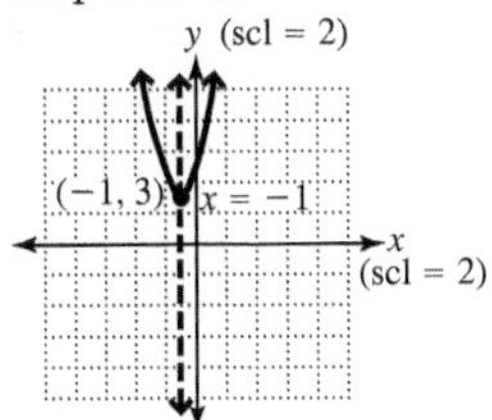

43. ellipse

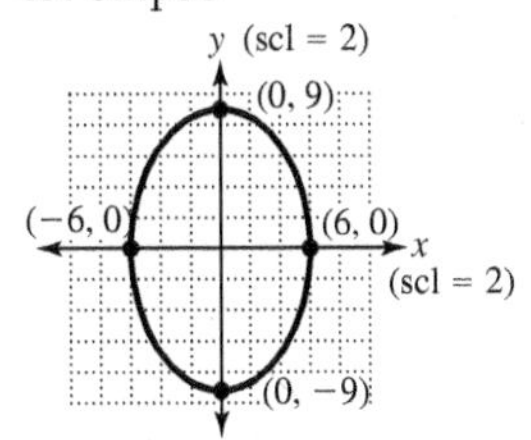

45. hyperbola

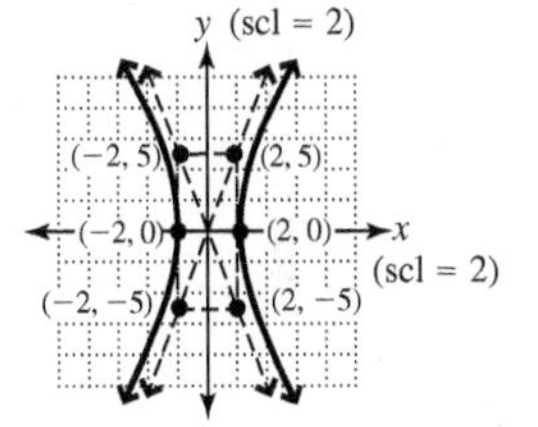

47. parabola

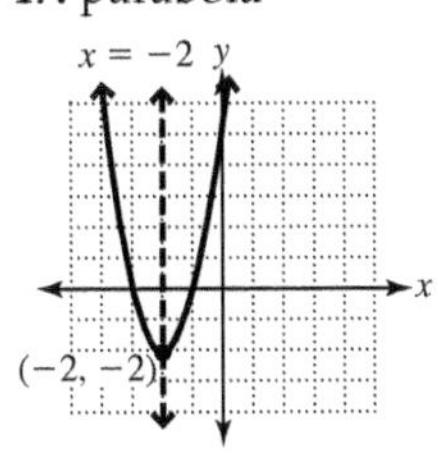

49. hyperbola

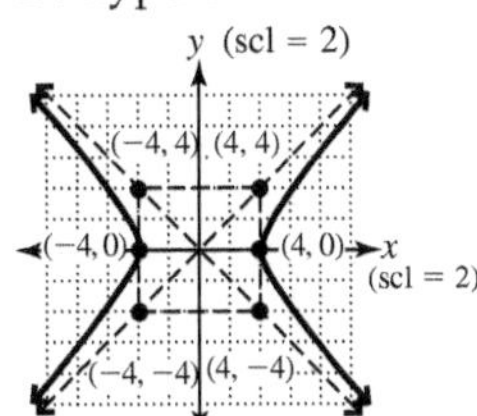

51. ellipse

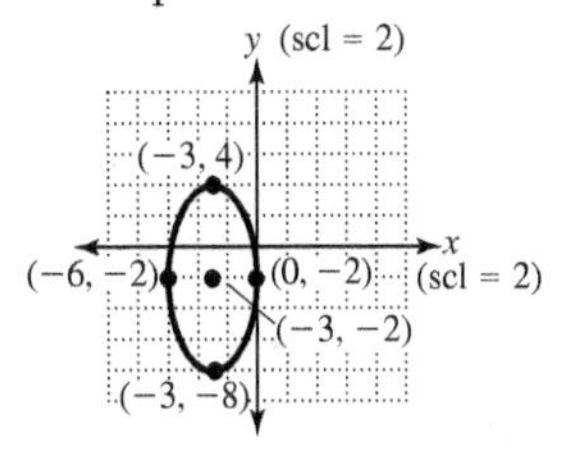

53. a. Yes, the sailboat will clear the bridge. The height of the bridge at the center is 20 feet, and the boat's mast is only 18 feet above the water. **b.** The bridge is 32 feet wide at the base of the arch. **55.**

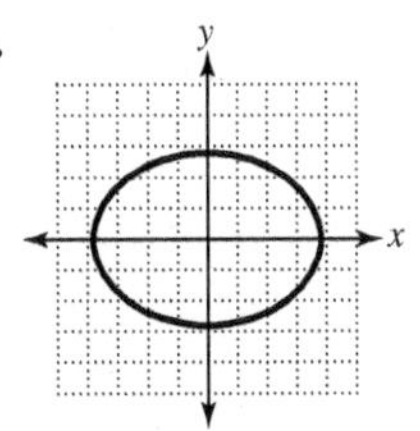

57. $\frac{x^2}{2.25} + \frac{y^2}{4} = 1$ **59.** 3 units

Review Exercises: **1.** $y = -3x^2 + 6$ **2.** $(1, 2)$ **3.** $(2, -3)$ **4.** $(-3, 4)$ **5.** $-4, \frac{2}{3}$ **6.** ± 3

Exercise Set 11.3

Prep Exercises **1. a.** 0, 1, 2, 3, or 4 solutions are possible.
b. Answers may vary.

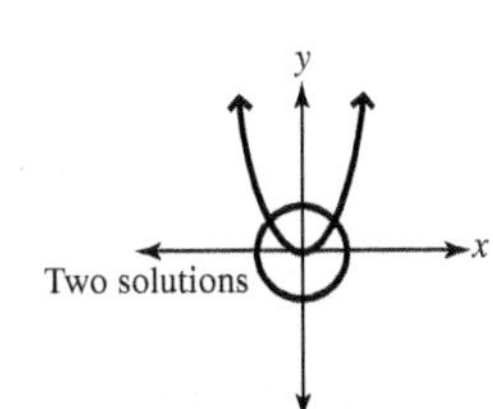

2. a. 0, 1, or 2 solutions are possible.
b. Answers may vary.

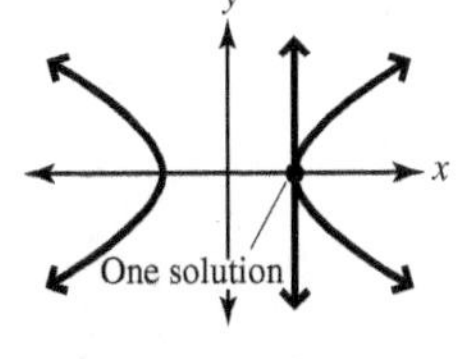

3. Substitution because the first equation has a linear term and is easy to solve for y. **4.** Elimination because all variables are squared. **5.** The graph of the first equation is a line, and the second is an ellipse; so the system could have 0, 1, or 2 solutions. **6.** No. If both graphs are centered at the origin, then because of the symmetry, there could be only 0, 2, or 4 solutions.

Exercises **1.** $(3, -4), (1, 0)$ **3.** $(-2, 0), (1, 9)$ **5.** $(-4, -3), (3, 4)$ **7.** $(2.6, -1.8), (1, 3)$ **9.** No solution **11.** $(0, 3), (1, 5)$ **13.** $(3, 2), (2, 3)$ **15.** $(4, 2), (4, -2), (-4, -2), (-4, 2)$ **17.** $(3, 2), (3, -2), (-3, -2), (-3, 2)$ **19.** $(2, 1), (2, -1), (-2, -1), (-2, 1)$ **21.** $(-6, 0), (3, 3), (3, -3)$ **23.** No solution **25.** $(2, 4), (-2, 4)$ **27.** $(4, 0), (-4, 0)$ **29.** $(2, 2), (-2, -2)$ **31.** $(5, 12), (-1, 0)$ **33.** $(2, 2), (2, -2), (-2, -2), (-2, 2)$ **35.** $(-7, 3), (-7, -3), (2, 0)$

37. Answers may vary, but one possible system is $\begin{cases} x + y = 4 \\ x^2 + y^2 = 1 \end{cases}$.

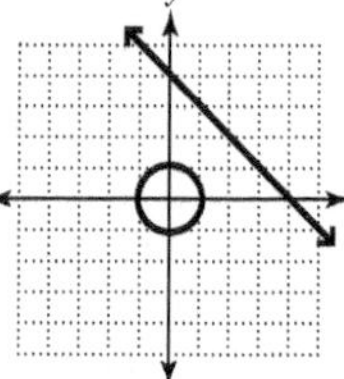

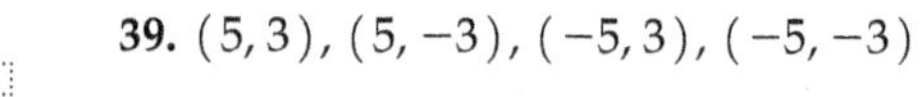

39. $(5, 3), (5, -3), (-5, 3), (-5, -3)$

41. Length: 18 in.; width: 8 in. **43.** At equilibrium, there should be 400 chairs at $72 per chair.

45.

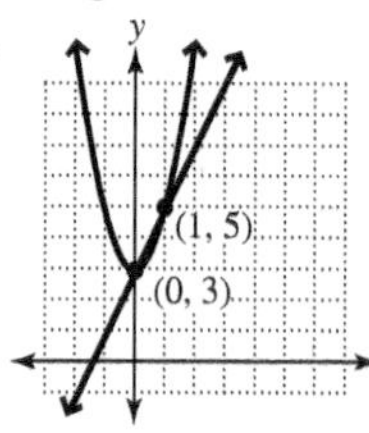

47.

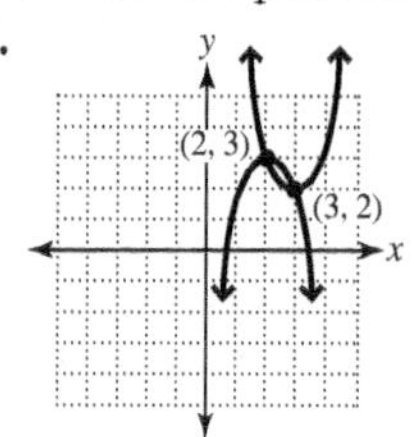

Review Exercises:

1. yes **2.** no **3.**

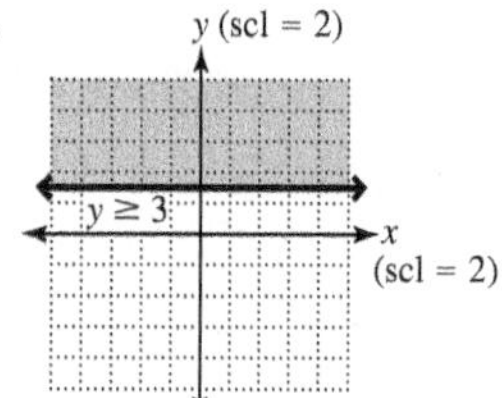

4.

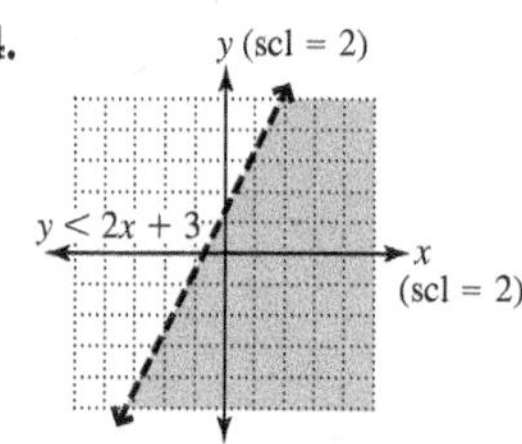

5.

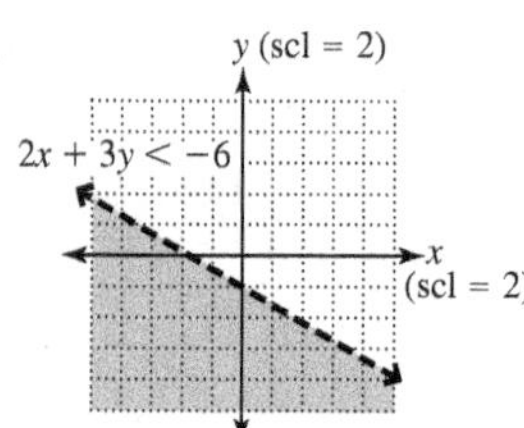

6.

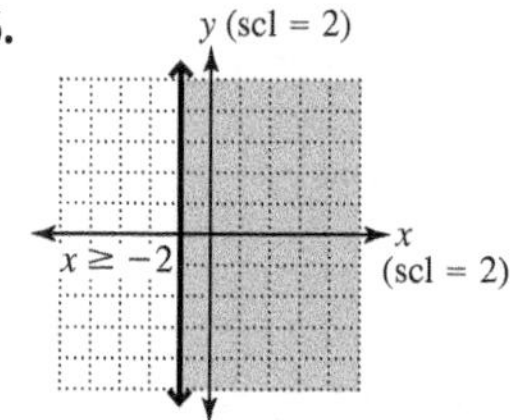

Exercise Set 11.4

Prep Exercises **1.** $9x^2 + 4y^2 = 36$; the ellipse is drawn dashed. **2.** The boundary curve is the parabola whose equation is $y = x^2 + 3$; the parabola is drawn solid. **3.** Draw a dashed hyperbola defined by $\frac{x^2}{9} - \frac{y^2}{16} = 1$. Then choose test points not on the curve to determine the shaded solution region. **4.** The solution region for the system is the intersection of the solution regions of the inequalities. **5.** The graph of $x^2 + y^2 < 1$ is the region inside the circle with radius 1. The graph of $x^2 + y^2 > 4$ is the region outside the circle with radius 2. Thus, there is no common region of solution.

Exercises

1.

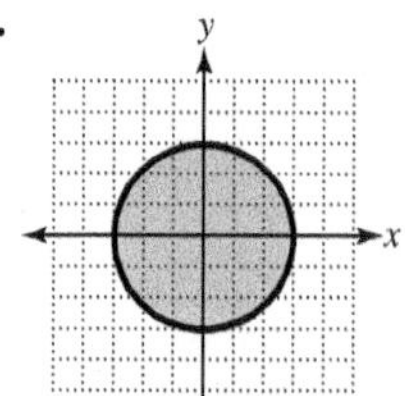

3.

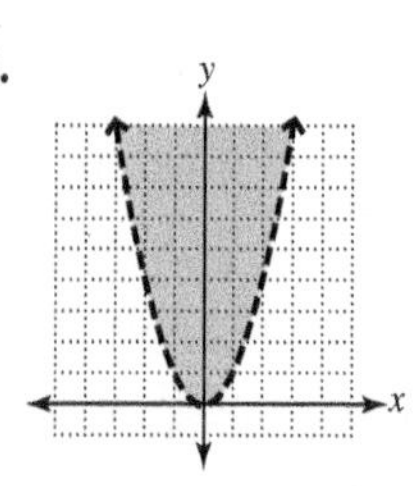

5.

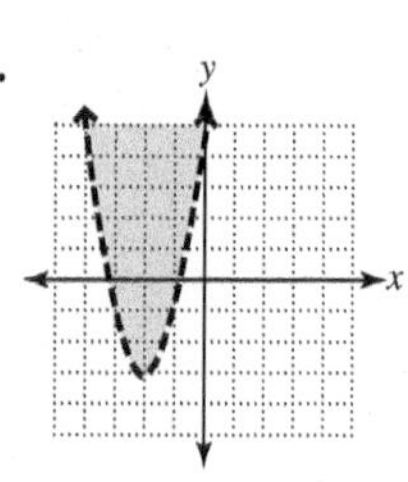

7.

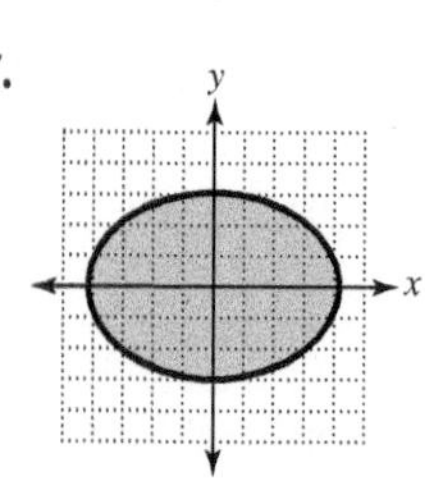

9.

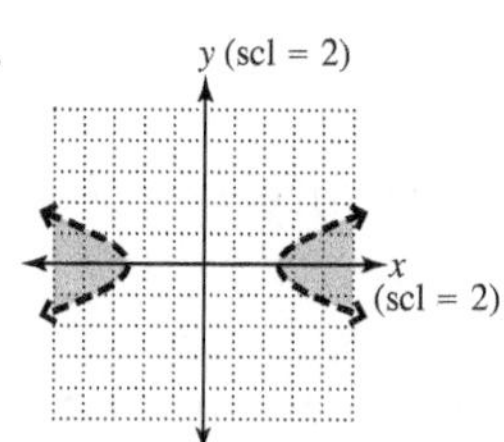

11.

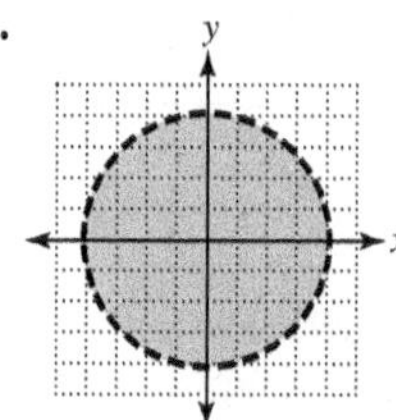

13.

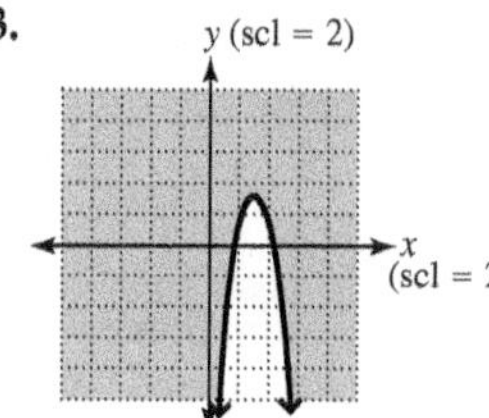

15.

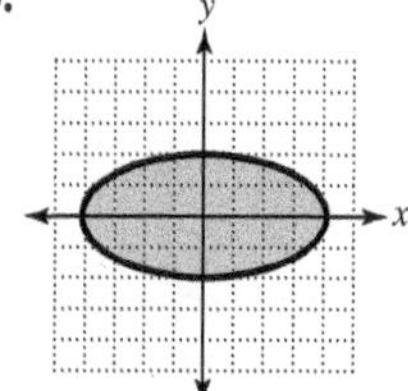

17.

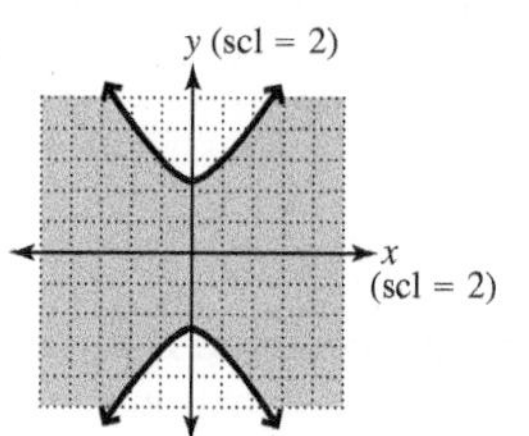

19.

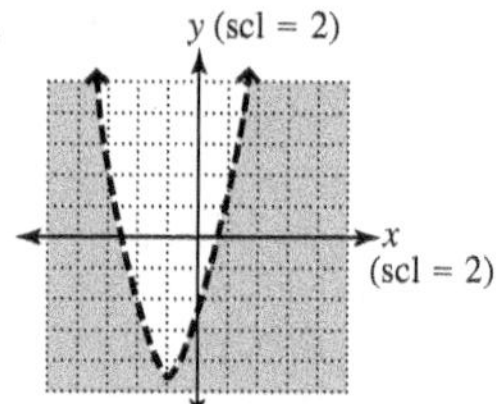

21.

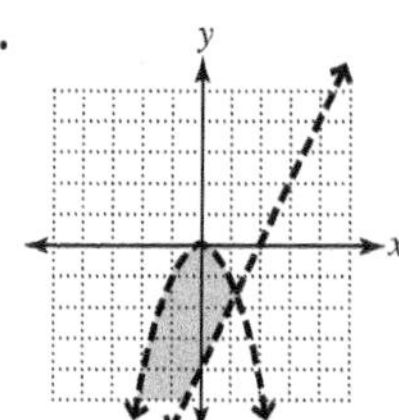

23.

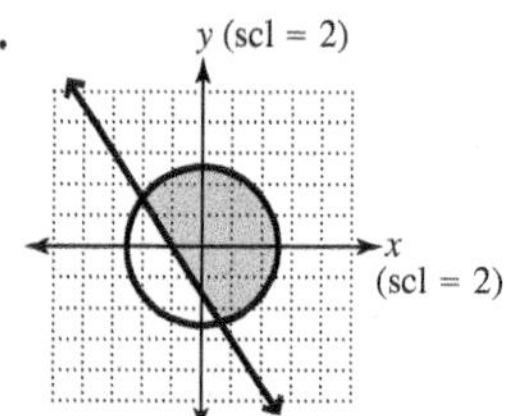

25.

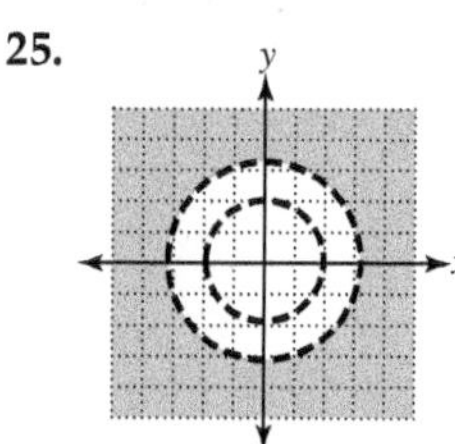

27.

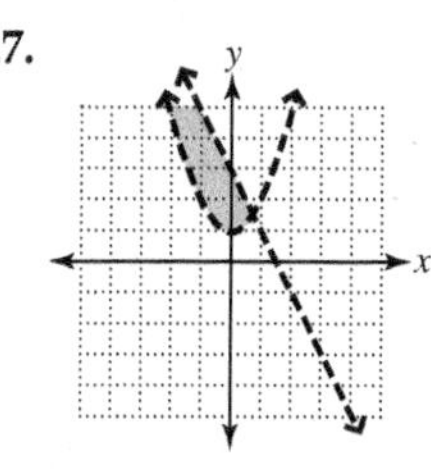

29.

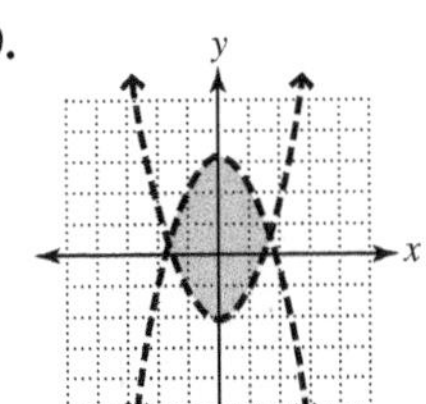

31.

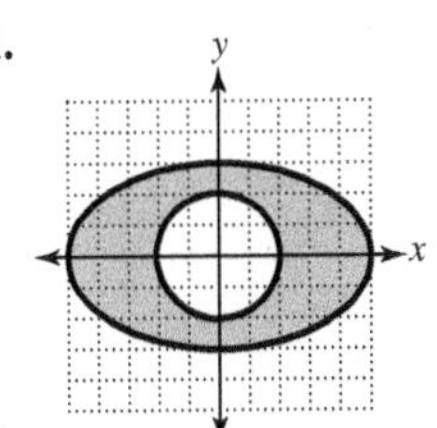

33.

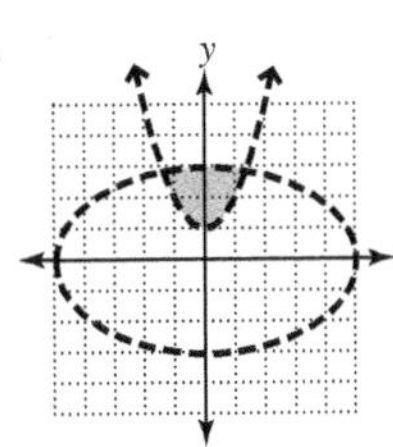

35.

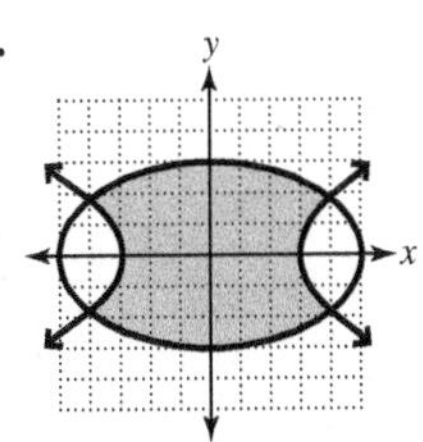

37.

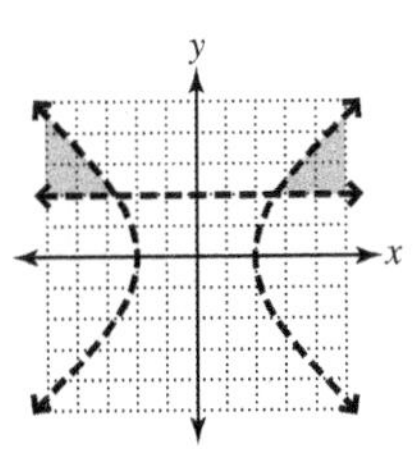

39.

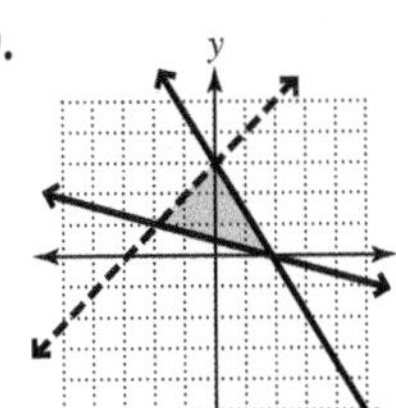

41.

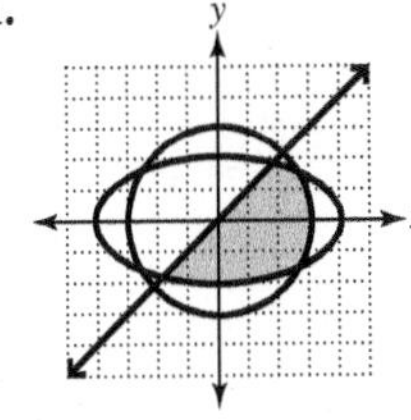

43.

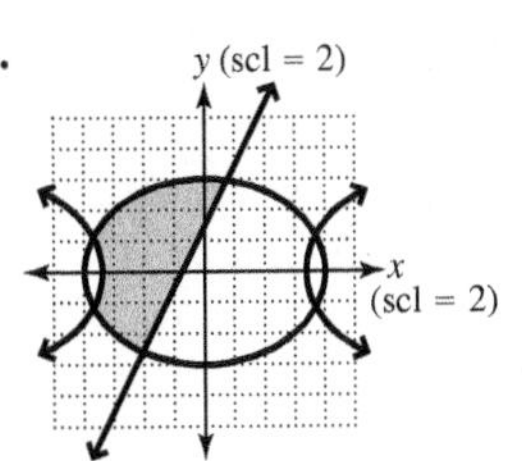

45.

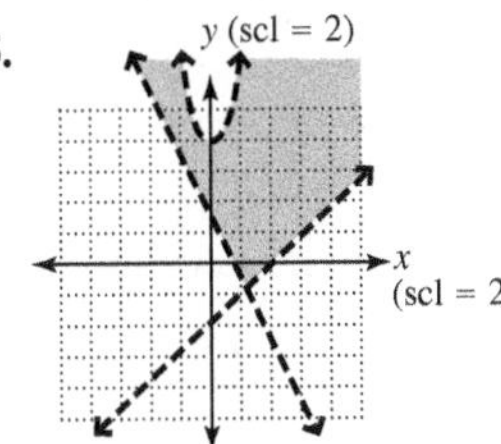

Review Exercises: **1.** 38 **2.** 312 **3.** −84 **4.** −1, 5, 15, 29 **5.** $-1, \frac{1}{4}, -\frac{1}{7}, \frac{1}{10}$ **6.** 67, 69, 71

Chapter 11 Summary and Review Exercises

11.1 Definitions/Rules/Procedures cone; hyperbola; (h, k); $a > 0$; $a < 0$; $y = k$ Exercises

1. Opens upward; vertex: $(3, -5)$; axis of symmetry: $x = 3$

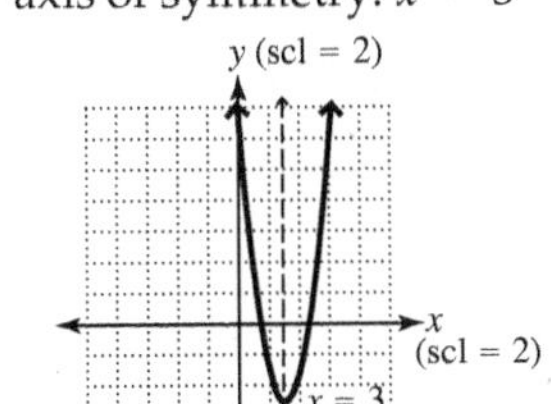

2. Opens downward; vertex: $(-2, 3)$; axis of symmetry: $x = -2$

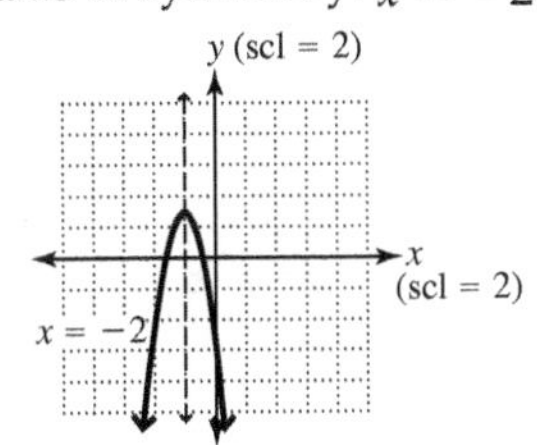

3. Opens left; vertex: $(4, 2)$; axis of symmetry: $y = 2$

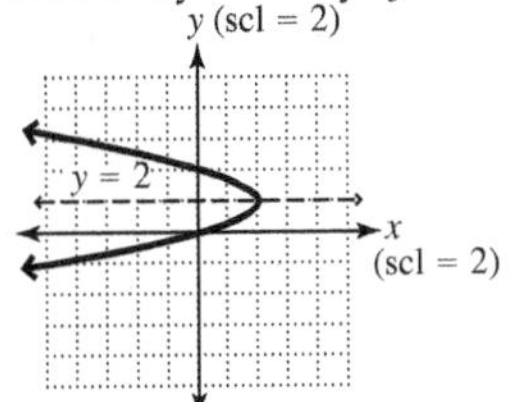

4. Opens right; vertex: $(-2, -3)$; axis of symmetry: $y = -3$

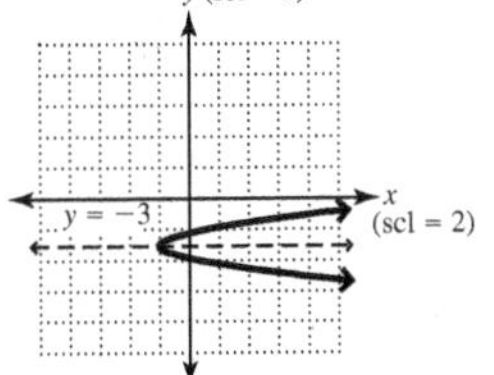

5. Opens right; vertex: $(-1, -3)$; axis of symmetry: $y = -3$

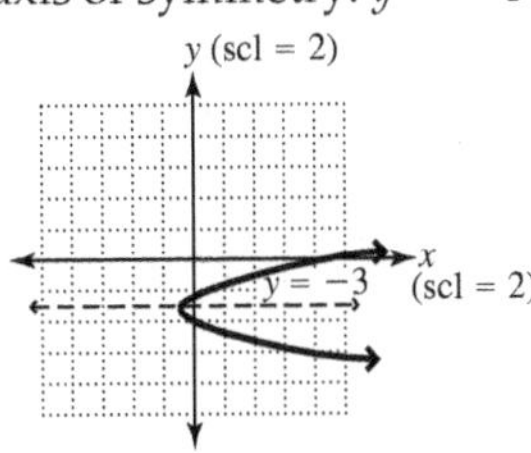

6. Opens right; vertex: $(-14, 2)$; axis of symmetry: $y = 2$

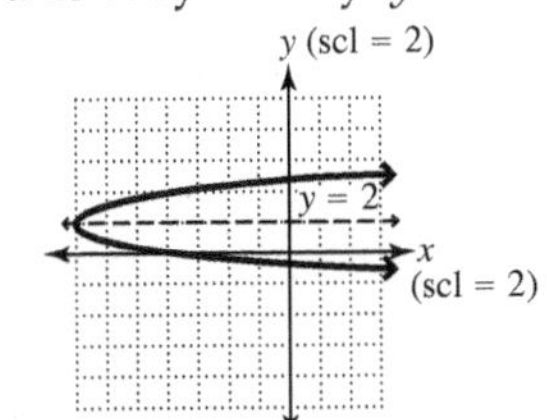

7. Opens left; vertex: $(4, -1)$; axis of symmetry: $y = -1$

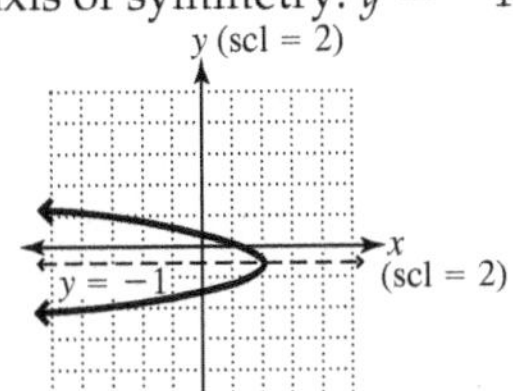

8. Opens left; vertex: $(3, -2)$; axis of symmetry: $y = -2$

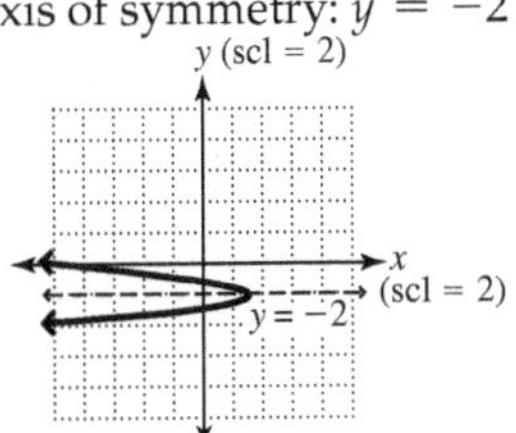

9. **a.** 144 ft. **b.** 1 sec. **c.** 4 sec. Definitions/Rules/Procedures $\sqrt{(x_2 - x_1)^2 + (y_2 - y_1)^2}$; $\left(\frac{x_1 + x_2}{2}, \frac{y_1 + y_2}{2}\right)$

Exercises **10.** 5, $\left(-3, -\frac{1}{2}\right)$ **11.** $4\sqrt{5}$, $(0, -1)$ Definitions/Rules/Procedures equally distant; center; center of a circle; any point on the circle; $(x - h)^2 + (y - k)^2 = r^2$; complete the square Exercises **12.** 10

13. Center: $(3, -2)$; radius: 5

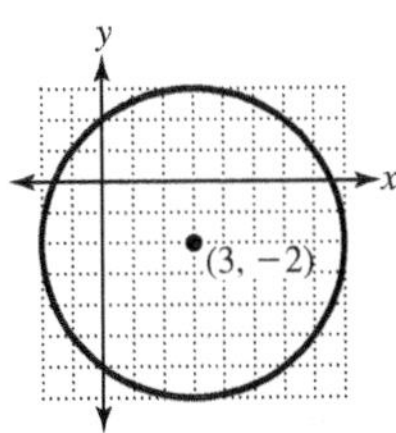

14. Center: $(-5, 1)$; radius: 2

15. Center: $(2, -4)$; radius: 3

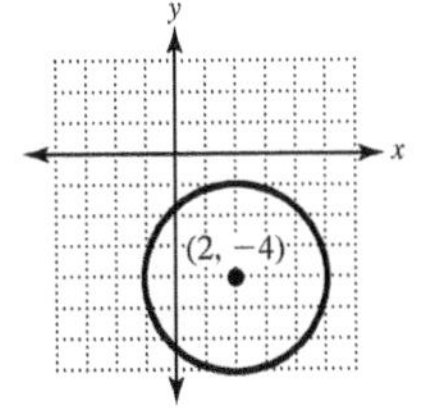

16. Center: $(-5, -1)$; radius: 2

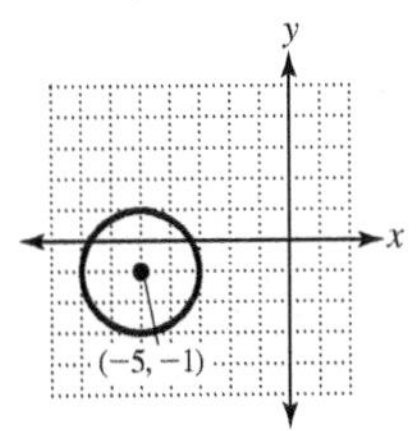

17. $(x - 6)^2 + (y + 8)^2 = 81$ **18.** $(x + 3)^2 + (y + 5)^2 = 100$ **19.** $x^2 + y^2 = 400$ **20.** $(x + 6)^2 + (y - 8)^2 = 225$

11.2 Definitions/Rules/Procedures sum; two fixed points; $\frac{x^2}{a^2} + \frac{y^2}{b^2} = 1$; $\frac{(x-h)^2}{a^2} + \frac{(y-k)^2}{b^2} = 1$; a; b

Exercises **21.**

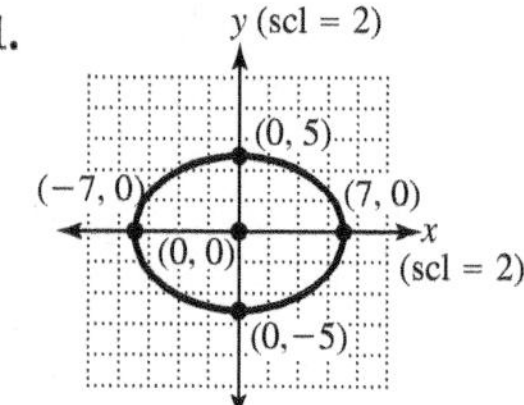

22.

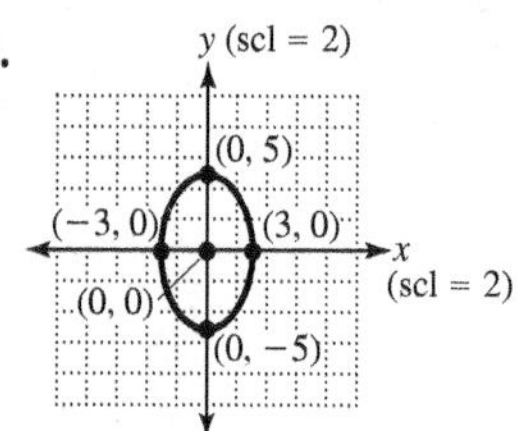

23.

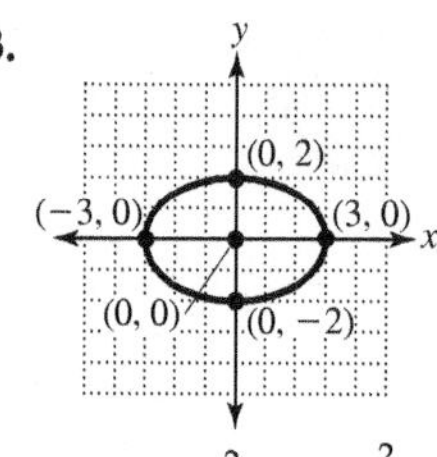

24.

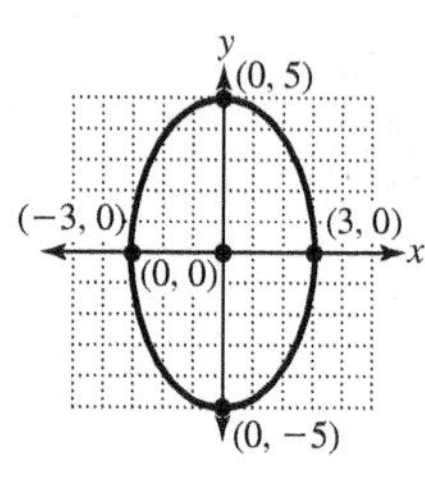

25.

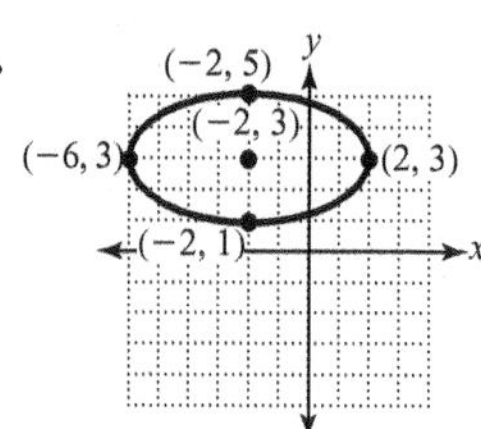

26.

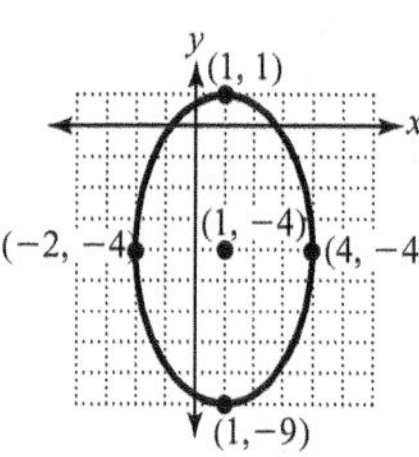

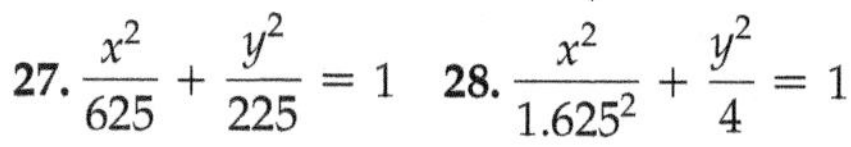

27. $\frac{x^2}{625} + \frac{y^2}{225} = 1$ **28.** $\frac{x^2}{1.625^2} + \frac{y^2}{4} = 1$

Definitions/Rules/Procedures difference; two fixed points; $\frac{x^2}{a^2} - \frac{y^2}{b^2} = 1$; $\frac{y^2}{b^2} - \frac{x^2}{a^2} = 1$; $(a, 0)$; $(-a, 0)$; y; $(0, b)$; $(0, -b)$; x; (a, b), $(-a, b)$, $(a, -b)$, and $(-a, -b)$; fundamental rectangle; intercept; asymptotes

Exercises **29.**

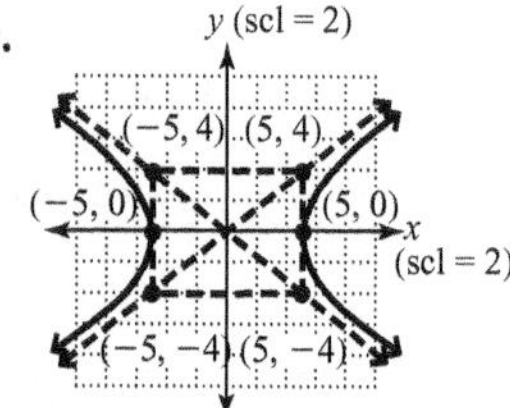

30.

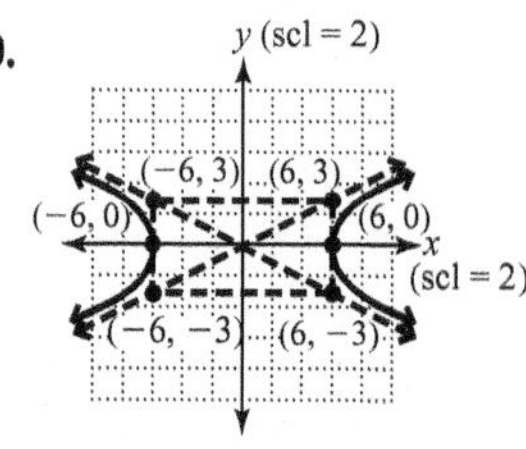

31.

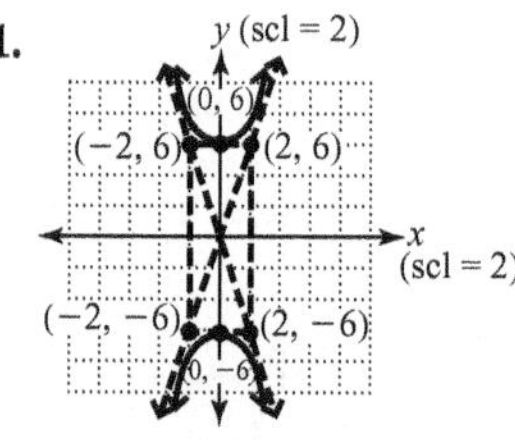

32.

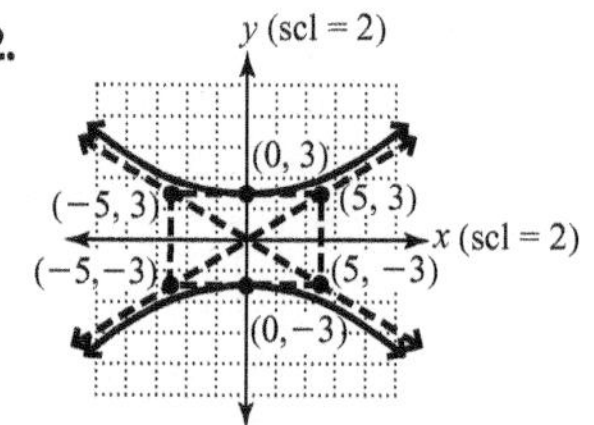

11.3 **Definitions**/Rules/Procedures nonlinear; substitution; substitute; elimination Exercises **33.** $(2, 5)$, $(-1, -1)$
34. $(-1, -4)$, $(4, 1)$ **35.** $(-73/37, -35/37)$, $(1, 5)$ **36.** $(8, 0)$, $(-8, 0)$ **37.** $(3, 4)$, $(3, -4)$, $(-3, -4)$, $(-3, 4)$
38. $(4, 2)$, $(4, -2)$, $(-4, -2)$, $(-4, 2)$ **39.** $(2, 2)$, $(-2, 2)$ **40.** $(2, 3)$, $(-2, 3)$ **41.** $(8, 5)$, $(8, -5)$, $(-8, -5)$, $(-8, 5)$
42. 8 ft. by 10 ft. **43.**

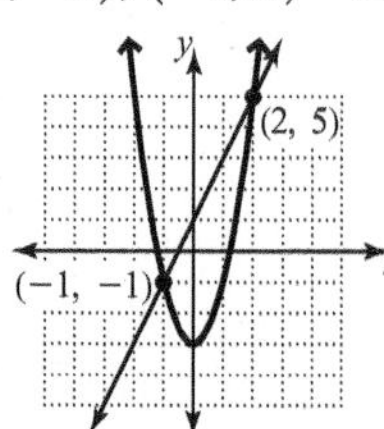

44.

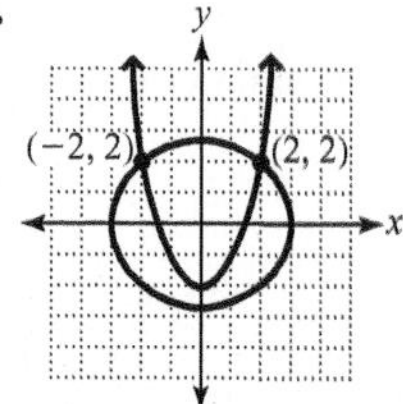

11.4 Definitions/Rules/Procedures equation; solid; dashed; ordered pair; the region containing that ordered pair
Exercises **45.**

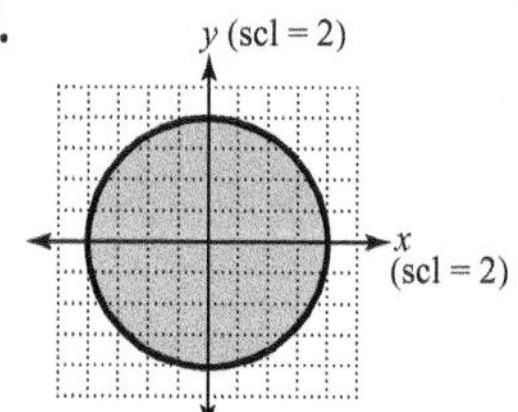

46.

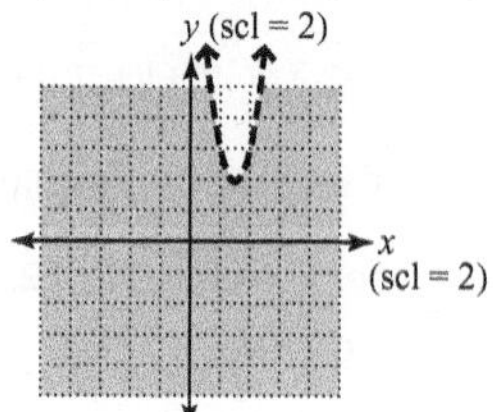

47.

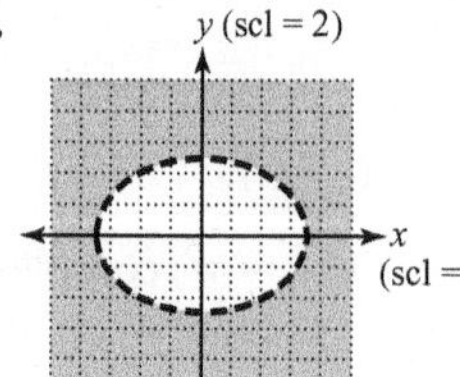

48.

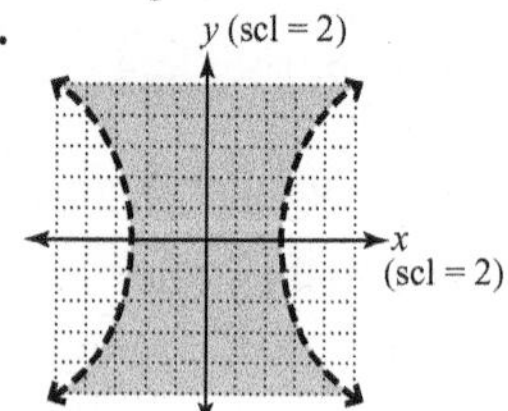

Definitions/Rules/Procedures the solution sets of the individual inequalities Exercises **49.**

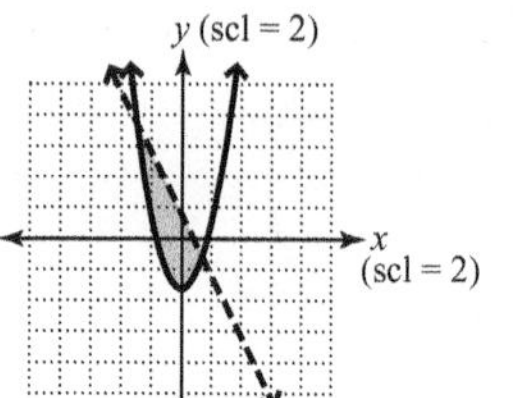

50.

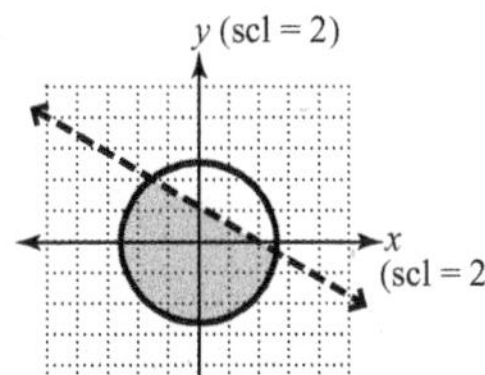

51.

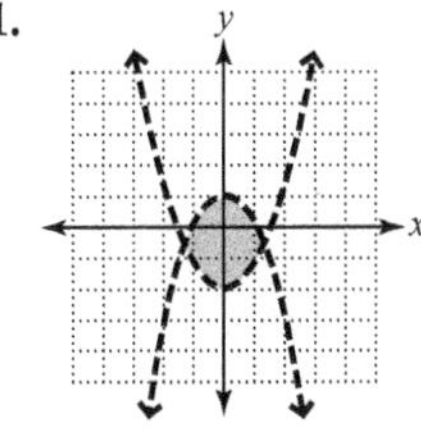

52.

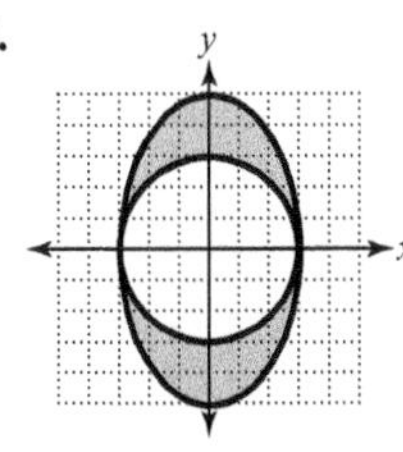

53.

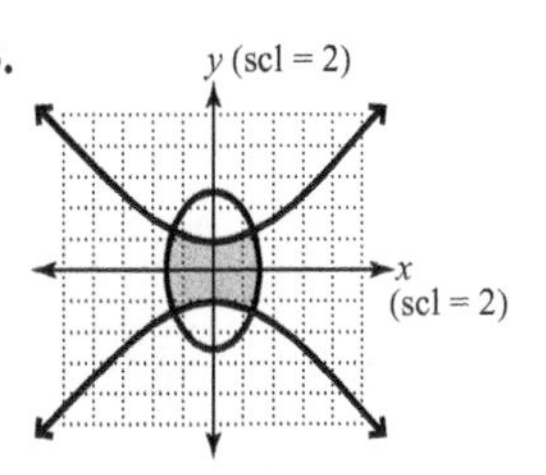

54.

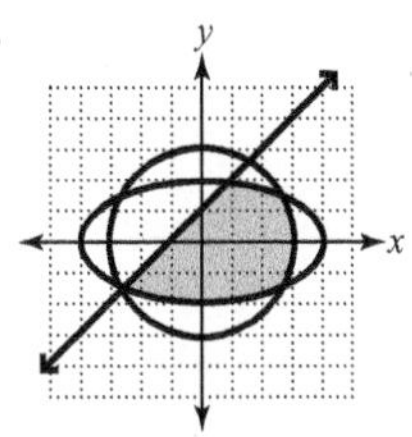

Chapter 11 Practice Test

1. Opens upward; vertex: $(-1, -4)$; axis of symmetry: $x = -1$ [11.1]

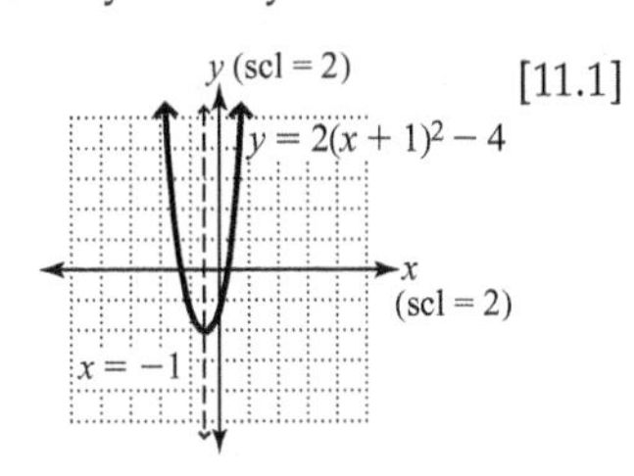

2. Opens left; vertex: $(1, 3)$; axis of symmetry: $y = 3$ [11.1]

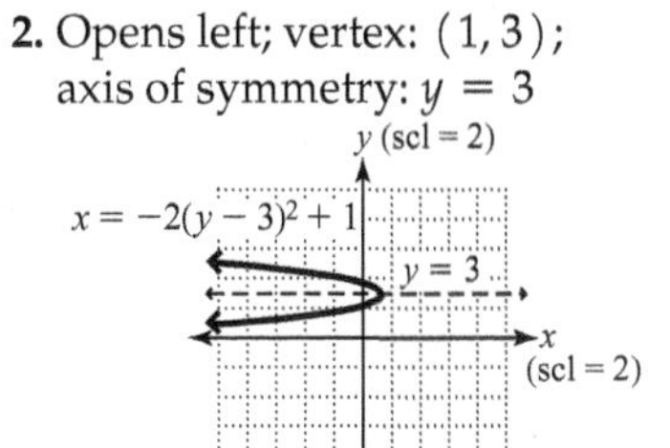

3. Opens right; vertex: $(-7, -2)$; axis of symmetry: $y = -2$ [11.1]

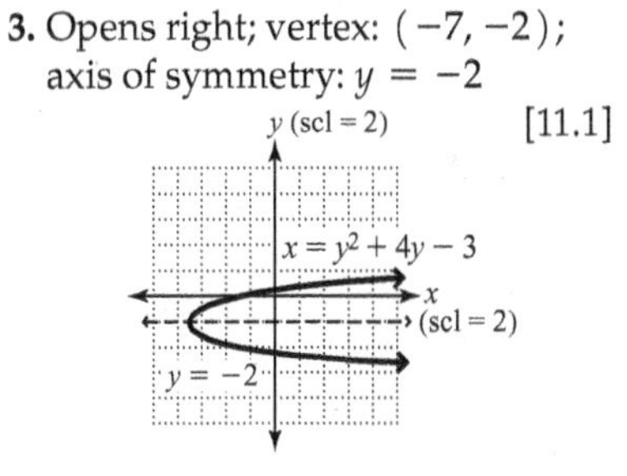

4. $2\sqrt{5}$, $(4, -4)$ [11.1]

5. Center: $(-4, 3)$; radius: 6 [11.1]

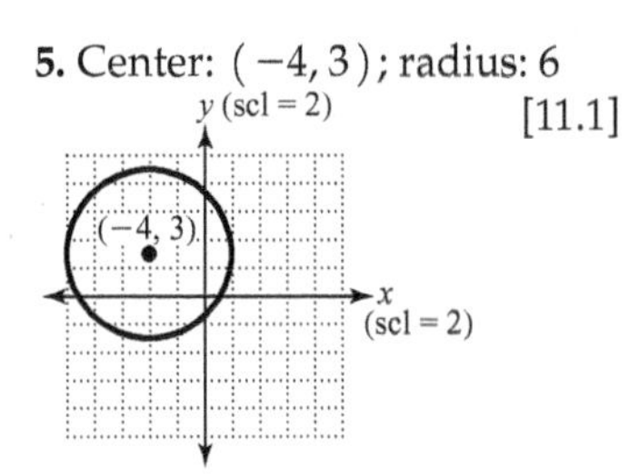

6. Center: $(-2, 5)$; radius: 3 [11.1]

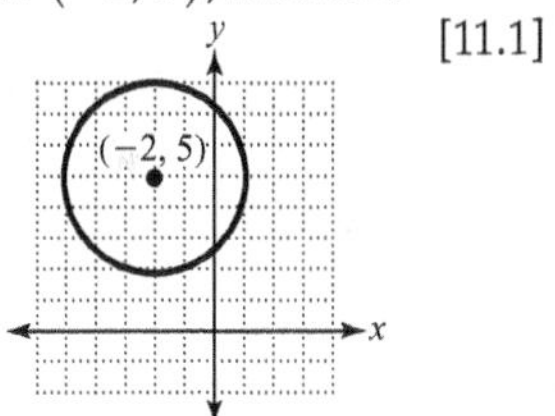

7. $(x - 2)^2 + (y + 4)^2 = 100$ [11.1]

8. [11.2]

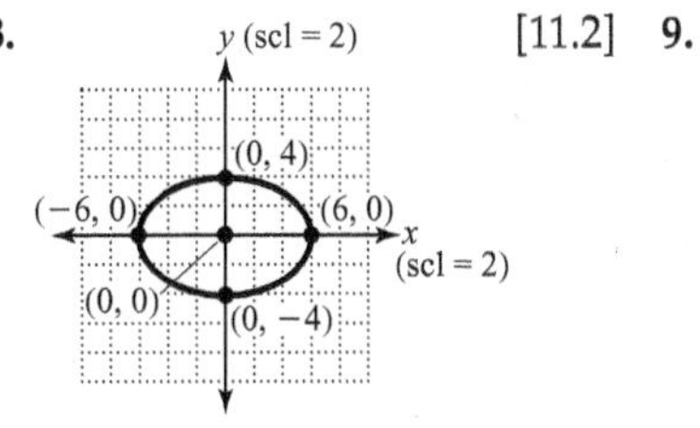

9. [11.2]

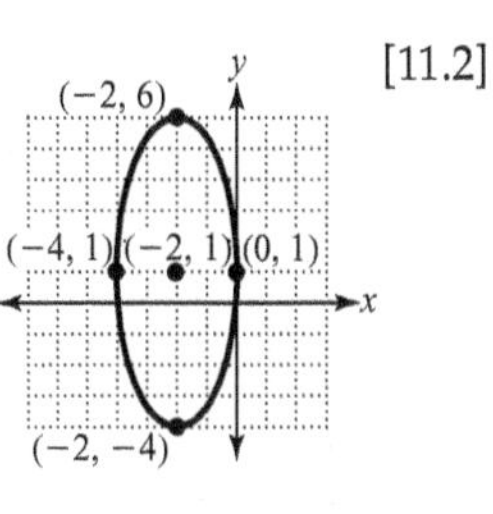

10. [11.2]

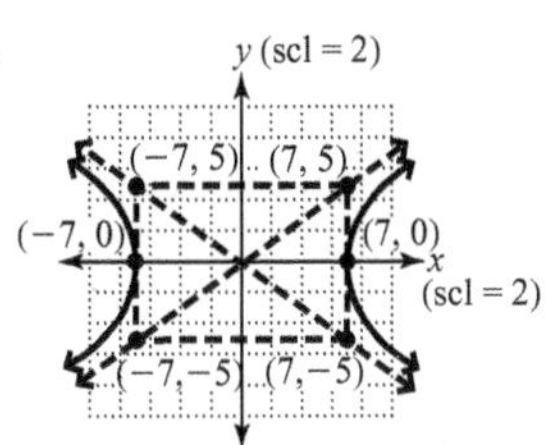

11. [11.2]

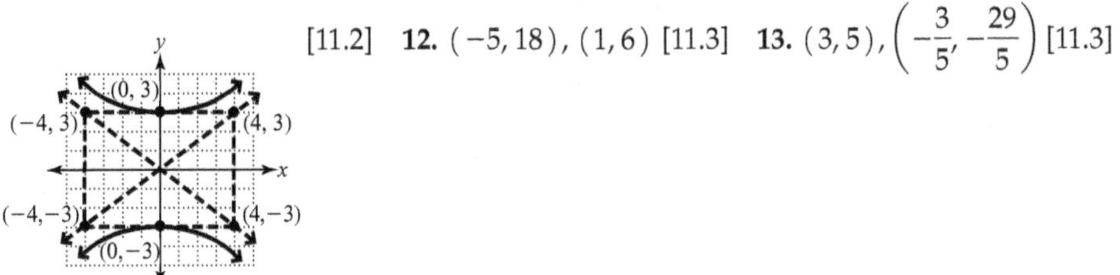

12. $(-5, 18)$, $(1, 6)$ [11.3] **13.** $(3, 5)$, $\left(-\frac{3}{5}, -\frac{29}{5}\right)$ [11.3]

14. $(2, 3)$, $(2, -3)$, $(-2, -3)$, $(-2, 3)$ [11.3] **15.** $(3, 2)$, $(3, -2)$, $(-3, -2)$, $(-3, 2)$ [11.3] **16.** [11.4]

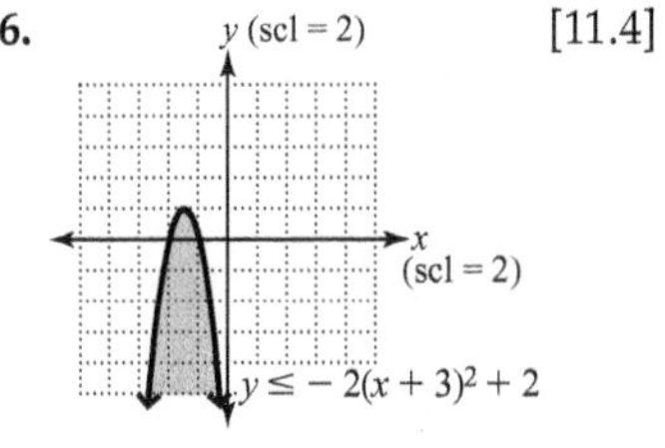

17. 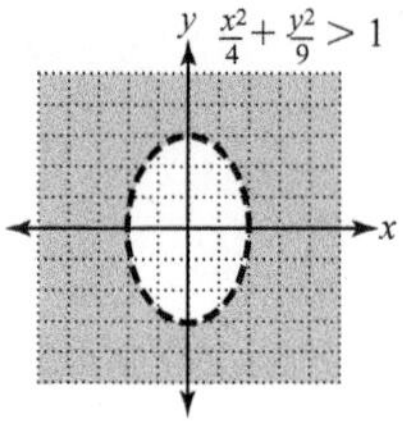

[11.4] **18.** 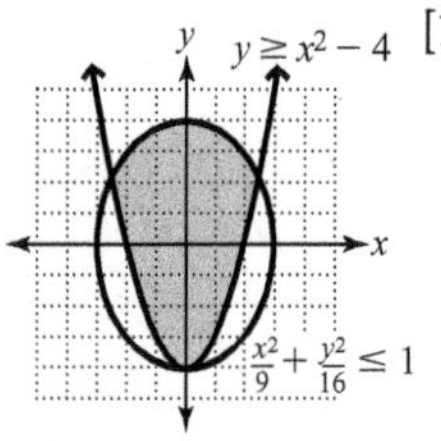

[11.4] **19.** Height is 18; distance across the base is 12. [11.1]

20. $\frac{x^2}{9.5^2} + y^2 = 1$ [11.2]

Chapters 1–11 Cumulative Review Exercises

1. false **2.** true **3.** true **4.** $-8; -5$ **5.** $i.$ **6.** $d = \sqrt{(x_2 - x_1)^2 + (y_2 - y_1)^2}.$ **7.** $9\frac{1}{16}$ **8.** $\frac{n^6}{125m^9}$ **9.** $2i\sqrt{5}$ **10.** $(x + y)(a + b)$ **11.** $(k^2 + 9)(k + 3)(k - 3)$ **12.** $-4, \frac{2}{3}$ **13.** 52 **14.** $\frac{1 \pm i\sqrt{3}}{4}$ **15.** $-\frac{4}{3}$ **16.** 1.5 **17.** $\frac{1}{8}$ **18. a.** 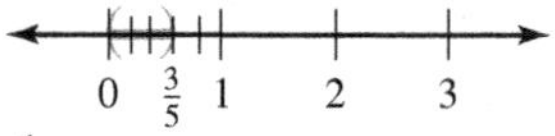

b. $\left\{m \mid 0 < m < \frac{3}{5}\right\}$ **c.** $\left(0, \frac{3}{5}\right)$ **19. a.** $x^2 + 2x$ **b.** $-x^2 + 2x + 2$ **c.** $2x^3 + x^2 - 2x - 1$ **d.** $\frac{2x + 1}{x^2 - 1}; x \neq \pm 1$ **e.** $2x^2 - 1$ **f.** $4x^2 + 4x$ **20.** $\log_5 125 = 3$ **21.** $5 \log_5 x + \log_5 y$ **22.** $(x - 2)^2 + (y - 4)^2 = 169$

23. 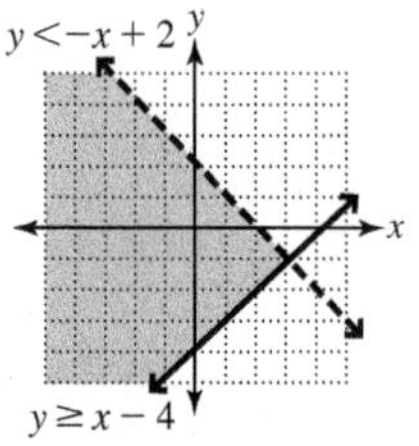

24. 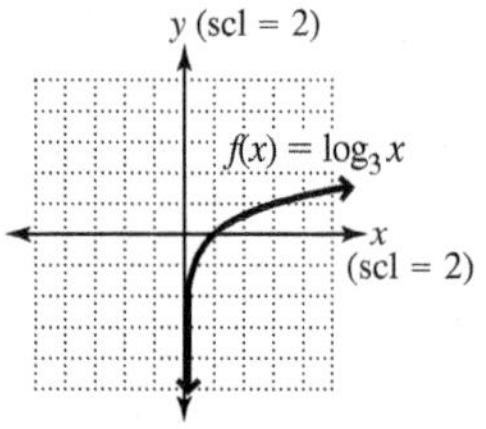

25. 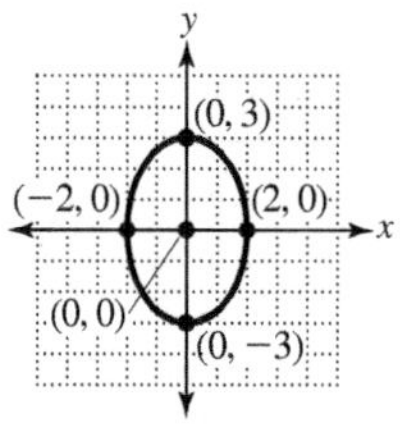

26. a. $h = \frac{V}{\pi r^2}$ **b.** $\frac{15}{r^2}$

27. 2.9 lb. of tuna and 3.4 lb. of shrimp **28.** \$1500 at 5%, \$2500 at 6%, \$1000 at 3% **29.** 6.9 **30.** $\frac{x^2}{5625} + \frac{y^2}{400} = 1$

Glossary

Absolute value: A number's distance from zero on a number line.

Additive inverses: Two numbers whose sum is 0.

Arithmetic sequence: A sequence in which each term after the first is found by adding the same number to the previous term.

Arithmetic series: The sum of the terms of an arithmetic sequence.

Augmented matrix: A matrix made up of the coefficients and the constant terms of a system. The constant terms are separated from the coefficients by a dashed vertical line.

Binomial: A polynomial containing two terms.

Binomial coefficient: A number written as $\binom{n}{r}$ and defined as $\frac{n!}{r!(n-r)!}$.

Binomial theorem: For any positive integer n, $(a+b)^n = \binom{n}{0}a^n + \binom{n}{1}a^{n-1}b + \binom{n}{2}a^{n-2}b^2 + \binom{n}{3}a^{n-3}b^3 + \cdots + \binom{n}{n}b^n$.

Circle: A set of points in a plane that are equally distant from a central point. The central point is the center.

Coefficient: The numerical factor in a monomial.

Common difference of an arithmetic sequence: The value of d is found by $d = a_n - a_{n-1}$, where a_n is any term in the sequence (except the first) and a_{n-1} is the previous term.

Common logarithms: Logarithms with a base of 10. $\text{Log}_{10}x$ is written as $\log x$. Note that $\log 10 = 1$.

Complementary angles: Two angles are complementary if the sum of their measures is 90°.

Complex conjugate: The complex conjugate of a complex number $a + bi$ is $a - bi$.

Complex number: A number that can be expressed in the form $a + bi$, where a and b are real numbers and i is the imaginary unit.

Complex rational expression: A rational expression that contains rational expressions in the numerator and/or denominator.

Composition of functions: If f and g are functions, then the composition of f and g is defined as $(f \circ g)(x) = f[g(x)]$ for all x in the domain g for which $g(x)$ is in the domain of f. The composition of g and f is defined as $(g \circ f)(x) = g[f(x)]$ for all x in the domain of f for which $f(x)$ is in the domain of g.

Compound inequality: Two inequalities joined by either *and* or *or*.

Conditional linear equation in one variable: A linear equation with exactly one solution.

Conjugates: Binomials that are the sum and difference of the same two terms.

Conic Section: A curve in a plane that is the result of intersecting the plane with a cone—more specifically, a circle, an ellipse, a parabola, or a hyperbola.

Consistent system of equations: A system of equations that has at least one solution.

Constant: A symbol that does not vary in value.

Constant Function: A function of the form $f(x) = c$, where c is a real number.

Contradiction: An equation that has no solution.

Cubic equation in one variable: An equation that can be written in the form $ax^3 + bx^2 + cx + d = 0$, where a, b, c, and d are real numbers and $a \neq 0$.

Cubic Function: A function of the form $f(x) = ax^3 + bx^2 + cx + d$, where a, b, c, and d are real numbers and $a \neq 0$.

Degree of a Monomial: The sum of the exponents of all variables in the monomial.

Degree of a Polynomial: The greatest degree of any of the terms in the polynomial.

Direct variation: Two variables y and x vary directly if $y = kx$. If y varies directly as the nth power of x, then $y = kx^n$, where k is the constant of variation.

Discriminant: The discriminant is the radicand, $b^2 - 4ac$, in the quadratic formula.

Domain: A set containing initial values of a relation; its input values; the first coordinates in ordered pairs.

Ellipse: The set of all points the sum of whose distances from two fixed points is constant.

Equation: Two expressions set equal to each other.

Equation quadratic in form: An equation is quadratic in form if it can be rewritten as a quadratic equation $au^2 + bu + c = 0$, where $a \neq 0$ and u is a variable or an expression.

Exponential function: If $b > 0, b \neq 1$, and x is any real number, then the exponential function is $f(x) = b^x$.

Expression: A constant; a variable; or any combination of constants, variables, and arithmetic operations.

Factored form: A number or an expression written as a product of factors.

Factorial notation: For any natural number n, the symbol $n!$ (read "n factorial") means $n(n-1)(n-2) \ldots 3 \cdot 2 \cdot 1$. $0!$ is defined to be 1, so $0! = 1$.

Finite sequence: A function with a domain that is the set of natural numbers from 1 to n.

Function: A relation in which each value in the domain is assigned to exactly one value in the range.

Geometric sequence: A sequence in which every term after the first is found by multiplying the previous term by the same number, called the common ratio, r, where $r = \frac{a_n}{a_{n-1}}$.

Geometric series: The sum of the terms of a geometric sequence.

Greatest common factor (GCF) of a set of terms: A monomial with the greatest coefficient and degree that evenly divides all of the given terms.

Hyperbola: The set of all points the difference of whose distances from two fixed points remains constant.

Identity: An equation in which every real number (for which the equation is defined) is a solution.

Imaginary number: A number that can be expressed in the form $a + bi$, where a and b are real numbers, i is the imaginary unit, and $b \neq 0$.

Imaginary unit: The number represented by i, where $i = \sqrt{-1}$ and $i^2 = -1$.

Inconsistent system of equations: A system of equations that has no solution.

Inequality: Two expressions separated by $\neq, <, >, \leq,$ or $\geq$.

Infinite sequence: A function with a domain that is the set of natural numbers.

Intersection: For two sets A and B, the intersection of A and B, symbolized by $A \cap B$, is a set containing only elements that are in both A and B.

Inverse functions: Two functions f and g are inverses if and only if $(f \circ g)(x) = x$ for all x in the domain of g and $(g \circ f)(x) = x$ for all x in the domain of f.

Inverse variation: Two variables y and x vary inversely if $y = \frac{k}{x}$. If $y = \frac{k}{x^n}$, then y varies inversely as the nth power of x, where k is the constant of variation.

Irrational number: Any real number that is not rational.

Joint variation: If y varies jointly as x and z, then $y = kxz$, where k is the constant of variation.

Like radicals: Radical expressions with identical radicands and identical root indices.

Like terms: Variable terms that have the same variable(s) raised to the same exponents, or constant terms.

Linear equation in one variable: An equation that can be written in the form $ax + b = c$, where a, b, and c are real numbers and $a \neq 0$.

Linear equation in two variables: An equation that can be written in the form $Ax + By = C$, where A, B, and C are real numbers and A and B are not both 0.

Linear function: A function of the form $f(x) = mx + b$, where m and b are real numbers.

Linear inequality in one variable: An inequality that can be written in the form $ax + b < c, ax + b > c, ax + b \leq c$, or $ax + b \geq c$, where a, b, and c are real numbers and $a \neq 0$.

Linear inequality in two variables: An inequality that can be written in the form $Ax + By > C$, where A, B and C are real numbers and A and B are not both 0. Note that the inequality could also be $<, \leq,$ or $\geq$.

Logarithm: If $b > 0$ and $b \neq 1$, then $y = \log_b x$ is equivalent to $x = b^y$.

Matrix: A rectangular array of numbers.

Minor of an element of a matrix: The determinant of the remaining matrix when the row and column in which the element is located are ignored.

Monomial: An expression that is a constant or a product of a constant and variables that are raised to whole number powers.

Multiplicative inverses: Two numbers whose product is 1.

Multivariable polynomial: A polynomial with more than one variable.

***n*th root:** The number b is an nth root of a number a if $b^n = a$.

Natural logarithms: Base-e logarithms are called natural logarithms, and $\log_e x$ is written as $\ln x$. Note that $\ln e = 1$.

Nonlinear system of equations: A system of equations that contains at least one nonlinear equation.

Numerical coefficient: The numerical factor in a term.

One-to-one function: A function f is one-to-one if for any two numbers a and b in its domain, when $f(a) = f(b)$, $a = b$ and when $a \neq b$, $f(a) \neq f(b)$.

Polynomial: A monomial or an expression that can be written as a sum of monomials.

Polynomial equation: An equation that equates two polynomials.

Polynomial equation in standard form: $P = 0$, where P is a polynomial in terms of one variable written in descending order of degree.

Polynomial function: A function of the form $f(x) = ax^m + bx^n + \cdots$ with a finite number of terms, where each coefficient is a real number and each exponent is a whole number.

Polynomial in one variable: A polynomial in which every variable term has the same variable.

Prime number: A natural number with exactly two different factors, 1 and the number itself.

Quadratic equation in one variable: An equation that can be written in the form $ax^2 + bx + c = 0$, where a, b, and c, are real numbers and $a \neq 0$.

Quadratic function: A function of the form $f(x) = ax^2 + bx + c$, where a, b, and c are real numbers and $a \neq 0$.

Quadratic inequality: An inequality that can be written in the form $ax^2 + bx + c > 0$ or $ax^2 + bx + c < 0$, where $a \neq 0$. Note: The symbols $<$ and $>$ can be replaced with $\leq$ and $\geq$.

Radical equation: An equation containing at least one radical expression whose radicand has a variable.

Radical function: A function containing a radical expression whose radicand has a variable.

Radius: The distance from the center of a circle to any point on the circle.

Range: A set containing all values that are paired to domain values in a relation; its output values; the second coordinates in ordered pairs.

Rational exponent: An exponent that is a rational number.

Rational expression: An expression that can be written in the form $\frac{P}{Q}$, where P and Q are polynomials and $Q \neq 0$.

Rational function: A function expressed in terms of rational expressions.

Rational inequality: An inequality containing a rational expression.

Rational number: Any real number that can be expressed in the form $\frac{a}{b}$, where a and b are integers and $b \neq 0$.

Relation: A set of ordered pairs.

Row echelon form: An augmented matrix whose coefficient portion has 1's on the diagonal from upper left to lower right and 0s below the 1's.

Scientific notation: A number expressed in the form $a \times 10^n$, where a is a decimal number with $1 \leq |a| < 10$ and n is an integer.

Sequence: A function list whose domain is $1, 2, 3, \ldots, n$.

Series: The sum of the terms of a sequence.

Set: A collection of objects.

Slope: The ratio of the vertical change (change in y) to the horizontal change (change in x) between any two points on a line.

Solution: A number that makes an equation true when it replaces the variable in the equation.

Solution set: A set containing all of the solutions for a given equation.

Solution for a system of equations: An ordered set of numbers that makes all equations in the system true.

Square matrix: A matrix with an equal number of rows and columns.

Subset: If every element of a set B is an element of set A, then B is a subset of A.

Supplementary angles: Two angles are supplementary if the sum of their measures is 180°.

System of equations: A group of two or more equations.

Term: A number or the product of a number and one or more variables raised to powers.

The set of integers: $\{\ldots, -3, -2, -1, 0, 1, 2, 3, \ldots\}$

The set of natural numbers: $\{1, 2, 3, \ldots\}$

The set of whole numbers: $\{0, 1, 2, 3, \ldots\}$

Trinomial: A polynomial containing three terms.

Union: For two sets A and B, the union of A and B, symbolized by$A \cup B$, is a set containing every element in A or in B.

Variable: A symbol varying in value.

x-intercept: A point where a graph intersects at the x-axis.

y-intercept: A point where a graph intersects at the y-axis.

Index of Applications

Index

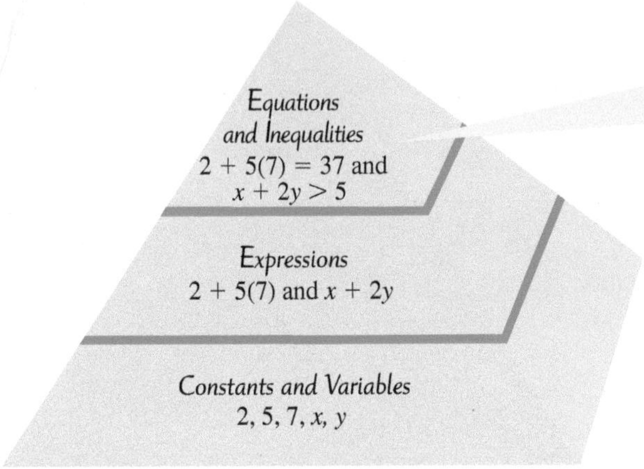

Equations and Inequalities
$2 + 5(7) = 37$ and $x + 2y > 5$

Expressions
$2 + 5(7)$ and $x + 2y$

Constants and Variables
$2, 5, 7, x, y$

Equations and Inequalities relate expressions using symbols such as $=, <, >, \leq,$ or $\geq$. We **solve** or **graph** equations and inequalities. To solve, we find every number that makes the equation or inequality true when substituted for the variables. To graph, we draw a representation of every possible solution either on a number line (for one-variable equations and inequalities) or on a grid called the *rectangular coordinate system.*

Text Overview

The text is divided into two parts:

Part I: *Chapters 1–4: Linear*

Chapter 1 forms the expression foundation for solving and graphing linear equations and inequalities (Chapters 2–4). An introduction to functions appears at the end of Chapter 3. The linear portion of the course culminates in Chapter 4 with systems of linear equations and inequalities.

Chapters 5–11: Nonlinear

Chapters 5 and 6 explore the expression foundation for solving equations by factoring. Solving equations by factoring is discussed at the end of Chapter 6.

Chapter 7 explores the expression foundation for solving equations containing rational expressions. Those equations are solved at the end of the chapter. Rational functions are also introduced.

Chapter 8 explores the expression foundation for solving equations containing rational exponents and radicals. Those equations are solved at the end of the chapter. Radicals are also needed for complex numbers. Radical functions are also introduced.

Chapter 9 explores methods for solving any quadratic equation or an equation that is quadratic in form. Quadratic functions and nonlinear inequalities are also explored.

Chapter 10 explores exponential and logarithmic functions.

Chapter 11 explores the equations related to the conic sections and finishes with a discussion of nonlinear systems of equations and inequalities.

The following list outlines the development of equations and inequalities in this text:

Chapter 2 Solving Linear Equations and Inequalities in One Variable

Graphing Solution Sets for Linear Inequalities in One Variable

Solving Compound Inequalities

Solving Equations and Inequalities Involving Absolute Value

Chapter 3 Graphing Linear Equations and Inequalities in the Rectangular Coordinate System

Writing Equations of Lines

Introduction to Functions

Chapter 4 Solving Systems of Linear Equations and Inequalities

Chapter 6 Solving Equations by Factoring

Introduction to Quadratic Functions

Chapter 7 Solving Equations Containing Rational Expressions

Introduction to Rational Functions

Chapter 8 Solving Radical Equations

Introduction to Radical Functions

Chapter 9 Solving Quadratic Equations

Quadratic Functions

Chapter 10 Exponential and Logarithmic Functions

Solving Exponential and Logarithmic Equations

Chapter 11 Equations of Conic Sections

Solving Nonlinear Systems of Equations and Inequalities

The Algebra Pyramid:
A Guide to the Development of Algebra Topics

Constants and variables are the foundation of algebra. The following diagram summarizes the entire complex number system (all constants). These constants, along with variables such as x or y, are the basic building blocks that give meaning to the expressions, equations, and inequalities.

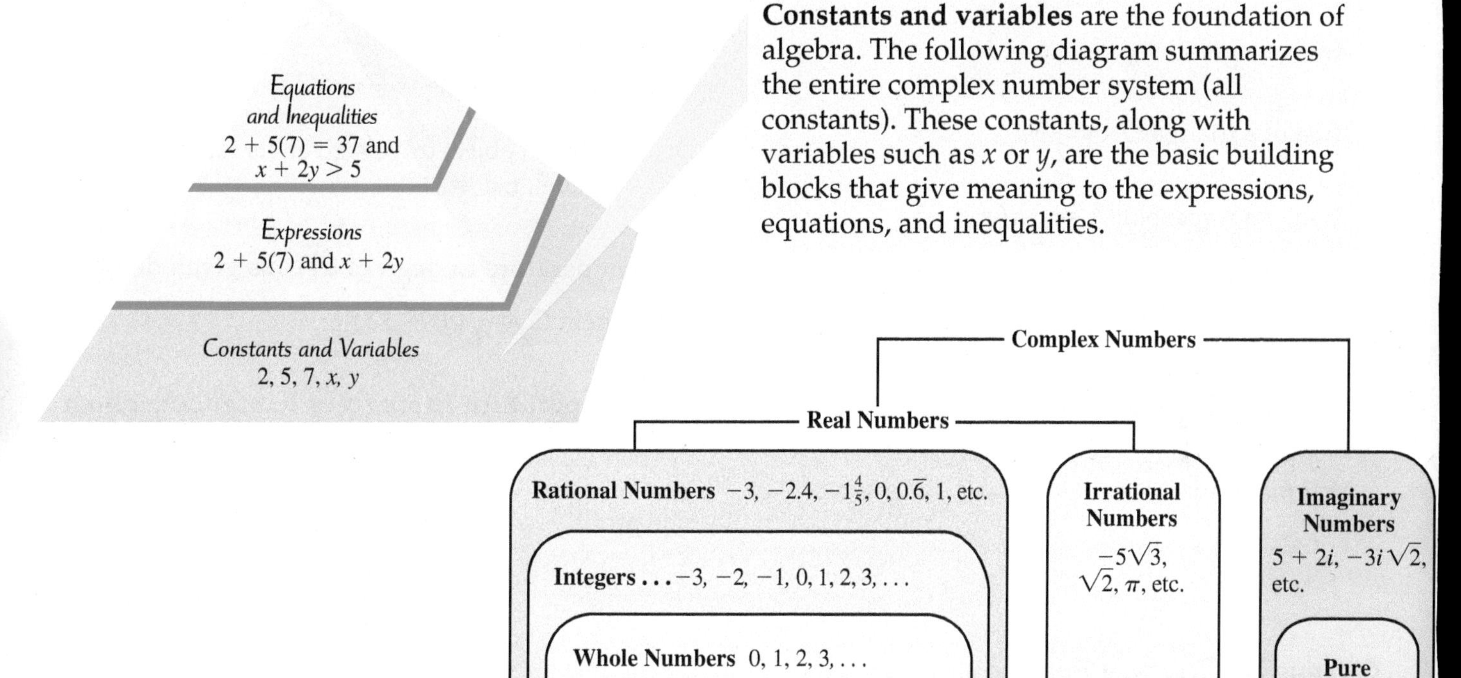

Expressions relate constants and variables using symbols such as operations of arithmetic to describe a calculation. We **evaluate** or **rewrite** expressions. To evaluate, we replace variables with numbers and perform the calculation. To rewrite, we use rules to write an alternative (and usually simpler) form.

The following list outlines the development of expressions in this text:

Chapter 1	Order of Operations
	Translating Word Phrases to Expressions
	Evaluating and Rewriting Expressions
Chapter 5	Exponent Rules and Scientific Notation
	Adding, Subtracting, Multiplying, and Dividing Polynomials
Chapter 6	Factoring Polynomials
Chapter 7	Simplifying, Adding, Subtracting, Multiplying, and Dividing Rational Expressions
	Simplifying Complex Rational Expressions
Chapter 8	Simplifying, Adding, Subtracting, Multiplying, and Dividing Radical Expressions
	Rational Exponents
	Rationalizing Numerators and Denominators of Radical Expressions
	Adding, Subtracting, Multiplying, and Dividing Complex Numbers

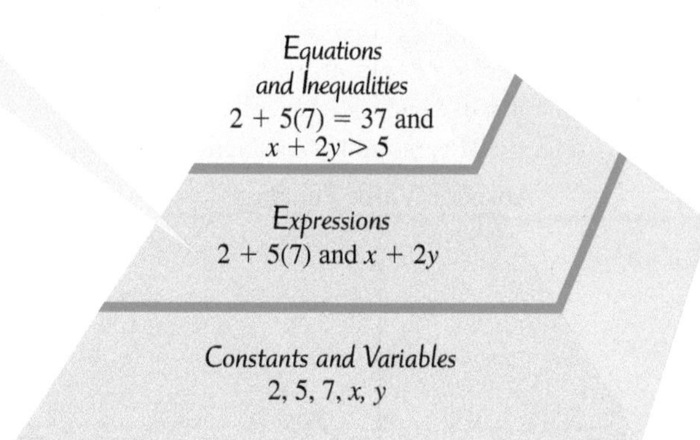

Useful Formulas

Perimeter of a rectangle: $P = 2l + 2w$

Circumference of a circle: $C = \pi d$ or $C = 2\pi r$

Area of a parallelogram: $A = bh$

Area of a triangle: $A = \frac{1}{2}bh$

Area of a trapezoid: $A = \frac{1}{2}h(a + b)$

Area of a circle: $A = \pi r^2$

Surface area of a box: $SA = 2lw + 2lh + 2wh$

Volume of a box: $V = lwh$

Volume of a pyramid: $V = \frac{1}{3}lwh$

Volume of a cylinder: $V = \pi r^2 h$

Volume of a cone: $V = \frac{1}{3}\pi r^2 h$

Volume of a sphere: $V = \frac{4}{3}\pi r^3$

Distance, d, an object travels given its rate, r, and the time of travel, t: $d = rt$

The temperature in degrees Celsius given degrees Fahrenheit: $C = \frac{5}{9}(F - 32)$

The temperature in degrees Fahrenheit given degrees Celsius: $F = \frac{9}{5}C + 32$

The profit, P, after cost, C, is deducted from revenue, R: $P = R - C$

Graphs of Common Functions

Linear Functions

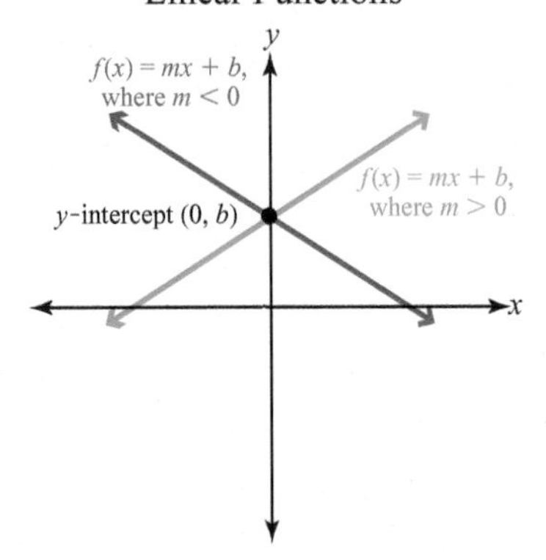

Quadratic Functions

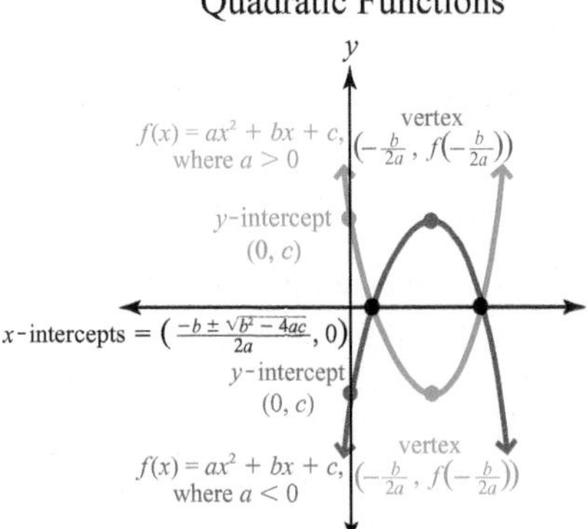

Cubic Functions

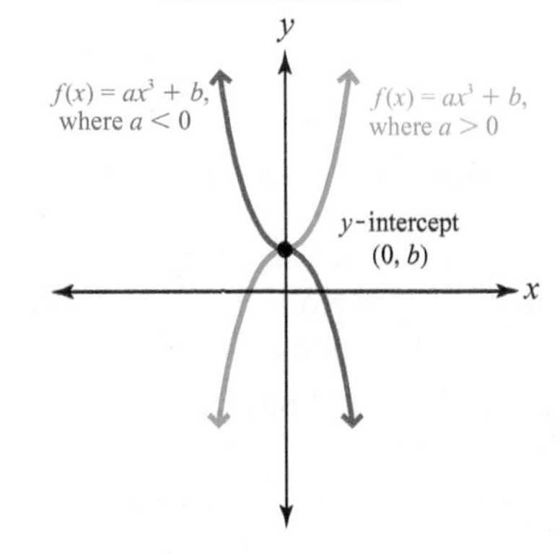

Absolute Value Functions

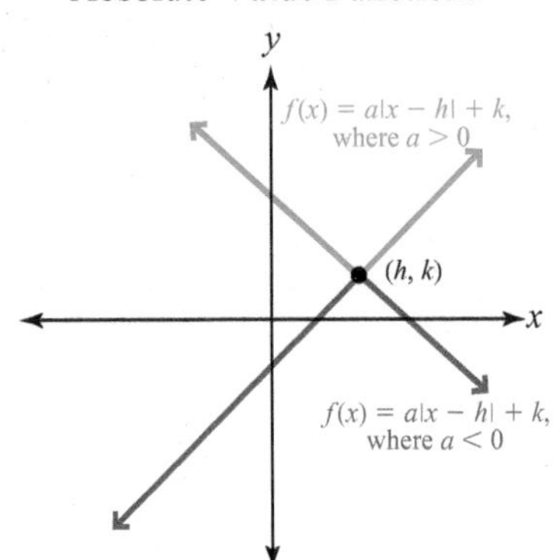

Radical Functions

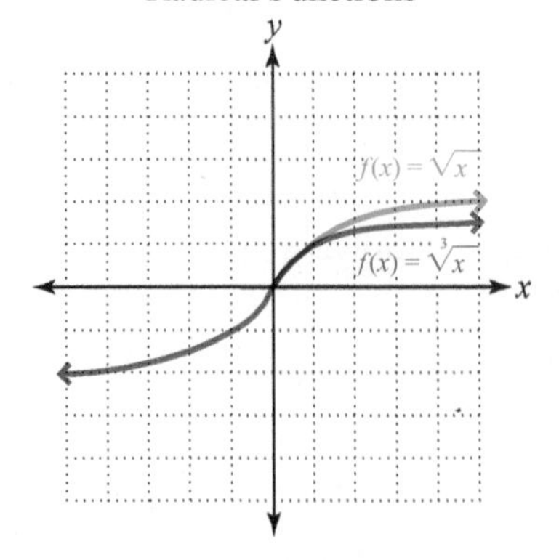

Exponential and Logarithmic Functions

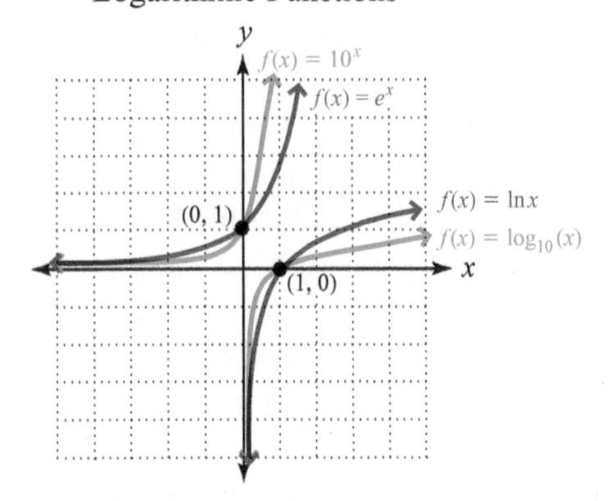

Notes

Notes

Notes

Notes

Notes

Powers and Roots

n	n^2	n^3	$\sqrt{n}$	$\sqrt[3]{n}$	$\sqrt{10n}$	n	n^2	n^3	$\sqrt{n}$	$\sqrt[3]{n}$	$\sqrt{10n}$
1	1	1	1.000	1.000	3.162	51	2,601	132,651	7.141	3.708	22.583
2	4	8	1.414	1.260	4.472	52	2,704	140,608	7.211	3.733	22.804
3	9	27	1.732	1.442	5.477	53	2,809	148,877	7.280	3.756	23.022
4	16	64	2.000	1.587	6.325	54	2,916	157,464	7.348	3.780	23.238
5	25	125	2.236	1.710	7.071	55	3,025	166,375	7.416	3.803	23.452
6	36	216	2.449	1.817	7.746	56	3,136	175,616	7.483	3.826	23.664
7	49	343	2.646	1.913	8.367	57	3,249	185,193	7.550	3.849	23.875
8	64	512	2.828	2.000	8.944	58	3,364	195,112	7.616	3.871	24.083
9	81	729	3.000	2.080	9.487	59	3,481	205,379	7.681	3.893	24.290
10	100	1,000	3.162	2.154	10.000	60	3,600	216,000	7.746	3.915	24.495
11	121	1,331	3.317	2.224	10.488	61	3,721	226,981	7.810	3.936	24.698
12	144	1,728	3.464	2.289	10.954	62	3,844	238,328	7.874	3.958	24.900
13	169	2,197	3.606	2.351	11.402	63	3,969	250,047	7.937	3.979	25.100
14	196	2,744	3.742	2.410	11.832	64	4,096	262,144	8.000	4.000	25.298
15	225	3,375	3.873	2.466	12.247	65	4,225	274,625	8.062	4.021	25.495
16	256	4,096	4.000	2.520	12.649	66	4,356	287,496	8.124	4.041	25.690
17	289	4,913	4.123	2.571	13.038	67	4,489	300,763	8.185	4.062	25.884
18	324	5,832	4.243	2.621	13.416	68	4,624	314,432	8.246	4.082	26.077
19	361	6,859	4.359	2.688	13.784	69	4,761	328,509	8.307	4.102	26.268
20	400	8,000	4.472	2.714	14.142	70	4,900	343,000	8.367	4.121	26.458
21	441	9,261	4.583	2.759	14.491	71	5,041	357,911	8.426	4.141	26.646
22	484	10,648	4.690	2.802	14.832	72	5,184	373,248	8.485	4.160	26.833
23	529	12,167	4.796	2.844	15.166	73	5,329	389,017	8.544	4.179	27.019
24	576	13,824	4.899	2.884	15.492	74	5,476	405,224	8.602	4.198	27.203
25	625	15,625	5.000	2.924	15.811	75	5,625	421,875	8.660	4.217	27.386
26	676	17,576	5.099	2.962	16.125	76	5,776	438,976	8.718	4.236	27.568
27	729	19,683	5.196	3.000	16.432	77	5,929	456,533	8.775	4.254	27.749
28	784	21,952	5.292	3.037	16.733	78	6,084	474,552	8.832	4.273	27.928
29	841	24,389	5.385	3.072	17.029	79	6,241	493,039	8.888	4.291	28.107
30	900	27,000	5.477	3.107	17.321	80	6,400	512,000	8.944	4.309	28.284
31	961	29,791	5.568	3.141	17.607	81	6,561	531,441	9.000	4.327	28.460
32	1,024	32,768	5.657	3.175	17.889	82	6,724	551,368	9.055	4.344	28.636
33	1,089	35,937	5.745	3.208	18.166	83	6,889	571,787	9.110	4.362	28.810
34	1,156	39,304	5.831	3.240	18.439	84	7,056	592,704	9.165	4.380	28.983
35	1,225	42,875	5.916	3.271	18.708	85	7,225	614,125	9.220	4.397	29.155
36	1,296	46,656	6.000	3.302	18.974	86	7,396	636,056	9.274	4.414	29.326
37	1,369	50,653	6.083	3.332	19.235	87	7,569	658,503	9.327	4.431	29.496
38	1,444	54,872	6.164	3.362	19.494	88	7,744	981,472	9.381	4.448	29.665
39	1,521	59,319	6.245	3.391	19.748	89	7,921	704,969	9.434	4.465	29.833
40	1,600	64,000	6.325	3.420	20.000	90	8,100	729,000	9.487	4.481	30.000
41	1,681	68,921	6.403	3.448	20.248	91	8,281	753,571	9.539	4.498	30.166
42	1,764	74,088	6.481	3.476	20.494	92	8,464	778,688	9.592	4.514	30.332
43	1,849	79,507	6.557	3.503	20.736	93	8,649	804,357	9.644	4.531	30.496
44	1,936	85,184	6.633	3.530	20.976	94	8,836	830,584	9.695	4.547	30.659
45	2,025	91,125	6.708	3.557	21.213	95	9,025	857,375	9.747	4.563	30.882
46	2,116	97,336	6.782	3.583	21.148	96	9,216	884,736	9.798	4.579	30.984
47	2,209	103,823	6.856	3.609	21.679	97	9,409	912,673	9.849	4.595	31.145
48	2,304	110,592	6.928	3.534	21.909	98	9,604	941,192	9.899	4.610	31.305
49	2,401	117,649	7.000	3.659	22.136	99	9,801	970,299	9.950	4.626	31.464
50	2,500	125,000	7.071	3.684	22.361	100	10,000	1,000,000	10.000	4.642	31.623